P9-CAB-777

Biology

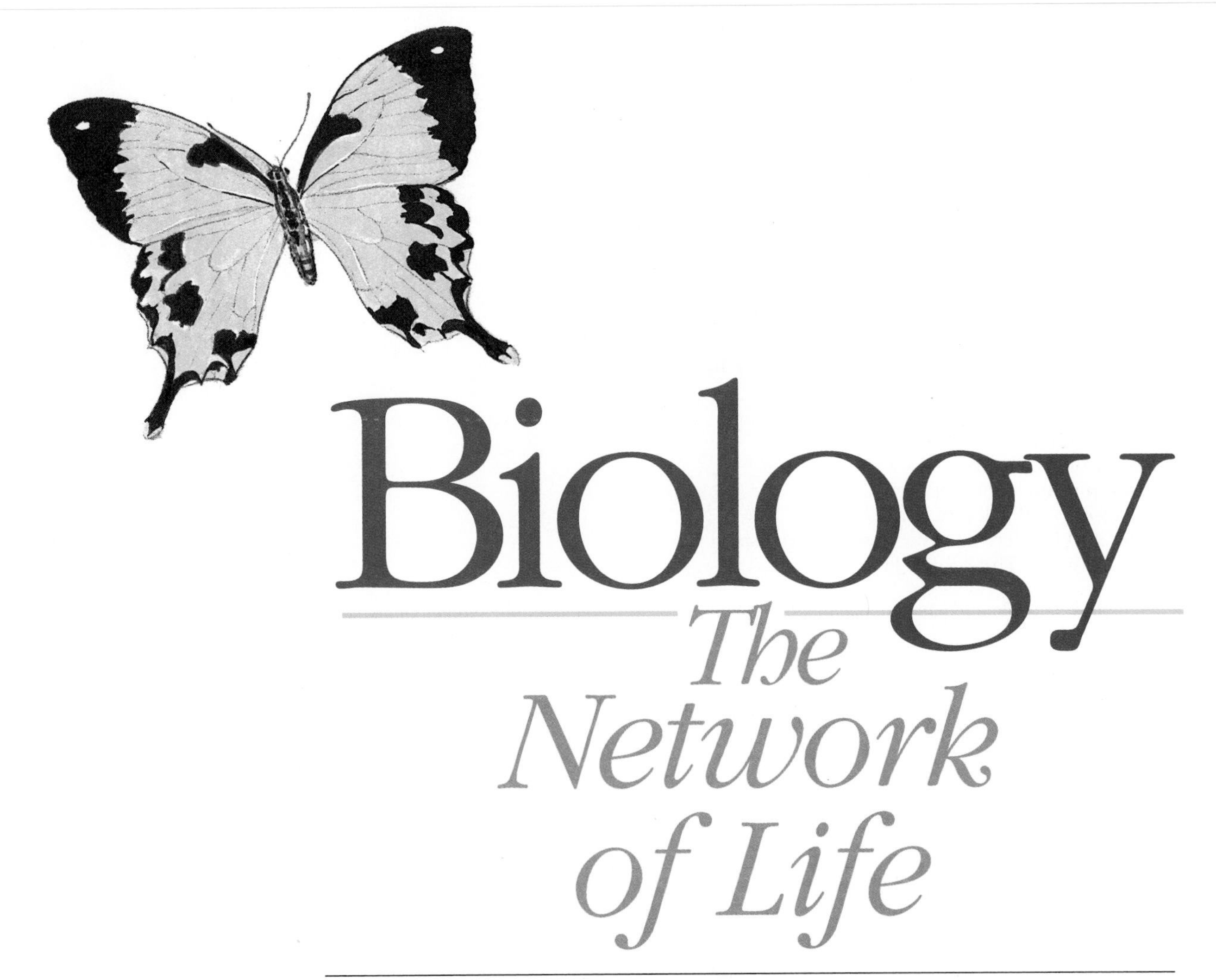

Biology

The Network of Life

Michael C. Mix

Paul Farber

Keith I. King

Oregon State University

Sponsoring Editor: Glyn Davies
Development Editor: Rebecca Strehlow
Project Editor: Nora Helfgott
Design Supervision and Cover Design: Lucy Krikorian
Text Design: Edward Smith Design, Inc.
Cover Photos: Jim Trotter/SUPERSTOCK (front); Kim Taylor/Bruce Coleman, Inc. (back)
Photo Researcher: Karen Koblik
Director of Production: Kewal K. Sharma
Production Assistant: Jeffrey Taub
Compositor: Black Dot Graphics
Printer and Binder: Arcata Graphics/Kingsport
Cover Printer: The Lehigh Press, Inc.

For permission to use copyrighted material, grateful acknowledgment is made to the copyright holders on pp. C1–C9, which are hereby made part of this copyright page.

BIOLOGY: THE NETWORK OF LIFE

Library of Congress Cataloging-in-Publication Data

Mix, Michael C.
Biology: the network of life / Michael C. Mix, Paul Farber, Keith I. King.
p. cm.
Includes index.
ISBN 0-673-39869-2 (student edition)
ISBN 0-673-52200-8 (teacher edition)
1. Biology. I. Farber, Paul Lawrence, 1944- . II. King, Keith I. III. Title.
[DNLM: 1. Biology. QH 308.2 M685b]
QH308.2.M58 1992
574—dc20
DNLM/DLC
for Library of Congress 91-20850
CIP

91 92 93 94 9 8 7 6 5 4 3 2 1

Brief Contents

Detailed Contents

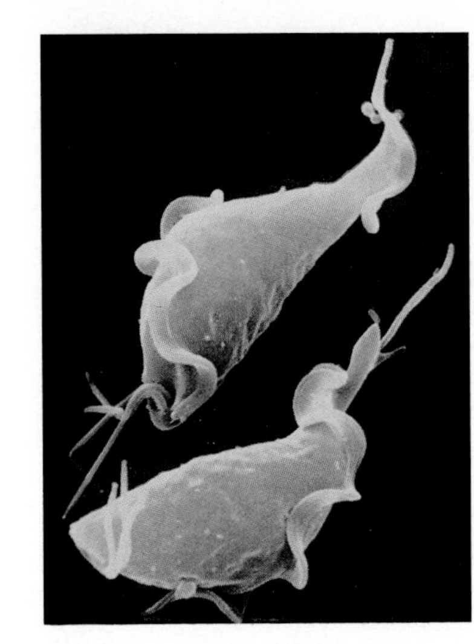

Detailed Contents

Detailed Contents

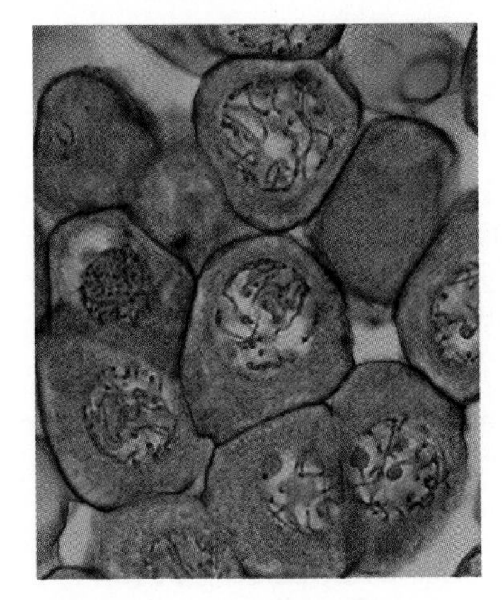

Detailed Contents

Detailed Contents

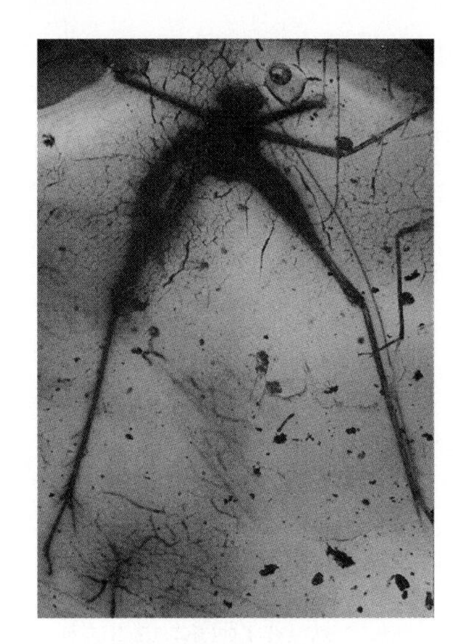

Detailed Contents

Detailed Contents

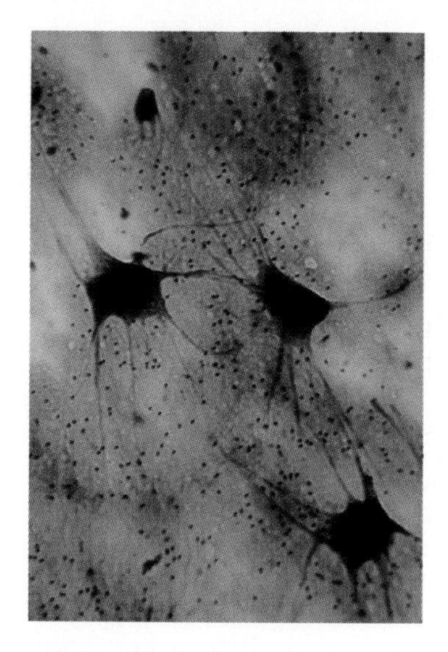

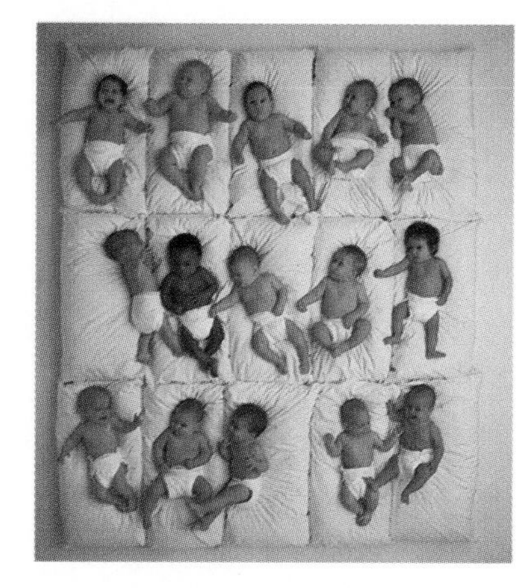

Detailed Contents

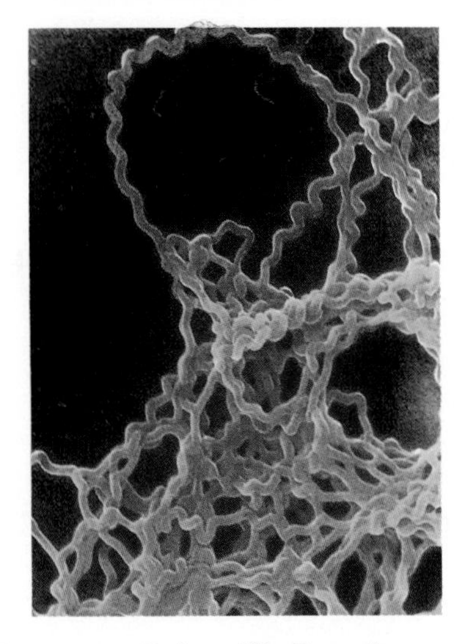

Preface

Biology and the teaching of biology are undergoing profound changes as we approach the twenty-first century. Recent reports from the largest and most prestigious scientific societies in the United States have provided guidance for transforming both textbooks and college courses. For years, too many biology textbooks and courses have emphasized facts, vocabulary, and memorization, and there has been little effort to expose students to scientific process and science as a way of thinking or knowing. Consequently, many students have not developed a sense of the importance of science, scientific thinking, or the excitement that characterizes studies of the living world.

Biology: The Network of Life concentrates on knowledge that an educated person requires for understanding the living world and making informed decisions related to that world. Scientific inquiry and scientific process are the conceptual foundations of our book. *Scientific inquiry* refers to the ways in which scientists investigate problems. *Scientific process* is a broader concept. Historically, science has been presented as a body of knowledge concerning the natural world. Using an elegantly simple but wholly imaginary "scientific method," scientists allegedly added bits of information to an ever-enlarging picture of nature. Several factors ultimately led to a fundamental alteration of this static view. These included the scientific revolutions of the early twentieth century, research done in the history and philosophy of science in response to those revolutions, and recent investigations into the sociology of science. What has emerged is a view of science as a process: a dynamic activity based on scientific inquiry—making careful observations, collecting relevant information, conducting experiments, and constructing hypotheses and broad explanations (theories).

The scientific process allows us to acquire a deep understanding of the natural world. Since the process of science is dynamic, a historical perspective is important for understanding current concepts, hypotheses and theories, and future biological research. Far from being based on a simple, "cookbook" scientific method, the scientific enterprise is stunningly creative, yet rigorously exact. Our study of biology goes far beyond the catalog of facts, which is merely an outer shell. By emphasizing the *process* of science, it becomes possible to understand science as a way of thinking that all citizens can use in problem solving.

Consistent with the new goals for teaching biology, Unit I begins with the origin and maturation of Earth (Chapter 2). After setting the stage, we describe the characters—the organisms that inhabit Earth—examine the different habitats of Earth, and learn how the physical world preconditions the biological world (Chapters 3–6). How do these organisms interact? How do humans affect those interactions? How do long-term physical changes influence those interactions? These questions are addressed in Chapters 7–11.

In Units II and III, we focus on this question: What explains the appearances, functions, and survival capacities of organisms? In the short term, organisms look and function as they do because they resemble their parents. The study of genetics explains why and how this occurs (Chapters 12–21). In the long term, organisms have characteristic traits because they are members of species populations that have evolved through time (Chapters 22–28). The theory of evolution explains how living organisms change in response to their environments and to each other.

Unit IV is concerned with how organisms function. It begins with the basic process of energy capture and use (Chapter 29). In three chapters on the world of plants (Chapters 30–32), the forms and functions of plants are related to their evolution and the environments in which they survive. Chapters 33–40 emphasize human biology. The enormous strides made in human biological research will become evident from reading these chapters. We describe what is now known about human organ systems that control our every activity, how the human immune system works, and where research is heading in medicine with special emphasis on infectious diseases, cancer, cardiovascular diseases, and AIDS. Chapter 41 looks to the future to underscore the reality that biology, like all exciting science, is constantly changing.

TO THE STUDENT

This textbook does not attempt to give you biology "once and forever." Rather, it will prepare you to understand the ever-widening scope of knowledge about the living world. Every week brings some interesting and potentially significant development in biology that may affect your life. As an educated citizen, you will probably be called on to discuss and even vote on issues that have a biological dimension, and to make intelligent choices you will need to stay informed. For example, you might be asked to take a position on the release of genetically engineered organisms in your state or on the teaching of evolution in your schools. We hope that our text will provide you with the background necessary to become scientifically curious and literate on biological issues. Yet it will not replace the excitement of a hands-on experience you will receive in a laboratory or the thrill of being out in the field on a gorgeous spring day. The photographs in the text are beautiful, but they cannot substitute for the experience of actually observing tide pools or birds in the wild or flowers poking up through snow or a cell dividing under a microscope. Perhaps, in the final analysis, this text can be a guide for identifying and understanding interesting and important areas of biology that will affect you throughout your life and help you develop a continuing appreciation for science.

LEARNING AND TEACHING TOOLS

Each chapter opens with an outline and a set of reading questions to establish a conceptual framework for the student. Within chapters, key terms are printed in boldface where they are defined. Four types of enrichment essays provide deeper insights into the nature of science: "Focus on Scientific Inquiry," "Focus on Scientific Explanation," "Focus on Science and Society," and "Focus on Science and Technology." At the end of each main section, a brief in-text summary highlights key ideas. Chapters conclude with a concise summary, review questions, essay and discussion questions, and a list of references and recommended readings. Numerous illustrations and tables complement the narrative. The book also includes an appendix on the classification system used, an extensive glossary, and a thorough index.

ANCILLARIES

Instructor's Manual by the authors. The manual is available free to adopters. It provides an index to appropriate images on the laser disk, plus suggested lecture outlines, lists of key concepts, and lecture demonstrations. The *Instructor's Manual* also includes 150 transparency masters.

Study Guide by Elizabeth Godrick of Boston University. The guide contains chapter overviews, learning objectives, concepts in review, key terms (with page numbers from the text for reference), and self-tests featuring matching, true/false, multiple-choice, and short-answer questions.

Laboratory Manual by Bill Tietjen of Bellarmine College. All lab experiments have been carefully chosen and class tested. Art is included for each exercise, helping to clarify the experiment.

Test Bank by Ken Saladin of Georgia College. The test bank consists of 2,500 multiple-choice, true/false, matching, and sentence completion questions.

Testmaster. The test bank is available to adopters in a computerized form for your IBM or Macintosh.

Acetate Transparencies. A comprehensive set of 125 four-color acetates of art and photomicrographs from the text is available free to adopters.

The HarperCollins Biology Encyclopedia Laser Disk. The Biology Encyclopedia Laser Disk, produced in conjunction with Nebraska Interactive Video, Inc., offers the latest in visual technology. It contains transparencies, micrographs, slides, and film and video footage. Over 1,500 images were provided by Carolina Biological Supply. The laser disk allows instant access to any image or footage, frame-by-frame or moving, simply by pushing a few buttons on a hand-held remote. The disk enhances the principles of biology covered in the text much more effectively than transparencies or videos.

Student Environmental Action Guide. The Earthworks Group and HarperCollins have joined with the Student Environmental Action Coalition to bring your students a handbook of the environmental movement on campuses around the country. It contains a series of strategies through real campus examples for approaching the administration, the community, political leaders, student leaders, and one's own personal habits to achieve positive change. Examples include population control, transportation, water conservation, and newsletter publication.

Two Minutes a Day For a Greener Planet by Marjorie Lamb, a veteran reporter on environmental affairs. This book provides easy, practical answers to what all of us can do to save the Earth. It gives suggestions for individual action, on a small scale, that can make a big impact on our planet's future.

Harper Dictionary of Biology by W. G. Hale and J. P. Margham, both of the Liverpool Polytechnic Institute. The dictionary contains 5,600 entries, which go far beyond basic definitions to provide in-depth explanations and examples. Diagrams illustrate such concepts as genetic organization, plant structure, and human physiology. The dictionary covers all major subjects (anatomy, biochemistry, ecology, etc.) and also includes biographies of important biologists.

The Biology Coloring Book, Anatomy Coloring Book, Physiology Coloring Book, Botany Coloring Book, Zoology Coloring Book. An exciting new approach to learning biology. Coloring provides an enjoyable and effective means of learning the fundamentals of biology. Participation by the reader, through creative coloring, provides significant learning reinforcement. The text accompanying each coloring plate provides supportive explanatory material and leads the reader through the plate in a step-by-step manner. Furthermore, when finished, the colored plates provide an excellent review that the reader has helped create.

Writing About Biology by Jan A. Pechenik, professor of biology at Tufts University. This brief but straightforward guide includes sections on writing lab reports, essays, term papers, research proposals, critiques and summaries, and in-class essay examinations. It also includes special sections on effective note taking, how to give oral presentations, and how to prepare applications for summer and permanent jobs in biology. Appendices listing commonly used abbreviations for lengths, weights, volumes, and concentrations are also featured.

REFERENCES

The following articles and reports are important to everyone interested in improving biology education.

American Association for the Advancement of Science. 1989. *Biological and Health Sciences: Report of the Project 2061 Phase I. Biological and Health Sciences Panel.* Washington, D.C.: AAAS Publications.

———. 1989. *Science for All Americans: A Project 2061 Report on Literacy Goals in Science, Mathematics, and Technology.* Washington, D.C.: AAAS Publications.

American Society of Zoologists. 1984–1990. *Science as a Way of Knowing.* Cosponsored by the American Society of Naturalists, the Society for the Study of Evolution, the Biological Sciences Curriculum Study, the American Institute of Biological Sciences, the American Association for the Advancement of Science, the Association for Biology Laboratory Education, the National Association of Biology Teachers, the Society for College Science Teachers, the Ecological Society of America, and the Genetic Society of America. All related materials were published in a special issue of *American Zoologist,* once each year from 1984 to 1990.

National Academy of Sciences. 1989. *On Being a Scientist.* Washington, D.C.: National Academy Press.

National Research Council. 1990. *Fulfilling the Promise: Biology Education in the Nation's Schools.* Washington, D.C.: National Academy Press.

ACKNOWLEDGMENTS

This book grew out of our 20 years of teaching university biology and history of science courses. Along the way we have been aided by many people from numerous institutions. A comprehensive list of acknowledgments would be unreasonably long and unavoidably incomplete. However, this project could not have been initiated or completed without help from the following individuals.

Bonnie Roesch was the first to recognize the novel dimensions of our project, and she has been a tireless supporter. Her continuous encouragement was inspiring, and we will always feel indebted to her. Rebecca Strehlow tactfully but rigorously used her editor's pencil on several chapter drafts; the end product was immeasurably improved by her efforts. Likewise, Marilyn Henderson edited all of the first drafts, sparing us considerable embarrassment once they fell into editors' and reviewers' hands. Glyn Davies skillfully guided the entire project, and we appreciated his hard questions and open mind. Karen Koblik was indefatigable in pursuing the beautiful photographs that grace this book. Bill Davis offered early, enthusiastic support for our ideas. We also thank Jim Winton and the many reviewers listed below who offered constructive criticism and shared their ideas for improving the book. They, too, deserve great credit for bringing this book to its final state.

John Adler, Michigan Technological University

Bonnie Amos, Angelo State University

Kay Antunes de Mayalo, American River College

Al Avenoso, University of Houston

Amy Bakken, University of Washington

Cecilio R. Barrera, New Mexico State University

Barry Batzing, State University of New York at Cortland

Robert A. Bell, Loyola University

Kristen Bender, California State University at Long Beach

Linda Berg, University of Maryland

Charles J. Biggers, Memphis State University
Richard G. Bjorklund, Bradley University
Richard Blazier, Parkland College
Antonie Blockler, Cornell University
James Botsford, New Mexico State University
William R. Bowen, University of Arkansas at Little Rock
Clyde Brashier, Dakota State College
J. N. U. Brown, University of Houston
Gloria Cadell, University of North Carolina at Chapel Hill
S. Dan Caldwell, Georgia College
John Campbell, Northwest College
Nina Caris-Underwood, Texas A&M University
Galen E. Clothier, Sonoma State University
David Davis, University of Alabama
Thomas Davis, University of New Hampshire
Linda Dion, University of Delaware
Lee C. Drickamer, Southern Illinois University
Marvin Druger, Syracuse University
Dorothy C. Dunning, West Virginia University
Andres Durstenfeld, University of California at Los Angeles
Thomas C. Emmel, University of Florida
Stanley H. Faeth, Arizona State University
Michael Fine, Virginia Commonwealth University
Herman S. Forest, State University of New York at Geneseo
Sally Frost, University of Kansas
Rita Ghosh, Indiana State University
David C. Glenn-Lewin, Iowa State University
Judith Goodenough, University of Massachusetts at Amherst
Kenneth Goodhue-McWilliams, California State University at Fullerton
Thomas Gorham, California Lutheran University
Nels H. Granholm, South Dakota State University
Thomas Griffiths, Illinois Wesleyan University
Gilbert F. Gwilliam, Reed College
James Habeck, University of Montana
Marcia Harrison, Marshall University
Thomas Herbert, University of Miami
David Hicks, Manchester College
Linda Margaret Hunt, Notre Dame University
David T. Jenkins, University of Alabama at Birmingham
C. Weldon Jones, Bethel College
Leonard Kass, University of Maine
Elizabeth Keith, University of Mississippi
Kenneth Klemow, Wilkes University
Bob Kosinski, Clemson University
John Krenetsky, Metropolitan State College
Karen Kurvink, Moravian College
John M. Lammert, Gustavus Adolphus College
Howard Lenhoff, University of California at Irvine
Ronald Lindahl, University of Alabama
Michael Lockhart, Northeast Missouri State University
Ruth Logan, Santa Monica College
Raymond Lynn, Utah State University
David Mark, St. Cloud State University
James Marker, University of Wisconsin at Platteville
Gayton C. Marks, Valparaiso University
Samuel Maroney, University of Virginia
Joyce Maxwell, California State University at Northridge
Helen C. Miller, Oklahoma State University
Neil Miller, Memphis State University
Charles Mims, University of Georgia
Herbert L. Monoson, Bradley University
Walter A. Morin, Bridgewater State College
Keith Morrill, South Dakota State University
Alexander Motten, Duke University
Tom Nye, Washington and Lee University
Clifford Night, East Carolina University
Robert W. O'Donnell, State University of New York at Geneseo
Lowell P. Orr, Kent State University
Elizabeth Painter, Colorado State University
Robert A. Paoletti, King's College
Patricia Pearson, Western Kentucky University
Richard Peifer, University of Minnesota

Herbert B. Posner, State University of New York at Binghamton
Greg Rose, West Valley College
Monica Rudzik, Westminster College
William Rumbach, Central Florida University
Douglas Sampson, Emory University
Lawrence C. Scharmann, Kansas State University
Joan Schuetz, Towson State University
Richard Search, Thomas College
David Senseman, University of Texas at San Antonio
Linda Simpson, University of North Carolina at Charlotte
Daryl Smith, University of Northern Iowa
Kingsley Stern, California State University at Chico
Charles L. Stevens, University of Pittsburgh
Lewis Stratton, Furman University
Gerald Summers, University of Missouri
Marshall Sundberg, Louisiana State University
Daryl Sweeney, University of Illinois at Urbana
Linda C. Twining, Northeast Missouri State University
Patrick K. Williams, University of Dayton
William Wissinger, St. Bonaventure University
John L. Zimmerman, Kansas State University

Many people have provided valuable aid on parts of this book. Foremost among these is Bill Winner, who wrote the initial drafts of Chapters 30–32 and supervised the creation and development of the accompanying art. He was also a valued source of information and ideas used in Unit I. Past and present colleagues in the Department of General Science at Oregon State University helped shape the general biology course over the past two decades. Some of our ideas originated in work done in the course by Jack Lyford, Henry Van Dyke, Dennis McDonald, Larry Forslund, Bruce McCune, and Patricia Muir. The teaching assistants in our course gave us excellent feedback on our ideas for teaching biology. Susie Bratsch helped us translate ideas into images and provided many sketches that led to finished illustrations in this book. Barbara Moritsch gave us advice on several art pieces. Administrators at Oregon State supported our ideas for developing a contemporary nonmajors general biology course. Fred Horne, dean of the College of Science, was especially helpful.

Jerry Kling, of Oregon State, provided assistance related to computer literacy and served as a consultant on material cycling in soils. Darrell L. King, of Michigan State University, provided photographs and ideas about deciduous forests and sand dune succession. La Verne D. Kulm, of Oregon State, supplied photographs and also reviewed sections of Chapter 9.

Several distinguished scientists and scholars graciously answered our questions about how they would like to see biology develop in the coming decades. We thank Ernst Mayr, Ledyard Stebbins, David Hull, E. O. Wilson, Marvin Druger, and Linus Pauling for their time and thoughts.

The production of this book also involved considerable assistance from others. Karla Russell, of the General Science Department, typed and edited drafts, balanced deadlines, and was otherwise helpful in ways too numerous to list but that are easily understood by anyone who has the good fortune to work where there is a superior office staff. Laura Mix Kohut, now of Morrison & Foerster, provided the legal assistance necessary in such a large project and even succeeded at making us understand what some of it meant. Leslie Mix typed drafts of some of the early material. Many remarkably talented and patient individuals at HarperCollins brought this book to completion; they include Nora Helfgott, project editor, and Teresa Delgado, art director. Bruce Emmer assisted as copyeditor, and David Fox provided input during art production.

Finally, we recognize our wives, Marilyn Henderson, Vreneli Farber, and Roberta King. They suffered longest and most, yet were unfailing in their encouragement and support.

The students we have had the good fortune to have in classes during the past 20 years were, in a real sense, the source of inspiration for this project. To all biology students—past, present, and future—we dedicate this book.

Michael C. Mix
Paul Farber
Keith I. King

UNIT I

The Sphere of Life

Earth is a sphere of life. Where did it come from? How has it developed and matured in the 4.5 billion years since it was formed? What kinds of relationships exist between Earth and its inhabitants? Scientists have long been interested in answering these questions. Their answers, grounded in scientific hypotheses and theories, provide the framework for Unit I.

Life appeared around 3.5 billion years ago. An astounding number of microorganisms, fungi, plants, and animals now live on the planet. Throughout its history, Earth has undergone continuous change. Originally hostile to life, it now consists of many different regions—forests, grasslands, mountains, deserts, lakes, streams, and oceans—that have unique characteristics and inhabitants. These regions are united through physical, chemical, and biological processes. Attempts to understand these processes have driven ecological research for most of this century. As a result, we are better able to assess Earth's complex environments and the factors involved in their regulation.

CHAPTER 1

Introduction: Studying the Living World

Biology is the scientific study of life. It is an immense field that traces its origins to different activities that occurred in the past. Early workers studied how the animal body functioned, examined hereditary patterns and nutritional needs of domestic plants and animals, and attempted to describe and name all living organisms of Earth. Today, biology is a diverse subject held together by certain unifying concepts. General methods of investigation are used to acquire knowledge of the living world, and a set of broad theories integrates bodies of that knowledge. Biological science is studied, taught, and used in institutions such as schools, government laboratories, museums, corporation research centers, and private foundations. Biology has both immediate personal interest for each of us and profound social importance. By understanding the network of life, we can anticipate and comprehend change and we can influence the world in which we live.

What is biology? To most people, it has a very broad meaning and includes a great variety of topics. Students beginning their study of this subject have been exposed to biological phenomena throughout their lives. Even the most casual observer has probably noticed the beauty, diversity, and complexity of the living world. The sight of birds flying in the early morning sun and the exquisite geometric patterns of microscopic foraminiferan shells like that shown in Figure 1.1 are strikingly beautiful. The different kinds of organisms that can be found in a bucket of sand collected at the beach, from a small area of a forest floor, in a pond, or even from a puddle of water in a field are fascinating to everyone, from grade school students to trained biologists (see Figure 1.2).

By studying organisms, biologists have been able to expose the most intimate details about the complex operations of plants and animals. Popular magazines and television programs bring these details to the general public. In many cases, the level of knowledge is truly incredible. For example, the exact molecular causes of certain types of cancer are now understood because of results from research completed during the past few years. Also, studies of life have revealed complex interactions among living organisms and the environments they inhabit. We now understand that individual organisms, including humans, are not isolated entities. Rather, they have vital ties to other organisms and to other environments, and their fate depends on the continued health of both.

A

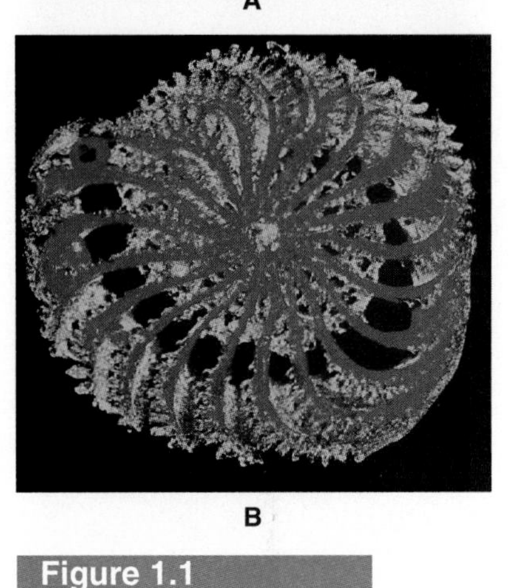

B

Figure 1.1

The beauty of nature: (A) a snowy egret fishing; (B) a microphotograph of a formaniferan.

BIOLOGY

The word **biology** originated in 1800; literally, it means "the study of life." At that time, scientists used the term in reference to studying the functions of living organisms, a field we now call *physiology*. During the twentieth century, biology has come to refer to the *scientific study of life*. Today, biologists study all organisms' functions, their appearance, their habitats, their interactions with one an-

Figure 1.2

As this illustration demonstrates, the diversity in a relatively small space can be impressive.

Table 1.1 Selected Member Societies of the American Institute of Biological Sciences

American Bryological and Lichenological Society	Association of Ecosystem Research Centers
American Fern Society	Association of Systematics Collections
American Ornithologists Union	Association for Tropical Biology
American Physiological Society	Botanical Society of America
American Phytopathological Society	Ecological Society of America
American Society of Agronomy	Entomological Society of America
American Society of Animal Science	Mycological Society of America
American Society of Mammologists	Organization of Biological Field Stations
American Society of Naturalists	Phycological Society of America
American Society of Parasitologists	Poultry Science Association
American Society for Photobiology	Society for Conservation Biology
American Society of Plant Physiologists	Society for Industrial Microbiology
American Society of Plant Taxonomists	Society of Mathematical Biology
American Society of Zoologists	Society of Nematologists
Animal Behavior Society	Weed Science Society of America

A partial listing of the members of the American Institute of Biological Sciences suggests the vast scope of present-day biological study.

other and with the environment, and their changes over time. The range of topics biologists study is immense, and for that reason, it is difficult to provide an overview that encompasses the entire field. The list of some of the member organizations of the American Institute of Biological Sciences in Table 1.1 reflects the scope of modern biology.

Although biology is a scientific field, it holds an intrinsic fascination and importance for everyone. Knowing the names of wildflowers we see along the roadside, interacting with even the most common of animals in nature, and feeling awe when we are lucky enough to observe a whale breaking the surface on a calm ocean enriches our lives (Figure 1.3). Rarely does a week go by without some interesting and potentially significant development in biology that may affect your life. Educated citizens are called on to discuss and even vote on issues that have a biological dimension, and to make intelligent choices, it is necessary to be well informed.

A

B

Figure 1.3

The living world has scenes that fascinate people of all ages. (A) Can we forget our earliest awareness of nature? (B) A humpback whale surfacing.

Though extremely broad, the study of biology does not incorporate everything known or written about living organisms because life can be viewed from numerous perspectives beyond the scientific. Artists and writers, for example, can express personal ideas about the beauty of organisms or about a landscape (see Figure 1.4) that are quite different from the ideas that a scientist considers.

The Origin of Biology

How did biology arise as a scientific discipline? The general study of living beings traces its roots back to the ancient Greeks, when it was known as **natural history.** Aristotle, a famous Greek philosopher, in the fourth century B.C. wrote a book titled *History of Animals* in which he described 500 types of animals and what he knew of them. Aristotle's book was one of the first classics of natural history.

The goals of natural history changed over the centuries. The Romans, whose civilization succeeded that of the Greeks in the second century B.C., were a practical people. Their natural history writers, like the famous Pliny (Figure

Figure 1.4

The Snake Charmer by the French painter Henri Rousseau conveys a sense of the exotic by featuring plants that are imaginary.

Figure 1.5

Pliny at work, as illustrated in a fifteenth-century manuscript.

1.5), concentrated on practical issues like the proper maintenance of horses, the recognition of medically valuable plants, and the treatment of domestic animals. In the Middle Ages—the period between the collapse of the Roman Empire in the fifth century A.D. and the Renaissance of the fifteenth century—natural history reflected the moral lessons of religious writers, as can be seen in Figure 1.6, which shows an illustration from a famous medieval work on animals. After the Rennaissance, the goals of natural history were to describe and classify all the "natural" products on Earth, that is, all plants, animals, and minerals. Starting in the Renaissance, natural history books were often illustrated, sometimes with magnificent but fanciful woodcuts, and later with beautiful and precise ones such as that shown in Figure 1.7.

Although the ambitions of natural scientists—to describe and classify natural objects—were important, modern biology focuses on many other fascinating questions about organisms. However, until relatively recently, efforts to answer most of these questions were made by scientists in other disciplines, such as medicine and agriculture. For example, until the nineteenth century, questions about the structure and function of organisms were studied in medicine, not natural history. Early physicians made great contributions to what we now know as biology. Similarly, detailed examination of hereditary traits was of considerable interest to people who raised plants and animals for food or pleasure. This same group was also interested in plant and animal diseases and nutrition. Agriculture is thus also a distant but honored ancestor of modern biology.

Figure 1.6

Illustration from *Physiologus,* a medieval collection of stories about animals. The woodcut is from a late edition (1587).

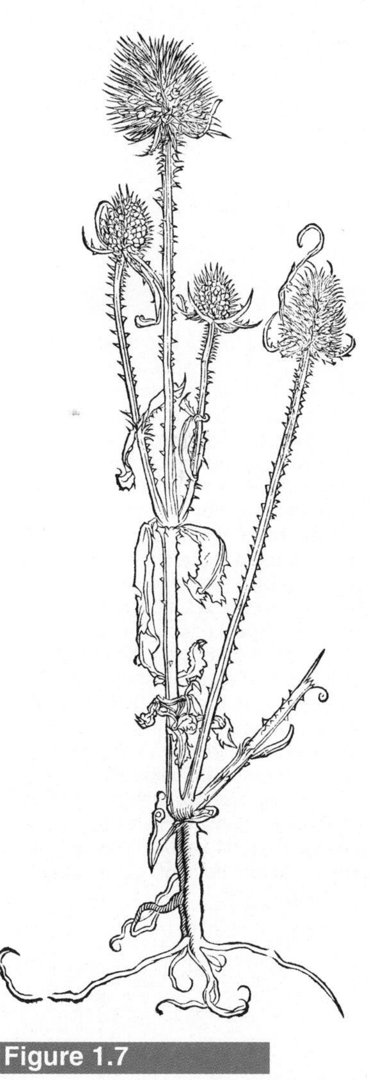

Figure 1.7

Renaissance natural history books had beautiful and accurate illustrations, like this one of teasel (*Dipsacus fullonum*) from Brunfels's *Herbarum vivae eicones* (1530).

Biology Today

Today biology encompasses different fields of research, each with its own perspective, methods, and goals. What holds the diverse field of biology together? That is, what gives it unity? Biology, like other intellectual disciplines, is held together by its methods, its major ideas, and its institutions.

Methods of Biology

Biology is unified in its *methods* of acquiring knowledge of the living world. Although science writers frequently refer to a formal set of procedures called "the scientific method," a careful assessment of the activities of scientists, both living and dead, indicates clearly that there is no single "scientific method" in biology or in any other science. Biologists, like all scientists, are interested in different questions and use methods that are appropriate for answering the questions at hand. However, certain distinct features of scientific research make it a powerful approach for learning about the natural world.

At the most basic level, biologists *observe* nature, directly or with instruments, as illustrated in Figure 1.8, in order to answer certain questions. Early naturalists simply wanted to know which plants and animals were present in different parts of the world. Today biologists ask deeper questions, such as, How are birds able to navigate hundreds of miles to arrive at specific winter feeding grounds, or what effect will increased acid levels have on fish in mountain lakes? To answer such questions, scientists study nature in search of regularities, mechanisms, and processes. They use logical arguments to relate particular phenomena to more general patterns. Sometimes when scientists do not understand or cannot interpret established patterns or changes in such patterns, they develop tentative hypotheses for explanation. Further studies are then conducted to test

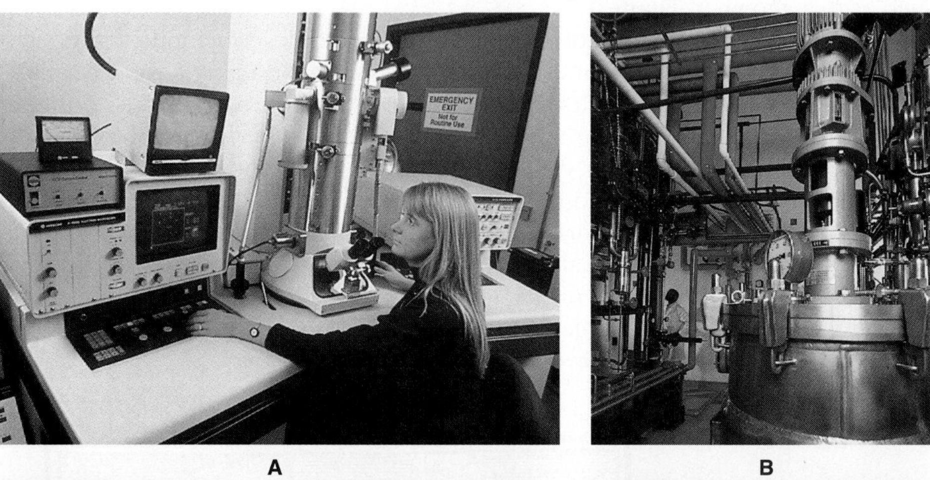

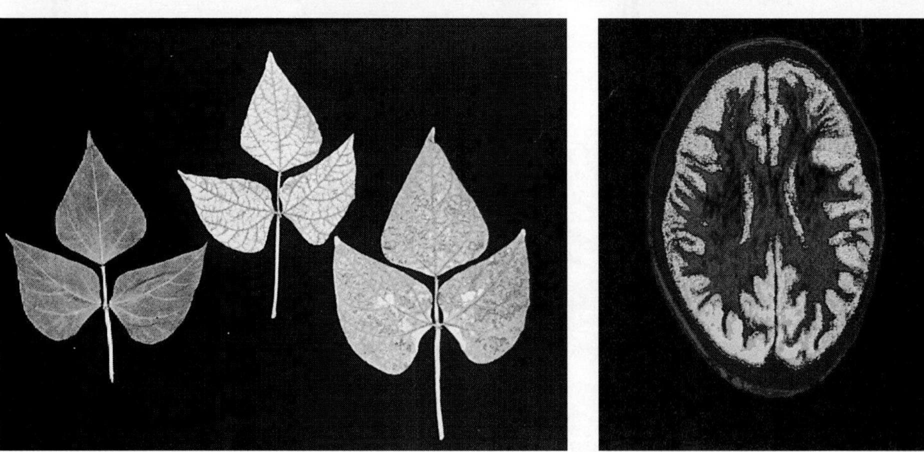

Figure 1.8

Progress in biology has been increasing rapidly, partly because advanced technologies have led to the development of new, powerful tools. (A) An electron microscope is used to magnify materials over 100,000 times. Electron micrographs (EMs) are photographs of such magnified objects, and many of them appear in this book. (B) Huge machines are now used in making products from genes, bits of genetic material that dictate structure and function in all organisms. (C) Special techniques allow biologists to determine the effects of pollutants on plants in their natural environment; a normal leaf (left) and two leaves damaged by air pollution (right). (D) Sophisticated instruments such as a magnetic resonance imaging (MRI) scanner are used to create vivid images of human organs. MRI images of the brain are used to evaluate normal and abnormal functions of this incredibly complex organ.

Figure 1.9

In various places, the remains of extinct fossils can be found which are similar to contemporary forms. In this photo, it can be seen that fossil scallops resemble modern scallops.

the accuracy of these hypotheses. For example, in recent years a number of scientists have voiced concern over what they believe is a worldwide decline in amphibian populations. This is an especially difficult issue because few data are available on the sizes of normal amphibian populations. Therefore, one of the first steps will be to conduct amphibian population studies. If amphibian populations are declining, what might be the cause? Since many of the suspected sites of amphibian decline are at high altitudes, scientists hypothesize that exposure to increased levels of ultraviolet light may be responsible. A lot of research will be necessary to test the accuracy of this hypothesis.

The questions biologists ask are conditioned by the social context. For example, two of the most compelling questions that drive research today, in terms of money spent and numbers of researchers involved, are tied to problems that concern many citizens: How can diseases such as AIDS and cancer be combated, and how can the quality of our environment be maintained or improved? These two issues are responsible for the existence of gigantic research programs, in the United States and in other countries.

Understanding the context in which science is carried out is an important component in understanding biology, for it partly explains why certain subjects attract attention. The social dimensions of biological research are discussed throughout this text, and various scientific methods are described. Experiments, the formulation of hypotheses and theories, and different levels of explanation and understanding are also considered in many chapters.

Ideas of Biology

Biology, as a scientific discipline, shares a set of methods for obtaining knowledge of the living world. Equally important, biology is a scientific discipline because individual bits of information about organisms are related to larger bodies of information and to more general ideas about life. On the broadest level, biologists formulate scientific theories, which explain why the world is as we see it and guide research in further understanding it.

The *theory of evolution* provides the most general framework in which all life can be understood. Many similarities among modern organisms are reflections of a shared ancestry. So too are the comparable physical characteristics of living organisms and fossil remains of extinct organisms (see Figure 1.9). The astonishing diversity of life on Earth today is a product of evolutionary forces that molded populations over millions of years. Even the pattern of geographical distribution among plants and animals can be understood through knowledge of their evolutionary history.

There are other unifying sets of ideas in biology that relate large bodies of information. Cell biology, ecology, genetics, and molecular biology all focus on fundamental aspects of life. These fields are based on broad, unifying ideas that are compatible with evolution and can be interpreted in evolutionary terms. Nevertheless, biology is far from being a completely systematic and unified science. In this sense, it is no different from most other sciences, such as physics or chemistry.

Institutions of Biology

Biology as a discipline is recognizable more mundanely by its institutional settings. Where do biologists perform their research today? Because of the wide range of subjects they study, biologists are found throughout the world and throughout society's institutions. Yet before biology became widely studied early in the twentieth century and before governments began to support biological research on a large scale, examinations of the living world were extremely restricted. The few researchers who did investigate the living world were typically found in one of three places: the museum, the field, or the laboratory.

Natural history was primarily a museum subject in the eighteenth and nineteenth centuries. Preserved specimens were examined with the object of classifying and describing them. Relatively few individuals studied organisms in their natural habitats. They either collected specimens to be sent back to museums or, more rarely, made field observations on organisms' behavior, life stages, or distribution. Governments supported such work on a limited scale, mostly in cases where it held potential commercial value. For example, in the nineteenth century, the British were interested in making observations on plants that could be raised on plantations and exported from their colonies in Asia. For the most part, however, natural history was an amateur undertaking; that is, it was done not for pay but rather for the love of the subject.

Laboratory work in the nineteenth century was performed mostly by medical researchers. Because it was impractical to use human subjects, animals were often studied. Claude Bernard was one of the great researchers of the nineteenth century who conducted experiments on animals. At that time, society was less sensitive to animal suffering, and many of these experiments were, from a modern point of view, rather appalling and probably could have been conducted with considerably less discomfort to the subjects. Nevertheless, such studies contributed enormously to our understanding of animal body functions. Agricultural research also yielded knowledge about both plants and animals, especially in the areas of heredity, growth, and nutrition.

Given the large number of biologists today, it is difficult to appreciate that until comparatively recently, studies in the biological sciences were done by a small group of people. In the eighteenth century, for example, individuals who were seriously interested in natural history were able to correspond with all other such individuals in the world! Today it would be extraordinary to maintain communications with all the specialists in any subdiscipline, such as human physiology.

The explosion of biological research is distinctly a twentieth-century phenomenon. Universities and other centers of higher learning have extensive research faculties; the federal government maintains large research operations—the National Institutes of Health, near Washington, D.C., alone contain more biological scientists than the entire English-speaking world had at any one time in the nineteenth century—and government agencies, such as the Environmental Protection Agency, fund extensive research programs. Private foundations, such as the Salk Institute, and numerous corporations and businesses are also major contributors to our growing knowledge of the living world.

BIOLOGY: THE NETWORK OF LIFE

Biology is no longer the exclusive domain of the few who can afford the pleasure of studying the fascinating living world at their leisure. Today, it is a subject of great significance for all members of our society. The current attention focused on biology is due not only to its inherent interest but also to the enormous practical value of the knowledge gained. This knowledge deepens our appreciation and is also a key to control. Those of us who wish to influence the quality of our environment, the state of health of the people we care about, or the supply and quality of food we consume depend on knowledge provided by the biological sciences.

The current biological knowledge is obviously too extensive to encapsulate in a single book. In *Biology,* we have concentrated on concepts and knowledge that we believe are interesting and important to educated members of society. Because biology is a subject that is growing rapidly, the organizing theme of this book is *scientific inquiry* and the *scientific process.*

As explained in the preface, biology is not simply a collection of facts or a systematic arrangement of concepts. It is a dynamic activity based on **scientific inquiry**—a way of answering questions that requires careful observations, the collection of relevant information, experiments, and the construction of hypotheses and broad explanations (theories). The **scientific process** has enabled humans to acquire knowledge of the living world. However, this knowledge is in a state of constant change. In some fields of biology, it has become progressively deeper. In other areas, the questions that scientists ask have changed, often in response to societal and technological issues. Each of the four units of *Biology* focuses on basic dimensions of life, explores how we acquired our current state of knowledge, and indicates directions of future research. This book will introduce you to contemporary biology and to science as a way of learning, and will give you a sense of the importance of biology in modern society. We hope that it will also provide the basis for a lifelong interest in biology.

CHAPTER 2

Origins: Setting the Stage

Chapter Outline

Reading Questions

1. What are the current hypotheses on the origin of the universe?
2. How was the planet Earth formed?
3. How was Earth transformed from a barren, hostile planet into one that could support life?
4. How did the first living systems develop from the organic soup?
5. How did prokaryotic and eukaryotic cells originate and evolve?

From the earliest written records, it is evident that our ancestors were curious about the universe they observed (see Figure 2.1). Over the millennia, humans have attempted to explain the origin of Earth and the universe. Until the seventeenth century, explanations of Earth's history were based on mythical or religious ideas. All known ancient cultures developed creation myths grounded in beliefs of sacred idols, mysterious forces, or supernatural spirits. For example, an ancient Babylonian creation epic called the *Enuma elish* described the beginning as a "separation of land and sky." Later, concepts about Earth's origin and age were derived from the Old Testament. For example, in the fourth century A.D., Saint Augustine, based on his careful studies of biblical family trees, estimated that Earth was 6,000 years old. In 1654, James Ussher, an archbishop and biblical scholar, calculated that Earth was created at 9 A.M. on Sunday, October 23, 4004 B.C.

However, since the seventeenth century, Western cultures have tried to understand the natural world in terms of fundamental scientific laws. Such laws are based on objective measurements of natural phenomena, not statements about supernatural events or processes. Scientific laws provided the fertile ground from which modern cosmological sciences have sprung and matured. By the beginning of the nineteenth century, geologists (scientists who study Earth's history) had formulated a significant conclusion—the planet Earth was far older than the biblical 6,000 years. How old is Earth? How was it formed? How old is the universe? How did it arise? We consider these and other questions related to origins in this chapter.

UNDERSTANDING ORIGINS

Scientists have developed **hypotheses**—explanations derived from careful observations and supported by results from experiments and other evidence—about the probable events involved in the origins of the universe, Earth, and the life that came to exist on Earth. Scientific hypotheses are associated with levels of certainty. For some of the hypotheses described in this chapter, supporting evidence is limited for a number of reasons, and thus the hypotheses may be considered to have low levels of certainty. For other hypotheses, more concrete experimental evidence exists, and they have higher levels of certainty. In general, the lower the level of certainty, the more likely it is that the hypothesis will change as new evidence becomes available. Figure 2.2 conveys the relative levels of certainty associated with hypotheses discussed in this chapter.

Few hypotheses in science are associated with absolute or near-absolute certainty. The history of science warns scientists never to be uncritically confident, for the past is littered with discarded hypotheses that once were universally accepted. A hypothesis that has stood the test of time and is continuously confirmed by new evidence generally takes the form of a **law** or, if it is more complex, a **theory** (for example, the cell theory, discussed in Chapter 13, or the theory of evolution, discussed in Chapter 22).

A complete understanding of hypotheses examined in this chapter requires a depth of knowledge of principles from the physical sciences that is beyond the reach of this book. It is possible, however, to grasp current ideas about origins of the universe and Earth by acquiring some basic information from those areas (as well as putting to work a lively imagination!). Most of the hypotheses concerned with origins are supported by techniques and data that have been used to determine the age of various objects (for example, meteorites, rocks, and fossils). By

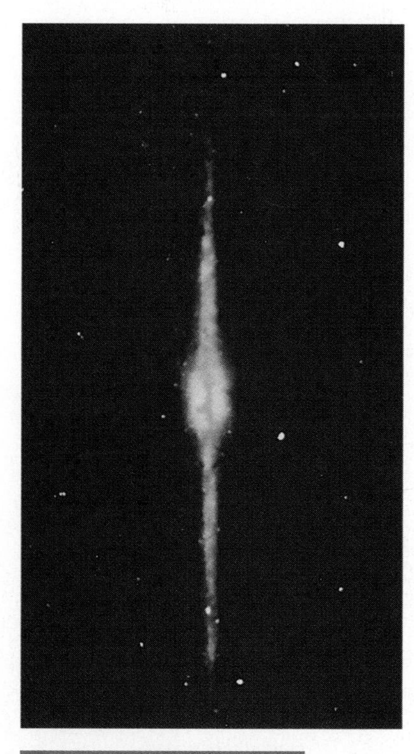

Figure 2.1

Humans throughout history have been fascinated by stars appearing in the evening sky. Most stars we see are in the Milky Way galaxy, one of billions in the universe and the home of our planet, Earth.

Figure 2.2

Scientific hypotheses are associated with levels of certainty, depending on the strength of supporting evidence. Various hypotheses described in this chapter have different levels of certainty as indicated in this figure. The big bang theory has a relatively low level of certainty because it is very broad and inclusive and may change as new experimental results are obtained.

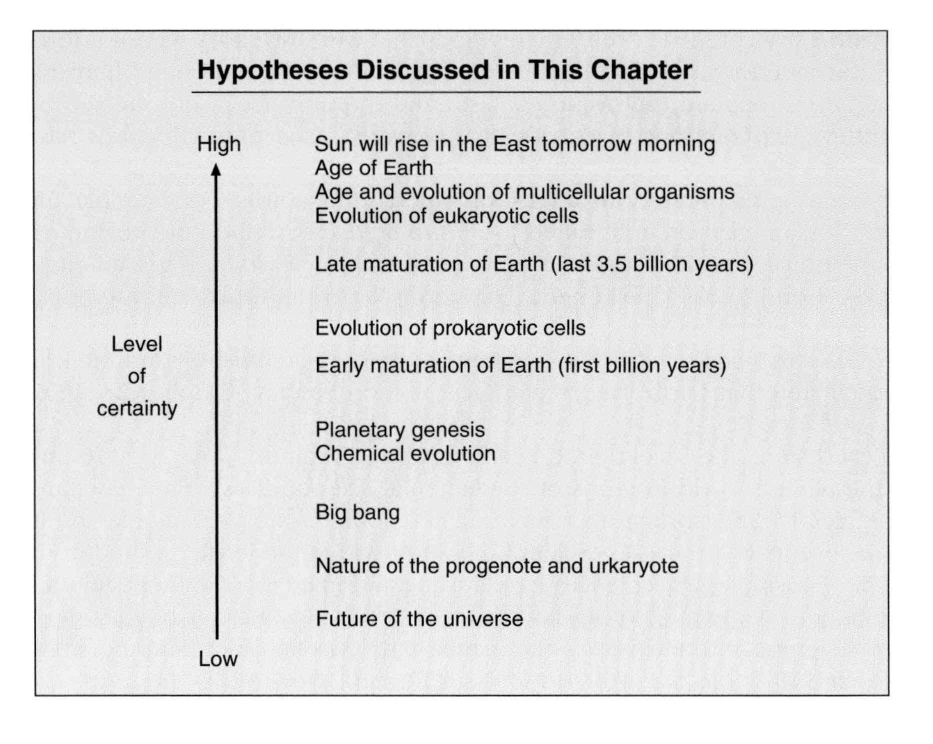

analyzing the chemical composition of such objects, scientists are able to determine their age with great accuracy. Figure 2.3 describes the relationships in the universe. Figure 2.4 indicates the sequence of events described in this chapter.

Humans have always been interested in understanding the universe that surrounds them. Prior to the seventeenth century, explanations about the origin of the universe and of Earth were based on supernatural phenomena. Beginning in the seventeenth century, naturalists attempted to understand and explain natural events and processes through the formulation and testing of scientific hypotheses.

Figure 2.3

The universe is infinite space with no known boundaries. Within the universe are billions of galaxies, such as the Milky Way, and within galaxies are stars, solar systems, and planets.

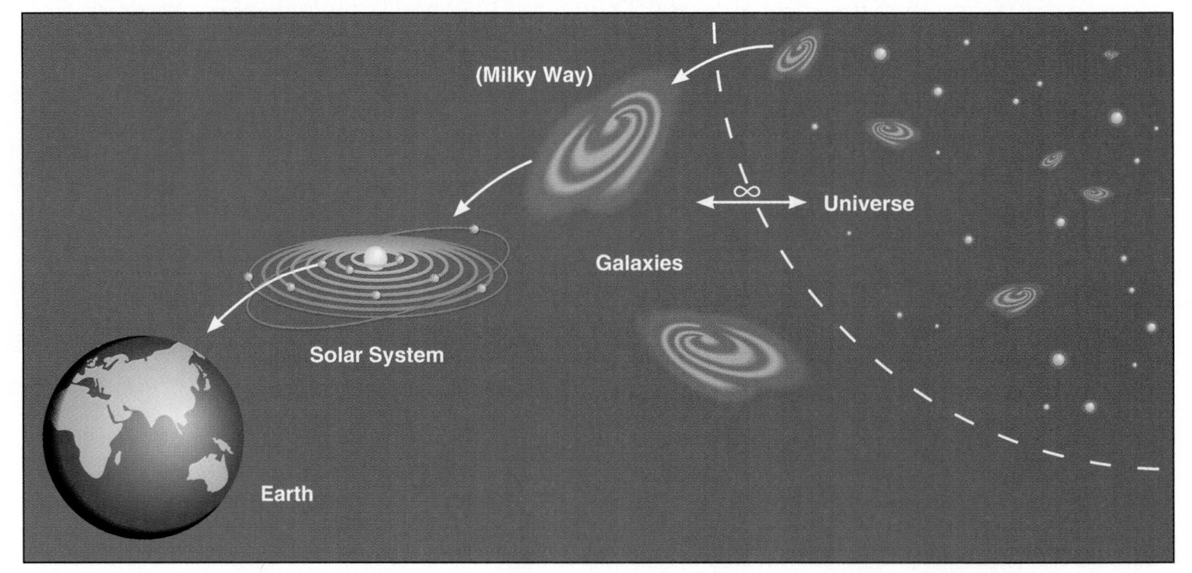

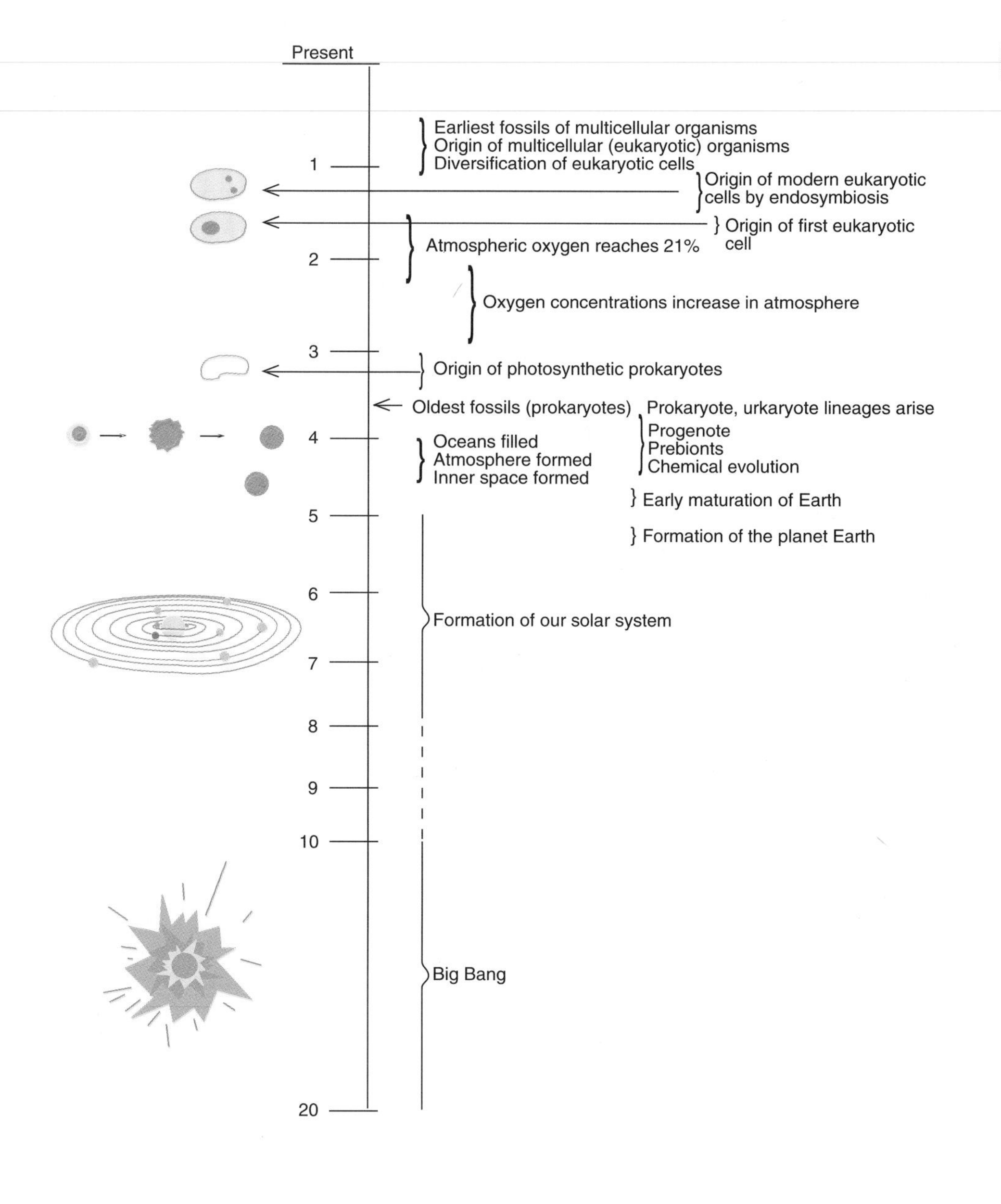

Figure 2.4

The history of the universe is thought to have begun 10 to 20 billion years ago with the "big bang." Subsequently, over a period of several billion years, galaxies and solar systems formed. Our solar system exists in one of the spiral arms of the Milky Way galaxy and was formed about 5 billion years ago. Because of its position relative to the sun, Earth was the only planet in our solar system that developed stable environmental conditions conducive to the origin and evolution of life.

ORIGIN OF THE UNIVERSE

A Key Observation

By the early twentieth century, astronomers had formed a partial picture of the universe. It was filled with galaxies, stars, and smaller bodies arrayed in vast areas of space. But where did these objects come from, or how had they originated? Contemporary ideas on **cosmology** (the study of the origin and evolution of the universe) date from less than a century ago. In the late 1920s, Edwin P. Hubble, using giant telescopes at Mount Wilson and Mount Palomar in California, made a remarkable discovery: all galaxies in the universe were moving away from each other at incredible speed! A fascinating consequence of this rapid expansion is that light from some of the stars you may observe in the sky tonight was emitted thousands or millions of years ago and is just now reaching Earth. What did Hubble's stunning observation mean? Most scientists felt it indicated that the universe was expanding from a highly concentrated initial state.

The Big Bang Theory

Cosmologists continued to study the expansion phenomenon for many years. The rate of expansion was estimated, and by extrapolating backward, it was possible to determine the approximate time, 10 to 20 billion years ago, when all matter and energy were originally present at a single point in space, no larger than a dust particle. In 1955, a major theory about the origin of the universe emerged from these calculations. According to the **big bang theory**, the universe originated from an infinitely dense, infinitely hot point in space called a *singularity*. Some 10 to 20 billion years ago, the singularity exploded, an expansion of space began, and ultimately, matter formed.

By understanding the behavior and properties of radiation and matter and knowing that light travels at a definite speed, scientists have been able to create a picture of the universe that begins 10^{-43} second (this number is a decimal point followed by 42 zeros and then a one) after time zero. No hypotheses permit an assessment of anything preceding that event.

The First Second of the Universe

In the first fraction of a second after the big bang at time zero, the universe was contained within a space smaller than an atom (10^{-28} centimeter). Neither matter nor particles existed, only energy. The embryonic universe had begun a brief but extremely rapid period of expansion, and its temperature, initially billions of degrees Celsius (°C), began to decrease.

Within the next fraction of a second, the universe increased to the size of a grapefruit, but its temperature was still too high for matter to exist. A split second later, the inflationary period ended and the universe became filled with exotic classes of newborn particles created from energy. These first particles were the precursors of more familiar particles, neutrons and protons, which came into existence a moment later.

Galaxy Formation

During the first 500,000 years after time zero, the temperature of the universe decreased enough for the simplest atoms, hydrogen and helium, to form. Once atoms existed, the formation of matter could begin. Small concentrations of matter formed seedlike clusters that ultimately ballooned into vast galaxies or clusters of galaxies. Heavier elements, such as the calcium in your bones and the iron in your red blood cells, were created later in stars within the galaxies. Our own galaxy, the Milky Way, was created several billion years after the big bang and is one of billions of galaxies that now fill the universe. Thus the history of the universe since time zero consists of an ongoing expansion phase during which the forces, matter, particles, planets, stars, and galaxies existing today were created from energy that was initially present at the big bang.

The foregoing description of the origin of the universe reflects current hypotheses from astronomy and physics. Just as for other hypotheses in science, supporting evidence comes from complex experiments. However, many of the assumptions rely on highly theoretical constructs with only limited opportunities for testing.

ORIGIN OF OUR SOLAR SYSTEM

Origin of the Sun

Proceeding from the big bang, less speculative hypotheses about the origin of our solar system are supported by research results. Cosmologists believe that the formation of our sun and the system of planets that encircle it, our *solar system,* followed a developmental path that may not be uncommon in the universe.

Our solar system is hypothesized to have originated 5 to 10 billion years ago in a fog of gas and dust that existed in an arm of the Milky Way galaxy. The purported sequence of events in the creation of our solar system is traced in Figure 2.5. The sun began its life when an ancient cloud of hydrogen, helium, and other elements became large enough to be condensed and compressed by gravity. After its initial formation, the sun began to get hotter. As the young sun continued to increase in size, it accumulated more matter from the galactic haze. Eventually, great quantities of these substances were spewed into surrounding orbits, and a slowly rotating disk formed. This huge disk, billions of miles across, was the forerunner of our solar system.

Planetary Genesis

The raw materials necessary for forming planets—hydrogen and helium gases and "dust" containing calcium, silicon, uranium, aluminum, iron, carbon, sulfur, phosphorus, and oxygen—were present in the disk surrounding the young sun. At some point, these substances started to coalesce into solids, beginning with pebble-sized objects, that grew larger and larger. As the temperature of the sun rose, a few of these developing bodies increased in size until they formed small embryonic planets that continued to obtain additional materials by gravitational attraction.

About 5 billion years ago, our solar system entered its final phase of development. At that time, the sun began emitting scouring blasts of solar wind, blowing gases out into space. This reduced the smaller inner planets essentially to rock with no surrounding gases. The giant outer planets, farther from the sun, retained more of their primordial gases. This gave them greater size.

The sun, shown in Figure 2.6, is the major driving force of our solar system. It is a star that derives its enormous energy from the conversion of hydrogen to helium. We recognize this energy as sunshine that illuminates and warms the planet we inhabit. At 4.6 billion years of age, the sun has consumed about 50 percent of its fuel, so its life is nearly half over.

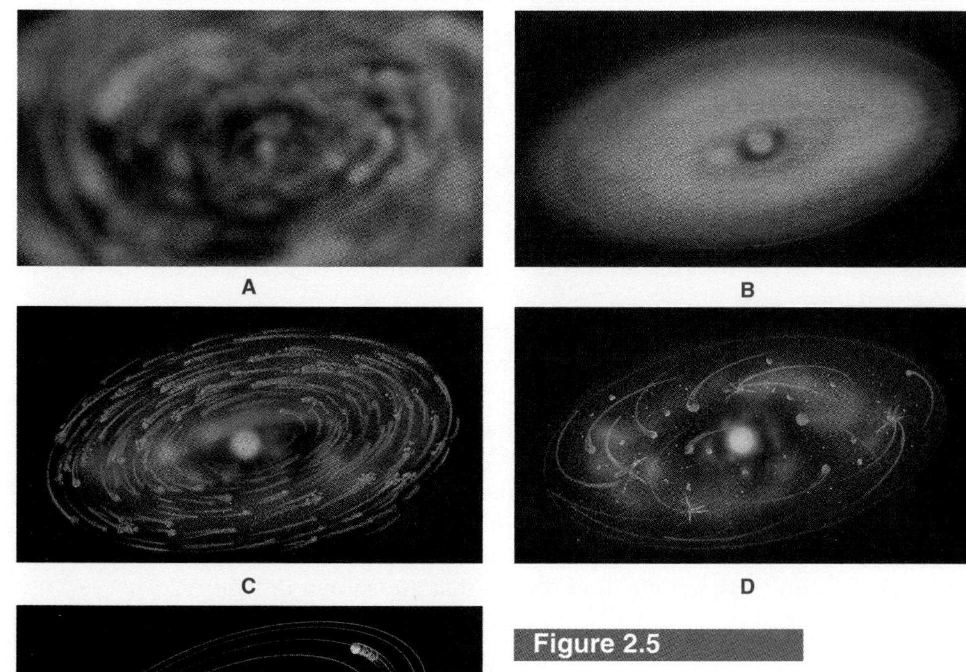

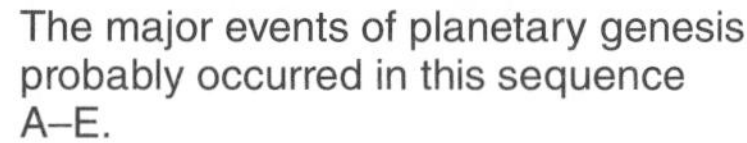

Figure 2.5

The major events of planetary genesis probably occurred in this sequence A–E.

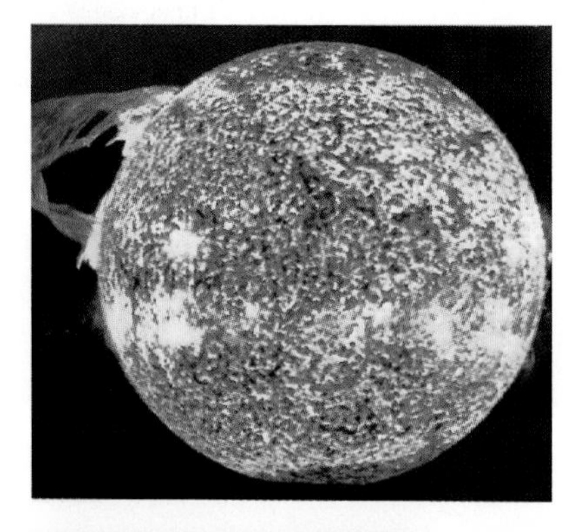

Figure 2.6

The sun is the source of radiant energy and heat, two forms of energy that maintain Earth's environments and the life that inhabits the planet.

Figure 2.7

The blue planet—Earth—as it appears from outer space.

Understanding the creation processes that occurred in the universe taxes the intellect and imagination. To comprehend the origin, nature, and future of the universe, it is necessary to traverse an expanse of time from mathematical singularity—infinite density at time zero—to forever, the length of time that the universe may continue to expand. It began with ultimate simplicity when matter did not exist and all forces and forms of energy were interchangeable. In its present stage, the universe consists of billions of galaxies and trillions of stars spread over immeasurable space. These galaxies are continuing to rush apart, the distance between any particular pair doubling every 10 billion years. How long will this expansion of space continue?

Cosmologists are uncertain about what ultimately lies ahead. The universe may eventually collapse, in which case another big bang may occur someday, or it may simply continue to expand, slowly dissipating its energy, and finally fade away, its cold ashes forever spreading throughout the expanse of space. However, the universe is now considered to be in its youth, and its demise is not expected for several billion years.

The origin of our solar system within the Milky Way galaxy may have followed a developmental path that has been taken in other galaxies in other universes at other times. After the evolution of our solar system, the stage was set for one planet, third in distance from the sun, to continue its development under the creative forces of nature (Figure 2.7).

The universe began 10 to 20 billion years ago at the moment of the big bang. Subsequently, billions of galaxies were created from the energy and matter that had originally been present at a single point in space. Our solar system—the sun and its planets—arose in an arm of the Milky Way galaxy 5 to 10 billion years ago as a result of the sun's activities.

DEVELOPMENT OF THE PLANET EARTH

Early Earth

How did the beautiful blue planet known as Earth change after its origin nearly 4.6 billion years ago? Knowledge from several fields of science, including geology, atmospheric science, chemistry, and biology, has been used in forming and testing scientific hypotheses about the maturation and development of Earth into its present form.

Geologists hypothesize that early Earth was a bare, cold, rocky planet with no oceans or atmosphere. As time passed, Earth warmed as a result of gravitational compression and radioactive decay of heavy elements deposited at the time of its origin. Consequently, the heated interior of the planet became molten, and layers of different densities formed. More than 4 billion years ago, as Earth began to cool, this density differentiation resulted in the formation of internal zones as very heavy materials sank toward the planet's center and lighter elements accumulated near the surface. This process produced the major divisions of Earth's inner space—the core, mantle, and crust—that are shown in Figure 2.8.

Earth's Physical Structure

Geologists believe that Earth's core consists of hot, molten metals, primarily iron and nickel, with smaller amounts of other chemical elements. The planet's thick mantle constitutes about 80 percent of its volume. The thin outer crust has two components, one consisting of blocks that form the continents and the other of blocks that constitute the ocean floors.

Early Landmasses

Early continental landmasses were not anchored at specific locations for great lengths of time. Rather, they moved restlessly on shifting, brittle slabs or mantle plates. Study of the continuous but glacially slow movement of these blocks is referred to as *plate tectonics*. As described in Figure 2.9, it is now known that plate movements have occurred throughout Earth's history, from the time when the first crustal blocks formed to the present day. The ceaseless wanderings and interactions of the plates resulted in earthquakes, mountain building, and volcanoes. These forces had a profound effect on the geological history of Earth and on biological evolution. Movements of these plates continue to have an impact on geographical areas now occupied by humans (see Figure 2.10).

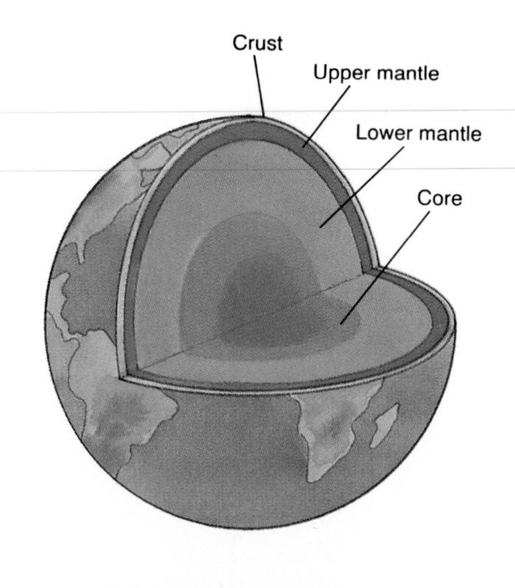

Figure 2.8

A cutaway view of Earth showing its interior and exterior structure. The hot molten core is surrounded by a thick mantle. The lower mantle and upper mantle are 1,300 and 300 miles deep. The thin outer crust is composed of continental plates (those supporting the continents) and oceanic plates (those supporting the oceans).

The Planet Matures

Several factors influenced the development of early Earth. Foremost among these were the sun and physical (i.e., geological) changes.

The Nature of Energy in Sunlight

The sun emits *electromagnetic radiation,* which consists of beams of light, or **photons**, of different wavelengths, as shown in Figure 2.11. The wavelength of a photon is related to its energy: the shorter the wavelength, the greater the energy of the photon. The high-energy photons called gamma rays and X-rays are very harmful to living systems because they can alter biological molecules. However, these types of solar radiation rarely reach Earth's surface. Ultraviolet light is also damaging to biological molecules, and this type of radiation penetrated to the surface of early Earth. Infrared radiation emitted by the sun, and also from young Earth as it cooled, has low energy levels, as do microwaves and radio waves. Photons with characteristic wavelengths in the visible light spectrum are perceived as distinct colors because they affect light receptor cells in the eye differently. The

Figure 2.9

Continental and oceanic plates have moved slowly throughout Earth's history. Since existing in a single colossal landmass called Pangaea 200 million years ago, the major continental plates migrated to positions shown in this figure.

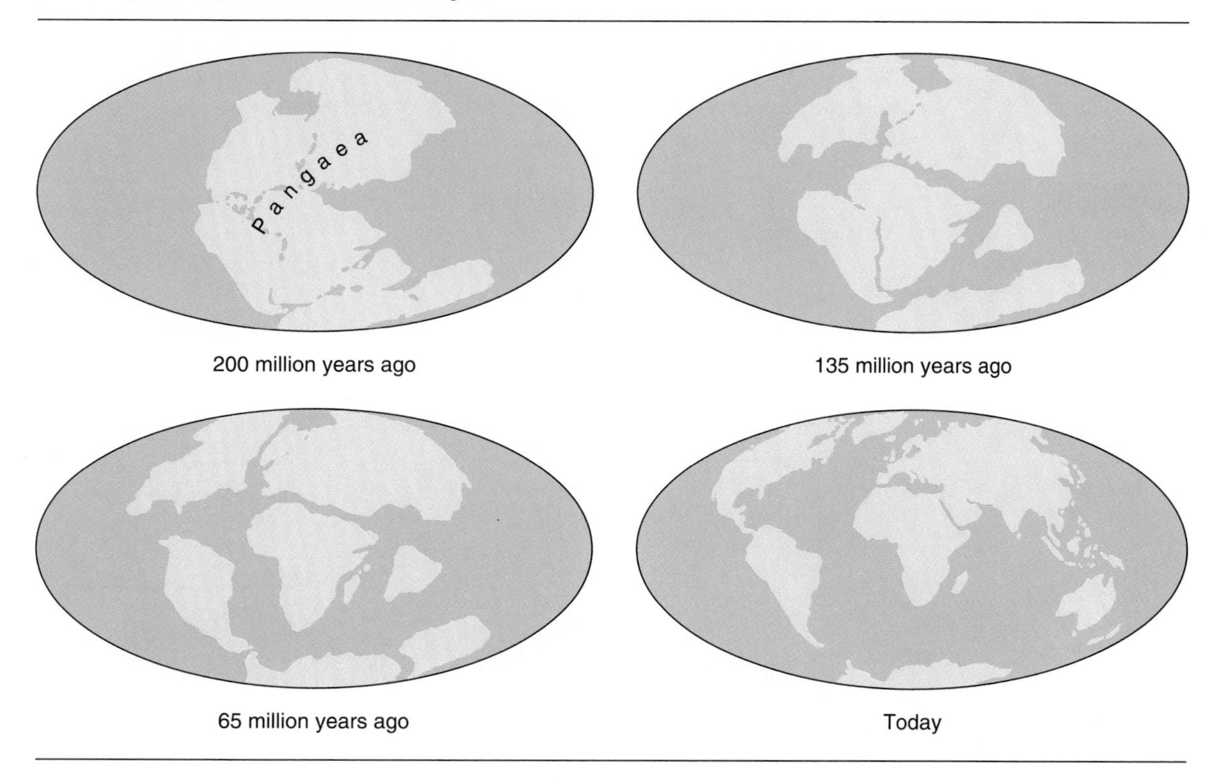

Figure 2.10

When continental or oceanic plates shift position, major changes can occur in areas supported by the affected plates. The San Francisco earthquake of 1989 (A) and the continuing eruption of the Kilauea volcano in Hawaii (B) were consequences of plate movements.

shortest wavelengths are seen as violet or blue colors, and the longest visible photons appear red. Radiation of wavelengths longer than those of visible light does not directly alter biological molecules.

Formation of the Early Atmosphere

Approximately 4 billion years ago, the first volcanoes penetrated through Earth's crust and vented steam and volatile gases that formed the primordial atmosphere. This early atmosphere is now hypothesized to have consisted of water (H_2O), carbon monoxide (CO), carbon dioxide (CO_2), nitrogen (N_2), hydrogen sulfide (H_2S), and hydrogen (H_2). Little or no oxygen gas (O_2) is thought to have been present at this time. Thus the gaseous composition of early Earth's atmosphere was dramatically different from our present atmosphere, which consists primarily of N_2 and O_2. In addition, lightning and intense concentrations of ultraviolet radiation, significant energy sources, may have played important roles in the later origin of life forms. Characterized by erupting volcanoes; lightning; vicious winds and heat; the absence of solid earth, oceans, or lakes; and noxious gases in the atmosphere, early Earth was an incredibly hostile environment.

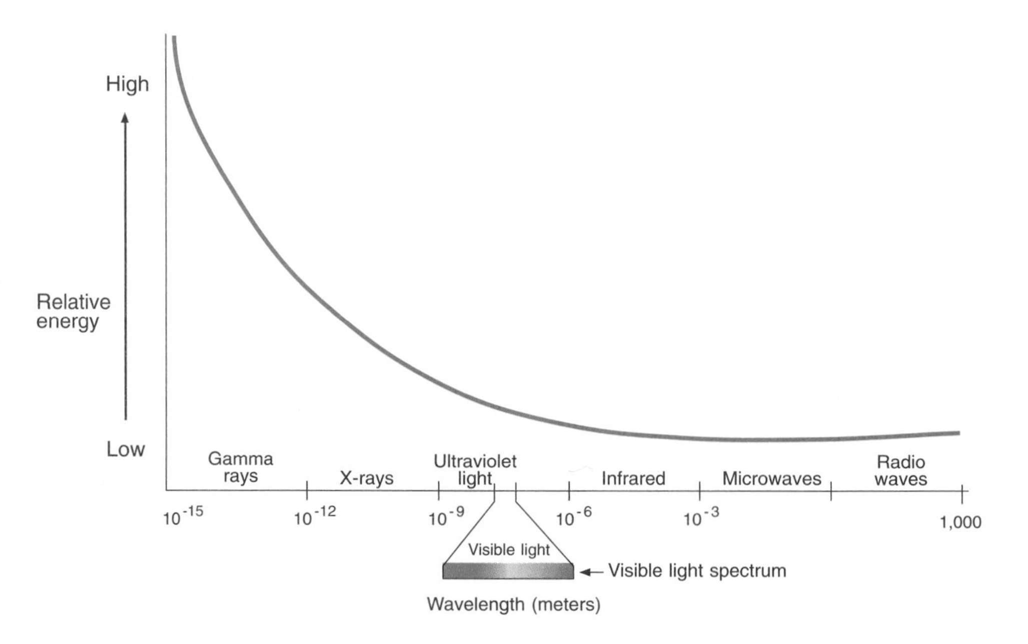

Figure 2.11

Sunlight consists of different photons, each with a characteristic wavelength and energy. Photons with short wavelengths (gamma rays, X-rays, and ultraviolet light) can alter biological molecules and are therefore harmful to living organisms. Photons with longer wavelengths do not cause direct damage to molecules but produce low levels of heat.

Figure 2.12

During the first billion years after its formation, Earth was characterized by erupting volcanoes, lightning, poisonous gases in the atmosphere, and little, if any, water on its surface. As the planet matured, it cooled and became covered with water, gases in the atmosphere changed, and the first life forms developed.

Water

As the early atmosphere cooled, water vapor condensed into rain that fell continuously for thousands of years and began to fill the ocean basins. Carbon dioxide, water vapor, and other gases in the atmosphere prevented the escape of infrared radiation that was being emitted from Earth's surface; eventually this trapping process led to a balancing of the young planet's energy budget. Were it not for this energy retention within the early atmosphere, uncontrolled cooling would have occurred, water would have become ice, and Earth would have become a frozen wasteland, bathed in faint light as it orbited the young sun.

During the early period of Earth's development, precipitation containing dissolved CO_2 would have been weakly acidic. Thus acid rain is not a new phenomenon. As these acid waters flowed over the ancient landscapes, they are thought to have dissolved carbon-containing carbonates. Carbon is of critical importance because it is the fundamental building block of all chemicals used and produced by living organisms; such carbon-containing compounds are called **organic chemicals**. After entering primordial oceans, the carbonates, along with other molecules, provided the raw materials from which the first life forms developed (see Figure 2.12).

> After Earth was created about 4.6 billion years ago, it slowly changed during a billion-year period. From a bare, cold, rocky planet, it became transformed into one that was warm and had solid landmasses, water, and a gaseous atmosphere.

ORIGIN OF LIVING SYSTEMS

Where, when, and how did life arise on Earth? More specifically, how did the first biological system originate from nonliving matter on Earth? How did this first system give rise to *cells,* the fundamental structural and functional units of all living organisms? These questions are among the oldest and the most difficult in biology. The antiquity of events makes it impossible to design experiments that duplicate or re-create circumstances that may have existed on Earth when life began. Nevertheless, beginning in the 1920s, relevant information from many fields of science was used in formulating hypotheses about the probable course of events that took place on Earth during the previous 4 billion years. These hypotheses have been greatly strengthened by research conducted during the past 40 years.

Four major scientific hypotheses deal with the origin of life on our planet. One involves the transport of living systems to Earth from somewhere in outer space and is of little interest to scientists since it does not explain the origin of *that* life. The other three hypotheses are concerned with the sequence of changes in chemicals that are thought to have been present on early Earth. The first of these proposes that there was a random, chance assembly of chemicals into a system that could be considered living. Most scientists have rejected this hypothesis because it lacks experimental supporting evidence. The second hypothesis, now most widely accepted, is that beginning about 4 billion years ago, life developed gradually through a series of continuous changes in chemical systems that existed on the primitive planet. Certain features of this so-called **chemical evolution hypothesis** suggested some experiments that were carried out, and their results offer support of varying strength for its primary ideas (see the Focus on Scientific Inquiry, "The Chemical Evolution Hypothesis").

A third hypothesis about the origin of organic compounds and life on Earth has recently been proposed. After viewing and studying underwater vents along the seams of oceanic plates, where superheated water, gases, and magma blend with cool salty water, as they probably did 4 billion years ago, some geologists and oceanographers believe that those sites may have been the major source of early organic compounds. Further, they feel that life could have originated at ancient vents where hydrogen-rich gases escaping from Earth's interior reacted with carbon-rich gases in the water. This hypothesis is also supported by the presence of unique bacteria at similar sites today (see the Focus on Scientific Inquiry). One related finding of considerable interest is that organisms existing at the vent sites do not depend on the sun as their source of energy since these sites are shrouded in complete darkness. Energy that currently supports the assemblage of life at these sites originates from the heat and chemicals emitted from Earth's interior.

The Chemical Evolution Hypothesis and the First Prebiont

The purported sequence of changes in the chemical evolution hypothesis can be summarized as follows:

1. Formation and organization of molecules present in the early oceans into an "organic soup" (though it was more accurately a "chemical brew," we use the traditional term, *organic soup*)
2. Development of systems capable of arranging and using organic molecules for life processes such as self-replication and energy capture
3. Origin of the first primitive cell

The broad outlines of this theoretical sequence are generally accepted by the scientific community, although numerous details are still subject to debate.

Formation of the Organic Soup

By 4 billion years ago, Earth's crust had begun to solidify and its surface had cooled sufficiently for naturally formed, stable, complex organic molecules to exist. The first organisms for which fossil evidence is available are nearly 3.5 billion years old. These fossils are apparently the remains of ancient bacteria. Therefore, it can be concluded that the time required for the formation of the organic soup and the appearance of the first primitive cells was approximately 500 million years.

How did the early oceans become an organic soup? During the time before organisms were present on Earth—the *prebiotic phase*—conditions prevailed that were favorable for the synthesis of organic compounds. There was an abundance of simple chemicals in the atmosphere that could serve as building blocks for larger organic chemicals. A variety of energy sources existed, including electrical discharges (lightning), extremely reactive ultraviolet radiation from the sun, visible light, and heat. Such energy could promote the combination of simple chemicals into more complex chemicals in the atmosphere that would become incorporated into rain and enter the warm early oceans. Alternatively, organic compounds may have come from outer space, since these chemicals are known

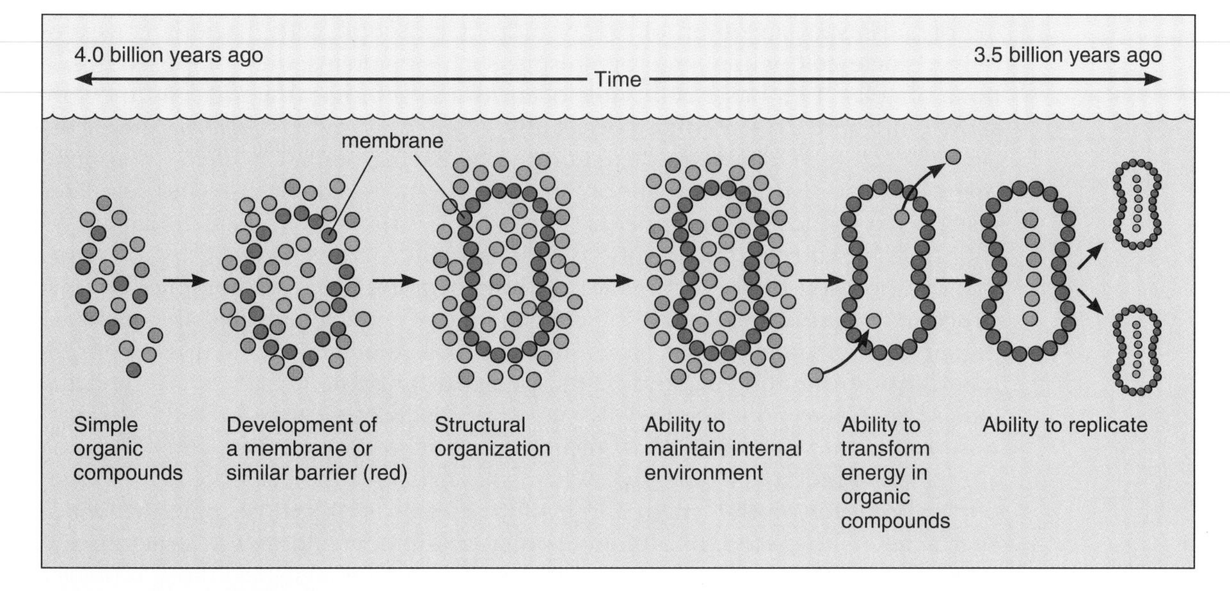

Figure 2.13

A summary of events hypothesized to have led to the development of the first life form on Earth, beginning 4 billion years ago.

to be widespread in the universe and have been found deep inside meteorites that collided with Earth. Finally, there was no free (uncombined) O_2 present to react with (oxidize) and destroy the newly formed organic compounds. Thus it is hypothesized that stable organic compounds originated as a result of these circumstances. Elaborate experiments have been done that offer strong support for this sequence of events, if assumptions about the nature of the early atmosphere are correct.

Development of Organized Forms from the Organic Soup

Prebiological systems, referred to as *prebionts,* which developed next, are thought to have originated as aggregates of organic molecules that became capable of arranging and using chemicals in the soup for processes we associate with life. These hypothesized prebionts, shown in Figure 2.13, may have evolved through the following sequence of events: development of a membrane, or similar barrier, that separated the interior and exterior environments of the "individual"; structural organization; ability to maintain an internal environment that differed from the surrounding external environment; competence in converting chemical energy available in compounds present in the organic soup into energy that could be used for maintenance; and finally, self-replication, or the ability to reproduce.

The exact nature of such prebionts is, of course, not known. Some interesting studies suggest that these early aggregates of chemicals were first formed within stable microscopic droplets. No abrupt conversion of prebionts to organisms that could be classified as living is thought to have occurred.

The first prebiological system probably arose from the rich, warm, shallow seas of early Earth. This first prebiont is hypothesized to have been surrounded by a membrane that separated its internal environment from the environment around it. It may also have been able to use energy and to reproduce.

Origin of Cells

Cells are the fundamental structural and functional units of life. Cells carry out activities that enable organisms to develop, grow, survive, and reproduce. All living organisms are made up of one or more microscopic cells; bacteria consist of only a single cell, while large organisms, such as humans, are composed of billions of cells. Two basic cell types occur in the biological world, and they are shown in Figure 2.14. **Prokaryotic cells**, commonly called *bacteria,* are of the simplest cell type. They are distinguished by having neither a *nucleus* (a membrane-enclosed structure that contains genetic material—DNA—and regulates many cell activities) nor *organelles* (microscopic structures in the cell that are responsible for carrying on specific functions such as converting sugars to energy that can be used by the organism). **Eukaryotic cells** have a complex internal structure and are able to carry out a wide range of cellular activities. In contrast to prokaryotes, eukaryotic cells have a true nucleus, genetic material that is concentrated inside the nucleus, and many specialized organelles. They are typically much larger than prokaryotic cells. Most organisms other than bacteria are composed of eukaryotic cells. How did prokaryotic and eukaryotic cells originate after the first prebionts appeared? What is the nature of the evidence used to support hypotheses on the origin and evolution of cells? It will become evident in later chapters that modern biology emphasizes understanding the structure and activities of molecules within cells. Using very powerful analytical tools, scientists have generated great amounts of data that have been used in defining the relationships between different types of cells. Much of what follows is supported by results from experiments conducted during the past decade.

Origin of Self-replicating Systems

The development of a mechanism for self-replication was a critical step in the origin of cells, since the ability to reproduce is a hallmark of life. Once self-replicating, ordered systems had developed, rapid progress in a continuous evolution toward a more highly ordered cell became possible.

Biological evolution, the process of changes that occur in populations of living organisms over long periods of time, and the beginning of definite *lines of descent* (for example, an individual of one type of cell could give rise to an indi-

Figure 2.14

All organisms on Earth are composed of one of two types of cells. Bacteria are prokaryotic cells, a type of cell that has no nucleus or organelles. Most other organisms (including humans) consist of eukaryotic cells, a more complex cell type that has a nucleus and numerous organelles.

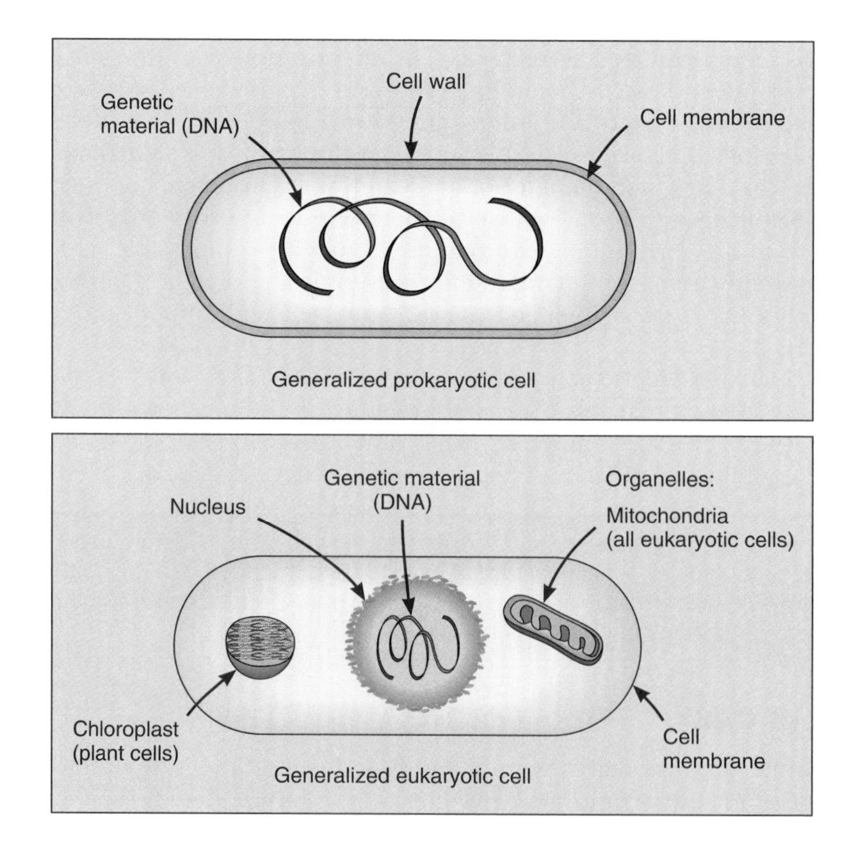

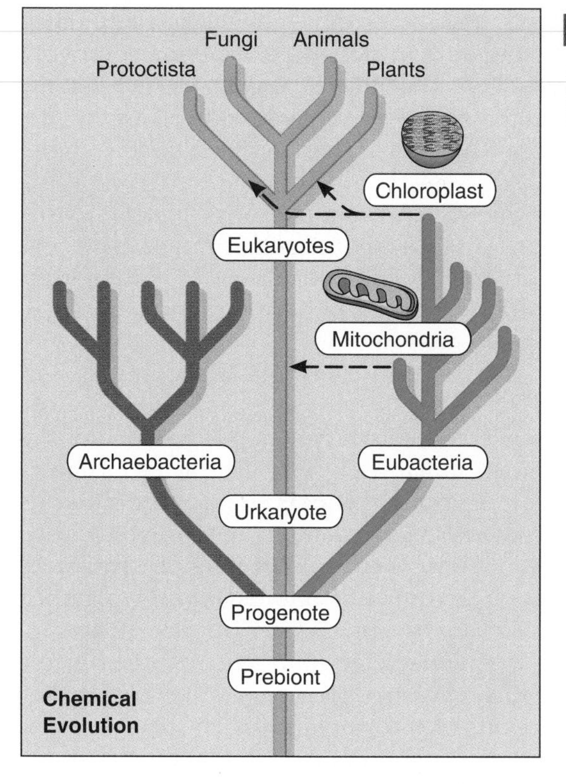

Figure 2.15

A phylogenetic chart describing the origin and evolution of major life forms on Earth.

vidual of a slightly different type of cell, which, with the process continuing over thousands or millions of years, eventually results in the evolution of a new cell type) are thought to have begun at this stage in Earth's history. Figure 2.15 illustrates current ideas about how different cells and organisms arose and evolved on Earth.

The point at which a "living" cell evolved from a prebiont remains speculative. The escalation from a self-replicating, ordered system (a prebiont) to even the simplest cell is immense and must have included many small developmental steps and possibly one or more precellular stages. Eventually, however, living cells did appear. Scientists hypothesize that both eukaryotic and prokaryotic cells arose independently from a common precellular stage of chemical organization called a **progenote** (after the term *progenitor*). That is, the progenote was the ancestor of both cell types. Little is known about the hypothesized progenote since it left no fossils or any other traces of its existence. Thus there is no direct evidence that the progenote ever existed. However, based on a sound biological principle—more complex forms evolve from simpler forms—it seems likely that such a transitional form did exist and gave rise to the two major cell lines. Nevertheless, the origin of cells remains the area of greatest scientific ignorance in the field of origins.

Origin of Prokaryotic Cells

According to the idea now most widely accepted by scientists, two separate ancestral lines of prokaryotic cells evolved from the progenote between 3.5 and 3.8 billion years ago. One branch gave rise to the **archaebacteria**, primitive prokaryotes that inhabited harsh environments where little or no oxygen was present. Archaebacteria that exist on Earth today are found in similar environments—extremely salty, hot, or acidic, with or without oxygen. The second branch gave rise to **eubacteria**, more complicated prokaryotes that could live in diverse environments. Most modern prokaryotes are eubacteria. The oldest fossils are about 3.5 billion years old and resemble modern archaebacteria.

A critical function of all living cells is the ability to obtain energy by converting chemical substances from one form to another. Basically, high-energy forms are converted to low-energy forms, and the energy liberated by these transformations is used by the cell to carry out its activities. What were the sources of energy used by these first prokaryotes? The most likely source was high-energy organic molecules that were present in the primordial organic soup. Thus the first primitive cells are thought to have had a "free lunch" that provided them with the energy they required.

In time, the rich organic soup in the vast early seas became depleted. What happened when the days of the free lunch ended? Life on Earth might have entered the abyss of extinction were it not for the evolution, at least 2.5 billion years ago, of a new chemical process in certain types of eubacteria that provided a fresh and continuous source of high-energy molecules.

A New Supply of Energy

The new process, *photosynthesis,* has been appraised by scientists as one of the most important evolutionary developments in Earth's history. The steps of photosynthesis are described in Figure 2.16. **Photosynthesis** yields a continuous source of energy because photosynthetic cells use radiant energy from the sun to convert the raw materials of CO_2 and H_2O into energy-rich sugars, the source of energy for most organisms. Oxygen, a by-product of photosynthesis, played a central role in shaping the life forms that were to appear on Earth later. The sugars from photosynthesis served to replace the depleted nutrients from the organic soup and were used by the organisms that continued to appear and evolve during the next 2 billion years. The life functions of most organisms existing on Earth today are directly or indirectly dependent on the photosynthetic energy-transforming process carried on by modern green plants.

The Importance of Oxygen

As the O_2 produced during photosynthesis began to accumulate, it had a major impact on existing organisms and on Earth itself. Free O_2 was probably toxic to many of the prokaryotes that first inhabited the planet. Thus as the concentration of O_2 increased from 0.0001 percent to about 21 percent 1.5 to 2 billion years ago, most oxygen-intolerant prokaryotes became extinct. However, new prokary-

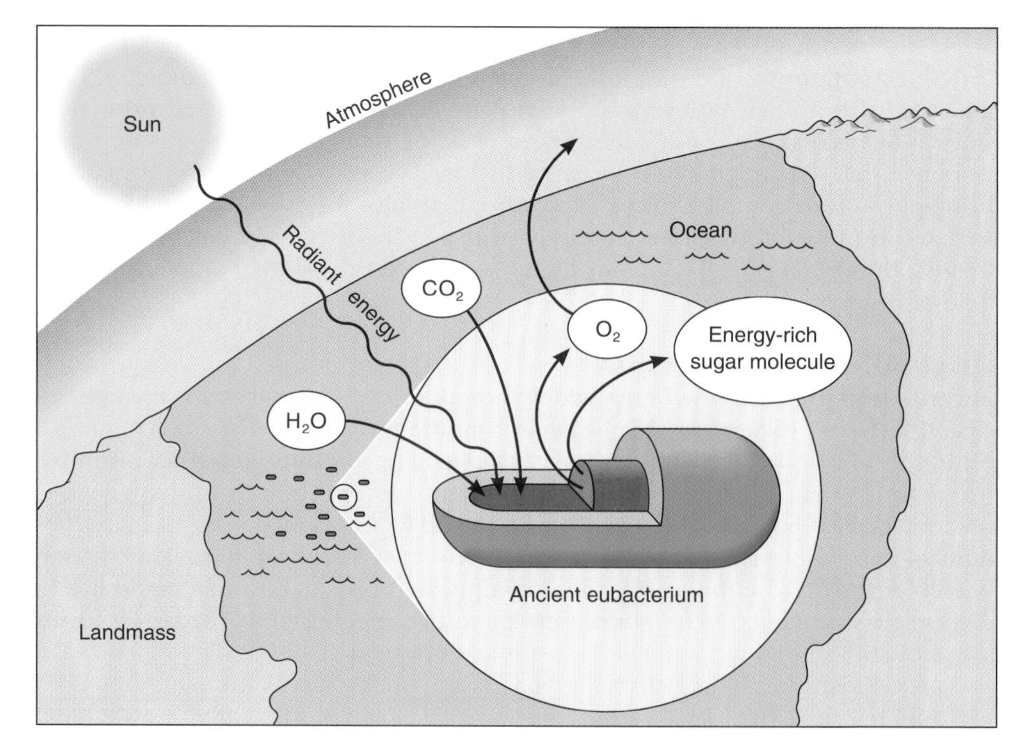

Figure 2.16

Photosynthesis is a chemical process in which radiant energy from the sun is used to convert carbon dioxide and water into energy-rich sugars with oxygen released as a by-product. Today, cells of plants and algae carry out this process. Scientists believe that photosynthesis evolved in eubacteria more than 2.5 billion years ago.

otes evolved that were not only able to tolerate O_2 but could actually use it in reactions that yielded great quantities of energy. Another significant development occurred as O_2 diffused into the outer atmosphere, where a portion of it was converted into the gas ozone (O_3). *Ozone* forms a layer in the outer atmosphere that absorbs ultraviolet radiation emitted by the sun and protects life from the harmful effects of this highly reactive energy source. Thus photosynthesis provided two of the original conditions thought essential for life to arise on ancient Earth—free O_2 and protection from ultraviolet radiation.

Despite its primitiveness in comparison with modern eukaryotic cells, prokaryotic life must be judged a marvelously successful series of biological advances. Perhaps the most significant measure of prokaryotes' accomplishments is that they were the only inhabitants of Earth for at least 2 billion years. During that long interval, the prokaryotes evolved all of the significant chemical systems and processes used by modern cells to carry on life. They helped transform Earth from a violent, barren landscape into a fertile planet capable of accommodating the other life forms that began to appear 2 billion years ago. As a result of their great success in evolving chemical processes, such as photosynthesis, and inhabiting all areas of the world, the first prokaryotes paved the way for the appearance of more complex cells.

Origin of Eukaryotic Cells

Most organisms present on Earth today are composed of eukaryotic cells. These cells all have a similar structure, use the same systems for carrying on chemical reactions necessary for normal life, and employ identical mechanisms for processing and preparing their genetic material for transmission to the next generation. Because of these similarities, biologists believe that eukaryotic cells evolved from a single, common precursor cell. But what was the nature of this ancestral cell, and how did the complicated eukaryotic cell evolve?

Eukaryotic cells first appear in the fossil record about 1.5 billion years ago. Until the 1960s, it was assumed that the ancestral cell was a prokaryote (that is, eukaryotic cells evolved directly from a prokaryotic cell line). However, a more complex hypothesis was then formulated that traces the emergence of the eukaryotic cell line from the progenote (see Figure 2.15). According to the hypothesis, the first eukaryote to evolve from the progenote was the **urkaryote**. Just as for the progenote, we have no direct evidence of the urkaryote. However, analyses of the molecular structure of prokaryotic and eukaryotic cells from hundreds of different organisms suggest that the urkaryote lineage appeared at about the same time as the prokaryote lineages.

The **endosymbiotic hypothesis** (see Figure 2.15) envisions modern, eukaryotic cells (those that evolved between 1 and 1.5 billion years ago) arising as a result of unions between at least two types of eubacteria that came to live within ancestral eukaryotic cells. Two organelles found in modern eukaryotic cells are of particular importance in this hypothesis. *Mitochondria* are found in all eukaryotic cells. They function in providing energy for the cell. *Chloroplasts* function in photosynthesis and are found only in eukaryotic plant and algae cells. Where did these organelles come from? Molecular biologists have found that the appearance and molecular structure of eukaryotic mitochondria and chloroplasts are very similar to ancient prokaryotic cells. Mitochondria and chloroplasts also behave as independent organisms when it comes time for the cell to reproduce. These findings offer strong support for the endosymbiotic hypothesis.

Evolution is explored in Chapters 22–28. For now, we simply point out that the evolution of organisms is strongly affected by factors present in their surrounding environment. Organisms that can cope with such factors may survive and produce offspring. What major factors influenced the development and evolution of living systems during the first 4 billion years of Earth's history? From examining the various hypotheses, it seems evident that the earliest living systems and cells, in order to survive and reproduce, must have been able to exist in ex-

The Chemical Evolution Hypothesis

Focus on Scientific Inquiry

The sequence of developments whereby simple chemicals gave rise to complex organic compounds that eventually developed into organized chemical-biological systems is termed *chemical evolution.* The basic ideas about these events were formulated independently in the 1920s by A. I. Oparin of the Soviet Union and J. B. S. Haldane of Great Britain. However, not until the 1950s did scientists begin to conduct experiments that supported the chemical evolution hypothesis and allowed for its widespread acceptance. What types of experiments could provide useful information about events that occurred 4 billion years ago?

By the 1950s, geological evidence, based on analyses of ancient rocks and sediments, offered some support for Oparin's and Haldane's model of an early atmosphere that contained hydrogen, water vapor, methane (a simple organic compound), and nitrogen-containing ammonia. Also, the energy sources present on early Earth were assumed to be electricity (lightning), ultraviolet light, and heat. According to the chemical evolution hypothesis, the first step in the origin of life was the creation of complex organic molecules from simple compounds that existed in Earth's atmosphere at that time. Logical questions were an outgrowth of this information. How were complex chemicals formed? Was it possible to design a relevant experiment that could demonstrate the construction of organic compounds from simple chemicals? Could existing energy sources, together or separately, have been capable of causing the formation of complex molecules? In 1953, Stanley L. Miller conducted a classic experiment that not only provided answers to these questions but also caused a major expansion of research efforts that further explored and explained chemical evolution.

Miller, at the time a graduate student at the University of Chicago working under the direction of Harold C. Urey, designed an experiment that attempted to simulate conditions on early Earth. He set up an airtight apparatus in which warmed water vapor containing a mixture of hydrogen, methane, and ammonia (components of the early atmosphere) was circulated past electrical discharges (used to mimic lightning) (see Figure 1). These gases were circulated for one week; then the contents of the experimental system were removed and analyzed. The researchers found that an astonishing number of organic compounds, including amino acids that have great biological importance in cellular activities, had been formed.

Miller's experiment provided the first evidence that one of the stages of the Oparin-Haldane hypothesis, organic compound synthesis, could have occurred. Since 1953, many additional experiments have been carried out using different mixtures of gases and other energy sources, such as ultraviolet light and heat. All have yielded similar results.

It is worth emphasizing here that scientific hypotheses are continually modified as new information becomes available, and the chemical evolution hypothesis is no exception. Since Miller's experiment, new knowledge about the early atmosphere has led scientists to revise their earlier view of its probable composition. It is now thought that the early atmosphere contained gases known to be emitted during the present-day outgassing of volcanoes. These include water, carbon monoxide, carbon dioxide, nitrogen, hydrogen sulfide, and hydrogen. Experi-

Figure 1 This figure depicts the type of apparatus Miller used, superimposed on a drawing depicting the conditions on primitive Earth that were similar to those in the apparatus. The results of his experiments offer support for the chemical evolution hypothesis.

ments using those gas mixtures have also yielded positive results for organic compound synthesis.

Hundreds of different organic compounds have been formed in the many primitive Earth simulation experiments conducted during the past 40 years, including representatives of all important molecules found in cells today. The assortment of experimental conditions in which atmospheric chemical synthesis has been demonstrated enables scientists to conclude that the first step in chemical evolution could have occurred. Although there is good evidence available for all the ideas discussed in this chapter, hypotheses about these events must be considered tentative pending further study. Details about many of the major steps are sketchy or missing, and there is no way of knowing if the laboratory early Earth simulations are representative of what actually took place on the primitive planet.

Are there other scientific hypotheses concerned with the origin of life on Earth? In fact, a controversial hypothesis emerged during the 1980s. As described in this chapter, some scientists believe that life may have originated in deep ocean vents where tectonic plates came together or volcanoes existed. They postulate the following sequence of events:

1. Water seeped into cracks and came in contact with molten rock lying under the plates.
2. The water became superheated, reacted with the rocks, and extracted carbon, nitrogen, oxygen, hydrogen, and sulfur, the chemicals required to make organic molecules.
3. Organic molecules formed.
4. Primitive cells developed.

What type of evidence would be required to support this "vent origin of life" hypothesis? What questions could be formulated to guide research on this hypothesis? Could a Miller-Urey type of experiment be conducted? What types of organisms exist at such sites today? Two key findings have been that hydrogen sulfide gas (H_2S) streams from the vents, and bacteria presently exist near these sites that grow in H_2S in laboratory experiments. Such bacteria are considered to be ancient. Could they have evolved at these vents billions of years ago?

What about water temperatures at the vents? They have been reported to be extremely hot, up to 350°C. These high temperatures have figured in some intense debates between scientists favoring the two major hypotheses (life originating in shallow seas or deep vents). In a disputed experiment, it was reported that bacteria collected from deep vent sites thrived at temperatures of 250°C. Another scientist was unable to confirm this finding and later published a paper in a prestigious journal that called the original finding an "experimental mistake." In 1988, Stanley Miller (yes, the same Stanley Miller) and a colleague reported results from an experiment that indicated that organic molecules are unstable and decompose rapidly at temperatures of 350°C. From this they concluded that life was highly unlikely to have originated at vent sites. A counterargument to this conclusion is that newly formed organic molecules could have moved rapidly into cooler water immediately after their synthesis. But then, how could "organisms" be created in the absence of an energy source to promote further chemical reactions? (Chapter 9 provides more information on life present at deep ocean water sites.)

The debates will undoubtedly continue. It is likely that complete answers to many questions will remain elusive given the antiquity and complexity of events associated with the origins of life on Earth. It must be understood that scientists are unable to offer total or absolute truths. Rather, their views are often stated in terms of the quality and quantity of evidence available and the degree to which it supports the hypothesis or idea being considered. Obtaining such evidence is the primary objective of scientific inquiry.

Figure 2.17

Plants and animals had become well established on Earth by 600 million years ago. These are pictures of two of the fossils from that period. (A) Fish and (B) sea pen.

tremely harsh environments with no O_2 present. Millions of years later, their descendants were able to use energy sources (molecules) present in the organic soup. Over the next 2 to 3 billion years, successful organisms and their offspring were presumably able to use different energy sources (for example, sunlight) once the organic soup was depleted, survive in the presence of O_2, and finally use O_2 for more efficient energy production in the cell.

The appearance of the modern eukaryotic cell ignited an explosion of great diversification among single-celled organisms, a phase that lasted 600 to 800 million years. By then, the opportunity existed for the evolution of organisms composed of aggregates of many cells that could become specialized to carry on a wide variety of specific functions.

Origin of Multicellular Organisms

Multicellular forms of life were abundant on Earth 600 million years ago. Fossils from that period reveal relatively complex organisms such as those shown in Figure 2.17. Many of the major groups of plants and animals present on Earth today are represented in the fossil record from that period. However, because of the general absence of fossils from the period between 600 million and 1.5 billion years ago, little is known about how multicellular organisms arose from unicellular (single-cell) eukaryotes. It is believed that multicellular evolution resulted from two different processes that occurred in eukaryotic cells: cellular aggregation, in which cells clumped together, and membrane formation, which resulted in single cells becoming divided into many cells. It is clear from the fossil record that once multicellularity evolved, there was a rapid increase in the appearance and diversity of new life forms on Earth.

Two cell types evolved from a single ancestral life form called a progenote. *Prokaryotic* cells are of the simplest cell type. They appeared first on Earth and flourished for over 2 billion years. During that period, they changed conditions on Earth in a way that favored the development and evolution of more complex eukaryotic cells. Multicellular organisms evolved from single-celled eukaryotes about 1 billion years ago.

THE FIVE KINGDOMS OF LIFE

For convenience and to permit order where there would be turmoil, biologists place modern organisms into five major groups according to certain characteristics that they possess. This process is known as *classification,* and the major groups are called *kingdoms.* The five kingdoms are named Prokaryotae (prokaryotes), Protoctista (unicellular eukaryotes and multicellular algae), Fungi, Plantae (plants), and Animalia (animals). This classification system recognizes the profound differences between the two major types of cells. Bacteria have been placed into their own kingdom, the Prokaryotae. The other four kingdoms include organisms made up of one or more eukaryotic cells and are further distinguished by their modes of nutrition, structure, reproduction, and development. Classification and the types of organisms found on Earth are considered in Chapters 3 and 4.

Multicellular organisms, constructed of eukaryotic cells, appeared on the planet less than 1 billion years ago and now seem to dominate it. However, that is an illusion related to size—microscopic prokaryotes outnumber eukaryotic organisms by at least 10^{43} to 1! The prokaryotes, though no longer the sole inhabitants of our planet, continue to play a critical role in maintaining life on Earth.

Summary

1. The big bang theory proposes that the universe began 10 to 20 billion years ago with a colossal explosion at a single point in space where all matter and energy originally existed in one form. During the first 500,000 years after the big bang, matter, mass, and particles were created from energy, and galaxies began to form in the universe.
2. Our solar system, consisting of the sun and its orbiting planets, formed 5 to 10 billion years ago as a result of matter becoming condensed and compressed. Planets formed when very small bodies began to accumulate vast amounts of solar material that later coalesced. The physical development of the planet Earth was completed between 4 and 5 billion years ago.
3. From its original rocky state, Earth matured into its present form through the effects of several processes that led to the formation and trapping of heat, the separation of internal substances into layers, the creation of landmasses, the development of an atmosphere, and the formation of warm, shallow oceans.
4. Life on Earth is hypothesized to trace its origins to chemicals that originated in the early oceans, creating an organic soup. These chemicals ultimately became organized into self-replicating systems that later evolved into simple cells. It is now thought that the first cells evolved from the progenote, a precellular stage of chemical organization, between 3.5 and 4 billion years ago.
5. The earliest cells were prokaryotes. They dominated Earth for 2 billion years. Eventually, they depleted the energy supply provided by the organic soup, and a new biological process, photosynthesis, evolved. The products of photosynthesis, sugars and oxygen, ultimately altered conditions on Earth and enhanced the evolution of new, more complex eukaryote cells.
6. The first multicellular organisms arose from unicellular eukaryotes about 1 billion years ago. These early life forms gave rise to all the plants and animals found on Earth today.

Review Questions

1. What are the relationships between hypotheses and levels of certainty?
2. How do scientific hypotheses, laws, and theories differ?
3. How does the big bang theory explain the origin of the universe?
4. Describe the sequence of events that led to the formation of Earth.
5. What factors influenced the early development of Earth 4 to 4.5 billion years ago?
6. How does the chemical evolution hypothesis explain the origin of the first prebiological system on Earth?
7. What factors led to the formation of the "organic soup" on early Earth?
8. What were the properties of the hypothesized prebiont?
9. Describe the differences between prokaryotic cells and eukaryotic cells.
10. How did photosynthesis and changes in atmospheric oxygen concentration influence the evolution of life on Earth?
11. How does the endosymbiotic hypothesis explain the origin of eukaryotic cells?

Essay and Discussion Questions

1. Why doesn't science use supernatural events or processes to explain or interpret natural phenomena?
2. Is it possible that a biological system, such as a prebiont or a progenote, could arise from nonliving materials on Earth today? Explain.
3. Figure 2.2 indicates the relative levels of certainty associated with hypotheses

discussed in this chapter. Make a list of other scientific hypotheses you are familiar with and describe their relative levels of certainty. What criteria can be used to make such judgments?
4. Over the years, scientists have held different opinions on the division between living and nonliving objects on Earth. Given the current ideas about the origin of life on Earth, is the distinction between living and nonliving valid? Explain.
5. If life is discovered on a distant planet, what features would you expect it to share with Earth's living beings? What differences would you expect?

References and Recommended Reading

Ben-Avraham, Z. 1981. The movement of continents. *Scientific American* 69:291–299.

Black, D. C. 1991. Worlds around other stars. *Scientific American,* 264:76–82.

Cloud, P. 1988. *Oasis in Space: Earth History from the Beginning*. New York: Norton.

Ferris, T. 1988. *Coming of Age in the Milky Way*. New York: Morrow.

Hawking, S. W. 1988. *A Brief History of Time: From the Big Bang to Black Holes*. Toronto: Bantam.

Kabnick, K. S., and D. A. Peattie. 1991. *Giardia*: A missing link between prokaryotes and eukaryotes. *American Scientist* 79:34–43.

Loomis, W. F. 1988. *Four Billion Years: An Essay on the Evolution of Genes and Organisms*. Sunderland, Mass.: Sinauer Associates.

Margulis, L. 1982. *Early Life*. Boston: Science Books International.

Margulis, L., and D. Sagan. 1986. *Microcosmos: Four Billion Years of Microbial Evolution*. New York: Summit Books.

Morgan, J. 1991. Trends in evolution. In the beginning . . . *Scientific American* 264:116–125.

Oparin, A. I. 1953. *The Origin of Life*. New York: Dover.

The once and future universe. 1983. *National Geographic* 163:704–749.

The planets: Between fire and ice. 1985. *National Geographic* 167:4–51.

CHAPTER 3

The Diversity and Classification of Life: Kingdoms Prokaryotae, Protoctista, and Fungi

Chapter Outline

Reading Questions

1. How great is the diversity of life on Earth?

2. What hierarchies are used to organize the various life forms on Earth?

3. What are the five kingdoms of life, and how do they differ?

4. How is each species assigned a unique name?

The enormous diversity of species present in tropical rain forests has recently been verified by a number of studies. A **species** consists of all the organisms of a particular kind that have the ability to interbreed and produce offspring. Alwyn H. Gentrey of the Missouri Botanical Gardens found nearly 300 species of trees per hectare in a forest near Iquitos, Peru, and stated:

> I conclude that the ever-wet forests of Upper Amazonia may be the world's richest in tree species. . . . Indeed, it is hard to imagine a more diverse forest than at Yanamono where there are only twice as many individuals . . . as species in a 1-hectare patch of forest, with 63% of species represented by single individuals and only 15% of species represented by more than two individuals.

Other recent studies in tropical Amazonia have determined that the world's largest inventories of birds, butterflies, amphibians, reptiles, and mammals occur in these rain forests.

The extent of species diversity on a worldwide basis remains unknown, but Edward O. Wilson of the Museum of Comparative Zoology at Harvard University has concluded that about 1.7 million species have been described, including approximately 250,000 flowering plants, 47,000 vertebrates, and over 750,000 insects. Estimates of total potential diversity, considering all biota of Earth, range from 5 to 30 million species!

SYSTEMATIC ORGANIZATION OF THE DIVERSITY OF LIFE

The systematic organization of this complex diversity of life was given its modern form in the eighteenth century, when the Swedish naturalist Carolus Linnaeus (1707–1778) developed a hierarchial classification system. He divided all known living organisms into two *kingdoms,* plants and animals. These kingdoms were each subdivided into sequentially less inclusive groups, from larger to smaller, which he named *class, order, genus,* and *species* (see the Focus on Scientific Explanations, "Classification"). An analogy can be used to illustrate the increasing specificity of information in this system. If a similar classification scheme were to be used to locate the place where a certain student lives, we might expect the following organization: *kingdom,* Earth; *class,* country (say, United States); *order,* state (California); *genus,* town (Eureka); *species,* address (555 North Student Lane).

Other levels have since been added to the Linnaean system. These include the *phylum* between the kingdom and class and the *family* between the order and genus. The use of terms or names from the two smallest groups, genus and species, constitutes a **binomial system** that provides an individual name for each species.

Figure 3.1

The lodgepole pine has the scientific (genus and species) name *Pinus contorta.*

The Binomial System

Binomial nomenclature is used today for naming all species. For example, the genus *Pinus* contains all the various pine trees of the world; the species *contorta* is limited to the lodgepole pine (Figure 3.1). By convention, genus names always begin with a capital letter and species names begin with a lowercase letter. Both are italicized or underlined when used as a scientific name (for example, *Pinus contorta*).

Linnaeus, who invented the binomial system, obtained the names he gave to organisms from the Latin language, a practice that continues today. Commonly, the names refer to some characteristic of the species. However, the surname of an individual is sometimes used in naming a species. Often such individuals have identified or studied the new species, but not always. For example, the scientific name of the cutthroat trout is linked with the famous Lewis and Clark expedition that first explored the western United States. As described by Meriwether Lewis in his *Journals of the Lewis and Clark Expedition* on June 13, 1805:

These trout are from 16 to 23 inches in length, precisely resemble our mountain or speckled trout in form and the position of their fins, but the specks on these are of a deep black instead of the red or gold of those common in the United States. These are furnished with long teeth on the pallet and tongue and have generally a small dash of red on each side behind the front ventral fins; the flesh is of a pale yellowish red, or when in good order, of a rose red.

The scientific name given this trout first described by Lewis is *Oncorhynchus clarki* after William Clark. Cutthroat trout with slightly different characteristics inhabit a number of distinct habitats throughout the West. Although there is some disagreement among scientists, each of the 14 different forms is designated by a different **subspecies** name. For example, *Oncorhynchus clarki clarki* is the cutthroat trout that lives in western coastal streams and migrates to the ocean. *O. clarki lewisi* inhabits areas of the inland Pacific Northwest, *O. clarki bouvieri* is the Yellowstone species, and *O. clarki pleuriticus* inhabits the Colorado River.

Both genus and species names are required to identify a given species. A common term can be used in naming two different species. For example, *Bison bison* is the North American bison, but *Enophrys bison* is the buffalo sculpin, a small (38-millimeter) marine fish living over shallow reefs along the Pacific coast (Figure 3.2).

A

B

Figure 3.2

The same name has been used for more than one species. For example, the term *bison* is used in the scientific name of both the North American bison, *Bison bison,* shown in (A), and the small marine buffalo sculpin, *Enophrys bison,* shown in (B).

The Kingdoms of Life

For a time, Linnaeus's two-kingdom system worked well enough to classify the organisms that had been identified in the early eighteenth century. However, as better microscopes brought the world of microorganisms into view, the two kingdoms, Plantae and Animalia, were no longer sufficient for classifying the huge number of diverse and unique microscopic organisms (Figure 3.3). Many microscopic species were found to have characteristics of both plants and animals, and it became increasingly difficult to place them comfortably in either of the two kingdoms. This problem prompted a number of scientists to suggest the creation of a third kingdom.

For microscopic organisms, John Hogg proposed the kingdom Protoctista in 1861, and in 1866, Ernst Haeckel suggested creating the kingdom Protista. The classification system advanced by Haeckel separated the blue-green algae and bacteria (prokaryotes) from the nucleated protists (eukaryotes) and placed them in the group Monera within the kingdom Protista. Other classification systems were proposed during the late nineteenth and early twentieth centuries, but Haeckel's three-kingdom system predominated.

In 1938, Herbert Copeland proposed a four-kingdom system. In addition to the plants and animals, he elevated the monerans to kingdom status and in 1956 accepted Hogg's term, Protoctista, as the kingdom name for the eukaryotic pro-

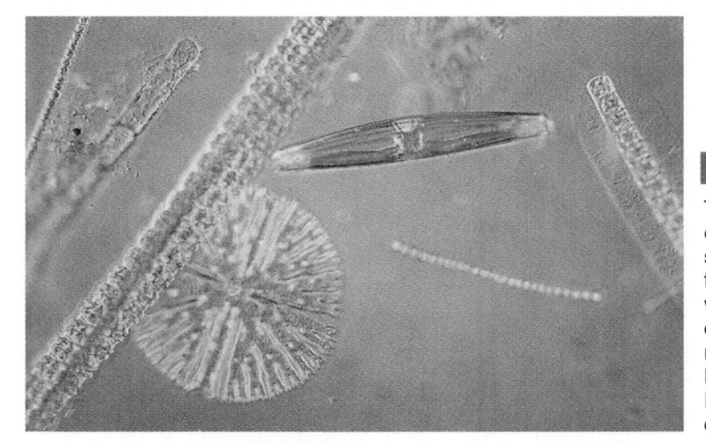

Figure 3.3

The compound microscope enabled early naturalists to study a new universe of life forms that could not be seen with the unaided eye. Consequently, a third kingdom, named Protoctista by John Hogg in 1861 and Protista by Ernst Haeckel in 1866, was created.

Figure 3.4

As depicted in the eighteenth century, all beings on Earth were linked in a scale that went from most perfect (humans) to nonliving minerals.

tists. The names applied to the two kingdoms, Monera and Protoctista, have varied over the past half century. However, the conceptual separation of life into four kingdoms remained unchanged until 1957, when Robert Whittaker proposed to elevate the fungi to kingdom rank, thus creating a fifth kingdom. Whittaker's five kingdoms were Monera, Protista, Fungi, Plantae, and Animalia, and this system has been adopted by most biologists since the 1960s.

In this book, we use Whittaker's five kingdom system as modified by Lynn Margulis and Karlene Schwartz in their 1988 edition of *Five Kingdoms: An Illustrated Guide to the Phyla of Life on Earth*. Their five kingdoms are Prokaryotae, Protoctista, Fungi, Plantae, and Animalia. However, they concede that their system, like all its predecessors, is imperfect, stating: "Our system has the advantage of defining the three multicellular kingdoms precisely, but the disadvantage of grouping together as protoctists amoebae, kelps, water molds, and other eukaryotes that have little in common with one another."

Information in this textbook concentrates on the major groups of living organisms in the five kingdoms that contribute significantly to our understanding of biological science (see also the appendix). The classification groups (taxa) now used by scientists are kingdom, division (in the kingdom Prokaryotae only), phylum, class, order, family, genus, and species; each of these groups may be further broken down into subcategories (subkingdoms, subphyla, and so on).

Biologists today refer to organisms as "higher" or "lower," "simple" or "advanced," and "primitive" or "complex." These are not value-laden terms and are not used to degrade the humble sponge and glorify the proud peacock. In part, the use of such terms is traditional, from a time when it was believed that all living organisms could be arranged on a single scale called the **chain of being** as depicted in Figure 3.4. This scale began with minerals, moved to plants and animals, and finally reached what were obviously the most "perfect" animals, humans. Although we no longer hold such ideas, terms like *simple* and *advanced* do distinguish between organisms that have a relatively simple organization and those that have many complicated organ systems. *Primitive* and *complex* do not mean that the primitive species have given rise through evolution to advanced forms but rather that their organization is less developed. *Higher* and *lower* similarly reflect levels of organization. Since in the evolution of organisms, more complex, advanced, higher forms generally came into being later than simple, lower, primitive ones (not the ones alive today, but their evolutionary ancestors), these terms also reflect some aspects of ancestry, but only in a very general way.

Life on Earth is classified into the kingdoms Prokaryotae, Protoctista, Fungi, Plantae, and Animalia. Other hierarchies include the phylum, class, order, family, genus, and species, each of which is less inclusive than the preceding level. The genus and species names combine to give each kind of organism a unique scientific name.

KINGDOM PROKARYOTAE

The kingdom Prokaryotae includes all known organisms constructed of cells that have neither an organized nucleus nor, with few exceptions, specialized cellular organelles (intracellular structures). It contains some species with characteristics similar to the earliest life forms that are thought to have occupied Earth. Today, they continue to participate in the cycling of many mineral elements that are required by other organisms within the biosphere. However, some prokaryotes are detrimental because they cause diseases in eukaryotic organisms from the other four kingdoms of life.

Prokaryotes are either *autotrophs* ("self-feeders") or *heterotrophs* ("other-feeders"). **Autotrophic bacteria** can synthesize complex organic molecules from simple molecules. Some can use light as an energy source for this synthesis and are known as **photosynthetic bacteria**, while others can use energy-rich molecules like hydrogen sulfide or methane: these are known as **chemosynthetic bacteria**. **Heterotrophic bacteria** obtain their nourishment from organic molecules formed by autotrophs and other heterotrophs. Bacteria participate in the decay of organic matter from organisms in all five kingdoms.

Research completed in the early 1980s indicates that there are two distinct types of prokaryotes. Consequently, Margulis and Schwartz separate the prokaryotes into two *subkingdoms,* the Archaebacteria and the Eubacteria.

Subkingdom Archaebacteria

The subkingdom Archaebacteria includes only two phyla, one containing *methanogenic bacteria* and the other *halophilic* ("salt-loving") and *thermoacidophilic* ("heat-acid-loving") bacteria. **Methanogenic bacteria** are unusual organisms that use simple organic molecules other than carbohydrates (sugars) or *proteins* as food sources. **Proteins** are a diverse group of organic molecules that are composed of chains of amino acid molecules built from carbon, hydrogen, oxygen, and nitrogen atoms. In Chapter 7 we will see that methanogenic bacteria are very important in the biogeochemical cycling of carbon, for they chemically

Figure 3.5

When substances dissolve in water, protons can separate from the water molecules. This disassociation of water molecules produces hydrogen ions (H^+) and hydroxide ions (OH^-). When acids and bases dissolve in water, acids release hydrogen ions, but bases combine with hydrogen ions. The concentration of H^+ and OH^- in solutions can vary by 100 trillion or more times. To reduce these concentrations to a manageable range of numbers, negative mathematical logarithms are used. The resulting pH values range on a scale from 0 to 14, corresponding to concentrations of $H^+ = 10^0$ to 10^{-14} and $OH^- = 10^{-14}$ to 10^0. At a standard temperature of 25°C, a solution would be neutral at pH 7 when both H^+ and OH^- concentrations are 10^{-7}. Most solutions with pH values below 7 are acidic, whereas those above 7 are basic.

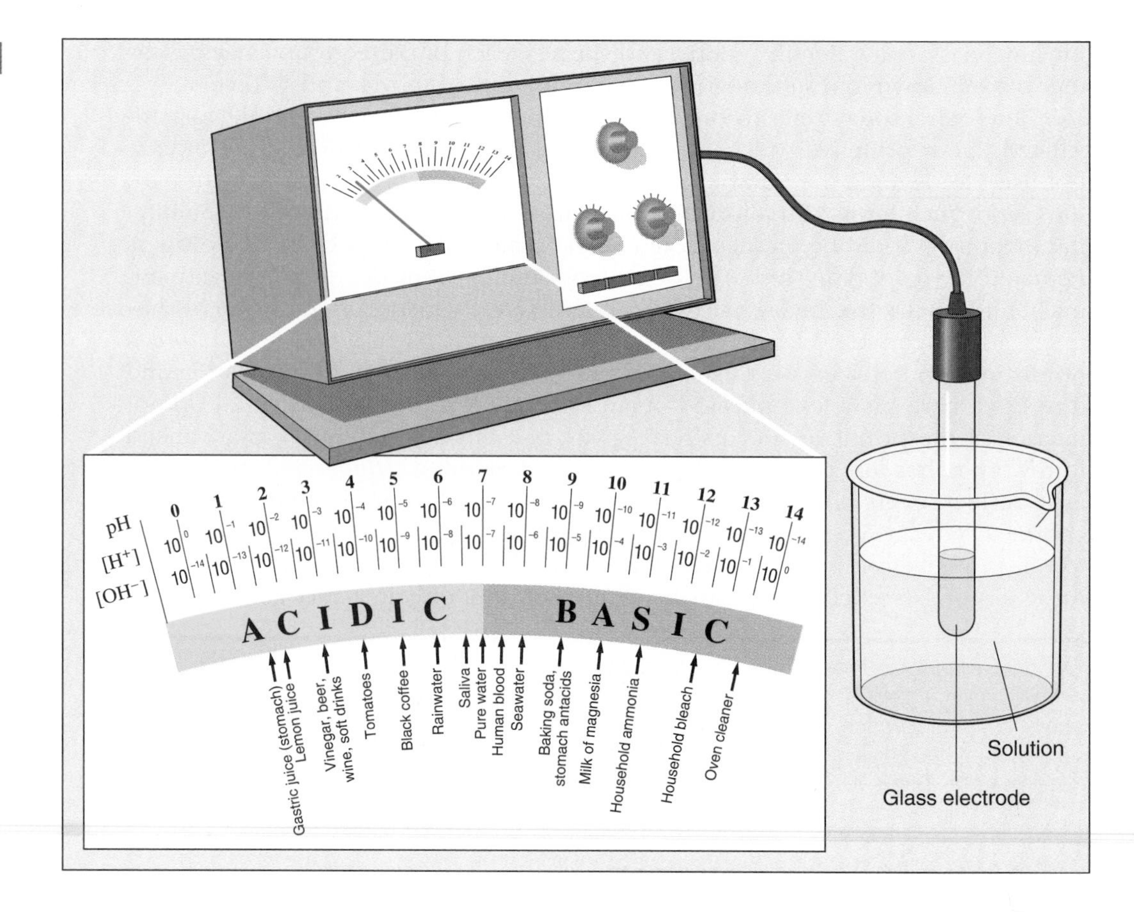

reduce carbon dioxide and *oxidize* hydrogen to produce methane, a colorless, odorless, flammable gas. **Oxidation** is a process in which an atom or molecule loses one or more electrons, and **reduction** is the process whereby an atom or molecule gains one or more electrons.

Halophilic and thermoacidophilic bacteria can exist in environments that extend their habitable ranges to nearly unbelievable limits. Some live in hot springs where temperatures may approach 90°C, and others live in environments where extreme pH ranges of 1 or 2 are common (see Figure 3.5). Representatives of the thermoacidophilic archaebacteria are found in the hot springs of Yellowstone National Park (see Figure 3.6).

Figure 3.6

Yellowstone National Park is located on a geological "hot spot." Surface water seeps into the ground, where it becomes superheated; as it rises back to the surface, the heated water carries mineral compounds that have dissolved in it.

Subkingdom Eubacteria

This subkingdom includes all "true" bacteria (eubacteria) and is partitioned into three different *divisions* based on the type of cell wall present in the bacteria. The first division contains only one phylum, which consists of species lacking rigid cell walls. The bacteria in this phylum cause some types of pneumonia in mammals, including humans.

The other two divisions have been distinguished historically by a contrasting affinity for a specific biological staining agent known as *Gram stain* (named after the Danish microbiologist Hans Christian Gram). **Gram-negative bacteria** have a cell wall that incorporates little of the Gram stain, causing them to appear light pink. In contrast, **gram-positive bacteria** have cell walls that retain greater amounts of the stain, causing the cells to develop a deep purple color.

Gram-negative bacteria include many of the photosynthetic cyanobacteria; the spirochetes that cause syphilis, yaws, and other pathogenic diseases; and one group of nitrogen-fixing bacteria, which are able to convert atmospheric nitrogen into a form that plants can use (discussed in Chapter 7), including species in the genus *Rhizobium* found in the root nodules of legumes such as alfalfa, beans, and lupine.

Gram-positive species include the fermenting bacteria that emit lactic acids or hydrogen sulfide (marsh or rotten-egg gas) as end products of their chemical activities. Also among the gram-positive bacteria are species capable of nitrogen fixation and some species in the genus *Streptomyces*. These bacteria are important because they produce antibiotics such as streptomycin.

The kingdom Prokaryotae contains all organisms composed of prokaryotic cells. This kingdom has two subkingdoms, the Archaebacteria and the Eubacteria, each of which has one or more divisions, and each division contains a number of phyla.

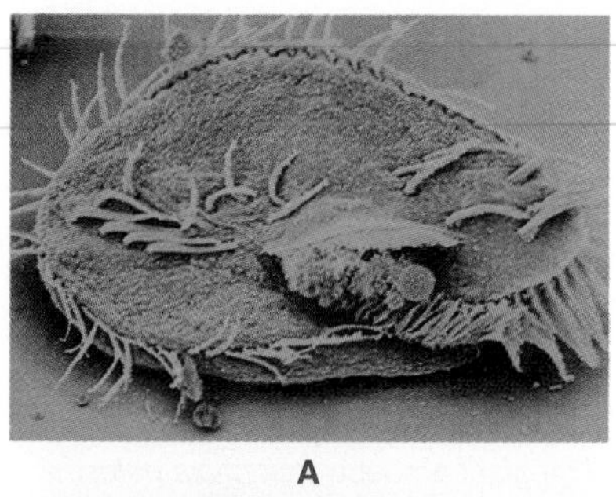

A

B

Figure 3.7

The kingdom Protoctista includes such diverse species as single-celled ciliated protozoans (A) and kelps (B). Kelps are giant brown algae that live in coastal waters.

KINGDOM PROTOCTISTA

The kingdom Protoctista contains four major groups of eukaryotic organisms: single-celled protozoans, unicellular algae, multicellular algae, and slime molds. Table 3.1 shows that this kingdom is defined by exclusion. All members have characteristics that exclude them from the other four kingdoms, but they are constructed of nucleated cells, respire aerobically (that is, they require oxygen), and most have **undulipodia**—specialized structures used in mobility—at some stage of their life cycle. However, they do not develop from a **blastula** (the hollow ball stage of animal development) or an **embryo**. It is a kingdom of extreme body forms, for it encompasses organisms as diverse as protozoans and giant kelps (Figure 3.7). This diversity is reflected in the myriad of methods of reproduction and development, from the simple cell division of an amoeba to the sexual reproduction of kelp (the fertilized egg germinates in response to light, rootlike rhizoids grow downward to become the holdfast, and a leaflike blade grows upward toward the surface of the ocean).

Table 3.1 Key to the Five Kingdoms

			Kingdom
1.	(a)	Cells without nucleus →	Prokaryotae
		or	
	(b)	Cells with nucleus →	2 (go to number 2)
2.	(a)	Undulipodia absent, absorptive nutrition, form spores →	Fungi
		or	
	(b)	Undulipodia usually present, photosynthetic or ingestive nutrition →	3
3.	(a)	Ingestive nutrition, blastula present during development →	Animalia
		or	
	(b)	Ingestive or photosynthetic nutrition, blastula absent →	4
4.	(a)	Photosynthetic nutrition, embryo present during development →	Plantae
		or	
	(b)	Ingestive or photosynthetic nutrition, embryo absent →	Protoctista

Keys designed to classify living organisms provide for choices between sets of characteristics. In this key, the absence of a nucleus (choice 1a) separates the kingdom Prokaryotae from the other four. The absence of undulipodia, absorptive nutrition, and the presence of spores (choice 2a) separates the kingdom Fungi. Ingestive nutrition and blastula formation are characteristics of organisms in the kingdom Animalia (choice 3a). The presence of an embryo (choice 4b) separates the kingdom Plantae from the kingdom Protoctista.

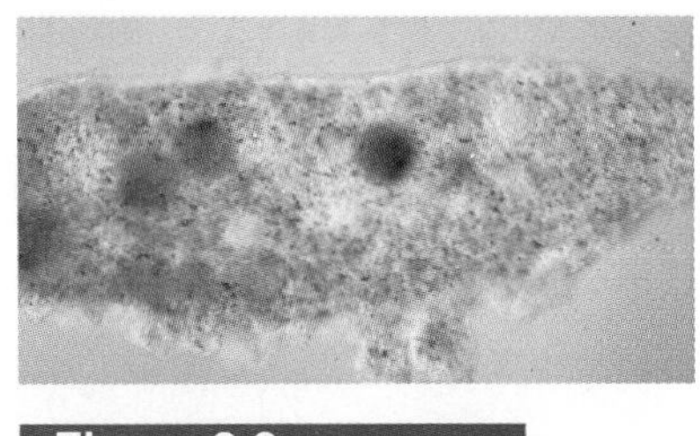

Figure 3.8

The giant amoeba *Pelomyxa palustris* may be the most primitive of all eukaryote life forms. This species has multiple membrane-bound nuclei but none of the other organelles that are found in all other eukaryotes.

Some scientists have recently suggested that various members of this kingdom are so distinctive that they could justify the creation of at least 20 separate kingdoms. Others have suggested that 40 to 50 phyla should be developed for all of the different varieties of organisms. Margulis and Schwartz list 27 phyla to accommodate this diverse assemblage of organisms.

The Giant Amoeba

The very primitive giant amoeba, *Pelomyxa palustris,* is the only species in the first phylum of the kingdom Protoctista (Figure 3.8). These amoebas are large, multinucleated cells with no chromosomes or specialized cellular organelles. They obtain energy from methanogenic bacteria that reside inside them. Giant amoebas inhabit mud at the bottom of freshwater ponds, where they contribute to the degradation of organic molecules and the cycling of carbon within the biosphere.

Protozoans

The protozoans classified by Haeckel in his kingdom Protista are placed in the kingdom Protoctista by Margulis and Schwartz. They have created seven phyla that include these "animallike," generally unicellular organisms. The four major phyla contain the following general types of eukaryotic organisms: complex, single-celled amoebas; flagellated protozoans; ciliated protozoans; and foraminifera. The undulipodia known as **flagella** and **cilia** are hairlike structures used in mobility.

Single-celled Amoebas

This phylum includes all free-living freshwater, marine, and soil amoebas, as well as those that are parasites of animals. Amoebas lack undulipodia and move by forming specialized, temporary membrane extensions called **pseudopodia** (Figure 3.9). They do not reproduce sexually but rather produce new cells through the asexual process of simple cell division (discussed in Chapter 13). Many members of this phylum form tough, enclosed, resting stages called **cysts** at some stage of their life cycle. This allows them to withstand adverse conditions for long periods of time. The intestinal parasites that cause amoebic dysentery germinate from such resistant cysts within the digestive tracts of their mammalian hosts, including humans.

Flagellated Protozoans

Members of this phylum are free-living, *mutualistic,* and *parasitic* species. **Mutualism** refers to a relationship in which two dissimilar organisms live together in an intimate association that benefits both. By contrast, in **parasitism**, only one of the organisms benefits while the other is generally affected in a negative way. An example of a mutualistic flagellate is *Trichonympha,* which inhabits the digestive systems of dry-wood termites (Figure 3.10). Bacteria associated with these flagellates digest the wood eaten by the termites. Both the termites and the flagellates benefit by deriving energy from the digested wood products. Some flagellated protozoans cause severe human parasitic diseases, including a form of sleeping sickness in Africa and Chagas' disease, which affects millions of people in Central and South America (Figure 3.11). In temperate North America, the parasitic flagellates are represented by the genus *Giardia,* which includes a number of forms that inhabit many seemingly pristine mountain lakes and streams. *Giardia* causes severe digestive tract problems when taken in by a suitable mammalian host, such as a human. The potential presence of this flagellate in all wilderness areas has made it necessary for campers routinely to boil all water used for personal consumption.

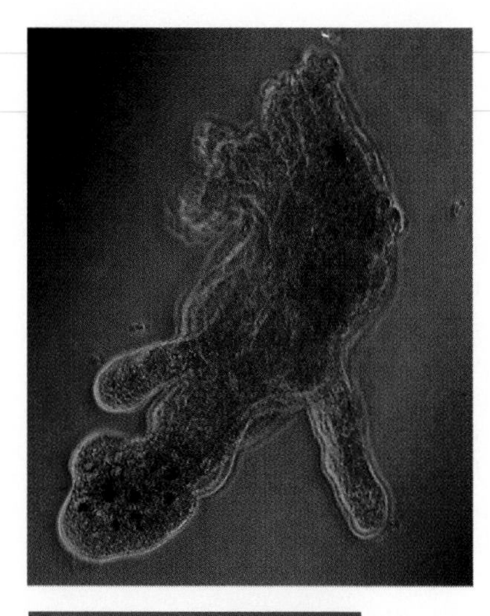

Figure 3.9

The amoeba *Amoeba proteus* moves by using pseudopodia. These "false feet" form from temporary extensions of the cell body, allowing the amoeba to engulf its food.

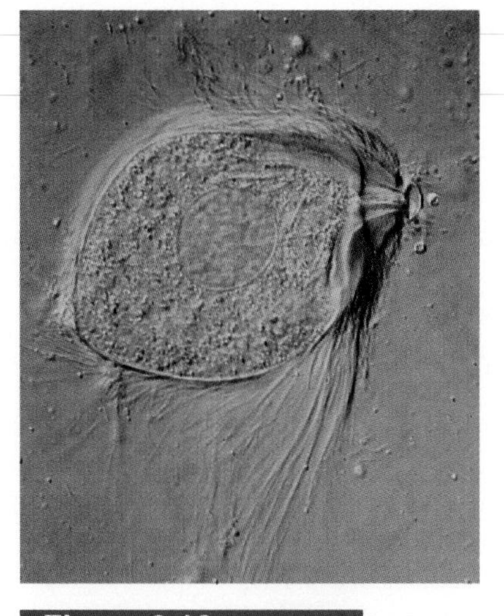

Figure 3.10

The flagellate *Trichonympha* sp. has up to 14 undulipodia known as flagella. Trichonymphs live as symbionts in the gut of dry-wood termites, where they digest cellulose in the wood fibers consumed by the termites.

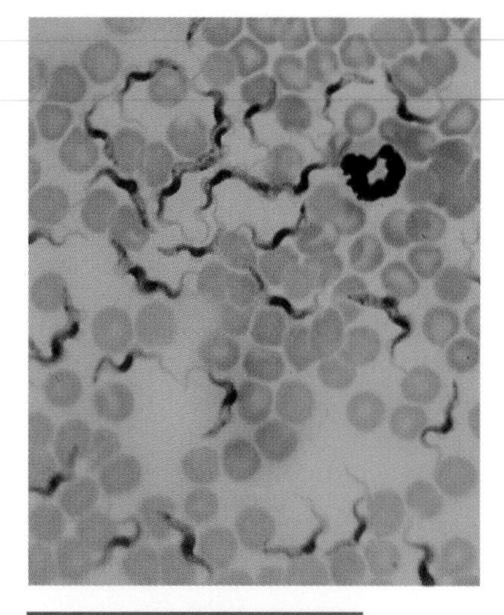

Figure 3.11

Some flagellated protozoans, including these trypanosomes, are parasitic and cause human diseases such as African sleeping sickness.

Ciliated Protozoans

This phylum contains species that are among the most complicated single-celled organisms. Members of this phylum have many specialized structures including cilia for mobility, **trichocysts** for capturing food and defense, **micronuclei** and **macronuclei**, and **contractile vacuoles** for expelling excess water from their cells (Figure 3.12). They are also distinguished by their use of **conjugation**, a variety of sexual reproduction that involves the "mating" of two cells, with an actual exchange of nuclear material, to produce new individuals. Members of this phylum live in aquatic environments, and more than 8,000 freshwater and marine species have been described. Most samples of water collected near the bottom of any pond will contain ciliates. They are very efficient "swimmers" that feed on bacteria and other microscopic creatures as they move through the water column. Few ciliates cause human diseases, but they have been of interest for decades to cell biologists because they are large, complex, and easy to grow in laboratory environments.

Foraminifera

Some species of foraminifera (forams) are the most aesthetically pleasing protoctists. All forams are marine species, and most live attached to various surfaces or to other organisms. However, some are free-living members of the plankton community (floating or drifting microscopic organisms in oceans, ponds, and lakes) and are important members of marine food chains. Forams are known as the "shelled protoctists" because they all have **tests**—external, hard coverings composed of mineral and organic molecular complexes. Great numbers of **filopodia**, needlelike pseudopodia, protrude from pores in the tests and are employed for mobility and feeding on algae, ciliates and other microscopic forms of life.

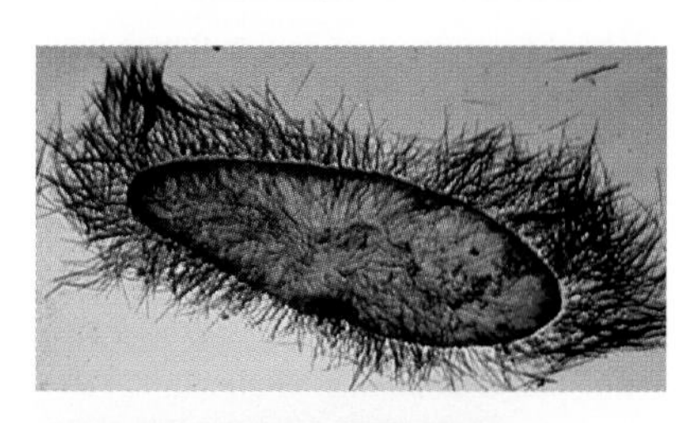

Figure 3.12

When living *Paramecium* species are stained, their trichocysts explode. In living paramecia, trichocysts function as darts that can be used to capture the small prey on which they feed.

Other Protozoan Phyla

The three remaining phyla of protozoans cover one group of small marine plankton known as **actinopodes** and two groups, **sporozoans** and **microsporidians**, that are parasites of higher vertebrates. One sporozoan genus is *Plasmodium,* which contains the parasites that cause malaria.

Unicellular Algae

Unicellular algae are separated into a number of protoctist phyla, the most important being the *dinoflagellates.* Microscopic **dinoflagellates** have nuclei bound to intercellular membranes and two undulipodia, a longitudinal one and a traverse one that circumscribes the cell within a groove in the cell wall or test. The tests of most dinoflagellates consist of **cellulose** (structural carbohydrate) plates embedded in the cell membrane and encrusted with cilia. Dinoflagellates often undergo an explosive increase in numbers during the summer months, resulting in "blooms" that cause "red tides" in near-shore marine waters. An inevitable consequence of red tides is a suspension in the harvesting of clams, mussels, and oysters in affected areas. The dinoflagellate responsible, *Gonyaulax tamarensis,* produces a neurotoxin that is concentrated by filter-feeding shellfish. Since this toxin affects humans (effects range from a tingling of the ears to respiratory paralysis and death), contaminated shellfish is unfit for consumption until the red tide dissipates (Figure 3.13).

Figure 3.13

Summer blooms of the dinoflagellate *Gonyaulax tamarensis* can produce red tides in near-shore ocean waters. When filter-feeding shellfish feed on *Gonyaulax,* they may accumulate high levels of the neurotoxin concentrated in the unicellular algae. If mussels are then eaten by humans, the neurotoxin they contain can be lethal.

The amount of photosynthesis carried out by dinoflagellates is surpassed only by the marine diatoms. The phylum of the diatoms includes marine and freshwater species that are very important in aquatic food chains. Each one of the nearly 10,000 species of diatoms displays a unique pattern in its test, an extraordinary example of the variety present in nature (Figure 3.14).

Figure 3.14

Every diatom species has a unique test pattern. Photosynthetic diatoms are very important in aquatic systems because they are able to convert sunlight into a form of chemical energy that is stored in organic molecules.

More unicellular algae are found in two other phyla. **Euglenoid flagellates** possess flagella for motility. *Euglena gracilis,* a representative species, is often used in studies of cell structure and function (Figure 3.15). **Golden algae** have flagella and tend to have tests made primarily of silicon. Golden algae can also exist in aggregates of cells, which suggests that they may have given rise to the true multicellular algae that arose later.

Figure 3.15

Flagellated unicellular algae, including *Euglena,* are important photosynthetic organisms in aquatic systems.

Multicellular Algae

The greatest degree of "organized cell aggregation" (approaching the level of a multicellular organism) in the kingdom Protoctista is seen in the phyla of brown, red, and green algae. These phyla include the major aquatic **producers** (organisms capable of capturing light energy and making complex organic substances from simple inorganic molecules) of freshwater and marine environments.

Brown Algae

About 1,500 species of brown algae are found within and below the intertidal zone of rocky coasts in temperate continental regions. The larger forms, some attaining a length of nearly 100 meters, are known as **kelp** (Figure 3.16). Kelp dominates these intertidal habitats and forms the base of some food chains described in Chapter 7.

Brown algae are a varied group, ranging in size from microscopic forms composed of single filaments to the large kelps. They also have a number of characteristics that enable them to survive in harsh intertidal areas. Some species have hollow sacs that are filled with water, which helps to prevent drying during exposure to air at low tides. In other species, these sacs are inflated with air and serve as float bladders, bringing the algae up near the surface and into the light during high tides, allowing them to carry on photosynthesis throughout the entire tidal cycle. Finally, kelps have anchoring structures called **holdfasts** that anchor them to the rocky shore and allow them to survive in the pounding surf (Figure 3.17).

Red Algae

Red algae also reside in rocky intertidal areas along temperate coasts; however, their distribution is more cosmopolitan since they can occupy both sandy and rocky tropical beaches (Figure 3.18). There are about 4,000 species of red algae, making them the most diverse multicellular algae phylum. Because some become

Figure 3.16

Bullwhip kelp (*Nereocystis luetkeana*) forms dense beds in offshore ocean waters up to 80 meters deep. Its hold-fasts often give way in the pounding surf, and masses of kelp wash up on Pacific beaches from Alaska to Baja California. *N. luetkeana's* float bladders and long stalks, up to 40 meters in length, resemble the whips used by "bull teamsters," hence their common name, bullwhip kelp.

Figure 3.17

Sea palms are common brown algae that inhabit many outer rocky coasts. The breaking surf of their habitat poses a great danger to people who walk there.

Figure 3.18

Coralline red algae give both color and texture to tidepools found in the rocky intertidal zones of most oceans.

encrusted with calcareous materials composed of calcium carbonate, these algae add a splash of color and texture to many intertidal pools exposed at low tide along rocky, temperate coastlines.

Green Algae

The green algae are contained in two phyla, the gamophytes and the chlorophytes. **Gamophytic green algae** reproduce sexually when their amoeboid gametes (sex cells), which do not have undulipodia, fuse during conjugation. These algae are important members of freshwater communities, and we study them further in Chapter 9. Microscopic gamophytes include **desmids** and some common **filamentous algae** that produce "pond scums" during summer blooms (Figure 3.19).

Chlorophytic green algae live in both freshwater and marine environments and are distinguished from gamophytes by their undulipodate gametes,

Figure 3.19

The scums that form on the surfaces of ponds during summer are usually due to the growth of green algae.

CLASSIFICATION

Focus on Scientific Explanations

Classification is the orderly arrangement of information. It has many practical values because it permits the retrieval of that information and can organize it to reflect relationships that exist among the objects classified.

A survey of different cultures shows that all peoples have constructed classification systems of natural objects. Some are complex, and others are simple. Distinctions are made in some languages that are not possible in others. Attempts to classify organisms date back to the beginnings of science. The earliest scientific classifications were fairly simple and reflected the relatively small number of plants and animals known at the time. Aristotle, for example, described only 500 animals in his writings.

As a response to new information during the Renaissance, many "herbals" were published. These books, which were beautifully illustrated, as shown in Figure 1, attempted to describe all the plants and animals then known. They were for the most part arranged in a simple style, usually alphabetically, although some herbalists attempted to devise a system of classification.

During the seventeenth century, classification became the central issue among naturalists, and two general types of classification systems were developed: *artificial* and *natural*. An **artificial system** is a classification system that is used primarily for retrieving information and makes no claims about the relationships among the objects classified. By contrast, the main purpose of a **natural system** is the construction of a classification that reflects an order existing in nature. Natural systems can also be used for information retrieval, but often they are not as convenient as artificial systems. In the seventeenth century, Joseph Pitton de Tournefort, one of France's most famous botanists, organized 10,146 species of plants into 698 genera, many of which are still recognized today. Tournefort believed that God created plants according to a plan and that by careful observation, a trained scientist could "intuitively" come to know the plan. His system was a natural classification system, and for him it was a picture of God's arrangement.

An English contemporary of Tournefort, John Ray, did not believe that a natural system was possible. He didn't doubt that God had a plan, but he thought that people had no access to it. Ray described 18,000 species of plants and went on to classify animals as well. He was able to arrange these numerous organisms into simple systems called keys that allowed someone to identify a specific plant or animal. Ray did not claim that his keys reflected anything other than convenient systems for identification and organization.

As might be expected, there has been considerable debate over the decades between proponents of artificial and natural systems. In the eighteenth century, Linnaeus constructed an artificial system of classification that became widely used throughout the world. He also proposed using a binomial nomenclature, which brought some order into the then-chaotic realm of naming organisms. Until Linnaeus, there was no universal method of naming organisms, and one plant might have 25 different names given at different times by different writers!

In Paris, Linnaeus's contemporary and rival, Georges-Louis Leclerc de Buffon, constructed a natural system of classification of animals and encouraged others to work on a natural system for

Figure 1 This beautiful water lily graced Brunfels's famous herbal, published in 1530. It was one of the first of many Renaissance herbals that contained plants drawn from nature.

plants. Buffon's system was based on more than 20 years of research, but it encountered a problem that plagued most of the writers who constructed natural classifications—disagreement over the criteria. Tournefort had earlier said that naturalists could intuitively see the relationships among organisms. Although many agreed with this point of view, when it came to elaborating a system, many naturalists saw different patterns of organization, resulting in no common agreement over what constituted the natural system.

Debates persisted throughout the nineteenth century. Some experts hoped that comparative anatomy of animals would uncover the natural criteria; others felt that animal behavior would provide the key. In botany, a wide variety of plant parts were proposed as determinants.

Darwin's theory of evolution (discussed in Chapter 22) provided a new perspective from which to approach classification. He had stated in his writings that species of animals or plants often resembled other species because they descended from a common ancestor. He went on to suggest that classification should be based on evolutionary relationships.

Darwin's idea quickly caught on; in practice it has often been problematic because virtually no historical record remains for many plants and animals. Moreover, organisms that share similarities are not always related directly but rather may have evolved the similarities independently.

Today, a variety of tools are used to assess the evolutionary relationships among organisms. In addition to adult and embryo morphology, scientists can use information from biochemical analysis, biogeography, fossils, geology, and behavior. Scientists who attempt evolutionary classifications still disagree as to whether emphasis should be placed on the point of divergence between groups or on the significance of the divergence in terms of structural features. The systems that stress the branching off of groups are called *cladistic,* and those derived from the alternative point of view, which emphasizes the usefulness and significance of evolutionarily derived characteristics, are called *phylogenetic*.

When information on evolutionary relationships is insufficient, taxonomists resort to simple keys for part or all of the classification system. There are also scientists who recognize the intellectual value of evolutionary classification systems but nonetheless construct artificial keys for practical purposes. For example, a field guide to birds is much more convenient for the early morning bird-watcher than a multivolume treatise that attempts to organize the avian world into a phylogenetic scheme.

Still other scientists reject evolutionary classifications and believe that a more rigorous system can be devised by using the number of similarities that exist among organisms. These **phenetic systems** tabulate traits, and through computer manipulation they can generate lists of groups that share the most traits. Generally, the scientists who have been attracted by phenetic classification are those who work with organisms with uncertain evolutionary histories.

Figure 3.20

Slime molds are protoctists that have some fungal characteristics during certain stages of their life cycle. For example, they may develop spore-forming reproductive structures that mature and release spores. After dispersal, these spores germinate, forming a new generation of slime molds like the one shown here growing on a dead Douglas fir tree.

which fuse during conjugation. Freshwater chlorophytes, along with gamophytic green algae and cyanobacteria (blue-green algae), are the floating producers present in lakes and ponds. In marine environments, chlorophytes add bright green colors that contrast with the drab brown algae and deep red algae also present in the intertidal areas. Many unique chlorophytes occur in these marine habitats, characterized by the following distinguishing forms: stiff or flexible filaments similar to those of freshwater species; thin, bright green sheets, as in the sea lettuce, *Ulva*; and, thick, spongy growths of dark green "fingers," as in the genus *Codium*.

Other Protoctists

Myxomycota and the other remaining protoctist phyla include the peculiar group of organisms collectively known as *slime molds* (Figure 3.20). They are separated into different phyla based on their body types. Some tend to be cellular, some form nets, and some form large cellular masses at certain stages in their life cycles. A few of these unusual organisms share many characteristics with species found in the kingdom Fungi, but they still produce undulipodiated cells, thereby retaining their membership in the kingdom Protoctista.

The kingdom Protoctista contains single-celled protozoans, unicellular and multicellular algae, and slime molds. This kingdom includes the more primitive eukaryote life forms that have in common aerobic respiration and undulipodia during some life stage.

KINGDOM FUNGI

The kingdom Fungi encompasses eukaryotic, spore-forming organisms that lack undulipodia. It consists of five phyla that contain more than 100,000 species. Most fungi are **saprophytic**, obtaining their nutrients through absorption of breakdown products from dead plants and animals. However, a number of parasitic species cause many diseases in plants and a lesser number in animals.

Most new fungi arise directly from asexual spores called **conidia**. Conidia are produced in fruiting bodies (asci, sporangia, or basidia) of mature fungi. Conidia give rise to structures called **hyphae**, which usually become multicellular when connecting "cross walls," called **septa**, develop in the hyphae (Figure 3.21). These hyphae form mats of tissue called **mycelia** that constitute the body of the fungus. Within the life cycle of most fungi, sexual reproduction also occurs through conjugation and results in production of other types of spores (ascospores or basidiospores) from which new hyphae can develop.

Zygomycotes

The phylum of the zygomycotic fungi includes black bread molds in the genus *Rhizopus*. These molds were a problem in food preservation before mold inhibitors were routinely added to bakery products. Some of the zygomycotic fungi are parasitic, but most are saprophytic (Figure 3.22).

Ascomycotes

In this phylum, bladder fungi are distinguished by microscopic reproductive structures called **asci** that appear as tubular spore sacs filled with ascospores. This phylum includes certain edible products such as morels and truffles, as well as the yeasts used in many fermenting processes, such as those used for producing beer and wine and leavening bread (Figure 3.23). There are a few parasitic forms in this phylum, including the pathogenic species that causes Dutch elm disease, which has decimated most populations of North American elm trees.

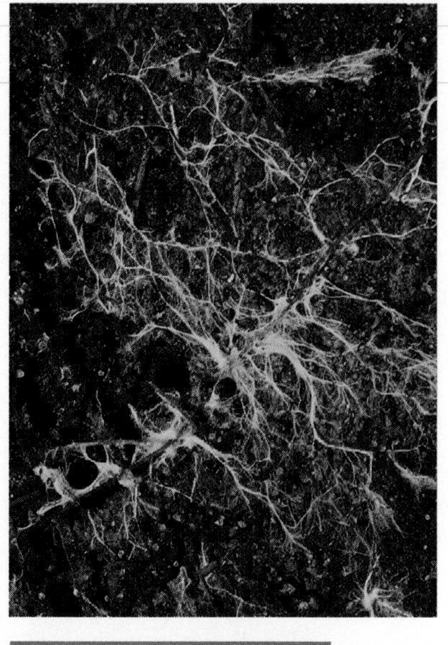

Figure 3.21

Most fungi consist of hyphae that form mats of mycelia within the soil, rotting logs, or other vegetation in which they grow. Microscopic magnification of fungal hyphae reveals networks such as those shown here.

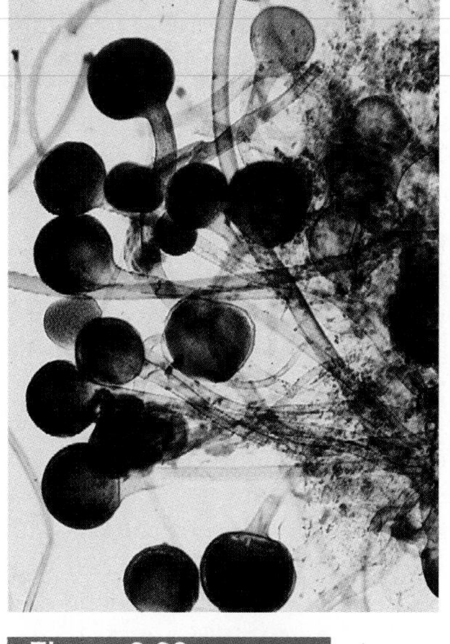

Figure 3.22

Black bread mold, belonging in the genus *Rhizopus,* forms stalks on which sporangia develop spores. If dispersed to suitable substrates, these spores will germinate and grow into new fungi.

Figure 3.23

The ascomycotes are a diverse group of fungi that include the edible yellow morels shown here (*Morchella esculenta*).

Basidiomycotes

The most familiar fungi are basidiomycotes. This phylum contains smuts, puffballs, and mushrooms (Figure 3.24). Most of a living basidiomycote actually exists underground or within the rotting log on which it may live. The mushroom protruding aboveground is the spore-producing, reproductive structure.

Many species of zygomycotes, ascomycotes, and basidiomycotes form mutualistic associations with the roots of trees and other plants that are called *mycorrhizae.* The mycorrhizal relationship between roots and fungi is beneficial to the tree species, for it increases the rate at which mineral nutrients liberated by the fungi are transported from the soil into the roots. Fungi benefit by receiving carbohydrates produced by the tree.

Deuteromycotes

The fourth phylum of fungi includes species that have lost the ability to reproduce sexually or never were sexual; for this reason, they have often been classified as the **fungi imperfecti**. This group is defined as a phylum on the basis of structural form. Deuteromycotes include the genus *Penicillium,* from which the important antibiotic penicillin is obtained, and also a number of parasitic genera including those that cause root rot in wheat and athlete's foot in humans. Most scientists consider fungi in this phylum to have evolved from the ascomycotes or basidiomycotes.

Lichens

Lichens make up the last phylum of fungi. Like the Deuteromycotes, they are classified by structural characteristics or form. Lichens are classic *symbionts* (participants in a close living association), for their bodies consist of two different species, one a fungus (most are ascomycotes) and the other a species of cyanobacteria or chlorophytic algae (Figure 3.25). Lichens play important roles in many natural communities.

A

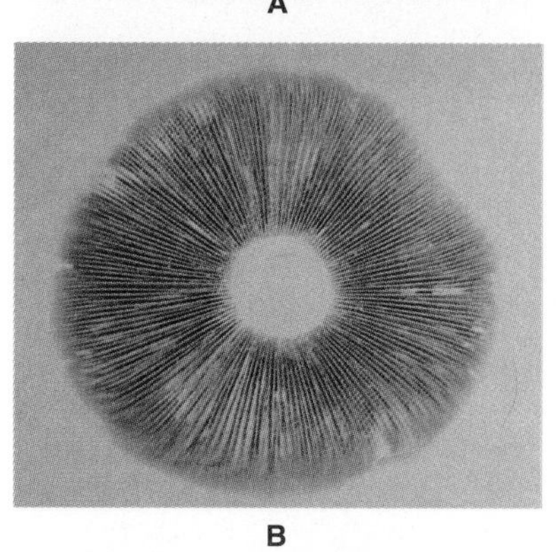

B

Figure 3.24

(A) Basidiomycotes include familiar puffballs that produce spores. (B) The spore print of a mushroom is often used to identify a species (*Agaricus campestris*).

Figure 3.25

Lichens are classified with fungi because of their growth form. However, they are composed of both a fungus and a cyanobacterium, or green alga, forming a symbiotic relationship that is beneficial to both organisms.

The kingdom Fungi includes eukaryotic organisms that have no undulipodia. Most fungi are saprophytic, but some are important parasites in both plants and animals, including humans. They all produce reproductive spores, some through asexual reproduction and others through sexual processes. Usually only the spore-forming structure is visible, for the bulk of a fungus grows within the soil or decaying vegetation.

Summary

1. Estimates of the total species diversity in the biosphere range from 5 to 30 million, but only 1.7 million species have been described. Linnaeus developed a binomial system of nomenclature that uses a unique genus and species name for identifying organisms.
2. All the biota of Earth are classified into five kingdoms: Prokaryotae, Protoctista, Fungi, Plantae, and Animalia.
3. The prokaryotes are separated into two subkingdoms, the Archaebacteria and the Eubacteria, which are further partitioned into a number of divisions based on the staining properties of their cell walls. These divisions are each broken down into a number of phyla. Archaebacteria phyla include methanogenic, halophilic, and thermoacidophilic forms. Eubacteria include autotrophs, heterotrophs, important nitrogen-fixing forms, and pathogenic forms.
4. Organisms classified in the kingdom Protoctista are aerobic eukaryotes that have undulipodia during some stage of their life cycle. They include both single-celled and multicellular forms that range in size from the microscopic protozoans and algae to the large marine kelps.
5. The kingdom Fungi contains saprophytic or parasitic organisms composed of hyphae that form mats of mycelia and asexual reproductive structures that produce conidia. Most fungi can also reproduce sexually.

Review Questions

1. What is a scientific name? How does it differ from the common name of an organism?
2. How is the hierarchy of a division used in the kingdom Prokaryotae?
3. Why is the kingdom Protoctista said to be defined by exclusion?
4. Describe the various types of undulipodia and how they are used in the classification of protoctists.
5. Which types of algae are you likely to find in the intertidal zone along rocky coasts?
6. How do cyanobacteria differ from green algae?
7. How do fungi obtain their food?
8. Which phyla of fungi are most useful to humans, and in what ways?
9. What are lichens? Why are they classified with the fungi?

Essay and Discussion Questions

1. Scientists believe that large-scale destruction or alteration of natural environments (for example, tropical rain forests) threatens the diversity of life. Why might the loss of species diversity be a serious problem?
2. Why are some species in the kingdom Prokaryotae thought to be most similar to the first organisms to evolve on Earth?
3. Construct an artificial system of classification.

References and Recommended Reading

Ahmadjian, V. 1982. The nature of lichens. *Natural History* 91:30–37.

Corliss, J. O. 1984. The protista kingdom and its 45 phyla. *BioSystems* 17:87–126.

Gentry, A. H. 1988. Tree species richness of upper Amazonian forests. *Proc. Natl. Acad. Sci., USA* 85 (January):156–159.

Krogmann, D. 1981. Cyanobacteria (blue-green algae): Their evolution and relation to other photosynthetic organisms. *BioScience* 31(2): 121–124.

Margulis, L., and K. Schwartz. 1988. *Five Kingdoms: An Illustrated Guide to the Phyla of Life on Earth*. New York: Freeman.

Smith, A. 1980. *The Mushroom Hunter's Field Guide*. Ann Arbor: University of Michigan Press.

Vidal, G. 1984. The oldest eukaryotic cells. *Scientific American* 250:48–57.

Whittaker, R., and L. Margulis. 1978. Protist classification and the kingdoms of organisms. *BioSystems* 10:3–18.

Woese, C. 1981. Archaebacteria. *Scientific American* 244:92–122.

CHAPTER 4

The Diversity and Classification of Life: Kingdoms Plantae and Animalia

Chapter Outline

Reading Questions

1. What are the characteristics used in classifying plants?
2. What are the characteristics used in classifying animals?
3. What is the basis for classifying tetrapod vertebrates?
4. What is a taxonomic classification system? A functional classification system? How do they differ?

KINGDOM PLANTAE

Plants have been the major occupants of terrestrial landscapes for more than 400 million years, and today nearly 500,000 species are known to inhabit Earth. The actual number is certainly much larger. Plants are eukaryotic organisms that are distinguished by a life cycle featuring an **alternation of generations** in which two different structural forms, *gametophytes* and *sporophytes,* alternatively produce one another through time. The **gametophyte** is the gamete-producing stage of plants, and the **sporophyte** is the spore-producing phase shown in Figure 4.1. Through the process of *mitosis,* gametophytes produce true gametes—sperm and egg cells—that unite during fertilization, develop into multicellular embryos, and ultimately mature to become sporophytes. Through the process of *meiosis,* sporophytes produce spores that develop into gametophytes. Details of these processes are presented in Chapter 13.

Plants are classified into one of two major groups, *nonvascular bryophytes* or *vascular tracheophytes.* **Vascular tracheophytes** develop vascular tissues—vessels and tubes—that transport water and minerals obtained from their environment and sugar produced during photosynthesis from one region to another within the plant. **Nonvascular bryophytes** have no vascular tissues and required substances simply *diffuse* from the environment into their tissues. **Diffusion** is a physical process in which molecules in solution move from regions of high concentration to low concentration. All the phyla in the kingdom Plantae, as well as one typical genus from each phylum, are identified in the appendix.

Bryophytes

The bryophyte phylum includes about 24,000 species that are separated into three classes with somewhat unappealing names: *liverworts, hornworts,* and *mosses* (see Figure 4.2). Nonvascular bryophytes are considered to be the most primitive plants because the sporophyte is inconspicuous and nutritionally dependent on the gametophyte, which dominates the life cycle. The gametophyte stage develops from spores and appears as a small leafy structure with rootlike **rhizoids** that are composed of elongated single cells or multicellular filaments. Female gametophytes produce eggs that are fertilized by undulipodiated sperm furnished by male gametophytes. The sperm must swim to the egg, an awkward requirement for land plants that limits bryophyte distribution to environments that are wet during some part of the year. Zygotes result from the union of egg

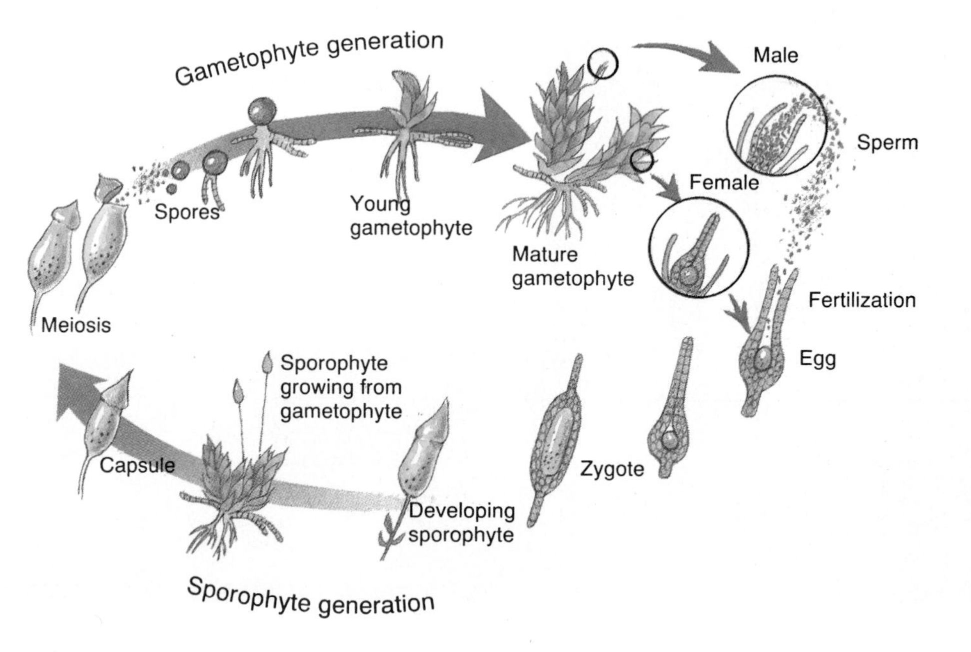

Figure 4.1

Plant life cycles have an alternation of generations. The sporophyte generation of a moss produces spores through the process of meiosis. Spores germinate and develop into the gametophyte generation, which produces sperm and eggs. These gametes unite during fertilization and form a zygote, which will grow into a mature sporophyte.

and sperm and develop into embryos that grow and develop into the sporophyte phase of the bryophyte life cycle.

Most bryophytes are terrestrial organisms, although some are aquatic. A few species inhabit the splash zone of intertidal habitats along rocky coasts, but none are considered to be strictly marine organisms. They are important members of many different terrestrial communities.

> Bryophytes are photosynthetic nonvascular plants. This phylum includes mosses, liverworts, and hornworts. All plants, including the bryophytes, have an alternation of generations. In this type of life cycle, a gametophyte generation produces gametes that after fertilization develop into a sporophyte generation, which produces spores that germinate and grow into gametophytes.

Tracheophytes

Vascular tracheophytes include the most advanced species of plant life. The more conspicuous sporophyte generation is characterized by the presence of specialized groups of cells and intercellular substances that form vascular conducting tissues.

Spore-forming Tracheophytes

Four phyla (the whisk ferns, clubmosses, horsetails, and true ferns; see Figure 4.3) include primitive tracheophytes that do not form *seeds*. **Seeds** are structures that contain the embryos of more advanced plants. Among the seedless tracheophytes, the ferns are by far the most widespread on Earth at the present time.

Ferns have undulipodiated sperm, and like the bryophytes, they are restricted to habitats that occasionally have sufficient moisture to allow fertilization. Such moist conditions occur in a number of different biomes, permitting ferns to become members of many plant communities. A few aquatic ferns form symbiotic associations with nitrogen-fixing cyanobacteria, which leads to their involvement in the nutrient cycles of many aquatic systems.

Seed-forming Tracheophytes

There are two types of seed-forming tracheophytes: those that have a "naked" seed, the *gymnosperms,* and those with seeds enclosed within certain plant structures, the *angiosperms*. Seed-forming tracheophytes are the dominant plant forms on Earth today.

A

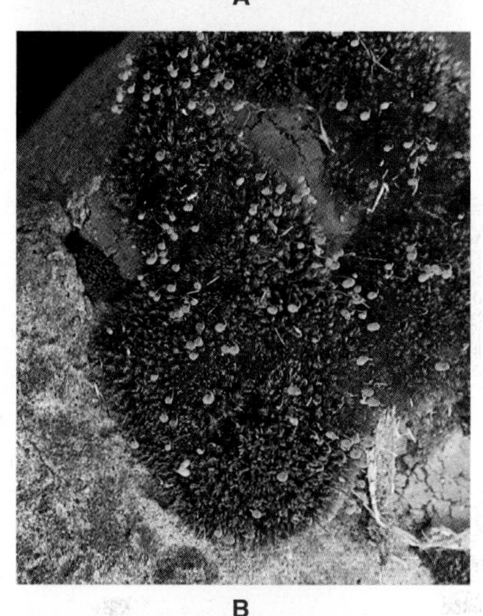

B

Figure 4.2

Liverworts (A) and mosses (B) have no vascular tissue, which results in their having a low growth form.

Figure 4.3

Clubmosses (A), horsetails (B), and true ferns (C) are tracheophytes that do not produce seeds.

A B C

A

B

C

Figure 4.4

Living gymnosperms include (A) cycads, small shrubby plants that have a cursory resemblance to pineapple; (B) gentophytes, like the cone-bearing, seed-producing *Welwitschia mirabilis* that grows in the deserts of southern Africa; and (C) conifers, including the largest plant known, the giant sequoia tree.

Gymnosperms

Gymnosperms are separated into four different phyla: *cycads, ginkgos, gnetophytes,* and *conifers* (see Figure 4.4). **Conifers** are the most common gymnosperms and are presently represented by about 50 genera of *softwood* species, including those that are harvested by the timber industries in North America. **Softwood** is a common term applied to lumber from conifers. Commercially important species include pines, Douglas fir, true fir, larch, spruce, and redwoods (see Figure 4.5).

Angiosperms

All plants that develop a *true flower* are classified as **angiosperms**. The angiosperms vary with respect to their number of floral parts, but all **flowers** are reproductive structures that contain both male and female reproductive parts, and their seeds develop within fruits. More than 300 families of angiosperms are separated into two major classes, monocots and dicots.

Monocots **Monocots** develop from germinated seeds that have a single embryonic leaf (called a cotyledon) and grow into plants with parallel leaf venation and floral parts that occur in threes or multiples of three, as shown in Figure 4.6. The vascular tissues in cross sections of monocot stems appear as scattered bundles of cells.

Monocots include most species that are cultivated as food crops by humans. The grass family is by far the largest and most important; it includes rice, wheat, barley, oats, other grains, and the grasses on which most animal *herbivores* feed (see Figure 4.7). **Herbivores** are organisms that feed on plants or other primary producers. Larger members of the grass family include corn, sugarcane, and bamboo. The largest monocot species belong to the pineapple, banana, and palm families.

Many spring wildflowers are monocots in the lily, iris, and orchid families (see Figure 4.8). The orchid family has the largest number of flowering plants, over 20,000 species, most of which grow in tropical or subtropical regions of the world.

Dicots **Dicots** develop from germinated seeds that have two cotyledons, grow into plants with netted leaf venation, and produce flowers with parts in fours or fives or multiples of four or five. The vascular tissue in the dicot stem forms a circular pattern when viewed in cross section.

Figure 4.5

Douglas fir is an important lumber-producing conifer in the forests of western mountains.

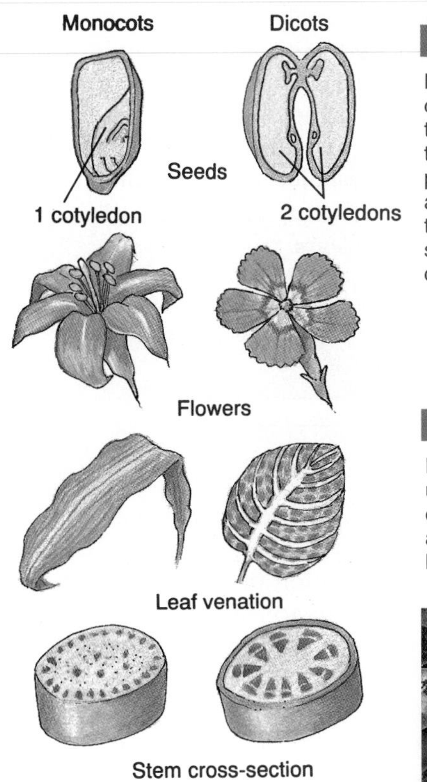

Figure 4.6

Monocots and dicots can be distinguished in a number of ways. Monocots have one cotyledon and dicots have two. The floral parts of monocots occur in threes or multiples of three, those of dicots in fours or fives or multiples of fours or fives. Vascular bundles in monocots form a parallel pattern of leaf veins; dicot bundles form a netted pattern. In monocot stems, vascular tissue forms a scattered pattern in cross section; in dicots it forms a circular pattern.

A

Figure 4.7

Rice (A) is a monocot in the grass family. Grains make up an important part of the human food supply throughout the world. Bananas (B) and date palms (C) are among the largest monocots. They are an important human food source in some regions.

B

C

Nearly 70 percent of the 250,000 known species of flowering plants are dicots. They vary in form and habitat, ranging from deciduous trees, such as maples and elms, to herbaceous (nonwoody) ground cover species found on the floor of old-growth forests, to the giant saguaro cactus, a hallmark species in the Sonoran Desert of southern Arizona (see Figure 4.9).

Figure 4.8

Many early wildflowers, including these yellow fawn lilies, are monocots.

A

B

Figure 4.9

The dicots include more species than any other terrestrial plant class. Their diversity ranges from the saguaro cactus that lives in the hot southwestern deserts (A) to many of our fruits, like the peaches shown here (B).

Dicot species that have been adapted for human cultivation include peas, beans, alfalfa, and clover—all members of the pea family; lemons, oranges, and grapefruit in the citrus family; and the diverse rose family, which includes roses, berries, cherries, peaches, almonds, pears, and apples among its 3,000 species.

The aster family features over 19,000 species, some of which are among the most evolutionarily specialized plants. This largest, most diverse dicot family includes sunflowers, ragweed, zinnias, asters, sagebrush, lettuce, and dandelions.

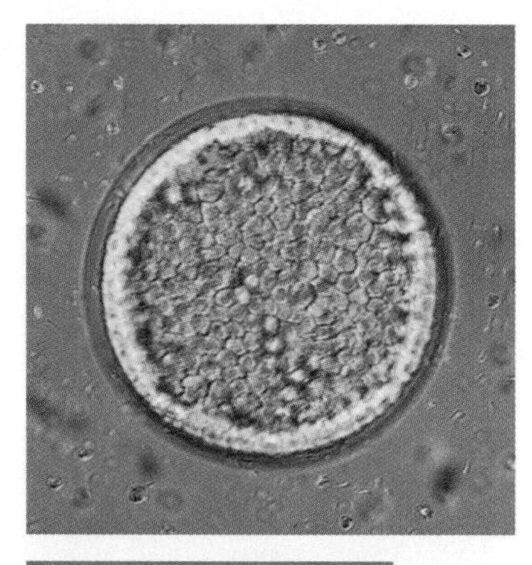

Figure 4.10

The formation of a blastula, like that of a sea urchin shown here, is an important characteristic used in classifying the animals.

> The tracheophytes include whisk ferns, clubmosses, horsetails, true ferns, gymnosperms, and angiosperms. They all produce spores during their life cycle, but only gymnosperms and angiosperms form seeds. Cycads, ginkgos, gnetophytes, and conifers comprise four phyla of gymnosperms. All angiosperms produce true flowers. They are separated into two classes, the monocots and the dicots.

KINGDOM ANIMALIA

The classification of the animal kingdom is based to a large extent on patterns of early embryonic development. This kingdom includes all heterotrophic eukaryotes in which a **zygote** is produced from the fertilization of an egg by a sperm. Animal zygotes undergo successive cellular divisions to form a hollow ball of cells, the *blastula,* shown in Figure 4.10. Further development of the blastula proceeds when certain surface cells begin to move inward, forming the **blastopore**, an opening that will develop into the mouth in some animal species and into the anus in others. These migrating surface cells become a cell layer known as the **endoderm**. As they continue to migrate inward, some cells move to the area between the developing endoderm and the outer layer of cells, the **ectoderm**. These "middle cells" form the **mesoderm**. At this stage, the original zygote has reached the embryonic stage of development known as the **gastrula**, shown in Figure 4.11. The inward-moving cylinder of endoderm eventually unites with the ectoderm on the other pole of the gastrula, thus forming the animal's primitive gut.

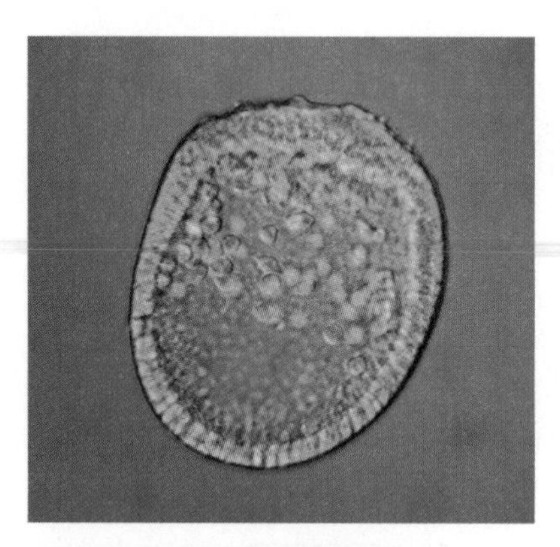

Figure 4.11

During sea urchin development, a blastopore appears as the blastula develops into a gastrula, shown here. In echinoderms and chordates, this opening is destined to become the anus, but in all the other animals in which it forms, it becomes the mouth.

One of the major characteristics that is used to separate animal phyla is the fate of the blastopore. Animal phyla in which it becomes the mouth are known as *Protostomes* ("first mouth"), and those in which it becomes the anus are called the *Deuterostomes* ("second mouth").

The symmetry of developing animal embryos follows one of two general patterns, as described in Figure 4.12: a primitive radial type or the more advanced bilateral form. Phyla in which animals develop bilateral symmetry are separated into three groups based on the degree of development of a body cavity within the mesoderm, known as the **coelom**, as shown in Figure 4.13. The

A

B

Figure 4.12

The radial symmetry of a green sea anemone *Anthropleura xanthogrammica* (A) is quite different from the bilateral symmetry of a rough-skinned newt *Taricha granulosa* (B).

most primitive bilateral phyla do not develop a coelom and are known as the **acoelomates**. The second group of phyla develop a body cavity between the mesoderm and the endoderm that is often called a false coelom; hence they are termed the **pseudocoelomates**. Animals in the most advanced bilateral phyla form a true coelom within the mesoderm and are called the **coelomates**. Most internal organ systems of coelomates develop and are suspended within the coelom. The appendix lists the phyla in the kingdom Animalia and gives an example of a genus from each.

Primitive Animal Phyla

Table 4.1 indicates that the 33 phyla of living animals are separated according to their pattern of embryonic development and the presence or absence of organ systems. The most primitive animal phyla are those that have only a tissue level

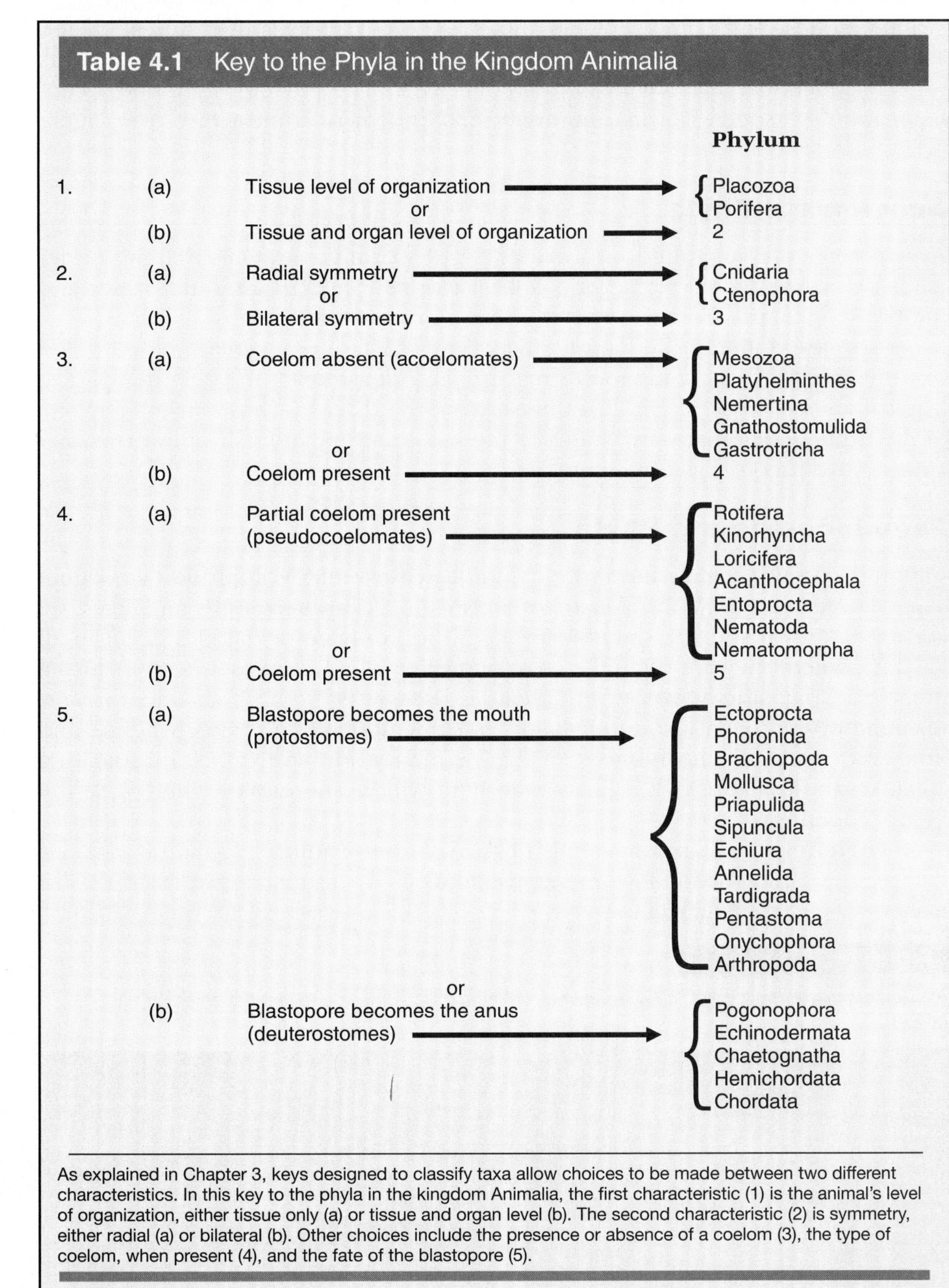

Table 4.1 Key to the Phyla in the Kingdom Animalia

			Phylum
1.	(a)	Tissue level of organization →	Placozoa, Porifera
		or	
	(b)	Tissue and organ level of organization →	2
2.	(a)	Radial symmetry →	Cnidaria, Ctenophora
		or	
	(b)	Bilateral symmetry →	3
3.	(a)	Coelom absent (acoelomates) →	Mesozoa, Platyhelminthes, Nemertina, Gnathostomulida, Gastrotricha
		or	
	(b)	Coelom present →	4
4.	(a)	Partial coelom present (pseudocoelomates) →	Rotifera, Kinorhyncha, Loricifera, Acanthocephala, Entoprocta, Nematoda, Nematomorpha
		or	
	(b)	Coelom present →	5
5.	(a)	Blastopore becomes the mouth (protostomes) →	Ectoprocta, Phoronida, Brachiopoda, Mollusca, Priapulida, Sipuncula, Echiura, Annelida, Tardigrada, Pentastoma, Onychophora, Arthropoda
		or	
	(b)	Blastopore becomes the anus (deuterostomes) →	Pogonophora, Echinodermata, Chaetognatha, Hemichordata, Chordata

As explained in Chapter 3, keys designed to classify taxa allow choices to be made between two different characteristics. In this key to the phyla in the kingdom Animalia, the first characteristic (1) is the animal's level of organization, either tissue only (a) or tissue and organ level (b). The second characteristic (2) is symmetry, either radial (a) or bilateral (b). Other choices include the presence or absence of a coelom (3), the type of coelom, when present (4), and the fate of the blastopore (5).

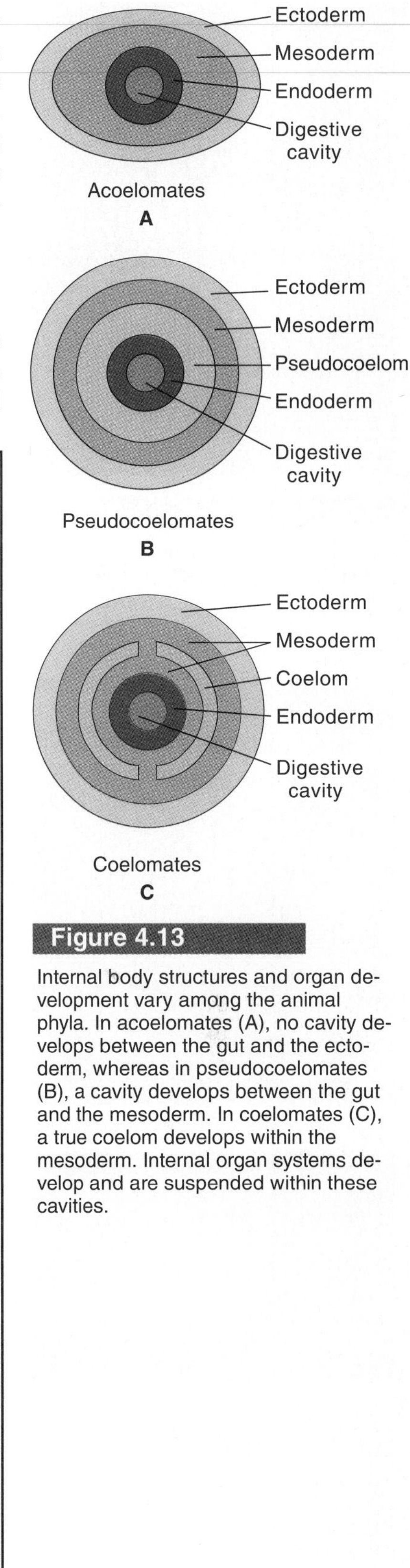

Figure 4.13

Internal body structures and organ development vary among the animal phyla. In acoelomates (A), no cavity develops between the gut and the ectoderm, whereas in pseudocoelomates (B), a cavity develops between the gut and the mesoderm. In coelomates (C), a true coelom develops within the mesoderm. Internal organ systems develop and are suspended within these cavities.

Figure 4.14

The yellow candle sponge is in the Porifera, one of the most primitive of the animal phyla.

of organization (that is, they have no true organs). They are represented by living animals in two phyla: **Placozoa**, which has only a single species, *Trichoplax adhaerens,* a microscopic animal found in seawater and on the walls of marine tanks in laboratories and aquariums; and **Porifera**, which includes all the sponges. The sponges shown in Figure 4.14 have tissues that are composed of two cell layers separated by **mesenchyme**, a gelatinous layer in which *spicules* and *amoebocytes* can be found. **Spicules** are skeletal elements composed of calcium carbonate or silicates produced by **amoebocytes**, cells that can move through the cell layers with their pseudopodia. Sponges are filter feeders that remove particles and plankton from the water.

Radially Symmetrical Phyla

The two phyla of animals with radial symmetry are the **Cnidaria** and the **Ctenophora**. Most cnidarians are marine and include the familiar true jellyfish, sea anemones, and corals. Hydras and a few other species live in fresh water. The marine corals are responsible for the production of some of the largest structures produced by any organism, the coral reefs. The phylum Ctenophora contains about 90 species of marine organisms known as comb jellies (see Figure 4.15).

A

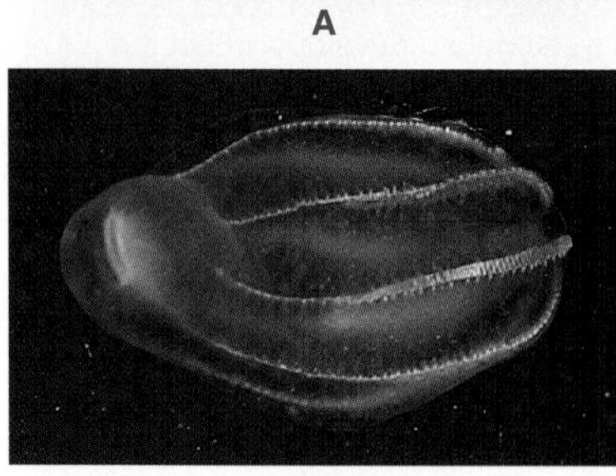

B

Figure 4.15

The hydra in the phylum Cnidaria (A) and the comb jelly in the phylum Ctenophora (B) represent the two radially symmetrical animal phyla.

Acoelomate Phyla

Acoelomates are separated into five phyla of wormlike animals that are either parasitic or free-living. Nemertine ribbon worms found in intertidal zones and the brown planaria from freshwater streams are among the most common free-living species. Some of the parasitic flatworms, such as liver flukes, blood flukes like the schistosomas, and tapeworms, are very detrimental in many developing tropical and subtropical countries because of the diseases they cause. Members of the other acoelomate phyla are for the most part small freshwater or marine animals seldom observed in nature.

Pseudocoelomate Phyla

Important pseudocoelomate phyla include **Rotifera** and **Nematoda** (see Figure 4.16). Rotifers are common aquatic organisms in freshwater ponds and lakes. Of the 2,000 known species, only 50 are marine. Free-living nematodes occur in all habitats, where they are often the most numerous of all animals. There are also a number of parasitic nematodes that cause diseases in both plants and animals, resulting in significant economic losses of some agricultural crops, wildlife, and domestic animals. The number of known nematode species is approaching 100,000, but it is considered likely that there will be more than 1 million at the final accounting.

Loricifera is the most recently described (1983) animal phylum. It includes microscopic, spiny-headed, wormlike pseudocoelomates that live in clean, coarse marine sands or fine gravel.

Figure 4.16

(A) Rotifers live in most aquatic environments. (B) Nematodes are found in most environments on Earth.

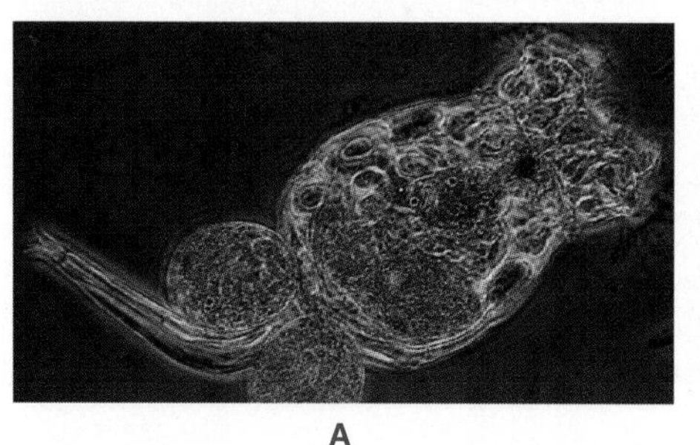

A

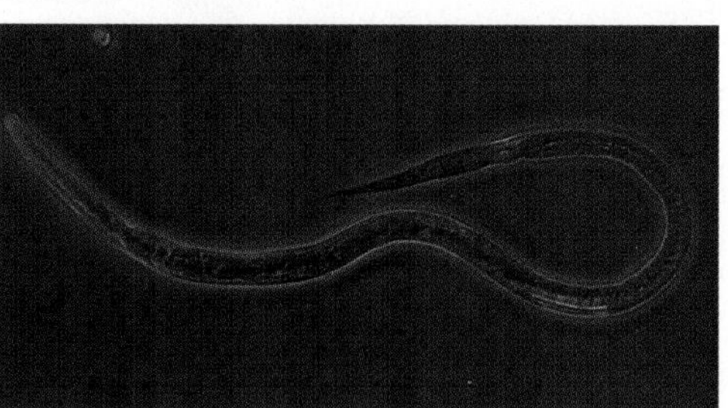

B

Coelomate Protostomes

The most primitive coelomate animals occupy 12 protostome phyla. In these animals, the blastopore forms the mouth of mature adults. Many of these phyla include marine, wormlike organisms rarely seen by most biologists. **Mollusca** and **Arthropoda** are the two protostome phyla that contain organisms of the greatest economic importance.

Annelids

Segmented worms in the phylum **Annelida** are among the best-known protostomes, the most familiar being the common earthworm or night crawler. Species in this phylum are distributed from the tropics to the antarctic seas, where they exist as scavengers or predators in terrestrial, marine, and freshwater habitats (see Figure 4.17).

Figure 4.17

Annelid worms live in both marine and terrestrial areas. (A) This marine worm, *Nereis vexillosa,* reaches about 15 centimeters in length and lives in mussel beds along the rocky Pacific coast. (B) Earthworms or night crawlers are terrestrial annelids.

Mollusks

The phylum **Mollusca** is composed of about 110,000 species separated into eight classes, of which three (Gastropoda, Bivalvia, and Cephalopoda) are the best known and of the greatest importance to humans (see Figure 4.18). **Gastropods** occur in aquatic and terrestrial environments and include the land snails and slugs familiar to gardeners in moist climates, both temperate and tropical. **Bivalves** include most of the marine and freshwater organisms known as "shellfish." Fish they are not, but they do include many economically important species of clams, oysters, scallops, and mussels that are cosmopolitan in both their distribution and their appeal as food to humans. **Cephalopods** are a class of mollusks that includes the largest and most complex protostome coelomates. The giant squids are known to reach lengths greater than 18 meters and weights in excess of 1,000 kilograms. Other cephalopods include the cuttlefish, the octopus, and the nautilus.

Arthropods

The phylum **Arthropoda** is the largest animal phylum, containing 18 classes distributed among three subphyla: **Crustacea**, **Chelicerata**, and **Uniramia**. As a group, the crustaceans predominate in many aquatic environments. Crayfish, barnacles, shrimp, crabs, and lobsters contribute to the bounty that humans harvest from rivers and oceans (see Figure 4.19). The chelicerates include marine horseshoe crabs, sea spiders, and the class of terrestrial arachnids containing mites, spiders, and scorpions. Uniramians include centipedes, millipedes, and insects. Insects are the most numerous terrestrial animals. Many insect species have aquatic stages in their life cycle. The class **Insecta** is partitioned into more orders than any other animal class, and each order contains numerous families,

Figure 4.18

Mollusks are a diverse coelomate protostome phylum. Garden snails (A), mussels (B), and octopuses (C) represent three of the eight classes in this phylum.

A

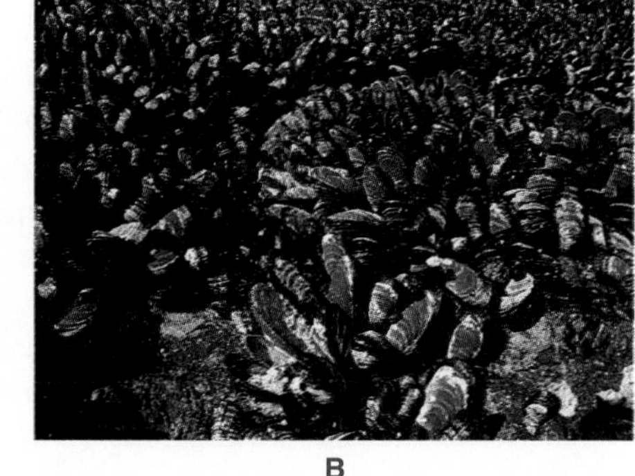

B

C

A B C D E F

Figure 4.19

Arthropoda, the largest animal phylum, is divided into three subphyla: Crustacea, Chelicerata, and Uniramia. Of the arthropods shown here, the ghost crabs (A) and crayfish (B) are in the first subphylum, marine horseshoe crabs (C) and spiders (D) in the second, and centipedes (E) and insects represented by the dragon fly (F) are in the third.

many with vast numbers of genera and species. About 750,000 species of insects have been described, but intensive collections in tropical forests indicate that their total number, on a world basis, may exceed 30 million!

> Sponges have only a tissue level of organization, jellyfish and comb jellies have radial symmetry, and the other animal phyla are distinguished by their coelom development and the fate of the blastopore.

Coelomate Deuterostome Phyla

Five phyla of deuterostomes contain the most familiar animals that inhabit Earth. They all have a true coelom, and the blastopore of developing embryos becomes the anus in adults. **Echinodermata** is by far the best known of the four primitive deuterostome phyla. It includes the sea stars (starfish) and sea urchins, shown in Figure 4.20, that live in the intertidal areas along rocky coasts. This phylum is unique among the advanced phyla, for the adults have radial symmetry, a trait that is characteristic of more primitive jellyfish and other cnidarians.

The phylum **Chordata** contains the most advanced animals. Members of this phylum all have a **dorsal** hollow nerve cord (located near the back), **pharyngeal gill slits** (openings that allow water to flow through the mouth, over the gills, and out the gill slits) during some stage of their life, and a **notochord**—an internal, cartilaginous rod. The chordates are divided into four major subphyla: **Tunicata**, **Cephalochordata**, **Agnatha**, and **Gnathostomata** (see Table 4.2).

Tunicates and Cephalochordates

Tunicates have a notochord only during their motile, larval stage, and adults live as sessile (attached) organisms in marine environments. **Lancelets** are

Table 4.2 Key to the Phylum Chordata

1.	(a)	Brain absent →	2
		or	
	(b)	Brain present →	3
2.	(a)	Notochord present in larvae only →	Subphylum Tunicata (tunicates)
		or	
	(b)	Notochord present in larvae and adults →	Subphylum Cephalochordata (lancelets)
3.	(a)	Jaws absent →	Subphylum Agnatha (lampreys)
		or	
	(b)	Jaws present →	Subphylum Gnathostoma, 4
4.	(a)	Paired jointed appendages absent →	Superclass Pisces, 5
		or	
	(b)	Paired jointed appendages present →	Superclass Tetrapoda, 6
5.	(a)	Internal skeleton composed of cartilage →	Class Chondrichthyes (sharks and rays)
		or	
	(b)	Internal skeleton composed of bone →	Class Osteichthyes (bony fishes)
6.	(a)	Poikilothermic →	7
		or	
	(b)	Homeothermic →	8
7.	(a)	Scales absent →	Class Amphibia (frogs and salamanders)
		or	
	(b)	Scales present →	Class Reptilia (lizards, snakes, and turtles)
8.	(a)	Teeth absent, feathers present, hard-shelled egg →	Class Aves (birds)
		or	
	(b)	Teeth present, feathers absent, young nourished with milk →	Class Mammalia (mammals)

This key to the phylum Chordata separates the taxa by the presence or absence of the following characteristics: brain (1), notochord (2), jaws (3), paired jointed appendages (4), internal skeleton (5), temperature control (6), scales (7), and teeth, feathers, and nourishment of young (8).

A

B

Figure 4.20

Sea stars (A) and red urchins (B) are in the phylum Echinodermata. Adult echinoderms have radial symmetry, but they develop from bilateral larvae.

cephalochordates that have a notochord and dorsal nerve cord in both their larval and adult stages. They are primitive, marine, fishlike chordates that live buried in sand. They feed on plankton delivered to their mouth by cilia-induced currents (see Figure 4.21).

Figure 4.21

Tunicates (A) and lancelets (B) are members of primitive chordate phyla.

A B

Agnaths

The parasitic lampreys are living relics of the primitive agnaths, the jawless fishes. Their notochord and pharyngeal gill slits persist throughout life, and their dorsal nervous system includes a very simple brain. By parasitizing and killing large numbers of fish, lampreys have had a great negative impact on fisheries in the Great Lakes.

Gnathostomates

Superclass Pisces

The superclass **Pisces** includes the two classes of true fish: **Chondrichthyes**, the cartilaginous sharks and rays, and **Osteichthyes**, the bony marine and freshwater fishes. In adult sharks, the notochord is replaced by vertebrae that are composed of cartilage. The pharyngeal gill slits persist throughout life, serving as openings for water that flows through the mouth and over the gills. Sharks have paired **pectoral fins** located behind the head and an asymmetrical tail (see Figure 4.22A).

Members of the class Osteichthyes have an ossified (bony) skeleton composed of a skull, a vertebral column, and ribs. About 25,000 species of bony fishes live in marine and freshwater systems. Bony fishes have several adaptations for living in aquatic environments, including a swim bladder that functions in depth control, paired pectoral and pelvic fins, and a symmetrical tail fin (see Figure 4.22B). The gill slits of bony fishes consist of a single opening covered by an **operculum** that protects the gills and allows oxygen-containing water to enter the mouth and flow over the gills in a controlled fashion.

Superclass Tetrapoda

Tetrapoda is composed of the other four familiar vertebrate classes, **Amphibia**, **Reptilia**, **Aves**, and **Mammalia**. Organisms in these classes are characterized by the presence of bilateral symmetry, increasing **cephalization** (the concentration of specialized nervous tissues in the head), and four limbs used for locomotion. The amphibians and reptiles are **poikilothermic** in that they have no internal mechanisms that control their body temperature, but the birds and mammals are **homeothermic** because they have such mechanisms.

Amphibians Amphibians' transition from aquatic to terrestrial life was never completed, and they all must return to aquatic environments to reproduce (see Figure 4.23). Amphibians include salamanders, newts, frogs, and toads, many of which have existed with few changes for over 150 million years. They all have smooth, moist skin and gills at some life stage. Most species develop lungs as adults, like the frog shown in Figure 4.24.

Reptiles Reptiles were the first truly terrestrial vertebrates. Certain characteristics acquired over millions of years of evolution enabled them to inhabit terrestrial environments. These characteristics included skin with protective scales, internal fertilization, and the **amniote egg**, with a waterproof, leathery shell and a fluid-filled membrane, shown in Figure 4.25. The last two developments freed

Figure 4.22

(A) The leopard shark is a member of the class Chondrichthyes. (B) Members of the class Osteichthyes, the bony fishes, live in both fresh and marine water. Some class members, such as this sturgeon, resemble ancestors that swam in rivers 400 million years ago.

A

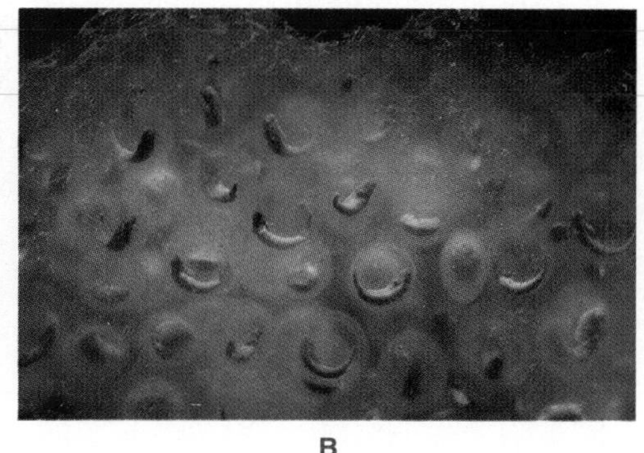

B

C

Figure 4.23

The life cycle of an amphibian like this salamander starts with sexually mature adults. Salamander eggs are fertilized within the female after she picks up a sperm packet that was deposited by a male on a stick, a leaf, or the ground. After internal fertilization, eggs are laid in fresh water (A) where they develop (B). Young larvae (C) resemble their parents and grow to adult size through successive metamorphosis.

Figure 4.24

This adult bullfrog is the largest frog species in the New World. As tadpoles, bullfrogs live in ponds and lakes and feed on algae, but as adults they are carnivores, feeding mainly on insects.

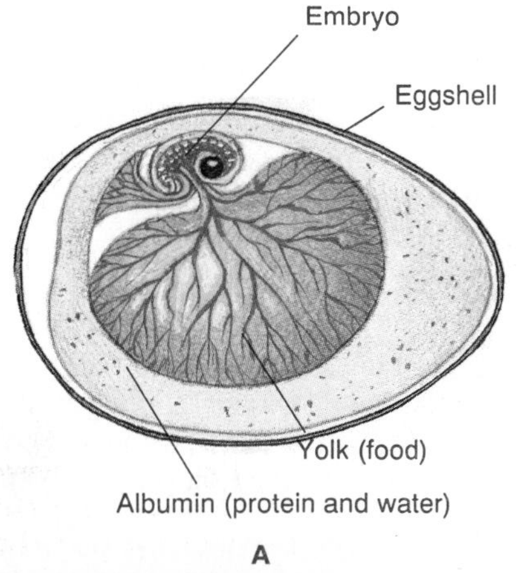

A

B

Figure 4.25

Reptiles are terrestrial poikilotherms that have been liberated from aquatic environments because of their amniote (shelled) egg (A). With a waterproof covering and an internal water supply, reptiles–like this pilot blacksnake with 14 eggs she just laid—have been able to inhabit most regions of Earth (B).

reptiles from any dependence on water to complete their reproduction. Living reptiles include lizards, snakes, turtles, alligators, and crocodiles.

Aves Birds evolved from reptiles millions of years ago. All birds retain reptilian-like scales on their feet, and their feathers are actually highly modified scales. Unlike reptiles, modern birds produce amniote eggs that have hard, waterproof, calcium-impregnated shells; they have no teeth; and their forelimbs have evolved into wings. There are 19 orders of birds containing more than 8,600 species (see Figure 4.26). With their highly evolved structures enabling them to fly, they dominate the airspace they share with arthropod insects and mammalian bats.

A B

Figure 4.26

Unique features of birds include feathers, wings, and calcareous shelled amniote eggs. Each of these birds, young screech owls (A) and a cardinal (B), represents one of the 19 bird orders.

A

B C

Figure 4.27

The class Mammalia includes species with diverse reproductive strategies that range from the primitive egg-laying platypus (A) to the marsupial opossum (B) to the more advanced placental rodent, a beaver (C).

Mammals Unique characteristics of the class Mammalia include the presence of hair, functional red blood cells without nuclei, and female mammary glands. There are two subclasses of mammals: the primitive **Prototheria**, consisting of the egg-laying, duck-billed platypus and the spiny anteater, and **Theria** ("true mammals"), which contains all other species (see Figure 4.27). Theria includes both *marsupials* and *placental mammals* and is divided into 17 orders containing 4,600 species that inhabit most aquatic, terrestrial, and aerial environments (see Figure 4.28). The **marsupials** have internal fertilization and early development, but later development occurs in an external pouch, where the embryo feeds on milk provided through the nipple of a mammary gland. The **placental mammals** complete their development within the female uterus, nourished by the **placenta**, an organ jointly produced by the mother and the developing embryo.

With the exception of the echinoderms, the most complex lower animals are coelomate protostomes. The echinoderms and chordates are coelomate deuterostomes. Reptiles, birds, and mammals are the only truly terrestrial chordate phyla, and the latter two are the only homeothermic ones.

Figure 4.28

The 17 mammalian orders are represented by the four shown here. Dolphins are marine mammals (A), eastern gray kangaroos are marsupials (B), elephants are the largest terrestrial mammals (C), and the pig-tailed macaque represents the primates (D).

A B

C

D

Table 4.3	A Comparison of the Taxonomic Classification of the Black Oak and the Domestic Sheep.	
	Black Oak	**Sheep**
Kingdom:	Plantae	Animalia
Phylum:	Tracheophyta	Chordata
Subphylum:	Spermatophyta	Vertebrata
Class:	Angiospermae	Mammalia
Subclass:	Dicotyledoneae	Theria
Order:	Spindales	Artiodactyla
Family:	Fagaceae	Bovidae
Genus:	*Quercus*	*Ovis*
Species:	*velutina*	*musimon*

TAXONOMY

Taxonomy (or systematics) is the science of classifying organisms. The different levels in the classification system are called **taxa**. From the time of Linnaeus, taxonomy has been an important component of descriptive biological sciences. The taxonomic classification of newly discovered organisms and modification of the status of other species continues today as new information becomes available. An example of the taxonomic classification of a typical plant and a typical animal is presented in Table 4.3.

You share the same taxa with the sheep from the kingdom level through the class Mammalia, but divergence occurs at the order level. You are classified in the order Primate, family Hominidae, and genus and species *Homo sapiens,* as shown in Figure 4.29.

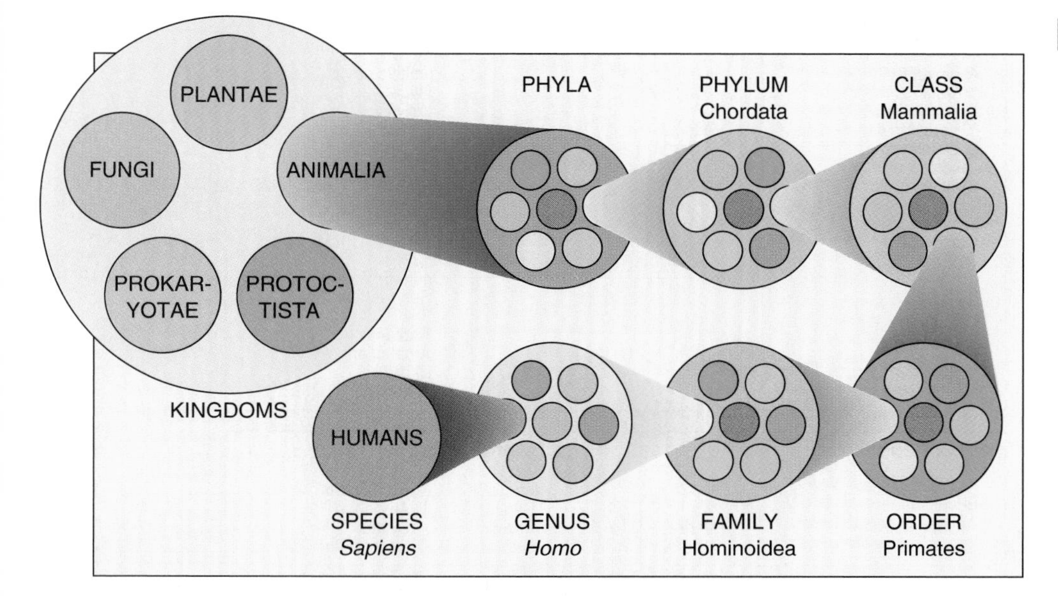

Figure 4.29

The taxonomic hierarchy used to classify the five kingdoms. Each of the kingdoms includes a number of phyla, each phylum a number of classes, each class a number of orders, each order a number of families, each family a number of genera, and each genus a number of species. Thus we all belong to kingdom Animalia, phylum Chordata, class Mammalia, order Primates, family Hominidae, genus *Homo,* and species *sapiens.*

A FUNCTIONAL CLASSIFICATION SYSTEM

The five kingdoms of life can also be classified by an alternative *functional classification system*. The basis for such a system is the mode of nutrient acquisition in each of the major groups as described in Figure 4.30. **Functional classification systems** emphasize processes that occur among organisms, and they are becoming important in both terrestrial and aquatic ecology.

Producers

Cyanobacteria in the kingdom Prokaryotae, algae in the kingdom Protoctista, and the entire kingdom Plantae are all classified as **primary producers**. Through photosynthesis, they combine the energy-poor molecules of carbon dioxide and water and convert them into energy-rich organic molecules (sugars) using sunlight as the energy source and releasing oxygen as a by-product. Most other organisms on Earth depend on the products formed during photosynthesis by these primary producers.

Decomposers

Archaebacteria and eubacteria in the kingdom Prokaryotae and the entire kingdom Fungi can be classified as **decomposers**. They all produce digestive enzymes that reduce complex organic molecules into a form that the organisms can absorb. The free-living species of this group are saprophytic since they digest nonliving organic matter. Pathogenic forms have the ability to digest and destroy living cells, and in some ways they might be considered consumers. However, for convenience, all of these organisms are classified as decomposers because they do not ingest nutrients but rather absorb essential energy-rich organic molecules.

Consumers

The protozoan species in the kingdom Protoctista and all members of the kingdom Animalia are considered to be **consumers** because they ingest complex energy-rich organic molecules. Herbivores consume primary producers, and carnivores feed on other animals or protozoans.

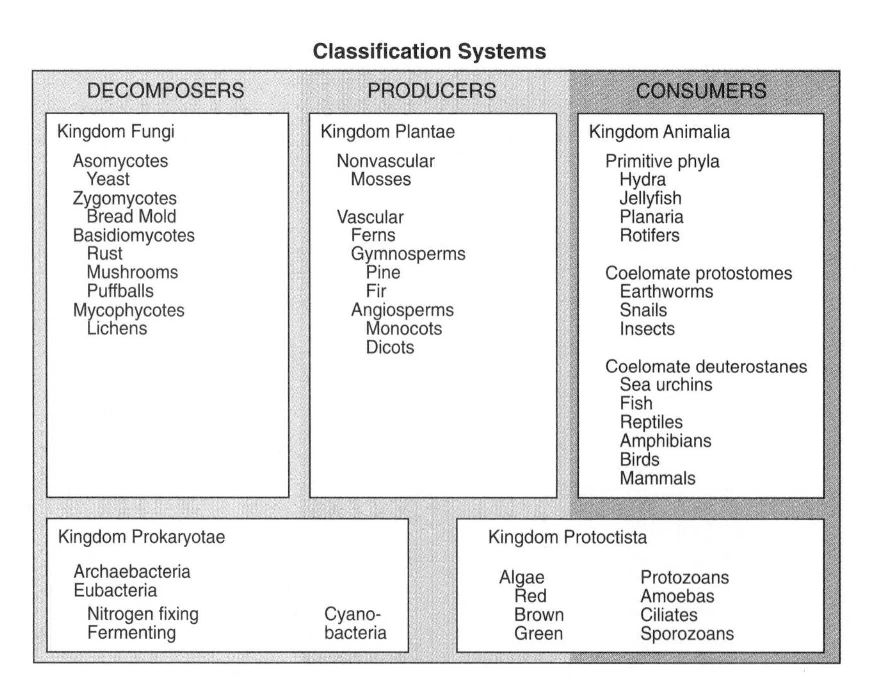

Figure 4.30

This classification system distinguishes organisms in the five kingdoms in two different ways. First, organisms are grouped with respect to their taxonomy within their kingdom. Second, in the functional system, organisms of a kingdom are regrouped depending on their mode of nutrient acquisition. Decomposers include all of the fungi and some prokaryotes, producers include all of the plants and the prokaryote cyanobacteria, and protoctist algae, and consumers include all of the animals and the non-photosynthetic protoctists.

Taxa are different levels of the classification system. Kingdom, phylum, class, order, family, genus, and species are the major taxa. The relative importance of each taxon differs among the five kingdoms, depending on species diversity. A functional classification system of producers, decomposers, and consumers is a useful alternative in the study of ecology.

Summary

1. All species in the kingdom Plantae exhibit alternation of generations. Sporophyte plants produce spores that germinate and grow into gametophytes, which in turn produce gametes that unite during fertilization to produce new sporophytes.
2. The nonvascular bryophytes are liverworts, hornworts, and mosses; the vascular tracheophytes group all of the higher forms of plant life. Ferns have undulipodiate sperm that restrict them to environments that are moist during their period of sexual reproduction.
3. Gymnosperms and angiosperms are two major types of living tracheophytes. Four phyla of gymnosperms contain the cycads, ginkgos, conifers, and gnetophytes, respectively, whereas all true flowering plants—those that produce seeds enclosed within fruits—are in a single phylum.
4. Angiosperms are divided into two classes: monocots, which have a single cotyledon, leaves with parallel venation, and floral parts in threes or multiples of three; and dicots, which have two cotyledons, netted leaf venation, and floral parts in fours or fives or multiples of four or five.
5. The phyla in the kingdom Animalia are segregated according to their pattern of embryonic development and the presence or absence of organ systems. Sponges have tissues composed of two cell layers separated by mesenchyme, whereas jellyfish and sea anemones are characterized by radial symmetry.
6. Acoelomates are represented by the planaria and ribbon worms, pseudocoelomates by rotifers and nematodes. Two developmental forms of coelomates, the protostomes and the deuterostomes, make up the remaining animal phyla. Annelids, mollusks, and arthropods are characteristic protostomes, and echinoderms and chordates are representative coelomates.
7. The brain, notochord, jaws, and paired jointed appendages are characteristics used to differentiate organisms in the phylum Chordata. The four classes of tetrapods are amphibians, reptiles, birds, and mammals.
8. Producers, decomposers, and consumers are the major categories in a functional classification system of all the organisms in the five kingdoms of life. The basis of this system is the mode of nutrient acquisition.

Review Questions

1. What are the major components in the life cycle of a moss?
2. What are the two major types of tracheophytes, and how are they different?
3. How do ferns and gymnosperms differ?
4. What characteristics are used to separate monocots from dicots?
5. How do sponges differ from members of the radially symmetrical phyla?
6. How does a pseudocoelom differ from a coelom?
7. What characteristics are typical of annelids, mollusks, and arthropods?
8. Which phyla are classified as coelomate deuterostomes? What characteristics result in their inclusion in this group?
9. What are the characteristics that separate classes in the superclass Tetrapoda?
10. What is a functional classification system?
11. How do taxonomic and functional classification systems differ?

Essay and Discussion Questions

1. What other, alternative classification systems might be devised?
2. Give some reasons for classifying organisms.
3. What are some characteristics that distinguish plants from animals?
4. How do plant and animal life cycles differ?

References and Recommended Reading

Hickman, C., L. Roberts, and F. Hickman. 1982. *Biology of Animals*. 3d ed. St. Louis: Mosby.

Little, L. 1980. *The Audubon Society Field Guide to North American Trees: Western Region*. New York: Knopf. Also see the other volumes in this series, including *Birds: Eastern Region* and *Western Region*; *Butterflies, Fish, Whales and Dolphins*; *Mammals*; *Reptiles and Amphibians*; *Seashore Creatures*; *Trees: Eastern Region*; and *Wildflowers: Eastern Region* and *Western Region.*

Raven, P., R. Evert, and S. Eichhorn. 1986. *Biology of Plants*. 4th ed. New York: Worth.

CHAPTER 5

The Biosphere

Chapter Outline

Conceptual Beginnings

Focus on Scientific Inquiry: Glacial Lake Missoula and the Spokane Flood

World Climate

Climates of North America

Focus on Scientific Explanations: The Properties of Water

World Biome Development

Reading Questions

1. What factors led to the formation of the biomes on Earth?

2. How is the curvature of Earth related to its climates?

3. Why do the climates differ across North America?

4. What are some predictive models of biome formation?

The **biosphere** consists of a thin envelope of air, water, and land in which all known life forms exist. The term *biosphere* dates from 1875, when the Austrian geologist Eduard Suess considered the interactions of several "envelopes" of Earth in his book on the development of the Alps.

In this chapter, we will explore how energy from deep within the planet caused the continents to form numerous alignments in past eons, how sculpting forces have created the present surface features of the biosphere, and how forces in our solar system contribute to the march of the seasons and the development of world climates. Integrating these developments will help you understand why distinct biomes exist in different areas of Earth. **Biomes** are large regions with characteristic assemblages of plants and animals that change from the polar to the equatorial latitudes.

CONCEPTUAL BEGINNINGS

The conceptual integration of living organisms with the **hydrosphere** (all the water on Earth's surface), the **lithosphere** (Earth's rigid crustal plates), and the **atmosphere** (the gaseous envelope surrounding Earth) was first established in the 1920s by the eminent Russian scientist Vladimir Ivanovitch Vernadsky. Life, he contended, is "intimately associated with the structure of the Earth's crust, it is involved in the mechanism of the latter, and in this mechanism it is responsible for the most important functions, without which the mechanism could not exist." In other words, he thought that living organisms of the past had been involved in shaping and maintaining the stability of habitable environments throughout the ages of Earth.

Energy Sources

Long-term changes in the biosphere through time were due to two energy sources. Tectonic movements of the continental plates were driven by energy derived from deep within Earth before and during the formation of Pangaea, 230 to 280 million years ago. These forces created the present alignment of the continents and will continue to rearrange them throughout the geologic future. The **geological time scale**, shown in Figure 5.1, depicts the eons, eras, periods, and epochs of the 4.6-billion-year history of Earth.

Different assemblages of organisms inhabited the continents as they drifted apart following the breakup of Pangaea 180 million years ago. Throughout this long period, most life forms have depended on energy derived from the sun. However, as shown in Figure 5.2, this energy is not evenly distributed over the face of the planet, for Earth is a sphere, and because of its curvature, the sun's rays are spread over a greater area near the poles than they are near the equator. Thus energy—light and heat—is more concentrated at the equator and becomes more diffuse toward the poles. This difference is responsible for the development of world climates and for major differences in the distribution of life within the biosphere.

Sculpting Forces

The topography (surface features) of Earth consists of coastal plains, elevated plateaus, canyons, and mountain ranges of varying age. These surface features have resulted from the interactions of major lithospheric, hydrospheric, and atmospheric forces throughout the vast expanse of geologic time. When the continental blocks of Pangaea broke apart during the Mesozoic era, they drifted up and over the oceanic blocks, producing massive uplifting forces that created many of the major continental mountain ranges (see Figure 5.3). Climatic conditions varied during these mountain-building episodes, and for extended periods of time, large areas of the continents in the Northern Hemisphere were covered with massive sheets of ice. These periodic ice ages had profound effects on the structure

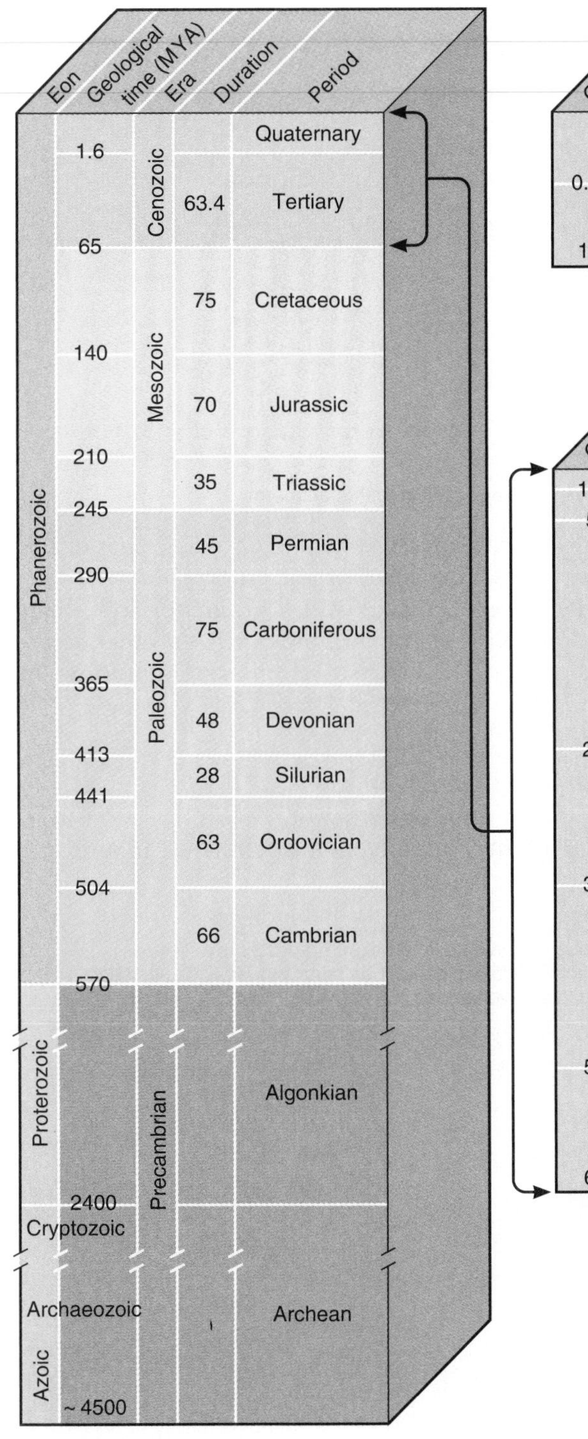

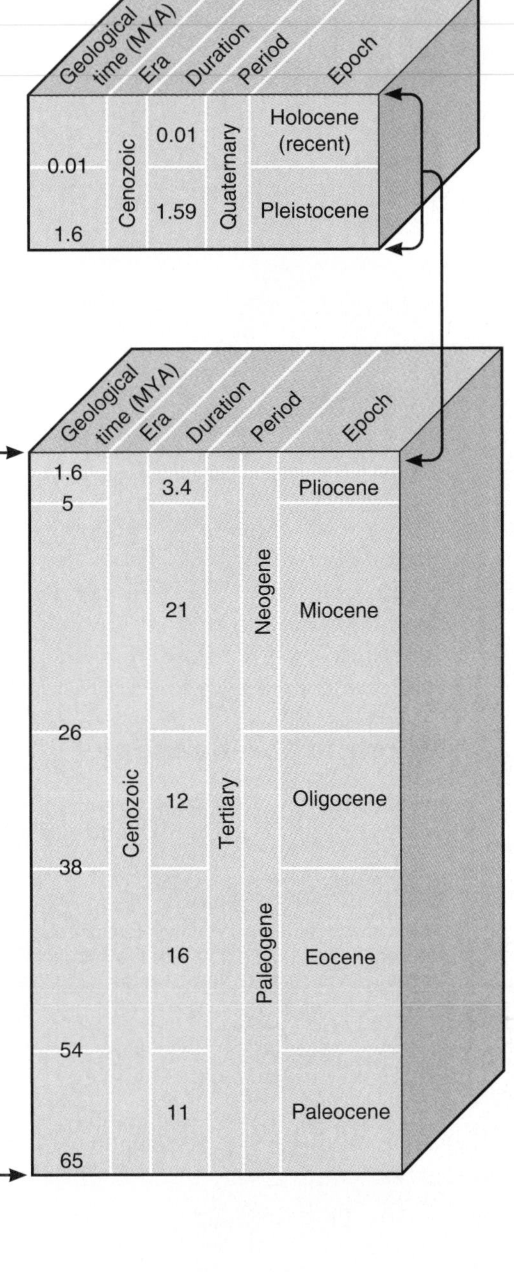

Geological Time Scale

Figure 5.1

The geological time scale chronicles the past eons, eras, periods, and epochs of the history of Earth. Most of these divisions are based on the ages that have been determined for major rock strata. The time scale used (MYA) is millions of years ago.

of mountain terrains. The last Great Ice Age occurred during the Pleistocene epoch, and lasted from 20,000 to 12,000 years ago. As seen in Figure 5.4, this ice age caused dramatic changes in Earth's surface. In North America, Pleistocene glaciers left many reminders of their colossal forces, including the Great Lakes in the Midwest and the alpine valleys and lakes in the Rocky Mountain and Sierra Nevada ranges (see Figure 5.5A). The receding glaciers even changed the direction of some river flows. When massive glacial dams melted, the enormous deluges of water they released scarred large landscapes in certain areas (see the Focus on Scientific Inquiry, "Glacial Lake Missoula and the Spokane Flood."

Figure 5.2

Solar energy strikes Earth directly at the equator and at oblique angles as latitude increases, which results in less energy per unit surface area.

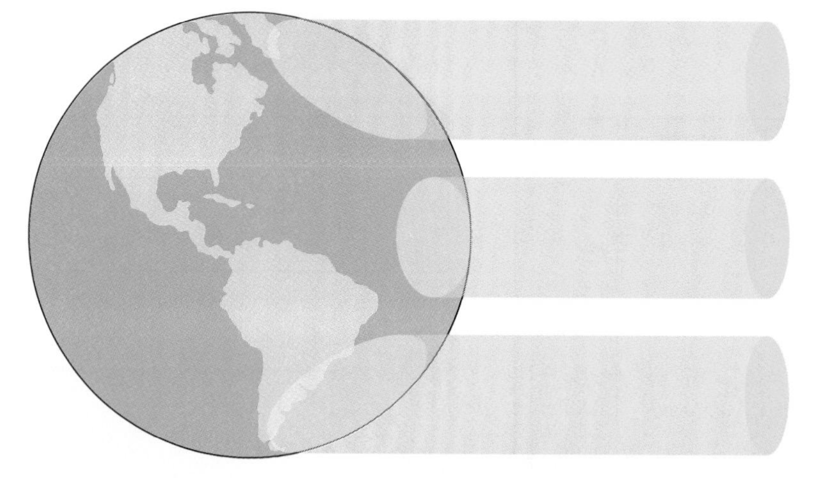

Variation in area covered by solar beams carrying equal amounts of energy

Not all areas of North America were affected by the ebb and flow of the glaciers, but other forces have shaped vast expanses of ancient terrains. The Colorado River, and others before it, carved canyons over a kilometer and a half deep, exposing rocks that are 1.7 billion years old (Figure 5.5B). In other areas, wind and water combined in an erosion process that sculpted unique forms in varied landscapes (Figure 5.5C).

March of the Seasons

Living organisms inhabit environments from polar regions to the equator, from below sea level in Death Valley to about 6 kilometers in the highest mountains,

Figure 5.3

When continental and oceanic plates collide, organic sediments are dragged down beneath the continental plate along with the subducting oceanic plate. This results in massive uplifting forces that can create new mountain ranges and volcanoes.

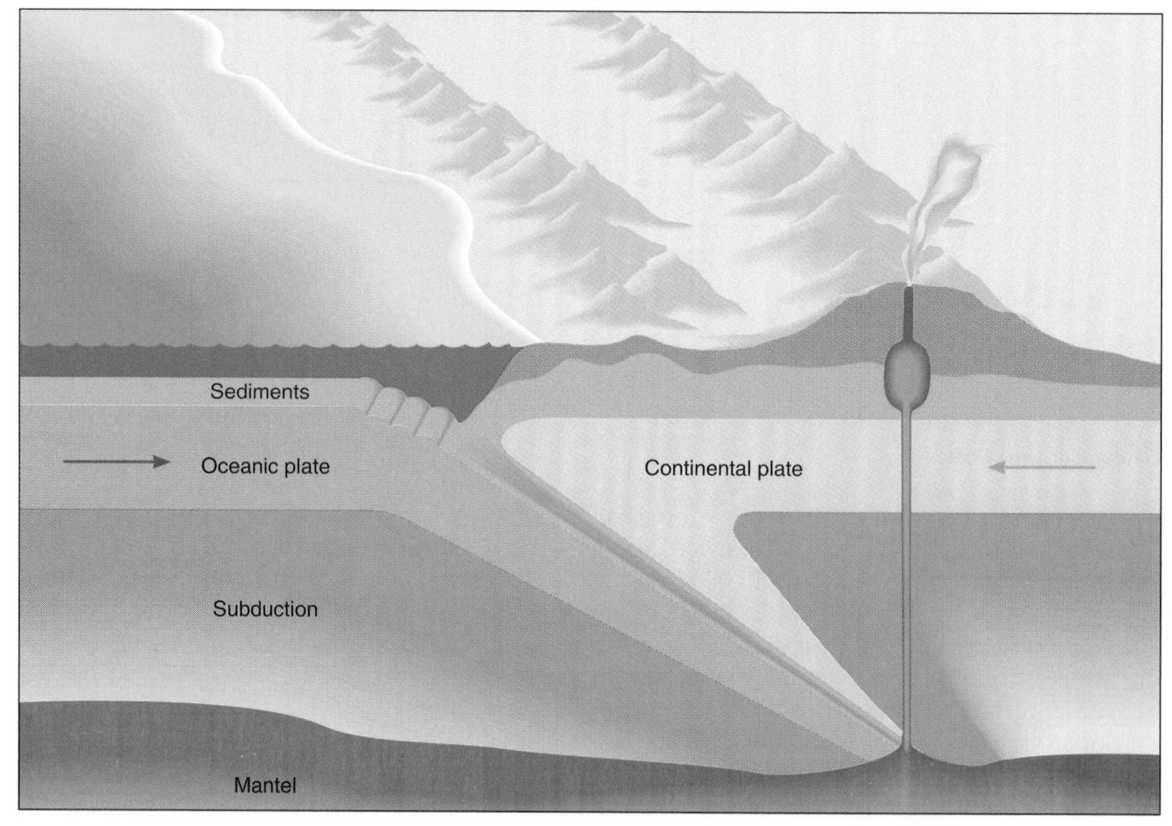

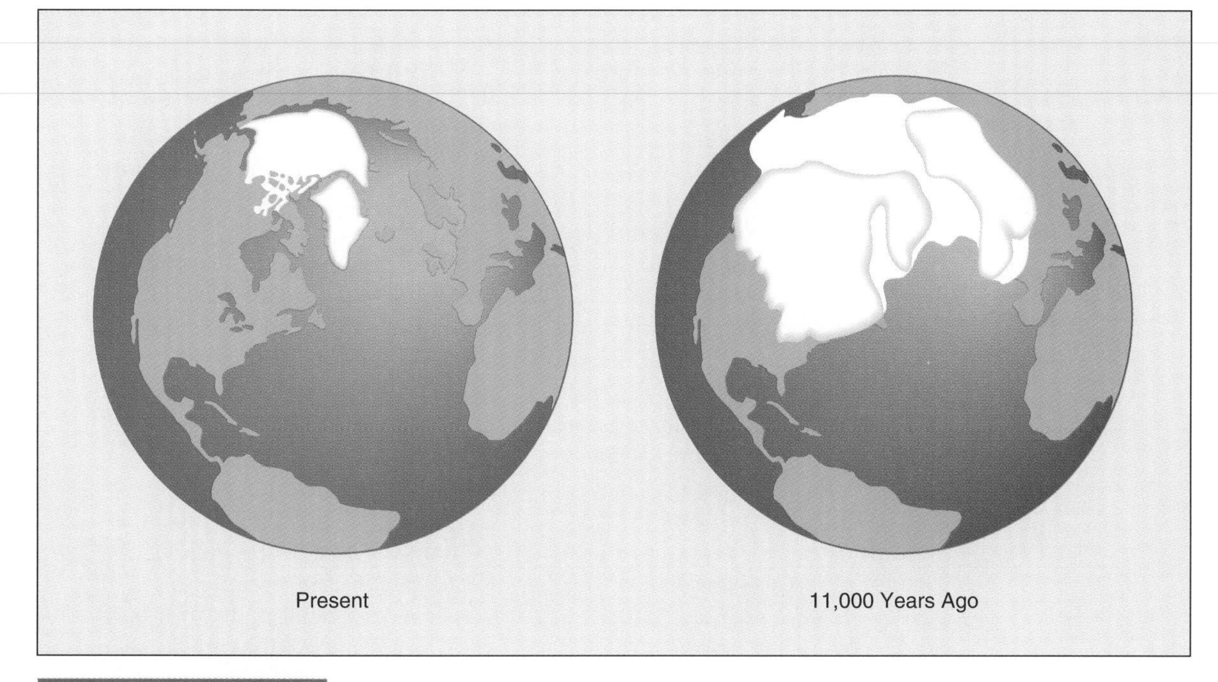

Figure 5.4

Comparing the present world map with a map of 11,000 years ago dramatizes the extent of glaciation at the end of the last ice age.

and from the oceans' surfaces to their deep trenches. The inhabitants of these different regions have adapted to tremendous variations in day length, temperature, and available moisture that result from Earth's planetary relationships within the solar system. The term **adaptation** can be used in both an ecological and an evolutionary context. In ecology it is the process of adjustment of an individual organism to environmental stresses. In its evolutionary context, adaptation can result from the processes that modify species over time, improving their reproductive and survival efficiencies.

Earth orbits the sun once each year and makes one complete rotation on its axis each day. However, the axis of its rotation is inclined from its orbital plane, causing the Northern Hemisphere to tip toward the sun during summer (see Figure 5.6 on page 72.). The first day of summer (the summer solstice) occurs on June 21, when the vertical rays of the sun strike Earth at the Tropic of Cancer, resulting in 24 hours of continual sunlight in the far northern latitudes. As summer progresses to autumn, the days become shorter in the Northern Hemisphere, and

Figure 5.5

(A) This lake was formed in the depression made by mountain glaciers during the last ice age. (B) Rock strata of the Grand Canyon represent all of the major eons, eras, periods, and epochs listed on the geologic time scale from the present to the Early Precambrian. (C) Ancient terrains in the Badlands of South Dakota show wind and water erosion.

A B C

Focus on Scientific Inquiry

Glacial Lake Missoula and the Spokane Flood

The Roaring Twenties produced many innovations in the science and culture of the Western world. New hypotheses were being advanced in many fields of science, and great controversies often developed as their merits were debated in the scientific literature. Early in the decade, J. Harlen Bretz at the University of Chicago published a paper in the *Journal of Geology* titled "The Channeled Scablands of the Columbia Plateau." In this 1923 paper, Dr. Bretz advanced the hypothesis that a large portion (40,000 square kilometers) of the eastern half of the state of Washington had been eroded by a catastrophic deluge, creating what he termed "channeled scabland" (Figure 1). His paper was received with less than enthusiastic acceptance, and for decades many geologists would give it no credence at all. This "Spokane flood" controversy focused on a central question: What could have been the source of the immense volume of water required to produce this massive erosion?

By 1930, Dr. Bretz had located this water source. It was known as Glacial Lake Missoula (see Figure 2). This enormous lake resulted from a regression of the late Pleistocene Cordilleran ice sheet that had covered southern British Columbia, eastern Washington, northern Idaho, and western Montana. The glacier came to rest against the Bitterroot Mountains in the panhandle of Idaho, blocking the Clark Fork of the Columbia River with an ice dam over 600 meters high. The lake behind the ice dam inundated the valleys of western Montana, creating an island lake of glacial meltwater half the size of Lake Michigan. This ice dam advanced and regressed a number of times, and with each regression, massive flooding occurred in eastern Washington, creating the "channeled scablands."

Evidence supporting Bretz's hypothesis can be found today. On the mountains east of Missoula, Montana, 240 kilometers upriver from the site of the glacial ice dam, strandlines of Glacial Lake Missoula reach an elevation of

Figure 1 Dry falls in the Grand Coulee of the Columbia River was formed during the great Spokane floods.

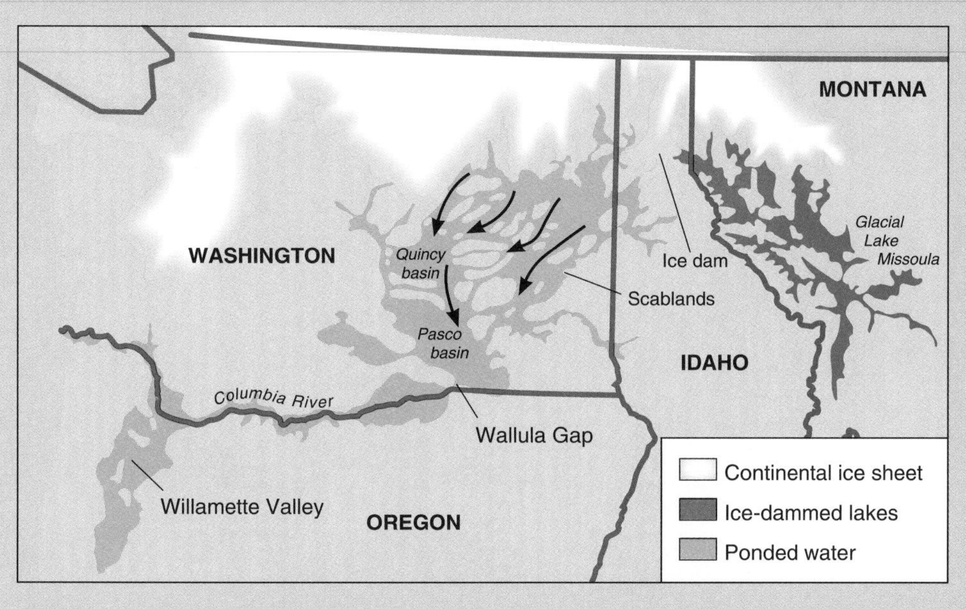

Figure 2 A great flood occurred when the water in Glacial Lake Missoula was released as an ice dam on the Clark Fork of the Columbia River melted.

1,280 meters. The 2,100 cubic kilometers of water released when the Cordilleran ice sheet made its final retreat 12,000 years ago performed the last sculpting of the channeled scablands of eastern Washington, terraced the Columbia River Gorge, and filled the 520-square-kilometer Willamette Valley of western Oregon, to a depth of 20 meters, with soils from eastern Washington, northern Idaho, and western Montana.

Figure 5.6

The "march of the seasons" results from the 23½degree tilt of Earth's axis from the plane of its orbit around the sun. In the fall, the oblique angle of sunlight increases in the Northern Hemisphere as the axis of rotation tips away from the sun. This reduces the amount of solar energy received per unit area and causes the cooler temperatures that lead to winter. As Earth's axis tips toward the sun, the oblique angle of light decreases, which increases the amount of solar energy per unit area and leads to spring and summer.

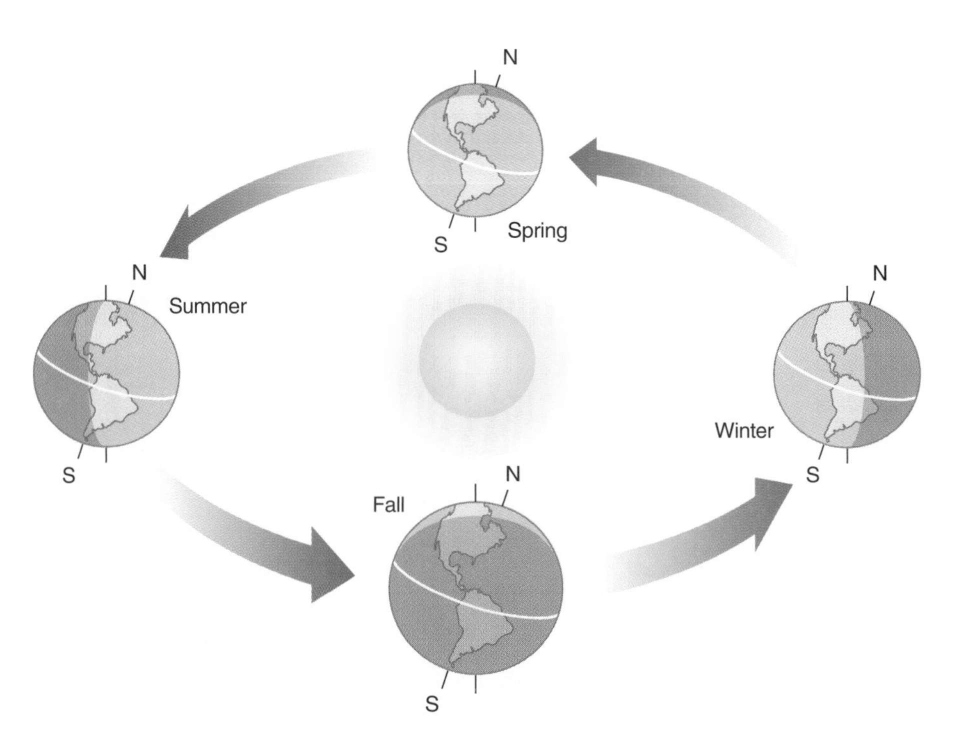

fall begins with the autumnal equinox on September 23. On this date, the vertical rays of the sun fall on the equator, and throughout the entire biosphere, day and night last 12 hours each.

As fall slips into winter in the northern latitudes, summer arrives in the Southern Hemisphere. On December 22, the winter solstice occurs, when the sun's rays are vertical on the Tropic of Capricorn. Far northern arctic latitudes are cast into continual darkness, but the southern latitudes bask in the warmth of full summer. While Alaskans brace for subzero temperatures and blizzards move over the northern Great Plains, revelers spend New Year's Day on sunny beaches in Sydney and Rio de Janeiro. Spring occurs on March 21 (the vernal equinox), when once more the vertical rays of the sun fall on the equator, and day and night are again equal.

The source of the seasons is related to the tilt of the planet. The differential heating from the equator to the polar latitudes creates forces with far-reaching effects over the entire biosphere. This variation in energy generates both the climates of the world and the major currents that circulate in the great oceans.

Life on Earth was hypothesized to play a role in producing and maintaining habitable environments in the 1920s. Sunlight and energy from deep within the planet affect both the development and the maintenance of the biosphere. Topography, the tilt of the planet, and Earth's orbital variations all interact in creating climates and seasons.

WORLD CLIMATE

The variations in energy received in different areas of the globe cause winds in the atmosphere that deliver moisture to the continents and establish the major circulation patterns in the oceans. The polar latitudes receive only about one fifth of the total annual sunlight received in the tropics. Temperate latitudes receive about twice the solar energy at the summer solstice as they do at the winter solstice. Why doesn't this uneven distribution of heat lead to even greater temperature extremes than are observed? Do global mechanisms redistribute this uneven concentration of energy?

Atmospheric Cells

In 1735, George Hadley, an English scientist, attempted to explain the prevailing wind patterns of the Atlantic Ocean. He knew that the tropics receive more energy from the sun than the polar regions and also that as air warms, it rises. This movement of air creates wind. From these observations, he developed a model of Earth showing an *atmospheric cell,* a circular zone where air would rise in the tropics, move north, and then descend in the colder polar regions. Surface winds generated in these cells would blow from the poles to the equator; however, in the Northern Hemisphere they would be deflected westward due to the counterclockwise rotation of Earth (see Figure 5.7A). Hadley's explanation required that all winds blow from east to west, but that is not the case. The trade winds blow from east to west, but the westerlies blow from west to east, and in the equatorial doldrums and horse latitudes, winds are rare. These latitudes were a bane to the sailors in their clipper ships, for in these doldrums, the air moves vertically, up at the equator and down in the horse latitudes, and there is little surface movement. To explain these observed wind patterns, a new three-cell model, shown in Figure 5.7B, was developed many years later. Air movement within these multiple atmospheric cells accounts for all major wind patterns.

Ocean Circulation

Atmospheric cells and their associated winds are responsible for the large surface circulation patterns in the oceans that are described in Figure 5.8. **Coriolis forces** are associated with Earth's rotation and cause circular currents in large ocean basins in the Northern Hemisphere to rotate clockwise and those in the Southern Hemisphere to rotate counterclockwise. These ocean currents transport about 30 percent of the excess equatorial heat north or south, while the atmosphere transports 70 percent. Consequently, the oceanic currents, in combination with solar heating and surface winds, contribute to continental climates.

Figure 5.7

(A) Hadley's single-atmospheric-cell model reflected his hypothesis that air becomes heated and rises near the equator and cools and descends near the poles. This would cause a pressure differential resulting in surface winds that would blow from the poles to the equator, and the rotation of Earth would deflect them westward at the surface and eastward in the upper atmosphere. (B) In the later three-cell model, a combination of the decreasing circumference of Earth from equator to pole and its rotation from west to east (counterclockwise) results in the formation of three atmospheric cells in each hemisphere. After air rises near the equator, creating the low pressure of the doldrums, it descends at about 30° north or south latitude, creating the high pressure that causes surface winds that blow toward the equator. This pressure gradient causes the northeast and southeast trade winds. The descending air between 30° and 35° north or south latitude produces little if any surface wind in an area known as the horse latitudes. A second cell forms when air rises once again, causing low pressure at about 65° north or south latitude. This air descends at the horse latitudes and produces a surface wind known as the westerlies. Polar easterlies form from the cycle of the third atmospheric cell.

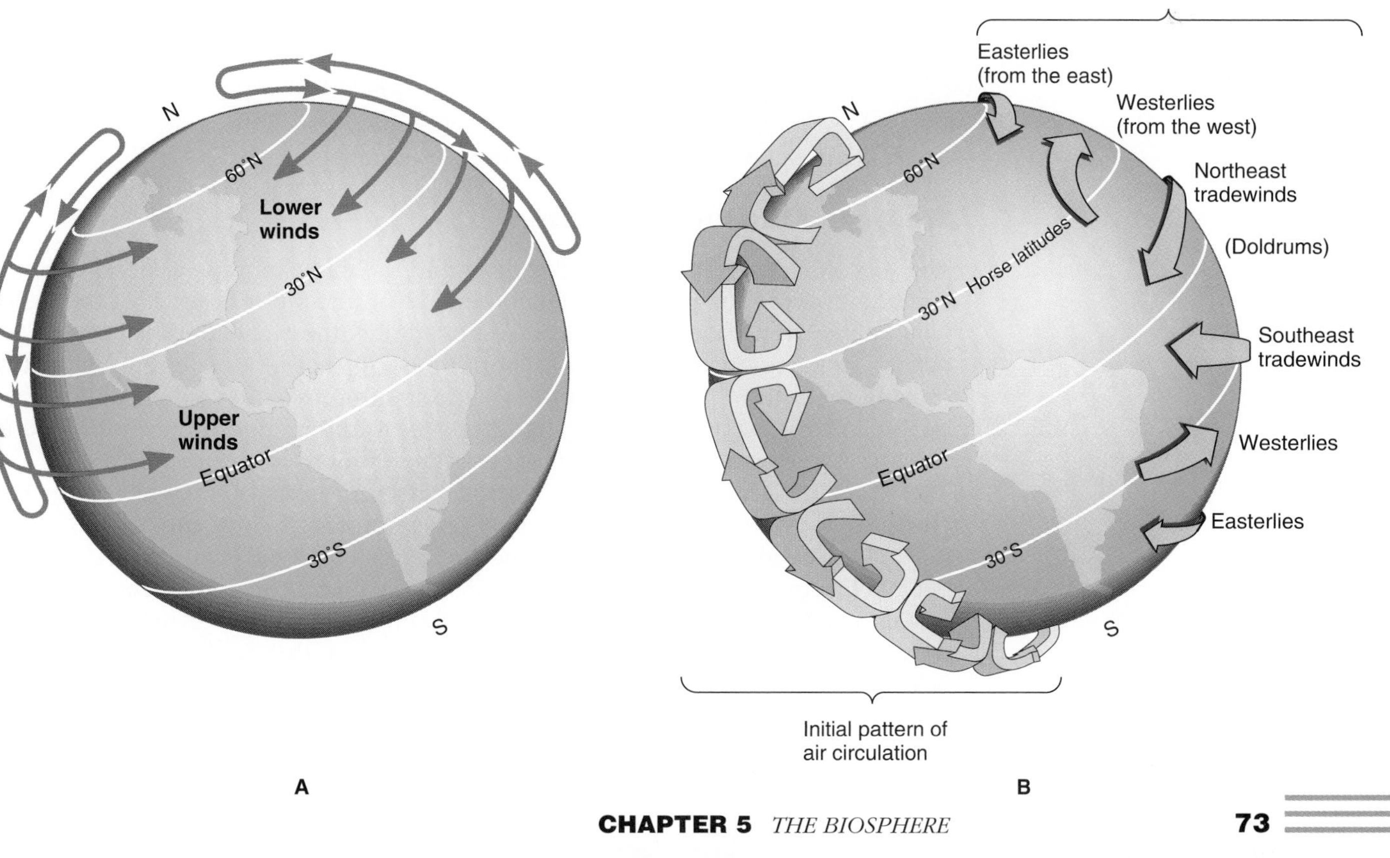

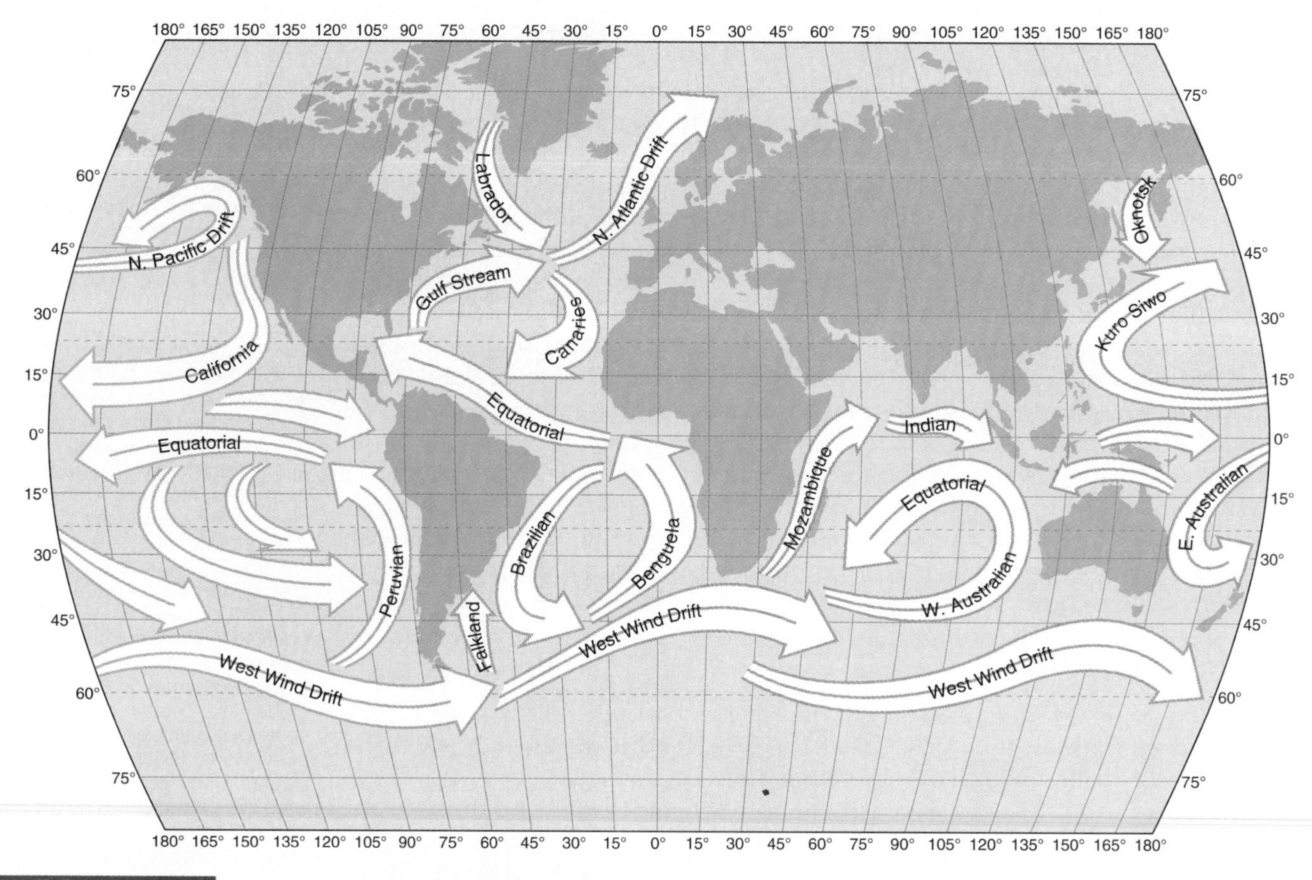

Figure 5.8

The North Pacific gyre (a spiral system of ocean currents) brings cold water in the California current that moderates climates of the west coast of North America. The gulf stream and the North Atlantic drift of the North Atlantic gyre tempers climates of the eastern seaboard and western Europe.

Three atmospheric cells in each hemisphere produce wind patterns that vary with latitude. These winds and Coriolis forces create the major ocean currents. Both the winds and the ocean currents transport excess equatorial heat poleward.

CLIMATES OF NORTH AMERICA

Climates vary greatly throughout the world, but most of them are represented somewhere in North America. These include the western maritime (ocean-influenced), mountain, and central and eastern continental climates.

Western Maritime Climates

The high capacity of oceans to store heat and the circulation patterns of currents combine to produce profound climatic effects on the western coast of the major continents. From about 40° north or south latitude, the circulation patterns in the oceans produce onshore flows of warm, moist air that moderate these climates, causing them to be milder than their latitude alone would predict. These maritime climates owe their special characteristics to the unique properties of water (see the Focus on Scientific Explanations, "The Properties of Water").

In North America, maritime climates sustain forests of ancient origin. The coastal redwoods of northern California, shown in Figure 5.9A, are remnants of a forest that covered much of North America in past geologic eras, when the continental climate was warmer and moister than it is today. The cool onshore breezes of summer bring moisture in the form of fog, but storm cells of winter become water-laden as they sweep over the warm waters of the Pacific Ocean, bearing heavy rains to these ancient forests.

A short distance north of the California-Oregon border, the effect of increasing latitude lowers the average annual temperature below the tolerance limit

Figure 5.9

(A) Redwood forests exist where cool, moist marine air flows onshore from the Pacific Ocean. (B) Sitka spruce forests grow from southern Oregon to coastal Alaska, replacing the redwoods in these higher latitudes.

of redwoods. Here they are replaced by Sitka spruce, a tree that dominates the coastal fringe from southern Oregon to southern Alaska (Figure 5.9B). The spruce forests of coastal Alaska persist to 60° north latitude. They could not exist that far north were it not for the moderating effects of the ocean.

Mountain Climates

Maritime coastal forests of the West would extend far inland were it not for the result of tectonic forces operating in the past and the present. These forces have been responsible for the development of major western mountain chains. The Coast, Cascade, and Sierra Nevada ranges and the Rocky Mountains all have major modifying effects on the climate of central North America (Figure 5.10A on page 78).

When winter storm cells surge over the west coast, they encounter these mountain ranges, and air is forced to rise. As air flows up and over mountain summits, cooling occurs, moisture condenses, and clouds release tremendous volumes of water in the form of rain or snow, depending on elevation and season (see Figure 5.10B on page 78). When rain forms, heat energy is released, warming the air mass, and as it descends the eastern mountain slopes, compression and further warming occur. This warm, descending air mass tends to absorb moisture, and precipitation decreases, forming a **rainshadow** east of these major mountain ranges. As moisture decreases in the rainshadow, forests are replaced by open woodlands, which give way to grasslands and deserts.

Central and Eastern Climates

The climates of central and eastern North America are often modified by maritime influences, but to a lesser extent than western and mountain regions. Weather systems spawned in the Gulf of Mexico bring summer rains to the Midwest. However, most major winter storms in the eastern half of North America originate in the Arctic, then travel south and east across the Great Plains, creating a continental climate with much greater fluctuations in temperature and moisture than are found in true maritime climates.

These continental climatic conditions are tempered on the eastern seaboard, where the waters of the Gulf Stream, a warm Atlantic Ocean current, moderate some of the extremes. The Gulf Stream extends its moderating effects beyond North America to the island provinces of eastern Canada, southern Greenland, Iceland, the isles of Great Britain, and much of northwestern Europe.

Western maritime, mountain, and central and eastern continental climates occur west to east across North America. Proximity to oceans, high mountain ranges, and latitude all contribute to climate.

THE PROPERTIES OF WATER

Focus on Scientific Explanations

Water is the most abundant substance in the biosphere. The oceans and freshwater lakes and rivers cover nearly 72 percent of Earth's surface. The heat balance of the planet allows water to exist in three physical states: liquid, solid, and gaseous.

Water molecules, as described in Figure 1, are composed of one atom of oxygen and two atoms of hydrogen, forming a polar molecule. This molecular polarity results from an electron charge variation in different regions, produced by the bond angles of the hydrogen atoms to the oxygen atom. The electron region nearest the hydrogen atoms has a positive charge, and the farthest region has a negative charge. The molecular polarity of water allows it to dissolve many substances. Other polar molecules, including acids, bases, salts, and many organic compounds, are soluble in water; for this reason, water is known as the *universal solvent* and is the major component of all life forms.

Another unique feature of water is the change in density that occurs as it changes state. Liquid water exists between 0°C and 100°C at standard pressure. However, when water begins to cool, the density increases until it reaches a temperature of 4°C. As cooling continues, the density decreases, and ice forms at 0°C. This property of water allows ice to float, which is significant indeed in the cold climates, because if ice were more dense than water, it would sink to the bottom of lakes and rivers when it formed, causing them to overflow and killing benthic (bottom-living) life forms. Other environmental effects of this property of water will be considered in Chapter 9.

The **specific heat** of water is exceeded only by that of liquid hydrogen and ammonia. It requires one calorie of heat to raise one gram of water 1°C; thus the specific heat of water is 1. If the oceans increase or decrease a number of degrees in temperature, a massive amount of heat energy is absorbed or released. This high **heat capacity** of water leads to the stabilizing effects oceans have on marine-influenced climates.

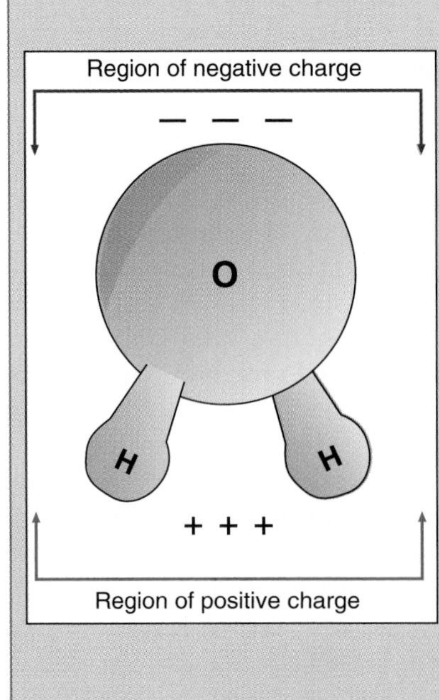

Figure 1 In a water molecule, an oxygen atom lies between two hydrogen atoms. The angle of attachment of the hydrogen atoms to the oxygen atom is known as the bond angle. In a water molecule, the bond angle is 105°. Since the bond angle is less than 180° and the protons of the hydrogen atoms carry a positive charge, a water molecule will have a region of relative positive charge and, on the opposite end, a region of negative charge. This charge difference is responsible for many of the solvent properties of water.

Figure 2 Water can exist in three different states on Earth: liquid (water), solid (ice), and gas (water vapor).

The surface waters of Earth have a high reflectivity. The different colors (wavelengths) of sunlight are differentially absorbed and reflected by water, resulting in the vivid blue color Earth projects into space, making it distinctive among the planets. When surface water vaporizes into a gaseous state, a large amount of energy is required, known as the **latent heat of evaporation**. When ice changes to water, energy known as the **latent heat of melting** is required (Figure 2). The release of latent heat can have profound effects on air temperature, and it is responsible, in large part, for the development of rainshadows.

Figure 5.10

(A) Mount Whitney is the highest mountain in the Sierra Nevada range. (B) The eastward-flowing air from the Pacific Ocean loses much of its moisture as it flows up and over the west-facing slopes of the Sierras. Deserts occur in rainshadows east of the Sierra Nevada range.

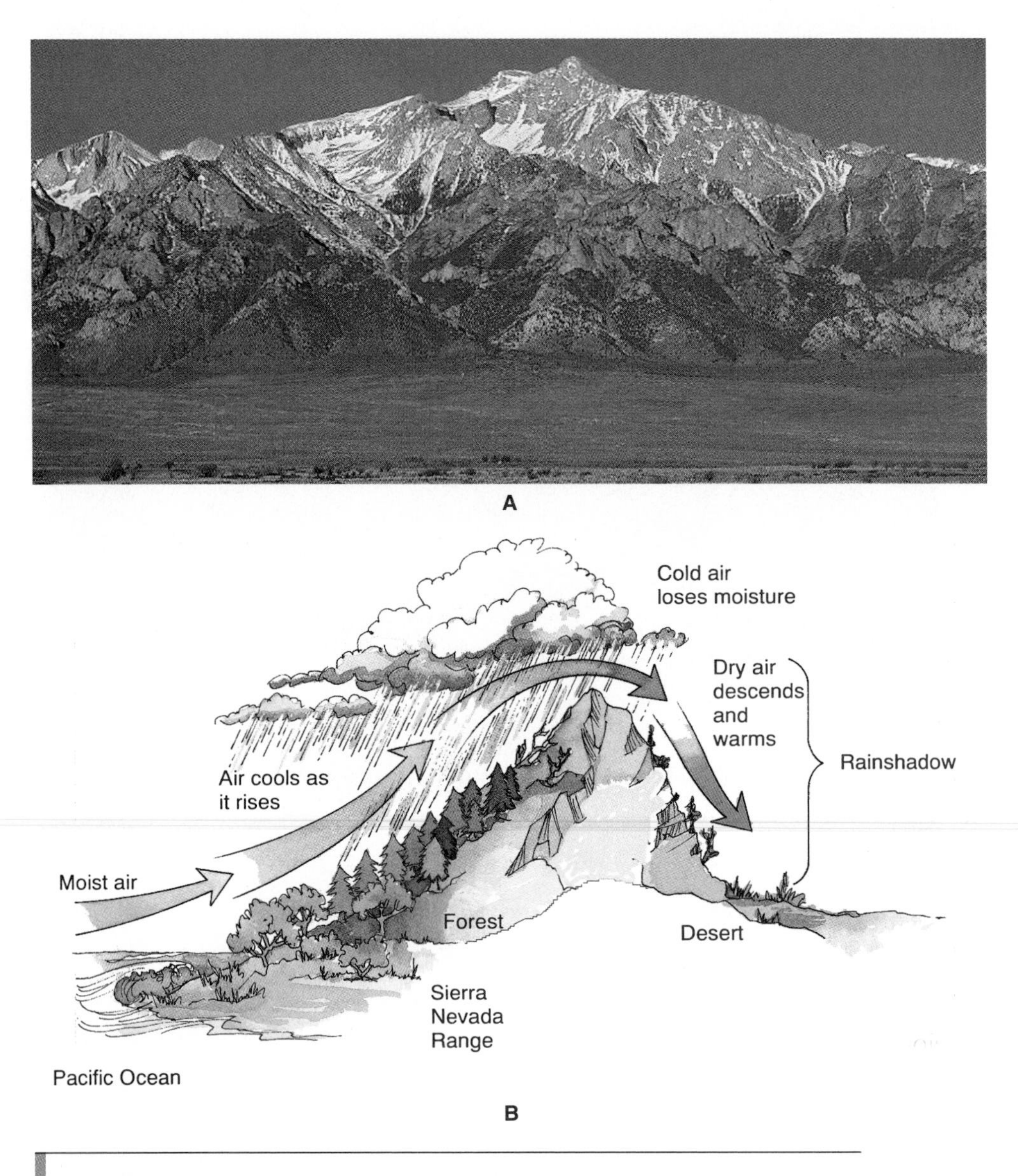

WORLD BIOME DEVELOPMENT

Variations among climates of the world result in a number of biomes, each with a unique assemblage of organisms. World biomes differ markedly between polar and equatorial regions.

Biome Formation

Variations in solar energy and moisture have led to the formation of a general latitudinal pattern of biome distribution that is described in Figure 5.11. This pattern began to be recognized in maps drawn during expeditions early in the sixteenth century. It is verified today by photographs from space, which show six major biome types: tropical forests, deserts, temperate forests, grasslands, taiga, and tundra.

Tropical forests are found in the equatorial regions of continents, between the Tropic of Cancer and the Tropic of Capricorn. *Deserts* tend to be located in the doldrums of 30° north and south latitude, where air in the Hadley cells descends. *Temperate forests* occur in latitudes of the westerlies, between 5° and 10° north and south of the 40th parallel. *Grasslands* develop in these temperate regions in areas devoid of major mountain systems and where distinct wet and dry seasons occur. The *taiga* begins at about 50° north latitude and ends at the lower limit of the permafrost, where the *tundra* begins. The taiga and tundra are characteristic of only the Northern Hemisphere, because there are no landmasses

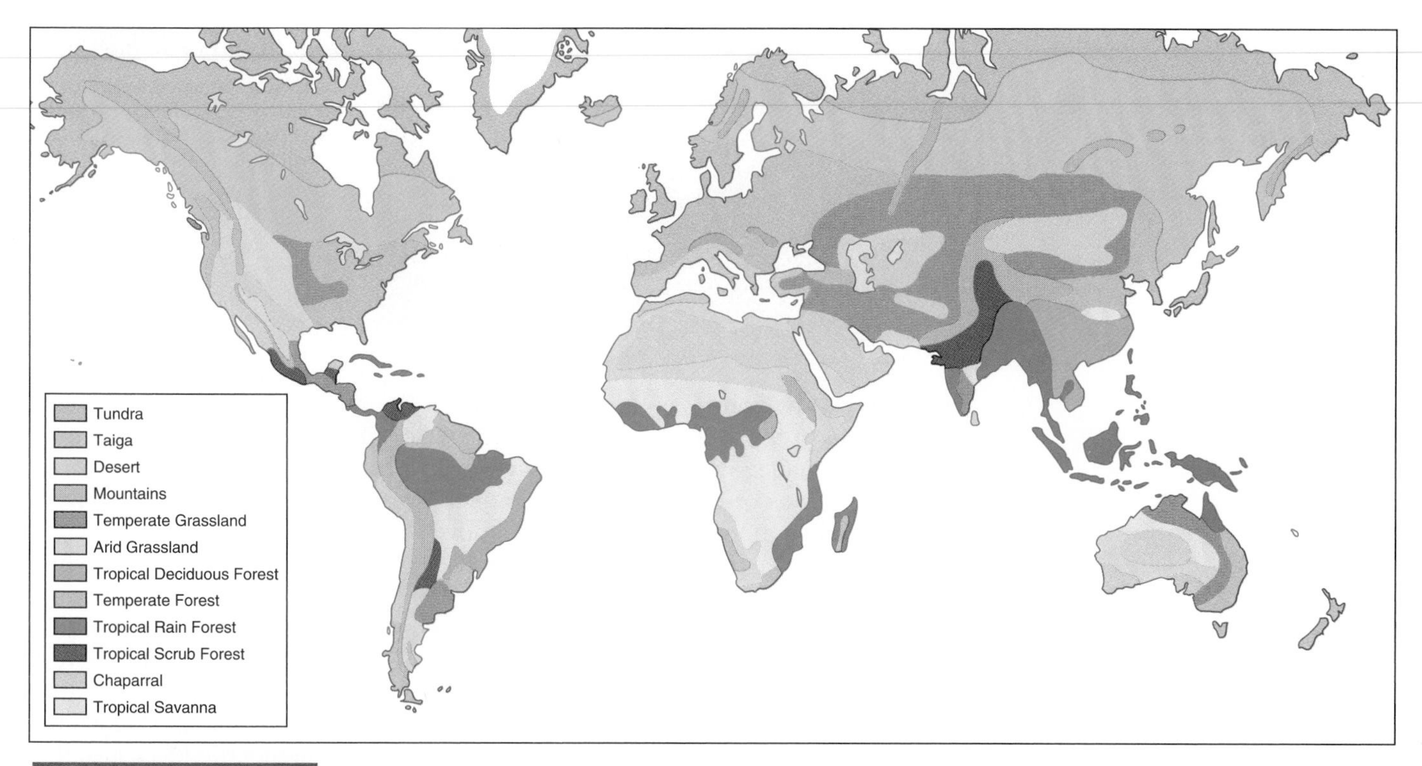

Figure 5.11

World biomes developed in response to latitude, proximity to oceans, topography, and other factors that generate world climates. Greater expanses of grasslands, boreal forests, and tundra occur in the Northern Hemisphere than in the Southern Hemisphere because there is insufficient landmass in southern latitudes for those biomes to be created.

of sufficient size in the Southern Hemisphere where these types of biomes could develop.

It becomes evident on inspection of a world biome map that the general pattern of latitudinal biome distribution is modified by topography, proximity to oceans, and continental size. These modifying factors change the distribution of some major biomes and lead to the development of other minor biomes. Minor biomes include *scrub forests* and *savannas* in the tropics and *chaparral* in the Mediterranean climates. We will consider the characteristics of the major and minor biomes in greater detail in Chapter 6.

Different biomes result from different climates, which are in turn caused by variations in the amount of available solar energy and moisture, which vary with latitude. However, topographic changes associated with increased elevation can produce similar differences in the general distribution of life forms. **Life zones** (the nature of a habitat and the plants and animals that live there) found in mountain regions are the direct result of *altitudinal zonation,* which can, for example, result in the development of **alpine zones** (similar to tundra biomes) on equatorial mountains.

Altitudinal Zonation

The Personal Narrative of the Travels to the Equinoctial Regions of America (1799–1804), by Alexander von Humboldt and Aimé Bonpland, includes numerous descriptions of their ascents of a number of high mountains, most in the Andes of Venezuela. From his journals, it is evident that Humboldt began to develop an understanding of how the pattern of life was related to altitude. Early in his journey with Bonpland, while still in the Canary Islands off the west coast of Africa, Humboldt described the ascent of the volcano Tenerife: "We beheld the

Figure 5.12

Early life zone studies were conducted by C. H. Merriam in the San Francisco Peaks of Arizona, shown here.

plants divided by zones, as the temperature of the atmosphere diminished with the elevation of the site." Humboldt's concept of **altitudinal zonation** can be defined as a change in community structure as a function of changing elevation. That is, the types of organisms found at sea level will be much different from those found on a mountain 2,000 meters above sea level.

This idea was developed further by C. Hart Merriam in the 1890s while studying the distribution of vegetation on San Francisco Peaks in Arizona (Figure 5.12). Merriam developed a *model* in which organisms present at specific latitudes and elevations (life zones) could be predicted by changes in temperature alone. A **model** is often used in attempts to describe or predict a cause-and-effect relationship in nature. In Merriam's model, a single *variable* (a characteristic that changes), temperature—the cause—was used to predict the life zones that would be found at any particular latitude or elevation—the effect. Although Merriam had an impact on ecological thought in the early 1900s, other models have appeared since that are more strongly predictive because they include more variables.

Distribution of Biome Vegetation: A Predictive Model

At present, a new model with high predictive value is emerging. Examination of this model offers insight into the type of research now being conducted on biomes. The first part of this model states that vegetation mass, structure, and *leaf area index* can be predicted from the *hydrological balance* in different biomes. **Leaf area index (LAI)** is the ratio of total leaf surface area to the total ground surface area covered by the leaves (that is, LAI = total leaf surface area ÷ total ground surface area covered). For example, if 1 square meter of ground is covered with 1 square meter of leaf surface area, LAI = 1, but if the same area of ground is covered with 3 square meters of leaf surface area, LAI = 3. Thus the greater the amount of foliage covering the ground, the larger the LAI.

The **hydrological balance (HB)** is computed using a variety of data, including the amount of precipitation that penetrates the canopy and reaches the ground surface and the total loss of water from evaporation and by the foliage itself. In general, there is a direct correlation between HB and LAI. A high HB is predictive of a high LAI, and a low HB is predictive of a low LAI.

Another component of the model uses minimum yearly temperatures to predict the *physiognomy* in each biome as described in Figure 5.13A. **Physiognomy** refers to the types of vegetation present in an area as described in Table 5.1. Physiognomy is affected by low temperatures associated with both freezing and **supercooling**, which occurs when water cools rapidly and fails to freeze at 0°C. When water cools slowly, ice crystals form, and freezing occurs at 0°C. These ice crystals may damage the cells of plants. However, when supercooling occurs, crystals do not form, and plants are not harmed at temperatures below freezing.

When both physiognomy and the LAI were computed for the biomes of Earth, the map shown in Figure 5.13B resulted. Compare the maps in Figure 5.13 with Figure 5.11. How accurate are the predictions made by this model?

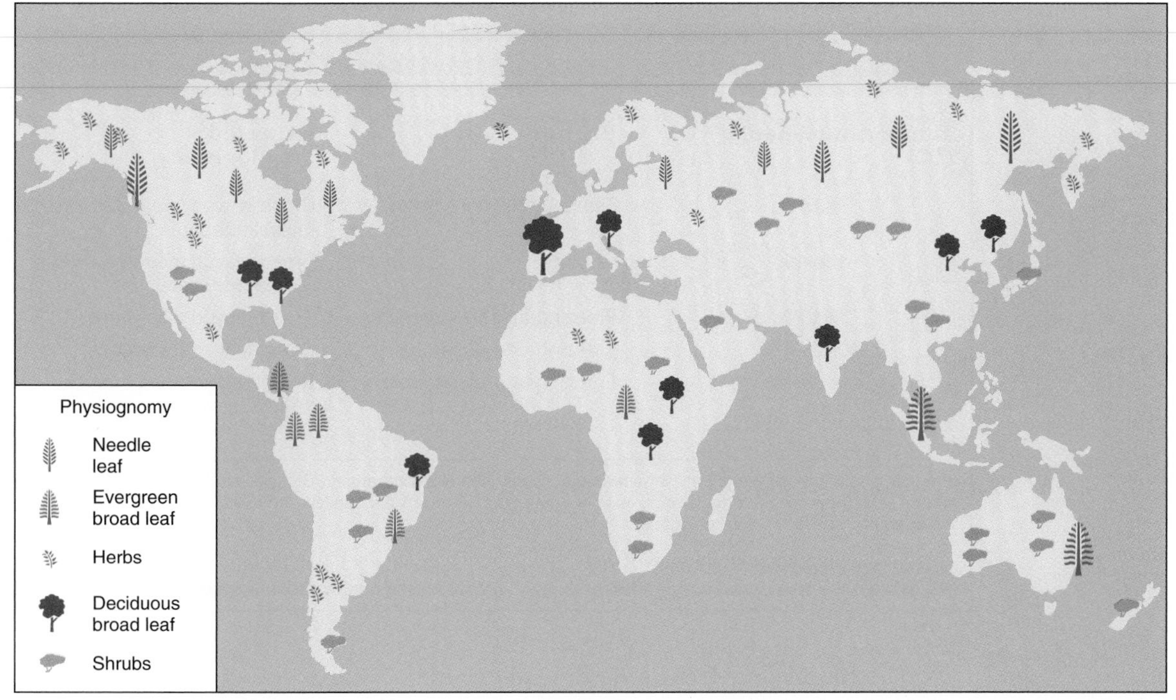

A

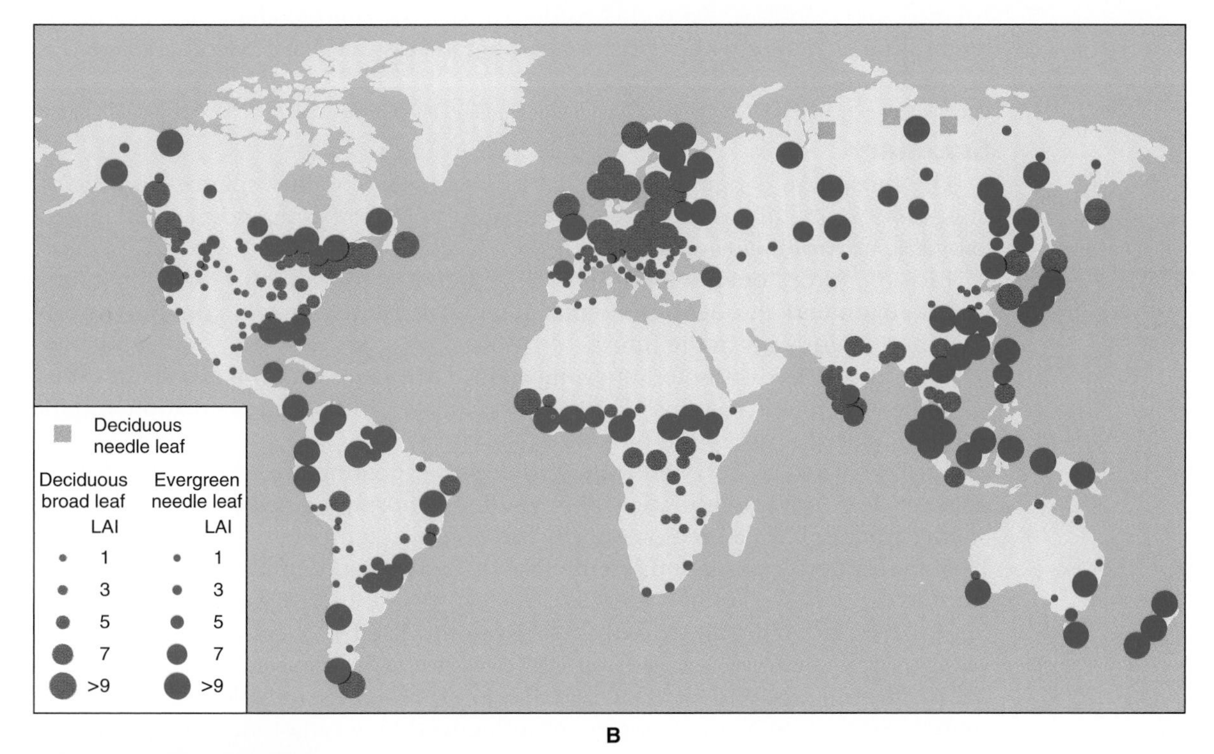

B

Figure 5.13

(A) Minimum yearly temperatures were used in creating this map of vegetation physiognomy for the world. (B) Hydrological balances were used to predict the leaf area index within world vegetation physiognomy regions. The LAI varies from 1 to more than 9. How does the distribution of these values seem to relate to latitude?

Variations in sunlight, temperature, and moisture produce latitudinal patterns of biome distribution. An increase in altitude results in changes in life zones. A highly predictive model that explains changes in vegetation between the biomes uses minimum temperatures and hydrological balance to predict leaf area index and physiognomy.

Table 5.1 Minimum Temperatures and Expected Physiognomy

Temperature Range (°C)	Phenomenon	Expected Physiognomy
> 15	Temperature not limiting	Broad-leaved evergreen when rainfall adequate
–1 to 15	Chilling	Broad-leaved evergreen when rainfall adequate
–15 to 0	Freezing and supercooling	Broad-leaved evergreen
–40 to –15	Freezing and supercooling	Broad-leaved deciduous
< –40	Freezing and supercooling	Evergreen and deciduous needle-leaved (coniferous)

Predicted dominant physiognomy is based on minimum yearly temperatures. Broad-leaved trees can be evergreen, as in the tropics, or deciduous, as in the temperate zones. Most conifers are evergreen, but some species, such as larch, are deciduous.

Source: After Woodward (1987), pp. 97–107.

The biomes present on Earth result from the interaction of altitude, latitude, and proximity to large oceans because they all affect both temperature and water balance. In Chapter 6, we will consider further the unique characteristics of different biomes, along with representative organisms that have become adapted to live within them.

Summary

1. The biosphere is composed of the hydrosphere, the lithosphere, and the atmosphere. Living organisms have contributed to the creation and maintenance of habitable environments on Earth.
2. Tectonic forces deep within Earth have been responsible for changing continental alignments throughout the planet's history, and they will continue to produce new alignments in the future.
3. Solar energy is distributed unevenly over Earth's surface. This results in differential heating that is responsible for world climates and ocean circulation and the development of major world biomes.
4. Earth's tilt on its axis and differential global heating are responsible for the seasons. The movement of air within Hadley atmospheric cells transports excess energy from equatorial regions north and south to temperate and polar regions. This atmospheric circulation, coupled with Earth's rotation, causes major ocean currents.
5. Moisture-laden air flowing onshore from the Pacific Ocean loses much of its water to precipitation as it rises up and over the major mountain ranges of western North America, and rainshadows develop on their eastern slopes. Climates in central and eastern North America are influenced by weather systems spawned in the Gulf of Mexico and the Arctic and by the Gulf Stream.
6. Major world biomes include tropical and temperate forests, deserts, grasslands, taiga, and tundra. Latitude has a major effect on the location of world biomes. Life zones similar to those of the major biomes develop in areas with increasing topographic elevation.
7. Older life zone models are not as strongly predictive as the current model using leaf area index and vegetation physiognomy.

Review Questions

1. What are the three major components of the biosphere?
2. How are tectonic forces related to the present continental alignments on Earth?
3. How is the uneven distribution of

solar energy over Earth's surface related to its climate and ocean circulation?
4. Which orbital properties of Earth are related to the "march of the seasons," and how does each affect it?
5. How do Hadley cells explain the existence of the horse latitudes, trade winds, and the doldrums?
6. Which topographic features are related to rainshadows?
7. Which factors are responsible for the development of the major biomes of Earth?
8. What is altitudinal zonation, and what are its causes?
9. How is a leaf area index computed?

Essay and Discussion Questions

1. How have energy sources, sculpting forces, and the march of the seasons caused the development of each major biome present on Earth?
2. How do oceans affect maritime climates in North America?
3. What factors combine to produce altitudinal life zones on mountains similar to the different biomes associated with increasing latitude?
4. What is the value of predictive models, such as the LAI physiognomy model? How might they be used by scientists?

References and Recommended Reading

Allen, J. E., and B. Burns. 1986. *Cataclysms on the Columbia*. Portland, Oreg.: Timber Press.

Balandin, R. K. 1982. *Vladimir Vernadsky, Outstanding Soviet Scientist*. Moscow: MIR Publishers.

COHMAP (Cooperative Holocene Mapping Project) Members. 1988. Climatic changes of the last 18,000 years: Observations and model simulations. *Science* 241:1043–1052.

Foukal, P. 1990. The variable sun. *Scientific American* 262:34–41.

Merriam, C. H. 1898. Life zones and crop zones of the United States. *Bureau Biological Survey* 10:1–79.

Overpeck, J., et al. 1989. Climate change in the circum-North Atlantic region during the last deglaciation. *Nature* 338:553–557.

Rowell, G. 1989. The John Muir Trail: Along the high, wild Sierra. *National Geographic* 175:467–493.

Smith, L. S. 1986. *Elements of Ecology*. 2d ed. New York: HarperCollins.

Woodward, F. I. 1987. *Climate and Plant Distribution*. Cambridge: Cambridge University Press.

CHAPTER 6

World Biomes

Chapter Outline

Forests of the Torrid Zone
Temperate Forests
Grasslands of the World
The Taiga
The Tundra
The Deserts
The Minor Biomes

Reading Questions

1. What environmental factors give rise to Torrid Zone forests?
2. Why do western coniferous forests exist at the same latitude as eastern deciduous forests ?
3. Which climatic conditions are responsible for growth of grasslands?
4. Why do most deserts occur at 30° north or 30° south latitude?
5. In what ways do savannas differ from temperate grasslands?

The journals written by intrepid naturalists of past centuries provide insightful descriptions of the environments encountered during their expeditions to regions of the world unknown to Europeans. These expeditions penetrated all major biomes of Earth, from the torrid jungles of the tropical forests to the frozen expanses of the arctic tundra. Excerpts from a few of these journals will be used in this chapter as introductions to the different biomes, for they provide a sense of the impressions that these naturalists had when first exploring these unique regions of Earth.

FORESTS OF THE TORRID ZONE

> We descended slightly from an elevated, dry, and sandy area to a low and swampy one; a cool air breathed on our faces, and a mouldy smell of rotting vegetation greeted us. The trees were now taller, the underwood less dense, and we could obtain glimpses in to the wilderness on all sides. The leafy crowns of the trees, scarcely two of which could be seen together of the same kind, were now far away above us, in another world as it were.

This description of a tropical rain forest at the mouth of the Para River is from *The Naturalist on the River Amazons* by Henry Walter Bates, which describes his explorations of Amazonia with Alfred Russel Wallace in the late 1840s.

The **Torrid Zone** is Earth's surface between the Tropic of Cancer and the Tropic of Capricorn, divided by the equator. It receives a greater amount of solar radiation than any other part of the planet. Tropical and subtropical forests are found in this zone.

Tropical Forests

Tropical forests (see Figure 6.1) occur in the equatorial portion of the Torrid Zone, where more than 240 centimeters of annual rainfall combines with an average annual temperature greater than 17°C to create the most productive forests on Earth. Many lowland areas, below 1,000 meters in elevation and within 10° to 15° of the equator, receive up to 450 centimeters of rain a year, creating environmental conditions that support a dense, stratified rain forest characterized by high species diversity among both plants and animals. It is thought that over 50 percent of all species on Earth are native to tropical forests. An example of this great diversity can be found in the book *Forest Environments in Tropical Life Zones: A Pilot Study*. It lists 860 tree species for the small Central American country of Costa Rica alone!

The distribution of animal life mirrors the stratification in these rain forests. There are an aerial community above the forest, including birds and bats; a canopy community with birds, fruit bats, squirrels, and monkeys; a subcanopy community of climbing animals; and a community on the ground that includes large and small mammals, reptiles, and birds. Insects are present in all strata throughout the forest.

Subtropical Forests

Subtropical forests tend to develop in northern and southern portions of the Torrid Zone and at higher elevations on mountains within rain forests. They result from a decrease in total available moisture or a seasonal distribution of rain. Tropical seasonal forests include semievergreen rain forests and deciduous seasonal forests.

Figure 6.1

Tropical forests include the greatest number of different species per unit area of any terrestrial biome, and they probably contain more than half the species living on Earth today.

Torrid Zone Forest Nutrients

Forests of the Torrid Zone have soils with unusual properties. The interactions of excessive rain, warm humid conditions, and the biota (all organisms present in a given area) combine either to recycle soluble nutrients quickly or to remove them from the system. In these forests, most of the nutrients reside within living organisms, not in the soil as they do in other biomes. Nutrients in the soil are the "life blood" of temperate biomes, but in tropical forest biomes, high rates of decomposition and nutrient cycling are the driving forces that contribute to the forest stability through time.

Torrid Zone forests occur between the Tropics of Cancer and Capricorn. They include tropical rain forests, semievergreen rain forests, and seasonal deciduous forests. Most nutrients are found within living organisms present in Torrid Zone forests rather than in the soils.

TEMPERATE FORESTS

> Virginia doth afford many excellent vegitables and living Creatures, yet grasse there is little or none, but that groweth in lowe Marishes: for all the Countrey is overgrowne with trees, whose dropings continually turneth their grasses to weedes, by reason of the ranckness of the ground which would soone be amended by good husbandry. The wood that is most common is Oke and Walnut, many of their Okes are so tall and straight, that they will beare two foote and a halfe square of good timber for 20 yards long. . . . There is some Elme, some black walnut tree, some Ash: of Ash and Elme they make sope Ashes.

This description of the forests that greeted the first European settlers in Virginia is from *A Map of Virginia*, written by Captain John Smith in 1612.

The **temperate forest biome** occurs in Earth's Temperate Zones, where increasing latitude results in greater seasonal extremes, with lower average temperatures and less precipitation than occur in biomes of the Torrid Zone. In North America, these environmental conditions have produced three major forest types: *deciduous forests* of the eastern region, *coniferous forests* of the Pacific Northwest, and *western montane forests* in mountainous areas of the West.

Deciduous Forests

Extensive **deciduous forests** have developed in only three temperate regions of Earth, all in the Northern Hemisphere. They are found in central Europe, eastern Asia, and eastern North America, where variations in temperature and precipitation correlate directly with increasing latitude and altitude.

In deciduous forests of North America, precipitation ranges from 75 to 250 centimeters per year, while temperatures vary from summer highs of 38°C to winter lows of –30°C. These extremes create a temperate forest biome characterized by four distinctive seasons, as shown in Figure 6.2. This seasonality results in large variations in the amount of available moisture. There is a high probability that water will be unavailable to the plant community for more than half the year, during the cold months, in northern regions of deciduous forests. Thus growth is limited to late spring, summer, and early fall, and the total amount of forest growth is far less than that of tropical forests.

When the lengthening days of early spring release the frozen grip of winter, the floor of a deciduous forest blooms with a profusion of flowers. To survive, these flowering plants must complete most of their annual growth and reproduce before oak, hickory, maple, beech, and other tree species grow leaves, closing the forest canopy and reducing light intensities on the forest floor by 95 to 99 percent. The growth of these tree species is supported to a large extent by rains spawned in the Gulf of Mexico and carried northeast by the many storms and hurricanes of summer.

As the days begin to shorten in late summer, growth slows in deciduous forests, and the deep green summer foliage is replaced by the brilliant colors of

Figure 6.2

Leaves of deciduous trees emerge in spring (A), and growth continues through summer (B) and into early fall, when the leaves become ablaze with color, die, and fall to the forest floor (C). Fallen leaves enrich the soil and prepare trees for a dormant period in winter (D).

fall. As fall gives way to winter, the trees lose their leaves, and the northern forests become snow-covered and dormant through the cold months of winter.

Leaves and other litter accumulate and decompose on the forest floor, producing rich soils. These soils contain far greater concentrations of nutrients than soils in the tropics. However, the rates of decomposition and nutrient cycling are much lower because of dormant periods during winter when they lie frozen.

The extensive deciduous forests that greeted colonists in the early settlement of eastern North America have been reduced in size by more than 80 percent. During the westward migration, an increasing human population cleared the land of trees and converted most of this magnificent forest biome into land for agricultural production, leaving only a few ancient stands undisturbed.

Coniferous Forests

Evergreen rain forests, composed of a variety of conifer (cone-bearing) species of spruce, fir, pine, and hemlock, blanket the coastal plains and mountains of the Pacific Northwest. They range from sea level to an elevation of about 1,800 meters in the Coast Range of Oregon and extend north along the Pacific coast to southern Alaska, where their upper elevation limit drops to about 900 meters. The southern portion of this **coniferous forest biome** occurs at the same latitude occupied by deciduous forests in eastern North America. Eastern deciduous forests result from summer rains and cold, frozen winters. In contrast, coniferous forests in the Pacific Northwest are influenced by onshore flows of winter storms that bring up to 400 centimeters of rain per year. As a result, western conifers have extended growth periods into the wet winter season, in which temperatures are often mild.

Redwood and Sitka spruce forests, characteristic of the fog zone along coastal lowlands, are replaced by forests of Douglas fir and hemlock as elevation increases in the Coast and Cascade ranges of Oregon, Washington, and British Columbia (Figure 6.3A). These forests are replaced at even higher elevations by mountain hemlock and the true fir (Figure 6.3B).

The rainshadow on eastern slopes of the Cascade Range is associated with a sharp decline in moisture that results in a rapid change of species. Fir and hemlock on the west side of the mountains are replaced by species of pine and ju-

A

B

Figure 6.3

(A) In coniferous forests of the Pacific Northwest, inland from the coastal plain, as the elevation increases, the fog zone is left behind and the spruce forest gives way to western hemlock. (B) At higher elevations, hemlock is replaced by mountain hemlock and true firs.

niper in the rainshadow on the east side. Open forests of mature ponderosa pine result from interactions of low-moisture, well-drained, volcanic soils and frequent fires, which burn all material on the forest floor and prevent other plants from becoming established.

Western Montane Forests

Montane forests occur in the mountains of western North America. The term *montane* refers to cool, moist, high areas of mountains where evergreen trees grow. These forests occupy a life zone above the coastal coniferous forests and end at the *timberline* in the alpine zone (Figure 6.4). The timberline varies from 4,200 meters in the southwestern forests to about 3,500 meters in Glacier National Park, near the Canadian border.

During the Pleistocene glaciation that ended 11,000 years ago, much of northern North America lay under massive continental glaciers, and in mountains of the West, alpine glaciers developed as far south as California. These alpine glaciers retreated when a general global warming brought the last glacial period to a close. In their wake, the melting glaciers created a diverse landscape that supports a variety of **habitats** (local areas occupied by an organism). These distinctive habitats include mountain lakes, bogs, meadows, and, depending on elevation and degree of exposure to solar energy, steep mountain slopes covered with a variety of tree species.

One feature of montane forest zones is the occurrence of large areas covered by coniferous forests composed of nearly pure stands of a single tree species. For example, western larch occurs in virtually pure stands in northwestern Montana, and in northern Idaho and Washington, western white pine forests cover large areas. However, a variety of tree species occur in much of this mountainous region. Open Douglas fir forests, present on warm, dry, south-facing slopes, contrast sharply with cool, moist, dark forests of pine, grand fir, and spruce present on north slopes. These variations result in distinct habitats, and in their transition zones, complex communities develop that support many different resident and transient populations of insects, birds, and mammals.

Many montane forests include several types of deciduous trees. One species in particular, quaking aspen, is the most widely distributed tree in North America. It became the major species to inhabit many forests that formed in the wake of receding glaciers. In montane forests of Colorado and Utah, it remains a relic of a past age (Figure 6.5).

Figure 6.4

The timberline occurs at about 3,500 meters elevation in Glacier National Park. The alpine zone above the tree line has habitats similar to those found in the arctic tundra biome.

Figure 6.5

The quaking aspen has a broader distribution in North America than any other tree species.

Temperate forest biomes of North America consist of eastern deciduous, western coniferous, and western montane forests. Distinctive climatic and topographic conditions contribute to the development of these different forest types.

GRASSLANDS OF THE WORLD

> Having for many days past confined myself to the boat, I determined to devote this day to amuse myself on shore with my gun and view the interior of the country lying between the river and the Corvus Creek. . . . this plain extends back about a mile to the foot of the hills one mile distant . . . and it is intirely occupyed by the burrows of the *barking squiril* hertefore described; this anamal appears here in infinite numbers and the shortness and virdue of grass gave the plain the appearance throughout it's whole extent of beautiful bowling-green in fine order. It's aspect is S. E. a great number of wolves of the small kind, halks and some pole-cats were to be seen. I presume that those animals feed on this squirril. . . . this senery already rich pleasing and beautiful was still farther hightened by immence herds of Buffaloe, deer Elk and Antelopes which we saw in every direction feeding on the hills and plains. I do not think I exagerate when I estimate the number of Buffaloe which could be compre[hend]ed at one view to amount to 3000.

This is a description of the grassland prairie encountered on Monday, September 17, 1804, by Meriwether Lewis, from the Lewis and Clark *Journals* (see Figure 6.6).

Grassland biomes of the world comprise about one fourth of the total area covered by vegetation. They are found in a number of different latitudinal zones and, depending on elevation, are interspersed within tropical and temperate forest biomes. From the pampas of South America to the outback of Australia or the tallgrass prairie of North America, grasslands vary in structure and species composition.

All plants need moisture to carry out photosynthesis (see Chapter 2). In most biomes, this moisture comes in the form of precipitation, which then leaves the community in four ways: by direct surface runoff, in groundwater, by direct evaporation, or through **transpiration**, the loss of water vapor from plants. **Evapotranspiration** refers to the loss of water by direct evaporation plus transpiration.

Major factors that influence the development of grasslands are the intensity and duration of solar energy and the amount of precipitation. The evapotranspi-

A

B

Figure 6.6

Prairie dogs (A) and bison (B) are two herbivores that were present in great abundance on shortgrass prairies traversed by Lewis and Clark on their epic journey west early in the nineteenth century.

ration rate in grasslands is a significant measure of the growth potential of the grasses. Surface runoff and loss to groundwater are difficult to measure, for they vary with degree of slope and soil type. However, all grasslands develop in areas where the amount of precipitation occasionally decreases sharply, causing major changes in the dominant vegetation.

Grasslands exist where droughts occur in 30- to 50-year cycles. The degree of moisture decline varies in different areas of the globe, leading to the development of various types of grasslands. Average annual rainfall can vary from 100 centimeters in the tallgrass prairies of Illinois to 25 centimeters in shortgrass prairies in the rainshadows of the Rocky Mountains. Some trees may live and grow for a time, but forests do not develop in these grasslands because in the long run a drought will occur and trees will die.

Trees do occur in grasslands, but only in drainage basins at the bottoms of hills and along streams where there is sufficient moisture to support them through periods of drought. Some drought-resistant tree species will occur singly or in scattered clumps on the tropical savannas of eastern Africa and northern Australia and in the velds of southern Africa and northern South America. In these areas, secondary disturbances—immense herds of grazing animals, wildfires, and activities of humans—also affect tree density and distribution (Figure 6.7).

North American Grassland Development

The major grasslands of North America occur in areas without large mountain ranges. On these rolling plains, persistent winds develop during hot summer months and cause an increase in the evapotranspiration rate far above that found in deciduous forests. This high rate of water loss reduces the amount of precipitation that percolates through the soil and thus reduces the amount lost to groundwater. As a result, the quantity of soil nutrients removed is small, and grassland soils develop a layer of rich organic matter.

North American Prairies

Tallgrass prairies, including some areas east of the Mississippi River, are among the most productive lands in the world. Historically, they were covered with grasses, but today they constitute the major portion of the "corn belt," where corn and soybeans form the base of an agricultural economy.

Vast expanses of *mixed* and *shortgrass prairies* originally extended from Texas north to Saskatchewan and west to the eastern slopes of the Rocky Mountains. Over the millennia, these prairies withstood the advance and retreat of at least three major glaciations, extensive seasonal fires, droughts, and grazing by herbivores, including migrating herds of up to 50 million bison (Figure 6.8A).

Figure 6.7

The most extensive savannas on Earth are found in Africa; they support numerous species of herbivores as well as the carnivores that feed on the herbivores.

Figure 6.8

Extensive grasslands of the prairie states associated today with cattle once supported vast numbers of large herbivores including bison, deer, elk, and pronghorn antelope (A). Vast areas of shortgrass prairie biomes have been plowed and modified for producing wheat and other grains harvested with modern combines (B).

However, after the westward migrations, farmers converted these grassland biomes into the breadbasket of the nation and, to some degree, the entire world. Today, combines are used to harvest vast crops of wheat, and cattle graze in areas where herds of deer, elk, pronghorn antelope, and bison once migrated in uncounted numbers (Figure 6.8B).

Grasslands occur in areas where droughts occur in 30- to 50-year cycles, which limits the establishment of forests. Average annual rainfall varies from 100 centimeters in tallgrass prairies to 25 centimeters in shortgrass prairies. Tallgrass, mixed, and shortgrass prairies have rich soils in which much of the world's food supply is now grown.

THE TAIGA

> *August 17th.* Rainy weather, made East coming on the River about 3½ Miles, horrid Roads. . . . The mountains confining this Valley are here on both sides lofty spacious & majestic bellying into promontories alternately advancing into the valley on both sides & their Red sides (for they are here partly composed of Redish stone & earth) appearing through a Thick Coat of wood & when viewed as we now see them through a Summer shower droping from a black pending cloud with all the phenomenon of the beautiful Variations of the Rain Bow, is realy Grand & Beautiful—here we find some sizeable Birch Trees, but not big enough to make a small Canoe—two kinds of the aspin Poplars & Liards but of stinted Growth.

This passage is from *A Journal of a Voyage from Rocky Mountain Portage in Peace River to the Sources of Finlays Branch and North West Ward in Summer 1824* by Samuel Black for the Hudson's Bay Company. It describes a valley in the northern forest of north-central British Columbia, Canada.

Figure 6.9

Boreal forests are the most extensive forest type in North America. They include woodland areas that form the habitat for woodland caribous shown here.

The **taiga** is the most northern coniferous forest biome. It forms a continuous belt across Eurasia and North America, beginning at the southern limit of the winter arctic front, near 50° north latitude, and extending to the southern limit of the summer arctic front, between 60° and 70° north latitude. The taiga is a relatively young biome, formed in the wake of receding glaciers of the last ice age. Species diversity in the taiga is the lowest of any forest biome, with dominant species being four types of coniferous trees: spruce, pine, fir, and larch. Deciduous species of alder, birch, and poplar are interspersed within these coniferous forests (Figure 6.9).

Many lakes and rivers are found in the valleys across this great expanse of taiga. These aquatic environments provide habitats for many migratory and resident species of birds and mammals. The large northern bears, wolves, and moose are mammals adapted to this unique biome (Figure 6.10).

Figure 6.10

Moose, the largest species in the deer family, is also adapted to the taiga.

Cool and cold northern *boreal* forests of the North American taiga biome are composed of two general forest types: *closed-canopy forests* and *open woodlands*. **Closed-canopy forests**, where spruce and larch shade understories of

Figure 6.11

Black spruce covers great expanses of the taiga.

small shrubs and moss-covered ground, extend from Alaska to Newfoundland (Figure 6.11). These forests are seldom limited by moisture, and water shortages are rare, occurring only on some open, south-facing slopes.

Extreme variations in solar radiation over the course of a year create a climate of cool summers and extremely cold winters. At some sites, seasonal average monthly temperatures can vary by 90°C. Within the taiga, the number of days with an average temperature of 10°C or greater ranges from 120 days per year in the south to 30 days in the north and seems to set a limit on the amount of vegetation present.

Permafrost, the subsurface layer of permanently frozen ground, is a major factor defining the distribution of trees in boreal forests. In summer, only a thin surface layer of soil thaws, allowing growth of grasses and shrubs. Permafrost, however, forms a barrier that prevents the growth of tree roots. Permafrost may also be responsible for the development of treeless areas within **boreal woodlands** of the northern taiga. In these woodland areas, white and black spruce grow where permafrost does not occur, and deciduous trees are restricted to margins along watercourses where deeper soils have developed from past sedimentation. The ground cover of these open woodlands includes mosses and lichens, which provide food for woodland caribou, the dominant herbivores inhabiting this biome.

The taiga of North America consists of both closed-canopy forests and open woodlands. The number of days with average temperatures of 10°C or greater sets a limit on the amount of taiga vegetation. Permafrost may be responsible for treeless areas within the boreal woodlands.

THE TUNDRA

> On the 9th. . . a large musk bull was shot, and his flesh was found excellent—the skeleton will be preserved. A short time after midday on the 10th we arrived here having been five days coming from the coast, during some of which we were 14 hours on foot and continually wading through ice cold water or wet snow which was too deep to allow our Exquimaus boots to be of any use.
>
> The latter part of our journey if not the most fatiguing was by far the most disagreeable. . . . Our principal food was geese, partridges, and lemmings. The latter being very fat and large were very fine when roasted before the fire or between two stones. These little animals were migrating northward and were so numerous that our dogs as they trotted on, killed as many as supported them, without any other food.

This passage is from a letter to Sir George Simpson, governor in chief of the Hudson's Bay Company, by John Rae at Kendall River Provision Station, describing part of his exploration of the High Arctic tundra in the northern Northwest Territories, Canada, in June 1851.

The **tundra**, treeless arctic plains, like the taiga, is a biome of the Northern Hemisphere. It does not exist in Antarctica, where terrestrial life is limited to a few meadows that develop during the short southern summer. The Northern Hemisphere has an extensive area of arctic tundra that is commonly divided into the *Low Arctic* and *High Arctic,* as shown in Figure 6.12.

The Low Arctic

The **Low Arctic** begins as a treeless zone at the edge of boreal woodlands and continues north to areas where low temperature and available moisture limit the growth of vegetation. Plants growing on the tundra are well adapted to low temperatures, and during the continuous light of summer, their growth rates approach those of similar temperate species. However, the growing season is short, beginning with the melting of snow in spring and ending with the first frost in late July or early August. With 24 hours of light each day in early summer, the total amount of solar energy approaches that of temperate regions, but the limited length of the arctic summer allows only a few centimeters of upper permafrost to thaw. This limited amount of soil is usually saturated with water in the Low Arctic summer, and tundra species develop shallow root systems.

The High Arctic

The **High Arctic** tundra is found at higher latitudes and in some areas is considered a desert, for the annual precipitation may be as low as 5 centimeters. These far northern latitudes are areas of high atmospheric pressure produced by descending air from the northern air cell. In many areas, if moisture does not limit vegetation, the permafrost does, leading to a scattered distribution of a few hardy species of moss, lichens, herbs, and dwarf willows.

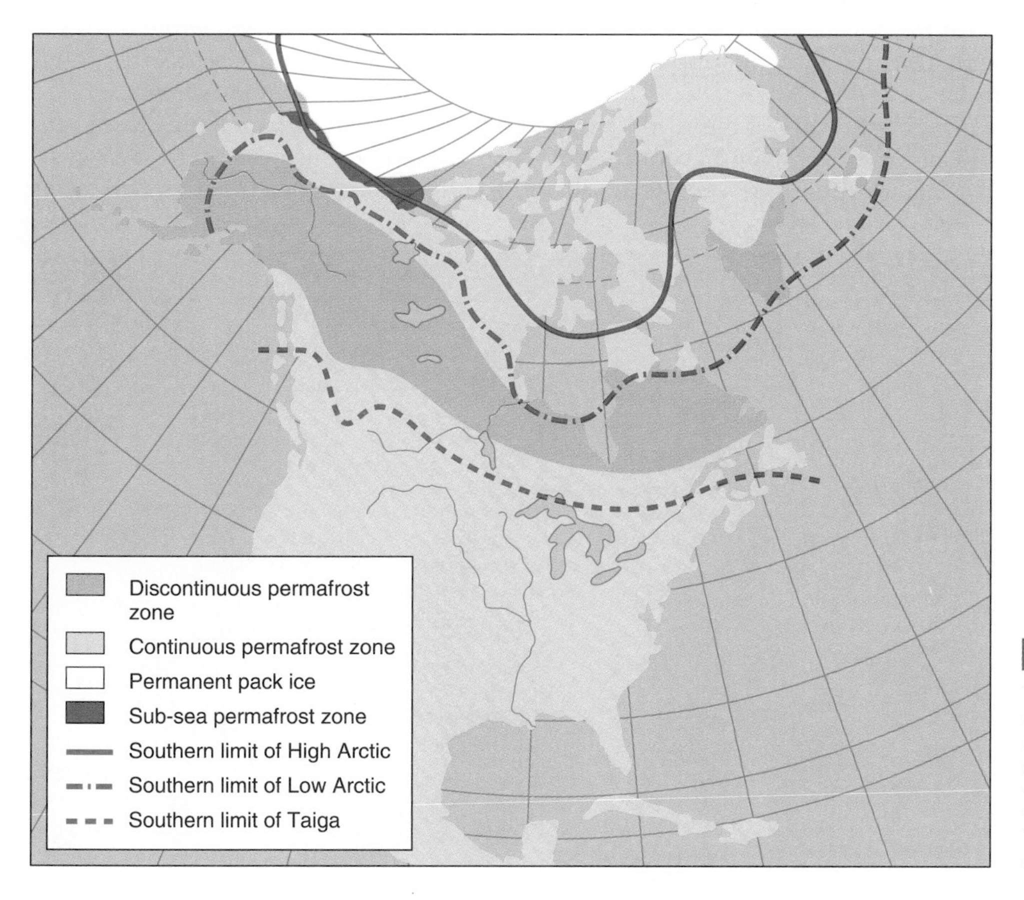

Figure 6.12

The Low Arctic tundra includes the tundra from just north of the boreal woodlands, where the dominant plant species are woody shrubs, to the southern edge of the High Arctic tundra, where nonwoody plants predominate. The High Arctic tundra ends at the edge of the permanent ice sheet.

A

B

Figure 6.13

The tundra supports resident populations of willow ptarmigan (A) and musk-oxen (B).

Permanent High Arctic residents include a few species of birds, including the ptarmigan (Figure 6.13A) and the snowy owl. Mammals include the musk-ox (Figure 6.13B), arctic hare, arctic fox, and lemmings. These resident species are joined by large herds of migrating mammals and immense flocks of migratory birds, including many species of waterfowl that breed and rear their young in the long days of arctic summer. Caribou migrate from their winter ranges in boreal forests and woodlands to graze on spring and summer foilage of the tundra. During these migrations of several hundred kilometers, cows give birth to young that grow and develop rapidly during the High Arctic summer.

Taiga and tundra biomes exist only in the Northern Hemisphere. The tundra consists of Low and High Arctic regions. Animals of the tundra consist of both resident and migratory species.

THE DESERTS

> *January 28.*—We were favored with another charming morning, mild and without a breeze. Following an Indian trail down White Cliff valley, we soon came to a projecting rock, beneath which were walled partitions, with remnants of fires, showing signs of having been recently occupied. . . . The valley was covered with dense groves of cotton-wood, beneath which flowed the prettiest brook we have found since leaving Pueblo creek. . . . The stream, turning southerly, appeared a short distance below to join a wide arroyo from the north, called Big Sandy. . . . The weather is spring-like. Vegetation begins to conform to that of Rio Gila. Canotias are mingled with cedars upon the dry arroyos, and mezquites with cotton-wood upon the flowing streams. Numerous varieties of cacti also abound, and from the huge Echino cactus of Wislizenus, to humbler mammillaria.

This descriptive passage is from *Reports of Explorations and Surveys, to Ascertain the Most Practicable and Economical Route for a Railroad,* written in 1853 and 1854 by A. W. Whipple, of the Corps of Topographical Engineers, on the route near the 35th parallel in the desert of Arizona (see Figure 6.14).

The **deserts** of the world are found in arid regions, where total annual precipitation is less than 25 centimeters. These deserts occur in western regions of the major continents, near 30° north and south latitude, where the descending air

Figure 6.14

The campsite depicted in the lithograph *Bivouac January 28,* from W. W. Whipple's report of his surveys, shows vegetation types he encountered near the 35th parallel in the Arizona desert.

of equatorial and temperate air cells creates high atmospheric pressures and low humidity.

The general circulation patterns of ocean currents contribute to the formation of deserts along western continental margins, where cold water originating in polar seas flows toward the equator, while eastern coasts are influenced by warm water originating in equatorial seas and flowing toward the poles. An example of this can be seen in South America, where on the west coast, the Atacama Desert lies on the Tropic of Capricorn, whereas a tropical rain forest occurs at this latitude in Brazil. The same conditions prevail in Australia, where extensive deserts occur on the western edge of the continent, but at the same latitude on the east coast, subtropical forests predominate (refer to Figure 5.11).

The largest deserts occur in the Northern Hemisphere because greater continental landmasses exist near the 30° latitude. The Sahara in Africa is the largest desert on Earth. It begins on the west coast and traverses the entire continent and the Arabian Peninsula.

Deserts also occur in western North America, where descending air from the equatorial and temperate air cells combines with the climatic effects produced by the cold offshore California Current, resulting in an arid climate. At the same latitude on the east coast, subtropical climates in Georgia and Florida result from the warm waters of the Gulf Stream.

Rainshadows of the Sierra Nevada and Cascade ranges extend western deserts beyond 45° north latitude, creating both cold and hot deserts that are distinguished by average yearly temperatures (see Figure 6.15A). **Cold deserts** are found within the Great Basin of southeastern Oregon, southern Idaho, western Utah, and the northern three fourths of Nevada. The elevation of this desert region is about 1,220 meters, with numerous mountain ranges reaching 2,440 meters (Figure 6.15B).

The Great Basin desert covers an area of about 410,000 square kilometers, making it the largest desert in the United States. Most drainage basins in this desert area have no outlet; all precipitation is stored as surface water, enters the groundwater, is transpired by plants, or evaporates from the surface. This results in a continual increase in salt concentration in these desert soils, which causes many desert lakes to have very high salinities (Figure 6.16).

The Great Basin desert can be divided into two different portions based on the major vegetation present in each. Sagebrush dominates the northern portion, and in the southern part, shadescale predominates. Other plants occur within these two desert areas, but trees are limited to areas of drainage at the base of rimrocks and along intermittent streams.

Hot deserts include the Mojave, Sonoran, and Chihuahuan, each differing in latitude and vegetation type. The Mojave Desert covers the south-central portion of California and the northwestern quarter of Arizona. The Sonoran Desert is

Figure 6.15

(A) Deserts of the western United States include the northernmost cold Great Basin desert and the hot Mojave, Sonoran, and Chihuahuan deserts of the Southwest. (B) The Steens Mountains rise from the Great Basin desert in southeastern Oregon to an elevation of 2,967 meters.

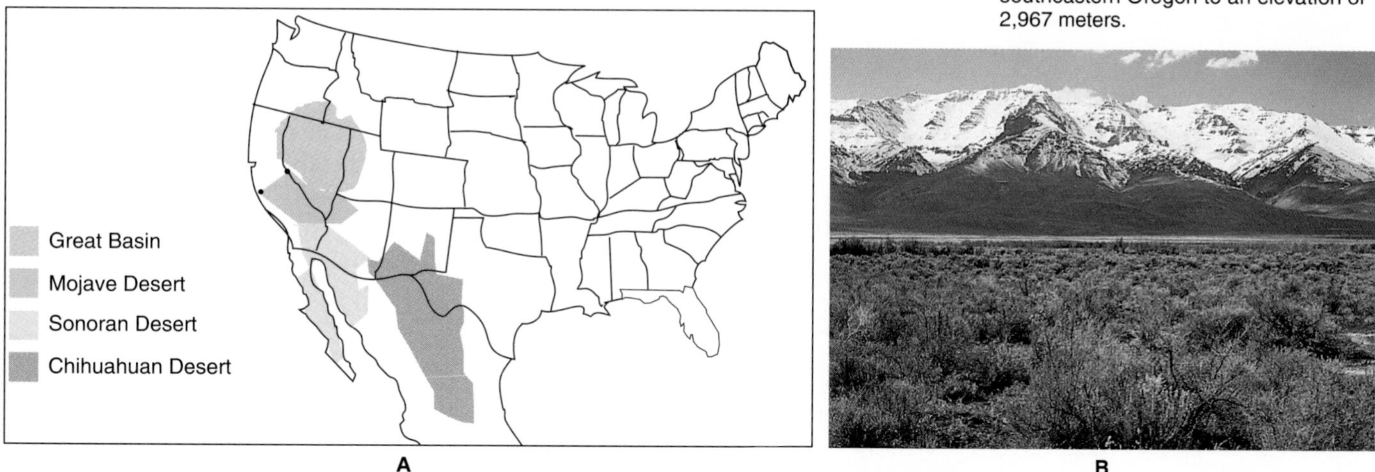

Figure 6.16

Many areas in western deserts occupy drainage basins with no outlets to oceans. As water evaporates from lakes in these areas, salts are left behind, which causes them to increase in salinity.

in the southeastern corner of California and southwestern Arizona and extends into Baja California and western Mexico, enclosing most of the Sea of Cortez. The Chihuahuan Desert occurs in central Mexico, with a northern extension across southern New Mexico and into western Texas.

The moisture regime is quite different among these four deserts. The Great Basin desert receives moisture from winter storms that move in from the Pacific Ocean. Some of this winter moisture also falls in the Mojave and Sonoran deserts but does not reach far enough east to affect the Chihuahuan Desert.

The southern deserts receive summer moisture from thundershowers spawned in storm systems that develop in the Gulf of Mexico. These summer thundershowers reach north into the Chihuahuan and Sonoran deserts but not into the Mojave or the Great Basin. Thus only the Sonoran Desert receives both summer and winter moisture, creating a unique assemblage of desert plants (Figure 6.17).

A

B

Figure 6.17

(A) Agavaceae, Yucca, and Joshua trees grow in the Mojave Desert where most of the annual moisture falls in the winter. The Sonoran Desert (B), characterized by creosote bush, ocotillo, and the giant saguaro cactus, receives moisture in both summer and winter.

Both cold and hot deserts occur in western North America. The Great Basin, Mojave, and Sonoran deserts receive winter moisture, and the Sonoran and Chihuahuan deserts receive summer moisture. Thus only the Sonoran receives both summer and winter moisture, which results in a distinctive plant community.

THE MINOR BIOMES

Tropical Savannas

Savannas of the world occur in a tract across Africa south of the Sahara, in areas in northern Australia, and in regions south of the Amazon basin and east of the Andes in South America. **Savannas** are modified tropical grasslands characterized by alternating wet and dry seasons that produce areas typical of grasslands. During the wet season, grasses develop rapidly, but the dry season reduces them to brown expanses. This type of climate supports variable densities of scattered thorn trees and brush, creating a habitat quite different from that of typical grasslands found in temperate regions.

Seasonal moisture cycles in savannas are reflected in a cyclic appearance of vegetation, which causes herbivores to follow predictable patterns of migration. Wildebeests of the Serengeti Plain are typical of this cycle of life (Figure 6.18). Giraffes, gazelles, zebras, and other herbivores inhabit these grasslands, accompanied by a variety of carnivores, including jackals, hyenas, and lions.

Chaparral

Most *chaparral* biomes are located within arid zones, near 30° north or south latitude. They occur in regions near the Mediterranean Sea, near the tip of South Africa, and on the west coast of North and South America. The proximity of

Figure 6.18

Vast herds of wildebeests migrate across the Serengeti Plain of East Africa in response to wet and dry seasonal cycles that are characteristic of tropical savannas.

Figure 6.19

Chaparrals occur in latitudes where deserts are usually found but where maritime climates moderate the effects of high temperature. Extensive foliage of the chaparral may become tinder dry in the hot summers and can burn readily, destroying homes and other structures.

these areas to major oceans subjects them to maritime climatic effects that modify the desert, producing a typical chaparral biome. **Chaparral** areas receive significant winter rains, but the summers are typical of the desert latitudes in which they occur.

Vegetation in the chaparral typically consists of a number of shrub species, with a few species of grass interspersed among them. The shrubs produce a considerable amount of living mass that is subjected to frequent burning, an environmental characteristic that many homeowners in southern California can verify (Figure 6.19).

Summary

1. Tropical forest biomes may contain more than 50 percent of all species on Earth. Tropical forest soils are usually nutrient-poor, and most nutrients are stored in organisms living in these biomes.
2. Deciduous forests occur in eastern North America, where wet summers are prevalent and the growing season is limited. In the west, coniferous forests predominate because summer precipitation is minimal, but growth occurs all year. Most of the original ancient deciduous forests in eastern North America were cleared and the arable lands converted to agricultural production.
3. Species of plants change with increasing elevation in western coniferous forests. Coastal plains and western slopes are carpeted with redwood, pine, spruce, and Douglas fir. True fir, hemlock, and cedar predominate at higher elevations, and on the eastern slopes, pine and juniper form open woodlands.
4. Quantities of moisture define the transition from forests to grasslands and between short- and tall-grass prairies. Grazing animals, wildfires, and human activities are secondary disturbances that contribute to the continued existence of grasslands.
5. The taiga is the youngest and most northern coniferous forest biome. Its species diversity is the lowest of any forest biome. The presence of permafrost is a major factor in defining tree distribution in the northern forest.
6. The tundra biome occurs only in the Northern Hemisphere and is composed of both Low Arctic and High Arctic areas. During the continuous light of summer, growth rates in some Low Arctic species approach those of similar temper-

ate species. Much of the High Arctic area often receives so little precipitation that many experts consider it a desert. The tundra provides many migratory species with habitats conducive to their reproduction and resources necessary for the early growth of their young.

7. Major deserts occur in areas near 30° north and south latitude, where descending air of the equatorial air cells creates high pressure and low humidity. The four major North America deserts are the Great Basin, Mojave, Sonoran, and Chihuahuan. Variations in total sunlight and the frequency of precipitation are major factors contributing to the development of deserts.

8. Minor world biomes include tropical savannas and chaparrals. Savannas are modified grasslands that have cyclic wet and dry seasons. They are characterized by the presence of widely dispersed trees, and in Africa they support vast numbers of animals. Chaparrals occur in arid zones, but they are influenced by their proximity to oceans and are characterized by shrub species.

Review Questions

1. How do tropical rain forests differ from temperate deciduous forests in terms of important nutrients?
2. Why are coniferous forests found in the Pacific Northwest at the same latitude where deciduous forests occur in the northeastern states?
3. What factors are responsible for the rich soils formed in grasslands?
4. What factors are related to the formation of grassland biomes?
5. Why are taiga and tundra biomes not found in the Southern Hemisphere?
6. How does permafrost affect the development of Low and High Arctic areas in the tundra biome?
7. How do growth rates of plants in arctic biomes compare with rates for similar species in the temperate biomes? What accounts for the differences?
8. Why do major deserts occur near 30° north and south latitude?
9. What are some minor biomes? Where do they occur? What factors influence their development?

Essay and Discussion Questions

1. What factors may have been responsible for the great species diversity that characterizes tropical rain forest biomes?
2. As you travel north from the Torrid Zone to tundra biomes, what changes would you expect in species diversity? Why?
3. Why are most grassland biomes located in interior areas of continents?
4. If weather patterns are disturbed or change naturally, what will happen to the current distribution of biomes? Construct some possibilities.

References and Recommended Reading

Barbour, P. L., ed. 1986. *The Complete Works of Captain John Smith* (1580–1631). Vol. 1. Chapel Hill: The University of North Carolina Press.

Brown, J., and B. Maurer. 1989. Macroecology: The division of food and space among species on continents. *Science* 243:1145–1150.

Brown, L. 1985. *Grasslands*. New York: Knopf (and other volumes in the Audubon Society Nature Guide series, including *Atlantic and Gulf Coasts, Deserts, Eastern Forests, Pacific Coast,* and *Western Forests*).

Emmons, L., and F. Feer. 1990. *Neotropical Rainforest Mammals: A Field Guide*. Chicago: University of Chicago Press.

Holdridge, L. R., W. C. Grenke, W. H. Hatheway, T. Liang, and J. A. Tosi. 1971. *Forest Environments in Tropical Life Zones: A Pilot Study*. Elmsford, N.Y.: Pergamon Press.

Kelley, D. 1972. *Edge of a Continent*. New York: Crown.

McNaughton, S., R. Ruess, and S. Seagle. 1988. Large mammals and process dynamics in African ecosystems. *BioScience* 38:794–800.

National Geographic Society. 1976. *Our Continent: A Natural History of North America*. Washington, D.C.

Nichol, J., and C. Newton. 1990. *The Mighty Rain Forest—in Association with Worldforest 1990*. London: Newton Abbot.

Norse, E. 1990. *Ancient Forests of the Pacific Northwest*. Washington, D.C.: Wilderness Society/Island Press.

Sinclair, A., and M. Norton-Griffiths. 1979. *Serengeti: Dynamics of an Ecosystem*. Chicago: University of Chicago Press.

Trimble, S. 1989. *The Sagebrush Ocean: A Natural History of the Great Basin*. Reno: University of Nevada Press.

CHAPTER 7

Ecology

Chapter Outline

Origins of Ecology

Early Concepts in Plant Ecology

Early Concepts in Animal Ecology

Focus on Scientific Inquiry: Succession on the Sand Dunes of Lake Michigan

Ecosystems

The Ecological Niche

Modern Ecology

Reading Questions

1. What discoveries led to the development of early ecological concepts?
2. What causes ecological succession?
3. In what ways are trophic levels related to communities?
4. What are ecological pyramids, and how are they used?
5. Which environmental factors move through biogeochemical cycles?
6. What is an ecological niche?

Ecology is defined as the scientific study of interrelationships that exist between organisms and their environments. Although the term *ecology* did not appear in the scientific literature until the late nineteenth century, for convenience we use it throughout this chapter.

Within an environment, organisms are greatly affected by both **biotic factors**, such as other living plants, animals, and microbes, and **abiotic factors**, various physical, chemical, and temporal (time) components (see Figure 7.1). Major *physical factors* are the force of gravity and energy in the form of light and heat. Important *chemical factors* are water and all the nutritional elements required by living organisms. *Temporal factors* consist of normal changes that occur throughout the life of an organism and gradual environmental changes over long periods of time.

ORIGINS OF ECOLOGY

The roots of modern ecology follow many paths back through the history of science, and ecological concepts can be found in the work of many early biologists. For over a century before the work of Charles Darwin, naturalists believed not only that God had created each species but also that he had created a "balance of nature." In other words, animal and plant numbers remained roughly constant due to the forethought of the creator, who planned just the right balance of animals to plants, predator to prey, and so on. Linnaeus, in his overview of classification, popularized this view of a natural balance that used the forces of reproduction and destruction to maintain an equilibrium of animal and plant numbers. Darwin proposed mechanisms to explain his ideas of the balance of nature. He believed that different kinds of organisms that lived in an environment remained in balance due to the "forces of nature." His balance was a dynamic

Figure 7.1

(A) All organisms are affected by factors present in their environment. Physical factors include energy and gravity; chemical factors include water and all essential nutrients; biotic factors include all other living organisms; and temporal factors consist of changes in environments that occur during the life of the organism and those that have occurred over very long time periods. (B) The organisms living in this lake are affected by energy (sunlight), gravity (which will pull them to the bottom if they are not mobile), chemicals in the water, and changes that occur as the lake slowly fills with sediments.

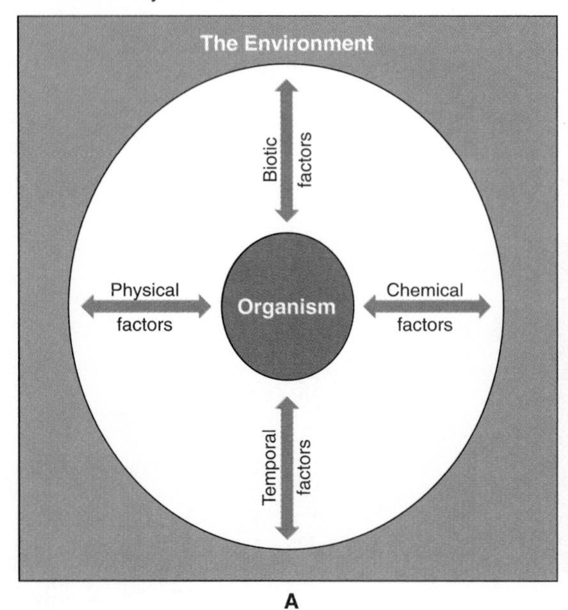

A

B

one and could change due to these same forces of nature. But what underlying factors created this balance? What processes led to environmental stability or change?

Many important discoveries during the nineteenth century shed light on such questions and led to the formulation of significant new ecological concepts. Ultimately, these new principles, derived from careful scientific investigations, led to a deeper understanding than the simple balance-of-natural-forces view of ecology. Most of the early supporting studies for the new ideas involved plants. For example, chemical factors in the environment were found to be related to plant growth rates. It was also determined that not all chemical elements were required by plants in equal amounts, and further, different elements could limit plant growth at different concentrations.

These and other findings led to the development of numerous central principles in ecology. Examination of the development of these concepts offers insight on how the discipline of ecology matured and reached its present status.

Different schools of ecology have attracted varying numbers of ecologists since Darwin. In one school, adherents consider the biota to be an ecosystem and climate, soil, sediments, and other abiotic factors to be external influences. They regard *populations* as the fundamental ecological level, because **populations** can be defined explicitly as a group of individuals in the same species occupying a specific area. In addition, the factors that cause populations to change, including competition and predation, can be isolated and studied. This **population-community school** garnered support from Darwin's theory of natural selection (discussed in Chapter 23) because the population is the ecological level at which natural selection operates. However, population changes induced by natural selection originate with individual organisms. Individuals live and die, and they may or may not leave offspring, depending on a multitude of environmental factors.

Another school of ecology considers the ecosystem to consist of both the biotic and the abiotic components. Through time, this **ecosystem-centered school** developed a functional view of the interactions among the biota and between it and abiotic environmental components. The ecosystem-centered school became partitioned into a number of different "subschools," each headed by different prominent ecologists during the late nineteenth and early twentieth centuries. The views of many of these ecologists will be developed further in this chapter and in Chapter 8.

Darwin thought that the dynamic "balance of nature" resulted from changes in the "forces of nature." Supporters of the population-community school held that populations were the fundamental hierarchy of ecology; proponents of the ecosystem-centered school considered interactions between abiotic factors and the biota within an ecosystem to be equally important.

EARLY CONCEPTS IN PLANT ECOLOGY

Early concepts in terrestrial ecology emerged from studies of the relationships among plants and between plants and their environments. These studies set the stage for later studies in terrestrial animal ecology.

Limiting Factors

In 1840, an agricultural chemist, Justus von Liebig, conducted studies designed to evaluate successful plant growth. He first established that certain inorganic salts of magnesium and other elements were essential for adequate growth. Then Liebig discovered that if there were insufficient quantities of a salt in the environment, growth of the plant would be limited regardless of the concentrations of other important nutrients. These conclusions led to the formulation of Liebig's "law of the minimum." It was later amplified to accommodate the idea that there were optimal amounts of each essential nutrient, and if those amounts were not present, growth would be limited. This new, revised concept became known as the **law of limiting factors**.

Finally, it was postulated that nutrient availability changed as environments became modified through the passage of time. Such fluctuations would be expected to influence plant growth and subsequently to lead to the establishment of new plant populations with different nutritional requirements.

Succession

Near the end of the nineteenth century, two American plant ecologists, F. E. Clements and H. C. Cowles, developed the major ecological principle of *succession,* which explained more clearly why communities change over time. They postulated that natural systems resulted from a gradual **succession** of different plant-animal associations through time. If an area of "new" ground came into existence (for example, through volcanic outflow or sand dune formation), **primary succession** would occur. Clements believed that a progression of different species would come to inhabit the new site. The sequence would begin with early *pioneer species* that would slowly change the original microenvironmental (small local habitat) conditions. Changes in concentrations of specific elements or the amount of shade available are examples of microenvironmental alterations. These first modifications would then enable other species with different environmental requirements to invade the original site. Sequential succession would continue until a permanent community of organisms called the **climax community** existed. This stable climax community would continue to perpetuate itself unless some major disturbance occurred and the environment-organism equilibrium was disrupted. If that happened, the site would revert back to an earlier stage, and **secondary succession** would begin. The communities present on the site would progress again through a number of predictable types and in time would culminate in a new climax community.

In Clements's opinion, the climax community developed much like an organism, resulting in an "organic entity" within each community. Cowles shared Clements's idea of dynamic community succession, but he disagreed with the concept of a stable climax community as indicated by his statement that "a condition of equilibrium is never reached." He felt that succession was variable and could retrogress as well as advance to a specific climax community (see the Focus on Scientific Inquiry, "Succession on the Sand Dunes of Lake Michigan").

The Clementsian "school of organismic ecology" prevailed in many early American ecological writings, but a number of ecologists did not agree with its principal concepts. In 1913, W. S. Cooper, a student of Cowles's, described results from a study he conducted on Isle Royale in Lake Superior. He observed that succession did not occur uniformly throughout a community. Rather, it resembled a series of "braided streams" that, when woven together, would appear as a "mosaic" of various aged stands of trees, each representing a different stage of continuous change. He felt that such mosaics existed because of small, localized disturbances within the forest that were caused by diseases, insects, or windfalls. Much of Cowles's work and that of his students led to establishment of the discipline of *plant physiological ecology*. This field continues to be important today in determining how plants relate among themselves and to their environments.

Two early plant physiological ecologists who studied individual plant responses to environmental changes were L. G. Ramensky and H. A. Gleason. They were among the first to apply mathematical methods to studies of ecology.

Ramensky was a Soviet scientist who in 1924 developed a method for measuring plant abundance that could be used to quantify the percentage of an area that each species covered. To these data he applied statistical analyses, including tests that could be used to describe variations in the abundance of different plant species. Gleason was an American plant ecologist who, like Ramensky, emphasized variation in the response of single species populations. He thought that such variations resulted from changing *gradients,* or differences in concentrations, of important limiting factors as temporal changes occurred within the environment.

Gleason and Ramensky both felt that changes in the *spatial patterns,* or distributions of specific plant populations, actually determined the distribution of other species within the community. This contrasted with Clements's view, which was slanted toward the idea that temporal changes occurring during succession affected all organisms in the community. Current papers in modern ecological literature include elements of both points of view; however, the importance of spatial distribution of species within communities is at present receiving the greater amount of study.

Clements thought that a climax community developed as an organic entity and had characteristics of a superorganism. In contrast, Cowles thought that true equilibrium was never reached and hence that a stable climax community never developed. Later studies concentrated more on spatial patterns than on temporal factors when considering community succession.

EARLY CONCEPTS IN ANIMAL ECOLOGY

In Europe, ideas about plant ecology progressed well ahead of ecological concepts that incorporated communities of animals. However, in North America, two zoologists, S. A. Forbes and V. E. Shelford, began to formulate ecological principles that could be applied to animal populations.

Succession

In a famous paper, *The Lake as a Microcosm,* published in 1887, Forbes extended the equilibrium concepts inherent in the "balance of nature" hypothesis to include animals along with abiotic factors. In his paper, he suggested that a lake was "a little world within itself—a microcosm within which all of the elemental forces are at work."

In 1913, Shelford attempted formally to integrate animal communities into the grand design of succession that had previously been developed for plants. He concentrated on interactions within and among different plant and animal species and also on their relationships with abiotic factors in their environment. From this work, he created some of the first diagrams, such as that shown in Figure 7.2 on page 106, to describe food relations within a natural community.

The Law of Toleration

Shelford also expanded the concept of limiting factors to incorporate his belief that there were maximum, optimum, and minimum concentrations of abiotic elements present in the environment. Consequently, to survive, organisms had to be able to function within a range of concentrations for different elements. This principle became known as Shelford's **law of toleration** (see Figure 7.3, page 106.).

Trophic Levels

All organisms require a constant supply of energy in order to survive. However, laws of physics dictate that energy can be neither created nor destroyed. Given those constraints, a preexisting energy source is required to sustain life on Earth. But what is it, and where does it come from? The answer, of course, is that the sun is the ultimate source of energy used by most organisms, and solar energy is

Focus on Scientific Inquiry

SUCCESSION ON THE SAND DUNES OF LAKE MICHIGAN

Early concepts of *ecological succession* emerged through a detailed study of the changing vegetation on sand dunes at the southern end of Lake Michigan (Figure 1). This pioneering work formed the basis of a Ph.D. dissertation, by Henry C. Cowles of the University of Chicago, which was published in the *Botanical Gazette* in 1899. The area where Cowles first observed the sequence of succession events, which occurred when "new" sand was transformed into different communities with the passage of time, has been modified by human activities to such a degree that many of the relationships are no longer apparent. However, on the east shore of Lake Michigan, in areas of Sleeping Bear Dunes National Lakeshore or in areas of Ludington State Park, it is still possible to walk back in time.

If you begin your walk at the lake shore, you will be on the *lower beach,* an area of sand usually devoid of life due to the pounding waves of the storms of winter (see Figure 2). Walking landward, you will next cross the *middle* and *upper beaches,* areas of sparse vegetation with accumulations of driftwood. The upper beach ends with the first sand dunes, the foredunes.

A *pioneer community* exists where beach grasses invade the foredune; there they bind the sand and provide a base for the development of a community that includes insects such as grasshoppers and tiger beetles, spiders, and decomposers. This *beach grass community* continues to dominate the larger dunes that stretch inland. As time passes, organic matter gradually accumulates, soil forms, and *shrub communities* are able to invade and replace the beach grass community. Farther inland, the shrubs have changed the microenvironmental factors, allowing pine trees to survive and develop into a *pine forest community.*

Pines are *transitional species* because, as the forest canopy closes with the growth of the trees, pine seedlings are not able to germinate and grow. However, oak seedlings thrive in the shaded environment, and oak trees replace the pine as time passes. Within these pine and oak forests, the

Figure 1 Henry C. Cowles developed some of the early concepts of ecological succession from studies he conducted on the south shore of Lake Michigan.

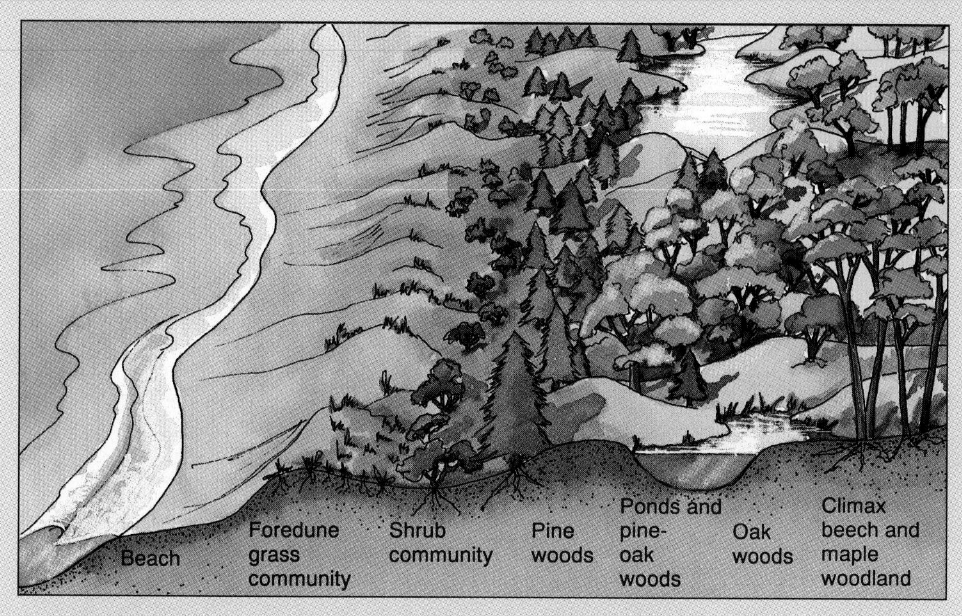

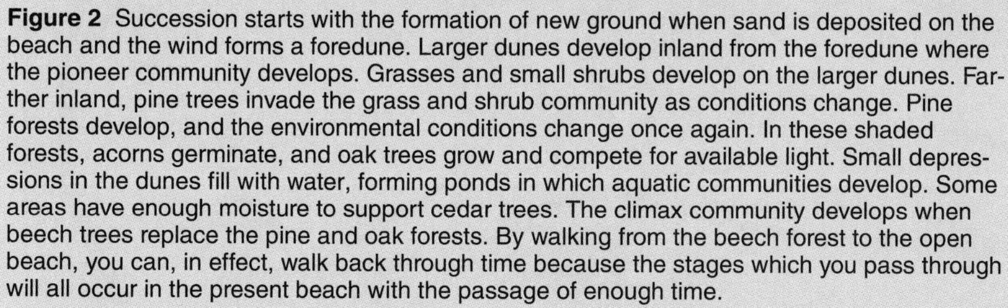

Figure 2 Succession starts with the formation of new ground when sand is deposited on the beach and the wind forms a foredune. Larger dunes develop inland from the foredune where the pioneer community develops. Grasses and small shrubs develop on the larger dunes. Farther inland, pine trees invade the grass and shrub community as conditions change. Pine forests develop, and the environmental conditions change once again. In these shaded forests, acorns germinate, and oak trees grow and compete for available light. Small depressions in the dunes fill with water, forming ponds in which aquatic communities develop. Some areas have enough moisture to support cedar trees. The climax community develops when beech trees replace the pine and oak forests. By walking from the beech forest to the open beach, you can, in effect, walk back through time because the stages which you pass through will all occur in the present beach with the passage of enough time.

dunes often form depressions in which water accumulates, allowing unique pond communities to develop. Some low areas remain wet enough for the pine to be replaced by cedar trees, and these *cedar bogs* can exist for long periods of time.

As the soil accumulates a deep humus layer, the oak forest finally gives way to the final stage of succession, a *climax beech forest community.* Thus by walking from the lake shore landward to the beech forest community, you traverse all successional community types that would develop on the beach. If, like Rip Van Winkle, you were to fall asleep on the beach, how long do you think you would have to slumber before you would awaken in a beech forest?

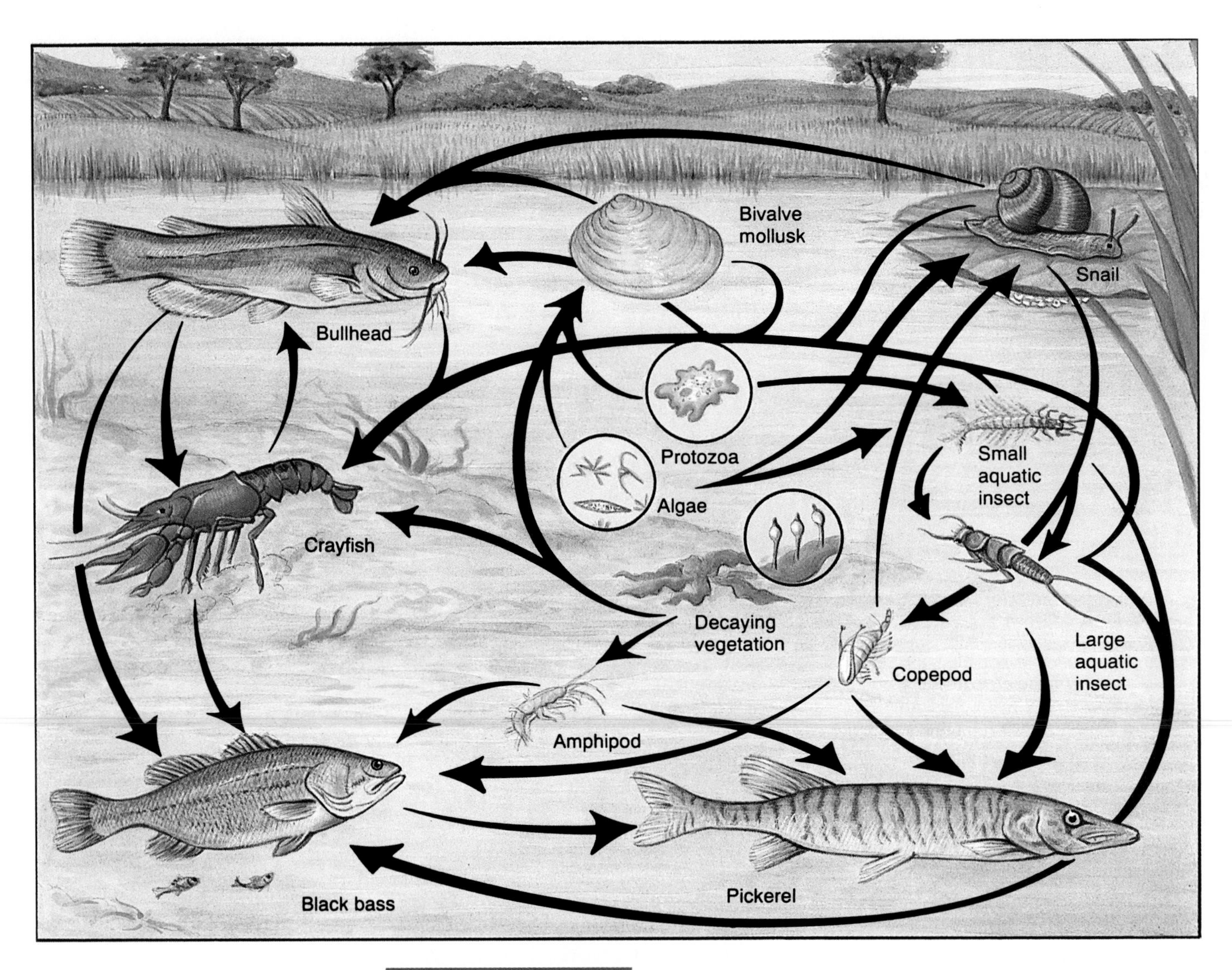

Figure 7.2

Energy enters Shelford's food relations diagram through photosynthesis by algae and flows as food through the system as indicated by the arrows. This diagram combines the flow of both energy and nutrients that are essential to the organisms present. The concepts of trophic levels, energy flow, and material cycling were refined and further developed by other scientists later in the twentieth century.

used by plants to build energy-rich molecules that can in turn be used by other organisms.

Once plants produce biological molecules that can be used as an energy source, how is that energy made available to other organisms in a community? The concept that food energy flows from one feeding level, or group of organisms, to another was formalized during the 1920s. Organisms that feed on others in different levels constitute specific **trophic levels**. Thus plants become the first trophic level and are fed on by animals in the second (herbivore) trophic

Figure 7.3

In Shelford's law of toleration, organisms were thought to function most effectively and be most abundant when abiotic factors were present in their optimum range. Decreases in function and abundance would occur if higher or lower concentrations of the abiotic elements were present in the organisms' environment.

Shelford's Law of Toleration

Minimum limit of toleration		Range of optimum	Maximum limit of toleration	
Absent	Decreasing ←	Habitat or center of distribution Great abundance	Decreasing →	Absent

level, and animals that consume herbivores (carnivores) represent the third trophic level. At each trophic level, the original quantity of food energy made available by plants becomes reduced, and by the time the highest trophic level is reached, very little of the original energy captured by the plants remains. This process is described as a *one-way flow of energy,* and it explains why life on Earth is dependent on a never-ending supply of energy. Like gasoline, once food energy is burned, it is converted to unusable forms, usually heat that ultimately radiates into space.

In contemporary ecology, a **food chain** describes the sequence of transfers of energy in the form of food from one trophic level to another. For example, a simple food chain would be grass-deer-cougar. **Food webs** depict complex interlocking series of food chains that exist in a community.

It also became recognized that *biogeochemicals*—naturally occurring chemicals—essential for life, such as phosphorus and carbon, were also transferred between trophic levels. However, unlike energy, there are finite amounts of available biogeochemicals in nature, and for life to endure, they must be continuously recycled through communities.

In 1935, Charles Elton developed the concept of *food cycles* in his book *Animal Ecology*. His **food cycles** attempted to incorporate both energy flow and biogeochemical cycles, as shown in Figure 7.4. Later the concept of food cy-

Figure 7.4

Energy entered Elton's food cycle through photosynthesis by aquatic algae and mosses (11, 14, and 15) and terrestrial plants (3). Algae become food for zooplankton (12, 13, and 16), which feed aquatic invertebrates (17); these are eaten by fishes (9), which in turn are eaten by birds and bears (7, 8, and 10). Terrestrial plants are eaten by insects (6) and birds such as the ptarmigan (4), which are then eaten by arctic foxes (5). Dung (18) and other wastes feed bacteria (1), which release mineral salts and fix nitrogen (2) used by the plants and algae. As the last few patches of snow melt, both the arctic fox and the ptarmigan replace their white winter color with brown fur or feathers that will be more adaptive during the coming summer.

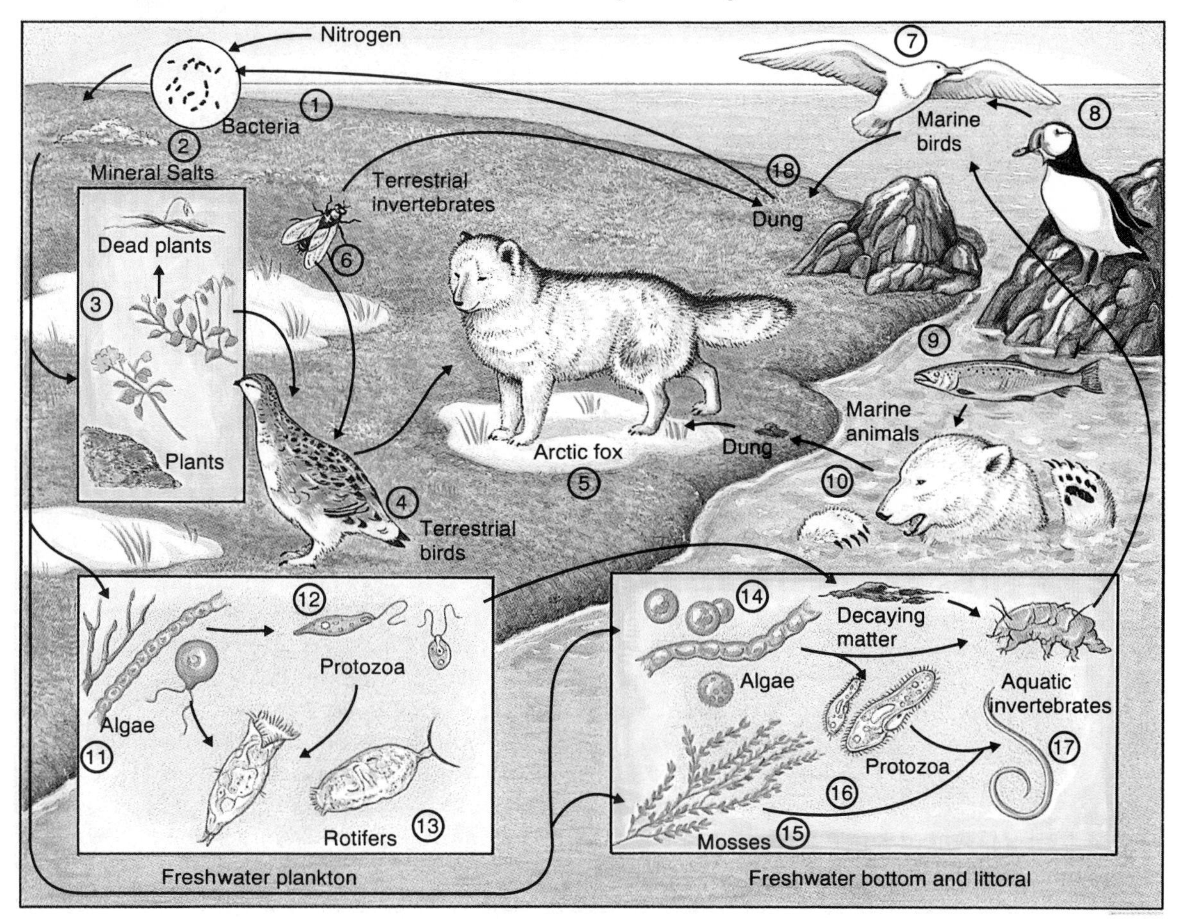

cles evolved further, and Elton's diagrammatic representations of functional communities were renamed *ecosystems*. The development of the concept of ecosystems has occupied many ecologists from the mid-1930s to the present, while other ecologists concentrated on individual organisms and how they function in their ecological role, or *niche,* in a community.

Early animal ecologists, including Shelford and Elton, developed the conceptual framework that led to concepts of energy flow and material flow between trophic levels in ecosystems.

ECOSYSTEMS

In 1935, A. G. Tansley defined an **ecosystem** as "not only the organisms-complex, but also the whole complex of physical factors forming what we call the environment." Today, ecosystems continue to be recognized as fundamental ecological units with both biotic and abiotic components.

Raymond Lindeman, a faculty member at Yale University, wrote a paper titled "The Trophic-Dynamic Aspects of Ecology" that was published in the journal *Ecology* in 1942. In it, he redefined an ecosystem as "the system composed of physical-biological processes active within a space-time unit of any magnitude, i.e., the biotic community *plus* its abiotic environment." Lindeman introduced many new concepts to the field of ecology, and the logic of his ideas suggested that he would have made further conceptual advances during his career. Tragically, however, Lindeman died at the age of 27.

Although the impact of Lindeman's short career is difficult to assess, he did make several important contributions to the emerging field of ecology. He described *trophic dynamics* as the transfer of energy within ecosystems. He also developed a process for measuring "biological efficiencies" by comparing amounts of energy used with amounts of growth occurring within each trophic level. This approach led to the creation of new methods that could be used to determine the relative productivity, as indicated by growth, between adjacent trophic levels. In one major section of his 1942 paper, he related trophic dynamics to changes that occurred during succession: "From the trophic-dynamic viewpoint, succession is the process of development in an ecosystem, brought about primarily by the effects of the organisms on the environment and upon each other, towards a relatively stable condition of equilibrium."

Finally, Lindeman related a concept of lake productivity, which states that there is a rapid increase in the numbers of organisms during early phases of aquatic succession, to a similar concept of succession in terrestrial ecosystems. He wrote, "It is equally apparent that the colonization of a bare terrestrial area represents a similar acceleration in productivity. In the later phases of succession, productivity increases much more slowly."

Ecosystem ecology flourished in the United States during the 1940s and 1950s. Howard T. Odum developed more advanced concepts about energy flow through ecosystems and illustrated relationships in ecosystem energetics through the creation of complex energy flow diagrams such as the one shown in Figure 7.5. Odum's system refined the traditional way of describing energy flow within ecosystems because it incorporated quantitative estimates of the fate of energy as it flowed between different trophic levels.

Odum's aquatic ecosystem in Silver Springs, Florida, received abundant amounts of energy from the sun. A fraction of this radiant energy was captured by the producers—plants and algae—through photosynthesis and stored as sugars or other plant materials in the form of **photosynthetic biomass**, the amount of living organisms. However, not all of the photosynthetic biomass was available to the *herbivores*. The producers' **gross production**, the total amount of new plant

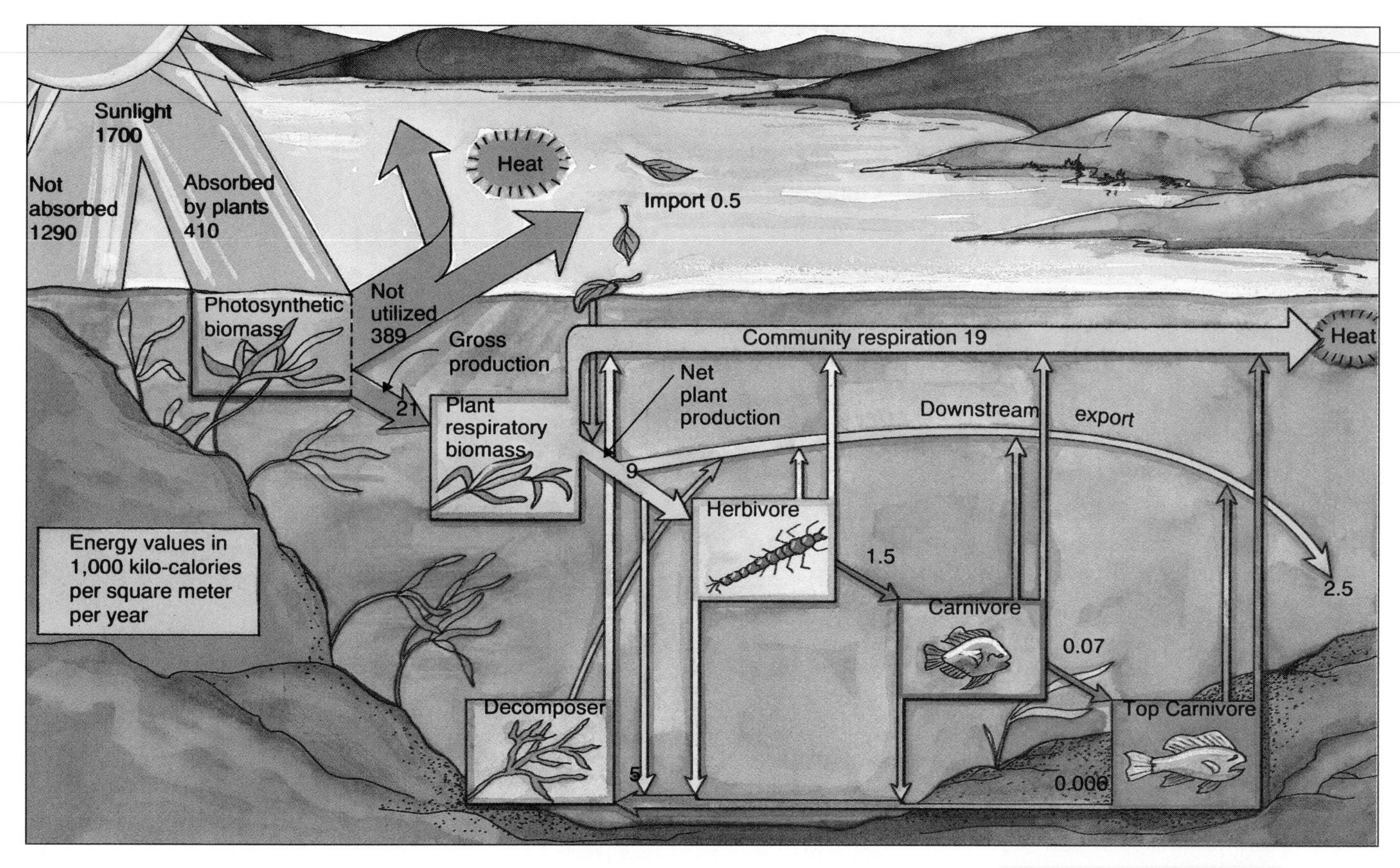

Figure 7.5

Howard T. Odum attempted to show both the direction and the amount of energy moving among trophic levels in Silver Springs, Florida. Of the 410,000 kilocalories per square meter per year absorbed by plants, only 6 kilocalories moved through the trophic levels and into the top carnivores. Ultimately, most of the energy left the ecosystem in the form of heat.

substances generated, was reduced by a quantity called the **plant respiratory biomass**, which represented the amount of energy required by the plants to conduct their normal activities. Consequently, only the **net plant production** remained available to the herbivores. In a similar fashion, energy received by the herbivores was reduced by their respiration. As a result, only a small fraction of the original energy that entered the system was available to the **carnivores** and even less to the **top carnivores**. Ultimately, **decomposers** broke down the biomass that was not passed up the food chain. All of the remaining energy finally departed the system in the form of low-grade heat created during metabolic activities of the decomposers or was exported as organisms that drifted downstream and entered other ecosystems.

Ecological Pyramids

In his 1935 book, Elton devised "pyramids of numbers" that were based on his observations of an oak forest and a small pond. He wrote:

> The animals at the base of a food-chain are relatively abundant, while those at the end are relatively few in numbers, and there is a progressive decrease in between the two extremes. . . . This arrangement of numbers in a community, the relative decrease in numbers at each stage in a food-chain, is characteristically found in animal communities all over the world, and to it we have applied the term "pyramid of numbers."

Howard Odum used Elton's pyramid concept in structuring his energy flow diagrams for Silver Springs, and Eugene Odum, his brother, applied it to pyramids of numbers, biomass, and energy flow, as shown in Figure 7.6. **Pyramids of energy** are used to describe rates of energy flow and can be graphically presented as kilocalories per area (square meters) per year. When energy flows from one trophic level to another, there is always less available to the next highest level because of the inescapable loss of some energy in the form of heat when organisms

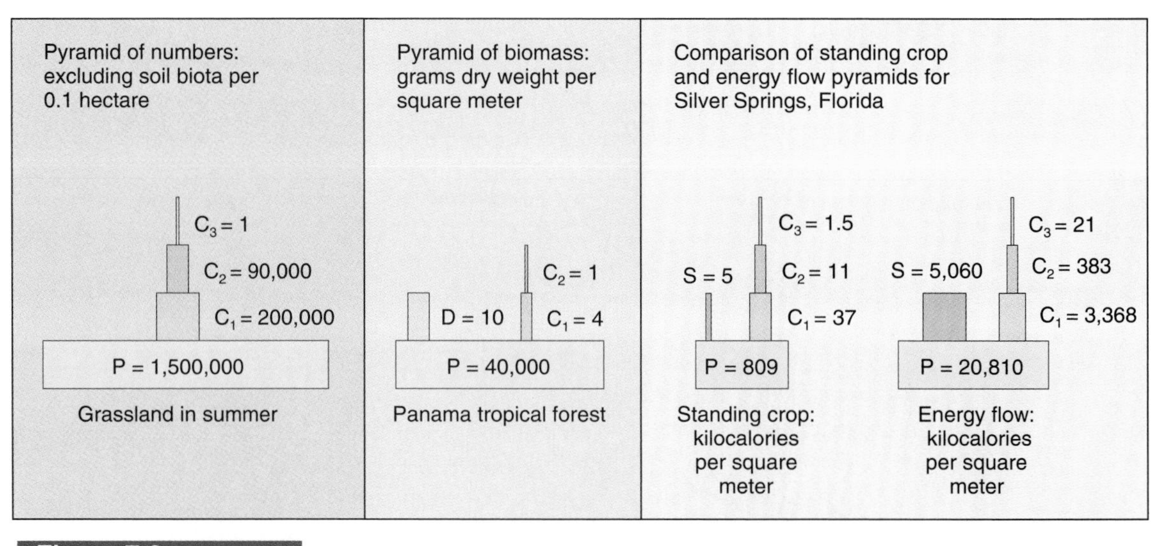

Figure 7.6

Eltonian pyramids of numbers, biomass, and energy were created by Eugene Odum and show the relative amounts of each present in different trophic levels. P = producers, C_1 = primary consumers (herbivores), C_2 = secondary consumers (carnivores), C_3 = tertiary consumers (carnivores), D = decomposers, and S = saprotrophs (decomposers). The energy flow data from Silver Springs show that only 16 percent of the energy available in producers was transferred to primary consumers and that less than 12 percent of that energy flowed to secondary consumers. Tertiary consumers represented only 5 percent of the energy present in the secondary consumers. Similar decreases occur in pyramids of numbers for grasslands in summer, biomass in tropical forests, and the standing crop of Silver Springs.

transform energy from one form to another (discussed in Chapter 29). As a result, energy pyramids are never inverted.

Pyramids of numbers are graphic representations of the density of organisms in each trophic level, expressed as numbers per area or volume. **Pyramids of biomass** depict measurements of the amount of living mass in each trophic level and are usually expressed as grams of dry weight per area or volume. Pyramids of numbers and biomass represent the **standing crop** (the number or amount of organisms present at any time) in each trophic level of an ecosystem. The standing crop is usually expressed in grams of dry weight per area or volume, or it can be converted into energy equivalents and expressed as kilocalories per area or volume. Because there is no time interval involved, pyramids of numbers and biomass can be inverted. For example, in a pyramid of numbers representing a population of bark beetles infesting a Douglas fir, the tree represents a single producer (P), but the herbivorous beetles (C_1) may number in the thousands (see Figure 7.7).

Biogeochemical Cycles

The early food cycle of Elton (Figure 7.4) actually illustrates several food webs and traces the flow of energy (food) through an arctic ecosystem. It also shows,

Figure 7.7

Douglas firs can be attacked by Douglas fir bark beetles when growing under stressful conditions, and thousands of beetles can infest a single tree and lay eggs under the bark. Hatched larva feed on nutrient-rich tree tissues and leave a network of tracks shown here. A pyramid of numbers representing these two trophic levels would be inverted. Would the pyramid of energy constructed from these two trophic levels be inverted?

in an uncomplicated way, the cycling of important "mineral salts," nitrogen, and "dung" by bacteria. Within Elton's Bear Island ecosystem, energy could flow through food webs, but only in one direction and in ever-decreasing amounts. In contrast, bacteria could release the "mineral salts" in decaying matter, making them available for the future growth of new plants and algae. Thus Elton described a recycling of minerals among producers, consumers, and decomposers.

It is now known that many biogeochemicals are cycled through ecosystems—carbon, calcium, potassium, phosphorus, oxygen, and other elements required by the biota. In general, cycling of these nutrients occurs through one of three types of *biogeochemical cycles.* Oxygen and nitrogen are cycled rather quickly through **gaseous cycles** in which the atmosphere and hydrosphere serve as the primary *reservoirs,* places where large amounts of an element are found. In contrast, cycling of common elements such as phosphorus, calcium, sulfur, magnesium, and potassium occurs slowly through complicated **sedimentary cycles** in which rocks, soil, or sediments act as reservoirs. Movement through sedimentary cycles may involve hundreds, thousands, or millions of years. The **hydrologic cycle** (water cycle) connects and drives the various sedimentary biogeochemical cycles that are critical to functioning ecosystems. That is, crucial elements are transported by water as it moves through its global cycle as shown in Figure 7.8.

Phosphorus Cycle

Most sedimentary nutrients move through similar cycles, so the movements of phosphorus can serve as a model to illustrate a sedimentary cycle. Phosphorus is particularly important because insufficient quantities of this element often limit

Figure 7.8

The hydrologic cycle describes the movement of water between oceans and landmasses. Solar energy causes evaporation from the water surface. Onshore flows of moist air result in a net moisture transfer landward, where the increased elevation of mountains causes air to cool and lose its moisture as precipitation. Water deposited over land evaporates; is stored in snowfields, glaciers, lakes, and the biota; or is transported back to the ocean in groundwater or surface streams and rivers.

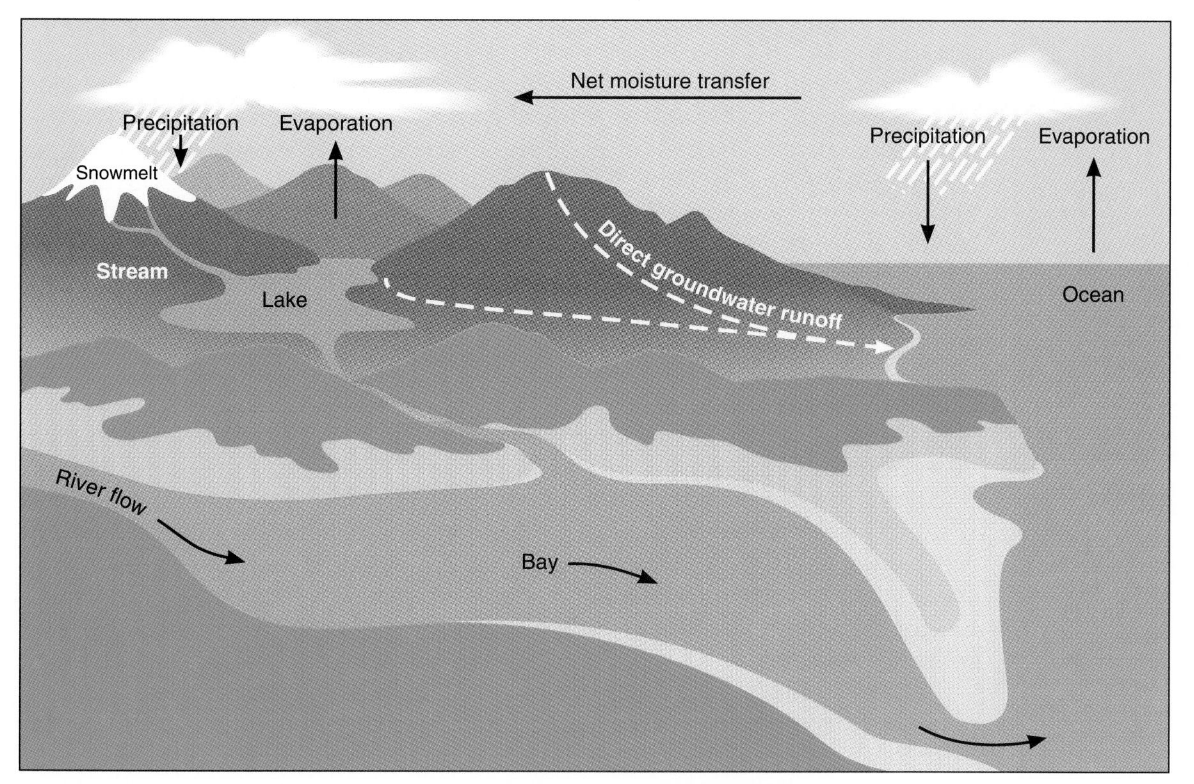

Figure 7.9

Inorganic phosphate (IP) enters terrestrial ecosystems when parent (sedimentary) rock breaks down during weathering. Once in soil, plants convert IP into organic phosphate (OP) in their growing tissues. Plants are either eaten by herbivores or are decomposed by fungi and bacteria when they die. OP from animals is also degraded by decomposers. Once returned to the soil, OP can be taken up once again by plants or enter aquatic systems, where it cycles through both freshwater and marine communities. Some OP can move from oceans landward in migrating fish, like salmon, or in the droppings (guano) of birds that feed in marine ecosystems. However, most OP in the ocean is lost to deep ocean sediments when marine organisms die and sink to the bottom. There the OP is destined to be compressed and uplifted during some future eon to become "new" parent rock.

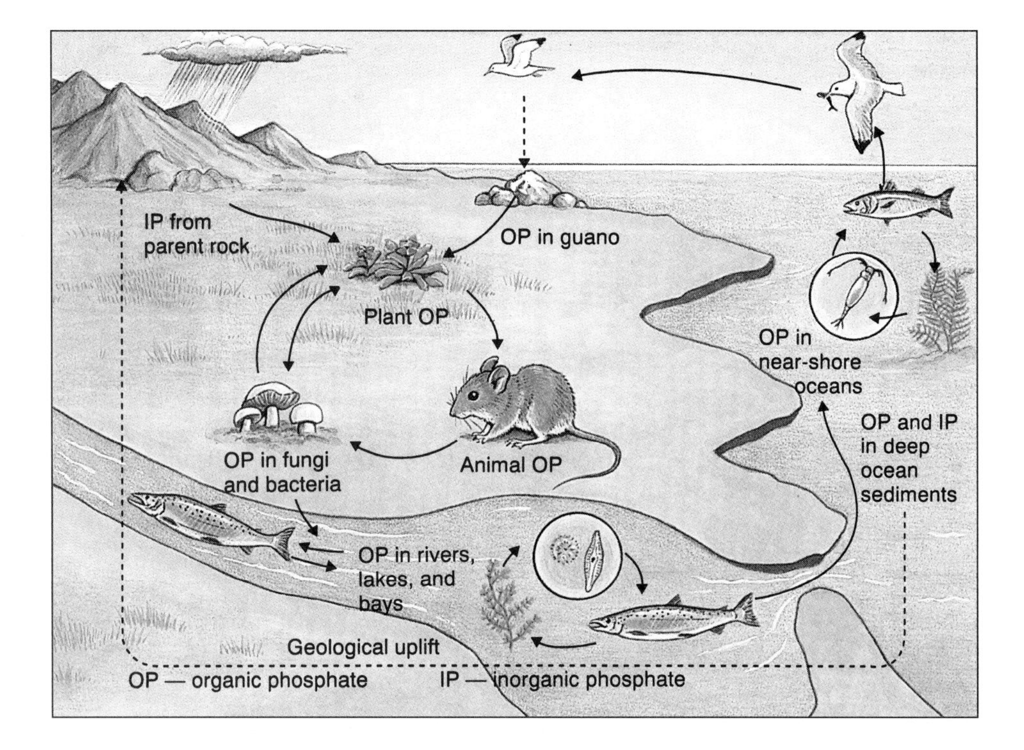

the growth of producers in many aquatic and some terrestrial ecosystems (see Figure 7.9). Phosphorus contained in rocks is liberated by erosion as phosphate (PO_4^{-3}). This inorganic phosphate (IP) becomes available to plants and algae in ecosystems. These producers incorporate PO_4^{-3} into an organic molecule to create organic phosphate (OP), the form present in organisms. Phosphate can cycle through numerous food webs, moving from plants to animals, ultimately to decomposers, and then back to other plants.

Phosphate and other biogeochemicals can be routed through many different cycles. As ecological succession occurs in terrestrial environments, varying amounts of phosphates (important chemical forms of phosphorus) are removed by surface runoff and transported to rivers and lakes, where they cycle through aquatic food webs. Eventually, those phosphates are transported downstream and enter oceans. Once there, phosphates may be cycled through marine food webs, returned to land in the form of bird guano, or even conveyed back to headwater streams in the bodies of salmon returning from the ocean to spawn (Figure 7.10). Once spawning is completed, the salmon die, and through decomposition they replenish the stream or river system with phosphates and other nutrients. Phosphates may also accumulate in the bodies of marine organisms, which sink to the ocean floor when they die. If this path is followed, deep ocean sediments serve as a reservoir from which the phosphates may not escape for millions of years.

Figure 7.10

Phosphates present in ocean food chains are concentrated in the bodies of adult salmon as they feed in offshore marine communities. When they return to their home rivers and streams, they mate, spawn, and die. Bacterial and fungal decomposition of these large fish release phosphates that can once again cycle in the food webs of these freshwater ecosystems.

Carbon-Oxygen Cycle

All biological systems, cells, and tissues that have ever existed on Earth have been primarily constructed with carbon atoms. Oxygen (O_2) is primarily confined to a gaseous cycle, and carbon is involved in both gaseous and sedimentary cycles as shown in Figure 7.11A. In the gaseous phase, carbon contained in carbon dioxide (CO_2) is cycled through the hydrosphere and atmosphere. However, the major carbon reservoir is calcium carbonates of sedimentary rock strata. Erosion of rocks allows "new" carbon to enter ecosystems, where it can be used by living organisms.

The two major metabolic processes of life, photosynthesis and cellular respiration (Chapter 29 provides complete details of these two processes), link the hydrologic and carbon-oxygen cycles as indicated in the following equation:

$$CO_2 + \text{water} + \text{energy} \underset{\leftarrow \text{ cellular respiration}}{\overset{\text{photosynthesis } \rightarrow}{=\!=\!=\!=\!=\!=}} \text{sugars} + O_2$$

The reactions of *photosynthesis,* which are dependent on solar energy, drive the equation to the right and lead to the production of energy-rich carbohydrate molecules (sugars), as shown in Figure 7.11B. *Aerobic cellular respiration* reactions release the energy stored in the sugar so that it can be used by the organism, driving the equation to the left. Relative to carbon cycling, plants are able to extract carbon molecules, as CO_2, from the hydrosphere and atmosphere during photosynthesis. Cellular respiration occurs in all plants, animals, and decomposers and results in the return of CO_2 to the environment. Similarly, cellular respiration requires O_2 while photosynthesis releases O_2.

During some past geological periods (for example, the Triassic period, about 200 million years ago), cellular respiration by all of the existing biota occurred at lower rates than the photosynthetic reactions of plants. As a result, great quantities of carbon were extracted by plants from the hydrosphere and atmosphere. Eventually the masses of resulting plant material became stored in fossil fuel "sinks" such as coal, natural gas, oil, and oil shales through complex

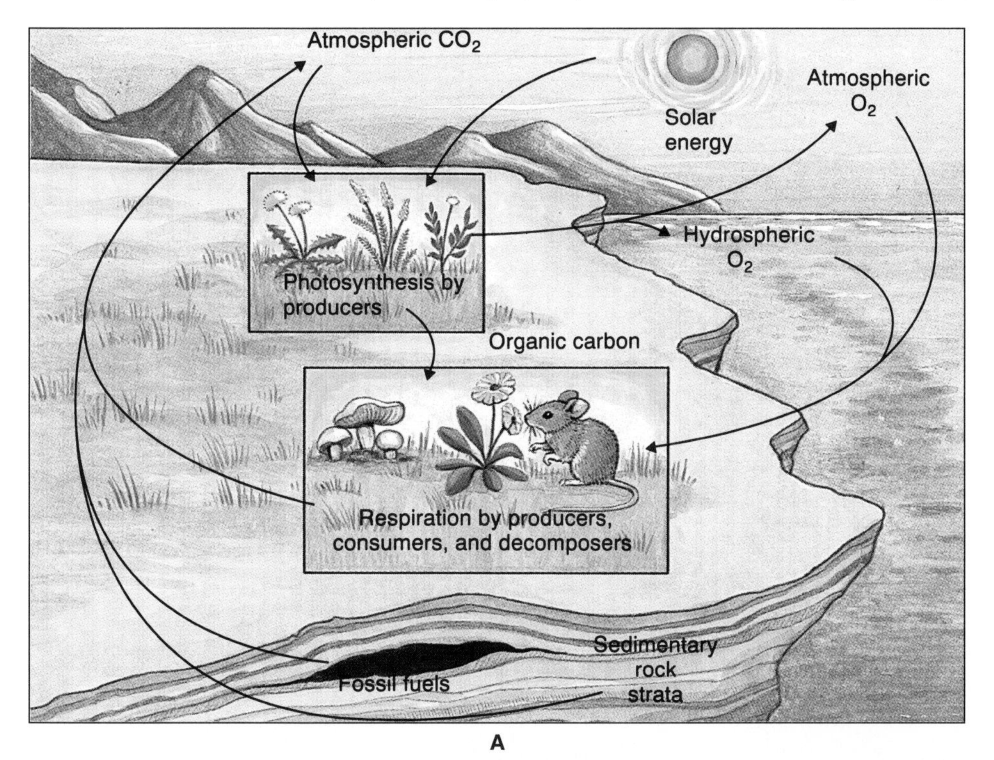

A

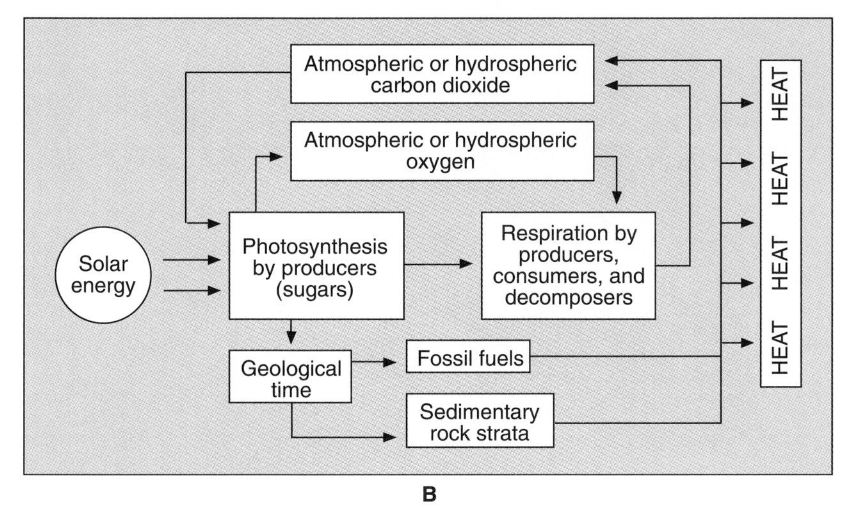

B

Figure 7.11

(A) Carbon and oxygen move through ecosystems in both short- and long-term cycles. Short-term cycles involve the extraction of carbon dioxide (CO_2) by plants during photosynthesis. Complex carbon compounds formed by photosynthesis are eaten by animals or degraded by decomposers, releasing CO_2 back into the atmosphere. Oxygen (O_2) is released during photosynthesis and used during the respiratory activities of plants, animals, and decomposers. Long-term cycles involve fossil fuels and sedimentary rock strata containing calcium carbonate. (B) The energy that drives all these cycles comes from the sun, and most of it is ultimately lost as heat.

Figure 7.12

Carbon was extracted from the atmosphere during the Triassic period and stored in fossil fuel "sinks" like the coal in the mine shown here. When we burn this coal, the carbon is released, which increases the concentration of carbon dioxide in our atmosphere.

geological processes (Figure 7.12). Burning of these fossil fuels by humans is returning much of this carbon to the atmosphere, where it once again is available to plants (discussed in Chapter 11). Consequently, some of the carbon atoms present in the breakfast cereal you most recently consumed may have last cycled through biota that existed on Earth during the Triassic period! Thus a carbon atom now in your body may have once been transferred from the leaves of a primitive cycad to the muscles of a herbivorous *Triceratops* to the brain of a carnivorous *Tyrannosaurus* to a fossil fuel sink to an oil refinery to a tractor in an Iowa cornfield to the corn to your cornflakes and finally to you (see Figure 7.13).

Figure 7.13

Some of the carbon atoms present in your cornflakes could have last cycled through the biota of Earth during the Triassic period. These carbon atoms might have been extracted by photosynthetic activities of an ancient cycad and moved through a dinosaur food chain, where they became deposited in a fossil fuel sink. The extraction and refining of oil and the burning of tractor fuel could have released them as carbon dioxide in an Iowa cornfield, where a corn plant could have captured the carbon atoms. The processed cornflakes may have contained some of these ancient carbon atoms, which may now reside in your body tissues.

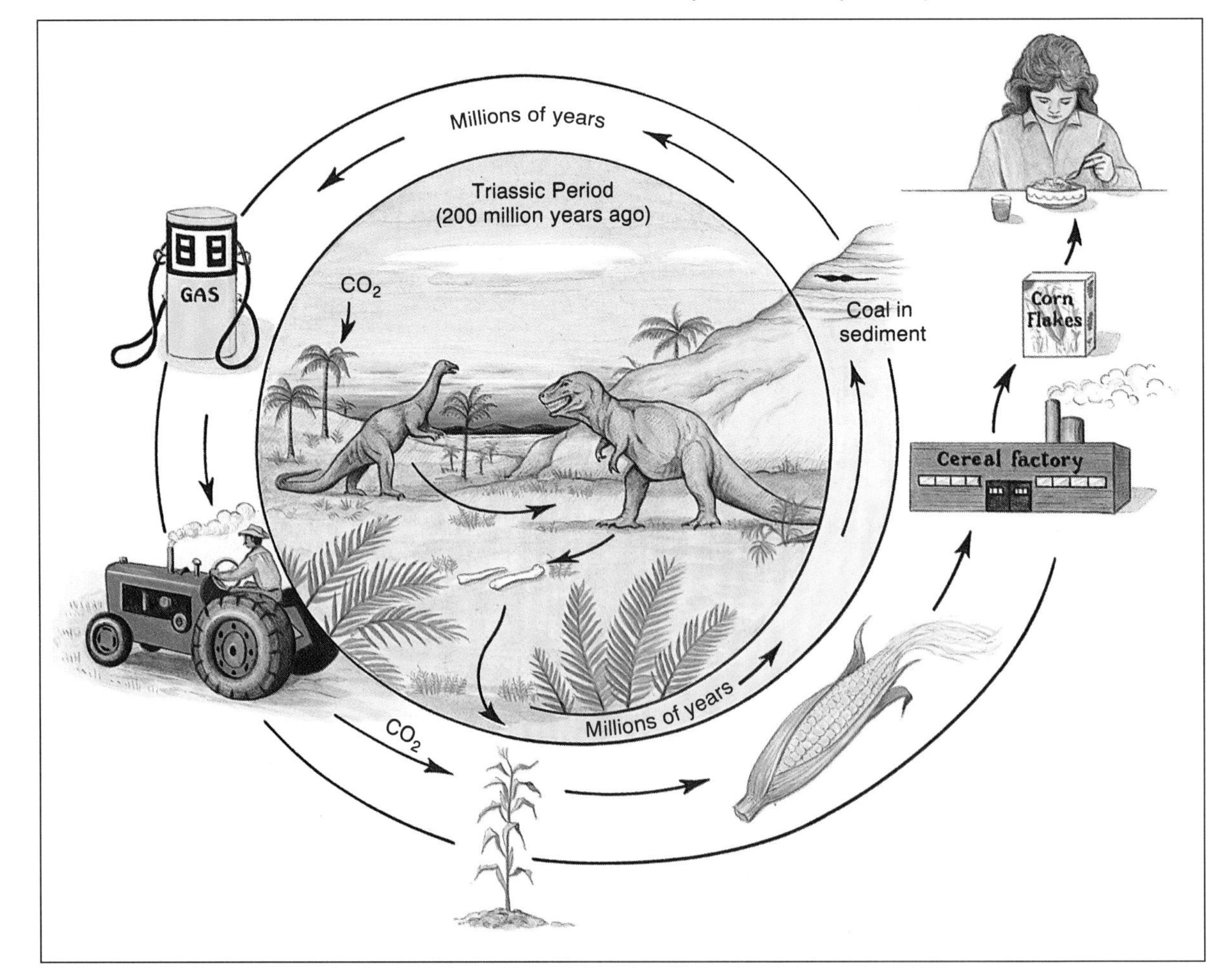

Nitrogen Cycle

The nitrogen cycle is probably the most complex biogeochemical cycle. In the biosphere, the element nitrogen occurs in three different states: dinitrogen molecules, nitrogen oxides, and reduced nitrogen. The atmosphere is composed of about 79 percent *dinitrogen molecules* (N_2), the most stable form of nitrogen. A number of different *nitrogen oxides* exist, but those of greatest biological importance are *nitrite* (NO_2^-) and *nitrate* (NO_3^-). Forms of *reduced nitrogen* include *ammonia* (NH_3), *ammonium* (NH_4^+), and various forms of *organic nitrogen* (ON), including that in organic chemicals that make up living organisms and different nitrogenous products excreted by animals.

Most plants absorb nitrogen from soil as NO_3^-, but once within their tissues, it changes to NH_4^+, which is then used by rapidly growing younger cells for building new plant tissues (see Figure 7.14). A few plant species are able to assimilate NH_4^+ directly from the soil, which effectively shortens the cycling time for nitrogen.

Organic nitrogen in plant tissues is consumed by animals and converted to animal ON. When death occurs, both plant and animal ON enters decomposer

A

Figure 7.14

(A) Inert atmospheric nitrogen (N_2) enters terrestrial ecosystems when nitrogen-fixing bacteria convert it to ammonium (NH_4^+). Some of these bacteria are free-living, and others live in the root nodules of legumes such as alfalfa shown here (B). Ammonium and ammonia (NH_3^+) can be converted to nitrite (NO_2^-) and nitrate (NO_3^-) forms of nitrogen by the process of nitrification. This process is accomplished by bacteria such as *Nitrosomonas* and *Nitrobacter,* which live free in the soil, not in nodules of legumes. Many plants can use both the ammonia and nitrate forms of nitrogen for their growth processes, converting the nitrogen compounds into organic nitrogen (ON) that is used by animals for their growth and body maintenance. Nitrogenous waste products from both plants and animals are processed by decomposers, and through ammonification, the modified nitrogen enters the ammonia pool in the soil. Other bacteria present in the environment can convert nitrate forms of nitrogen into N_2 by the process of denitrification, thus completing the nitrogen cycle.

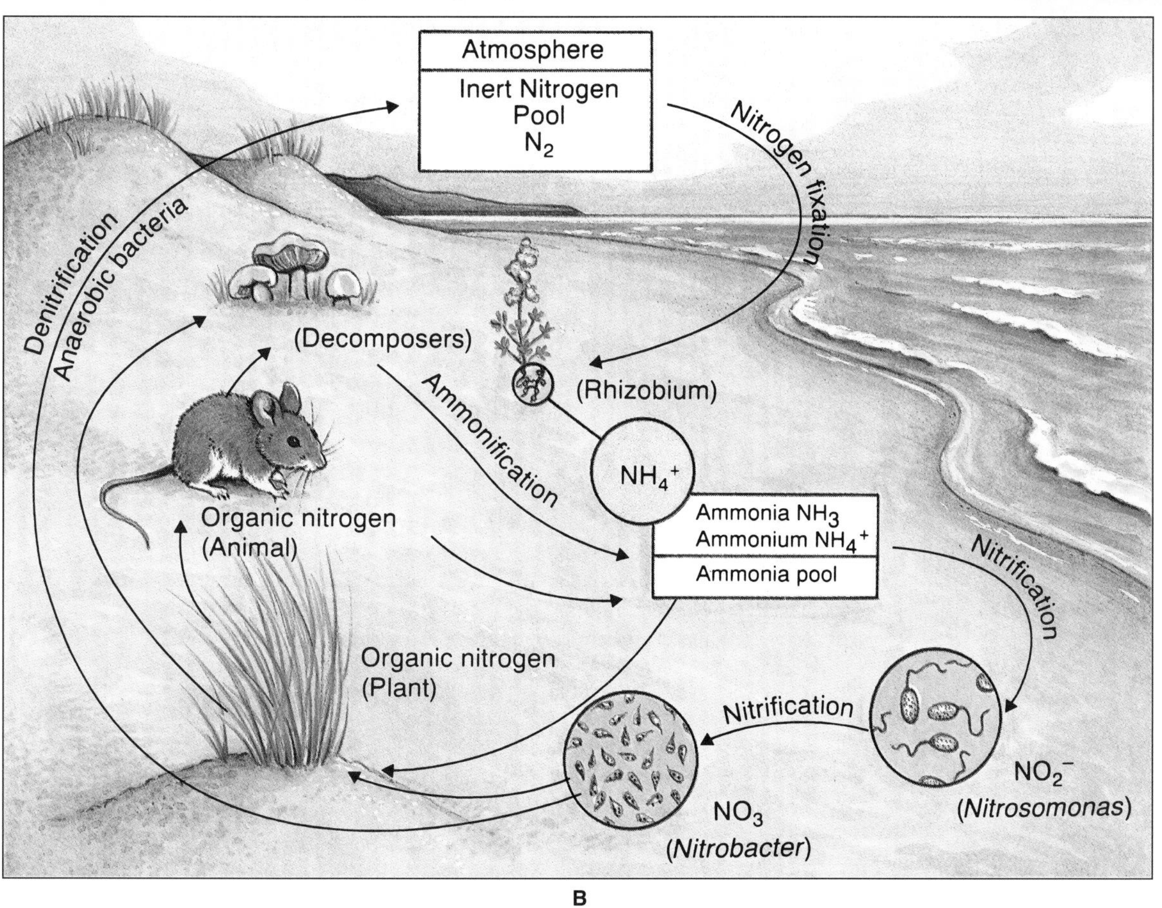

B

food chains, where fungi and bacteria convert it to ammonia through the process of *ammonification*. Soil bacteria in the genus *Nitrosomonas* convert NH_4^+ to NO_2^-, and *Nitrobacter* species convert NO_2^- into NO_3^-, thus making nitrogen available once again for new plant growth.

Within oxygen-limited terrestrial and aquatic environments, about 25 different genera of anaerobic bacteria convert NO_3^- into inert forms of nitrogen, including the atmospheric form N_2, through a process known as *denitrification*. However, more than 70 bacterial genera, including some free-living forms and those like *Rhizobium* that inhabit the root nodules of plants in the pea family (Leguminosae), convert atmospheric N_2 into ammonia by a process called *nitrogen fixation*.

In terrestrial ecosystems, nitrogen-fixing organisms are critically important because they change biologically inert nitrogen into ammonia, thus making it available to producers. Eukaryotic cells are not capable of carrying out nitrogen fixation. Therefore, most eukaryotic life forms on Earth are dependent on the nitrogen-fixing reactions carried on by free-living and nodule-associated prokaryotes. Exceptions are crops grown using modern nitrogen-based fertilizers, which are produced by using fossil fuels, usually natural gas.

When water is present in a soil environment, NH_3 is converted into NH_4^+, which becomes bound to soil particles. This serves to retard nitrogen loss by runoff or through groundwater movement. However, some soil bacteria can convert NH_4^+ into NO_2^- and NO_3^- in a process known as *nitrification*. These oxide forms of nitrogen are both very soluble in water. Consequently, they can be removed from the ecosystem during surface runoff or groundwater transport. These processes tend to reduce the amount of available nitrites and nitrates, often causing them to become major limiting macronutrients in terrestrial ecosystems.

Lindeman and the Odum brothers applied concepts from physics to quantify the flow of energy and material cycling within ecosystems. Organic life forms actively participate in the cycling of materials such as carbon, oxygen, nitrogen, and phosphorus. The solar energy that drives the ecosystems of Earth is converted to many useful forms, but ultimately it is all dissipated as heat.

THE ECOLOGICAL NICHE

The ecosystem was not the only concept used to describe ways in which communities of organisms lived together. Another fruitful approach is to consider a population as the fundamental ecological unit and determine which biotic and abiotic factors affect its abundance and distribution. Joseph Grinnell at the University of California was one of the first ecologists to use the term *niche,* in a 1917 paper titled *The Niche-Relationships of the California Thrasher.* He summarized the factors that defined the distribution of this secretive bird (Figure 7.15) as follows:

> These various circumstances, which emphasize dependence upon cover, and adaptation in physical structure and temperament, thereto, go to demonstrate the nature of the ultimate associational niche occupied by the California Thrasher. This is one of the minor niches which with their occupants all together make up the chaparral association. It is, of course, axiomatic that no two species regularly established in a single fauna have precisely the same niche relationship.

Grinnell conceptualized the thrasher's *niche* as a unit defined by its collective behaviors in association with elements from its physical structure.

Elton (1927) expanded the niche concept to include other factors present in the community. He reasoned, "Animals have all manner of external factors acting upon them—chemical, physical, and biotic—and the niche of an animal means its place in the biotic environment, *its relations to food and enemies*."

Figure 7.15

California thrashers live in chaparral foothills and dense shrubs in parks and gardens west of the Sierra range and south into Baja California. Grinnell first used the term *niche* to describe the factors that defined the distribution of this bird species.

The emerging concept of the species niche was subjected to mathematical interpretations during the 1920s and 1930s by V. Volterra, A. J. Lotka, G. F. Gause, and numerous other ecologists. Grinnell's general idea that only one species can occupy a single niche was further expanded by Volterra and Gause, and it became known as the Gause-Volterra principle or, more descriptively, the *competitive exclusion principle.* Today, a **niche** is considered to be the ecological role a species plays within a community. Competition between two similar species at niche boundaries became the focal point for one school of ecology, and it continues today as a focus for many scientific studies.

A niche is defined as the role a species plays in a community. It includes the behaviors of individuals in a species and the influences of other biotic and abiotic factors that affect its life.

MODERN ECOLOGY

Two general schools of ecology have been described in this chapter. One is centered on the ecosystem and the other on the species population (niche). Both continue to present viable approaches for new studies of natural environmental systems.

Chapters 8 and 9 will describe the ecology of terrestrial and aquatic ecosystems. Much of what we know today about these natural communities of organisms can be traced to the seminal studies and principles generated through the pioneering efforts of Liebig, Clements, Cowles, Lindeman, Gause, the Odums, and others. Developing the concepts of limiting factors, succession, trophic levels, ecosystems, ecological pyramids, biogeochemical cycles, the one-way flow of energy, and the ecological niche represent stunning intellectual achievements. They remain today as the cardinal principles of ecology.

Summary

1. Ecology is the scientific study of interrelationships between organisms and other biotic and abiotic factors in their environment. Early ecological concepts included the idea of a *balance of nature*. Darwin and other naturalists proposed mechanisms to help explain this perceived balance.
2. Liebig discovered that an insufficient concentration of only one required salt could limit the growth of plants in an area. This idea later became the *law of limiting factors*.
3. Clements and Cowles presented different views of how ecological succession occurs. Clements thought that terrestrial succession led to a climax community with an "organic entity" similar to that of an organism. Cowles thought that succession could both advance and retrogress and that natural communities never reach equilibrium.

4. Gleason and Ramensky advanced the idea that the spatial pattern of plant population distributions was a result of their differential responses to changing gradients of important limiting factors over time.
5. Shelford integrated animal communities into the grand design of succession. He transformed the law of limiting factors into the *law of toleration* when he included minimum, optimum, and maximum concentrations of abiotic elements required by all organisms.
6. The ecosystem concept was developed during the 1930s and 1940s to include both the biotic community and its abiotic environment. Lindeman and Odum attempted to quantify ecosystem functions by measuring the energy flow between trophic levels. They determined basic biological efficiencies by relating this flow to the growth that occurred within each trophic level. They also applied the concepts of trophic dynamics to changes that occur during succession.
7. Plants, herbivores, and carnivores represent three different trophic levels. The amount of energy present at a given trophic level is always less than that present in the previous (lower) trophic level but greater than the energy that will be available to the next (higher) trophic level. Food chains represent one possible route of energy flow through an ecosystem; food webs represent all possible routes.
8. Ecological pyramids of energy, numbers, and biomass are graphical representations of the trophic structure of an ecosystem.
9. Elton's food cycles illustrate food webs, energy flow, and the cycling of important materials in an ecosystem. Nutrients are supplied to ecosystems through gaseous, sedimentary, and hydrologic biogeochemical cycles.
10. The niche is the ecological role that a species plays within a community.

Review Questions

1. What contribution did Darwin make to early ecological concepts?
2. What is the significance of the law of limiting factors?
3. How did Clements and Cowles differ in their views on ecological succession?
4. What contributions did Shelford and Elton make to early animal ecology?
5. What are trophic levels, and how do they affect the flow of energy and material cycling in ecosystems?
6. Which environmental factor can be accurately represented by ecological pyramids?
7. What is the hydrologic cycle, and how does it operate?
8. What are the major sinks for carbon, oxygen, nitrogen, and phosphorus?
9. In the nitrogen cycle, which processes produce nitrogen compounds that are useful to plants?
10. What is a species' ecological niche?

Essay and Discussion Questions

1. Energy flows through an ecosystem only once, whereas materials can cycle through numerous times. What will happen if Earth's energy source is extinguished? Why? What human activities might disrupt biogeochemical cycling?
2. What might be some reasons that early concepts of plant ecology were developed before those of animal ecology?
3. Which one of the concepts described in this chapter do you feel was probably the most important in advancing the field of ecology? Why?

References and Recommended Reading

Cowles, H. 1899. The ecological relations of the vegetation of the sand dunes of Lake Michigan. *Botanical Gazette* 27:95–117, 167–202, 281–308, 361–391.

Darwin, C. 1859. *On the Origin of Species*. London: Murray.

Elton, C. 1935. *Animal Ecology*. London: Sidgwick & Jackson.

Hoekstra, T. W., T. F. H. Allen, and C. H. Flather. 1991. Implicit scaling in ecological research. *BioScience* 41:148–154.

McIntosh, R. 1987. *The Background of Ecology*. Cambridge: Cambridge University Press.

Odum, E. 1971. *Fundamentals of Ecology*. 3d ed. Philadelphia: Saunders.

Odum, H. 1957. Trophic structure and productivity of Silver Springs, Florida. *Ecological Monographs* 27:55–112.

O'Neill, R., D. DeAngelis, J. Waide, and T. Allen. 1986. *A Hierarchical Concept of Ecosystems*. Princeton, N.J.: Princeton University Press.

Smith, L. 1986. *Elements of Ecology*. 2d ed. New York: HarperCollins.

Stauffer, R. 1957. Haeckel, Darwin, and ecology. *Quarterly Review of Biology* 32:138–144.

CHAPTER 8

Terrestrial Ecosystems

Chapter Outline

Reading Questions

1. What are the basic components of biotic ecosystem models?
2. What are the basic components of functional ecosystem models?
3. How do major terrestrial ecosystems differ in their patterns of energy flow, material cycling, population densities, and succession?
4. What types of factors regulate these ecosystem processes?

Ecosystem concepts for terrestrial communities have largely been developed by integrating results from many environmental studies that focused on different space and time scales. Ecosystems have commonly been viewed as complex systems that change over time. Within these systems, environmental regulatory factors often affect different trophic levels during each successional stage. Early successional stages are usually dominated by producers, and their growth is largely dependent on available amounts of water and nutrients. However, in later successional stages, herbivores play a major role in regulating the growth of producers through their grazing activities. As succession continues, the complexity of the entire system increases. The degree of complexity, as indicated by such factors as *species diversity*—the number of different species present in an area—is considered to be of major importance in sustaining an ecosystem when external disturbances occur.

ECOSYSTEMS AND COMMUNITIES

Ecosystems have long been described in *biotic* terms by considering each species to be a member of a different trophic level in a community. An alternative approach has emerged in which ecosystems are perceived as being composed of different *functional elements* that can be identified and subjected to analysis, as described in Figure 8.1. The simpler, traditional **biotic ecosystem models** have emphasized the *amount* of energy and materials moving up through different trophic levels. The more recently conceived **functional ecosystem models** are concerned with the *rate* at which energy and materials move through the system.

Biotic Ecosystem Models

Until recently, most descriptions of terrestrial ecosystems have had a biotic emphasis, in which species populations are considered to operate in fixed trophic levels within constraints imposed by the abiotic environment. Under this system of organization, producers, herbivores, carnivores, and decomposers take center stage, and populations belonging to the various trophic levels become the focal point of most studies. Following Lindeman's original ideas, published in 1942, it has commonly been assumed that energy flows sequentially from producers to

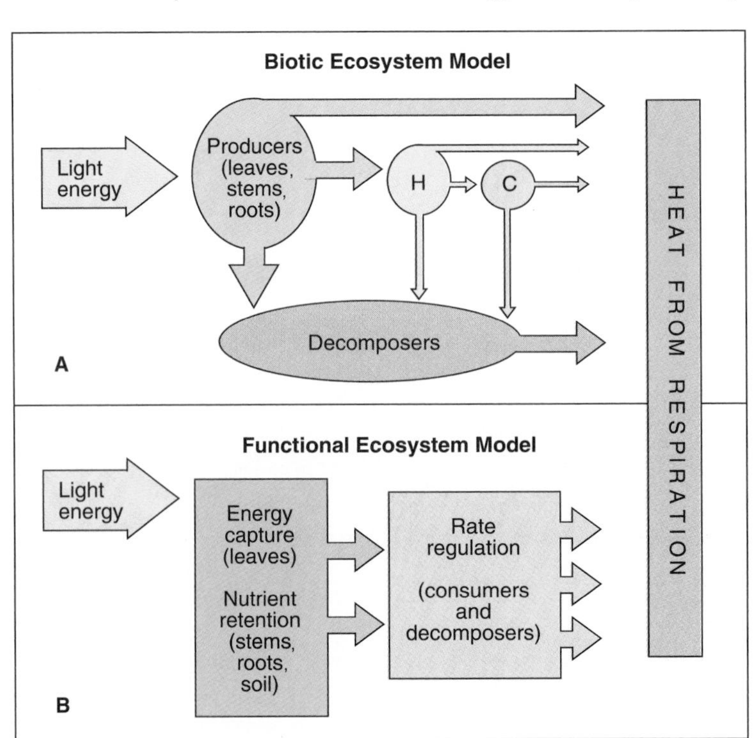

Figure 8.1

Ecosystems can be described by both biotic and functional models. (A) The biotic ecosystem model is based on the flow of energy through trophic levels. Energy enters the ecosystem through photosynthesis by producers; flows to herbivores (H), carnivores (C), and decomposers; and is ultimately lost from the ecosystem as heat. (B) In the functional ecosystem model, the hierarchy includes components of energy capture, nutrient retention, and regulation of the rate at which energy flows to consumers, decomposers, and ultimately out of the ecosystem as heat. Material cycling is not shown in this figure.

herbivores to carnivores to decomposers (as described in Chapter 7). Contemporary ecology research has uncovered drawbacks in using this simple model to define energy flow in ecosystems. We shall consider some of these drawbacks.

Trophic Levels

Producers

Producers in terrestrial ecosystems are easily identified: they are all the plants that carry on photosynthesis. However, in most plants, such as a maple tree, only a relatively small fraction of living biomass, usually leaves and the surface layer of stems, actually participates in photosynthetic activities; that is, only a small fraction of the plant is actually a producer. The nonphotosynthetic plant biomass—woody stems and roots—interacts with the soil and serves as a reservoir for key nutrients in terrestrial ecosystems.

Herbivores, Carnivores, and Decomposers

Energy and nutrients generally move from producers to herbivores, carnivores, and decomposers. However, assigning a species to a single trophic level is rarely an accurate way to describe its role in an ecosystem. Consider the case of North America grizzly bears (see Figure 8.2). To the relief of nature lovers and wilderness hikers, grizzlies feed most often as herbivores. Of the many types of plants they eat, they consume only one plant genus, *Equisetum* (horsetails), throughout their entire North American range. Various species of horsetails populate different sections of their range, and grizzlies feed on them during all seasons. Horsetails are usually found in association with grasses and sedges, which the bears also consume. In coastal and mountain habitats, cow parsnip, clover, and dandelions supplement the bear's spring and early summer diet. The animals also consume a variety of other plants, including shrub genera that produce huckleberries and blackberries, during late summer and fall.

The question of trophic level placement for grizzlies becomes more complex because they are also carnivores. Grizzly bears are opportunistic and will

Figure 8.2

(A) The grizzly bear *(Ursus arctos)* is classified as an omnivore. In the coastal regions of the grizzly bears' range, salmon become an important component of their diet. (B) Grizzlies range from central Nevada north to Alaska and in the Mackenzie District of the Northwest Territories in Canada.

A

Range of grizzly bears

B

feed on any living fish or mammal they are able to kill, as well as on most carrion (dead, decaying flesh). In the northern portion of their range, they have been observed to steal carcasses of animals killed by wolf packs or by other predators. Herbivores such as caribou, elk, moose, and small rodents, such as ground squirrels and lemmings, make up significant portions of grizzly bears' spring diet. In the coastal areas of their range, they consume large quantities of migrating salmon, and here they may also feed on the beached carcasses of marine mammals (whales, walruses, and seals), some of them herbivores and others carnivores. In which conventional trophic level does a grizzly bear belong? The term **omnivore**, a consumer of both plants and animals, living or dead, is often used, but food webs describing energy flow in far northern ecosystems become quite confusing when they include grizzly bears. Looking ahead, however, if grizzlies are considered as elements in a *functional system* (review Figure 8.1), they are simply classed as *consumers,* which influence the system by affecting the rates of energy flow and the *flux,* or movement, of materials.

Trophic Guilds

In developing his concepts of trophic dynamics, Lindeman used the community of Cedar Bog Lake as a model. His trophic concepts had their roots in an unpublished paper written by his mentor, G. E. Hutchinson. Hutchinson considered the productivity of any trophic level to be the rate at which energy (food) was advanced from the previous trophic level; he felt that trophic structure could be defined in terms of a linear chain consisting of eaters and the eaten. However, during the half century of ecological research since these concepts originated, scientists have come to realize that few, if any, heterotrophic organisms feed exclusively at any one trophic level.

Currently, trophic levels are no longer considered to be connected in a series to form a chain, and species are not necessarily assigned to a single trophic level. Instead, species are assigned to **trophic guilds**, which are defined as a group of species that exploit the same class of trophic resources in a similar way. Trophic guilds are comparable to the traditional concept of trophic levels; that is, they consist of producers, herbivores, carnivores, and decomposers. However, membership in trophic guilds is fluid, and most organisms, including grizzly bears, are thought to belong to more than one guild, depending on their feeding habits. Use of the trophic guild concept allows modern ecologists to define and quantify more precisely the flow of energy and nutrient cycling in the ecosystems they study.

The biotic model of an ecosystem categorizes organisms into different trophic levels identified as producers, herbivores, carnivores, and decomposers. However, since most animal species can belong to more than one trophic level, the new concept of trophic guilds was formulated. By focusing on trophic guilds, researchers can more accurately measure energy flow and material cycling in ecosystems.

The study of populations has been a key element in expanding our knowledge of ecosystems, particularly in understanding succession. Currently, there is great emphasis on identifying factors that influence the size of populations within ecosystems.

Population Parameters

Populations of organisms affect one another over time. However, within each community, only a small portion of each species population is present at any given time. These local interbreeding populations represent a lower tier within the *biotic hierarchy*—a level of organization within the biosphere—called a **deme**, described in Figure 8.3.

Research has shown that changes within demes constitute a major driving force of succession. This observation has led to a number of important questions

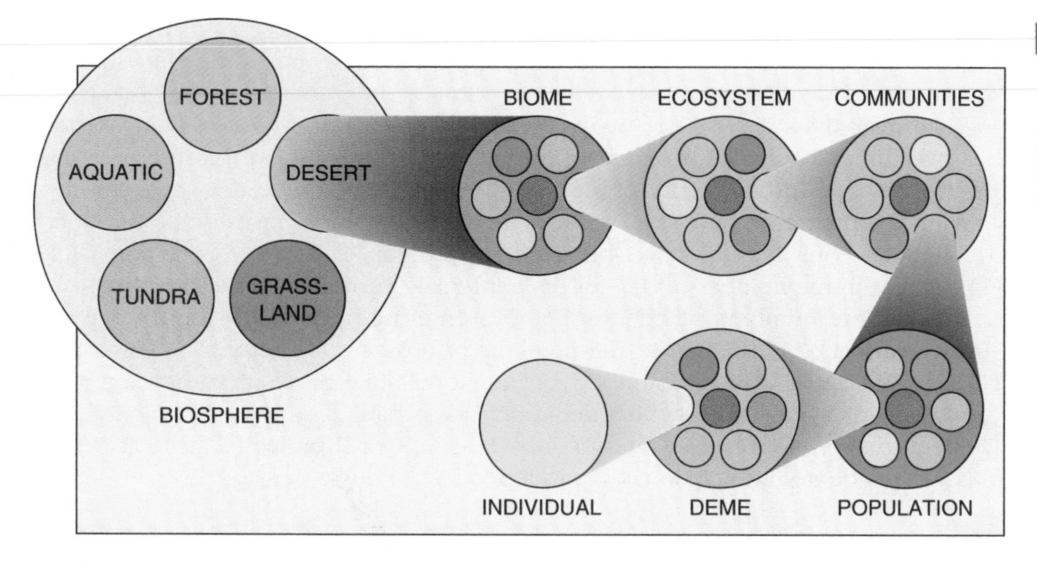

Figure 8.3

In the biotic ecosystem model, an ecological hierarchy exists in which there are different levels of organization. The biosphere is the highest level, followed by biomes, ecosystems, communities, populations, demes, and individuals.

that are now guiding research that could lead to a better understanding of terrestrial ecosystems. What are the relationships between different demes and other important components of an ecosystem? What factors regulate the demes in a population? Are those factors biotic, abiotic, or both? The search for answers to these and other pertinent questions presents major challenges for ecologists today and tomorrow. These questions and some of the partial answers already obtained are addressed in the remainder of this chapter.

Deme Density

The **density** of a deme is usually expressed as the number of individuals of the same species in a given area. Examination of a specific case provides an opportunity for some interesting analyses relative to deme density. Table 8.1 summarizes results from a trapping study conducted in Point Pelee National Park, in southern Ontario, Canada, that determined the density of white-footed mice *(Peromyscus leucopus)* and meadow voles *(Microtus pennsylvanicus)* in four different *microhabitats* (small, specialized habitats) during 1978 and 1979. A trapping system covered an area of about 3.4 hectares in each of the four microhabitats. Thus in the grassland microhabitat, the density of *Peromyscus* in 1978 is expressed as 10 per 3.4 hectares.

The data in Table 8.1 show that white-footed mice are adapted to a wider range of microhabitats than the meadow voles. (Why can this be concluded?) In general, the four microhabitats each represent a sequential stage in a succession-

Table 8.1 Deme Density of Two Species in Different Microhabitats

		MICROHABITAT			
Species	**Year**	**Grassland**	**Old Field**	**Sumac**	**Forest**
Peromyscus	1978	10	11	31	59
(White-footed mouse)	1979	11	39	84	121
Microtus	1978	97	26	0	5
(Meadow vole)	1979	147	30	0	2

The number of different individuals of *Peromyscus* and *Microtus* captured in four adjacent microhabitats, each 3.4 hectares in size, in Point Pelee National Park, Canada, during 1978 and 1979.

Source: After Morris (1987).

al series that begins with the grassland and climaxes with the forest. Therefore, it appears that meadow voles may not be adapted for surviving in the later successional stages in Point Pelee Park.

Terrestrial succession can be understood in terms of changes in deme densities over time. Therefore, it is important to identify the factors that regulate the *rate of change* in the density of a deme. Densities of a deme can increase, decrease, or remain the same over a given time period. What causes deme densities to change? Four factors have been identified, as illustrated in Figure 8.4: (1) changes in the numbers of live births per female per unit of time (**natality**), (2) changes in the numbers of new individuals entering from demes in other areas (**immigration**), (3) changes in the numbers of deaths per unit of time (**mortality**), and (4) changes in the numbers of individuals that move away from an area (**emigration**). The rates at which deme densities change are shown in Figure 8.5. Elements that influence these rates of change include environmental factors that are either dependent on deme density or independent of deme density.

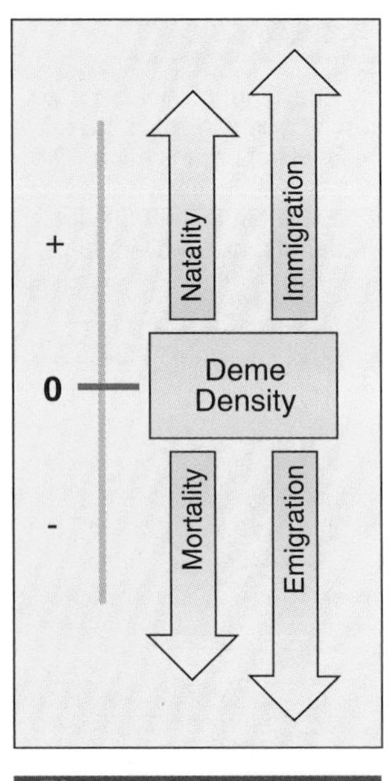

Figure 8.4

Natality and immigration of individuals increase deme density, while mortality and emigration of individuals decrease deme density.

Density-dependent Factors

Density-dependent factors vary directly with changes in the deme density (that is, as the factor increases, so does the deme density). These factors include nutrients, food, cover, *intraspecies interactions* (among individuals of a species within a deme or between members of the same species from other demes), and *interspecies interactions* (between demes of different species). Important intraspecies interactions that affect the densities of animal demes are direct competition for space, food, and mates; those that affect plant densities are nutrients, moisture, space, and light. Critical interspecies, density-dependent interactions include parasitism, predation, and competition.

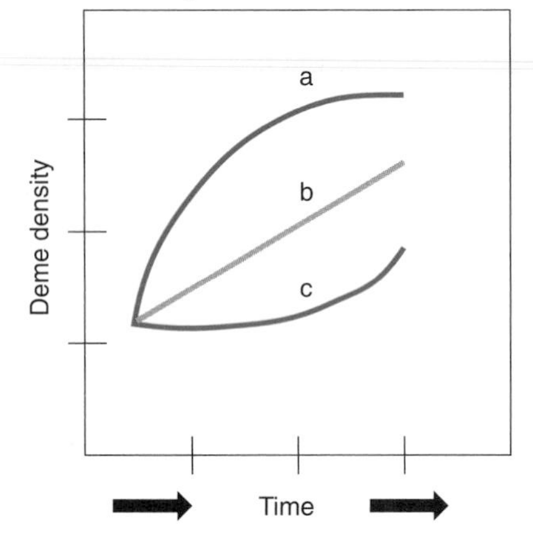

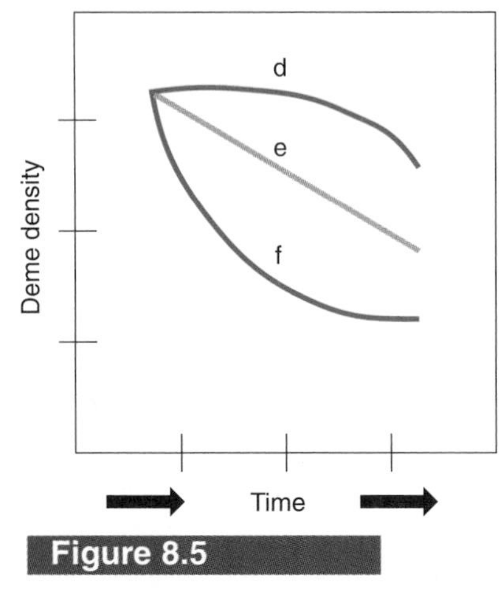

Figure 8.5

Deme densities can expand at rates that are increasing (a), constant (b), or decreasing (c), or they can be reduced at increasing (d), constant (e), or decreasing (f) rates.

Density-independent Factors

Density-independent factors are usually abiotic. Severe climatic conditions, including droughts, extended periods of unusual freezing temperatures, flash floods, or volcanic eruptions, can alter the rate at which a deme's density would change (Figure 8.6). However, the amount of change in these cases would not depend on the deme's density.

Population Density Controls

It has been difficult for ecologists to measure the separate effects of density-dependent and density-independent factors in changing animal population densities. To illustrate the problems, consider the following classic study, based on an examination of historic records and data of the Hudson's Bay Company. From these data, Charles Elton graphed the yearly totals of fur pelts taken from arctic snowshoe hare and their major predator, the Canada lynx, over a 100-year period (see Figure 8.7). He made two assumptions about the data. First, the geographical

Figure 8.6

Deme densities can be affected by abiotic factors that range from small, localized changes in temperature or available moisture to catastrophic regional events such as the eruption of Mount St. Helens in southwestern Washington. Not only are demes decimated, but entire communities, both terrestrial and aquatic, can be destroyed.

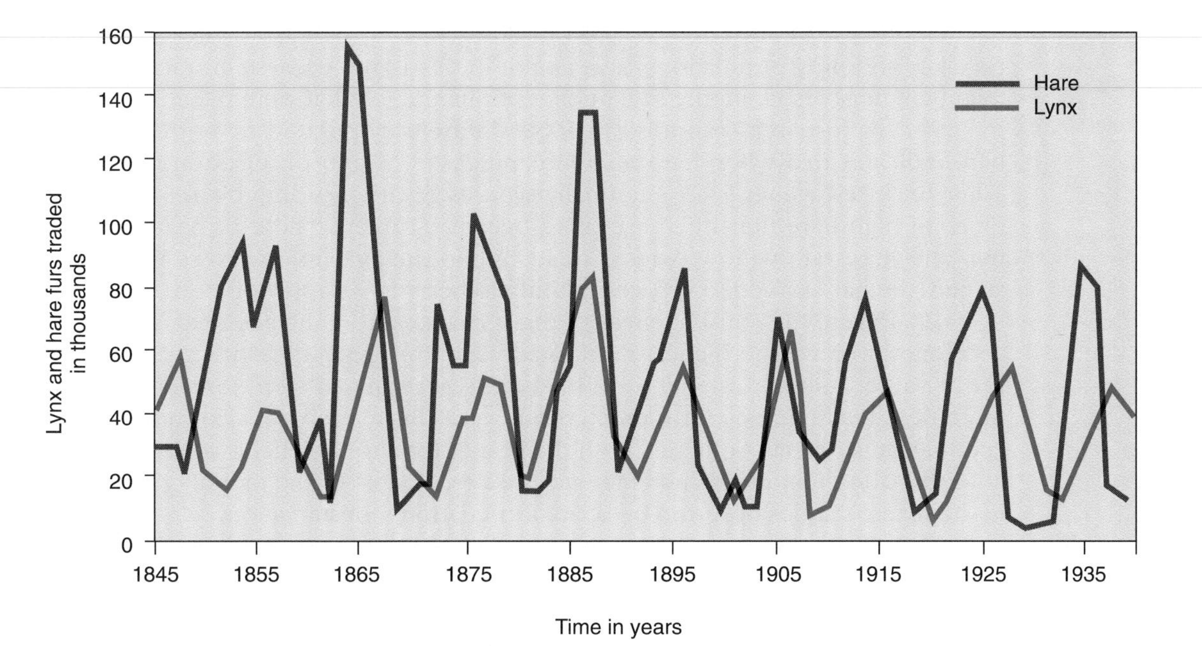

Figure 8.7

Graphs drawn from Elton's data (A) suggested to him that the density of prey such as hares (B) allowed the density of predators such as lynx (C) to reach such high levels that increased predation caused a crash in prey density, which in turn led to a subsequent crash in the density of predators. These cyclic oscillations occurred about every ten years. Recent research has shown that Elton's explanation was incomplete.

area from which the pelt data were collected remained constant, and second, the graph represented the approximate densities of the two species over the 100-year period. Inspection of the graph revealed an apparent cyclic oscillation in lynx and hare densities that was a function of time. Hare numbers peaked at intervals of about ten years, and each peak was followed by a rapid decline, the result of massive mortalities. The graph of the lynx pelt data revealed similar oscillations, but in most decades, the lynx peak numbers lagged a few years behind those of the hare. What would be a logical hypothesis to explain this pattern?

Elton's observations led him to form a concept of **intrinsic control** between these two species. When hare numbers were high, lynx numbers would also increase. Ultimately, after heavy predation, hare numbers crashed, followed by a dramatic decline in the numbers of lynx. This attractive, simple hypothesis was accepted by ecologists for more than 30 years. Like any hypothesis, accumulating corroborative evidence increases confidence in its validity, but dissenting observations can undermine this confidence. Such contradictory observations lay ahead, and as a result, the "hare-lynx intrinsic control hypothesis" entered a phase of scientific limbo.

In the early 1960s, populations of snowshoe hares that inhabited an island devoid of lynx were observed to fluctuate in ten-year cycles! Thus Elton's explanations about oscillations in lynx and hare numbers came under suspicion. The fluctuations apparently could not be caused exclusively by intrinsic interactions of predator and prey populations, at least not in this island ecosystem. Factors other than lynx must also limit the size of hare populations. What were they?

Recent research (reported in 1988) in areas populated by both lynx and hare indicates that when hare numbers increased, a shift occurred in their diet. Their preferred winter food is normally older twigs from swamp birch, but during times of high hare density, these twigs became scarce because of heavy browsing, and hares ate willow twigs instead. When the available willows were consumed, the hare turned to spruce and buffalo berry as a supply of food.

What were the consequences of these changes from their normal diet? Birch was found to have the highest nutritional value for hares, the other twig species lower value. As high-quality food become scarce, hares were forced to spend more time feeding away from protective cover. This altered behavior had two related negative effects on hares. Both the search for and the effects of lower-quality food made them more vulnerable to predation. Most of the mortality observed in the recent studies occurred because of predation, not starvation.

Thus it now appears that a modification of Elton's original hypothesis is in order. It cannot be rejected entirely because the oscillations in numbers of hare do seem to influence the numbers of lynx. Nevertheless, the decline in hare numbers in the 1988 studies resulted directly from changes in the quantity and quality of the animals' food. Increased lynx predation occurred only because of the poor nutritional state of the hare. If the lynx had not been present, as was the case in the island study in the 1960s, hare mortality would still have occurred at the same level due to starvation but would have been spread over a longer period of time. Is it possible to make confident conclusions about the controlling factors operating on these two populations? Are they density-dependent or density-independent? The quality and quantity of available twigs was related to the density of hare, as was the density of lynx when they were present. However, the presence of birch and willow twigs was a function of ecological succession. Both birch and willow are species that invade areas shortly after fires or other density-independent disturbances. Thus in this complicated case, population density controls can operate at a number of levels in the biotic hierarchy and in a number of different time frames.

The hare-lynx story constitutes a typically complex but fascinating example of the scientific search for truth in the field of ecology. In another sense, this case study illustrates the transition being made between an earlier time, when ecology was based on a few elementary principles, to the present, when it has become clear that little in ecology is as simple as it might initially seem.

Deme density is determined by both density-dependent and density-independent factors. Ecological succession results in modifications of habitats that are populated by animal demes and leads to further changes in their density.

Functional Ecosystem Models

Functional ecosystem models of energy flow represent a modern, alternative conceptual approach to the more traditional concepts of the biotic model. The central assumption in functional ecosystem theory is that the existing structure of complex ecosystems results from different rates of energy flow and material cycling during each successional stage. Thus the major controlling factors that operate during each successional stage affect the availability of nutrients and energy for a certain period of time. However, modifications of those factors can occur through the successional history of a site and must also be recognized and factored into the functional ecosystem model.

When viewing ecosystems from a functional perspective, organisms are placed into one of two general categories: (1) energy capture and nutrient retention and (2) rate regulation (see Figure 8.1). Consumers and decomposers are important because they regulate rates of energy flow through the system. Thus ground squirrels, lemmings, snowshoe hare, elk, moose, lynx, grizzly bears, fungi, and bacteria are important in "functional ecology" because they all influence the *rate* of energy flow and the flux of nutrients and other materials. Functional models also incorporate population-regulating factors derived from studies grounded in the biotic ecosystem model.

Applying functional perspectives to ecosystem research is now guiding studies of terrestrial biomes and aquatic systems. A primary objective of most modern ecological studies is to obtain abundant, precise numerical data that can be analyzed and used to explain different aspects of ecosystem processes. We shall describe some specific case studies for two reasons: (1) to illustrate how components of the functional ecosystem model are influenced by regulating factors and (2) to highlight experimental approaches used in contemporary studies of terrestrial ecosystems.

MAJOR TEMPERATE TERRESTRIAL ECOSYSTEMS

Different terrestrial ecosystems exist within the major biomes described in Chapter 6. For each biome, the number of distinct terrestrial ecosystems varies with latitude, longitude, elevation, and proximity to oceans or other large bodies of water. Forest biomes dominate the northern, eastern, and many western and mountainous regions of North America. The central portion of the continent is covered with grasslands, and deserts occur in the southwest.

Temperate Forest Ecosystems

In North America, a large-scale biogeographical pattern exists for total organism diversity. For example, in Figure 8.8, transects (lines used to chart organism distribution across a given area) A and B show that the number of different tree species decreases from a maximum of 180 in the high plateau south of the Appalachian Mountains to only 10 species near the timberline in northern Canada and Alaska.

As you can see on the transects, change in species diversity is largely associated with increasing longitude from central North America to the Appalachian plateau (transect A), while from central Canada to the northern timberline, increasing latitude has a greater effect (transect B). The complex patterns in the west are related to the presence of major mountain ranges.

What factors might account for this pattern of tree species diversity? Recent research correlates the number of tree species with the **average annual evapotranspiration (AAET)**, which is defined as the average amount of water evaporated from the soil plus that transpired from vegetation, per year. When AAET increases, tree species diversity decreases, and when AAET decreases, tree species diversity increases. About 76 percent of the variation in tree species numbers across North America can be explained by AAET; the remaining 24 percent is related to topography and proximity to oceans. Generally, greater plant species diversity is correlated with higher levels of energy capture in natural ecosystems. Thus AAET can be a significant factor in regulating energy capture in certain ecosystems.

Longitude, latitude, climate, topography, and elevation determine the type of forest—coniferous or deciduous—but within each forest type, local modifications among microhabitats determine the characteristics of a specific forest. For example, the forest biome east of the Cascade Mountains in Oregon includes a number of different species of evergreen trees that are affected by various factors. However, elevation is a primary regulating element, and at higher elevations, ponderosa pine is replaced by lodgepole pine, which can thrive at higher altitudes.

Figure 8.8

Tree species diversity in North America decreases from east to west and from south to north. Along transect A, the change in species diversity is primarily associated with increasing longitude, whereas along transect B, species diversity decreases with increasing latitude. The red lines deliniate areas, each labeled with the number of species that grow there.

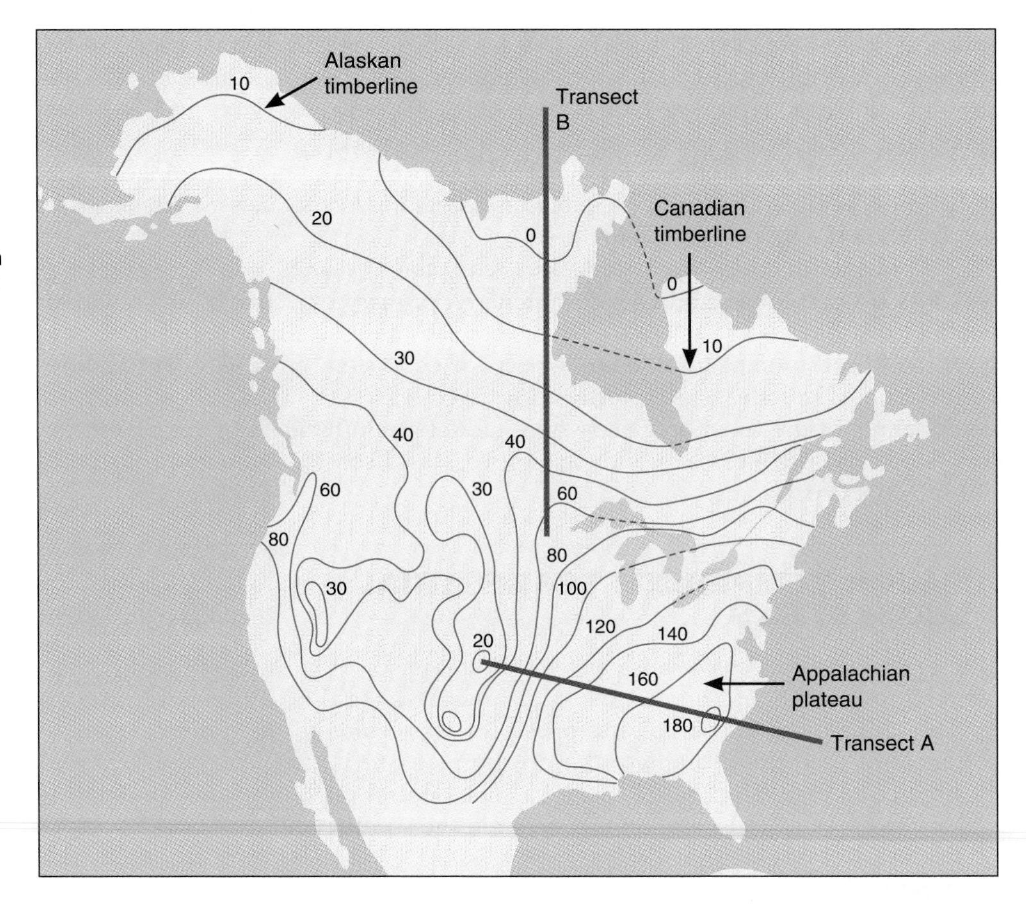

A different controlling factor regulates parasitic bark beetle populations in pine forests. Ponderosa and lodgepole pine forests are often infested with species of bark beetles. The larvae of these beetles consume nutrient-transporting tissues located under the bark, and if their density reaches high levels, they can kill the trees. One important species, mountain pine beetles *(Dendroctonus ponderosae),* infests both lodgepole and ponderosa pine trees (see Figure 8.9A). One major problem ecologists face is identifying the factors that trigger large increases in populations of bark beetles, which result in massive pine tree mortalities over large forested tracts.

In 1985, forest ecologists conducted a study of an epidemic population of beetles infesting lodgepole pine in eastern Oregon. First, they established experimental plots and added nitrogen fertilizers to the forest floor. Then they compared the growth rates of trees on nitrogen-supplemented plots with growth rates on plots that received no additional nitrogen. The results indicated that nitrogen fertilizers stimulated the growth of trees and that greater numbers of beetles were required to kill these more vigorous trees. In areas where tree growth exceeded 100 grams of new wood per square meter of foliage, the beetles were unable to penetrate the bark because they were "pitched out" (see Figure 8.9B). Adult beetles are "pitched out" when the flow of tree sap exceeds their ability to penetrate past the sap flow and into the nutrient layers under the bark. When the sap is then exposed to air, it dries to form pitch and entombs the encroaching insect.

What about the amount of available nitrogen on the unfertilized plots? Was it adequate to support growth efficiencies capable of retarding beetle attacks? What conditions in the pine forest ecosystem would influence levels of available nitrogen? What options are available to forest managers that would help control epidemics of bark beetles? Such questions are typical of those raised by ecologists who address modern problems involving forest ecosystems. The answers can only be determined by carefully designed experiments in pine forest ecosys-

Figure 8.9

Trees in both lodgepole and ponderosa pine forests can be attacked by bark beetles. When sufficient numbers of mountain pine beetles infest a pine tree, it dies (A). However, if environmental conditions are optimal for the tree, they can produce enough sap to "pitch out" invading beetles. When red turpentine beetles infest ponderosa pines, whitish pitch tubes form where the beetles bore through the thick bark and into the sap (B).

tems. Some studies have been completed on bark beetles, and they provide insights into the interesting and complex relationships of ecosystem components that may initially seem to be unrelated.

Under natural conditions, low available nitrogen levels most likely result from extended periods of drought. During drought years, trees are stressed in a number of ways. Most significant, soil moisture decreases while soil temperature increases. These conditions cause a reduction in rates of nitrogen fixation and ammonification by soil bacteria and cyanobacteria. The drought conditions also place added stress on pine trees by reducing the amount of water available for photosynthesis, cell activities, and growth. The combined effects of these processes reduce the overall vigor of pine trees. As a result, they can no longer produce sufficient quantities of sap to protect themselves from beetle attacks.

In turn, prolonged drought has a positive effect on the parasitic beetles. Local beetle demes have access to vastly increased food supplies in the form of stressed trees. As a result, they are released from the major density-dependent factor, food, that normally controls the size of their regional population. Thus during drought years, their numbers can explode and lead to epidemics in pine forests.

Results of these studies appear to offer limited options for forest managers interested in maximizing tree growth while preventing bark beetle epidemics in pine forests. The best approach may be to increase the amount of nitrogen available to trees during drought years. This could be accomplished by fertilizing forests with expensive commercial fertilizers or by thinning the affected stands, which would reduce the competition for nitrogen among trees. By removing some trees, those remaining might be vigorous enough to withstand beetle attacks. However, thinned stands are more vulnerable to damage by high winds.

These studies of interacting functional ecosystem components—beetles and trees, consumers and energy capturers—have related numbers of bark beetles to concentrations of available nitrogen at the ecosystem level. In drought years, beetles play a major role in regulating the rate of total energy flow and material cycling in the forest by increasing the mortality of pine trees.

Species diversity in forest ecosystems is strongly affected by AAET. Characteristics of forest microhabitats are also influenced by abiotic factors such as elevation and biotic factors such as parasites. Studying the relationships between parasitic bark beetle populations and pine species offers insights into the complexity of regulating factors in forest ecosystems.

Grassland Ecosystems

Primary production, or rates of energy capture, of the North American central grassland region differs between *shortgrass* and *tallgrass* communities. Research has shown that the average **aboveground net primary production (ANPP)** in both the shortgrass and tallgrass prairies is correlated directly with the amount and distribution of precipitation (see Figure 8.10).

Figure 8.10

The relationship between aboveground net primary production (ANPP) and average annual precipitation was determined by taking measurements from 100 major land resource areas across the central grassland region of North America. The graph shows that locations with greater precipitation have a higher ANPP.

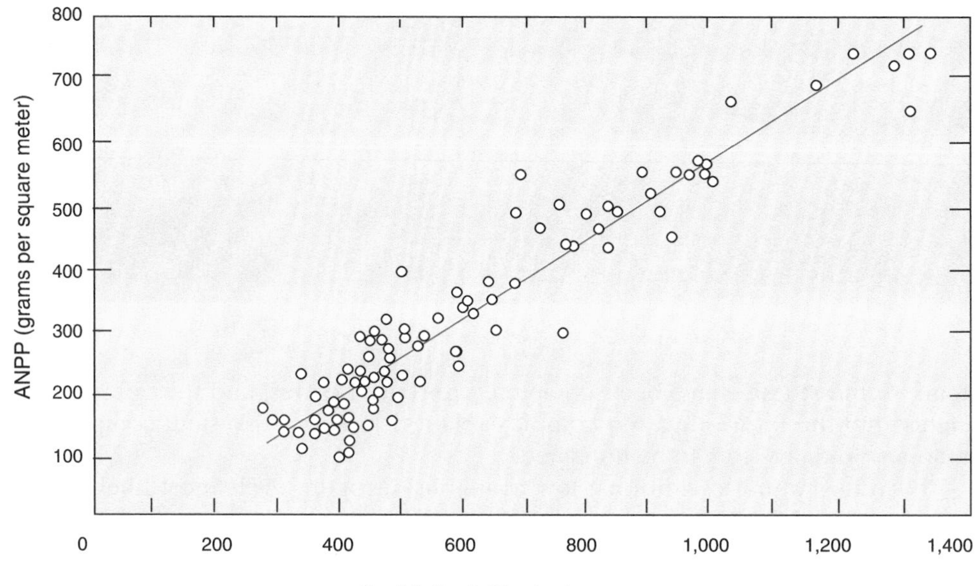

Major controlling factors affecting ANPP include water, carbon dioxide, nitrogen, and sunlight. Increased solar radiation leads to accelerated growth in the spring when light levels and day lengths are maximal. Plant growth rates change with longitude from east to west, but the total sunlight energy required for growth decreases with increasing latitude, as shown in Figure 8.11. For example, an ANPP **isopleth** (a line on a graph or map that connects points of equal or corresponding values) of 300 grams per square meter extends from northeastern North Dakota to southwestern Texas in years of average precipitation. During years of below-normal precipitation, the isopleth shifts eastward and extends from north-central Minnesota to southeastern Texas. In what ways does this range of annual precipitation affect community structure in these prairie ecosystems? Do other abiotic factors contribute substantially to community structure and function?

Prairie grassland ecosystems are largely dependent on adequate moisture, but soil type and nitrogen availability also influence grassland ecology. In fact, all

Figure 8.11

(A) Isopleths of ANPP range from 150 to 600 grams per square meter in the central grassland region of the United States during years of average precipitation. (B) During drought years, production decreases to ANPP isopleths of 100 to 500 grams per square meter.

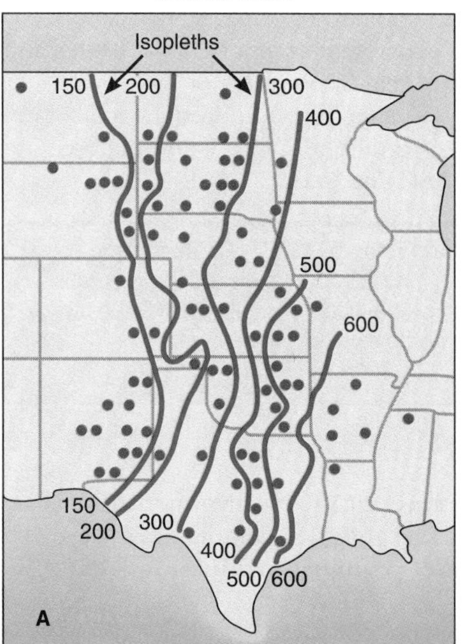

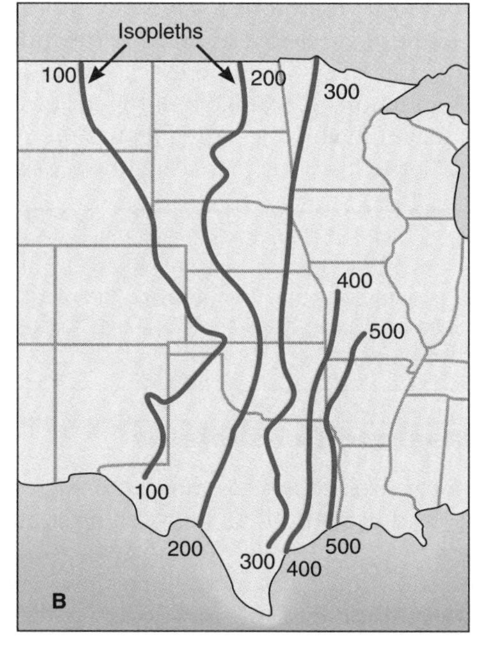

three factors are interrelated. One hypothesis suggests that ANPP in dry prairie ecosystems is less for fine-textured soils with high water-holding capacity (loamy soils containing clay, silt, sand, and organic matter) than for coarse-textured soils with low water-holding capacity (sandy soils). Also, rates of ammonification and nitrification are dependent on specific soil moisture and temperature combinations.

Variations in prairie community structure are influenced not only by annual precipitation and soil texture but also by factors that affect the amount of water available to the plant community over time. One such factor is the *rate of moisture loss*. Surface runoff, groundwater export, surface evaporation, and transpiration from vegetation all contribute to moisture loss from prairie communities.

Other factors that can affect the structure and function of grassland communities are grazing animals and wildfires. By consuming grass and carrying out their normal activities, grazing animals influence rates of energy flow and material cycling in the ecosystem. Wildfires remove surface litter and dead foliage, allowing increased amounts of light to reach the surface. Although some nitrogen is lost in smoke when wildfires sweep the prairies, the loss of surface litter results in elevated soil temperatures that stimulate nitrogen fixation and ammonification in the critical weeks of early spring when moisture is present.

Recent research implicates prairie wildfires as a positive secondary factor in established grasslands. The major impact of such fires is to increase the demand for materials essential for perpetuating prairie grassland communities. In effect, fires serve to accelerate the recycling of important nutrients and minerals. This type of research results has led some ecologists to advocate "let it burn" policies that have proved to be controversial (see the Focus on Scientific Explanations, "Yellowstone: Drought, Fires, and Carrying Capacity").

Grassland ecosystems illustrate a general ecological principle that is related to *scale of analysis*. As the scale becomes smaller, in moving from biome to ecosystem to community to population to deme to individual, the total number of variables that may influence an observed pattern becomes greater. Thus when the scale reaches a single plant on a prairie, an enormous number of variables can affect it. Each species' niche can be influenced by several levels in both biotic and functional ecosystem models: those that determine the type of biome, those that specify the ecosystem, those that regulate the rate at which energy flows and materials cycle within the community, those that determine the density of populations, and ultimately those that control the growth and reproduction of individual prairie plants.

A key measure of the condition of grassland ecosystems is ANPP, which is determined primarily by precipitation. ANPP isopleths are lines that describe a specific level of grassland productivity. In low-rainfall years, they shift from west to east, reflecting the influence of moisture on ANPP. Prairie grassland ecosystems are also affected by soil types, nitrogen availability, and the rate of moisture loss. All of these may be strongly influenced by grazing animals and wildfires.

Desert Ecosystems

To anyone who has visited a desert, it comes as no surprise that water availability is the chief factor that influences desert ecology. The amount of moisture and its geographical distribution throughout the year play leading roles in regulating the rates of energy flow and material cycling in desert ecosystems.

In the vast, hot deserts of southwestern North America, two species of plants are most common. The creosote bush *(Larrea divaricata)* and white burr sage *(Ambrosia dumosoa)* compete in both space and time in many desert communities (see Figure 8.12 on page 135). Research has shown that these two shrubs, both *perennials* (plants that live for many successive years), respond to

Focus on Scientific Explanations

YELLOWSTONE: DROUGHT, FIRES, AND CARRYING CAPACITY

During the summer of 1988, dramatic events in Yellowstone, our nation's oldest national park, resulted in ecological changes on a scale never before observed in this region. Hot, dry weather during the summer, coupled with a low snowpack formed during the previous winter, had significantly reduced the grassland productivity in low-elevation areas of the park. Precipitation during June, July, and August was only 36 percent of normal levels, and temperatures in June were 5°C higher than normal, leading to the development of high winds. Collectively, these conditions caused a 50 percent reduction in grass production on the summer wildlife range.

Historical factors also played a role in the story that was to unfold. From its establishment in 1872 until 1972, the National Park Service's forest management policy was to extinguish all natural fires within park boundaries. This management strategy, in conjunction with natural forest succession, resulted in a century during which the park experienced no fires of significant size. Consequently, extensive tracts of mature lodgepole pine forest characterized the Greater Yellowstone Area (GYA), and great quantities of accumulated deadwood lay on much of the forest floor.

In August, relative humidities were as low as 6 percent, and winds up to 96 kilometers (60 miles) per hour swept through the park. Thus the stage was set for the most extensive forest fire in the park's history (Figure 1). As shown in Figure 2, the August-September infernos burned about 45 percent of the area within Yellowstone Park. Figure 3 reveals that after these extensive fires, the park appeared as a mosaic of black, severely burned areas and brown, less burned area, interspersed with green, unburned forests.

What effect did these fires have on the park's famous wildlife? Predators such as black bears, grizzly bears, and mountain lions are dependent on populations of elk, deer, and other herbivores. Ultimately, then, the effects of the fire on predator populations are linked to the fire's impact on herbivore populations, with elk being the most important. During the summer of 1988, some 31,000 elk inhabited the park. After the fires, surveys of burned areas within the park revealed that less than 1 percent of the elk had been killed. However, the fires, combined with existing drought conditions and an extremely harsh winter that lay ahead, set the stage for a massive winterkill of the northern Yellowstone elk herd, the largest migratory elk herd in the world.

The northern herd, consisting of about 22,500 elk in 1988, typically spends most of the year grazing within Yellowstone Park. When winter comes, the population migrates northward to a range that straddles the park boundary, as

Figure 1 Extensive fires burned huge areas in Yellowstone National Park during the summer of 1988.

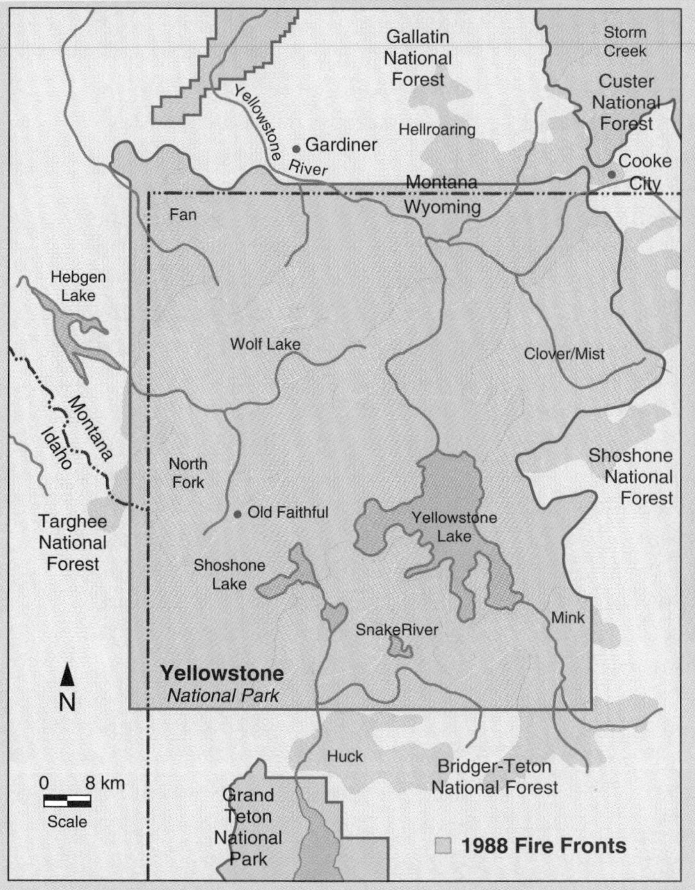

Figure 2 This map shows the major fire fronts in Yellowstone National Park during the summer of 1988.

Figure 3 This photograph shows the extensive areas that were burned during the 1988 Yellowstone fires.

Figure 4 The winter range of the northern elk herd includes parts of Yellowstone National Park and regions north of the park boundary.

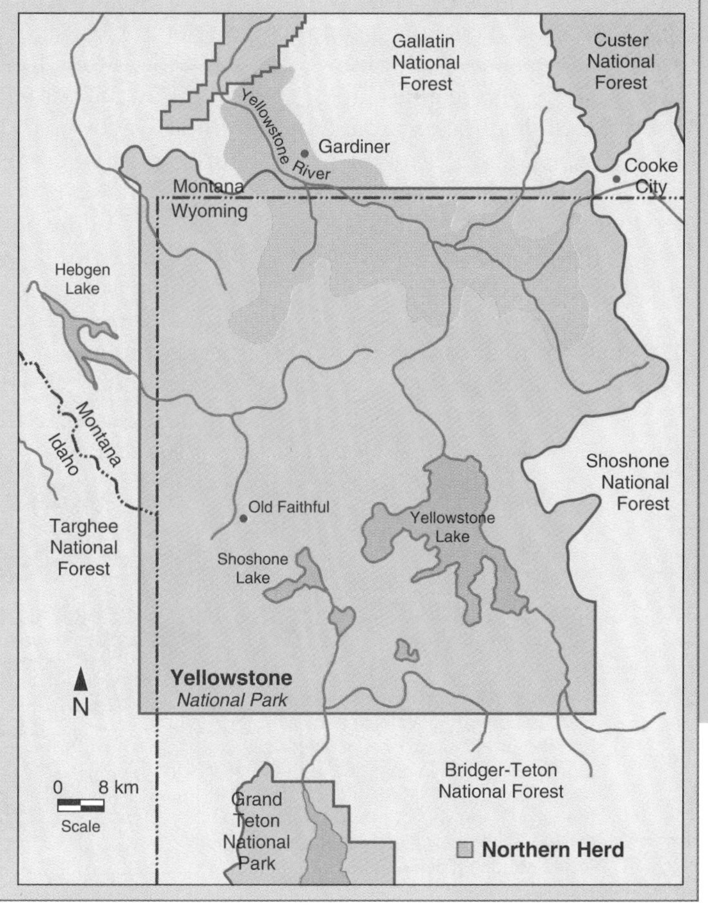

seen in Figure 4. The 1988 fires burned extensive, overlapping regions of the northern summer and winter ranges within park boundaries. Because of this loss, about 50 percent of the northern elk herd was forced to migrate out of the park four to six weeks earlier than usual in search of food.

The 1988 drought and fires, a deep snowpack during the winter of 1988–1989, and bitterly cold temperatures during the last two weeks of February 1989 resulted in high mortalities in the northern elk herd. The National Park Service estimates that 4,400 to 5,500 elk died of "natural" causes during the winter of 1988–1989. This mortality, added to the 2,350 elk harvested during a late season hunt near Gardiner, Montana, reduced the size of the population by about 35 percent.

Focus on Scientific Explanations

Wildlife biologists carefully analyze mortality data to determine answers to a number of important questions. For example, was mortality evenly distributed among all age classes, or were certain age classes of elk more likely to die? Did the loss due to hunting affect the age classes in the same way as the winterkill mortalities? What age classes are important for reproduction? Hunters generally harvest elk from the 3- to 8-year-old age class, the group most responsible for reproduction in the population. In contrast, surveys of dead elk on the northern winter range indicated that only about 17 percent of the carcasses were in this age class. Thus the northern elk herd appears to have the potential for a full recovery. However, new troublesome questions have emerged.

The poor condition of the northern elk herd during the winter of 1988–1989 resulted in a universal clamor for feeding the stressed elk. This humanitarian request seems to constitute a simple and effective solution, but how does it mesh with realities imposed by the natural environment? Ecologists recognize a number of problems related to artificial winter feeding programs for wild big game animals such as elk, including these:

1. Winter feeding can sustain excessive elk population sizes that will crash if feeding is terminated—that is, if the populations reach a size that cannot be supported by the natural environment.
2. Elk, at excessive densities, degrade the natural plant community, which can in turn affect the entire ecosystem.
3. Winter feeding can create a long-term dependency on feeding locations that may alter or stop natural migrations to normal winter ranges.
4. The concentration of large numbers of animals in a limited area results in a high risk of disease in the herd.
5. The usual pattern of winter mortality—the loss of a certain percentage of young, old, and diseased animals—that normally regulates the population size does not occur, and excessive elk numbers result.

For these reasons, the Montana Fish and Game Commission decided not to feed elk from the northern Yellowstone herd.

Can the natural summer and winter ranges provide long-term support for an elk population of the size that existed in the summer of 1988? Will the Yellowstone fires of 1988 result in a greater or smaller summer range in the future? After this major fire, natural succession is expected to take place over the next 300 years and result in a mature forest of lodgepole pine, subalpine fir, and Engelmann spruce, depending on elevation. Previous research indicates that during early successional stages, wildlife range increases greatly as grasses and shrubs become established after the fire. However, as the forest begins to grow and mature, the canopy closes, and the amount of food for herbivores decreases (see Figure 5). Consequently, the **environmental carrying capacity (ECC)** of Yellowstone Park can be expected to change over time. If we consider the ECC as being the density of an animal deme that can be sustained during each stage of succession, then the total number of elk that both the summer and winter range can support will change as the forest undergoes natural succession. In the short term, the ECC of the summer range is expected to be high. (Why?) However, wildlife ecologists are now considering the answers to some difficult questions: Will the ECC of the summer range within the park greatly exceed the ECC of the winter range? If so, what will be the consequences for the northern elk populations? Will winterkill continue to be the major density-dependent regulating factor for the northern elk herd in Yellowstone Park in the foreseeable future? The answers to these and other questions will evolve in the coming decades.

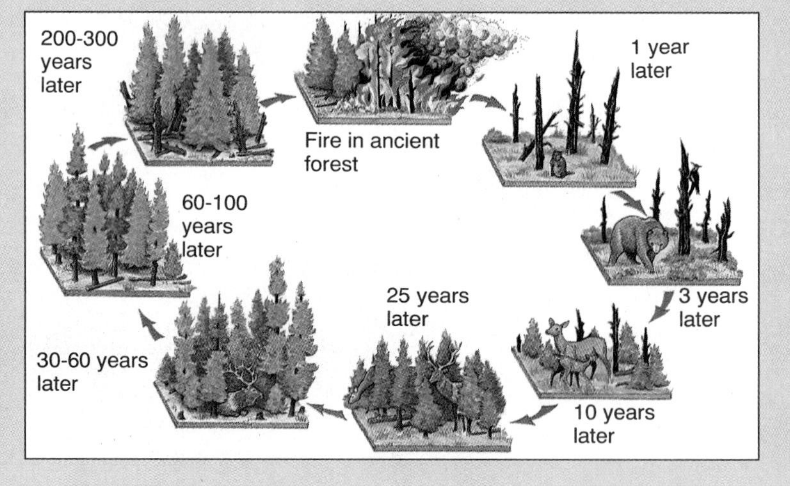

Figure 5 After the major fire in the GYA, secondary successions will begin, and the progression of successional stages will result in changes in the environmental carrying capacity for elk in the area. Food will be more abundant in the early stages than later, when the forest canopy closes.

Figure 8.12

The growth pattern of the creosote bush (A) is quite different than that of the burr sage (B). Competition for water results in the uniform distribution shown here for the creosote bush. Burr sage tends to grow much closer together if water is available.

the harsh environment in different ways over the course of time. Burr sage is deciduous and loses its leaves each year, but the creosote bush is evergreen. Burr sage usually forms a continuous pattern of distribution; it is spread more or less evenly over the desert. In contrast, creosote bushes form a dispersed or clumped pattern. The dispersed pattern of the long-lived creosote bush results from competition for water among individuals that occurs during the critical period when they become established. Over time, certain individual creosote bushes will have a competitive advantage in obtaining water. This leads to the elimination of some plants in the area, and the resultant dispersed pattern eventually appears. In areas of older, established creosote bushes, competition for water is reduced due to the spacing arrangement of the surviving plants.

In contrast, burr sage is a short-lived deciduous shrub that can respond quickly to short-term climatic changes. When water is available, it grows close together, but as water becomes limiting, the plants compete, and only a few will survive. They are able to propagate numerous offspring because they produce many seeds under favorable conditions. During periods of water stress in areas where both burr sage and creosote bush exist, competition occurs between the two species and among burr sage individuals, but not among the older creosote bushes.

In areas of the Sonoran Desert, many cacti cannot survive until after early successional plants, including species of palo verde, become established. Why is it essential for palo verde to become established in an area before cacti can survive? Basically, cacti and other desert perennials cannot germinate and grow except in the shade of mature palo verde or other pioneer plant species. The microenvironment under these "nurse plants" is enriched with leaf and twig litter, and the palo verde also provides some protection from the hot sun (see Figure 8.13).

Surprisingly, it has been shown that plant productivity in hot deserts can approach that of a young coniferous forest if the data are expressed as ANPP per amount of living plant foliage. The differences in total annual productivity between desert communities and young coniferous forests are due to differences in the length of the growing season. In coniferous forests, trees grow throughout the entire year, but in deserts, growth occurs only for a short time, after rare periods of rainfall.

Consumers in deserts are adapted for tolerating long periods of water stress through specializations in their form, function, and behavior. Many desert rodents have the most efficient kidneys among mammals, excreting virtually no water in their urine. Also, desert mammals, reptiles, and birds behave in ways that limit their loss of water. For example, many desert animals, like those shown in Figure 8.14, forage during the early morning or night and spend the hot days resting in burrows or other areas removed from the sun.

Figure 8.13

Palo verde often acts as a nurse plant that provides the shade and nutrients necessary for the growth of cacti such as the saguaro shown here.

Rates of decomposition in desert ecosystems are controlled primarily by moisture, with lesser contributions coming from secondary biotic factors. Bacterial decomposition is limited to periods of high soil moisture following rains. Thus for most of the year, little nutrient addition occurs through bacterial decomposi-

Figure 8.14

Many animals are adapted to live in deserts. Sidewinder rattlesnakes (A) feed on small desert rodents, and the greater roadrunner (B) feeds on small snakes, lizards, insects, and scorpions.

tion. This moisture-limited bacterial decomposition in the soil intensifies the importance of an indirect form of decomposition that is carried out by the insect community. Research has shown that termites consume up to 50 percent of the leaf fall in the Chihuahuan Desert and over 90 percent of the deadwood on the ground in the Sonoran Desert. Since they harbor bacteria in their digestive systems, they contribute bacterial decomposition products, in their excrements, to these harsh, water-stressed environments.

Water availability regulates virtually all aspects of organism growth, distribution, and survival in desert ecosystems. The structure of plant communities and the existence of different species are largely determined by adaptations related to obtaining water. Desert animals have many unique structural, functional, and behavioral adaptations that enable them to survive in hot, dry conditions. Nutrient cycling and decomposition are also largely regulated by moisture.

In this chapter, we have identified some of the new ecological concepts—trophic guilds, demes, functional ecosystem models—that now guide research. Ecological studies of terrestrial ecosystems will likely continue to be conducted from both biotic and functional viewpoints. Both approaches have been fruitful in the past, but the functional model of determining environmental factors that control the rate of energy flow and the movement of materials within ecosystems now appears to be the more promising approach for future studies of terrestrial ecosystems.

Summary

1. Ecosystems can be understood from either a biotic or a functional viewpoint. In a biotic ecosystem model, organisms are partitioned into trophic guilds according to the trophic levels from which they obtain their energy at any particular time. In a functional ecosystem model, organisms are considered to be important as regulators of both the rate of energy flow and the cycling of materials.
2. Changes that occur in demes, local interbreeding species populations, are now thought to constitute the major driving force of succession. The densities of demes are affected by both density-independent and density-dependent environmental factors. Density-independent factors are usually abiotic, whereas density-dependent factors are either biotic or abiotic.
3. The interactions of arctic snowshoe hare, a prey species, and Canada lynx, a predator species, involve a number of regulating factors, both density-dependent and density-independent.
4. Variations in precipitation and average annual evapotranspiration (AAET) are the major environmental factors that determine biome types across North America. AAET is influenced by topography and proximity to large bodies of water.
5. The case studies of beetles and pine species, grassland communities, and desert ecosystems illustrate the complex biotic and abiotic regulatory factors that operate on different ecosystems and affect rates of energy flow and material cycling.

Review Questions

1. In what ways do biotic and functional ecosystem models differ?
2. What trophic levels are used in biotic ecosystem models?
3. What are trophic guilds? Why are they now preferred over trophic levels?
4. What is a deme, and where does it fit into a biotic hierarchy?
5. How do density-dependent and density-independent factors affect deme densities?
6. What factors regulate population levels of arctic hare?
7. What environmental factors affect tree species richness in North America?
8. How does available moisture relate to the ability of bark beetles to kill pine trees?
9. How is ANPP related to primary production of grassland ecosystems?
10. How do the growth habits of the burr sage and the creosote bush differ? Why?

Essay and Discussion Questions

1. Compare and contrast the value of studying ecosystems from both a biotic and a functional viewpoint.
2. Describe a deme that exists in your area, and list some density-dependent and density-independent factors that may affect it.
3. What lessons can be drawn from the history of ideas concerning the population densities of the Canadian lynx and arctic snowshoe hare?

References and Recommended Reading

Barbour, M., and W. Billings, eds. 1988. *North American Terrestrial Vegetation*. Cambridge: Cambridge University Press.

Burns, T. 1989. Lindeman's contradiction and the trophic structure of ecosystems. *Ecology* 70:1355–1362.

Caswell, H. 1988. Theory and models in ecology: A different perspective. *Ecological Modeling* 43:33–44.

Currie, D., and V. Paquin. 1987. Large-scale biogeographical patterns of species richness of trees. *Nature* 329:326–327.

Diamond, J., and T. Chase, eds. 1986. *Community Ecology*. New York: HarperCollins.

Lindeman, R. L. 1942. The trophic-dynamic aspect of ecology. *Ecology* 23:399–418.

Ludwig, J., G. Cunningham, and P. Whitson. 1988. Review: Distribution of annual plants in North American deserts. *Journal of Arid Environments* 15:221–227.

McIntosh, R. 1989. Citation classics of ecology. *Quarterly Review of Biology* 64:31–49.

Morris, D. W. 1987. Tests of density-dependent habitat selection in a patchy environment. *Ecological Monographs,* 57(4):269–281.

O'Neill, R., R. DeAngelis, J. Waide, and T. Allen. 1986. *A Hierarchical Concept of Ecosystems*. Monographs in Population Biology 23. Princeton, N.J.: Princeton University Press.

Roughgarden, J., R. May, and S. Levin, eds. 1989. *Perspectives in Ecological Theory*. Princeton, N.J.: Princeton University Press.

Sala, O., W. Parton, L. Joyce, and W. Lauenroth. 1988. Primary production of the central grassland region of the United States. *Ecology* 69:40–45.

Van Cleve, K., F. S. Chapin III, C. T. Dryness, and L. A. Viereck. 1991. Element cycling in taiga forests: State-factor control. *BioScience* 41:78–88.

CHAPTER 9

Aquatic Ecosystems

Chapter Outline

Reading Questions

1. What are the principal types of freshwater ecosystems?

2. What is the nature of regulatory mechanisms that operate in different freshwater ecosystems?

3. What are the major classes of marine ecosystems?

4. What factors account for the differences among various marine ecosystems in their capacities for supporting life?

Aquatic ecosystems are found within both freshwater and marine bodies of water. **Freshwater ecosystems** comprise *lotic* systems, characterized by free-flowing streams and rivers, and *lentic* systems, which are ponds and lakes. **Marine ecosystems** include the oceans of Earth, bays, and estuaries (areas where fresh water from rivers mixes with salt water of the ocean).

FRESHWATER ECOSYSTEMS

The extension of ecological concepts to aquatic systems has occurred concurrently with, but often isolated from, applications to terrestrial ecology. There have been several reasons for this disconnected advancement. Perhaps the most important has been the publication of relevant research in separate scientific journals, a practice that tended to isolate the two fields of ecology. However, many of the cardinal ecological concepts have now been applied to both broad areas of ecology. These concepts include Forbes's microcosms, Shelford's laws of toleration, Lindeman's trophic dynamics, and the Odums' ecosystem energetics (see Chapter 7).

Many conceptual differences between aquatic and terrestrial ecology originated because of the dissimilarities between the two physical environments. The availability of water and abiotic factors that regulate evapotranspiration are critical elements in defining the structure of terrestrial ecosystems. Obviously, neither of those are serious limiting factors in aquatic communities. Instead, amounts of light energy and nutrient availability are the critical limiting factors in most aquatic ecosystems.

Lotic Ecosystems

Major environmental factors that influence free-flowing lotic ecosystems vary greatly between small headwater streams (those at the highest elevations in drainage basins) and huge rivers such as the Columbia or the Mississippi (see Figure 9.1). Water volumes and **gradient variations** resulting from the slope of the drainage basin create corresponding changes in the communities that exist within streams and rivers of different size.

Classification of Lotic Ecosystems by Size

In the stream and river classification system now widely used, small headwater streams are categorized as *first-order streams. Second-order streams* originate when two first-order streams meet. *Third-order streams* arise at the junction of

A

B

Figure 9.1

Streams and rivers are classified by orders ranging from 1 to 12. First-order streams originate at the top of drainage basins, usually from melting snow in the deep shade of a forest community. They flow into larger streams (A). The order of streams and rivers continues to increase until they reach the size of huge rivers like the Columbia River shown here (B).

two second-order streams, *fourth-order streams* at the junction of two third-order streams, and so on up to *twelfth-order rivers,* which are the size of the Columbia or the Mississippi by the time they reach the ocean.

Aquatic ecosystems exist in both freshwater and marine environments. While different ecological principles have been used to explain terrestrial and aquatic ecosystems, some concepts, such as functional ecosystem models and trophic guilds, have been applied to both. Free-flowing lotic ecosystems are classified according to gradiant variation and size.

A Functional Model of Lotic Ecosystems

What approaches are used by modern ecologists in studying aquatic ecosystems? A useful conceptual model for describing energy flow in lotic systems developed in 1977 by K. W. Cummins is diagramed in Figure 9.2. This model allowed aquatic ecologists to identify and investigate the major routes through which nutrients move in biota that inhabit lotic ecosystems. It helped provide answers to the question, How do different organisms survive in lotic ecosystems? Cummins's model consists of a functional classification of the community that is based on

Figure 9.2

Functional models of lotic ecosystems have been useful for designing ecological studies. Aquatic insects are classified on the basis of their mode of nutrient acquisition. Fallen leaves and other terrestrial materials enter small-order streams, where bacterial and fungal degradation begin. Shredders feed on the decomposing leaves, a major component of coarse particulate organic matter (CPOM). Insect fecal pellets and other organic matter form fine particulate organic matter (FPOM), which flows downstream. Collectors feed on the FPOM, and predators feed on both shredders and collectors. Farther downstream, as light penetration increases, diatoms and other microproducers and mosses, algae, and aquatic plants—the macroproducers—grow and become food for increasing numbers of scrapers and other grazers.

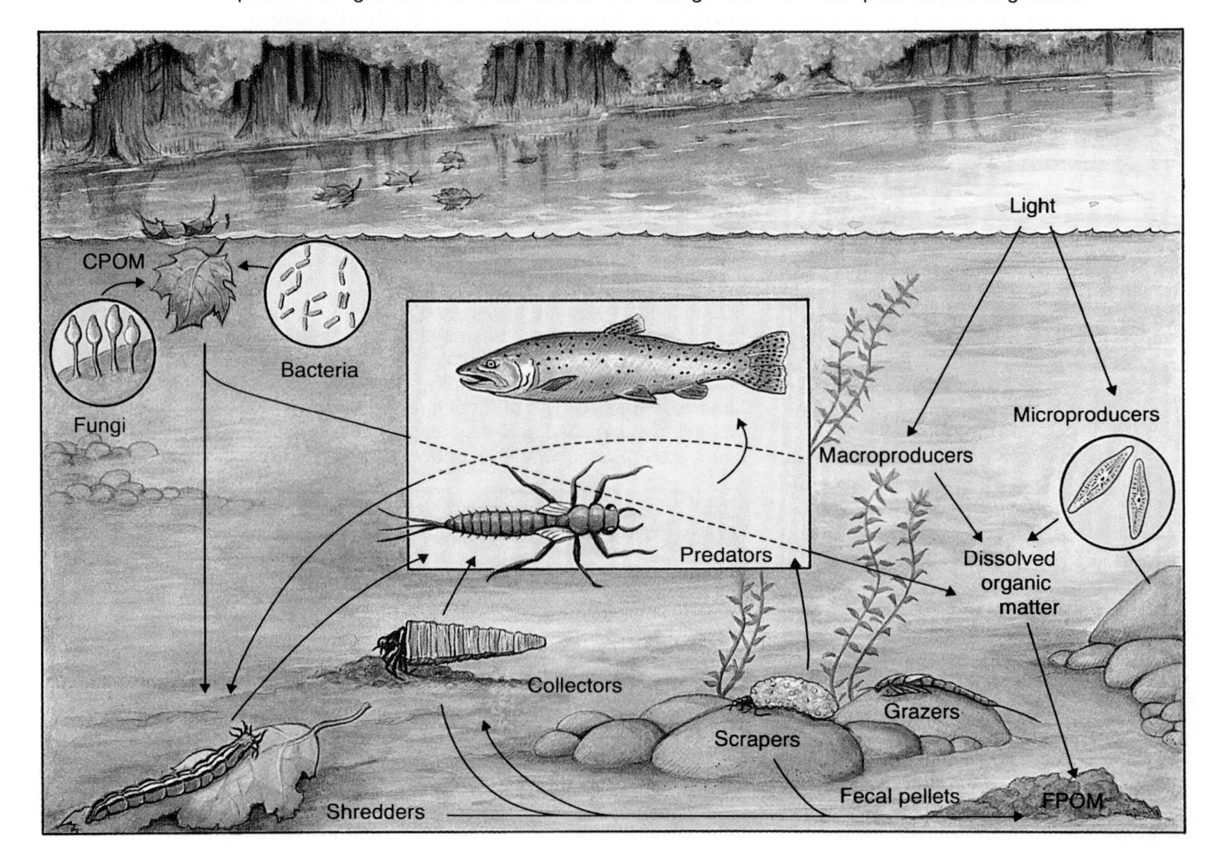

the niches of organisms within the trophic structure of streams and rivers. It includes the light energy used in primary production by *microproducers* (diatoms) and *macroproducers* (mosses and algae) living in the stream and terrestrial products (for example, leaves) and plant debris that enter the stream. Rate-regulating decomposers are bacteria and fungi, and the consumers consist of aquatic insect shredders, collectors, scrapers, and numerous species of insect predators and fish.

Two different components form the base of food chains in lotic systems of small and large sizes. Small headwater streams usually occur in steep terrain, near the head of drainage basins. Their **riparian zones** (the areas along the banks that are influenced by water flow) are bordered by terrestrial producers such as deciduous willows, birch or alder trees, or conifers like pine or fir trees. These trees often grow out over the stream and provide deep shade on the stream's surface. In shaded streams, photosynthesis by aquatic plants and algae is minimal, so primary production in these aquatic environments is low. However, the overhanging foliage contributes substantial quantities of organic matter to the streams in the form of leaves, bark, twigs, and occasionally larger parts of trees. Thus in headwater streams, most aquatic food chains begin with this *coarse particulate organic matter* (CPOM).

Insects familiar to anyone who has walked along a stream include mayflies, stoneflies, caddisflies, and dragonflies. These are adults that have emerged to breed; they will lay eggs in the stream that will hatch and become the next generation, as shown in Figure 9.3. Most stages in the life cycle of aquatic insects occur in streams or rivers.

Cummins classified the aquatic insects that consume CPOM as **shredders**. Shredders generally obtain most of their nutrients from digesting the bacteria and fungi that first colonize leaves and other organic debris as it enters the stream. Shredder fecal pellets (undigested foods that pass through their digestive system) constitute much of the *fine particulate organic matter* (FPOM) that is taken up by insect **collectors**. The importance of other biotic elements in Cummins's model varies in lotic systems of different sizes. The changes in energy and nutri-

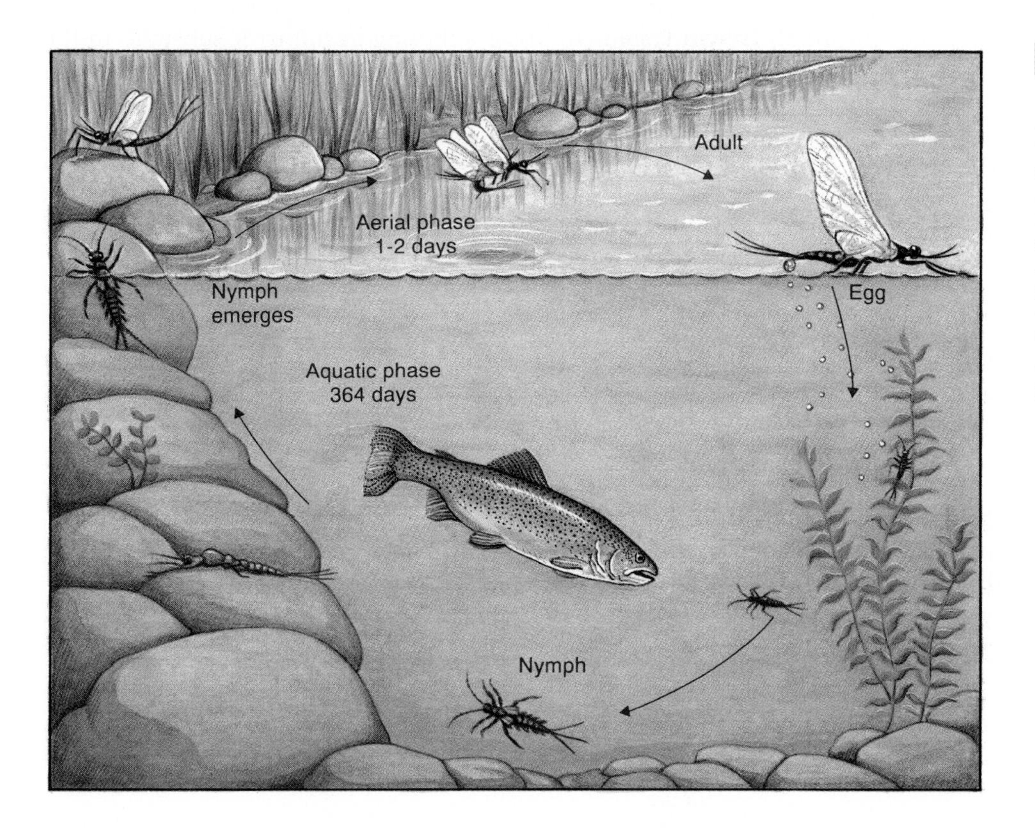

Figure 9.3

Mayfly life cycles are typical of aquatic insect species. Adult females deposit their eggs at the surface of the water after mating. The eggs hatch into tiny nymphs that begin to feed and grow. Mayflies progress through a number of successively larger nymphal stages. After about one year, mature nymphs crawl out of the water onto rocks along the shore, and adults emerge from the last nymphal stage. The adults form mating pairs that reproduce and complete the life cycle.

ent flow that occur as water moves toward the ocean are described in Figure 9.4, beginning with streams and ending with large-order rivers.

Small-Order Streams

One important indication of how factors influence energy flow in aquatic systems is the *production to respiration ratio* (P/R). A P/R of greater than 1 indicates that productivity—the amount of new organic material produced by photosynthesis—exceeds the amount of organic material used in the activities (respiration) of all organisms in the community. A P/R of 1 symbolizes a precise balance, and if P/R is less than 1, the level of community productivity alone is inadequate to sustain the total ecosystem.

The community structure of small streams is influenced to a large extent by the presence of CPOM. Because of low light levels due to overhanging plants, primary production within the stream provides little food for invertebrate **grazers**, a group composed primarily of insect nymphs or larvae that consume algae on rocks or other surfaces. The "herbivore" trophic guild consisting of all shredders, collectors, and grazers provides enough energy for only a small biomass of predators, as shown in Figure 9.4. With little photosynthesis occurring in the stream, and because of the high levels of community respiration due to CPOM received from terrestrial riparian plants, the P/R would be less than 1.

Mid-order Rivers

As smaller streams combine to form larger fourth-, fifth-, and sixth-order rivers, width and depth increase, but gradient and current velocity decrease. Rivers are usually distinguished from streams between the third and fourth orders.

Riparian vegetation does not cover entire expanses of mid-order rivers, so greater amounts of sunlight are received at their surfaces. Decomposition activities in small-order streams farther up a drainage basin result in increased concentrations of carbon dioxide, nitrogen, phosphorus, and other nutrients that move downstream. Consequently, the conditions required for plant growth and photosynthesis are often ideal in order 4, 5, and 6 rivers. Because of small quantities of terrestrial-generated CPOM and increased levels of primary production by vascular plants and **periphyton** (plants or algae adhering to the rock substrate in the bottom of the stream), significant changes occur in the proportions of herbivore types. The relative number of collectors remains about the same as in small-order streams, but grazers increase and shredders decrease in number and importance. The herbivore community in mid-order rivers supports about the same percentage of predators. With increased photosynthesis, the P/R increases and becomes greater than 1. Thus organisms in these mid-order rivers generate more biomass than they consume, and the excess organic products can be exported to the larger rivers into which they flow.

The most diverse biological communities in lotic systems are normally found in mid-order rivers. These larger lotic systems provide a wide range of habitats for different forms of collectors, grazers, a few shredders, and insect predators. In turn, this abundance of insect fauna can support a greater variety of fishes, including many warm-water species in rivers at low elevations, and sculpin and trout species in rivers at higher elevations. Mid-order streams and rivers are the most highly productive for trout species. Many "free stone rivers," those flowing over gravel and rocks, like the river shown in Figure 9.5, on page 144, provide the best trout fishing because they include areas of maximum primary productivity. This leads to greater insect diversity and numbers and hence to larger trout populations.

Large-Order Rivers

Large rivers, in the orders from 7 through 12, become deep and wide. Because of depth and an increased load of light-filtering organic matter and inorganic sedi-

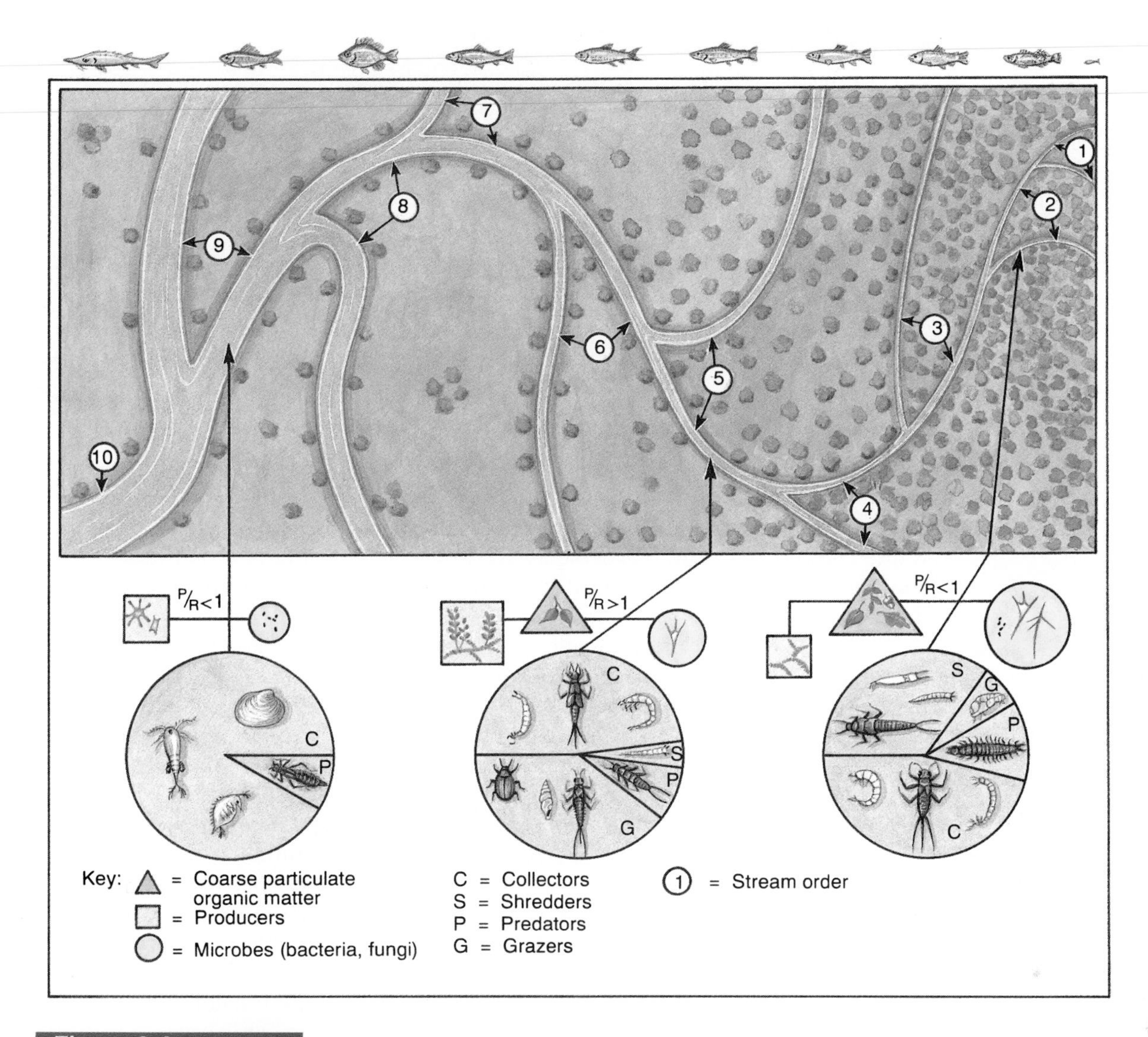

Figure 9.4

When first-order streams flow together, they form second-order streams, which combine to form third-order streams. This pattern continues, eventually forming tenth-order or larger rivers. The insect community changes as the stream order changes. Tree canopies cover many first-, second-, and third-order streams. In these streams, the insect community is represented by large proportions of shredders and collectors, few grazers, and only about 10 percent predators. Farther down the drainage basin, the streams become rivers, and the insect community changes. Here grazers increase and shredders decrease in abundance, but the collectors and predators remain in about the same proportions as in the smaller streams farther up the drainage basin. In tenth-order and larger rivers, the insects are usually represented by collectors and about 10 percent predators. The production-to-respiration ratios (P/R) are usually less than 1 in smaller-order streams because the tree canopy restricts light penetration to the water surface, limiting in-stream photosynthesis. As the canopy opens up in mid-order streams and rivers, P/R becomes greater than 1 due to increasing light penetration, resulting in greater amounts of in-stream photosynthesis. The depth of large rivers prevents light penetration to depths where aquatic plants and benthic algae could live, which results in a P/R of less than 1.

ments received from upstream, the light energy required for photosynthesis does not penetrate the entire water column in these systems (see Figure 9.6). Since primary production is limited because of insufficient light, the consumer community, now composed of only collectors and predators, uses energy at rates higher than it is fixed during the total primary production. This results in a P/R of less than 1 in most large-order rivers. Consequently, large-order rivers are not productive areas for aquatic algae or plants; the insect population consists mainly of collectors, few predators thrive, and the fish tend to be species like sturgeon or catfish that feed on the organic sediments present on the river bottom.

Figure 9.5

The rock substrate of a free stone river maximizes the habitat for aquatic producers and insect life, which can support large fish populations.

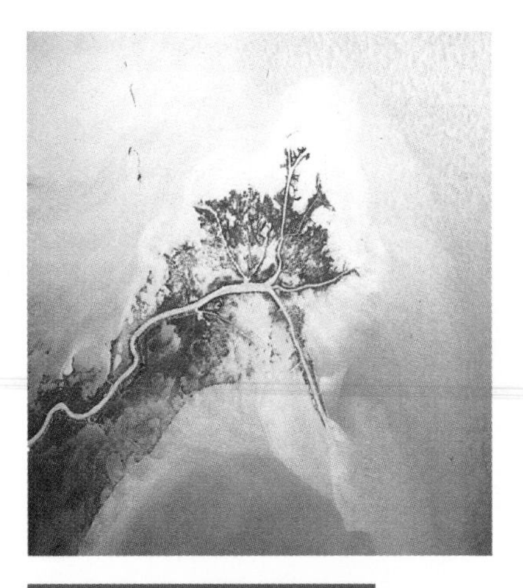

Figure 9.6

This classic view of the Mississippi River shows an example of a "crow foot" delta. The distribution of river-borne sediments is clearly evident in the Gulf of Mexico.

Drift

A unique feature of lotic systems stems from the fact that water flows ever seaward. The moving water carries dissolved materials, suspended particulates—including CPOM and FPOM—and living insects. These collectively constitute the **drift**.

The life cycle of most aquatic insects begins when the adults emerge, mate, and then fly upstream to lay their eggs. The eggs hatch in the stream and develop into nymphs. They drift downstream as they grow and increase in size. This drift usually occurs at night, and it is related to a number of self-sustaining features of the stream community. Normal spring floods that occur in most drainage systems without dams scour the bottoms of streams and remove significant portions of the flora and fauna. However, since adult insects fly upstream to lay their eggs, these flood-stressed systems are repopulated on an annual basis. Most trout species also swim upstream to spawn. Thus when the eggs hatch, young fish occupy an area populated with small nymphs, which are their most important insect food source. As both fish and insects grow larger, they tend to move downstream together.

Drift not removed during its seaward movement will ultimately settle and be deposited in estuaries near the mouth of the river system. This organic material adds to the biological richness of these unique estuarine ecosystems.

Energy flow in lotic ecosystems can be defined using a functional model. The principal organism components of the model are producers; insects that are classified as either collectors, shredders, predators, or grazers; and microbes. Other significant elements of the model include light energy received, different types of organic matter, and P/R. All of these change as a function of increasing stream order.

Lentic Ecosystems

Lentic ecosystems consist of standing bodies of fresh water, including all lakes and ponds. Four major zones of life exist in lentic ecosystems as shown in Figure 9.7. The **littoral zone** includes the area of transition between the riparian zone along the shore and water extending to a depth of about 10 meters. It is usually the most productive zone and contains the greatest diversity of aquatic life. The **limnetic zone** consists of the portion of the lake, excluding the littoral zone, in which sufficient light energy penetrates to depths where photosynthesis can occur. The **profundal zone** is the deeper portion of the lake where light does not penetrate, and the **benthic zone** is the bottom of the lake, where sediments accumulate and most bacterial decomposition occurs.

Lentic systems differ fundamentally from lotic systems because the water does not move in a constant-elevation gradient-induced flow. However, water in

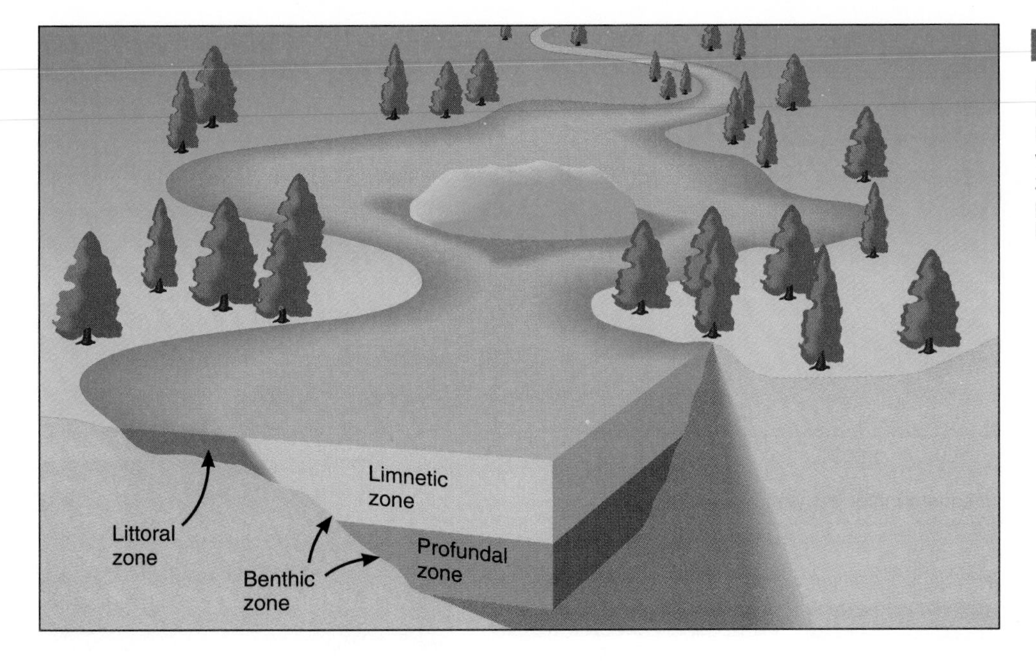

Figure 9.7

Life zones present in lakes and other lentic systems include the near-shore littoral zone; the limnetic zone, through which light will penetrate; the benthic zone, or bottom of the lake; and the profundal zone, which is too deep for light to penetrate.

lentic systems can move, depending on such factors as the amounts of water received from the drainage basin and the rates of discharge to groundwater, streams, or rivers. In addition, wind-generated waves can circulate surface water to variable depths, depending on the wind velocity and duration. Finally, surface water actually sinks to the bottom of a pond or lake when it reaches its maximum density at 4°C. (See the Focus on Scientific Explanations in Chapter 3.) This occurs in lakes where water temperatures cool in the fall before ice forms and again in the spring when the water warms after the ice melts.

Primary production in temperate lakes has a marked seasonal aspect, with photosynthesis being highest in the late spring, after the surface water reaches its greatest density and sinks. As the cold surface water sinks, it is replaced by deep water from near the lake bottom in a process called **turnover**. The water brought from the depths is rich in nitrates, phosphates, and other nutrients because of bacterial decomposition of organic material that occurred on the lake bottom throughout the winter. The nutrient-laden bottom waters circulate to the surface in late spring, and sunlight continues to increase until the summer solstice.

During the summer, primary productivity will often become limited by either decreased concentrations of available nitrogen or phosphorus. However, the total amount of primary productivity is dependent on the growth of algae and aquatic plants, which is in turn related to the quantities of available nutrients and a number of biotic factors.

Trophic Classification of Lakes

In temperate regions, lakes are often classified on the basis of their relative *net productivity* (P_n), the amount of organic material produced by plant photosynthesis minus the amount used by plant respiration. Figure 9.8A describes this classification system. **Oligotrophic lakes** like the one shown in Figure 9.8B are the least productive, with P_n ranging from less than 0.1 to about 10 grams of carbon per square meter per year ($gC/m^2/y$). **Mesotrophic lakes** (Figure 9.8C) range from about 10 to 70 $gC/m^2/y$, and **natural eutrophic lakes** (Figure 9.8D) from 70 to 400 $gC/m^2/y$. **Cultural eutrophic lakes** can produce 1,000 or more $gC/m^2/y$ because excessive nutrients generated by human activity are introduced into the system.

The productivity of oligotrophic lakes is usually nutrient-limited. Mesotrophic lakes are more productive and often contain larger populations of

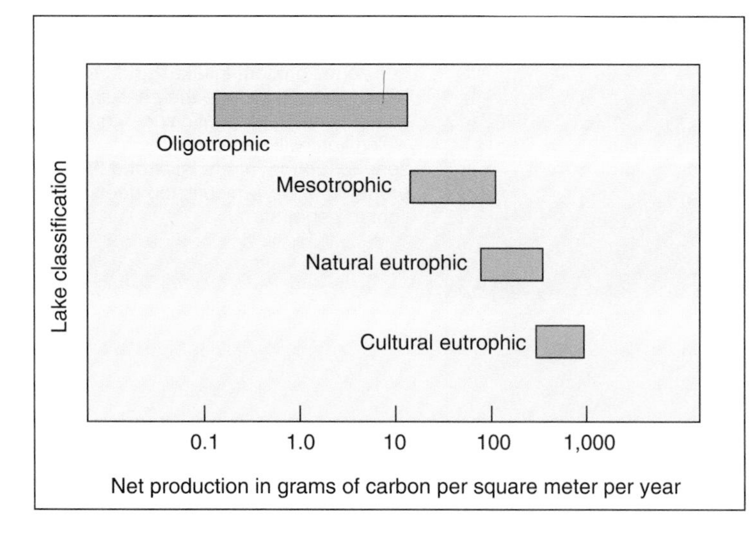

A

B

C

D

Figure 9.8

Lakes are often classified on the basis of net production (A), measured in grams of carbon per square meter per year. Oligotrophic lakes (B) are the least productive, ranging from about 0.1 to 10 grams, mesotrophic lakes (C) range from 10 to 70 grams, natural eutrophic lakes (D) from about 70 to 400 grams, and cultural eutrophic lakes from 400 to 1,000 grams or more.

Figure 9.9

When eutrophication occurs in lentic ecosystems, the concentration of oxygen in the water may decrease, resulting in the death of many fish like those shown here.

fish. Natural eutrophic lakes can accommodate the growth of algae and aquatic plants to such levels that the herbivore community cannot consume them all. The excess eventually becomes organic matter that is decomposed by bacteria. This decomposition is an aerobic process that may lead to depletion of oxygen in the lake water. Occasionally, oxygen levels become so low that fish can no longer survive. Massive fish mortalities caused by oxygen depletion can occur in natural eutrophic lakes, and they are very common in cultural eutrophic lakes (see Figure 9.9).

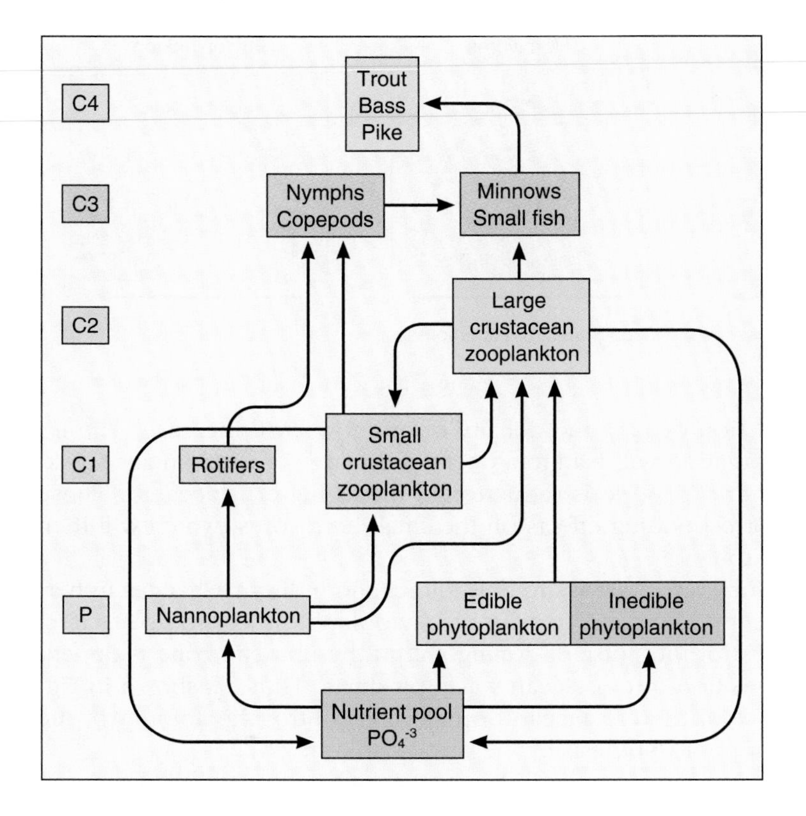

Figure 9.10

Five trophic guilds can be used as a functional ecosystem model for the organisms comprising the food web of a lake. Nutrients and energy are fixed during photosynthesis by the producers, which are consumed by first-order consumers, including rotifers, and both small and large crustacean zooplankton. Large zooplankton (second-order consumers) feed on small zooplankton as well, but when the large zooplankton are young, they can be eaten by mature small zooplankton species. Demes at the C1 and C2 level become food for insect nymphs, copepods, minnows, and other small fish (including the young of larger fish species). Third-order consumers in level C3 are consumed by mature trout, bass, or pike.

Biotic Regulation of Lentic Ecosystems

Recent research in the field of limnology (the study of lentic ecosystems) has revealed some factors that profoundly affect lake productivity, material cycling, and food web interactions. In most lakes, phosphate (PO_4^{-3}) concentrations limit primary production. Thus factors that regulate PO_4^{-3} availability control and shape the lentic community. Low nitrate concentrations can also be limiting, but if sufficient PO_4^{-3} is present, nitrogen-fixing cyanobacteria like *Oscillatoria* or *Anabaena* will often "bloom," which leads to increases in the available nitrate concentrations.

A functional model of the organisms in a lake food web can be described using five trophic guilds as illustrated in Figure 9.10. The *producers* (P) are first, consisting of nannoplankton (microscopic algae between 2 and 20 micrometers in diameter) and larger edible and inedible phytoplankton. **First-order consumers** (C1), the herbivores, include rotifers, and small and large crustacean zooplankton. However, large crustacean zooplankton also feed on phytoplankton, small crustacean zooplankton, and the rotifers, requiring that they be placed in two trophic guilds, C1 and **second-order consumers** (C2). **Third-order consumers** in C3 trophic guilds include insect nymphs and other small anthropods, small minnows, and young trout, bass, or pike that feed on the C1 and C2 trophic guilds. Finally, these minnows and small fish are eaten by **fourth-order consumers** (C4), larger trout, bass, or pike, depending on the lake system.

Cascading Trophic Interactions

A number of studies have shown that changes in the relative abundance of organisms in different trophic guilds can regulate the rate of PO_4^{-3} cycling, which in turn serves to regulate the entire community through a series of interrelated reactions. The discovery of these related effects has given rise to a theory of **cascading trophic interactions (CTI)**. Like all theories, it allows scientists to make predictions that can be tested in carefully designed experiments.

To illustrate the theory, assume that the number of trout, bass, or pike is depleted by human harvest with nets, traps, or hook and line. What will happen?

Figure 9.11

When the deme densities of fish in level C4 decreases, deme densities in guild C3 increase, and biomass and production in level C2 decreases. This reduces the feeding pressure on herbivore demes of level C1, allowing their biomass and production to increase. As a result, increased feeding by herbivores leads to decreases in biomass and production of producers of the P level. Thus the effects of increasing numbers of predator fish cascade down through the entire food web.

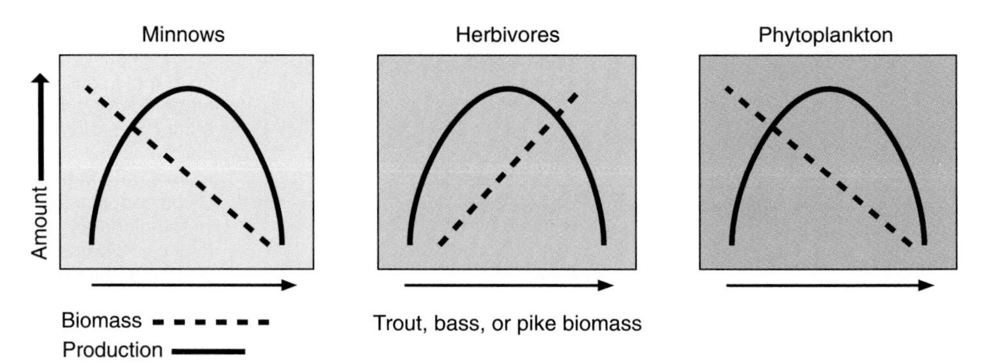

Why? If such a reduction occurs, the density of minnows will increase. The increased minnow population will lead to a decrease in large crustacean zooplankton, insect nymphs, and copepods (aquatic gill-breathing crustaceans). These changes will have corresponding effects on the small herbivores, which will then affect the phytoplankton.

The effects of such interactions are both direct and indirect. Changes in herbivore densities directly affect the density of phytoplankton. An indirect effect also occurs since most of the nutrient cycling within the limnetic zone is dependent on the decomposition of crustacean waste products. Thus, as shown in Figure 9.10, the densities of the crustacean populations control the amount of the nutrients available for photosynthesis by the phytoplankton.

Figure 9.10 and the three graphs in Figure 9.11 summarize results from several studies concerned with the CTI theory. They indicate that maximum production rates for the minnows, herbivores, and phytoplankton occur when levels of trout, bass, or pike biomass are intermediate. It is of great significance that the biomass of the lower trophic guilds was found to change in concurrence with predictions derived from the CTI theory. This affirmation indicates that the theory has power and can be used to guide aquatic ecologists in designing further experiments to learn more about lentic systems.

Lentic ecosystems consist of different zones that are defined in terms of light energy received and productivity. These factors are greatly influenced by depth, temperature, and available nutrients. A functional model consisting of five trophic guilds describes energy flow in lentic ecosystems. The theory of cascading trophic interactions explains the system of effects that follow if one of the trophic guilds changes significantly.

Eutrophication Abatement

The CTI theory and an examination of Figure 9.10 suggests a possible management strategy for lakes suffering from eutrophication. What might be done to prevent the explosive growth of aquatic algae and plant populations? If excessive algae blooms occur, the addition of large predatory fish (trout, bass, or pike) may offer a viable solution. Why? Because increasing their density would lead to a reduction in the number of minnows, and that effect would cascade through the food web. The most desired effect would be increased densities of herbivore species that would consume much of the excess plankton productivity. This strategy, however, would be appropriate only for lentic systems that contained reasonable levels of PO_4^{-3}. In cultural eutrophic lakes, the excessive amounts of nutrients would place the system beyond the regulatory capacity offered by the cascading trophic interactions.

Figure 9.12

Many lakes were formed in the wake of receding alpine glaciers as the last ice age ended These lakes are usually oligotrophic.

Beavers and Lake Succession

Ironically, the ultimate fate of all lakes is that one day they will cease to exist. A number of distinct successional patterns occur in different natural lentic systems. For example, one type of succession may occur in a lake formed in the wake of a receding glacier (see Figure 9.12). With the passage of time, the combination of

Figure 9.13

(A) By 1977, lower Crabtree Lake in the Cascade Mountains of western Oregon was nearing the end of its aquatic succession because it was nearly filled with sediments. (B) In 1981, a family of beavers moved into the Crabtree Lake area and constructed a dam. Consequently, succession was reversed, and a "new lake" was formed.

an accumulation of organic matter entering the lake from the watershed along with an excess of lentic biomass production will result in the lake's eventually filling with sediments.

A different type of succession can occur in lakes similar to that shown in Figure 9.13A. The outline of the original lake can be identified in the photograph. Trees now grow in areas that once were part of the lake. If you were to walk from these trees toward the lake, you would be passing through time, from the oldest shrub communities to a younger boggy wetland area to a still younger marshy area and finally to the present lake. The lake system in the photograph is in the last successional stages of its existence. In time, the lake will disappear, and terrestrial succession will begin. In the distant future, the area will be covered by a climax forest that is characteristic of its elevation and biome.

However, the time line of lake succession is not always predictable. Unanticipated forces in the environment can lead to a surprising alteration in the predicted progression of succession. In 1981, the lake shown in Figure 9.13A became occupied by a family of beavers, which constructed a dam that they continued to expand during the next few years. The dam blocked the outflow of water from the lake, and rather quickly, the lake enlarged and water flooded the riparian habitats, killing many of the shrubs and trees, as shown in Figure 9.13B. What would you predict will be the ultimate fate of this "new" lake?

MARINE ECOSYSTEMS

Marine and freshwater ecosystems differ in a number of significant ways. Most notably, a high concentration of dissolved salts (salinity) is the major parameter that affects the occurrence, distribution, and abundance of organisms in marine environments. Also, the great ocean depths act as "sinks" for important nutrients. In most areas of the oceans, nutrients locked in sediments are slowly cycled through geological time scales.

Marine systems are categorized as either *neritic* or *oceanic* (see Figure 9.14). **Neritic waters** include a *littoral zone* that occurs in near-shore waters adjacent to the coasts, bays, and estuaries and waters extending over a portion of the continental shelf. **Oceanic waters** comprise all the other areas of marine ecosystems.

Oceanic Ecosystems

Oceans are classified into four vertical zones because environmental conditions vary so greatly with increasing depth that distinct ecosystems with very different structures and functions exist. The **euphotic zone** is defined by the depth of light penetration and occurs from the surface down to a maximum depth of about 200 meters. The **bathyal zone** occurs between 200 and 1,500 meters, the **abyssal zone** from 1,500 to 6,000 meters, and the **hadal zone** from 6,000 down to 11,000 meters in deep ocean trenches.

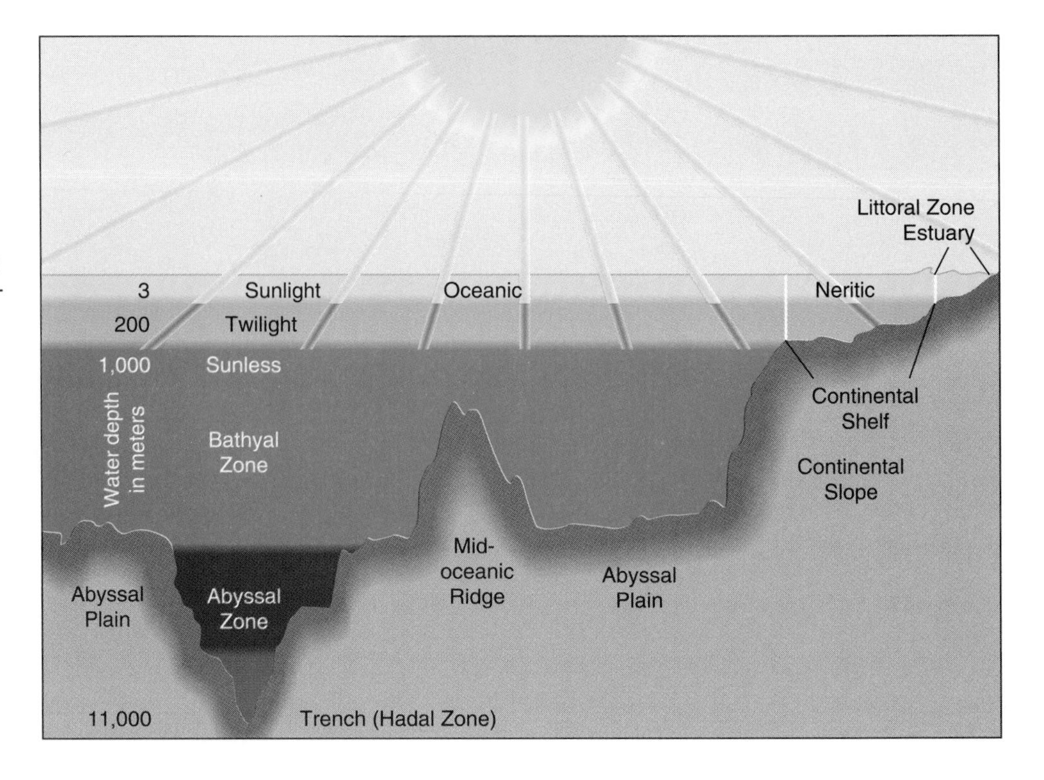

Figure 9.14

Marine life zones vary with distance from shore and depth. The zones include near-shore neritic waters over continental shelves and littoral zones along the shore, bays, and estuaries. Oceanic life zones consist of the upper regions where light will penetrate and the sunless bathyal, abyssal, and hadal zone of the deep oceans. Ocean topography includes abyssal plains, mid-oceanic ridges, the continental slope, and the continental shelf.

Upwelling

The topography of the ocean floor directly affects both the movement of deep ocean currents and nutrient circulation. Gravity results in organic nutrients' being deposited in the depths of the oceans. Only in specific areas do conditions allow these nutrients to be transported back to the surface waters. The vertical transport of nutrients from ocean depths to surface waters is called **upwelling**.

In general, three different current systems are capable of causing upwelling, as shown in Figure 9.15. In areas where divergent currents separate surface waters, deeper ocean waters move to the surface. The presence of mid-oceanic ridges on the ocean floor causes deep ocean currents to be forced up and over them, which brings deeper waters to the surface. Some oceanic ridges emerge above the surface and create islands like the Hawaiian chain. Finally, **continental upwelling** occurs where the combined effect of surface winds and Earth's rotation toward the east causes a net offshore movement of surface water on many western continental coasts. As this water moves offshore, it is replaced by cold, deep ocean water that moves up the continental slope, across the continental shelf, and into the near-shore surface waters. All these types of upwelling result in the transfer of nutrient-rich deeper waters to the surface layers of the oceans.

In addition to the three types of upwelling, a mechanism exists for the vertical transport of nutrients in far northern and southern ocean areas that is similar to turnover in temperate lakes. In the cold polar regions, ocean surface water reaches its maximum density just before it freezes. The temperature of greatest density varies with salinity, but when it is most dense, the seawater sinks and is replaced by deep, nutrient-laden water from the ocean floor.

The global regions where upwelling and vertical transport of deep ocean water occur are the areas of highest marine productivity. These are the sites of major modern marine fisheries and the whaling industry of the past. The salmon, tuna, and halibut fisheries off the west coast of North America are located in areas of continental upwelling, as are the anchovy fisheries off western South America and major fisheries along both coasts of Africa.

Nutrient Deserts and Coral Reefs

Most areas of the deep oceans are not in upwelling zones. These expanses are nutrient-poor, for there is no mechanism to cycle any nutrients present on the

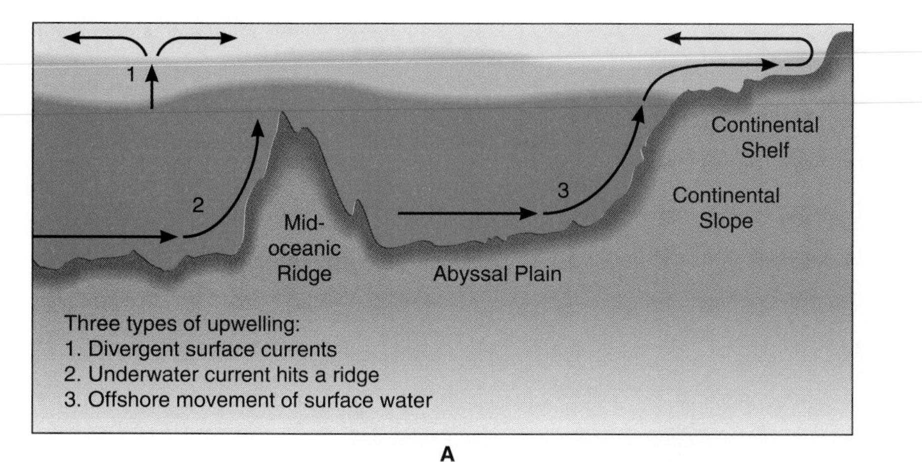

A

= upwelling

B

Figure 9.15

(A) Upwelling occurs as a result of (1) divergent surface currents, (2) underwater currents reflected off mid-oceanic ridges, and (3) the offshore movement of surface water, which is then replaced by deep water from the abyssal plain, which moves up the continental slope and along the continental shelf to the near-shore surface areas. Upwelled water enriches the surface zones with important nutrients released by bacterial degradation of organic matter in the ocean benthic environments. (B) This world map shows the locations of global upwelling and other major regions of marine community productivity.

bottom back into the euphotic zone. Such areas are identified as *nutrient deserts* where few organisms live. Exceptions to these nutrient-poor ocean areas are the newly discovered deep-ocean-trench ecosystems (see the Focus on Scientific Inquiry, "Serendipity and the Deep").

Coral reefs in shallow tropical and subtropical seas are also unique marine systems. They have an enormous diversity of life that is dependent on the nutrients cycled with continental upwelling or wave action circulation (see Figure 9.16).

A Marine Food Web

Much of the whaling industry of past decades was centered in the upwelling regions. Heavily hunted areas included the North Pacific from Alaska to the northern islands of Japan, south and west of Greenland in the North Atlantic, and the nutrient-rich waters north of Antarctica in the southern oceans.

Most large whales feed on **krill**, small crustaceans that are present in unimaginably large numbers in some marine systems. Krill feed on zooplankton, which feed on phytoplankton, whose growth is dependent on nutrients brought to the surface in these areas of upwelling (see Figure 9.17A). As shown in Figures 9.17B–C, excessive harvests of blue whales led to decreases in total catch and average length of females, an increase in the proportion of sexually immature whales, and a dramatic decrease in catch per unit effort, a measure of how long it takes to find and kill each whale. The sharp declines in populations of the great blue, the fin, and other large whales have led most nations to adhere to interna-

Figure 9.16

(A) Coral reef communities are important ecosystems in warm, shallow ocean areas. (B) Most coral reefs are found in tropical and subtropical ocean areas where they are usually associated with the continental shelf or with islands of archipelago systems.

A

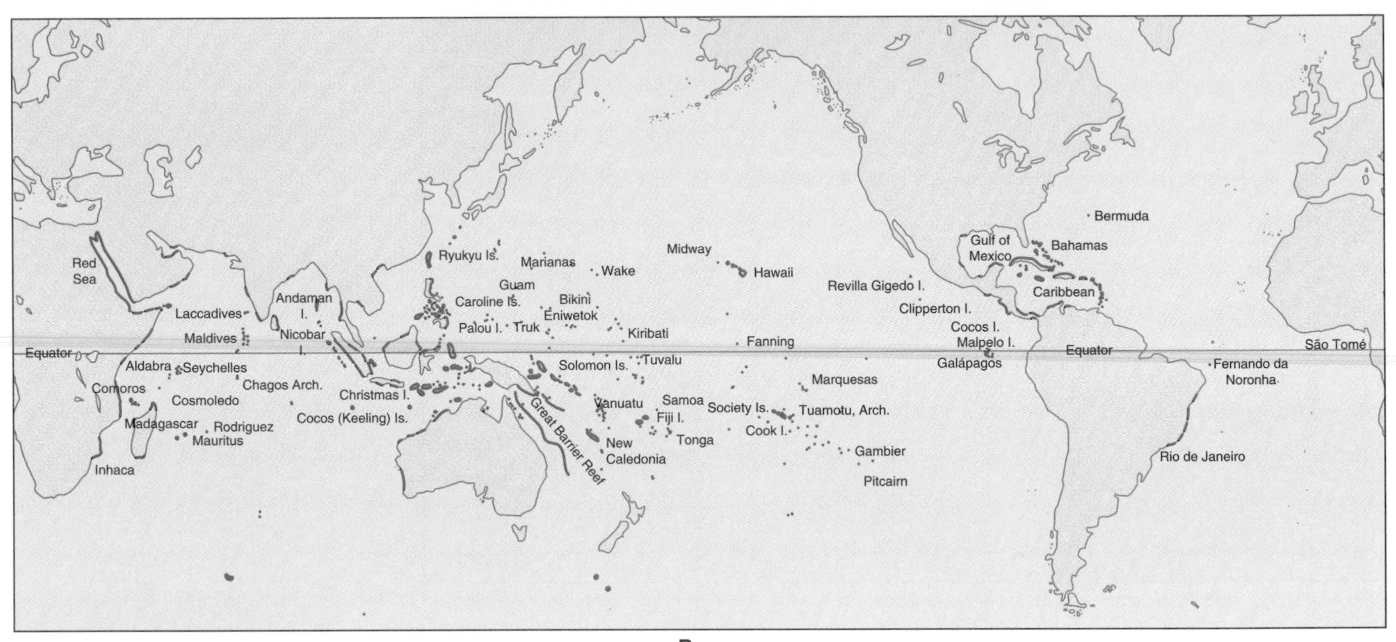

B

Figure 9.17

(A) In a food web that occurs in southern oceans, upwelling enriches the surface waters with important nutrients used by phytoplankton that supports a thriving zooplankton herbivore guild, which in turn sustains immense numbers of krill, the main food for the great blue whale and other large whales. Other guilds in the food web include fishes, penguins, and seal populations. (B) Between 1934 and 1965, the blue whale catch varied, but the catch per unit effort declined. (C) Between 1934 and 1964, a greater percentage of the more abundant younger whales were killed when adults became scarce after excessive harvesting. No data were available for 1942–1947.

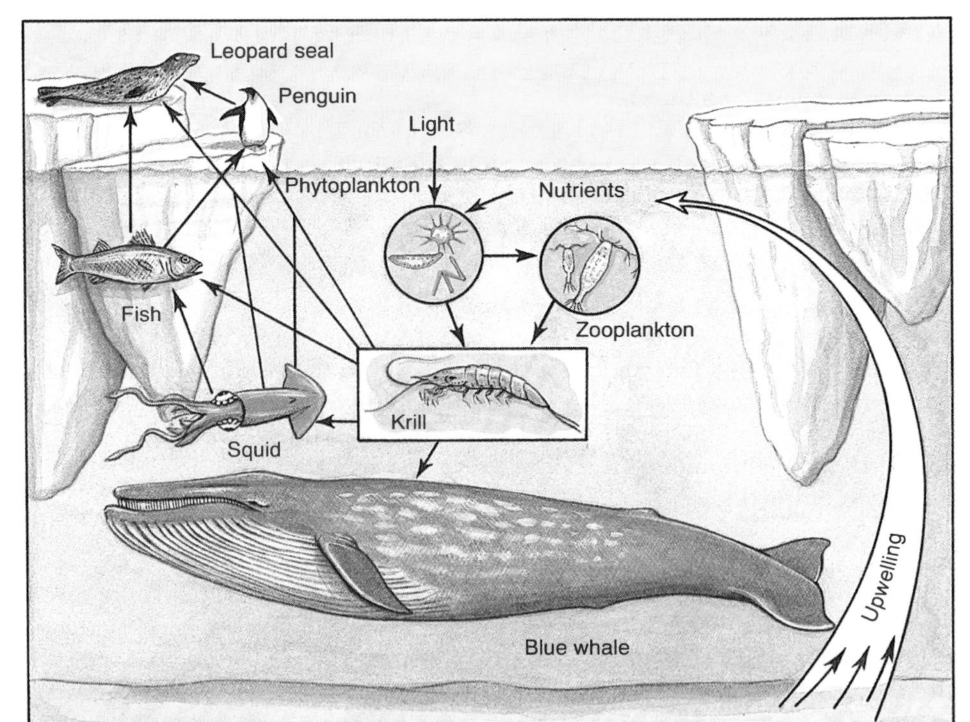

A

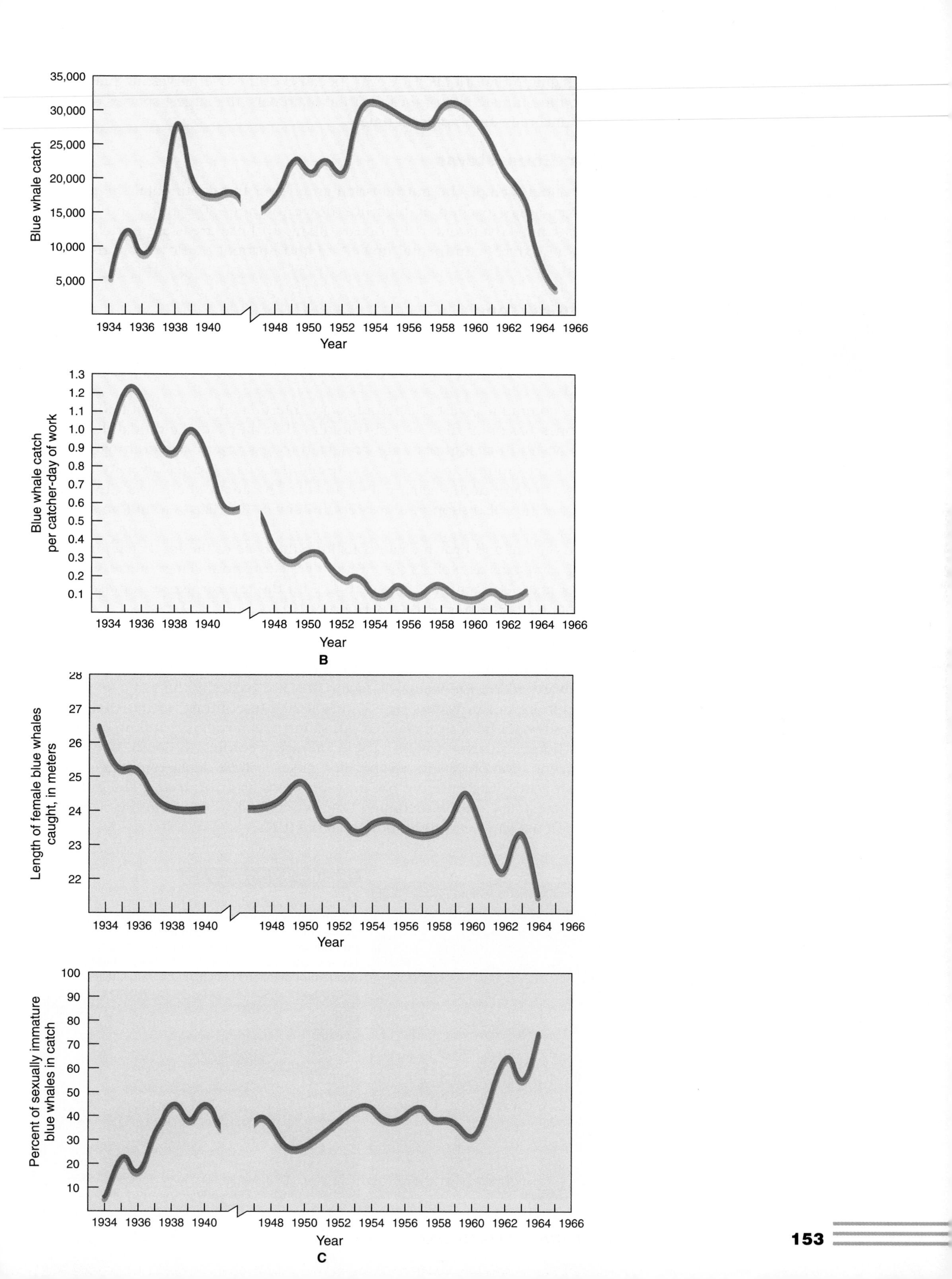
35,000
30,000
25,000
20,000
15,000
10,000
5,000
Blue whale catch
1934 1936 1938 1940 1948 1950 1952 1954 1956 1958 1960 1962 1964 1966
Year
1.3
1.2
1.1
1.0
0.9
0.8
0.7
0.6
0.5
0.4
0.3
0.2
0.1
Blue whale catch per catcher-day of work
1934 1936 1938 1940 1948 1950 1952 1954 1956 1958 1960 1962 1964 1966
Year
B
28
27
26
25
24
23
22
Length of female blue whales caught, in meters
1934 1936 1938 1940 1948 1950 1952 1954 1956 1958 1960 1962 1964 1966
Year
100
90
80
70
60
50
40
30
20
10
Percent of sexually immature blue whales in catch
1934 1936 1938 1940 1948 1950 1952 1954 1956 1958 1960 1962 1964 1966
Year
C

Figure 9.18

Common intertidal organisms are sea stars and anemones.

tional whaling moratoriums. However, because of the overharvesting practices of the past, the ultimate fate of these marine mammals, the largest species on Earth, remains uncertain at the present time.

Neritic Ecosystems

Neritic waters include the nutrient-rich areas over the continental shelf, littoral zones from areas reached by high tide down to about 30 meters, and bays and estuaries where rivers enter the ocean. Many of these areas are enriched from two major sources, continental upwelling and nutrient outflow from rivers and estuaries.

The Littoral Zone

The littoral zone on both coasts of North America hosts a number of ecological systems that support rich diversities of marine life (see Figure 9.18). They include the subtidal areas offshore, open beaches, and the rocky intertidal region around headlands.

The communities established in rocky intertidal areas are stratified into different zones based on their aspect (the direction they face) and tidal depth. Organisms on north-facing rocks are subjected to less solar drying during low tide than those on south-facing rocks. As a result, there is a greater diversity of life in the moist, shaded, north-facing microhabitats. The areas where water remains in pools at low tide also support a greater diversity of life than is found on adjacent, open rock faces.

Red, green, and brown algae constitute the base of the food chain in the intertidal ecosystems as well as the subtidal areas offshore. These algae support a variety of marine organisms, culminating with the top carnivores represented by marine otters and harbor seals (see Figure 9.19).

Bays and Estuaries

Bays differ from estuaries in many ways. A bay is a quiet arm of the ocean extending landward where the water characteristics are basically the same as those in the near-shore ocean waters (see Figure 9.20). In contrast, estuaries are areas where fresh water from rivers mixes with saline ocean waters that flow upbay during high tides. Estuarine waters vary in salinity according to depth, because the high concentration of salts in ocean water makes it more dense (and thus heavier) than fresh water. The entering ocean water flows as a *salt wedge* along the bottom of the estuary on the incoming tide, while fresh water from the river flows seaward on the surface as described in Figure 9.21.

Figure 9.19

The marine otter shown here represents mammals that feed at the top of marine food chains. They also eat crabs, sea urchins, scallops, abalone, and fish.

Figure 9.20

Marine bays are quiet landward extentions of the ocean where the water characteristics are very close to those of the near-shore ocean waters. They are important nursery areas for marine organisms, and many, like this one in the Bay of Islands, New Zealand, have resident populations of fish, crabs, and shellfish.

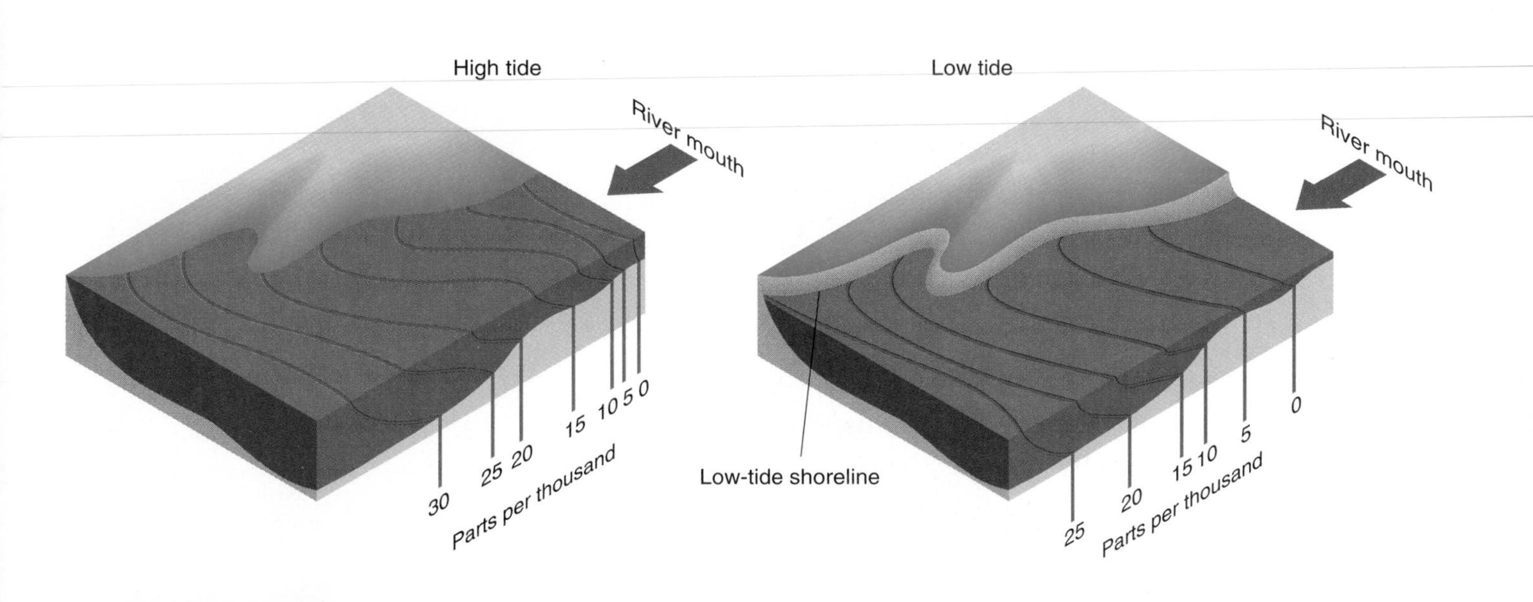

Figure 9.21

The salt water that enters a marine bay or estuary during an incoming tide is more dense than the fresh water flowing seaward from the river. This results in the formation of salinity gradients from the bottom to the surface and from the ocean to the river mouth. Thus the salinity of bays and estuaries varies during the day between high tide and low tide and seasonally with fluctuations in the volume of river flow.

Marine bays are common along the west coast of North America, and their estuarine portions occur far inland. The location of the major area where fresh and salt water mix in these estuaries varies seasonally, depending on the volume of river flow. East-coast rivers develop more extensive estuaries and wetlands of coastal marshes, mangrove swamps, and deltas along major rivers in the southern states (see Figure 9.22).

Marine ecosystems are incredibly diverse with respect to size and productivity. Depth and upwelling are two major factors that determine the capacity of a marine ecosystem to support life. Neritic ecosystems are generally the richest biologically of marine ecosystems because they receive an abundance of light and nutrients. They support some of the most diverse communities on Earth.

HUMAN INTERVENTIONS

Bays, estuaries, and wetlands are important feeding habitats for many species of birds and major nursery areas for numerous species of marine life (see Figure 9.23A). However, being close to major areas of human habitation, these fragile

Figure 9.22

Mangrove swamps are shallow tidal ecosystems found in tropical and subtropical regions.

Focus on Scientific Inquiry

SERENDIPITY AND THE DEEP

Many important scientific discoveries have occurred by chance or accident. Such *scientific serendipity* is not the result of a special aptitude for making fortunate accidental discoveries, as was the case in Horace Walpole's 1754 tales from Ceylon, "Three Princes of Serendip." Rather, it results from critical observations by trained, open-minded scientists who are able to evaluate the unexpected in the light of new possibilities. In 1984, such an observation was made on the floor of the Pacific Ocean over 2,000 meters below the surface.

Figure 1 The *Deep Submergent Vehicle*, nicknamed *Alvin*, suspended from the support research vessel, *Atlantis II*, is often used in research involving deep ocean biology and geology.

A team of oceanographers led by LaVerne Kulm, from Oregon State University, was conducting a multidisciplinary research program designed to investigate the process of subduction-induced mudstone and sandstone sediment formation. These sediments accumulate where the Juan de Fuca oceanic plate is forced downward into Earth's mantle when the North American continental plate rides up and over it *(subduction)*. When deep ocean sediments accumulate at the bottom of the continental slope, they are scraped along and piled up *(off-scraping)* as the continental plate of North America moves westward. While these sediments accumulate, they have a water content of about 70 percent, but as they become compressed during off-scraping, this decreases to about 30 percent through a process known as *pore fluid venting* (see Figure 5.3 in Chapter 5). In the summer of 1984, these processes were studied off the coast of Oregon using the *Deep Submergent Vehicle*, referred to as *Alvin*, shown in Figure 1.

Early in a sequence of 15 dives into the perpetual darkness of the deep ocean floor, one of the scientists thought he saw tube worms illuminated by the lights on *Alvin*. If he had, it would represent the first such finding for the subduction zone along the North American plate. Deep sea communities of tube worms, giant clams, and associated bacteria were discovered in 1977 at hydrothermal vent systems in the Galápagos Rift. However, the Galápagos systems are supplied with copious quantities of hot water and dissolved hydrogen sulfide (H_2S). No H_2S was detected during the 1984 dives, although methane (CH_4) was collected in the venting pore fluids in concentrations up to five times greater than background levels. Also, water temperatures at these vents were cool, only 0.3°C higher than the surrounding water. Videotapes taken during the dive confirmed the presence of tube worms, and in 7 of the 15 dives, other populations of tube worms, large clams, small clams, crabs, and predatory fish were documented, as shown in Figure 2.

The clams and tube worms collected off Oregon have been classified in the same genera as

A

Figure 2 (A) Giant tube worms, fish, clams, and an unknown crab were photographed through the viewing port of *Alvin* in the subduction zone off the coast of Oregon. (B) This drawing shows details of the animals' appearance.

B

those that occur in the Galápagos Rift; the crab species remains unidentified. Similar species of the large clams have been found in fossil strata on the islands off Barbados, which date back to the Oligocene, 30 million years ago. Until the discovery of deep ocean communities these fossil shells were thought to have been from clams that lived in shallow Oligocene seas. However, the discovery of deep ocean communities indicates that these ancient clam populations have survived for millions of years. Further, the vent communities must have occupied tectonic areas of the deep ocean floors for a very long period of time.

The 1984 dives of the *Alvin* established that a new type of venting process, not previously described at a tectonic boundary, was occurring in the Oregon subduction zone. The serendipitous discovery of deep ocean communities also raised an interesting question: How were these organisms able to obtain energy from a surrounding environment that was totally dark? Since sunlight penetrates to only a depth of 200 meters, was some other energy source sustaining life in these deep ocean regions? In 1986, Kulm's group formulated a new hypothesis, which was published in the journal *Science.* "We hypothesize that clams and tube worms found in the cool vent areas of the Oregon subduction zone have successfully adapted to another type of energy metabolism, that is, the capacity to utilize dissolved CH_4." Specifically, they deduced that the unusual benthic communities are sustained by chemosynthetic bacteria that are able to use CH_4 as an energy source.

Focus on Scientific Inquiry

Until the late 1970s, scientists believed that solar-driven photosynthesis and bacterial chemosynthesis at terrestrial hot springs were the only energy-transforming processes that sustained life on Earth. Then, in 1977, it was discovered that giant clams from the Galápagos site were supported by energy from a new source. These clams feed on symbionic chemosynthetic bacteria that live in their gills. Here is what occurs in this system: (1) H_2S is used by these bacteria to support their activities through a complex series of chemosynthetic reactions. (2) The clams provide a site of attachment for the bacteria at the H_2S vents. (3) The bacteria are used as food by the clams. Is the CH_4 present at the ocean floor vents in the subduction zone off Oregon used in a similar way to sustain life? If so, a new energy source would be added to the list.

To test the new hypothesis about CH_4 as an energy source, Kulm's group developed a *benthic barrel* that would sample the gases discharged into vent water when placed over the deep ocean clam beds (see Figure 3). This barrel was used successfully during dives in 1987, 1988, and 1990. During the 1987 dives, only CH_4 was detected at all sites sampled, but in 1988, both CH_4 and H_2S were recorded at a different sampling site. Chemical analysis of the samples obtained with the benthic barrel indicate that the CH_4 and H_2S are of *biogenic origin* (that is, they are produced from organic matter decomposition within the accumulating sediments). In contrast, the H_2S and CH_4 venting in the Galápagos Rift is of *primordial origin,* which means these gases were created when the planet was young and are still being released from Earth's interior.

Kulm and his colleagues plan to continue their efforts to determine the biological role of CH_4 (and H_2S) in the communities at the deep ocean sites. They have already discovered that these communities are important indicators of active tectonic zones—areas on the ocean floor where sediment accumulation, off-scraping, and pore fluid venting are occurring. The age and distribution of the vent communities are the best indicators of these marine geological events in deep ocean areas; the presence of clams and tube worms is correlated with areas of active fluid venting that occurs when sediments are compressed to form mudstone and sandstone.

Research on future dives will attempt to determine how widespread these communities are along the subduction zone and if CH_4 acts as an energy source for the bacteria on which these organisms apparently feed. Some marine biologists in Kulm's group think that CH_4 will not prove to be a new energy source for this community, and they expect to find new, undetected levels of H_2S at these sites off the Oregon coast, but only future research will resolve these questions. However, the serendipitous discovery of these communities has added to our biological knowledge, and it has also extended our understanding of geological events involved in these tectonically active deep ocean areas.

Figure 3 This benthic barrel was developed to sample gas exchange in the community after being placed over the clam beds on the ocean floor.

A B

Figure 9.23

(A) Great blue herons are important predators that feed on fish, shrimp, and other aquatic life present on or in exposed mudflats during low tide in marine bays and estuaries. (B) Biologically rich areas of marine bays and estuaries have been lost in the past due to filling. The "new land" is often very valuable real estate because of its proximity and access to the water. Uses of such filled areas are varied, but they can include sites for ocean front development as shown here in Ocean County, New Jersey.

ecosystems have been seriously affected by disturbances, including agricultural, industrial, and municipal wastes and draining and filling for housing, industrial, and recreational uses, as shown in Figure 9.23B. The scope of the loss was described in a federal survey, which determined that about 11 million acres of wetlands were converted to other uses in the lower 48 states between the mid-1950s and the mid-1970s. Most of this loss was due to drainage or diking, the conversion of tidelands into agricultural lands. What will be the ultimate cost of these alterations to the affected estuarine systems?

We now turn our attention from ecosystem structure and function as we attempt to assess some of the effects human activities are having on natural systems. Increasing environmental impact from human activities has been correlated directly with the increase in human population density. Parameters that influence human population growth over time are the major topics in Chapter 10. In Chapter 11, we will consider the extent to which human activity is changing the global climate and how such changes will likely affect the biota present in all biomes of Earth.

Summary

1. Aquatic ecology includes studies of lotic and lentic freshwater systems and estuarine, neritic, and open-ocean marine systems. The amount of light energy and nutrient availability are the most critical limiting factors in aquatic systems.
2. The order of a stream increases by one unit when two streams of the same order meet. Insect shredders in headwater stream communities use CPOM, whereas insect collectors use FPOM as major energy sources. As the stream order increases, shredders are replaced by grazers that feed on periphyton. Only mid-order streams and rivers have a P/R greater than 1. Drift not removed during its seaward movement will settle in estuarine sediments, adding to their biological richness.
3. Turnover increases the productivity of temperate lakes by recharging surface waters with nutrients twice each year. Lakes are classified as oligotrophic, mesotrophic, natural eutrophic, and cultural eutrophic, depending on community productivity.
4. A functional model of lake food webs assigns the organisms to a number of trophic guilds, depending on their food habits. Changes in these guilds can change the rate of PO_4^{-3} cycling, which can lead to a modification of the trophic structure and can regulate the function of the entire community. All lakes are temporary because, given enough time, natural successional events will cause them to be replaced by terrestrial ecosystems.
5. Cold arctic and antarctic waters and areas along some continental margins are the most productive marine environments. In these areas, mechanisms recharge the euphotic zone with nutrients liberated by the decomposition of organic matter on the ocean floor. Most other ocean areas have no mechanism for this type of nutrient transport and are therefore considered to be nutrient deserts.
6. Estuaries are among the most productive aquatic areas, but their proximity to major areas of human habitation has resulted in the significant degradation of many of these important natural ecosystems.

Review Questions

1. What are the classes of aquatic and marine ecosystems?
2. How are lotic ecosystems classified?
3. How does Cummins's functional model explain energy flow in lotic ecosystems of different sizes? What are the types and roles of different classes of insects in the model? What is the significance of P/R in the model?
4. What is drift? Why is it important?
5. What factors influence lentic ecosystems?
6. How are lakes classified? What are some important regulating factors in lentic ecosystems?
7. What does the theory of cascading trophic interactions explain?
8. How are marine ecosystems classified?
9. What is upwelling? Where does it occur? Why is it important?
10. Why are neritic ecosystems generally richer biologically than oceanic ecosystems?

Essay and Discussion Questions

1. Assume that you are an environmental consultant responsible for preparing an impact statement. Predict the direct impact of logging (removing trees from) watersheds of first- and second-order streams. Also project the effects on larger streams and rivers into which these streams flow.
2. How would you rate the vulnerability of the following aquatic ecosystems to disruption by human activities (from most sensitive to least sensitive): an order 2 stream, a small bay, coral reefs, the Mississippi River, a small lake, Lake Superior, Chesapeake Bay, the open ocean between Hawaii and the mainland? What criteria would you use in making such a list?
3. Some ecologists believe that it is easier to conduct "good" studies of aquatic systems than of terrestrial ecosystems. Do you agree? What factors might be considered in coming to such a conclusion?

References and Recommended Reading

Baker, M., and W. Wolff, eds. 1987. *Estuarine and Brackish-Water Sciences Association Handbook*. Biological Surveys of Estuaries and Coasts. Cambridge: Cambridge University Press.

Barnes, R., and R. Hughes. 1988. *An Introduction to Marine Ecology*. 2d ed. Oxford: Blackwell.

Carpenter, S., and J. Kitchell. 1988. Consumer control of lake productivity. *BioScience* 38:764–769.

Carpenter, S., J. Kitchell, and R. Hodgson. 1985. Cascading trophic interactions and lake productivity. *BioScience* 35:634–639.

Cummins, K. W. 1977. From headwater streams to rivers. *American Biology Teacher* 5:305–312.

Cummins, K. W., M. A. Wilzbach, D. M. Gates, J. B. Perry, and W. B. Taliaferro. 1989. Shredders and riparian vegetation. *BioScience* 39:24–30.

Kulm, L. D., E. Suess, J. C. Moore, B. Carson, B. T. Lewis, S. D. Ritger, D. C. Kadko, T. M. Thornburg, R. W. Embley, W. D. Rugh, G. S. Massoth, M. G. Langseth, G. R. Cochrane, and R. L. Scamman. 1986. Oregon subduction zone: Venting, fauna, and carbonates. *Science* 231:661–666.

Northcote, T. 1988. Fish in the structure and function of freshwater ecosystems: A "top-down" view. *Canadian Journal of Fisheries and Aquatic Science* 45:361–379.

Smith, R. 1986. *Elements of Ecology*. 2d ed. New York: HarperCollins.

Spencer, C. N., B. R. McClelland, and J. A. Stanford. 1990. Shrimp stocking, salmon collapse, and eagle displacement—cascading interactions in the food web of a large aquatic ecosystem, *BioScience* 41:14–21.

Suess, E., B. Carson, S. D. Ritger, J. C. Moore, M. L. Jones, L. D. Kulm, and G. R. Cochrane. 1985. Biological communities at vent sites along the subduction zone off Oregon. *Biology Society of Washington Bulletin* 6:475–484.

Turner, R. E., and N. N. Rabal. 1991. Changes in Mississippi River water quality this century. *BioScience* 41:140–147.

CHAPTER 10

Human Populations and the Environment

Chapter Outline

Reading Questions

1. What factors have influenced the size of human populations throughout history?

2. What factors may act to limit human population growth?

3. In what ways have human populations affected the environment?

4. What types of technological solutions to biological problems have resulted in new environmental problems?

Growth in the number of humans during the twentieth century has resulted in a global population that now exceeds 5 billion people. Our effects on the ecosystems of Earth are now greater than the effects of all human generations that lived before this century! In this chapter, we explore the history of human population growth and project some repercussions of future population growth on global environments into the next century.

EARLY HUMAN SETTLEMENT

For millennia, the environmental impact of *Homo sapiens* was related to the increasing densities of various human populations that inhabited Earth. The success of the human species at producing offspring often resulted in population densities that exceeded the food-producing capacity of the environment. This led to famine or the emigration of part or all of the affected population to less used areas.

Certain ancient human populations led an existence that enabled them to obtain food from a dependable source. They did this by correlating their movements with those of certain animals. Many species of large herbivores, such as the North American bison and caribou and the wildebeests of Africa, evolved a survival strategy that allowed their population densities to exceed what any given habitat could long support. When such areas became devoid of important nutrients, the herds migrated, following the "march of the seasons" to greener fields (Figure 10.1). Early human populations often developed a similar nomadic lifestyle. They followed these migratory herds on their annual treks or settled along their routes of migration and exploited them as food resources during specific periods of each year.

The oldest permanent human settlement sites date from approximately 11,500 years ago. They appeared between the end of the Pleistocene glaciation and the early interglacial Holocene epochs, a time when the climate began to warm. What factors led to the change from a nomadic existence to life in a village? How did survival become dependent on cultivating food resources as opposed to simply harvesting available food? Was the development of agriculture a cause or an effect of settlement?

These questions cannot be answered completely because information is lacking. The story is complicated by the need to consider that human history is influenced by cultural as well as biological factors. A unique attribute of our species's evolution was the development and application of **culture**. In the broadest sense, culture includes our languages, our social structures, our value systems, and our development of tools and their use in agriculture. Human cul-

Figure 10.1

Many species of large herbivores have evolved strategies that include extensive migrations on an annual basis. The wildebeests in East Africa move in such patterns, following more favorable forage conditions as the seasons change.

ture can be viewed as an adaptive trait, but many anthropologists are skeptical that adaptation is all that is involved in culture. In recent years, human culture has been treated as a partly independent factor of human evolution constrained, but not fully determined, by biological limits.

In this text, we are concerned primarily with the physical and biological aspects of human population effects. However, it is important to understand that societies at different times have chosen to limit or expand their size for reasons that are not strictly related to biological concerns. Whatever choices societies make, they are still limited by biological factors. A society may decide to quadruple in size, but if it lacks food or other necessary resources, its decision cannot be implemented. Similarly, if millions of people want to drive to work every day, they may at some point have to face the realization that the air that they breathe is contaminated by this action.

One of the physical factors hypothesized to have been important in the development of permanent human settlements is global climate change. Because of warming climates, native flora and fauna flourished in many areas and may have induced humans to settle there. However, as the villages' population densities increased, environmental damage often began to occur in the surrounding areas used for hunting and gathering food. As the food supplies diminished, human options were probably limited; the population could starve, emigrate, or attempt to increase the production of some important plants and animals that were used for food. This last choice may have been the impetus for the development of agriculture. Early farming allowed settlements to exist for long periods of time, but available evidence indicates that decreased soil fertility and changing climate, coupled with ever-increasing population densities, led to the collapse of most prehistoric villages.

Early human populations were nomadic because they obtained much of their food from killing and eating migratory animals. Approximately 11,000 years ago, humans began to settle into villages. The reasons for this change are not known with certainty, but cultural, biological, and physical factors may have been involved. Expanding populations led to environmental damage that ultimately forced these early villages to be abandoned.

HUMAN POPULATION GROWTH THROUGH TIME

Growth of global human populations was very slow for more than 15,000 years; then, during the seventeenth century, it began to grow at a very rapid rate. If this high growth rate continues, the global human population could double to 10 billion within your lifetime.

Prehistoric Human Population Growth

By the end of the Pleistocene glaciation, the human species had radiated from its cradle of evolution in eastern Africa to inhabit most of the major continents of Earth. The estimated world population numbered about 5 million people, who generally subsisted in small hunter-gatherer tribes. Because of the application of primitive agricultural techniques, humans were able to multiply, and the world population increased slowly for the next 9,500 years. About 2,000 years ago, the rate of population growth slowed and remained relatively constant for the following 1,000 years. During this millennium, few major advances occurred in agriculture. Periodic famines during times of drought were common. Although very impressive "high" civilizations like those of the Greeks, Romans, and Arabs existed, most people lived in appalling conditions during this time. Poor sanitary practices resulted in epidemics of typhus, smallpox, plagues, and other serious diseases. These diseases caused an increase in the mortality rate, and as it approached the birth rate, world population growth approached zero.

Figure 10.2

(A) An arithmetic graph of human population growth during the past 17,000 years. (B) The history of human population growth expressed as a logarithmic plot.

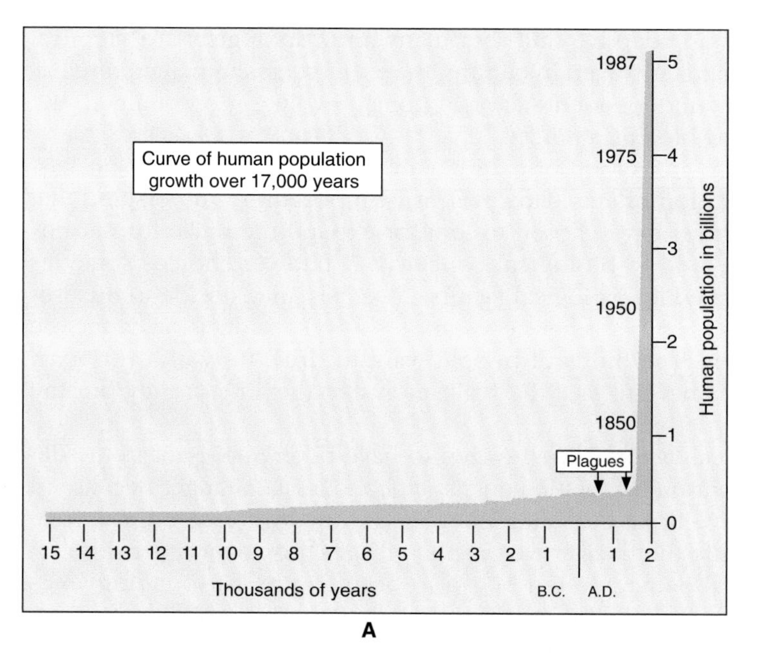

A

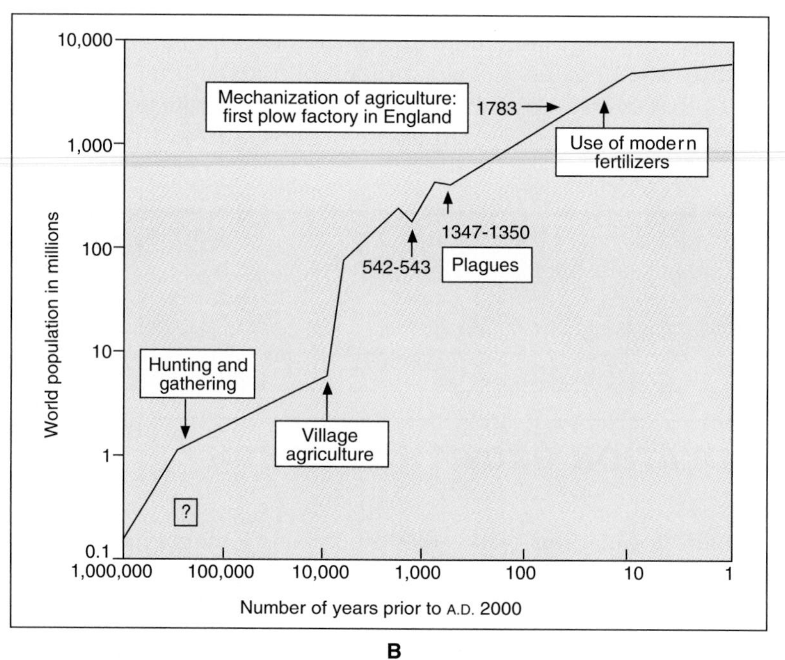

B

Historic Human Population Growth

In Figure 10.2A, population data have been graphed on an arithmetic scale, but this approach fails to divulge details about the nature of human population growth over long periods of time. However, if the arithmetic data are converted to logarithmic numbers, the resultant graph, shown in Figure 10.2B, is much more revealing. The logarithm of a number is the power to which 10 must be raised to produce the number; for example, $100 = 10^2$, so $\log(100) = 2$. This logarithmic plot shows more clearly many of the major events that occurred throughout the history of human population growth.

Village agriculture allowed early human populations to enlarge by increasing the amount of available food (see Figure 10.3). Figure 10.2B describes this general trend and also reflects some other significant historical events, including the effects of two major disease plagues. These plagues of 542–543 and 1347–1350 caused a temporary decrease in the density of the human world population, but they had little long-term effect on population growth.

Figure 10.3

Early village life may have developed due to advances in crop production and animal husbandry.

The total human population stood at about 500 million by the year 1650, when the rate of growth began to increase. A number of factors are likely to have contributed to this increase. European populations began to emigrate, especially to Africa, Oceania (Australia and the Pacific islands of New Zealand, Melanesia, Micronesia, and Polynesia), and the New World (North and South America). Nutrition began to improve because of new developments in agriculture. The death rate began to decrease, not yet due to medical advances, but most likely due to declines in the severity of diseases. Public health and sanitation advances in the nineteenth century extended the human life span, and by the century's end, medicine had developed into a scientific art that began to alter life expectancy dramatically.

The global human population had increased to 1.6 billion by 1900, and the rapid, exponential nature of its growth continued worldwide into the late 1960s. *Exponential growth* (also called *geometric growth*) occurs when the growth rate of a population can be described mathematically by the formula 10^x, where x (the exponent) is some computed number that defines the rate of growth. Exponential growth still characterizes populations of many developing regions and it is expected to continue into the foreseeable future (see Table 10.1). The data from Table 10.1, when graphed, indicate a great disparity in growth rates between the developing and developed regions of Earth (see Figure 10.4).

Table 10.1 Human Population Trends, 1900–2100

	POPULATION (MILLIONS)					
	1900	**1950**	**1985**	**2000**	**2025**	**2100**
Developing regions (total)	1,070	1,681	3,657	4,837	6,799	8,748
Africa	133	224	555	872	1,617	2,591
Asia[a]	867	1,292	2,679	3,419	4,403	4,919
Latin America	70	165	405	546	779	1,238
Developed regions (total)	560	835	1,181	1,284	1,407	1,437
Europe, USSR, Japan, Oceania[b]	478	669	917	987	1,062	1,055
Canada, United States	82	166	264	297	345	382
World total	1,630	2,516	4,837	6,122	8,206	10,185

[a]Excludes Japan.
[b]Includes Australia and New Zealand.

Source: Merrick, T. W. (1986), p. 16.

Figure 10.4

Human population growth trends differ greatly between developed and developing regions of Earth.

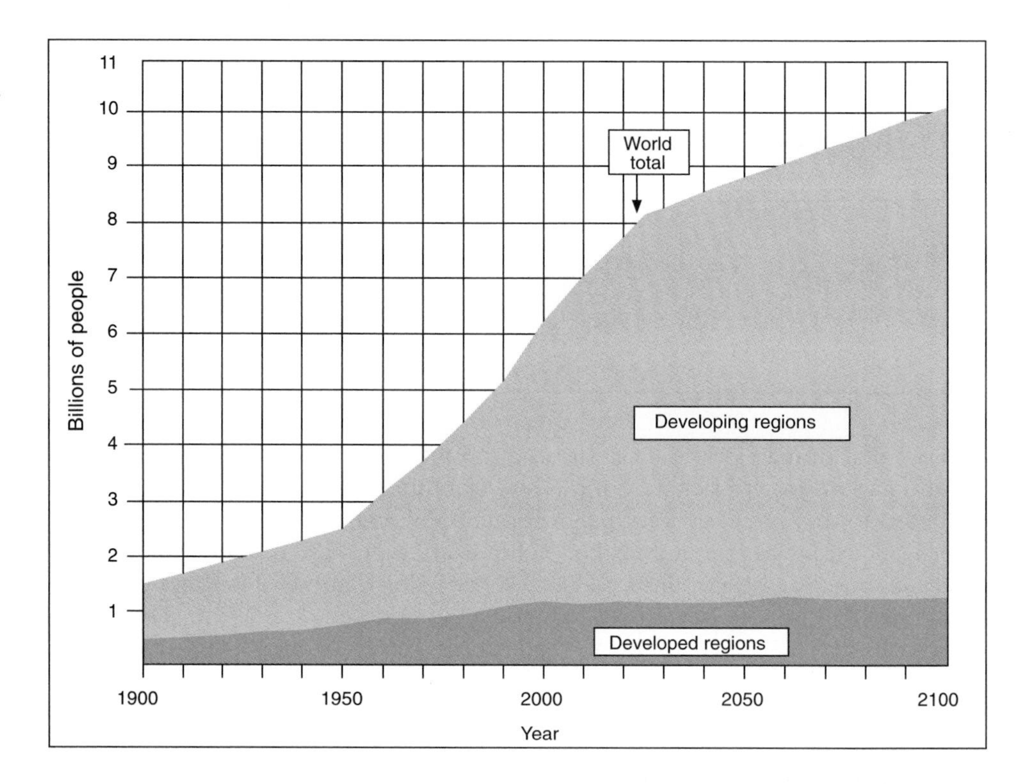

Around 11,000 to 12,000 years ago, the world held about 5 million people. Through the next 10,000 years, the human population increased slowly, limited largely by food and disease. By 1650, some 500 million humans inhabited Earth, and at that time the population began to increase sharply because of an increased food supply and a decline in mortality caused by disease. This accelerated rate of growth continued until recently.

Modern Human Population Growth

In the late 1960s, the world rate of human population growth began to decrease (see Figure 10.5). However, many countries in developing regions of the world continue to increase their populations at high rates. For example, the growth rate throughout most of Africa is expected to exceed 3.0 percent per year until the end of this century. A growth rate of 3.0 percent translates to a **doubling time** (the number of years it takes a population growing at a given rate to double in

Figure 10.5

The average annual rate of human population growth for developing and developed regions and the entire world is expected to decrease.

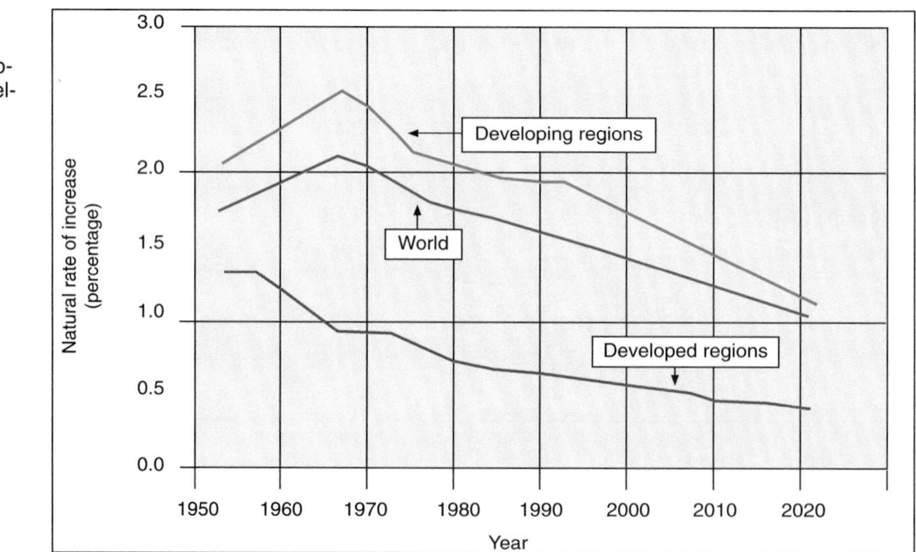

Table 10.2 Human Density in Developed and Developing Regions, 1900–2100

	POPULATION (MILLIONS)					
	1900	**1950**	**1985**	**2000**	**2025**	**2100**
Developing regions	14	22	48	63	88	114
Developed regions	10	15	21	23	25	26

Values are expressed in numbers of people per square mile (2.59 square kilometers). Developed regions are Europe, the USSR, Japan, Oceania, Canada, and the United States. Developing regions are Africa, Asia, and Latin America. These data are graphed in Figure 10.6.

size) of 23 years. Thus the total population in Africa was projected to double between 1989 and 2012. Other developing areas where high population growth rates persist include ten Middle Eastern Arab nations and Belize, Guatemala, Honduras, and Nicaragua in Central America.

Differences in the densities of humans in developing and developed regions of Earth have also increased in relation to their changing populations. Past, present, and expected densities are shown in Table 10.2.

The land area of the developing regions will remain constant at about 76,700 square kilometers and the developed regions at about 33,700 square kilometers. If the rates of population increase remain constant in the developing regions, their density will become greater than 58 people per square kilometer by the year 2000, and by 2100 it will have increased to 114 people per square kilometer (see Figure 10.6). Is this likely to occur? Will some factor intercede to reduce this potential population explosion?

In an even more extreme case, the rate of natural increase now approaches 4.0 percent in some developing countries, the most prominent being Kenya in eastern Africa. In mid-1990, the growth rate in Kenya was 3.8 percent, which computes to a doubling time of 18 years. By contrast, in mid-1990, France, a developed country covering an area similar to that of Kenya, had a rate of natural increase of 0.4 percent and a doubling time of 175 years. The expected change in density between these two countries is shown in Table 10.3. In 1990, the population density of France was greater than twice that in Kenya, by 2015 they will be equal, and by 2037, the density in Kenya will be twice that in France. Are there factors that will prevent this extreme population growth from occurring in Kenya?

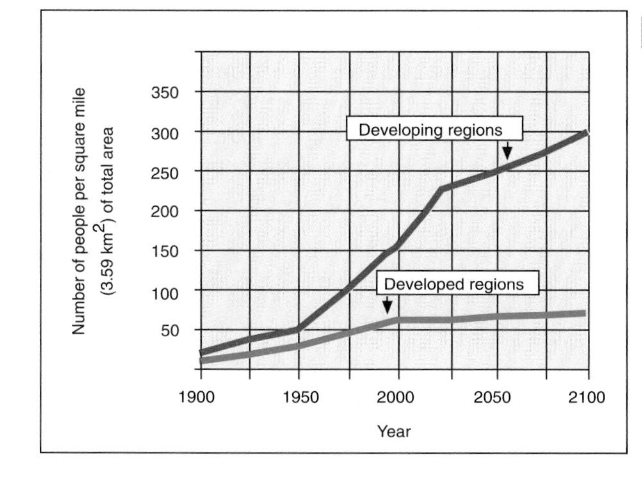

Figure 10.6

The density of humans is expected to continue increasing in the developing regions of Earth but to plateau in the developed regions. (Data are from Table 10.2.)

Table 10.3 Projected Population Density in Kenya and France

	1990	2008	2026	2044
Kenya	42	84	168	336
France	104	113	121	131

The rate of increase *r* in 1990 was 3.8 percent in Kenya and 0.4 percent in France. If these rates remain constant, Kenya's population density will double every 18 years, while the population density of France will double in 175 years, or in the year 2165. Values are expressed in number of people per square kilometer.

Malthusian Limits to Human Population Growth

The emergence of early human populations from the Rift Valley in East Africa set the stage for the development of a species that would radiate into all of the major biomes of Earth. We evolved out of the dim past, were "fruitful and multiplied," and our numbers now exceed 5 billion. That number will likely double in the next four decades. Factors that have contributed to this phenomenal growth include the development of agriculture, the industrial revolution, advances in medicine and public health, the green revolution of the 1960s, and the genetic revolution developing today. The effect of these factors on the population growth rate is illustrated in Figure 10.2B.

In 1798, Thomas Robert Malthus published the first edition of *An Essay on the Principle of Population* in England. He based his essay on two assumptions:

> First, That food is necessary to the existence of man. Secondly, That the passion between the sexes is necessary and will remain nearly in its present state. . . . Assuming then, my postulata as granted, I say, that the power of the population is indefinitely greater than the power in the earth to produce subsistence for man.

Malthus was aware of the disparity between the rate of population growth and the rate at which the food supply could be increased. "Population, when unchecked, increases in a geometrical ratio. Subsistence increases only in an arithmetical ratio." He concluded:

> The only true criterion of a real and permanent increase in the population of any country is the increase of the means of subsistence. . . . Famine seems to be the last, the most dreadful resource of nature. The power of population is so superior to the power in the earth to produce subsistence for man, that premature death must in some shape or other visit the human race.

Malthus assumed that further increases in human populations would greatly exceed the ability of agriculture to feed them. In this belief, he did not anticipate the impact of the industrial revolution that was dawning throughout Europe. Evolving mechanization during the nineteenth and twentieth centuries allowed agricultural production to keep pace with the increasing human population in most Western countries. Nevertheless, numerous famines did occur in less developed regions, but migrations of significant numbers of people to the Western Hemisphere often reduced their impact.

Revolutions and Malthus

The *green revolution* of the 1960s was built on the selective propagation of high-yielding varieties of cereal grains such as wheat and rice. These new varieties helped especially in feeding the burgeoning populations in Asia and Latin America, for when they were treated with modern chemical fertilizers and pesticides, crop yields often increased two- or threefold. In many respects, the green revolution of the past three decades has allowed the world population to surpass 5 billion people. We have now entered the next revolution, the *genetic revolution,* in which advanced techniques of modern science and technology are further amplifying the production of plant and animal varieties.

Malthus, therefore, was wrong in his predictions, not only because of his assumptions about the rate of population growth but also in his assumptions about the growth rate of food production. However, even today, will the "power of the population," in time, exceed the ability of agriculture to supply subsistence for humans? That is, will our human population ultimately grow to a size that exceeds the ability of Earth's ecosystems to feed us? Of equal importance, why does a world that produces enough food to feed the entire human population have such extensive famine? The answers to this question will have to come through cultural, political, and social actions, not from biological research.

In 1798, Malthus described a numerical relationship that showed a great difference between the rate at which human populations grew and the rate at which food was produced. He felt that unless human population growth was slowed, mass starvation was inevitable. However, during the nineteenth and twentieth centuries, technologies were developed that increased rates of food production in the world to a level that kept pace with population growth.

HUMAN IMPACTS ON DRINKING WATER

Early environmental problems caused by our distant ancestors were related primarily to the degradation of agricultural areas adjacent to developing villages, towns, and cities and to effects caused by inadequate disposal of human wastes. Human feces and kitchen wastes accumulated in or near the inhabited areas and created squalid conditions. Such sites were ideal breeding grounds for vermin involved in the transmission of infectious diseases, including rats and fleas infected with the bacteria that caused the famous plagues of the fourteenth century. However, in attempting to rid their villages of garbage, early city dwellers often exchanged one environmental problem for another. In general, they dumped their wastes into the nearest aquatic system, usually the river on which the city was located. This often led to a reduction in fish populations and to water that became unsafe for drinking. Contaminated waters were also the major sources of typhus, cholera, and many diarrheal diseases. Clearly, early planners did not triumph in their attempts to create cleaner, safer villages and cities.

In developed regions, water pollution continued to be a major environmental problem well into the twentieth century, and it remains a critical problem in many developing regions. However, with the application of modern sewage treatment and disposal, most developed countries have now succeeded in reducing the degrading effects of human municipal and industrial wastes in their aquatic systems.

TECHNOLOGICAL SOLUTIONS TO MODERN ENVIRONMENTAL PROBLEMS

Technological advances in agricultural practices have led to huge increases in world crop production, which have sustained the exponential growth of the human population. However, many of these new technologies have created a new set of environmental problems.

Environmental Consequences of the Green Revolution

The success of the green revolution resulted, in part, from the development and distribution of new plant varieties. However, these new crops required the ex-

Figure 10.7

Modern agriculture in developed regions often uses large areas for monoculturing more valuable crops like the wheat being harvested here.

penditure of considerably more energy in field preparation, pumping irrigation water, manufacturing synthetic fertilizers and pesticides, and applying them to fields and forests. These new agricultural methods allowed vast areas to be planted with single-species crops known as **monocultures**. These areas, with their uniform crops, essentially became expansive tracts of unlimited food supplies for classic agricultural pests including insects, soil nematodes, and other destructive species (see Figure 10.7).

Pesticides are any substances that kill unwanted or harmful organisms. Pesticides developed before and during World War II were adapted for agricultural use and significantly reduced the loss of crops attributed to many pests. One class of *insecticides* (substances that kill insects), the chlorinated hydrocarbons that include the well-known DDT, proved to be nearly 100 percent effective against many insect pests when first used. Also, the application of *herbicides* (substances that kill plants) reduced competition from weeds. The promiscuous application of these pesticides provided the foundation for increasing crop production and managing pests. This approach persisted well into the 1960s, when it suddenly became apparent that the concentrations of pesticides required for acceptable pest control had severe environmental consequences.

Pesticide Resistance

After a decade or so of use, it became necessary to use stronger pesticide concentrations because a small number of pests survived each application. These survivors then gave rise to new populations that could tolerate pesticide applications in ever-increasing numbers. Eventually, certain pest populations developed **resistance** to a number of pesticides, some of which were used at remarkably high concentrations, and the ability to manage their numbers slipped from control (see Figure 10.8).

Pesticide Effects in Nontarget Species

DDT and other chlorinated hydrocarbon insecticides were effective because they have two important characteristics. They are *broad-spectrum insecticides,* which means that they kill most insect species, and they have very low water solubilities (that is, very little DDT will dissolve in water). The low water solubility of DDT allowed it to remain effective for a long time because rains would not wash it from the plant foliage and it was less likely to be broken down by natural mechanisms. Thus it would often continue to kill many insects, including those that were not pests, for days or weeks after application.

In the absence of chemical insecticides, insect pests are normally subjected to the assaults of insect predators, parasites, and disease organisms. Unfortunately, many of these beneficial insects are more sensitive to synthetic insecticides than crop pests. Consequently, when chlorinated hydrocarbons such as DDT were used, these natural systems of control were squandered.

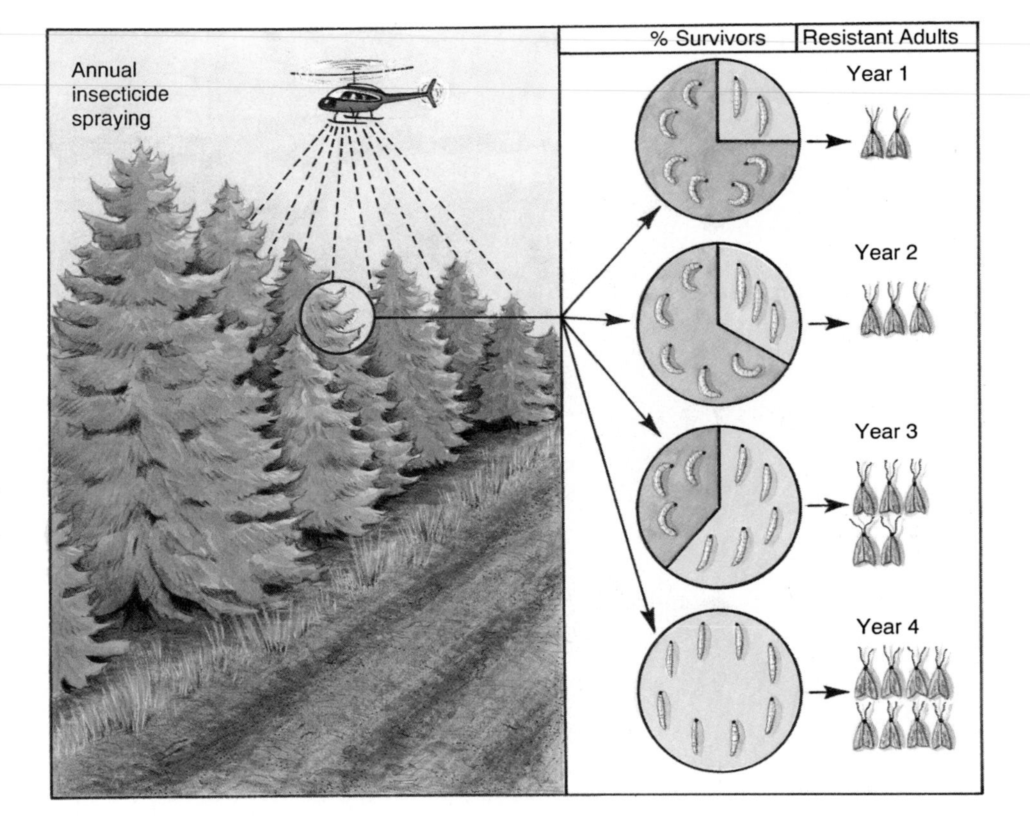

Figure 10.8

Spruce budworm larvae feed on the new foliage of many coniferous tree species. Spraying forests with DDT caused high budworm mortality, represented by the curved larvae. However, a few individuals that were resistant to the insecticide, represented by the straight larvae, survived. After several generations, some budworm populations were highly resistant to the insecticide, and either much larger doses or different insecticides had to be used if desired insect mortality was to be achieved.

Biological Magnification

Bird populations make up another important element of natural insect pest control. By the early 1960s, evidence was accumulating that several bird populations were being adversely affected by high concentrations of DDT. Terrestrial raptors (birds of prey) including owls, hawks, ospreys, and eagles, and many marine predators, such as cormorants and pelicans, were among the species being affected by increasing environmental DDT concentrations (see Figure 10.9). Why were birds so vulnerable to DDT?

In large-scale spraying operations, DDT was applied to crops and forests (see the Focus on Science and Technology, "DDT, Spruce Budworms, and the Insect Bomb"). Scientists now know that once sprayed into the environment, DDT moved from one trophic guild to the next, being stored in the lipids (fats and oils) of each organism. As a result, body burdens of DDT increased within each successive trophic guild in the affected food chains, a process known as **biologi-**

A

B

Figure 10.9

Ospreys (A) and brown pelicans (B) feed at the top of aquatic food chains by eating fish. They were vulnerable to biological magnification of DDT and other chlorinated hydrocarbon insecticides. Consequently, their population densities declined largely because of reproductive failures caused by insecticides.

Figure 10.10

Food chain concentrations of insecticides like DDT occur as a result of its low water solubility and its high affinity for oils and fats. The DDT concentrations in aquatic plants may be very low, but when the insecticide passes to each successive trophic guild, concentrations increase until, at the top of a food chain, they may be high enough to harm top-level carnivores.

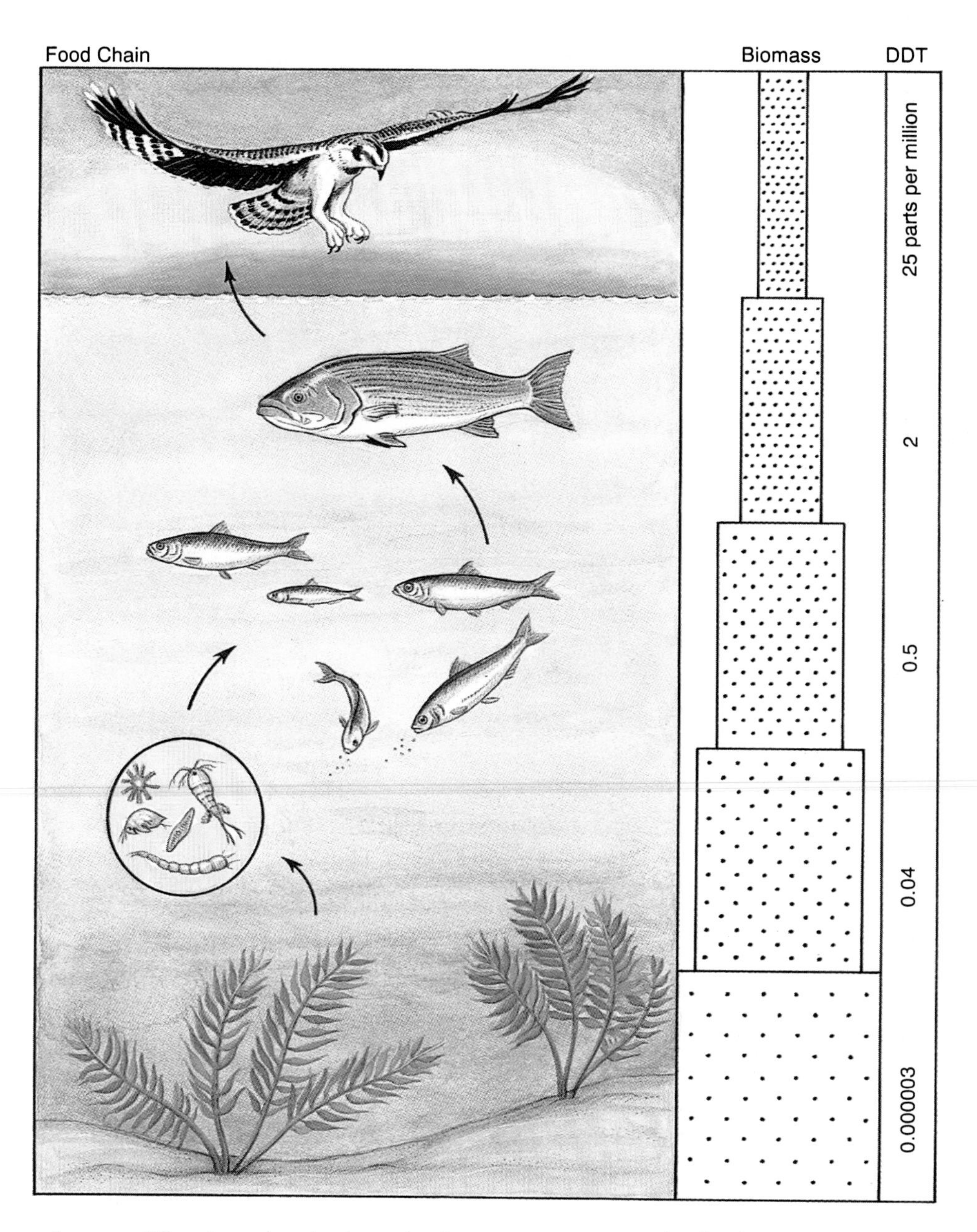

cal magnification that is described in Figure 10.10. The highest concentrations generally occurred in birds feeding at the top of DDT-fouled food chains. These birds, most notably the peregrine falcon, developed a condition known as the thin-eggshell syndrome. When body burdens of DDT reached high levels, females produced eggs with abnormally thin shells. Consequently, these fragile eggs were often broken by the parent birds during nesting. This resulted in low hatching success and sharp declines in the population densities of many birds of prey.

Finally, it was discovered that DDT had been distributed over the entire globe by wind transport mechanisms. Thus after massive spraying operations conducted over two decades, even polar bears in pristine arctic regions were found to have measurable quantities of DDT in their flesh.

As a result of all these negative environmental effects—biological magnification as well as persistence, toxicity to nontarget species, and worldwide transport—the U.S. Environmental Protection Agency banned the use of DDT in the United States in 1972 and prohibited the use of other chlorinated hydrocarbon insecticides by 1975. Nevertheless, DDT continues to be used in tropical regions of the world to control malaria-spreading mosquitoes. For example, 19 million kilograms of DDT are now used for malaria control each year in India. About 80 percent of it is used for mosquito control in domestic houses and cattle sheds in rural areas. Concern is building about significant DDT contamination of stored grains. By eating these stored, contaminated foods, people may raise their tissue DDT levels to nearly twice the currently acceptable U.S. standard. However, in these countries, administrators have decided that the severe economic and personal costs of malaria override the environmental and human consequences of the DDT contamination.

Contamination of water with human wastes and related effects were early environmental problems associated with population growth. In the twentieth century, technological solutions to biological problems resulted in new environmental impacts. Many of these environmental consequences were related to the extensive use of chemical pesticides, especially DDT.

The Future of Agriculture

The success of the green revolution has altered world agriculture in many positive ways. Untold millions of people have avoided famine, and new varieties of rice and wheat are major staples throughout much of the developing world. In many areas, including China and India, more than half of the available cropland consists of monocultures of these two grain crops. However, too much of a good thing can often have negative economic and environmental consequences. For example, global rice production now often exceeds demand. As a result, prices have fallen to very low levels, and many farmers in both developed and developing countries are left with excess crops that have market values less than their costs of production.

Monoculturing of grains leads to a dependency on high-energy agriculture, including synthetic fertilizers and pesticides, irrigation, and high transportation costs. The potential for crop failure is especially high in drought-prone regions, and in some areas, the use of monocultures may have already reached the point of diminishing returns.

Research in many developing countries is once again focusing on *multiple cropping,* an approach that offers several economic and environmental advantages. By growing a number of different interspersed crop species, the demand for pesticides often decreases, and farmers are placed in a more stable economic position should the price of a single crop fall on the world market. Other innovative agricultural technologies hold the promise of reducing energy requirements and dependence on expensive chemical pesticides and synthetic fertilizers.

Malthusian Population Limits and the Future

Many important questions about the quality of human life in the future remain open. Will modern agricultural practices and the emerging genetic revolution contribute adequate food resources for the ever-increasing human population density, or will the predictions of Malthus ultimately prevail? What will be the social, cultural, and environmental costs of a continuously expanding human population? Will fertility control be applied on a worldwide basis, and if so, will it be successful in slowing human population growth?

Modern methods of human fertility control include procedures that prevent fertilization or implantation, and in some countries, abortion if implantation has occurred. Fertilization prevention can be achieved through the use of mechanical, chemical, surgical, and behavioral contraceptive techniques (discussed in Chapter 37). Mechanical and chemical mechanisms are also used to prevent implantation. These methods have ethical considerations and thus are applied in varying degrees among the different human cultures.

In developing regions of the world, contraceptive use is highest in eastern Asia and portions of Latin America, where up to 75 percent of the women report

DDT, Spruce Budworms, and the Insect Bomb

Focus on Science and Technology

The use of DDT was not restricted to agricultural crops in the post–World War II years, for it also seemed to hold great promise as a new method of insect pest control in forest management. From the late 1940s through the mid-1960s, DDT was used extensively to suppress epidemic populations of spruce budworm in the forests of the northern, northwestern, and northeastern United States and the maritime provinces in eastern Canada (see Figure 1).

Aerial application of DDT was used over more than 3.6 million hectares of national, private, and state forests in the Pacific Northwest and the Rocky Mountain regions of the United States from 1949 through 1958 (see Figure 2). Despite these efforts, by 1961 the budworm epidemic had extended into a number of states, and large-scale spraying programs were projected to continue into the 1960s and 1970s. However, data had begun to accumulate indicating that DDT and other chlorinated hydrocarbon insecticides caused environmental damage, including food chain magnification. A new question challenged scientists: What could be done to alleviate the harmful effects of DDT?

In the early 1960s, federal officials began to search for replacement insecticides that would not concentrate in food chains. One such insecticide was Zectran. Early field tests indicated that Zectran would be the most effective replacement for DDT because it satisfied a primary criterion: when applied to the forest, it produced the greatest mortality of western spruce budworms.

In early tests, a number of problems were discovered concerning the delivery system. The spray systems developed for DDT were not highly effective when adapted to Zectran. Basically, smaller drops of Zectran were more effective in killing insects and resulted in less waste than the larger drops produced by the old delivery system. Thus it was necessary to develop a technological system that would produce spray droplets between 5 and 50 micrometers in diameter. (One micrometer is one millionth of a meter.) To accomplish this, the Equipment Development Center of the U.S. Forest Service constructed an "insect bomb" the main feature of which was a Freon gas-charged pressure-and-nozzle system that would produce an insecticide mist of the required droplet size when the liquid Freon changed into a gas (see Figure 3). When sprayed from the aircraft, the insecticide would act like mist from a modern insect fogger and penetrate throughout

Figure 1 (A) Spruce budworms are small insects whose larvae feed on the needles of Douglas fir, spruce, and other tree species, causing damage and, in severe infestations, death (B).

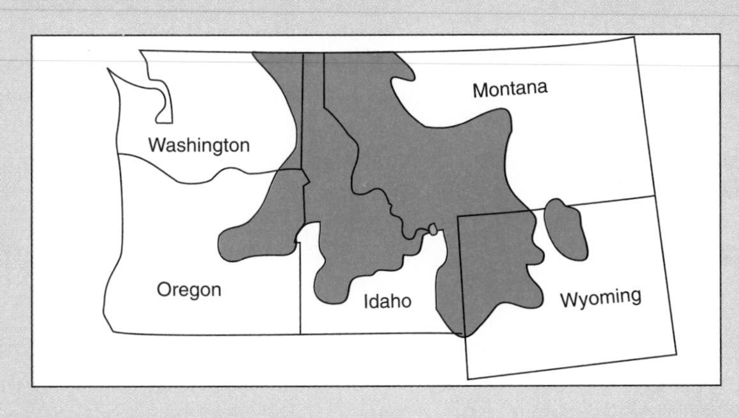

Figure 2 Aircraft were used to spray large forested tracts in the areas of spruce budworm distribution shown here. These spraying operations attempted to control population density in these Northwest states from the late 1940s through the mid-1960s.

the forest foliage. Unfortunately, like DDT, this elaborate system had little permanent effect in controlling the spruce budworm. And another unpleasant surprise lay ahead.

This brief history of attempts to eradicate the spruce budworm illustrates the unpredictable nature of results from the use of a new technology. Chemical technology created DDT and other chlorinated hydrocarbon insecticides. By the middle of the twentieth century, these chemicals were viewed as holding great promise for making major reductions in agricultural and forest productivity losses due to insect pests. However, biological magnification was an unanticipated result and led to the abandonment of these chemicals.

Despite this setback, technology was able to replace DDT oil-based systems with Zectran linked to Freon gas, thus reviving the great promise. Ironically, in 1974, only two years after DDT was officially banned, an article appeared in the scientific journal *Nature* that first warned of the devastating atmospheric effects of Freon and other chlorofluorocarbons (discussed in Chapter 11). Most notably, chlorofluorocarbons have been implicated in reducing the concentrations of stratospheric ozone, a gas that shields Earth from harmful ultraviolet radiation emitted from the sun.

A common theme in environmental science is that for every technological advance, unforeseen problems may compromise apparent benefits. During the 1940s and 1950s, no one could predict the harmful effects of DDT. In the late 1960s, few scientists considered that any potential problem would develop from spraying millions of forested acres with liquid Freon, for at that time it was thought to be an inert gas. What environmental problems will develop in the future from the application of our present technology? Is the ultimate solution for every environmental problem the development of a new technology?

Figure 3 After DDT was banned for use in the United States, other insecticides were substituted in new attempts to control the spruce budworm. Many of these new pesticides required the development of complex pressure-spraying equipment as shown here.

using some method. In general, the burden of fertility reduction rests with the women of the world at the present time. It is widely believed that significant control will result only if more women in developing regions employ some method of contraception for in many cultures throughout the world, men seldom use contraceptive measures. However, this is unlikely to occur unless women in the developing areas are liberated from their heavy workload and given a form of security other than that derived from their children. It has been estimated that women do up to two thirds of the physical work and produce 60 to 80 percent of the food in many developing countries, including Asia, Africa, and parts of Latin America.

Women and children frequently represent the largest proportion of the very poor, landless peoples of the world. Often their only security is in numbers. Unless this cycle is broken, through the education of females and agrarian reforms, population growth rates will likely continue at high levels. If the females among the poor are educated and allowed to participate in the security of land ownership, their dependence on large numbers of children will likely be diminished.

Many important questions about human population growth in the future cannot be answered at the present time. Major uncertainties concern food production, the quality of human life, and social and cultural dilemmas related to controlling population growth.

Malthus's postulate "that the passion between the sexes is necessary and will remain nearly in its present state" retains some validity today. However, although we now have the technical means to prevent conception, the question remains if we will develop the determination, on a global basis, to apply them. What is the upper limit of human population density that the ecosystems of Earth can support? Will new food-producing revolutions unfold in the future? Can we, as modern humans, control our fertility in time, and what will be the economic and environmental consequences if we do not?

Summary

1. The development of agriculture by early human cultures allowed the human population to expand from 5 million to over 5 billion in only 10,000 years. Major human migrations resulted in the habitation of all the major biomes on most continents of Earth.
2. The world population increased exponentially from the late 1800s until the late 1960s, when the rate of growth began to decrease. However, the total population continues to increase and will double in about 40 years if growth continues at the present rate.
3. Malthus assumed that the need for food and the continuing passion between the sexes would lead populations to outstrip the food supply, and famine would visit the human race. However, the agricultural, industrial, and genetic revolutions have allowed human population densities to exceed the limits that Malthus predicted.
4. The propagation of monocultures has led to a reliance on pesticides, fertilizers, and other high-energy processes in agriculture. Pesticide resistance, biological magnification, and nontarget species mortality are some negative side effects of pesticide use.
5. If significant population control is to occur, women in the developing regions must have greater security beyond that derived from their children. This security could include increased access to education and participation in land ownership.

Review Questions

1. What was the relationship between early human culture and food supply?
2. Describe the pattern of prehistoric human population growth.
3. What factors affected human population growth until the modern period?
4. Describe the current pattern of human population growth.
5. What is the significance of Malthus's ideas on human population growth, published in 1798?
6. What was the green revolution? How did it affect human population growth?

7. Describe some of the consequences of using DDT to control insects.
8. What changes may be in store for agriculture in the future?
9. What do experts predict about human population growth in the future?

Essay and Discussion Questions

1. What predictions would you make about the size of the world's human population in the year 2500? What factors might be considered in making such predictions?
2. What types of technological "revolutions" might occur in the future that would allow for expanded human population growth? What might be the consequences of such revolutions?
3. What are the lessons to be learned from the DDT experience?

References and Recommended Readings

Ehrlich, P., and A. Ehrlich. 1990. *The Population Explosion.* New York: Simon & Schuster.

Goudie, A. 1990. *The Human Impacts on the Natural Environment.* 3d ed. Cambridge, Mass.: MIT Press.

Caldwell, J. C., and P. Caldwell. 1990. High fertility in sub-Saharan Africa. *Scientific American* 262: 118–125.

Krebs, C. 1985. *Ecology: The Experimental Analysis of Distribution and Abundance.* 3d ed. New York: HarperCollins.

Lewin, R. 1988. A revolution of ideas in agricultural origins. *Science* 240:984–986.

Managing Planet Earth (special issue). 1989. *Scientific American* 261 (3).

Merrick et al. 1986. World population in transition. *Population Bulletin* 42:16.

Steinhart, P. 1991. Beyond pills and condoms. *Audubon* 1:22–25.

Torrey, B. B., and W. W. Kingkade. 1990. Population dynamics of the United States and the Soviet Union. *Science* 247:1548–1552.

Watt, K. 1982. *Understanding the Environment.* Newton, Mass.: Allyn & Bacon.

World Resources Institute, International Institute for Environment and Development, and United Nations Environment Programme. 1988–1989. *World Resources,* New York: Basic Books.

World Resources, 1990–1991. New York: Basic Books.

CHAPTER 11

Global Climate Change

Chapter Outline

Reading Questions

1. How has Earth's climate changed throughout its history? What factors account for past changes?
2. What factors are influencing global climate at the present time?
3. What are the hypothesized consequences of global climate change?
4. How have various types of pollution affected forest biomes?
5. What types of human activities have affected global carbon sinks? What are the predicted consequences?

Environmental problems have resulted from anthropogenic (human-caused) activities and are now known to affect the biosphere at the local, regional, and global levels. The effects that lead to disruptions in terrestrial and aquatic biomes on a global scale are anticipated to have the greatest impacts in the future. Most environmental scientists believe that changes in world climate will be the most important environmental problem facing Earth's inhabitants in the twenty-first century.

To comprehend present and future global climate changes, it is first necessary to understand how similar changes occurred since Earth's atmosphere formed about 3 billion years ago. Records from the distant past have allowed scientists to study the natural sequences of climate change that have occurred throughout geologic history. From resulting data they have been able to formulate hypotheses about what is happening on Earth now and to make predictions about what the future may hold. In this chapter, we concentrate on those hypotheses and what they tell us about the future of our planet. Though there are many uncertainties, the predicted effects are diverse and generally serious. It is a complex but critically important story for all educated citizens.

TIME SCALES AND CLIMATE CHANGES

Global climate has changed over the past decades, centuries, thousands of years, and millions of years. These changes have resulted in major biome shifts over Earth's surface as glacial and interglacial periods followed one another.

Hundred-Million-Year Intervals

Earth's climate has fluctuated throughout its history. **Paleoclimatic indicators** (records that indicate types of ancient climates) contained within rock strata, lake sediments, ancient coral reefs, and continental ice accumulations have revealed that ice age cycles began 2.5 billion years ago. Other major ice ages occurred at 1 billion, 700 million, 450 million, and from 300 million to 250 million years ago. There have also been numerous minor ice ages. What caused these irregular glacial cycles? Did similar factors lead to the different ice ages? An abundance of scientific data has been used in developing hypotheses about global climate changes in the past and in the future.

Million-Year Intervals

A large body of data dating back to the ice ages of the Pleistocene documents a series of glacial cycles that began about 30 million years ago. At that time, a number of factors intensified a general cooling trend. Those factors, in combination with tectonic forces centered in Antarctica, caused the formation of a massive ice cap over the South Pole.

Orbital Factors

Three properties of Earth's orbit are thought to contribute to the alternating climatic cycles that are responsible for very long glacial and interglacial periods (see Figure 11.1). First, the present tilt of Earth's axis of rotation is 23.5°. However, evidence suggests that it shifts every 41,000 years to a new tilt, between 22° and 24.5° (see Figure 5.6 in Chapter 5). Second, during every 95,000-year period, Earth's orbit around the sun changes from circular to elliptical and back to circular. The third orbital factor is a change in the time of year when Earth comes closest to the sun. Figure 11.1 shows that at present, on the northern summer solstice (June 21), Earth is at its greatest distance from the sun. However, 11,000 years ago, it was at its closest point on this date. All of these factors have affected Earth's climatic history.

Figure 11.1

Regular changes in Earth's orbital properties are responsible for long-term changes in climatic cycles. These changes include variations from the present 23.5° tilt of Earth's axis of rotation to between 22° and 24.5°, a shift in Earth's orbit from a circular to an elliptical orbit around the sun with a period of about 95,000 years, and a variation in the time of year when Earth is closest to the sun.

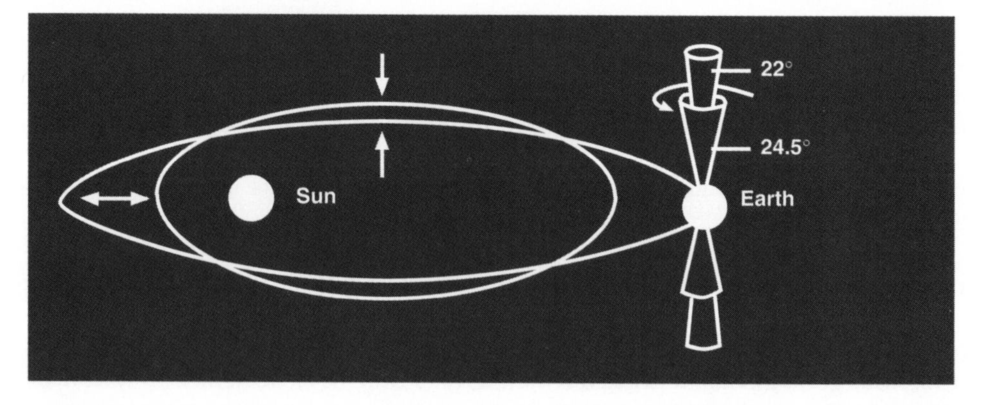

Tectonic Factors

Plate tectonics, the movements of Earth's continental blocks, constitute another factor that may have influenced long-term cooling and warming patterns in the distant past. One recent hypothesis addresses the effects of uplifting of two major plateaus in the Northern Hemisphere. The mountainous plateaus of western North America and the Tibetan Plateau in southern Asia were thrust up during the past 45 million years. These rising landmasses created large weather cells that deflected both the prevailing west-to-east flows of surface air and the jet stream northward. The deflected jet stream would have developed an arc, with a north-to-south orientation that caused cold arctic air to flow into the continental interiors of both Asia and North America. This change may have led to a general period of cooling that culminated in the Pleistocene glaciation.

Other aspects of tectonic activity are also thought to be implicated in the shorter glacial cycles of the Pleistocene. Atmospheric warming probably occurred when volcanic activity, in association with the continental plate movements responsible for creating plateaus, injected massive quantities of gaseous and particulate matter into the atmosphere. These volcanic gases and ash could have altered Earth's energy balance by increasing the concentration of carbon dioxide (CO_2) and amplifying the atmospheric *greenhouse effect,* as described in Figure 11.2. The **greenhouse effect** occurs when concentrations of CO_2, water vapor, and other gases increase and absorb more of the sun's longer (infrared) wavelengths that are radiated from Earth's surface. The effect of this process is to warm the atmosphere. Other important greenhouse gases include sulfur dioxide (SO_2), methane (CH_4), various nitrous oxides (NO_x), ozone (O_3), and chlorofluorocarbons (CFCs). The source (where these substances come from), effects (how they alter atmospheric chemistry), and sinks (biospheric areas capable of storing large amounts of them) differ for each of these gases, but they all contribute to the greenhouse effect (see Table 11.1).

As the atmosphere warmed during an interglacial cycle, more surface water evaporated, and the water vapor load of the atmosphere increased. This atmospheric water vapor would also have absorbed more heat energy radiating from Earth's surface, thus augmenting the general warming trend.

Volcanoes may have been one factor that combined with others to bring the last ice age to an end. However, volcanoes have also been implicated in atmospheric cooling trends that have varied in length from a few years or decades to a century or two. Cooling resulted from injections of SO_2 into the stratosphere during volcanic eruptions. Once present in the atmosphere, SO_2 was converted into tiny droplets of sulfuric acid. Such droplets are very shiny and, as their numbers increased, more incoming solar energy was reflected by Earth's atmosphere. This reduction in incoming solar energy would have resulted in a global cooling trend.

Disappearances of numerous early human cultures have been correlated with periods of intense volcanic activity. If volcanic aerosols, very fine particles

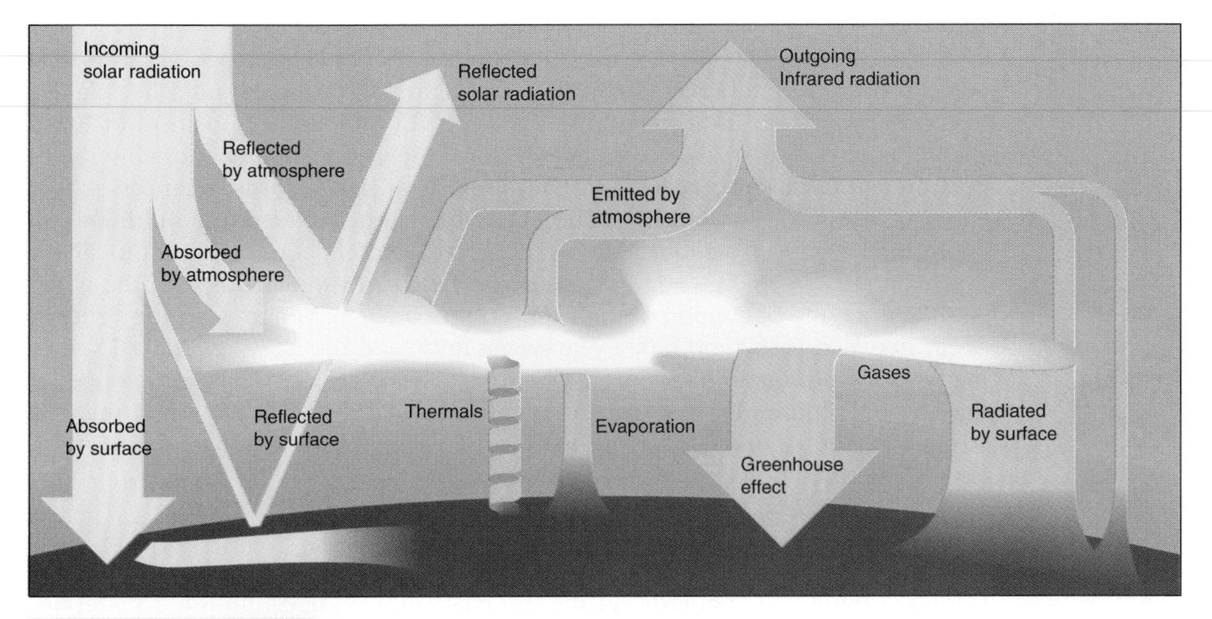

Figure 11.2

Water vapor, CO_2, and other greenhouse gases generally do not inhibit incoming visible and near-infrared wavelengths of sunlight from reaching Earth's surface. The hydrosphere and lithosphere absorb some of this incoming energy but also reflect part of it. The reflected wavelengths are radiated back into the atmosphere as longer-wavelength infrared energy. Greenhouse gases absorb infrared energy more efficiently than visible or near-infrared wavelengths, which results in an increase in atmospheric warming. Thus as the concentration of greenhouse gases increases, atmospheric warming also increases. Widths of arrows indicate relative amounts of energy.

injected into and suspended in the upper atmosphere, led to precipitous drops in the average global temperature, they may have contributed to the demise of certain human cultures by shortening the growing seasons, reducing rainfall, or lowering winter temperatures.

In order to comprehend current changes in global climate, it is helpful to understand natural patterns of change that have occurred in the past. Evidence indicates that Earth's climate has changed regularly during the past 2.5 billion years. These regularities have been primarily related to Earth's orbital properties and tectonic factors associated with wind, volcanoes, and atmospheric changes.

Table 11.1 Sources, Sinks, and Effects of Greenhouse Gases

Pollutant	Primary Source	Sink	Effects
Sulfur dioxide (SO_2)	Burning of fossil fuels	Atmosphere; removed by acid depositions	Soil and water acidification
Nitrous oxides (NO_x)	Burning of fossil fuels and agricultural effluent	Atmosphere; removed by acid depositions	Soil and water acidification
Methane (CH_4)	Cow and termite digestion products; melting permafrost	Atmosphere	Trapping of infrared energy (20 times more effective than CO_2)
Ozone (O_3)	Result from conversions of NO_x and hydrocarbons	Atmosphere; some removed by global forests	Damage to forests and crops
Chloroflurocarbons (CFCs)	Foam expansion, refrigerants, and solvents	Atmosphere	Breakdown of O_3 in the stratosphere

Figure 11.3

Alpine glaciers are presently retreating, as are arctic and antarctic ice caps. As the ice melts, the volume of water on Earth increases, resulting in a general rise in sea level.

The Present

Calculations of orbital relationships between Earth and the sun and other planets in our solar system indicate that the present tilt and orbital position of Earth should result in the smallest average difference between summer and winter temperatures in Earth's recent history. On the basis of past records, scientists would expect that summers would now be rather cool and winters quite mild. During cool summers, the accumulated snow and ice of winter should not melt totally. If this occurred, an enlargement of existing icefields and snowfields over the globe would herald the beginning of the next ice age. If so, maximum glaciation should occur between 5,000 and 6,000 years from now, and the next interglacial warming peak should begin in about 10,000 years. However, all available evidence indicates that alpine glaciers and arctic and antarctic icefields are retreating, not advancing (see Figure 11.3). Are there errors in the orbital calculations or in scientists' interpretations, or are there other explanations?

HUMAN-INDUCED CLIMATE CHANGES

Human activities, particularly during the past century, have modified the atmosphere by increasing the relative concentrations of greenhouse gases. Burning fossil fuels, depleting temperate and tropical forests, increasing desertification, and certain modern agricultural and forestry practices have combined to increase significantly the atmospheric concentrations of CO_2 (see Figure 11.4). Because these activities are expected to continue, this trend is likely to persist into the next century.

The photosynthetic activities of living plants and algae extract CO_2 from the atmosphere or hydrosphere, and the activities of all organisms return CO_2 to the air or water through cellular respiration (see Chapters 2, 7, and 29). Minor temperature variations have a relatively limited effect on the photosynthetic rate but cause major changes in respiration rates. Thus cellular respiration of fungi and bacteria, the major decomposers, varies directly with changes in temperature. With increasing environmental temperatures, the rate of organic decomposition is expected to increase while the global photosynthetic rate remains relatively constant. What effect will this have on the atmospheric concentration of CO_2?

Anaerobic decomposition takes place in areas where oxygen is limited. Methane is a major by-product of this type of decay, and if temperatures rise, the amount of atmospheric methane should increase because of a general increase in the rate of decomposition. At present, the global atmospheric concentration of methane is 1.68 parts per million, but it is increasing at a rate of nearly 1 percent per year.

Analyses of polar ice cores from Vostok station, Antarctica, and others from Greenland indicate that atmospheric methane concentrations began increasing significantly between 100 and 200 years ago, resulting in a doubling of the atmospheric load of this gas. These data correlate well with increasing CO_2 concentrations in the atmosphere. When coupled with the effects of increasing CFC concentrations and the depletion of stratospheric ozone, these changes in atmospheric chemistry are now predicted to cause an increase in average global temperature between 1.5°C and 4.5°C by 2030.

General circulation models (GCMs) are computer models used to make predictions about long-term changes in global conditions. GCMs that incorporate greenhouse warming data predict that the general rise in temperature will not be uniform over Earth's surface. Rather, temperatures in the subpolar and temperate regions may rise between 5°C and 10°C, and in the polar regions they may increase by as much as 20°C. How significant would a 4.5°C average global temperature rise be in the next 38 years? What would be the effects of a 20°C temperature rise in far northern and southern latitudes? Is the greenhouse effect induced by human activities inhibiting the general cooling expected? Is Earth entering its predicted next ice age or not?

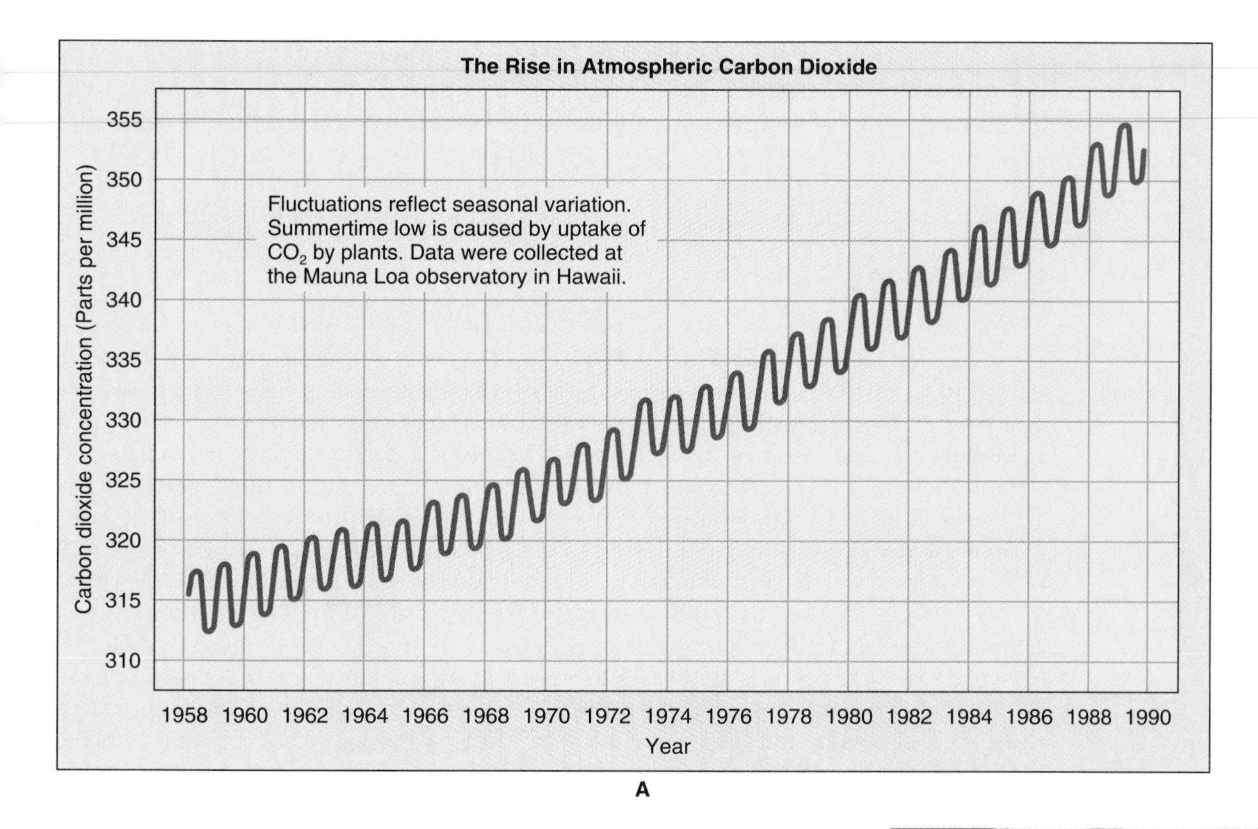

A

B

C

D

Figure 11.4

(A) Data collected at the Mauna Loa climate observatory in Hawaii indicate that the atmospheric concentration of carbon dioxide has increased by about 40 parts per million since 1957. Seasonal variations in these data reflect the increased photosynthesis by producers during Northern Hemisphere summers. The burning of Douglas fir forest debris after logging (B); trophical forest clearing and burning (C); and industrial pollution from a carbon black plant in Transylvania, Romania (D) all contribute to the ever-increasing concentration of atmospheric CO_2.

During the past century, human activities have caused changes in concentrations of greenhouse gases in the atmosphere. As a result, normal biological cycles related to photosynthesis, respiration, and decomposition have been modified quantitatively. Scientific analyses indicate that the long-term global consequences of these changes will include depletion of stratospheric ozone and increased warming.

THE ECOLOGICAL EFFECTS OF GLOBAL TEMPERATURE INCREASE

During the peak of the last ice age, 18,000 years ago, average surface temperatures for both land and oceans are estimated to have been 5°C cooler than at present. The last major ice age cycle ended about 6,000 years ago, when summer temperatures throughout the continental interiors of Europe, Asia, and North America were 2°C to 4°C higher than at present. If Earth has indeed entered its next cooling period, and it took 6,000 years for the temperature to decline 2°C to 4°C, what might happen if global average surface temperatures increase 4.5°C in the next 38 years? Major predicted effects include an increase in the average global sea level and severe negative impacts on Earth's biota.

The Rising Tide

A global rise in sea level has occurred during the past century. Ocean levels are 10 to 20 cm higher over Earth's surface, and they continue to rise at an average rate in excess of 1 millimeter per year. Between 25 and 50 percent of this sea level rise can be attributed to thermal expansion of the ocean waters caused by global warming, because when water is heated, its density decreases but its volume increases. The remaining portion is thought to be due to the melting of terrestrial glaciers and floating icebergs. If the greenhouse effect continues unabated, Earth's oceans are expected to continue to rise, and salt water will flood important estuarine habitats and many low-lying terrestrial areas.

Recent ice core data indicate that most, if not all, of the Greenland ice sheet melted during a major interglacial period more than 100,000 years ago. This evidence correlates closely with data from studies of ancient coral reefs, which indicate that during the same period, the average sea level was 6 meters higher than at present. Much more of Greenland's ice sheet is thought to have melted during that earlier interglacial period than during the present phase. It is postulated that this was due to a number of combined orbital and tectonic factors that resulted in greater insulation and warmer temperatures.

What effects would an ever-increasing rise in sea levels have on major centers of human population that are located along the coasts of most continents? The effect will likely vary with latitude, but one recent computer model predicts that in Florida and in other Gulf states, vast areas will be flooded and beaches will disappear or re-form inland. In addition, there will be an increase in the frequency and severity of hurricanes, which will cause further direct damage and extensive saltwater intrusions into many freshwater systems.

Warming Effects on Biomes

The increasing thermal effects linked to accumulating atmospheric greenhouse gases are elevating the average surface temperatures of both marine and terrestrial areas of Earth. This warming is expected to produce changes in general circulation patterns of the atmosphere and oceans (see Chapter 5). These changes could produce markedly different rainfall distributions that may result in variations in global soil moisture (see Figure 11.5). When coupled with increases in average annual temperatures, major climatic conditions will likely be altered within and between the biomes of Earth.

The response of communities within different biomes will vary with increasing latitude. However, major effects will probably occur in all biomes as each species responds to rapidly changing environmental conditions. Plant communities will suffer the greatest effects because plants are affected directly by changes in available moisture, and unlike animals, plants cannot move.

Shifts in Forests

Temperate and boreal forests will probably be affected to a greater degree than tropical and subtropical forests. One computer model that assumes a doubling of CO_2 levels predicts that today's environmental conditions in temperate forests of beech, birch, hemlock, and sugar maple in eastern North America will change and cause these forests to shift north as far as 500 kilometers. In other words, if the eastern beech forests now growing from southern Canada to Florida are to survive, they would need to "migrate" (colonize a new area) to the area between

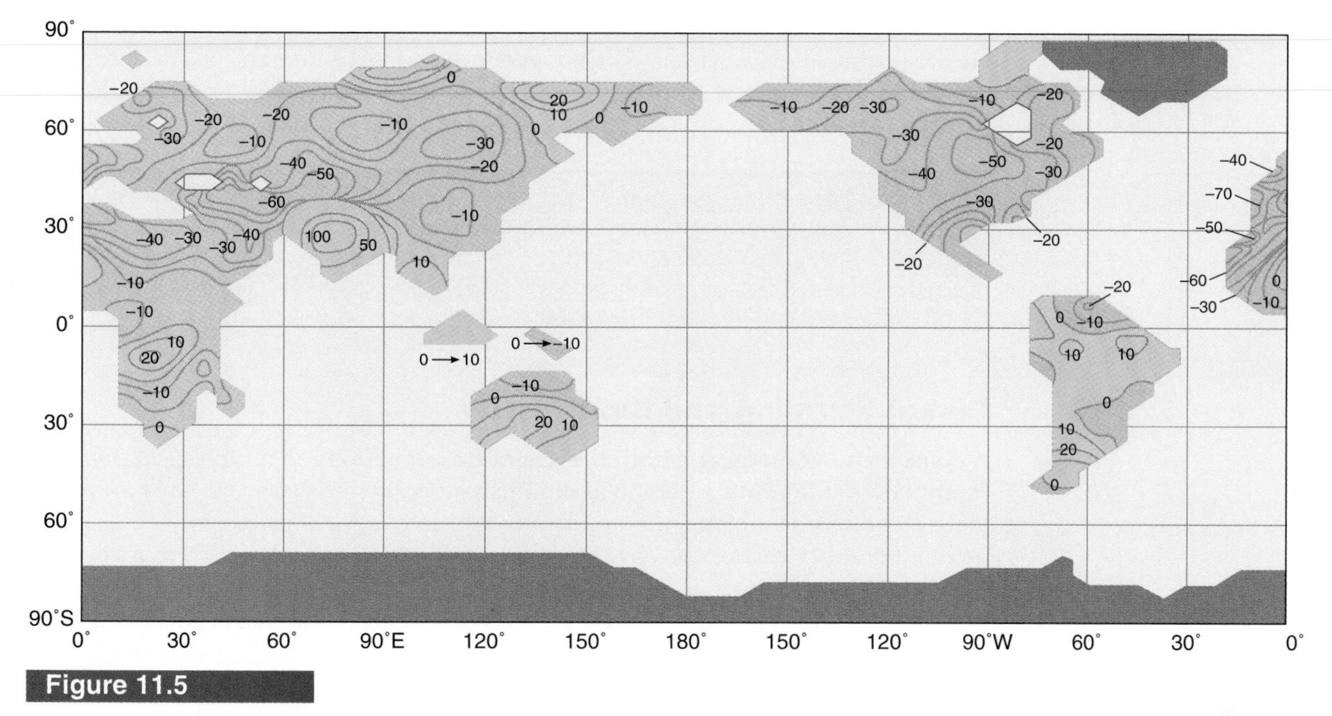

Figure 11.5

This map, generated from a computer model, shows the percentage change in soil moisture that could result from a doubling of atmospheric CO_2. The variations correlated with changing latitudes have important ecological ramifications.

northern New England and the shores of Hudson Bay. Can beech trees migrate this far this fast?

With the retreat of Pleistocene glaciers, developing forests in eastern North America migrated at a rate of about 20 kilometers per century as temperatures rose an average of 3°C to 5°C over a 10,000-year period. It is clear that if the temperature rises by an equal amount in only one century, the eastern deciduous forests cannot migrate at an equal rate, and this forest type may undergo a rapid decline. The old trees could survive for a time, but the production of new saplings would be reduced from current levels.

The coniferous forests of western North America will also undergo significant changes as their temperature and moisture regimes respond to greenhouse warming. Some species may migrate to higher elevations, but the available area in the higher reaches of the western mountains is limited, and for some trees, the proper climatic conditions may exist only farther north in the boreal forest and tundra regions of Alaska and Canada.

Forest migrations are expected to occur throughout most of the Northern Hemisphere. Forests now growing in northern Europe and the great expanses of boreal forest in the Soviet Union will likely join this parade.

Shifts in Grasslands and Deserts

With moisture and temperature maximums shifting northward, major environmental changes are also expected in the shortgrass and tallgrass prairie regions of central North America. The growing season in the corn belt will likely become drier and hotter, and normal summer rains may fail to fall. The vast areas of wheat now growing in the original shortgrass prairie regions may shift from the northern Great Plains into north-central Canada. The deserts of the western United States are also likely to expand northward.

The amounts of moisture now captured and stored in the deep winter snowpacks of the Cascade, Sierra, and Rocky Mountain ranges may decrease. Such a decline would severely limit the amount of water available to the major population centers in the southwestern states. What will be the fate of Los Angeles, Phoenix, and other large southwestern cities as their sources of water decrease during shorter winters and longer, even hotter summers? Will California's Central Valley continue to provide its agricultural bounty? Will the aqueducts of Arizona run dry?

> Slight increases in global temperature over time may result in significant changes in global sea levels. Increasing temperatures and sea level and atmospheric changes are expected to have major impacts on Earth's biomes. The species composition of forests and grasslands will change, and the geographical locations of most biomes will shift. Some biomes will shrink while others expand.

Global Warming and Biodiversity

A consensus is emerging within the scientific community that biological diversity (numbers of different species) will diminish globally in response to the continued global warming resulting from anthropogenic activities. This loss in diversity will be in addition to that caused by the continued loss of habitat related to growth of the global human population. The decrease in species diversity due to global warming will not be uniform over the surface of Earth, and major effects are predicted for the arctic biomes.

In his book *Biodiversity* (1988), E. O. Wilson develops numerous reasons why the loss of biodiversity is important to all humans. Among them he states: "We have come to depend completely on less than one percent of the living species for our existence, the remainder waiting untested and fallow." What potential treasures will be lost when these uncounted organisms enter the abyss of extinction? Will a new important food source have been among them? Will we lose a potential cure for some devastating human disease? Do the biota with which we share the biosphere have any right to continued existence? At what point will the loss of biodiversity begin to affect human populations directly?

Changes in Arctic Biomes

The warming trend will be greater at higher latitudes where extreme effects on numerous arctic tundra species are predicted. Many species of migratory birds could be affected when rising sea levels inundate their nesting sites. The massive insect blooms that occur in the Arctic each spring may appear much earlier, long before the birds arrive for their nesting season. Numerous species depend on these insects for feeding their young. Will migratory bird species be able to adjust their life cycles to these new conditions?

Higher average temperatures, earlier seasonal changes, and melting of the permafrost will have major effects on many resident and migratory animal species and on seasonal progression in the growth of different plant species. If these predicted scenarios occur in the future, some species may not have the genetic, behavioral, or physiological flexibility to accommodate such radical changes. If not, will they face extinction?

Changes in Temperate and Tropical Biomes

Global warming will also affect species in temperate and tropical regions. Tropical diseases from equatorial regions may spread into the subtropical and southern temperate regions. Parasitic diseases such as malaria, schistosomiasis (snail fever), and onchocerciasis (river blindness) may be able to enter large human population centers that are presently located north and south of the areas in which they now occur. Malaria still kills about 5 million people per year worldwide. How many may die if global warming brings more rain, warmer temperatures, and malaria-carrying mosquitoes into today's disease-free subtropical regions?

With increasing insolation, the hot winds blowing out of East Africa will increase the rate of water evaporation as they cross the Indian Ocean. As global warming continues, monsoons sweeping across south-central India may carry twice the amount of rainfall of current levels. What impact will these torrential rains have on the immense human populations of India and Pakistan? Will their effects on rain forests increase or decrease species diversity in these tropical regions?

Changes in Aquatic Systems

In the Great Lakes basin, the effects of a global temperature rise are predicted to include warming of habitats in streams and lakes, shrinking areas acceptable to salmon and trout species, and increasing habitats for less valuable warm-water fish species. Habitat changes will include decreases in wetland and littoral areas as less precipitation occurs in the basin and more water is lost to evaporation. These changes will adversely affect shallow fish-spawning and nursery areas. The general effects will probably include geographical shifts of entire fish associations that will result in changes in the relative abundance of different species. Salmon and trout species will move farther north within the basin.

The impact these changes will have on the southern Great Lakes basin sport fishery will likely be significant. Will warm-water species capture the interest of the sport fishing industry that the cold-water salmon and trout species do at present? If not, what will be the total economic consequences in affected communities?

> Scientists hypothesize that biodiversity will be reduced on a global scale if abnormal global warming occurs. The effects of this loss cannot be predicted with certainty, but they will likely be significant for the human population. While all terrestrial biomes will be disrupted, arctic biomes will probably undergo the greatest change. Changes in aquatic ecosystems will ultimately change the species composition of most communities.

REGIONAL EFFECTS OF HUMAN ACTIVITY

The environmental effects of human activities are not uniform over all inhabited areas of Earth. Factors that contribute to these variations in effects include differences in population density, variable levels of technological development, and climatic variations.

Acid Deposition

Acid deposition is a process in which acidic substances are delivered from the atmosphere to Earth's surface. The major precursors of these acids enter the atmosphere as emissions of SO_2 and NO_x. Chemical reactions in the atmosphere convert these precursors into sulfuric and nitric acids. As described in Figure 11.6, these acids arrive on surfaces in three forms, *wet deposition* as acid precipitation, *occult deposition* directly from fog or cloud droplets, and *dry deposition* of particles.

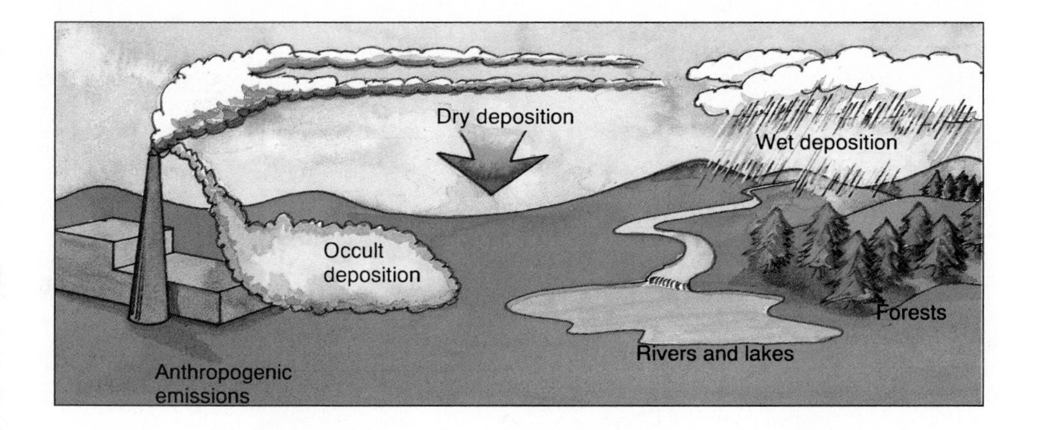

Figure 11.6

Major pathways of acidic compounds from anthropogenic emissions include occult deposition directly from fog or cloud droplets, dry deposition of acidic particles, and wet deposition as acidic precipitation. Both terrestrial and aquatic ecosystems have been seriously affected by acid deposition.

Emission Sources

Substances that contribute to total atmospheric acidity originate from both natural and anthropogenic sources. Globally, SO_2 emissions from industrial processes are estimated to be about equal to those from natural sources. However, biogenic (natural) sulfur emissions are greater in the tropics, while anthropogenic sulfur is greatest in Europe and North America. Major biogenic sulfur emissions include large quantities of dimethylsulfide produced by marine phytoplankton. In addition, hydrogen sulfides and organic sulfur compounds are emitted from swamps, mud flats, and other anaerobic environments.

Major anthropogenic sources of both SO_2 and NO_x include coal-fired electrical generating plants and, for NO_x especially, exhaust from internal combustion engines (see Figure 11.7). By far the greatest amounts of SO_2 come from burning coal that has a high sulfur content. These emissions are associated with high atmospheric acid concentrations on a regional basis. In North America, the regions of highest atmospheric acid concentrations are downwind from major power-generating plants located in midwestern and northeastern states and the southeastern Canadian provinces, as shown in Figure 11.8. More than 75 percent of the United States' SO_2 emissions originate east of the Mississippi River.

Ecological Effects

Ecological effects of acidic depositions include measurable increases in the acidity of lakes and streams, observed deterioration in the quality of coastal waters, and increased groundwater contamination. Acidic depositions have also resulted in major declines in aquatic communities. Many affected lakes no longer contain fish. Finally, acid deposition has tentatively been identified as one of the contributing factors in forest diebacks in affected regions of the United States and Europe.

Surface Water Acidification

The degree to which acid deposition results in the acidification of surface waters depends to a large extent on the *buffering capacity* (ability to neutralize acids) of the water. Alkalinity (the amount of hydroxide ion, OH^-) is the most useful measure of the *acid neutralizing capacity* (*ANC*) of receiving waters because alkaline substances in the water will neutralize a certain quantity of acid. The geology of a watershed and its hydrologic characteristics (flow rates, flow paths, and storage capacity) determine the buffering capacity of its lentic systems.

Figure 11.7

This is an example of a coal-fired, electrical-generating plant.

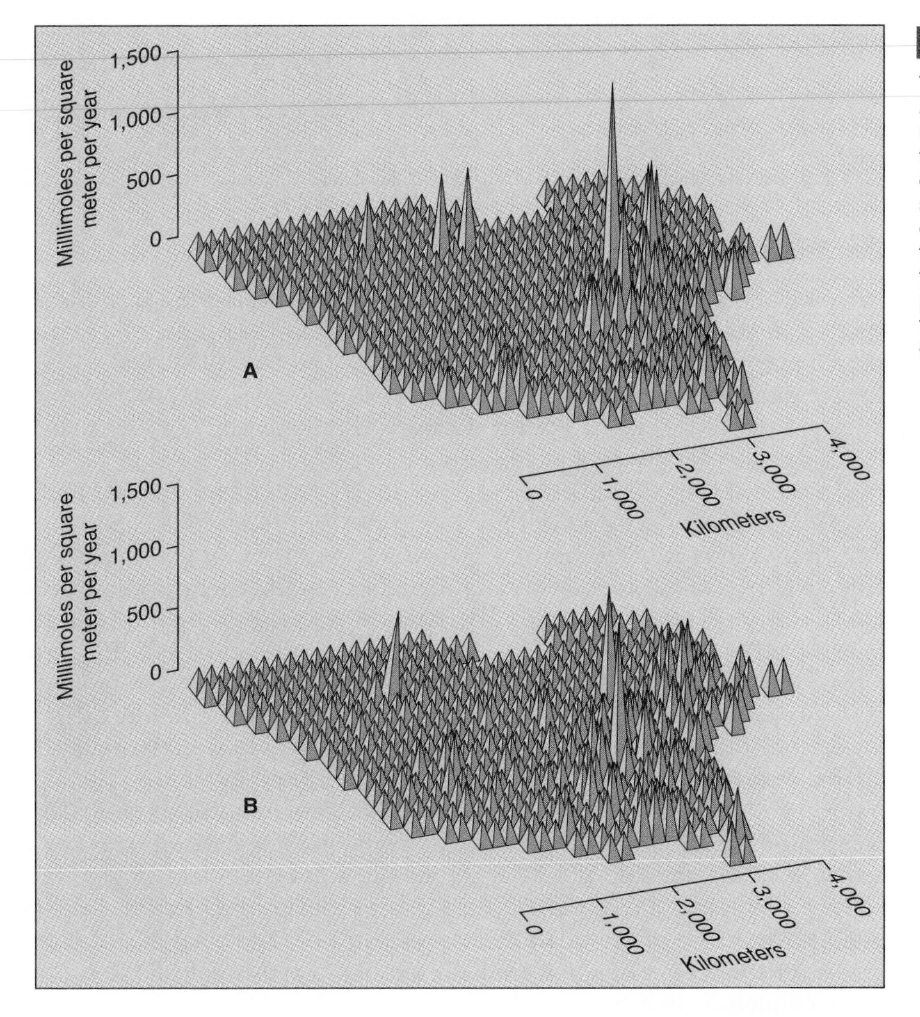

Figure 11.8

These maps show portions of Canada and the United States south from 60° north latitude. They have a grid structure of 1° longitude by 1° latitude per cell, and each cell represents 9,435 square kilometers. The upper map (A) describes emissions of sulfur dioxide; the lower map (B), nitrogen oxides. At a finer resolution, the major concentration peaks indicate that emissions come from point sources, such as coal-fired electricity-generating plants.

When acidic depositions fall on a watershed, the magnitude of their effects is generally determined by the ANC of the system. Lakes with low alkalinity are more sensitive to acidification than lakes with higher alkalinity. If wet, occult, and dry acidic depositions from anthropogenic sources exceed the ANC of susceptible lakes, they become acidic.

Biotic Effects

When the acidity of surface waters increases, biotic effects often include declines in the populations of salmon, trout, and other cold-water fish species. Changes in species diversity are among the first effects of aquatic acidification. Studies of the effect of increasing acidity suggest that mortality during early life stages is responsible for declines in fish population in acidified lakes.

Recent studies estimate that nearly 500 lakes (18.7 percent of the total number) in the Adirondack Mountains in northern New York State have become acidified as a result of acidic depositions from anthropogenic sources. These lakes once supported large recreational fisheries, but due to increasing acidification, the golden days have passed. The demise of these fish populations has resulted in projected losses in excess of $13 million per year, excluding the "loss of amenity" costs to the surrounding communities.

Acidification of surface waters has also resulted in losses or declines of Atlantic salmon in rivers in Nova Scotia and other areas in eastern Canada, Norway, and Sweden. The total dollar cost of these losses is unknown, but with continued input of SO_2 and NO_x from coal-fired power plants and other anthropogenic sources, it will continue to increase.

Acid depositions trace their origin to gaseous sulfur and nitrogen compounds emitted into the atmosphere from various natural and industrial processes. Lakes that receive acid precipitation may or may not become acidic, depending on their buffering capacity. Fish populations have been eliminated from lakes that have become acidified.

Air Pollution and Forest Decline

In some temperate forests of Europe and North America, a number of tree species in stands adjacent to or downwind from major industrial centers began showing symptoms of decline and dieback in the late 1970s and early 1980s. In **forest decline**, trees develop yellow leaves or needles, and tree growth per unit of ground area decreases. **Forest dieback** is the mortality of select species within a forest stand or even of an entire stand.

Forest decline and dieback are rapidly becoming global in scope. Major forest declines have occurred in New Hampshire and New York, and in Vermont, a significant dieback of mature red spruce has occurred on the upper slopes of an area called Camels Hump. Profound declines and diebacks have occurred in the forests surrounding the Mexico City basin, in Bavarian forests in Germany, in numerous other forests in central Europe, and in the subalpine forest of central Japan.

In general, sites of forest decline and dieback contain trees that are dying for reasons other than normal successional or density-related mortality. Studies in Germany suggest that a network of effects, described in Figure 11.9, are combining to produce forest decline and dieback. This hypothesis indicts numerous compounds common to the atmospheric emissions generated by many industrial activities. These include SO_2, NO_x, ammonia, and ozone. For some of these atmospheric pollutants, the pathways leading to reduced tree growth have been well established; those of the others are at present only speculative. For example, surface-level concentrations of ozone are known to cause damage to the needles in many coniferous tree species.

Soil Acidification

Certain atmospheric pollutants are thought to damage forest stands principally by their effects on soil. Depositions of sulfur, nitrate, and ammonium have been shown to alter soil chemistry. In these chemically altered soil environments, tree roots absorb ammonium rather than nitrate, a process that diminishes the uptake of magnesium (Mg). Trees suffering from reduced Mg uptake often compensate by mobilizing Mg reserves in old needles, causing those needles to turn yellow. If normal tree growth is to continue, Mg concentrations must somehow be maintained since Mg is an essential element for all plants. Acidification also alters a tree root's ability to obtain the necessary ratios of other required nutrients.

Studies in forests of the Camels Hump region of Vermont support the German results. They also indicate that existing atmospheric concentrations of nitrogen-containing compounds can *overfertilize* the forest because trees can take up these substances from the air through their leaves or needles (**foliar uptake**). These compounds include nitrate, ammonium, and, at some sites, gaseous nitric acid (HNO_3). Foliar uptake of excess nitrogen compounds induces trees to grow rapidly, at rates that cause Mg to become limiting. Liebig's law of the minimum (see Chapter 7) seems to apply in these circumstances, because the lack of one required element (Mg) seems to limit tree growth. Thus uptake of atmospheric nitrogen-based pollutants can promote canopy growth, which in turn exhausts the tree's reserves of Mg. To compound the problem, in soils that have become acidic, essential elements are unable to move freely into a tree's roots because they become bound to the soil. Thus acidic soils prevent adequate supplies of Mg (and other elements) from entering the roots.

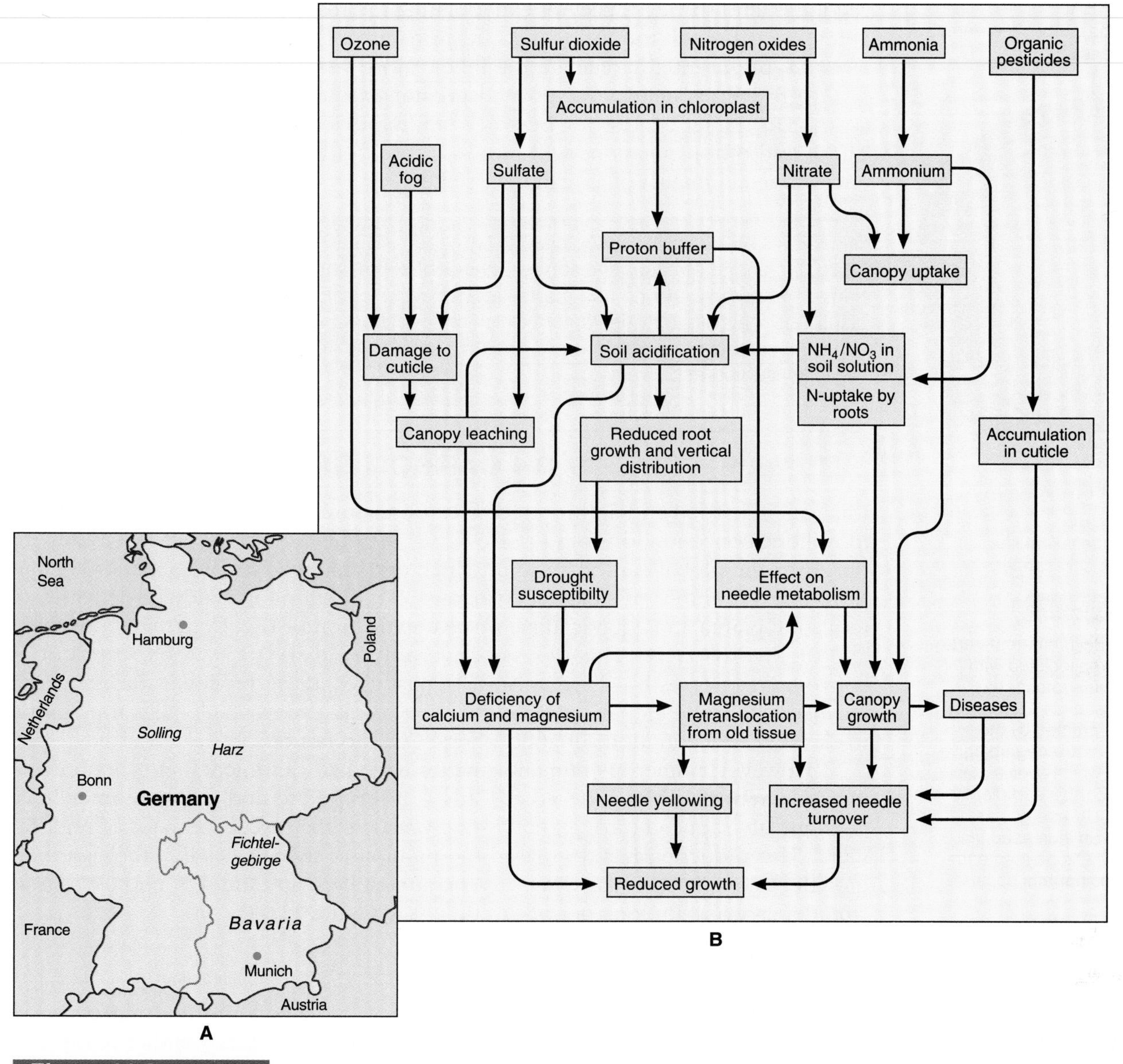

Figure 11.9

(A) One study of forest decline was conducted in the Fichtelgebirge, in northeastern Bavaria. (B) This schematic diagram relates important air pollutants to their effects in soil and plants. The total effect of the pollutants is to reduce plant growth, which can lead to forest decline.

Although other atmospheric pollutants are involved, SO_2 and NO_x are primarily responsible for causing soil acidification. Acidification induces various nutrient imbalances that create stress on trees. Also, nitrates are removed from the soil and transferred to groundwater, where they are often inaccessible to plants. The resultant nutrient imbalance leads to reduced growth, forest decline, and eventually forest dieback.

Ozone Effects on Forests

Increasing concentrations of O_3 in the lower portion of the troposphere (the near-surface, weather-generating layer of the atmosphere) are adding a stress to many temperate forests, and, as we will see later, stratospheric ozone concentrations are related to the amount of ultraviolet light reaching Earth's surface (see Figure 11.10). Recent laboratory studies indicate that photosynthesis and growth rates are reduced in many tree species when they are exposed to O_3 levels now common in many areas of the United States. These studies were conducted on

STRATOSPHERE
CFCs degrade ozone, allowing more UV to pass through.

60° N

Equator

60° S

TROPOSPHERE
Ozone damages plant life; UV increases skin cancer.

Figure 11.10

The presence of ozone in different parts of the atmosphere is associated with different effects. In the stratosphere, ozone is beneficial because it absorbs harmful UV radiation emitted by the sun. At present, CFC's are degrading stratospheric ozone, which allows more UV energy to reach Earth's surface. Increased exposure to UV energy has been correlated with an increased incidence of skin cancer. Ozone in trhe troposphere, however, is harmful because it can damage plants.

seedlings, and the legitimacy of extrapolating the results to a living forest is debatable. Nevertheless, the following questions are of great economic importance: Does O_3 preferentially damage the taller, more valuable trees in the forest? Are faster-growing species affected to a greater extent by O_3 than slower-growing types? Do the effects differ between multispecies and single-species stands? These and other questions will require more field studies before the complete environmental impact of O_3 can be assessed.

The forests of the world have proved to be especially susceptible to various forms of pollution. Increased atmospheric concentrations of sulfur- and nitrogen-containing gases have been correlated with tree mortalities in several parts of the world. Acidified soils disrupt normal tree activities and diminish chances for long-term survival. Increased ozone levels are suspected to have negative effects on trees, and this problem is being carefully studied at present.

HUMAN IMPACTS ON CARBON SINKS

Changes in natural and anthropogenic carbon sources are causing increases in atmospheric concentrations of CO_2. However, the effect of this increase is partly reduced by the global carbon cycle in which dynamic mechanisms extract CO_2 and store it in a number of sinks (see Chapter 7). Global photosynthetic activities of terrestrial plants and aquatic algae involve extraction of CO_2 from the atmosphere and the hydrosphere. This CO_2 is converted to sugars and can be stored in or converted to tissues by all organisms. In past ages, these processes maintained a balance within the global carbon cycle. As a result, excessive global warming due to elevated concentrations of atmospheric CO_2 did not occur.

However, numerous human activities have had detrimental effects on these natural **carbon sinks**. The great forests of temperate and tropic biomes have been greatly reduced through harvesting practices that exceeded rates of replacement. This reduction has decreased the number of living trees that through their growth removed carbon as CO_2 from the atmosphere and through photosynthesis stored it as wood for long periods of time (see Figure 11.11).

Tropical Deforestation

Large-scale tropical **deforestation** continues to accelerate in the Amazon basin south of the Amazon River because of the construction of all-weather roads. Nations with jurisdiction over this tropical forest biome include Brazil, Venezuela, Colombia, Ecuador, and Peru. The major areas of deforestation have occurred in the Brazilian states of Acre, Mato Grosso, and Rondônia. For example, the state of

Figure 11.11

(A) Students in Kenya, Africa, planting trees to reclaim deforested areas. (B) An "urban forest" created on Earth Day 1990.

A

B

Rondônia is at present undergoing the greatest amount of deforestation in its tropical forests (see Figure 11.12). Rondônian deforestation has been continuing since the early 1970s, but with increased access provided by new all-weather roads, it has now reached devastating proportions, as indicated by Figure 11.12C. To a large extent, this deforestation correlates directly with new frontier expansion policies. These policies have resulted in a surge of human migrations from overpopulated cities in Brazil into these rain forests, which has led to the clearance of vast areas of the forest for planting subsistence crops and grazing cattle.

The loss of these vast tracts of tropical forests will obviously reduce this global carbon sink to some degree. However, Amazonia is not the only tropical area where deforestation is occurring. Areas of Africa, India, and Oceania are also undergoing major deforestation. It has been estimated that on a global basis, tropical forests are disappearing at an annual rate of 11 million hectares, an area larger than the state of Ohio. Because of the acceleration in deforestation rates, the extent of denuded areas is expected to increase rapidly in the near future.

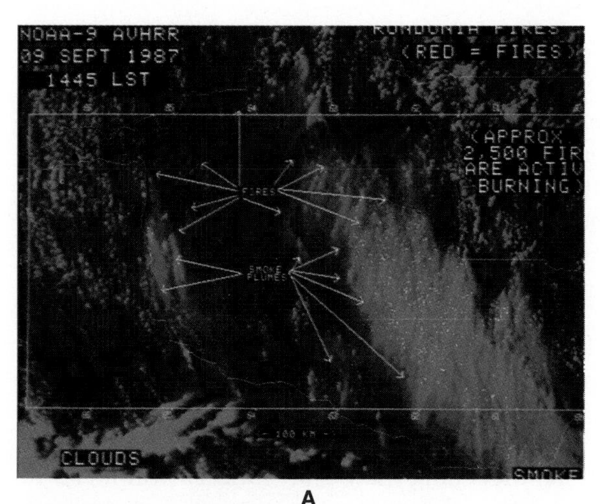

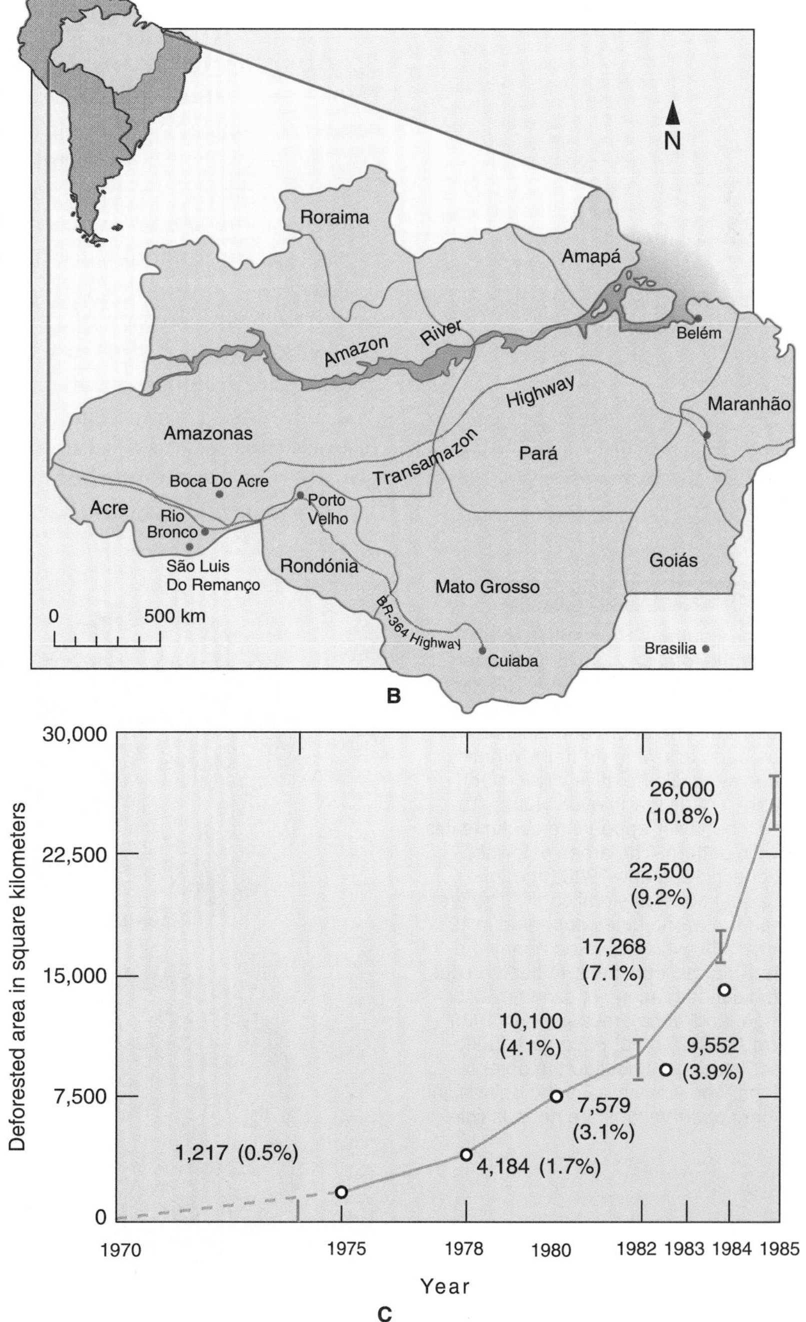

Figure 11.12

(A) This view from space was generated from data transmitted by the NOAA-9 weather satellite as it passed over the Brazilian state of Rondônia. Clouds are labeled at the lower left, and fires and smoke plumes can be seen in the center of the picture. These fires are the result of massive efforts to clean and burn tropical rain forests in this region. (B) This map of the northern states of Brazil shows highway BR-364, which runs through Rondônia. The access provided by this all-weather road opened up the tropical forest to new population migrations, which led to the forest being cleared and burned. (C) This graph shows the rate of deforestation in Rondônia from 1970 to 1985. Note the rapid increase since the early 1980s. The brackets at the data points for 1982, 1984, and 1985 indicate a range of more or less than 1,000 square kilometers.

Marine Carbon Sinks

The other major global carbon sink consists of marine algae and diatoms that extract CO_2 from ocean waters. When these organisms die, the carbon they contain eventually sinks, forming carbonate deposits on the ocean floor. Human activities may also be decreasing the amount of carbon entering this sink as a result of marine pollution and stratospheric ozone depletion.

Marine Pollution

The pollution of near-shore waters with municipal garbage, industrial effluent, and major oil spills is likely to have an adverse effect on the primary producers in these aquatic life zones. Worldwide, the magnitude of marine pollution is unknown; thus the extent of decrease in CO_2 extraction rates is also unknown. However, some scientists believe that this loss is significant because deposits from past ages contain immense amounts of carbon. Hence in the past, major quantities of CO_2 must have been removed by primary producers inhabiting these areas.

Stratospheric Ozone Depletion and CO_2 Sinks

Depletion of stratospheric ozone has resulted from increases in stratospheric CFC concentrations and concentrations of some other anthropogenic and naturally occurring aerosols. Ozone depletion has become especially apparent in the Southern Hemisphere. A hole in the ozone layer there appears seasonally over Antarctica and seems to be enlarging (see Figure 11.13). As O_3 concentrations decrease, greater quantities of high-energy ultraviolet (UV) light, consisting of wavelengths normally blocked by stratospheric O_3, enter the troposphere and penetrate to the surface of Earth. These higher surface levels of UV energy are expected to lead to an increase in the incidence of skin cancer since this high-energy form of sunlight is known to induce damaging changes in skin cells.

Increased stratospheric concentrations of CFC and other aerosol compounds may also indirectly affect marine phytoplankton because as concentrations decrease, the amount of UV light reaching the ocean surfaces will increase. These increased levels of UV light may have a negative impact on the phytoplank-

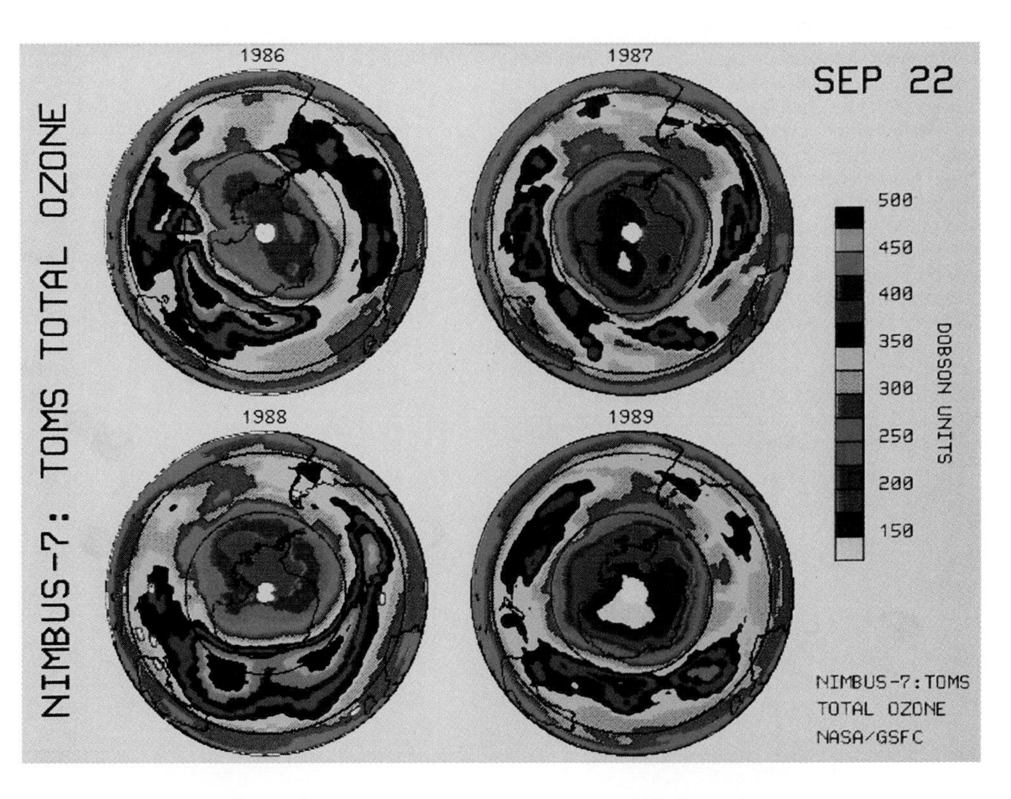

Figure 11.13

These Southern Hemisphere plots show the total ozone distribution for September 22, 1986–1989. The lowest recorded ozone level was during October 1987. The depth of the ozone hole has been following an approximately two-year cycle (relatively low in odd years, not as low in even years). The 1989 behavior appears to continue this cyclic pattern. The areas of lowest ozone in 1986 and 1988 (purple shades) covered significantly less area than the ozone holes observed in 1987 and 1989, which covered nearly the entire Antarctic continent. In addition, the high concentrations of ozone (red, orange, and white), measured in 1986 and 1988 were not observed during 1987 and 1989 outside the polar regions. The white area is the polar night where observations are not possible.

ton community. If elevated levels of UV light reduce the photosynthetic rates of marine phytoplankton, it could reduce their efficiency as a major carbon sink by decreasing the rate at which CO_2 is removed from the hydrosphere and, ultimately, the atmosphere as well.

In the past, atmospheric levels of CO_2 and ozone remained relatively constant. As a result of human activities during the past century, atmospheric CO_2 concentrations have been increasing and global carbon sinks have been reduced. Stratospheric ozone concentrations have also decreased, which has apparently led to elevated levels of harmful UV radiation reaching Earth's surface.

IS GLOBAL GREENHOUSE WARMING UPON US?

Are the effects of anthropogenic atmospheric changes now being reflected in a change in average global temperature? (See Figure 11.14.) Is global greenhouse warming now occurring? These questions may seem rhetorical, but if the answers are affirmative, as most researchers believe, they will have enormous importance for the policies formulated by governments throughout the world (see the Focus on Scientific Inquiry, "Interdisciplinary Teams"). Since the major air pollutants involved in global climate change are associated with the production and uses of energy sources (oil, gas, coal), any regulations aimed at their reduction would require changes in our patterns of energy use and lifestyles. What would be the effects of mandated reductions in energy use? How will our industrial societies accommodate these reductions? Will people stop driving cars?

Throughout the geologic past, the best predictor of climate variations has been the global sea level. When warming occurred, the sea level rose, and when cooling resulted in glaciation, the sea level dropped. At present, the sea level is rising. In the past, when global sea levels rose, concentrations of CO_2 increased. The atmospheric level of CO_2 is now also increasing.

At the present time, it is not clear whether or not Earth's temperature is rising or if the recent temperature increases are within the normal range of natural

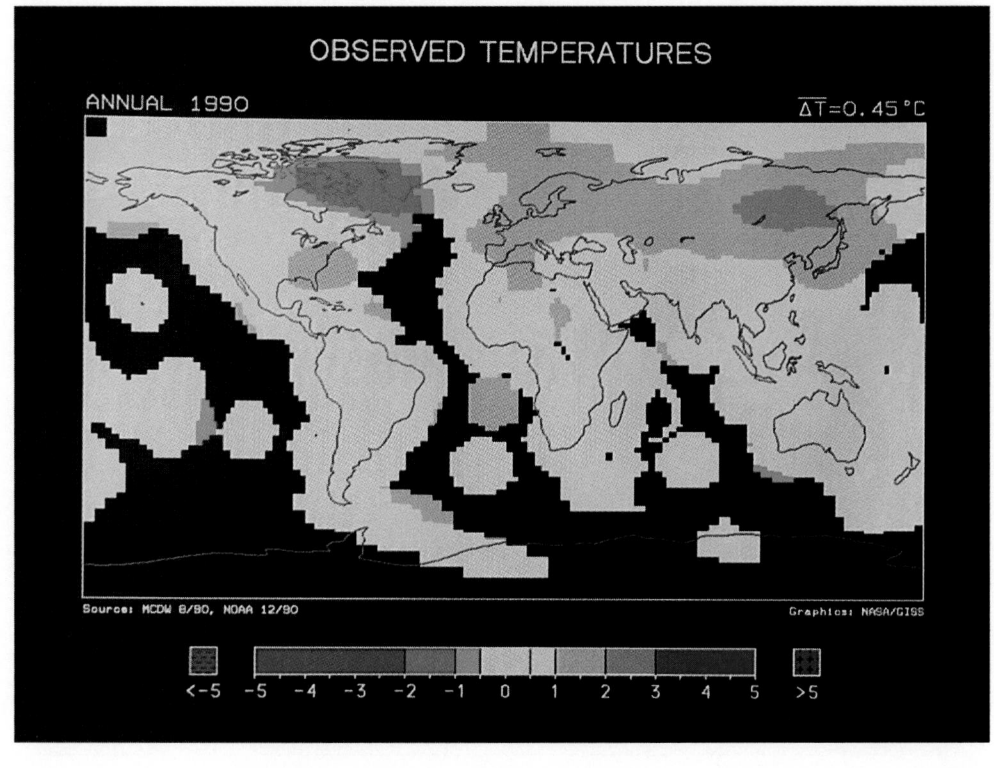

Figure 11.14

Recent data from 2,000 weather stations show patterns of temperature change. Average temperatures in 1990 were compared with average temperatures from the reference period of 1951-1980. Colors represent deviations from the reference temperatures, which are shown in green. Except for two areas, most regional temperatures increased. The global average temperature in 1990 was 0.45°C higher than the 30-year reference average.

Focus on Scientific Inquiry

INTER-DISCIPLINARY TEAMS

As we have seen, the climate of Earth has been subject to continual change. Temperatures have varied throughout geological history, and these changes have had dramatic biological effects. The current period of global warming is so different because of the time scale involved. From our understanding of the history of our planet, we know that species come and go, that biomes have changed over time, and that climates have been radically different in the past. But these events involved such great lengths of time as to be irrelevant to us as individuals. It is difficult to think of them as bearing on our lives or our families' lives. Who worries about catching a trout or a salmon in 10,000 years? Now, however, with the high probability of rapid global changes suddenly upon us, it becomes necessary to think of foreseeable effects on the lives of people who are of concern to us. Although some of the changes might be welcome (longer growing seasons in Minnesota, for instance), the combined effects of predicted changes described in this chapter are ominous.

Facing such a potentially disruptive set of changes, the scientific and various political communities now realize that the issues are so great that conventional responses will most likely be inadequate. Tremendously complex ocean systems, atmospheric systems, chemical systems, ecological systems, and physiological systems are involved, and for each, our ignorance greatly exceeds our knowledge. The effects of global climate change are the result of interactions among these imperfectly understood systems.

Scientists have faced difficult intellectual challenges in the past. Individual scientists have been able to survey what was known and propose new theories or solutions. However, during the past 100 years, the production of new information in biology (and indeed, in all sciences) has accelerated to such a degree that it is now impossible for any single scientist to have a grasp of "all the facts." Solutions to problems now often come from groups of scientists working together. Instead of a single author writing a scientific paper or book, multiple authors are the rule. We now realize that understanding global climate change and its effects falls beyond the capabilities of high-power scientific group research as we know it. It is clear that international scientific cooperation, involving the coordinated efforts of many national agencies, will have to be employed to achieve meaningful results. In the past, only wartime efforts could bring about such large-scale cooperative scientific efforts, and even then, they usually involved only the physical sciences.

Global climate change and its effects are issues that require the integration of physical, earth, and biological sciences. Programs like the International Geosphere-Biosphere Program sponsored by the International Council of Scientific Unions hope to be able to coordinate long-term studies that may provide some answers to the questions posed throughout this chapter. In the United States, the National Science Foundation (NSF), the National Aeronautics and Space Administration (NASA), the United States Geological Survey (USGS), the National Oceanic and Atmospheric Administration (NOAA), the Department of Energy, and the Office of Naval Research are taking part in the international effort. If these efforts are even moderately successful, they will not only mark a new level of scientific investigation but will also produce interdisciplinary research teams that will probably tackle other, seemingly intractable scientific problems. We may well be in the middle of a scientific revolution that future undergraduate students will read about and even experience for the rest of their lives.

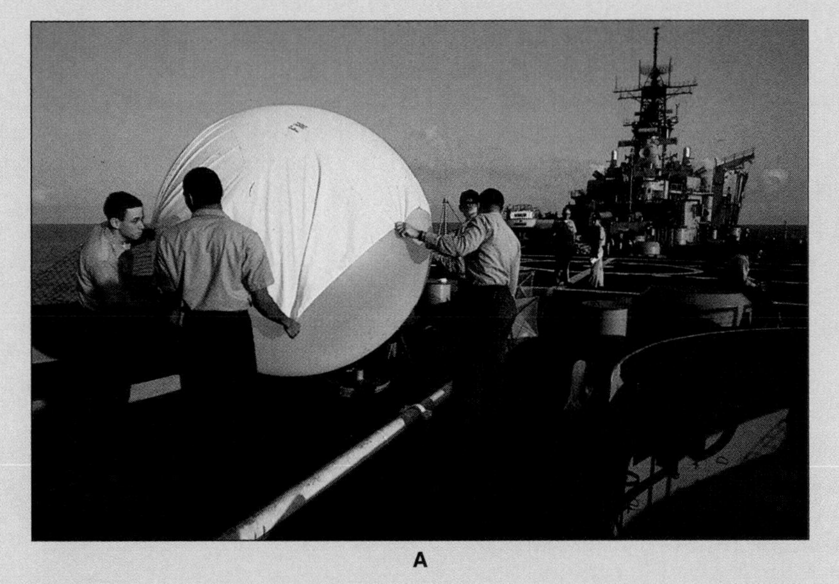

A

B

Figure 1 (A) Crew members of the USS *Iowa* prepare to release a weather balloon to gather information about wind velocities. (B) An NOAA scientist analyzing data at the Forecast Systems Laboratory at Boulder, Colorado.

variations. What will be the consequences if the implementation of new energy production and use policies is delayed until the temperature rise clearly exceeds this "variation noise"? Will it be too late to alter the total global greenhouse warming effects?

When one considers the potential environmental problems likely to be of major importance in the twenty-first century, none compares to the potential effects of continued global warming. How will the biota of Earth and our ever-increasing human population cope with the effects brought on by global warming?

Summary

1. Anthropogenic activities affect the biosphere at the local, regional, and global levels. Earth's climate has fluctuated on different time scales as a result of both orbital and tectonic factors.
2. The greenhouse effect, resulting from atmospheric increases of carbon dioxide, methane, and compounds of nitrogen and sulfur, may be causing global rises in both temperature and sea levels that may lead to major biome changes. Scientists predict that these effects will occur in variable magnitudes that increase with latitude. The northern biomes will be affected to the greatest degree.
3. Anthropogenic activities contributing to global warming include the direct addition of greenhouse gases to the atmosphere, forest destruction, and marine pollution.
4. Acids arrive on surfaces from the atmosphere in wet, occult, and dry depositions. The sources of these depositions are related to fossil fuels burned for transportation and energy production. Acid rain degrades aquatic systems and is implicated in significant forest declines and diebacks in many areas downwind from major industrial centers.
5. Increased stratospheric CFC concentrations are thought to be destroying Earth's protective ozone shield, which will result in an increased intensity of harmful UV light received at the surface.

Review Questions

1. What are paleoclimatic indicators? How are they used in studies of global climate change?
2. What types of long-term regularities characterized global climate changes of the past? What factors are thought to have been responsible for these cycles?
3. What is the greenhouse effect? What causes it?
4. How should natural factors be changing global climate at the present time?
5. What types of human activities have modified the normal patterns of global climate change? How has the expected pattern been modified?
6. What types of ecological effects are expected to occur if global temperatures increase?
7. How are biomes expected to be affected by increasing global temperatures?
8. How will biodiversity be affected by global warming?
9. What is acid deposition? What are its causes? What are its effects?
10. How has air pollution affected forests throughout the world?
11. What effects have soil acidification and abnormal ozone concentrations had on forests?
12. Why have vast tracts of tropical forests disappeared?
13. How have marine pollution and stratospheric ozone depletion affected carbon sinks? What are some of the possible consequences of these actions?

Essay and Discussion Questions

1. Some ecologists believe that changes in temperature serve as a single integrative measure of human-caused change. What might be the basis for coming to this conclusion? Do you agree? Explain.
2. Assume that you are a political leader (say, the head of the U.S. Environmental Protection Agency) responsible for taking steps to reduce global warming. What recommendations would you and your task force make? Do you think they would be implemented? Why?
3. If predicted global temperatures increase in the future, how might different biomes of Earth be changed after 100 years? After 1,000 years?

References and Recommended Reading

Anderson, J. G., D. W. Toohey, and W. H. Brune. 1991. Free radicals within the Antarctic vortex: The role of CFCs in Antarctic ozone loss. *Science* 251:39–46.

Ausubel, J. H., and H. E. Sladovich, eds. 1989. *Technology and Environment.* Washington, D.C.: National Academy Press.

Booth, W. 1989. Monitoring the fate of the forests from space. *Science* 243:1428–1429.

Corson, W. 1990. *The Global Ecology Handbook: What You Can Do About the Environmental Crisis.* Boston: Beacon Press.

Curtzen, P. J., and M. O. Andrede. 1990. Biomass burning in the tropics: Impact on atmospheric chemistry and biogeochemical cycles. *Science* 250:1669–1677.

Energy for Planet Earth (special issue). 1990. *Scientific American* 263(3).

Fearnside, P. 1989. Extractive reserves in Brazilian Amazonia. *BioScience* 39:386–393.

Goudie, A. 1990. *The Human Impacts on the Natural Environment.* 3d ed. Cambridge, Mass.: MIT Press.

Houghton, R., and G. Woodwell. 1989. Global climate change. *Scientific American* 260:36–34.

Irwin, J. 1989. Acid rain: Emissions and deposition. *Archives of Environmental Contamination and Toxicology* 18:95–107.

Jones, P., and T. Wigley. 1990. Global warming trends. *Scientific American* 264:84–91.

Managing Planet Earth (special issue). 1989. *Scientific American* 261(3).

Matthews, S., and J. Sugar. 1990. Under the sun: Is our world warming? *National Geographic* 178(4):66–99.

Meisner, J., J. L. Goodier, H. A. Regier, B. J. Shuter, and W. J. Christie. 1987. An assessment of the effects of climate warming on Great Lakes basin fishes. *Journal of Great Lakes Research* 13:340–352.

Molingreau, J., and C. J. Tucker. 1988. *AMBIO* 17:49–55.

Mooney, H. A., B. G. Drake, R. J. Luxmoore, W. C. Oechel, and L. F. Pitelka. 1991. Predicting ecosystem responses to elevated CO_2 concentrations. *BioScience* 41:96–104.

Peltier, W., and A. Tushingham. 1989. Global sea level rise and the greenhouse effect: Might they be connected? *Science* 244:806-810.

Repetto, R. 1990. Deforestation in the tropics. *Scientific American* 262:36–47.

Schneider, S. 1989. The greenhouse effect: Science and policy. *Science* 243:771–781.

Schoeberl, M. R., and D. L. Hantmann. 1991. The dynamics of stratospheric polar vortex and its relation to springtime ozone depletions. *Science* 251:46–52.

Schulze, E. 1989. Air pollution and forest decline in a spruce (*Picea abies*) forest. *Science* 244:776–783.

Schwartz, S. 1989. Acid deposition: Unraveling a regional phenomenon. *Science* 243:753–762.

Silver, C. S., and R. S. DeFries. 1990. *One Earth, One Future: Our Changing Global Environment.* Washington, D.C.: National Academy Press.

White, R. 1990. The great climate debate. *Scientific American* 263:36–43.

World Resources Institute, International Institute for Environmental Development, and United Nations Environment Programme. *World Resources, 1988–1989.* New York: Basic Books.

World Resources, 1990–1991. New York: Basic Books.

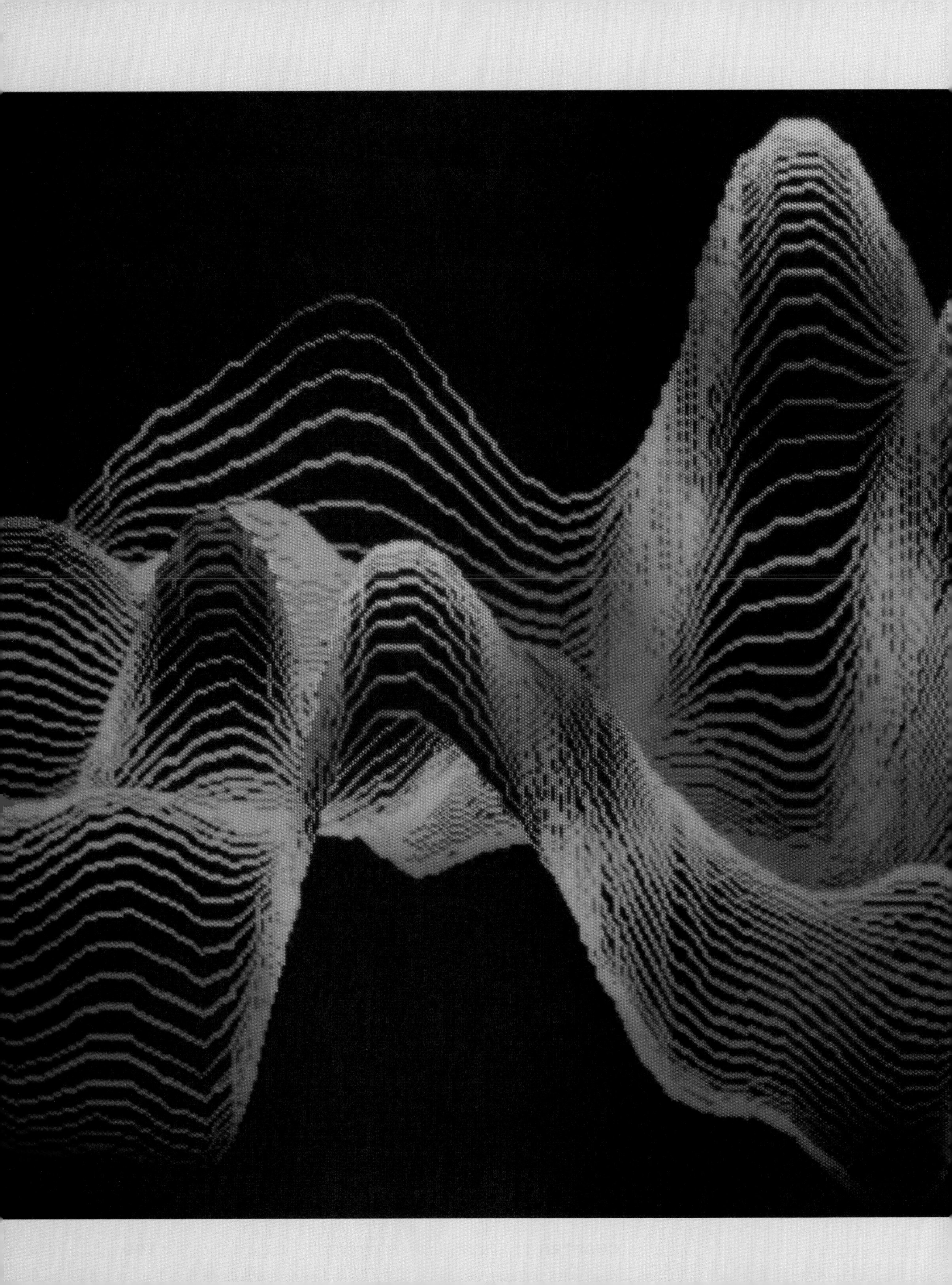

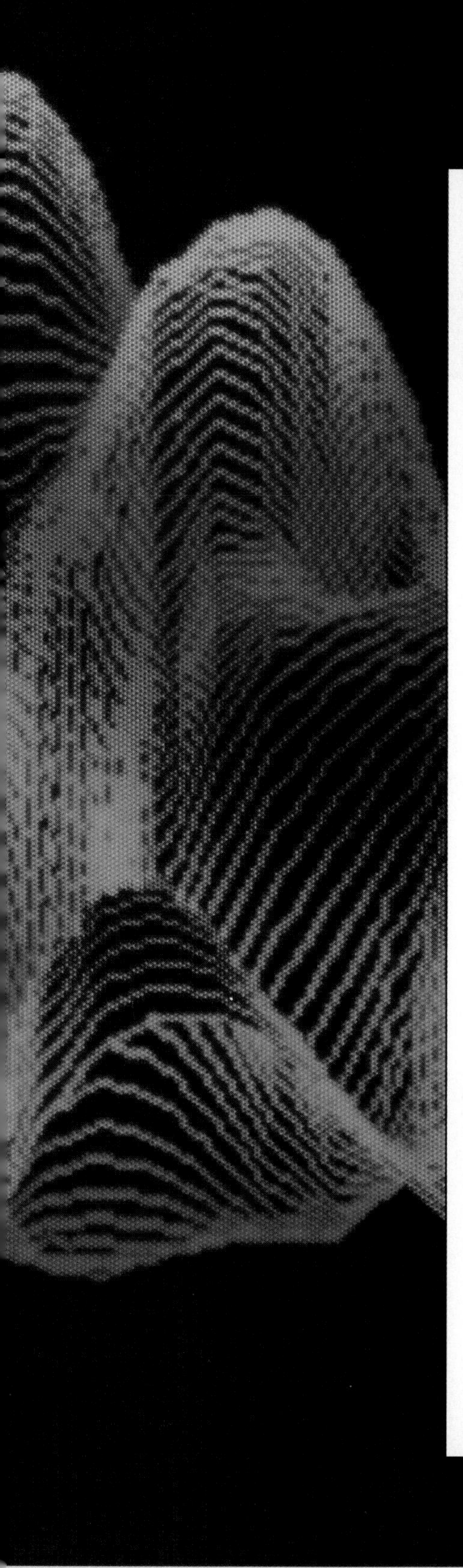

UNIT II

The Language of Life

What enables organisms to produce offspring with all the characteristics of their parents and their species? From the earliest written records, we know that this question has always interested our human forebears. For thousands of years, ideas about reproduction were based primarily on commonsense observations and philosophic beliefs. Not until the end of the nineteenth century did biologists develop an adequate scientific perspective on reproduction and heredity. Unit II describes scientific accomplishments that led to an understanding of how hereditary information—the language of life—is transmitted between generations and expressed in offspring.

During the first half of the twentieth century, scientists attempted to identify the basic genetic material. Between 1910 and 1925, they determined that hereditary information is transmitted on chromosomes. During the next 25 years, DNA was recognized as the genetic material. In 1952, the structure of DNA was described, and during the following decade, scientists defined the underlying processes that make it the language of life for all organisms on Earth.

Knowledge about DNA has been extended over the past 40 years. We have learned the hereditary basis of many normal human traits and human genetic disorders. Organisms with new capabilities and new biological products have been created by applying genetic technologies. Now new frontiers are being approached, and decisions must be made about extending the use of these technologies directly to humans.

CHAPTER 12

Foundations of Genetics

Chapter Outline

Reading Questions

1. What were the major pre-Mendelian explanations of heredity?
2. What are Mendel's two laws?
3. How did Mendel's laws differ from earlier ideas on heredity?
4. Why were Mendel's laws not accepted in his own lifetime?

Genetics is the scientific study of heredity, the transmission of characteristics from parents to offspring. Genetics explains why offspring resemble their parents and also why they are not identical to them. Genetics is a subject that has considerable economic, medical, and social significance and is partly the basis for the modern theory of evolution. Because of its importance, genetics has been a topic of central interest in the study of life for centuries. Modern concepts in genetics are fundamentally different, however, from earlier ones. In this chapter, we consider what people thought about heredity in the past and see why modern ideas replaced older ones.

EARLY IDEAS ABOUT REPRODUCTION

From earliest recorded times, people have realized that although individuals die, the different "kinds" (what we now call species) of plants and animals persist because individuals reproduce offspring similar to themselves. Since humans reproduce sexually, and since the study of life until recent times has emphasized humans and organisms closest to humans, most theories of reproduction have been formulated in terms of sexually reproducing organisms.

Greek Ideas About Reproduction

The greatest naturalist of ancient times, Aristotle, in the fourth century B.C. wrote an extensive treatise on reproduction. Like most scientists before the eighteenth century, he thought that only animals reproduce sexually. He concentrated primarily on animals in his studies because of their importance for understanding the biology of humans. Therefore, he described heredity in terms of sexual reproduction.

In his treatise *On the Generation of Animals,* Aristotle stated that males and females made different contributions to what become their offspring. Reflecting the social values of his day, Aristotle claimed that the male's contribution was more important than the female's. He held that women were inferior to men and should not be allowed any place in politics, nor should they be given the same education as men. A woman's place, from this view, was basically in the household. In reproduction, according to Aristotle, the female primarily supplied "matter." This was acted on by an "active spirit" contained in the male's semen, which molded and formed the future offspring.

If all went well in the process, the normal result would be a healthy male child that resembled his father. If conditions were suboptimal (due to the male's not being vigorous), the offspring would be a male that displayed some of his mother's traits. In the worst case, where the male was inadequate, the result was a monstrous, deformed offspring who lacked the ability to reason clearly, had less virtue, and did not even have "proper" sexual organs. In other words, it was a female.

How could such a great mind believe these disparaging and ridiculous views on the female role in reproduction, which we find so objectionable? Aristotle merely stated ideas that seemed "obvious" to him; that is, they were so widely held at the time that apparently no one stopped to consider if they were true. Although other Greek, and later Roman, writers occasionally differed with Aristotle's views, for the most part his theory of heredity lasted for almost 2,000 years.

Aristotle's theory of heredity reflected the culture of his time. He held that the female contributed matter, whereas the male contributed an active spirit that molded the matter into the future offspring.

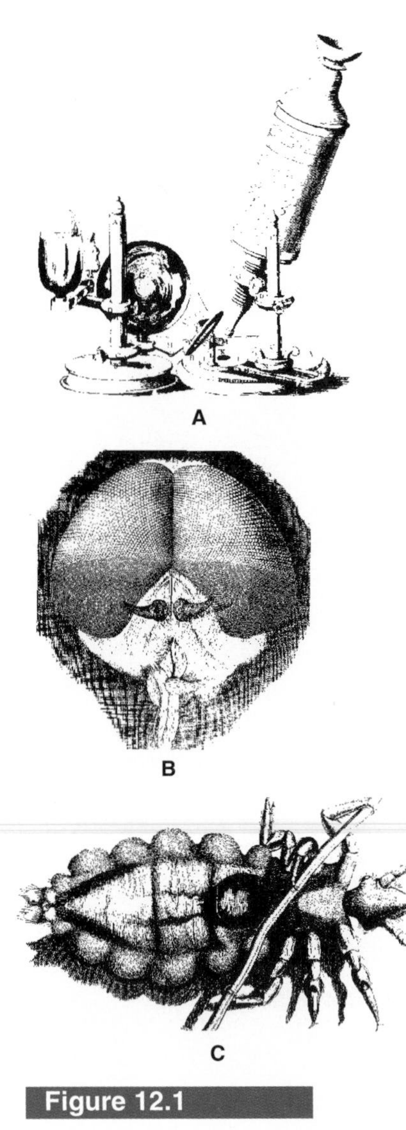

Figure 12.1

These illustrations are from Robert Hooke's book *Micrographia,* published in 1665, which popularized the microscope by reproducing a set of interesting observations. Hooke used this compound microscope (A) for his observations. Among the many objects he viewed were the head of an insect (B) and the body louse (C), an all-too-frequent companion in the seventeenth century.

The Scientific Revolution and Sex

Aristotle's view was finally challenged in the 1600s because of a major shift in the way nature was conceptualized and because of new scientific discoveries. The seventeenth century was the age of scientific revolution, and with it came the notion that nature could be regarded as a wondrous mechanism created by God. The study of nature was an expression of piety, and contemplation of God's creation was thought by scientists to lead to a greater appreciation of his power. Central to the scientific revolution was the metaphor of the world as a machine. Philosophers and scientists tried to explain all phenomena in terms of pieces of matter following the laws of physics, a worldview known as **mechanical philosophy**.

From this new perspective, Aristotle's theory of reproduction (indeed, most of Aristotle's scientific writings) looked hopelessly out of date and misguided. What could an "active spirit" possibly mean in a world that was like a big windup clock?

A new set of discoveries also called into question Aristotle's position on reproduction. The seventeenth century saw the invention of numerous scientific instruments, including the *microscope*. It created a sensation, for when it was trained on even the most mundane objects (moldy cheese, pond water, onion skin, fleas, the ever-present body lice), an entirely new world of microstructure was revealed (see Figure 12.1).

Researchers soon focused their microscopes on plants and animals and exposed their microstructure. A study of the reproductive organs of mammals led to the discovery of what appeared to be eggs in the female ovaries. Since eggs were known to be one of the main reproductive products in other vertebrates, the discovery of mammalian eggs led some scientists to believe that all animal life began in eggs.

Even more exciting was what careful observations with the microscope revealed about the metamorphosis of caterpillars into butterflies. To a number of microscopists, it appeared that all parts of a butterfly were actually already present in the caterpillar stage, merely folded up and ready to expand. These and similar observations led to the formulation of the idea of **preformation**, which states that successive generations are all preformed and encapsulated in previous generations like nested boxes or Russian matrioshka dolls (see Figure 12.2).

Today this idea appears rather strange. Why was the theory of preformation attractive to major thinkers of the scientific revolution? One reason was that it did resolve a number of issues associated with viewing the world as a machine. For example, although machines are rather marvelous, they were never seen reproducing. Put a pair of gerbils in a cage for a period of time, and one will soon see baby gerbils. Put two watches in a box, and no baby watches will issue forth, not

Figure 12.2

These Russian matrioshka dolls can fit inside one another (smaller into the next larger) to become nested into a single unit. Similarly, the idea of preformation described all future generations as being contained in the first living beings.

even cheap imitations. The preformation theory, by conceiving of all generations as being preformed, explained how a mechanism like an animal could produce similar mechanisms. The theory also fit nicely with religious ideas of the day. Because scientists who took part in the scientific revolution believed that God had created the world, they easily accepted the idea that God had encapsulated all generations of humankind within Eve and generations of all animals and plants within the first creations. They held that slowly, over hundreds of years, the entire course of history unfolded according to a divine plan. Preformation, therefore, not only explained a complex biological problem but also tied together science and religion.

Preformation was a highly popular scientific position. However, some naturalists expressed dissatisfaction over the emphasis on the female's role in reproduction. If all generations of living organisms are encapsulated in ovaries or their equivalents, then the male's role was reduced, at best, to merely stimulating the egg to begin development. A few scientists presented an alternative idea that sperm contain "little candidates for life" (*homunculi*) that develop after being introduced into the uterus. This view, although it saved male pride, didn't catch on, and for almost 100 years the classic preformation theory based on the unfolding of eggs (called *ovism*) dominated writings on reproduction.

Preformation reflected the mechanical philosophy of science. It explained heredity as part of an overall divine plan wherein successive generations were preformed and encapsulated in previous generations like a set of nested boxes. The site generally thought to contain future generations was the female ovary.

The Beginning of Modern Genetics

Belief in the preformation theory broke down in the late eighteenth century for several reasons. The most important was that many scientists had come to believe that although science and religion were compatible, science should not rely on the supernatural to explain the physical or biological world. Instead, they thought it more fruitful to attempt to understand the world by searching for physical and biological laws of nature and then seeking to test and verify them. Since religious and supernatural concepts cannot be physically tested, they were considered inappropriate for scientific investigation. Although religious ideas may provide an overview of the significance of the world, they have no role in designing scientific investigations.

Preformation ran into observational problems as well. The late eighteenth century was a time of great reform in agriculture and animal husbandry. Part of that movement consisted of vast breeding programs intended to improve the quality of cattle, sheep, corn, potatoes, and fruit. The information collected by these practical efforts showed farmers and scientists that traits from *both* the male and the female were passed on to their descendants.

An additional problem for preformation was a general recognition in the eighteenth century that plants reproduce sexually. The study of plant reproduction, moreover, demonstrated the complexity of the process. The view that the embryo was somehow preformed in the egg made no sense in light of this information.

By the nineteenth century, scientists had concluded that preformation was not an adequate explanation for inheritance. Unfortunately, they did not agree on a good working theory. Most scientists believed that both the male and the female contributed to the offspring and that the hereditary material was blended. This view, illustrated in Figure 12.3, was called the **blending theory of inheritance**. However, scientists were unable to describe rigorous, predictable laws that governed inheritance. Considerable research was conducted in this area since the problem had considerable practical application. The impressive research

Figure 12.3

These imaginary plants illustrate the blending theory of inheritance. Hereditary material from each parent "blended" or "mixed" to produce intermediate offspring. The blending of a particular trait is seen when red and white parents produce pink offspring and when tall and short parents blend to produce an offspring of average height. The mixing of different traits is illustrated by the thorns of one parent and the wide leaf shape of the other parent, resulting in an offspring that has both thorns and wide leaves.

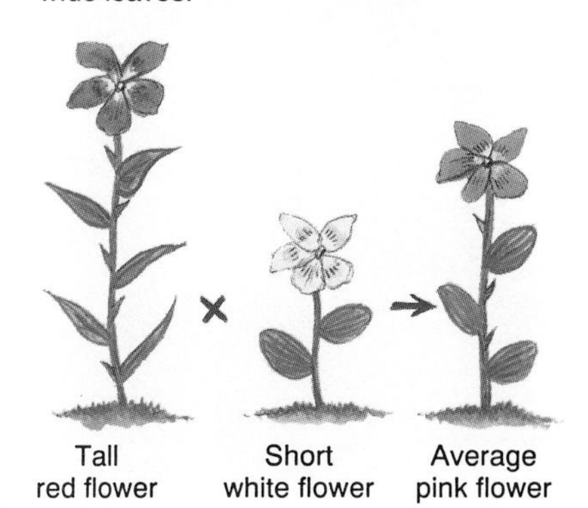

Figure 12.4

Hugo de Vries was a Dutch botanist whose research on genetics led to wide acceptance of the idea that genetic traits were inherited in discrete units rather than through blending in the offspring. His work was the first to call attention to Mendel's famous 1866 publication.

Figure 12.5

The evening primrose was one of de Vries's most important experimental plants.

on animal and plant breeding had literally revolutionized farming, and in countries like Great Britain, the threat of famine was eliminated for the first time in history. If predictable laws of heredity could be discovered, this agricultural revolution could be extended to other organisms. In addition, Darwin's theory of evolution had stimulated research on inheritance (discussed in Chapter 22).

Around the end of the nineteenth century, a new way of looking at inheritance and a set of careful experiments fundamentally altered existing concepts of heredity. The central figure in this turning point was a Dutch botanist, Hugo de Vries (see Figure 12.4). From his studies with the evening primrose (Figure 12.5), he collected evidence that a blending of hereditary material did *not* occur. Instead, it appeared to him that individuals inherited from each parent specific traits that were then passed on to later generations in unblended, discrete units. He also claimed that he had discovered a law that described and predicted the inheritance of these specific traits. De Vries was actually engaged in a project larger than genetics. He hoped to reformulate Darwin's theory of evolution in such a manner as to make it a more experimental theory.

The publication in 1900 of de Vries's experiments marked a turning point in genetics between the idea of blending inheritance and the idea of discrete units of inheritance. Unfortunately, de Vries is usually not credited with the innovation; instead, the honor goes to Gregor Mendel (1822–1884), a relatively obscure church figure who lived in what is now Brno, Czechoslovakia (then part of the Austro-Hungarian Empire). Thirty-four years before de Vries published the results of his experiments, Mendel (Figure 12.6) had published in the proceedings of the Natural History Society of Brno a paper that clearly stated the law that de Vries had "discovered." In fact, de Vries may have used Mendel's paper in formulating his ideas. In any case, de Vries's publication called attention to Mendel's paper,

A

Figure 12.6

(A) Gregor Mendel's simple and perceptive experiments gave rise to the first genetic principles that later became his laws of segregation and independent assortment. (B) Mendel's "laboratory" was a simple garden at his monastery.

B

which had been ignored until then. The recovery of what came to be known as *Mendel's laws* eventually overshadowed de Vries's work, and today few people recognize his importance in reorienting our understanding of heredity.

> The decline of preformation theory resulted from new observations and new criteria for studying living organisms. In 1900, de Vries popularized the idea of discrete units of inheritance and rediscovered Mendel's laws of heredity.

MENDEL'S LAWS

Although Mendel's research had little influence in its own day, his experiments, conducted with 10,000 plants over a period of eight years, were so carefully designed, his data so intelligently analyzed, and his conclusions so well expressed that his 1866 paper has received much belated honor. It is, indeed, a model scientific paper. What did Mendel's research tell us about heredity?

Mendel's Experiments and the First Law

Mendel was searching for laws that regulated the transmission of characteristics from generation to generation, and he chose as his experimental subject the garden pea plant (*Pisum sativum*), which has easily recognizable traits. Figure 12.7 illustrates the seven traits he selected for his study: (1) shape of ripe seeds (round or wrinkled); (2) color of cotyledon or seed leaf (yellow or green); (3) color of seed coat (gray or white), which is always associated with flower color (violet or white); (4) shape of ripe pod (inflated or constricted); (5) color of unripe pod (green or yellow); (6) flower position (axial or terminal); and (7) stem length (long or short).

The pea plant Mendel chose not only had easily recognizable traits but also had a flower structure that permitted him to control fertilization carefully. The plant has male and female organs on the same flower, and both the structure of the flower and the development of its male and female gametes (pollen and eggs) are such that the plant normally fertilizes itself. These features, shown in Figure 12.8, allowed Mendel to control mating reliably. He could allow the plant to self-

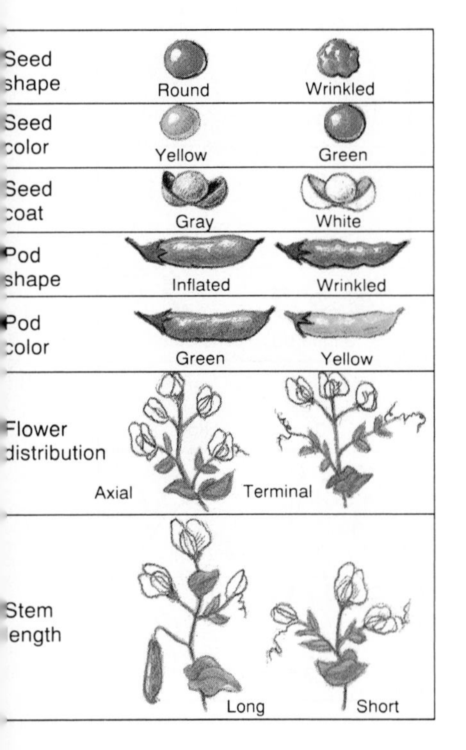

Figure 12.7

The seven pairs of pea plant traits that Mendel studied.

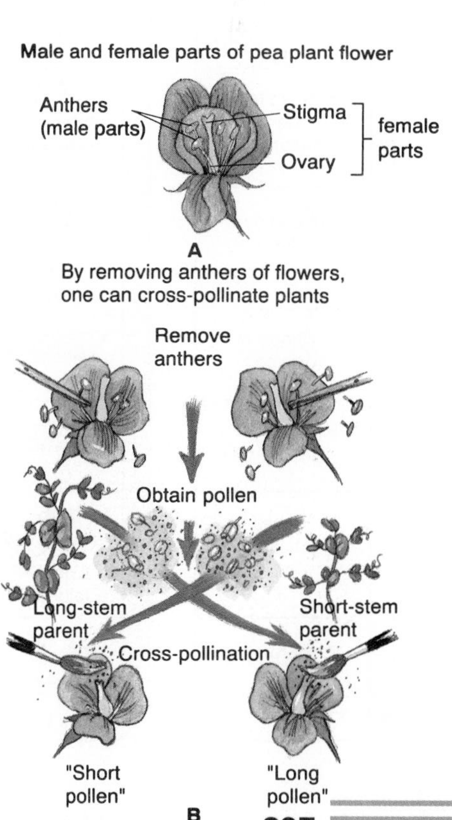

Figure 12.8

(A) A cross section of the pea plant flower shows the location of its male and female parts. (B) By removing the anthers of flowers, one can cross-pollinate plants.

pollinate, or he could artificially cross-pollinate by removing the male pollen-bearing part of the flower (anthers) and then dusting the end of the female part (stigma) with pollen from a chosen plant.

The self-pollinating nature of the plant also made it possible for Mendel to acquire and use **pure-breeding lines**, plants that when self-pollinated or cross-pollinated with a plant from the same line produce only offspring that are identical to the parent or parents. He could therefore plan experiments with individual plants that were known to display particular traits.

Mendel's Monohybrid Crosses

Mendel set out to determine what happened to traits when various crosses were made. He began his work by focusing attention on pure-breeding lines that differed by only a single trait. Such crosses, involving two forms of a trait (for example, a long stem and a short stem), are called **monohybrid crosses** (see Figure 12.9). The offspring of the crosses between parents having different forms of the trait he called **hybrids**.

Mendel found that his hybrids did *not* show blended inheritance. Offspring of a plant with a long stem crossed with a short-stemmed plant did not have medium stem length. Instead, one of the two traits, in each of the seven pairs of traits, was always expressed. He called these **dominant traits**. The traits that did not show up in the first generation of offspring (called the first filial generation and represented as F_1) he called **recessive traits** because they appeared to recede or disappear in the hybrid. For example, as shown in Figure 12.9, when Mendel crossed a plant that had white seed coats with a plant that had gray seed coats, the resulting F_1 plants did *not* yield seeds with light gray seed coats; they all had gray seed coats. Thus the trait for gray seed coats was dominant over the

Figure 12.9

The results of Mendel's first monohybrid crosses.

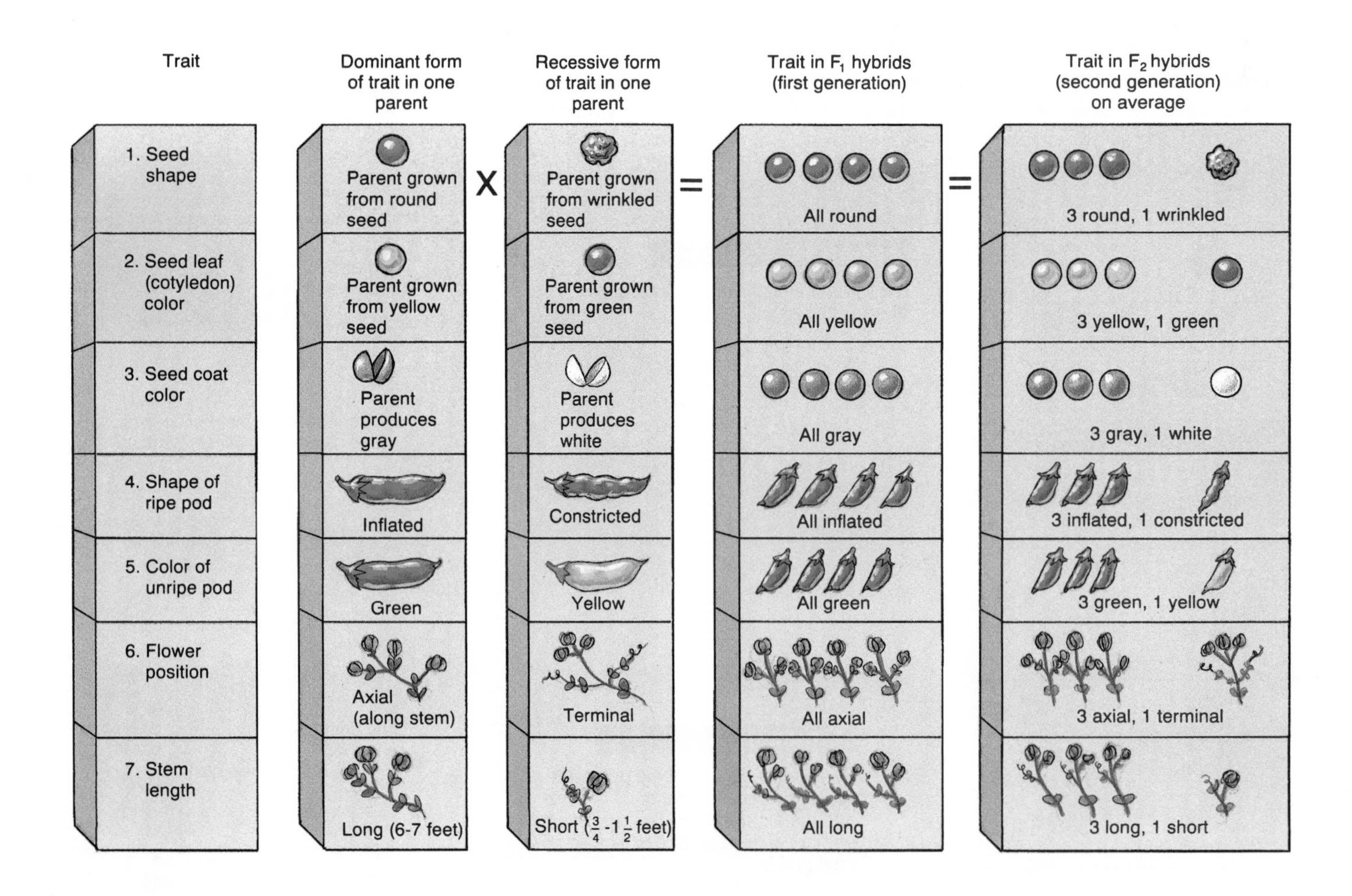

recessive trait, white seed coats. Later researchers would discover that dominance of one trait over another is not universal and that traits do not always exist in pairs. Our modern understanding of Mendel's traits will be discussed in Chapter 13.

Were there any other patterns that could be uncovered? Mendel's monohybrid crosses revealed even more interesting results than the existence of dominant and recessive traits. When he allowed his F_1 hybrids to self-pollinate and then examined the resulting second filial generation (F_2), he found some with the dominant trait and some with the recessive trait. By keeping careful records of the number of each, he showed that the traits appeared in a ratio of about three dominant to one recessive. Clearly, the hereditary material had not been blended together in the F_1 hybrids. Rather, Mendel concluded, the F_1 hybrids received **hereditary factors** from each of its parents, but the expression of the recessive factor was masked by the dominant factor. The recessive factor, although not expressed in the F_1, had not ceased to exist and could be passed on to the F_2.

Mendel conducted experiments to determine the hereditary makeup of these F_2-generation plants. Plants that showed the recessive trait produced only offspring with the recessive trait when allowed to self-pollinate. The plants with a dominant trait in the F_2, however, were not all the same. When he allowed them to self-pollinate, Mendel discovered that one third produced offspring with the dominant trait only, and subsequent generations raised from their seeds continued to do the same. However, the other two thirds of the F_2 that displayed the dominant trait gave rise to dominant and recessive in the same ratio as the offspring of the F_1 plants, 3:1. How could this result be explained? Mendel reasoned that they must be like the F_1 hybrids. The F_2 generation, then, was complicated. The F_2 plants exhibiting the ratio of three dominant to one recessive were actually one recessive to one dominant to two hybrid.

Mendel's First Law: The Law of Segregation

Mendel concluded from his results that the two alternative forms of a trait were determined by a pair of hereditary factors. He held that each individual pea plant had a pair of factors for each trait. The pair of factors could be the same, a state we now call *homozygous,* or the pair could consist of the two alternative factors, a state we now call *heterozygous*. Future chapters will discuss some of what scientists have discovered about these factors and how much more complicated the situation is than it first appeared to Mendel.

The generalizations that Mendel drew from his set of monohybrid crosses are still considered valid. He explained that the appearance of the F_1 generation of hybrids was due to the combination of a dominant factor inherited from one of the parent plants and a recessive factor inherited from the other parent plant. It made no difference which gamete, the egg or pollen, carried the factor. But why were these regularly observed ratios of dominant and recessive traits not present in the F_2? A brief look at the possible combinations of gametes will explain why.

The checkerboard-like diagram in Figure 12.10 is called a **Punnett square**

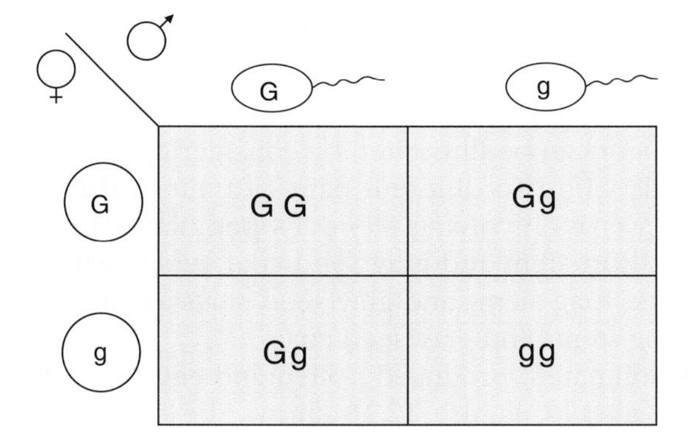

Figure 12.10

The Punnett square is a convenient modern method used to tabulate all the possible combinations of gametes in fertilization. The larger circles represent eggs, and the little circles with tails represent sperm or pollen. The letters inside the circles depict factors for the trait being studied. Capital letters stand for dominant; lowercase letters stand for recessive. Each square represents one of the possible combinations.

Figure 12.11

A Punnett square illustrating the ratio Mendel obtained in a monohybrid cross, involving gray seed coat and white seed coat plants. The F_1 heterozygote yields a 3:1 ratio in the F_2.

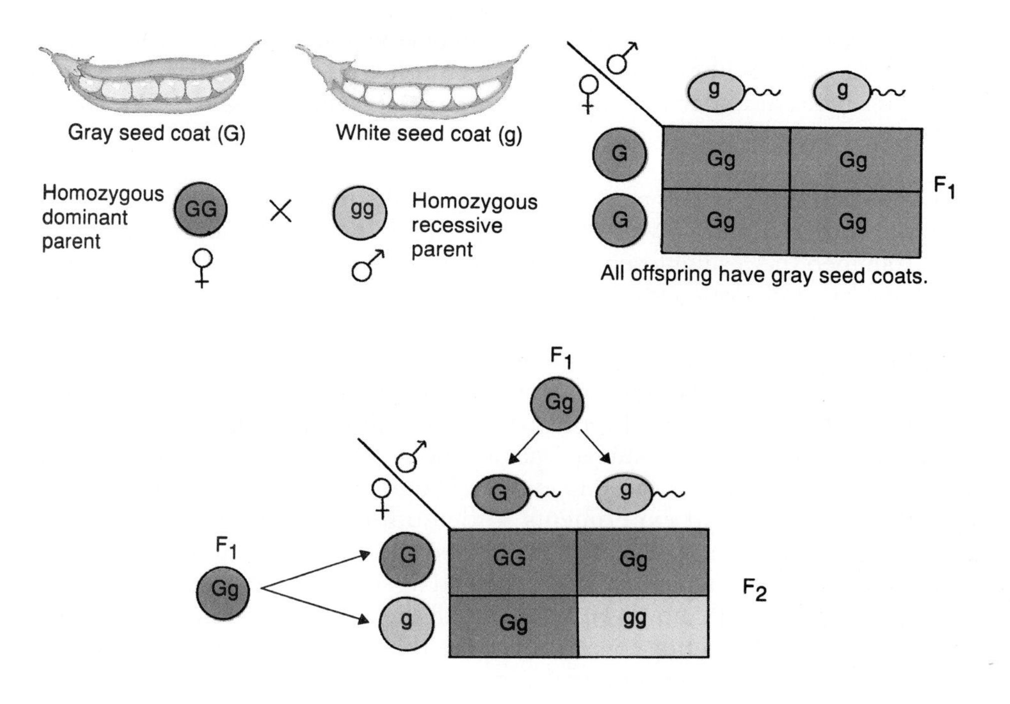

and is a convenient, modern way to illustrate the possible combinations of gametes. By convention, each pair of factors can be represented by a pair of letters, a capital letter standing for the dominant factor, and a lowercase letter standing for the recessive factor. Figure 12.11 illustrates Mendel's explanation of his data. He suggested that the dominant and recessive factors in the F_1 plant separated in the course of egg and pollen formation and that these factors randomly reunited in fertilization to produce the F_2. The separation of parental factors is known as Mendel's **law of segregation**. This law directly contradicted the then-current idea of blending inheritance.

Mendel experimented with pairs of traits in pea plants. He found that for each pair of traits, one was dominant and one was recessive. He believed that each form of the trait was caused by a pair of factors, one from each parent. Mendel's law of segregation states that during gamete production, these factors separate, and only one member of the pair enters a particular gamete. Fertilization randomly brings the pairs of factors together again and determines the type of trait in offspring.

Mendel's Second Law

Mendel's experimental organism had seven true-breeding traits, and his first law was discovered by considering plants that differed in a single trait. What occurred when two or more traits were considered? Mendel did extensive experiments and found that not only did his law of segregation always hold when more than one trait was studied, but the traits were transmitted independent of one another—a discovery now referred to as Mendel's second law, the **law of independent assortment**.

The law of independent assortment can be illustrated by considering one of Mendel's crosses involving individuals with two different pairs of traits, a **dihybrid cross**. Figure 12.12 describes a cross involving Mendel's pea plants with two pairs of traits: long stem versus short stem and gray seed coat versus white seed coat. As can be seen in the figure, long stems and gray seed coats are dominant, and in the F_1 the plants have long stems and gray seed coats.

These F_1 dihybrids, when self-pollinated, produced pollen and eggs in four different combinations: LG, Lg, lG, and lg. Figure 12.13 shows the eggs and

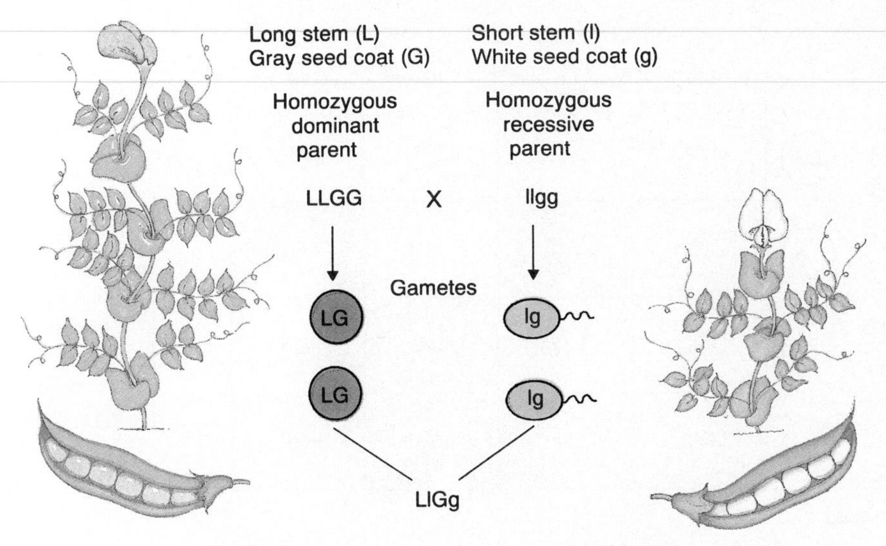

Figure 12.12

One of Mendel's crosses that illustrates the law of independent assortment. This cross involves two traits: long stem (L) versus short stem (l) and gray seed coat (G) versus white seed coat (g).

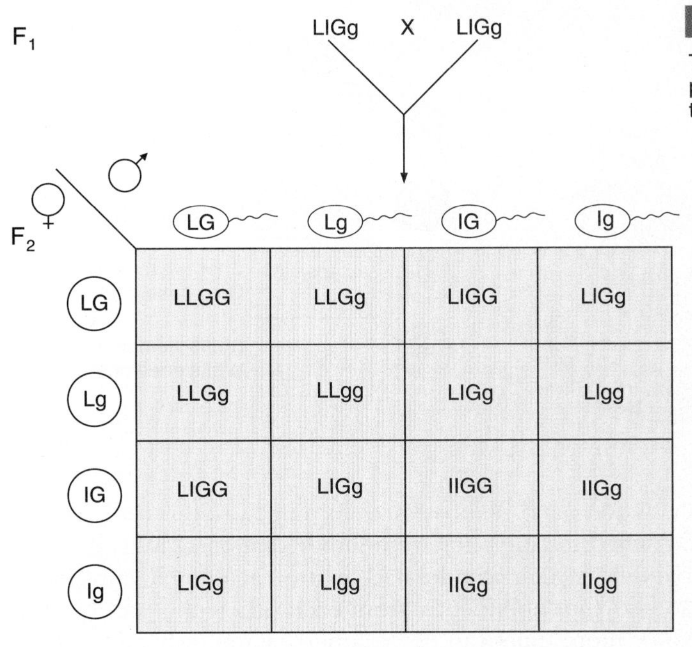

Figure 12.13

The dihybrid cross offers nine possible combinations of traits in the F_2.

pollen produced by the F_2 and their possible combinations. Nine different combinations are possible. Simple observation, however, will not distinguish all nine. Because of dominance, only four different combinations of factors are observed, as indicated in Figure 12.14.

Mendel was able to determine that the factors were sorting independently by making a *test cross,* in which he mated F_1 dihybrid pea plants with plants that were known to be homozygous recessive for both traits. Figure 12.15 illustrates the results of his test cross. Since the homozygous test-cross parent could contribute only double recessive factors to the eggs and pollen, the ratio of traits seen in the offspring reflects the ratio of different pairs of factors in the eggs and pollen produced by the dihybrid parent. If the two traits were assorted independently, then four combinations are possible and would be observed in a ratio of 1:1:1:1. If the traits were not independent, some other ratio would have been observed.

Another way of showing independent assortment was to consider the ratios of each trait in the F_2 dihybrids separately (that is, long stem with gray seed coats

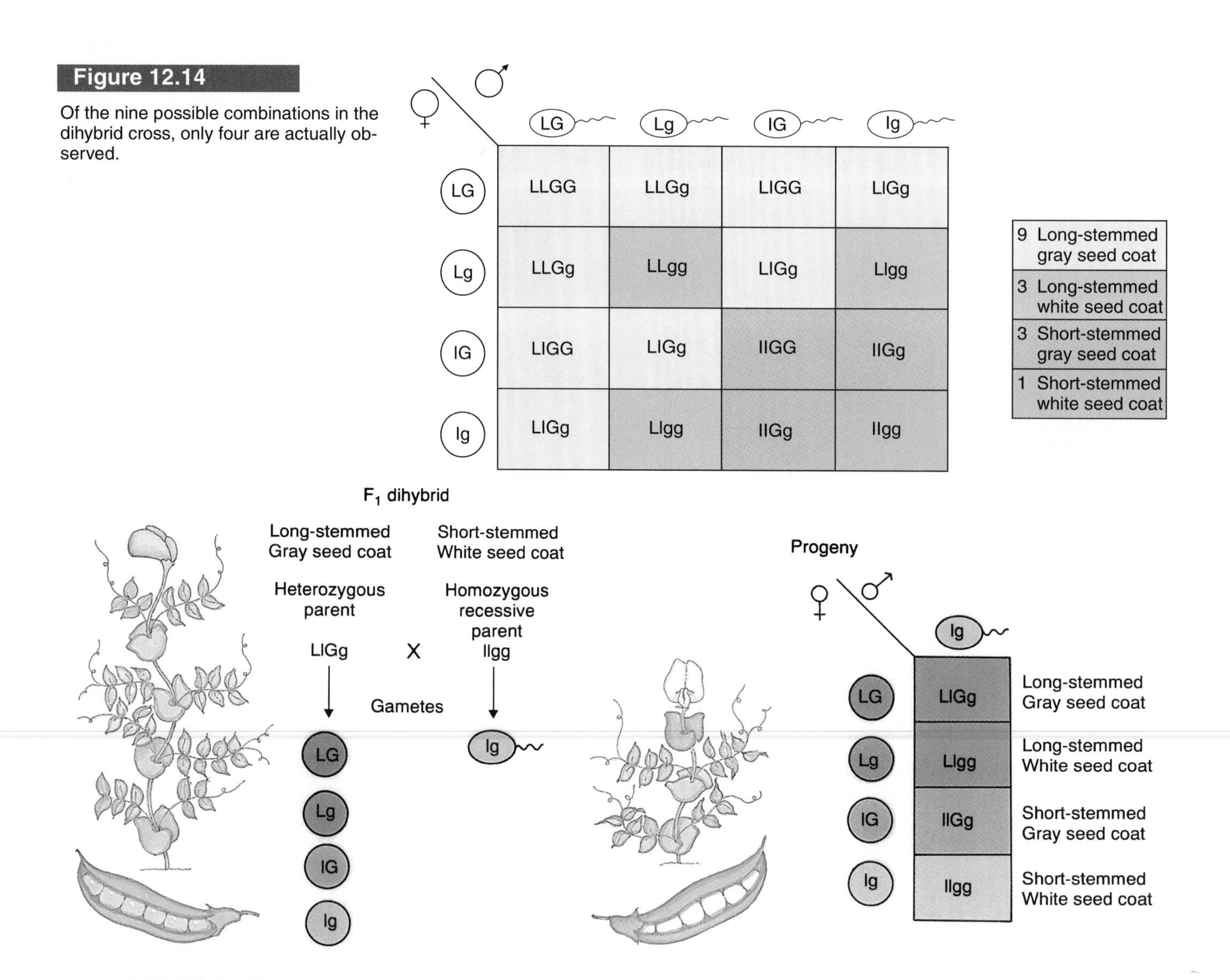

Figure 12.14

Of the nine possible combinations in the dihybrid cross, only four are actually observed.

Figure 12.15

Mendel's test cross demonstrated the law of independent assortment.

and short stem with white seed coats). Ratios for each trait turned out to be 3:1. Students who are familiar with statistics may remember that the multiplication law of mathematical probability (which Mendel had learned at Vienna University) predicts that for traits that assort independently from each other, the proportions from a cross involving two or more traits can be obtained by multiplying the proportions expected for each trait alone. So in this case, the 3:1 ratios (3 long stem, 1 short stem) × (3 gray seed coats, 1 white seed coat) = 9 long stem with gray seed coats to 3 long stem with white seed coats to 3 short stem with gray seed coats to 1 short stem with white seed coat.

Mendel's law of independent assortment states that the factors responsible for two or more traits are inherited independently.

MENDEL AND LATER WORK IN GENETICS

When we read Mendel's paper today, it is difficult to imagine that it was generally ignored in the years after 1866. Why was such an important paper overlooked for more than 30 years? In part, it went against the common wisdom that hereditary material was blended. It is difficult for scientists, as it is for other professionals, to give up a widely held position unless compelling evidence is brought forth. In Mendel's case, that evidence didn't appear until the turn of the century when de

Vries and two other biologists, Carl Correns and Erich von Tschermak, rediscovered and confirmed the laws that now bear Mendel's name.

Another reason Mendel's laws did not receive a better reception was that the use of mathematics—ratios, statistics, and counting—was not widely practiced in biology at that time. In the 1860s, biology was largely a descriptive science. Even when experiments were conducted, as in physiology, the work was largely qualitative rather than quantitative (see the Focus on Scientific Explanations, "Understanding Nature").

Mendel's laws were not accepted for many years for a third reason. When other scientists attempted to verify Mendel's results by testing other organisms, they did not always get the simple ratios that Mendel's laws predict—nor did Mendel when he experimented with certain other plants. Although Mendel's laws are valid, they are not observed for all traits in all crosses in all species. We will see why in Chapters 14 and 15.

After finishing his famous experiments, Mendel's life became increasingly busy with professional duties (he was elected the abbot of his monastery), and he was not able to pursue his own research. In 1900, however, after the rediscovery of his work, a very extensive research program in genetics began. Other researchers not only verified Mendel's laws but also expanded the knowledge of genetics far beyond anything Mendel could have imagined.

Summary

1. Genetics is the scientific study of heredity.
2. Theories of reproduction have changed dramatically throughout history. Aristotle held that females contributed matter and that males contributed an active spirit that molded the matter into the future offspring.
3. The mechanical philosophy of the scientific revolution replaced Aristotle's worldview. The invention of the microscope made possible new observations that, combined with the mechanical view of nature, led to the formation of the theory of preformation, which stated that all successive generations are encapsulated in previous generations.
4. Preformation theory was discarded because scientists chose to explain nature in terms of physical and biological laws and because observations seemed to contradict it. A blending theory of inheritance replaced the preformation theory.
5. Blending inheritance, however, did not give rise to rigorous, predictable laws that governed heredity. Dissatisfaction with the blending theory led to a new way of looking at heredity, which was the belief that traits are transmitted in discrete, unblended units.
6. Hugo de Vries carried out experiments to demonstrate the idea of discrete units of inheritance. His discoveries, published in 1900, called attention to the earlier work of Gregor Mendel.
7. Mendel's famous paper of 1866 describes his experiments on pea plants. He chose true-breeding lines and selected seven traits to study. His monohybrid crosses yielded hybrids that displayed dominant traits in the F_1 generation, but the recessive traits reappeared in the F_2. Mendel's experiments also showed that the traits were inherited in the F_2 in a ratio of three dominant to one recessive. However, two of the dominants were hybrids, so he concluded that the F_2 consisted of one dominant, two hybrids, and one recessive.
8. Mendel postulated that a pair of factors existed for each trait. If an individual has a pair of like factors, it is called *homozygous*. If the individual has the two alternative factors, it is called *heterozygous*.
9. In the production of eggs and pollen, the two factors that determine a trait in the parent separate and then recombine in the F_2. The separation of factors is known as Mendel's first law, the law of segregation.
10. Mendel discovered that the traits he studied were transmitted independent of one another. This independence is known as Mendel's second law, the law of independent assortment.

UNDER-STANDING NATURE

Focus on Scientific Explanations

Biologists use different methods to identify the laws of nature. For hundreds of years, studies of the biological world focused on describing and naming organisms. *Natural history,* as the subject was called, had as its goal creating a catalog of all living beings. Some natural historians also hoped that when all plants and animals were discovered, described, and cataloged, a natural order would reveal itself. The assumption was that God had a plan and that the plan would be recognizable once all the pieces were assembled. Other natural historians were skeptical that humans had the ability to discern God's plan. Rather, they held that the best people could hope for would be to devise some pattern to organize the vast amount of information (a brief discussion of the history of classification can be found in Chapter 3).

A method that proved to be very helpful in extending the knowledge gained from the description of organisms was the **comparative method**. By comparing the anatomy (structure) of various animals, scientists were able to discover similarities and differences that led them to classify organisms in new ways. It gave them a sense of what was fundamental to particular groups, and it suggested relationships among various life processes. Georges Cuvier, perhaps the greatest comparative anatomist who ever lived, was able to construct a complete classification system based on the comparative anatomy of different organ systems.

The **experimental method** didn't enter the biological sciences until quite late. Although William Harvey performed some astonishingly insightful experiments in the seventeenth century on the circulation of blood in animals (described in Chapter 34), it wasn't until the late eighteenth century that experiments began to be used as a tool of investigation in biology. When first introduced, experimentation was employed primarily for the investigation of the function of the body, that is, physiology. The method was so powerful, however, that soon it was employed in most scientific areas. Today experimentation is almost synonymous with the biological sciences.

With the experimental method came the use of **quantitative methods**. Experimenters measured, counted, and calculated. For example, the famous French scientists Antoine Lavoisier and Simon Laplace showed that the amount of CO_2 given off by both a candle burning and a mouse breathing was proportional to the amount of heat produced by each. From that experiment they concluded that breathing and combustion involve the same chemical process.

Mendel was one of the first to employ **statistical methods** to analyze experimental results. Statistics refers to the mathematical analyses and interpretations of data (numerical facts) that enable a scientist to make an objective appraisal about the reliability of the conclusions based on the data obtained. The key to generating powerful conclusions is to make as many observations as possible and to obtain a large data set. For example, Mendel could be relatively confident about his 3:1 ratios, not because in every cross he obtained exactly a 3:1 ratio but because he made many crosses and had hundreds of offspring data that always resulted in about a 3:1 ratio. Modern biological investigations are carefully designed to provide data that can be organized and subjected to specific statistical tests that indicate whether or not an original hypothesis should be rejected.

Figure 1 Lavoisier and Laplace used an ice calorimeter to measure the amount of heat given off in respiration and combustion. This engraving from their paper of 1783 shows a cutaway illustration of their basic apparatus.

11. Mendel's work wasn't appreciated in his own lifetime because it didn't fit with accepted ideas, it was not properly verified using other species, and the use of quantitative methods was novel in biology. After the rediscovery of his laws, however, active research in genetics began.

Review Questions

1. What is genetics?
2. What was Aristotle's theory of reproduction?
3. Why was preformation theory ultimately rejected?
4. Huge de Vries believed that the idea of blending inheritance should be replaced by what?
5. What is a monohybrid cross? A dihybrid cross?
6. Why did Mendel conclude that the two alternative forms of the traits he was studying were determined by a pair of hereditary factors?
7. What is the difference between a homozygous pair and a heterozygous pair of factors?
8. What is the law of segregation?
9. What is the law of independent assortment?
10. Why was Mendel's work neglected for so long?

Problems

1. What ratios among the F_1 generation would you predict would result from the following monohybrid crosses with Mendel's pea plants: (a) yellow pod (YY) × green pod (yy)? (b) $F_1 \times F_1 \rightarrow F_2$?
2. Construct a Punnett square that illustrates the possible gamete combination in a dihybrid cross using two of the seven traits Mendel studied.
3. What sort of test cross could Mendel have used to determine that the F_1 plants in his monohybrid crosses were heterozygous?

Essay and Discussion Questions

1. Aristotle was unaware of the value judgments he brought to his theory of reproduction. What assumptions that we hold today might influence our picture of nature?
2. What might you have thought if Mendel had written to you describing his experiments with *Pisum* and describing other experiments with a different plant that did not yield the same ratios? What advice might you have given him? Why?

References and Recommended Readings

Allen, G. E. 1975. *Life Science in the Twentieth Century*. New York: Wiley.

Corcos, A., and F. Monaghan. 1985. Role of de Vries in the recovery of Mendel's work. *Journal of Heredity* 76:187–190.

Dunn, L. C. 1965. *A Short History of Genetics*. New York: McGraw-Hill.

Gasking, E. 1967. *Investigations into Generation, 1651–1828*. Baltimore: Johns Hopkins University Press.

Olby, R. 1966. *Origins of Mendelism*. New York: Schocken Books.

Roger, J. 1963. *Les sciences de la vie*. Paris: Colin.

Ross, D. 1964. *Aristotle*. Oxford: Oxford University Press.

Stern, C., and E. R. Sherwood, eds. 1966. *The Origin of Genetics: A Mendel Source Book*. New York: Freeman.

CHAPTER 13

Cells: Structure, Function, and Reproduction

Chapter Outline

Reading Questions

1. What does the cell theory explain?
2. What are the major eukaryotic cell structures and their functions?
3. What is mitosis? What does it accomplish?
4. What is meiosis? What does it accomplish?
5. What is the role of fertilization in sexual reproduction?

A major unifying set of ideas in biology is the **cell theory**, which states that living organisms consist of individual *cells* and that the basic processes that define life are carried out by cells or groups of cells. For many years after the theory was originated, researchers were convinced that the cell theory would be *the* unifying theory in biology. However, the theory of evolution that emerged a few decades after the cell theory had even broader applications. Nonetheless, the importance of the cell theory cannot be underestimated, for it provided, and continues to provide, a key to understanding the functions of living organisms.

In Chapter 12, we examined the early development of genetics, a story that continues in Chapters 14–21. In this chapter, we consider the origin of the cell theory, provide an overview of cell structure, function, and reproduction, and consider an important cellular process, *meiosis,* that is essential for understanding genetics. More detailed information about certain cell structures and their functions appears in subsequent chapters.

THE CELL THEORY

How do major scientific theories originate? The origin of the cell theory is interesting because it demonstrates the complexity that often characterizes the birth of new scientific ideas. An understanding of the fundamental importance of cells, how they are organized, and how they function began to develop about 150 years ago. To a large extent, this knowledge grew out of medical studies on the human body, but results from this research were extended to include all living organisms. In part, the theory developed as a result of advances in technology. Since most cells cannot be seen with the naked eye, it wasn't until the invention of the microscope in the seventeenth century that the cell theory could begin to unfold. Even then, the early microscopes were imperfect instruments that often greatly distorted any object being examined. Early scientists, like the famous Robert Hooke in the seventeenth century, had glimpsed structures like the cell walls shown in Figure 13.1, but they had no idea of the significance of such observations. Although many interesting discoveries were made with these early microscopes, development of the cell theory was delayed until the invention of high-quality lenses in the 1820s and 1830s.

The origin of the cell theory, however, depended on more than observations made using a good microscope. The theory was formulated because a number of scientists in the first half of the nineteenth century were trying to find something like the cell. Several French and German medical researchers were convinced by then that the key to understanding the human body was to consider it to be a large, organized chemical machine. This seems to be an unusual idea, for when we look in the mirror, we certainly don't see a machinelike entity, nor do we imagine ourselves and our friends to be bags of chemicals. However, powerful ideas in science have not always been intuitively obvious. Scientists like Henri Dutrochet in France and Theodor Schwann in Germany also felt that a new way of looking at how bodies function was necessary if they were to understand them. Not only did they think that living organisms were made of organized chemicals, but they also recognized that these chemicals must be contained within some type of small basic unit. Unfortunately, Dutrochet's microscope was not adequate to see these units, but two German scientists, Matthias Schleiden (working with plants) and Schwann (working with animals), saw many different kinds of cells through their microscopes. In 1838 and 1839, they published separately their idea that the bodies of animals and plants were composed of cells. Thus the cell theory was first articulated.

Figure 13.1

Using a simple microscope in the seventeenth century, Robert Hooke observed plant cell walls and made these drawings.

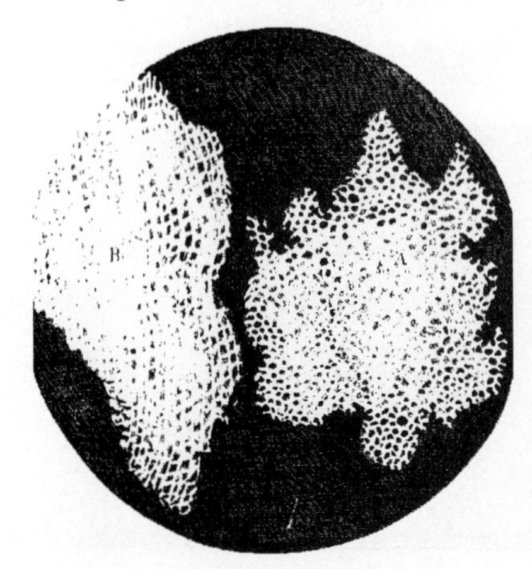

Maturation of the Cell Theory

Throughout the rest of the nineteenth century, new discoveries and theoretical advances contributed to an extension of the cell theory. In 1855, Rudolf Virchow stated that new cells could originate only by the division of preexisting cells and not from nonliving substances. The corollary statement, that all existing cells can trace their origins to ancestral cells, was added in 1880. Throughout this period, descriptions were published of some of the structures observed inside of cells and the wondrous events that occurred when a cell divided into two apparently identical cells. It was also discovered that during **sexual reproduction**, hereditary factors are transferred from one generation to the next by the passage of genetic material carried in specialized sex cells.

By the end of the nineteenth century, there was a broad, but shallow, understanding of cell structure and function as a result of the earlier descriptive studies. Biologists began to penetrate ever further into the cell's interior to discover the ultimate secrets of its internal structures and functions. In the twentieth century, they were aided in these endeavors by the development of new, more powerful microscopes and advances made in the fields of biochemistry and molecular genetics.

The Modern Cell Theory

Today the cell theory has been extended and modified to reflect advances in knowledge gained from sophisticated studies conducted during recent decades. It is now clear that all eukaryotic cells share many basic similarities. They are similar in their chemical composition and in many structures they possess. They also share similar mechanisms for using energy, building structural components, and synthesizing genetic material and transmitting it to the next generation. An acknowledgment of these similarities, along with the view that activities of organisms are a function of the activities and interactions of cells, has been incorporated into the modern cell theory.

The cell theory has had a long and rich history that is not yet over. Technological innovations have taken the initial ideas far beyond what was originally described, but not beyond the original inspiration to understand the functioning of living organisms in terms of their chemical and physical components. The pioneers of the cell theory also believed that it would provide for a unification of all our knowledge of living organisms. However, the theory of evolution was even more comprehensive and actually incorporated the cell theory into its framework.

The concept that all organisms are composed of cells emerged as a result of research using microscopes and the formulation of powerful new ideas derived primarily from medical studies of the human body. These activities led to the formation of the cell theory, a broad, unifying set of statements about the nature of cells and of life.

THE ESSENCE OF CELLS

The two basic cell types, prokaryotic and eukaryotic, were introduced in Chapter 2. Additional details about the more complex eukaryotic cell structures and functions are provided in this chapter.

Cell Structure

An examination of cell types from organisms in different kingdoms reveals a bewildering array of sizes (eukaryotic cells average about 10 micrometers in size; see Figure 13.2) and shapes. However, there is a basic pattern in the structure of all living cells that is apparent when viewed through a microscope. Although many cells in multicellular plants and animals have the potential to operate independently, they do not do so under normal circumstances. Rather, most cells are combined with other cells to form tissues and organs that perform specific functions within the body of an organism.

Figure 13.2

Cells differ greatly in size.

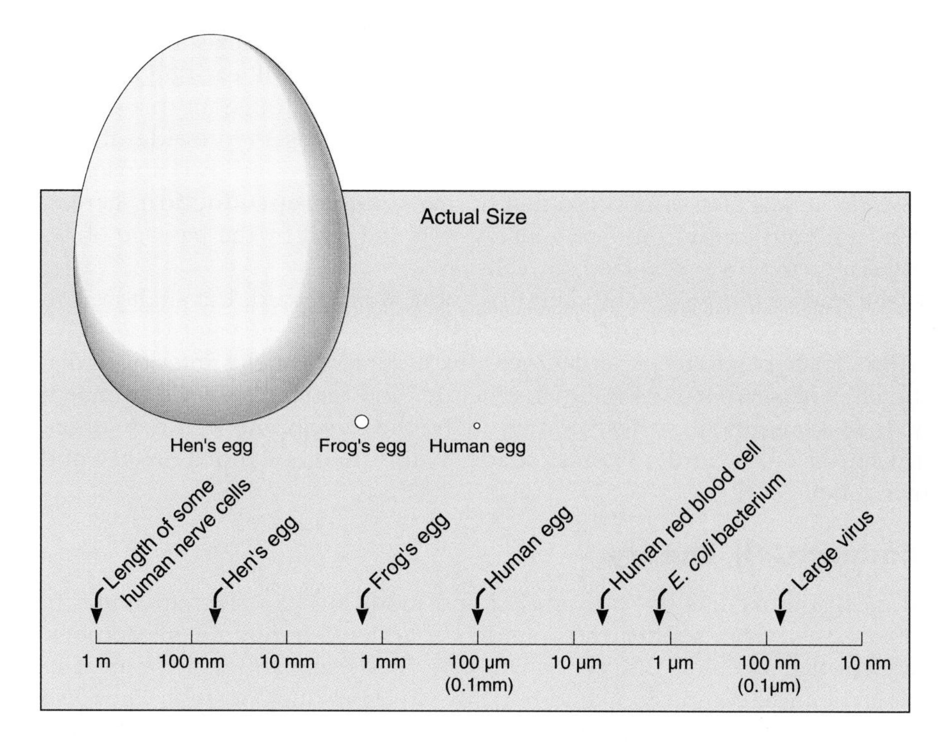

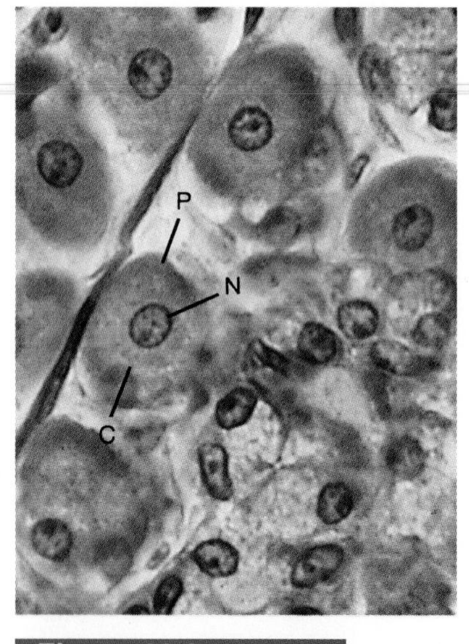

Figure 13.3

Most eukaryotic cells, such as these parietal cells of a cat's stomach, consist of a nucleus (N) and cytoplasm (C) surrounded by a plasma membrane.

Visualizing Cells

The compound (light) microscope has made it clear that eukaryotic cells are discrete units bound by a *plasma membrane* that encloses a prominent *nucleus,* which is surrounded by *cytoplasm* (see Figure 13.3). What exactly exists in the nucleus and the cytoplasm? Since a light microscope can magnify objects clearly only 1,000 times, that question could not be answered until the 1950s, when an exciting new instrument, the **transmission electron microscope (TEM** or **EM)**, was created that could magnify objects over 100,000 times. Later came the **scanning electron microscope (SEM)**, which makes it possible to view a cell three-dimensionally at these high magnifications.

Many kinds of microscopes and other instruments are now used to study and visualize cells. For cells to be visualized with most microscopes, including the EMs, they must first be treated and processed by special techniques. Basically, cells are fixed for viewing by placing them in chemicals that preserve their structure. Next, they are embedded in a medium (for example, wax or plastic) that can be sliced into ultrathin sections. Finally, these sections are stained or coated with some substance that makes their structure more visible. Photographs of cells and cell parts, using different microscopes and other techniques, appear in this chapter and later parts of the text.

Eukaryotic Cells

Figures 13.4 and 13.5 illustrate the detailed structure of animal and plant cells. Once the TEM was developed, it became possible to examine the internal architecture of the cell. A beautiful collection of fibers and structures awaited the first investigators. Within the interior matrix of each cell are a number of microscopic structures called *organelles* along with various fibers collectively termed the *cytoskeleton*. These fibers are important in maintaining cell shape and creating cell movement. In most cells, each internal **organelle** is bound by a membrane or is part of a membrane system and is specialized to conduct a specific function in the cell. All eukaryotic cells contain organelles, although there are variations in type and number among the cell types. Cell biologists have also been able to study these structures in isolation and learn about their functions. The results of these studies made it apparent that cells have a compartmental organization that

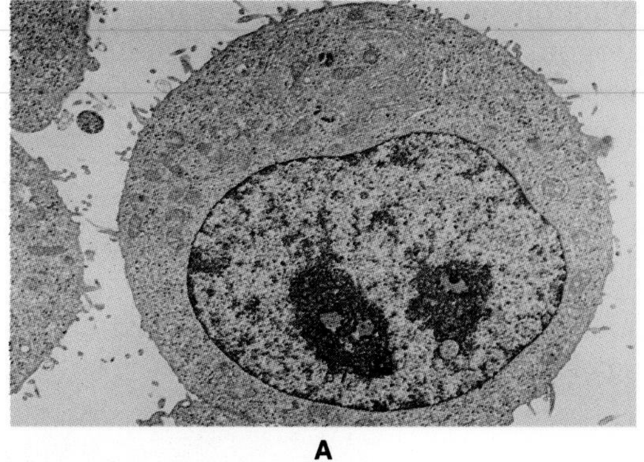

A

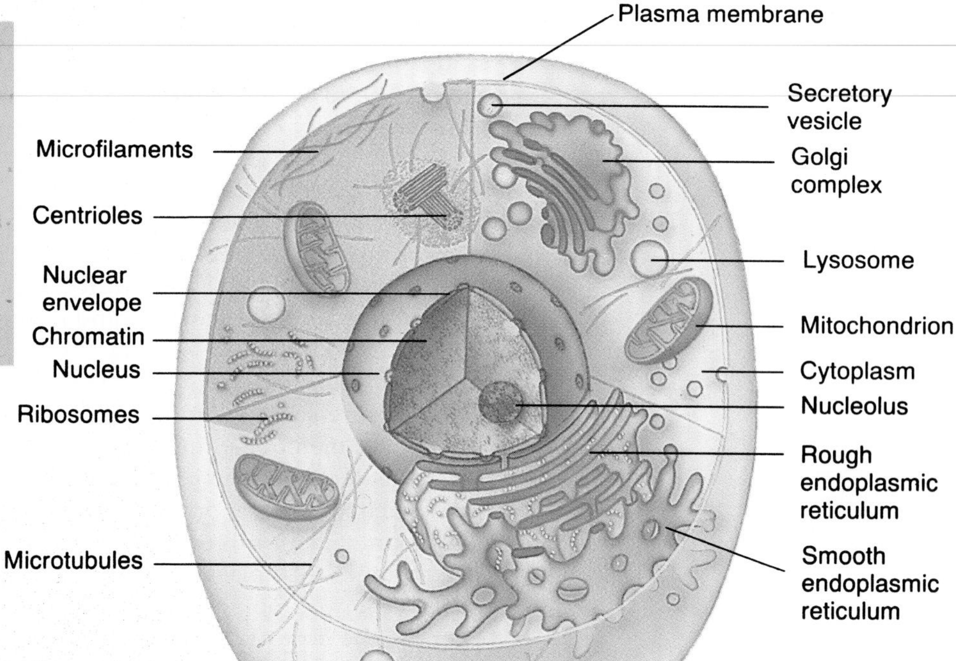

B

Figure 13.4

(A) An electron micrograph (EM) of an animal cell showing the nucleus and cytoplasm. (B) Animal cells contain some or all of the structures shown in this figure, depending on their function.

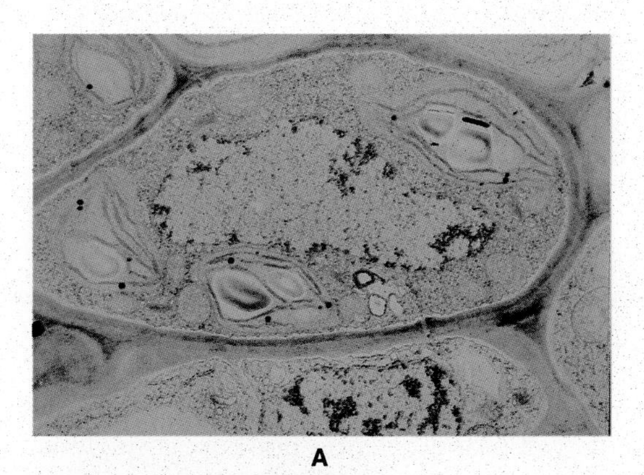

A

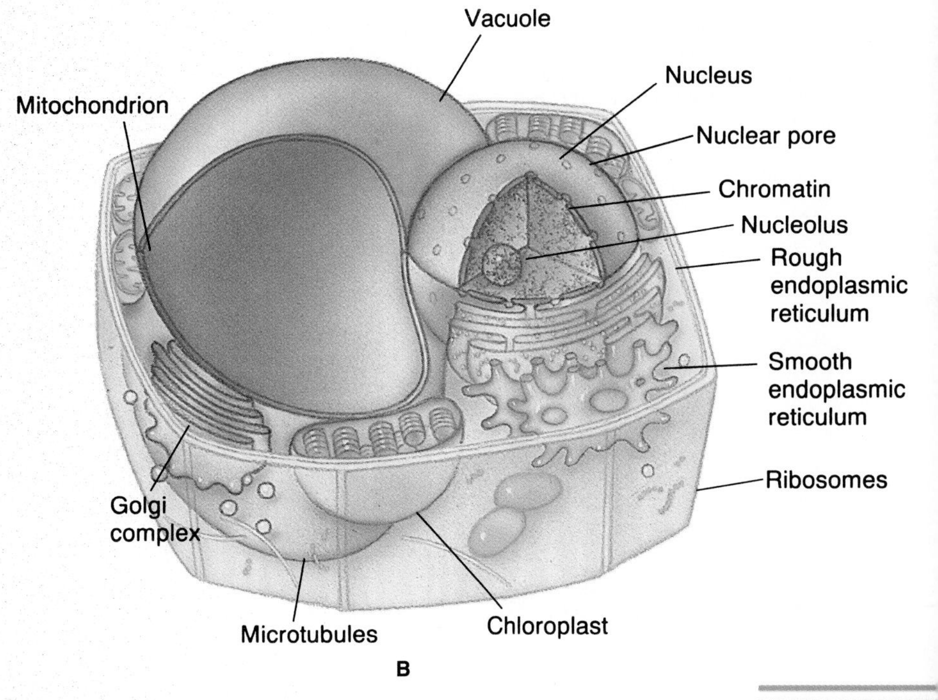

B

Figure 13.5

(A) An EM of a plant cell. (B) Plant cells (and some protoctists) contain some or all of the structures found in animal cells. They also feature cell walls, chloroplasts, and vacuoles.

is reflected by discrete organelles carrying on specific but separate functions in the cell. The cell's operations are analogous to an orchestra's; individuals (organelles), playing certain instruments (carrying out a specific function), together create a single piece of beautiful music (make a finely tuned, efficient cell).

Basic Cell Structures

Eukaryotic cells generally consist of a plasma membrane that encloses cytoplasm and a nucleus. Figure 13.5B reveals that plant cells differ from animal cells in being surrounded by a **cell wall**—a tough, extracellular matrix composed of *cellulose* and other substances—and having chloroplasts and vacuoles.

Plasma Membrane

All cells are surrounded by a **plasma membrane**. As described in Figure 13.6, biological membranes are very complex structures composed of highly organized large molecules (macromolecules), mostly proteins and lipids (fats and cholesterol). Lipids are the primary structural component of membranes, and their two-layer organization is associated with the dynamic properties and functions of membranes. The plasma membrane is responsible for a variety of operations critical to the well-being of the cell it surrounds. Its most important function is to control the flow of chemical substances, such as water, nutrients, and waste products, into or out of the cell. Since each cell has exact requirements for its internal chemical composition, precise regulation by the plasma membrane is required for maintaining stable operating conditions within the cell.

Cytoplasm

The interior of a cell contains a gelatinous material called the **cytosol** that consists primarily of chemical substances dissolved in water. Within this medium, a variety of specialized cellular organelles exist. **Cytoplasm** refers to all of the substances and organelles inside the plasma membrane, except for the nucleus.

Nucleus

The **nucleus** is a large spherical structure that is surrounded by a double-layer membrane called the **nuclear envelope** (see Figure 13.7). The inner membrane encloses the nucleus itself. In many cells, the outer membrane branches and connects with an intracellular membrane system called the *endoplasmic reticulum*. Each cell usually contains a single nucleus, although there are some interesting exceptions. For example, mature human red blood cells do not have a nucleus, and each muscle cell in your bicep contains many nuclei. The nuclear envelope, like the plasma membrane, regulates the movement of materials into and out of

Figure 13.6

(A) An EM showing a plasma membrane (arrows). (B) All cells are enclosed by a plasma membrane, a complex structure that regulates the movement of substances into and out of the cell.

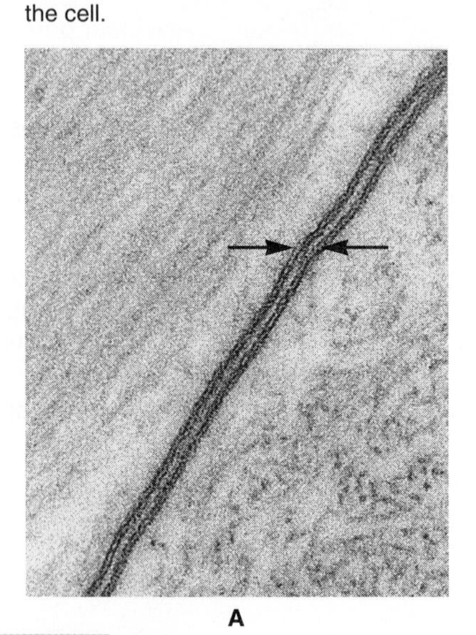

A

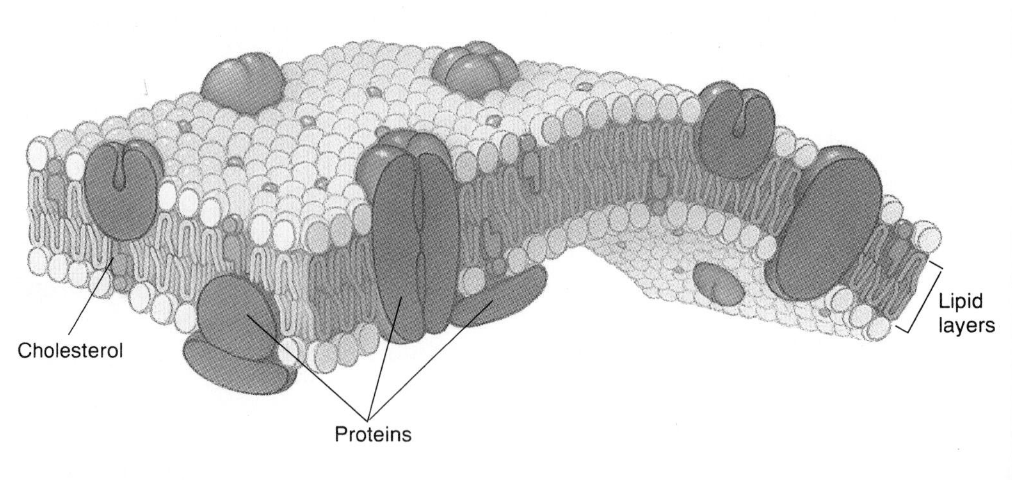

B

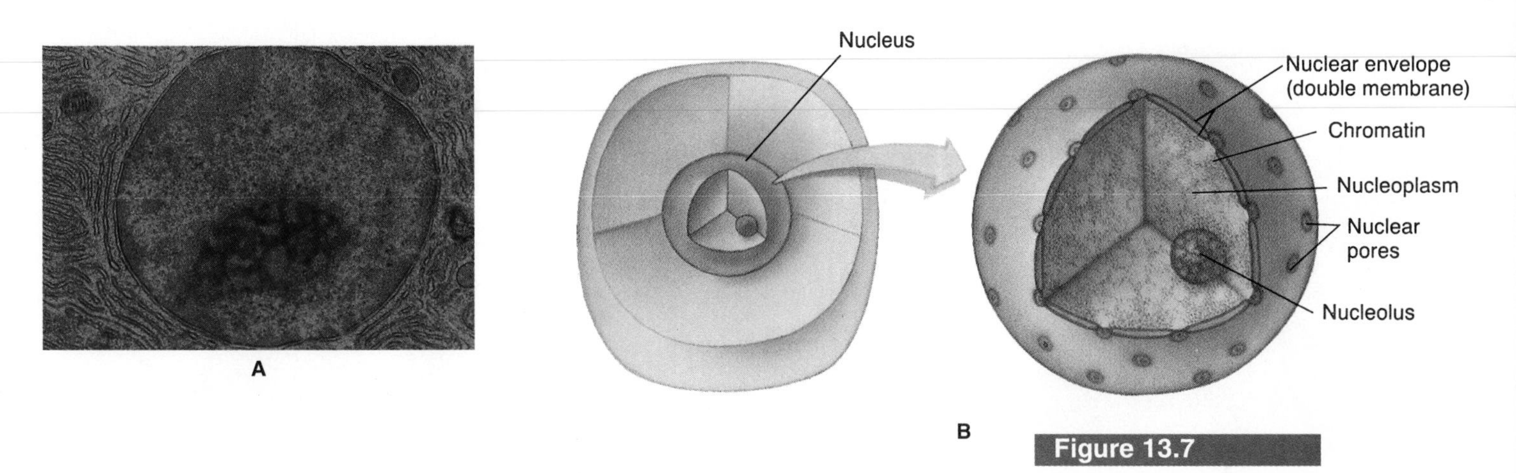

Figure 13.7

(A) An EM of a nucleus. (B) The nucleus is a complex structure enclosed by a nuclear envelope (membrane) that is riddled with nuclear pores. These large openings regulate the movement of substances in and out of the nucleoplasm. Chromatin is scattered genetic material that forms into chromosomes prior to cell division. The nucleolus is a specialized structure involved in protein synthesis.

the nucleus. Large **nuclear pores** are apparent in the surface of the membrane, and evidence suggests that they are involved in the transport of substances between the nucleus and the cytoplasm.

The interior of the nucleus is referred to as **nucleoplasm**. A prominent structure seen in the nucleus of most cells is a **nucleolus** ("little nucleus"). Its primary function is related to *protein synthesis,* which is described in Chapter 16. Mechanisms for the control and regulation of most cell functions reside within the nucleus. The underlying basis for most of these regulatory activities lies in the production of unique proteins that cause specific effects in different cells.

Cellular Organelles

Eukaryotic cells have certain organelles in common. Both plants and animals possess an endoplasmic reticulum, Golgi complexes, and mitochondria. However, only plant cells and photosynthetic protoctists contain chloroplasts and, often, vacuoles.

Endoplasmic Reticulum

The **endoplasmic reticulum (ER)** is a large network of interconnected membranous tubules where proteins and lipids are synthesized in the cell (see Figure 13.8). **Rough ER** is studded with small structures called **ribosomes** that play a major role in the synthesis of proteins. Some proteins synthesized by the cell are required for building and maintaining organelles or for participating in activities within the cell, but others may be released for use by other cells in the organism.

Figure 13.8

(A) An EM of endoplasmic reticulum. (B) The endoplasmic reticulum (ER) is a network of membranes that branch off the nuclear envelope but are continuous with it. Ribosomes are scattered on the surface of rough endoplasmic reticulum, where they participate in protein synthesis. Smooth endoplasmic reticulum has no ribosomes. It conveys proteins produced on the rough ER to the Golgi complex. Lipids may also be synthesized in smooth ER.

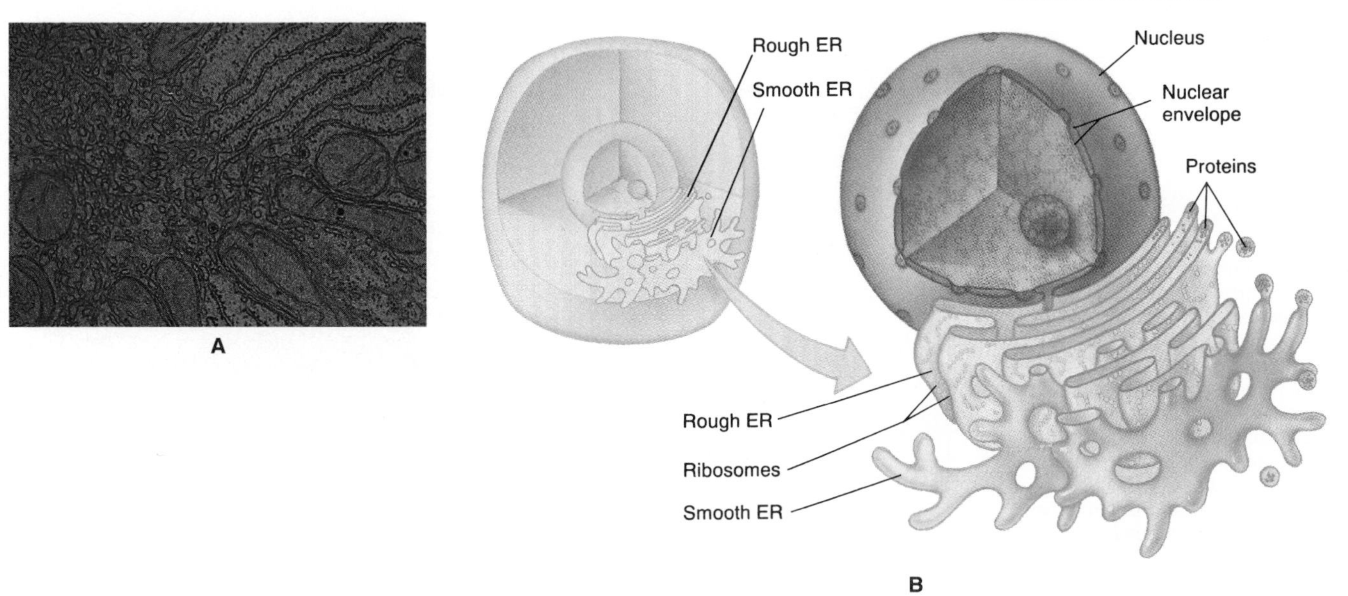

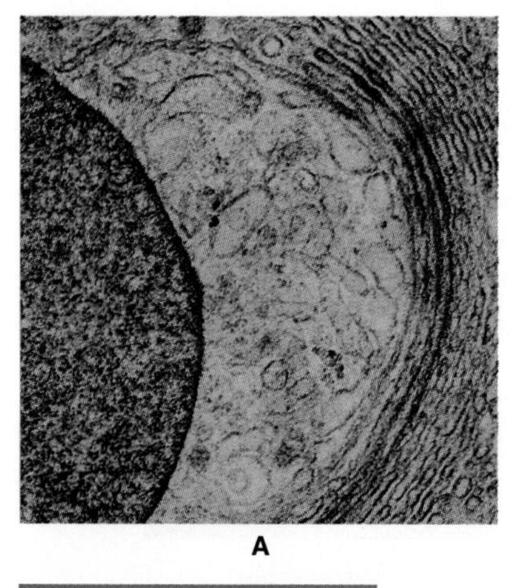

A

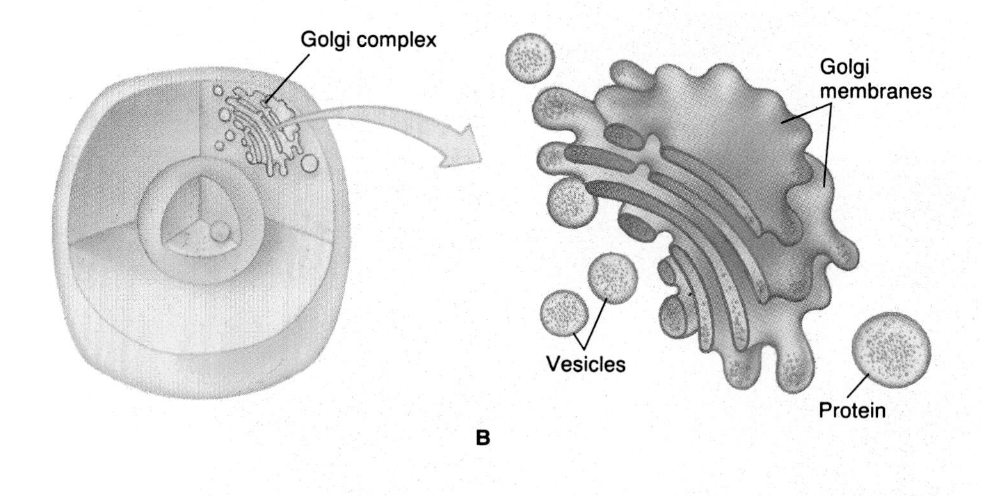

B

Figure 13.9

(A) An EM of a Golgi complex. (B) Golgi complexes are networks of flattened membranes. Their primary function is to process and move proteins produced within the cell. Transported proteins are enclosed in membranous vesicles.

Smooth ER, which does not have ribosomes on its surface, is the site where lipids are synthesized. It also receives proteins from the rough ER and incorporates them into small membrane-bound sacs called **vesicles**.

Golgi Complex

Proteins synthesized on ribosomes attached to the rough ER are transported from the ER to a **Golgi complex**, an organelle that consists of a series of flattened membranes (*Golgi sacs*; see Figure 13.9). It is generally believed that Golgi complexes are responsible for modifying, sorting, and transporting proteins within the cell so that they arrive at their proper location. Proteins that are to be exported from the cell are packaged in vesicles that move to the plasma membrane and are expelled to the cell's exterior. Figure 13.10 describes the relationships among membranous organelles.

Mitochondria

Cells obtain energy by converting a simple sugar, glucose, into another energy-rich substance that can be used by the cell (described in Chapter 7). This transformation of glucose occurs primarily in **mitochondria**, large organelles that are encircled by a double membrane (see Figure 13.11). The outer membrane surrounds the mitochondria, and the inner membrane has numerous infoldings that are the sites of energy generation. Most cells contain numerous mitochondria; the more active the cell, the larger the number. For example, muscle cells are filled with mitochondria, whereas a fat cell may have only one.

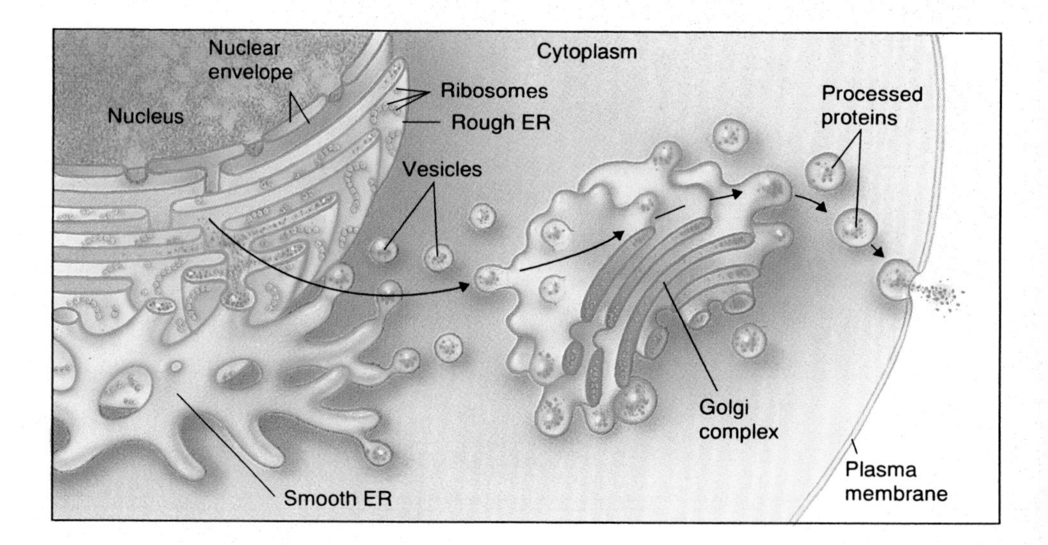

Figure 13.10

This diagram describes the structural relationships among the various membranous organelles in eukaryotic cells. Macromolecules responsible for initiating protein synthesis move from the nucleus to the rough endoplasmic reticulum, where synthesis occurs. From there, newly produced proteins move in sequence through smooth ER, the Golgi complex, and finally the plasma membrane, from where they leave the cell. Proteins are transported from one organelle to another in membranous vesicles.

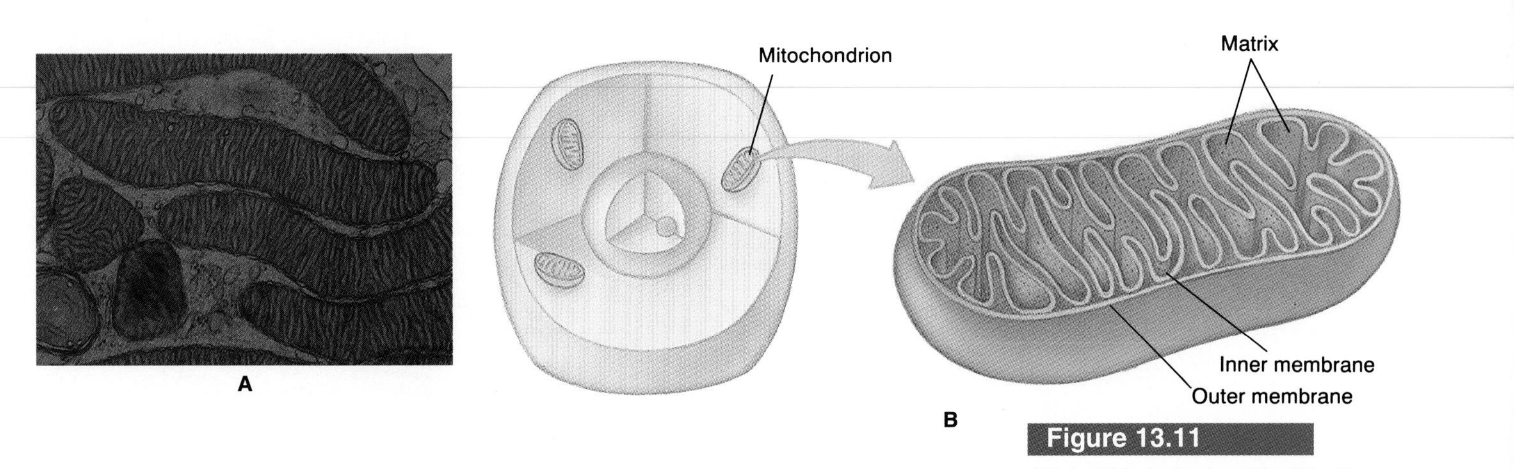

Figure 13.11

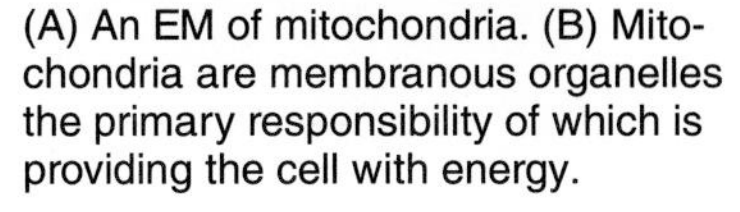

(A) An EM of mitochondria. (B) Mitochondria are membranous organelles the primary responsibility of which is providing the cell with energy.

Chloroplasts

Chloroplasts, shown in Figure 13.12, usually contain a green pigment called **chlorophyll** that is critical in carrying out photosynthesis, the process whereby radiant energy is used to produce sugars (explained in Chapter 2). Chloroplasts have a double outer membrane and an extensive internal membrane system, which is where the reactions of photosynthesis take place.

Vacuoles

Most plant cells contain at least one large **vacuole**, a small cavity that is surrounded by a single membrane. These vacuoles store substances such as water, sugar, proteins, and sometimes pigments and may also act as reservoirs for harmful waste products created by the plant cell.

The Cytoskeleton

Within the cytoplasm is a delicate weblike structure called the **cytoskeleton** (see Figure 13.13). This microscopic internal skeleton gives animal cells their shape, encloses and anchors organelles, and permits movement in certain cells (for example, muscle cells). The principal components of the cytoskeleton are *microtubules* and *microfilaments.* **Microtubules** are made of globular, beadlike subunits organized into hollow, cylindrical tubes that form a skeletal network of fibers in the cytoplasm of many cells. Microtubules have a variety of roles in cells of living organisms, such as control of cell shape and the movement of different materials within the cell. **Microfilaments** are thin fibers composed of globular protein subunits. They function in moving organelles around the cell and in contraction movements in cells specialized for that activity.

Figure 13.12

(A) An EM of a chloroplast. (B) Plant cells and some protoctists have chloroplasts, large organelles that contain chlorophyll, a pigment capable of transforming radiant energy into organic molecules during photosynthesis.

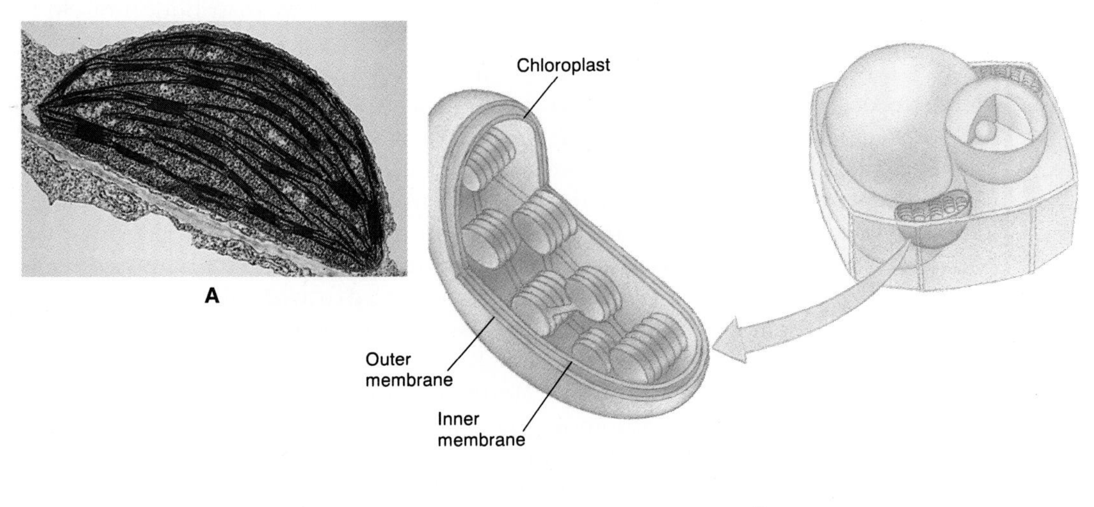

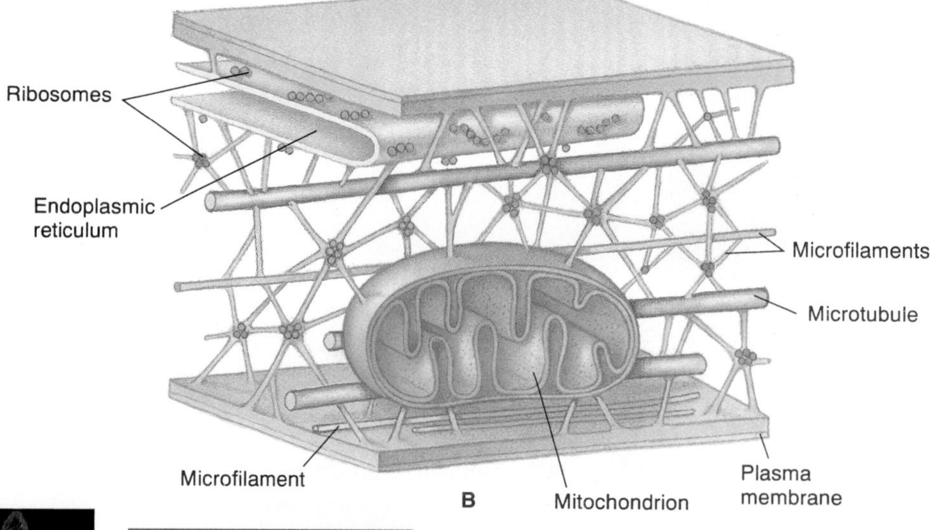

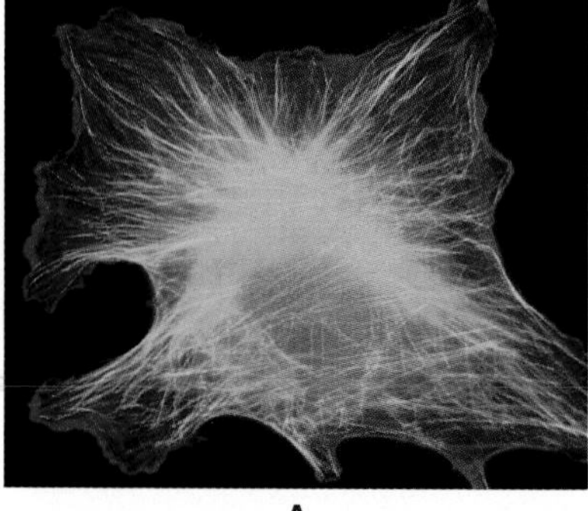

Figure 13.13

(A) The cytoskeleton is an internal cell support system composed of microtubules and microfilaments. In this photograph of a cell, microtubules appear as green fibers and microfilaments are red. (B) Microtubules have a hollow, tubular structure. They control cell shape, act as guides for the movement of various substances inside the cell, and play a key role in cell division by manipulating chromosomes. Microfilaments have a fibrous structure that provides the cell with structural reinforcement. The cytoskeleton in a cell has an organized structure as shown in this figure.

All eukaryotic cells have a basic structure consisting of an outer plasma membrane that encloses a nucleus and cytoplasm. Within the cytoplasm are various organelles—endoplasmic reticulum, Golgi complexes, mitochondria, chloroplasts, and vacuoles—that carry out specific cellular functions. The numbers and kinds of organelles may vary in different cells of the same organism and between cells of organisms from different kingdoms.

CHROMOSOMES

Chromosomes are filamentous structures composed of protein and tightly coiled *DNA,* the genetic material. Sets of specific numbers of chromosomes are found within the nuclei of all eukaryotic species but are usually visible only when a cell divides. During most of a cell's life, its chromosomes exist in a dispersed state called **chromatin**. Chromatin consists of uncoiled DNA and proteins, loosely distributed throughout the nucleoplasm. The patchy distribution of chromatin within the nucleus results in a characteristic microscopic appearance.

For convenience in discussing certain concepts, biologists have developed terms to describe chromosomes and their relationships with one another. These are presented in summary form in Figure 13.14, along with related terms used in genetics, using one pair of chromosomes. Each chromosome pair consists of two **homologs**, one contributed by the female parent (the maternal chromosome) and one by the male parent (the paternal chromosome); together the two homologs make up a **homologous pair of chromosomes**. A cell with two complete sets of homologous chromosomes is called **diploid**, abbreviated as $2n$ where n = the number of different chromosomes in one set. Sex cells (gametes) or cells with only one homolog of each chromosome type are **haploid** and are signified as n. To extend this example to humans, each of our body cells is diploid ($2n$) and contains 23 homologous pairs of chromosomes, or 46 chromosomes in total; one homolog of each pair came from each parent. Our sperm or egg cells (n) contain only 23 chromosomes, one of each type of homolog.

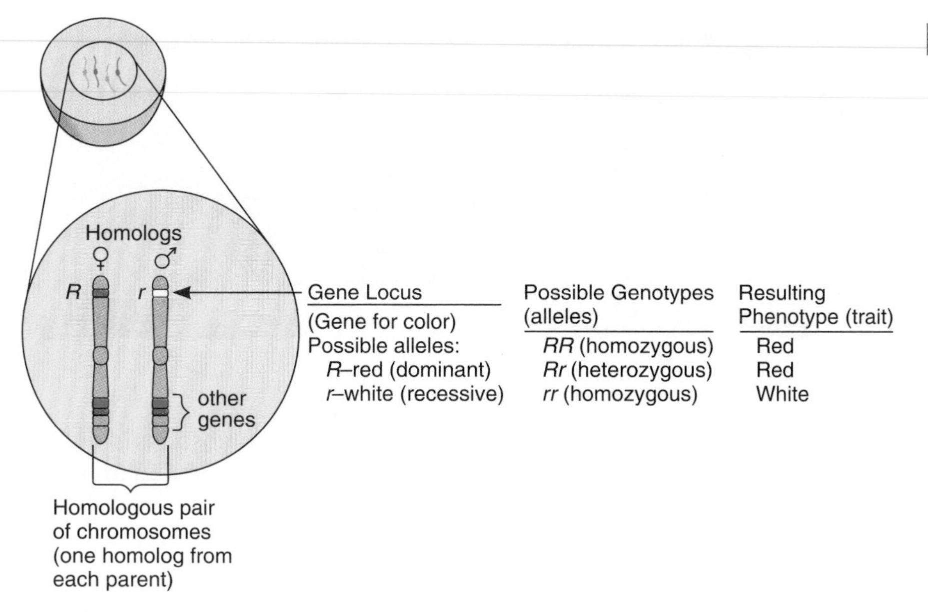

Figure 13.14

This diagram gives terms used in describing the relationships between chromosomes and also the chromosomal structures and locations that are important in genetics. Chromosomes are not visible until after they have been duplicated prior to cell division. The duplicate chromosome is not shown in this figure.

Human chromosomes contain 50,000 to 100,000 bits of genetic information, called **genes**, which are composed of DNA. The term *gene,* the fundamental hereditary unit, was introduced in 1909 and replaced hereditary *character* or *factor.* Each homolog of the homologous pair contains one or more genes for the same trait (for example, flower color) at a specific **locus**, or identical location, along its length. A single gene may have alternate forms that are called **alleles**. For example, a gene for flower color, a single trait, may have different alleles that result in red, white, or purple colors, which are alternate forms of the trait. The specific alleles or genes contained in a cell (or an individual) are called its **genotype**. The physical trait that occurs as a result of a specific genotype is referred to as the **phenotype**.

All sexually reproducing species have a characteristic number of homologous chromosome pairs in their cells. For example, fruit flies have 4 pairs; corn, 10 pairs; humans, 23 pairs; ducks and pumpkins, 40 pairs; and crayfish, 50 pairs. Obviously, complexity is not related to the number of chromosomes a species possesses. The relationship between the two members of each homologous chromosome pair is interesting and fundamental in conveying hereditary information between generations in sexually reproducing organisms. One member of each pair of homologous chromosomes came from each parent. Therefore, each parent contributed exactly one half of the chromosomes present in cells of the offspring. In humans, 23 chromosomes came from the mother and 23 from the father.

ASEXUAL REPRODUCTION

As stated in the cell theory, a fundamental characteristic of cells is an ability to reproduce themselves by a process called *cell division.* Cell division is a type of **asexual reproduction**, a process of replication in which a single cell or individual gives rise to two or more identical (or nearly identical) daughter cells or parts. How do cells reproduce? What does cell division achieve?

Prokaryotic Cell (Bacterial) Division: Binary Fission

Bacterial cells, shown in Figure 13.15, have the following characteristics: (1) they are enclosed by a plasma membrane that is surrounded by a cell wall, (2) they have no internal membranes or membrane systems, (3) their ribosomes are scattered throughout the cytosol, (4) they have a diffuse *nucleoid,* not a well-organized nucleus, and (5) they have a single, circular chromosome. Bacterial cells produce "offspring" by **binary fission**, a process in which one cell divides

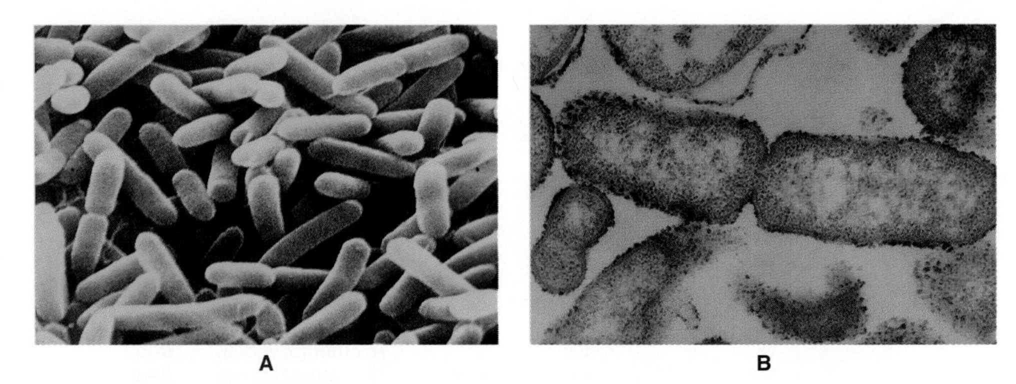

Figure 13.15

(A) A scanning EM of bacteria (prokaryote) cells. (B) An EM showing a bacterium undergoing binary fission.

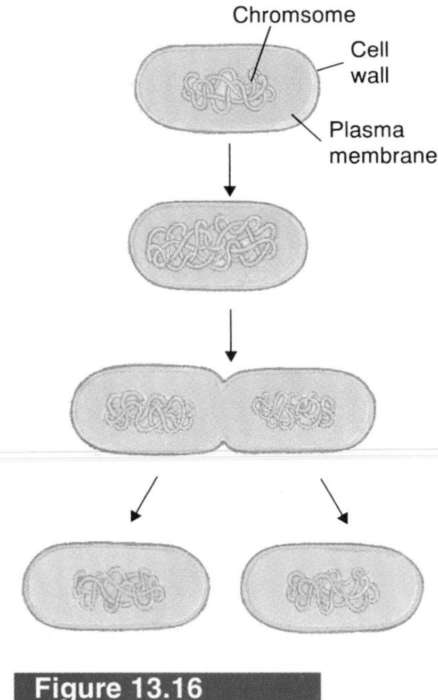

Figure 13.16

Bacterial cells reproduce by a type of cell division called binary fission.

into two cells (see Figure 13.16). In summary, the chromosome is duplicated before cell division, the two chromosomes separate, the plasma membrane wedges inward, the cell wall forms along the same plane, and the parent cell divides into two daughter cells. Binary fission is a bacterium's means of reproduction, since the daughter cells are independent new "offspring."

Eukaryotic Cell Division: Mitosis

Eukaryotic cells divide by *mitosis,* a process in which each daughter cell receives the same number and kind of chromosomes as the parent cell. **Mitosis** refers to the nuclear division and **cytokinesis** to the division of a *parent cell* into two *daughter cells*. Mitosis is divided into stages, shown in Figure 13.17, for convenience in describing and understanding the process. We describe below a general case of animal cell mitosis. Be aware, however, that there are a great number of variations in the process in eukaryotic organisms from different kingdoms.

For clarity in describing the sequence of events during mitosis, biologists have given names to the phases. However, the process is continuous with no distinct borders. In cells that will divide, the first and longest stage of mitosis is called **interphase**. Some scientists do not consider this stage to be part of mitosis, but for clarity we include it as the first phase. During interphase, each chromosome is duplicated, and the cell acquires enough cellular materials—nutrients and structural proteins—to survive division into two daughter cells.

Following interphase, **prophase** occurs, characterized by changes in the form of the chromosomes. During prophase, the chromosomes shorten in appearance and thicken to the extent that they can be seen with the light microscope. At this stage, each of the two duplicated chromosomes is said to consist of two identical **sister chromatids** that are held together at their **centromeres**, or constricted regions. As the chromosomes become visible, small cylindrical structures called **centrioles**, which are composed of microtubules, appear in animal cells and move to opposite ends (poles) of the cell. Soon after this occurs, the nuclear envelope and nucleolus disappear, and the centrioles generate *spindle fibers* (specialized microtubules) that attach to the centromeres of the sister chromatids. At the end of prophase, the sister chromatids become aligned at the equatorial plane of the cell.

After prophase, the cell enters **metaphase**, a stage when sister chromatids are aligned at the center of the cell. During **anaphase**, the paired chromatids separate at their centromeres and appear to be pulled toward the poles of the parent cell as the spindle fibers shorten. This stage marks the point at which the duplicated set of chromosomes of the parent cell becomes separated into two sets of identical chromosomes received by each of the two daughter cells.

During the last stage of mitosis, **telophase**, daughter nuclei form, nuclear envelopes and nucleoli appear, and chromosomes disappear. Finally, cytokinesis occurs, resulting in two new daughter cells that are exact (or nearly exact) duplicates of the original parent cell.

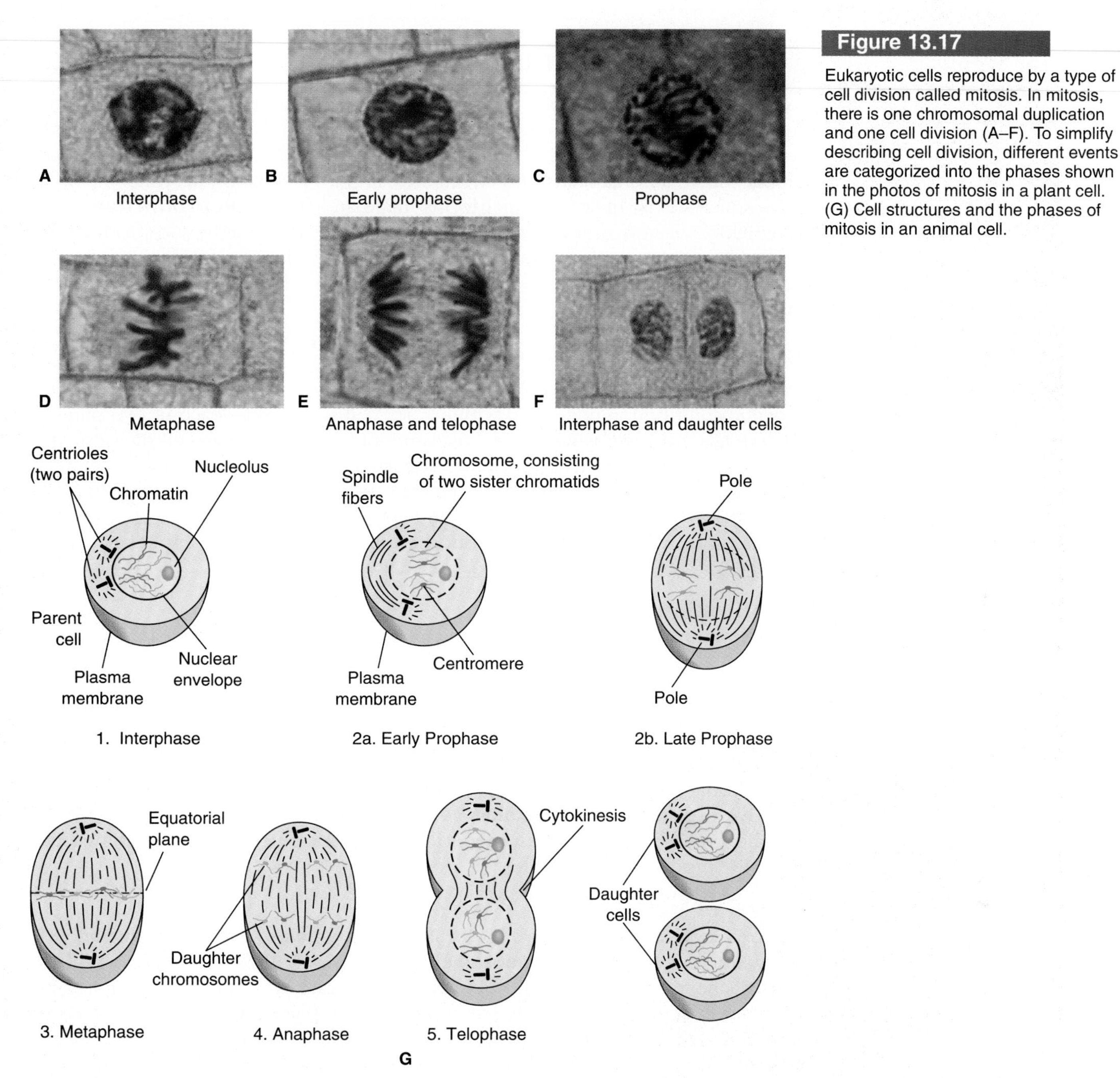

Figure 13.17

Eukaryotic cells reproduce by a type of cell division called mitosis. In mitosis, there is one chromosomal duplication and one cell division (A–F). To simplify describing cell division, different events are categorized into the phases shown in the photos of mitosis in a plant cell. (G) Cell structures and the phases of mitosis in an animal cell.

What kinds of cells divide by mitosis? What is accomplished? The answers may become evident in thinking about these questions as they relate to humans. How are cells replaced in our body as they wear out and die? For example, an average red blood cell lives approximately 120 days, and millions of these cells die every hour. How are wounds or injuries repaired? How does a fertilized egg (one cell) become a baby (trillions of cells) in about nine months? Mitosis occurs in cells involved in growth, repair, development, and cell replacement. It is estimated that in adult humans, about 1 trillion mitoses occur every 24 hours to provide new cells to replace those that normally die in performing their functions. Thus it has tremendous importance in the lives of all eukaryotic organisms.

Many cells have the ability to divide by a form of cell division. In bacteria, binary fission gives rise to new individuals. Eukaryotic cells divide by mitosis, a process characterized by one chromosomal duplication, one nuclear division, and one cell division. The resulting daughter cells are identical to the original parent cell.

SEXUAL REPRODUCTION

Sexual reproduction is not a term that usually sends people in search of a dictionary. However, since hundreds of different reproductive processes occur in protoctists, fungi, plants, and animals, it is difficult to formulate a definition applicable to all organisms. In general, **sex** implies a situation in which members of a species exist in two distinct forms, male and female. **Reproduction** is the origination of new organisms from preexisting ones. The ability to produce offspring by sexual reproduction has advantages, yet places great demands on organisms (see the Focus on Scientific Explanations, "The Significance of Meiosis and Sexual Reproduction").

Meiosis

Most eukaryotic animals reproduce sexually, a process that typically requires a genetic contribution contained within a sex cell supplied by each of two parents. In animals, **meiosis** is a process consisting of two consecutive cell divisions that results in the formation of four **gametes**, specialized cells that contain the haploid number of chromosomes. Interesting questions to consider are why meiosis is necessary and why gametes must be haploid rather than diploid. (Meiosis also occurs in other eukaryotic cells, but the following discussion applies to animal cells.)

Fertilization

Fertilization is a process in which gametes unite to form a *zygote* that develops into a new offspring. The outcome of fertilization is the restoration of the normal, diploid number of chromosomes ($n + n = 2n$). Thus the goal of meiosis is to reduce the chromosome number by half so that fertilization of gametes will result in a constant chromosome number in a species.

Haploid gametes—*sperm* produced by males and *eggs* by females—are generated from meiotic nuclear divisions of gamete-producing cells present in reproductive organs, which are called *gonads* in animals. Meiosis begins with the duplication of each chromosome, followed by two nuclear divisions, and ends with the production of gametes.

Stages of Meiosis

Meiosis is described in Figure 13.18. During *interphase,* gamete-producing cells grow and develop, and chromosomes exist only as diffuse chromatin. While in this phase, each chromosome becomes duplicated, and thus the chromosome number is doubled. Meiosis is now ready to take place. Since there are two cell divisions during meiosis, roman numerals are appended to indicate whether the events occur during the first or second division.

Meiosis I

The first stage of meiosis, *prophase I,* begins when the chromatin coalesces and the sister chromatids become visible. During this stage, duplicated homologous chromosomes undergo **synapsis**, or align themselves lengthwise with each other, and their chromatids become intertwined. This intertwining is associated with a process, *crossing-over,* that can introduce genetic variation in sexually reproducing organisms. During **crossing-over**, chromatids of the duplicated homologs frequently break, exchange, and become reattached to the other homolog. Crossing-over is described in greater detail in Chapter 14. Prophase I concludes as all chromatid pairs become aligned at the center of the cell.

The continuing meiotic process advances to *metaphase I* when the paired homologous chromosomes become aligned at the equatorial plane of the cell. The orientation of the homologous pairs of chromosomes is random, and there is a 50-50 chance that any pair of sister chromatids will go to a particular pole during the next phase. Metaphase I ends when the spindle fibers begin to shorten.

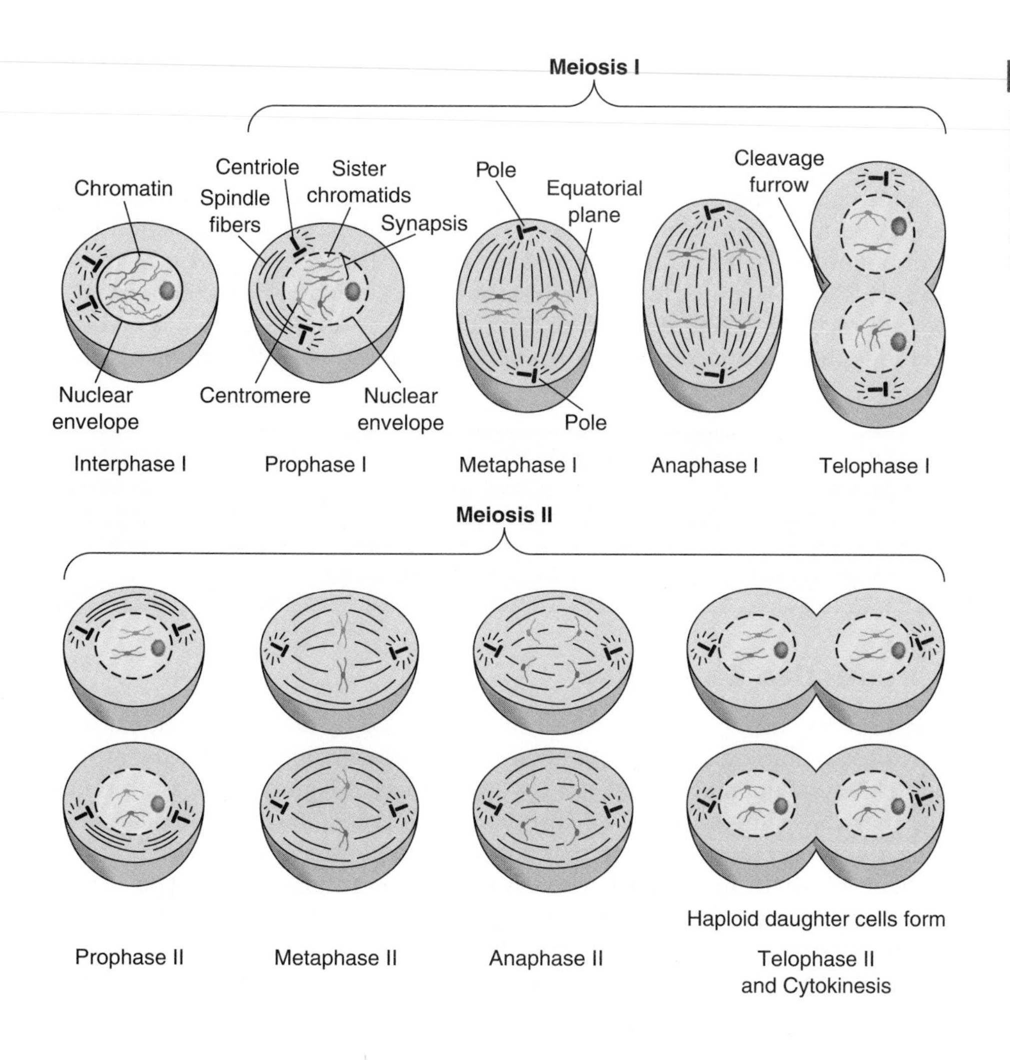

Figure 13.18

Specialized eukaryotic cells produce haploid cells (gametes in animals, spores in plants) by a cell division process called meiosis. In meiosis, there is one chromosomal duplication and two cell divisions, termed meiosis I and meiosis II.

During *anaphase I,* spindle fibers separate the homologous chromosomes, which then move toward opposite poles of the cell. This is a key event in creating new genetic combinations in sexually reproducing species. New combinations of genes or chromosomes, created by any process (including crossing-over), is termed **genetic recombination**. The original homologous pairs of chromosomes are *independently assorted* in anaphase I since each duplicated maternal and paternal chromosome has a 50-50 chance of ending up in one of the new daughter cells. Consequently, the gametes produced can have any possible combination of maternal and paternal chromosomes (for mathematicians, the number of combinations possible = 2^n, where n = the haploid number of chromosomes. For humans, this number would be 2^{23} = 8.4 million).

The first meiotic division concludes during *telophase I,* when the separated homologous chromosomes cluster at each pole and cytokinesis takes place. Since each chromosome is composed of two sister chromatids sharing a single centromere, the total chromosome number has been reduced by half. The two daughter cells each harbor an equal number of chromatids, but each contains different genetic information since all of the original homologs separated independently during this first division.

Interkinesis

In most species, daughter cells created during the first meiotic division enter **interkinesis**, a period when the chromosomes fade and new nuclear envelopes form. The amount of time cells spend in this state varies with the species and sex of the reproducing organism, but for most cells, it is a relatively short period. No chromosomal duplication occurs during interkinesis.

Meiosis II

The second meiotic division begins with *prophase II.* The newly formed nuclear envelopes of the two daughter cells disappear as the chromosomes re-form. This second prophase ends when sister chromatid pairs move to the center of each daughter cell. *Metaphase II* is a static phase during which the chromatid pairs are aligned at the equatorial plane of each daughter cell. The attached spindle fibers begin to contract, and metaphase II ends. During *anaphase II,* the sister chromatids are separated when their centromeres divide. The result is that each pole will receive, in *telophase II,* one set of chromosomes. A second cytokinesis concludes the meiotic process, forming four haploid cells that may mature into gametes.

Meiosis is a complex mechanism used by sexually reproducing, eukaryotic animals to create gametes. It consists of one chromosomal duplication in a parent cell followed by two cell divisions. In completing meiosis, a single diploid ($2n$) parent cell will produce four haploid (n) gametes. Because of the random distribution of sister chromatids during meiosis, each gamete represents a new genetic combination.

Reproductive Strategies

In the universe of reproductive strategies employed by animals, the most familiar is classified by biologists under the colorless heading of "basic bisexual reproduction." This fundamental animal strategy, described in Figure 13.19, is a process in which haploid gametes from two distinct sexual types, male and female, unite in fertilization to produce a zygote that develops into a single offspring.

Deviations from the basic animal model occur among the large variety of animals that exist. Two unusual examples may suffice to illustrate the complexities encountered in categorizing reproduction. There is a type of parasitic worm that follows the common pattern until shortly after fertilization is completed. At

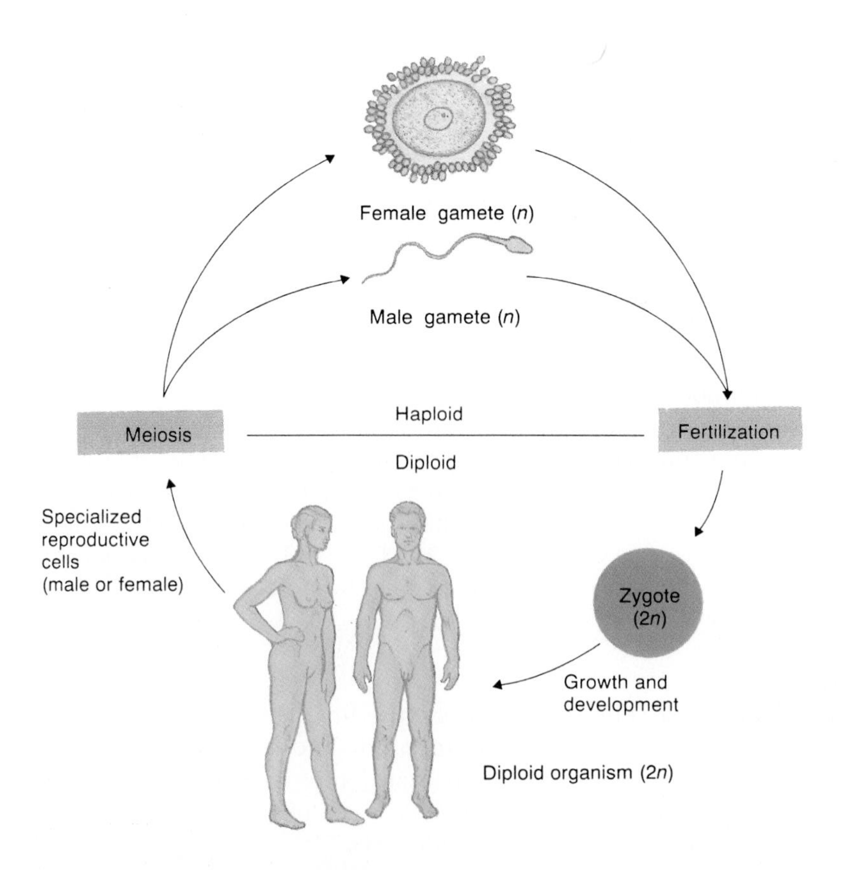

Figure 13.19

In the basic animal reproductive strategy, haploid gametes are produced by two distinct sexes, male and female. When gametes unite in fertilization, a diploid zygote is formed that develops into an offspring resembling the adult.

that point, the fertilized egg, rather than developing into a single offspring, always separates into many identical zygotes. A second example involves a crustacean species that employs "gonad transplants" in its reproduction! In this species, a young male crab uses "hypodermic injection" to insert a few of its cells into a female. The injected cells eventually form a *testis,* the male gonad, which then produces the sperm used by the female to fertilize her eggs. Thus the female produces both types of sex cells, yet technically the sperm cells came from a male. The list of reproductive strategies used by animals is almost endless if all deviations from the basic bisexual form are included. Most, however, follow a general pattern in which meiosis and fertilization are of central importance.

As described in Figure 13.20, plants employ a basic strategy that is fundamentally different from that used by animals. Two distinct phases, haploid and diploid, occur in the life cycle of a plant species (described in Chapter 4). In a single species, each phase results in the production of a life form that is often remarkably different from the other. As in animals, the diploid plant produces haploid cells by meiosis. However, these haploid cells are **spores**, not gametes; that is, they can germinate directly into one form of the plant. Since fertilization does not occur, this form is haploid. The haploid plant will produce haploid cells, some of which mature into gametes. Under proper conditions, fertilization—the fusion of two of these haploid gametes—will occur and lead to the development of the alternate, diploid form of the plant.

Mosses, familiar to many of us as the green carpeting on trees, logs, forest floors, and roofs, have a conspicuous haploid form with an obscure diploid form. In contrast, the fern that most of us recognize is a diploid, spore-producing form. The haploid phase of the fern, which produces gametes, is inconspicuous and generally overlooked. All other plants have reproductive strategies that involve variations of the same haploid-diploid alternation theme.

Animals and plants use a vast number of reproductive strategies. The importance of this diversity can best be understood in terms of species survival. Each

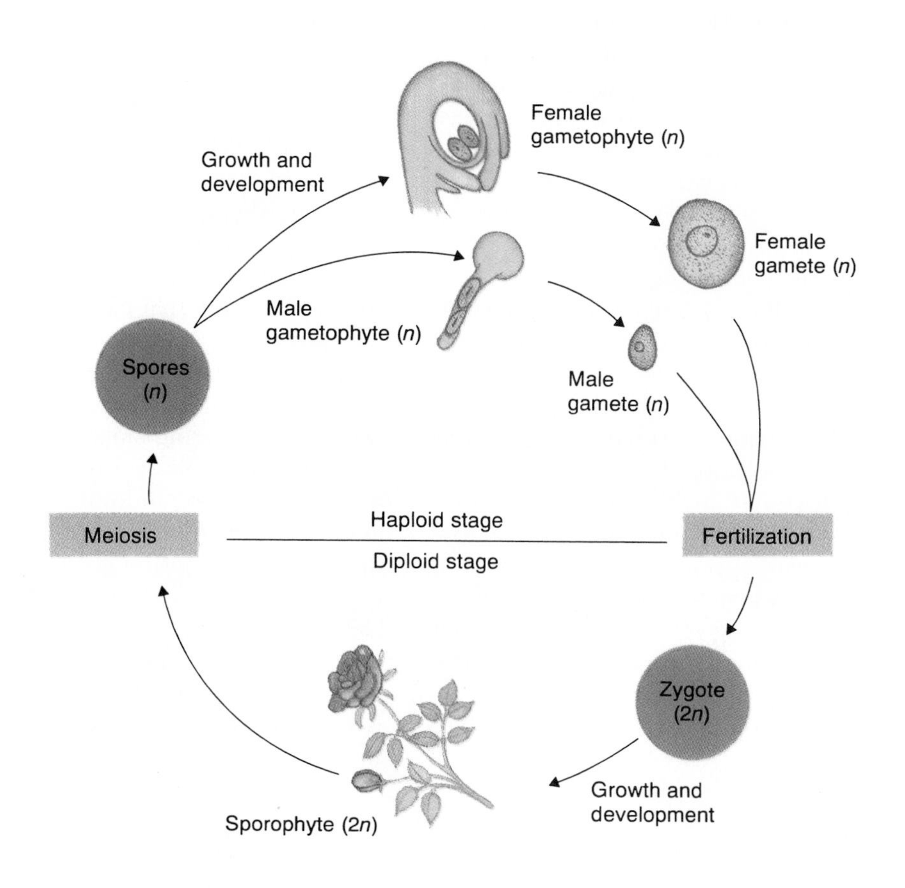

Figure 13.20

A plant life cycle consists of two distinctly different forms, one which is haploid and the other diploid. The haploid plant produces gametes by mitosis, and the diploid plant produces spores by meiosis. After fertilization, the resulting zygote develops into a diploid plant form. Spores grow directly into the haploid plant form. This type of life cycle is known as alternation of generations.

The Significance of Meiosis and Sexual Reproduction

Focus on Scientific Explanations

Understanding the ultimate nature of sex and sexual reproduction requires, first, full comprehension of the process of meiosis and the significance of its final accomplishment, the production of discrete reproductive cells containing new arrangements of genetic information. Second, it requires an appreciation of the significance of fertilization, the process that restores the normal chromosome number for the species being perpetuated. Though somewhat abstract, this is an intriguing example of the types of inquiries that scientists make about biological processes.

Given the complexities involved, biologists have posed a number of questions about the wisdom of a dependence on meiosis and sexual reproduction for producing offspring. There are a number of obvious *disadvantages* in these operations, among them reliance on an extremely complex mechanism, meiosis, for producing gametes; the frequent occurrence of errors (for example, inaccurate duplication or separation of chromosomes) during gamete formation; and the effort required to locate a mate and establish circumstances that can bring about fertilization. These costs cause scientists to ask: Why bother with this waste of energy and time? What possible advantages outweigh the costs of using this elaborate system of reproduction?

The conventional answer to those questions has employed reasoning along the following lines. Together, meiosis and sexual reproduction, including fertilization, create opportunities for generating novel genetic combinations and producing offspring that are genetic originals. Variation can be introduced in two ways. (1) Genetic recombination is any process that generates new gene or chromosome combinations. In most organisms, this occurs in meiosis I when the homologs are randomly assorted; consequently, each gamete is almost certain to be distinctive in the type of information it contains. It also occurs when crossing-over during meiosis increases the variety of genetic information in gametes produced. (2) In *outcrossing,* offspring are produced from gametes contributed by two different parents.

Genetic recombination due to crossing-over and independent assortment of homologs in meiosis I is thought to be the more fundamental aspect of sex since many reproductive strategies used by different plants and animals retain this feature but do not preserve outcrossing. However, there are questions that are very troublesome for this line of reasoning. For example, why is it advantageous to break up genetic combinations once they are established and presumably successful?

The traditional view is now being reexamined by scientists who have studied new evidence about sex and sexual reproduction in the biological world. Of fundamental importance has been the discovery that genetic recombination is virtually universal, occurring in all organisms including those, such as bacteria, that usually reproduce asexually. In addition, it is now known that genetic recombination occurs in a wide variety of ways, at astounding rates, in bacteria and other prokaryotes. Recombination tends to randomize genetic information in organisms; the greater the randomization, the greater the variety of offspring that can be produced. For bacteria, this may be an advantage that is related to creating new variations quickly. Because of recombination, bacteria, which may produce a new generation in minutes or hours, can quickly adjust to new environmental conditions. For example, this allows them to become resistant to certain drugs in relatively short periods of time.

Can recombination and the resulting randomization of genetic material convey such marked advantages to plants and animals? In comparison with bacteria, genetic recombination must have less importance in long-lived organisms that are dependent on meiosis and sexual reproduction because rapid adjustments to changing conditions are not possible. While it is clear that meiosis and sexual reproduction do produce genetic variation, and this may be important in the long run, many biologists now believe that their preeminent effect is to serve as stabilizing mechanisms. According to this modern hypothesis, the randomization of genetic information is minimized as a consequence of the elaborate events required for a successful completion of meiosis and sexual reproduction. Thus the long-term genetic stability of a sexually reproducing species is felt to be of greater importance than any short-term advantage arising from variations resulting from recombination.

An intriguing extension of this *stabilization hypothesis* has recently been published. In simplified form, it has been proposed that recombination and outcrossing provide opportunities for repairing DNA that has become altered in gamete-producing parental cells (the *repair hypothesis*). Since transmitting damaged DNA or deficient chromosomes to a new generation would probably be very harmful, these mechanisms counteract such an effect by providing for repair, replacement, or elimination of the damaged goods.

The true significance of meiosis and fertilization to sexually reproducing organisms is not yet clear. However, the old view that they conveyed unique advantages for creating new variations in offspring has come under attack. Reexamining accepted ideas like these is a marked characteristic of all good science.

method of reproduction, no matter how bizarre it may seem to us, has permitted some members of a population to leave viable offspring that were able to sustain the species throughout the course of its existence. This is the most important criterion used to judge a successful reproductive strategy. These concepts are further explored in Unit III.

MENDEL REVISITED

Understanding meiosis and fertilization makes it possible to look back and reevaluate Mendel's studies and the laws of genetics that bear his name. Using pea plants and mathematical analyses, but with no knowledge of genes or chromosomes, Mendel was able to describe regular patterns of inheritance that had not been identified previously. By luck or design, he chose traits that occurred in one of two phenotypic forms (for example, stems were either long or short, and seeds were either yellow or green). When his experiments were completed, he was able to make the following inferences:

1. Pea plants possessed two discrete hereditary factors (alleles) for each trait.
2. One member of the pair expressed itself, or was dominant, in the presence of the other, recessive factor.
3. When gametes were formed, the two factors separated independently and the resulting sex cells contained only one factor for each trait.
4. Factors for each of the characters were sorted randomly into the gametes.
5. The factors were recombined when fertilization occurred.

When Mendel's work was rediscovered in the early part of this century, his conclusions were formally summarized as laws and named in his honor.

The mechanisms of meiosis and fertilization are now understood, and thus it is possible to correlate Mendel's observations and laws with events that occur during these two processes. His discrete pairs of factors, now called genes (or alleles), are carried on separate chromosomes, with each homolog of a homologous pair carrying one allele for a trait. The two alleles separate as a function of chromosome separation during meiosis (anaphase I) and return to the diploid state during fertilization. All the axioms from Mendelian genetics are derived from these two core principles. Thus Mendel's law of segregation is based on the fact that the two alleles determining a trait separate, or segregate, during meiosis and each gamete contains only one of the alleles. Mendel's law of independent assortment states that the alleles of one gene pair separate independently, or randomly, from alleles of another gene pair during meiosis. In Mendel's pea plants, for example, the alleles for plant height sorted independently from the alleles for seed color because they were on separate chromosomes. Once these central principles were defined and understood, they provided a starting point for the "golden age of genetics" that was to follow.

Summary

1. During the nineteenth century, the cell theory proposed that all living organisms were composed of cells and that all cells arose from preexisting cells.
2. The structure and function of cells became an important area of research. New technologies, such as the electron microscope, provided scientists with the opportunity to examine cells at extremely high powers of magnification. As a result, previously unknown internal cell structures were identified, and their functions were determined using other techniques.
3. Eukaryotic cells consist of a nucleus and cytoplasm enclosed by a plasma membrane. Within the cell cytoplasm are small organelles that carry out specific

activities and a cytoskeleton that provides support and allows for movement. Major organelles found in plant and animal cells include endoplasmic reticulum, Golgi complexes, and mitochondria. Plant cells also contain chloroplasts and vacuoles.
4. Chromosomes in the nuclei of cells contain genetic information in the form of DNA. Each sexually reproducing species has a characteristic number of chromosomes. Offspring receive half of their chromosomes, and thus half of their genetic information, from each parent when fertilization occurs.
5. Binary fission and mitosis are types of cell division that give rise to two cells that are identical to the parent cell.
6. Meiosis is a process that gives rise to gametes in animals. In meiosis, there is one chromosome duplication followed by two cell divisions. The resulting gametes contain the haploid number of chromosomes. Two events in meiosis, independent assortment and crossing-over, are the source of much of the genetic variation that occurs in sexually reproducing species.

Review Questions

1. Summarize the major ideas of the cell theory.
2. What are the general functions or roles of the plasma membrane, the nucleus, and the cytoplasm?
3. What are the specific functions of the following cellular organelles: rough and smooth endoplasmic reticulum; Golgi complexes, mitochondria, chloroplasts, microtubules, and microfilaments?
4. Explain the difference between genes and alleles.
5. Assume that a flower with alleles *RR* or *Rr* is red and one with *rr* is white. How many genes determine flower color? How many different alleles are there for flower color? What is the genotype of a white flower? What is the phenotype?
6. Humans have 23 pairs of homologous chromosomes. How many homologs came from the mother? What is the haploid number of chromosomes for humans? How many chromosomes does a human intestinal cell contain? A human sperm cell?
7. What is asexual reproduction?
8. What is mitosis? Why is it important in eukaryotic cells?
9. What is sexual reproduction?
10. How are gametes produced in sexually reproducing animals?
11. How does the basic animal life cycle differ from a plant life cycle?

Essay and Discussion Questions

1. Why is the cell theory classified as a unifying scientific theory with broad application?
2. Technological factors were important in the development of the cell theory. What sorts of technological advances can you imagine that would be considered trailblazing today?
3. One of the key ideas in the early cell theory was that the difference between living and nonliving objects was in the *material organization* of living beings. Is this an adequate distinction today? Why?
4. How do biologists explain the common structures and organelles that are present in the cells of all eukaryotic organisms?
5. What are the advantages of being multicellular?

References and Recommended Readings

American Society of Zoologists. 1989. Science as a way of knowing: VI. Cell and molecular biology. *American Zoologist* 29:483–817.

Baker, J. R. 1948–1955. The cell-theory: A restatement, history, and critique. Five parts were published over seven years in the *Quarterly Journal of Microscopical Sciences.*

Bernstein, H., H. C. Byerly, F. A. Hopf, and R. E. Michod. 1985. Genetic damage, mutation and the evolution of sex. *Science* 229:1277–1280.

Blackwelder, R. E., and B. A. Shepard. 1981. *The Diversity of Animal Reproduction*. Boca Raton, Fla.: CRC Press.

Fredrick, J. F., ed. 1981. Origins and evolution of eukaryotic intracellular organelles. *Annals of the New York Academy of Sciences* 361:1–512.

Hughes, A. 1959. *A History of Cytology*. London: Abelard-Schuman.

Margulis, L., and D. Sagan. 1986. *Microcosmos: Four Billion Years of Microbial Evolution*. New York: Summit Books.

McIntosh, J. R., and M. P. Koonce. 1989. Mitosis. *Science* 246:622–628.

Murray, A. W., and M. W. Kirschner. 1991. What controls the cell cycle? *Scientific American* 264:56–63.

Penny, D. 1985. The evolution of meiosis and sexual reproduction. *Biological Journal of the Linnean Society* 25:209–220.

Prescott, D. M. 1988. *Cells: Principles of Molecular Structure and Function*. Boston: Jones & Bartlett.

Scientific American. 1985. A series of articles on cell structure and function appeared in various issues in 1985.

CHAPTER 14

Chromosomes and Heredity

Chapter Outline

Reading Questions

1. What kinds of knowledge led to the formation of the chromosome theory?
2. How does the chromosome theory explain genetic inheritance?
3. What kinds of experiments proved the chromosome theory to be correct?
4. What are the bases for non-Mendelian patterns of inheritance?
5. What are the genetic consequences of chromosomal alterations?

ORIGIN OF THE CHROMOSOME THEORY OF HEREDITY

By the dawn of the twentieth century, it had become clear that Mendel's studies provided a theoretical foundation for understanding the heredity of certain traits in plants and animals. Within this framework, it was possible to make and test predictions about the inheritance of at least some traits in some organisms. Many interesting questions followed. Where did Mendelian factors, which later became known as *genes* (the term we will use in most of this chapter), reside? Since gametes are the only direct link between generations, a partial answer to that question was evident, but the more difficult part of the question remained unanswered. Where in the sperm and egg cells were genes found? What structure or structures were they associated with?

As microscope technology improved in the late 1800s, a new science, **cytology**, the study of cells, generated much information that led to a greater understanding of cell structure and function. Microscopes had revealed that sperm and egg cells consisted of a nucleus and cytoplasm. Were genes found in one or both of these cell parts? The cytoplasm seemed to be an unlikely location. There was very little of this substance in sperm cells compared to egg cells, and by then it was believed that both sexes made equal contributions in producing offspring. The nucleus, however, was approximately the same size in both gametes; thus most attention became focused on this structure. Where in the nucleus were genes found?

Key Discoveries

Several discoveries during the last half of the nineteenth century were relevant to the unanswered question. Certain threadlike structures within the nucleus were found to strongly absorb special dyes or stains that made them easily visible for microscopic studies (see Figure 14.1). These structures, called *chromosomes*, could now be examined by fixing cells, staining them, and observing them through the microscope. Using these methods, cytologists acquired some basic knowledge about the behavior of chromosomes during meiosis. Although the exact connection between meiotic events and the hereditary transmission of traits remained undefined, there seemed to be some relationship between chromosomes and genes.

Formulation of the Chromosome Theory

In 1883, it was postulated that chromosomes carried the units of heredity, but this concept required further elaboration before it could be considered a theory. By the first years of the twentieth century, scientists were in hot pursuit of the biological basis of heredity. Many people were closing in on the answer, but the entire matter was suddenly resolved by two precise and detailed papers, one written in 1902 and the other in 1903, by William S. Sutton. The earliest of his papers, which described results from experiments on grasshopper chromosomes, included the first clear demonstration that chromosomes exist within the cell as sets of distinguishably different pairs of like (homologous) chromosomes, one having come from each parent. He concluded with this statement: "I may finally call attention to the probability that the association of paternal and maternal chromosomes in pairs and their subsequent separation during the reducing division [meiosis]. . . . may constitute the physical basis of the Mendelian law of heredity." The 1903 paper expanded on this hypothesis and included additional data on the random orientation of paired chromosomes on meiotic spindles, thus accounting for the independent segregation of separate pairs of the factors (genes) previously described by Mendel. Sutton also proposed that Mendel's re-

Figure 14.1

By the end of the nineteenth century, improved microscopes had made it possible to study chromosomes. What was their function in heredity?

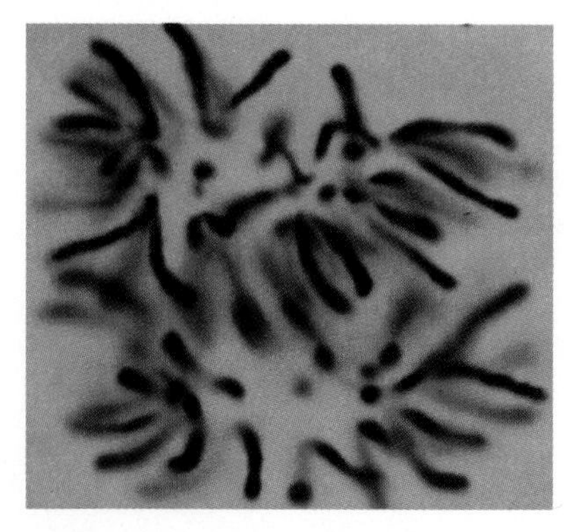

sults could be explained if genes were located on chromosomes. Further, he speculated that each chromosome must carry many genes.

With Sutton's brilliant conceptual breakthrough, the hereditary basis of segregation and independent assortment and their relation to chromosomes was explained. However, many years would pass before Sutton's hypothesis became widely accepted. The primary obstacle for many scientists was lack of experimental evidence that proved that genes were located on chromosomes or that chromosomes contained numerous genes. Ultimately, the formalized statement that inheritance patterns may be generally explained by assuming that genes are located on chromosomes became known as the **chromosome theory** of heredity.

Thus by 1903, a mechanism that explained Mendelian inheritance had been proposed, and the field of *genetics* (named in 1900) was ready to explode. Sutton's hypothesis served as a guide for much of the genetics research that occurred during the next two decades.

Following Mendel's classic experiments, there was great interest in learning where genes were located in the cell. Based on a brilliant analysis of results from his own studies, Sutton advanced the idea that numerous genes were located on chromosomes.

CONFIRMATION OF THE CHROMOSOME THEORY

Sutton's papers concluded one phase of genetics history, but additional pieces were necessary to complete the puzzle of chromosomes and heredity. In fact, the field of genetics was in a general state of confusion during the first decade of the 1900s. Many scientists were not aware of Sutton's papers or did not understand them. Also, many investigators had difficulty moving away from Mendel's simple model of inheritance. Although Sutton had formally associated chromosomes with genes, chromosomes could not be related to specific traits in a one-to-one fashion, as inferred from Mendel's studies.

Unanswered Questions

Recall that Mendel studied seven specific hereditary traits, each determined by one pair of genes. His experiments with peas always resulted in the offspring ratios expected if independent assortment and segregation occurred. From his work, it could reasonably be concluded that each chromosome carried a gene for only one trait. However, when scientists conducted new investigations with additional traits or with other plants or animals, the offspring ratios obtained often failed to conform to the common Mendelian ratios.

The primary difficulty, recognized by Sutton and others, was accounting for the fact that in most organisms there were obviously more hereditary traits than there were chromosomes. It became essential to conduct experiments that answered certain questions. Specifically, what exactly did chromosomes have to do with heredity? Were genes contained within chromosomes, and were several genes located on each chromosome, as Sutton suggested? How could the inheritance of hundreds of traits be explained? Did Mendelian laws apply only to pea plants? Definitive answers to these questions emerged as a result of research conducted in an unlikely setting—a small laboratory at a major university in New York City that was jammed with half-pint milk bottles full of flies and filled with the aromas of bananas and ether.

Fruit Flies Yield Their Secrets

The key figures during the next phase of developing ideas and hypotheses in hereditary were Thomas Hunt Morgan (Figure 14.2) and his brilliant group of

Figure 14.2

Thomas Hunt Morgan, shown here late in his career, and his students confirmed the chromosome theory of heredity.

graduate students (Alfred Sturtevant, Calvin Bridges, and Hermann Muller) at Columbia University. In 1904, they began the research in the "Fly Room" that eventually defined the relationships between chromosomes and heredity.

For their experiments, Morgan selected the tiny, red-eyed fruit fly, *Drosophila melanogaster,* which proved to be an ideal research animal and has since been widely studied in laboratories around the world (see Figure 14.3). The use of fruit flies offered many advantages over using other animals: they are small, only 2 to 3 millimeters in length; they are cheap to raise and feed (banana gelatin); they are easy to maintain in small bottles and simple to examine (after being anesthetized with ether); they reproduce frequently, furnishing a new generation every two weeks; and hundreds of offspring are produced in each genera-

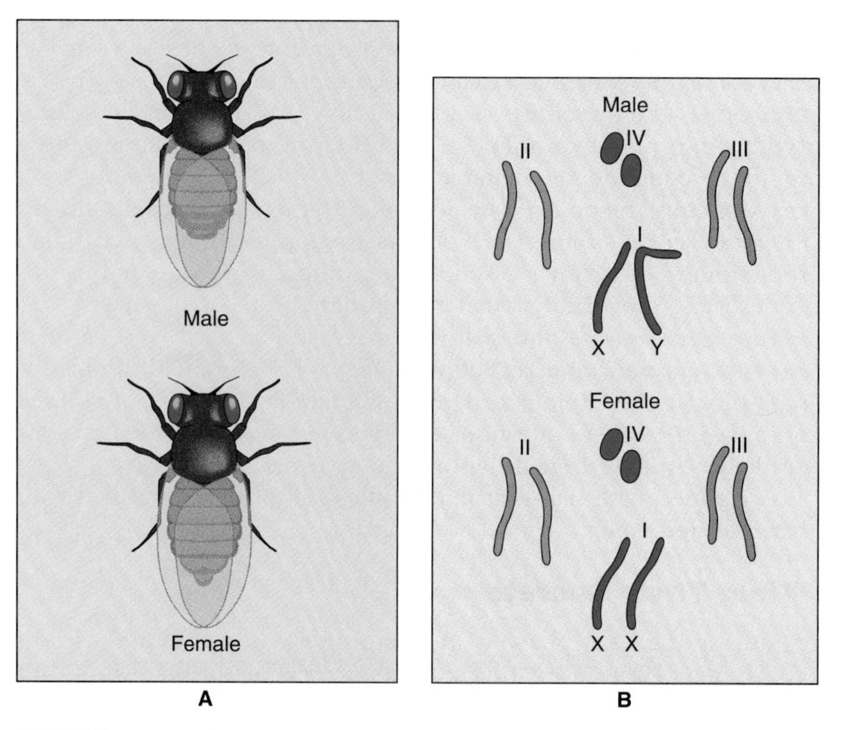

Figure 14.3

Morgan's research group used fruit flies (*Drosophila melanogaster*) in their genetics studies. (A) The natural eye and body coloration and the relative size of males and females are shown. These flies proved to be ideal experimental animals because of their small size, the ease of maintaining them in the laboratory, and their ability to produce many generations of offspring in a short period of time. The cells of *Drosophila* have four pairs of homologous chromosomes (B)—three pairs of autosomes (designated as Group II, III, and IV) and one pair of sex chromosomes (Group I, XX, or XY).

tion. Experiments that would have taken Mendel years with his pea plants could be completed within months using fruit flies. Another major advantage is that *Drosophila* has only four pairs of chromosomes.

Drosophila Chromosomes and Sex Determination

There are two basic types of chromosomes in the cells of fruit flies and most other animals. A single pair of **sex chromosomes** play a role in determining the sex of an individual; all other chromosomes are known as **autosomes**. Examination of *Drosophila* chromosomes reveals the presence of one pair of sex chromosomes and three pairs of autosomes. The relation between the sex chromosomes and sex type had been described for many insects by 1909. Sex determination in *Drosophila* was determined a year later.

In X-Y sex determination, which occurs in many organisms including humans, the larger of the sex chromosome pair is referred to as the X chromosome and the smaller, the Y chromosome. In 1910, the model for sex determination illustrated in Figure 14.4 was described. Females contain two X chromosomes, but males have one X and one Y; thus with respect to sex chromosomes, females are XX and males are XY. During meiosis, the male produces two types of sperm, one type bearing X chromosomes and the other type bearing Y chromosomes. The female produces eggs that all have an X chromosome. Fertilization by an X-bearing sperm produces a female (XX), by a Y-bearing sperm, a male (XY). Equal numbers of males and females are produced because of the segregation of X and Y chromosomes during meiosis. This documentation of the link between sex determination and chromosomes offered strong support for Sutton's hypothesis that chromosomes were the basis of inheritance.

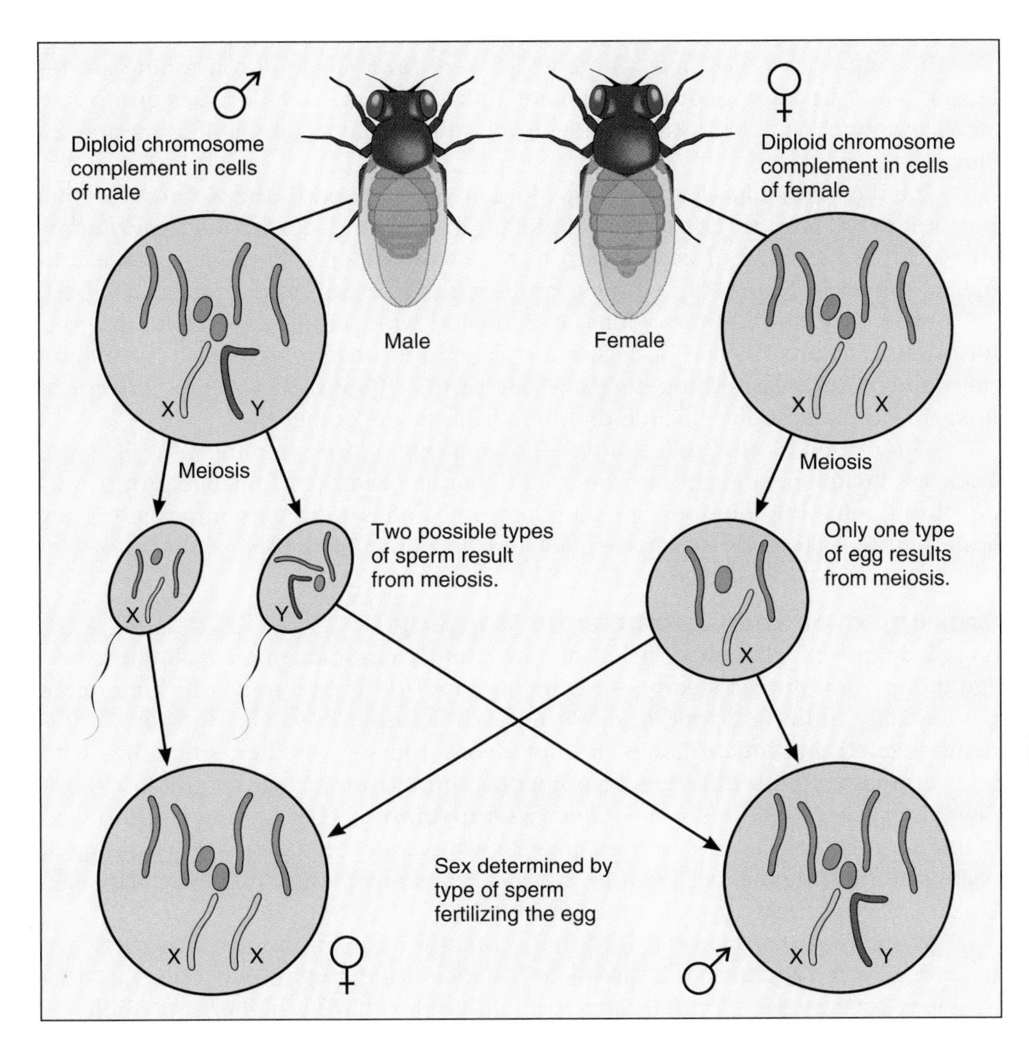

Figure 14.4

In organisms with X-Y sex determination, females have two X chromosomes and males have one X and one Y chromosome. During meiosis, sperm cells receive either an X or a Y sex chromosome, whereas all eggs produced carry an X chromosome. The sex of the resulting offspring depends on whether the egg is fertilized by an X- or a Y-bearing sperm.

Abnormal Fruit Flies

Morgan's group was enthusiastic about detecting fruit flies with unusual traits. A **mutation** occurs when a gene undergoes a permanent structural change that results in a new form of an expressed trait. The possessor of such traits, which are almost always recessive to the normal trait, is called a *mutant*. Analyses of mutants played an important part in the development of many fundamental concepts in classical genetics. Early in 1910, one such mutant appeared in the form of a white-eyed male fruit fly. This fly became the most famous fruit fly in the history of science, and the subsequent studies it generated constitute a classic story in the history of genetics (see the first Focus on Scientific Inquiry, "The White-eyed Fruit Fly").

Linkage

Experiments with white-eyed male fruit flies yielded a critical finding: the gene for a specific trait (eye color) was shown to be part of the X chromosome. Other unusual mutant traits, such as yellow body and miniature wings, were soon discovered also to be linked to the X chromosome. Thus it became apparent, as Sutton had hypothesized, that several traits were determined by genes located on a single, specific chromosome. Sutton had also postulated that "all of the [alleles] represented by any one chromosome must be inherited together." Could this hypothesis now be tested in the Fly Room?

Morgan and his students continued their search for mutant flies, and by 1915, they had identified 85 different mutant traits. Further, as Sutton had suggested, many of these traits, and the genes that caused them, tended to be inherited together. **Linkage** refers to the concept that specific traits tend to be inherited together because their associated genes are arranged in linear fashion along the same chromosome. All genes present on a single chromosome are referred to as a **linkage group**. How many linkage groups did Morgan's group find for *Drosophila*? Not surprisingly, they found four, equal to the haploid number of chromosomes.

Many of the traits described by Morgan were linked to the X chromosome, but others fell into three separate linkage groups, and numerous experiments showed that the traits of each group were usually inherited together. For example, as shown in Figure 14.5, if flies with three mutant recessive traits (black bodies, purple eyes, and dumpy wings) were mated with standard flies heterozygous for dominant traits (brown bodies, red eyes, and normal wings), all three mutant traits generally appeared together in some of the offspring. Thus the concept of linkage was extended to include the three fruit fly autosomes.

The results of Morgan's studies showing that a large number of traits were associated with specific chromosomes and that the number of linkage groups was equal to the haploid number of fruit fly chromosomes were very important. They added substantial credibility to the evolving chromosome theory.

Crossing-over and Chromosome Mapping

As just described, all genes carried on one fruit fly chromosome constitute a linkage group, and Morgan's early experiments showed that the traits they determine were usually inherited together. Within a short time, however, further studies demonstrated that linked genes did not always remain together, since flies with traits connected with different homologous chromosomes *did* appear in experimental crosses; such flies were called **recombinants**. This finding could be explained by the effects of chromosomes crossing over, breaking, and rejoining at some point during meiosis (see the second Focus on Scientific Inquiry, "Linkage, Linkage Groups, and Chromosome Mapping").

Morgan's group hypothesized that a precise correlation must exist between recombination frequency and the linear distance separating the linked genes; that is, genes separated by long distances would tend to have a higher probability of

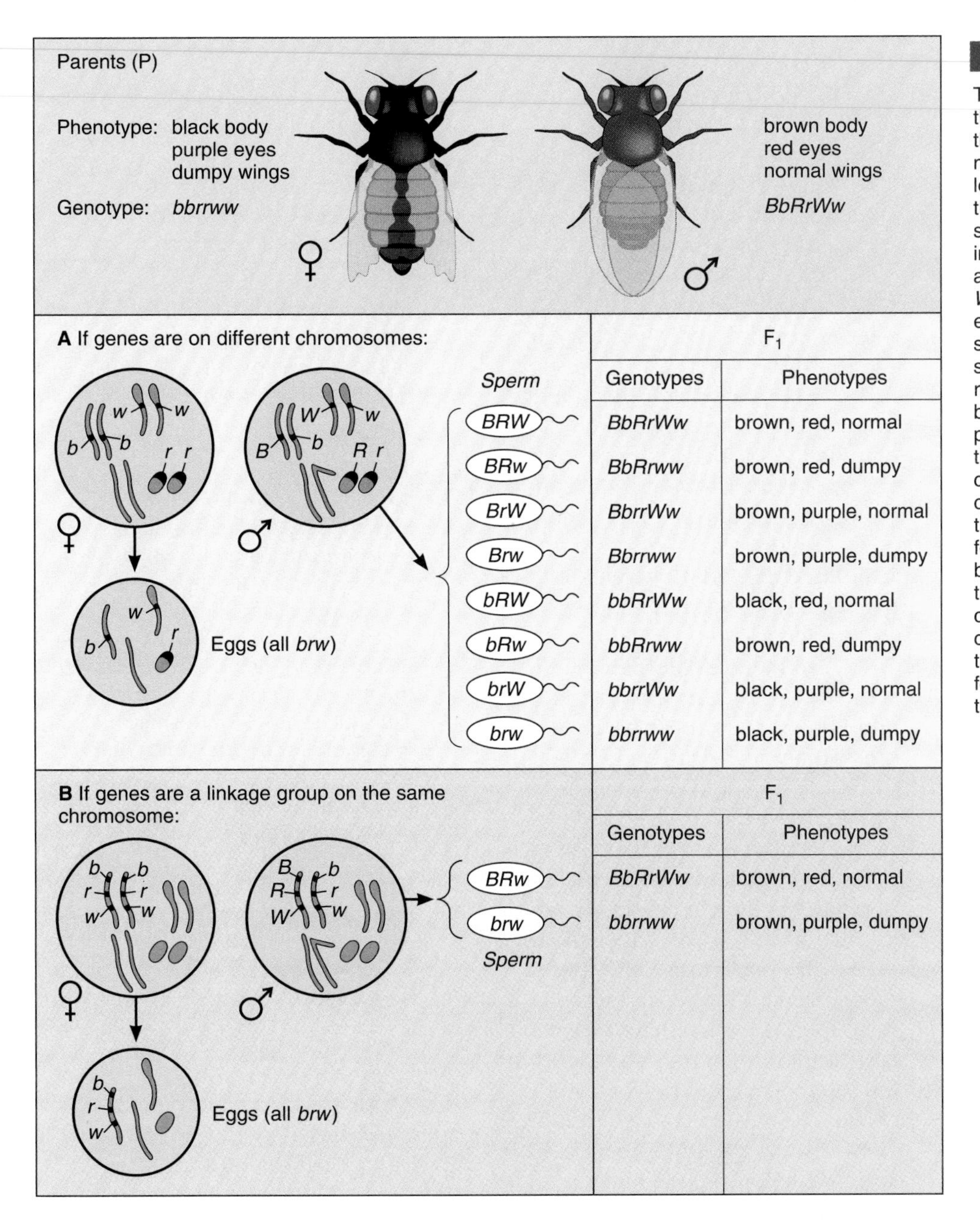

Figure 14.5

This figure compares the inheritance of three *Drosophila* traits when genes for the trait are located on different chromosomes and genes for the traits are located on the same chromosome. The three traits involve body color, wing shape, and eye color. The normal, dominant alleles and their related phenotype are as follows: *B* = normal brown body, *W* = normal wing, and *R* = normal red eyes. The corresponding mutant, recessive alleles and the phenotype observed in flies that are homozygous recessive for the trait are as follows: *b* = black body, *w* = "dumpy" wing, and *r* = purple eye color. (A) If genes for the three traits are located on different chromosomes, then the F_1 from a test cross between a female with the mutant traits and a male that is heterozygous for all three traits will include all possible phenotypes. (B) If genes for the three traits are located on the same chromosome, then the F_1 from a test cross between a female with the mutant traits and a male that is heterozygous for all three traits will include only the two parental phenotypes.

crossing over than genes that were close neighbors on their chromosome. After conducting further experiments with their fruit flies and obtaining additional data, Morgan's group was able to create **chromosome maps**, which showed both the location of linked genes and their exact linear order along the length of the chromosome. Final confirmation of their *Drosophila* maps required more sophisticated techniques that did not become available until much later. A remarkable aspect of their maps, which were constructed using an indirect, mathematical approach, was that their gene sequences proved to be entirely correct, although the distances they proposed turned out to be somewhat less precise.

Sturtevant's mapping study (see the second Focus on Scientific Inquiry), in particular, offered solid evidence that genes were arranged in a linear sequence along the chromosome. It also provided the theoretical foundation for constructing chromosome maps for many species besides *Drosophila*. It remained for another of Morgan's graduate students, Calvin Bridges, to conduct experiments that overwhelmingly convinced the scientific community of the accuracy of the chro-

The White-Eyed Fruit Fly

Focus on Scientific Inquiry I

"In a pedigree culture of *Drosophila* which had been running for nearly a year through a considerable number of generations, a male appeared with white eyes. The normal flies have brilliant red eyes" (see Figure 1). So reads the first paragraph in T. H. Morgan's classic paper, "Sex-limited Inheritance in *Drosophila*," which was published in 1910. He speculated that a mutation within a single egg resulted in elimination of the factor (gene) for red eye color. The new character, white eyes, and the "sport" (Morgan's term for the white-eyed male) that carried it, were exhaustively analyzed in different genetic experiments.

Like Mendel, Morgan made standard crosses involving P, F_1, and F_2 flies. In the first parental cross, the white-eyed male was mated with his red-eyed sisters. The F_1 consisted of 1,237 red-eyed offspring along with 3 white-eyed males. The unexpected appearance of white-eyed males in the F_1 was thought to have been "due evidently to further sporting," and these flies were ignored for the rest of the paper (they may have been early emerging F_2 flies contained in the same bottles). The red-eyed F_1 hybrids were then crossed, and the F_2 consisted of the following offspring: 2,459 red-eyed females, 1,011 red-eyed males, 782 white-eyed males, and no white-eyed females. Finally, the white-eyed male was crossed with some of his F_1 daughters, producing 129 red-eyed females, 132 red-eyed males, 88 white-eyed females, and 86 white-eyed males ("the four classes of individuals occur in approximately equal numbers"). How could these experimental results be interpreted?

Figure 1 Two fruit flies, one with normal red eyes, the other a white-eyed mutant.

Morgan described a hypothesis to account for the results. Note that his hypothesis was formulated *after* the first experiments were completed.

> The results just described can be accounted for by the following hypothesis. Assume that all of the spermatozoa of the white-eyed male carry the "factor" for white eyes *W* and that half of the spermatozoa carry a sex factor *X* that the other half lacks; that is, the male is heterozygous for sex. Thus the symbol for the male is *WWX* and for his two kinds of spermatozoa *WX* and *W*. Assume that all of the eggs of the red-eyed female carry the red-eyed factor *R* and that all of her eggs (after reduction) carry one *X* each; the symbol for the red-eyed female will therefore be *RRXX*, and that for her eggs will be *RX* and *RX*.

The relevant phenotypes, genotypes, and gametes produced, using Morgan's terminology in his paper, are shown in Table 1. Since a hypothesis can be used to make predictions, Morgan then conducted several additional crosses to test predictions made by his hypothesis. The predictions and results of the experiments are summarized in Table 2. In each experiment, his prediction was confirmed.

This famous paper—the description of a white-eyed male, the results of various test crosses, the formulation and testing of a hypothesis, and its contribution of convincing support for Sutton's hypothesis that genes were carried by chromosomes—occupies a prominent spot in the library of genetics historians. Yet it had major defi-

Table 1 Phenotypes, Genotypes, and Possible Gametes of Fruit Flies Used in Morgan's 1910 Experiments

Generation	Phenotype	Genotype	Possible Gametes
P	Red-eyed female	*RRXX*	*RX*
	White-eyed male	*WWX*	*WX, W*
F_1	Red-eyed female	*RWXX*	*RX, WX*
	Red-eyed male	*RWX*[1]	*RX, W*[2]
F_2	Red-eyed female	*RRXX* or *RWXX*	
	Red-eyed male	*RWX*	
	White-eyed male	*WWX*	
Other	White-eyed female	*WWXX*	

[1]Males were assumed to have only one sex chromosome, an X.
[2]Morgan assumed that only two types of sperm were produced by the red-eyed males, *RX* or *W*, "otherwise the results will not follow."

Morgan's symbols are used in this table: *R* = red eye factor (allele), *W* = white eye factor, and *X* = sex chromosome.

Table 2 Results of Morgan's Experiments

PREDICTION 1: Offspring from a cross between a white-eyed female and a white-eyed male should "be white [eyed], and male and female in equal numbers."
Result of Experiment 1: All offspring were white-eyed.

PREDICTION 2: If the two types of red-eyed F_2 females are crossed with white-eyed males, "there should be four classes of individuals in equal numbers."
Result of Experiment 2: The predicted ratios (25 percent each of red-eyed females, white-eyed females, red-eyed males, and white-eyed males) were obtained.

PREDICTION 3: If F_1 females heterozygous for white eyes are crossed with white-eyed males, "the four combinations [of offspring] last described" will occur.
Result of Experiment 3: Offspring consisted of approximately 25 percent each of red-eyed females, white-eyed females, red-eyed males, and white-eyed males.

PREDICTION 4: If F_1 red-eyed males (*RWX*) are crossed with white-eyed females, "all the female offspring should be red-eyed, and all the male offspring white-eyed."
Result of Experiment 4: All females were red-eyed and all males were white-eyed.

Morgan's experiments were designed to test predictions derived from his hypothesis that the factor for white eye color is associated with the X chromosome.

ciencies, since one of its primary assumptions was incorrect. Can you identify the problems with the hypothesis set forth by Morgan?

Recall that sex in *Drosophila* is determined by X and Y chromosomes (females are XX and males are XY). Morgan incorrectly assumed that females were XX and males simply had one X chromosome (expressed as X or X0). Further, even though he considered eye color alleles to be "associated" with the X chromosome, his hypothesis and experimental interpretations were based on the idea that eye color genes and sex chromosomes were *inherited independently*. This led to complications in explaining the results of some additional experiments not discussed in this essay.

Within months, it became clear to Morgan that all his experimental results made sense if he assumed that the eye color alleles were actually transmitted *as part of the X chromosome*. Also, by then he had become aware that male fruit flies have an X and a Y chromosome. Thus in 1911, he formulated and published a modification of his original hypothesis that proved to be entirely correct.

> The experiments on *Drosophila* have led me to two principal conclusions: First, that sex-limited [now called *sex-linked*] inheritance is explicable on the assumption that one of the material factors of a sex-limited character is carried by the same chromosomes that carry the material factor for femaleness [that is, the X chromosome]. Second, that the "association" of certain characters in inheritance is due to the proximity in the chromosomes of the chemical substances [genes] that are essential for the production of those characters.

Figure 2 summarizes key aspects of Morgan's experiments in light of his modified hypothesis.

Based on his data, Morgan hypothesized that the allele for white eyes was recessive and was carried on the X chromosome. Figure 2A shows that the gene for eye color occurs at a specific location (locus) on the X chromosome but that there is no corresponding gene on the Y. Since male fruit flies have only one X chromosome, they carry only one allele for eye color. The two alleles for eye color, carried by the X chromosome, are shown as *R* for the dominant red color and *w* for the recessive white color. According to Morgan's modified hypothesis, gametes of the P_1

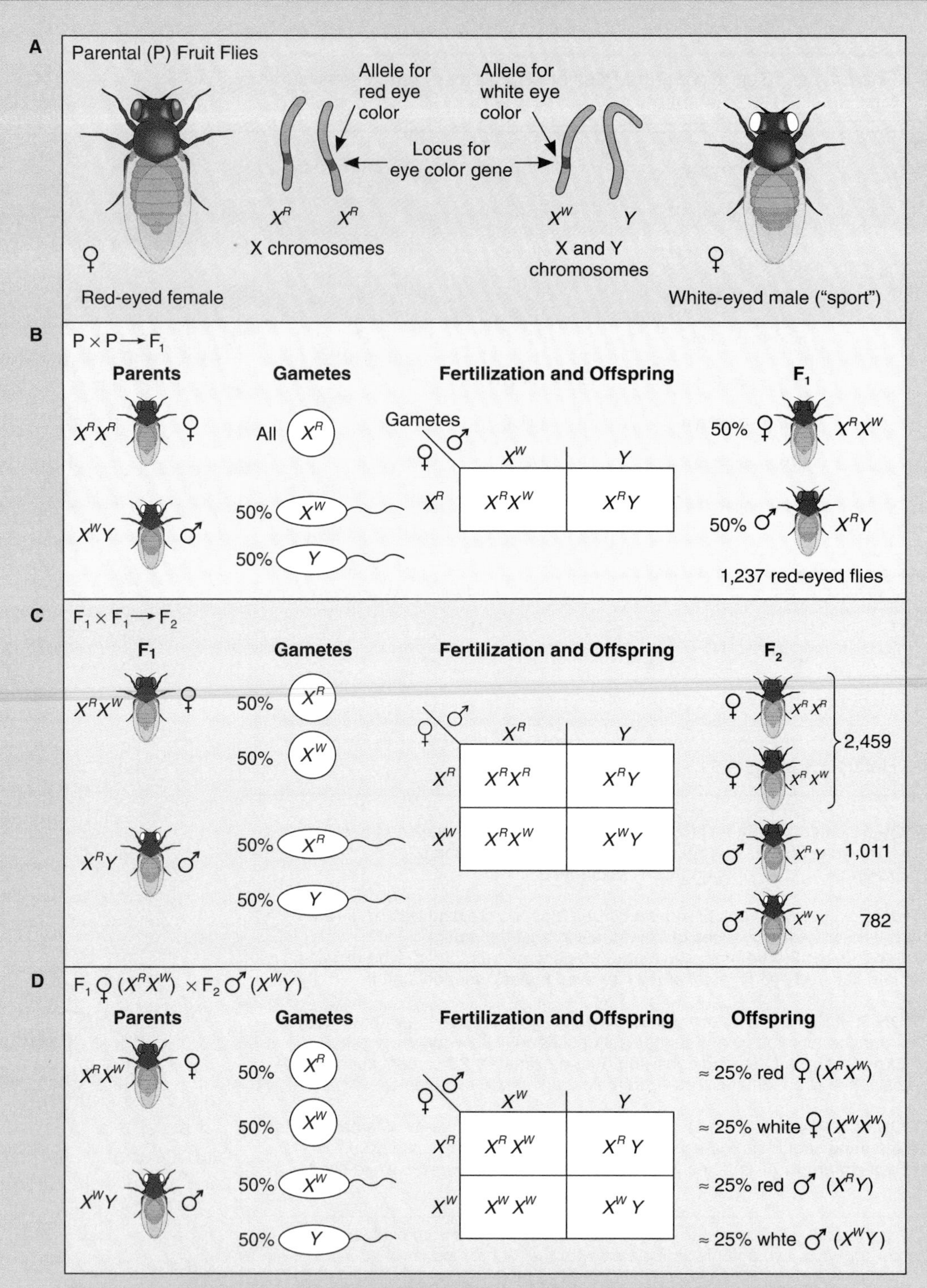

Figure 2 A summary of some of Morgan's experiments and results. *X* and *Y* = sex chromosomes, *R* = the dominant allele for red eye color, *w* = the recessive allele for white eye color; X^R indicates that the X chromosome carries the allele for red eyes; X^W indicates that the X chromosome carries the allele for white eyes. The Y chromosome has no locus for eye color genes. (A) These are the phenotypes, sex chromosomes, and alleles for eye color in Morgan's experimental fruit flies in the P generation. (B) In Morgan's first preliminary experiment (P × P), homozygous red-eyed females were mated to the white-eyed male. The females could produce gametes containing only X^R. The sperm of the male could contain either a Y chromosome or an X^W. A Punnett square is used to describe the offspring produced by fertilization of the parental gametes. The phenotypes and genotypes (and ratios) of the F_1 are also shown. (C) In Morgan's second preliminary experiment, F_1 females were mated with F_1 males. (D) The results of Morgan's third prediction were confirmed by this experiment.

male contained either an X chromosome with a white-eye allele or a Y chromosome, which had no eye color allele. Gametes from red-eyed P_1 females all contained an X chromosome with an allele for red eye color. Genotypes of the F_1 are shown in Figure 2B. All F_1 flies were red-eyed, but 50 percent of the females would have been expected to be heterozygous, or carriers of the white-eye allele.

The possible gametes formed by F_1 flies are shown in Figure 2C. The F_1 flies were mated, and the results are shown in Figure 2D. The F_2 females were all red-eyed, as predicted by the hypothesis. Morgan also predicted that an equal number of males would be produced with approximately 50 percent having red eyes and 50 percent white eyes. As can be seen, the actual numbers were considerably lower, especially the number of white-eyed males. It is now known that in any such experiment, there is always a lower number of white-eyed offspring produced than would be predicted. This is due to the death of a significant number of flies with the white-eye allele during embryonic development, which suggests that some other (unknown) characteristic may also be affected.

These experiments were repeated by numerous investigators, with the same results. They provided strong support for the hypothesis that the gene for this trait was carried on the X chromosome and thus for Sutton's genes-on-chromosomes hypothesis. A valid hypothesis enables scientists to make predictions. In this case, Morgan tested many predictions by conducting experiments summarized in Table 2. (Students are encouraged to construct diagrams similar to those of Figure 2 for the crosses described in Table 1.) These experiments served to verify his hypothesis, since no other interpretation for the white-eye phenomenon could be seriously considered after a careful examination of the results.

The relatively simple picture Morgan had of white and red eyes in his fruit flies has since become exceedingly complex as a result of modern studies using new experimental tools. It is now known that genes at 18 separate loci on different chromosomes act together to produce the normal red eye color. If any of these genes undergoes a mutation, different eye colors, including the classic white eyes, will result. Some of the other eye colors are coffee, carrot, buff, apricot, honey, purple, and spotted. Sex determination in *Drosophila* has also been found to depart from the original X-Y model. Sex-determining genes are present on both the X chromosome (*X*) and the three pairs of autosomes (*A*), but not on the Y chromosome, although it is required for male fertility. It is the *balance* between *X* and *A* that determines the sex of any fruit fly. In normal flies, an *XX* genotype has an *X:A* ratio of $2X{:}2A = 1$ (that is, both are present in diploid numbers; it will have two X chromosomes and two of each autosome) and will be a female. An *XY* genotype has an *X:A* ratio of $1X{:}2A = 0.5$ and will be a male. In cases where nondisjunction occurs, some interesting phenotypes are produced. Individuals with $2X{:}3A = 0.67$ have traits of both males and females; $1X{:}3A = 0.33$ have an extreme male phenotype; and $3X{:}2A = 1.5$ are extreme females. These type of experimental observations seem to support a **balance model** of sex determination in *Drosophila*.

Morgan's first experiments initially led to the formulation of an interesting hypothesis. Subsequently, new experiments provided convincing evidence to support his modified hypothesis. His studies on the white-eyed fruit flies, though conducted far in the past, are still considered an elegant example of the experimental scientific method.

LINKAGE, LINKAGE GROUPS, AND CHROMOSOME MAPPING

Focus on Scientific Inquiry II

Genes belonging to linkage groups are located on the same chromosome. Thus they do not assort independently, and recombinants would not be expected to appear in offspring from *Drosophila* test crosses (refer to Figure 14.5). Nevertheless, experimental crosses with fruit flies revealed that recombination of genes for traits carried on the same chromosome *did* occur. How could this be explained?

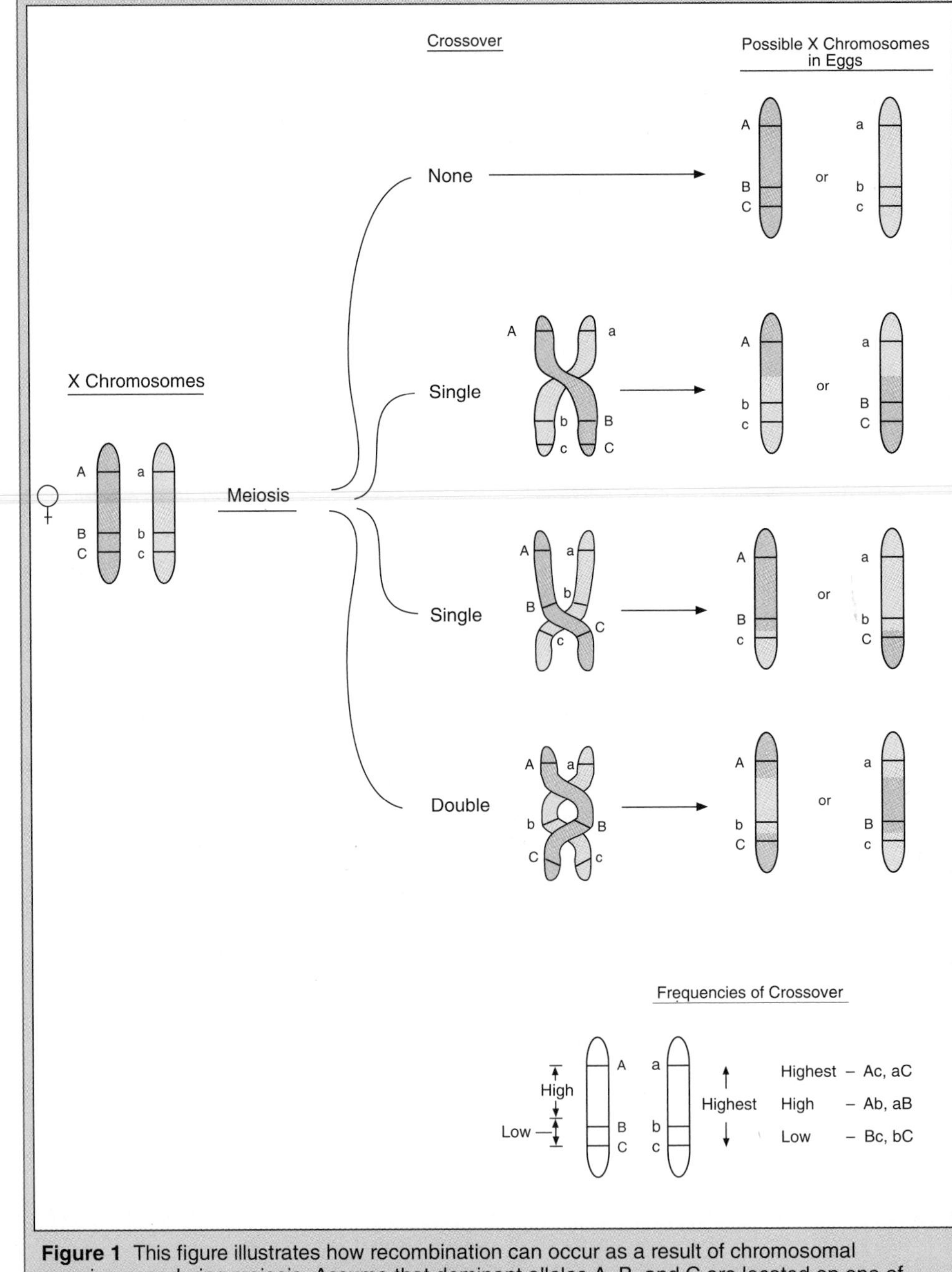

Figure 1 This figure illustrates how recombination can occur as a result of chromosomal crossing-over during meiosis. Assume that dominant alleles A, B, and C are located on one of the X chromosomes of a female *Drosophila* and recessive alleles a, b, and c on the other. If crossover does not occur, the three traits they affect will be inherited together (that is, no recombination will occur). In cases where crossovers occur, gametes may have new combinations of alleles (for example, A, b, and c or a, b, and C) and consequently, new combinations of traits—recombinations—may appear in offspring. The frequency of recombination is directly related to the distance between two genes on the chromosomes. Thus only a small number of recombinations would be expected for traits determined by alleles B and c, but a high number of recombinations would be expected for alleles A and c.

Morgan proposed that an exchange of chromosome segments must occur in meiosis (during crossing-over) that resulted in a separation of linked genes along the same chromosome. Figure 1 summarizes Morgan's ideas on genes and the effects of crossing-over. Since the linked genes would be found on different chromosomes after crossing-over, recombination could occur. Morgan further suggested that the "strength" of a linkage between two genes or traits depended on the distance between those genes on the chromosome. If that were true, the frequency of recombination between linked genes would provide an index for determining the distance between genes and, after careful anaylsis, the gene sequence on the chromosome. Chromosome maps showing the location of those genes and the relative distances between genes could then be drawn. If recombination seldom occurred for two traits, then according to this hypothesis, the genes for these traits would lie close to each other on the chromosome. Genes separated by long distances would be expected to recombine more frequently.

In 1913, one of Morgan's graduate students, A. H. Sturtevant, tested this hypothesis by completing another of the classic studies in the history of genetics. He experimented with the six sex-linked traits (traits determined by genes located on the X chromosome), which are recessive to the dominant, normal characteristic: yellow body color, white eyes, eosin eyes, vermilion eyes, miniature wing, and rudimentary wing. He soon realized that white and eosin eye color were caused by two alleles of one gene, and vermilion eye color was due to a second gene. Thus he was actually working with *five* genes, not six. Sturtevant obtained thousands of offspring in making every possible test cross involving the five traits. His general experimental approach is described in Figure 2, and selected data from his studies are presented in Tables 1 and 2.

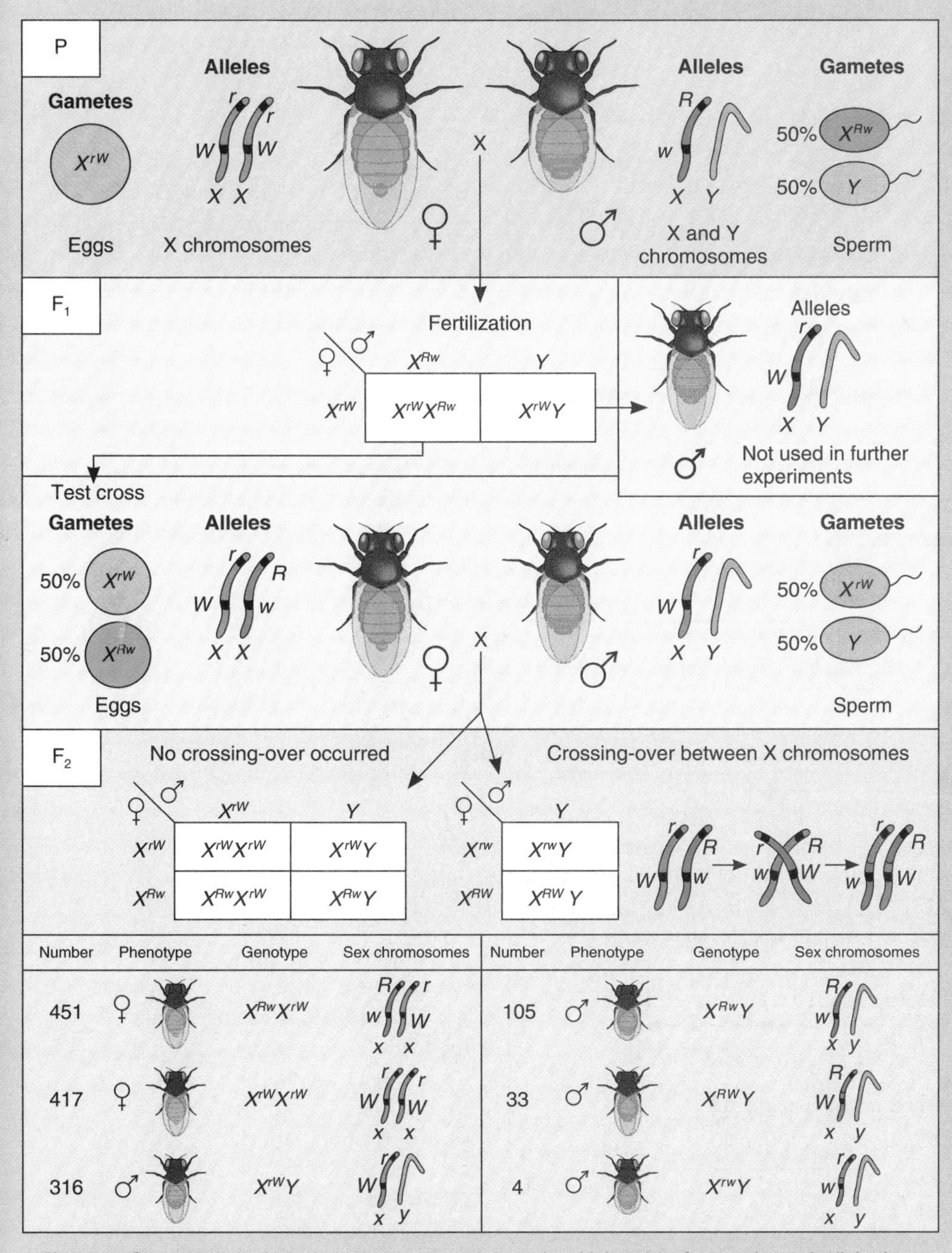

Figure 2 Sturtevant carried out numerous experiments in which pairs of recessive traits linked to the X chromosome in *Drosophila* were studied. His basic approach was to first cross females (P) homozygous for one normal dominant trait (in this example, normal long wings) and for one recessive trait (vermilion eyes) with males (P) that had the opposite combination of traits on their X chromosome (in this example, rudimentary wings and red eyes). The resulting F_1 females, which were heterozygous for both traits being studied, were then test-crossed with males that had normal wings and vermilion eyes to produce the F_2 (W = normal wing, w = miniature wings, R = red eyes, r = vermilion eyes.) Numbers of F_2 flies with the different traits were then counted. Those data are included in Table 1.

Focus on Scientific Inquiry II

Table 1 Results of Sturtevant's Experiments with Traits Linked to the X Chromosome in *Drosophila*

Pairs of Traits Studied	Proportion of Crossovers	Percentage of Crossovers
Yellow body and white eyes	214/21,736	**1.0**
Yellow body and vermilion eyes	1,464/4,551	**32.2**
Yellow body and rudimentary wings	115/324	**35.5**
Yellow body and miniature wings	260/693	**37.6**
White eyes and vermilion eyes	471/1,584	**29.7**
White eyes and rudimentary wings	2,062/6,116	**33.7**
White eyes and miniature wings	406/898	**45.2**
Vermilion eyes and rudimentary wings	17/573	**3.0**
Vermilion eyes and miniature wings	109/405	**26.9**

The proportion of crossovers represents the number of flies in which recombined traits appeared/total number of flies with no recombined traits.

Table 2 Sturtevant's Reasoning

(a) The distance between *Y* and *V* = 32.2 (the percentage of crossovers between *Y* and *V*). These can be placed arbitrarily as the first genes on the X chromosome.

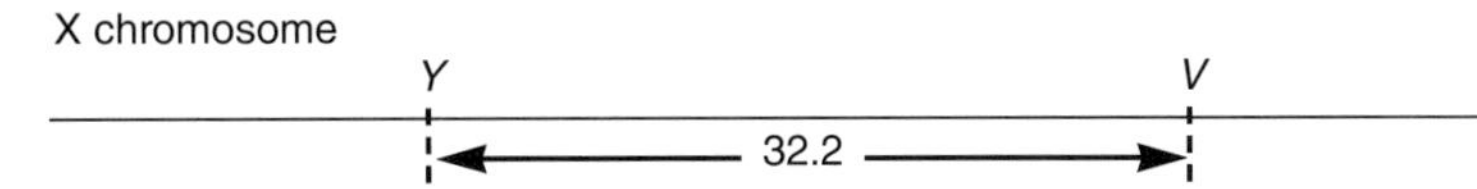

(b) The distance between *Y* and *R* = 35.5. But is *R* to the right or to the left of *Y*? If it lies to the left of *Y*, then the distance between *R* and *V* would be approximately 32.5 + 35.5 = 68. If *R* is found to the right of *Y*, then the distance between *V* and *R* would be about 35.5 - 32.5 = 3. The data in Table 1 show that the actual distance, as indicated by the percentage of crossovers, between *V* and *R* is 3. Therefore, *R* lies to the right of *Y*.

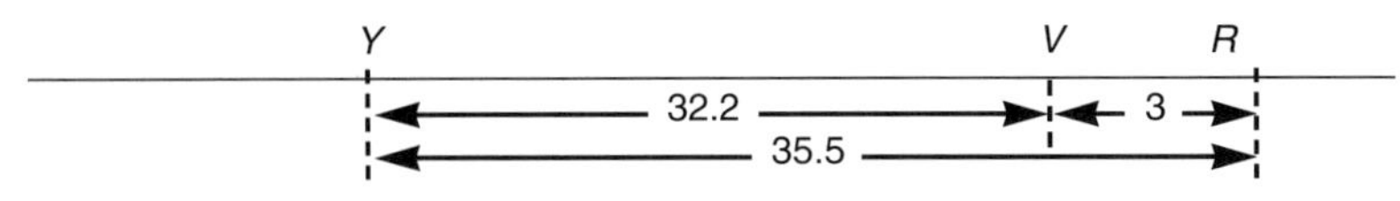

(c) This is Sturtevant's complete map, published in his paper in 1913.

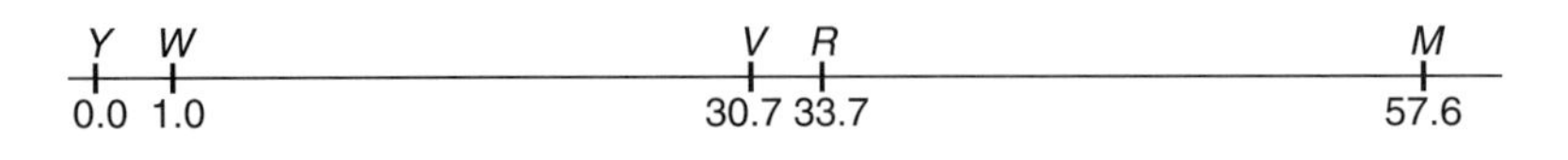

Sturtevant used this type of reasoning in analyzing his data and mapping the X chromosome for the five genes studied in his experiments. *Y* = yellow body, *W* = white eyes, *V* = vermilion eyes, *M* = miniature wings, and *R* = rudimentary wings.

Examination and analysis of this data set make it possible to tentatively answer some questions about distances separating the five genes on the X chromosome. For example, which two genes appear to lie very close to one another? Why? Which two genes appear to be separated by the greatest distance? These answers provide the basis for beginning to create a chromosome map for the X chromosome. Logical analysis of the other recombination frequencies (expressed as "percentage of crossovers" in Table 1) leads to the correct gene sequence and a measure of the distance between each of the genes. Sturtevant created the map shown in Table 2 mostly from the data shown here. The map units were defined as the equivalent of a 1 percent recombination frequency (that is, the percentage of crossover).

There appear to be some inconsistencies between the data and his map. The most obvious are the distances indicated between *Y*, *W*, and *M*. The map indicates greater distances (57.6 and 56.6) than shown by the data (37.6 and 45.2). Sturtevant explained that the lower than expected number of crossovers involving these genes was probably due to double crossovers, which would result in a reduction in the percentage of crossovers between these genes (see Figure 1).

Sturtevant concluded his magnificent paper by stating that the results "form a new argument in favor of the chromosome view of inheritance, since they strongly indicate that the factors investigated are arranged in a linear series at least mathematically." In fact, this study, and others conducted by members of Morgan's group during the next three years, completed the picture showing the relationships of chromosomes, genes, and inheritance.

mosome theory. His studies, published in 1916, also centered on crosses of red- and white-eyed fruit flies, but they involved extremely complex aspects of abnormal chromosome behavior (nondisjunction) that are beyond the scope of this book.

End of the Golden Age of Classical Genetics

Morgan and his ingenious coworkers (including some students who stayed to work with him after receiving their graduate degrees) continued their increasingly complex studies for several more years. Finally, in 1915, they published the results of their studies in a book, *The Mechanism of Mendelian Inheritance,* which interpreted Mendel's laws in terms of the chromosome theory. The book was a milestone in the history of genetics and closes the story of the chromosome theory of heredity. Though a few critics remained skeptical, most of the scientific community accepted the general conclusion that genes and chromosomes were the bases of heredity. Thomas Hunt Morgan left Columbia for California Institute of Technology in 1928. He received the Nobel Prize for his *Drosophila* work in 1934. For the next quarter century, the breadth and depth of knowledge about genes, chromosomes, and inheritable traits continued to expand as the field of genetics moved toward new frontiers.

T. H. Morgan and his students used fruit flies in conducting genetics experiments. They were able to show that specific traits were associated with the X chromosome. They also demonstrated that genes for many traits existed on separate chromosomes. Further, the sequence of genes and their relative distance from one another on a chromosome could be determined experimentally. Their research led to confirmation of the chromosome theory.

IMPORTANCE OF THE CHROMOSOME THEORY

The period that began with Mendel and ended with the time-honored studies of Morgan and his students was full of remarkable accomplishments. Beginning with only a dim outline of how traits were transmitted from parent to offspring, scientists established these three fundamental principles:

1. Genes are the fundamental hereditary units that determine specific traits in an organism.
2. Chromosomes are the carriers of genes.
3. The basis of Mendel's laws—segregation and independent assortment—lies in the behavior of chromosomes during the events of meiosis.

In the process of establishing these concepts through hypothesis generation, scientific experiments, and data analysis, many misconceptions about heredity were laid to rest. Chromosomes were shown to sort independently during meiosis, but not genes (except when separated by crossing-over) unless they occur on different chromosomes. Genes on the same chromosome constitute a linkage group, and they normally segregate into the same gamete; since they are on the same chromosome, they do not sort independently. The old hypothesis that a chromosome and a Mendelian hereditary factor (gene) were one and the same became modified to state that a chromosome contains several genes and that these genes are located in a linear order along its length. This picture of genes as parts of chromosomes provided a logical basis for understanding the normal mechanism of genetic transmission of different traits in sexually reproducing organisms. Equally important, the chromosome theory united the disciplines of genetics and cell biology. Now that many of the questions about chromosomal inheritance had been answered, new challenges focused on understanding the mechanisms and roles of chromosomes and genes in inheritance.

PATTERNS OF INHERITANCE

The most familiar and easily understood patterns of gene transmission fall into the category of *Mendelian inheritance,* in which traits normally exist in one of two forms (tall versus short, red versus white) and are determined by a single pair of genes (*Tt* and *Rr*) with complete dominance (tall over short, red over white). It is a very useful model for beginning to understand genetics. However, even a cursory examination of common plants and animals, including humans, makes it clear that surprisingly few traits appear to be inherited in the classic Mendelian pattern. Most flowers seem to come in a variety of colors, and humans are not either tall or short. Rather, a remarkable range of shapes, colors, heights, and other features characterizes the inheritable traits of all species. We will examine briefly a few patterns of inheritance that differ from the single-gene, all-or-none Mendelian concepts.

Incomplete Dominance

Mendel's F_1-generation peas always looked like one of the two parents because all their traits were determined by genes with dominant and recessive alleles. However, most flowering plants fail to produce simple Mendelian ratios in standard crosses. For example, as shown in Figure 14.6, when red and white four-o'clock flowers are crossed, the F_1 are all various shades of pink. When F_1 pinks are crossed, the F_2 consists of an approximate 1:2:1 ratio of three phenotypes—1 red to 2 pink to 1 white. This type of inheritance, in which heterozygotes show a range of intermediate phenotypes between homozygous parents, is called **incomplete dominance**. Mendelian rules, however, can explain these non-Mendelian results. A single pair of alleles is involved, segregation and independent assortment occur for the red and white alleles, and there is a 1:2:1 F_2 genotypic ratio. The difference occurs because neither allele is dominant, and as a result, the heterozygous form, some tone of pink, is intermediate between the two parents.

Multiple Alleles

A single gene may have many different or **multiple alleles** for a particular trait. For many traits, the number of alleles for a given gene in a homologous pair of chromosomes is two, one on each chromosome. Although each individual member of a species may have only two alleles, several different alleles may exist in

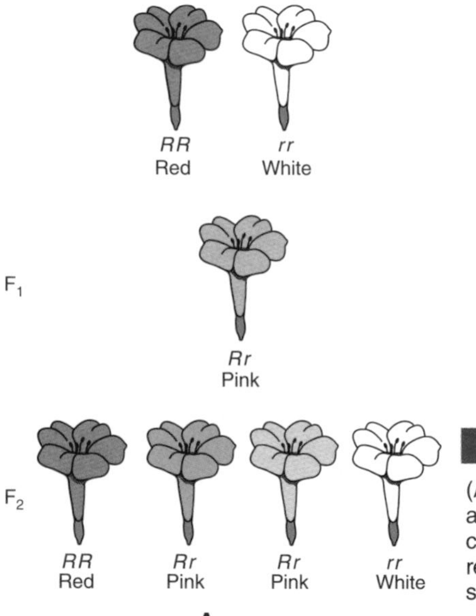

Figure 14.6

(A) The color of four-o'clock flowers is determined by a gene pair with two alleles, *R* and *r,* that have incomplete dominance. Flowers with genotype *RR* are red and *rr* are white, but flowers with genotype *Rr* are some shade of pink. (B) Photo of four-o'clock flowers.

the population. A familiar example of a human trait determined by a gene with multiple alleles is blood type. Each of us has two genes for this trait, but our blood type is determined by the two alleles present on homologous chromosomes: alleles AA or AO = blood type A, BB or BO = type B, AB = type AB, and OO = type O.

Polygenic Inheritance

Many traits show a great range of phenotypic variation. Such traits include human height, the color of wheat kernels, and possibly the weights of whales (not easily determined!). If the heights from a random sample of 100 students are recorded, a typical result might show that the shortest measured 5 feet and the tallest 6 feet 4 inches. The heights of the other 98 would form a continuum between the two extremes. A variable phenotypic trait, such as height, that is determined by multiple genes (not alleles) is said to be governed by *polygenic inheritance.*

Polygenic inheritance involves a number of genes that have a small but cumulative effect, with or without dominance or with partial dominance, in determining a trait. A classic example of this concept was demonstrated for kernel color in wheat by Herman Nilsson-Ehle in 1909 (see Figure 14.7). In a certain wheat variety, kernel color is determined by the cumulative action of two pairs of genes, each with two possible alleles: a partly dominant allele for red and another for white. Kernels with four red alleles are dark red in color, and those with four white alleles are white. Kernels with different combinations, such as two white and two red or three red and one white, are intermediate in color.

ALTERATIONS IN GENETIC MATERIALS

In sexually reproducing organisms, the normal mechanisms of inheritance follow a precise trail that begins with the events of meiosis and ends with fertilization, the formation of a zygote, and the development of new offspring. This pathway leads to the orderly, largely predictable transmission of chromosomes, genes, and genetic traits between generations.

Occasionally, however, deviations in normal genetic processes lead to abnormal combinations or a modification in the structure of genes or chromosomes within the cells that produce gametes. A permanent structural change within the hereditary material is called a *mutation*. The end result of a mutation—the effect on the offspring—can vary dramatically: there may be no effect, new variants (offspring with new forms of a trait) may appear, or premature death may occur before or after birth. A change that results in death is classified as a **lethal muta-**

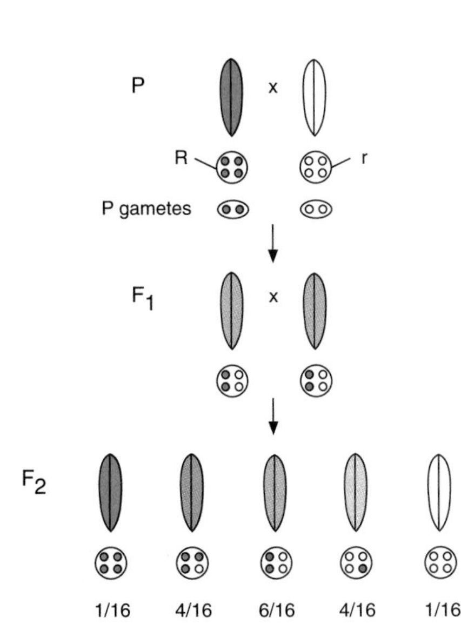

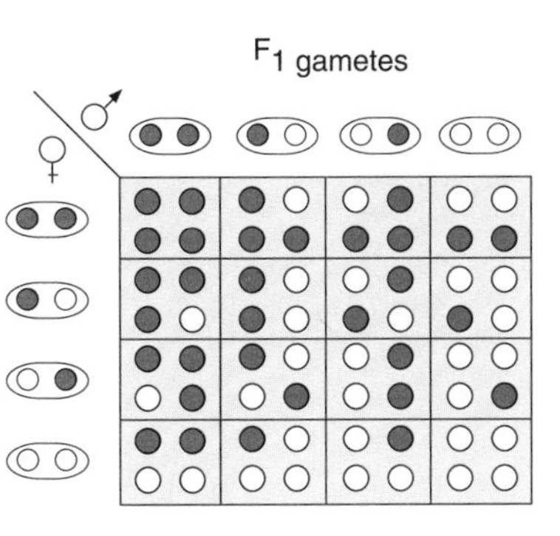

Figure 14.7

Nilsson-Ehle found that kernel color in a certain strain of wheat was determined by two gene pairs, each with two alleles. One gene pair came from each parent. Wheat with dark red kernels had four alleles for red color (designated *R*), and white kernels were found in wheat in which all four alleles were *r*. If dark red-kernel wheat was crossed with white-kernel wheat, the F_1 were all medium red. $F_1 \times F_1$—diagrammed in the Punnett square—yielded an F_2 with kernel colors in the ratios indicated in the figure.

tion. Although most mutations are considered harmful, the creation of new genetic combinations and new variations is also a vital process in the evolution of organisms. Two basic levels of mutation are recognized, chromosomal mutations and gene mutations.

Chromosomal Mutations

Chromosomal mutations occur when segments of chromosomes, entire chromosomes, or even complete sets of chromosomes are involved in genetic change. The effects of chromosomal mutations are caused primarily by new arrangements of the chromosomes. They are generally harmful in animals but may produce interesting new variations in plants. Chromosomal mutations can be classified as aberrations either in chromosome structure or in chromosome number.

Changes in Chromosome Structure

The origin of most chromosomal aberrations can be traced back to their behavior during prophase I of meiosis because at that time, homologous regions of chromosomes have a powerful pairing affinity. During this phase, breakage of chromosomes frequently occurs, and the broken ends of such chromosomes are very likely to attach to other broken ends. The consequences of such behavior are loss, rearrangement, or duplication of chromosomes or certain segments of chromosomes. Such chromosomal changes are described in Figure 14.8.

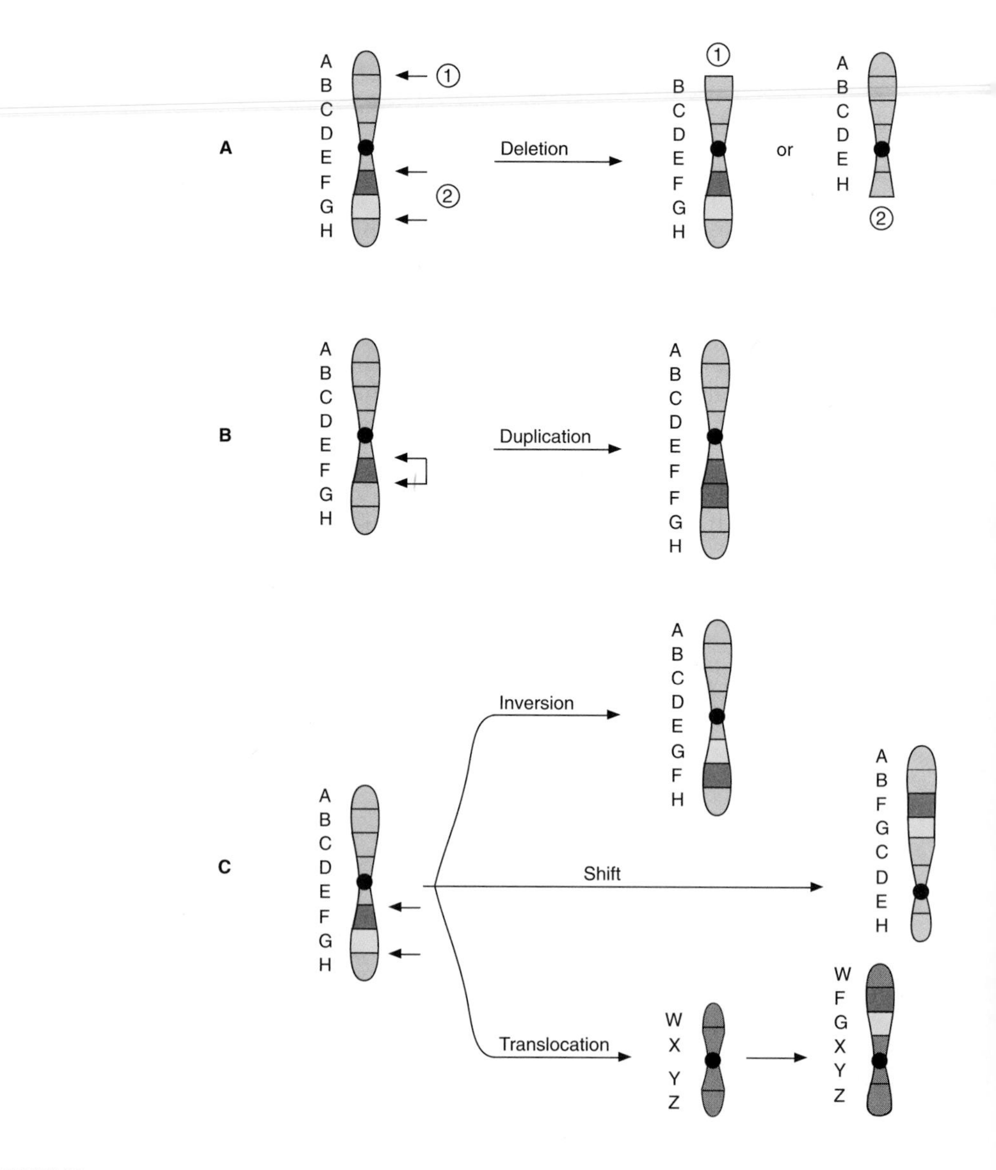

Figure 14.8

Several types of changes in chromosome structure may occur during meiosis. In this figure, changes in one chromosome from a homologous pair are described. Capital letters indicate different regions of the chromosome. (A) Deletions occur when breaks develop (at sites indicated by the arrows) and portions of the chromosome are detached. (B) Duplications (arrows) result in extra copies of genes or larger regions of chromosomes. (C) If a section of a chromosome breaks (arrows), it may become inverted in the same part of the chromosome (inversion), become incorporated into a different part of the same chromosome (shift), or become integrated into a different chromosome (translocation).

Deletions
The loss of a segment of a chromosome after breakage is called a **deletion**. The size of the deletion can range from a small section of a gene to a portion of a chromosome that contains several genes. If the same deletion occurs in both members of a homologous chromosome pair, it is almost always lethal. Studies of deletions indicate that most of the genetic information carried on chromosomes is required for an organism to develop and function normally. Even single deletions are frequently harmful because for some traits, this upsets a specific balance of genes. Deletions of certain regions of a chromosome often create a characteristic phenotype. For example, in *Drosophila,* notched wings result from one particular deletion.

Duplications
Duplications result in an increase in the amount of genetic material carried by a chromosome. Unless a large number of genes or a major segment of the chromosome is duplicated, the potential for a negative effect is apparently quite small. Indeed, there are some indications that duplications may, in some instances, actually supply additional genetic material that leads to an increase in the diversity of gene functions. Thus duplications can be a source of new genetic variation.

Inversions, Shifts, and Translocations
When a portion of a chromosome breaks, it may rotate 180° and become reinserted at the same position in the chromosome (**inversion**), become inserted in a different region of the same chromosome (**shift**), or become inserted in a nonhomologous chromosome (**translocation**). These types of mutations can also occur during the complex chromosomal gymnastics connected with crossing-over (discussed in Chapter 13). Chromosomal inversions, shifts, and translocations all have the potential to generate unequal meiotic products (gametes) and, consequently, are usually harmful to any zygote produced.

Changes in Chromosome Number
In the normal flow of events during meiosis, paired homologous chromosomes separate during anaphase of meiosis I and paired chromatids separate during anaphase of meiosis II. Occasionally, **nondisjunction**, the failure of certain chromosomes to separate during either of these phases, results in the production of gametes with abnormal numbers of chromosomes (see Figure 14.9). Changes of this type, which involve a fraction of the chromosomes in a complete chromosome set (one chromosome in Figure 14.9), result in offspring that have cells with an abnormal number of chromosomes. Such cells or individuals are called **aneuploid**. Inspection of Figure 14.9 shows that the gametes produced will contain one extra chromosome or be deficient by one chromosome. Assuming fertilization of these abnormal gametes with a normal sperm or egg, the resulting cells of the aneuploid zygote will contain one chromosome too many (in which case it is called trisomic) or one too few (monosomic), compared to the normal diploid number. In humans and other animals that have been carefully studied, aneuploidy usually results in spontaneous (natural) abortion of the fetus. If the offspring survives, it usually suffers from a variety of symptoms caused by the abnormal concentration of genetic material.

Normal diploid organisms have two multiple sets, $2n$, of the monoploid, n, number of chromosomes in their cells (technically, the term *haploid* refers to the n number of chromosomes found in a gamete; *monoploid* refers to the n number of chromosomes in any cell). However, organisms can be produced that have $3n$, $4n$, or even a greater number of complete monoploid sets of chromosomes. Such organisms are called *polyploids*.

Breeders of plants have a long history of developing new varieties and species of flowers, fruits, and grains using methods that involve the creation of

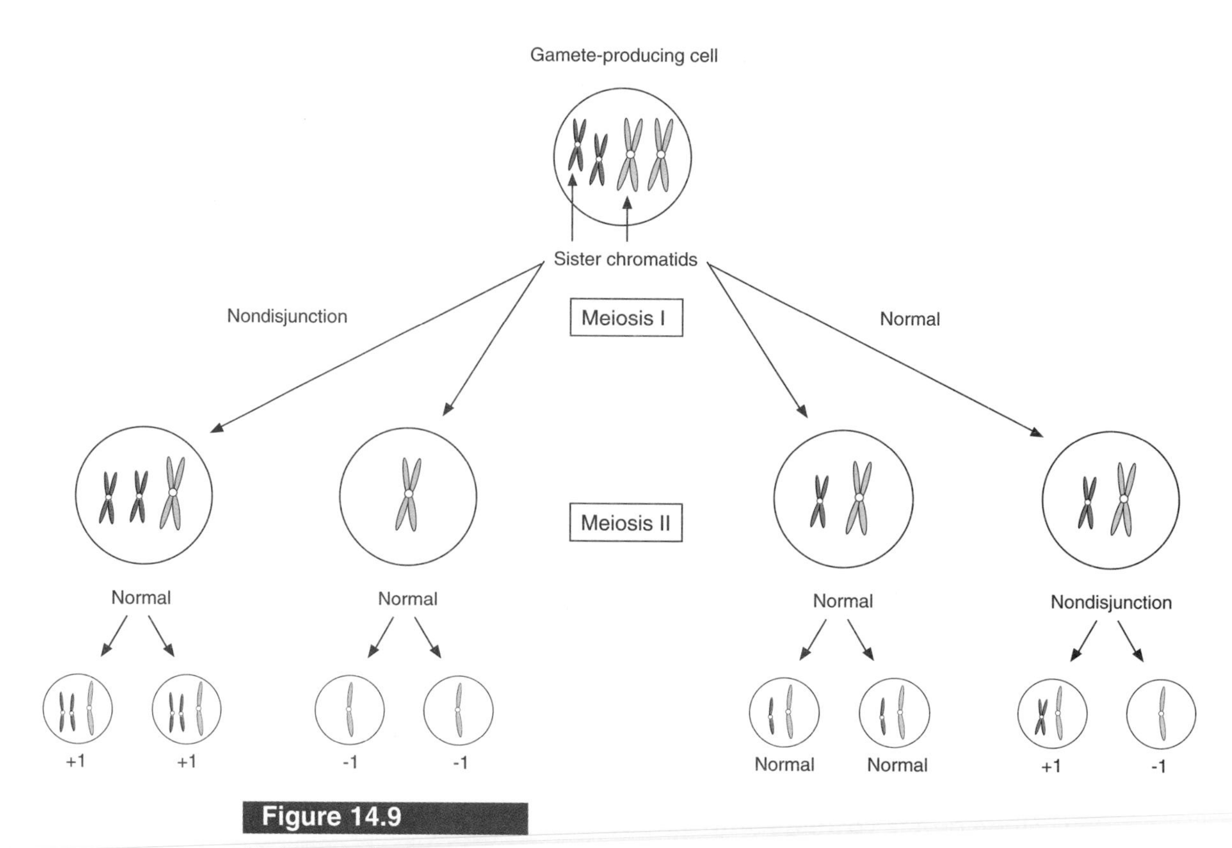

Figure 14.9

During meiosis, sister chromatids may fail to separate normally during either the first or second division. As a result of this nondisjunction, gametes produced will have either one more or one less than the normal chromosome number. If one of these irregular gametes participates in fertilization, the resulting zygote will also have the same abnormal chromosome number as the gamete (+1 or –1).

polyploid forms. For example, New World cotton has 26 pairs of chromosomes and is a polyploid of Old World cotton, which has 13 pairs. Polyploidy is of minor importance in animal genetics since possession of multiple sets of chromosomes is usually incompatible with survival.

Gene Mutations

Single genes can mutate from one allelic form to another. The white eyes of the fruit fly in Morgan's studies were caused by a gene mutation. Most **gene mutations** are probably harmful to a species since they alter a stable, successful genetic constitution. However, gene mutations that occur in gamete-producing cells of animals are a source of new genetic material and may have a positive aspect, since they can increase variation within a population.

Once predictions made by the chromosome theory were proved correct, it was possible to explain non-Mendelian patterns of inheritance. Chromosome and gene mutations, and their consequences, also became understood.

CLOSE OF AN ERA

The age of classical genetics slowly came to an end during the 1930s. During the first third of the twentieth century, enormous progress had been made; the reality of genes, their location on chromosomes, and their vital roles in determining hereditary traits had become clearly established. T. H. Morgan and many others had successfully employed multicellular organisms in complex breeding experiments and mathematical analyses to derive the first genetic principles. Eventually, however, such methods proved limiting, further advances became difficult, and new questions about the nature of the hereditary material were being asked.

Summary

1. Following the rediscovery of Mendel's work, there was great interest in establishing the cellular location of Mendelian factors (genes) and the mechanisms that accounted for the determination of genetic traits.
2. The chromosome theory of heredity, conceived early in the twentieth century primarily by William Sutton, stated that genes were located on chromosomes. Mendel's laws and the transmission of genetic traits were thought to be related to events that occurred during meiosis.
3. Thomas H. Morgan and his group of graduate students conducted experiments that eventually proved the chromosome theory to be correct. By using an ideal experimental organism, the fruit fly (*Drosophila melanogaster*), they carried out studies of mutants with unusual traits such as white eyes. Their research demonstrated conclusively that these traits were due to genes carried on the X chromosome. Thus certain genes were shown to be located on specific chromosomes, and the chromosome theory was confirmed.
4. Through studies involving linked traits, crossing-over of chromosomes during meiosis, and uncomplicated mathematical analyses, it became possible to create maps of fruit fly chromosomes. Such maps serve to identify the locations of specific genes on chromosomes and also to determine their sequence and relative distances from one another.
5. Different patterns of inheritance were identified in the first third of the twentieth century. It also became recognized that chromosomes and genes undergo mutations that give rise to new genetic variations that are usually harmful.

Review Questions

1. Why did early cytologists believe that there was a link between meiosis and chromosomes and genes?
2. How does the chromosome theory explain genetic inheritance?
3. Why are fruit flies (*Drosophila*) excellent experimental animals for genetics studies?
4. Explain the mechanism of X-Y sex determination.
5. How was it shown that genes for certain traits of *Drosophila* were located on the X chromosome?
6. What types of experiments and results confirmed the chromosome theory?
7. What is meant by Mendelian inheritance?
8. What is meant by non-Mendelian inheritance? Give three examples and explain why they are non-Mendelian.
9. Why are chromosomal mutations more likely to be harmful than gene mutations?

Problems

1. Using Nilsson-Ehle's wheat variety, what offspring ratios would be expected if one of the two parent plants had three alleles for red kernels and one for white and the other had two red alleles and two white alleles?
2. Assume that a gene pair for wing length in *Drosophila* is found on one of the autosomes and that the allele for long wings (*L*) is dominant over the allele for short wings (*l*). What genotype and phenotype ratios would be expected in the F_1 if both parents were heterozygous for wing length? If one was heterozygous and the other had short wings?
3. From Morgan's eye color experiments, what would be expected in the F_1 if a white-eyed female was crossed with a red-eyed male?
4. Construct a chromosome map for alleles *A, B, C, D,* and *E,* given the following percentage rate of crossovers: *A* and *B* = 12, *A* and *C* = 18, *A* and *D* = 2, *A* and *E* = 22, *B* and *C* = 30, *B* and *D* = 14, *B* and *E* = 10, *C* and *D* = 16, *C* and *E* = 40, *D* and *E* = 24.

Essay and Discussion Questions

1. When the chromosome theory was proposed, no one had actually seen a gene. Genes were a theoretical concept that were hypothesized to exist and to determine specific traits in organisms. Later research verified the existence of genes and their structure. Were scientists justified in postulating the existence of genes and a hereditary role for these as yet unseen factors? Why?

2. Mendel proposed two simple laws of genetics. Subsequent investigation showed the situation to be much more complex for most traits in most organisms. To what extent are we justified in believing in the validity of so-called Mendelian inheritance?

3. The chromosome theory developed through theoretical ideas, experimental investigations, and careful observations. Evaluate the relative importance and roles of each of these elements in creating hypotheses or theories.

4. Confirmation of the chromosome theory depended on a group of scientists who had laboratory facilities, graduate students who did research, and time. What were the possible sources of funds for all of the space, equipment, supplies, and investigator salaries? Why were funds made available for supporting the type of research described in this chapter?

References and Recommended Readings

Allen, G. E. 1978. *Thomas Hunt Morgan: The Man and His Science*. Princeton, N.J.: Princeton University Press.

Crowe, J. F. 1991. Antedotal, historical and critical commentaries on genetics. Our diamond anniversary. *Genetics* 127:1–3.

Dunn, L. C. 1965. *A Short History of Genetics: The Development of Some of the Main Lines of Thought, 1864–1939*. New York: McGraw-Hill.

Lederman, M. 1989. Research note: Genes on chromosomes: The conversion of Thomas Hunt Morgan. *Journal of the History of Biology* 22:163–176.

Moore, J. A. 1986. Science as a way of knowing: Genetics. *American Zoologist* 26:583–747.

Morgan, T. H. 1910. Sex-limited inheritance in *Drosophila*. *Science* 32:120–122.

Sturtevant, A. H. 1913. The linear arrangement of six sex-linked factors in *Drosophila,* as shown by their mode of association. *Journal of Experimental Zoology* 14:43–59.

———. 1965. *A History of Genetics*. New York: HarperCollins.

Sutton, W. S. 1902. On the morphology of the chromosome group in *Brachystola magna*. *Biological Bulletin* 4:24–39.

———. 1903. The chromosomes in heredity. *Biological Bulletin* 4:231–251.

Suzuki, D. T., A. J. F. Griffiths, J. H. Miller, and R. C. Lewontin. 1986. *An Introduction to Genetic Analysis*. New York: Freeman.

CHAPTER 15

The Birth of Modern Genetics

Chapter Outline

Reading Questions

1. What questions guided genetics research between 1920 and 1955?
2. What types of organisms were used to answer questions about DNA?
3. What is molecular biology?
4. What were the key studies in identifying the genetic material?
5. How did Watson and Crick determine the structure of DNA?
6. How is DNA replicated?

NEW FRONTIERS

After the first third of the twentieth century, the focus of genetics research gradually turned to new problems as more was learned about the elemental mechanisms of genetics. Mendel had established that hereditary factors (genes) were responsible for determining specific traits. Later, Sutton postulated that genes were located on chromosomes, a theory that was verified by T. H. Morgan and his students.

New Questions

Following these achievements, new questions emerged that defined the direction of genetics research. The basis of many of these questions concerned the chemical identity of the hereditary material. Exactly what molecules in the cell constituted genes and chromosomes? The research conducted to answer this question and the knowledge uncovered by numerous studies are described in this chapter and Chapter 16.

New Developments

During the 1930s, several developments changed the science of genetics. In the history of genetics from Mendel forward, progress has often been accelerated when scientists turned to new organisms for their experiments. Most of the early work, including Mendel's, was performed on plants. Later, Morgan and his colleagues advanced the field of genetics by using their famous fruit flies.

By the 1930s, finer degrees of genetic resolution became the goal of scientists. To accomplish this goal, bacterial cells and later viruses were recruited as experimental subjects (see Figure 15.1). The most obvious advantage of using microorganisms was that they were simple and easy to work with. However, in modern studies, an underlying rationale for using such organisms is that all species evolved from a common ancestral cell line (see Chapter 2) and have been found to have the same mechanisms for genetic operations. Thus the discovery and understanding of a genetic molecule or process in a plant, fruit fly, or bacterium leads to at least a tentative perception of how similar mechanisms operate in more complex organisms, such as humans.

Bacteria and viruses have major advantages over the eukaryotic plants and animals that had been used in earlier studies. Their genetic materials are simpler and more easily accessible, culturing expenses are lower, and they multiply at extraordinary rates, giving rise to large populations in a short period of time. Perhaps the greatest gain was that experiments could be completed within days, or even hours, instead of years (Mendel's peas) or months (Morgan's *Drosophila*).

Other developments that aided progress in genetics included new methods, techniques, and scientific instruments. These advances made it possible to conduct more complex experiments and to analyze results with greater precision.

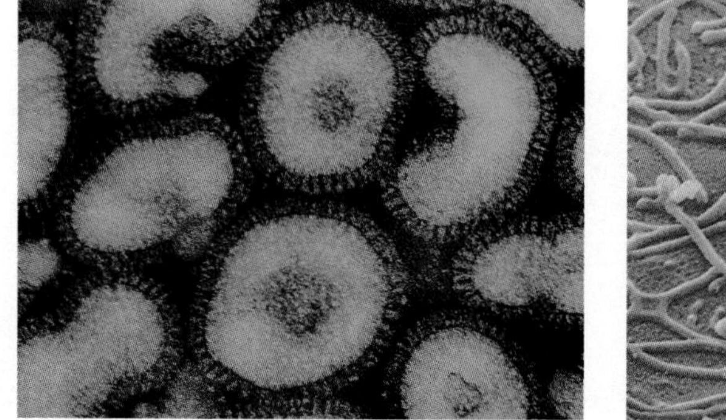

A

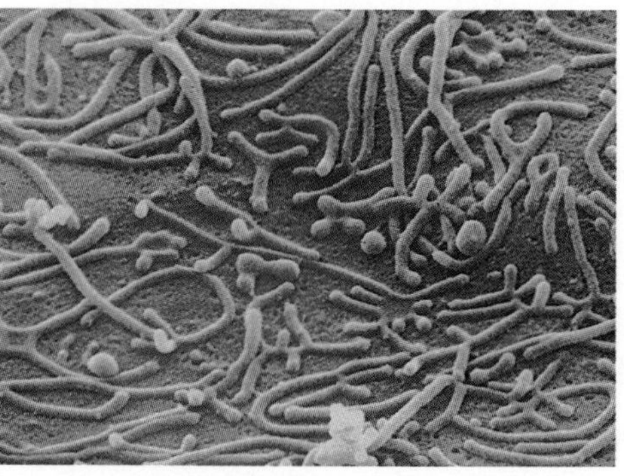

B

Figure 15.1

Studies using viruses (A) and bacteria (B) accelerated progress in genetics research during the twentieth century.

The development that had the greatest impact on genetics—indeed, it resulted in the transformation of biology—was the merging of physics, biochemistry, microbiology (the study of microorganisms, such as bacteria), and genetics to create a new field of science that came to be known as **molecular biology**. During the 1930s, the great interest in comprehending genes at the molecular level required experimental approaches, methods, and expertise that were within the domains of physicists and chemists. Fortunately, many such scientists had become interested in understanding biological processes at the level of atoms and molecules and had immigrated to biology. Their classical studies on the molecules involved in heredity provided a basis for creating the field of **molecular genetics**.

In the period following Morgan's verification of the chromosome theory, geneticists turned to microorganisms as experimental subjects. A new scientific discipline, molecular biology, arose from an influx of scientists from the fields of physics, biochemistry, microbiology, and genetics who had an interest in understanding life processes.

CELLS AS CHEMICAL FACTORIES

To understand the investigations described in this chapter, it will be helpful to learn some basic concepts about the nature of chemicals that exist within cells. The incredible variety of activities carried out by different cell types can ultimately be understood as the result of numerous chemical processes that take place within the cell. To sustain their functions, cells are always engaged in producing, using, and recycling chemicals.

Two classes of molecules exist within cells. At least 750 types of small chemicals (for example, sodium and calcium) are found in most cells, along with thousands of large *macromolecules*. The *small chemicals* include salts, trace elements, and molecules used to construct larger molecules. Most of the **macromolecules** involved in the chemical processes of organisms occupy one of the following categories: *carbohydrates* (sugars), *lipids* (fats), *proteins,* and *nucleic acids*. These macromolecules are all formed by linking specific types of small molecules into long chains. Proteins and nucleic acids are the subjects of this chapter and Chapter 16.

Proteins

Proteins of every organism on Earth are composed from an alphabet of 20 different small, building-block molecules called **amino acids** (see Table 15.1). Amino

Table 15.1 The 20 Amino Acids Used to Construct Proteins

Alanine (ALA)	Glycine (GLY)	Proline (PRO)
Arginine (ARG)	Histidine (HIS)	Serine (SER)
Asparagine (ASN)	Isoleucine (ILE)	Threonine (THR)
Aspartic acid (ASP)	Leucine (LEU)	Tryptophan (TRP)
Cysteine (CYS)	Lysine (LYS)	Tyrosine (TYR)
Glutamic acid (GLU)	Methionine (MET)	Valine (VAL)
Glutamine (GLN)	Phenylalanine (PHE)	

The proteins in all living organisms on Earth are constructed of combinations of these 20 amino acids.

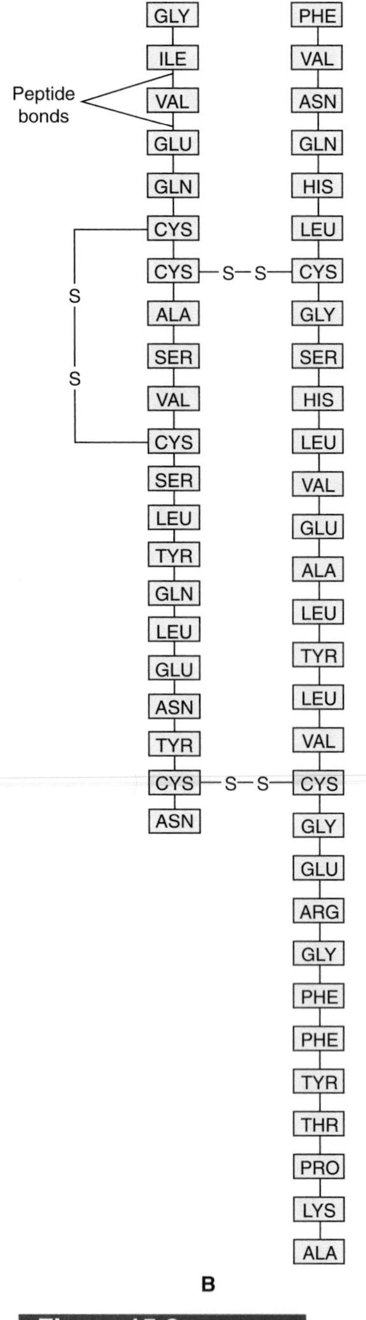

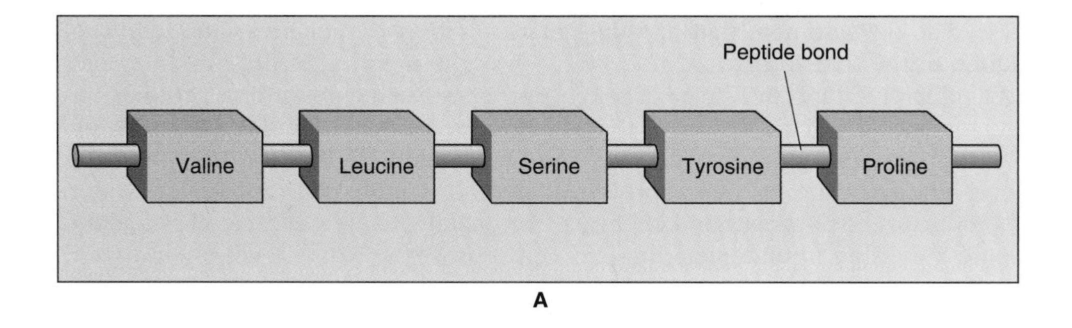

Figure 15.2

Proteins are made of chains of amino acids that are attached to each other by peptide bonds (A). Most individual proteins contain 50 to over 1,250 amino acids. Insulin (B) is a protein molecule composed of two chains, 21 and 30 amino acids long, held together by bonds between sulfur atoms.

acids are attached to each other by a form of chemical linkage called a **peptide bond** (Figure 15.2A); two amino acids are joined together by a single peptide bond. Proteins, such as insulin, diagramed in Figure 15.2B, differ from one another because each has a distinctive sequence of amino acids. There are no predictable patterns or rules in the amino acid sequence of proteins; the 20 amino acids never occur in equal amounts, and not all may be included in any single protein. Most proteins range in size from about 50 amino acids (insulin) to 1,250 amino acids (gamma globulin, an antibody protein).

A central theme in biology is that the specific structures and functions of each cell type are determined by the proteins it contains. There are more than 50,000 different kinds of protein in the cells and tissues of your body. Individual cells may contain hundreds of different proteins or only a few. Proteins are the most abundant macromolecules, accounting for over 50 percent of the dry weight of living cells. Proteins have enormous structural and functional diversity and are engaged in a variety of biological roles.

In higher organisms, proteins serve as structural units, carry out specific biochemical reactions within cells (enzymes), transport substances (hemoglobin in blood, for example, transports oxygen), defend cells (antibodies), and regulate processes (hormones). In addition, diverse biological molecules and products are made of proteins, including muscle fibers, feathers, antibiotics, and mushroom poisons.

Enzymes

Enzymes are *catalysts* that participate in most chemical reactions within cells and are usually composed of proteins. The concept of enzyme action is described in Figure 15.3. Like other catalysts, enzymes greatly accelerate the rate at which chemical reactions occur, but they are not used up during the course of the reactions. A single enzyme molecule may enter into thousands of reactions every second without being modified or degraded. Most enzymes catalyze only a single, specific chemical reaction within cells.

Given the great quantity and complexity of reactions carried out in the cell, it is not surprising that the vast majority of macromolecules present in most eu-

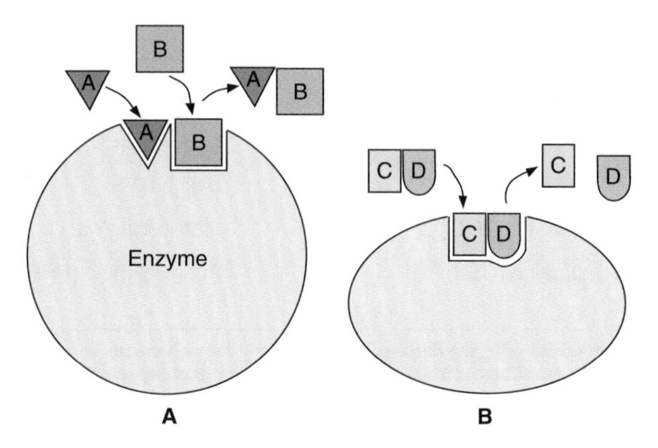

Figure 15.3

Enzymes are protein catalysts that participate in thousands of chemical reactions within the cell. Enzyme reactions can join molecules together (A) or split molecules apart (B).

karyotic cells are enzymes. It has been estimated that a typical animal cell contains 1,000 to 4,000 different enzymes, each of which catalyzes one chemical reaction or a set of closely related reactions.

Nucleic Acids

Two types of **nucleic acids**, **DNA** (deoxyribonucleic acid) and **RNA** (ribonucleic acid), exist in cells, and both are associated with crucial activities, which we will describe. Both DNA and RNA have the same fundamental structure; each is a linear unit composed of varying numbers of four different basic molecules called **nucleotides**. The structures of different nucleotides are nearly identical except for the fact that each is distinguished by the presence of a specific nitrogen-containing **base**.

In DNA, the four bases and their common abbreviation are adenine (A), guanine (G), cytosine (C), and thymine (T); RNA is made of A, G, C, and uracil (U). Figure 15.4A shows structural diagrams of the four DNA bases and uracil. Nucleotides with A and G are classified as *purines* because of their double-ring structure; those with C, T, and U are *pyrimidines,* having single rings. DNA and RNA are technically polynucleotides, and each nucleotide consists of a phosphate molecule, a sugar (called deoxyribose in DNA—hence the formal name—and ribose in RNA), and one base (see Figure 15.4B).

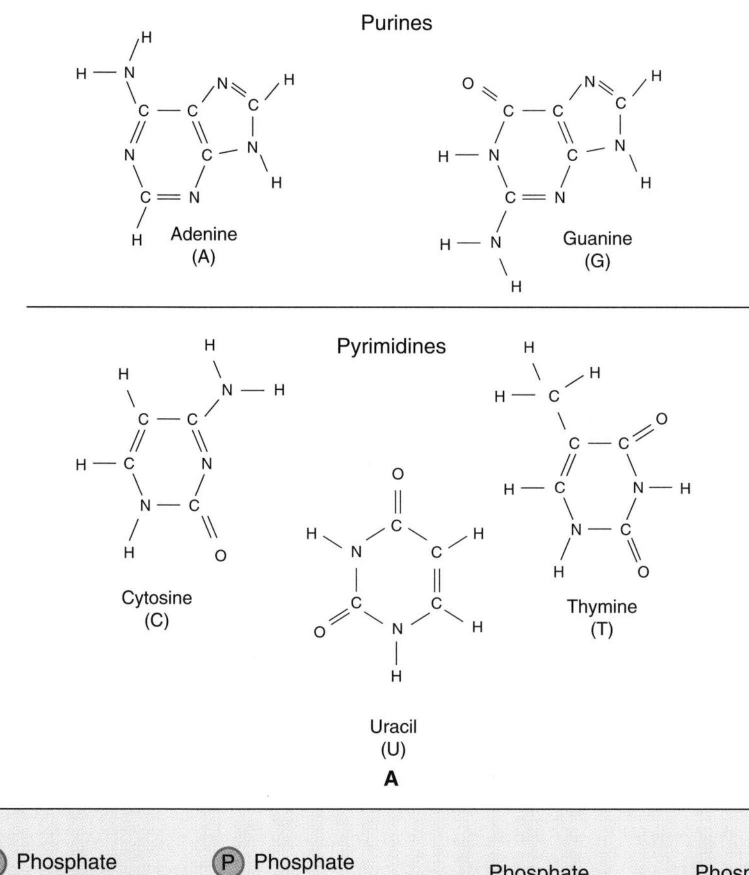

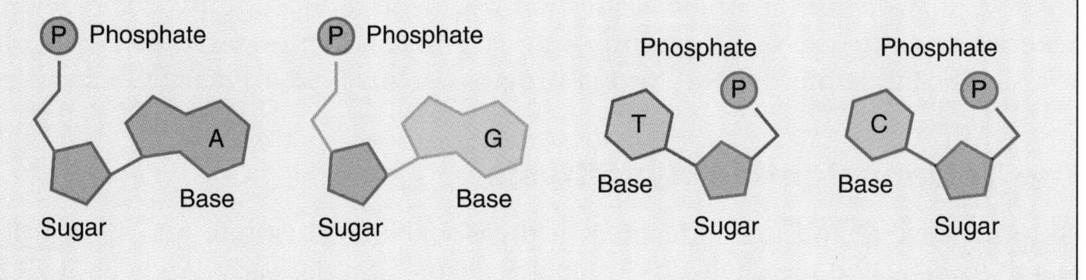

Figure 15.4

The primary structural units of nucleic acids are five nucleotides that each consist of a unique, nitrogen-containing base. (A) The bases used in the construction of DNA are adenine (A), guanine (G), cytosine (C), and thymine (T). RNA is composed of A, G, C, and uracil (U). (B) Each nucleotide contains one base, a phosphate molecule (P), and a sugar molecule. In DNA molecules, the sugar is deoxyribose, and in RNA molecules, it is ribose.

A wide variety of chemicals are found inside cells. Two types of macromolecules are especially important in cellular activities. Proteins are composed of amino acids and have a great diversity of structure. Proteins are used in building cellular structures and performing various functions, depending on the type of cell. Nucleic acids, RNA and DNA, are linear molecules constructed of nucleotides.

THE DISCOVERY OF DNA AND ITS ROLE IN LIVING ORGANISMS

The quest to uncover the repository of hereditary information spanned a century, from the 1860s to the 1960s. In looking back, the road traveled by scientists on this journey, except for one major detour, appears to have been long but relatively straight.

Several experiments conducted during this period were extraordinarily beautiful in their design and outcome. Since biology is, in part, an experimental science, we will describe some of these elegant experiments in this chapter for readers to gain an understanding of how science progresses.

A Chemical in the Nucleus

The revolution in molecular biology traces its origin back to the time when chemists were trying to identify the different chemicals present in the nucleus of cells. The discovery of DNA is credited to a Swiss biochemist named Johann Friedrich Miescher, who worked on the problem from the 1860s until 1874 and then again, shortly before his death, in 1895. His source of experimental material would not have been the first choice of many scientists. He analyzed pus, largely composed of white blood cells, which he obtained by washing out bandages taken from surgical patients. After analyzing these cells, Miescher found that the nuclei contained an unknown organic chemical with an unusually high amount of phosphorus. He and his colleagues named the new chemical *nuclein* (see Figure 15.5). He later continued his studies using sperm from salmon. Not only was this material probably more pleasant to work with, but sperm were an especially rich source of nuclein, since few other chemicals are present. Miescher published his results on nuclein, but they were vague and were severely criticized. Also, he was unable to offer a viable hypothesis about its role in the cell.

Figure 15.5

In Miescher's early studies to identify chemicals inside cells, special stains revealed a previously unknown chemical he called *nuclein.* In this photo, the green substance is Miescher's nuclein.

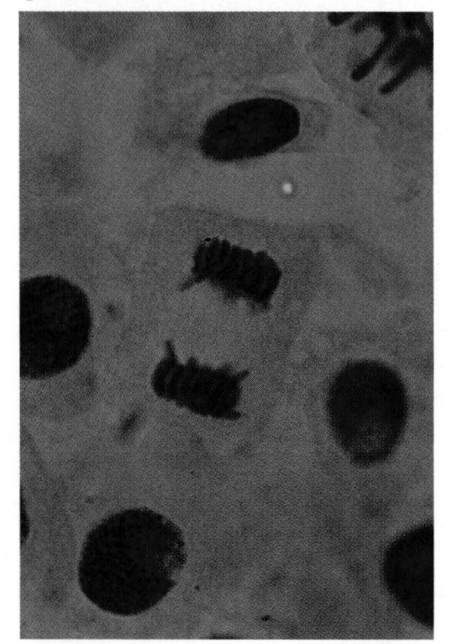

Based on work by Miescher and others, nineteenth-century scientists formed opinions about the possible nature of nuclein. Some felt, with little supporting evidence, that nuclein and chromatin (known to be the substance of chromosomes) were identical.

Further advances were made during the next three decades as a result of new research in biochemistry. The previously controversial suggestion that nuclein and chromatin were the same substance was shown to be true. This finding, coupled with the assumption from genetics that chromosomes carried the hereditary material, made the chemical analysis of nuclein highly significant. Through the use of new biochemical techniques, nuclein was found to consist of DNA and protein. This critical discovery can be seen as a very promising beginning toward understanding the ultimate chemical nature of chromosomes. The number of possible answers for the original question had been reduced to two. The genetic material was either DNA or protein—but which? The trail was now to grow cold, as a significant detour delayed further progress in this field for many years.

The Tetranucleotide Hypothesis

Phoebus Levene and his colleagues at the Rockefeller Institute for Medical Research investigated the chemistry of nucleic acids from the early 1900s until 1940.

During the first decade of this work, they were responsible for many of the chief discoveries that helped form a clearer picture of nucleic acid structure.

By his great successes and distinguished reputation, Levene came to dominate the field of nucleic acid research. Unfortunately, his analyses suggested to him that equal amounts of the four nucleotides were present in the DNA of all organisms. The implication of this *tetranucleotide hypothesis* was that DNA was simple and repetitive. Given the complexity and great diversity of genetic traits, it seemed obvious that a static DNA molecule could not be the hereditary blueprint. Rather, it was assumed that proteins, because of their remarkably varied structure, must somehow be responsible for defining the exceptional variations of hereditary traits. This view, in contrast with most unconfirmed hypotheses, became an unchallenged paradigm of biochemistry. Consequently, little headway was made until this hypothesis was finally abandoned because of results from later experiments.

In hindsight, it is easy to understand that the uncritical acceptance of the tetranucleotide hypothesis, and its implications, led to a period in the 1920s and 1930s when little was accomplished in determining the chemical identity of the genetic material. This case illustrates how premature acceptance of a hypothesis retards the normal processes of questioning (modifying, rejecting, or forming new hypotheses) and experimenting that generate progress in science.

The "Transforming Principle"

It often happens in science that research conducted in one area provides unexpected insights into another, seemingly unrelated field. One such event revolves around work with certain strains of a bacterium, *Diplococcus pneumoniae* (pneumococcus), that causes pneumonia in humans.

In the 1920s, it was shown that two basic *Diplococcus* varieties existed. The first type was an encapsulated form that was surrounded by a sugar coat and produced smooth, shiny, rounded colonies when grown in culture; this became known as the S form. A second type did not have a sugar coat, and its colonies appeared rough rather than smooth; this was called the R form. The form that causes pneumonia is the S form. Exposure to the R form causes no negative effects.

The Griffith Experiment

Fred Griffith was a microbiologist interested in learning more about the effects of different strains of pneumococcus on humans. In 1928, he published a paper describing some curious observations that remained puzzling for many years.

Griffith's experiments are described in Figure 15.6. Standard results were first obtained when he injected mice with the two forms of *Diplococcus:* mice injected with the S form developed pneumonia and usually died, while those injected with the R form suffered no ill effects. In a second set of experiments, he first killed the S form of bacteria by applying gentle heat. Not surprisingly, the heat-killed S bacteria did not cause pneumonia. However, a startling result occurred when he mixed the live R forms with the heat-killed S form. Mice exposed to this combination contracted pneumonia. Further, blood samples collected and analyzed from those mice revealed the presence of live S bacteria that had caused the disease! When those bacteria were grown in culture, they produced the smooth colony characteristic of the S variety. What had happened? How could these results be explained? To Griffith, the results indicated that the R bacteria had mysteriously obtained something, a *transforming principle,* from the heat-killed S form. The harmless R form had somehow been transformed into the deadly S type of pneumococcus.

Griffith did not attempt any further experiments, and there is no indication that he related his findings to a transfer of hereditary material between the dead and living bacteria. He was killed in his lab during a bombing of Britain in 1941, and the puzzle he created was not solved until 1944.

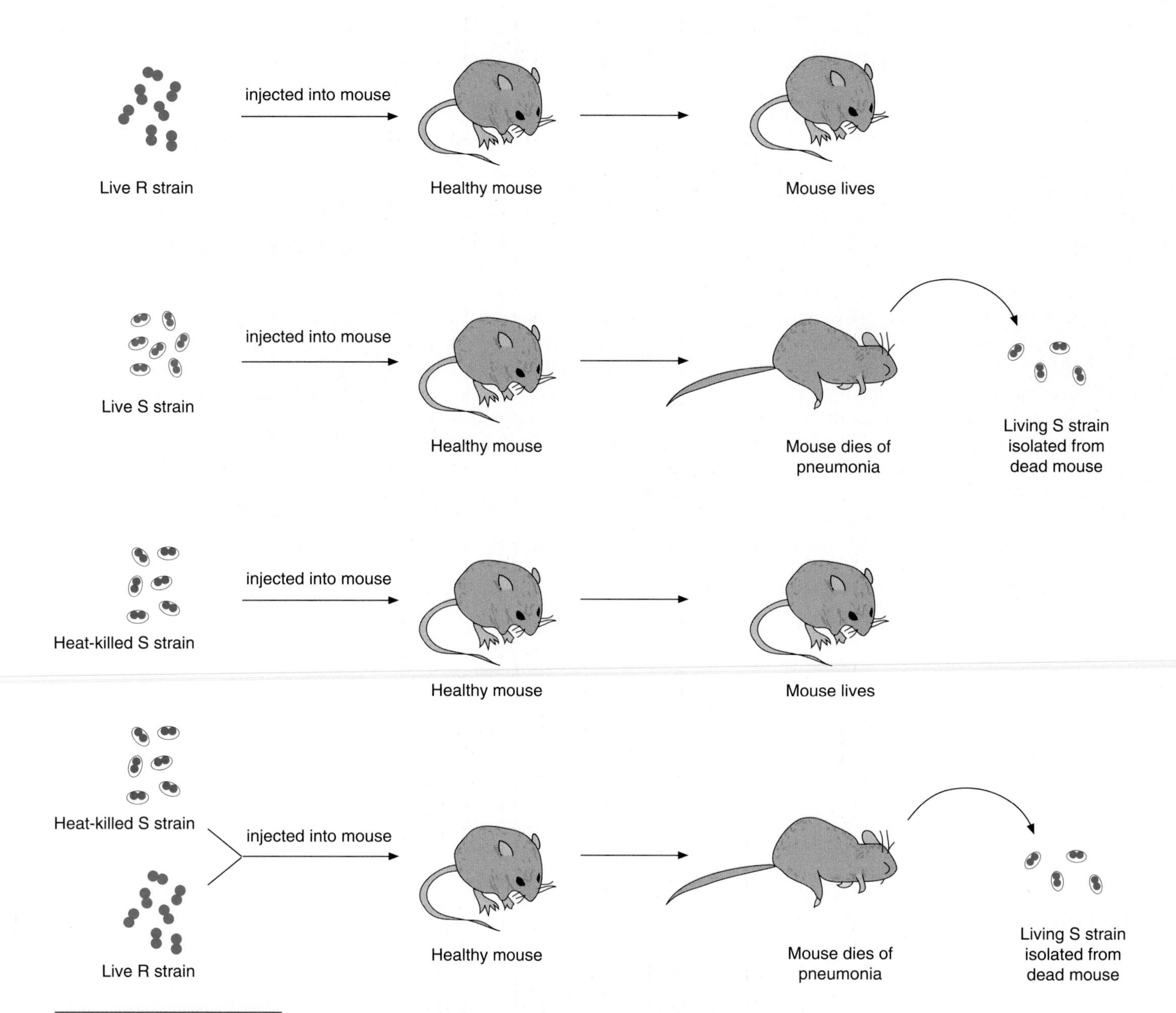

Figure 15.6

Beginning in the 1920's, key experiments were conducted using two strains of *Diplococcus* bacteria, a smooth (S) form and a rough (R) form. In Griffith's first experiment, mice were injected with either the S or R *Diplococcus*. Those injected with the R strain were unaffected, but those receiving the S strain died of pneumonia. In later experiments, mice injected with heat-killed S bacteria lived. However, if those S forms were first mixed with live R bacteria and then injected into a mouse, it died and was found to contain live S forms.

The Studies of Avery, Macleod, and McCarty

Soon after Griffith's publication in 1928, several other scientists confirmed his results. Also, transformations were produced under normal cell culture conditions without the involvement of mice. In the cell culture system, living R bacteria placed in test tubes containing the heat-killed S form became transformed into the encapsulated S pneumococcus. In a clever 1933 study, a method was devised to shatter the cells of heat-killed S bacteria by alternately subjecting them to freezing and heating. The different chemicals obtained from inside the cells were then separated from the cell debris using a special filtration technique. The transforming substance was found to be contained in the group of chemicals isolated from inside the cell. In other words, this extract alone was found to transform harmless R bacteria to the virulent S form.

Much of the work just described was done in the laboratory of Oswald T. Avery, where the transforming principle was the center of attention for over a decade. Beginning in 1935, Avery and his colleagues initiated experiments to try to isolate the active principle and to identify its chemical composition. A long period elapsed before a positive outcome was achieved. However, when the results were finally published in 1944, they left little doubt about the identity of the transforming principle.

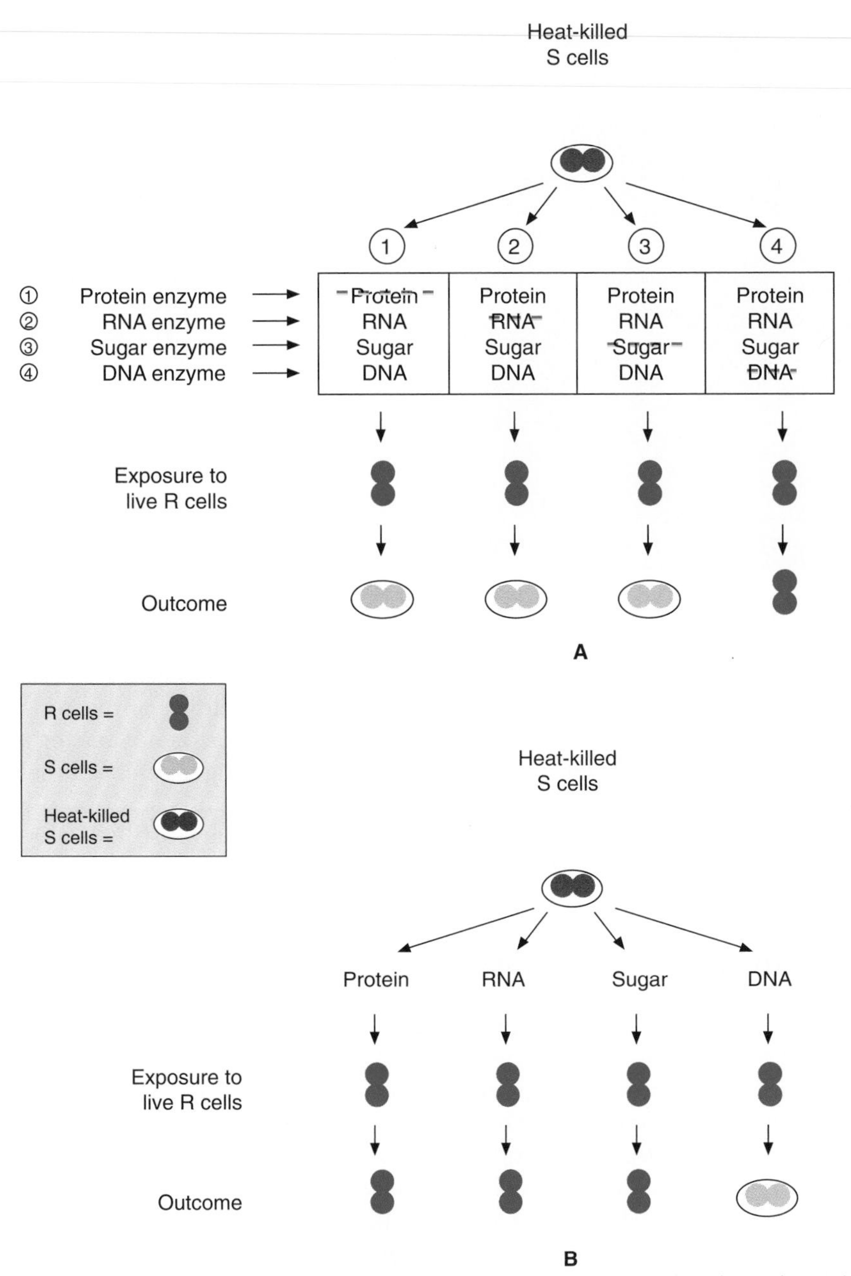

Figure 15.7

In a classic scientific paper published in 1944, Avery, Macleod, and McCarty summarized the results of experiments they conducted with *Diplococcus.* (A) Using suitable enzymes, they sequentially inactivated proteins (1), RNA (2), sugars (3), and DNA (4) from heat-killed smooth (S) cells, while leaving the other molecules intact. All groups of molecules, each containing one inactive type, were then introduced separately into live rough (R) cells; only those receiving inactive DNA failed to become transformed into S cells. (B) In a second approach, live R cells were exposed to purified molecules extracted from S cells. Only those R cells receiving DNA became transformed to S cells. The results of these experiments seemed to answer the question about the chemical identity of the genetic material.

The famous 1944 publication by Avery, Colin Macleod, and Maclyn McCarty combined results from several studies they conducted over many years using the smooth and rough pneumococcus strains. The complete set of results seemed to provide a clear answer about the chemical nature of the transforming principle.

Their initial tactic involved an indirect process of elimination as described in Figure 15.7A. First, proteins, RNA, and sugars were eliminated as candidates as the active material. This was accomplished in different studies where the chemical extract from inside S cells was attacked by specific enzymes. Proteins were chopped apart, yet transformation still occurred. The same result was obtained for RNA and sugars. Thus none of these chemicals could be the transforming principle. However, when an enzyme that split DNA apart was introduced into the extract, transformation no longer took place! What did this suggest?

The second approach, described in Figure 15.7B, involved direct attempts to identify the transforming material by using chemical analytical methods. Using this technique, R cells were exposed to purified protein, RNA, sugar, and DNA extracted from S cells. Only cells exposed to the DNA were transformed. The authors stated that the active material was "a highly polymerized and viscous form of sodium desoxyribonucleate," that is, a form of DNA.

Figure 15.8

(A) Bacteriophages are viruses that infect and kill bacteria. In this electron micrograph, T2 phages appear as black ovals outside and inside an infected bacterium. (B) A diagram of a T2 phage.

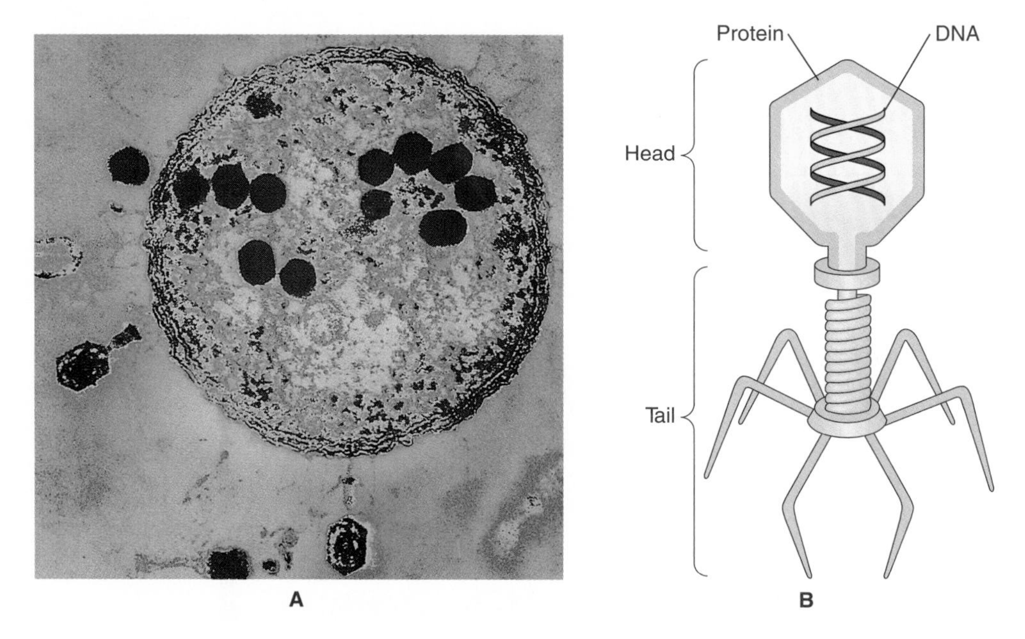

The obvious conclusion from these powerful studies was that DNA was the transforming principle. However, the impact of this report was delayed because of the absence of a solid connection between biochemical studies on bacteria and genetics at that time. For example, it was unclear whether or not bacteria even had chromosomes. Nevertheless, many researchers no longer doubted that DNA was, in fact, the genetic material. More conservative scientists adopted a cautious attitude and awaited further evidence. To some of them, it still seemed possible that proteins might have a hereditary role in other organisms or in other circumstances.

The Hershey and Chase Experiment

By 1952, many molecular biologists were using viruses to learn more about DNA. Viruses that infect bacteria, called **bacteriophages** or simply **phages**, were special favorites (see Figure 15.8). Viruses occupy a twilight zone between living and nonliving forms. They depend on host cells for all phases of their life cycle. Phages have an intricate structure, but chemically they are very simple, consisting primarily of a protein coat that surrounds an inner core of DNA. When a phage encounters a bacterium, it attaches itself and injects DNA into the bacterial cell. The protein portion of the phage remains outside the bacterium, where it appears as a "ghost." The bacterium, unable to distinguish between phage DNA and its own DNA, reproduces phages according to instructions contained within the phage DNA. Eventually, the new phages cause the bacterial cell to rupture, thus releasing a new set of viruses that can then infect other bacteria. This entire cycle, from DNA injection to bacterial rupture, takes only a few minutes. Human viruses operate in much the same way when they infect our cells.

In a brilliant experiment published in 1952, Alfred Hershey and Martha Chase used T2 phages—a type of phage that infects the common bacterium, *Escherichia coli,* which is a normal inhabitant of the intestines of mammals, including humans—to establish beyond doubt the identity of the genetic material. The Hershey-Chase experiments are described in Figure 15.9. The phages were first grown in culture media containing radioactive types of either phosphorus (^{32}P) or sulfur (^{35}S). Since viruses depend on the media for these elements, the radiolabeled phosphorus and sulfur were taken up and incorporated into their proteins or nucleic acids as they were being synthesized. Radioactive phosphorus and sulfur each have unique characteristics that make it possible to trace their fates. Since DNA contains phosphorus but not sulfur, the presence of ^{32}P indicated that DNA was present. Proteins were identified by the ^{35}S label, since they contain sulfur but no phosphorus.

After labeling, phages were introduced into a bacterial culture, and the cycle

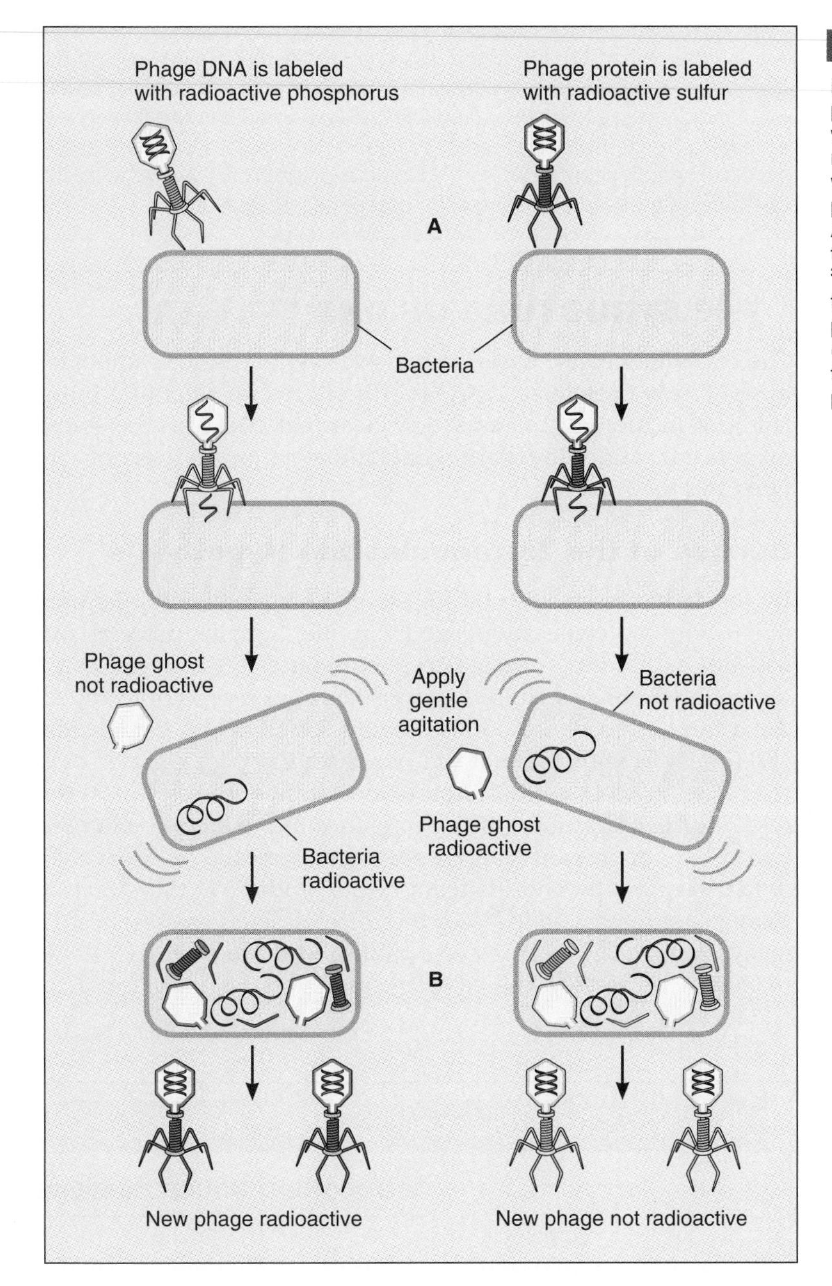

Figure 15.9

Hershey and Chase's famous phage experiment finally settled the question of whether protein or DNA was the genetic material. (A) Phage DNA was labeled with radioactive phosphorus (^{32}P) and proteins with radioactive sulfur (^{35}S). After labeling, the phages infected bacteria. By measuring radioactivity from ^{32}P and ^{35}S, it was found that most of the phage DNA entered the cell but the protein remained with the phage ghost. (B) Progeny phages were found to contain ^{32}P-labeled DNA derived from the parental phages, but almost no ^{35}S.

of infection and phage reproduction was allowed to follow its normal course. By determining which chemical was present in the bacteria after a replication cycle was completed, the true identity of the genetic material would be revealed. Unfortunately, a slight snag developed when at first both ^{32}P and ^{35}S seemed to be incorporated into the bacteria. The researchers then discovered that the phage ghosts (protein coats) remained attached to the bacteria, so a method to remove the ghosts had to be devised. Gentle agitation in a blender was found to detach the ghosts from the bacteria. The resulting mixture was then centrifuged, forcing the bacteria to the bottom while the phage ghosts remained at the top. The ^{32}P label, indicating DNA, was found inside the bacterial cells, whereas the ghost layer contained protein, as indicated by the presence of ^{35}S. Further, phage progeny produced by bacteria originally infected by ^{32}P labeled viruses contained 30 percent or more of the parental phage phosphorus but almost no radioactive sulfur.

This elegant experiment marked the end of any remaining scientific skepticism as to the identity of DNA as the genetic material (at least in this group of viruses). For this and other studies on phages, Hershey received the Nobel Prize in 1969.

A central question had guided research in molecular biology during the first half of the twentieth century: what was the chemical identity of the genetic material? The two leading candidates were proteins and DNA. Due primarily to the flawed tetranucleotide hypothesis, proteins were assumed to be the hereditary substance in the 1920s and 1930s. Then, through a series of classic experiments, DNA was finally identified as the chemical responsible for transmitting genetic information.

THE STRUCTURE OF DNA

The experiments just described answered the question about whether the genetic material was protein or DNA (see the Focus on Scientific Inquiry, "Experimental Methods in Biology"). It was now clear that DNA was the source of genetic information transmitted between generations or, in the case of viruses, between the virus and the host cell.

Demise of the Tetranucleotide Hypothesis

By the 1940s, molecular biologists were aggressively pursuing answers to questions about molecules associated with life. Initially, proteins were of greater interest, but as evidence began to accumulate in favor of DNA as the hereditary molecule, emphasis shifted toward learning more about this complex chemical. What kind of structure could account for all of the genetically determined traits and functions with which it was associated?

The strength of the tetranucleotide hypothesis was seriously weakened as results of experiments on the transforming principle emerged. Edwin Chargaff was deeply impressed with the findings presented by Avery, Macleod, and McCarty in 1944, and he and his team began studies on the structure of nucleic acids. They made careful analyses of the nucleotide composition of DNA obtained from many organisms. Their results, published during the late 1940s, are summarized in Table 15.2. Review the table, examine the data, and formulate any conclusions

Table 15.2 Base Composition of DNA in Several Species

		COMPONENT NUCLEOTIDE BASES (PERCENT)			
Organism	**Cells Analyzed**	**Adenine**	**Thymine**	**Guanine**	**Cytosine**
E. coli	—	26.0	23.9	24.9	25.2
Diplococcus	—	29.8	31.6	20.5	18.0
Yeast	—	31.3	32.9	18.7	17.1
Wheat germ	—	27.3	27.1	22.7	22.8
Sea urchin	Sperm	32.8	32.1	17.7	18.4
Herring	Sperm	27.8	27.5	22.2	22.6
Rat	Bone marrow	28.6	28.4	21.4	21.5
Human	Thymus	30.9	29.4	19.9	19.8
	Liver	30.3	30.3	19.5	19.9
	Sperm	30.7	31.2	19.3	18.8

Summary of results from Chargaff's experiments measuring the nucleotide base composition of DNA from several species.

that are supported by this extremely interesting data set. Do the data support or contradict the tetranucleotide hypothesis? Why?

Two conclusions that were drawn from Chargaff's data struck at the heart of the tetranucleotide hypothesis. First, the DNA in all species does not contain equal amounts of the four bases, and second, the base composition of DNA differs from one species to another. These conclusions effectively demolished whatever support remained for the obsolete tetranucleotide hypothesis. It had became obvious that DNA is complex and highly variable in structure.

It was also apparent that the base composition of DNA from different cells of the same species was characteristic and constant. Finally, the data indicated a consistency in the base composition of all DNAs that became known as *Chargaff's rule:* the amount of adenine is always approximately equal to the amount of thymine (A = T), and the amount of guanine equals the amount of cytosine (G = C). At the time, no hypothesis could explain this observation. (The data in Table 15.2 indicate that the A-T, G-C amounts are not absolutely equal because techniques used at that time did not permit perfect measurements.)

In Pursuit of DNA's Structure

By the early 1950s, interest in DNA was in full blossom. No one doubted the genetic role of DNA any longer, and attention became directed toward determining the mechanism whereby DNA accomplished the transfer of information. Exactly how did DNA operate? To answer this question, the fundamental structure of the molecule would have to be determined.

Scientists from many fields, including many members of the first generation of molecular biologists, were drawn to the challenge of accomplishing this difficult task. The cast of characters that became involved in this venture included many celebrated scientists of the period, located primarily in the United States and England. Yet success was finally achieved by two young, ambitious scientists, James D. Watson and Francis Crick, who had not yet acquired strong reputations. The story of their discovery of the structure of DNA in 1953 is fascinating and has been the subject of numerous books and at least one film.

Watson, a young Ph.D. from the United States, and Crick, a brilliant Ph.D. candidate, came to share a lab at Cambridge University in Cambridge, England, in 1951 (Figure 15.10). Both were interested in determining the structure of DNA,

Figure 15.10

In 1953, using unconventional methods, Watson (left) and Crick (right) identified the fundamental molecular structure of DNA, one of the most important accomplishments in the history of biology.

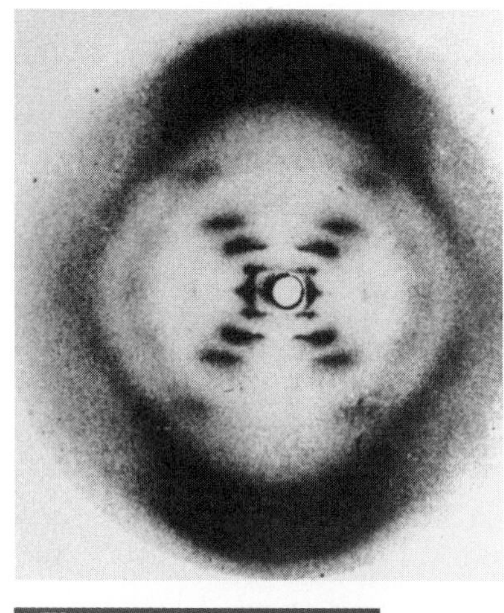

Figure 15.11

X-ray crystallography was a powerful tool used in identifying the structure of DNA. This X-ray diffraction photograph of DNA, taken by Franklin, was apparently used by Watson and Crick in determining the structure of DNA.

and each brought with him unique skills that were useful in attacking the problem. Their approach did not involve conventional experiments. Rather, they used a combination of strategies, including an application of existing principles and assumptions from quantum physics; the construction of models with cardboard cutouts, wire, and sheet metal; and guesswork. In addition, they exploited vital information obtained from others.

Uncovering the three-dimensional structure of DNA presented substantial technical challenges. Not even large molecules, such as DNA, could be visualized with existing microscopes. Special instruments and techniques were used to acquire data that could provide hints about the structure of DNA. The most important clues were obtained using a technique called *X-ray crystallography*. In this method, a crystal of the purified chemical of interest is exposed to X-rays. The X-rays are diffracted in specific ways characteristic of the crystal's structure. The resulting pattern is then captured on photographic film (see Figure 15.11). Examination and analysis of the photograph by expert crystallographers provided details about the precise structure of the crystal DNA.

Neither Watson nor Crick had exceptional expertise in crystallography, yet they fully realized the importance of obtaining X-ray diffraction pictures of DNA. Meanwhile, a major effort to obtain such pictures was being made at King's College in London by Maurice Wilkins and Rosalind Franklin, both acknowledged crystallography experts, who were also interested in determining the structure of DNA (Figure 15.12). In early 1953, Watson obtained a print of one of Franklin's best photographs. The events surrounding that acquisition, the subsequent use Watson and Crick made of the picture, and their failure to acknowledge its critical importance in their success remain controversial. It later became apparent, however, that Franklin and her co-workers had been very close to determining the structure of DNA.

Within a week after obtaining Franklin's picture, Watson and Crick began a feverish month of model building that ended on March 7, 1953, when a full DNA model had been pieced together. It conformed perfectly with the X-ray measurements and the requirements imposed by quantum physics and chemical principles. Their accomplishment, elucidating the three-dimensional structure of DNA, has been described as one of the greatest achievements of twentieth-century biology.

In 1962, Watson, Crick, and Wilkins received a Nobel Prize for their work on DNA. Franklin would probably have shared in this honor had she not died of cancer before the award was bestowed (these prizes are awarded only to living scientists).

Figure 15.12

Rosalind Franklin was an expert in X-ray crystallography. One of her diffraction photographs of DNA proved crucial in determining its structure.

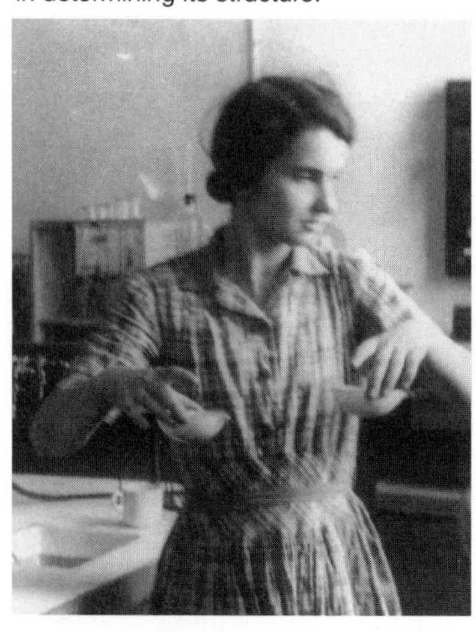

The Watson-Crick DNA Model

The DNA molecule described by Watson and Crick in their historic paper is shown in Figure 15.13A. It consists of two strands twisted into the shape of a **double helix**, much like a ladder twisted about its long axis. The backbone of each strand (the upright of the twisted ladder) consists of long chains of nucleotides, bonded sugar and phosphate groups each with one of the four bases (A, T, G, or C) attached and projecting inward. The two strands are held together (appearing as rungs of the twisted ladder) by hydrogen bonds, a weak type of chemical bond, which join the bases from one strand to the bases from the other strand. Also, the strands are *antiparallel* because they run in opposite directions, relative to their sequence of bases (see Figure 15.13B). The key to the Watson-Crick model is that the bases always pair in the same way, A with T (or T with A) and C with G (or G with C), partly because of hydrogen-bonding requirements. Three hydrogen bonds can form between G and C, but only two between A and T. The base pairs (A-T, G-C) are called *DNA complementary base pairs*. Note that the underlying basis for Chargaff's rule—equal amounts of A and T and of C and G in DNA—is now evident.

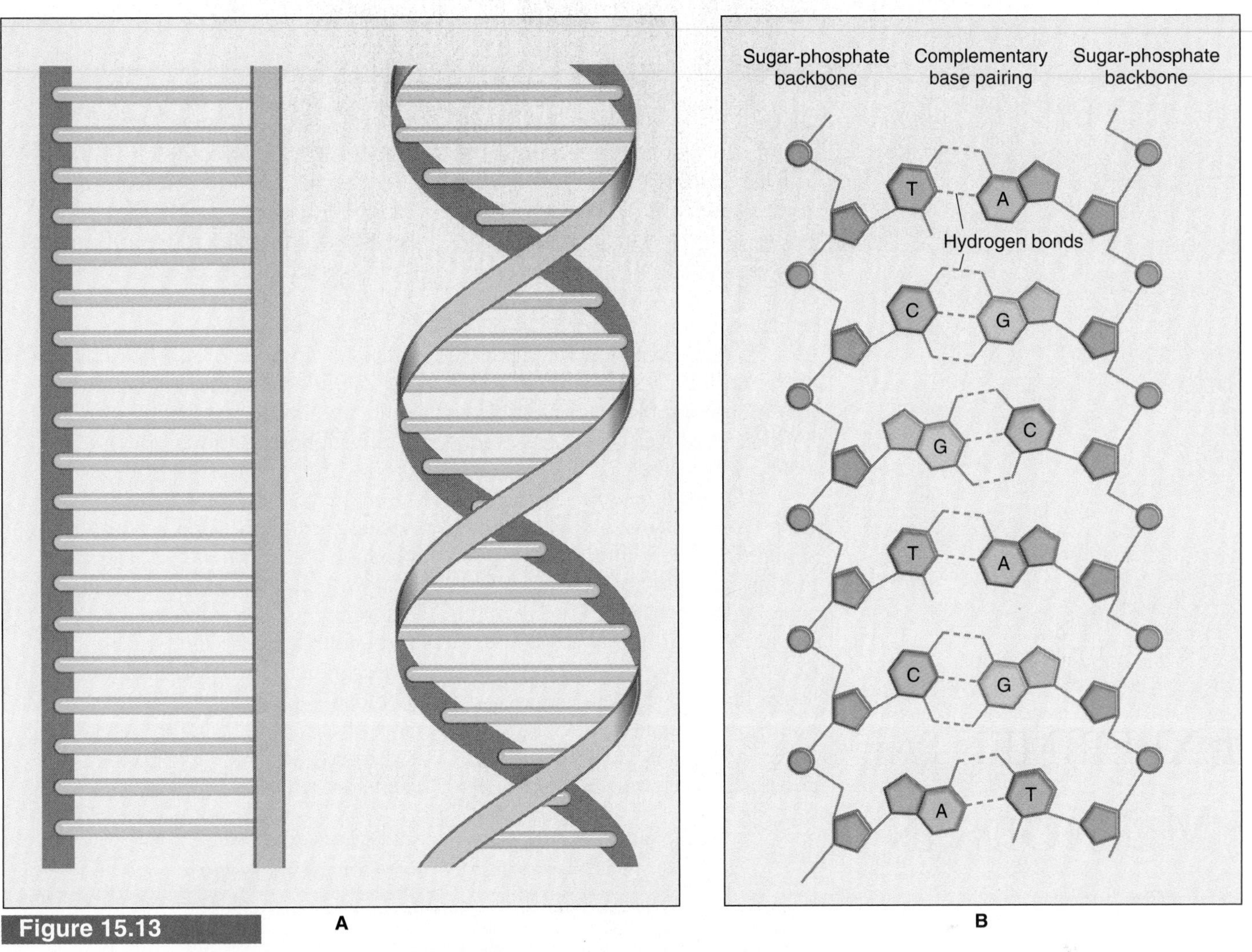

Figure 15.13

(A) As described in Watson and Crick's historic 1953 paper, the basic structure of DNA is a double helix in which two strands (rails of the ladder), connected by hydrogen bonds (steps of the ladder), are twisted around each other. (B) Each DNA strand consists of a long chain of different nucleotides. The two strands are held together by weak hydrogen bonds that exist between complementary base pairs of each strand.

Several features of the Watson-Crick DNA model are of interest. The two antiparallel strands are not identical to each other either in the base sequence or in composition. Rather, they are complementary to each other with A always opposite T and G opposite C. Any linear sequence of bases can exist in a DNA molecule; there are no restrictions except that the composition of one strand is determined by the sequence of the other because of complementary base pairing. This last point is related to the most fascinating feature of the Watson-Crick model; it hints at how chromosomes—and DNA—become duplicated before cell division. Indeed, near the end of their paper is a sentence, apparently put in at the insistence of Crick, that is considered one of the coyest statements in the literature of science: "It has not escaped our notice that the specific pairing we have postulated immediately suggests a possible copying mechanism for the genetic material."

After DNA was identified as being the genetic material, the next challenge facing molecular biologists was to determine its structure. After an exciting race, James Watson and Francis Crick described the chemical structure of DNA in 1953. The DNA molecule has the shape of a double helix, with two chains of nucleotides held together by hydrogen bonds formed between complementary base pairs.

EXPERIMENTAL METHODS IN BIOLOGY

Focus on Scientific Inquiry

Experimentation in science consists of a set of procedures designed to accomplish certain goals (Figure 1). The experimental methods used in biology are certainly among the most powerful tools available to modern science for addressing questions and investigating problems. These methods can be categorized into three main groups, depending on the purpose of the experiments: manipulation for extending observations, deciding between two hypotheses, and testing a hypothesis.

Manipulation for extending observations is the oldest and most straightforward experimental method. Until 200 years ago, researchers interested in the living world relied primarily on simple powers of observation. They looked to see what species of plants and animals were found in different regions, described them, and classified them. In attempting to understand how the human body worked, they dissected bodies, observed the parts, and tried to figure out what each did.

Slowly the idea of manipulating nature to "see what would happen" attracted a number of scientists, particularly those interested in physiology (the study of function). For example, the most famous naturalist of the first half of the eighteenth century, René Réaumur, was interested in digestion. He thought that it might be associated with chemical reactions, so he let a chicken swallow a small sponge, which he later retrieved. He then showed that the gastric juices obtained from the sponge reacted chemically with several different substances. The Griffith experiment on the two forms of *Diplococcus* is an example of this sort of experiment, as are Avery's successive attempts to find the transforming principle.

Figure 1 A young scientist assisting in an experiment.

Deciding between two hypotheses was a method first used with great success in the physical sciences. In the eighteenth century, naturalists attempted to apply this method to problems of the living world. One of the most elegant applications was by the Rev. Stephen Hales, whose book *Vegetable Staticks* is a classic in the development of physiology. One of the central tasks he tackled in that book was to demonstrate whether or not plants circulate their sap in the same way that animals circulate blood. After discovering the circulation of blood in animals, many naturalists assumed that a similar circulation must exist in plants. Thus the two hypotheses were (1) that plants circulated sap in the same way that animals circulated blood and (2) that they did not. Hales demonstrated that such circulation does not exist in plants.

The Hershey and Chase experiment was a brilliant example of an experiment that would settle whether protein or DNA was the genetic material. Since the protein and the DNA were marked with radioactive atoms and were the only possible agents involved in the experiment, the results were unequivocal.

Testing a hypothesis has become the principal experimental method in modern biology. However, there are many different ways of testing an explanatory idea. The formulation of an observable prediction or set of predictions is basic to most approaches. In Chapter 14, the famous white-eyed fruit fly experiments were described. We examined Morgan's derivation of predictions and his experiments and how his experiments were designed to test their accuracy.

One of the things that makes science such an exciting field is that each problem a scientist faces is new and usually calls for novel experimental designs to solve it. Rarely are two problems solved in the same way, so scientists have to be very creative.

Another important feature of science is that it is dynamic. Scientists don't just manipulate the environment (stuffing sponges down chicken throats) or test hypotheses and rest contentedly. Scientists perform experiments to attempt to understand the world from a particular perspective (for example, digestion as a chemical process) or to try to solve some general problem (for example, nature of the genetic material). Occasionally, a "classic" experiment convinces scientists that a particular hypothesis is very powerful and has great predictive value. However, experiments usually lead to other experiments by extending or modifying hypotheses. In the course of an experiment, fresh results may lead an investigator to reformulate initial ideas. Negative results might lead the investigator to reject the basic hypothesis and formulate a new one. In general, experimental research is a process that goes from one hypothesis to another in a branching fashion.

Although experimental methods often have different goals, methods, and procedures, they do have certain common features. The most important of these are controls, isolation of variables, concern with sample size, and analysis of data (usually statistical).

A *control* is a basic feature of experiments, a sample that is not manipulated. The control describes what happens in an untreated case. Without a control group, there is no way to know if what occurred did so because of the manipulation or because it would have happened anyway. For example, if scientists were interested in determining the effect of zinc on plant growth, groups of plants would receive different concentrations of zinc. The control in this experiment would be a group of plants that received no zinc.

Isolation of variables is often the trickiest part of experimental design. In trying to explain a certain phenomenon, it is always difficult to determine if more than one factor is operating, either to produce the effect or to influence it. Scientists try to simplify their experiments as much as possible so as to evaluate single influences (variables). Several instances in this chapter demonstrate how one can easily be misled by not considering other variables. The tetranucleotide hypothesis, for example, by focusing on only one aspect of DNA (the claim of equal *amounts* of four bases), overlooked the most interesting feature of DNA, its highly variable structure.

Sample sizes of an adequately large number are necessary to ensure that the experimental result obtained is real and not due to chance. For example, in Griffith's experiments, many mice, not just one or two, were exposed to the two different bacteria to determine which form, S or R or both, actually caused pneumonia. If only one or two mice were tested and developed pneumonia, there might be some uncertainty about the cause. If two groups of 50 mice each were exposed to each type of bacteria, and none (or one) in one group and 48 or 49 of the second group developed pneumonia after exposure, a final conclusion about the cause of pneumonia could be made with some confidence.

Analysis of data is required to determine whether or not the results obtained in an experiment are "significant." Statistical analysis makes use of complex mathematics. Such analysis provides researchers with objective infor- mation that enables them to make decisions about the meaning of their data and to formulate conclusions. A result is *statistically significant* if it is shown by mathematical analysis to be associated with a low probability of error (or uncertainty).

For example, let's assume that in Griffith's experiment, two groups of 50 mice were used. The experimental group would have been exposed to the combination of live R bacteria and heat-killed S bacteria, and the control group would not have been exposed to either. What hypothesis would be tested? Statisticians demand a concise statement of the hypothesis to be tested and require an accounting of all possible outcomes. To satisfy these criteria, scientists create a **null hypothesis** (H_0), which is a statement of "no difference," and an **alternative hypothesis** (H_A), which is accepted if H_0 is rejected. In this case, the H_0 would be that exposed mice would get pneumonia at the same rate as the control mice. When the experiment was conducted, most of the exposed mice developed pneumonia (say, 41/50) compared to none of the controls (0/50). Should the H_0 be *accepted* ("no difference") or *rejected* ("significant difference")? Statisticians ask, Is it likely that the observed difference (41/50 versus 0/50) could occur in mice sampled randomly from a population? If it could occur quite often, there may be no basis for rejecting the H_0. If it could almost never occur, the H_0 would be rejected and the H_A would be accepted. The next question asked is, How small a probability should be required for rejecting the null hypothesis? This is determined by statistical analysis, and the accepted level for rejection is 5 percent; that is, there is less than a 5 percent chance that if you rejected the H_0, you would be wrong. In our example of Griffith's study, the results would be statistically significant, the H_0 would be rejected, and the alternative hypothesis ("significant differences between the two groups") would be accepted.

Experiments are often repeated by different scientists to confirm a significant finding from an original study. If the results of an experiment are found to be both statistically significant and repeatable, the hypothesis is strengthened or confirmed. What if, in Griffith's study, only one or two out of 50 exposed mice had developed pneumonia? Such a result would be *not statistically significant* and would not have supported rejection of the null hypothesis. When this type of result is obtained, new experiments may be conducted, and the hypothesis may be modified or even abandoned.

DNA REPLICATION

Recall that prior to mitosis and meiosis, chromosomes are duplicated during interphase. The basis of chromosome duplication is the **replication**, or synthesis, of DNA. A new question now occupied researchers: How was DNA replicated in the nucleus?

The Watson-Crick Hypothesis

In a second paper that followed soon after the first, Watson and Crick proposed that by breaking the weak hydrogen bonds, the two parent (original) strands of the double helix could unwind and serve as templates for constructing new daughter strands. The two new strands of DNA, each consisting of a parent combined with its daughter, would be identical. This type of replication, described in Figure 15.14, is called *semiconservative* because each new DNA molecule contains one parent strand and one newly synthesized daughter strand.

Shortly after this paper appeared in 1953, Watson and Crick developed new interests, and their celebrated collaboration ended. In the years since, Crick has become established as a dominant intellectual force in the field of experimental molecular biology. Watson became director of the prestigious Cold Spring Harbor laboratories on Long Island, New York, and has also written several important books on molecular biology (see also Chapter 21). Both have written delightful books that are recommended for anyone interested in their accounts of the DNA story (Watson published *The Double Helix* in 1968, Crick *What Mad Pursuit* in 1988).

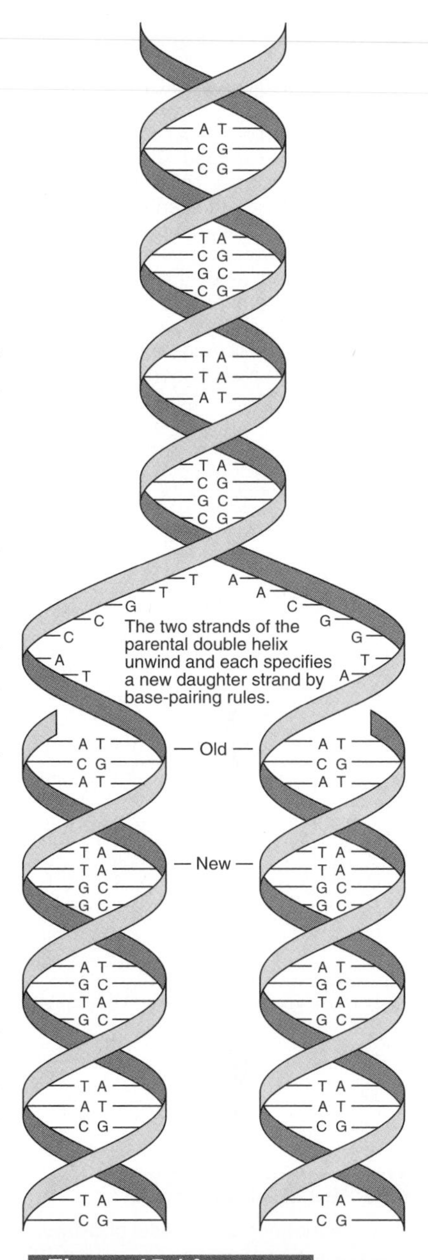

Figure 15.14

This simple model describes the basic process of DNA replication. The two parent strands separate, and each serves as a template for forming a new daughter strand. This type of replication is called semiconservative because each new DNA molecule consists of one "old" (parent) strand and one "new" (daughter) strand.

The Replication Process

In 1958, scientists confirmed the essential correctness of Watson and Crick's 1953 outline for semiconservative DNA replication. Further studies completed since that time have provided full details about how DNA is synthesized.

DNA replication is described in Figure 15.15. It begins when hydrogen bonds linking the base pairs are broken and the two parent strands of DNA partially unwind. **DNA polymerases**, enzymes that copy DNA by joining complementary bases, then move along each opened strand, bonding appropriate base pairs together. The parent base sequence serves as a template for determining the precise arrangement of the new daughter chain. Regular complementary base pairing occurs as the replication proceeds; A pairs with T and C with G.

Several features of the replication process are noteworthy. Prior to replication, the cell synthesizes the proper number of nucleotides that will be required in the construction of new DNA strands. Replication begins at specific positions, or regions, in the inner portion of the DNA chain (never at the end) and then proceeds in both directions away from the *initiation site*. The number of initiation sites ranges from one in small bacterial chromosomes to hundreds or thousands in the comparatively huge eukaryotic chromosomes. During replication, DNA polymerases race along the two DNA strands in opposite directions. Along one parental strand, the "leading" strand, the chain is copied continuously and smoothly. Replication proceeds in the opposite direction along the other "lagging" strand. Here numerous short segments of the new DNA strand are produced simultaneously by many polymerases and then pieced together and joined to the template strand by other enzymes. In eukaryotes, replication progresses at a rate of about 50 base pairs per second.

Complementary base-pairing rules dictate the organization of the new strand formed from the parental template. Thus each daughter strand is identical to the original parental strand that did not serve as its template. Consequently, the two new double helices are identical to each other and also to the original parent DNA molecule. Since DNA is the intergenerational genetic link, the in-

Figure 15.15

A more detailed model of DNA replication. Replication begins when DNA parent strands partially unwind, and the initiation site is exposed. DNA polymerases then add complementary bases to each parent strand, creating a daughter strand. Along the leading strand, bases are added continuously until completion; along the lagging strand short segments are added at several points and then pieced together. The newly replicated DNA molecule consists of one parent strand and one daughter strand.

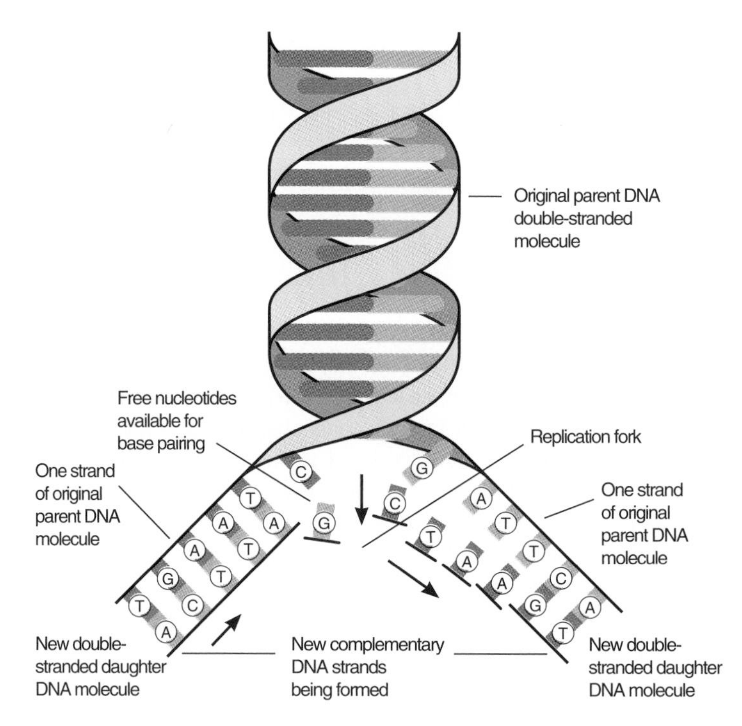

tegrity of the base sequence must be maintained through correct base pairing during replication. To ensure the accuracy of newly formed DNA molecules, cells of eukaryotes and prokaryotes have "proofreading" molecular mechanisms that pinpoint and correct errors in base pairing. In *E. coli,* replication errors, after correction, occur at an amazingly low frequency—one mistake per every billion pairs bonded!

DNA replication occurs before cell division and the transmission of genetic information between generations. The basic process begins when the two parent strands of DNA open, followed by the construction of two daughter strands, by complementary base pairing, using the parent strands as templates. The end products of this semiconservative replication process are two identical DNA molecules, each consisting of one parent strand and one new daughter strand.

Figure 15.16

A computer-generated model of DNA.

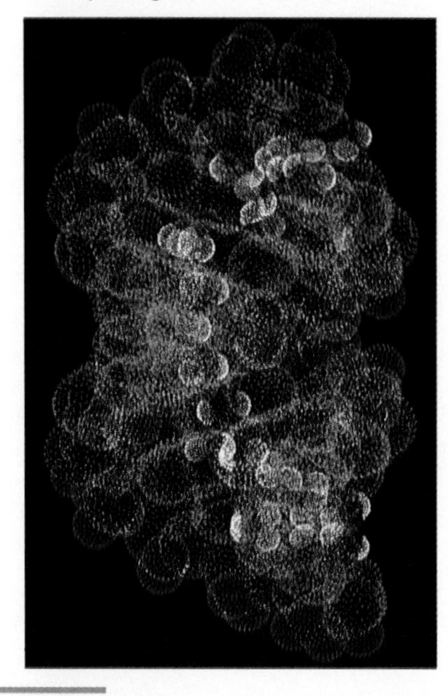

Finally, most of what is known about DNA replication has been derived from studies of bacteria. Considerably less is understood about DNA synthesis in eukaryotes. Nevertheless, cells of all organisms appear to produce copies of their DNA in essentially the same manner (Figure 15.16). The universal possession of such a critically important biochemical mechanism is consistent with the concept of unity in biology, that all life holds in common features that developed early in the history of life on Earth.

Summary

1. Following confirmation of the chromosome theory, the next fundamental problem to be addressed in genetics was determining the molecular substance associated with genes and chromosomes. The leading candidates were protein and DNA.
2. In experiments using bacteria that cause pneumonia, Griffith discovered that a "transforming principle" could convert one strain of bacteria to another.
3. Several ingenious experiments were conducted to identify the substance that could transform bacteria. Ultimately, it was determined that DNA, not protein, was the genetic material.
4. The next problem that generated intense interest was determining the structure of DNA. Using nontraditional scientific methods, Watson and Crick discov-

ered that DNA had the structure of a double helix. Attention then shifted to the process of DNA replication.

5. Watson and Crick proposed that parent DNA strands served as templates for producing new daughter strands, a process termed semiconservative replication. That hypothesis was later confirmed.

Review Questions

1. Why are microorganisms used in genetics studies?
2. What are the roles of proteins in cells?
3. What was the significance of Miescher's studies?
4. What is the tetranucleotide hypothesis? What did it contribute to identifying the genetic material?
5. What was concluded from Griffith's *Diplococcus* experiments?
6. How did the studies by Avery, Macleod, and McCarty and by Hershey and Chase prove that DNA was the genetic material?
7. What was the significance of Chargaff's experiments?
8. What methods did Watson and Crick use in determining the structure of DNA?
9. How is DNA constructed? What features of DNA structure are important in its replication?

Essay and Discussion Questions

1. Why is the Hershey-Chase experiment considered a classic in science?
2. Is it possible that another tetranucleotide-type hypothesis—one that restrains progress in a field—could appear today? Why?
3. What might explain the emergence of the field of molecular biology in the first half of the twentieth century?

References and Recommended Readings

Avery, O. T., C. M. Macleod, and M. McCarty. 1944. Studies on the nature of the substance inducing transformation of pneumococcal types. *Journal of Experimental Medicine* 79:137–158.

Crick, F. 1988. *What Mad Pursuit: A Personal View of Scientific Discovery.* New York: Basic Books.

Davis, B. D. 1980. Frontiers of the biological sciences. *Science* 209:78–89.

Gribbon, J. 1985. *In Search of the Double Helix: Quantum Physics and Life.* New York: McGraw-Hill.

Griffith, F. 1928. The significance of pneumococcal types. *Journal of Hygiene* 27:113–159.

Hershey, A. D., and M. Chase. 1952. Independent functions of viral protein and nucleic acid in growth of bacteriophage. *Journal of General Physiology* 36:39–56.

Judson, H. F. 1979. *The Eighth Day of Creation: The Makers of the Revolution in Biology.* New York: Simon & Schuster.

Magasanik, B. 1988. Research on bacteria in the mainstream of biology. *Science* 240:1435–1438.

McCarty, M. 1985. *The Transforming Principle: Discovering That Genes Are Made of DNA.* New York: Norton.

Moore, J. 1986. Science as a way of knowing: Genetics. *American Zoologist* 26:583–747.

Stent, G. S., ed. 1980. *The Double Helix. A Norton Critical Edition.* New York: Norton.

Watson, J. D. 1968. *The Double Helix: A Personal Account of the Discovery of the Structure of DNA.* New York: Athenum.

Watson, J. D., and F. H. C. Crick. 1953. Genetical implications of the structure of deoxyribonucleic acid. *Nature* 171:964–967.

———. 1953. Molecular structure of nucleic acids: A structure for deoxyribose nucleic acid. *Nature* 171:737–738.

CHAPTER 16

The Nature of Genetic Information

Chapter Outline

The Genetic Role of DNA

Protein Synthesis

A Closer Examination of Some Particulars

Focus on Scientific Explanations: Ancestors

The Language of Life

Reading Questions

1. Which early studies provided insights into the genetic role of DNA?
2. What is the relationship between metabolic pathways and genes?
3. How are proteins synthesized in eukaryotic cells?
4. What is the genetic code?
5. What is the explanation for the existence of exons and introns?

The wondrous discoveries recounted in Chapter 15 opened rich new pastures of scientific research in molecular biology. In the 1950s and 1960s, queries about DNA were always at the heart of these investigations. Foremost among the questions was, What information is actually transmitted from generation to generation? There were other questions of interest. How is DNA connected with the formation of a new individual, with all of the characteristics of its parents, and with properties of its species? How does DNA regulate the universe of activities carried on in cells that is necessary for life? The search for answers to these last two queries continues as we approach the twenty-first century.

THE GENETIC ROLE OF DNA

Even before DNA's structure and its replication mechanism were described in the 1950s, earlier research had provided scientists with major clues about the nature of DNA and how it functioned in the cell.

One Gene, One Enzyme

In 1902, Archibald Garrod, an English physician, published a paper in which he stated that a human disease (alkaptonuria) was inherited as a Mendelian recessive trait. Garrod continued his research on human hereditary diseases caused by recessive genes, and in 1908 the results were published in a classic book titled *Inborn Errors of Metabolism*. What disorders are caused by such recessive genes? Garrod described several, including **alkaptonuria**, a disorder characterized by the person's urine turning black after it is exposed briefly to air. Though undoubtedly horrifying to parents changing diapers, alkaptonuria has no harmful effects on babies. Later in life, however, it commonly results in a form of arthritis (swelling of joints). Garrod also discussed **albinism**, the condition of having no pigment in certain cells, which results in milky-colored skin, white hair, and pinkish eyes (see Figure 16.1), and *phenylketonuria,* or *PKU*. **PKU** is distinguished by abnormally high levels of an amino acid, phenylalanine, in the blood of newborn babies and can result in severe mental retardation. Garrod hypothesized that each of these inherited disorders was due to a missing enzyme. Further, he made a bold proposal that a single gene pair specified the enzyme and that homozygous recessives lacked the particular gene associated with the enzyme.

Several aspects of Garrod's work are fascinating. In common with Mendel, his studies were scientifically premature. Recall that at this time (1902–1909), when Garrod was proposing how a gene acted in the cell, Sutton had recently articulated the chromosome theory of heredity, Morgan was just beginning the fruit fly studies that led to confirmation of the chromosome theory, and nothing was known about the identity of the genetic material. It is also interesting that Garrod's studies dealt with humans, a species rarely encountered in the genetics story that unfolded over a 90-year period, from Mendel to Watson and Crick. Finally, Garrod's book reveals that the field of biochemistry was greatly advanced at that time in comparison to genetics.

Nearly three decades after Garrod's work, in the late 1930s, George Beadle and Edward Tatum conducted a classic study that employed a simpler genetic system. They used a fungus, the pink bread mold (*Neurospora*) shown in Figure 16.2A, that could be grown easily and rapidly in the laboratory on a medium containing only a few simple substances. The fungal cells could synthesize all required amino acids, vitamins, and other molecules from the basic chemicals contained in the simple growth medium. In their experiments, described in Figure 16.2B, Beadle and Tatum first exposed this microorganism to X rays and ultraviolet light, agents known to induce mutations in genes. After exposure, they found mutant *Neurospora* that could survive only if the simple growth medium was augmented with certain biochemicals. Some of the mutants required supple-

Figure 16.1

Albinos lack pigments in their hair, eyes, and skin. This phenotypic effect is caused by an "inborn error of metabolism" associated with a recessive gene.

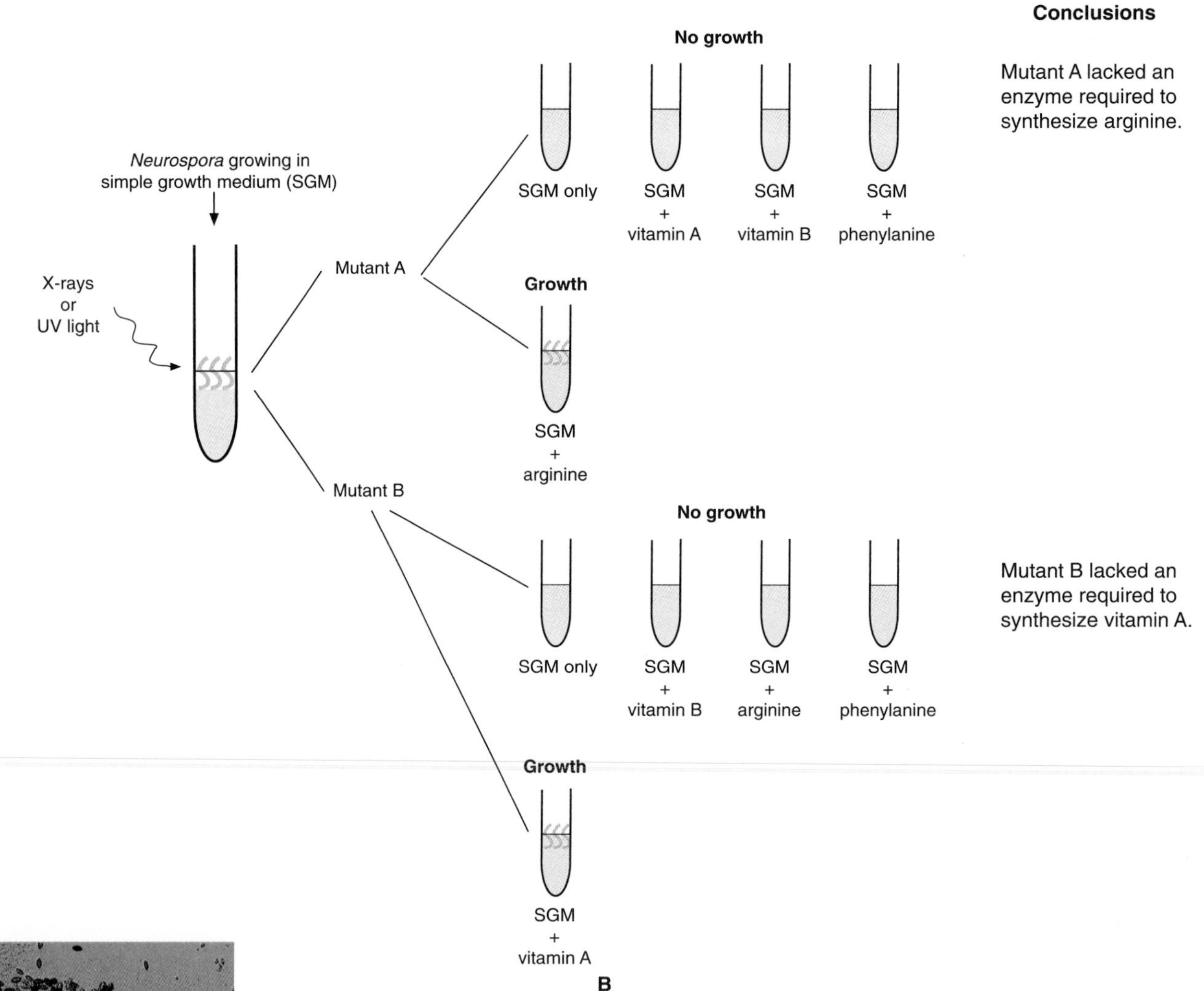

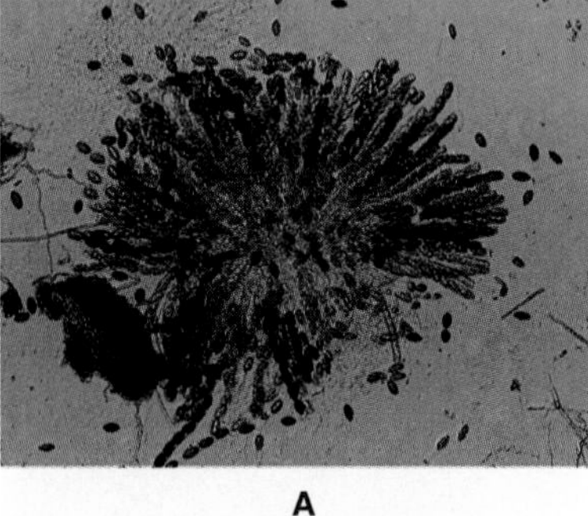

Figure 16.2

(A) *Neurospora* is a simple fungus that proved useful in early genetics studies designed to identify the role of genes in organisms. (B) Beadle and Tatum induced mutations in *Neurospora* by exposing it to radiation that caused mutations. In analyzing the effects of these mutations, they found that affected *Neurospora* required a specific chemical, such as arginine or vitamin A, be added to the simple growth medium for growth to occur. Beadle and Tatum concluded that mutants lacked a necessary enzyme for synthesizing the required substances and that genes were responsible for producing these enzymes. These experiments gave rise to the one-gene, one-enzyme hypothesis.

mental amino acids such as arginine; others required vitamins or other nutrients. In each case, the need for a specific substance was shown to be due to a missing enzyme that, if present, would have promoted synthesis of the required substance. In 1941, Beadle and Tatum concluded that each biochemical reaction in *Neurospora* was catalyzed by a single enzyme and that each enzyme was specified by one gene. This concept became known as the **one-gene, one-enzyme hypothesis**. For their discoveries, they received the Nobel Prize in 1958.

Two early investigations provided important information about the possible role of genes in cells. Between 1902 and 1908, Garrod reported that certain human hereditary disorders appeared to be associated with recessive genes. Individuals who were homozygous recessive for a specific gene lacked an enzyme and were unable to process phenylalanine normally. In the late 1930s, Beadle and Tatum studied the effects of gene mutations in *Neurospora.* They found that mutations resulted in missing enzymes. This finding led to the one-gene, one-enzyme hypothesis.

Metabolism

Having described, in a general way, the results of two historically important research projects, we now introduce metabolism in order to analyze the full significance of these studies. Within a living cell, thousands of different molecules are continuously synthesized from simple precursor molecules while others are being broken down or modified. **Anabolism** refers to chemical reactions involved in the synthesis of molecules (for example, the synthesis of proteins). **Catabolism** includes all chemical reactions by which a cell breaks down molecules (for example, sugar broken down into water and carbon dioxide). **Metabolism** refers to *all* chemical reactions involved in the synthesis and degradation of molecules within a living cell. A wide variety of biochemical reactions enable cells to carry on activities that include the breakdown of energy-rich molecules and the synthesis of cellular components such as structural proteins, enzymes, hormones, and DNA. Cellular metabolism is involved in every aspect of an organism's structure and function.

A **metabolic pathway** is a complete sequence of chemical reactions in which a molecule becomes progressively modified. Some metabolic pathways consist of only a few reactions; others may involve hundreds of stepwise reactions. Each reaction is specifically catalyzed by a single enzyme. The basic types of metabolic reactions include **synthesis** (simple substances *A, B, C,* and so on are combined to form a final product, *A* + *B* + *C* Æ *X*), stepwise **modification** of a chemical (*M* Æ *N* Æ *O* Æ *P* and so on), or **breakdown** into different products (*X* Æ *Y* + *Z*); each arrow represents a chemical modification. Each reaction typically requires the presence of a specific enzyme. Given this background, how can Garrod's work, which centered on missing enzymes, be interpreted? How can this same reasoning be applied to Beadle and Tatum's studies?

A small part of a metabolic pathway is shown in Figure 16.3. Phenylalanine normally undergoes a series of modifications that ultimately lead to much of it

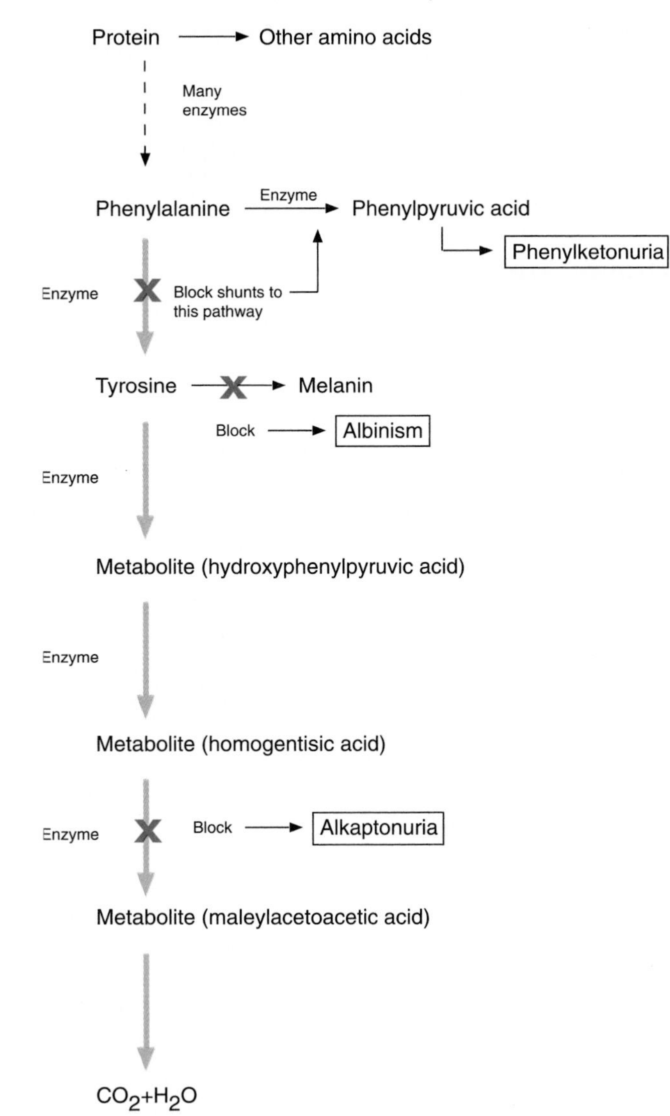

Figure 16.3

Metabolic pathways describe the sequence of chemical reactions in which a molecule becomes modified. This partial metabolic pathway describes both the normal breakdown of the amino acid phenylalanine and the phenotypic effects caused if specific enzymes are missing at different parts of the pathway.

being broken down into carbon dioxide and water. As indicated in Figure 16.3, some of it is converted into tyrosine, another amino acid, which is then transformed into **melanin**, the pigment that gives color to our skin, hair, and eyes. What happens if one of the enzymes of this pathway is not present or not functional? Garrod accurately described three possible consequences:

1. A substance would increase in concentration since it would not be broken down by an enzyme (alkaptonuria).
2. A needed compound would not be produced in the absence of a required enzyme (albinism).
3. Excesses of unwanted substances would be produced from compounds that would accumulate if no enzyme was present (PKU).

In PKU, the major pathway—phenylalanine converted to tyrosine—is blocked, and a minor pathway—phenylalanine converted to phenylpyruvic acid—becomes dominant. This leads to elevated levels of phenylalanine in the blood and phenylpyruvic acid in the urine; both of these substances have been implicated in causing mental retardation in untreated individuals. Fortunately, once the basis of this disorder was understood, screening tests for detecting phenylalanine and its **metabolites** (products of metabolism) in urine of newborn babies were developed and are now required by law in most states. An infant who tests positive is treated by being placed for a few years on a special diet low in phenylalanine to aid in intellectual development.

Metabolism refers to the breakdown and synthesis of chemicals within the cell. A metabolic pathway includes all of the reactions in which a single chemical is modified. Each step in a metabolic pathway usually requires a specific enzyme. If all components of a pathway are known, it is sometimes possible to determine the underlying cause of a genetic disorder and also to devise a treatment.

Genes and Proteins

As a result of the pioneering metabolic studies of Garrod, Beadle and Tatum, and many others, the one-gene, one-enzyme hypothesis came to be widely accepted. As more became known about genes and proteins and their structure, the original hypothesis was extended. It later became known as the **one-gene, one-polypeptide hypothesis**; a **polypeptide** is a chain of amino acids that may be smaller than a complete protein.

More recently, because of the discovery of *introns* and *exons* in DNA (described later in this chapter), the concept of a *gene* was again modified. A *gene* is now defined as all of the DNA sequences necessary to produce a single protein or RNA product. Thus the earlier form of the hypothesis can now be restated as *one-gene, one-protein or RNA-product.*

In the early 1950s, it was generally assumed that genes were small segments of DNA and that they gave rise to specific proteins in the cell, but how was the genetic information in genes turned into proteins? This question came to dominate research in the 1950s and beyond.

The General Idea

Watson and Crick had ushered in a new era of molecular biology with their description of DNA's structure in 1953. They, and many other scientists, now entered a decade of fruitful research that focused on the central questions of genes and proteins.

What was known about DNA and protein synthesis in 1953? First, it was understood that DNA is located on chromosomes within the nucleus, and second, it was known that protein synthesis occurs outside the nucleus, in the cytoplasm. RNA, the other nucleic acid, had also been identified and described. Whereas

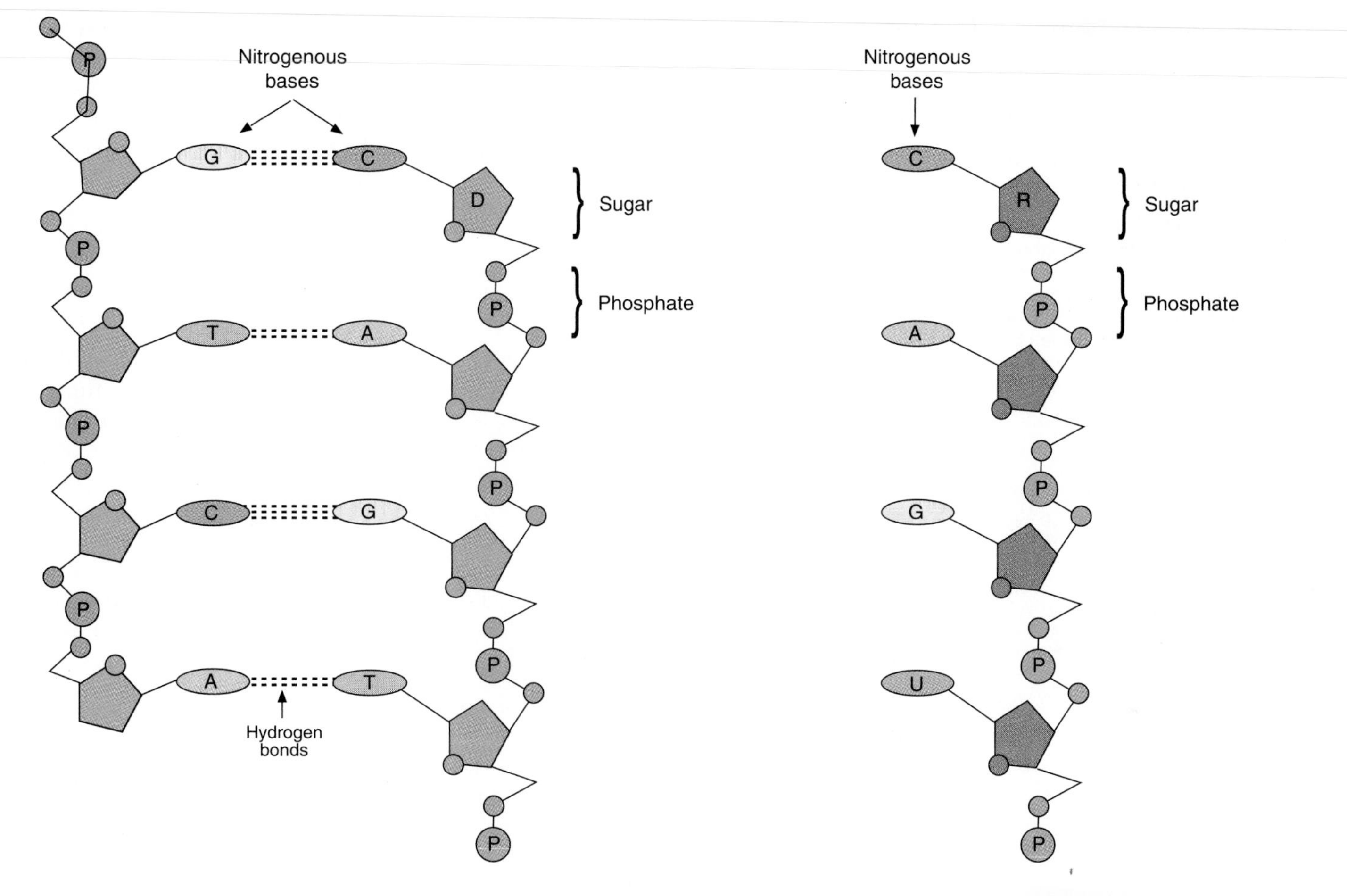

Figure 16.4

A comparison of DNA and RNA shows basic similarities. They are both composed of linear chains of nitrogenous bases, sugars, and phosphate groups (P). There are three principal differences in their structure: the sugar molecule is deoxyribose (D) in DNA and ribose (R) in RNA; DNA is double-stranded, but RNA is single-stranded; and both contain four nitrogenous bases, but RNA has uracil instead of the thymine found in DNA.

DNA molecules are large and double-stranded, RNA molecules are much smaller and single-stranded (see Figure 16.4). Since there could be no direct transfer of information from DNA confined to the nucleus to proteins in the cytoplasm, scientists postulated that some intermediate molecule must be involved in the transfer of genetic information from DNA to protein.

From the time the structure of DNA was described in 1953, the leading candidate for the intermediate information transfer molecule was RNA. Results from several studies supported this view. The cytoplasm of cells actively engaged in protein synthesis was found to contain large amounts of RNA. Other experiments, such as the one summarized in Figure 16.5, demonstrated that RNA was synthesized in the nucleus and hours later appeared in the cytoplasm. Also, it was easy to visualize the possible synthesis of single-stranded RNA with one strand of a DNA molecule serving as a template.

For historical flavor on how answers to the gene-protein question were pieced together, consider this later quote by Francis Crick: "Jim [Watson], you might say, had it first. DNA makes RNA makes protein. That became then the *general idea*." In a famous 1958 article, Crick elaborated on the **general idea**. It basically consisted of two principles. The first was the **sequence hypothesis**, the concept that the sequence of bases in DNA and RNA molecules specified the sequence of amino acids in proteins. The second principle, articulated by Watson, became famous as the **central dogma** summarized in Figure 16.6. Essentially, the central dogma described the flow of genetic information as DNA to RNA to protein. It also asserted that information flow in the reverse direction, from protein to RNA to DNA, was not possible.

As conceived in 1953, the central dogma made these predictions: first, DNA strands serve as templates for the production of either complementary DNA molecules (replication) or RNA molecules (*transcription*); next, the RNA molecules move from the nucleus to the cytoplasm, where they serve as templates for the sequence of amino acids in proteins (*translation*). The general idea turned out to be mostly accurate. Two decades later, however, it was discovered that certain viruses called *retroviruses* could direct the synthesis of DNA from RNA. Thus the concept of a forbidden path of genetic information flow proved to be incorrect.

Figure 16.5

(A) To show that RNA is synthesized in the nucleus and then moves into the cytoplasm, cells were first exposed to a "pulse" of RNA nucleotides labelled with radioactive molecules. The labelled molecules were incorporated into RNA molecules within 15 minutes, and special sensitive film, which is exposed by radiation emitted from the labelled nucleotides, was used to reveal their location. For a brief time the dots, indicative of exposed film from radioactive RNA, were found only in the nucleus. (B) One to two hours later, they were found in the cytoplasm, indicating that they moved from the nucleus to the cytoplasm.

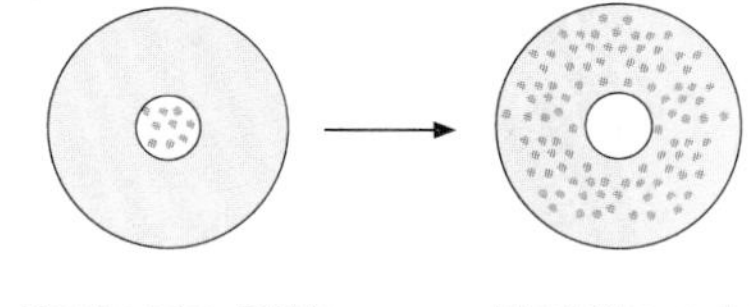

After the pulse, RNA is found in the nucleus.

A

After 1-2 hours, the RNA had moved out to the cytoplasm.

B

PROTEIN SYNTHESIS

DNA constitutes the information that is transferred between generations of organisms. Proteins make up the active labor force carrying out the biological functions of the cell. To maintain the integrity of a cell's activities, mechanisms are necessary for ensuring the continuous production of proteins, each with an exact sequence of amino acids. As described by the one-gene, one-protein hypothesis, a gene encodes instructions for building a specific protein. This is analogous to a recipe that encodes instructions for making a certain type of cookie. As predicted by the general idea, the linear sequence of amino acids in each protein is specified by DNA, and translation is carried out by RNA. Furthermore, part of only one strand of the DNA molecule—the gene—possesses the information used in making a protein. How does this remarkable process take place inside eukaryotic cells?

Transcription

Figure 16.7 is an overview of the events in protein synthesis in a eukaryotic cell. When a cell requires a specific protein, a gene becomes activated, and the stretch

Figure 16.6

The central dogma postulated that the flow of genetic information proceeded from DNA to RNA to protein but not in the reverse direction indicated by the dashed lines.

Figure 16.7

Genes in eukaryotes are usually expressed at some point through the synthesis of proteins. DNA is transcribed, in the nucleus, into messenger RNA (mRNA), which then moves out into the cytoplasm through nuclear pores. Once in the cytoplasm, mRNA becomes associated with ribosomes and is translated through the production of amino acid chains that form proteins.

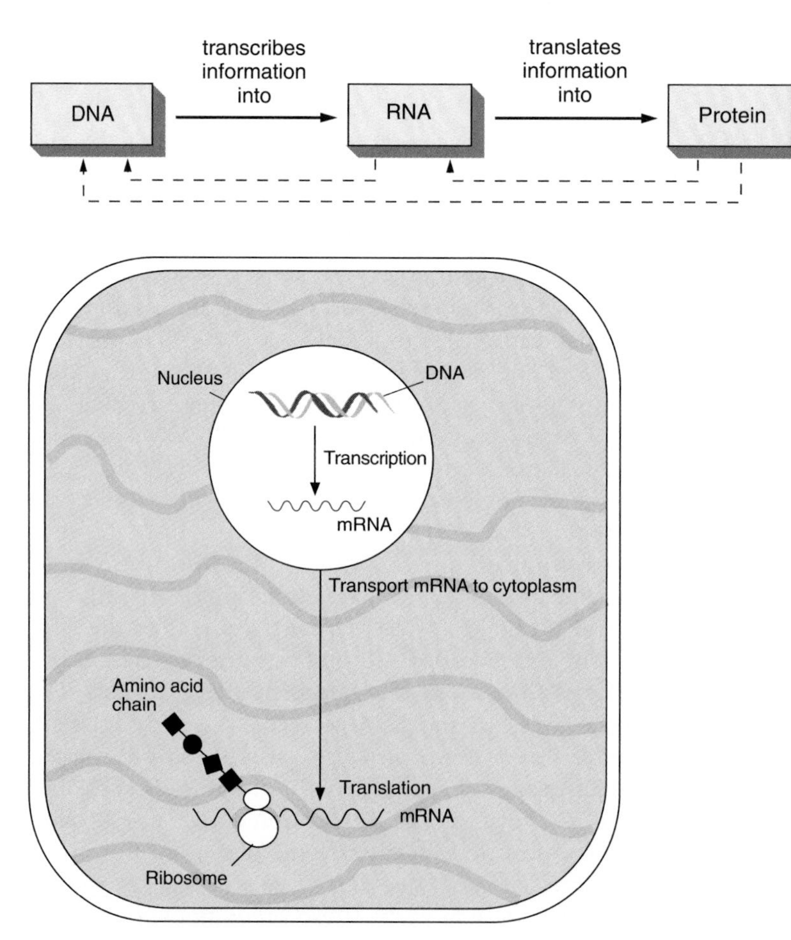

of DNA that constitutes the gene makes an RNA copy of itself. This is the initial step in the transfer of information from DNA. Essentially, the information-bearing RNA serves as a "messenger" from DNA and is therefore called **messenger RNA (mRNA)**.

RNA synthesis from DNA is called **transcription** because information coded in the sequence of DNA nucleotides is copied (transcribed) into RNA nucleotides. Transcription begins when an enzyme called **RNA polymerase** attaches to a specific site on the DNA molecule and causes it to separate partially into two strands; only the strand containing the gene is transcribed. Numerous other enzymes also participate in transcription.

Messenger RNA is assembled as the polymerase moves along the DNA strand, adding RNA nucleotide bases—according to complementary base-pairing rules G-C, C-G, T-A, A-U (recall that uracil replaces thymine in RNA)—to the DNA base sequence being transcribed, as shown in Figure 16.8. Nucleotides continue to be added one at a time until the RNA polymerase reaches a specific DNA nucleotide sequence that acts as a signal to terminate the transcription. The single strand of mRNA is then released from the DNA, and the two DNA strands are rejoined.

In addition to mRNA, two other types of RNA that participate in protein synthesis are transcribed from DNA. These are called **transfer RNA (tRNA)** and **ribosomal RNA (rRNA)**, and their roles will be described shortly. The three types of RNA are constructed using the same complementary base-pairing procedure. Once synthesized, the RNA molecules leave the nucleus and enter into activities related to protein synthesis. Thus there are two general types of genes associated

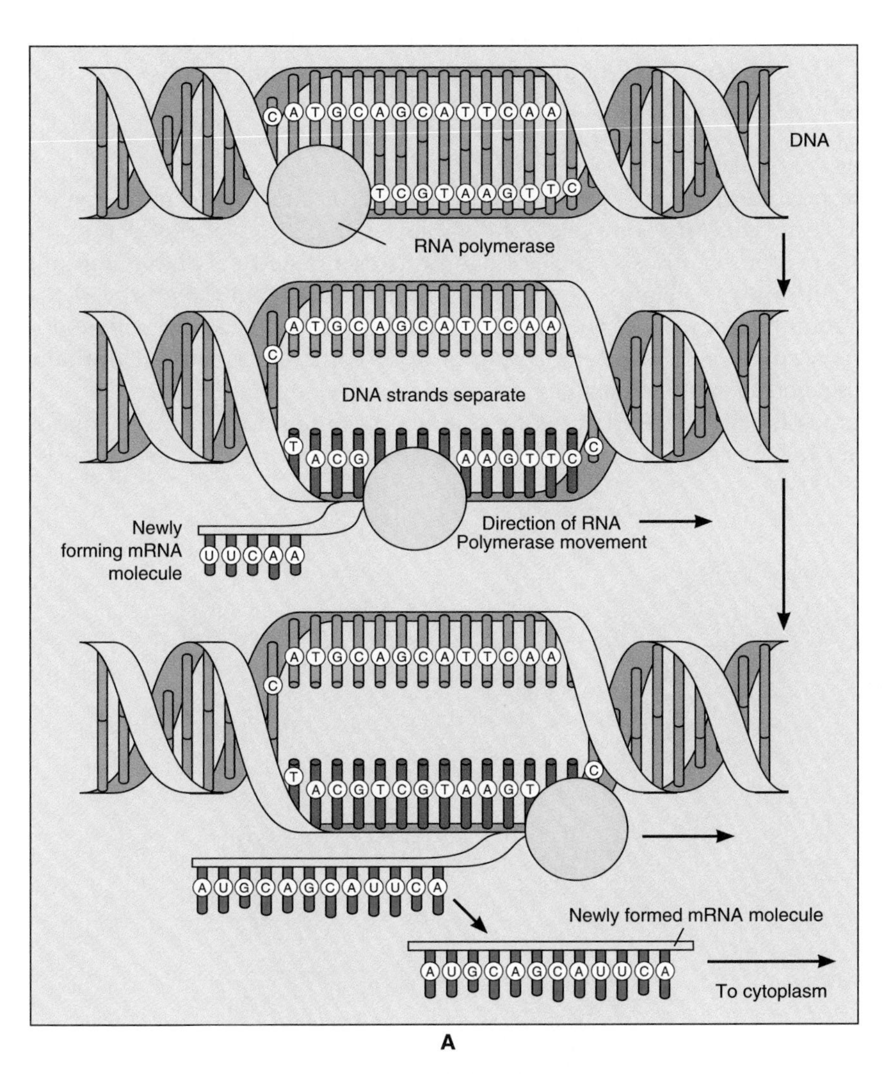

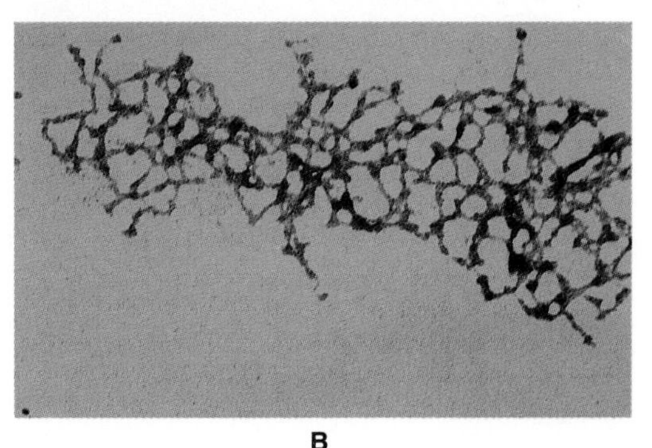

Figure 16.8

(A) DNA is transcribed into RNA primarily through the actions of an enzyme called RNA polymerase. This enzyme first separates the double-stranded DNA; the gene is then transcribed into a strand of RNA by adding base pairs that are complementary to those of the DNA template. Thus the base sequence of the mRNA strand is complementary to the base sequence of the gene. (B) In this electron micrograph, DNA appears as an axial fiber and newly synthesized RNA is coming off laterally.

with protein synthesis. **Structural genes** are those that code for proteins, and **nonstructural genes** code for tRNA and rRNA.

Translation

Translation is the process of producing a protein whose linear amino acid sequence is derived from the mRNA codon specified by a gene. In this operation, the genetic information of the nucleotides is translated into the language of amino acids and proteins. These proteins are synthesized on small organelles called ribosomes.

Codons

The discovery that a sequence of DNA nucleotides coded for the synthesis of a specific protein with its precise array of amino acids led to a new question. How were sequences of DNA nucleotides translated into sequences of amino acids? Clearly, it was not possible for one DNA nucleotide to specify one amino acid in a one-to-one fashion, since there are 20 of the latter but only 4 of the former (A, T, G, and C). A pair of nucleotides would expand the coding possibilities but would still be inadequate since only 16 (4^2) different amino acids could be accommodated. Thus scientists postulated that each amino acid was dictated by a code of three nucleotides, a prediction that was accurate. Although there are only 20 amino acids in proteins produced by living organisms, a triplet code can specify 64 unique "words" (4^3). In fact, it is now known that several different triplets code for the same amino acid.

Each mRNA nucleotide triplet that codes for an amino acid is called a **codon**. For example, the DNA sequences TAC and TCC stipulate, via complementary base pairing, mRNA codons reading AUG and AGG. Since most proteins contain between 50 and 1,250 amino acids, each mRNA strand consists of between 50 and 1,250 codons.

Ribosomes

Ribosomes appear as small particles in the cytoplasm or attached to rough endoplasmic reticulum. They are composed of complex aggregations of rRNA and proteins. During protein synthesis, each ribosome is composed of a large subunit and a small subunit (see Figure 16.9). The larger subunit contains three rRNA molecules, ranging in size from about 120 to 4,500 nucleotides. The smaller subunit contains a single long rRNA molecule of approximately 1,800 nucleotides. When not engaged in synthesis activity, the two subunits separate.

Ribosomes have a grooved structure that enables them to provide a temporary abode for mRNA as its message is being deciphered during protein synthesis.

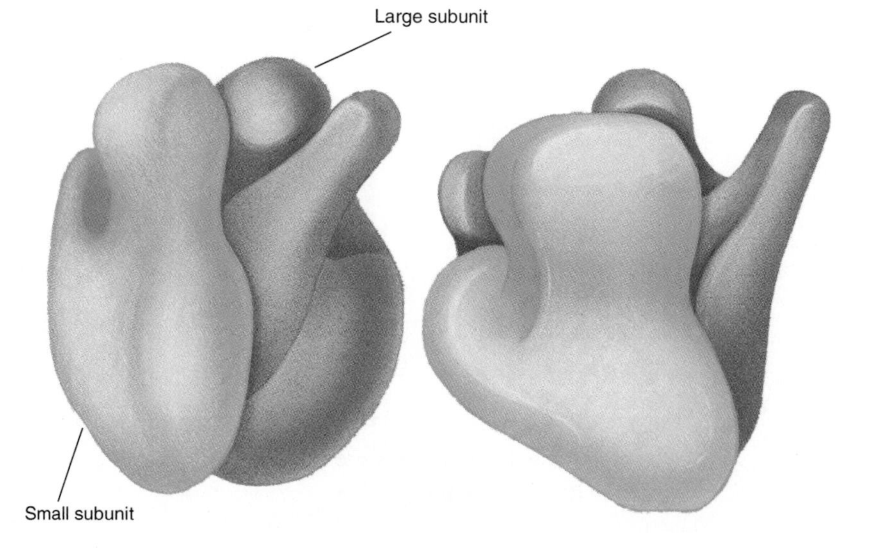

Figure 16.9

During protein synthesis, mRNA is translated on ribosomes that consist of a small and a large subunit with a specific three-dimensional structure.

In cells actively engaged in protein synthesis, clusters of several ribosomes may read the same mRNA strand to increase the amount of protein being manufactured. Ribosomes perform several critical functions in protein synthesis. These include correct alignment and reading of the mRNA strand, proper orientation of the tRNAs bearing their amino acids, and accurate handling of the growing protein chain as synthesis progresses.

Positioning of mRNA

As mRNA moves over the surface of the ribosome, successive codons are brought into position for ordering the respective amino acids of the protein being synthesized. How are amino acids correctly matched with a codon?

tRNA Adapters

Individual amino acids are brought into position by specific tRNA molecules with which they are chemically linked, and are aligned with mRNA. These tRNA "adapter" molecules must recognize the information in the mRNA codon so that the appropriate amino acid is added to the protein at the correct location in the linear sequence. How might this be accomplished?

Two attributes of tRNA account for its ability to serve as the adapter molecule in protein synthesis: each tRNA recognizes one amino acid and one specific codon. Transfer RNA molecules are short (70 to 80 nucleotides long) and have a complex, three-dimensional structure (see Figure 16.10). Several sites exist that are associated with specific functions (see Figure 16.10B). The amino acid attachment site has the same base sequence, ACC, in all tRNAs, but this is apparently not critical in the linking process between tRNAs and their amino acids. Rather, appropriate amino acids are attached to their tRNAs by the action of specific en-

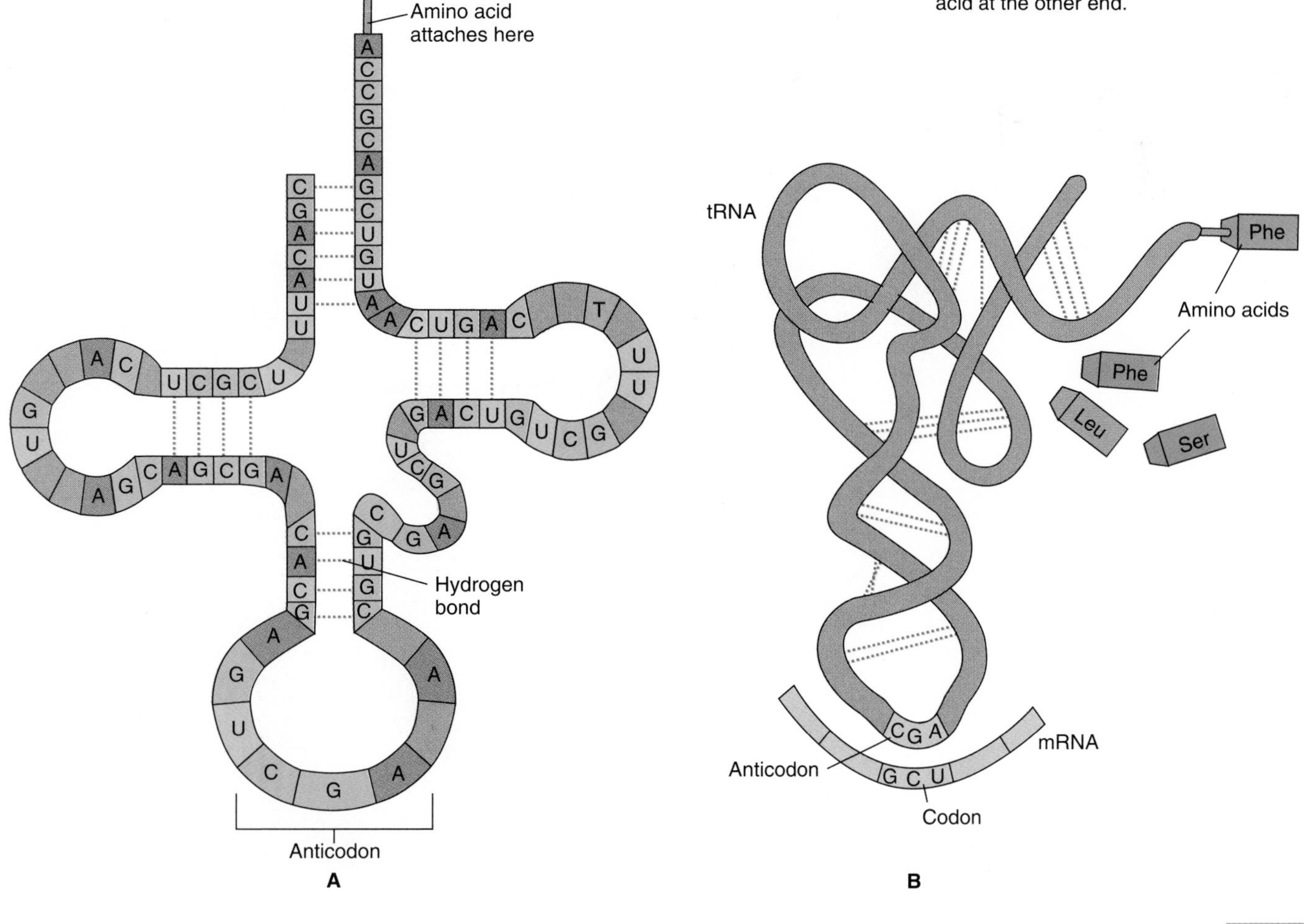

Figure 16.10

(A) Transfer RNA molecules have a two-dimensional cloverleaf shape but are considerably more complex, as indicated by their three-dimensional structure. Weak hydrogen bonds formed between complementary base pairs account for the shape of tRNA molecules. (B) Transfer RNA molecules have two functions, indicated by their structure and nucleotide composition. The tRNA anticodon at one end pairs with the complementary mRNA codon, and each tRNA molecule can carry one amino acid at the other end.

zymes. The precise mechanism by which a tRNA recognizes "its" amino acid is currently the subject of intense study. The tRNA site that can recognize the complementary base pairs of a specific codon is called the **anticodon**; its bases are complementary to those of the codon. As each tRNA brings its amino acid to the site of the growing chain on the ribosome, it binds to the complementary mRNA codon and transfers its amino acid to the protein being synthesized.

Constructing a Protein

Figure 16.11 describes the synthesis of proteins inside a cell. Protein synthesis consists of three complex stages: initiation, elongation, and termination. All protein chains begin with an *initiation* ("start") codon, AUG, which also specifies a form of the amino acid methionine if it is located elsewhere in the mRNA strand. In addition, several other interacting molecules called initiation factors must be present before synthesis can begin.

Once mRNA is positioned and initiation takes place, the ribosome moves

Figure 16.11

This figure traces the major events of protein synthesis. Three types of RNAs—mRNA, tRNA, and rRNA—are synthesized from genes (DNA) in the nucleus and enter the cytoplasm. rRNAs becomes ribosomal structural subunits, and tRNAs transport amino acids to ribosomes, the site of protein synthesis. A structural protein encoded by a gene is transcribed from DNA to mRNA, which becomes attached to ribosomes and is then translated into the protein. Amino acids are brought to the ribosome by tRNA and placed in the appropriate sequence specified by the mRNA. When the protein is assembled, it is released from the ribosome.

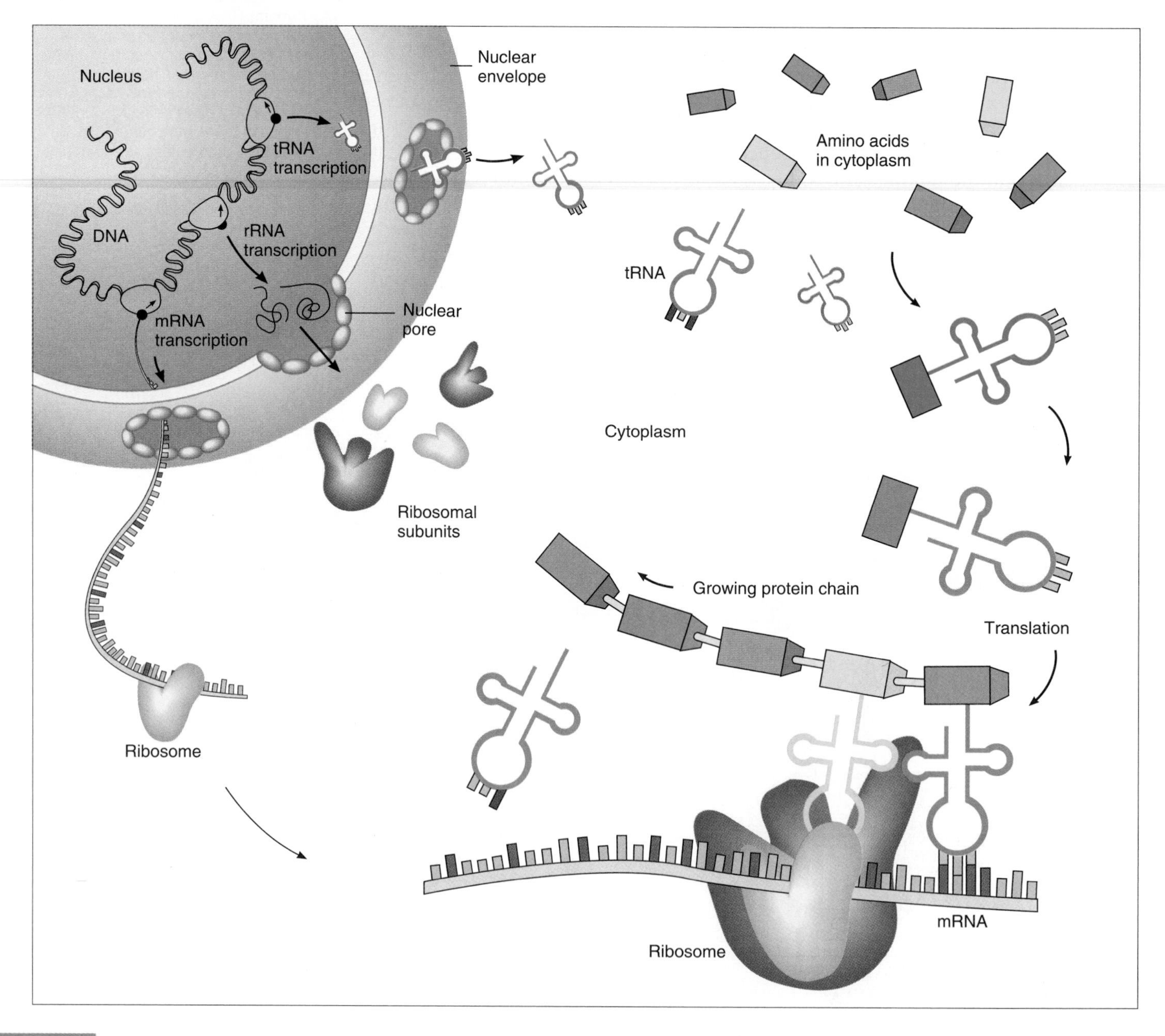

along its length one codon at a time. As each codon is exposed, a tRNA with the appropriate anticodon moves into place. The amino acid it carries is linked, by a peptide bond, to the growing protein chain during the *elongation* phase. This cycle is repeated until the complete protein has been synthesized. At that point, one of three codons, UAA, UAG, or UGA, serves as a *termination* ("stop") signal. Thus the sequence of codons specifies a precise linear sequence of amino acids in a protein and also contains signals indicating where synthesis is to begin and end.

When the genetic message has been completely translated, the finished protein is released from the ribosome. The new protein is then usually modified by enzymes and assumes a distinctive three-dimensional configuration, like the one shown in Figure 16.12, that has relevance to its function. The rules that govern the precise bending and folding of amino acid chains into functional proteins are not yet understood. However, this is a hot area of research, and through the use of new technologies, it has become possible to view and study precise "pictures" of protein structure.

Eventually, the mRNA that encoded the genetic message for producing the protein is destroyed by appropriate enzymes. This mechanism guards against the overproduction of a protein, which could have harmful effects on cell activities. The average life span of an mRNA molecule in eukaryotic cells is 6 to 24 hours, during which time it can direct the synthesis of thousands of protein molecules.

> Synthesis of a protein involves a sequence of reactions that begins in the nucleus where a segment of DNA—a gene—is transcribed into mRNA. Two other types of RNA, tRNA and rRNA, are also synthesized from DNA in the nucleus. Messenger RNA migrates to the cytoplasm, where translation occurs on ribosomes. Messenger RNA codons specify the exact type and sequence of amino acids that are used to build the protein defined by the gene.

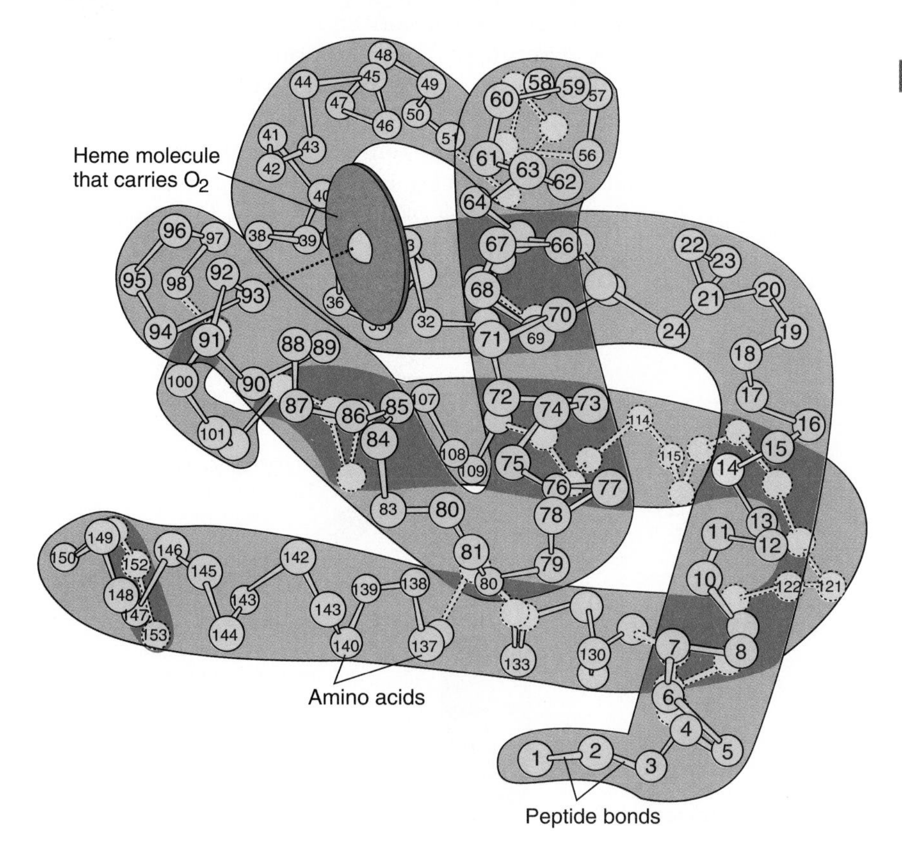

Figure 16.12

All proteins have a specific structure that is related to their amino acid sequence. The shape of the protein is usually related to its function. This complex molecule is myoglobin, a protein that is responsible for carrying oxygen to muscle cells. Numbered spheres represent amino acids, which are joined together by peptide bonds.

A CLOSER EXAMINATION OF SOME PARTICULARS

The description of protein synthesis in the preceding section provided a general review of the process. Protein synthesis is, of course, extremely complex. Molecular biologists have expended great amounts of time and money in learning more about all aspects of protein synthesis and the molecules involved. However, we will focus further discussion on answers to the following questions: Which codons specify which amino acids? What mechanisms control or regulate the events involved in protein synthesis? What happens when DNA is changed by mutations? What is now known about the "fine structure" of chromosomes?

Breaking the Genetic Code

After mRNA had been found to serve as the interpreter between DNA and the sequence of amino acids in specific proteins, there was great interest in determining how the nucleotide message (64 possible codons) was translated. What was the *genetic code* by which DNA, through mRNA base triplets, dictated specific amino acids? Breaking this genetic code is considered to be one of the greatest achievements of modern biochemistry.

The problem was solved by employing all of the usual molecules necessary for normal protein synthesis (tRNAs, amino acids, and ribosomes) along with mRNAs that were produced synthetically. The base sequence of synthetic mRNAs could be specified, and analysis of experimental results in which they were used enabled investigators to determine the amino acid indicated by each codon.

In 1961, Marshall Nirenberg and Heinrich Mathaei conducted the first experiment, described in Figure 16.13, that consisted of using mRNAs composed only of uracil ("poly-U"); thus the codon read UUU. When incorporated into the experimental system, a protein was produced from poly-U that contained only phenylalanine! What did this mean? Clearly, it meant that UUU codes for phenylalanine. Codons consisting of pure strings of the other three DNA bases (AAA, GGG, and CCC) and their specified amino acids were also quickly determined.

The next series of experiments, using codons of more than one nucleotide, were obviously more difficult. Basically, synthetic mRNAs consisting of alternating bases (for example, . . . ACACAC . . .) were used. A second experiment used a different repeat of the two bases (for example, . . . AACAAC . . .). By logical analysis of the end product (strings of certain amino acids), it was possible to assign amino acids to specific codons. Most of this work was done by groups working in Nirenberg's lab at the National Institutes of Health in Bethesda, Maryland, or with Severo Ochoa at New York University.

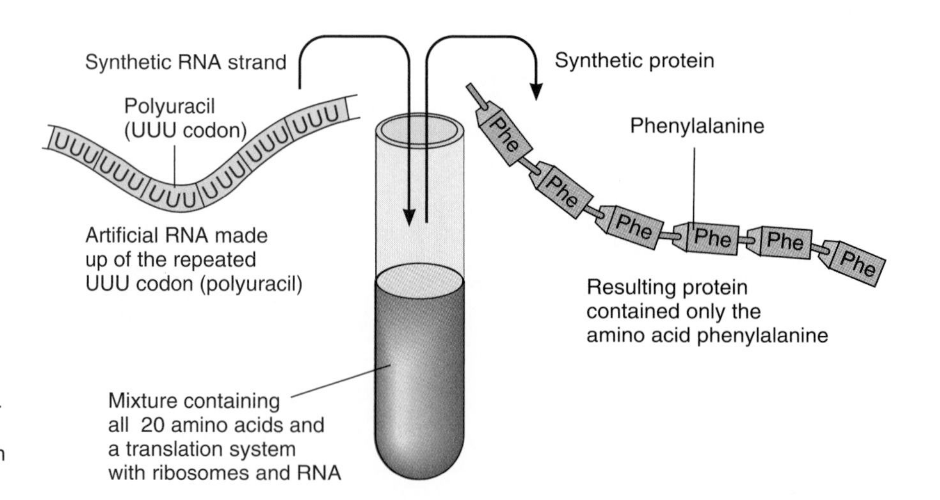

Figure 16.13

To determine the amino acid called for by a specific codon, an early experiment was conducted that used an artificial mRNA with a codon reading UUU. The only amino acid found in the protein produced was phenylalanine.

First letter of code	Second letter of code				Third letter of code
	U	C	A	G	
U	Phenylalanine	Serine	Tyrosine	Cysteine	U
	Phenylalanine	Serine	Tyrosine	Cysteine	C
	Leucine	Serine	STOP	STOP	A
	Leucine	Serine	STOP	Tryptophan	G
C	Leucine	Proline	Histidine	Arginine	U
	Leucine	Proline	Histidine	Arginine	C
	Leucine	Proline	Glutamine	Arginine	A
	Leucine	Proline	Glutamine	Arginine	G
A	Isoleucine	Threonine	Asparagine	Serine	U
	Isoleucine	Threonine	Asparagine	Serine	C
	Isoleucine	Threonine	Lysine	Arginine	A
	START or methionine	Threonine	Lysine	Arginine	G
G	Valine	Alanine	Aspartic acid	Glycine	U
	Valine	Alanine	Aspartic acid	Glycine	C
	Valine	Alanine	Glutamic acid	Glycine	A
	Valine	Alanine	Glutamic acid	Glycine	G

Figure 16.14

This table of the genetic code—the language of life—is in the form of a 64-word dictionary with each amino acid or start-stop code identified by a specific mRNA triplet codon. To determine the amino acid specified by a codon, read the three-letter sequence of any codon and determine where they intersect. For example, C (first letter) A (second letter) A (third letter) codes for the amino acid glutamine; UGG codes for tryptophan.

Other experiments eventually led to the complete deciphering of the **genetic code**, which is shown in Figure 16.14. Results from all this research produced the following information:

1. All codons consist of three nucleotides.
2. The code is "degenerate" in that many amino acids are specified by more than one codon.
3. Of the 64 codons, 61 code for specific amino acids.
4. One of the codons, AUG, codes for methionine but may also act as a start signal, depending on its location in the mRNA strand, and three of the codons—UAA, UAG, and UGA—are stop signals that terminate synthesis of the protein chain.

Control of Gene Expression

We now understand that genes are expressed through the production of specific proteins. **Gene expression** is defined broadly to include all the steps necessary to transpose a genotype to a phenotype. However, many proteins are rarely needed; some are synthesized only by certain types of cells; and others must be available only for a brief time in an organism's life. What happens when the protein product of gene expression is not needed? How does a cell "know" when a certain protein is required? How do cells regulate the expression of their genes?

Prokaryotes

A variety of genetic control mechanisms have been fully described in prokaryotes. In general, they are much simpler to describe than control mechanisms in eukaryotic cells. Therefore, we will focus our consideration of these complex activities on one of the best-understood models to illustrate the concept of gene regulation.

The bacterial lifestyle often requires the sudden production of a certain enzyme in response to a rapid change in the environment. If, for example, a new food source (say, a specific sugar) is encountered, it may be beneficial to a bac-

Figure 16.15

The *E. coli* genetic system responsible for breaking down the sugar lactose is called the *lac* operon. It is composed of three major components, a gene (*I*) that controls regulation; control sites (promoter, *P*, and operator, *O*), acted on by products encoded by the regulatory gene; and three structural genes (*Z*, *Y*, and *A*) that encode enzymes necessary for metabolizing lactose. Regulation of the *lac* operon system is controlled by a repressor protein encoded by the *I* gene that interacts with the operator region. In the presence of lactose, the repressor-lactose complex cannot bind to the operator; structural genes *Z*, *Y*, and *A* are turned on; transciption occurs; enzymes are produced; and lactose can be broken down.

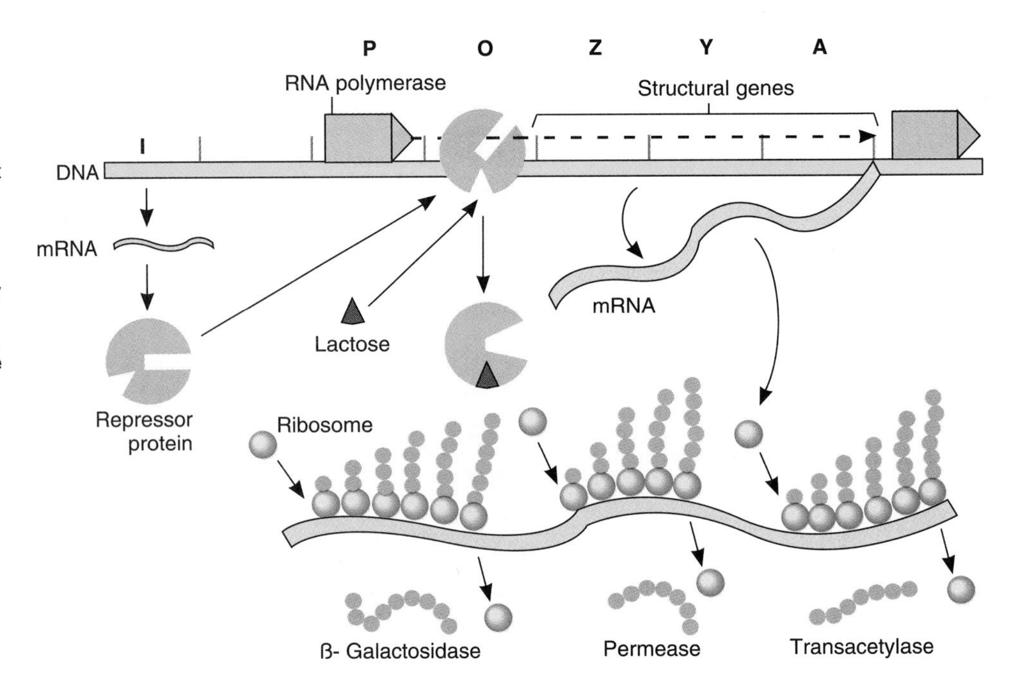

terium to be able to respond immediately. It does this by producing enzymes capable of converting the sugar molecules into forms that can be metabolized. However, because of an extremely small amount of space within its cell and limited cellular resources, a bacterium does not profit from having extra enzymes it may never use. Thus scientists faced a challenging question: Is there some mechanism by which bacteria control the production of enzymes (or other proteins) by using some sort of on-off switch? Two French scientists, Jacques Monod and François Jacob, received the Nobel Prize in 1965 for their classic studies, conducted in the 1950s, that helped answer this question.

What worthwhile information was available that could help put the puzzle together? From earlier studies, Monod and Jacob knew that the common bacterium *E. coli* produces a specific enzyme, b-galactosidase (b-gal), at high rates when the sugar lactose is present. For lactose to be used by the bacterium, it must first be broken down by b-gal into two smaller sugars, glucose and galactose. They also knew that the presence of lactose greatly accelerates the rate at which RNA polymerase binds to the gene (DNA) that codes for b-gal and initiates the synthesis of mRNA. The increased amount of the enzyme-coding mRNA leads to an expanded supply of b-gal.

From their own experiments and the work of others, a picture of the system emerged (see Figure 16.15). The gene for b-gal (gene *Z*) lies next to two other genes that code for enzymes that also have a role in *E. coli* lactose metabolism. One of the genes (*Y*) codes for permease, an enzyme that helps lactose enter the bacterium. The other gene (*A*) codes for transacetylase, an enzyme that removes modified lactose fragments that are not useful to the bacterium. A single strand of mRNA contains codons for expressing all three genes, and when lactose is available, equivalent amounts of the three enzymes are produced (from here on, *b-gal* refers to this three-gene complex). Now the central question could be addressed. How is the synthesis of b-gal turned on and off?

From their studies, Monod and Jacob developed the following hypotheses:

1. A certain molecule acted as a b-gal **repressor** when it was bound to a specific site on the DNA called the **operator** (*O*).
2. When the repressor was bound to the operator, RNA polymerase was prevented from attaching to the DNA and synthesizing b-gal mRNA.
3. The repressor protein molecule was encoded by a gene referred to as *I* (because it controlled *inducibility*).

4. Lactose would act as an **inducer** of b-gal synthesis by binding to the repressor molecule and preventing it from becoming attached to the operator.
5. Once the repressor was removed from the operator, the structural genes would be turned on, and mRNA and b-gal would be synthesized.
6. b-Gal mRNA synthesis began at a specific site called the **promoter** (P).
7. When the lactose was broken down and eliminated from the cell, the repressor would once again bind to the operator, and the b-gal genes would be turned off.

Through further investigations, all seven hypotheses were verified. This *E. coli* genetic system became known as the *lac* operon. An **operon** is a set of adjacent structural genes whose mRNA is synthesized in one piece, plus the adjacent regulating signals that affect transcription of the structural genes. Thus the ***lac* operon** includes three consecutive regions of DNA consisting of structural genes that code for enzymes needed to metabolize lactose (*Z, Y,* and *A*) and the adjacent **control site**, which features a promoter and an operator. A separate **regulatory gene** (*I*), located elsewhere on the chromosome, encodes a repressor protein that controls inducibility. The work that led to an understanding of the *lac* operon and the mechanisms associated with the regulation of the operon illustrate the efficiency of the systems that control prokaryotic gene expression. Many similar operons have been identified in other prokaryotes.

Eukaryotes

Gene regulation in multicellular organisms involves a complex collection of signals, many of which are not yet understood. Mechanisms of eukaryotic gene control apparently exist at each stage of protein synthesis, from transcription through translation. Experiments have revealed that multiple control processes exist at the transcriptional level. Available evidence also indicates that regulation of gene expression occurs after transcription in eukaryotes. However, major gaps exist in defining the mechanisms responsible for gene control at this level, and it remains an exciting research frontier.

In the early 1960s, the genetic code was determined through ingenious experiments. Meanwhile, a deep question began to drive genetics research: How is gene translation regulated in the cell? Monod and Jacob discovered how the *lac* operon regulates the production of enzymes involved in a specific metabolic pathway in the bacterium *E. coli.* Subsequently, many other prokaryotic gene regulation mechanisms have been described. Less is known about gene regulatory mechanisms in eukaryotic cells.

Mutations

Mutations are permanent alterations in DNA (see Chapter 14). They occur spontaneously, with no apparent cause, or after DNA is exposed to radiation (for example, ultraviolet light or X-rays) and certain chemicals and viruses. Mutations have always been the subject of considerable interest in genetics. If they occur in gamete-producing cells, they can be a source of new genes and, through translation, new hereditary traits (remember the white-eyed fruit fly?). Understanding gene expression allows us to cast a new light on certain types of gene mutations.

Gene mutations are generally described as changes in the normal sequence of nucleotide bases within a single gene. When one nucleotide or base pair is replaced by another—for example, a T replaces an A or a T-A base pair becomes a G-C pair—the change is called a **base substitution**. Such mutations may have no effect—no change in the amino acid sequence—or they may lead to the synthesis of an altered protein because of changes in the amino acid sequence. Mutations can also result when one or more base pairs are added to or deleted from a DNA strand. If the gene affected by this mutation codes for a protein, a **frameshift mutation** occurs, and the translation process is disrupted. Since the genetic code involves codons, the addition or deletion of any number that is not a multiple of three re sultsi nani nabil itytor ead thec odonsc orrectly (this analogy shows how a *frameshift mutation* "results in an inability to read the codons correctly"). These mutations frequently alter gene function and may therefore be harmful.

Chromosomal mutations affect many thousands of base pairs and may involve rearrangements, misalignments, additions, or deletions of large pieces of chromosomal DNA (see Chapter 14). Another class of apparent aberration involves "jumping genes" or **transposons**, which are DNA sequences capable of migrating and inserting into a different chromosomal site. Evidence suggests that transposons are not beneficial to the organism and may cause harmful effects characteristic of mutations. However, they may also be an important force in evolution by creating new combinations of DNA.

Since few mutations are beneficial, it is not surprising that cellular mechanisms have evolved that reduce the rate of, but do not eliminate, mutation. DNA polymerases, the enzymes primarily responsible for DNA synthesis, play a major role in identifying and deleting erroneous bases or base sequences in DNA and also in repairing any damage at the site involved.

A Second Look at Chromosomes and DNA

Many important discoveries have occurred since the breaking of the genetic code and the elucidation of gene regulatory systems. Several of these have further clarified the enormous complexities associated with protein synthesis. Eukaryotic genes are now known to exist as discrete islands along the continuous DNA molecule and to be composed of a sequence of nucleotides along one of the DNA strands, parts of which are not transcribed. **Exons** are all sequences that are transcribed as RNA. The sequences between exons are called **introns** (see Figure 16.16). After transcription, introns are removed from the mRNA by enzymes before the mRNA migrates to the cytoplasm to be translated.

The discovery of exons and introns in 1977 caused a bit of a shock to molecular biologists. Until that time, it was assumed that a gene was a single, continuous interval of DNA that would code for a single protein. It is now known that in human and other mammalian cells, over 75 percent of "gene DNA" consists of introns. In light of this new information, the gene could no longer be thought of as a single stretch of DNA. Further, it became known that a small number of genes code for tRNAs and rRNAs, not proteins. This discovery led to the revised definition of a gene given earlier in this chapter.

A Look Back

Research in molecular biology has also yielded some fascinating details about DNA and its history (see the Focus on Scientific Explanations, "Ancestors"). The chromosomes of prokaryotes are not cluttered with collections of introns. Rather, many of their protein-coding genes are arranged in operons, and little nontranslated DNA is found within or between these operons.

The simplest eukaryotes, such as yeast, also contain little excess DNA. However, chromosomes of complex plants and animals contain an enormous load of "surplus" DNA in the form of introns and other apparently nontranslated segments. Protein-coding genes (exons) occupy only 1 to 20 percent of the total DNA in the cells of higher organisms, and active transcription units are usually separated by great distances. What is the significance of this great difference in exon-to-intron DNA ratio between lower and higher organisms?

Hypotheses concerned with the origins of life on Earth offer an explanation for the disparity in amounts of nonfunctional DNA in prokaryotes and eukaryotes. As explained in Chapter 2, scientists now believe that three cell lines—ar-

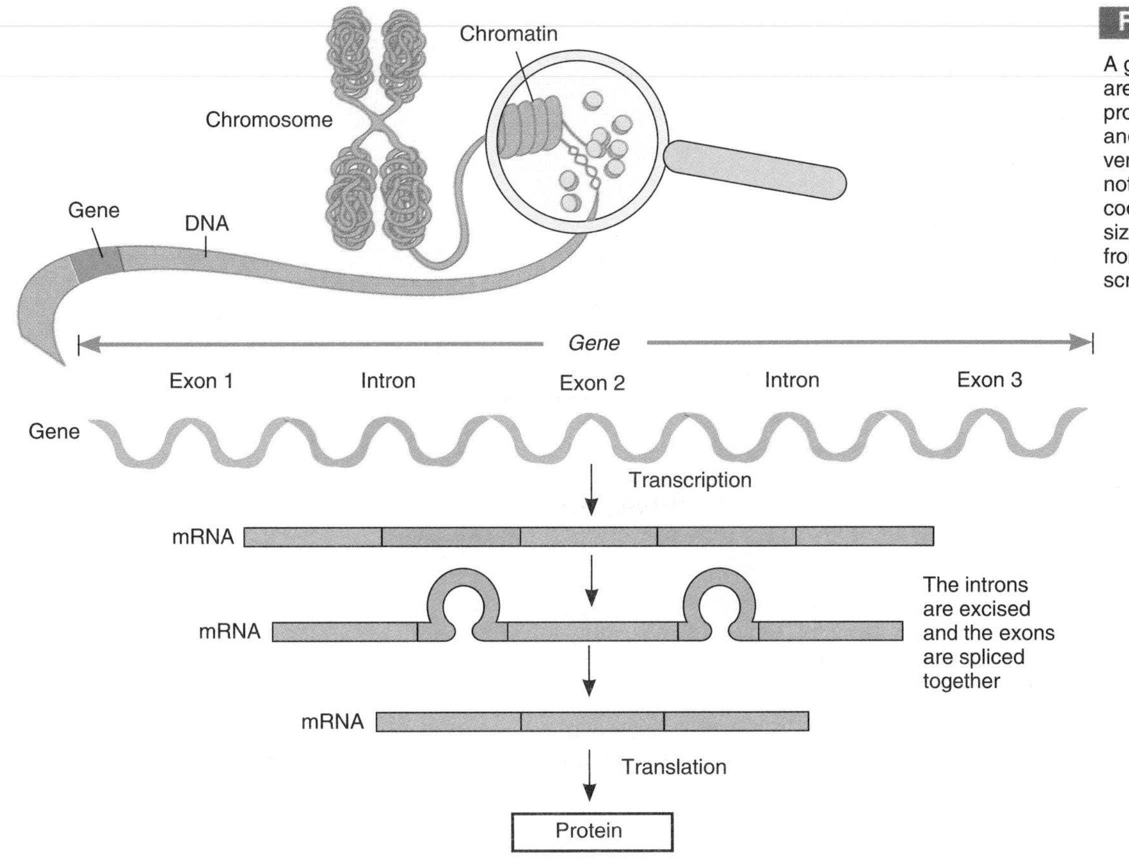

Figure 16.16

A gene consists of exons that are *ex*pressed as part of a protein encoded by the gene and introns, which are *in*tervening sequences that are not expressed. For the encoded protein to be synthesized, introns are removed from the mRNA after transcription occurs.

chaebacteria, eubacteria ("common" bacteria), and eukaryotes—arose several billion years ago from a single ancestral cell type called the progenote.

Based on results from many studies, an interesting hypothesis has emerged (see Figure 16.17). The progenote is now postulated to have contained intron-rich DNA. Thus the earliest cells that descended from the progenote would also

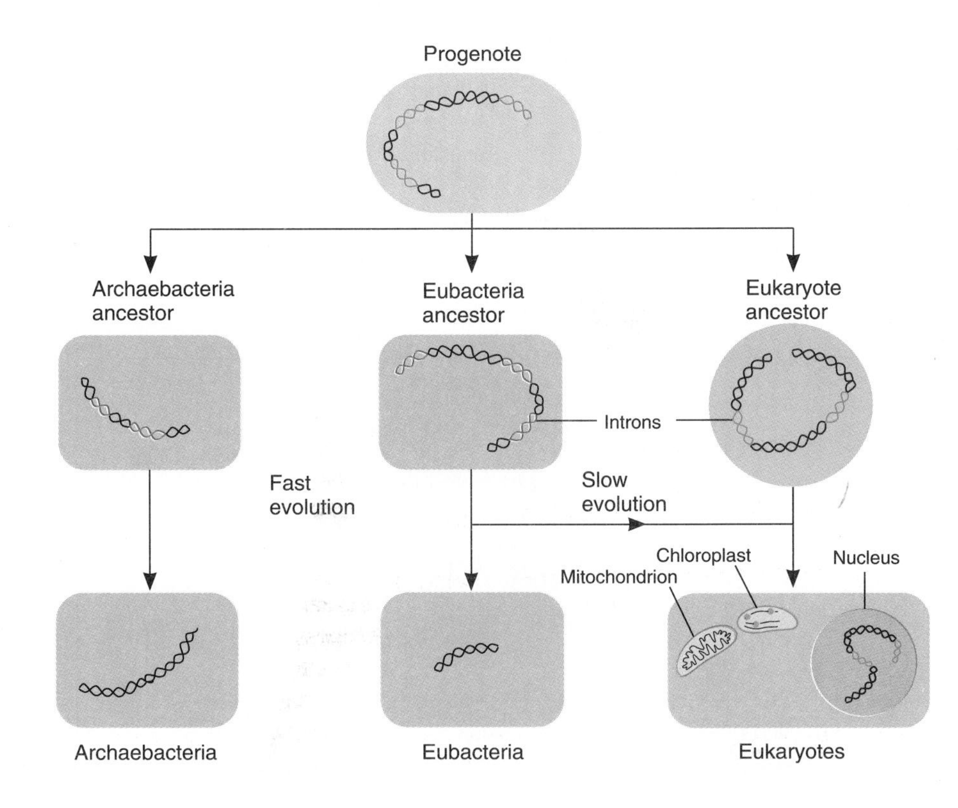

Figure 16.17

All life is hypothesized to have originated from a cell type called the progenote. The progenote gave rise to three cell lines that evolved into primitive bacteria (archaebacteria), modern bacteria (eubacteria), and modern plants and animals (eukaryotes). The progenote probably contained intron-rich sequences. Because of their short life cycles and rapid evolution, archaebacteria and eubacteria have eliminated the "useless" introns, whereas such DNA remains in eukaryotic cells.

Focus on Scientific Explanations

For over a century, biologists have been aware that there are great similarities in structure and function among widely dissimilar organisms. This view is expressed in the phrase "unity in diversity." Diversity refers to the millions of species that have inhabited Earth since life evolved.

Unity was first observed in the common anatomical structures of different plants and the similar structures present among animals. Recently, as the highly sophisticated techniques of molecular biology have been applied to studying lower organisms, it has become clear that common molecular mechanisms are found throughout nature. For example, all organisms synthesize proteins and transmit genetic information in essentially the same way; such fundamental mechanisms are known as *biological generalizations*. Thus some scientists hypothesize that molecular mechanisms that exist in all organisms trace their origins to the progenote, the ancestor of all living species. The molecules and processes involved have been highly conserved (that is, they have not changed significantly in species over time), and they are widely shared by prokaryotes and eukaryotes, including humans, at the cellular and molecular levels. Two specific examples illustrate this point. A molecule used by yeast cells for "mating" has likewise been found to be a component of certain sex hormones of higher organisms. Also, an insulinlike molecule has been found in fungi and yeasts that functions as a cellular signal, much as insulin does in mammals.

Interest in this area increased dramatically during the 1980s, pri-

ANCESTORS

Figure 1 The latest hypotheses on the origin and nature of the first living organisms to have appeared on Earth and the evolution of life are reflected in this diagram (BYA means billions of years ago). The hypothesized first organism to develop from the organic soup was capable of carrying out metabolic activities and reproducing, both through RNA molecules. This RNA world inhabitant gave rise to the breakthrough organism, which used DNA as its genetic material. Finally, the progenote evolved with a metabolism based on proteins (enzymes); this organism is thought to have been the ancestral cell to all modern life forms on Earth (discussed in Chapter 2).

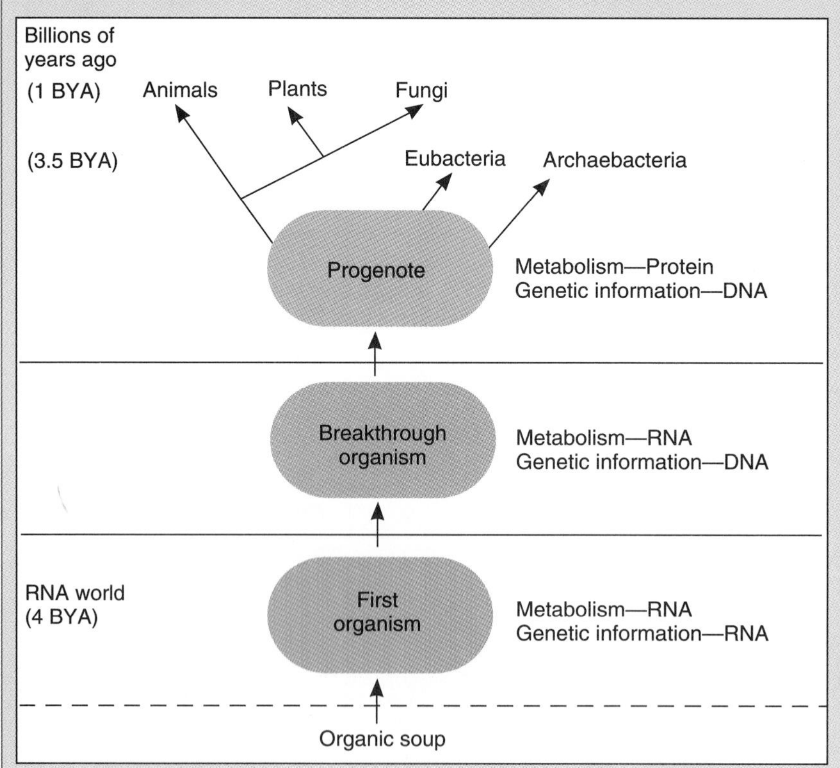

marily because of two developments. First, in a 1985 report (*Models for Biomedical Research: A New Perspective*) from the National Academy of Sciences, a suggestion was made to enlarge the scope of biological generalizations by creating a computerized information system that incorporates new data and information gained from studies of all organisms. Analysis of such information may reveal fundamental biological patterns that exist throughout the living world.

At the heart of this approach is the widely held belief that undiscovered organizational principles can be demonstrated from studies of much different or simpler organisms. We have examined a number of such cases in Chapters 14, 15, and 16. The fruit fly (*Drosophila*) was used to derive many important genetic concepts that are now known to be universal in nature. The universal genetic code was discovered first in bacteria and viruses but is now known to be shared by all living organisms on Earth.

Second, new knowledge about DNA, RNA, and protein synthesis has led to an opportunity to reconsider the organic soup and the earliest life forms that may have existed on Earth between 2 and 4 billion years ago. Recent research findings have revolutionized thinking about the earliest forms of life and led to the formation of new hypotheses. A startling discovery of a unique RNA led to the rejuvenation of interest in this area.

The 1989 Nobel Prize in chemistry was awarded to Sidney Altman of Yale University and Thomas Cech of the University of Colorado for their independent discovery, in 1983 and 1984, that "catalytic" RNA molecules can act as enzymes in some microorganisms. Specifically, catalytic RNA segments termed *ribozymes* snip the introns out of various pieces of tRNA and rRNA and rejoin the remaining pieces before they become functional in protein synthesis. Whether or not ribozymes "edit" mRNA in these organisms is not yet known. Since the discovery of this self-splicing RNA, many molecular biologists have come to believe that the first living systems to evolve from the organic soup must have used RNA for all of their chemical activities. Some extremely interesting models of this RNA world have appeared recently.

To be classified as living, an organism must be able to perform two primary functions: catalyzing chemical reactions required for life and storing and transmitting genetic information. As we have seen in this chapter, modern organisms accomplish these two general tasks through interactions among DNA, RNA, and proteins. How did the earliest organisms solve the two fundamental problems?

In 1989, an enchanting and hotly debated hypothesis, summarized in Figure 1, was published by Steven A. Benner and Andreas Tauer of the Swiss Federal Institute of Technology and Andrew Ellington of the Massachusetts General Hospital. Their hypothesis is based on analyses of metabolic pathways, genetic organization, chemical structure, enzymes that exist in modern organisms, and complex evolutionary considerations.

Basically, they proposed that the first "riboorganism" used only RNA for catalyzing reactions and encoding genetic information; DNA and proteins did not exist in the "RNA world." In the second stage of the origin of life, the "breakthrough organism" (this name is also controversial) appeared, with the following characteristics: it obtained energy and carried out other complex chemical reactions that were catalyzed by RNA, not protein enzymes, and it probably stored genetic information in DNA. Though the breakthrough organism did not possess standard protein-synthesizing capabilities, it may have possessed pieces of the translation machinery. The presence of all protein-synthesizing machinery, including ribosomes, and the use of protein enzymes in metabolic pathways first appeared in the progenote, the common ancestor for all forms of existing life. Thus, according to this hypothesis, the sequence in which the fundamental molecules of life originated, from first to last, was RNA, DNA, then proteins.

The intriguing RNA world hypothesis promises to occupy molecular biologists and theoreticians for some time. Not all scientists agree that events occurred in this postulated sequence or that they could have occurred so quickly (over the course of a billion years or so). Also, no one has yet determined how RNA arose from the organic soup. The beauty of the model is that it provides opportunities for biochemists to test some of its predictions about metabolism and structure using modern organisms. It will be intriguing to follow future developments in this area of biology.

have had DNA that was littered with introns. However, since bacteria reproduce very quickly, they are thought to have evolved more rapidly. As a result, they developed into streamlined protein-synthesizing units by eliminating the noncoding ("useless") regions of their DNA. In contrast, eukaryotic cells have evolved much more slowly and still retain the unproductive DNA inherited eons ago from the ancestral progenote.

Research on DNA has revealed that genes are composed of exons and introns. Prokaryotic genes are composed mostly of exons, whereas the genes of eukaryotes have an abundance of introns. Differences in the exon-to-intron balance between DNA of prokaryotes and eukaryotes are remarkable. Scientists have formulated a hypothesis that attempts to explain the contrast in terms of differential evolution of the two cell types from the ancestral progenote.

THE LANGUAGE OF LIFE

The path from Mendel to the present traverses an expanse of knowledge that began with the concept of hereditary factors and ended with the mechanism of gene expression. Despite overlying complexities, the genetic language of DNA turned out to be rather basic—three nucleotides = one codon = one amino acid—considering the possibilities. The simplicity of the language and its translation illustrates an interesting concept in molecular biology, the **rule of parsimony**. This rule is used to answer such questions as, Why only three nucleotides in a codon instead of four? Why are there only 20 amino acids used to construct the proteins of living organisms instead of 30, 40, or more? Essentially, the rule of parsimony dictates that the simplest solution is always the answer.

The genetic information transmitted from parent to offspring is basically a set of instructions on how to make specific proteins and RNAs. This process underlies many aspects of biology—human genetic traits, genetic disorders, high-technology uses of DNA, evolution, human physiology, cancer, AIDS—that are considered in later chapters of this book.

Summary

1. A desire to learn about the nature of genetic information guided research through much of the twentieth century. Studies of certain human disorders done early in the century by Garrod suggested that genes were responsible for producing enzymes in the cell. If enzymes were not present because of an absence of certain genes, severe consequences resulted.
2. Beadle and Tatum extended Garrod's results using another organism, the pink mold (*Neurospora*). They showed that radiation altered genes in such a way that the organism failed to produce certain enzymes. This finding led to the formulation of the one-gene, one-enzyme hypothesis, which ultimately became modified to one-gene, one-protein or RNA-product.
3. Watson and Crick postulated that DNA in the nucleus produced RNA, which then moved to the cytoplasm, where it guided the synthesis of specific proteins. This scheme, termed the *general idea,* was later confirmed by innumerable experiments.
4. DNA controls the production of proteins through two mechanisms. In transcription, RNA polymerase converts the sequence of nucleotides in DNA into messenger RNA (mRNA). Two other types of RNA, transfer RNA (tRNA) and ribosomal RNA (rRNA), are also produced from DNA. All of these RNAs then enter the cytoplasm to participate in protein synthesis. The production of proteins specified by DNA is a central feature of gene expression.
5. Protein synthesis occurs on ribosomes. Transfer RNAs bring specific amino acids to the site where they are joined to another amino acid in a sequence specified by the mRNA. The basis of correct matching is complementary base pairing.

In the genetic code, an mRNA three-base sequence called a codon is read by a corresponding anticodon on the tRNA. Amino acids continue to be added until the protein encoded by the DNA is completed.
6. Gene expression in prokaryotes involves regulation by complex systems of genes, proteins, and other factors called operons. The well-understood *lac* operon enables bacteria to respond to the presence of lactose in the surrounding environment.
7. Gene mutations cause changes in the normal nucleotide sequence of DNA. As a result, alterations in protein production that can be harmful to the organism may occur.

Review Questions

1. What was the significance of Garrod's studies of metabolism?
2. How did Beadle and Tatum's studies lead them to the one-gene, one-enzyme hypothesis?
3. What is metabolism? What is a metabolic pathway?
4. What was Watson's general idea?
5. To what extent were the sequence hypothesis and the central dogma correct in their predictions?
6. Which molecules are responsible for transcription? What are the roles of each?
7. Which molecules are responsible for translation? What are the roles of each in protein synthesis?
8. If a portion of a transcribed DNA base sequence reads AAACACGCATCGATC, what will be the resulting amino acid sequence?
9. How does the genetic code work?
10. Describe the operation and function of the *lac* operon.
11. Explain the relationships between mutations and the genetic code.
12. What are exons and introns? In which types of cells—prokaryotes, eukaryotes, or both—are they found? What explains this finding?

Essay and Discussion Questions

1. Why would scientists in the 1950s be driven to determine how DNA was turned into proteins?
2. Why might the central dogma have come to be labeled a dogma? Would you expect dogmas to be common in biological science? Why?
3. Which single experiment, idea, or hypothesis described in this chapter do you think might have been the most important in biology? Why?
4. Why has the definition of a *gene* changed repeatedly during the past century?

References and Recommended Readings

Beadle, G. W. 1959. Genes and chemical reactions in *Neurospora* (Nobel Prize lecture). *Science* 129:1715–1726.

Benner, S. A., A. D. Ellington, and A. Tauer. 1989. Modern metabolism as a palimpsest of the RNA world. *Proceedings of the National Academy of Science* 86:7054–7058.

Crick, F. H. C. 1958. On protein synthesis. *Symposium of the Society for Experimental Biology* 12:138–163.

———. 1970. Central dogma of molecular biology. *Nature* 227:561–563.

Darnell, J. E., Jr. 1985. RNA. *Scientific American* 253:68–78.

Doolittle, R. F. 1985. Proteins. *Scientific American* 253:88–99.

Doudna, J. A., S. Coufure, and J. W. Szostak. 1991. A multisubunit ribozyme that is a catalyst of and template for complementary strand RNA synthesis. *Science* 251:1605–1608.

Felsenfeld, G. 1985. DNA. *Scientific American* 253:58–67.

Garrod, A. 1908. *Inborn Errors of Metabolism*. Oxford: Oxford University Press.

Jacob, F., and J. Monod. 1961. Genetic regulatory mechanisms in the synthesis of proteins. *Journal of Molecular Biology* 3:318–356.

Judson, H. F. 1979. *The Eighth Day of Creation: The Makers of the Revolution in Biology*. New York: Simon & Schuster.

National Academy of Sciences. 1985. *Models for Biomedical Research: A New Perspective*. Washington, D.C.: National Academy Press.

CHAPTER 17

Human Chromosomes and Patterns of Normal Inheritance

Chapter Outline

Reading Questions

1. What is known about the human karyotype?
2. What types of DNA are found in human chromosomes?
3. Which human traits are determined by a single gene pair? By multiple alleles? By polygenes?
4. How is human sex determined?

Many basic principles of genetics have been described in earlier chapters. The studies of Mendel, de Vries, Morgan, and others provided a general understanding of how traits are passed between generations and the roles chromosomes and genes play in inheritance. The work of these researchers was of great importance because in using relatively simple organisms, they described many fundamental genetic principles, including segregation, independent assortment, sex-linked inheritance, linkage groups, and chromosome mapping.

These studies gave hope to scientists who had long believed that the scientific investigation of human heredity was potentially of great importance. As early as 1869, Francis Galton, a leading British scientist (and a first cousin of Charles Darwin), argued that hereditary factors were of major significance in determining intelligence. His research had been conducted by using mathematical analyses to evaluate information about human "genius," and although his conclusions were controversial, they provided a starting point for the study of human genetics. Galton's results suggested to him that certain human traits were inherited, but they did not provide any knowledge about how these traits were inherited.

Despite occasional stimulating discoveries, the history of human genetics reveals that this field did not enter the mainstream of scientific research until the middle of the twentieth century. One of the primary impediments facing scientists interested in studying human genetics was that humans are not ideal organisms for genetics studies or analyses. Also, the simple Mendelian model of single gene pairs determining certain traits was generally inadequate for explaining inheritance in humans.

Mendelian patterns of inheritance do, of course, occur in humans. Traits *are* determined by genes carried on chromosomes that segregate and undergo independent assortment. Genes are expressed in terms of proteins synthesized, which are in turn responsible for determining phenotypic characters. Deeper levels of understanding have resulted from the explosive growth of knowledge in human genetics during the 1980s. Thus while it is convenient to use the classical genetic principles as a departure point for understanding human genetics, new information makes it possible to obtain a more sophisticated appreciation of the enormous complexities involved.

HUMAN CHROMOSOMES

The existence of chromosomes in the nuclei of cells has been recognized for over a century, and their role in transmitting genetic information was first described by Sutton in 1902. All species have a unique **karyotype**, a specific number of chromosomes, each with a characteristic size, shape, and staining pattern. However, correctly determining this set of features for humans proved difficult. Not until 1956 was the human karyotype accurately determined by two Swedish **cytogeneticists**, scientists who study chromosome structure and behavior using the methods of cell biology and genetics. Joe Hin Tjio and Albert Lavan reported that the human karyotype consists of 46 chromosomes (23 pairs), not 48 as had long been thought.

The primary difficulty in determining a karyotype is that chromosomes are not ordinarily visible in the cell except during cell division, either meiosis or mitosis (see Chapter 13), when they condense after being duplicated. Special techniques described in Figure 17.1 are used by cytologists to reveal condensed chromosomes microscopically in certain cells. Appropriate cells (usually white blood cells) are removed from the body and grown in culture dishes containing a substance that stimulates cell division. After about 72 hours, the cells are exposed to *colchicine,* a chemical that stops cell division at metaphase. At this time, the cells are placed in a solution that causes them to swell and allows the chromatids to separate. After swelling, the cells are gently squashed and split open on a microscope slide, and the chromosomes are then stained and allowed to dry. After

Figure 17.1

To determine a karyotype, chromosomes can be viewed in a cell during cell division. After extraction from the cell, chromosomes are stained, photographed, and arranged according to their size and the position of the centromere.

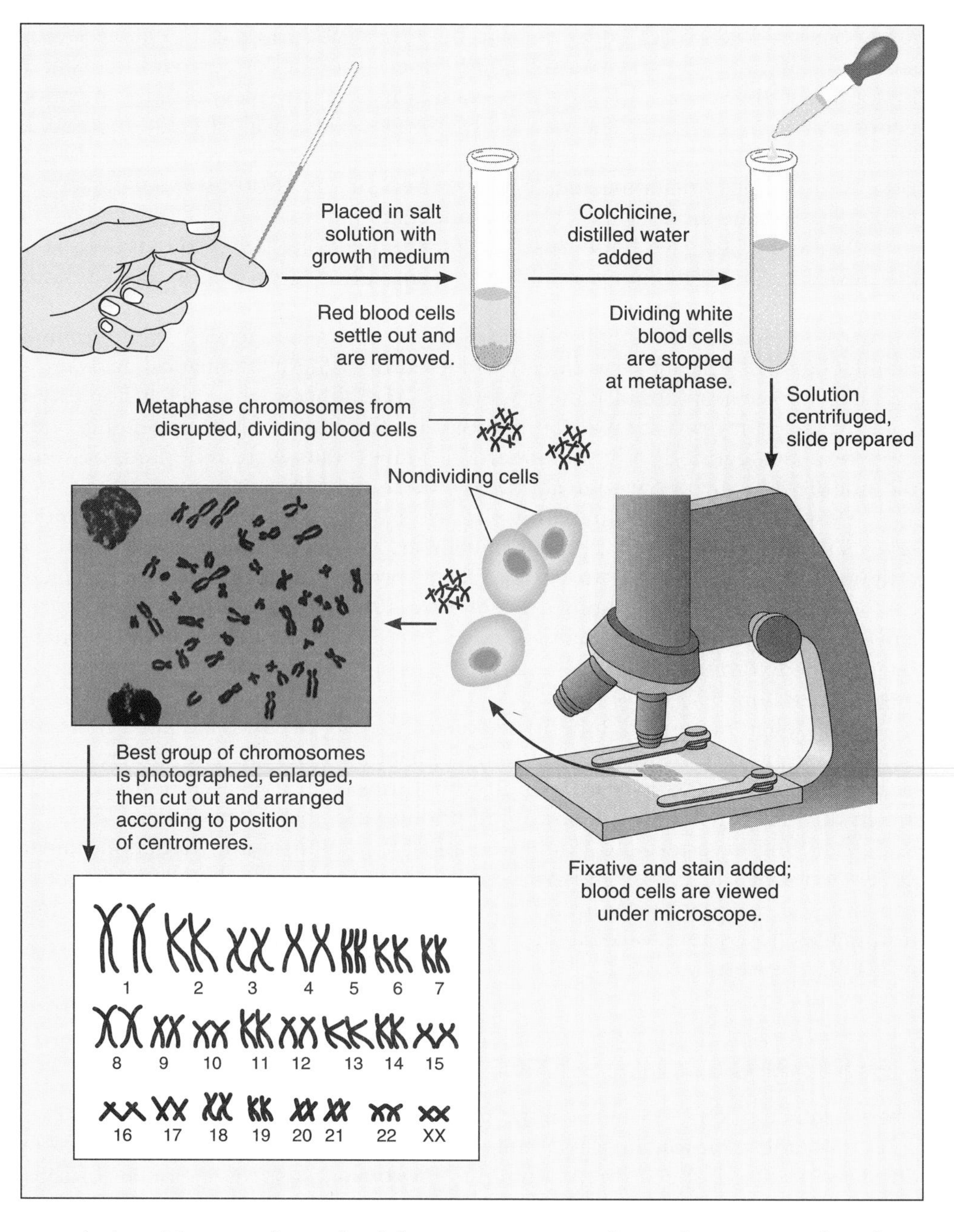

completing this procedure, the "chromosome spread" can be examined under a microscope. To prepare a karyotype, photographs are taken, enlargements are made, and pictures of metaphase chromosomes are cut out and placed in a conventional sequence according to their size. The amount of genetic information each chromosome contains is roughly related to its dimensions.

Human male and female karyotypes have 23 pairs of chromosomes. The first 22 pairs of the chromosomes are identical in males and females and are called *autosomes.* The unnumbered X and Y chromosomes are called *sex chromosomes;* cells of males contain an X and a Y chromosome, whereas those of fe-

Research on human genetics lagged far behind studies of the genetics of other, simpler organisms. The human karyotype was not accurately described until 1956, when it was determined that human cells contain 23 pairs of chromosomes. Normal males and females each have 22 pairs of autosomes and one pair of sex chromosomes. The autosomes are identical in both sexes, but females have two X sex chromosomes, males one X and one Y.

males have two X chromosomes. By convention, the karyotype of a normal male is represented as 46,XY (46 total chromosomes of which one is an X and one a Y) and a female is 46,XX.

Chromosome Nomenclature

To facilitate communication, scientists studying genetics worldwide have developed standard terminology to classify human chromosomes.

Position of the Centromere

Before 1970, stains used in karyotyping gave chromosomes a uniform dark color. As a result, all human chromosomes were classified according to two variables, length and the position of the *centromere* (middle, off-center, or near the end of a chromosome). Based on the position of the centromere, the short (upper) arm of a chromosome is designated by the letter *p* and the long (lower) arm by the letter *q*.

Chromosome Bands

After 1970, a number of special staining techniques were developed that significantly advanced the study of human chromosomes. Now each human chromosome can be further distinguished by a unique series of bands along its length (see Figure 17.2). A **band** is a section of the chromosome that differs from other areas because of its lighter or darker staining intensity. These bands are revealed using quinacrine mustard or related stains (Q bands), Giemsa stain (G bands), or heat pretreatment of chromosomes followed by Giemsa staining that results in a banding pattern that is the reverse (R bands) of Q or G bands. Bands with variable staining patterns are found in a few autosomes, especially near the centromere, and the Y chromosome.

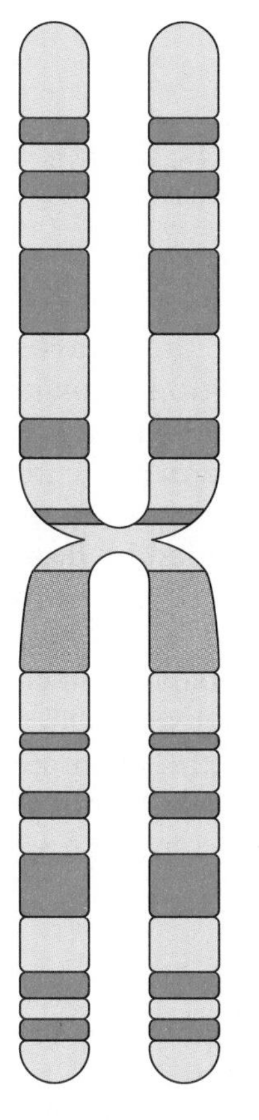

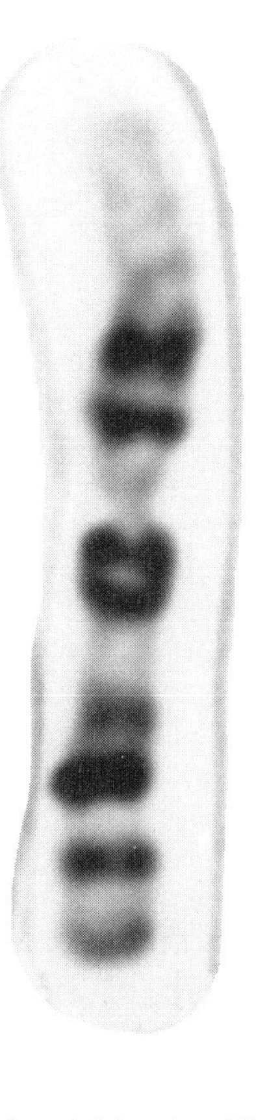

Positive R bands
Negative or pale-staining Q and G bands

Positive Q and G bands
Negative R bands

Variable bands

Figure 17.2

Each chromosome has distinctive banding patterns that show up after they are stained with different chemicals. Specific bands are used as landmarks to locate and identify whole chromosomes and regions of chromosomes. This is human chromosome number 1.

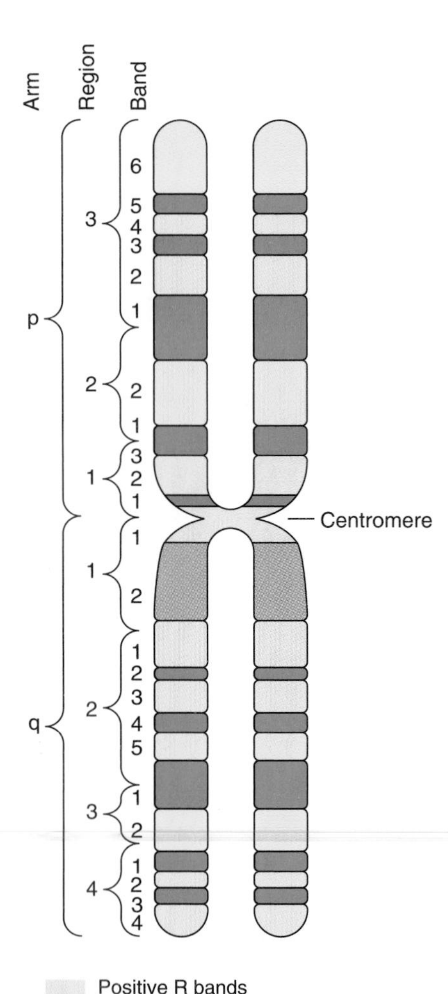

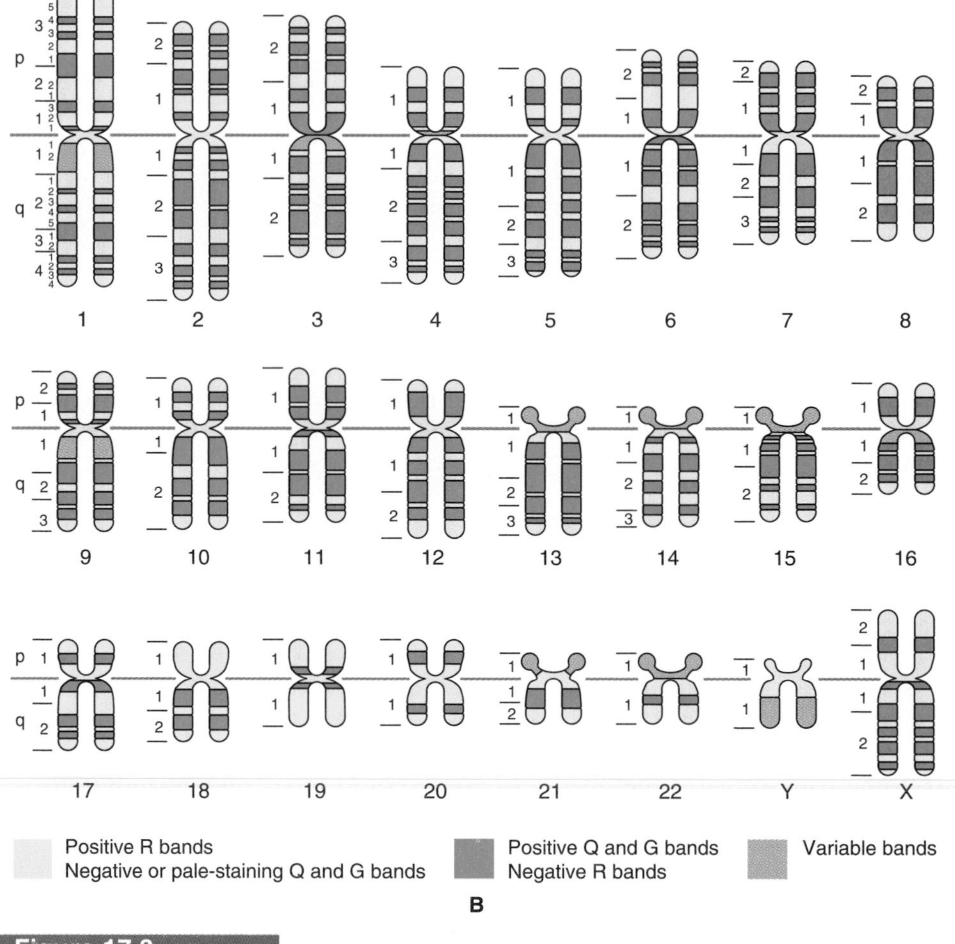

Figure 17.3

The bands of a chromosome and their position relative to the centromere provide the basis for a numbering system that is used to create a map of the chromosome. The details about the bands and the numbering system for human chromosome 1 (A) apply to the band maps for all human autosomes and sex chromosomes (B).

The chromosome arms are divided into regions separated by **landmarks**, defined as consistent features—position of the centromere, position of major bands, or the ends of chromosome arms—that can be used in identifying a specific chromosome. As shown in Figure 17.3, geneticists divide each arm of a chromosome into large areas or regions that are numbered consecutively from the centromere (for example, p1) and then into smaller subareas identified by **band numbers**. Use of this procedure allows for the precise location of an area of interest in a chromosome. For example, 1p22 refers to chromosome 1, the short (upper) arm, region 2, band number 2. Recent improvements in banding techniques now permit using **subbands** for even more precise chromosomal location; subbands are added as a decimal to the band number (in the example, 1p22.3). Distinctive banding patterns have allowed geneticists to make standardized maps of bands of human chromosomes that are used widely in modern genetics studies.

The Human Genome

A **genome** refers to the complete collection of genes in one organism or in a chromosome set. The human genetic material is immense and complex. Each of the 46 chromosomes in a human cell is thought to contain a single long strand of DNA. The total length of the DNA from all chromosomes in a single cell is about 2 meters and consists of 3 to 3.6 billion nucleotide base pairs. Sequences of these base pairs make up genes, and genes are ultimately responsible for directing the structure and function of cells through protein and RNA synthesis. From these data, it is estimated that the total number of structural genes in humans is between 50,000 and 100,000.

DNA in the Human Chromosome

At the molecular level, human chromosomes share common characteristics with chromosomes of all eukaryotic organisms. There are several classes of DNA in humans and other multicellular plants and animals (see Figure 17.4). In eukaryotes, relatively few protein-coding genes are probably solitary or appear only once in the genome, although this is not known with certainty. Most protein-coding genes are apparently members of **families**, which consist of two or more genes that encode similar proteins. The genes in any given family are often clustered together on a chromosome, although in some cases, they are widely distributed throughout the genome. What is the nature of gene families? The genes in any given family are thought to have descended from one ancestral gene. According to this view, some genes of early eukaryotes were replicated but never separated by cell division processes. Like unwelcome relatives who come to visit but end up staying, these duplicated genes accumulated continuously over time. Today, in modern eukaryotes, they are represented as clusters of genes that are similar to the ancestral gene. Because of gene mutation, many of these genes are no longer functional—they are not transcribed—and are called **pseudogenes**. Others encode proteins that are different from the original protein specified by the ancestral gene. These new proteins have been put to different uses by organisms throughout their evolutionary history. Thus gene duplication is thought to have provided new opportunities for expanded and specialized gene functions.

Eukaryotic chromosomes contain vast stretches of DNA that have no known function. Some of this DNA represents unused family genes, and the remainder consists of *spacer DNA* that apparently serves only to connect remote islands of structural genes. Sprinkled throughout the connecting DNA are regions known as *repetitive DNA* that contain repeated sequences of varying lengths. These sequences are classified as *simple repeated DNA sequences* of five to ten nu-

Figure 17.4

Chromosomes of eukaryotes contain relatively few solitary structural genes, which occupy only about 2 to 4 percent of the DNA strands. The rest of the chromosome consists of nonstructural genes and vast stretches of repetitive DNA sequences of varying length that have no known function.

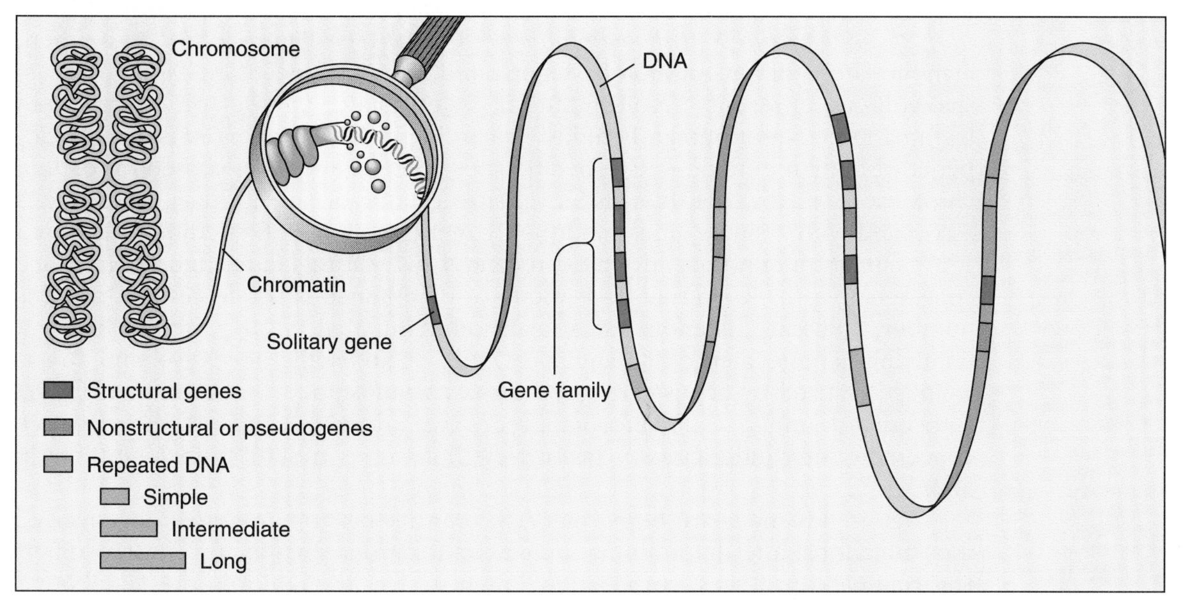

cleotides; *intermediate repeated DNA sequences* of 150 to 300 nucleotides; and *long interspersed repeated DNA sequences* of 5,000 to 6,000 nucleotides. The intermediate and long repeats may not contain units that are exactly the same.

In humans and other mammals, many of the **simple repeated DNA sequences**, which make up about 60 percent of the total DNA in humans, appear to be concentrated near the centromeres of chromosomes. These simple sequences may have some structural or organizational role involving the centromere.

Intermediate repeated DNA sequences account for about 40 percent of the total DNA. They are widely distributed throughout the genome and are not concentrated in clusters. Virtually nothing is known about the possible functions of this abundant form of human DNA. One intriguing possibility is that some of them may represent transposons that are associated with chromosomal rearrangements (see Chapter 16).

Long interspersed repeated DNA sequences make up about 1 to 2 percent of the total DNA in the genome. A few of their sequences may encode proteins since they contain numerous stop codons. Conversely, the abundance of stop codons may indicate that no proteins can ever be synthesized from this DNA.

After staining with special dyes, each human chromosome appears with a unique series of bands and subbands. These bands, together with centromere position, are used in creating a standard blueprint of each chromosome. The human genome is thought to contain between 50,000 and 100,000 genes. Most genes are thought to belong to families, and few single genes are apparently found on chromosomes. Much of the DNA in chromosomes consists of repetitive sequences of varying length, most of which have no known function.

MENDELIAN INHERITANCE IN HUMANS

Mendelian inheritance refers to traits that are inherited in accordance with Mendel's laws of segregation and independent assortment and also show dominance or recessiveness as Mendel defined them. Such traits are generally determined by a limited number of genes, most commonly one pair. What is known about classic Mendelian inheritance in humans?

In 1966, Victor A. McKusick published the first edition of a historic book titled *Mendelian Inheritance in Man: Catalogs of Autosomal Dominant, Autosomal Recessive, and X-linked Phenotypes* (*MIM*) (see Figure 17.5). Since then, eight updated editions of *MIM* have appeared, the most recent in 1990. This massive volume represents a gold mine for geneticists or any person interested in learning more about human heredity. According to McKusick, *MIM* is an encyclopedia of human genetic *loci*—specific sites where genes are located on certain chromosomes. However, it is also a valuable catalog of human traits (phenotypes associated with a single locus) and their genetic basis. The first edition identified 1,487 human phenotypes, the most recent, 4,937. Of this latter figure, McKusick considers that for 2,656 traits, the mode of inheritance is "quite certain." Of this number, 1,864 are listed as autosomal dominant (that is, they are caused by a dominant gene located on an autosome), 631 as autosomal recessive, and 161 as X-linked (caused by genes carried on the X chromosome). For 2,281 traits (1,183 autosomal dominant, 923 autosomal recessive, and 175 X-linked), the evidence concerning their inheritance is incomplete but strong enough to include them in *MIM*.

Rapid advances are being made in identifying human genes, discovering their functions, and determining their exact locations on specific chromosomes. For example, our largest chromosome, number 1, is thought to represent about 6

Figure 17.5

Since 1966, Victor McKusick has published nine editions of an extremely valuable book that describes all human traits known or suspected to be inherited according to Mendelian rules.

percent of the total human genome. At present, 159 gene loci (thought to be about 1 percent of the total genes on this chromosome) have been identified and mapped. A very small human autosome, chromosome 21, contains about 2 percent of the human DNA but has been one of the most intensively investigated regions of the genome. So far, 23 gene loci have been assigned to this chromosome.

Autosomal Dominant and Recessive Traits

The question of whether a trait (or an allele) is dominant or recessive is generally much more complex in humans than for the simple alternative phenotypes (for example, round or wrinkled peas) first studied by Mendel. In those cases, the dominant allele is expressed in the phenotype of individuals with either the homozygous or heterozygous genotype. A recessive allele is expressed only when it is homozygous. However, in many carefully studied human traits, the distinction between dominant (*H*) and recessive (*h*) alleles is not this perfect. Complications arise because in heterozygotes (*Hh*), each allele frequently forms a gene product, usually a structural protein or an enzyme. In these cases, there may be either incomplete dominance or *codominance* (equal expression of both alleles), or some other mechanism may be involved. As a result, different levels of the protein may be present in each of the three genotypes (*HH, Hh,* or *hh*), and this variation may result in three different phenotypes. For traits determined by more than one gene pair, it is easy to see how complex the genetic analysis can become.

Some Human Traits Determined by a Single Gene Pair

What sorts of human traits are determined by a single gene pair? We invite you to check out your phenotype as you proceed in reading this section.

A relatively limited number of observable human traits are clearly determined by a single gene pair, and attempts to list them are not without peril. For example, *tongue rolling,* the ability to curl the tongue into a longitudinal trough, has long been cited in biology texts as an example of a dominant Mendelian trait in humans since it was first described in 1940 by A. H. Sturtevant (of fruit fly fame). However, several studies of human groups have shown conclusively that it is not due to a single gene pair but must involve a variety of complex factors. To illustrate this point, nonrolling parents ("homozygous recessive" according to the old model) often produce tongue-rolling offspring, and people have been known

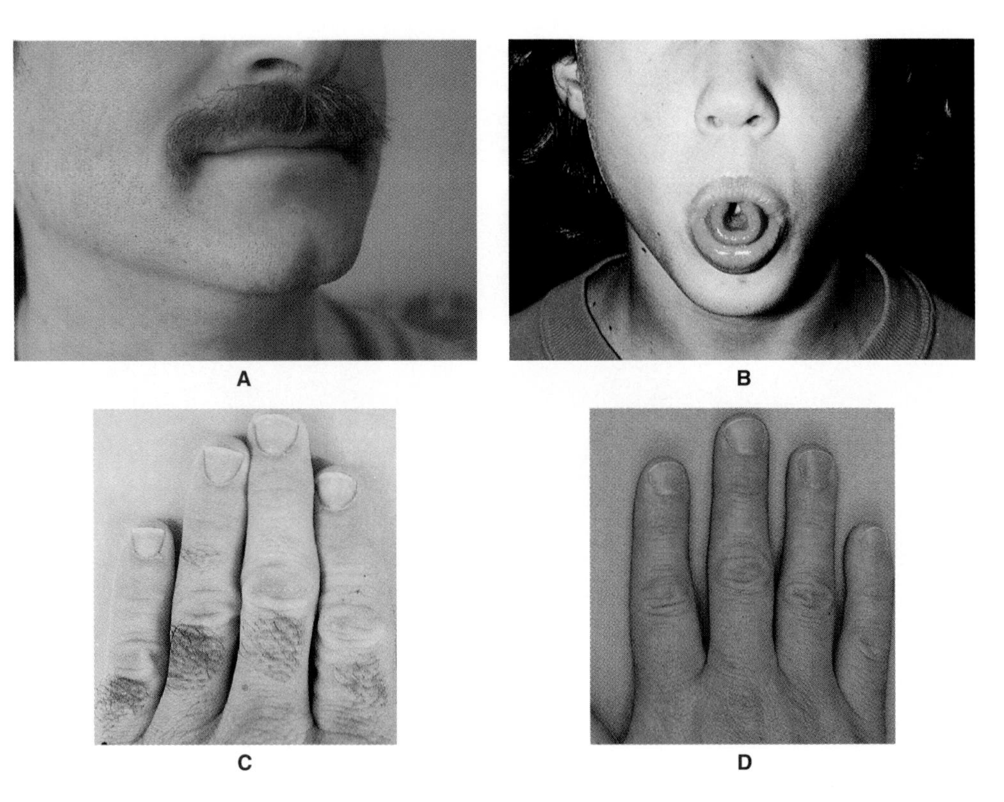

Figure 17.6

A few observable human traits are thought to be determined by a single gene pair including (A) chin fissure, (B) tongue rolling, (C) middigital hair, and (D) bent little finger .

to learn tongue rolling through practice. Nevertheless, tongue rolling remains in many biology and genetics texts as an example of a human trait determined by a single gene pair.

Figure 17.6 shows some human traits commonly assumed to be determined by a single gene pair. Some caution is recommended in evaluating these traits, since most are not simply expressed as two alternative forms. There is often **variable expressivity**, which means that the trait appears differently in people that apparently have the same genotype. There may also be **incomplete penetrance**, meaning that the trait is not evident in all individuals who have the relevant genotype. Finally, more than a single gene pair may eventually prove to be involved in determining a trait, or environmental or sexual influences may also affect the gene's expression.

For some traits, reference is made to a particular race. The underlying reason for this is that geneticists have found that many traits, particularly genetic disorders, are strongly associated with race. To identify and evaluate such associations, data collected on hereditary traits are categorized according to this standard. In many cases, there are extensive data sets for Caucasians but incomplete data for members of other races. Thus reference to Caucasians reflects the reliability of the data and does not indicate that the traits are not found in individuals of other races.

The inheritance of a *free earlobe* is thought by some researchers to be due to a dominant gene. People with the homozygous recessive genotype have earlobes attached directly to their heads. Since several ear-related traits are known to be controlled by many genes, some geneticists doubt that lobe attachment is controlled by a single gene pair. Nevertheless, inheritance of earlobes seems to fit a simple Mendelian recessive-dominant model involving a single gene pair.

The presence of *hair on the middle segment* of the four fingers may be due to a dominant allele. Homozygotes for the recessive allele have no hair on these finger segments, even though they may have abundant hair elsewhere.

The *ability to taste PTC* (phenylthiocarbamide) has long interested human geneticists. It also illustrates that genes can influence some rather bizarre sensory traits in humans, since PTC is a chemical that is never normally encountered in substances consumed by humans. To most people (the 70 percent with a dominant allele), PTC tastes quite awful, while to the rest it is tasteless. Evidence suggests that the gene for PTC tasting is located on chromosome number 7. In the North American Caucasian population, approximately 20 percent are homozygous tasters (genotype *TT*), 50 percent are heterozygous tasters (*Tt*), and 30 percent are homozygous nontasters (*tt*). However, the simple tasting-nontasting classification may be misleading, since among tasters there is considerable variation in taste quality reported, indicating there may be more than one gene pair involved.

A *bent little finger* (camptodactyly) is a minor hand abnormality characterized by an inward curvature of the little finger. It is due to a dominant allele and has both incomplete penetrance and variable expressivity.

A *mid-chin fissure* is the most common type of chin dimple and is caused by a dominant allele. It is a nice example of how a single gene pair can affect skeletal development. A number of features associated with this trait illustrate the complexities of "simple" inherited human traits:

1. It shows variable expressivity, ranging from a small dimple to a deep fissure.
2. It has nearly complete penetrance in males, but only about 50 percent of females with the allele have a fissure. Thus expression of the trait is *sex-influenced*—affected by the gender of the individual bearing the allele.
3. It has a *variable age of onset,* most commonly appearing in childhood or early adulthood.
4. It is most common among people of German extraction, with 20 percent of males and 10 percent of females having the trait.

Male-pattern baldness is the common form of hair loss in Caucasians and is another example of a sex-influenced trait caused by a gene that is fully expressed in males but less so in females. The hairline begins a progressive recession in young adult males as hair begins to fall out at the sides near the front of the scalp. Eventually, a bald spot appears at the rear of the crown, and finally, in about a third of affected males, only a fringe of hair remains along the lower crown of the head. It is estimated that at least 50 percent of Caucasian males and a much smaller number of Caucasian females are affected by such hair loss. Females experience much less hair loss than males and do not become bald unless they are homozygous for this trait. Thus, as with chin fissures, male-pattern baldness is sex-influenced, shows variable expressivity, and has a variable age of onset.

Some Human Traits Determined by Multiple Alleles and Polygenes

Most familiar human traits are determined by *multiple alleles*—the case in which there are alternative forms of a gene (alleles) that map to a single locus—or by *polygenes*—many genes at different loci. In *polygenic inheritance,* the numbers, chromosomal locations, and degree of expression of all the different genes are usually not known. Also, external environmental factors often influence the expression of such genes. Some of these traits are listed in Table 17.1.

ABO Blood Groups

The ABO blood group system was discovered in 1900 and was proved to be genetically determined in 1911. Its early discovery was probably connected with its clear pattern of Mendelian inheritance. The ABO blood group is one of the most completely understood human traits associated with multiple alleles; it is controlled by a single gene locus with three alleles. Even the site of the gene locus is known; it is located on chromosome 9q34. A, B, and O refer to the presence of specific molecules found on the surfaces of red blood cells. Blood types are linked to three alleles, designated I^A, I^B, and i. All humans have two alleles that determine which types of molecules will be present in their red blood cells. Humans

Table 17.1 Some Human Traits Determined by Multiple Alleles or Polygenic Inheritance

Trait	Pattern of Inheritance
ABO blood groups	Multiple alleles with codominance
Eye color	Polygenic with incomplete dominance
Fingerprints	Polygenic, influenced by external factors
Ear structure	Many features polygenic, development related to gender

with genotype I^AI^A or I^Ai have blood type A, genotypes I^BI^B and I^Bi are type B; I^AI^B results in type AB, and individuals with genotype ii have blood type O.

Eye Color

Human eye color is a trait that was poorly understood by classical geneticists. In 1907, an interesting research paper by Gertrude and Charles Davenport, titled "Heredity of Eye-Color in Man," appeared in the journal *Science*. The appeal of a simple Mendelian model had led to an assumption that eye color "in man is inherited as an alternative character" (that is, there were two human eye colors, brown and blue). The intent of the 1907 study was to test the hypothesis that human eye colors were inherited according to Mendelian predictions. The basic approach used was to study eye color in individuals from a number of families. After examining the data, the authors concluded that "blue eye-color is recessive to brown" and that "when two recessive individuals are mated *inter se* they throw [produce] only the recessive type." There was also a consideration of "imperfect [blue] owing to the presence of specks or patches of pigment—the 'gray' or 'hazel' color." However, the major conclusions of the paper—two primary eye colors, with brown being dominant over blue—became incorporated into textbooks and was therefore accepted as correct until relatively recently. The inadequacy of this simple model becomes obvious upon examining any large group of humans: you can see all sorts of black, brown, green, and blue eye colors.

Recent studies have led to the conclusion that eye color is a polygenic trait. There are no true blue, green, or black pigments in the human eye. Rather, eye color is mostly determined by the amount of *melanin* (a single brown pigment that also determines skin and hair color) present in the iris. A current model, described in Figure 17.7, proposes that at least two separate genes, each with two incompletely dominant alleles, control eye color. Each dominant allele results in the production of a certain amount of melanin. If there is little pigment (no dominant alleles), the eyes appear blue because of the way light is scattered by the sparse amount of pigment. The amount of pigment increases progressively with an increase in the number of dominant alleles. As a result, humans have eye colors that may be shades of green (one dominant allele), light brown (two), brown (three), and dark brown or black (four). This model, along with meiosis, also explains how two parents with light-colored eyes can have children with dark eyes.

Fingerprints

One of the most carefully studied human traits determined by polygenic inheritance is fingerprints. Three basic types of fingerprints occur in humans (see Figure 17.8). All humans have unique sets of permanent fingerprints, which are formed by the twelfth week of embryonic development. The three basic patterns are thought to be determined by a complicated polygenic system of dominant, incompletely dominant, and recessive genes. However, since not even identical

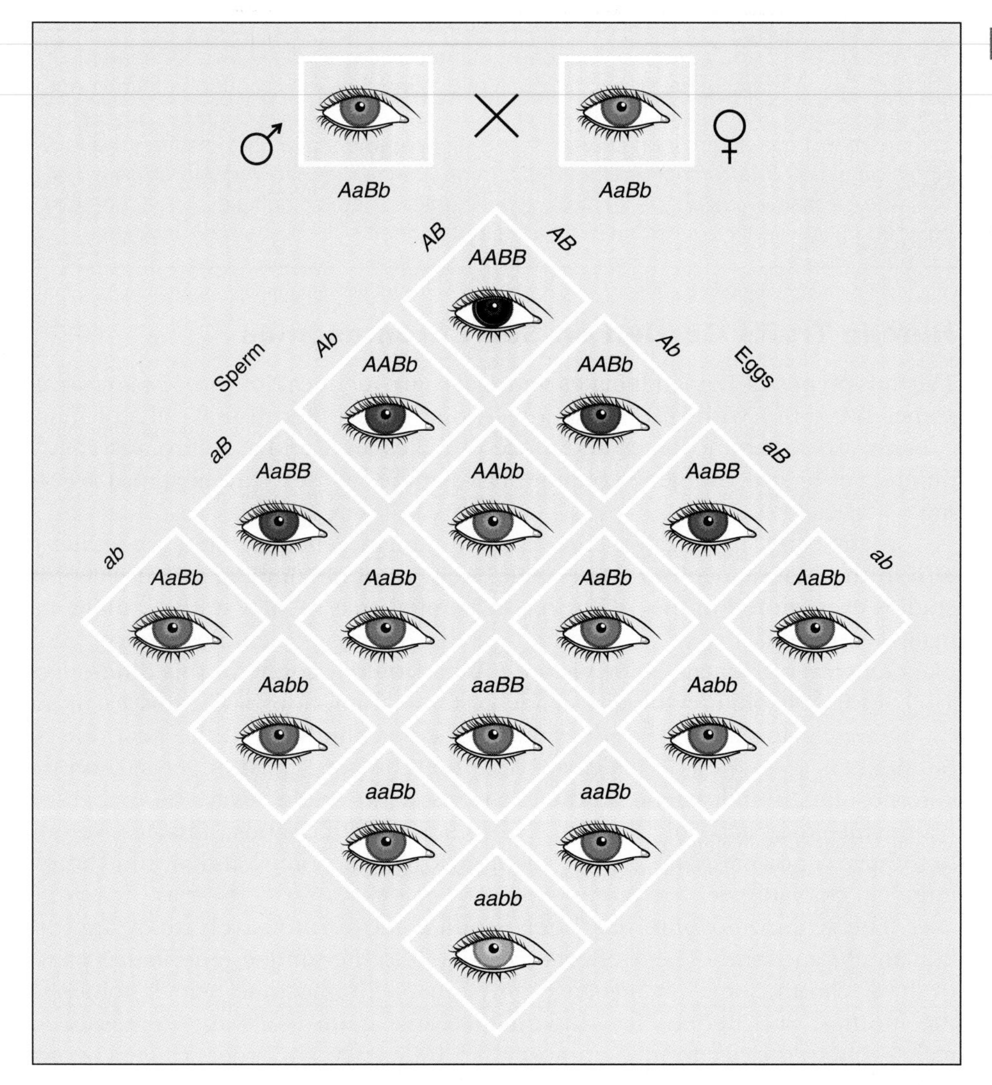

Figure 17.7

Human eye color is thought to be determined by polygenic inheritance. One model proposes that at least two genes, each with two incompletely dominant alleles, are responsible for producing melanin. Eye color is determined by the quantities of this pigment produced.

twins have the same fingerprints, external factors in the uterine environment must play some role in their formation during pregnancy.

Ear Structure

A final example of an odd trait determined by polygenic inheritance is ear structure. Like fingerprints, every human has a unique pair of ears. The number of genes and chromosomes involved in the creation of an ear is unknown but is thought to be large, given the impressive variety of ear shapes and sizes. Also, it is clear that many structural features of the human ear are under different genetic control and developmental regulation in males and females. The net effect of these differences is that males tend to have bigger ears than females, even though their heads are of a similar size. The significance of the ear size difference is unknown.

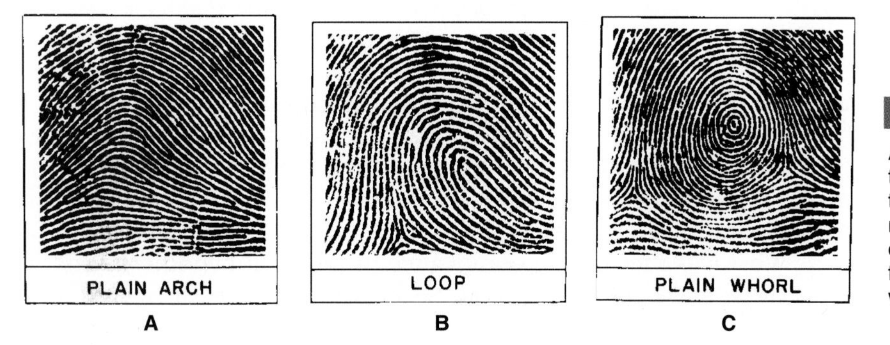

Figure 17.8

All humans have unique fingerprints that are determined by complicated interactions of polygenes and environmental factors during embryonic development. The three basic types of fingerprints—arch (A), loop (B), and whorl (C)—are shown.

The genetic basis and pattern of Mendelian inheritance are now known for nearly 5,000 human traits. A majority are due to autosomal dominant genes, about one third to autosomal recessive genes, and 5 to 10 percent are caused by X-linked genes. Several human traits, including hair on fingers, free earlobes, and a mid-chin fissure, involve a single gene pair. Many single-gene traits can be modified due to variable expressivity, sex, age, and other factors. ABO blood type is determined by multiple alleles, and eye color and fingerprints are examples of traits determined by polygenic inheritance.

Human Traits Carried on Sex Chromosomes

The karyotypes of human females and males are identical for the 22 pairs of autosomes but differ with respect to sex chromosomes. Recall that cells of females contain two X chromosomes and those of males an X and a Y chromosome. The differences in size and gene content between the two sex chromosomes are striking.

Many early investigators interested in human genetics noted a peculiar pattern in the inheritance of certain traits. A Swiss ophthalmologist (physician who studies eyes) noted in the 1870s that in families that he had studied, many more males than females were color-blind (see the Focus on Scientific Inquiry, "Molecular Genetics of Human Color Vision"). What could account for this difference? In 1911, E. B. Wilson, a cytologist, made certain assumptions and came to an interesting conclusion about the inheritance of color blindness. If the gene for color blindness is recessive and males are XY, then the gene must be located on the X chromosome. A faulty gene on the X chromosome would always be expressed in males since they had only the one X chromosome. In contrast, females were likely to have a normal gene on one of their X chromosomes and would therefore rarely be color-blind.

The X chromosome has been studied more intensively than any other human chromosome. These efforts are related to the unique inheritance patterns of the X chromosome—males have only one X chromosome, which comes from the mother, and hence an X-linked recessive gene is always expressed in a male—and to the occurrence of more than 100 genetic disorders caused by genes carried on this chromosome.

X Chromosome Inactivation

In 1949, M. L. Barr, a Canadian cytologist, observed a dark-staining structure in the nucleus of cells from female cats that was not present in cells from male cats. Barr referred to the structure as "sex chromatin," and it is now called the **Barr body** (see Figure 17.9). Subsequently, Barr bodies were found in nuclei from human females, and it was also shown, in studies of individuals with abnormal numbers of sex chromosomes, that the number of Barr bodies present in a cell nucleus was always one less than the number of X chromosomes. Thus cells of normal females (46,XX) have one Barr body, males (46,XY) have none, 47,XXX females have two, and 47,XXY males have one. What could explain this peculiar but consistent finding?

Figure 17.9

Barr bodies, which appear as dark-staining objects in the nucleus, represent X chromosomes that became inactivated during embryonic development.

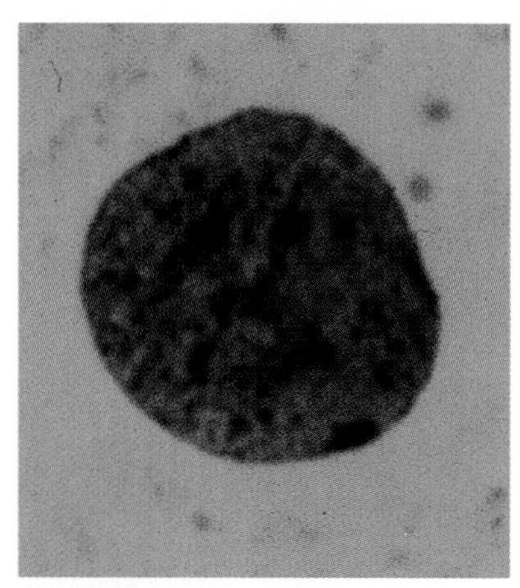

In 1959, it was proposed that the Barr body was an X chromosome with highly condensed, dark-staining DNA. Two years later, Mary F. Lyon, a British cytologist, formulated a hypothesis of **X chromosome inactivation** (the Lyon hypothesis), which stated that (1) the dark-staining X chromosome (Barr body) was genetically inactive; (2) either of the X chromosomes (that is, the one contributed by either the mother, X_m, or the father, X_p, could be inactivated in any given cell; (3) inactivation of the X chromosome took place early in embryonic development; and (4) any descendants of the cell would have the same inactivated X chromosome as the parent cell.

What was the genetic significance of the Lyon hypothesis, and what predictions did it make? Also, how could it be verified? First, it offered an explanation for the phenomenon of **gene dosage compensation** in which both males and females produce the same amount of X gene protein products. The hypothesis made a number of predictions, one of which was inaccurate. If gene dosage compensation were complete and perfect, 47,XXY males and 45,X0 females, assuming no inactivation of the single X chromosome, should be normal. In fact, they are abnormal. This suggests that some other genetic factor remains to be discovered, not that the hypothesis is incorrect. The most important prediction of the Lyon hypothesis was that cells of females are mosaics for X-linked genes; cells with an inactivated X_m chromosome would express the gene from X_p, and vice versa. Several studies have verified this prediction and thus confirmed the Lyon hypothesis. For example, an enzyme called G6PD has F and S forms. In females that are heterozygous for the enzyme, cells produce either G6PD-F or G6PD-S, but not both.

The Lyon hypothesis continues to be extended as new studies provide more detailed information. For example, it was recently discovered that inactivation is not entirely random if one of the X chromosomes is structurally abnormal (for example, if part of it has been deleted). In that situation, the defective X chromosome is always inactivated. From studies of mice, it is now understood that inactivation may not be a permanent condition. For reasons unknown, the proportion of cells with reactivated X chromosomes seems to increase with age. This finding raises many questions, the most important of which may be whether reactivation is an example of a general deterioration of gene control that is related to the aging process.

Genes Carried by the X Chromosome

It is well known that many genetic disorders trace their origins to defective genes carried on the X chromosome. Much less is known about normal traits carried by this chromosome, although certain inferences can be made from the effects of X-linked genetic disorders. For example, various forms of muscular dystrophy occur as a result of mutated genes on the X chromosome; therefore, some genes on that chromosome must be responsible for the normal development or functioning of muscle cells in humans.

In common with all chromosomes, the X chromosome contains genes that encode structural proteins and enzymes that are important for normal operations of the organism. Genes for blood-clotting factors and normal red-green color vision are also carried by the X chromosome.

Genes Carried by the Y Chromosome

Despite major progress since the 1950s, certain aspects of human sex determination are still not entirely understood. The X-Y basis of sex determination in humans and other mammals has been known since 1923; females have two X chromosomes (XX) and males have one X and one Y (XY). However, until the late 1950s, an open question remained unanswered: Which sex chromosome, the X or the Y, possesses the gene or genes responsible for governing the sexual destiny of a developing embryo? Some discoveries in the late 1950s and early 1960s seemed to point to the answer. Studies of individuals with abnormal numbers of sex chromosomes revealed that 45,X0 humans developed ovaries (female reproductive organs) and 47,XXY developed testes (male reproductive organs). What did this indicate? Consistent with these findings, it was hypothesized that the Y chromosome possessed a gene that dictated development of a male.

In 1987, David Page and his colleagues at the Whitehead Institute for Biomedical Research in Cambridge, Massachusetts, reported the results of studies that enabled them to pinpoint the location of a gene on the Y chromosome whose presence always resulted in the development of a male. The gene was called the **testis-determining factor (TDF)**, and it is located in Yp11.2. The TDF gene was also found on the Y chromosomes of other species of mammals, including monkeys, dogs, horses, and goats. Recently, a specific gene has been

MOLECULAR GENETICS OF HUMAN COLOR VISION

Focus on Scientific Inquiry

The story of understanding human color vision involves many elements of science and constitutes a rich and stunning success when considered in its entirety. The basis of human color vision has involved some of the world's greatest scientists, including Sir Isaac Newton and John Dalton. Newton's contribution dates back to the seventeenth century, when he analyzed the nature of colors in light using prisms. He found that "white light" is composed of a continuous series of colors that can be separated using a prism, as shown in Figure 1. By the eighteenth century, it was recognized that in most humans, all colors from the light spectrum were produced by mixing three principal colors (trichromacy)—red, green, and blue—together in different combinations. In 1774, Dalton, the father of atomic theory, read a paper before the Literary and Philosophical Society of Manchester (England) in which he described red-green color blindness in himself and one of his two brothers. Rather than being able to see the universe of color using the three principal colors, his view was limited to colors produced by combining only yellow and blue. Thus Dalton had diagnosed himself as being a dichromat with red-green "color blindness" (the term color-blind is a misnomer since very few people see no colors).

In 1802, Thomas Young, an early experimenter on light, hypothesized that trichromacy is a result of humans having three independent, light-sensitive mechanisms. This was, of course, long before even the most primitive principles of human genetics had been developed. From 1850 to 1900, studies were conducted using an instrument called the **anomaloscope** (which is still in use) that projects lights of different colors onto a screen. Individuals being tested are asked to adjust the ratio of different colors being projected to produce a specific color. Persons having many types of dichromatic color vision were identified in these studies; in each case, one of the three pigments (red, green, or blue) appeared to be missing. These results were consistent with Young's three-receptor hypothesis. Most of these early investigators recognized that males were much more likely to have anomalous color vision than females.

An Updated Hypothesis

The mechanism that allows humans to discriminate colors is based on three classes of color-

Figure 1 When reflected through a prism, white light is seen to be composed of different colors, each with a specific wavelength.

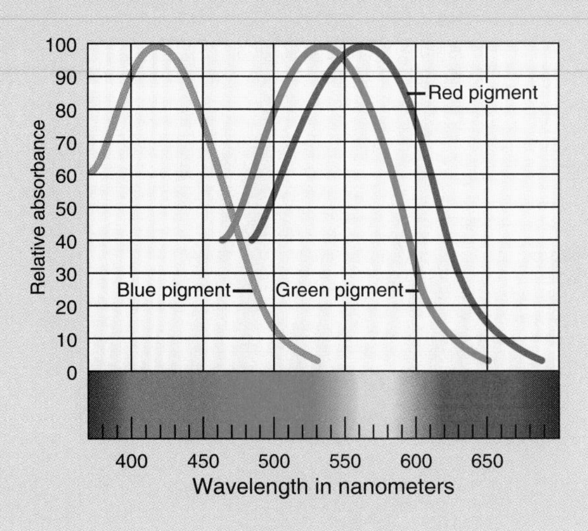

Figure 2 The eyes of humans possess three visual pigments—blue, red, and green—that respond to specific wavelengths of the visible light spectrum.

sensing cells, called **cones**, located in the retina of the eye. Each class of cells contains a pigment—red, green, or blue—that responds to different wavelengths of the visible light spectrum; the pigments are made of proteins. As shown in Figure 2, blue-sensitive pigment absorbs short wavelengths (420 nanometers in length), green-sensitive pigment absorbs intermediate wavelengths (530 nanometers), and red-sensitive pigment long wavelengths (560 nanometers).

Color-sensitive cones are one of two types of photoreceptors (light-sensing cells) in the retina of the eye. A fourth pigment, *rhodopsin,* enables us to see in dim light, but only in shades of gray, and is found in **rods**, the second type of photoreceptor. Light striking rods and cones stimulates the light-sensitive pigments and triggers a series of cellular reactions that ultimately results in nerve signals that are sent to the brain, where they are assembled into a color image.

By using techniques of molecular genetics, scientists gathered further information and gained new insights into human color vision. As a result, the following hypothesis was formulated: (1) three genes (or gene families) exist that encode blue, red, and green visual pigments; and (2) the three light-sensitive color pigments, plus rhodopsin, originated from a common ancestral pigment gene. This hypothesis was verified by research conducted at Stanford University, using advanced biological technologies. The results of these studies were published in 1988 and 1989 by Jeremy Nathans and coworkers.

Pigment Genes: Locations, Arrangements, and Variations

The three light-sensitive pigment genes are located on two different chromosomes. The blue pigment gene is present on chromosome number 7 between 7q22 and 7qter (*ter* refers to the terminal portion of the arm). The red and green pigment genes are found on the X chromosome in the Xq22-to-Xq28 interval. The gene encoding rhodopsin resides in chromosome 3 between 3q21 and 3qter.

The red pigment and green pigment genes are very closely related (98 percent similarity in DNA structure) and make up a family of repeated genes that exist as a head-to-tail array on the X chromosome described in Figure 3. A surprising research finding was the existence of considerable variation in the number of green pigment genes in males. Most individuals have two green pigment genes and one red pigment gene, or a 2:1 ratio, but 1:1 and 3:1 ratios are not uncommon. These differences result in red and green color vision variation, a condition that presently affects about 8 percent of Caucasian males and 1 percent of females. In carefully studied males with atypical red-green visualization, including different forms of color blindness, researchers have been able to correlate specific gene alterations that occur in meiosis with distinctive color variations. This very powerful result confirms the exact functions of specific red and green pigment genes in human males.

Evolutionary Aspects

There are a number of fascinating spin-offs from molecular studies of the human color vision system. The nucleotide sequences that code for the visual pigments have been used to determine the evolutionary relationships between the pigment genes. In this type of analysis, degrees of similarity (*homology*) between different genes (or gene sequences in different organisms) indicate how closely related they are in time, assuming that they originally evolved from a common ancestor. Genes (or organisms) with nearly identical DNA sequences are more closely related than those with more variation.

Analysis of this type of information has led to the formation of a model for the evolution of visual pigments that is summarized in Figure 4. The hypothesis indicates that the short-wavelength pigment (blue), long-wavelength pigments (red and green), and rhodopsin all originated from a common ancestral pigment gene at about the same time, between 500 million and 1 billion years ago. Since the red and green pigment genes are highly homologous, they apparent-

Focus on Scientific Inquiry

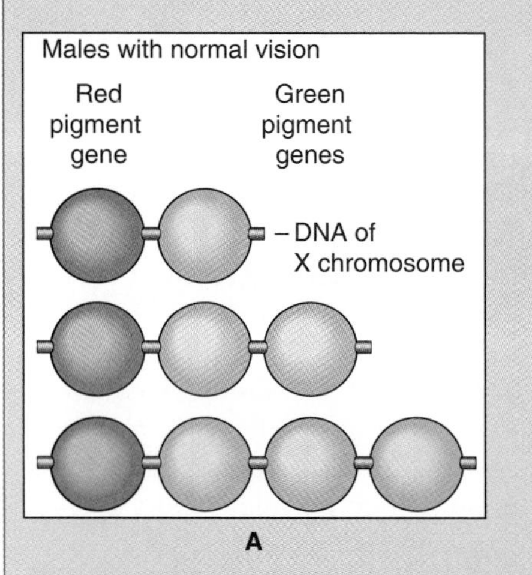

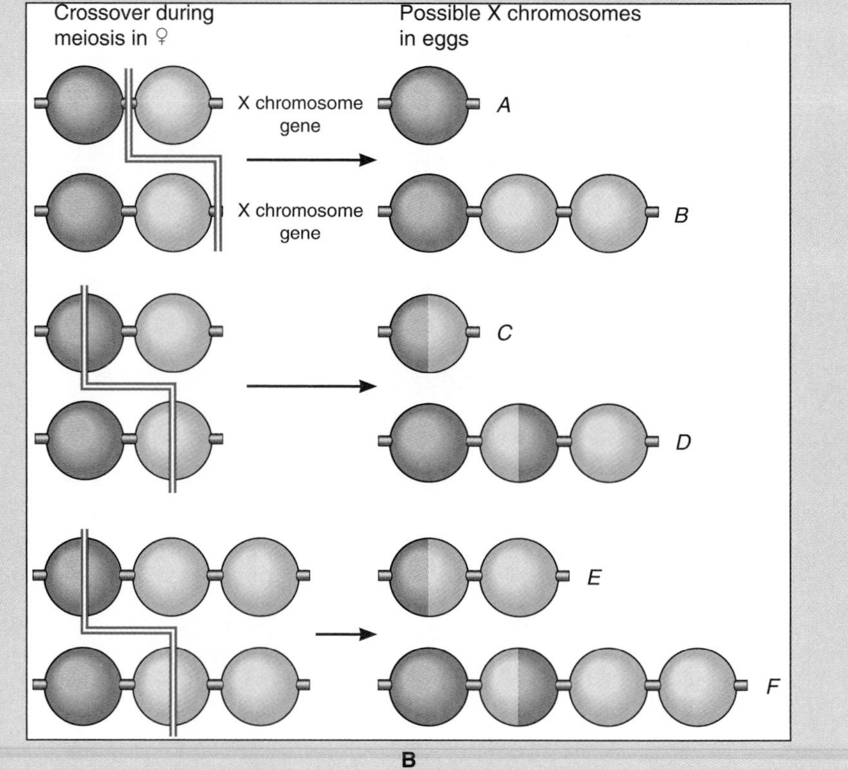

Figure 3 The genes encoding red and green visual pigments are located very close together on the X chromosome. Variations in the number of these genes lead to different red-green visualization. (A) Males with normal color vision were found to have one, two, or three green pigment genes. (B) The loss, partial loss, or addition of either a red or a green pigment gene results in different types of red-green color blindness. Male offspring resulting from the fertilization of eggs *A* and *C* would be deficient in green pigment genes; *C* and *E* would be deficient in red pigment genes. Consequently, all would have some variation of red-green color blindness. Males from eggs *B* and *F* would have normal red-green color vision.

ly diverged much more recently, approximately 30 to 40 million years ago.

The genetic process through which our existing visual pigment system arose—gene duplication, divergence, and shuffling of duplicated genes—is thought to be an important mechanism in the evolution of organisms since new genetic combinations can be produced in a relatively short period of time. The research on human visual pigments supports this hypothesis.

It seems possible that there is a potential for humans to develop even greater powers of color discrimination. The presence of multiple copies of green pigment genes has been correlated with an altered range of color visualization in affected individuals. Perhaps such variations in pigment genes indicate that a third X-linked visual pigment, sensitive to a new spectrum, could develop and enhance the visual capabilities of people who happen to inherit it. Would the world look the same to those with these new genes?

Figure 4 Molecular geneticists hypothesize that the human visual pigments evolved according to this model.

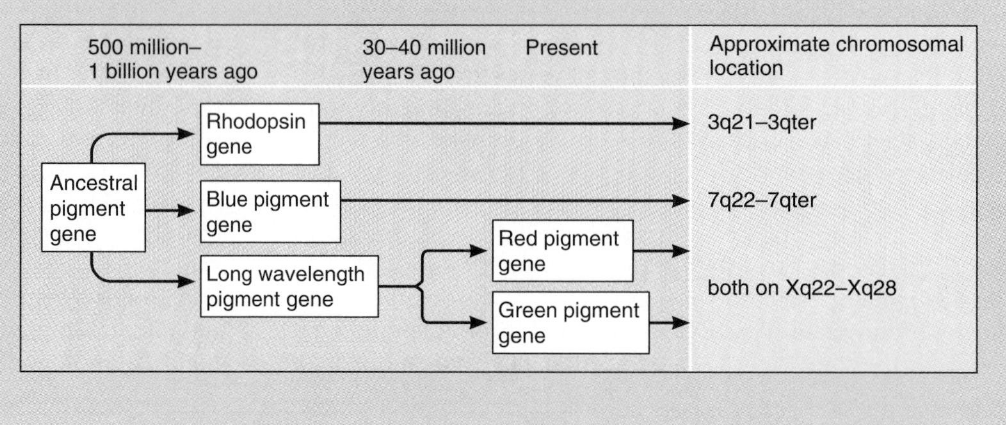

tentatively identified as being the TDF gene in humans; it is called *SRY*, sex-determining region of the Y chromosome.

The following picture of human sex determination has now emerged. The development of male and female embryos proceeds identically until about six or seven weeks after fertilization, when sexual differentiation occurs. At that time, a genetic signal initiates a cascade of biochemical events that leads to the development of a male. The genetic signal is hypothesized to be a protein encoded by *SRY* that regulates the expression of other genes. If the *SRY* gene is absent, the embryo develops into a female.

However, the story becomes more complicated because a nearly identical gene has also been found on the X chromosome. Its functional role, if any, in sex determination has not yet been defined. Thus in spite of the great excitement created by the discovery of the *SRY* gene, the complete story of sex determination in humans (and other mammals) has not yet been deciphered, and research on this exciting question continues.

A few other structural genes are located on the Y chromosome. Two or three Y genes encode specific proteins found on cell surfaces. Also, available evidence indicates that one or more genes affecting sperm production, stature (height), tooth size, and hair on ears are located on the Y chromosome.

The X chromosome has been more carefully studied than any other human chromosome. Many genes on the X chromosome determine traits that can be studied because of their pattern of inheritance in males and females. In females, only one of the X chromosomes remains active in cells, and the other is inactivated during embryonic development. Only a few genes are known to be carried by the Y chromosome, the most significant being the TDF (*SRY*) gene that determines the sex of offspring.

SOME FINAL THOUGHTS

It seems strange that the discipline of human genetics lagged far behind *Drosophila* genetics, microbial genetics, and the genetics of agricultural plants and animals. What were the reasons for this delay? A simple but incomplete answer is that human genetics was a much more complicated field (See the Focus on Science and Society, "Human Genetics and the Abuse of Science"). At first glance, the concepts of classical genetics probably seemed to have rather limited application in studying inherited human traits. Nevertheless, it is obvious that the basic genetic principles defined by Mendel, Morgan, and others—segregation, recombination, independent assortment, linkage, and chromosome mapping—all apply to the inheritance of human traits.

There are other reasons why human genetics did not emerge as a major scientific field until the 1960s. Most obviously, because of ethical considerations—humans are not used as research subjects in controlled genetics studies—long generation time, and few offspring, this species does not lend itself to the same experimental manipulations as fruit flies and bacteria.

Despite the beautiful pioneering work of Garrod described in Chapter 16, few researchers in the medical sciences saw a connection between genetics and human disorders. During the first half of the twentieth century, human genetics did not exist as a discipline in medical schools because the field was not considered fundamentally relevant to medicine. Thus most geneticists had appointments in biology departments or agricultural experimental stations where the organisms of major interest were not humans. What changed this situation? Events described in Chapters 15 and 16 had some role, but the development of improved cytologic techniques and the accurate determination of the human karyotype in 1956 were especially significant. Soon after, cytogeneticists discovered numerous abnormalities in human karyotypes (for example, 45,X0, 47,XXY, and

Focus on Science and Society

Human Genetics and the Abuse of Science

A review of the history of human genetics shows how science and human values can become intertwined. It also demonstrates how scientific research results can be abused by flawed analyses and unwarranted extrapolations.

In 1900, at the time of the rediscovery of Mendel's laws of segregation and independent assortment, biologists throughout Europe and the United States were debating the nature of human intelligence and the comparative influences of heredity and the environment. To many of these biologists, Mendelian genetics appeared to provide a valuable additional perspective in understanding human heredity. Some of these scientists were also concerned with implications of evolution for human heredity and believed that knowledge from both genetics and evolution could be combined to "improve the human race." These individuals called themselves *eugenicists* and conducted different types of research. For example, the American eugenicist (and famous geneticist) Charles Davenport established a well-funded station at Cold Spring Harbor on Long Island, New York, for "the experimental study of evolution." There he studied different families with the aim of improving humankind.

Davenport and others throughout the worldwide scientific community were genuinely concerned with what they perceived to be disturbing implications of evolution and genetics for the human population. They argued that harmful variations, normally removed by natural selection in other species, were "artificially" preserved by human society. In 1914, W. Grant Hague, in *The Eugenic Marriage,* expanded on this theme. Two quotations will serve to convey the flavor of eugenicists' concerns. Hague wrote that unless something was done immediately,

> every second child born in this country, in fifty years, will be unfit; and, in one hundred years, the American race will have ceased to exist. We mean by this that every second child born will be born to die in infancy, or, if it lives, will be incapable of self-support during its life, because either of mental degeneracy or physical inefficiency.

In calling attention to the potential harm of existing conditions, we find,

> any condition that fundamentally means race deterioration must be rendered intolerable. The prevalent dancing craze is an anti-eugenic institution, as is the popularity of the delicatessen store. No sane person can regard with complacency the vicious environment in which the future mothers of the race "tango" their time, their morals, and their vitality away. We do not assume to pass judgment on the merits of the dance; we do, however, emphatically condemn the surroundings.

In retrospect, it can be seen that these early researchers had a very incomplete view of human genetics. There is, of course, no such group as the "American race" Hague wrote about. These scientists did not appreciate that a deleterious allele could remain in a population in a heterozygous state. Even more misguided and of greater importance, the eugenicists also mixed morality with genetics. They did this by assuming that all physical, mental, and moral traits were inherited in accordance with simple Mendelian ratios. Therefore, insanity, epilepsy, alcoholism, criminality, and feeblemindedness were treated in the same fashion as eye color and albinism. Feeblemindedness in particular drew the attention of early eugenicists, as indicated by Figures 1 and 2.

The early eugenicists also mixed their political opinions into their scientific discussions. Since the post–Civil War years, the United States had been receiving millions of immigrants from southern and eastern Europe. These people came from different ethnic stocks than the predominately Anglo-Saxon people who colonized the United States and who represented the majority of the population. Many of the early eugenicists considered these immigrants, as well as the blacks already in the United States, as racially inferior and therefore worried about the "degeneration" of the American people. More enlightened eugenicists argued that only individuals with a family history of undesirable elements (mental retardation, insanity, criminality, alcoholism, and so on) should be barred from entry to the United States. Still, many eugenicists completely opposed the immigration of "inferior races." Similar racial and ethnic prejudices existed in European countries.

Today, we can look back with disbelief on some of the excesses of these naive views. However, it is important to realize that they had significant influence at the time. Eugenicists in the United States were active in campaigning for sterilization laws. Although most of the geneticists considered these laws premature, eugenicists considered them an important part of "racial hygiene." By 1931, as many as 30 of the 48 states had laws that permitted sterilization of the feebleminded, the insane, alcoholics, and rapists. By 1958, over 60,000 sterilizations had been performed for eugenic reasons. Although in later years many of these laws were

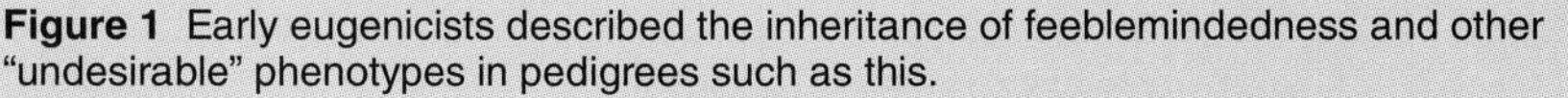

Figure 1 Early eugenicists described the inheritance of feeblemindedness and other "undesirable" phenotypes in pedigrees such as this.

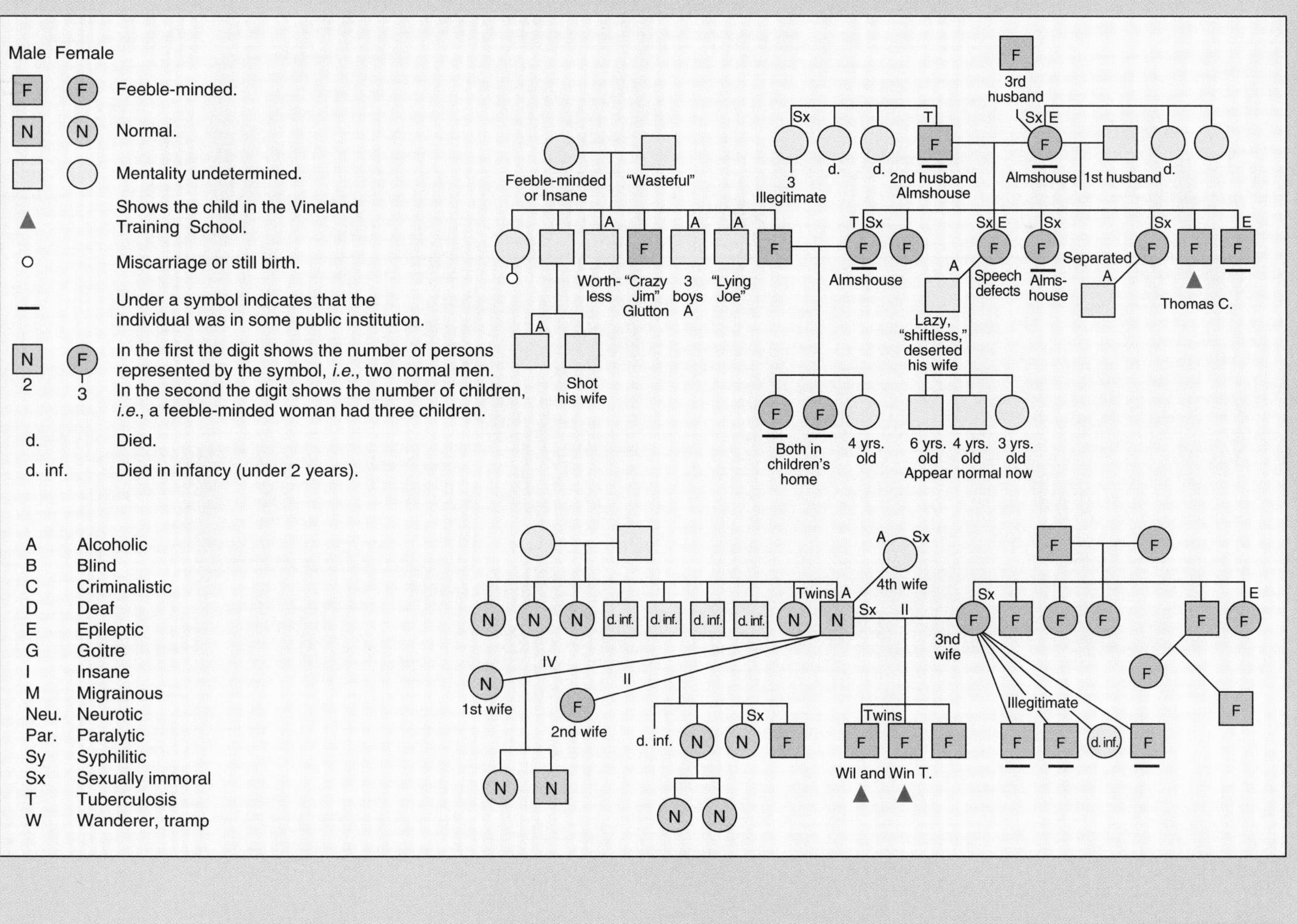

modified to protect individual rights and to reflect more "enlightened" genetics, most are still on the books today.

Similar to the sterilization laws, immigration laws passed early in the twentieth century were strongly influenced by the eugenics movement. The Immigration Restriction Act of 1924 limited the annual immigration from each European nation to 2 percent of that country's natives living in the United States during the 1890 census. That act was an attempt to keep the Anglo-Saxon Protestant majority from being "diluted," and it served as the template for future immigration legislation.

By 1930, American geneticists had come to realize that the early eugenics program was seriously flawed. They understood that politics and morality had been mixed with incomplete genetic knowledge, and for the most part, they repudiated their earlier program. They argued for a more "scientific" approach that distinguished cultural from biological factors and stressed the need for research on human genetics. In spite of repudiation by leading geneticists, the early American eugenics movement persisted. Not until the worldwide reaction of outrage to Nazi extermination of millions of people in the name of "racial hygiene" was the early American eugenics movement finally extinguished.

Figure 2 Eugenicists felt that the type of house a person occupied was related to qualities of the mind.

47,XXX) that were clearly the causes of specific disorders (some of these are described in Chapter 18). Thus, beginning in the 1960s, the central issue of medical implications, combined with increased funding for research, ushered in an era of explosive growth in human genetics, which is now a preeminent discipline in biology.

During the 1980s, scientists achieved much finer resolutions in all aspects of human genetics. The organizational features of our chromosomes are now well known. Details of the structure of chromosomes have permitted cytologists to draw precise maps based on constant banding patterns. Stunning progress has been made in unraveling the structure, types, numbers, and locations of genes. Finally, it is somewhat shocking to consider that the complete human genome can be compared to a hypothetical library holding 2 to 5 million books, of which only 50,000 to 100,000 contain pages that we know how to read.

Summary

1. The field of human genetics developed slowly because humans are complex, have long life spans, and could not be used in controlled genetics experiments. Until the 1960s, genetics researchers were much more interested in using organisms such as fruit flies, bacteria, and viruses that could help them answer questions in a relatively brief period of time.
2. The human karyotype was not correctly determined until 1956. Cells of normal females have 22 pairs of autosomes and one pair of X sex chromosomes; males have 22 pairs of autosomes and one X and one Y sex chromosome.
3. Chromosomes have a consistent structure. When stained with certain dyes, each chromosome has a unique set of bands and subbands that serves as a landmark that can be used to distinguish it from other chromosomes. These bands are also used as maps in determining the precise locations of genes on a chromosome.
4. Human and other eukaryotic chromosomes are characterized by enormous amounts of DNA that contain no known functional genes.
5. To date, almost 5,000 human phenotypic traits can be explained in terms of Mendelian inheritance. The genes that determine over 2,500 of these traits have been identified, and their locations on specific chromosomes are known.
6. Human traits thought to be determined by a single gene pair with complete dominance are free earlobes, hair on the mid-joint of fingers, the ability to taste PTC, a bent little finger, a mid-chin fissure, and male pattern baldness. A trait determined by multiple alleles is ABO blood types. Eye color, fingerprints, and ear structure are examples of polygenic traits.
7. Many genes are carried on the X chromosome, but only a few have been found on the Y chromosome; one, the *SRY* gene, determines sex of offspring.
8. Only one of two X chromosomes is genetically active in humans and other mammals. The second X chromosome of females is randomly inactivated during embryonic development. Consequently, the amount of gene products derived from the X chromosome is the same in both males and females.

Review Questions

1. What is a karyotype? How are karyotypes determined?
2. Describe the human karyotype.
3. What are chromosome bands? How are they used in human genetics?
4. Describe the human genome.
5. What different types of DNA are found in the nucleus? What is known about the function of each?
6. Explain differences between the genetic bases of these human traits: free earlobe, ABO blood type, eye color, and color blindness.
7. Explain the pattern of inheritance for sex-linked traits in humans.
8. What is X chromosome inactivation? Why is it significant?
9. What is the *SRY* gene? Where is it located? What does it do?

Essay and Discussion Questions

1. Using ideas from hypotheses related to the origin of life on Earth (Chapter 2) and

the RNA world, how might the existence of different DNAs in eukaryotic chromosomes be explained?

2. Why might studies of "normal" human, genetically determined traits be expected to be less fruitful than studies on human genetic disorders? Can results from one possibly have applications in the other? Explain.

3. Which areas of human genetics research do you feel should be emphasized (and hence receive funding) in the next decade? Why?

References and Recommended Readings

Badge, R. L. 1991. *SRY* and sex determination. *The Journal of NIH Research* 3:57–59.

Carson, H. L. 1986. Patterns of inheritance. *American Zoologist* 26:797–809.

Charlesworth, B. 1991. The evolution of sex chromosomes. *Science* 251:1030–1033.

Childs, B. 1986. Science as a way of knowing: Human genetics. *American Zoologist* 26:835–844.

Davenport, G. D., and C. B. Davenport. 1907. Heredity of eye-color in man. *Science* 26:589–592.

Hague, W. G. 1914. *The Eugenic Marriage: A Parental Guide to the New Science of Better Living and Better Babies.* New York: Review of Reviews.

Jenkins, J. B. 1990. *Human Genetics.* New York: HarperCollins.

Kevles, D. J. 1985. *In the Name of Eugenics: Genetics and the Uses of Human Heredity.* Berkeley: University of California Press.

Lyon, M. F. 1988. The William Allan Memorial Award Address: X-chromosome inactivation and the location and expression of X-linked genes. *American Journal of Human Genetics* 42:8–16.

McKusick, V. A. 1990. *Mendelian Inheritance in Man: Catalogs of Autosomal Dominant, Autosomal Recessive, and X-linked Phenotypes.* 9th ed. Baltimore: Johns Hopkins University Press.

Nathans, J. 1989. The genes for color vision. *Scientific American* 263:42–49.

Strickberger, M. W. 1986. The structure and organization of genetic material. *American Zoologist* 26:769–780.

Sutton, H. E. 1988. *An Introduction to Human Genetics.* 4th ed. Orlando, Fla.: Harcourt Brace Jovanovich.

Vogel, F., and A. G. Molusky. 1986. *Human Genetics: Problems and Approaches.* 2d ed. New York: Springer-Verlag.

CHAPTER 18

Human Genetic Disorders

Chapter Outline

Reading Questions

1. What is the importance of human genetic disorders?
2. What are pedigrees? How are they used in human genetics?
3. What is the genetic basis of sickle-cell anemia? What are the effects of this genetic disorder?
4. What are the genetic nature and effects of Huntington disease? Of familial hypercholesterolemia?
5. What is the function of prenatal diagnosis and genetic counseling?

During the 1980s, knowledge related to human genetics accumulated at a phenomenal rate. Methods and experimental approaches used in the so-called *new genetics,* a term introduced in 1979, allowed scientists to probe human chromosomes at a level that was not previously imaginable. Abnormalities in genes and chromosomes cause **genetic disorders**, a predictable set of consequences associated with a specific gene or chromosomal mutation.

Research designed to identify the causes of serious genetic disorders occupies a mainstream position in the field of human genetics. There are several important reasons for this: genetic disorders are present in a significant number of individuals, they are not curable and often not even treatable, they cause great emotional and financial stress on the families of affected individuals, and they place a considerable burden on health, social, and other community services. In addition to contributing to understanding the disease process, detailed knowledge about an abnormal condition will frequently provide insights into the underlying, but often poorly understood, normal processes that have been disrupted.

THE NATURE OF GENETIC DISORDERS

The causes of genetic disorders are rooted in modifications of normal DNA or chromosomes (or both) in tissues that produce sperm or egg cells. Some **congenital malformations** (deformities that occur during fetal development) seem to have a strong inherited element and may also be traced to altered genetic material. The genetic basis of most human genetic disorders is described in Figure 18.1. In offspring produced by fertilization involving a defective gamete, the altered DNA or chromosomes may result in abnormal concentrations of normal proteins, the absence of critical proteins, or the production of abnormal proteins. Such mutations can then lead to biochemical deficiencies or structural abnormalities of cells, tissues, or organs.

Recall that alterations in genetic material are called *mutations.* Two general categories of mutations are recognized: *gene mutations* occur when a single gene changes from one allelic form to another, and *chromosome mutations* occur

Figure 18.1

The basis of most human genetic disorders lies in the alteration or elimination of normal proteins because of mutated genes.

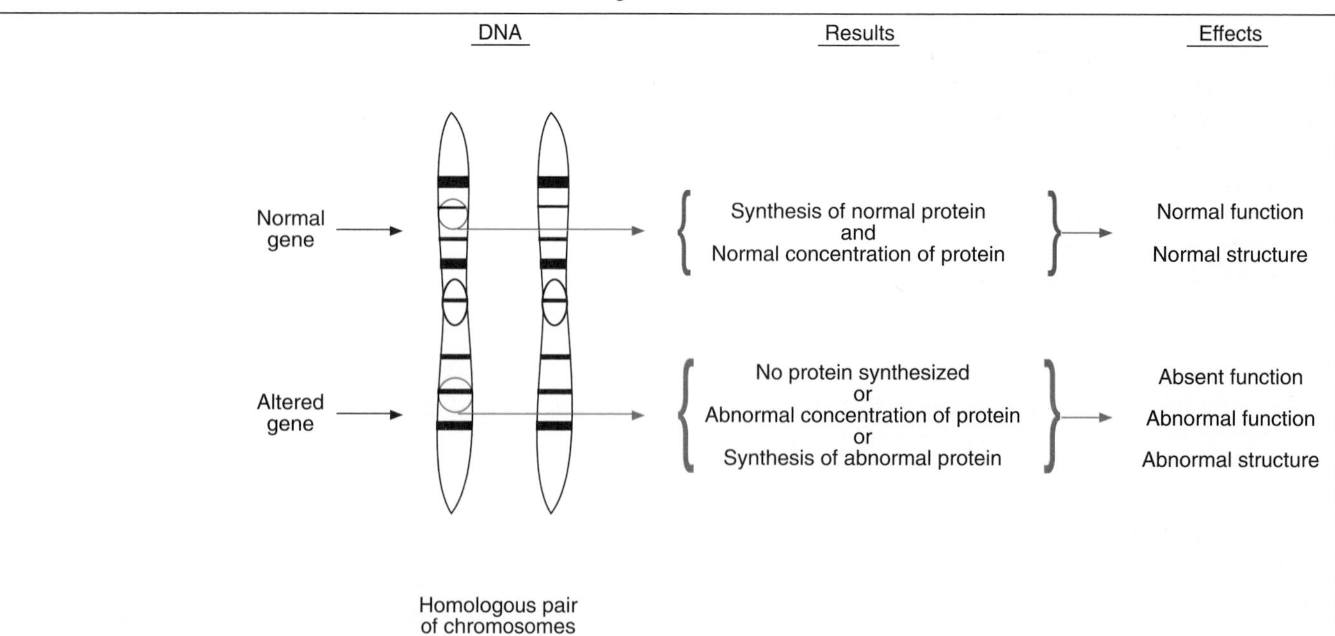

with changes in segments of chromosomes, whole chromosomes, or entire sets of chromosomes (see Chapter 14). By definition, chromosomal mutations affect more than one gene.

Gene Mutations

Single-gene defects cause many classical diseases that can often be followed through many generations of an affected family. Genetic disorders caused by single-gene mutations are classified as dominant or recessive, autosomal or sex-linked. Recessive disorders usually involve a defective enzyme or hormone, whereas dominant disorders are frequently characterized by an abnormality in a structural protein. All human organ systems (for example, the nervous, circulatory, skeletal, and immune systems) are affected by one or more genetic disorders. Most, if not all, metabolic pathways that depend on a variety of normal enzymes for processing DNA, RNA, lipids, carbohydrates, or proteins can also be disrupted by inherited disorders. More than 3,000 human genetic disorders caused by single-gene mutations have now been identified.

Genetic disorders caused by all mutations and their associated abnormalities occur in 2 to 5 percent of all live births and are an important cause of death in children under the age of 15 years. About 10 percent of the adult population may be affected by chronic disorders with a significant genetic component (for example, certain types of heart disease and cancer).

Chromosomal Mutations

A second group of inherited abnormalities are caused by chromosomal mutations that arise primarily as a result of imperfect meiosis during gamete formation, as described in Chapter 14. There are two principal types of chromosomal mutations, structural and numerical.

Structural mutations occur when there is a rearrangement of genes or sets of genes along specific chromosomes. Recall that sections of chromosomes containing many genes may break and do one of the following: (1) reattach to a different chromosome (an instance of *translocation*), (2) be removed from the chromosome (an instance of *deletion*), (3) be *inserted* somewhere into the middle of the same chromosome or a different chromosome (a type of translocation), or (4) be *inverted* and incorporated in reverse orientation in the same chromosome. A *duplication* results when multiple copies of genes are produced. Though usually harmful, this may, on rare occasions, have a positive effect by producing additional genetic material capable of performing new functions.

Numerical mutations increase or decrease the number of whole chromosomes without changing the structure of the individual chromosomes. The primary cause of numerical mutations is *nondisjunction,* the failure of paired homologous chromosomes to separate during meiosis, resulting in one gamete receiving both and the other gamete receiving neither of the affected chromosomes (see Figure 18.2). If the defective gamete participates in fertilization, an abnormal number of chromosomes will be found in every diploid cell of the resulting embryo. *Trisomy* occurs when a third copy of a homologous chromosome is present in cells. The condition in which cells lack one chromosome from a homologous pair is called *monosomy.*

It is estimated that chromosomal mutations occur in approximately one of every ten conceptions. About 90 percent of these cause such serious abnormalities that development ceases after a short period and a spontaneous abortion occurs, often before the mother is even aware that she is pregnant. Offspring that are born usually have distinctive structural and functional symptoms, a **syndrome** that is quite characteristic for each specific chromosome mutation. The biological consequences of such effects include premature mortality, infertility, and physical and mental handicaps.

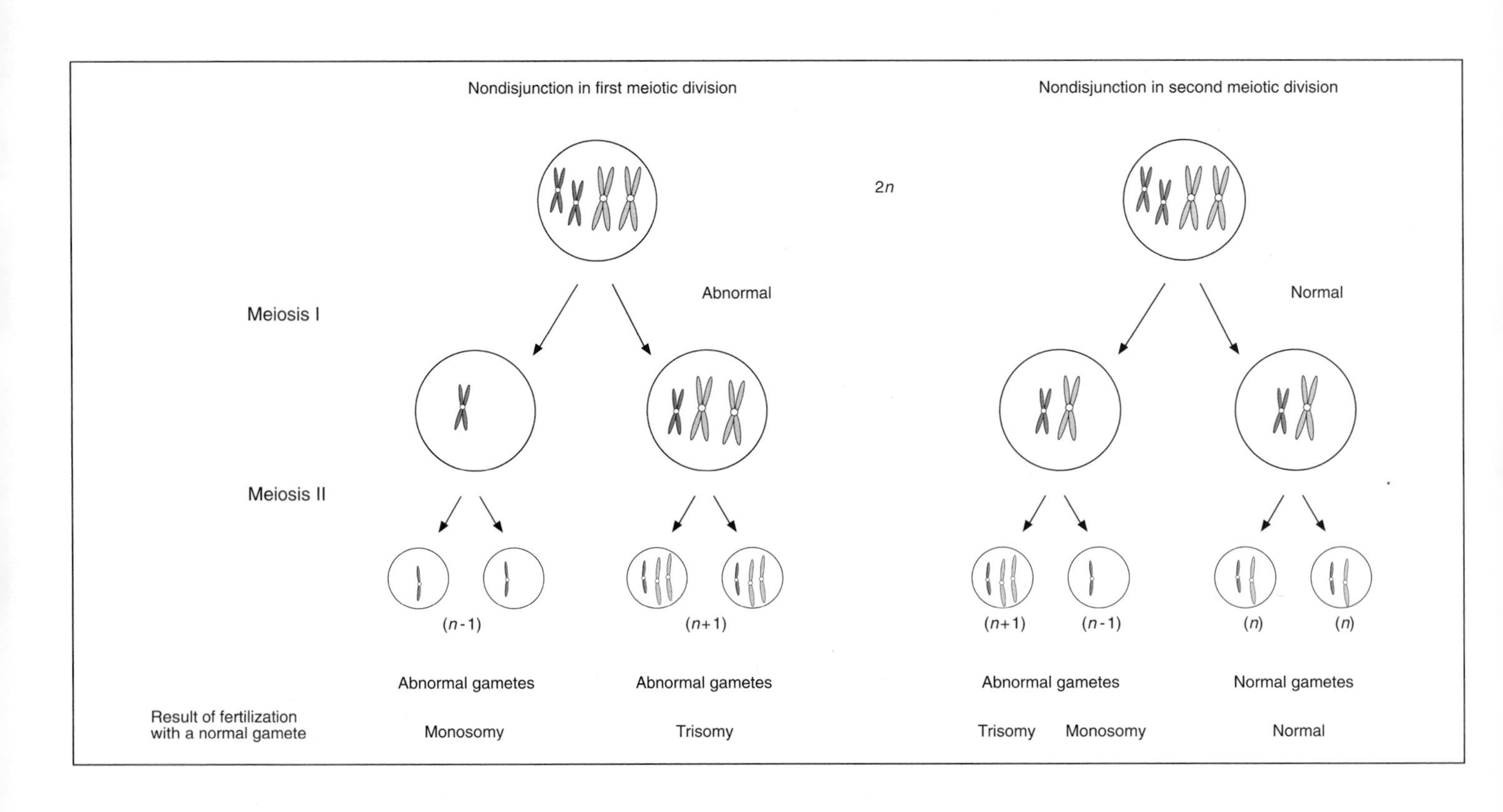

Figure 18.2

Nondisjunction of chromosomes may occur during either the first or the second division of meiosis and results in the production of eggs and sperm with abnormal numbers of chromosomes. If such gametes take part in fertilization, the resulting zygote will have one more chromosome than the normal diploid number (a condition called *trisomy*) or one less (a condition called *monosomy*).

For live births, the most frequent structural mutations involve chromosome numbers 13, 14, 15, 21, and 22; frequent numerical mutations include trisomies of autosomes 13, 18, and 21 and abnormal numbers of sex chromosomes, usually 47,XXY, 47,XXX, or a monosomy, 45,X0. The most common chromosome mutation is trisomy 21, which causes Down syndrome (see the Focus on Scientific Inquiry, "Down Syndrome", on pages 344–345). Chromosomal mutations occur in other chromosomes, but their devastating effects are not compatible with survival and so are rarely seen in live births. Figure 18.3 summarizes the quantitative effects of chromosomal mutations on human conceptions.

Human genetic disorders are caused by gene and chromosomal mutations and affect a significant number of humans. Mutant genes associated with more than 3,000 human genetic disorders have now been identified. Chromosomal mutations are relatively common, and their effects on the developing embryo are usually extremely harmful. At present, knowledge about the underlying genetic and molecular nature of genetic disorders is increasing rapidly, but there has been limited success in treating or curing most conditions.

HUMAN PEDIGREE ANALYSIS

Pedigrees are simple diagrams that are useful for tracing the inheritance of genetic disorders in families. For best results, it is necessary to have detailed records for as many family members as possible, spanning several generations, which indicate whether or not each individual has or had the trait of interest. Extensive and complete records permit construction of pedigrees that may reveal informative patterns in the way that traits are passed from one generation to the next.

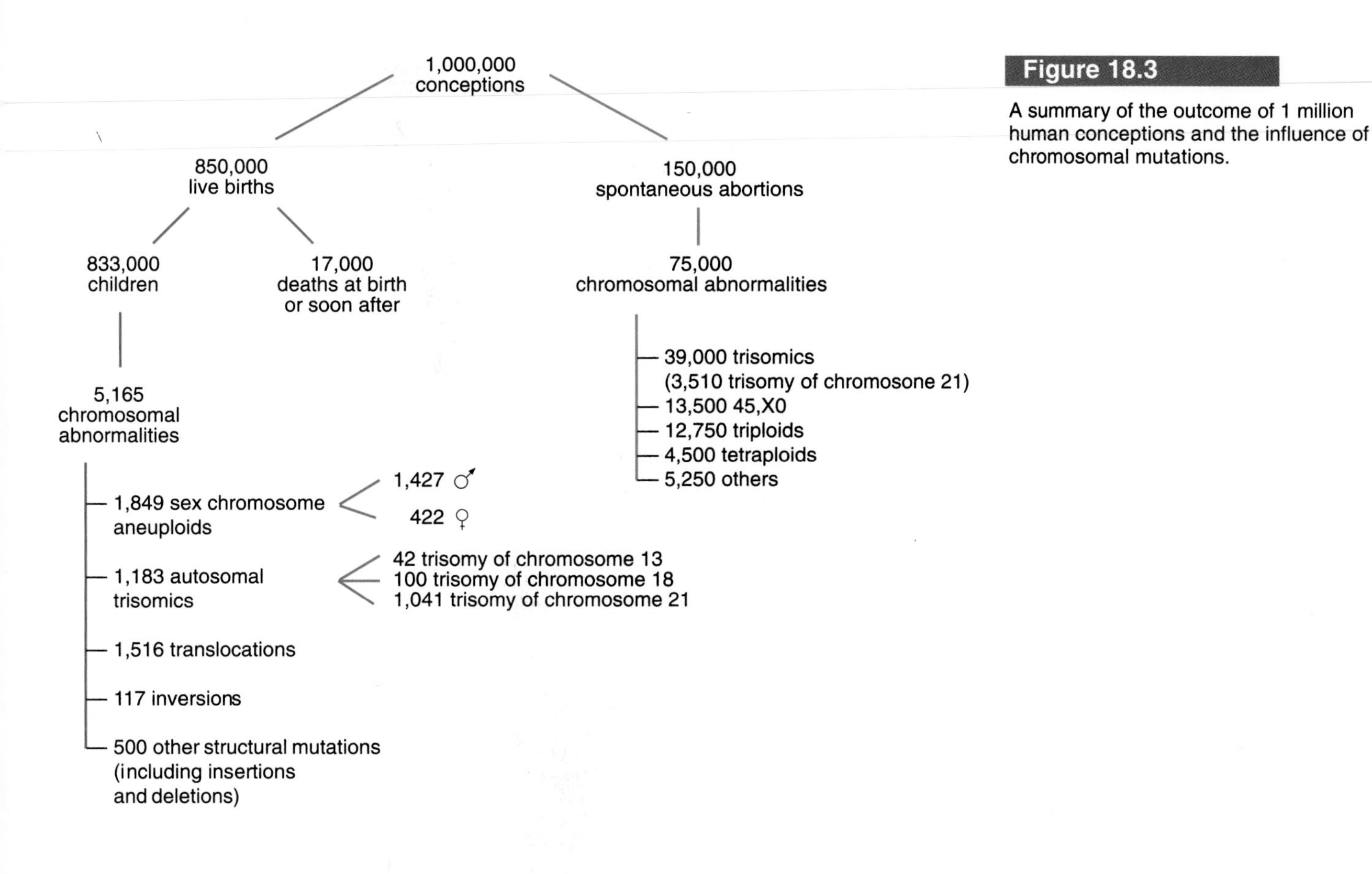

Figure 18.3

A summary of the outcome of 1 million human conceptions and the influence of chromosomal mutations.

Geneticists use standard symbols and conventions in diagraming pedigrees, as shown in Figure 18.4A. A pedigree can often determine the type of gene mutation responsible for the disorder being studied. What patterns of inheritance would be expected for each type of gene mutation? Researchers use established criteria to distinguish recessive, dominant, and sex-linked traits in families. *Autosomal dominants* can be transmitted through continuous generations in a family, affect both males and females, and are transmitted by both males and females to male and female offspring. Disorders caused by *autosomal recessives* tend to appear in alternate generations, but otherwise hereditary patterns are similar to those of autosomal dominants. A pedigree with many affected males in two or more family units connected through female relatives is evidence of *X-linked recessive* inheritance. If a trait is due to an *X-linked dominant,* all the daughters, but no sons, of an affected father will have the disorder. Figure 18.4B shows typical pedigrees that demonstrate the forms of inheritance involving various gene mutations.

Pedigrees are charts of family generations that show individuals affected by genetic disorders. Through logical analysis of pedigrees, it is usually possible to determine the type of gene mutation—dominant or recessive, autosomal or sex-linked—that is responsible for the disorder.

GENETIC DISORDERS OF HUMANS

In this section, we will look at a limited number of human genetic disorders to provide a general understanding of their underlying causes and effects. The relative frequency of most genetic disorders varies significantly from one human population or ethnic group to another. For example, sickle-cell anemia occurs in 1 to 2 percent of offspring born to black Africans but is extremely rare in other races. Cystic fibrosis is relatively common in offspring of northern Europeans but

Figure 18.4

(A) Human pedigrees are constructed using these symbols. (B) These pedigrees show typical patterns of inheritance associated with different types of gene mutations.

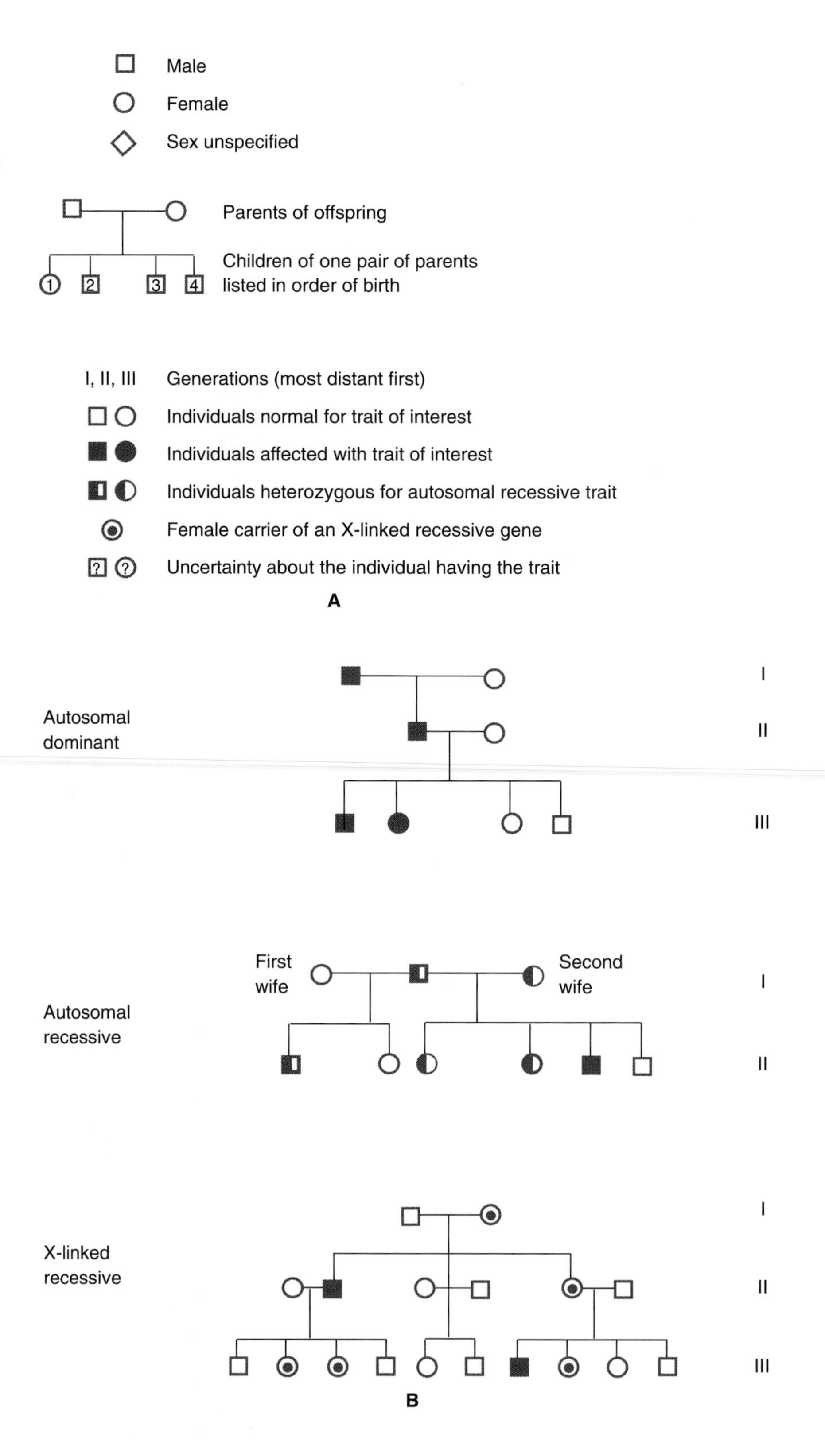

is very unusual in Orientals or African-Americans. The reasons for these differences are not always clear, but geneticists have speculated that they may often be related to some survival advantage conveyed by the allele in individuals who are heterozygous for the condition. Thus heterozygotes may be able to better withstand some environmental factor (for example, a disease) than individuals without the allele. Table 18.1 provides some information on the genetic disorders discussed in this chapter.

Table 18.1 Genetic Disorders Described in This Chapter

Disorder	Genetic Basis
Sickle-cell anemia	Single gene locus on chromosome 11 at 11p15.5
Sickle-cell disorder	Individuals that are homozygous recessive
Sickle-cell trait	Individuals that are heterozygous
Huntington disease	Single gene locus on chromozome 4 at 4p16.3; caused by a dominant allele
Familial hypercholesterolemia	Single gene locus on chromosome 19 at 19p13.2; multiple alleles with incomplete dominance
Muscular dystrophy	Single gene locus on the X chromosome at Xp21.2; sex-linked recessive allele
Down syndrome	Usually a trisomy of chromosome 21; rarely a translocation of chromosome 21

Site of Locus

Disorders Due to Autosomal Recessive Mutations

Hundreds of genetic disorders are caused by autosomal recessive mutations, but few are as well understood as sickle-cell anemia. **Sickle-cell anemia** is caused by an autosomal recessive gene and has great significance in the history of human genetic disorders. It is now known that there are two forms of sickle-cell anemia, and they are distinguished by name. *Sickle-cell disorder* occurs in individuals who are homozygous recessive for the gene. Heterozygotes have the *sickle-cell trait;* such individuals generally have milder anemia and are not as seriously affected as those with sickle-cell disorder. Sickle-cell disorder was the first disorder in which a clear relationship was established between a simple Mendelian recessive trait, a heritable gene mutation, and an abnormal gene product.

A syndrome referred to as *sickle-cell disease* was first described in 1910, but there was relatively little information about the disorder at that time. The following facts were known: it affected both male and female blacks; it often caused serious symptoms, including fever and extreme pain in bones and joints; and some of the red blood cells of affected individuals were sickle-shaped (see Figure 18.5A–B.). Severe *anemia* (a low number of red blood cells) often led to premature death. By examining pedigrees, it was soon shown that sickle-cell disease was transmitted as an autosomal recessive trait.

Subsequently, scientists discovered that red blood cells of affected individuals underwent a reversible change in shape called *sickling* when internal oxygen concentrations fell. Some suspected that this effect might be due to a structural flaw in the *hemoglobin* (Hb) *molecule,* an iron-containing protein of red blood cells that carries oxygen from the lungs to the internal tissues. What were the biochemical and genetic bases of such a defect? In 1949, Linus Pauling and his colleagues found that the sickling characteristic was in fact due to a structural aberration in hemoglobin molecules that was caused by an abnormal protein produced from a mutant gene.

Since Pauling's pioneering work, sickle-cell anemia has become one of the best-understood human genetic diseases. It is instructive to consider the exact nature of the underlying cause of this disorder. Normal hemoglobin, partly shown in Figure 18.5C, is a protein molecule containing 574 amino acids arranged in two chains each of two specific sequences, α and β. The gene that encodes these se-

Figure 18.5

(A) Photo of a normal red blood cell. Sickle-cell anemia is characterized by red blood cells that develop an abnormal shape in the presence of reduced oxygen concentrations (B). Normal hemoglobin is a protein composed of two α chains and two β chains. (C) Sickle-cell anemia is caused by a gene mutation that results in the production of abnormal β chains in the hemoglobin protein. The abnormal chain differs from the normal β chain by only one amino acid, having valine instead of glutamic acid. (D) The effects of sickle-cell anemia are complex and numerous. Many different tissues and organs are affected. (E) A higher percentage of individuals from tropical regions of the world have sickle-cell anemia than people who live elsewhere. This difference is thought to be related to its advantage in conveying resistance to malaria.

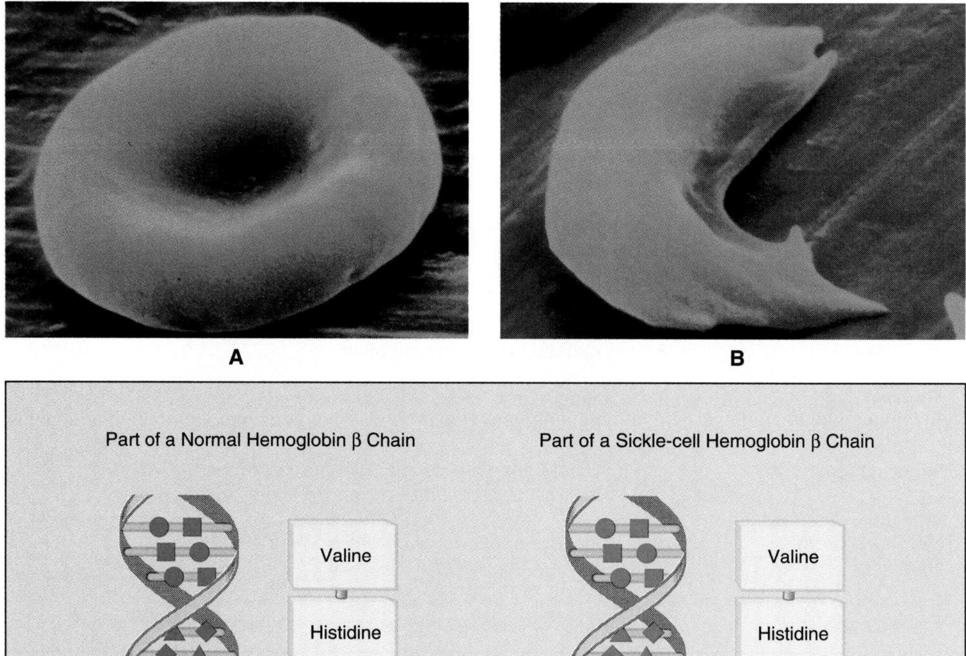

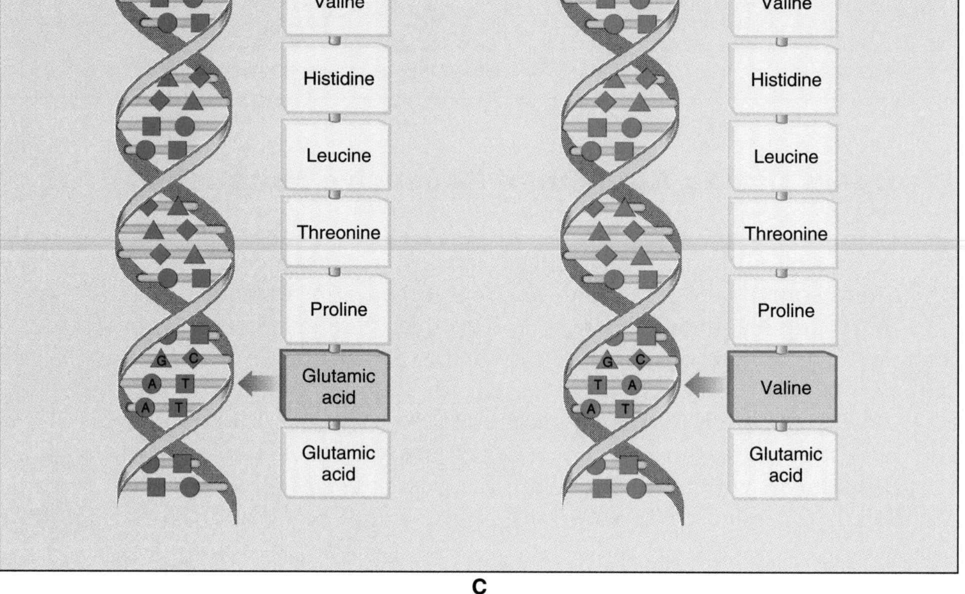

quences is located on the *p* arm of chromosome 11. As described in Figure 18.5C, sickle-cell hemoglobin differs by only one amino acid in each of the β chains, yet this single change alters the molecule to such an extent that it can become nonfunctional. Recall that three base pairs in DNA code for one amino acid. In normal hemoglobin, the DNA triplet CTT (cytosine-thymine-thymine) codes for glutamic acid, but in sickle-cell hemoglobin, the corresponding sequence reads CAT (cytosine-*adenine*-thymine), which codes for valine.

This would seem to represent the ultimate level to which the cause of a genetic disorder can be traced. The single change in one DNA base creates a defective hemoglobin molecule that under low-oxygen conditions forms a spiral, rigid, or fiberlike structure that causes the affected red blood cell to lose its normal shape and assume a sickled form. Figure 18.5D describes the cascade of effects that results from this sickling, including one of the hallmarks of the disease, anemia. Anemia occurs because sickled cells are susceptible to damage during circulation through blood vessels and are likely to be destroyed as a result. Although sickle-cell disorders are incurable, results of tests for an experimental drug treatment in 1990 and 1991 were very encouraging. The drug used, hydrourea, appeared to promote increased synthesis of normal hemoglobin. Further tests are being conducted to determine effective doses and to identify any potential long-term effects.

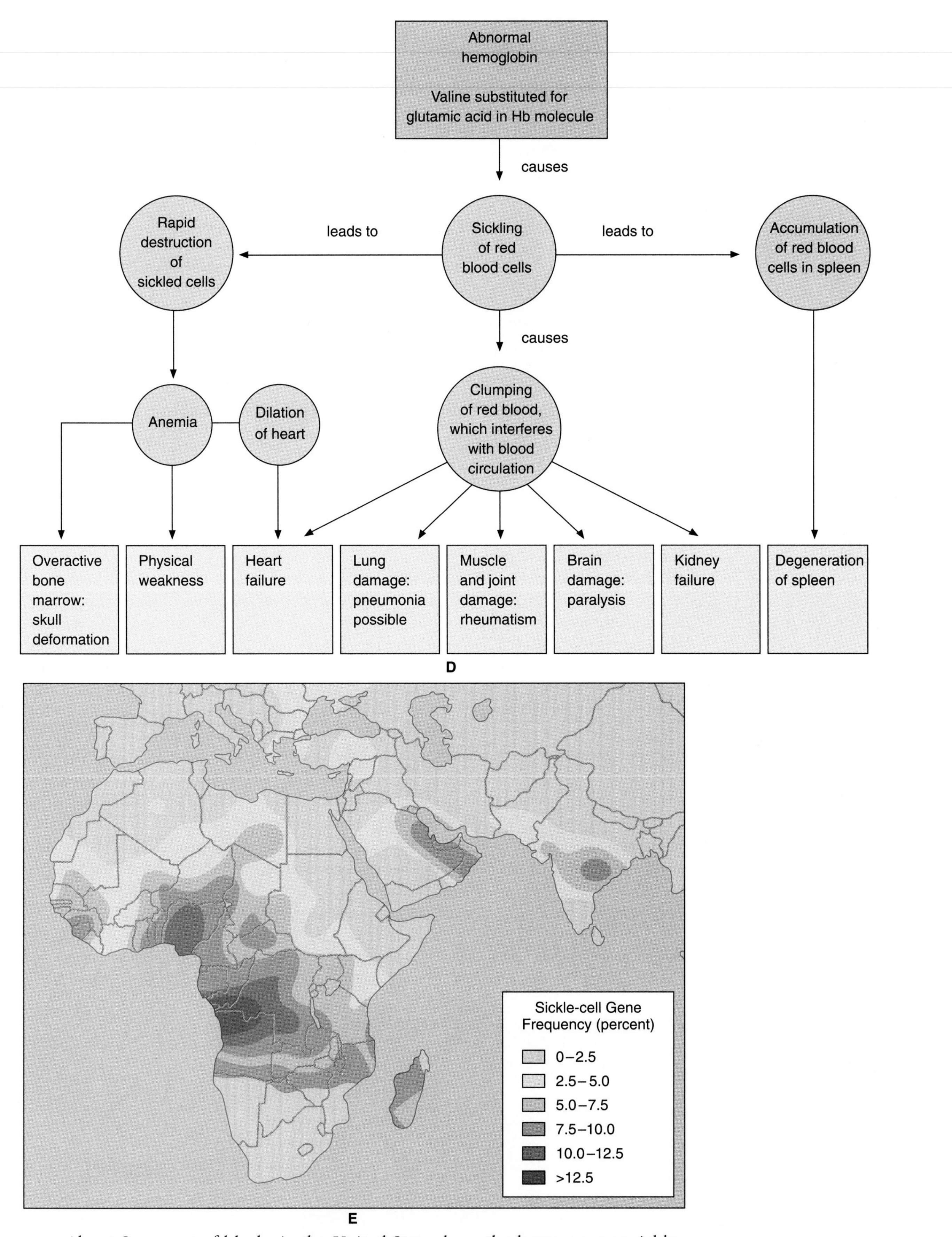

About 8 percent of blacks in the United States have the heterozygous sickle-cell trait. This frequency is apparently decreasing slowly. As indicated in Figure 18.5E, the mutant allele is also quite common in inhabitants of Africa, India, Arabia, and the Mediterranean area. Since alleles that cause such negative biological effects are relatively rare in human populations, why do certain human populations have these high allelic frequencies?

Scientists now know that heterozygotes with the sickle-cell trait are more resistant to malaria than those with normal hemoglobin. Thus in areas where malaria is present, there is a survival advantage associated with the allele that apparently outweighs the disadvantage related to the early deaths of a small number of homozygous individuals.

Sickle-cell anemia is a genetic disorder associated with a recessive gene. Homozygous individuals develop sickle-cell disorder, a dangerous condition that has direct effects on red blood cells and indirect effects on many other tissues and functions. Sickle-cell disorder results from a flaw in the structure of hemoglobin due to the replacement of a single amino acid in the β protein chains that make up part of this molecule. Heterozygotes suffer from sickle-cell trait, a less serious form of this hemoglobin disorder.

Disorders Due to Autosomal Dominant Mutations

Some of the best-understood and most serious genetic disorders of humans are caused by autosomal dominant mutations. We will examine two of these in detail, Huntington disease and familial hypercholesterolemia, to illustrate the nature of these types of genetic disorders.

Huntington Disease

Huntington disease (HD) was first described in 1872 by George Huntington, an American physician. Through the 100 years after its description, much was learned about HD, even though it occurs in a relatively small number of individuals (Figure 18.6). HD is a dominantly inherited genetic disorder that is considered one of the cruelest human genetic disorders for several reasons: it does not appear until middle age; it affects the nervous system, causing a long decline that leads to total incapacitation and death 10 to 20 years after the initial onset; children of a heterozygous, affected parent have a 50 percent chance of getting the disorder; and there is no treatment or cure at the present time. Though relatively rare, with an incidence of about 1 per 25,000 humans, primarily of British ancestry, it has been one of the most intensively studied human genetic disorders.

Most genetic mutations are expressed through the creation of abnormal phenotypes during prenatal development or shortly after birth. HD, however, lies at the other end of the scale; it is autosomal dominant and has a *variable age of onset,* with the first symptoms typically appearing between 30 and 40 years of age. No explanation for the middle-age onset is yet available.

The genetic basis of the disorder is a dominant allele, *H,* with *h* being the recessive allele. Individuals with the genotype *Hh* will develop Huntington disease if they live long enough. Since an affected parent can contribute either allele

Figure 18.6

Nancy Wexler and colleagues have studied Huntington disorder in members of a Venezuelan village for over 10 years. Here she is shown evaluating an individual from the village with HD.

to an offspring, the chances are 50 percent any children will have HD. Homozygous (*HH*) cases would require both parents to have HD; no such families have yet been studied.

The early symptoms of HD vary but often include clumsiness, forgetfulness, and depression. The effects soon escalate, with *chorea*—constant and uncontrollable body movements—being the most obvious. Gradually, an affected individual begins to stagger and fall, compulsively clench and unclench the fists, and experience rapid, uncontrolled movements of the arms and legs. In the later stages of HD, there is loss of memory and intellectual functions and a general physiological deterioration leading to death. These effects are all caused by the death of nerve cells in the brain.

Until the 1970s, opportunities for studying HD on a large scale were limited because of its low incidence. However, in 1972, a surprising discovery was reported—an extraordinary number of people living in a remote village on Lake Maracaibo, Venezuela, had HD. In 1976, a national commission on HD was established by Congress, and Nancy Wexler, now of Columbia University, was chosen as its executive director. The commission recommended in 1977 that a genetic study of the Venezuelan community be undertaken, and investigations began in 1979 under Wexler's direction. An extensive pedigree, consisting of over 7,000 individuals, has been traced out, and blood samples were obtained from about 1,500 of these people. Of the 7,000, more than 100 have HD and another 1,100 are *at risk* (that is, one of their parents had the disease). Why does such a large proportion of this population have HD? Apparently, five generations ago, when the population of the village was small, one woman had several children, including some who developed HD. These affected individuals may have constituted 5 to 10 percent of the population at that time. As the village population expanded, the HD gene continued to appear in each succeeding generation in approximately the same relatively high percentage.

HD illustrates some of the personal and social dilemmas created by technological advances. Until recently, parents who were at risk for HD were most likely to have already had children by the time they knew whether or not they would develop the disease. However, analysis of the blood samples obtained from the affected Venezuelan population led to a fascinating development. In 1983, James Gusella of Massachusetts General Hospital and his colleagues (including Nancy Wexler) discovered a *genetic marker* for HD (a segment of DNA that is almost always linked with the HD gene) using new techniques that are described in Chapter 19. The presence of the marker indicates, with a high level of certainty, that the person will develop HD. The gene marker was mapped to the terminal band of the short arm of chromosome number 4 (4p16). It is now possible, in some cases, to test people at risk and determine whether or not they have the HD marker. They must have relatively large families so that enough individuals can be tested to generate sufficient data to permit a confident interpretation of the results. Finally, in some cases, it is possible for researchers to test developing fetuses for the presence of the HD marker.

Exceedingly complex questions now arise for people at risk for HD. Should they have the HD diagnosis? Should they even consider having children? How do they organize their lives if they test positive for HD? Should a pregnant woman at risk have a prenatal test to determine if the developing fetus has HD? Should a fetus that tests positive for HD be carried to term if it will die from the disorder unless a cure is found? These questions focus on some very sensitive personal and societal issues that have yet to be addressed or resolved.

The biochemical basis of HD is not yet known. Is there a missing protein? An abnormal protein? Over- or underproduction of a protein? Until these questions can be answered, there are limited ways to address the problem of curing or preventing the disorder. There is some hope that in the future, the precise function of the affected allele can be determined. If researchers can trace the cause of

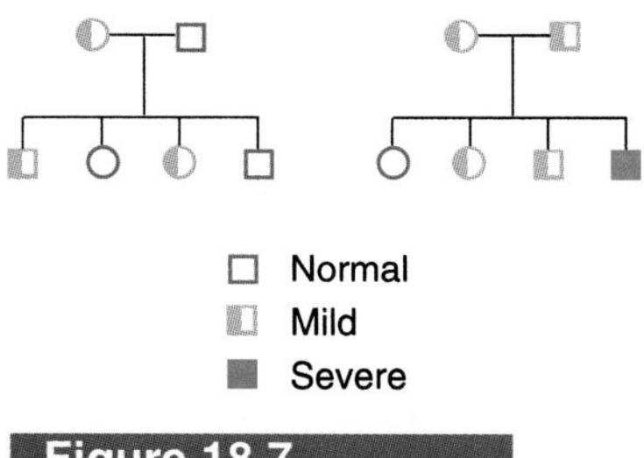

Figure 18.7

Two hypothetical pedigrees illustrate how FH is transmitted in humans.

HD to an aberrant protein—or to a normal protein that disrupts nerve function when present in abnormal concentrations (high or low)—then it may be possible to develop some countermeasures to prevent, cure, or treat the disease. At present, investigators in six major laboratories in the United States and England are collaborating to identify the gene. Progress has been slow, however, because of difficulty in determining the precise location of the gene.

Familial Hypercholesterolemia

Collectively, heart diseases are the most common disorders resulting from known genetic defects. A metabolic disorder, **familial hypercholesterolemia (FH)** ("having too much cholesterol in the blood"), is the most frequent cause of genetic heart disease. FH is caused by a mutant, incompletely dominant gene with multiple alleles that is located on chromosome 19. Heterozygous (*Cc*) individuals develop *atherosclerosis,* a blockage of arteries that supply the heart, that consequently leads to early *coronary heart disease* in their thirties and forties. For homozygous (*CC*) individuals, heart attacks occur in childhood and usually result in death. In most populations, the frequency of heterozygotes born is at least 1 in 500, making it the most prevalent Mendelian disorder in humans and one of the most serious. Homozygotes occur with a frequency of 1 in 1 million. Figure 18.7 shows two hypothetical pedigrees to illustrate how FH is transmitted in humans.

The underlying molecular cause of FH is understood in great detail. Like many genetic disorders, it involves an inborn error of metabolism, but of a special variety that is not due to an absent or a deficient enzyme.

Molecules of cholesterol (a type of fat) are required by every cell for forming cell membranes. Under normal circumstances, the liver synthesizes most of the cholesterol we require, with the remainder provided by the food we eat. Once in the blood, hundreds of molecules of cholesterol are packed into a complex molecule of **low-density lipoprotein (LDL)**. LDL serves to transport cholesterol to cells, which will take it inside if they have LDL receptor proteins that bind to LDL on their plasma membrane; that is, a specific cellular *receptor* is required for a molecule to enter the cell (see Figure 18.8). LDL receptors bind with LDL and usher it, along with its cholesterol cargo, inside the cell. LDL recep-

Figure 18.8

LDL cholesterol in the blood is normally removed by protein receptors on the plasma membrane of liver cells and is then processed internally. In familial hypercholesterolemia, LDL is not removed because of deficiencies in either the number, structure, or function of liver cell receptors.

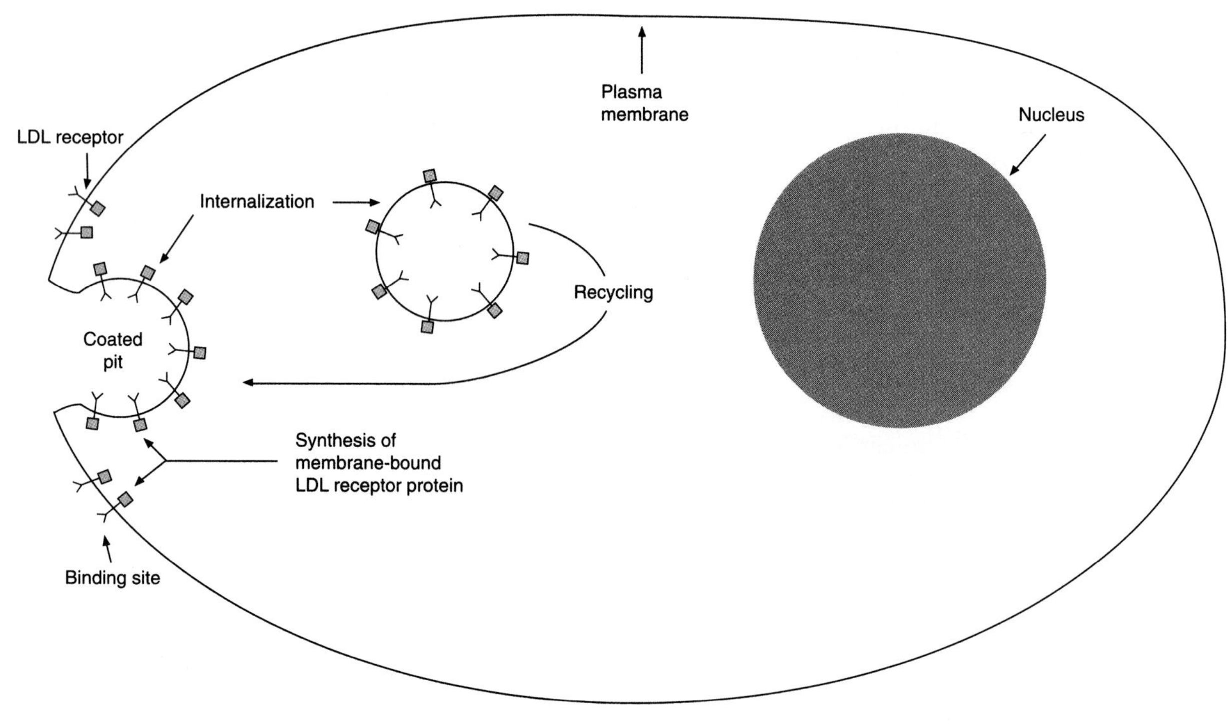

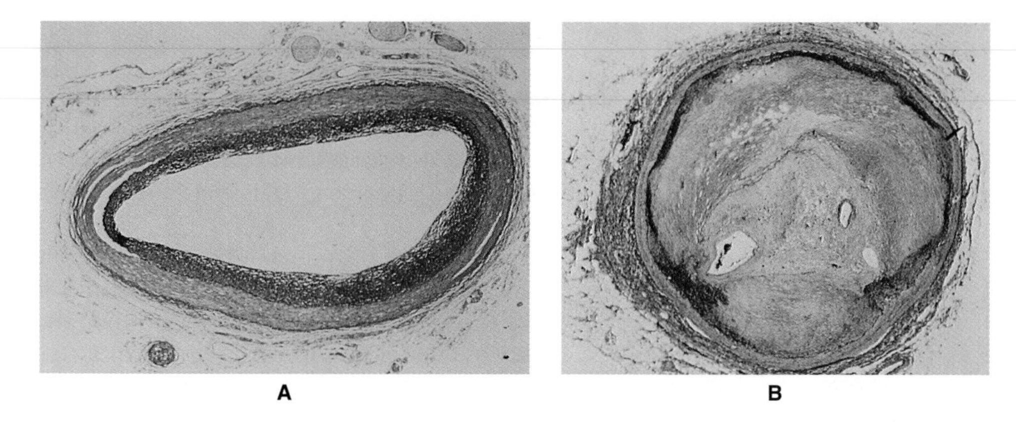

Figure 18.9

Continuous high levels of cholesterol in the blood eventually result in the formation of deposits in arteries that can block or inhibit normal blood flow and lead to a heart attack. (A) A normal artery from a 100-year-old woman who never had evidence of heart disease in her life. (B) This clogged artery has only three small openings but is otherwise blocked.

tor proteins are concentrated in regions called *coated pits*. Once inside the cell, the cholesterol is used by the cell or stored until required.

The genetic defects in FH affect the production of LDL protein receptors. Four alleles are known for the LDL gene locus: the normal allele encoding normal LDL receptor protein, an allele that encodes a receptor with a reduced binding capacity, an allele that encodes a receptor that cannot bind LDL, and an allele that encodes a receptor that cannot transport LDL. Heterozygotes have one normal allele and one of the three mutant alleles. The net result of all heterozygous conditions is that only a small amount of cholesterol can enter the cell. There are two negative consequences of this deficiency: (1) cells deprived of cholesterol, including liver cells, begin to make their own, and (2) LDL and cholesterol levels in blood serum increase since the excess is not taken up by other cells. The body's mechanism for coping with this excess involves depositing cholesterol in the walls of blood vessels, as shown in Figure 18.9; the result is atherosclerosis, which ultimately leads to heart attacks. Individuals homozygous for FH are severely deficient in LDL receptors and exhibit these effects during their youth. Heterozygotes have about half the normal number of receptors, and it takes longer for the damage to occur. In these individuals, coronary heart disease usually becomes evident when they are 40 to 50 years old.

Unlike Huntington disease and many other genetic disorders, FH does not occur more frequently in specific ethnic groups. Also, scientists know the precise molecular effects of the mutant dominant allele. As a result, they have devised methods of treatment for heterozygous individuals who are at risk for premature heart diseases associated with FH. Today, drugs are used that accelerate excretion of excess cholesterol from the body. Though not a perfect solution because of side effects such as nausea and diarrhea, the medication, along with lifestyle modifications that limit cholesterol intake in food and increase regular exercise, has allowed heterozygotes to improve their chances of survival beyond middle age.

Huntington disease is a genetic disorder caused by an autosomal dominant gene. It usually appears in middle-aged individuals and causes a gradual decline in nervous system function that leads to death. Presently, there is no treatment for HD. Familial hypercholesterolemia is caused by an incompletely dominant gene that affects normal cholesterol processing in the cell. FH is the most common human genetic disorder and the leading cause of genetic heart disease. Homozygous individuals rarely reach their teens. Heterozygotes commonly suffer from coronary heart disease by middle age. Drugs and diet are used to treat FH.

Disorders Due to X-linked Gene Mutations

Different mutant alleles and the resulting affected phenotypes can be traced to the X chromosome by their distinctive pattern of inheritance in families; they typically appear in males but only very rarely in females, unless dominant. More than

300 human genetic disorders are thought to be caused by defective alleles on the X chromosome.

Duchenne muscular dystrophy (DMD) is caused by an X-linked, recessive allele that appears in at least 1 in 3,500 males born; by comparison, hemophilia, perhaps the best-known X-linked disorder, occurs once in every 10,000 males. DMD is one of ten forms of muscular dystrophy but is by far the most common and one of the most severe.

DMD is a muscle-destroying disease that in its classic, childhood-onset form begins to affect boys by the age of 5. The muscle-wasting process begins in the legs and progresses upward to the rest of the body. Muscle cells actually die, but the nature of the responsible biochemical defect is unknown. Affected children are invariably in wheelchairs by age 12, and death related to the failure of muscles used in breathing commonly occurs before age 20. In some cases, mental retardation may also occur.

The gene associated with DMD is located at Xp21, is extremely large with at least 60 exons, and encodes a protein called *dystrophin*. By studying the dystrophin gene's structure from several hundred muscular dystrophy patients, scientists have found that exon deletions from the gene are responsible for the disease. The severity of the muscular dystrophies is apparently not related in a quantitative way to the fraction of the gene missing. Instead, it is related to the quantity and quality of the protein produced. If some or most of the key portions of the dystrophin gene are present at Xp21, a milder form of muscular dystrophy will result because some dystrophin (or a simpler form of it) is produced. If major portions of the gene are deleted, little or no dystrophin is produced, and severe DMD results. A complete understanding of the underlying mechanisms involved and the role of dystrophin in normal muscles awaits further research.

Research on Human Genetic Disorders

The genetic disorders examined in this chapter can be used as examples to illustrate the sequence of questions that guides contemporary investigations of human genetic disorders. This sequence of questions that follows generally reflects the degree of difficulty in acquiring answers.

The first step typically involves attempts to identify the residence of the gene causing the disorder. On which chromosome is the gene found? Exactly where on the chromosome is it located? The second phase consists of trying diverse approaches that might lead to the identity of the protein encoded by the gene. What is the protein? What is the amino acid sequence of the protein itself? What is the DNA sequence of the gene? Next, the function of the normal protein encoded by the unaltered gene is pursued if it is not already known, along with the effects of the modified or missing protein that results from the mutant allele. What is the role of the normal protein? Why does disease occur if the protein is modified or missing? The final step of the process leads to questions that have only rarely been answered. What can be done to treat the disorder? Can the disorder be prevented or cured? Ultimately, molecular biologists strive for answers to all of these questions for all genetic disorders. Only when they have answers will it be possible to counteract the effect of gene mutations in the human population. Gene therapy—replacement of defective alleles or, perhaps more likely, insertion of normal alleles—is also a distant goal. As we have seen, however, there have been some partial successes in treating genetic disorders for which the precise molecular deficiency has been elucidated.

PRENATAL DIAGNOSIS AND GENETIC COUNSELING

It is well beyond the scope of this book to describe the varying methods used by sociologists and economists to measure the costs of genetic disorders in human populations. Genetic disorders and congenital malformations appear in 2 to 5 percent of all live births. In developed countries, they account for up to 30 percent of children's admissions to hospitals and are a significant cause of childhood deaths. In the United States, it is estimated that over 1 million persons are hospitalized each year for hereditary or congenital disorders. Chronic disorders with a genetic cause are thought to occur in about 10 percent of the total adult population.

The costs of genetic disorders have two aspects. They place major financial strains on affected families and community health and social services. And since they primarily affect children, there is an emotional price that is incalculable. For these reasons, prospective parents at risk for a genetic disorder may often opt for *prenatal diagnosis,* a process used to determine the genetic status of the developing fetus. Genetic counselors explore various choices, including medical abortion, before such testing and assist in the evaluation of test results and any decision-making processes that follow. Parents unwilling to consider abortion may also choose to determine if the fetus has a genetic disorder. If the tests are positive, they can then become familiar with the care that will be required for the child and make other necessary plans.

Techniques of Prenatal Diagnosis

The field of prenatal diagnosis is growing rapidly, along with advances in molecular genetics. Many single-gene defects, such as Duchenne muscular dystrophy, and all chromosomal mutations can be identified in developing fetuses. In addition to classic genetic disorders, many common congenital malformations without known causes can be identified prenatally. Since they frequently appear in certain families, they are assumed to involve multiple genes or various environmental factors (or combinations of both). Many common structural malformations are related to abnormal development of the neural tube, the structure that gives rise to the brain and spinal cord. These *neural tube defects* can result in extremely severe abnormalities. Depending on the portion of the neural tube affected, effects range from a partial skull with no higher brain centers to abnormal spinal cord closure, a condition called *spina bifida,* that affects approximately 1 in 500 children born in the United States. Neural tube defects are first indicated by the presence of abnormally high concentrations of a protein called α-fetoprotein in the amniotic fluid surrounding the fetus. If this protein is found in prenatal tests, further tests are usually conducted.

Visualization Techniques

Two methods are used routinely to visualize a developing embryo inside the womb. **Ultrasound scanning** employs high-frequency sound waves to provide a profile of the structural features of the fetus such as that seen in Figure 18.10. Ul-

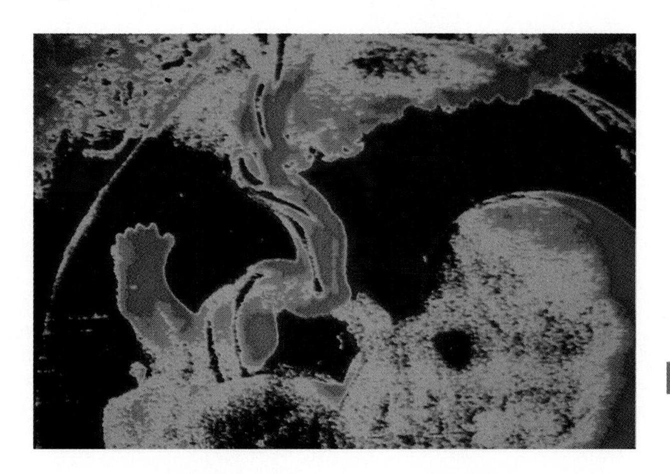

Figure 18.10

Image of a 10-week-old fetus as shown by ultrasound scanning.

trasound scanning can illustrate the orientation of the fetus and the location of the placenta prior to conducting further tests. It can also be used to confirm multiple fetuses and normal or abnormal brain development. Ultrasound scanning is generally considered safe for both the fetus and the mother.

Fetoscopy requires insertion of a fetoscope, a small instrument, into the amniotic sac through a small surgical opening in the mother's abdomen. Because it is an invasive surgical procedure, there is a risk associated with fetoscopy, and its use is limited. Its advantages over ultrasound scanning are that it can detect deformities of the limbs and can also be used to remove a fetal blood sample for evaluation.

Amniocentesis

Two principal methods, described in Figure 18.11, are used in prenatal diagnoses for genetic disorders: amniocentesis and chorionic villus sampling. The developing fetus is surrounded by fluid within the amniotic cavity. The amniotic fluid is produced by the fetus itself and contains various solutions, including fetal urine, and several kinds of fetal cells. **Amniocentesis** is a procedure in which a long, thin needle is inserted through the mother's abdominal wall and into the amniotic cavity. A small amount of fluid is removed and tested to identify possible ge-

Figure 18.11

Amniocentesis (A) and chorionic villus sampling (B) are procedures used in prenatal tests to determine the occurrence of chromosomal mutations or certain genetic disorders in a developing fetus.

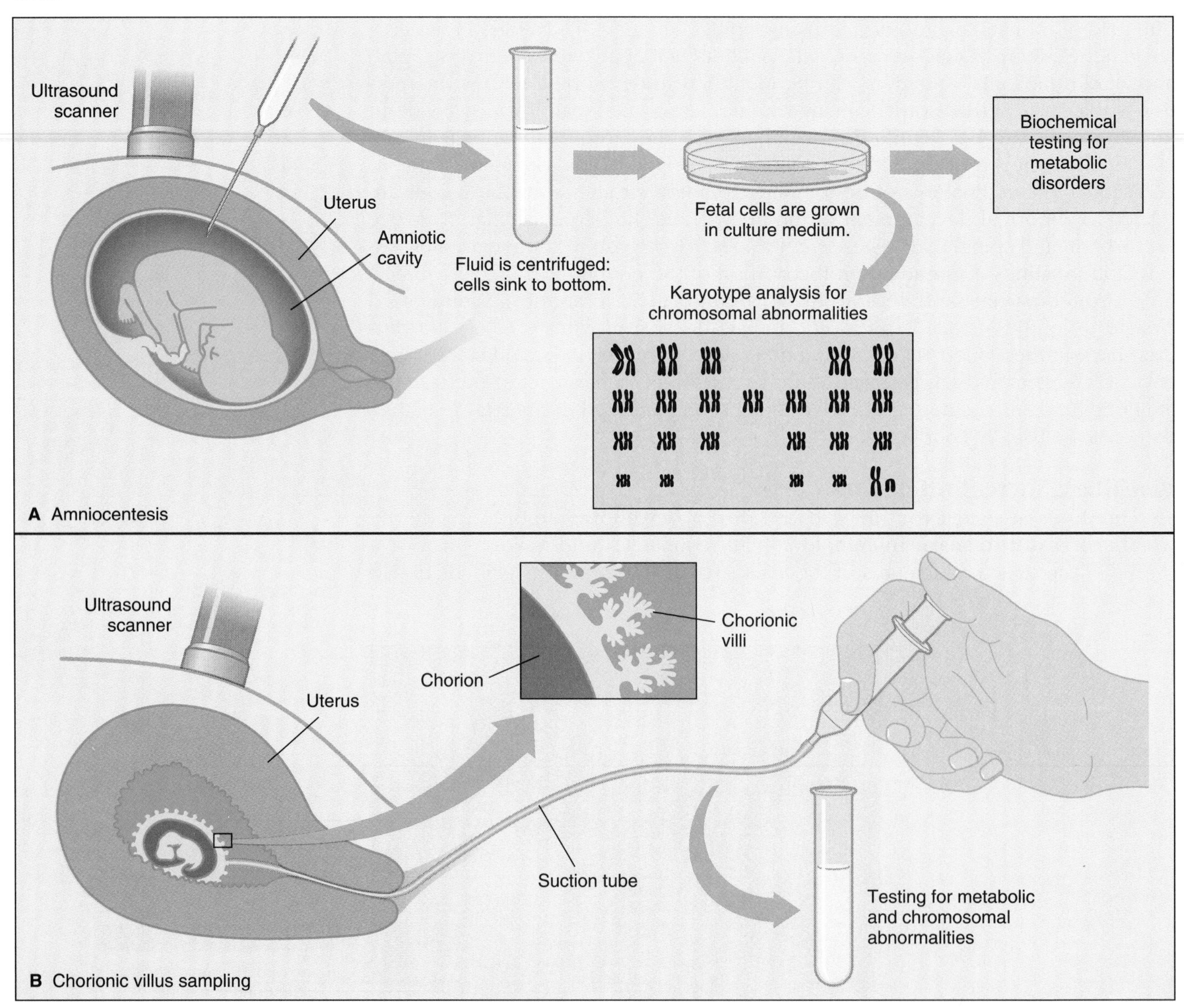

netic and biochemical defects. The extracted fluid can be analyzed to detect enzyme abnormalities that suggest certain developmental defects or genetic disorders. Fetal cells in the fluid must first be cultured (grown in an artificial medium) and later analyzed for any chromosomal abnormalities when enough cells have been produced.

Amniocentesis has revolutionized prenatal genetic counseling since 1980. It is usually performed during the fourteenth to sixteenth week of pregnancy and, when ultrasound is used to locate the fetal head and placenta, is considered to be reasonably safe for both the fetus and the mother. It is very accurate in detecting chromosomal mutations and about 50 genetic mutations expressed as inborn errors of metabolism. A disadvantage of amniocentesis is that a final analysis of the cells cannot be made until 17 to 20 weeks into the pregnancy.

Chorionic Villus Sampling

Chorionic villi are hairlike projections of the membrane that surrounds the embryo early in pregnancy. This membrane is a rich source of fetal cells until the tenth week of pregnancy, when it is replaced by the placenta. **Chorionic villus sampling (CVS)** consists of removing a plug of tissue from the villi with a small tube inserted through the cervix. Since there is an abundance of fetal cells in the collected tissue, analyses can begin immediately.

CVS offers certain advantages over amniocentesis. The sample is taken six to nine weeks into pregnancy, and results of chromosomal analyses and biochemical tests are available within days. A major drawback is that it is not suitable for determining neural tube defects since they cannot be detected until later in fetal development.

CVS was introduced in 1983. Its use was restricted by the U.S. Food and Drug Administration until suitable tests could be conducted to evaluate its safety. However, a 1988 study found that CVS does not present an elevated risk to mother or fetus when conducted by experienced physicians. Therefore, it may soon become more widely available and may eventually replace amniocentesis in testing for many genetic disorders.

Prenatal diagnosis involves various techniques that are used to examine a developing fetus and also to analyze its cells. Such procedures may enable a doctor to determine whether or not the fetus has or will develop a specific genetic disorder. Many, but not all, genetic disorders can be detected. Prenatal diagnosis and genetic counseling are usually recommended for parents who may be at risk for a specific genetic disorder.

Genetic Counseling

The American Society of Human Genetics offers the following definition of genetic counseling:

> Genetic counseling is a communication process concerning the risks of occurrence of a genetic disorder in a family. It involves an attempt to help the person or family comprehend the medical facts, appreciate the hereditary nature and recurrence risks in specific relatives, understand the options for dealing with the risk, choose the most appropriate course of action, and make the best possible judgment.

In general, prenatal testing is recommended for parents at risk for a genetic disorder. Any of the following criteria is indicative of risk:

1. The prospective mother is 35 or older or the father 50 or older, as increased age is correlated positively with chromosomal mutations.
2. The prospective mother is known or suspected to be a carrier of an X-linked genetic disorder.

Focus on Scientific Inquiry

Down Syndrome

The most common chromosomal mutation, occurring in about 1 in 700 to 800 live births in the general population, is a trisomy of chromosome 21 that causes **Down syndrome (DS)** (see Figure 1). The occurrence of trisomic DS is known to be related to the age of the mother: the risk of having a DS child is 1 in 50 in women over age 40. Recent evidence also suggests that older men are more likely to father DS offspring. Down syndrome is the leading cause of mental retardation in the United States. Affected children suffer from multiple physical and mental problems.

Reviewing DS offers an opportunity to examine how advances in scientific knowledge and scientific technology are used to extend the understanding of a genetic disorder. As more becomes known, particularly at the molecular level, it may become feasible to design methods for treating the effects of the genetic disorder. Ultimately, it may even become possible to prevent the effects caused by such disorders. Scientists are now investigating the molecular implications of having an extra copy of chromosome 21 and are trying to correlate these molecular effects with the DS phenotype.

Various records indicate that DS has occurred in children of all racial and ethnic groups throughout human history. The syndrome was formally recognized in 1866 when an English physician, John L. Down, published a comprehensive description of physical symptoms in certain mentally retarded patients. However, the cause of DS was not discovered until the late 1950s, when scientists observed that affected patients had three copies of chromosome 21 in their cells. Individuals with DS suffer from many anatomical and biochemical abnormalities (Figure 2). They have flattened facial features, epicanthic folds of the eyes, unusual creases on the palm, muscular flaccidity, and short stature. In addition to mental retardation, 40 percent of DS patients have congenital heart defects, many develop cataracts or other eye problems because of lens defects, and they are more susceptible to infections and are much more likely to develop leukemia than normal people. Most individuals with DS live a maximum of 30 to 50 years.

Figure 1 Down syndrome is most often caused by a trisomy of chromosome 21.

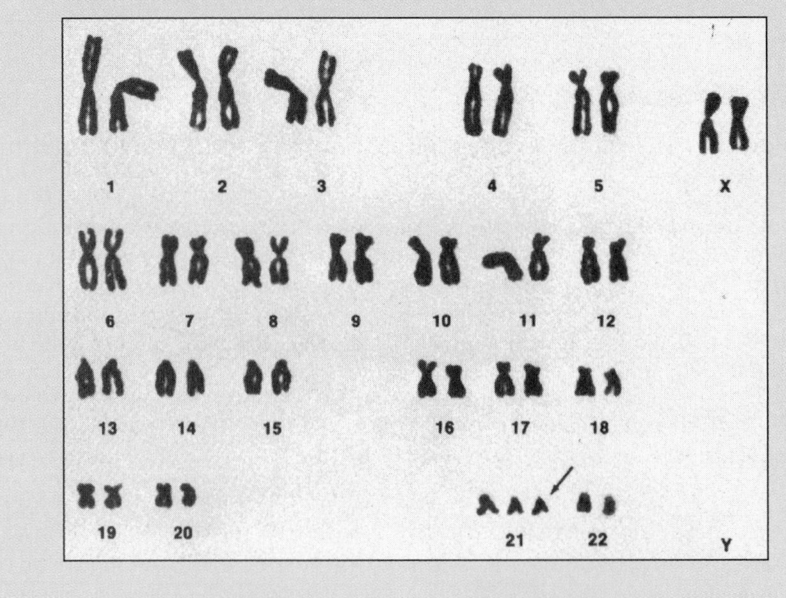

Figure 2 Children with Down syndrome have characteristic phenotypic features.

Recently, the number of families raising DS children at home, rather than committing them to an institution at birth, has increased. Studies indicate that DS children benefit significantly when allowed to remain in a home environment with caring and loving family members along with formal occupational and physical therapy. They show enhanced intellectual and physical development, achieve a higher degree of independence, and have an increased life expectancy.

Most chromosome 21 trisomies are a result of nondisjunction; however, 3 to 5 percent of DS individuals are a result of chromosomal translocations. This process involves the translocation of a major portion of chromosome 21 to another chromosome, usually 13, 14, 15, 22, or, in rare cases, the other 21. In translocation DS, one of the parents is generally a translocation carrier that is phenotypically normal but at increased risk of producing a DS offspring. Chromosome 21 is estimated to contain roughly 1,500 of the 50,000 to 100,000 genes in the complete human genome. Of that number, only a few have been precisely located and identified using high-resolution genetic-mapping procedures. Molecular biologists are particularly interested in answering the following questions: Which genes on chromosome 21 are implicated in producing the abnormal phenotype? Which genes are responsible for the harmful effects? What proteins are encoded by these genes? Why does having three copies of these genes lead to DS?

Investigations have now established that the DS phenotype is linked to a number of genes that lie within the 21q22 band of chromosome 21, as shown in Figure 3. Although speculative at the present time, ideas about the relationships among the genes identified, the proteins they produce, and the effects they cause have recently emerged.

The high risk for leukemia is thought to be associated with the *ets*-2 gene, which is known to be a cancer-causing gene (oncogene). Studies are under way to learn more about the expression of this gene in DS and normal individuals.

The *Gart* gene codes for three different enzymes that play a role in the synthesis of purines (recall that two purine bases, adenine and guanine, are found in DNA molecules). Individuals with DS have elevated purine levels in their blood serum. High purine levels are linked with a variety of problems, including mental retardation. Thus it has been hypothesized that the presence of a third *Gart* gene leads to a higher dosage of purines, which accounts for many of the problems that are characteristic of DS. An alternative hypothesis has recently been proposed. A gene has been described that codes for *S100 protein,* which is found in the nervous systems of all vertebrates. The highest levels of S100 are found in the brain, where it interacts with other brain proteins. It is possible that abnormal concentrations of S100 protein may also be responsible for the adverse nervous system effects of DS. One scenario for therapy in the future involves early detection of DS in the fetus followed by immediate attempts to regulate levels of the purines or proteins ultimately found to be responsible for the severe symptoms.

The *α-A-crystalline* gene codes for a protein of the same name that is a structural component of the lens of the eye. The abnormal expression of that protein may be related to the increased risk of cataracts and lens defects in DS.

Finally, the gene *SOD-1* codes for a protective enzyme that prevents damage to cells from reactive chemicals produced in certain metabolic pathways. Alterations in SOD-1 levels may account for the accelerated rate of aging in DS and may also contribute to mental retardation.

These hypotheses remain to be verified. Nevertheless, the supporting evidence is impressive, and these efforts offer a detailed view of ongoing research into an important human genetic disorder.

Figure 3 New techniques have been used to identify genes associated with Down syndrome. Scientists are now trying to determine how these genes cause the effects leading to the Down syndrome phenotype.

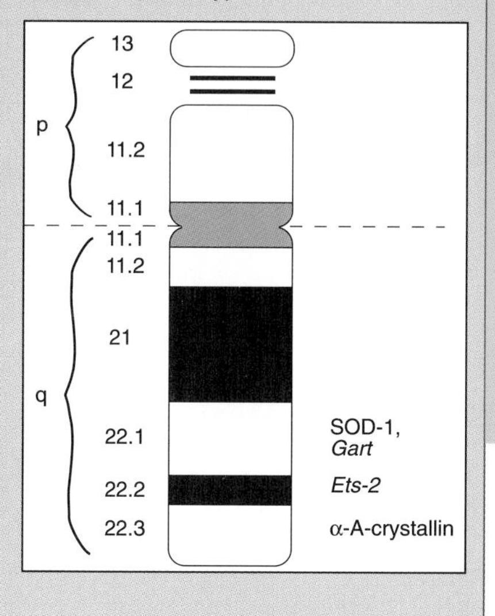

3. The parents have previously had a child with a chromosomal mutation or a neural tube defect.
4. Both prospective parents are known to be carriers of an autosomal recessive allele that causes a genetic disorder in homozygotes.
5. One parent is known to carry a defective autosomal dominant allele.

Ultimately, it is left to the individuals to make an informed decision once they have received all the relevant information about prenatal tests, obtained results from the tests, and considered the consequences of having a child with a genetic disorder. Science, scientists, and legitimate counselors do not abrogate the personal responsibilities of individuals in making their own reproductive decisions.

Summary

1. In the age of new genetics, much research on human genetics has been directed at determining the specific nature of inherited disorders. Genetic disorders are caused by gene mutations and chromosome mutations.
2. Gene mutations result in disorders that are classified as dominant or recessive, autosomal or sex-linked. Genetic disorders associated with recessive gene mutations are usually due to defective enzymes or hormones. Dominant disorders typically affect a structural protein. More than 3,000 genetic disorders are known to be caused by single-gene mutations.
3. Chromosomal mutations consist of structural changes in chromosomes or alterations of the normal number of chromosomes. Chromosomal mutations usually result in such serious abnormalities that development is arrested early in pregnancy and spontaneous abortion occurs.
4. Human pedigrees are used to determine the pattern of an inherited disorder. Evaluations of pedigrees may enable investigators to decide if a genetic disorder is caused by a recessive, dominant, or X-linked gene.
5. Sickle-cell anemia is a human genetic disorder associated with a recessive gene. Disorders caused by autosomal dominant genes include Huntington disease and familial hypercholesterolemia. Duchenne muscular dystrophy and milder forms of muscular dystrophy are associated with a large X-linked gene. The genetic bases for these disorders are well established, and at the molecular level, the phenotypic effects of the molecular modification are at least partly understood.
6. Individuals at risk for a genetic disorder have access to various prenatal diagnostic procedures that will enable them to learn whether or not the developing fetus has a specific disorder. These procedures include visualization techniques, amniocentesis, and chorionic villus sampling. All have advantages and disadvantages that must be considered before diagnosis proceeds. Genetic counselors are trained to interpret the results and inform parents of available options. Final decisions, such as continuation or termination of pregnancy, are left to the individuals at risk after they have considered all available options.

Review Questions

1. What are genetic disorders? What is their importance in the human population?
2. What types of gene and chromosomal mutations cause human genetic disorders?
3. Describe the pattern of inheritance that is typical for different classes of gene mutations.
4. What is the probable type of gene mutation that causes the genetic disorder present in the family pedigrees below?

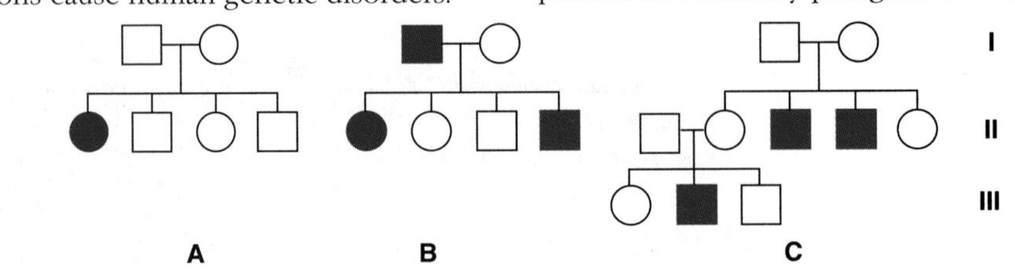

5. What is the genetic cause of sickle-cell anemia? What is the molecular basis of this disorder? What accounts for its effects?
6. What is the genetic cause of Huntington disease? What are the biological effects of this disorder? What are some of the social and family problems associated with HD?
7. Describe the cause and effects of familial hypercholesterolemia. What is the molecular basis of this disorder? How can its harmful effects be reduced?
8. What is the genetic nature of muscular dystrophy? Explain the underlying mechanism responsible for its effects.
9. Describe the techniques used in prenatal diagnoses for detecting genetic disorders. What are the advantages and disadvantages of each?
10. What is genetic counseling?

Essay and Discussion Questions

1. Why did the discovery of the affected Venezuelan population constitute a breakthrough for researchers studying Huntington disease?
2. What recommendations would you make to agencies that fund research on human genetic disorders relative to emphasis in the decade ahead? Which disorders should receive the most attention? Why? What are some criteria that might be used in making this decision?
3. Should insurance companies pay for prenatal diagnoses and genetic counseling? Why?

References and Recommended Readings

Chen, H. C. 1988. *Medical Genetics Handbook*. St. Louis: Green.

Heim, W. G. 1991. What is a recessive allele? *The American Biology Teacher* 53:94–97.

Jenkins, J. B. 1990. *Human Genetics*. 2d ed. New York: HarperCollins.

Marx, J. 1990. Dissecting the complex diseases. *Science* 247:1540–1542.

McKusick, V. 1990. *Mendelian Inheritance in Man: Catalogs of Autosomal Dominant, Autosomal Recessive, and X-linked Phenotypes*. 9th ed. Baltimore: Johns Hopkins University Press.

Myant, N. B. 1990. *Cholesterol Metabolism, LDL, and the LDL Receptor*. Orlando, Fla.: Academic Press.

Patterson, D. 1987. The causes of Down syndrome. *Scientific American* 259:52–60.

Roberts, L. 1990. Huntington's gene: So near, yet so far. *Science* 247:624–627.

Scriver, C. R., A. L. Beaudet, W. S. Sly, and D. Valle, eds. 1989. *The Metabolic Basis of Inherited Disease*. 6th ed. New York: McGraw-Hill.

Vogel, F., and A. G. Molulsky. 1986. *Human Genetics: Problems and Approaches*. 2d ed. New York: Springer-Verlag.

CHAPTER 19

Recombinant DNA and Genetic Technology

Chapter Outline

Recombinant DNA Technology

Gene Cloning

Biotechnology

Focus on Science and Society: Biotechnology Policies in the United States

Reading Questions

1. What is recombinant DNA?
2. What is gene cloning? Why are genes cloned?
3. How are genes cloned?
4. What are gene libraries?
5. What are some important biotechnology products?

The period following Watson and Crick's description of the molecular structure of DNA in 1953 was one of great scientific activity that produced much new information about this remarkable molecule. A constant stream of knowledge flowed from research laboratories throughout the 1950s, 1960s, and 1970s, providing information about the mechanism of DNA replication (1958), the complete genetic code used to specify proteins synthesized by cells (1966), and a variety of enzymes that participate in reactions involving DNA molecules. Enzymes in this category are *DNA ligase* (1967), *reverse transcriptase* (1970), and *restriction enzymes* (1970). Table 19.1 contains definitions of technical terms used throughout this chapter.

Molecular biologists also discovered unconventional processes for transferring genetic information through two different biological entities, *plasmids* (1965) and *viruses* (1960s and 1970s). Neither are considered to be true organisms in a biological sense because they cannot reproduce independently outside of the host cells they normally inhabit. Nevertheless, it became clear that plasmids and viruses were capable of transmitting genetic information from one species to another in ways that were eventually harnessed to have an enormous impact on biology and society.

The new methods and techniques developed in the 1970s provided deeper insights into the chemistry of DNA and genes. A spectacular accomplishment occurred in 1973 when DNA from two different organisms was spliced together—the first successful experiment in which DNA from two different species was "recombined." It later became possible to clone eukaryotic genes, carried by plasmids, inside of bacterial (prokaryotic) cells (1974) and to determine the exact DNA nucleotide sequences of entire genes and longer segments of DNA (1977). Companies were founded in the 1980s to take advantage of the new biological technologies to produce valuable substances such as vaccines and hormones.

Like several other major breakthroughs in biology, manipulating genes for human purposes has led to a number of issues and questions that are of concern to informed citizens. The new molecular tools gave rise to novel techniques referred to as **genetic engineering** and products known as *recombinant DNA*. For the first time, it became possible for molecular biologists to isolate genes from one organism, clone them in another, and have them expressed in bacteria, manufacturing the protein encoded by the gene. The pace of discovery in genetic engineering remains rapid, and useful products from this field continue to be developed.

In this chapter, we describe the biological processes and technologies developed by genetic engineers. We also examine the emergence of new industries based on these technologies and the important biological products that can be manufactured.

RECOMBINANT DNA TECHNOLOGY

DNA can be viewed as a master set of instructions that, once expressed, regulate the development, growth, functions, and aging of organisms throughout their lives. The instructions of DNA are translated through a genetic code into proteins, the workhorses of biochemistry (described in Chapter 16). The common feature of all life on Earth that makes genetic engineering possible is the genetic code; it is virtually the same for all organisms, from bacteria to humans.

In the two decades following the description of DNA's structure, new biological tools were developed for manipulating this molecule. Enzymes were discovered that could cut, splice, or replicate DNA, and it was found that plasmids and viruses could be used to transfer DNA between cells of different species.

Table 19.1 Terms Used in the Field of Genetic Engineering

Bacteria One-celled organisms; prokaryotes lacking a nucleus; capable of carrying on many chemical reactions (for example, protein synthesis) that are similar to those of higher organisms. *Strains* of bacteria are a group of organisms within a species that is characterized by some particular quality (for example, the rough and smooth strains of *Diplococcus pneumoniae* discussed in Chapter 15).

Bacteriophage Viruses that infect bacteria; commonly referred to as *phages.*

Biotechnology A broad field that employs an industrial technology based on the biological synthesis of important chemical compounds, especially proteins (for example, insulin), genetic engineering of plants and animals, and other, related technologies.

Clone A large number of *cells* or *molecules* identical to an ancestral cell or molecule. Also, a specific gene sequence, isolated and replicated.

Cloning vectors Small *plasmid, phage,* or animal *virus* DNA molecules used to transfer a DNA fragment from a test tube into a living cell. Cloning vectors are capable of multiplying inside of living cells.

Complementary DNA (cDNA) DNA copied from a messenger RNA (mRNA) molecule using an enzyme called *reverse transcriptase.* The DNA sequence is thus *complementary* to that of the mRNA.

DNA ligase An enzyme that connects two separate DNA molecules, end to end.

Escherichia coli (E. coli) The bacterium most widely used in recombinant DNA research; commonly found in the digestive tracts of mammals.

Expression vector A *plasmid* designed to permit expression (transcription) of a foreign gene inside a cell.

Genetic code The sequence of mRNA base triplets that specify the exact series of amino acids for a protein.

Genetic engineering The manipulation of genetic information (DNA or RNA) of an organism to alter the characteristics of that organism. The basic process consists of inserting foreign DNA carrying instructions for a valuable enzyme, hormone, or other protein into the DNA of some other organism so that the host organism makes the desired product at the same time as it makes its own proteins.

Genomic clone A fragment of genomic DNA from an organism.

Genomic DNA All DNA sequences of an organism.

Host cell A cell (usually a bacterium) in which a *cloning vector* can be propagated.

λ (lambda) phage A particular *bacteriophage* used extensively in gene cloning.

Library A set of cloned fragments together representing a part or all of the genome of a species.

Plasmids Small, circular DNA molecules found inside bacterial cells. Plasmids reproduce every time the bacterial cell reproduces.

Recombinant DNA (rDNA) A DNA molecule containing two or more regions of DNA from two different genes or species (for example, a fragment of human DNA spliced into plasmid DNA).

Restriction enzymes Enzymes that cut DNA molecules at specific nucleotide sequences.

Reverse transcriptase An enzyme purified from retroviruses that makes DNA from RNA.

Transduction The transfer of DNA (genes) from one bacterium to another by bacteriophages.

Transformation The insertion of any foreign gene into any organism.

GENE CLONING

Gene cloning is the central process of genetic engineering that is used to create large numbers of a specific gene, usually one that codes for a "desirable" protein. Such proteins are usually associated with human genetic diseases or have roles in normal biological functions. Genes from any organism, prokaryotic or eukaryotic, can now be cloned. Figure 19.1 is a simplified description of the major steps used in cloning genes.

The first step in the procedure is to obtain a desired gene for cloning. Standard techniques enable molecular geneticists to isolate genes directly from strands of DNA that have been removed from cells. Alternatively, indirect processes may be employed in which specific DNA sequences (genes) can be produced using messenger RNA as the template or by using a "gene machine," an

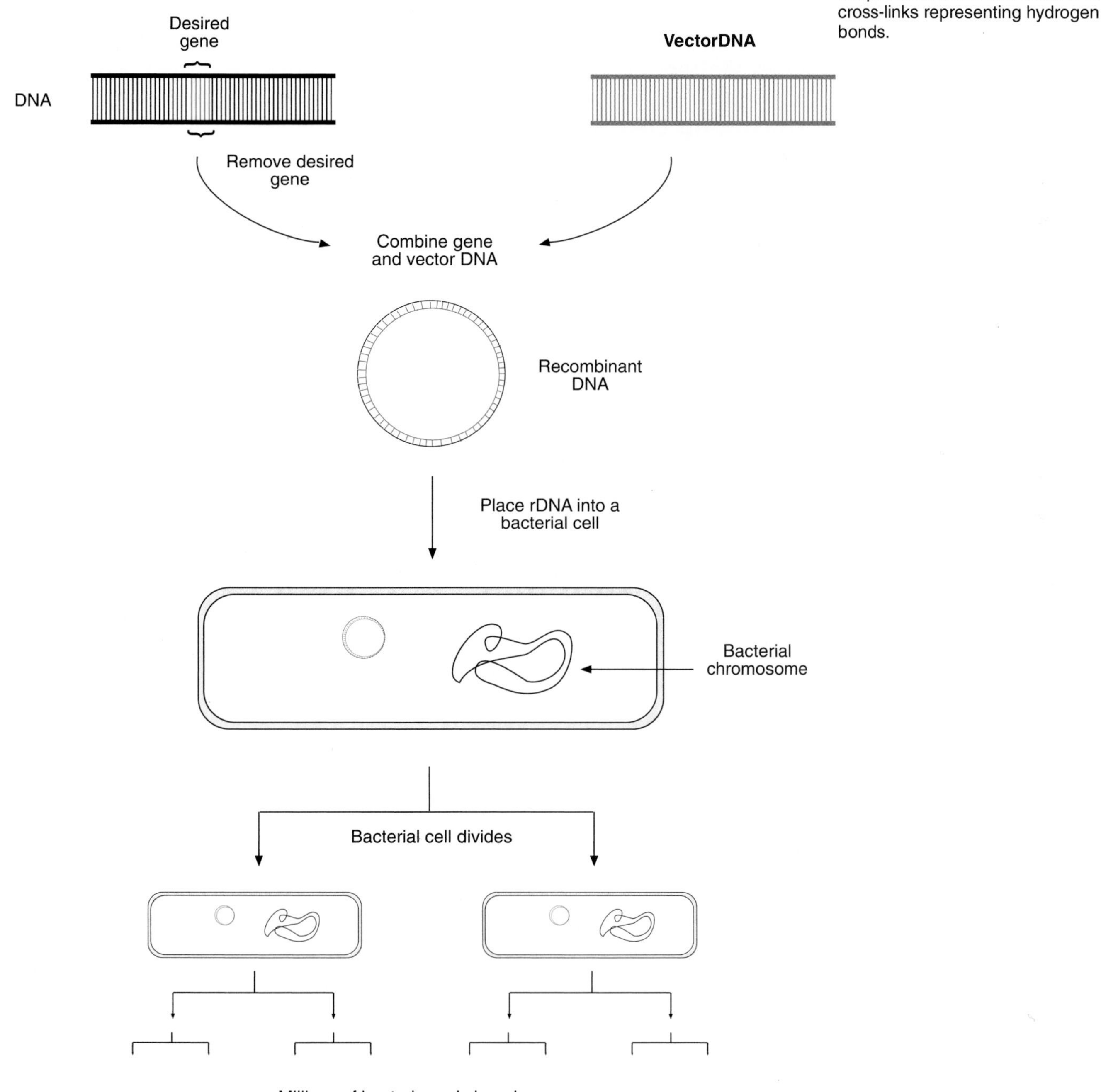

Figure 19.1

New genetic technologies are now used to make unlimited numbers of desired genes in a process called *gene cloning.* A gene of interest, most commonly one that encodes an important protein, can be identified and joined with the DNA of a vector. The resulting product—a combination of DNA from two different species or organisms—is known as *recombinant DNA.* This DNA is then introduced into a bacterial cell, usually *E. coli,* which is then placed in culture. After two or three days of bacterial reproduction, millions of *E. coli* and an equal or greater number of the gene of interest have been produced. The simplified DNA shown in the figures in this chapter consists of two strands with cross-links representing hydrogen bonds.

automated instrument that produces DNA sequences specified by a worker at a keyboard.

Since isolated genes cannot replicate themselves, a gene to be cloned must be inserted into the DNA of a suitable **cloning vector**, usually a plasmid or a virus, that can enter a living cell where replication can occur under appropriate conditions. The DNA formed by splicing DNA of the cloning vector with DNA of the foreign gene is referred to as **recombinant DNA (rDNA)**.

The rDNA is transferred into a host cell, where it will be replicated. The customary host cell is the bacterium *Escherichia coli,* although other cells are also used. Finally, the host cell is placed in an appropriate culture medium and allowed to multiply. Ultimately, millions of bacterial cells will be produced in a short period, along with an equal or greater number of cloned genes.

Every stage in the cloning "recipe" is fraught with technical difficulties. Although each step may seem straightforward, it is often necessary to repeat experiments a number of times. Tedium and frustration are constant adversaries, and failure is a frequent outcome. Diligence and creative approaches characterize efforts in this field. A few of the most common procedures used in rDNA technology are described next.

Obtaining DNA to Be Cloned

Recall that genes are specific sequences of DNA that encode proteins. One way to obtain relevant genes is to isolate them from the *genome* of the organism; that is, to identify and remove one gene from all of the DNA present in the chromosomes of a single somatic cell. This is a daunting task when one considers that the genome of an average mammal consists of approximately 1 billion nucleotide base pairs and that a typical gene contains 5,000 base pairs. To reduce the scope of this "needle in a haystack" problem, two general approaches have been developed for isolating small segments of DNA.

Genomic DNA Isolation and Restriction Enzymes

DNA molecules can be removed from cells and cut into small segments that can be manipulated and analyzed. Two basic methods are used to remove genomic DNA from cells and divide it into small segments (see Figure 19.2A). One employs mechanical techniques such as exposing cells and DNA to high-frequency sound (sonication), causing the rupture and release of fragmented DNA. The other uses a two-step approach. First, enzymes are used to digest membranes, which liberates the DNA. Second, *restriction enzymes* obtained from bacteria are used to cut the DNA into small pieces. The normal function of **restriction enzymes** in bacteria is to destroy ("restrict") foreign DNA attempting to gain entry into the cell. Through this mechanism, bacteria are protected against the insertion of DNA by viruses that infect and usually kill them.

Restriction enzymes recognize specific, short nucleotide sequences and cleave this DNA sequence at a specific *restriction site*. For example, *Bam*HI, a restriction enzyme from *Bacillus amyloliquefaciens,* specifically recognizes the DNA sequence GGATCC and cleaves it at a restriction site between the two guanines. Thus it recognizes the DNA sequence reading G↓GATCC and cuts between the two Gs, leaving G and GATCC. Approximately 200 restriction enzymes are now commercially available. By exposing a genome to suitable restriction enzymes, a set of fragments is obtained.

Complementary DNA (cDNA)

A second method for obtaining a gene of interest is to create a **complementary DNA (cDNA)** strand from a strand of messenger RNA (mRNA). This procedure is used when specific mRNA molecules can be isolated from cells that produce high levels of a certain protein. For example, a protein called *factor VIII* is necessary for proper blood clotting. It is the protein that most hemophiliacs are missing be-

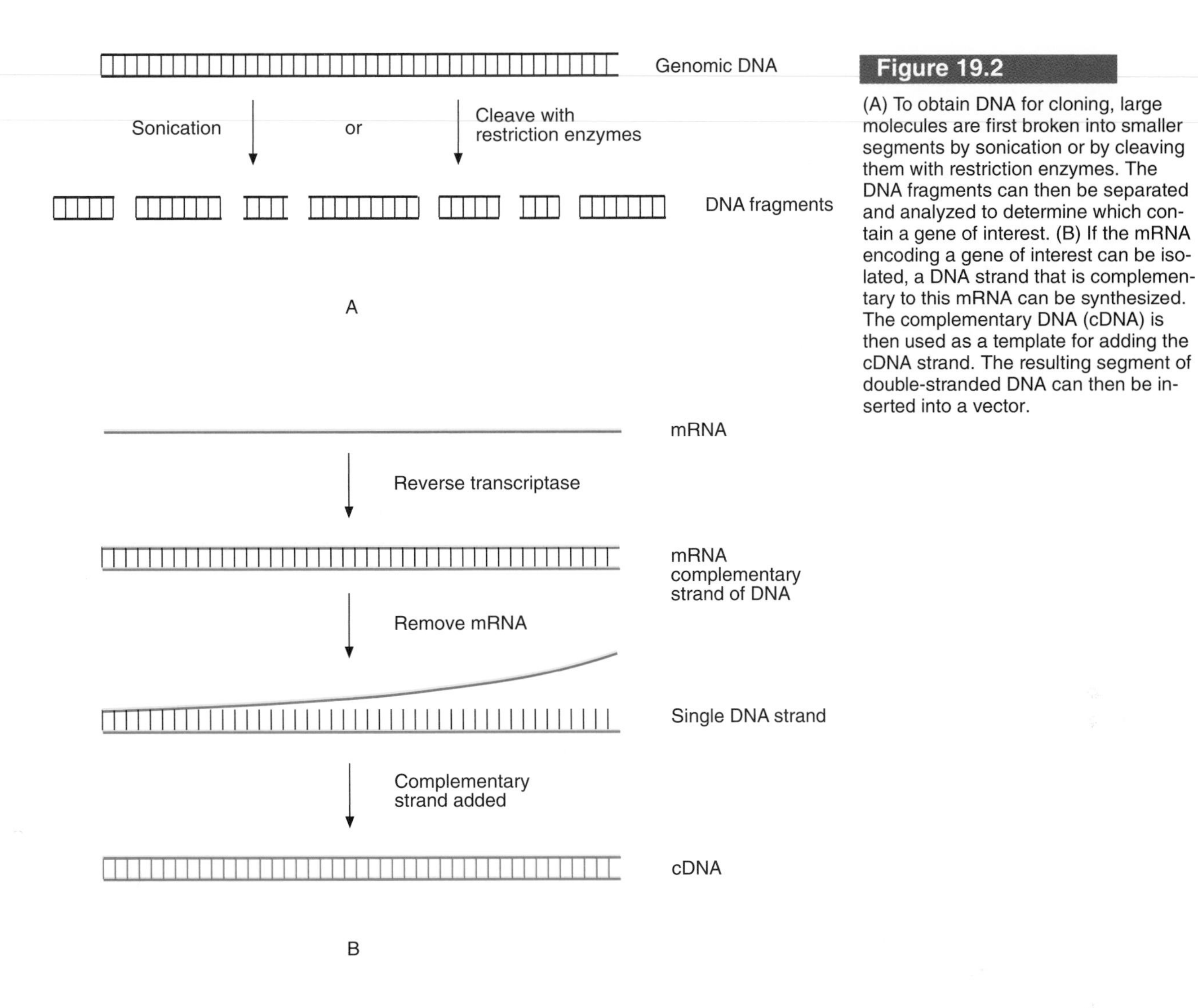

Figure 19.2

(A) To obtain DNA for cloning, large molecules are first broken into smaller segments by sonication or by cleaving them with restriction enzymes. The DNA fragments can then be separated and analyzed to determine which contain a gene of interest. (B) If the mRNA encoding a gene of interest can be isolated, a DNA strand that is complementary to this mRNA can be synthesized. The complementary DNA (cDNA) is then used as a template for adding the cDNA strand. The resulting segment of double-stranded DNA can then be inserted into a vector.

cause of a mutation in the relevant encoding gene. From several clever studies, high levels of this protein were found to be produced in the liver along with the key mRNA molecules directing its synthesis. By separating the factor VIII protein and the affiliated mRNA from other proteins and other mRNAs that were also present, it was possible to identify the gene encoding factor VIII. The major advantage of this method is that a "pure" DNA sequence can be produced that does not contain the extraneous DNA sequences (introns) that occur in genomic DNA. Once the mRNA is isolated and purified, it is used as a template to construct a cDNA strand using the unique enzyme **reverse transcriptase**, which synthesizes DNA from RNA (see Figure 19.2B). When the cDNA has been formed, the mRNA strand is removed, and a DNA strand that is complementary to the cDNA is then added, resulting in a typical double-stranded molecule. Once DNA has been obtained, the next step in gene cloning is to create an rDNA molecule that can be used for cloning cDNA or any of the thousands of fragments that can be obtained from genomic DNA.

Gene cloning is used to produce desired genes or DNA sequences. Two general types of DNA—genomic DNA and complementary DNA—are used in gene cloning. Genomic DNA can be removed from cells and cut into small fragments using restriction enzymes. Complementary DNA is produced from purified mRNA strands that code for a protein of interest. Once obtained, the DNA is spliced into a cloning vector, and the resulting rDNA is introduced into a bacterial cell for amplification.

Cloning Vectors

Almost any genomic or cDNA fragment can be cloned after it is integrated into a cloning vector forming rDNA. The rDNA molecule is inserted into a bacterial cell, where it is *amplified.* Two submicroscopic agents normally associated with bacteria, *plasmids* and *phages,* are commonly used as vector systems for cloning DNA fragments.

Plasmids

Plasmids are small, circular, double-stranded DNA molecules that reside inside prokaryotic cells such as bacteria (see Figure 19.3A). Plasmids replicate independently of the bacterial chromosome, although they are dependent on the host cell's DNA-synthesizing machinery. Plasmids are retained throughout cell division, and they generally encode proteins that are of some benefit to the bacterial host cells. For example, many plasmids (so-called R factors) contain genes that confer drug resistance on their bacterial host cells. Several strains of bacteria are now resistant to specific antibiotics such as ampicillin and tetracycline because they harbor plasmids with antibiotic-resistant genes. The fact that plasmids are able to confer drug resistance on a bacterial host has been exploited in identifying bacteria that contain such plasmids.

Figure 19.3B describes the insertion of foreign genomic or cDNA into a plasmid. The plasmid is opened through the action of a restriction enzyme, which cuts its DNA at a specific restriction site. The plasmid DNA and foreign DNA, which have been cleaved by the same restriction enzyme, are mixed together in the presence of the enzyme *DNA ligase,* which can bind them together to create an rDNA molecule. The manipulations described so far are carried out *in vitro* (in a test tube). For the rDNA molecules to be amplified, they are placed in the presence of a suitable bacterial strain. Only a very small percentage of the bacterial cells become *transformed,* meaning that they take up the plasmid containing the foreign DNA. The cells containing plasmid DNA must then be distinguished from those that do not; this is done by growing the bacterial cells in a medium containing the antibiotic whose resistance is encoded by the plasmid. Only transformed bacterial cells containing the plasmids survive; untransformed bacterial cells perish.

Phages

Bacteriophages, commonly known as *phages,* are viruses that infect bacteria (see Figure 19.4A). They are more complicated than plasmids in that they contain genes that encode proteins used in their own structures and replication. Like plasmids, however, they depend on a bacterial host cell to carry out their DNA replication. Each phage type will replicate only in a specific bacterial species or strain.

Using phages instead of plasmids as cloning vectors has three advantages: phages are easier to introduce into bacteria, larger foreign DNA fragments can be inserted into the phage DNA to create stable rDNA molecules, and phages have a greater efficiency for incorporating foreign DNA. The success for plasmids is usually 10 percent or less, whereas for phages it is greater than 50 percent.

Many phages have been used in gene-cloning operations, but one of the most frequently used has been the λ (lambda) phage (bacteriophage λ) that infects *E. coli.* The procedures for incorporating foreign DNA into the λ phage genome follow the same lines as for plasmids and are described in Figure 19.4B. The viral DNA is opened using specific restriction enzymes; foreign DNA is then introduced, and the two DNAs are joined by DNA ligase to form an rDNA molecule. The λ phage cloning vector then infects a bacterial cell where the rDNA may become integrated into the bacterial chromosome. In this phase, the rDNA will be replicated along with the bacterial DNA. Under other conditions,

Figure 19.3

(A) Plasmids, shown in this scanning electron micrograph, are small pieces of circular DNA that live in prokaryotic cells and replicate independently. (B) Segments of foreign DNA containing a gene of interest can be inserted into plasmid DNA to produce recombinant DNA (rDNA). In this example, both gene removal and opening of the plasmid vector are accomplished using the *Eco*RI restriction enzyme (G↓AATTC). The rDNA is then created by using the enzyme DNA ligase to join the complementary base pairs of the gene DNA and the plasmid DNA. The resulting rDNA is then introduced into a bacterial cell with the plasmid serving as a cloning vector. Bacterial cells will replicate the vector system, including the foreign gene.

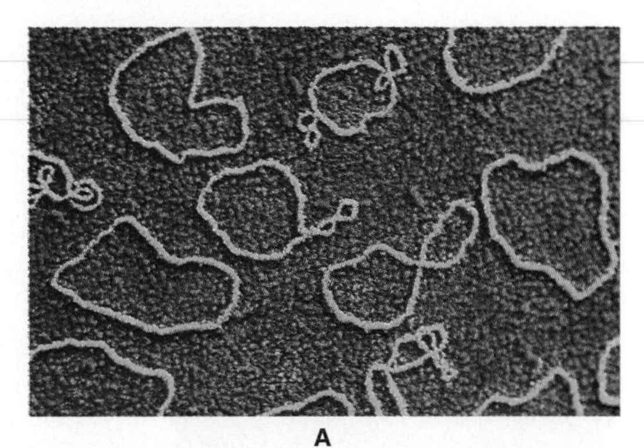

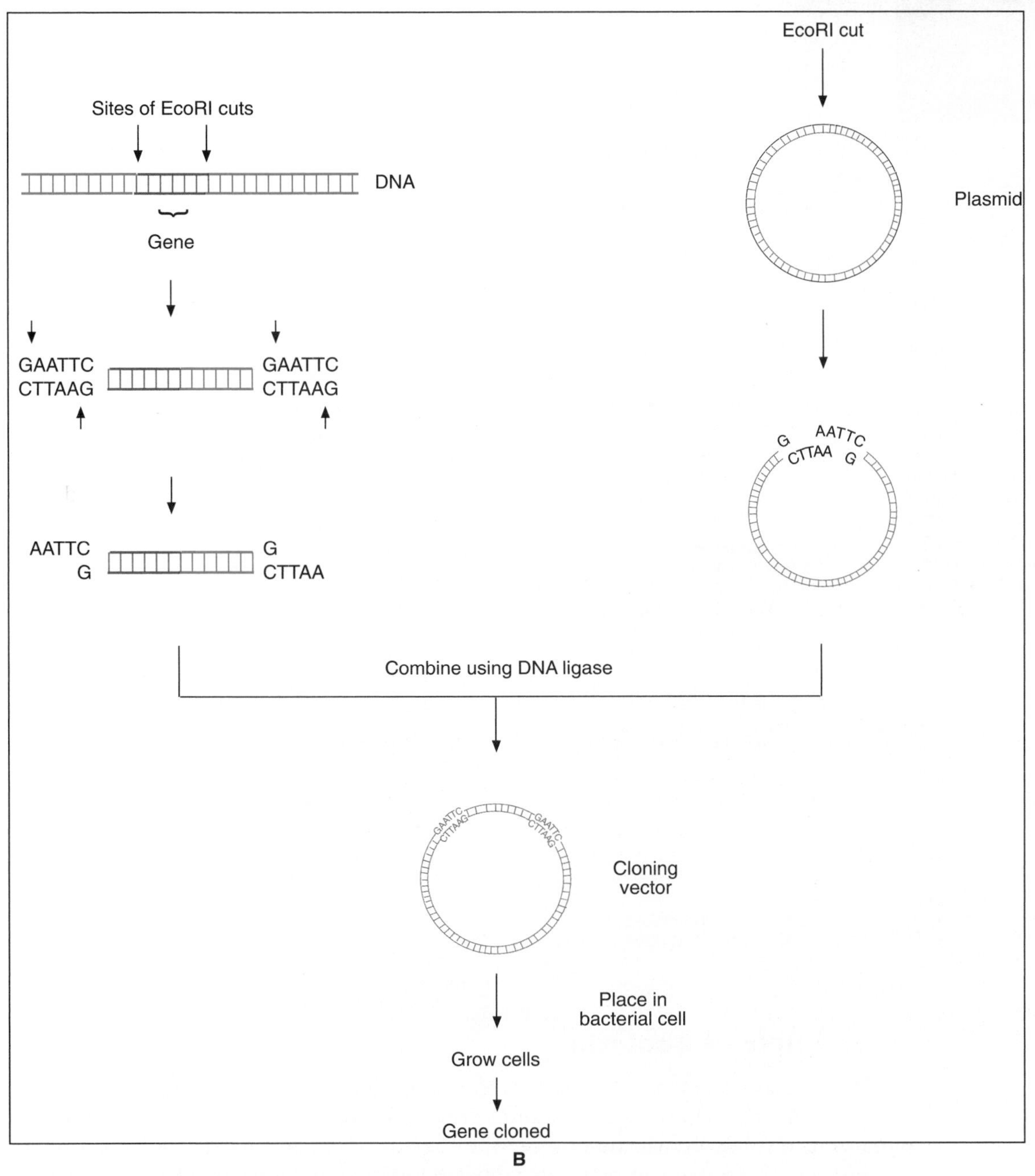

the λ phage DNA remains separated from the host chromosome and replicates autonomously. When this happens, the phage DNA replicates at a much greater rate, which causes death of the bacterial host but also results in greater yields of the cloned gene.

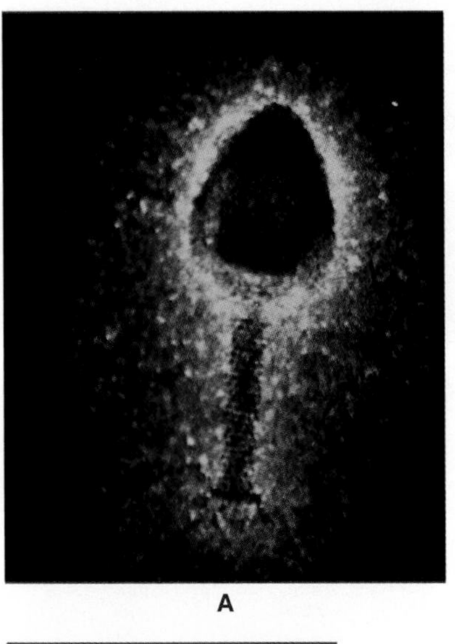

A

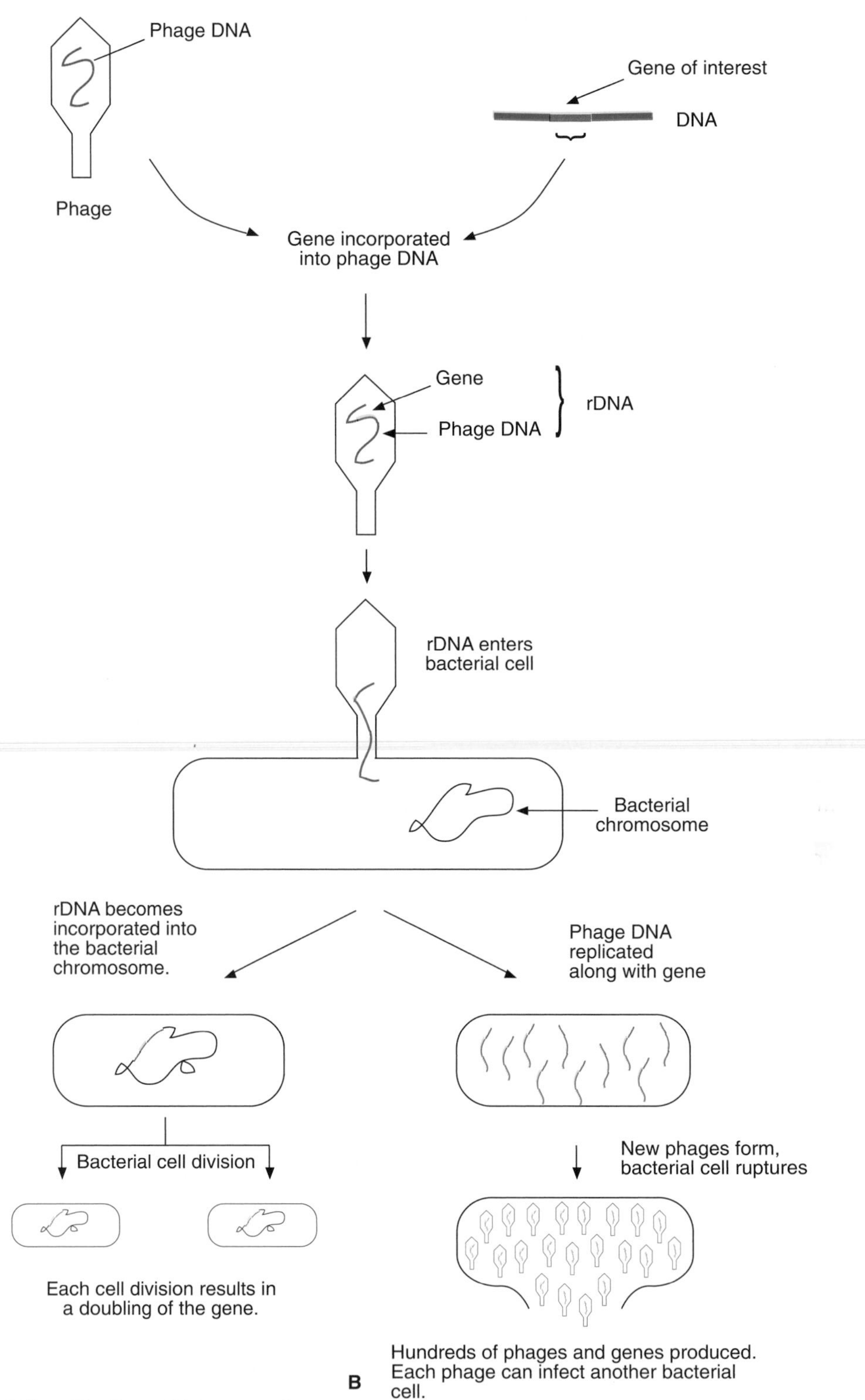

Figure 19.4

(A) Bacteriophages, shown in this scanning micrograph, are viruses that infect bacteria. The "head" of the phage contains DNA used to direct the synthesis of proteins and nucleic acids necessary to complete its life cycle. The "tail" is used for attachment to the bacteria and for inserting the viral DNA. (B) Phages can be used as cloning vectors to introduce rDNA into bacterial cells. Once inside a cell, the phage-rDNA system can follow one of two genetic routes. It may be incorporated directly into the bacterial chromosome and be replicated along with the bacterial chromosome whenever the bacterium divides. Alternatively, the phage and the gene it carries may begin replicating immediately without being integrated into the bacterial chromosome. In the latter case, new phages, each containing the gene of interest, are formed. The bacterial cellular machinery synthesizes the vector system proteins and DNA.

The Role of Bacteria

Bacteria are one-celled organisms that reproduce very rapidly by binary fission (dividing into two cells) after a brief growth period (see Figure 19.5). They live in practically every environment, but the species of greatest use to molecular biologists are those that can be easily cultivated in the laboratory by placing them on a suitable culture medium contained in a small dish. A single bacterial cell can give rise to tens or hundreds of thousands of identical cells (clones) in a single day. When a vector containing rDNA is inserted successfully into a bacterial cell, it is replicated, and hence the DNA fragment of interest is cloned.

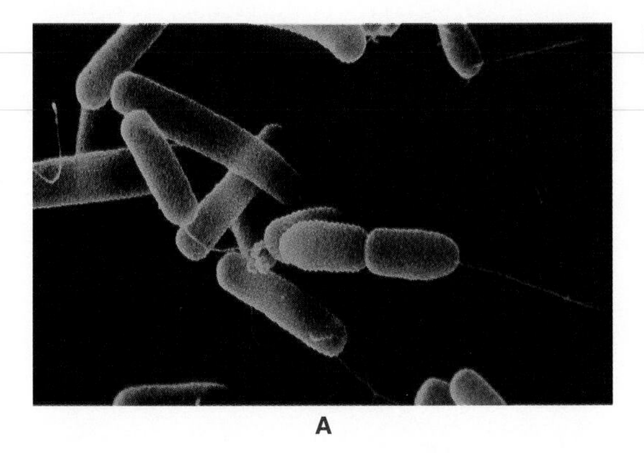

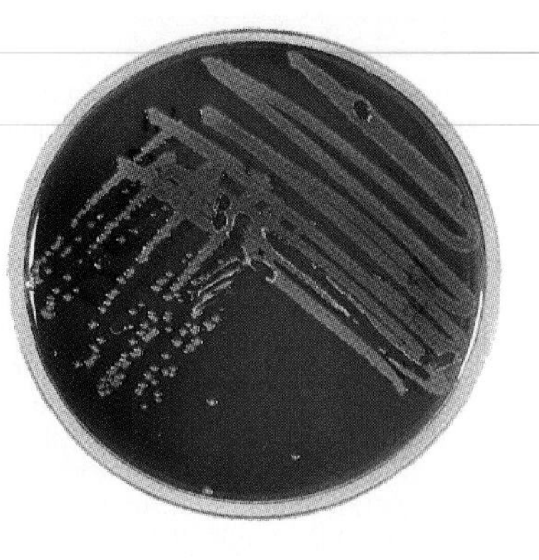

Figure 19.5

(A) Bacteria are single-celled prokaryotes that inhabit all types of environments. *E. coli* is widely used in recombinant DNA technology for synthesizing foreign genes and proteins of interest. (B) One advantage of using bacteria in scientific studies is that enormous numbers can be grown quickly in culture dishes in the laboratory. Each spot in the culture dish represents a colony consisting of millions of bacteria that originated from a single cell.

Bacteria have long been used by scientists probing the inner workings of cells. Much of the earliest information about cell biology and molecular biology came from studies of bacteria. Today, they play a critical role in gene-cloning technologies. The most popular bacterium in all of this work has been *E. coli,* a normally harmless inhabitant of mammalian digestive tracts. More is known about the molecular biology of *E. coli* than about any other species on Earth.

E. coli and some other bacteria are used to perform two functions in rDNA technology. First, they are used to clone genomic DNA or cDNA to obtain unlimited numbers of genes that can be used in further studies. Second, pure cultures, in which every bacterium contains a gene encoding a desired protein, can be created and put to work manufacturing huge quantities of that protein (see Figure 19.6). Plasmids that contain specific sequences to direct protein synthesis are called *expression vectors.*

Plasmids and phages are generally used as cloning vectors. Both normally inhabit bacterial cells, and both are dependent on the host cell for their replication. In gene cloning, foreign genomic DNA or cDNA is inserted into the DNA of the cloning vector, which is then introduced into a bacterial cell. In replicating the plasmid or phage, host bacteria also replicate the foreign DNA carried by these cloning vectors, thus producing large numbers of the desired DNA sequence. Such sequences may then be used to produce important proteins or to create gene libraries.

Gene Libraries

One of the major uses of rDNA technology is to create *gene libraries* for different species. In accordance with the procedure used to obtain the DNA, there are libraries of both genomic clones and cDNA clones. The number of "volumes" (DNA fragments or genes) required in the library to ensure that all gene se-

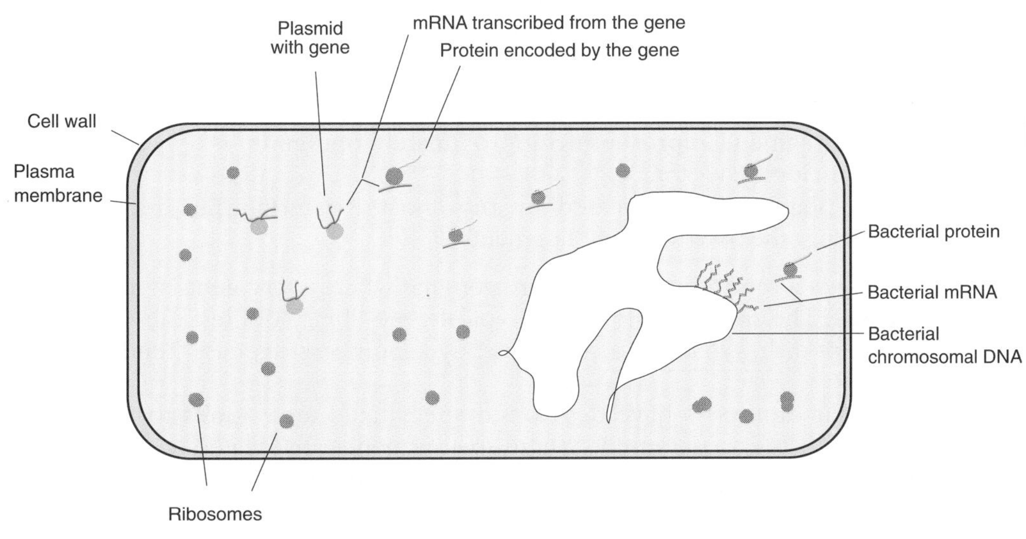

Figure 19.6

Bacteria, such as *E. coli,* are used as "protein-producing factories" in recombinant DNA technology. This schematic drawing shows the basic components of the system used for producing bacterial proteins and proteins encoded by a gene of interest that was obtained through gene cloning. The bacterium transcribes the gene carried by the plasmid vector, producing mRNA that is then translated on bacterial ribosomes. Under ideal conditions, the resulting foreign protein may account for 10 to 40 percent of the total protein synthesized by the bacterium.

Figure 19.7

To create a gene library, DNA from the entire genome of an organism is cut into segments using restriction enzymes; each segment carries one or more genes of interest. The genomic DNA is first combined with DNA extracted from a phage to create recombinant DNA, which is inserted into a phage vector and then into bacteria for cloning. After culturing, bacteria from each colony are broken apart, and the phages carrying the cloned genes are removed and stored together until further analyses are carried out to identify the genes. Together, all of the phages that contain all the genes in a genome are referred to as a *gene library.* Specialized techniques can then be used to identify the genes ("books") in the library.

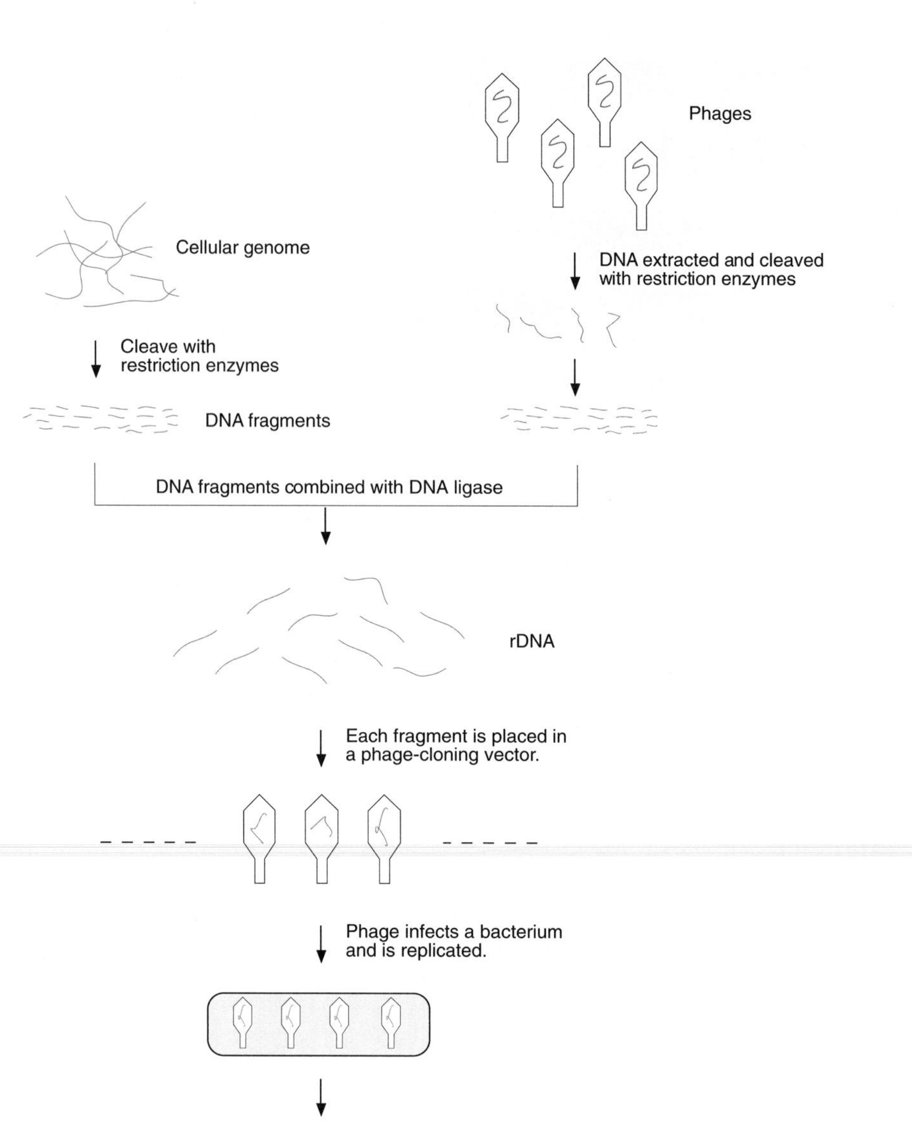

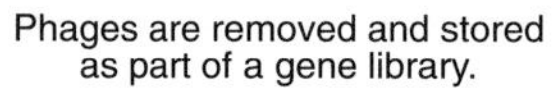

quences are represented is generally related to the size and complexity of the species. For *E. coli,* 1,500 fragments are necessary; for yeast, 4,600; fruit flies, 48,000; and humans, 800,000.

Researchers create a **gene library** for a particular species in the following sequence of events, described in Figure 19.7:

1. Genomic DNA is cut into thousands of fragments with appropriate restriction enzymes.
2. Each fragment, representing approximately one gene, is spliced into a phage-cloning vector.
3. The phages are replicated by the "host" bacteria, thus replicating (cloning) the inserted gene fragments.

Such a library is a repository of the entire genome where every gene of a species is represented. The space required for a typical gene library is a test tube or a set of small culture dishes that can be conveniently kept in a refrigerator pending further analysis.

Molecular biologists have developed specialized techniques to "screen" gene libraries, that is, to determine where specific genes are located in the pieces of DNA. This involves isolating a gene and identifying the protein it encodes. Once a gene has been isolated and cloned, it is possible quickly to determine its entire base sequence. As you might expect, the greatest emphasis is currently on research related to the human gene library, and remarkable progress has been made in isolating human genes of medical importance. Nevertheless, the human gene library is so immense, and many genes are so difficult to locate, that a complete reading of our library probably lies one or more decades in the future. This subject is explored further in Chapter 21.

BIOTECHNOLOGY

As you have learned, the discovery of the molecular tools, enzymes, and biological processes associated with genetic engineering and rDNA spanned two decades. Subsequently, the ingenious integration of these innovations led to the formation of new "bioindustries" that employ biological technology for manufacturing products. The first such company, Genentech, was founded in 1977.

The first part of this chapter described the underlying basis of one important biotechnology—an industrial technology based on the biological synthesis of important chemicals, usually proteins. A gene that encodes an important protein can be isolated, placed into a bacterium, and cloned. Subsequently, the gene can be manipulated to be expressed by a bacterium, resulting in the production of unlimited amounts of the protein.

Biotechnology is now beginning to have a significant impact in industrialized nations as its unique products, such as human insulin, become available. Some observers feel that genetic engineering promises a revolution beyond the scale of that generated by the computer. Certainly, the potential seems unlimited. In the remainder of this chapter, we will describe some general aspects and products of this biotechnology.

The Products of Biotechnology

The products developed and manufactured by biotechnology generally fall into two categories: (1) chemicals—primarily proteins—that can be manufactured by genetically engineered organisms and (2) genetically engineered organisms themselves that have acquired new capabilities.

Companies entering this field face many difficult technical and economic problems. Many are related to the large-scale production process necessary to manufacture commercial amounts of the product. Others are related to perplexing biological obstacles that are often encountered in trying to create suitable expression vectors capable of synthesizing commercial quantities of the key molecules.

Molecular Products

Modern bioindustries use complex, high-technology modifications of an ancient biological process called fermentation. *Fermentation* occurs when microorganisms synthesize useful products from nutrients and raw materials in the absence of oxygen. For thousands of years, humans have depended on this process to generate beer, wine, olives, pickles, cheese, soy sauce, and other substances that increase dietary diversity. Rather than using barrels or other simple containers, industrial operations employ very large scale systems in which microorganisms are cultured in huge reactor vessels maintained under carefully controlled conditions (see Figure 19.8).

Three general classes of proteins (and a few other chemicals) are synthesized by genetically engineered microorganisms: new molecules that have never been available, rare proteins that are difficult and expensive to obtain, and proteins that can be produced cheaply through biotechnology. So far, much effort has been directed at producing substances with human medical applications. These include insulin and other hormones, vaccines, blood products, immune regulators, antibodies, and antibiotics. Product development of proteins using ge-

Figure 19.8

Large reactor vessels are used in the industrial production of proteins by microorganisms carrying rDNA.

netic technology consists of three phases: creating the gene-vector-bacterium protein-producing system, large-scale production of the protein, and approval for marketing. The last phase requires extensive testing of the product, first in animals to establish its safety, then in clinical trials on human patients. Along the way, various applications must be submitted to and approval obtained from various federal agencies responsible for product safety. In the United States, final approval for marketing comes from the Food and Drug Administration.

In addition to pharmaceutical and medical companies, other industries involved in food processing, agriculture, chemicals, pollution control, and energy are exploiting these technologies to manufacture commercially important products. The production of vaccines and hormones by genetically engineered microorganisms serves to illustrate the importance of the new industry.

Vaccines

A **vaccine** is a substance containing a protein or a number of proteins called *antigens* that can be used to induce immunity against certain viruses, bacteria, and parasites. Mammals produce specific proteins called *antibodies* that react against antigens of disease-causing organisms and lead to the destruction of the harmful agent. At the present time, vaccination frequently consists of receiving an inoculation containing dead or weakened forms of the agent that causes the disease. This scheme is very effective, and children in the United States are routinely vaccinated against diphtheria, tetanus, polio, measles, rubella, and mumps by the time they are 5 years old. Many animal diseases are also controlled by vaccines.

Nevertheless, current vaccines pose several problems. A vaccine containing weakened microbes can occasionally cause the disease that it is supposed to prevent. Although the risk is slight (on the order of 1 in 100,000 inoculations), lawsuits by people infected have caused some pharmaceutical companies to stop making certain vaccines. Many vaccines are very expensive, and others are relatively ineffective. For some important diseases, no vaccines are available, and for others, the infecting virus has become resistant to the antibodies. Many disease organisms, particularly viruses, have multiple strains, and vaccines do not always immunize against all the different strains. That is why new flu vaccines may be required yearly.

A number of creative tactics are being employed by genetic engineers in developing new vaccines. The general approach has been to produce vaccines containing only small parts of viruses, not the entire virus itself. Genes that code for some surface antigens of viruses have been isolated, cloned, and incorporated into expression vectors where they produce the antigen that can be used in a vaccine. An alternative method involves removing genes that actually cause the disease from a virus, cloning these altered viruses, and then using them in inoculations. Antibody production will occur because virus antigens are present, but the disease will not since those genes are not present. Other approaches are also being investigated.

To date, vaccines for a limited number of viral diseases have been developed. A hepatitis vaccine was the first genetically engineered human vaccine approved for general use by the Food and Drug Administration. Other viral vaccines are now in the developmental stage, including those for influenza types A and B and herpes. Several vaccines against malaria, a disease caused by a protozoan, have been developed and are in various stages of clinical trials. Since malaria affects approximately 800 million people in the world, the development of an effective vaccine is crucial, but the test results obtained to date are very discouraging.

Hormones

Products from this class of proteins have been developed and marketed with a high degree of success. A number of commercial products are now available, including the protein *insulin,* a hormone required by humans for the normal processing of sugar in the diet. People with the disease *diabetes* lack the ability to produce normal insulin or normal levels of insulin, and many require injections of this hormone for survival. Until 1983, when genetically engineered human insulin became available (marketed as Humulin by Genentech), insulin was obtained from the pancreases of slaughtered pigs and cows. Animal insulins were very effective, being nearly identical to human insulin, but they could also cause allergic reactions in some individuals. This does not occur with pure human insulin.

Several companies are now marketing human growth hormone (hGH), which provides a cure for a certain form of dwarfism. Growth factors with structural similarities to hGH that facilitate wound healing have also been produced. In an interesting development, it was recently announced that hGH may soon be tested as a method for delaying the aging process.

Genetically Engineered Microorganisms in the Environment

Genetically engineered microorganisms may have direct application in the environment in four areas: mineral processing and recovery, enhanced oil recovery, wastewater treatment, and industrial cleanup and pollution control. In all of these cases but wastewater treatment, large-scale releases of altered microorganisms into the environment would be required.

Bacteria have been used to dissolve and separate certain metals (for example, copper) from low-grade ores; about 10 to 20 percent of the world copper supply comes from this technology. Such capabilities are believed to be under genetic control, but little is known about the biological processes involved. After conventional methods are no longer effective, oil can be recovered from depleted wells by injecting certain chemicals. Microorganisms can synthesize the chemicals that help to remove more oil. To date, progress has been relatively sluggish in these two areas.

Bacteria have long been used in municipal and industrial wastewater treatment facilities for breaking down solid wastes. Few of the relevant mechanisms in these processes are currently understood at the biochemical or genetic level. Nevertheless, significant efforts are being made to improve different parts of the breakdown processes by modifying bacteria using genetic technology.

Several companies have created mutant microbes that can break down such chemical pollutants as cyanides and phenols. Recently, genetic tools were used to develop bacteria with potentially important applications. Trichloroethylene (TCE), a liquid chemical used widely in industry, is one of the most significant environmental pollutants. The careless handling and storage of TCE over several decades has resulted in the widespread contamination of soils and groundwater in the United States. TCE is a suspected cancer-causing agent, and there is great concern about this chemical because groundwater is often the source of drinking water. Species of *Pseudomonas* bacteria have been identified that contain plasmids that can break down chemicals such as TCE. Would it be possible to isolate the gene or genes responsible for this capability and create a bacterial strain that could be used to break down TCE in the environment? Scientists isolated a fragment of DNA from *P. mendocina* that contained genes associated with TCE degradation. They then transferred this fragment into *E. coli* using appropriate vectors. In tests, it was found that the recombinant *E. coli* rapidly degraded TCE, reducing the concentration of this pollutant in water a thousandfold! Given the extent of TCE pollution in groundwater and in soils at contaminated waste sites, the use of these bacteria, or other such genetically altered microbes, to remove this substance from the environment may have tremendous potential.

Considerable additional research is necessary in these areas, and addressing environmental problems with altered microbes remains a speculative technology. By far, the greatest efforts to develop genetically engineered microorganisms to perform specific functions have been in the field of agriculture. Advancements in this area are considered in Chapter 20.

Focus on Science and Society

Biotechnology Policies in the United States

The United States government has long been involved in formulating policies related to biotechnology. The biotechnology issues include safety regulation, health-related activities, scientific research, environmental and agricultural developments, trade and property rights, and national defense.

Regulatory policies have evolved since the mid-1970s to keep pace with the explosive progress in recombinant DNA technology. The early concerns were centered on human safety in using engineered microorganisms in the laboratory. Today, the major issues are protection of the environment and human health associated with the release of altered biological materials.

In June 1986, the government published its *Coordinated Framework for the Registration of Biotechnology,* which assigned responsibilities for reviewing genetically engineered products and microorganisms to specific federal agencies. In most cases, the statutory authority comes from other, related acts (for example, the Environmental Protection Agency regulates under the Federal Insecticide, Fungicide, and Rodenticide Act; the Toxic Substances Control Act; and a few other antipollution statutes). This has often led to confusion, controversy, and legal actions. Nevertheless, there is presently no strong movement for creating new legislation that specifically covers the products of biotechnology. The responsibilities of the key agencies are outlined in Table 1.

The United States has made significant progress in developing a strong framework for defining and regulating products or microorganisms created by rDNA technology. In part, this is related to the widely held view that since biotechnology is closely associated with certain fundamental processes of life, the public's concern over related issues must always be addressed by scientists, industry, and/or the government.

In general, the regulatory thrust in the United States has been adaptive. The emphasis is toward monitoring developments rather than trying to create complex regulations that attempt to anticipate all potential problems that could arise. However, there is still no complete consensus about the value of this regulatory strategy, and two opposing views have emerged. On one side are people who agree with the flexible approach. They tend to support strong governmental oversight as a way to secure public trust in an industry sometimes viewed as potentially hazardous. To them, overly rigid regulations would impede progress and delay the enormous benefits biotechnology promises. On the other side are people who have criticized and legally contested the existing regulations, believing that the tightest possible restrictions are required to ensure that no unanticipated or undesirable out-

Table 1 Federal Agencies and Their Responsibilities in Regulating Biotechnology

Animal and Plant Health Inspection Service (APHIS) This part of the Department of Agriculture oversees broad categories of genetically engineered plants and animals and of microbes altered for agricultural purposes.

Environmental Protection Agency (EPA) This is the lead agency for monitoring all microbial pesticides and conducting required reviews of genetically engineered microorganisms prior to release into the environment. The primary criterion for approving a pesticide is that it will not cause "unreasonable adverse effects on human health or the environment." The data requirements for satisfying this criterion are detailed and complex.

Food and Drug Administration (FDA) This agency has authority over human and animal drugs—"biologics," including any "virus, serum, toxin, antitoxin, vaccine, blood, blood component or derivative, allergenic product or analogous product . . . applicable to the prevention, treatment, or cure of diseases or injuries," food or color additives, and medical devices. Obtaining approval of a new drug is normally a long and expensive process, averaging six to eight years and tens of millions of dollars. In the case of recombinant DNA products, approval time is often reduced because some information is already available.

National Institutes of Health (NIH) The NIH Recombinant DNA Advisory Committee (RAC) has primary responsibility for rDNA research. Its 1976 publication, *Guidelines for Research Involving Recombinant DNA Molecules,* assigned different categories of risk to certain types of experiments. Some types of studies were prohibited, and others had to employ special containment and safety procedures, depending on the level of risk. Since then, after the completion of millions of rDNA experiments, it has become clear that the early assessments were overly conservative. Many of the research risks either did not exist or were exaggerated. Today, the guidelines have been modified, and only 10 percent of the experiments conducted fall under guideline restrictions.

Occupational Safety and Health Administration (OSHA) This agency's primary biotechnological role is monitoring genetically engineered microbes in the workplace.

come occurs. As with other scientific issues that become political issues, the questions raised by rDNA technology are very complex and not always amenable to solution by rational analysis of experimental data.

It seems likely that informed citizens will continue to debate the lively topic of regulation for many more years. However, a 1987 publication (*New Developments in Biotechnology: Public Perceptions of Biotechnology*) indicates that biotechnology seems to enjoy a positive image. It presents data that show that the American public is generally supportive of biotechnology but also continues to see a need for government regulations and reviews by external scientific organizations to address areas of public concern.

Recombinant DNA encoding a protein of interest can be placed into bacteria that will then carry out translation and produce significant quantities of the protein. Using this technology, bioindustries have succeeded in producing vaccines, hormones, and many other products. The same technology has also been used to create genetically engineered bacteria that can be used in removing undesirable substances from the environment.

The Maturation of Biotechnology

Many of the scientists who made basic discoveries about DNA that eventually led to the birth of a new type of industry went on to collect the Nobel Prize for their achievements. As examples, a Nobel Prize was awarded for the discovery and use of restriction enzymes (1978), the creation of the first rDNA molecules (1980), and the development of powerful methods for elucidating the base sequence of DNA (1980). As the use of biotechnology has expanded, scientists and the federal government have formulated measures for regulating rDNA research activities and products manufactured by bioindustries (see the Focus on Science and Society, "Biotechnology Policies in the United States").

The significance of biotechnology cannot be overstated. It holds unlimited promise for the production of pharmaceuticals that are important for human health and chemicals that can be marketed as industrial commodities. At a different level, explored in Chapters 20 and 21, rDNA methodology can be used to modify multicellular organisms in ways that are now coming into focus.

Summary

1. During the three decades following Watson and Crick's report on the structure of DNA, scientists learned much about how DNA was replicated, the enzymes involved, and mechanisms that could be used to combine DNA from different species. These processes gave rise to a new field of science called genetic engineering and new products produced through recombinant DNA technology.
2. Genes from any organism that encode important proteins can now be isolated and cloned using the tools of molecular biology. A gene is first inserted into a cloning vector, creating recombinant DNA. This rDNA is then placed into a bacterial cell, where it will replicate in a way that produces thousands of copies of the gene.
3. Recombinant DNA technology is also used to create gene libraries that theoretically can include all the genes in a species' genome. Once a library has been established, specific genes can be studied, and in some cases, encoded proteins can be determined.
4. Once a gene has been isolated and amplified, copies can be introduced into bacterial cells, which then manufacture the protein specified. Industries now use this process to produce large quantities of commercially important proteins.
5. Products manufactured by biotechnology include hormones, vaccines, and immune system biochemicals. Genetically engineered microorganisms that may be used in mineral extraction, oil recovery, and pollution control have also been produced, although their use and regulation require further study.

Review Questions

1. What types of "biological tools" contributed to the development of rDNA technology?
2. Describe the general process of gene cloning.
3. How are desired DNA sequences obtained for cloning?
4. What is complementary DNA? How does it differ from genomic DNA?
5. What are restriction enzymes? How are they used in gene cloning?
6. What are cloning vectors? How are they used in gene cloning?
7. How are bacteria used in gene cloning?

8. How are gene libraries created? How are they used?
9. What are some important products now being manufactured using biotechnology?
10. What are the advantages of using biotechnology to create such products?
11. How are genetically engineered microorganisms now being used?

Essay and Discussion Questions

1. What types of biotechnology products should be emphasized for development in the future? Since the federal government funded much of the basic research that led to the elaboration of these technologies, should the general public have a voice in making such decisions, or should they be left to industry? Why?
2. Should individuals or bioindustry companies be allowed to patent any DNA sequences or genetically engineered organisms they create? Why?
3. How might gene libraries be used in the future?

References and Recommended Readings

Bains, W. 1987. *Genetic Engineering for Almost Everybody*. Harmondsworth, England: Penguin.

Brill, W. J. 1988. Why engineered organisms are safe. *Issues in Science and Technology* IV:44–50.

Drlica, K. 1984. *Understanding DNA and Gene Cloning: A Guide for the Curious*. New York: Wiley.

Glover, D. M. 1985. *Gene Cloning: The Mechanics of DNA Manipulation*. London: Chapman & Hall.

Holtzman, N. A. 1989. *Proceed with Caution: Predicting Genetic Risks in the Recombinant DNA Era*. Baltimore: Johns Hopkins University Press.

Murrell, J. C., and L. M. Roberts. 1989. *Understanding Genetic Engineering*. New York: Halsted Press.

Oehen, S., H. Hengartner, and R. M. Zinkernagel. 1991. Vaccination for disease. *Science* 251:195–198.

Olson, S. 1986. *Biotechnology: An Industry Comes of Age*. Washington, D.C.: National Academy Press.

Research and Education Association. 1982. *Genetic Engineering*. New York: Research and Education Association.

Sharples, F. E. 1987. Regulation of products from biotechnology. *Science* 235:1329–1332.

Watson, J. D., J. Tooze, and D. T. Kurtz. 1983. *Recombinant DNA: A Short Course*. New York: Freeman/Scientific American Books.

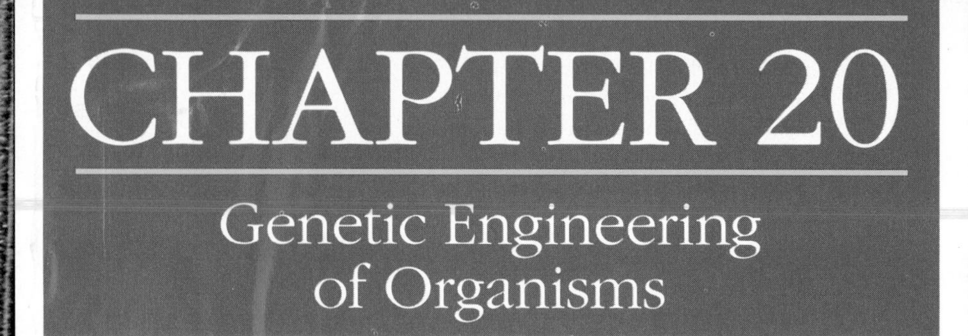

CHAPTER 20

Genetic Engineering of Organisms

Chapter Outline

Reading Questions

1. How might genetically engineered bacteria be used to improve plant growth and survival?
2. What methods have been developed for the genetic engineering of plants?
3. How has genetic engineering been applied to animals?
4. How are transgenic animals created?
5. What factors must be considered in releasing genetically engineered organisms into the environment?

Following the revolutionary developments in recombinant DNA technology and genetic engineering in the 1970s, interest turned to modifying microorganisms and multicellular plants and animals. Was it possible to use genetic technology to create new phenotypes with special capabilities that would benefit humans?

For the most part, attention has been focused on research with agricultural applications because of the enormous potential benefits for humankind. Astonishing changes in agricultural products are expected to occur during the 1990s.

Genetic technology can be used to improve agricultural products in two general ways. First, new biochemicals can be created, using "standard" recombinant DNA (rDNA) technology, that will enhance the growth or survival of agricultural commodities. Vaccines, drugs, antiviral substances, food additives, and growth hormones are examples of such chemicals. The importance of these products in agriculture cannot be overemphasized; for example, plant viral diseases cost U.S. farmers over $2 billion a year, and losses to animal viruses are even greater. There have already been some successes in this area. Antiviral treatments for serious viral diseases of economically important plants and animals have been or are being developed, and growth hormones have been used to help boost milk production in dairy cows (Figure 20.1).

A second approach involves altering the genetic material of agriculturally important microbes, plants, and animals to produce new individuals or strains with enhanced biological aptitudes. Although the potential value of such organisms is incalculable, many technical difficulties must be resolved before significant advances can be expected. Paramount among these difficulties are the complexities involved in identifying genes and gene products that determine traits of interest and creating stable changes in the DNA of genetically engineered plants and animals. Nevertheless, since the mid-1980s, there has been a rapid awakening in recognizing the potential value of genetic engineering, not only in agriculture but also in research on disease, environmental science, and fundamental molecular processes in plants and animals that are not yet understood.

Intense efforts are now being made to use recombinant DNA technology and other genetic technologies to create new traits or enhance existing traits in organisms used in food production.

GENETICALLY ENGINEERED BACTERIA

The growth of important crops is influenced in nature by microorganisms in a number of different ways, both positive and negative. How can bacteria be modified to produce desired effects on plant growth and survival? Different strategies

Figure 20.1

Since the 1960s, milk production has increased by 50 percent as a result of genetic processes.

have been developed. For beneficial traits, a single gene or gene cluster is typically added to the genome in order to improve the microbe in some way. For harmful bacteria (for example, those that cause a disease), one or more genes responsible for the undesirable trait may be removed. Examples of the genetic engineering of beneficial and harmful bacteria for agricultural uses will be considered.

Beneficial Bacteria

Certain bacteria (*Rhizobium* spp.) can supply particular crop plants with nitrogen, a critical nutrient more commonly provided by expensive fertilizers in the United States (see Figure 20.2). Plants cannot absorb nitrogen directly from the atmosphere and transform it into chemical forms they require (see Chapter 7). Industrial processes produce fertilizers with nitrogen in a usable form, and *Rhizobium* living in the soil in a symbiotic relationship with a limited number of plant species can also transform nitrogen into useful forms. *Rhizobium* forms nodules in the roots of legume plants such as peas, beans, peanuts, soybeans, clover, and alfalfa, providing nitrogen to the plant and receiving nutrients in return. Since Roman times, farmers have exploited this relationship in order to increase soil fertility.

Genetic engineers are attempting to identify and manipulate *Rhizobium* genes in order to create strains that will be more efficient in producing nitrogen or that will form associations with other plants. Since farmers spend over $1 billion a year just to fertilize corn, the advantages of extending nitrogen-fixing abilities to other plants are obvious.

Some bacterial species are beneficial because they protect plants from disease and from soils that are excessively acidic, salty, or polluted with toxic levels of heavy metals. Others are able to break down harmful pesticides or kill weeds that compete for nutrients. As scientists learn more about the underlying mechanisms involved in such functions, it may become possible to produce microbes with improved capabilities in performing these operations.

Harmful Bacteria

Bacteria and other microorganisms cause diseases of agricultural plants and animals and make life miserable in other ways. How can a harmful bacterial trait be neutralized by genetic engineering? A fascinating example involves a bacterium, *Pseudomonas syringae,* that produces frost injury in susceptible crop species. At-

Figure 20.2

(A) A scanning electron micrograph of *Rizobium* bacteria in a nodule. Species of *Rhizobium* are able to convert nitrogen from the atmosphere into forms that can be used by plants. (B) These bacteria can live in root cells of legume plants such as peas and soybeans. When rhizobia infect a root cell, a nodule is formed in which the bacteria live and form nitrogen compounds that pass into the plant's tissues from the roots.

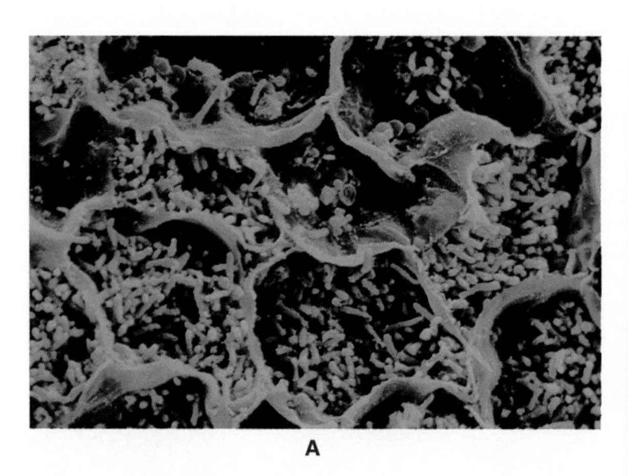

A

B

mospheric moisture condenses around dust particles, forming raindrops or snowflakes that then fall to the ground. This process is called *nucleation* because the dust particle provides a nucleus for droplet formation. Similarly, ice forms through nucleation around a variety of particles, including certain bacterial species that live on plants. *P. syringae* normally lives innocuously on many plants, including potatoes and strawberries. However, at temperatures just above freezing, the bacteria nucleate ice crystals from water vapor in the air, and the resulting damage kills millions of dollars' worth of these crops every year. What accounts for this property? Why don't all bacteria cause such damage in frost-sensitive plants?

Remarkably, the nucleation trait, called **ice plus (Ice$^+$)**, was found to be caused by the protein product of a single gene in *P. syringae*. Could genetic engineering be used to remove this gene? If so, would the new strain be able to colonize plants, survive, and grow in a natural environment? Figure 20.3 describes the action and effects of ice-nucleating and ice plus bacteria and summarizes results of some studies related to these bacteria. Scientists cloned the nucleation gene, then deleted certain sections from it, and finally exchanged the modified gene for the normal (Ice$^+$) gene in bacteria. The new strain, known as **ice minus (Ice$^-$)**, was then tested in laboratory and field studies by spraying it on young potato plants. From this research, completed in 1988, it was concluded that the Ice$^-$ strain colonized plants and survived as well as Ice$^+$ strains. In tests where an Ice$^-$ population first colonized a plant, Ice$^+$ populations remained low. These results suggest that introducing Ice$^-$ strains into fields where commercial crops are grown may reduce damage caused by mild frost, if these strains replace or reduce Ice$^+$ populations that cause nucleation. However, no final approval has yet been given for commercial applications of Ice$^-$.

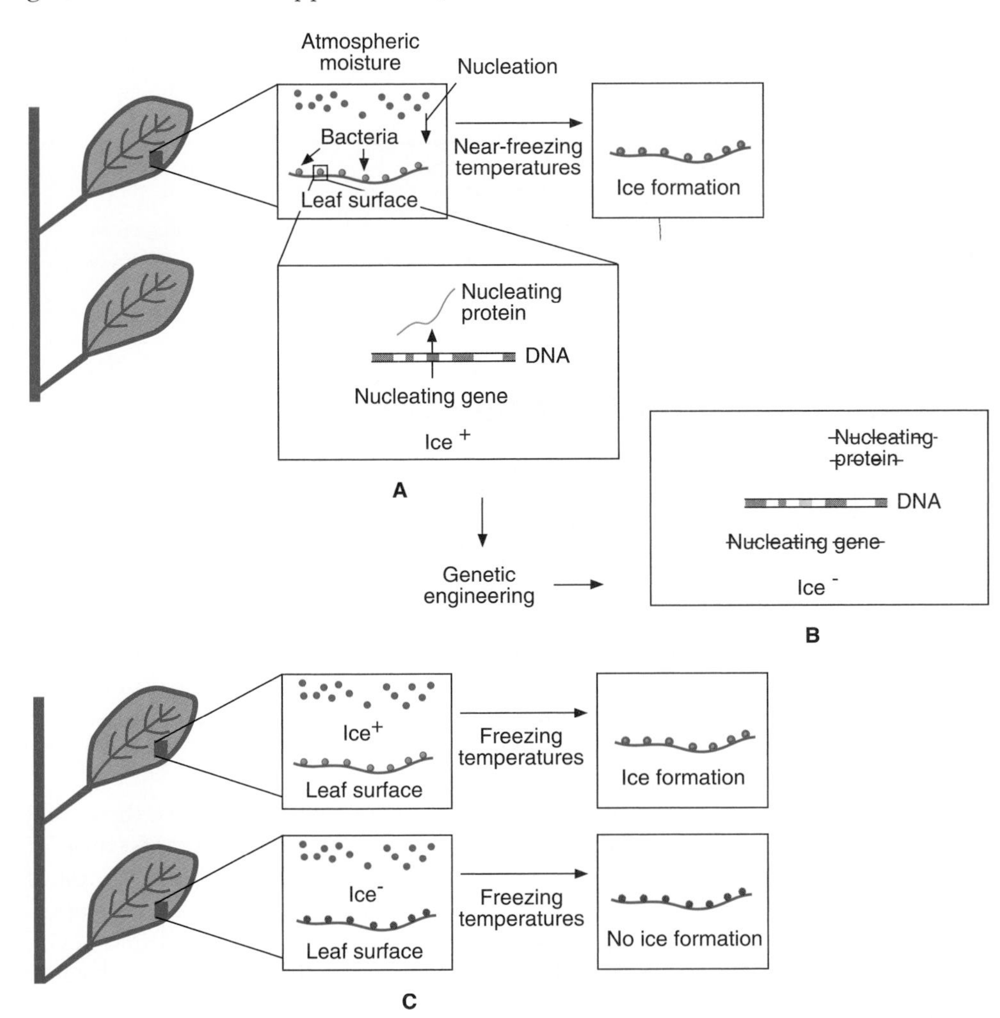

Figure 20.3

(A) "Ice plus" strains of *Pseudomonas* bacteria that live on crop plants can nucleate water vapor present in the atmosphere. At near-freezing temperatures, nucleation results in ice formation, which can result in frost damage or death of the plant. (B) By removing the nucleating gene from the bacteria, an "ice minus" strain is created that is no longer able to induce ice formation or cause this effect. (C) In field experiments, leaves covered with ice plus bacteria froze while those with the genetically engineered ice minus strain did not freeze.

Applying Genetically Engineered Bacteria

Despite the potential benefits, the idea of applying genetically engineered microorganisms for agricultural purposes has been controversial. For example, the first proposal to field test Ice$^-$ strains of bacteria in small, experimental fields of plants was submitted to the Recombinant DNA Advising Committee on September 17, 1982. The first authorized field studies were not conducted until April 24, 1987. In the intervening five years, there were numerous proposal revisions, lawsuits, unauthorized testing, charges of data falsification, fines for procedural violations, restraining orders, and other complications.

Why is there such controversy in this area? Answers to five questions provide a focus for considering the release of genetically engineered microorganisms. Will the altered microbe survive in the natural environment? Will it multiply? Will it spread to areas beyond the site of introduction? Can its altered genes be transferred to other species? Will the altered microbe or any species receiving its genes prove to be harmful? The general concern expressed by opponents of field testing with altered microbes is that dangerous new organisms might be released into the environment with a potential for causing harm. Proponents argue that a single or limited number of gene modifications are highly unlikely to result in the creation of a microorganism presenting any significant risk. No consensus on this subject has yet emerged among the public, although most scientists seem to favor the latter view.

Genetically engineered microbes are not now projected to have significant roles in agriculture. However, one area of promise involves the development of microbial biological control agents that kill disease-carrying insects such as mosquitoes. Also, ruminants (animals such as cows and sheep) require microorganisms in their digestive tracts to ferment the forage they consume when grazing. It may be possible to create microbes that would enhance the abilities of these animals to digest their food or even allow them to digest other plants that cannot be broken down normally.

Bacteria that affect plant growth and survival have been targeted for genetic engineering. Attempts are being made to extend nitrogen fixation capabilities to different bacterial strains and other plants. Harmful bacteria have been modified by removing genes responsible for negative effects on plants. Issues surrounding the release of genetically engineered bacteria have not yet been resolved.

GENETIC ENGINEERING OF PLANTS

Plants and animals have been reproductively manipulated since the dawn of the agricultural age. At the simplest level, plants with desirable "production traits" (such as juiciness in fruits, high yields, resistance to disease) have been bred for centuries in attempts to establish varieties that produce offspring well endowed with the valued feature. This approach has a long record of success. For example, yields of corn and wheat crops have increased continuously over the past half century. However, the conventional selection methods used by plant breeders typically require years to achieve the desired result, and there appear to be limits on further progress using these techniques. Since DNA governs the expression of these desired characteristics, genetic engineering may offer powerful opportunities for creating or amplifying production traits.

The genetic engineering of plants involves two basic approaches, those that effect genetic changes through cell fusion and those that involve the insertion or alteration of genes in cells. The second approach is similar to the methods of rDNA described for microorganisms in Chapter 19, with one important conceptual distinction. For microbes, genetic changes are made at the cellular level, whereas for crops, changes at the cellular level must usually be reproduced stably into every cell of the plant (and its offspring) if they are to be of value. The ultimate goal of these technologies is to increase productivity significantly through a variety of means.

Protoplasts

A number of methods have been developed to facilitate gene transfer in plants. A promising technique for producing new plants with genes and chromosomes from different species is described in Figure 20.4A. **Protoplasts** are plant cells that have had their outer cell wall removed by digestion with certain enzymes (Figure 20.4B). Protoplasts of some species have *developmental totipotency,* the potential to regenerate a whole plant from a single cell. In genetic engineering, foreign DNA can be introduced into these "naked" cells to create new plant varieties or even species. Different methods are used to insert DNA into protoplasts: (1) two protoplasts from two different plant species may be stimulated to fuse and form a *somatic-cell hybrid* in a process called **protoplast fusion**; (2) DNA may be introduced by a disease-causing cloning vector; or (3) most commonly, DNA segments are introduced into protoplasts by chemical, electrical, or mechanical procedures.

Genetically altered or unaltered protoplasts can be grown in media containing nutrients, vitamins, and plant growth hormones to form a mass of unspecialized cells called a *callus* (see Figure 20.4C). The callus is then induced to

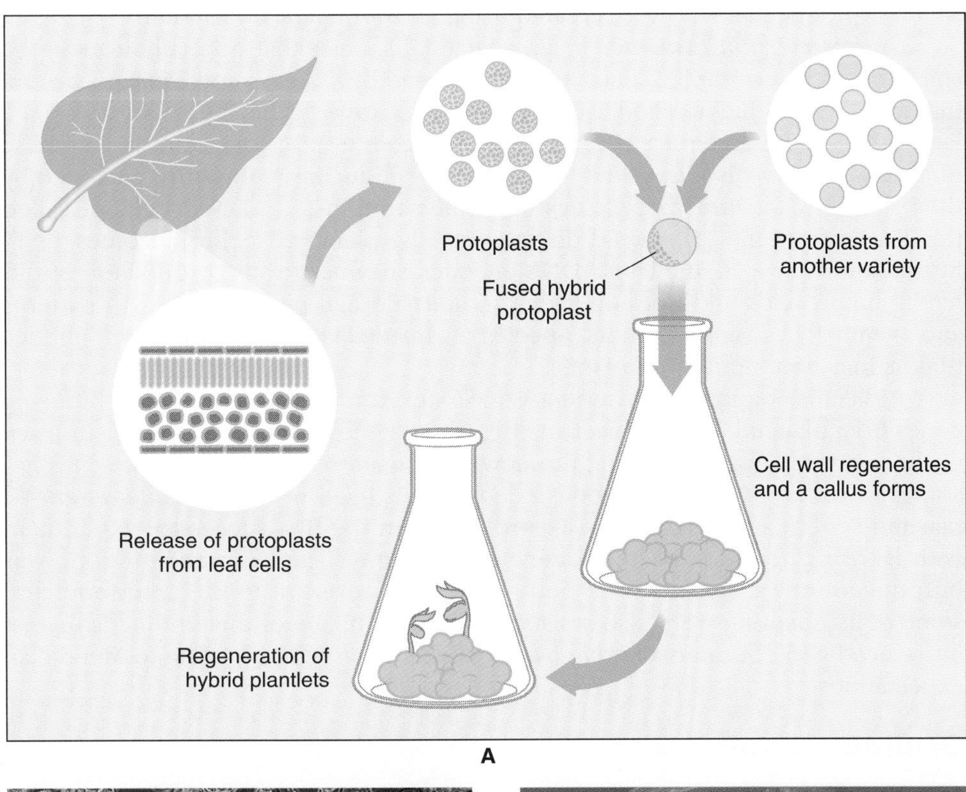

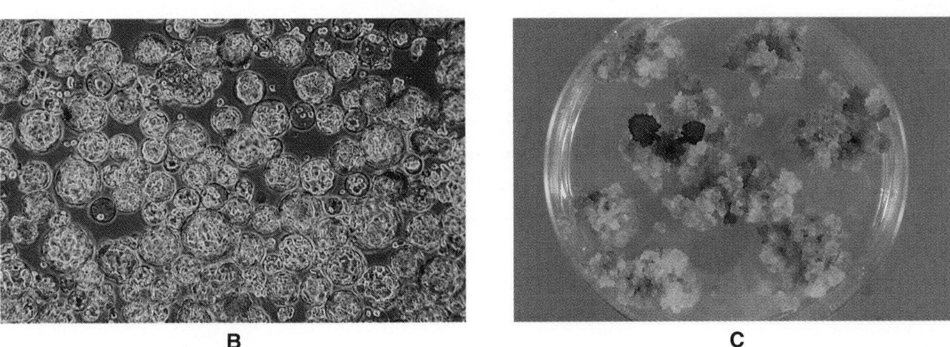

Figure 20.4

Protoplasts (plant cells without the surrounding cell wall) from different species can be isolated or fused with a protoplast from a different species and then cultured, as shown in the diagram (A). Protoplasts from separate species, like these tobacco leaf protoplasts (B), can be placed on culture medium and stimulated to grow into a callus (C) and ultimately into a plant.

develop into an entire plant by exposing it to appropriate hormones. Unaltered protoplasts and calluses of petunias, tobacco, tomatoes, and carrots have regenerated into whole plants. However, this process has only recently been used successfully for some of the most important crop species, such as legumes and cereals.

One curious plant, which resulted from the experimental fusion of potato and tomato protoplasts, illustrates the scatter-shot nature of the process. The hybrid was called a "pomato." In the best of worlds, the pomato would have had the tuber of a potato and the fruit of a tomato. Alas, it turned out to be an unattractive, amorphous vegetable with no delectable properties. However, this work had value because it demonstrated that protoplast fusion could be used to produce plants with new genetic combinations. Though the future importance of the protoplast fusion technique is uncertain, it could prove to be an important procedure in some cases.

Agrobacterium tumefaciens

For successful genetic engineering, plant geneticists must be able to insert genes into a plant's genetic material. Remarkable progress has been made in this area. Recombinant DNA methods can be used for cloning plant genes, although it has often been difficult to identify specific DNA segments harboring genes of importance. To introduce foreign DNA or genes, a unique, natural cloning vector has been used. Most **transgenic plants** (those containing genes from a different species) produced so far were created using an *Agrobacterium* system.

Agrobacterium tumefaciens is a bacterial species that normally lives in the soil. However, it can infect plant tissues exposed by wounds and cause a harmful tumorous enlargement called a *crown gall* that grows at the site of invasion (see Figure 20.5A). Research has revealed that the agent responsible for inducing these tumors is a large plasmid, called the **Ti** ("tumor-inducing") **plasmid**, harbored by the bacterium. *A. tumefaciens* has the ability to transfer a DNA segment called the *T-DNA* from the Ti plasmid directly into the plant cell's chromosomes, as illustrated in Figure 20.5B. The T-DNA encodes enzymes for the production of different amino acids and hormones that redirect normal plant cell growth patterns and eventually cause the tumors. Species of *Agrobacterium* can infect and induce tumors in a large number of plants.

Molecular biologists have been successful in removing crown gall–inducing genes from plasmid T-DNA, inserting beneficial genes from other species in place of those genes, and transferring the new combination of genes into the chromosomes of different plants. The basic techniques used in the *A. tumefaciens*–Ti plasmid system are described in Figure 20.5C. In the first successful experiment with *A. tumefaciens,* a bacterial gene conveying resistance to an antibiotic was introduced into a petunia. The petunia became resistant to the antibiotic, and some of its offspring were also resistant. The importance of this finding is that an introduced gene became a *stable trait* that was inherited according to Mendelian expectations.

Desirable Plant Traits

A number of other experiments have been conducted to examine the prospects for successfully introducing genetically engineered traits into plants that would lead to improvements in producing crops. What types of genetically determined traits would be associated with increased crop production? To date, four traits have received considerable attention—insect resistance, disease resistance, biological efficiency (for example, increased photosynthesis), and weed control.

Plants are constantly under attack from herbivorous insects. Billions of dollars are spent yearly on chemical pesticides to manage agriculturally important insect pests. Might some other control strategy be possible? Many years ago, a bacterium, *Bacillus thuringiensis,* was found to produce a protein that was lethal to the larvae of moths and butterflies and a few other insects that consume plants. This substance has no effect on other organisms, including beneficial adult insects, animals, or humans. The genes that encode the toxic protein were inserted into tomato, tobacco, and cotton plants. Field tests using the transgenic tomato and tobacco plants indicated a high level of insect control. In one experiment, tomato plants with the *B. thuringiensis* genes suffered no damage from insects while control plants without *B. thuringiensis* genes were destroyed by insects. These results indicate a promising commercial future for the *B. thuringiensis* system and perhaps for other insecticidal genes.

Genes for resistance against a broad spectrum of viruses have been successfully transferred into a range of plants. Transgenic tomato, tobacco, and potato plants have been created that are resistant to numerous plant virus diseases. Resistance to viruses could greatly increase the production of important crops such as wheat, corn, rice, and soybeans.

As all gardeners come to realize, weeds compete with economically important crops for space, nutrients, and other soil resources. To combat undesirable

A

Figure 20.5

(A) The bacterial species *Agrobacterium tumefaciens* can infect plant cells and cause them to grow at an uncontrolled rate, which leads to the formation of a tumor called a crown gall, shown here. (B) This effect is caused by a Ti plasmid in the bacterium that contains genes (T-DNA) that can be inserted into the plant's chromosomes, resulting in a transformed cell. When these plasmid genes are expressed, they cause abnormal plant growth. (C) Scientists use the Ti plasmid as a vector for introducing genes into the cells of plants. The tumor-inducing genes are removed and replaced by desirable genes using recombinant DNA technology. If successfully incorporated, the foreign genes may give the plant some new capability.

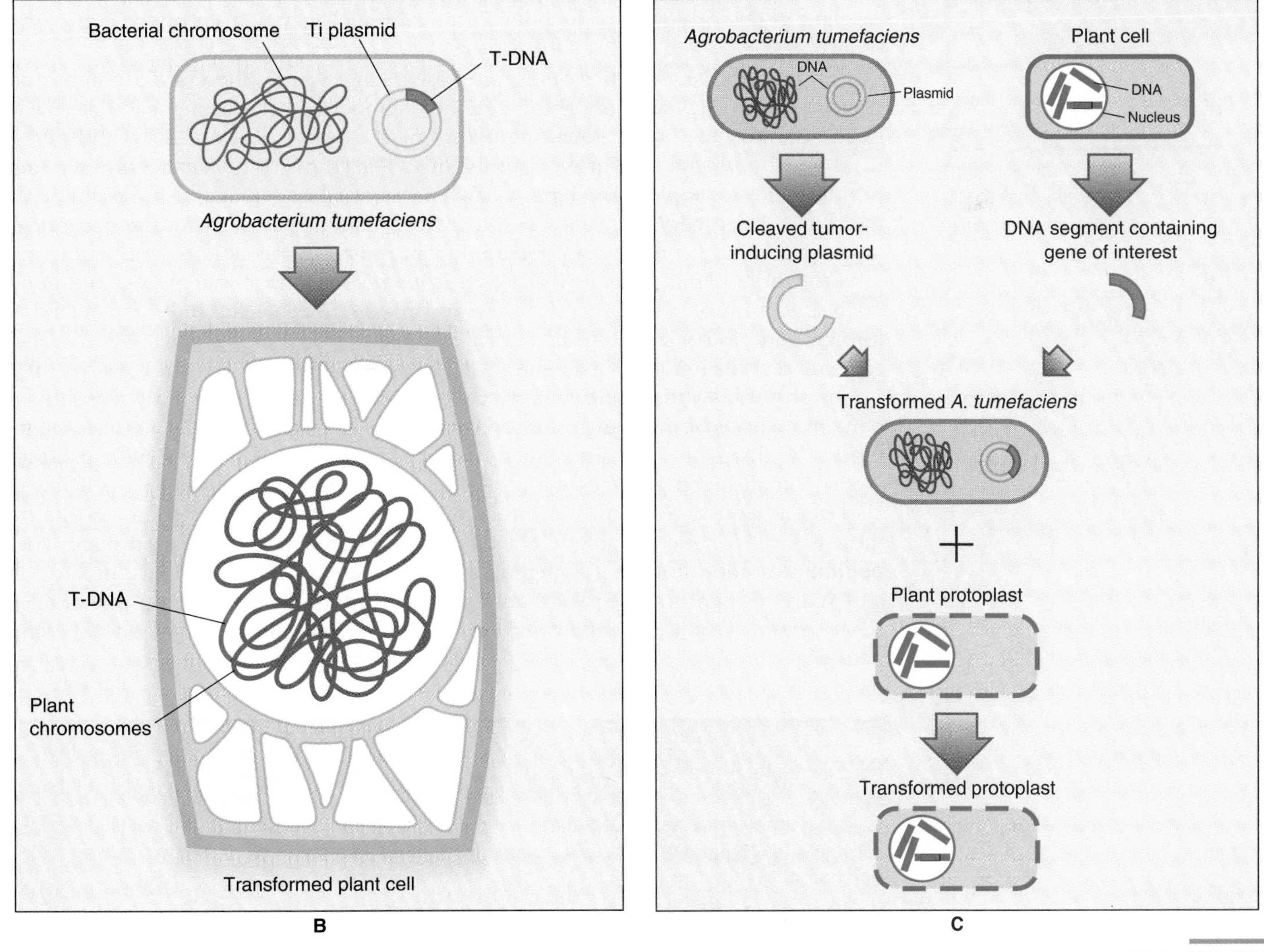

plants, chemical herbicides (substances that kill plants) are widely used in agriculture. Unfortunately, producing herbicides that kill weeds but not desired plants is an expensive proposition. Attempts to produce transgenic plants that are resistant to herbicides have had some success. For example, in 1990, fertile corn plants were produced that contained a gene making them resistant to a specific chemical herbicide (Bialaphos). Spraying herbicides over fields with resistant crops would allow for more effective and less costly weed control but might also promote the use of herbicide chemicals in agriculture, a controversial prospect. Although a few herbicide-resistant plants have already been developed, prospects for their production and use in the future are clouded by the herbicide chemical question.

Plant geneticists also hope to increase the photosynthetic proficiency of plants, which could result in increased crop yields without an increase in fertilizer requirements. Gene transfer in plants has become a fertile field of genetic engineering, and despite some nasty technical obstacles, the future seems very promising. Agricultural scientists now expect that many genetically engineered crops, including corn, soybeans, rice, cotton, sugar beets, and tomatoes, may reach the market by the mid-1990s.

Problems

A number of complex scientific problems have arisen in attempts to create new plants or plants with enhanced capabilities through genetic engineering. Many of the desirable properties, such as the ability to fix nitrogen, produce higher levels of plant growth hormones, or resist disease, heat, and cold, are determined by the interactions of many genes. For such traits, it is difficult to identify all of the relevant genes or gene products or to transfer such genes from one plant to another.

Even for characteristics determined by a single gene, there are a number of obstacles to overcome. Most of the genetic engineering achievements to date have been confined to *dicots,* flowering plants such as tomatoes, potatoes, and petunias. With *monocots,* the group that includes the plants of greatest economic importance (for example, cereal grains such as wheat, barley, oats, rice, and maize), there have been very few successes, for two primary reasons. With the exception of rice and maize, protoplasts obtained from most monocots have not regenerated into whole plants and the T-DNA–*A. tumefaciens* vector system does not generally infect monocots. Recently, however, researchers successfully transferred a foreign gene into maize ("Indian corn"), a major cereal crop, using T-DNA. The cloned, recombinant DNA (gene) was transferred by the plasmid vector into maize protoplasts that were subjected to an electric field. Only a few (usually 1 percent or less) of the transformed protoplasts regenerated, and none have been fertile. Mechanical procedures such as use of a "particle gun" (a device used to "shoot" small metal particles with DNA on their surface directly into a cell) have also been used successfully in introducing DNA into monocots. The results of these efforts are encouraging. Genes have been transferred to a cereal plant, and pending resolution of the low regeneration rates and fertility problems, neither of which are thought to be insurmountable, this important class of crop plants may soon become a full participant in the current blossoming of plant genetic engineering.

Plant protoplasts have been subjected to various genetic manipulations in efforts to create new traits. A system consisting of *A. tumefaciens,* its Ti plasmid, and the T-DNA it contains has been created to introduce genes into plants. Plant traits of interest to genetic engineers include resistance to insects and disease, increases in biological efficiency, and weed control. Despite some successes, several problems remain to be overcome.

GENETIC ENGINEERING OF ANIMALS

Commercially important animals in agriculture (cows, pigs, goats, sheep, and poultry), like plants, have benefited to some degree from the products of rDNA technology such as vaccines and growth hormones. Unlike plants, whole animals cannot be regenerated from a single somatic cell that has been extracted and manipulated genetically. Only a zygote, formed at fertilization by the fusion of gametes, normally has the capacity to develop into a whole animal. For a foreign gene to be incorporated into all cells of an animal (and their offspring), it must be introduced into a sperm or an egg cell or the zygote.

Although the esoteric genetic technologies have not yet been applied to farm animals on a broad scale, momentum is increasing. Also, a number of interesting reproductive technologies have already been developed that resulted in genetic improvements in livestock.

Reproductive Technologies

Reproductive technologies basically involve controlled breeding or the direct manipulation of sex cells. Success in these areas depends on the selection of animals that have desirable, heritable traits. For beef cattle, such traits include lean steaks, high birth weight, or high fertility. By successively breeding superior animals with respect to a heritable trait, there is a continuous improvement in the generations that follow. To illustrate the power of this approach, consider that during the past three decades, the average milk yield of cows has more than doubled, while the number of dairy cows has been reduced by more than 50 percent.

A somewhat more radical example involves turkeys such as those shown in Figure 20.6. In the United States, there has long been heavy selection for big-breasted commercial turkeys. This trait has become so pronounced that commercial turkeys can no longer breed naturally; the big-breasted males are physically incapable of mounting a female! *Artificial insemination,* the manual placement of sperm into the female reproductive tract, is now the sole means of producing commercial-grade turkeys.

In addition to artificial insemination, which makes use of sperm from exceptional males, a number of other reproductive technologies for livestock have been developed. Many of these involve increasing the number of eggs produced by females to form a greater number of embryos during each reproductive cycle. Ex-

Figure 20.6

Commercial turkeys have long been bred for the genetic trait of large breasts. As a result, they are no longer able to breed naturally, and artificial insemination is now used to produce all commercial turkeys in the United States.

cess embryos can be removed from one animal and transferred to another (perhaps in a different part of the world), and some are stored for future use. The most notable advances in this area took place in the late 1980s, when a prototype method for cloning cattle and sheep embryos was developed (see Figure 20.7B). The cloning method proceeds (ideally) according to the following sequence: (1) cells of developing embryos are separated at an early stage; (2) each of the separated cells develops into an identical embryo; (3) the embryos are placed into surrogate mothers, where they develop until birth. The early results are promising, but much further research is necessary before the cloning methods can be routinely applied in agriculture.

Transgenic Animals

Scientists began to develop techniques for inserting foreign genes into higher animals in the late 1970s. Not until the early 1980s were the first successful gene transplant experiments carried out.

The introduction of foreign genes into the germ cell lines (fertilized eggs or embryos) of mammals is one of the most significant advances in the revolutionary field of genetic engineering. Embryos that integrate genes or DNA from a different species into their chromosomes may subsequently develop into **transgenic animals** (see the Focus on Scientific Inquiry, "The Making of a Transgenic Mouse"). Transgenic mice have been used in a number of fruitful studies that have offered unprecedented insights into the ways in which DNA controls embryonic development and the functions of cells in the organism. The transfer of foreign genes into somatic (body) cells of adult organisms is not yet an established technology. For a transgenic experiment to be a success, the following events should occur: (1) genes inserted by the vector must be integrated into the host genome, (2) the encoded protein must be produced in specific tissues and be fully functional, and (3) any activities required to make the protein operate normally must take place (for example, molecular modifications or transport to a specific cellular site).

Figure 20.7

Using cloning technology, herds throughout the world can be improved (A). For example, superior cattle and sheep can be mated and the resulting embryo manipulated in such a way that it gives rise to 16 or 32 identical embryos (B). These embryos can be transplanted into surrogate mothers, where they develop normally.

A

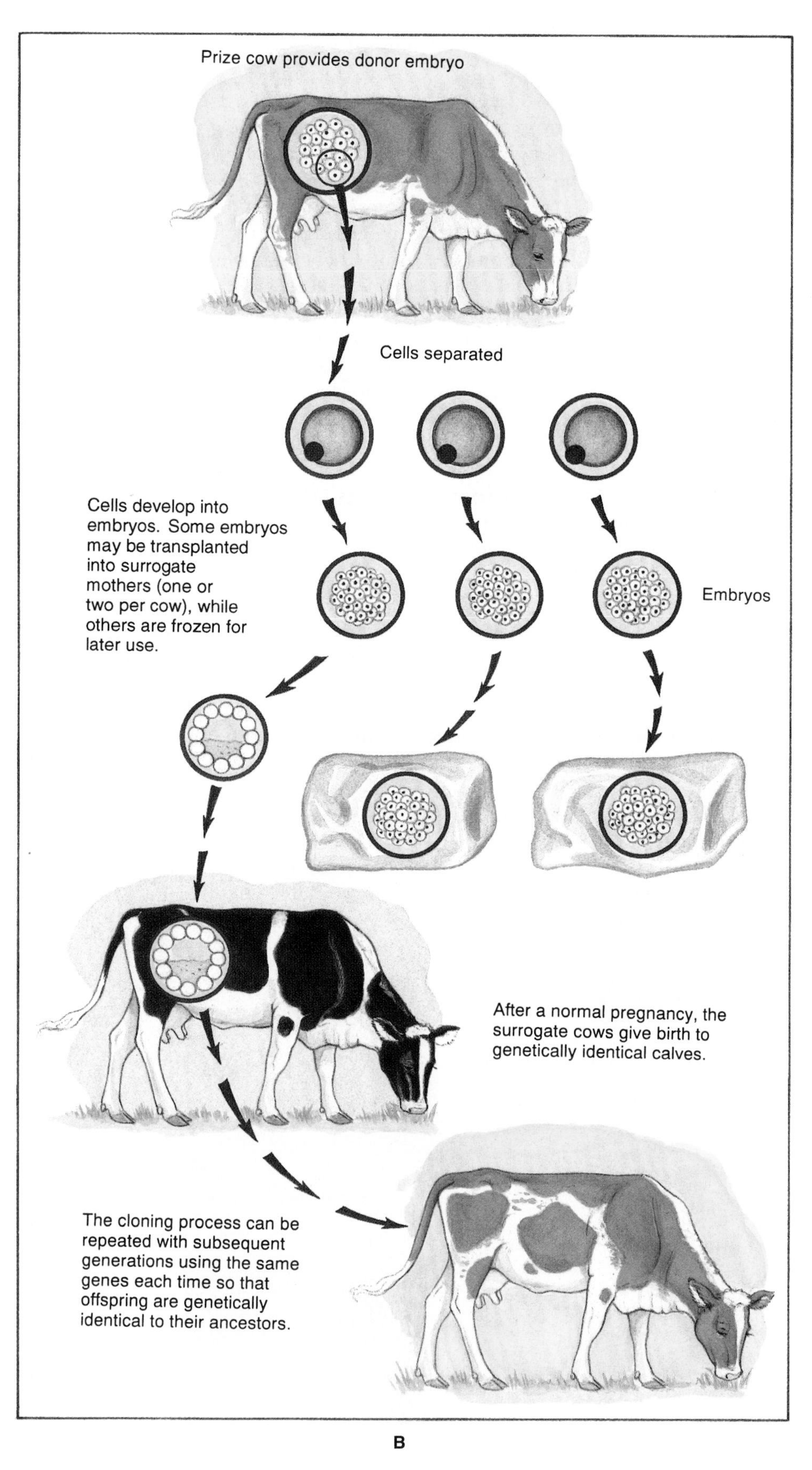
Prize cow provides donor embryo
Cells separated
Cells develop into embryos. Some embryos may be transplanted into surrogate mothers (one or two per cow), while others are frozen for later use.
Embryos
After a normal pregnancy, the surrogate cows give birth to genetically identical calves.
The cloning process can be repeated with subsequent generations using the same genes each time so that offspring are genetically identical to their ancestors.

B

Introducing Genes into Animals

Several experimental techniques were developed to introduce new genes into the chromosomes of mice, and recently these techniques have been extended to domestic animals. Most often, these genes were not of importance to the species but were selected because they, or their encoded products, could be identified after chromosomal integration into a host to determine if the process had worked.

Two techniques have been used for introducing genes into animals—virus vectors and microinjection. In using virus vectors, a foreign gene is incorporated into the viral genome, the virus is injected into an embryo, and if successful, the new gene becomes integrated into the host chromosome and the encoded protein is produced in the resulting offspring. *Microinjection* is the technique of inserting cloned genes or DNA strands generated by rDNA techniques into fertilized eggs or cells of an embryo using finely engineered instruments such as the one shown in Figure 20.8. It has been the most widely used method for creating transgenic animals. Using this procedure, between 100 and 30,000 foreign gene copies are introduced into the nucleus of a fertilized egg or into cells of a developing embryo. In a typical experiment, the injected zygote or embryo is implanted into a foster mother and allowed to develop. After birth, the offspring is subjected to specialized analyses to determine whether or not the microinjected gene became integrated.

Chromosomal Integration of Foreign DNA

Usually, 10 to 30 percent of microinjected eggs survive and develop into offspring, although higher percentages have been obtained in recent experiments. The proportion of offspring that have integrated the foreign gene has been highly variable, ranging from a few to as high as 40 percent.

Some generalizations can be drawn from animal studies conducted to date. (1) The foreign gene is usually integrated into only one of the two homologous chromosomes, (2) the number of genes integrated varies from a few to hundreds, and (3) the foreign gene is integrated into both somatic cells and gametes, eggs or sperm of the offspring. Consequently, the gene can be transmitted as a stable Mendelian trait.

Gene Expression in Transgenic Animals

In an ideal experiment, the integrated gene would be expressed in a predictable and tissue-specific manner. For example, if a foreign gene from a rat encodes an enzyme normally produced in the liver, certain levels of the same enzyme would be expected in a transgenic mouse receiving the gene. Further, synthesis of the

Figure 20.8

High-precision instruments, such as the microneedle in the photograph, are used to inject genes directly into the nucleus of an egg cell.

enzyme would be under host regulatory control and be produced by liver cells, not cells of the brain or muscle. That would indicate that the gene had been integrated and was operating normally. In the early studies, there was often very little evidence of regulated or tissue-specific expression of the inserted gene. However, these criteria have been satisfied in an increasing number of transgenic mouse experiments. Some examples of the successful expression of genes include insulin, a muscle protein, many immune molecules, and certain enzymes and hormones.

Applications of Genetic Technology

Introducing foreign genes and having them expressed in recipient mammalian hosts is rapidly becoming a routine procedure. In addition to mice, fertilized eggs of rabbits, sheep, and pigs have been microinjected with genes. Although transgenic technology seems to have more potential for improving plants than livestock, there is considerable optimism about future applications in animals.

There is also great promise for using transgenic technology to study both animal and human diseases, since many disorders trace their roots to the synthesis of aberrant proteins or the production of abnormal concentrations of normal proteins. Certain cancers and immune deficiencies are examples of diseases that lend themselves to study.

The ability to insert genes and follow their integration and expression constitutes an extraordinarily powerful tool for learning more about fundamental biological processes. Nevertheless, such a capability has raised concerns among certain sectors of the general public about scientists trespassing into areas where they may not belong. This subject will be investigated in Chapter 21.

Problems

Before genetic engineering of animals can be a reality, progress must be made in several areas. The regulation and expression of genes must be better understood. Ways must be devised to control where genes are inserted into chromosomes. The complexity of traits determined by multiple genes (probably the rule for most animal characteristics of economic importance) must be unraveled. Gene identification is a serious problem in the genetic engineering of mammals; of the 100,000 or so expressed genes, the protein products of only about 1,000 have been identified. Because of these information gaps, molecular manipulations are not expected to play a significant role in improving livestock for at least several years.

Different reproductive technologies such as artificial insemination and embryo cloning have been developed to increase the numbers of offspring produced by superior livestock. Serious efforts to use genetic engineering in improving agricultural animals are now under way. Transgenic animals that have incorporated genes from other species into their genome have been generated. Ultimately, the technology used in creating such animals may be employed for improving production traits in livestock.

CLONING OF WHOLE ORGANISMS

In the 1980s, a number of popular books were written that gave a new meaning to the term *cloning*. Rather than being limited to describing identical cells or molecules derived from a single ancestral cell or molecule, cloning was used to describe the production of humans from a single cell of a famous ancestor. These fictional clones were genetically identical to their illustrious father. This prospect confirmed the worst fears of certain people about genetic engineering and the road now being traveled by modern researchers. Educated citizens, however, with an ingrained skepticism about such "scientific" accounts will demand an answer

Focus on Scientific Inquiry

THE MAKING OF A TRANSGENIC MOUSE

By the early 1980s, molecular biologists faced an interesting challenge with important implications for research on human and animal diseases. The molecular tools and experimental techniques had become available for transferring genes between cells of different animal species. Yet before the first experiments could begin, careful thought had to be given to two important questions. First, which animal species would be the best candidates for use in the pioneering experiments? The answer to this question was relatively obvious. Rats and mice have long been used as experimental animals by biologists, and a wealth of relevant information on their reproductive processes, embryonic development, and genes was available. Second, which gene could be transferred? Perhaps the most important criterion for making this decision was, Which gene would make its presence known if it was expressed in the recipient animal? Richard Palmiter of the Howard Hughes Medical Institute in Seattle and his colleagues selected the gene for growth hormone. *Growth hormone* is required for normal growth; if insufficient quantities are produced by the pituitary gland, dwarfism results. Excessive levels cause abnormally large individuals. If genes for rat growth hormone (rGH) could be successfully introduced into fertilized mice eggs, integrated into chromosomes, and later expressed, recipient mice would be expected to be larger than experimental controls that did not receive the rGH gene because rats are larger than mice.

In what will probably become a classic experiment in the history of biology, Palmiter and his coworkers successfully microinjected copies of a *rat* gene encoding growth hormone into fertilized *mouse* eggs in 1982. The affected offspring were much larger than their normal littermates. The experimental procedures used are described in Figure 1. How could the rGH gene from a rat be introduced into the genome of a mouse?

To insert the rGH gene into fertilized mouse eggs, a *fusion gene* was constructed combining DNA segments from different organisms. First, recombinant DNA methods were used to clone the rGH gene, which consists of five introns and six exons. Second, an interesting question had to be addressed: How could the rGH gene be regulated during the experiment? Since little was known about normal regulation of the gene, was there some simple mechanism that could be devised that would allow researchers to turn the rGH gene on or off in mice that had the gene? An ingenious solution was found by fusing the regulatory sequence of a mouse metallothionine promoter to the rGH gene. *Regulatory sequences* are specific DNA sequences that control the production of a protein—in this case, the synthesis of metallothionine. *Metallothionine (MT)* is a protein that is expressed in all tissues but especially in the liver. It binds to metals and helps prevent metal poisoning. The MT gene is inducible (turned on) by exposure to heavy metals, such as cadmium or lead. Thus by injecting small quantities of such metals into experimental mice, the rGH gene would be expressed; when metal exposure was terminated, the MT gene would be turned off. Third, the rGH gene and the MT promoter were integrated into a special plasmid, pBR322, to create the fusion gene called a *MTrGH plasmid*.

About 600 copies of the fusion gene were microinjected into

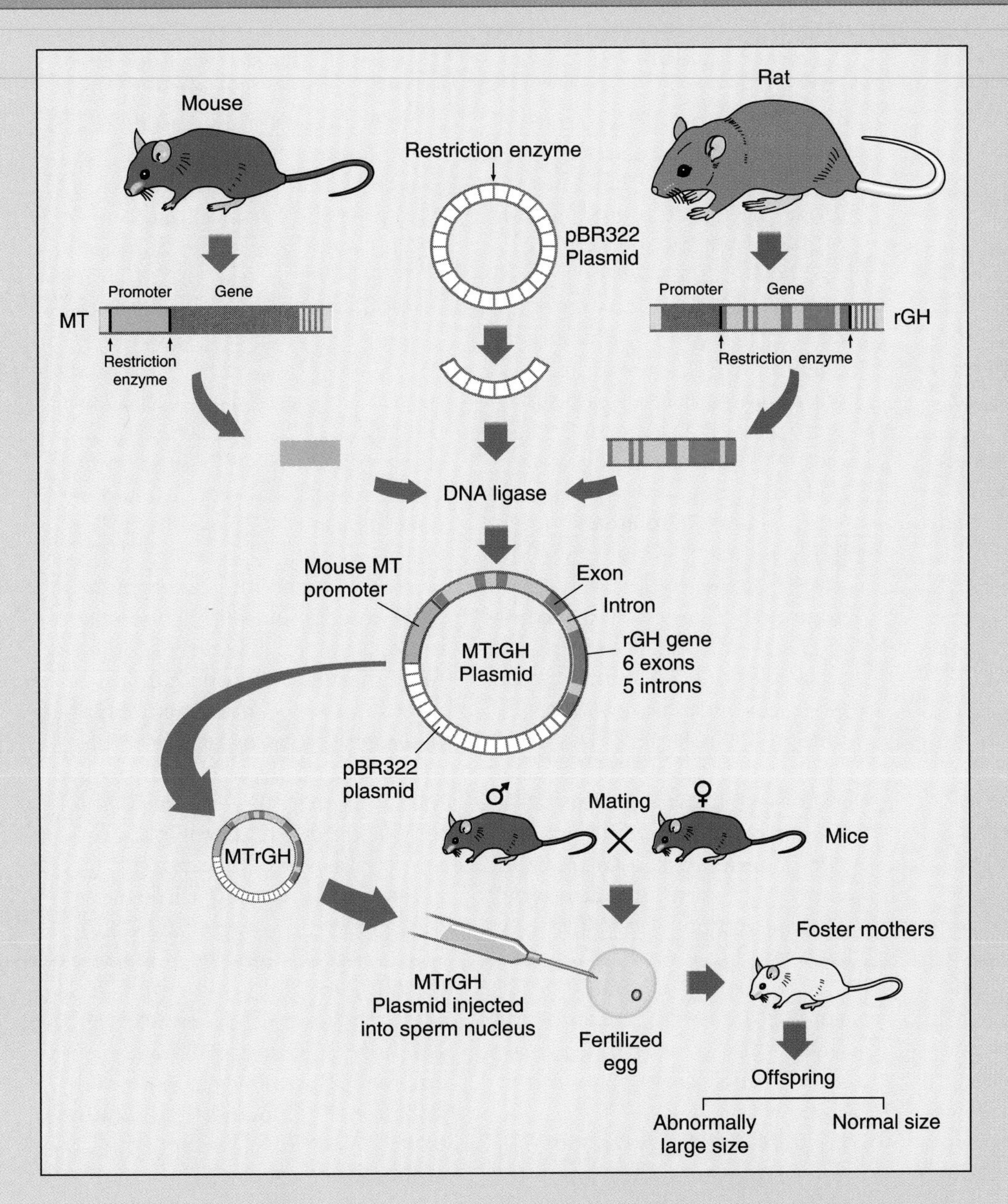

Figure 1 Summary of the genetic experiment involving mice and rats. A rat growth hormone (rGH) gene and the promoter region for a mouse metallothionine (MT) gene were removed from chromosomes by restriction enzymes and combined with a pBR322 plasmid to create a fusion gene called a MTrGH plasmid. After experimental mice were mated, copies of this plasmid were injected into the sperm nucleus just prior to fertilization. The resulting zygotes were placed in foster mothers to develop. Several of the offspring grew to an abnormally large size, indicating that the rGH gene functioned in these mice.

the mouse sperm nucleus after it entered the egg but before it fused with the egg nucleus to form a zygote. The mouse zygotes were then inserted into foster mothers. From 170 zygotes, 21 mice developed, of which 7 were positive for the rGH gene. Six of the seven grew to an abnormal size.

In a second, even more spectacular study published in 1983, the same group of scientists extended their earlier research by using *human growth hormone (hGH) genes*! They used comparable procedures to create an MThGH fusion gene and followed a similar experimental protocol. The results were very interesting, and some will be reviewed here.

Scientists often formulate questions in designing and evaluating their research. In evaluating results, new questions often suggest themselves, providing direction for future investigations. The brief summaries that follow relate some of the relevant questions and resulting conclusions. We leave it to you to think of new experiments that would be of value for extending the results of this study.

1. How many transgenic mice were produced? A total of 101 mice developed from approximately 1,400 injected eggs, of which 33 were transgenic (that is, positive for hGH DNA). Why was the percentage of offspring produced so low? It was concluded that since the injection

procedures involve trauma, implantation of the embryo into the placenta often failed to occur and therefore significant numbers of experimental eggs did not develop.

2. The number of MThGH genes in each transgenic animal varied from 0.9 to 455 copies per cell. From this it is obvious that there was great variability in the number of genes that became integrated.
3. Did transgenic mice grow to a greater size? Of the 33 transgenic mice, 23 were more than 18 percent larger than their littermates, and several were twice as large (see Figure 2). Therefore, enhanced growth in transgenic mice does not depend on GH from a closely related species or a hormone with close structural similarities. Rat and human GH differ in 67 out of 191 amino acids, yet growth patterns in the two studies were similar.
4. Was it possible to derive a *predictive relationship* (that is, to obtain a data set that is good enough to be able to predict the value for one variable, given a value for the other) between growth rate and the number of inserted hGH genes? There was no relationship between growth rate and the number of MThGH genes in the cell. Some of the largest animals had few copies, but of the ten that were not larger, most had fewer than three GH genes.
5. Was there a predictive relationship between the rate of growth and the amount of hGH hormone present in the blood? All of the larger transgenic mice had hGH in their serum, but only a rough relationship existed between growth rate and circulating hGH. Some animals with high hGH levels did not grow as well as others. One animal with substantial levels of hGH did not grow larger than normal at all. The largest transgenic mouse had only two integrated copies of the MThGH gene per cell and had a moderate level of hGH. Therefore, the answer to the question was no. However, expression of the foreign MThGH gene was undoubtedly responsible for the increased growth. Further, there was only a poor relationship between gene dosage and level of gene expression. In these types of experiments, variability may be related in a general way to the number of integrated genes, but the site on the chromosome where they are integrated may have a much more profound effect.
6. Is the MThGH gene in transgenic mice heritable? One transgenic male was bred to a nontransgenic female. The first litter contained seven pups, four of which had inherited the gene. A second litter showed the same results. Thus the trait could be inherited according to Mendelian expectations, since the genes were contained on only one of the two homologous chromosomes in the transgenic male. That is, half the offspring would be expected to have the trait.

There were many other significant aspects of this fascinating study. We have highlighted some of the results that were related to information considered in earlier chapters.

The techniques used in these experiments have since been used to create many more transgenic mice that have various human genes in their genome. For example, in 1990, it was reported that the DNA molecules containing the human genes encoding normal α and sickle-cell β hemoglobin protein chains were inserted into fertilized mouse eggs. As a result, a transgenic mouse line was established that synthesizes human sickle-cell hemoglobin. The transgenic methods have also been used successfully in experiments with other animal species. For example, the gene for the human blood-clotting protein, factor IX, was recently engineered into sheep. These transgenic sheep were found to secrete factor IX in their milk. This type of result is now becoming almost routine in genetic engineering.

Figure 2 Photograph of normal-sized mouse and abnormally large littermate that incorporated and expressed the gene for human growth hormone.

to the question, Can multicellular organisms, including humans, be "cloned" from *single cells?*

Gardeners and plant breeders have cloned plants by cutting and grafting various pieces onto recipient plants. Also, many plants reproduce asexually, which gives rise to identical progeny. Using the restricted definition—whole organisms from a single cell—some plants have been cloned by techniques already described. Thus far for plants, the answer to the question is frequently yes.

For higher animals, the answer to the question is, so far, generally no. In contrast to plants, very few animals reproduce asexually. To clone an animal—that is, to make an identical copy of one parent—it is necessary to remove the nucleus of a fertilized egg and replace it with the nucleus obtained from the individual to be cloned. Numerous experiments of this sort have been done with frogs (see Figure 20.9). When the transplanted nuclei came from very early frog embryos, identical adult frogs developed; that is, if the nuclei of eight cells from a single embryo were injected into eight separate eggs, eight identical frogs were produced. However, no frog has ever been cloned when the transplanted nuclei came from cells of an adult frog. Thus for frogs, it appears that once embryonic development progresses to a certain stage, the cells lose their ability to give rise to an entire organism.

Experiments done with mammals offer little reason to believe that whole-organism cloning will soon be possible. As usual, there are a number of technical problems. Mammalian eggs are so small that manipulations involving nuclei are extremely difficult without causing damage. Also, little is known about embryonic developmental pathways at the molecular level and the importance of nuclear or cytoplasmic factors. To date, no reproducible experiment has shown that a mammal can be cloned by transplanting the nucleus from any single cell. Thus for technical reasons alone, the likelihood of egocentric billionaires or demented dictators leaving clones of themselves is remote.

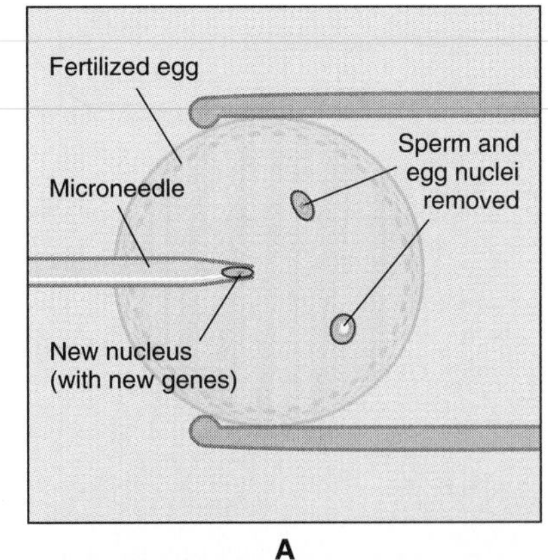

A

B

Figure 20.9

(A) In cloning experiments with frogs, the sperm and egg nuclei from a fertilized egg were removed and replaced by nuclei from frog embryos and adult frogs. The "new" fertilized egg divided until eight embryonic cells were produced. These eight cells were then separated and allowed to develop. (B) When nuclei that were obtained from a frog embryo were implanted into the egg, cloned (identical) frogs resulted. If the translated nuclei came from cells of an adult, no frogs developed.

RELEASE OF GENETICALLY ENGINEERED ORGANISMS

The fundamental question of whether or not to release new strains of genetically altered microorganisms, plants, and animals into the natural environment is still being debated. The case involving ice minus bacteria was considered earlier in this chapter. What about regulating the introduction of modified plants and animals that could provide important new food resources for humankind?

Progress in the science of producing such organisms has exceeded our ability to establish an appropriate decision-making process. Several nontechnical obstacles remain that must be addressed before genetically engineered plants and animals will become common. For example, biotechnology businesses feel that there is a critical need to set up an approval process that is straightforward and allows for the rapid change from field tests to full-scale market production of a new food commodity. Also, companies will insist on patent protection—court decisions have ruled that genetically engineered organisms can be patented—in order to recover their costs in developing new organisms. Yet problems in applying these legal decisions to all aspects of producing such organisms have not been resolved completely.

What about potential risks to the environment posed by genetically engineered organisms? In 1989, the Ecological Society of America published a special review article in the journal *Ecology* ("The Release of Genetically Engineered Organisms: A Perspective from the Ecological Society of America") that considers this question. Recommendations from this prestigious scientific society are expected to influence the development of regulations in this area. The major points expressed in the publication were these:

1. The authors "support the timely development of environmentally sound products, such as improved agricultural varieties, fertilizers, pest control agents, and microorganisms for waste treatment" that do not compromise "sound environmental management."
2. They "support and will continue to assist in the development of methods for scaling the level of oversight needed for individual cases according to objective, scientific criteria, with a goal of minimizing unnecessary regulatory burdens."
3. "Genetically engineered organisms should be evaluated and regulated according to their biological properties (phenotypes), rather than according to the genetic techniques used to produce them."
4. The issues that must be considered before releasing such organisms into the environment "include survival and reproduction of the introduced organism, interactions with other organisms in the environment, and effects of the introduced organism on ecosystem function."

The ESA also recommended procedures for regulatory approaches and suggested that instead of the present case-by-case review for each new product, it may become possible to use a generic approval process depending on the level of risk as determined by several criteria (for example, nature of the genetic alteration, genetic stability of the alteration, and survival under adverse conditions).

Decisions and policies relative to the release of genetically engineered organisms into the environment have not yet been established. The Ecological Society of America has published an important paper that attempts to focus the debate and move the decision-making process forward.

Progress in creating genetically engineered organisms with special attributes is expected to continue at a swift pace. This subject is frequently discussed in editorials and the science sections of most major newspapers and newsmagazines. We encourage you to follow new developments in this field by reading such articles.

Summary

1. Once techniques were developed for manipulating DNA and genes, scientists became interested in creating genetically engineered organisms with unique properties that could be beneficial to humans. Most attention has been paid to bacteria, plants, and animals that are important in food production.
2. Bacterial genetic engineering conducted in the past decade has focused primarily on introducing nitrogen-fixing genes into crops and reducing the harmful effects caused by bacteria that colonize economically important plants. Efforts are also being made to insert bacterial genes into plants that will improve their survival in marginal growing environments.
3. Beneficial genes can be incorporated into a plant species genome by protoplast fusion, an *A. tumefaciens*–Ti plasmid vector system, or other protoplast transformation systems. Most research has been directed toward improving economically important plants by conveying resistance to virus diseases and insect herbivores and reducing competition from unwanted plant species. Significant success has been achieved for insect resistance through insertion of a gene from the *B. thuringiensis* bacterium. Many transgenic plants have also been developed that demonstrate resistance to plant viruses.
4. So far, animals are somewhat less accommodating of the techniques of genetic engineering used for bacteria and plants. The production of commercially important animals through genetic engineering is not expected to occur for several years. Nevertheless, significant improvements have been made in livestock through the use of reproductive technologies such genetic selection, controlled breeding, and embryo cloning.

5. For both plants and animals, more knowledge is required about genes that control traits of interest, the products of such genes, the control of their expression, and methods of integrating them stably into the genome.
6. The debate about introducing transgenic organisms into the environment and their testing and regulation has not yet been resolved.

Review Questions

1. How is genetic engineering currently being used to improve agricultural products?
2. What approaches are being used to endow plants with nitrogen-fixation capabilities?
3. How do Ice^+ bacteria affect plants? How were Ice^- bacteria created? How does this strain affect plants?
4. Why is the release of genetically engineered bacteria into the environment controversial?
5. What are protoplasts? How are genes inserted into protoplasts? How has protoplast fusion been used in genetic engineering?
6. What are the relationships between *Agrobacterium tumefaciens,* Ti plasmids, and T-DNA? How is this system used in genetic engineering?
7. How has *B. thuringiensis* been used in genetic engineering?
8. What are some of the problems retarding progress in plant genetic engineering?
9. What types of reproductive technologies have been used successfully in animals?
10. What are the prospects for cloning multicellular plants and animals?
11. What criteria are used to determine success in creating transgenic animals?
12. What recommendations has the Ecological Society of America made regarding the release of genetically engineered organisms?

Essay and Discussion Questions

1. What are some possible explanations to account for the frequent failure of engineered traits to be transmitted to offspring?
2. How might transgenic animals be used for research on human diseases?
3. Design a transgenic plant or animal that might be economically important.
4. What recommendations about process, policies, or agency responsibilities would you make to guide decisions on releasing genetically engineered organisms into the environment?

References and Recommended Reading

Baskin, Y. 1990. Biotechnology: Getting the bugs out. *Atlantic* (June):40–47.

Cherfas, J. 1991. Transgenic crops get a test in the wild. *Science* 251:878.

Du Puis, E. M., and C. Geisler. 1988. Biotechnology and the small farm. *BioScience* 38:406–411.

Fox, J. L. 1990. Herbicide-resistant plant efforts condemned. *Biotechnology* 8:392.

Gasser, C. S., and R. T. Fraley. 1989. Genetically engineering plants for crop improvement. *Science* 244:1293–1299.

Knorr, D., and A. J. Sinskey. 1985. Biotechnology in food production and processing. *Science* 229:1224–1229.

Labeda, D. P. 1990. *Isolation of Biotechnological Organisms from Nature.* New York: McGraw-Hill.

Olson, B. H., and R. A. Goldstein. 1988. Applying genetic ecology to environmental management. *Environmental Science & Technology* 22:370–372.

Palmiter, R. D., G. Norstedt, R. E. Gelinas, R. E. Hammer, and R. L. Brinster. 1983. Metallothionein-human GH fusion genes stimulate growth of mice. *Science* 222:809–814.

Pirone, T. P., and J. G. Shaw, eds. 1990. *Viral Genes and Plant Pathogenesis.* New York: Springer-Verlag.

Pursel, V. G., C. A. Pinkert, K. F. Miller, D. J. Bolt, R. G. Campbell, R. D. Palmiter, R. L. Brinster, and R. E. Hammer. 1989. Genetic engineering of livestock. *Science* 244:1281–1288.

Raines, L. J. 1988. The mouse that roared. *Issues in Science and Technology* (Summer):64–70.

Ryan, T. M., T. M. Townes, M. P. Reilly, T. Asakura, R. D. Palmiter, R. L. Brinster, and R. R. Behringer. 1990. Human sickle hemoglobin in transgenic mice. *Science* 247:566–568.

Tiedje, J. M., R. K. Colwell, Y. L. Grossman, R. E. Hodson, R. E. Lenski, R. N. Mack, and P. J. Regal. 1989. The release of genetically engineered organisms: A perspective from the Ecological Society of America. Special feature: The planned introduction of genetically engineered organisms: ecological considerations and recommendations. *Ecology* 70:297–315.

Walden, R. 1989. *Genetic Transformation in Plants.* Englewood Cliffs, N.J.: Prentice-Hall.

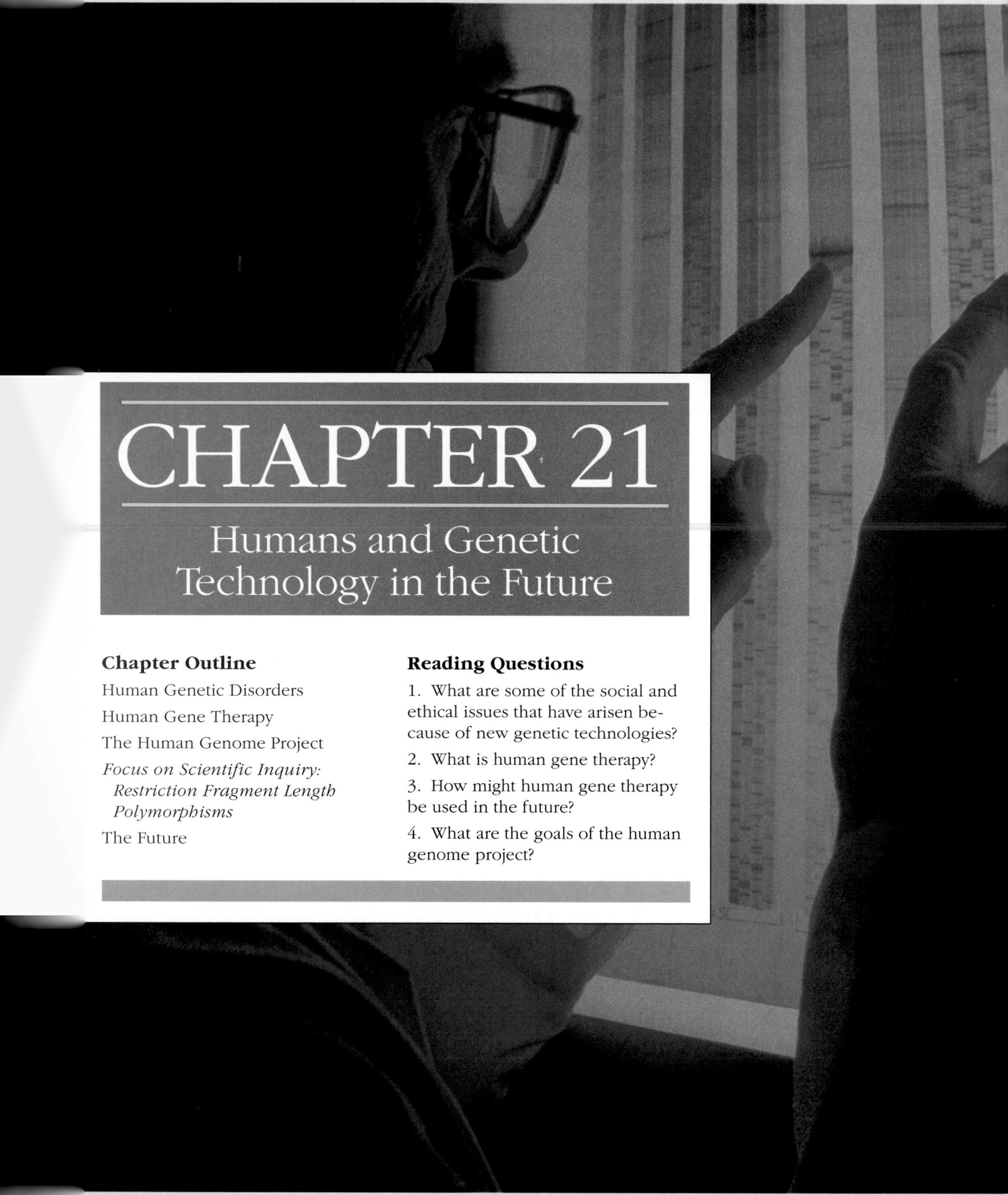

CHAPTER 21

Humans and Genetic Technology in the Future

Chapter Outline

Human Genetic Disorders

Human Gene Therapy

The Human Genome Project

Focus on Scientific Inquiry: Restriction Fragment Length Polymorphisms

The Future

Reading Questions

1. What are some of the social and ethical issues that have arisen because of new genetic technologies?
2. What is human gene therapy?
3. How might human gene therapy be used in the future?
4. What are the goals of the human genome project?

Research in molecular genetics has yielded insights into human biology that are beginning to be translated into actions at an applied level. Modern molecular tools and a clearer comprehension of the structural and functional properties and regulation of genes have led scientists to consider prospects for human applications.

The new technologies described in Chapters 19 and 20 require resolution of two general problems. First, extending many of the new genetic technologies to humans may proceed at a slow tempo because of complex technical problems. Many successes achieved with microorganisms, plants, and mice cannot be easily transferred to humans. Second, certain of the technologies, such as gene transfer, offer challenges to society in the form of ethical choices that have not yet been made. Clarification of society's view of such issues will require participation by scientists, politicians, and an informed public.

Genetic technology can be applied to humans in many ways. Applications include the diagnosis of genetic disorders, uses in gene therapy, and determining the location and nucleotide sequence of genes. We examine these and other topics in this chapter.

HUMAN GENETIC DISORDERS

Diagnosing human genetic disorders using new molecular techniques is now reaching an advanced state. Mutant alleles that cause a number of genetic disorders can now be identified in individuals before phenotypic effects become apparent. As a result, it has become possible to determine whether or not individuals *at risk* (that is, with a family history) for late-acting genetic disorders, such as Huntington disease, will develop the disorder later in their lives. Also, these methods have been adopted for prenatal diagnosis for an increasing number of genetic disorders including Duchenne muscular dystrophy, cystic fibrosis, retinoblastoma (eye cancer), and Huntington disease.

Other complex disorders that appear to have at least a partial genetic cause, such as certain cancers, diabetes, and cardiovascular disorders, may also become amenable to new DNA testing procedures. It is expected that such tests, when fully developed, will allow credible predictions of the risk for these and many other genetic disorders. This ability to predict the future accurately has severely challenged the ethics and values of people who may be directly or indirectly affected by the new genetic technologies. These individuals include scientists, physicians, politicians, parents and prospective parents, and all concerned citizens.

Societal Issues

A number of questions have been raised about the ethics of altering the fundamental molecules of life or using techniques from molecular biology to tell a person what the future holds. *Ethics* in this context is defined as a system or set of moral principles and values. The problem of applying ethical standards to current biological issues is, of course, that in pluralistic societies, there is a wide spectrum of opinion about what is or is not acceptable (Figure 21.1). That is certainly the case for genetic engineering in general and genetic diagnostic technology, gene therapy, and whole-organism cloning in particular.

A host of ethical questions arises from being able to determine who has or who will develop genetic disorders, especially late-onset disorders. Should individuals at risk for a genetic disorder be required to have a diagnostic test? For what purposes? Are there circumstances that make a difference? For example, individuals who are deficient in an enzyme called alpha-1-antitrypsin are more sensitive to certain substances (such as cotton dust) in the workplace than those who are not. Do companies have a right to require testing to reduce the risk of exposing susceptible employees and deny them higher-paying jobs associated with the

Figure 21.1

Besides being genetically unique, individuals have differing systems of values that influence their thinking about the uses of genetic technologies.

hazard? Conversely, can companies justify not requiring tests if without that information, employees might be placed at increased risk?

Do insurance companies have a right to obtain the results of genetic tests that can determine whether or not an individual has a high probability of having a gene associated with a disorder? If a family has a history of a genetic disorder, should insurance companies be able to mandate testing as a prerequisite for insuring individuals at risk? The relevance of this question is associated with the high costs of medical treatment for individuals with genetic disorders and the increased probability that they will die at an early age. Is it reasonable to withhold insurance from those at risk? From those who test positive? Should people who have no genetic disorders be required to pay higher insurance rates to subsidize those who do?

What about serious genetic disorders that appear in certain families and cause death soon after birth? For example, Tay-Sachs disease is a single-gene disorder inherited as an autosomal recessive (that is, both parents must carry the gene to produce a child with the disease). Affected infants progressively become severely retarded, paralyzed, and blind. Death occurs at age 2 or 3, and there is no effective treatment for the disease. Should physicians be responsible for requiring prenatal tests in pregnancies involving parents at risk for a genetic disorder (Figure 21.2)? If the offspring has a genetic disorder, should the parents be able to sue the doctor for negligence if the doctor did not recommend the test? Who should make those decisions?

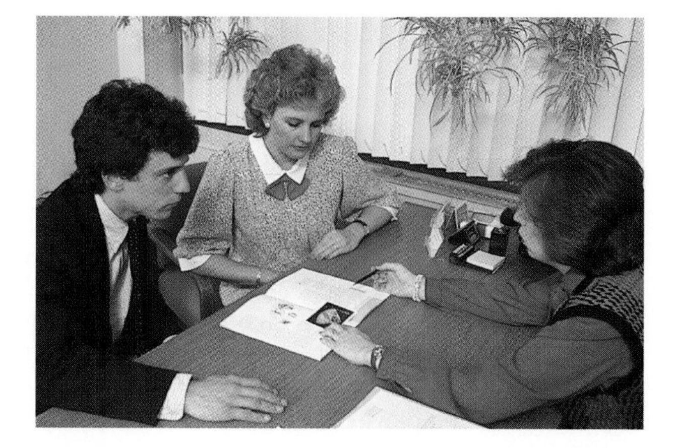

Figure 21.2

New procedures can be used in prenatal testing for an increasing number of genetic disorders. Prospective parents who are at risk for genetic disorder may or may not wish to take advantage of such tests.

Huntington Disease: A Model

What is a proper approach for using the new technologies to test individuals at risk for a genetic disorder? The ability to diagnose Huntington disease (HD), a late-onset, neurological disorder that is progressively debilitating and invariably fatal, has resulted in the evolution of an approach for addressing complicated problems that arise from administering predictive genetic tests. The HD test can be conducted at any time during an individual's life. Obviously, the decision to be tested or not is an enormously difficult one for a person to make. Such a highly personal decision can only be made by the individual at risk, usually in consultation with family members, since their blood samples are required to complete the test. An additional problem concerns children at risk for HD. So far, all institutions offering the test have chosen to restrict it to people age 18 and older who have undergone extensive counseling.

Since the diagnostic test became available in 1984, professional staffs conducting the diagnoses have recognized the importance of including the family at every level of the process. All test centers offer extensive counseling, both before and after testing, to enable individuals to deal with the results in the most positive way possible. Results to date indicate that such a highly personal approach, with a sensitive professional staff, is to be recommended as a model for other testing programs that will soon become available.

New genetic technologies may be applied to humans in the future. The ability to diagnose many genetic disorders has already reached a high level. This has aroused questions about regulating the use of such tests in our society and the difficult choices that individuals at risk face. Counseling programs developed for families at risk for Huntington disease provide a valuable model for other testing programs.

HUMAN GENE THERAPY

Gene therapy, the replacement, correction, or augmentation of existing genes, is a concept that emerged in the early 1970s when it became known that during some stage of their life cycles, tumor-causing viruses insert foreign genes into mammalian cells. When new laboratory techniques were developed during the age of recombinant DNA technology to perform this same feat, it seemed likely that forms of gene therapy could be devised for treating human genetic disorders (see Figure 21.3).

A great deal has been written in newspapers and magazines about gene therapy in humans; unfortunately, much of it has been based on a poor understanding of genetics. Also, there has been a tendency in such articles to oversimplify the technical aspects. The emphasis is frequently on some sensational angle such as being able to obtain a desired trait at a "gene supermarket." As a result, many people express considerable confusion and fear about the future use of gene therapy in humans.

There are two general types of gene therapy: **somatic cell therapy**, which targets genes in somatic (body) cells of the body, and the more controversial **germline therapy**, which consists of modifying genes in cells that produce sperm and eggs or in early embryonic cells. What progress has been made in the development of gene therapy? What does the future hold?

Technical Problems

To understand the future of gene therapy, it is important to appreciate some obstacles that must be overcome. In terms of total number and complexity, the technical problems are formidable.

Figure 21.3

Gene therapy consists of different strategies designed to modify the effects of a defective gene. In gene augmentation and gene correction, a normal gene is first placed into the DNA of a cell and then inserted into a tissue where it can be expressed. Gene replacement requires that the defective gene in a cell be replaced with the normal allele. If modified genes are placed in a somatic cell, they may reduce or eliminate effects caused by a defective gene but will not be transmitted to any offspring. If inserted into a germ cell that produces sperm or egg cells, modified genes may be transmitted to offspring but may not have a therapeutic effect.

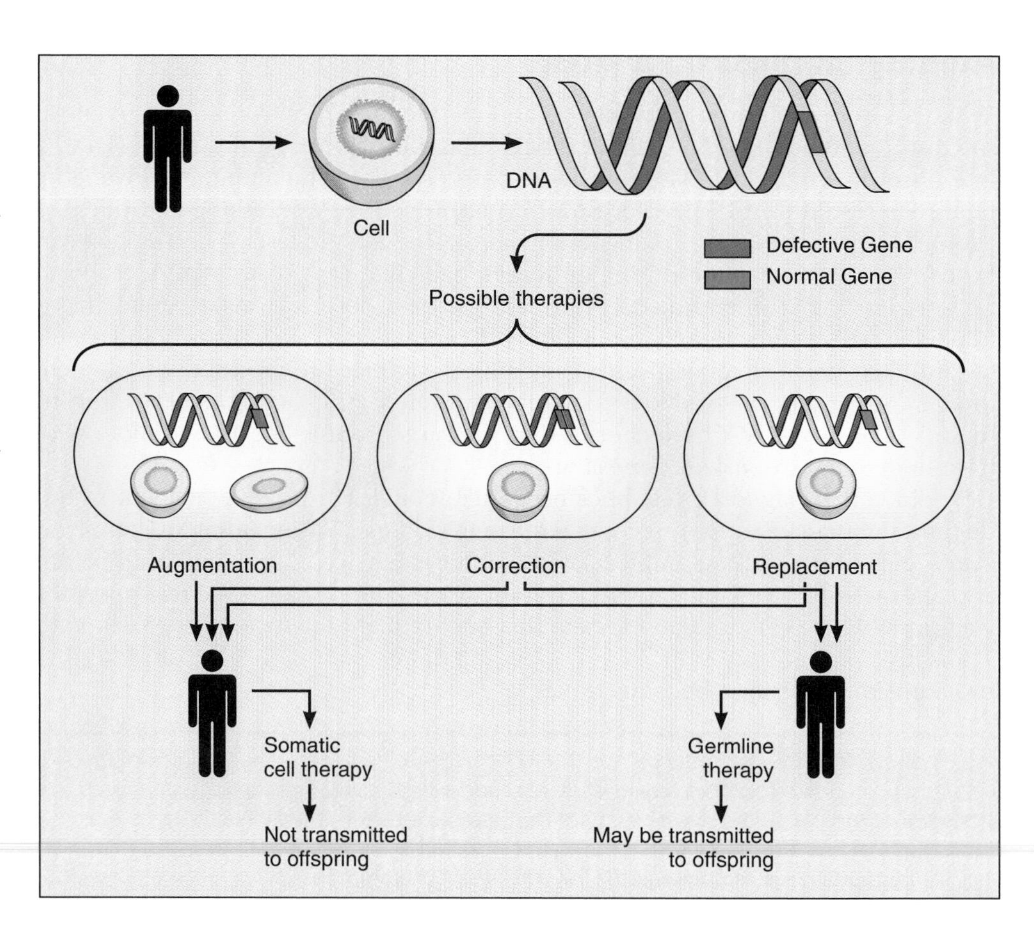

Genes and Gene Regulation

Researchers developing gene therapy approaches face two main obstacles: identifying key genes and their mode of regulation and inserting genes into appropriate chromosomal sites. For each single-gene disorder, all of the following will be necessary:

1. Identification of the normal gene site and nucleotide sequence that has been altered and the protein it encodes
2. Understanding of how the gene's expression is regulated
3. Creation of vectors or physical methods that can be used to transport the normal gene into cells where the gene can be expressed
4. Ensuring that the gene becomes stably integrated and expressed so that the protein product is present at levels sufficient to reduce or eliminate effects of the disorder

Gene Insertion

Since the early 1970s, the use of mouse and bird retroviruses as vectors for transferring normal human genes has been carefully investigated. **Retroviruses** are unusual viruses because their genetic material is RNA, not DNA. Once they infect a host cell, retroviruses are replicated through a mechanism that translates their RNA into DNA. This, of course, is the opposite of the normal translation of genetic information, DNA into RNA, hence the prefix *retro-*. Human retroviruses cause several important diseases, including some cancers and AIDS.

In the laboratory, retroviruses have several advantages as vectors compared to others that have been studied. They are easy to manipulate and have a demonstrated ability to transfer foreign genes into cells and have the genes expressed. However, retroviruses are sometimes unstable, it is impossible to control where the genes they carry are inserted, and even if they do become integrated, the genes are not always expressed. Recipient human cells must be dividing before the gene carried by a retrovirus can be integrated in their chromosomes. For example, bone marrow cells divide rapidly and constantly to produce blood cells and may be good candidates for retroviral vectors, depending on the purpose of the gene insertion. Because of the cell division requirement, these vectors cannot be used for introducing genes into the mature cells—brain, nerve, muscle, or other cells—that make up most human tissues.

Despite these problems, it is evident that retrovirus vectors will play a prominent role in future developments. Other viral vectors are now receiving increased attention. For example, **herpesviruses**, which normally live inside nerve cells (and cause such problems as cold sores and genital herpes in humans) can perhaps be modified to serve as vectors for inserting genes into nervous tissue. Physical methods, such as microinjection and high-velocity microprojectiles, are also being studied.

Though all of the difficulties have not yet been overcome, impressive progress has been made since the mid-1980s. Some forms of gene therapy now appear to offer great promise for ameliorating the effects of certain human genetic disorders.

Ethical Issues

The policy of the U.S. National Institutes of Health Recombinant DNA Advisory Committee (RAC) currently prohibits germline therapy experiments, but human somatic cell therapy research is taking place in laboratories throughout the country and the world. Most scientists and physicians view the projected uses of somatic cell therapies as extensions of accepted medical procedures, comparable to blood and marrow transfusions used for treating diseases. Also, few disagree with the justification that the use of somatic cell therapy is appropriate for treating a genetic disorder when no other treatment is available. Consequently, few specific objections have been raised based on ethical considerations.

In sharp contrast, many people, including ordinary citizens, scientists, politicians, and others, have questioned the wisdom of creating genetic alterations that can be transmitted to future generations. Is it acceptable or unacceptable to attempt to alter the germinal cells of an individual with a genetic disorder in such a way that the person can have children who will not have the disorder? Perhaps there would be general agreement that germline therapy might be allowed for single-gene disorders. However, most human traits are controlled by more than one gene, and little is known about the genetics of such traits.

With respect to humans, no clear consensus about germline therapies has yet emerged, in part because the research has not progressed to the point where serious concerns have arisen. However, because of the severe technical problems and a clear indication from most groups that they are not yet ready to accept experiments centered on human germline alterations, it seems unlikely that any research in this area will be approved in the near future.

Somatic Cell Gene Therapy

There are two general categories of somatic cell gene therapy, *gene substitution* and *gene augmentation* (see Figure 21.3). As originally conceived, it was thought that **gene substitution** could be used to remove and replace mutant genes responsible for genetic disorders. However, there have been no successes in creating such substitution therapies. Alternatively, would it be possible to "correct" the abnormal DNA sequences of mutant genes? This might be accomplished by using established gene transfer methods to introduce a normal DNA sequence that would recombine, through complementary base pairing, with the mutant sequence. If the normal gene were then expressed, the needed protein would be available and the effects of the disorder would be reduced or eliminated. Some progress has been made in gene correction techniques and their application to research on animals such as mice.

Gene augmentation involves techniques that will equip cells with added genes that can produce the missing protein associated with the genetic disorder. As described in Figure 21.4, two basic methods are being developed: (1) a "standard" technique in which competent genes are inserted into cells of an afflicted individual, followed by introduction of the altered cells into the appropriate tissue where they are normally expressed, and (2) the insertion of the needed gene into "donor" cells that are then placed back into the body, where they produce the protein. This latter approach has been referred to as a "drug delivery system," and it offers exciting prospects. How does the donor cell system work? As described in Figure 21.4, vectors containing a normal gene encoding a missing protein—an enzyme produced by the liver, for example—are inserted into nonliver cells such as bone marrow cells, which produce white blood cells, or endothelial cells, which line the inside of blood vessels. These donor cells are then trans-

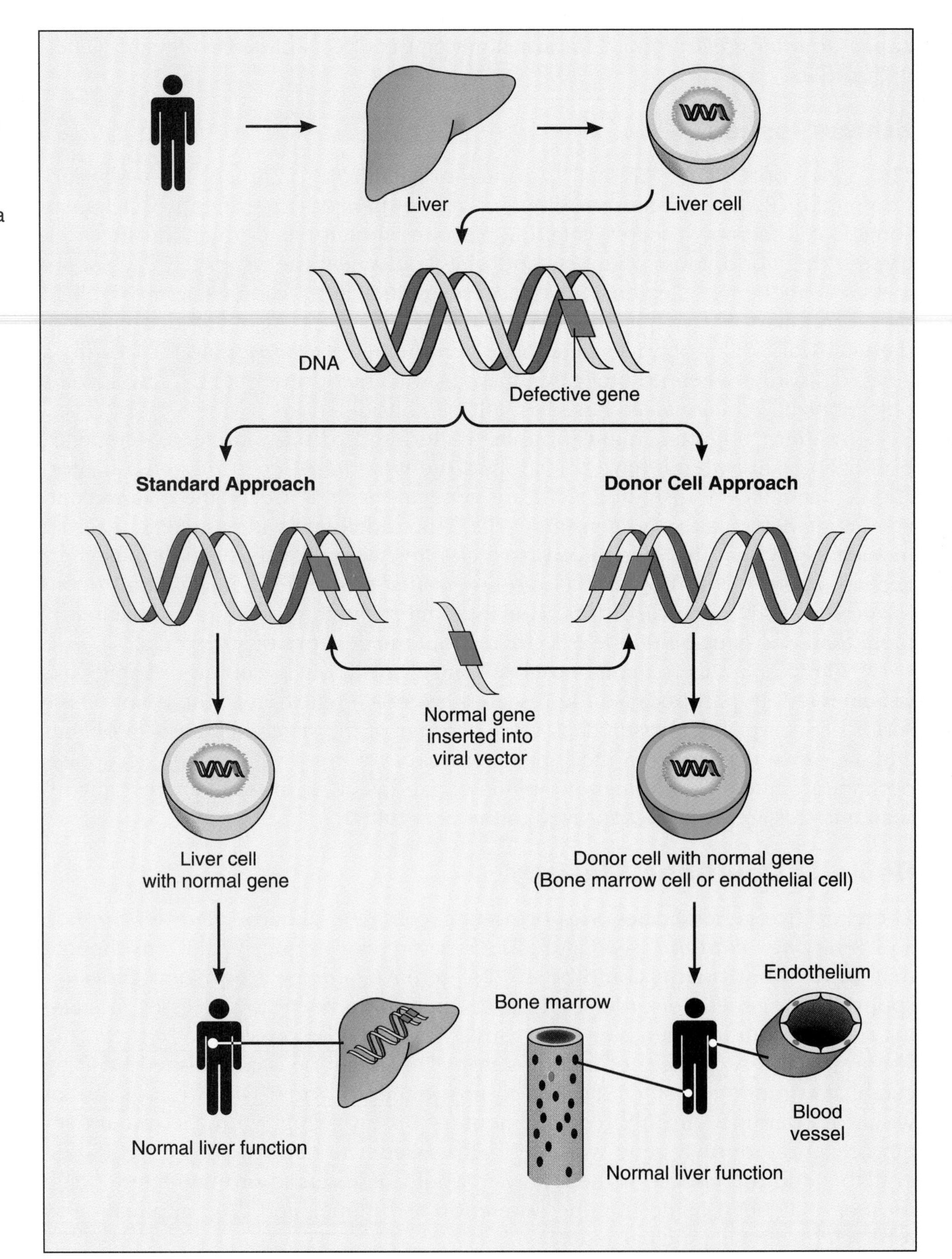

Figure 21.4

Two gene augmentation approaches have been developed. In the standard approach, a normal gene is inserted by a vector into a cell type that normally expresses the gene. That cell is then reintroduced into the tissue where it originated. In the donor cell approach, a normal gene is placed into a different cell type that usually has access to the bloodstream. When these cells are inserted into the appropriate tissue, they may produce and release the desired protein.

ferred into the body, where they become established, express the gene, and pump out useful quantities of the missing enzyme into the bloodstream. There have been notable accomplishments using both the standard and donor cell types of gene therapy.

Gene therapy consists of modifying or augmenting an abnormal gene associated with a genetic disorder. Although technical problems have delayed progress in this area, rapid gains are expected in the 1990s. Before certain gene therapies become an option for treating genetic disorders, society must address certain ethical issues that have not yet received thoughtful consideration.

Animal Models and a Case Study

For ethical and practical reasons, basic experimental research in gene therapy is not first conducted directly on humans. Given these constraints, how is it possible to make advances that may eventually lead to the treatment of humans? In general, new techniques and therapeutic approaches are developed and tested in the laboratory using cells obtained from various animals, including humans, which can be cultured (grown and amplified by cell division) in small containers. Subsequently, new methods are tested and analyzed using animals, primarily mice, rats, and rabbits. Once a technique that may have useful applications in gene therapy has been developed, it becomes necessary to conduct experiments on animals that have specific disorders that also occur in humans. Such a species is referred to as an **animal model**; that is, its disease serves as a model for learning about the nature of the condition in humans. The use of animal models for studying human diseases (for example, cancer) has a long and successful history. Only after exhaustive tests and demonstrations of safety will specific gene therapies ever be used on humans. Some very interesting research related to gene augmentation therapy demonstrates how this process works.

Figure 21.5

The Watanabe rabbit is an animal model used for studying familial hypercholesterolemia, the most common human genetic disorder.

Recall from Chapter 18 that familial hypercholesterolemia (FH) is a human genetic disorder characterized by high levels of cholesterol in the blood that often leads to premature death. The most common cause of FH is a deficiency in the number of protein receptors on liver cells that normally remove this substance from the blood. Is an animal model available for studying FH? What strategy might be developed for treating or curing FH? The answer to the first question is yes; the Watanabe rabbit has the same disorder (Figure 21.5). In the late 1980s, researchers provided information relevant to the second question by accomplishing the following:

1. Methods were devised for inserting foreign genes into the liver cells of various animals. This was notable because liver cells had previously been resistant to gene transfer.
2. Viral vectors were used to insert normal copies of the gene encoding the cholesterol receptor protein into cultured rabbit liver cells.
3. Once inside the rabbit cells, the cholesterol receptor gene produced functional receptor protein that removed cholesterol from the experimental system.

The next step will be to place the cells back into Watanabe rabbits in an attempt to cure the disease. For that to happen, the altered cells will have to become permanently established in the rabbit's body and continue to produce the receptor protein. Though these very exciting results have obvious implications for the treatment of FH, they represent only the first step in a long path leading to human clinical applications. What is required before similar types of gene therapy tests can be conducted on humans?

Authorization Procedures for Studies on Humans

Before conducting any experiments related to human gene therapy, scientists must obtain legal approval from several regulatory agencies. Research plans must first be reviewed by local institutional review boards and institutional biosafety committees. At the national level, approval is required from the RAC and the director of the NIH. In addition, the proposed experiment must be described in nonscientific language and published in the *Federal Register* so that members of the public can, if they wish, offer comments to the RAC and the NIH. The purposes of this review are to ensure that only qualified research teams conduct such tests, that all safety and ethical issues are fully considered and evaluated, and that the public has an opportunity to express its feelings.

The First Experiment

How does the authorization process work in real life? Early in 1988, a request to conduct the first human gene therapy experiment in the United States was submitted by W. French Anderson of the National Heart, Blood, and Lung Institute and Steven Rosenberg and R. Michael Blaese of the National Cancer Institute. Their proposal stemmed from earlier investigations on a promising new approach for treating cancer. Specific immune cells—white blood cells called *lymphocytes*—that kill "foreign" cells, including cancer cells, normally exist in the body. In early studies, lymphocytes that had infiltrated the tumor (*tumor-infiltrating lymphocytes,* or TIL cells) of seriously ill patients were removed surgically and induced to grow and multiply in the laboratory. When sufficient numbers of cultured TIL cells had been generated, they were reinfused into the patient. TIL therapy caused significant tumor regression in about 50 percent of the 25 patients tested who had advanced cancers and had failed to respond to other treatments.

Why was TIL therapy effective only in some of the patients? To try to answer this question, Anderson, Rosenberg, and Blaese wanted to insert a single-gene "marker" (for resistance to the antibiotic neomycin) into TIL cells using a retrovirus vector and follow their progress once placed back into a patient. Figure 21.6 illustrates the TIL marker research protocol. The marker would allow researchers to identify TIL cells with the gene by collecting blood samples periodically during treatment and then exposing all blood cells removed to neomycin; those with the marker would survive, while those without the resistance gene would perish. The marker would have no effect on either the TIL cells or their capacity for attacking cancer cells.

In accordance with the procedures described earlier, the proposal was approved by the various committees, but not without delays related to procedure (not all of the relevant preliminary data were made available to RAC members) and a large number of public hearings. A lawsuit alleging that the review process was not open to the public was eventually dismissed.

Finally, on May 22, 1989, a severely ill cancer patient received the first infusion of TIL cells that had been genetically altered to contain the foreign neomycin gene. Additional patients were treated during the following weeks. Results from the experiment were similar to those from earlier studies—some patients showed dramatic regression of their tumors, and others did not. The significance of the research lies in what was learned about the marked TIL cells introduced into the patient. They survived in the body, apparently attacked and killed tumor cells, and were unchanged by the inserted gene. The next step in this research was to seek approval to insert a gene for an active antitumor agent (a natural protein produced by the immune system) into TIL cells. Approval for such an experiment was granted in 1990, and results from new experiments may be available in 1991.

Though perhaps not technically a gene therapy experiment (the gene inserted into TIL was neutral, not therapeutic), this is considered the first such study in the United States. The TIL experiment opened the doors to further studies involving gene therapy. For example, the same group of researchers received approval to test the effects of a *tumor necrosis factor* (TNF) gene inserted into lymphocytes. TNF is a protein that shrinks tumors by cutting off their blood supply. The first patients were treated in 1990, and further tests are now under way.

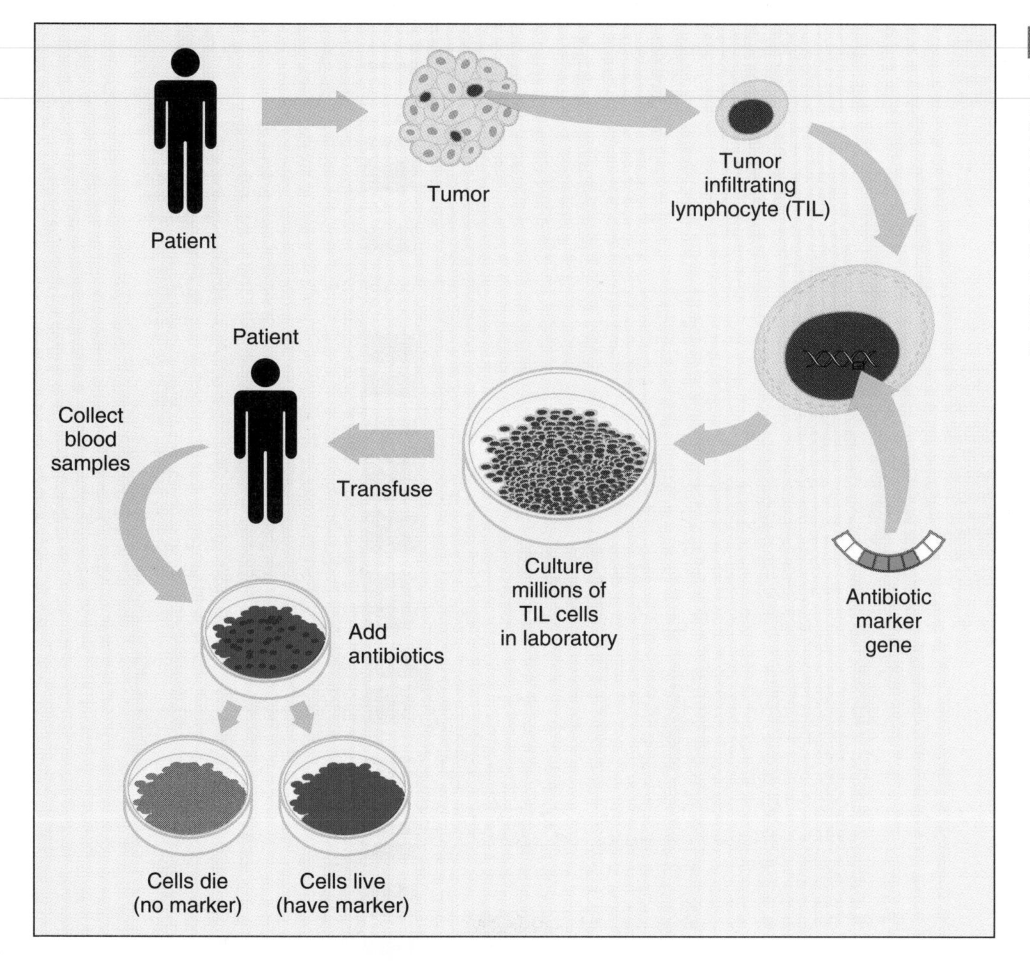

Figure 21.6

The first approved human gene therapy experiment was conducted to determine the fate of a type of lymphocyte cell that was known to react against tumor cells. Marker genes for antibiotic resistance were inserted into the lymphocytes, which were then transfused into the patient. By exposing lymphocytes that were collected periodically in blood samples to the antibiotic, it was possible to determine if they had survived and reacted against the tumor.

In 1990 and 1991, children with a rare, lethal immune deficiency known as ADA (because of a defective enzyme called *adenosine deaminase*) received transfusions of their own lymphocytes, which contained foreign ADA genes. One child, a 4-year-old girl, received cells on September 14, 1990. This is considered the first test of human gene therapy. Finally, in 1991, scientists succeeded in synthesizing normal dystrophin genes. (Recall that deletions of this gene result in Duchenne muscular dystrophy.) Gene therapy trials involving this gene are expected to begin soon. Clearly, an increasing number of human gene therapy experiments can be expected in the next five years.

Gene therapy has been successfully applied to animal models for human genetic disorders. From successful research on rabbits, which are animal models for familial hypercholesterolemia, it seems likely that similar therapies may soon be developed for humans. The first human gene transfer experiment was conducted in 1989. It consisted of introducing a marker gene into lymphocytes from cancer patients and then placing the marked cells back into the affected individuals. The first human gene therapy experiment, involving ADA genes, began in 1990.

THE HUMAN GENOME PROJECT

What is known about the location of genes on human chromosomes? In a monumental achievement, the first partial map of the human genome was published in the journal *Cell* in 1987. This **genetic linkage map** is based on the location of gene loci on each chromosome relative to the positions of specific markers (see the Focus on Scientific Inquiry, "Restriction Fragment Length Polymorphisms").

Figure 21.7

In 1990, a high resolution human genome map was published in the journal *Science*. This figure shows some of the details for one chromosome from the map, number 4. The three diagrammatic chromosomes represent, from left to right, (1) the percentage of chromosome segments for which the DNA sequence has been determined (white = 0%, yellow = 0.1%); (2) the percentage of gene loci determined in different segments (yellow = 0.1%, orange = 1%); and (3) the locations of specific gene loci. Note the location of the Huntington disease (HD) gene in the upper arm.

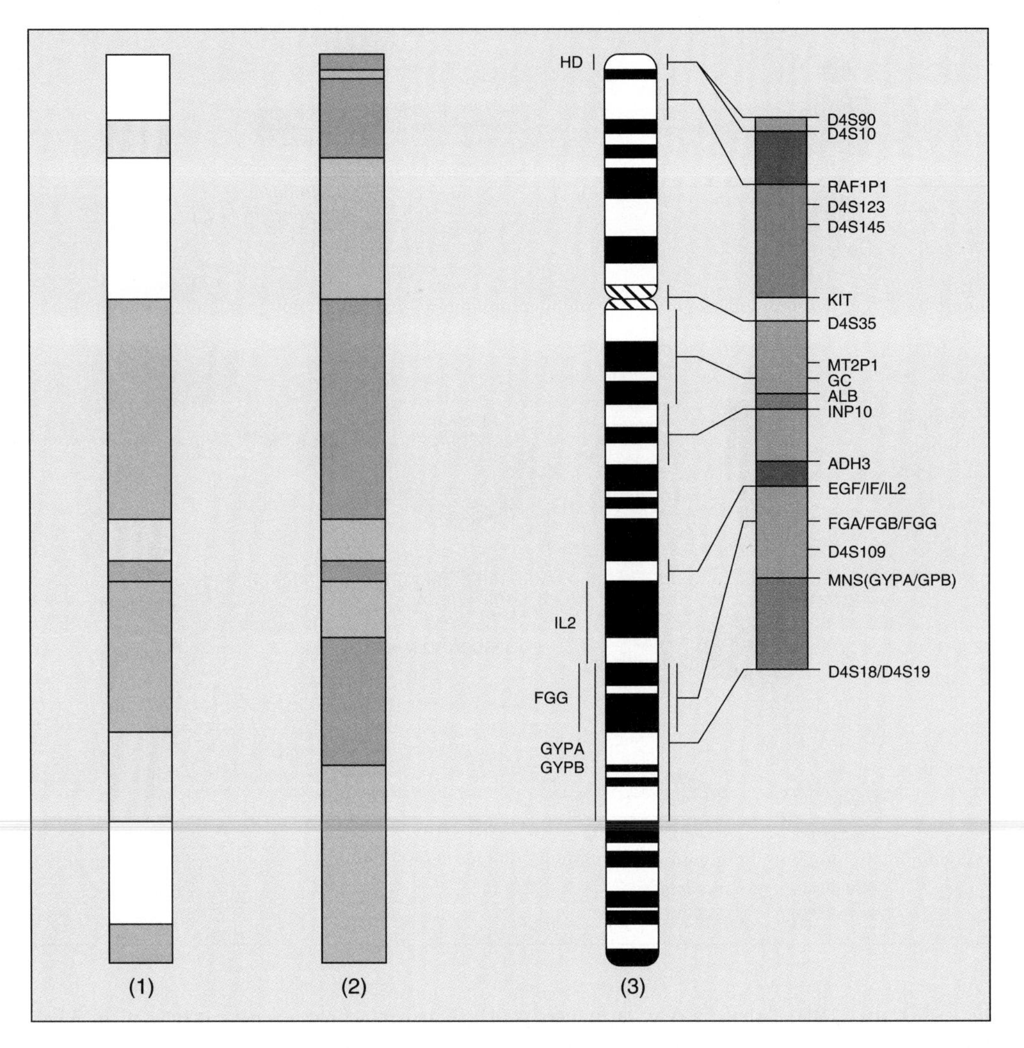

Each marker provides a distinct reference point on the chromosome, and together they serve as a primitive map for identifying the approximate location of various genes of interest. The blanks continued to be filled in and in October 1990, an updated human genome map was published in the journal *Science* (see Figure 21.7).

A question on an even grander scale was raised during the mid-1980s: Could a high-resolution physical map consisting of the exact sequence of nucleotides in the entire human genome be drawn? A **physical map** shows the actual location of genes and also the distances between genes as determined by constant, identifiable landmarks. Figure 21.8 shows the progression of human chromosome mapping. The crudest physical map uses chromosomal bands as landmarks. A map based on linkage markers represents a higher degree of resolution, and the ultimate physical map will be the complete nucleotide sequence of the entire genome.

In 1985, Charles De Lisi, director of the Department of Energy (DOE) Office of Health and Environmental Research, formally proposed a massive project designed to elucidate the complete sequence of the 3 billion nucleotide bases and 50,000 to 100,000 genes in the human genome. The enterprise became known as the *Human Genome Project,* and the original estimates for completion were $3 billion and 15 years.

The initial reaction of the scientific community was mixed. On the one hand, there was great enthusiasm related to gaining access to the "Holy Grail of biology," as one molecular biologist put it. On the other hand, there was concern about the magnitude and cost of what promised to be biology's first "big science" project (that is, costing billions of dollars and involving large, organized, interdis-

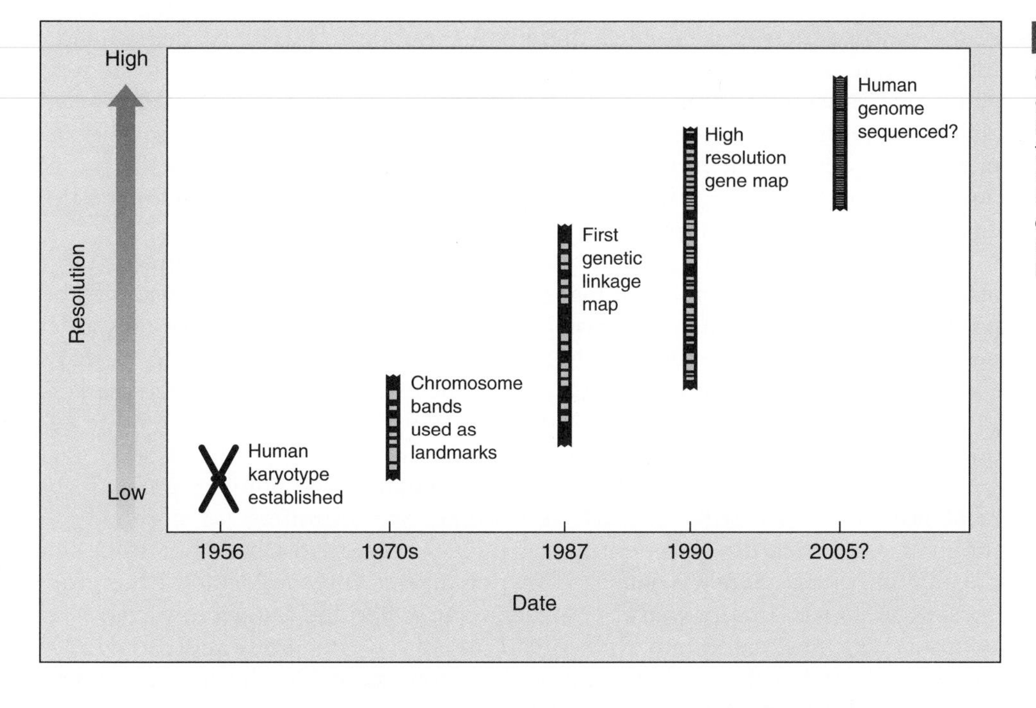

Figure 21.8

Our understanding of the human genome has progressed at a rapid pace because of the development of new technologies and advances in molecular biology. A primary goal of the Human Genome Project, with a target date of 2005, is to determine the exact nucleotide sequence of the entire human genome.

ciplinary teams of scientists) and the possible effects on traditional "small science." These concerns have not yet been completely dissipated, and funding of the project remains contentious. In 1989, the proposal was approved under NIH control (although the DOE will still play a research role). The official starting date was October 1, 1990, large-scale sequencing trials began in 1991, and the projected completion date is September 30, 2005, approximately 50 years after James Watson and Francis Crick published their seminal paper on the structure of DNA. The project is centered in the new Office of Human Genome Research at the NIH, and the same James Watson was named director in 1988 (Figure 21.9). Thus, as he put it, he will now complete his journey "starting with the double helix and going up to the double helical structure of man."

Project Research

How will the Human Genome Project proceed? From 1990 to 1995, emphasis will be on developing new technologies required to complete a physical map. Model organisms such as *E. coli,* yeast, *Drosophila,* and mice will be "sequenced" (that

Figure 21.9

Nobel laureate James Watson is the director of the NIH Human Genome Project.

is, the exact sequence of the nucleotide bases in their DNA will be determined). Interesting regions of the human genome—for example, those known to be near genes that cause important genetic disorders such as Huntington disease—will also be sequenced. During the next five years, technology for large-scale mapping will be developed, and some of the smaller human chromosomes will be sequenced. Finally, during the last five years of the project, sequencing of the entire genome will be completed.

Mapping the human genome will be the centerpiece of the project, but maps for other organisms, including economically important plants, will also be produced. The current human genome map contains the equivalent of one street marker for each 10 to 50 miles of DNA; a higher-resolution physical map, perhaps achievable as early as the mid-1990s, will reduce that distance between markers to 1 mile. The ultimate DNA sequence map will consist of the nucleotide sequences for each chromosome.

The precise timetable for the project will remain quite fluid since new technical advances can arise quickly and create opportunities for accelerating progress. One such development has already occurred. In December 1989, the prestigious journal *Science* named DNA polymerase the "molecule of the year" because of its role in a powerful new technique called the *polymerase chain reaction* (*PCR*). PCR came into widespread use in the late 1980s and early 1990s and was selected as the "major scientific development of 1989" in the same issue. Using PCR, short, unique pieces of DNA can be amplified, their sequences determined, and their precise chromosomal location established. In other words, single-copy DNA sequences—those that occur only once in the human genome—rather than linkage markers or other identifiers, can be used to define the basic landmarks on the physical map. These *sequence-tagged sites* (*STS*) represent a common language and a new approach that may move the timetable for the physical map forward at least five years and perhaps more.

Importance of the Project

The knowledge gained from the Human Genome Project will be of immediate importance in diagnosing single-gene disorders that affect our species. Progress can also be expected in learning about the underlying nature of enormously complex polygenic (or suspected polygenic) disorders—such as alcoholism, mental illnesses, certain cancers, and heart disease—that affect millions of people but are poorly understood at present. When the project is completed, genes associated with polygenic traits may become apparent as well as the underlying mechanisms that regulate their activities. As with single-gene defects, a basic understanding of the genetic defect should direct research toward methods for the treatment of various disorders. Certainly, many questions exist about the regulation of such genes, which seem to operate in some coordinated fashion to produce a given trait or genetic disorder. Farther down the road, the new information will also contribute to a greater understanding of gene control mechanisms in eukaryotes, human genome organization, normal and abnormal cellular growth and development, and evolutionary biology.

Besides being vital to the applied sciences, the project is of considerable importance to basic science. The genetic information in the genome represents the fundamental instructions for the origin, development, maintenance, and eventual demise of an organism. All questions related to these topics can conceivably be addressed from the conceptual framework provided by knowledge of the genome. However, deep knowledge of the genome does not guarantee that we will be able to understand complex processes such as the development of an organism from a fertilized egg. All the genes and their locations may be known, but key details for understanding how all of this information is processed may not be available. Nevertheless, the 1990s should be as fruitful for biological research as the previous decade, and almost certainly more spectacular.

Problems with the Project

Some of the problems of the Human Genome Project are related to technical uncertainties. For example, since only about 2 to 4 percent of the DNA in the human genome contains structural genes, some scientists question the wisdom of sequencing all of the DNA. Might there be a more effective strategy that could bypass sequencing the silent or "junk" DNA? Other scientists feel that new information might reveal unique functions associated with the silent DNA.

A more troublesome problem centers on the question of whether individuals or companies can copyright parts of the human DNA sequence. Historically, science has operated in an atmosphere of open communication in which research results are shared with others within a reasonable period of time. However, modern gene-cloning industries have imposed new realities on basic scientific research. Four factors are involved. First, the majority of leading molecular biologists now have some type of corporate link. Second, the courts have ruled that genetically altered microbes and multicellular organisms can be patented. Third, the potential profits related to resulting therapeutic products and their clinical applications are projected to be staggering. Fourth, molecular biology is expensive, and investments must be protected and recovered by industrial laboratories. There is considerable concern about the effects of these factors on the ways in which traditional science has operated. Currently, a number of open questions require very careful analysis. Is information about the human genome so important that it cannot be proprietary? Does science now condone withholding results based on economic considerations? And do individuals have a right to profit from research that is, in large part, funded by taxpayers?

Other countries are also active in studies of the human genome, although to a much lesser extent than the United States. The United Kingdom will spend about $20 million by 1991, other European Community members about $15 to $20 million per year, and the Soviet Union about $5 million per year. Political hackles have been raised in the United States by a concern that countries making only a minimal investment will reap the enormous rewards that will flow from the genome project. Some U.S. scientists and politicians feel that it may be appropriate to withhold access to data generated by the genome project from countries that do not "pay their fair share." It is not yet clear how this problem will be resolved.

The Human Genome Project is expected to enable scientists to learn the location of all genes in the human genome. Genomes of other organisms such as *E. coli* and mice will also be studied. Although support for the project is widespread, many questions about its costs, goals, and proprietary rights remain unresolved.

THE FUTURE

With respect to humans, the future applications of some new genetic technologies offer exciting prospects for benefiting humankind. For other technologies, the future remains uncertain. The molecular technologies we have described throughout this unit have revolutionized our understanding of many human diseases. The field of **DNA diagnostics**—the analysis of disease at the nucleic acid level—has progressed so rapidly that the time may be near when many genetic disorders, cancers, and viral diseases will be analyzed using rapid, automated, and inexpensive procedures. For some disorders, new technologies have already been used to determine the exact disease mechanism causing the phenotypic effect, and others will be identified shortly. Thus it may soon become possible to treat numerous disorders that diminish the quality of life or lead to premature death.

RESTRICTION FRAGMENT LENGTH POLY-MORPHISMS

Focus on Scientific Inquiry

Every so often in biology, a new method or tool appears that is so powerful that it transforms the way in which scientists are able to attack problems or opens entire new fields of research. In 1980, a theoretical paper by a group of Massachusetts Institute of Technology scientists described a procedure, *linkage analysis,* for locating the approximate sites of genes on chromosomes. In this case, *linkage* refers to the proximity of two or more markers (for example, genes or small segments of DNA) on a chromosome. This type of analysis traces its roots back to T. H. Morgan's fruit fly lab, where A. H. Sturtevant used linkage data to determine the relative positions of genes along the chromosome and subsequently created the first chromosome map (see Chapter 14). The proposed method involved the use of restriction enzymes that cut chromosomal DNA into short strands. Recall that each restriction enzyme (approximately 200 are available) cuts DNA at a specific recognition site composed of 4 to 12 consecutive base pairs. Analyses of the fragments produced can provide information that can actually be used in locating specific genes.

In 1983, linkage analysis moved from the realm of theory to reality when it was used to pinpoint the chromosomal location of the gene (but not the gene itself) for Huntington disease. Since then, linkage analysis has been used in the following ways in genetics research: (1) for identification of the location of genes causing other human genetic disorders, (2) in prenatal testing for serious heritable conditions caused by single genes, and (3) in highly precise pedigree analyses. It has also been applied in forensic tests of biological samples—most commonly blood and semen—in connection with criminal acts.

The methods used in linkage analysis are shown in Figure 1. First, DNA representing the entire genome is isolated from cells taken from an individual (for example, from white blood cells) and purified. The DNA is then digested with a specific restriction enzyme or a combination of enzymes to produce a series of small fragments. These fragments are separated according to their size using a process called *gel electrophoresis.* Using this technique, DNA fragments are placed in lanes on a gel and exposed to a charged electric field; smaller fragments move faster (and farther) than larger ones. When electrophoresis is completed, selected fragments can be exposed to gentle heat, which causes them to separate into single strands. The single strands are then transferred from the gel onto a nylon membrane, to which they become permanently bound. Once the fragments are bound, the membrane is placed in a medium containing small single-stranded segments of DNA called **probes** that are "labeled" with a radioactive element. Through complementary base pairing, a probe will bind with a specific nucleotide sequence on the fragment. Finally, X-ray film is placed over the membrane and exposed to the radioactivity emitted by the probe. The exposure results in a series of bands that can be visualized and analyzed.

The variable fragments generated are called *restriction fragment length polymorphisms* (*RFLPs*). **RFLPs** are defined as inherited variations in the size of DNA fragments produced when a defined piece of DNA is cut with a specific restriction enzyme. Of central importance to linkage anal-

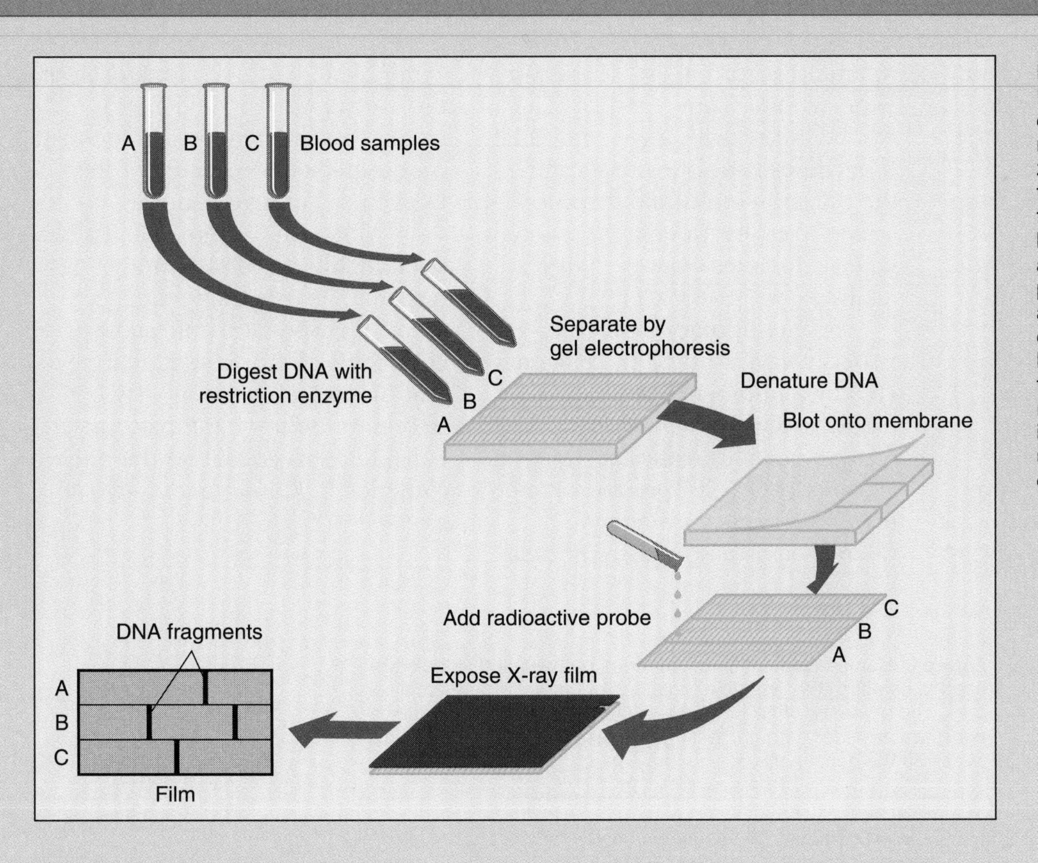

Figure 1 In linkage analysis, DNA taken from white blood cells is cut into small fragments using restriction enzymes. The fragments are then treated with various techniques so that they can be visualized. The number and size of the fragments produced reflect genetic variation as expressed by differences in the number and locations of enzyme recognition sites. The pattern of the resulting fragments is unique in all individuals and can be used to establish the identity of the DNA donor.

ysis is the high level of natural *genetic polymophisms,* or variations in the nucleotide sequences in a specific segment of an individual's DNA. These differences exist because of unique DNA sequences generated during countless human generations through the mechanisms of meiosis, crossing-over, recombination, gene mutation, and chromosomal deletions and insertions. If such variations affect exons (structural genes), there may be a loss of gene function, as in sickle-cell anemia. Conversely, if they occur in introns or in DNA regions between genes, they may simply result in the production or elimination of a recognition site. This results in a unique pattern of RFLPs.

The complexity of RFLP analyses depends on the combination of restriction enzymes and probes used. Probes covering multiple gene loci that yield RFLP patterns of more than 40 bands are now in use and result in a "DNA fingerprint" that is even more unique than a true fingerprint. How unique are you genetically? The chance of your having the same 40-band genetic fingerprint as someone else is about 1 in 10^{19}, compared to 1 in 10^{10} for your ordinary fingerprint.

In studying genetic disorders, an attempt is made to correlate the inheritance of a disorder with a distinctive RFLP *marker,* as shown in Figure 2. In other words, whenever individuals have the gene causing the disorder, they will also have a specific DNA marker (or more) nearby that is always inherited with the disease gene. Thus the

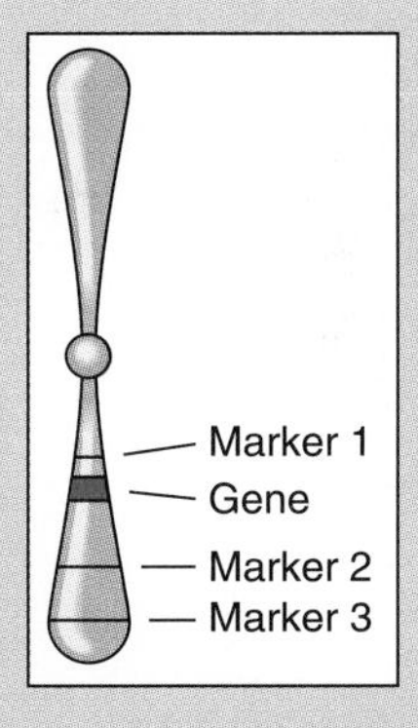

Figure 2 The degree of linkage between a specific RFLP marker and a gene for a disorder is indicated by how often they are inherited together. If they almost always tend to be inherited together, as is the case for the gene and marker 1, the sites of the RFLP locus and the gene lie very close to each other on the chromosome because they do not tend to become separated during crossing-over. If they are frequently not inherited together, as would be the case for the gene and marker number 3, the gene and the RFLP marker occupy locations more distant from each other and have a greater chance of crossing over.

marker indicates the presence of the disorder, and if such a marker is found on the chromosome, the individual either has or will ultimately develop the disorder. For analyzing genetic disorders, RFLP maps of patients with the disorder are compared with those of normal, unaffected people. A particular restriction site may always be present, or absent, from the RFLP profiles in patients (see Figure 3).

Alternatively, for disorders that tend to run in families, comparisons are made between the RFLPs of close relatives. RFLPs are inherited according to Mendelian expectations. Thus if a disorder can be linked with a specific RFLP, it becomes possible to determine the transmission of the disorder-linked RFLP between generations using principles from pedigree analysis.

Contemporary molecular geneticists have studied RFLPs generated by a number of restriction enzymes and have subjected the resulting data to complex statistical analyses. The 1987 genetic linkage map represents the linear sequence of common restriction enzyme sites identified along each of the human chromosomes. It is estimated that the genetic linkage map covers at least 95 percent of the DNA in the human genome.

Although linkage analysis has been highly successful in genetics research, its use in attempts to identify perpetrators of crimes is controversial. At present, it appears that further refinements may be necessary before RFLP patterns are considered to constitute definitive evidence in identifying guilty suspects.

Figure 3 Comparisons of RFLP markers between individuals are used to establish the probable locations of genes for genetic disorders. The presence of RFLP markers indicative of a genetic disorder is the basis for prenatal testing and also for testing individuals at risk for the genetic disorder.

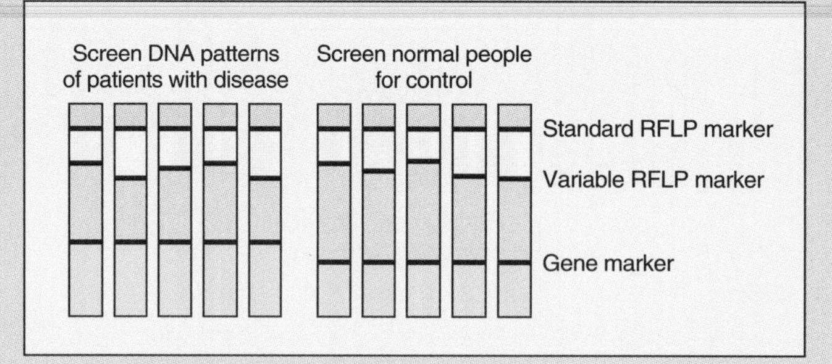

The Human Genome Project offers the supreme scientific challenge—to obtain the complete set of instructions responsible for our conception, development, maturation, and death. As Nobel laureate James Watson stated, "A more important set of instruction books will never be found by human beings. When finally interpreted, the genetic messages encoded within our DNA molecules will provide the ultimate answers to the chemical underpinnings of human existence."

Summary

1. New genetic technologies have emerged with the potential to benefit humans. It is already possible to diagnose a variety of genetic disorders through prenatal testing or analyzing chromosomes from individuals at risk for such disorders. In the future, the number of genetic disorders that can be diagnosed is expected to increase sharply.
2. The use or potential use of diagnostic technologies has led to a number of societal and ethical problems that have not yet been resolved.
3. Gene therapy—the replacement, alteration, or augmentation of existing genes—is now in the experimental stages. However, the potential applications of human somatic cell therapy have become evident in pioneering studies. The first successful human gene transfer experiment was conducted in 1989. The first human gene therapy experiment began in 1990.
4. Despite successes in experimental somatic cell gene therapy studies, many difficult technical problems remain for scientists to solve before gene therapy can be used routinely. Also, no consensus on how germline gene therapy will be used yet exists in the United States.
5. The Human Genome Project is the most expensive biology project ever conceived. The ultimate goal of the project is to determine the exact sequence of every nucleotide in the human genome by 2005. Researchers from many disciplines began studies in 1990 to achieve intermediate objectives such as completing a higher-resolution human genetic map than presently exists and sequencing the genomes of other organisms such as *E. coli, Drosophila,* and some plants and animals. Completion of the definitive nucleotide map will open doors for significant advances in both applied and basic sciences.

Review Questions

1. How is the ability to diagnose genetic disorders related to societal issues?
2. What are some of the social issues that we must confront as a result of new genetic technologies?
3. What are some of the technical problems that must be overcome before human gene therapy can be routinely used?
4. What are some of the ethical issues associated with human gene therapies?
5. What are the types of somatic cell therapy?
6. What are animal models? How are they used in biology research?
7. Describe the first experiment conducted on humans using a form of gene therapy.
8. Describe the types of genetic maps that are currently available for the human genome.
9. What types of research will be done under the umbrella of the Human Genome Project?
10. How are results of the Human Genome Project expected to be used?
11. What are some of the problems of the Human Genome Project?

Essay and Discussion Questions

1. Describe policies that you would recommend for (a) conducting germline therapy research and (b) applying the results of such research to humans.
2. What do you feel are the advantages and disadvantages of the Human Genome Project? Would you recommend that this project go forward, be modified, or be abandoned? Why?
3. James Watson has suggested that at least 3 percent of the Human Genome Project funds should go to support studies and analyses of the ethical and social aspects of the research. Do you agree or disagree? Why?
4. The accompanying hypothetical pedigree shows the inheritance of an RFLP that is linked with a late-acting genetic disease caused by a *dominant allele.* In

this case, the RFLP marker has several alleles that result in eight restriction fragments of particular lengths (here arbitrarily designated 1, 2, 3, and so on). Each one of the homologous chromosome pairs has one RFLP allele. Thus each individual in the pedigree has two RFLPs of varying lengths, one inherited from each parent. Analyze the pedigree and answer the following questions.

a. Which RFLPs characterize each individual in the pedigree? Males are represented by squares, females by circles; RFLPs for grandmother B are given as a "primer."

b. Which RFLP did E inherit from his father?

c. From which grandparents did J inherit her RFLPs?

d. Assume that grandmother B and parent E have the disease in question. Which RFLP serves as a marker for the disease?

e. Which children will later develop the disease?

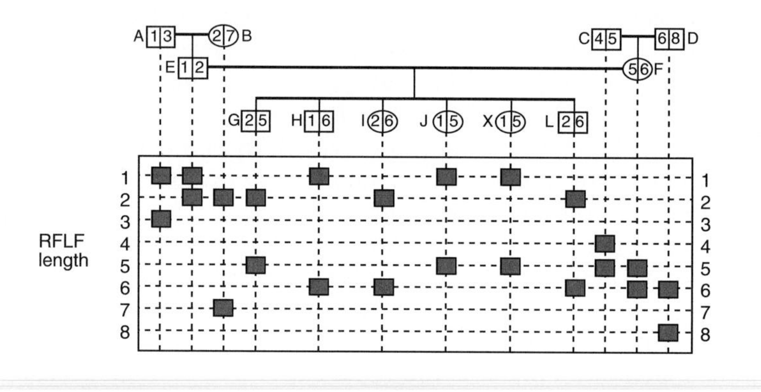

References and Recommended Reading

Bishop, J. C., and M. Waldholz. 1990. *Genome.* New York: Simon & Schuster.

Cavanaugh, M. L., M. I. Gradie, M. L. Mador, and K. K. Kidd. 1990. Mapping the human genome. *Science* 250:237–244.

De Lisi, C. 1988. The Human Genome Project. *American Scientist* 76:488– 493.

Friedmann, T. 1989. Progress toward human gene therapy. *Science* 244:1275–1280.

Guyer, R. L., and D. E. Koshland, Jr. 1989. The molecule of the year. *Science* 246:1543–1546.

Holden, C. 1991. Probing the complex genetics of alcoholism. *Science* 251:163–164.

Hull, R. T. 1990. *Ethical Issues in the New Reproductive Technologies.* Belmont, Calif.: Wadsworth.

Lewis, R. 1987. Genetic-marker testing: Are we ready for it? *Issues in Science and Technology* (Fall):76–82.

Neufeld, P. J., and N. Colman. 1990. When science takes the witness stand. *Scientific American* 262:46–53.

Olson, M., L. Hood, C. Cantor, and D. Botstein. 1989. A common language for physical mapping of the human genome. *Science* 245:1434–1435.

Roberts, L. 1989. Ethical questions haunt new genetic technologies. *Science* 243:1134–1136.

U.S. Congress, Office of Technology Assessment. 1988. *Mapping Our Genes: The Genome Projects: How Big, How Fast?* (OTA-BA-373). Washington, D.C.: U.S. Government Printing Office.

Watson, J. D. 1990. The Human Genome Project: Past, present, and future. *Science* 248:44–48.

White, R., and J. Laloeul. 1988. Chromosome mapping with DNA markers. *Scientific American* 258:40–48.

White, R., and C. T. Caskey. 1988. The human as an experimental system in molecular genetics. *Science* 240:1483–1488.

UNIT III

The Evolution of Life

Why do organisms appear, function, and behave as they do? How have they changed over time? Why are they able to survive in diverse environments? The theory of evolution answers these and other questions. It also provides a general framework that organizes all knowledge of the living world. Moreover, the theory places that knowledge within the context of a dynamic world that is constantly changing and helps us to understand that world better.

Charles Darwin formulated the theory of evolution to answer many interesting questions. Why do certain species resemble other species? What is the relationship between fossils and living organisms? Why are plants and animals distributed as they are over Earth? Darwin's answers to these questions were that organisms change in time, that we can understand mechanisms accounting for those changes, and that evolutionary change can explain the current appearance and distribution of organisms.

In the century since Darwin's death, the theory of evolution has been vastly expanded. Research from genetics led to a deeper understanding of evolutionary mechanisms, and the theory was able to explain many biological phenomena that were not well understood. Like all theories, it continues to change and to guide research.

CHAPTER 22

Darwin and the Origin of the *Origin*

Chapter Outline

Reading Questions

1. What types of information and material did Darwin collect on his voyage?
2. What questions did Darwin ask as a result of observations that he made on his journey?
3. What is the importance of Darwin's theory of evolution?
4. How did Darwin's ideas fit in with those at the time?

DARWIN'S VOYAGE

Shortly after he completed his studies in 1831 at the University of Cambridge, Charles Darwin received a letter from one of his professors, the Rev. John Stevens Henslow, informing him of an exciting opportunity. The British government was sending out a surveying ship to chart the coast of South America and to make various measurements at sea. The captain of the ship was looking for a naturalist to accompany him, and Henslow had suggested Darwin. The letter urged the young college graduate to follow up on the recommendation. Henslow knew that Darwin would be a reliable naturalist on the expedition, for he himself had helped to train him.

Darwin was excited by the prospect of the voyage, but his father, Robert Darwin, was initially opposed to the idea. He thought that it was about time for his son to settle down. Charles had already made an unsuccessful attempt at studying medicine in Edinburgh and had subsequently gone to Cambridge to prepare for the ministry. His father didn't think that a voyage would add to his son's qualifications for his chosen profession. Eventually, however, Charles was able to obtain his father's consent.

That voyage is very famous today. As the ship's naturalist, Darwin not only had the opportunity to make thousands of observations, but he also developed into a mature scientist (see Figure 22.1). Most important, he began to ask himself a set of questions that ultimately led him to formulate the theory of evolution.

The journey was long, dangerous, and at times difficult (he was dreadfully seasick at first), but it was clearly one of the greatest experiences of Darwin's life. The ship, HMS *Beagle* (Figure 22.2), left England two days after Christmas in 1831 and returned in October 1836 after circling the globe. A map of the voyage is shown in Figure 22.3.

Figure 22.1

Charles Darwin was 22 at the time of his voyage on the *Beagle*. Naturalists aboard British expedition vessels were expected to make observations and collections that were useful in several areas of science, including botany (plants), zoology (animals), paleontology (fossils), geography (exploration and mapping), ethnology (native peoples), and geology (minerals and land forms).

> Darwin's voyage around the world was an important part of his scientific education. It led him to consider questions that ultimately led him to formulate the theory of evolution.

Figure 22.2

This watercolor, owned by Darwin, shows HMS *Beagle* in the Murray Narrow, a small channel at the southern extremity of South America.

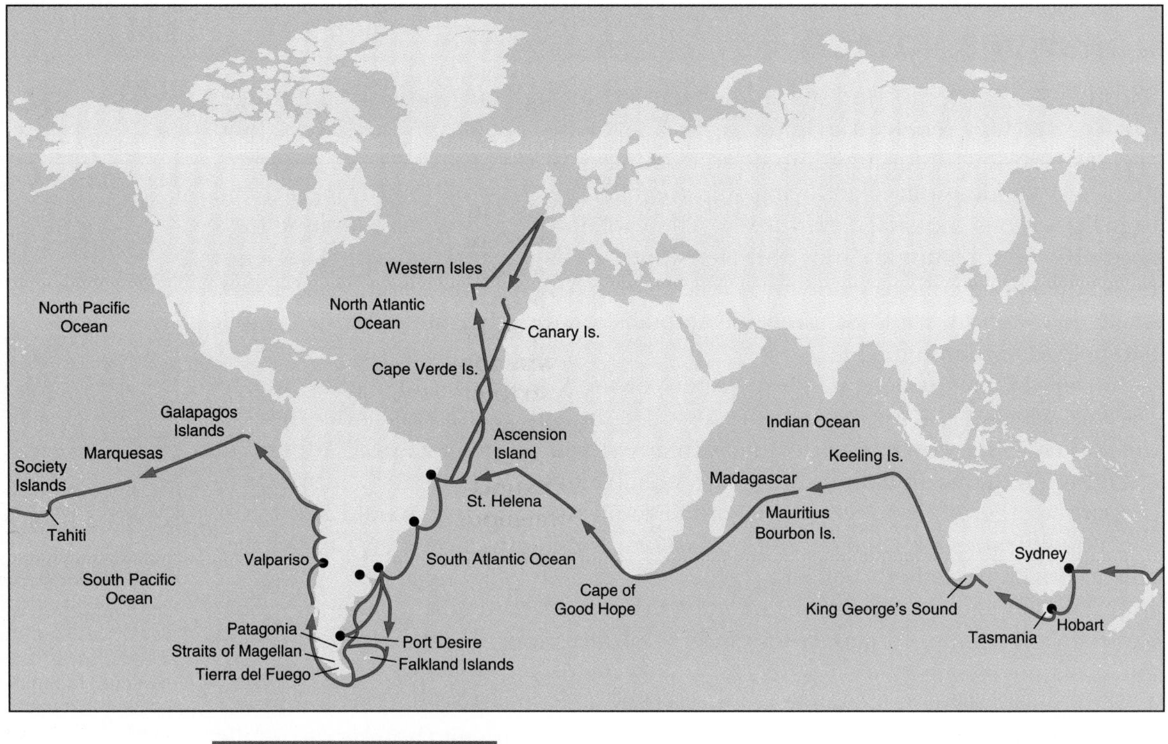

Figure 22.3

Darwin's voyage (1831–1836) proceeded from England to the coast of South America, across the Pacific to Australia, and back around the south of Africa to England.

Darwin's Observations and Questions

The expedition's activities were concentrated primarily on South America, but other areas—Australia, South Africa, and many oceanic islands—were also studied. Henslow's recommendation had been accurate, for Darwin had been well trained at Cambridge to look carefully at geological formations, plants, and animals. He assembled enormous collections, and it took years for scientists to examine them and publish descriptions of the new species he discovered. An illustration of some beetle specimens that Darwin observed is shown in Figure 22.4. Even more significant than the contribution Darwin made to the natural his-

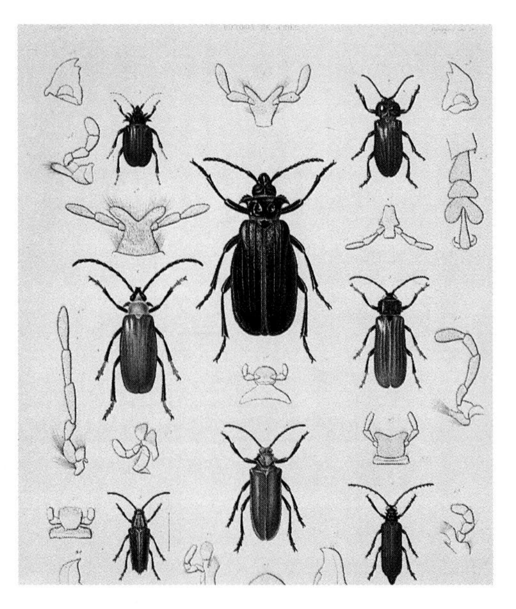

Figure 22.4

Darwin arranged to publish his observations after he returned to England. He contacted noted specialists to describe his collections and artists to draw them. This illustration contains beetles that Darwin saw in Chile.

tory of South America and other places he visited was what he did with his observations and collections. Darwin went well beyond being a collector by asking interesting questions about his data and what they meant.

Fossils

Fossils were among the most exciting finds Darwin made in South America. He was very impressed by the large number of fossils from extinct species. What was the significance of all these fossils? They suggested to him that the number of species that had become extinct must be enormous. Many of the fossils were of unusual extinct animals such as the giant ground sloth, shown in Figure 22.5, and giant armadillos. One of the things that struck Darwin about these discoveries was that although the bones were the remains of extinct animals, they were found in areas where similar animals still lived. Was there a relationship between the extinct forms and those presently living in the area? Other naturalists had noted similar relationships between the remains of extinct and living forms. For example, many Australian fossils were similar to contemporary Australian forms. The same relationship could be found in other areas of the world.

Biogeographical Patterns

Darwin also observed a similarity among organisms (living and extinct) in large geographical areas. Different regions of the world appeared to him to have a basic "character." Many South American animals and plants "looked South American" compared to European species. Darwin's view of a basic geographic pattern reflected a common idea in the nineteenth century: that the world could be divided into several large regions and that the plants and animals of those regions have distinct characteristics. Although some plant and animal groups are found throughout the world, many are centered in only one region. These large-scale distribution patterns are now called **biogeographical realms**. What particularly struck Darwin was that a definite pattern existed in localized regions within larger, general areas. As he traveled southward in South America, he noted that closely related species were often found in adjacent areas. For example, he discovered a previously unknown rhea (a large, flightless bird) and found that the common rhea, to which the new species was closely related, inhabited the adjacent area. This and similar examples suggested to him that there was a complex distribution of plants and animals, in both time (the fossils) and space (biogeographical patterns). What could explain these patterns?

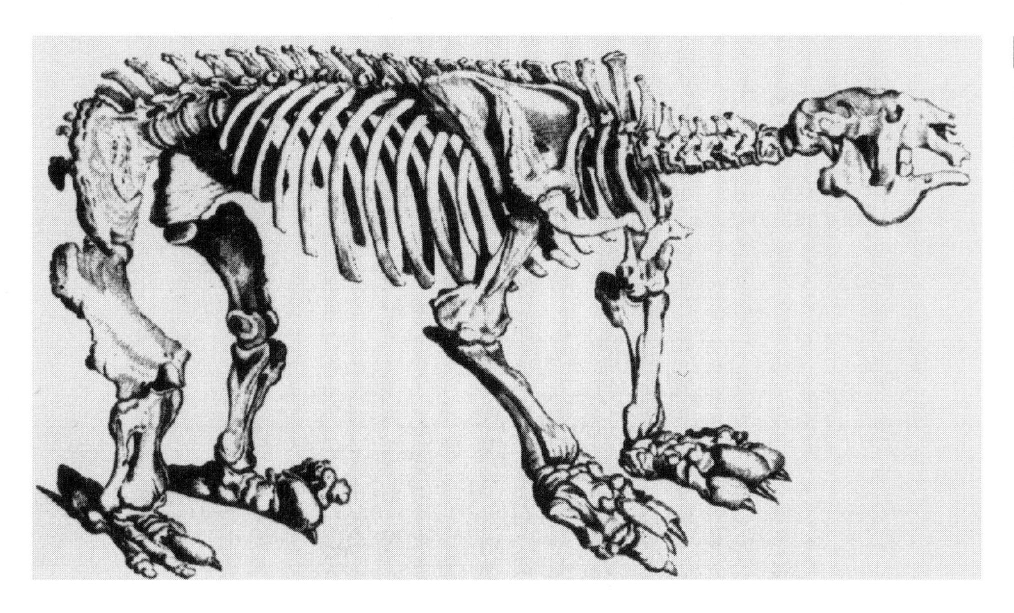

Figure 22.5

One of the fossils that impressed Darwin was the giant sloth. This fossil was found in an area that contained smaller living animals that resembled the giant fossil.

Figure 22.6

Mono Lake in California is a salt lake that resembles the one Darwin saw on his voyage. Salt-tolerant grasses grow along its shore.

Adaptation

In addition to being distributed in patterns, plants and animals are also remarkably well suited to the particular spot in which they are found. As a boy, Darwin had been taught that God had created each species, unchangeable and "perfectly adapted to its environment." On his voyage he was struck repeatedly by the wonderful adaptations of organisms. Adaptation in Darwin's day referred to how the structure and function of organisms were perfectly "designed" for survival in the specific environment in which they were found. Just as people design different machines for different jobs, so, too, had God designed different animals and plants to live in different places on the globe. For example, in what is now Argentina, Darwin visited a salt lake, like the one in Figure 22.6, and discovered that in spite of a high concentration of salt in the water, large numbers of animals survived along the shore. Similarly, he marveled at the unusual scissor-beak bird (*Rhynchops nigra*) (Figure 22.7), which has a long beak flattened laterally and an extended lower mandible, giving it an uncommon ability to catch fish. Toward the end of his voyage, however, Darwin made some observations that cast doubt on the concept that animals were perfectly adapted. In Australia, for example, he saw that some introduced species had replaced native species. This did not fit the concept of perfect adaptation, for if each species was indeed perfectly adapted to its own environment, then an introduced species should compete poorly in the "foreign" environment, and the local species should have the competitive advantage. Although Darwin came to realize that adaptation is never perfect, he nonetheless maintained his fascination with the exquisite adaptations found in nature and realized their importance for survival.

Oceanic Islands

> The natural history of these islands is eminently curious, and well deserves attention. Most of the organic productions are aboriginal creations, found nowhere else; there is even a difference between the inhabitants of the different islands; yet all show a marked relationship with those of America, though separated from that continent by an open space of ocean, between 500 and 600 miles in width. The archipelago is a little world within itself, or rather a satellite attached to America, whence it has derived a few stray colonists, and has received the general character of its indigenous productions. Considering the small size of these islands, we feel the more astonished at the number of their aboriginal beings, and at their confined range. Seeing every height crowned with its crater, and the boundaries of most of the lava-streams still distinct, we are led to believe that within a period, geologically recent, the unbroken ocean was here spread out. Hence, both in space and time, we seem to be brought somewhat near to that great fact—that mystery of mysteries—the first appearance of new beings on this earth.

This quotation from *The Voyage of the Beagle* (1845) reflects Darwin's fascination for the life that he found on islands in the ocean. The most famous of these islands, and the ones referred to in the quotation, are the **Galápagos Islands** (Figure 22.8), located in the Pacific near the equator about 600 miles off the coast of South America. Darwin was surprised by the number of indigenous species in the Galápagos; that is, they were native, not introduced from elsewhere. Despite being distinct species, it seemed to him that many of the island's plants and animals must be related to South American organisms. In descriptions of the Galápagos Islands, published when he returned to England, Darwin wrote that there were even differences in the animals among the islands. Why should there be such diversity in a cluster of small and remote islands? Moreover, what was the original source of the native plants and animals? The islands themselves had been formed by volcanoes during the present epoch, yet the flora and fauna were mostly unique to the Galápagos. In thinking about these fairly barren islands containing species found nowhere else, Darwin was led to ask, What is the origin of new species?

Figure 22.7

Darwin marveled at the beak of this *Rhynchops nigra*, which allowed it to catch fish. In *The Voyage of the Beagle*, he wrote: "I here saw a very extraordinary bird, called the Scissor-beak (*Rhynchops nigra*). It has short legs, web feet, extremely long-pointed wings, and is of about the size of a tern. The beak is flattened laterally, that is, in a plane at right angles to that of a spoonbill or duck. It is as flat and elastic as an ivory paper-cutter, and the lower mandible, differently from every other bird, is an inch and a half longer than the upper. . . . I saw several of these birds, generally in small flocks, flying rapidly backwards and forwards close to the surface of the lake. They kept their bills wide open, and the lower mandible half buried in the water. Thus skimming the surface, they ploughed it in their course."

Figure 22.8

The natural history of the Galápagos Islands fascinated Darwin. The main islands of the Galápagos feature a rugged terrain.

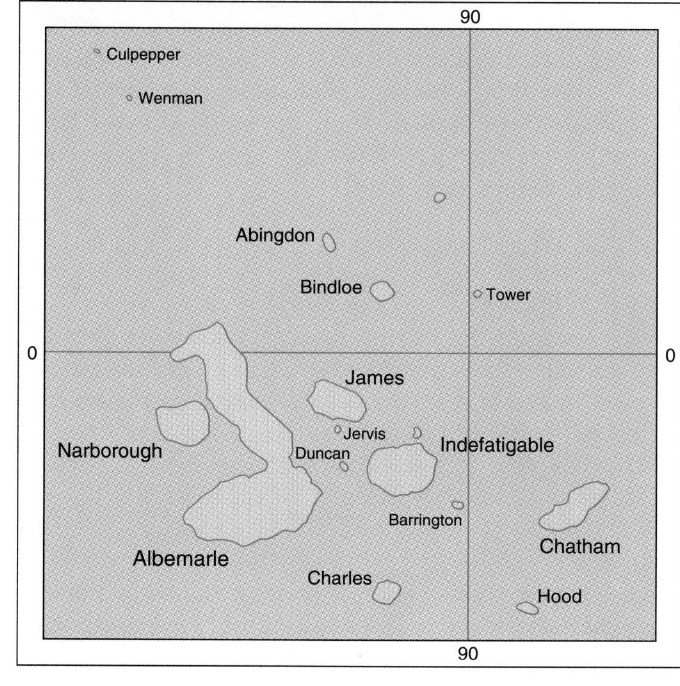

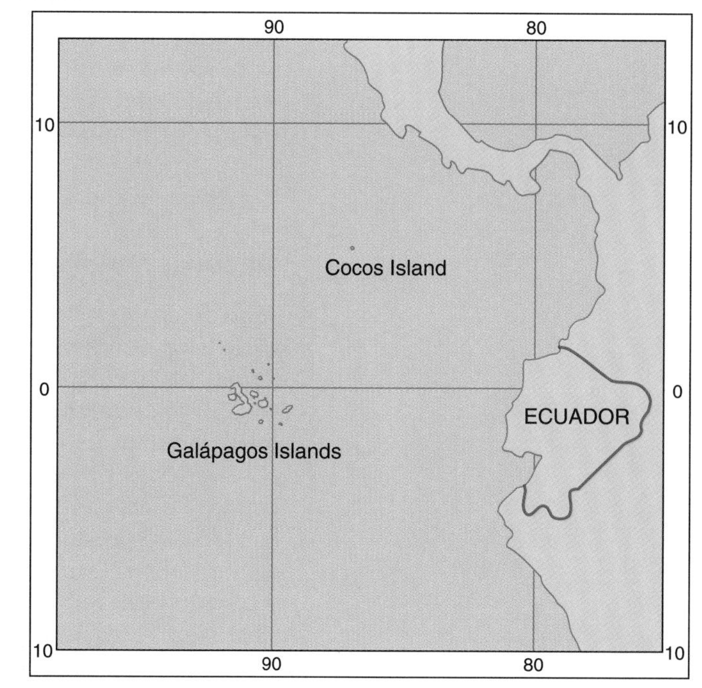

Darwin was struck by several critical questions during his voyage: What was the relationship of fossil to living forms? What was the cause of biogeographical patterns? How had animals and plants become adapted to their environment? What was the origin of plants and animals on the Galápagos Islands?

DARWIN'S COLLECTIONS AND THE CHANGE OF SPECIES

After Darwin returned from his voyage in 1836, he settled in London for several years to work on his collections and to arrange for the publication of their description. As part of this process, he consulted the leading British authorities in natural history. They were very impressed with the collections, which contained not only unusual specimens of unknown species but also interesting information on their distribution.

Questions About Darwin's Data

Discussions with preeminent naturalists led Darwin to consider some fruitful questions. For example, John Gould, a famous ornithologist (bird specialist), informed Darwin that his collection of Galápagos finches actually included several different species. Darwin was quite puzzled by this. Why should there be so many species of finches on a remote group of islands? Where had they come from?

For many months, he pondered these questions and others relating to fossils and geographic distribution. Finally he decided that if he made the unorthodox assumption that species could change over time, he could explain many of the unusual aspects of his data. It became evident why there were different species of finches: They were the descendants of an original species that had somehow become modified to fit new, different habitats on the islands. If species could change, many other questions could also be resolved. The relationship of fossils to present-day forms could be explained by postulating that the contemporary forms were related to the fossil forms by descent. That is, existing species that had strong similarities to fossils of extinct species were actually the distant descendants of those (or related) species. By studying the fossil record, it was possible to construct a **phylogeny**, a history of the evolutionary development of a species or of a larger group of organisms. Closely related species that were found in adjacent areas might be related to a common ancestor, or one species might be an offshoot of the other. The South American character of the Galápagos Islands might be explained if the original inhabitants had come from South America. The questions that could be generated and answered were endless. Some of these are illustrated in Figure 22.9.

Natural Selection

How do species change? Darwin reflected on this question for over a year trying to develop some hypothesis to explain it. He finally came up with a possible solution: *natural selection*. Natural selection is discussed in Chapter 23, but we can consider it here briefly. Darwin was aware that *variation* existed in plant and animal populations because he had studied how domestic breeders had created new varieties by selecting offspring with certain desired characteristics. For example, he was amazed by the great number of varieties of pigeons that pigeon fanciers had produced.

In 1838, Darwin read an essay by the Rev. Thomas Malthus (see Chapter 10), who believed that the rate of growth in human populations was geometric (that is, 2, 4, 16, 32, 64, . . .), whereas the rate of increase of food was arithmetic (that is, 2, 4, 6, 8, . . .). Consequently, Malthus concluded that humans would

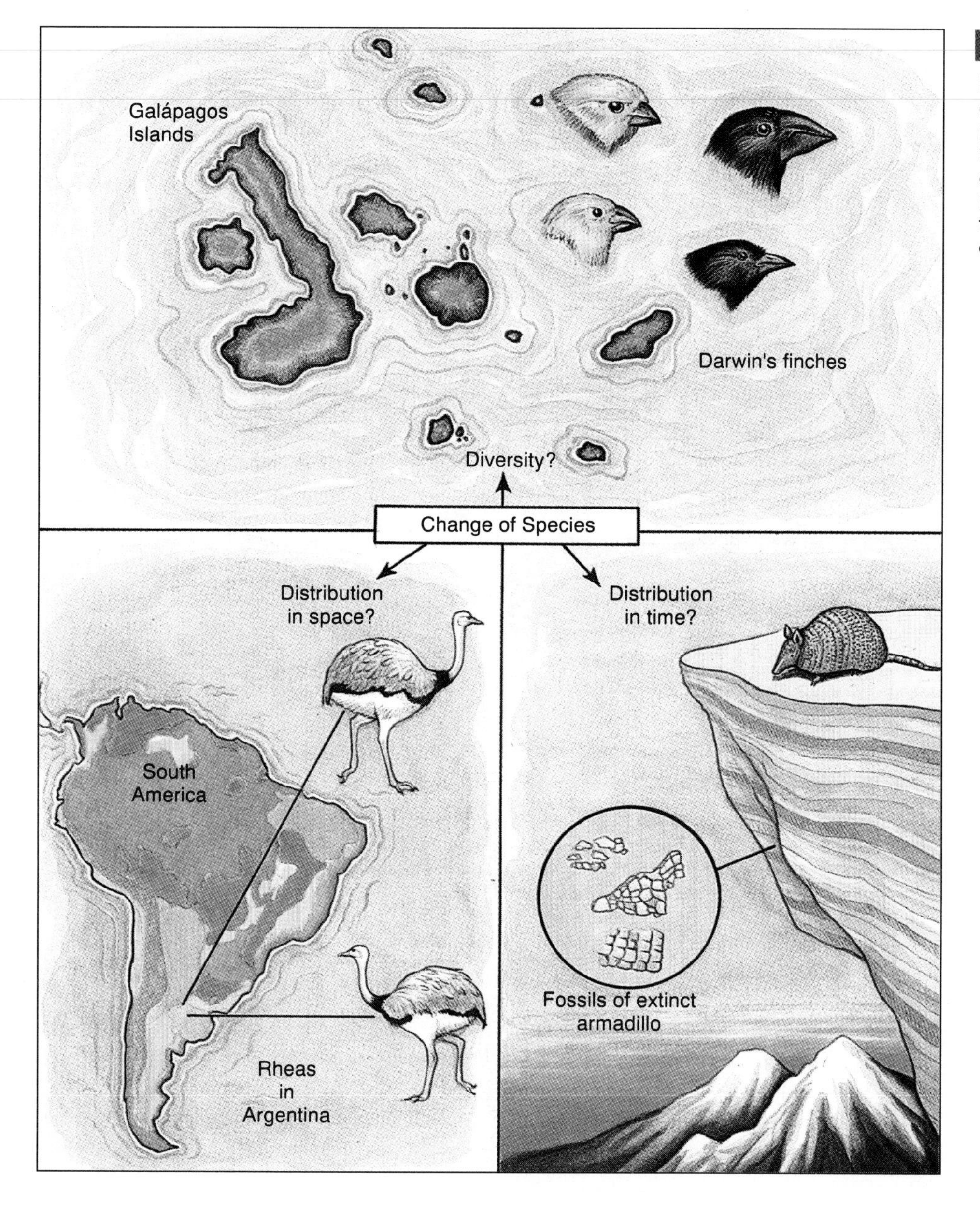

Figure 22.9

Darwin came to realize that if he accepted the idea that species change, many of the puzzling observations that he had made on his voyage could be explained. He could relate fossils to living forms, explain biogeographical distribution patterns, and understand the diversity of island organisms.

face a dramatic struggle for existence. Darwin combined these ideas of variation and population growth. He realized that animal and plant populations have tremendous potential for reproducing (consider how many eggs a frog lays in the spring, and see Figure 22.10), yet they appear to occur in roughly the same numbers over periods of time; therefore, only a few offspring must actually survive to reproduce and leave progeny. Further, he reasoned that those that were more "fit" would be more likely to survive and leave offspring than those less "fit." In other words, Darwin hypothesized that there would be a *natural selection*. The result of natural selection would be analogous to events in domestic breeding, where breeders select for mating only offspring displaying the traits that they are trying to enhance.

Natural selection differed from domestic selection in two important ways: no one was performing the selecting, and there was no predetermined goal in mind (for example, producing tomatoes that ripen early and have the diameter of hamburger buns). Also, domestic breeders had never created a new species, only new varieties of the same species. Darwin thought that given sufficient time, natural selection would produce new species. He thought that this would occur because as the environment gradually changed in time, new adaptations would be

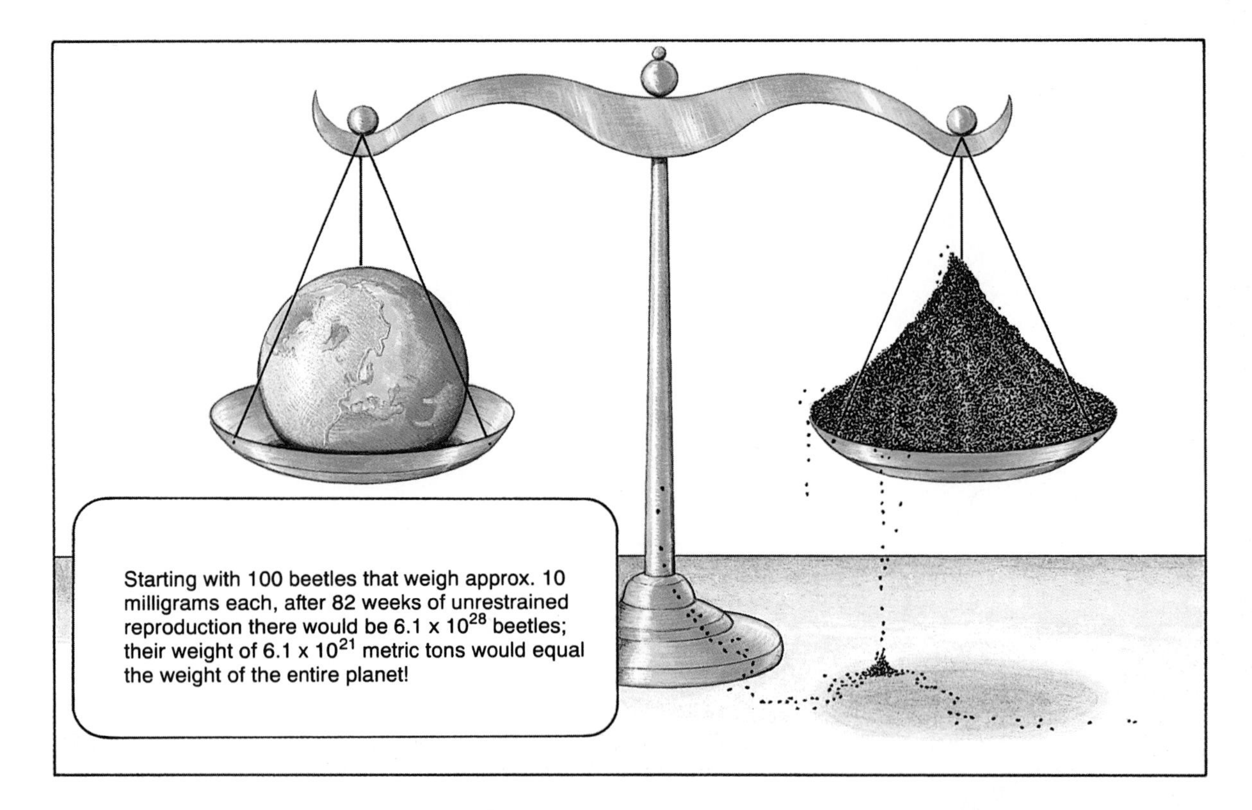

Figure 22.10

In his discussion of unrestrained growth, Malthus provided Darwin with the key to understanding how species change. If there are no checks, a population will increase and soon deplete its food sources. This drawing dramatizes how many beetles would exist if all offspring of a colony of 100 beetles lived and reproduced for 82 weeks.

favored in the new environment. Also, new ways of exploiting the environment might be developed by individuals with novel variations. In time, many new species could arise. Figure 22.11 provides a modern analysis of the Galápagos finches and the beaks that distinguish them and permit them to exploit resources from different habitats. It can be used to explain how the great *diversity* of life on Earth originated.

Darwin's hypothesis that natural selection resulted in the development of new species provided a new perspective on species. Rather than considering a species to be a "design" or blueprint for an organism, a species now had to be considered a population of individual organisms. The adaptation of a species to its environment was no longer just a part of the species' "design." Instead, adaptation was a *process* that resulted in survival in surrounding conditions. From this new perspective, a species can never be "perfectly" adapted, only relatively well adapted.

By hypothesizing that species changed over time, Darwin could solve all of the problems he addressed. His concept of natural selection explained how species changed.

THE THEORY OF EVOLUTION

Darwin's Theory

Of course, Darwin could not test his natural selection hypothesis directly (place a population of mice on an island and watch them for 100,000 years?), so he at-

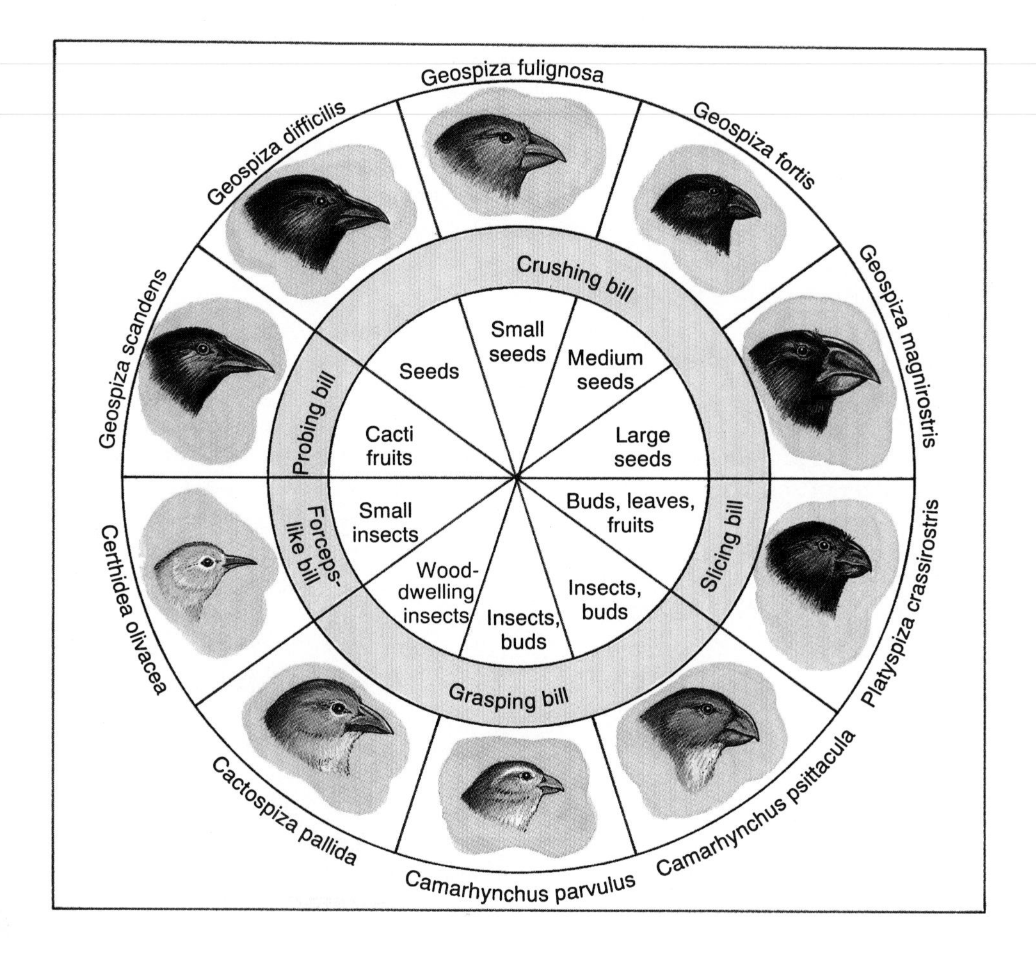

Figure 22.11

A result of natural selection is adaptation, as illustrated by the various beaks of Galápagos finch species. The diagram relates the beaks of some finches to their food supply.

tempted to test it indirectly by assuming that natural selection was operating. Then by considering the known facts of natural history, he attempted to determine if the expected patterns occurred or if they were consistent with what might be expected to occur. Darwin was a cautious investigator with indefatigable patience. He read vast numbers of books and journals, wrote to hundreds of specialists, made many observations, and conducted numerous experiments. As a result, he broadened his perspective and was finally able to construct a theory of evolution that not only explained the origin of species but also synthesized known facts from the biological sciences (see the Focus on Scientific Explanations, "Scientific Theories"). Moreover, the new theory pointed the way to areas of research that proved to be highly productive.

Darwin's explanation about the origin of species was a bold move in his day, but one that fit a general trend toward explaining nature in physical terms. Scientists wanted studies of the living world to be rigorous, like those of physics, chemistry, astronomy, and geology. To Darwin and many of his contemporaries, that meant that such studies should be based on careful observations of living objects and attempts should be made to discover the laws that governed them. Practitioners of the other leading sciences (such as physics) attempted to explain nature in terms of physical concepts and underlying mechanisms. Darwin's theory of evolution was in the spirit of modern science because it used physical and mechanical explanations to account for the phenomena of life.

The hypothesis of natural selection was able to explain more than the immediate problem of how species change. By considering how natural selection and the change of species were related to the major areas of biology, Darwin was able to formulate a general theory that explained the phenomena of life.

Scientific Theories

Focus on Scientific Explanations

Scientists use a variety of methods and processes to gain knowledge of the living world. There is no single, simple "scientific method," and what scientist do often resembles what scholars do in other areas of research. Let's consider what scientists do by looking at different *levels of inquiry*.

1. Observation

Observation provides much of the basic data of science. Observations are not, however, made randomly. On his voyage, Darwin made observations related to topics that interested him—indications of past geological change, unusual patterns in the distribution of organisms, and interesting adaptations of various plants and animals. On reading Darwin's account of his trip, *The Voyage of the Beagle,* one sees that he selected his material carefully. In describing South American fossils, for example, Darwin placed his observations in the context of the question "What is the relationship of fossils to living forms?" Scientists, like Darwin, make observations related to relevant questions or interests.

2. Generalization

An individual observation takes on greater significance if it can be related to some broader statement. Scientists look for patterns. In reviewing his observations, Darwin made many generalizations: he noted that closely related species were found in adjacent areas; he recognized a relationship of island fauna to nearby continents (such as the "African character" of the Cape Verde Islands); and after examining over 10,000 specimens of barnacles, he stated that practically all the external characters of a species are highly variable. Figure 1 shows some of the barnacles Darwin studied.

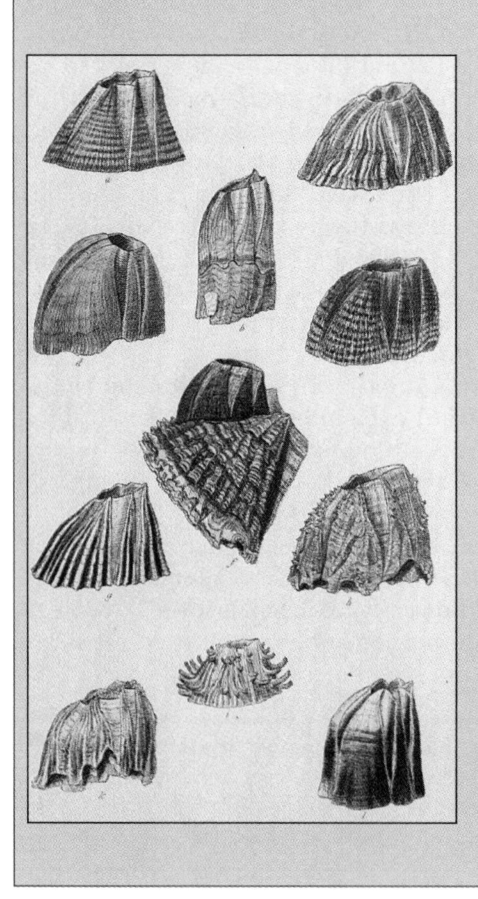

Figure 1 Darwin examined over 10,000 barnacles and came to appreciate the enormous variation that existed among them. This illustration is from Darwin's book on barnacles.

3. Explanation

What makes science interesting is the attempt to explain patterns, or even single events in nature. Initially, a calculated guess is made about the underlying reason to account for an observation or a natural pattern. These guesses are called *hypotheses,* and they are formulated in such a way that either they can be tested directly or indirect evidence can be gathered to support them.

Darwin formulated the hypothesis of natural selection to explain the change of species. He could not test it directly, so he compared what a scientist would expect to find in some different branches of natural history, if natural selection operated, with what was known to exist in those branches. For example, he reasoned that in examining the anatomy of the members of a taxon—say, a family—numerous anatomical similarities could be explained by descent from a common ancestor. The variations could be understood as the product of natural selection operating over time. His results were very convincing, and he went on to consider other areas of natural history.

Can testing, or gathering more evidence, make us believe that a hypothesis is true? Scientists are uncomfortable speaking about the truth of hypotheses. They have learned from long experience that hypotheses are often shown to be false or that observations can be better explained in some other way. Positive test results give us confidence to accept a hypothesis provisionally, while keeping an open mind. Negative results, however, may lead to the rejection or modification of a hypothesis. Some hypotheses turn out to be very good in that they repeatedly stand up to new tests. Natural selection, for example, has now been demonstrated in laboratory and field tests. When scientists have a great amount of confidence in a hypothesis, they call it a *law.* For example, the regularities that Mendel explained and then tested are called Mendel's laws (see Chapter 12).

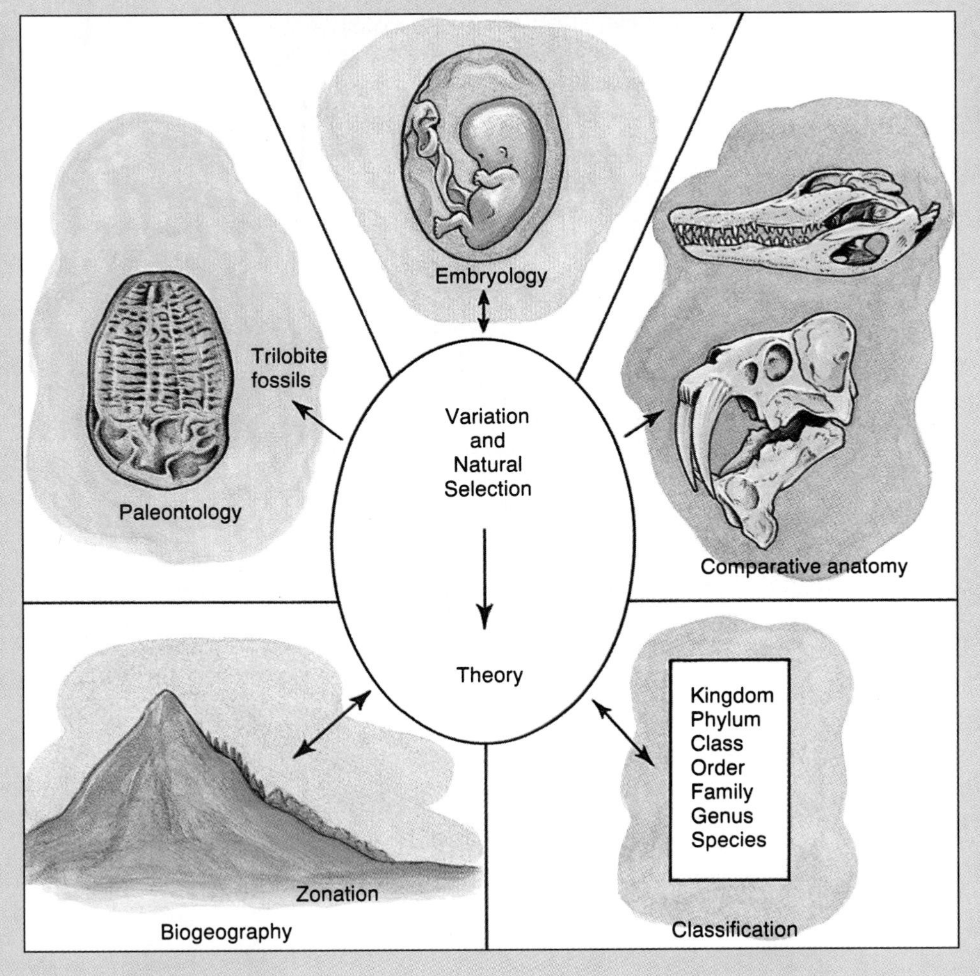

Figure 2 Darwin's hypothesis of natural selection led him to consider many areas of biology. In each case, it seemed that natural selection explained the facts involved and allowed interpretation from a new perspective. Ultimately, this led Darwin to formulate what has become the major synthesizing theory of biology, the theory of evolution.

Focus on Scientific Explanations

Darwin was searching for a hypothesis that he hoped would become a law that explained the origin of species. As it turned out, he did far more. He constructed a theory.

4. Theory

Darwin initially formulated a hypothesis dealing with the origin of species by natural selection. He could not test it directly, so he tested it indirectly by asking himself, "If natural selection did operate on populations, what patterns might we expect to find?" He applied this idea to the fossil record, biogeographical distribution, comparative anatomy (the comparison of animal structures), breeding experiments, embryology, classification—in short, to all the known areas of natural history at that time. In each case, his hypothesis not only seemed to hold but also served to explain related regularities. Using a few simple ideas (such as variation and natural selection), Darwin was able to provide an explanation for a vast quantity of data, and in so doing he synthesized these data in a coherent, unified formulation, as is indicated in Figure 2. A theory, like Darwin's theory of evolution, explains vast amounts of data with a relatively simple set of concepts and processes.

A theory does more than that, however. Theories are highly regarded in science, not only because their development represents an impressive intellectual achievement but also because they *direct research* by raising new questions. Much of the knowledge of genetics discussed in Chapters 14 and 15 was generated when scientists attempted to resolve the problems posed by Darwin's theory or its implications (see Figure 3).

Are theories true? As with hypotheses, it is inappropriate to label theories as true. Rather, if theories are successful in answering questions, unifying knowledge, and guiding research, scientists come to value them highly. Because a theory directs research, it is as much an approach (a plan of action, a research program) as it is an explanation, and it is therefore *dynamic* (it changes). A theory that never changes with time is not a very interesting or useful theory. Because of the changing nature of theories, a blind, unquestioning acceptance of theory (dogmatism) is unacceptable in science.

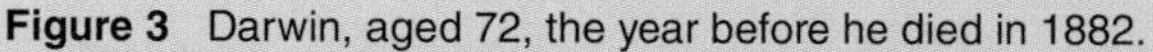

Figure 3 Darwin, aged 72, the year before he died in 1882.

Accepted Ideas on the Origin of Species in Darwin's Time

> All organized beings exhibit in themselves all those categories of structure and of existence upon which a natural system may be founded, in such a manner that, in tracing it, the human mind is only translating into human language the Divine thought expressed in nature in living realities. . . .
>
> The combination in time and space of all these thoughtful conceptions exhibits not only thought, it shows also premeditation, power, wisdom, greatness, prescience, omniscience, providence. In one word, all these facts in their natural connection proclaim aloud the One God, whom man may know, adore, and love; and Natural History must in good time become the analysis of the thoughts of the Creator of the Universe, as manifested in the animals and vegetable kingdoms, as well as in the inorganic world.
>
> Louis Agassiz, *Essay on Classification* (1858)

> He who believes in separate and innumerable acts of creation will say, that in these cases it has pleased the Creator to cause a being of one type to take the place of one of another type; but this seems to me only restating the fact in dignified language. He who believes in the struggle for existence and in the principle of natural selection, will acknowledge that every organic being is constantly endeavouring to increase in numbers; and that if any one being vary ever so little, either in habits or structure, and thus gain an advantage over some other inhabitant of the country, it will seize on the place of that inhabitant, however different it may be from its own place.
>
> Charles Darwin, *On the Origin of Species* (1859)

These quotations reflect the fresh approach that Darwin took in examining the problems of biology. In Darwin's time, the origin of species and the patterns found among living organisms were explained as part of God's plan. According to this view, each species was created at a particular time and was suited for a particular place. In 1858, Louis Agassiz (Figure 22.12), the leading zoologist in the United States, published the *Essay on Classification,* which attempted to

Figure 22.12

Louis Agassiz was the leading zoologist in the United States in the nineteenth century.

EVOLUTION AND CREATION IN U.S. PUBLIC HIGH SCHOOLS

Focus on Science and Society

High school science textbooks usually reflect beliefs of the scientific community. It is not surprising, therefore, that by the 1880s evolution was a topic included in new texts. By then the American scientific community had completely accepted the theory of evolution. Before the turn of the twentieth century, botany, geology, and zoology textbooks all contained serious discussions of evolution. After the turn of the century, curriculum reform resulted in the emergence of biology textbooks that combined the life sciences into one subject. They, too, continued the evolutionary thrust of earlier texts. In addition, the leading teachers' journal and the National Education Association stressed the value of teaching evolution as a unifying theme in biology.

By the 1920s, however, a strong antievolution crusade had emerged, with the goal of reversing the emphasis on evolution in biological science courses. This movement was not based on any change in the scientific community's commitment to evolutionary theory; instead, it reflected a number of social changes that had taken place in the United States.

One important factor was the tremendous expansion of the number of high school students. In 1890, only 3.8 percent of individuals 14 to 17 years of age were enrolled in high school. The number of students enrolling in high schools doubled each decade from 1890 to 1920, increasing from about 200,000 in 1890 to almost 2 million in 1920. This expansion brought many more high school students in contact with the theory of evolution and was a major factor in alarming the members of the general public who opposed the theory.

A second and equally important element was the rise of fundamentalism after the First World War (1914–1918) and its drive to combat "modernism." This entailed supporting a literal reading of the Bible and combating forces that fundamentalists believed undermined their traditional values.

William Jennings Bryan (Figure 1) was perhaps the most ardent crusader against teaching evolution in the public schools. He believed that government had the obligation to pass legislation to improve the social well-being of its citizens, and he had taken part in the campaign for prohibition as well as for numerous liberal reforms, such as the income tax and women's suffrage. He opposed the teaching of human evolution because he thought it would lead people away from God. Furthermore, he believed that it had contributed to the causes of World War I by allegedly encouraging in Germany a justification for an armed struggle among nations. He also feared that it would promote the exploitation of people by stressing the value of competition.

Although Bryan rejected the theory of evolution in general, he and the antievolution crusade of the 1920s focused their attack only on the teaching of *human evolution* and sought legislation that restricted it alone. Bryan's position was that church and state were separate and that therefore public schools should not teach *anything* concerning the origin of humans.

The antievolution campaign succeeded in introducing many bills in state legislatures to prohibit the teaching of human evolution. The first to pass was in Oklahoma (1923). The best-known bill was passed in Tennessee (1925), for it led to the famous "Monkey Trial" in which Bryan and the World's

Figure 1 William Jennings Bryan, the leading advocate of the anti-evolution crusade of the 1920s, is shown here as he stood up and made his first speech during the Scopes trial in 1925.

Christian Fundamentals Association (WCFA) confronted the American Civil Liberties Union (ACLU) and America's most famous defense attorney, Clarence Darrow. In that trial, a young science teacher, John Scopes, shown with Darrow in Figure 2, was accused of violating the newly enacted law, providing the occasion for an impassioned debate between two leading masters of rhetoric. Although Darrow succeeded in making the antievolution movement look dogmatic and ignorant, the jury convicted Scopes of violating the law. Darrow and the ACLU hoped to test the constitutionality of the law by appealing the decision to the Tennessee Supreme Court, but that court overturned the conviction on a technicality (without accepting the argument of the law's unconstitutionality), and so the defense could not take its case to the Supreme Court for a final judgment.

Some historians have described the Scopes trial as a turning point in the creation-evolution controversy and claim that the antievolution crusade was discredited as ignorant, intolerant, and backward-looking. A more accurate description would be to see the trial as marking the end of a short but stormy period of confrontation and the beginning of a truce. Since the Tennessee law had not been declared unconstitutional, other states could enact antievolution laws. Few did, however. For the next 30 years, no action was taken by any state regarding antievolution laws, although some local school boards did take action. The teaching of evolution in school nonetheless suffered because publishers who published high school texts tended to reduce or eliminate the topic of evolution in an effort to avoid controversy and to sell more books.

A turning point in the history

Figure 2 Clarence Darrow (left at table) was one of America's most famous defense attorneys when he defended John Scopes (second right at table) for teaching human evolution in 1925.

of evolution theory in the high school curriculum was reached in 1959. The launching of a space satellite, *Sputnik*, by the Soviet Union in 1957 had shocked the American educational community and had led to a serious effort to increase the quantity and quality of science taught in the schools. In 1959, the American Institute of Biological Sciences, financed by the National Science Foundation, sponsored the Biological Science Curriculum Study (BSCS). Many leading biologists who strongly believed that evolution should be given a prominent emphasis were brought into the process of planning and writing biology texts. The materials produced by the BSCS were widely used and strongly featured evolution.

Antievolutionists were quick to respond. However, this time their reaction was not to ban evolution from the schools but to demand "equal time" for teaching creation based on the biblical account in Genesis. This new approach was probably inspired by rules that regulate the media, which require equal time for political candidates and opposite sides of controversial public issues. The Creation-Science Research Center, the Creation Research Society, and other groups that either prepared teaching materials or worked to influence local and state political bodies mounted a strong campaign in support of this demand.

Science and teaching organizations, both on the state and national levels, launched a dual defense: to repeal the old antievolution laws and to fend off new equal-time legislation. The American Association for the Advancement of Science (AAAS), the National Academy of Science, and the National Science Teachers Association were active in these efforts, along with the ACLU. They were successful in overturning the old antievolution laws, but new equal-time laws were passed in some states.

In 1982, Federal Judge William Overton ruled that an Arkansas equal-time law was unconstitutional. He argued that its intent was to introduce particular religious beliefs into the curriculum and that "creation science" was not science. The same year, Judge Adrian Duplantier struck down a Louisiana equal-time law as an unconstitutional attempt to promote religion, a violation of the First Amendment of the Constitution which states, "Congress shall make no law respecting an establishment of religion." A Supreme Court decision in 1987 affirmed the Duplantier judgment, and for the present, it seems unlikely that any state law restricting the teaching of evolution will be acceptable. (For more information on the legal battle over creation and evolution, see Edward J. Larson's *Trial and Error: The American Controversy over Creation and Evolution*.)

place all natural history data into this religious framework. He argued that God had a complex plan of continuous creation and that by studying classification, biogeographical distribution, and the fossil record, the naturalist only translates into human language God's plan as revealed in nature. Like Darwin, Agassiz had traveled and read extensively—his personal library became the core of the library at Harvard University's Museum of Comparative Zoology—and he too wanted an explanation for the many phenomena and patterns he saw in nature. Agassiz's solution was to attribute all patterns directly to the Creator. This was not a modern scientific theory because it did not explain the world of natural history using a few basic scientific concepts, and it did not raise any useful scientific questions. Rather, life was explained as the result of an unknowable "divine intelligence" (see the Focus on Science and Society, "Evolution and Creation in U.S. Public High Schools").

Darwin's theory, which he published in 1859 as *On the Origin of Species by Means of Natural Selection, or the Preservation of Favoured Races in the Struggle for Life,* argued against using this form of explanation in biology. He did not argue against a belief in God or against a view that God might have created the world and its laws. Instead, he argued that such beliefs belonged in the realm of religion and theology, not in science. Darwin's position was that it was more useful to ask questions like "What are the causes of the origin of species?" and "How do species change?" than to attribute what we observe to a divine plan. Answering questions like "Why are there the distribution patterns that we observe?" with "Because it was God's plan" did not encourage research to solve scientific problems. It might lead to more descriptive work filling in details of the plan, but it would not extend our knowledge of the underlying laws of nature.

By struggling with the question "How do new species come into being?" Darwin was led to the hypothesis of natural selection and ultimately to a broad evolutionary theory. That theory in turn raised questions that led to research on the origin of variation, the laws of heredity, and the rates of evolutionary change. The hypotheses (and laws) invented to answer those questions have in turn led to dozens of other questions that have resulted in an enormous amount of valuable research.

By the third quarter of the nineteenth century, biologists had been convinced that they should limit their inquiry to questions that allow a physical answer. The results of this inquiry have been overwhelmingly successful, and it can hardly be surprising that today that bias is very strong among biological scientists.

Summary

1. Darwin's voyage in HMS *Beagle* provided him with a wealth of natural history data and started him thinking about a set of questions that led him to formulate the theory of evolution.
2. Darwin was especially struck with his data on fossils, biogeographical patterns, adaptations, and the relationship of island life to mainland life.
3. After his return to England, Darwin decided that he could explain many biological puzzles if he assumed that species change. He formulated the concept of natural selection to explain how species change.
4. Although his ideas were novel, they gave Darwin a key to understanding life on this planet.

Review Questions

1. What was the significance of Darwin's voyage?
2. What did Darwin find interesting about the fossils he saw on his trip?
3. What biogeographical patterns did Darwin observe?
4. What did adaptation mean in Darwin's day?
5. What struck Darwin about organisms living on islands?
6. How did the idea that species change answer the questions that Darwin was addressing?

7. Why is variation important for natural selection?
8. How does natural selection differ from domestic selection?
9. How did Darwin's ideas differ from Agassiz's?

Essay and Discussion Questions

1. It has often been pointed out that Louis Agassiz was familiar with all of the data that Darwin had and that he even later traveled and visited some of the locations that Darwin visited. Why was Agassiz unable to accept the theory of evolution?
2. In his day, Darwin was criticized for not providing "proof" of his theory. Was that a fair criticism? Why?
3. When Darwin proposed the theory of evolution, some naturalists indicated that it contradicted accepted laws in biology—for example, the law in anatomy stating that organisms were of such complexity that it was impossible for them to change. How might a scientist convinced of Darwin's ideas address this problem (a common one, it turns out, in the history of science)?

References and Recommended Reading

Appleman, P., ed. 1979. *Darwin*. New York: Norton.

Bowler, P. J. 1984. *Evolution: The History of an Idea*. Berkeley: University of California Press.

Kohn, D., ed. 1985. *The Darwinian Heritage*. Princeton, N.J.: Princeton University Press.

Larson, E. J. 1989. *Trial and Error: The American Controversy over Creation and Evolution*. Oxford: Oxford University Press.

Lurie, E. 1960. *Louis Agassiz: A Life in Science*. Chicago: University of Chicago Press.

Mayr, E. 1959. Agassiz, Darwin, and evolution. *Harvard Library Journal* 13:165–194.

Ruse, M. 1979. *The Darwinian Revolution*. Chicago: University of Chicago Press.

Sulloway, F. 1982. Darwin and his finches: The evolution of a legend. *Journal of the History of Biology* 5:1–53.

CHAPTER 23

The Modern Synthesis: Genetics and Natural Selection

Chapter Outline

The Genetics of Evolution

Forces of Change

Focus on Scientific Explanations: The Convergence of Ideas in Biology

Joint Action of Evolutionary Forces

Reading Questions

1. What is the modern synthesis?
2. What are the sources of variation in populations?
3. What is the significance of the Hardy-Weinberg law for evolution?
4. What type of selective processes affect the composition of gene pools?

By 1870, Darwin's **theory of evolution** had become widely accepted in the scientific community. It answered several critical questions and raised many new ones: If natural selection acted on small variations to create new species, then what regulates the inheritance of variations? What are the sources of variation? How is variation distributed in a population? For half a century after the general acceptance of the theory of evolution, much pioneering research addressed these and related questions. Results of this research led most scientists to believe that Darwin was correct about evolution in general but that he had overstated the role of natural selection. Research in the late nineteenth and early twentieth centuries focused on alternative mechanisms or processes that might also influence evolution. By the mid-1930s, new information and concepts derived from work done in the early part of this century had led several scientists to believe that a complete reformulation of the theory was in order. Ironically, that reformulation resulted in a return to Darwin's original perspective, emphasizing the role of natural selection.

The first attempt to articulate a reformulated theory was by Theodosius Dobzhansky in *Genetics and the Origin of Species* (1937), followed by Ernst Mayr's *Systematics and the Origin of Species* (1942), Julian Huxley's *Evolution: The Modern Synthesis* (1942), and George Gaylord Simpson's *Tempo and Mode in Evolution* (1944) (see Figure 23.1). All four books stressed the central role of natural selection and incorporated the results of new research in genetics. To distinguish the "new" Darwinian theory from Darwin's original theory, biologists call it the *modern synthesis*.

The **modern synthesis** is based on two assumptions: (1) that gradual evolutionary change can be explained by the action of natural selection on small genetic changes and recombination and (2) that the processes of species formation, **speciation**, and evolutionary change in taxonomic groups higher than species, **macroevolution**, are explainable in terms that are consistent with known genetic mechanisms. This chapter first considers the genetic changes that provide a source of variation on which natural selection operates and then explores some of the forces of change that cause populations to differ. The following two chapters describe speciation (Chapter 24) and larger-scale evolutionary change (Chapter 25).

The modern synthesis emphasizes the action of natural selection on small genetic changes and recombinations in explaining gradual evolutionary change and states that speciation and macroevolution can be explained in terms consistent with known genetic mechanisms.

Figure 23.1

The four chief architects of the modern synthesis: (A) Theodosius Dobzhansky, (B) Ernst Mayr, (C) Julian Huxley, and (D) George Gaylord Simpson.

A

B

C

D

THE GENETICS OF EVOLUTION

Genetic Variation

Darwin was acutely aware of variations in populations. He had, after all, examined over 10,000 barnacles and had observed the extensive variation in this invertebrate group. The modern synthesis has added to an appreciation of variations in the gross physical characteristics among individuals by providing an understanding of the basis of genetic variation in individuals and populations.

What are the sources of new variations? Hereditary mechanisms basically promote stability. The replication and translation of genetic messages result in offspring resembling parents in fundamentally important ways. However, chemical processes are not always absolutely reproducible. As described in Chapters 14, 15, and 16, the enormous chemical complexity of genetic systems results in many different kinds of errors or changes (inexact duplications, mutations, transposable elements, recombinations). These errors and changes are an important source of variation. Early supporters of the modern synthesis emphasized simple gene mutations—frameshift mutations and the more common point mutations.

A *point mutation* is a change in the sequence of nucleotide bases in DNA or RNA due to a base substitution or to the addition or deletion of a single base pair. These mutations can alter the amino acid sequence specified by the genetic code, with the result that the original message is modified. Altered sequences may have no serious effects, or they can cause a significant change in protein structure, as in the case of the base substitution that gives rise to sickle-cell anemia (see Chapter 18).

Frameshift mutations occur when the deletion or addition of a base pair shifts the reading of the triplet code along DNA by one unit, which ultimately leads to an entirely new amino acid sequence in a protein. Figure 23.2 illustrates point and frameshift mutations.

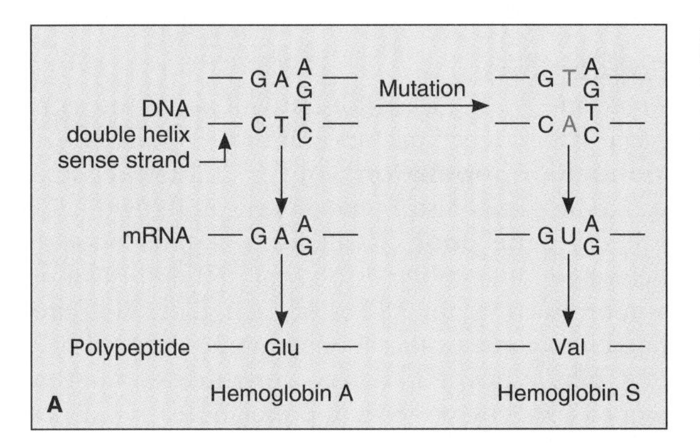

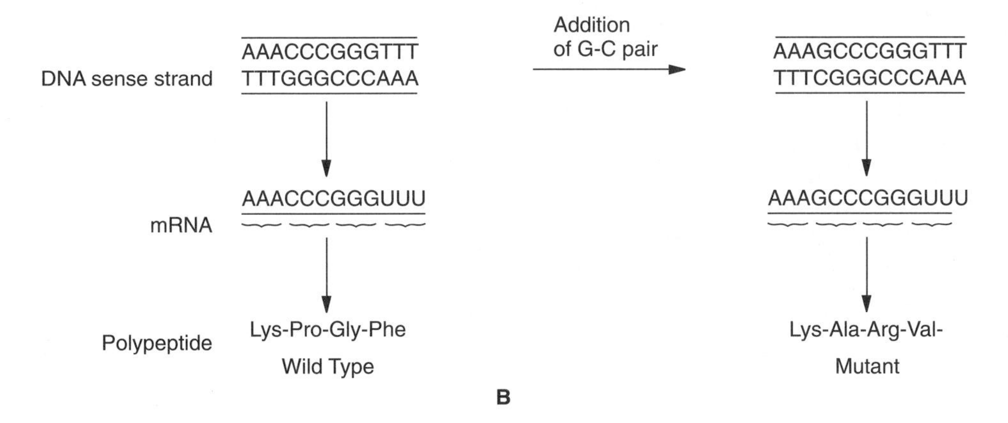

Figure 23.2

The modern synthesis initially stressed the two general forms of gene mutations, point mutations and frameshift mutations. (A) A base substitution (point mutation) is responsible for sickle-cell hemoglobin. (B) A frameshift mutation alters the amino acid sequence of a protein.

New genetic variation is not confined to gene mutations. Segments of DNA can "jump" or be transferred between chromosomes, and chromosomes can undergo a variety of changes during replication, which may result in deletions, alterations, or duplications of the genetic material. Although losses and alterations are usually harmful, duplication occasionally results in more robust forms. For example, in certain yeasts, chromosomal duplications have increased the capacity to produce particlar enzymes that improve their chance of survival in environments that are low in necessary nutrients.

Recall that duplication of whole chromosome sets, polyploidy, is common among plants (see Chapter 14). One third to one half of all flowering plants, including many familiar crop plants (for example, wheat, potatoes, and tobacco) and numerous common wildflowers, are thought to have originated through polyploidy. The changes that result from the loss, alteration, or duplication of chromosomes during replication, together with mutations, transpositions, and recombinations of genes, are important in generating variations in populations.

New genetic variation can arise by way of point mutation, frameshift mutation, transposition, recombination, and various structural changes in chromosomes during duplication. These changes provide the basis for variation in individuals.

Genetics of Populations

The study of genetics reveals a great potential for creating variation among the individuals of a population. Investigations of natural populations have confirmed that great genetic diversity does indeed exist in such populations. Attempts to understand the significance of this great storehouse of genetic differences within populations has raised fruitful questions. How are variable traits maintained, expressed, suppressed, or altered? The study of **population genetics**, which is concerned with the genetic constitution of populations and how they change, attempts to answer such questions. This field is of preeminent importance to the modern theory of evolution.

Populations

Populations are groups of individuals of the same species that live in the same place. Geneticists use several more specific terms to describe populations. A **Mendelian population** consists of interbreeding, sexually reproducing individuals. The genes distributed among all individuals of a population are collectively called a **gene pool**. All alleles in a gene pool are dispersed in individuals and determine the genotypes present in the population at any one time. Due to sexual reproduction, the alleles of each generation are sorted, shuffled, and incorporated into gametes that then combine in fertilization and lead to the formation of new genotypes. Population geneticists sometimes find it more convenient to emphasize gene pools than populations of individuals.

Genotype Frequencies: The Hardy-Weinberg Law

The essence of evolution is change, and population geneticists study how the frequency of genes and genotypes of populations changes. In this endeavor, they rely on an idealized model, just as physicists rely on ideal gas laws or models of perfectly spherical balls rolling down frictionless inclined planes. The model that serves as the touchstone in population genetics is called the **Hardy-Weinberg law** (see Figure 23.3). This law starts with an initially defined distribution of genotypes and then proceeds to predict the genotypic frequencies of succeeding generations. The law is based on the following assumptions: the population is very large; no mutation, selection, or migration occurs; and mating is random (that is, there is no selection of mates). In other words, the model describes a large, randomly mating Mendelian population that is not subjected to any selective pressures. By a simple mathematical calculation, this model predicts that

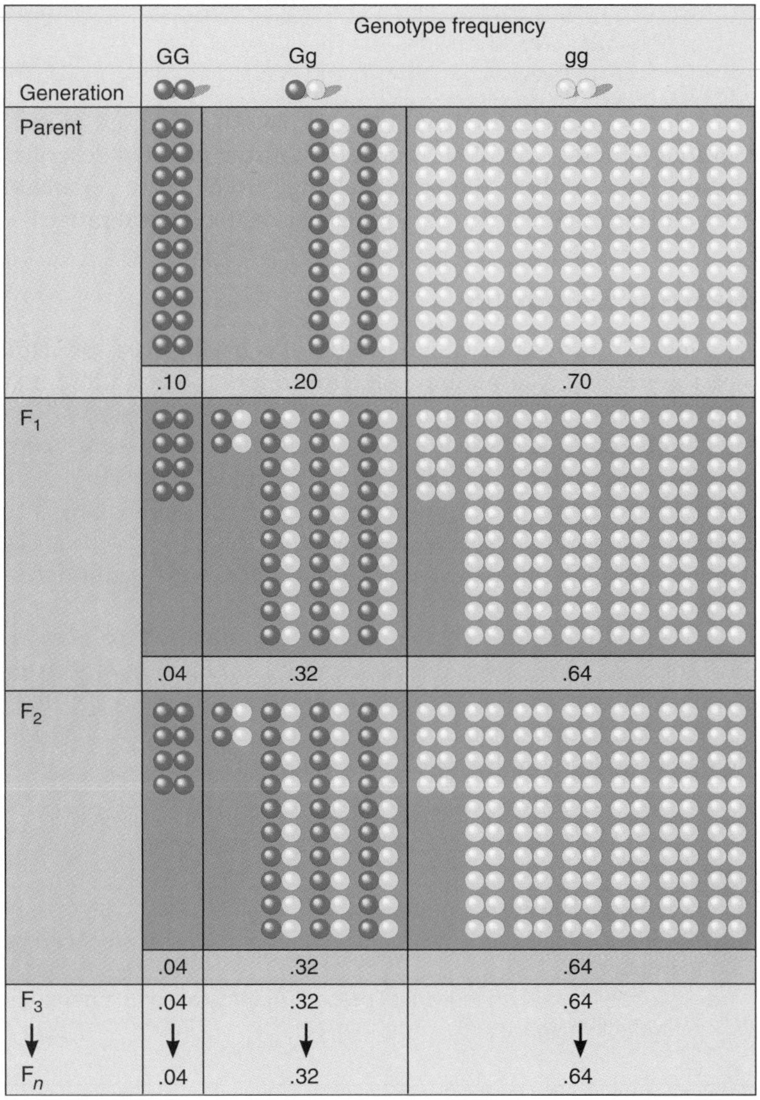

	Genotype frequency		
Generation	GG	Gg	gg
Parent	.10	.20	.70
F_1	.04	.32	.64
F_2	.04	.32	.64
F_3	.04	.32	.64
↓	↓	↓	↓
F_n	.04	.32	.64

Figure 23.3

The Hardy-Weinberg law predicts that in a large population where mating is random and no mutation, selection, or migration occurs, after one generation the frequencies of the genotypes will remain constant. In this illustration, the parent generation, shown in red, has genotype frequencies of .10 for GG, .20 for Gg, and .70 for gg. In one generation, the genotype frequencies become .04 for GG, .32 for Gg, and .64 for gg. All subsequent generations remain the same.

after a single generation, the frequency of each genotype will remain constant; that is, the frequencies will be in equilibrium.

This relationship was first described in nonmathematical terms in 1903 by the famous American geneticist William Castle, who was one of the first scientists to popularize the Mendelian laws in the United States. A few years later (1908), it was expressed independently in mathematical terms by G. H. Hardy and W. Weinberg, and since their writings called attention to the relationship, it is now called the Hardy-Weinberg law.

The Hardy-Weinberg law is useful because it defines the genetic stability in an idealized, undisturbed population. It thereby provides a standard of comparison with naturally occurring populations. By studying natural populations or experimental laboratory populations, which do change, biologists can compare them with an ideal, unchanging population and attempt to identify and analyze the factors that actually do cause changes in genotype frequencies.

The Hardy-Weinberg law provides a basis for understanding population changes. The law describes the genetic stability of an ideal large population in which no mutation, selection, or migration occurs and mating is random. Population geneticists use the law as a baseline for evaluating changes in genotype frequencies of natural populations.

FORCES OF CHANGE

Population geneticists recognize that four major, related forces of change can alter the frequencies in a population: natural selection, mutation, migration, and random genetic drift. The remainder of this chapter describes the operation of these forces and how they cause changes to occur in populations. In Chapter 24, we will learn that change in populations is the central process of evolution.

Natural Selection

The main thrust of Darwinian evolution is its emphasis on natural selection. In its most general sense, natural selection refers to the different survival and reproductive rates of beings that differ in one or more ways. Natural selection is a mechanical and statistical process that does not assume any sort of conscious selector, any value judgment concerning inherent worth, or any sort of predetermined goal. Natural selection can be the result of many different factors and, contrary to popular conceptions, can be discussed without reference to "a violent struggle for existence." It is also a concept that can refer to different kinds of objects and different levels of complexity.

As applied in the modern synthesis, **natural selection** is the differential reproductive success of genotypes, that is, their fitness. **Fitness** may be defined as the *net reproductive rate* (the average number of offspring produced by individuals) times the probability that the individuals will survive to reproductive age, as described in the following single equation:

$$w = n \times s$$

w = reproductive success (fitness)
n = net reproductive rate (average number of offspring produced per surviving individual)
s = probability that an individual present at the start of the generation will survive to reproductive age

The fitness of different genotypes in a population, that is, the **relative fitness**, can be expressed by dividing the net reproductive rate by the reproductive rate of the genotype with the greatest fitness. The following equation is used by population biologists:

$$F \frac{N_{ab}}{N_{xy}}$$

F_r = relative fitness
N_{ab} = net reproductive rate of genotype *ab*
N_{xy} = net reproductive rate of the genotype with the greatest fitness (*xy*)

Using this equation, 1 is the highest possible fitness (when $N_{ab} = N_{xy}$), and 0 is the lowest (when $ab = 0$).

Natural selection modifies the composition of genotypes in a population through an indirect process. Selection operates directly on phenotypes, the expression of genotypes in individual organisms. Since the phenotypic expression of genes is complex and often related to environmental factors, discussing selection of genotypes is correspondingly complicated. To get a grasp of the way geneticists study selection, we can consider a simple case involving natural selection operating on only two alleles at a single gene locus; then we can look at a more complicated example of natural selection creating an equilibrium involving more than one genotype.

Selection Against Deleterious Alleles

If an allele reduces the fitness of a phenotype, we might expect it to be selected against. For example, when an allele causes infertility or death before reproductive age, the fitness of the organism is 0. If the allele is recessive and has no effect on the fitness of heterozygotes, selection will operate only on recessive homozygotes. After several generations, the number of homozygous, recessive individuals will be reduced, but the allele, as shown in Figure 23.4, will never be eliminated from the population since some will always be conserved in the heterozygous state. For this reason, the ability to remove harmful or undesirable recessive alleles from a population (human, animal stock, or plant crop) is ultimately limited. If the harmful allele is dominant, however, it will be quickly eliminated since affected individuals will not reproduce. Nevertheless, new mutations may generate the allele at a low frequency in the population.

Balancing Selection

In contrast to reducing the frequency of or eliminating an allele, natural selection can also create a state of equilibrium involving more than one genotype. Such selection is called **balancing selection**, and the resulting existence within a population of two or more genotypes for a given trait is called **polymorphism**. For example, some heterozygous genotypes are more fit than either of the homozygous conditions; our discussion of sickle-cell anemia (see Chapter 18) was an example of balancing selection. Individuals homozygous for the allele are affected by a severe anemia that causes infant mortality. The heterozygotes have a milder form of the disorder but have a greater resistance to malaria, thereby increasing their fitness in environments where malaria is a health hazard. Figure 23.5 illustrates a different example of balancing selection, where different genotypes permit adaptation to different environmental conditions.

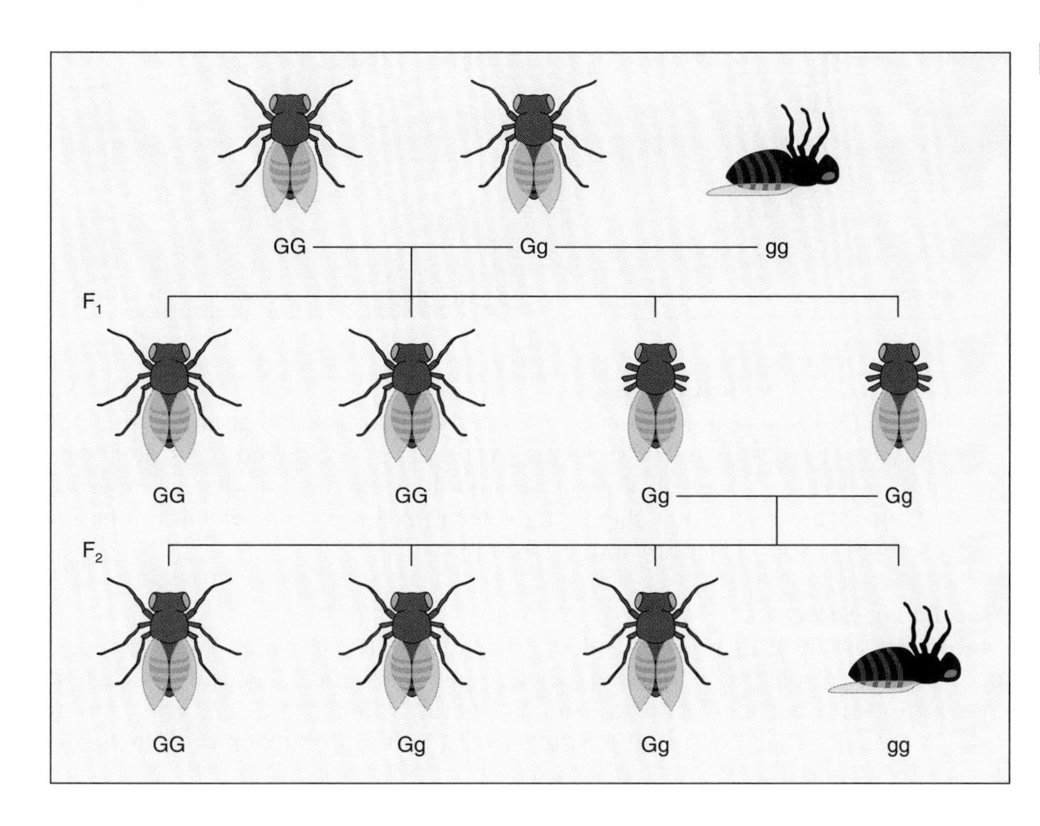

Figure 23.4

Selection against deleterious alleles may not lead to their elimination from a population if the allele is recessive and has no effect on the fitness of heterozygotes. As can be seen from this figure, the deleterious allele, g (which is lethal in the homozygous state), continues to appear in later generations because it has no deleterious effect on heterozygotes (Gg).

Figure 23.5

A classic example of balancing selection involves a land snail (*Cepaea nemoralis*) found near Oxford, England. This simplified illustration shows two forms of the snail's shell—the yellow, banded shell and the brown, unbanded shell. (A) Selection by thrushes in open grasslands favors the yellow snails, which are more difficult to distinguish than the darker brown ones. (B) Similarly, the darker shells are more difficult to distinguish among the leaves of the forest.

A

B

Other Patterns of Selection

The foregoing discussion of selection concentrates on the change in frequencies of genotypes. Several patterns of selection that are recognized in populations were discovered by studies of phenotypes and their interaction with the environment. These patterns include stabilizing selection, directional selection, disruptive selection, sexual selection, and coevolution.

Stabilizing Selection

Stabilizing selection refers to selection against extreme variants in a population, with the result that a "standard phenotype" is favored. This can occur when the relationship between an organism and its environment remains constant, at least in domains that are critical for the survival and reproduction of individuals.

Figure 23.6

An early example of stabilizing selection was recorded by Hermon C. Bumpus after observing English sparrows. In 1899, he published a classic paper on natural selection that described his research conducted during the previous year after "an uncommonly severe storm of snow, rain, and sleet." He studied a set of birds that was brought to the Anatomy Laboratory at Brown University after the storm. Some of the birds revived, and some died. Bumpus carefully measured and compared those that survived with those that perished. He observed that many of the birds with extreme dimensions (for example, the largest, smallest, heaviest, lightest, and those with the biggest and smallest heads) did not survive, whereas many with average dimensions did.

Stabilizing selection can be considered the usual state of affairs, in which a balance has been reached that promotes an optimum range of characteristics. Stabilizing selection does not, however, act to establish a set of uniformly homozygous genotypes.

Studies of variation in natural populations have shown that even where phenotypes are highly uniform, there are significant variations in the genotypes. Most populations possess a large reservoir of genetic variation. Some of the variation consists of small molecular variants that appear to be selectively neutral and cannot be detected in the phenotype. However, larger variations usually can be detected in populations, and they have potential importance for a population if, for example, the environment changes. This type of variation has been tapped by both scientists and breeders. They have altered the appearance of small groups of organisms by controlling their breeding and artificially selecting for certain desired characteristics that have scientific or commercial value, such as early ripening in tomatoes or exotic eye color in fruit flies. In nature, however, average individuals often appear to have a better chance of reproducing than those on the periphery of appearances. Almost a century ago, H. C. Bumpus illustrated this point by studying sparrows that died during a severe storm. He argued that surviving birds tended to be of average size, whereas those that died were larger or smaller (see Figure 23.6). Although more recent analyses of his data suggest that only female sparrows are subject to stabilizing selection for average size, Bumpus's classic study remains one of the standard examples of this pattern of selection.

Directional Selection

Directional selection shifts the mean (average) in a distribution of certain characteristics in response to a change in the environment. For example, if an area is becoming progressively cooler or drier, some traits, such as density of coat, behavior, or metabolic production of water, might be favored. Similarly, a population that migrates into a new environment that has recently undergone change (for example, due to glaciers or volcanic eruption) might be subjected to directional selection. One of the most famous examples of directional selection, the changes in the colors of peppered moths as a result of industrialization, belongs to this category (Figure 23.7).

Figure 23.7

Perhaps the most widely known example of directional selection, often used as a basic example of natural selection, concerns the peppered moth (*Biston betularia*), which has been carefully studied in England. These moths fly at night and pass the day resting on trees. (A) Until 1848, only the light form of the moth was known in England, and these moths were difficult to see against the background of lichen-encrusted trees. (B) During the nineteenth century, the environment in many parts of England was radically changed by industrialization. Smoke from factories killed the light-colored lichens and darkened the trunks of the trees that the moths rested on during the day. In 1848, a dark form of the moth was seen in Manchester, England, and the frequency of the dark form increased in polluted areas. Later studies showed that after installing new pollution controls, some areas have less pollution and, therefore, more light-colored lichens on trees. In those areas, the frequency of the light-colored moths has increased.

Manchester, 1850. There are six moths on this tree trunk. Which are better adapted to their environment the pale moths or the dark moths?

30 years later. There are six moths on this blackened tree trunk. Which are better adapted to their environment the pale moths or the dark ones?

Disruptive Selection

Disruptive selection (also called *diversifying selection*) acts to favor two or more traits simultaneously. This typically occurs when a population is subjected to separate selective pressures in different occupied areas. In Africa, for example, the Mocker swallowtail butterfly (*Papilio dardanus*) has a color pattern that mimics (that is, it has come to resemble another organism or object to which it is not related) other butterflies that have an offensive taste to their predators (see Figure 23.8). *Papilio dardanus,* however, is found over a wide range, and individuals mimic one of several different noxious butterflies! Populations of *Papilio* often have primarily one of the six patterns reflecting the distribution of the butterflies they mimic.

Another famous example, illustrated in Figure 23.9, involves certain bentgrass populations that grow in Wales (Great Britain) in areas that are heavily contaminated with toxic metals. Careful experimental tests have shown that the bentgrass growing on contaminated soil belongs to subpopulations that are genetically resistant to the contaminants, whereas the grass growing in adjacent

Figure 23.8

A well-known form of mimicry, called *Batesian mimicry*, occurs when a harmless species comes to resemble a poisonous, dangerous, or distasteful species. The Mocker swallowtail butterfly displays the result of disruptive selection and mimics a number of different noxious butterflies. (A) All the (1) male Mocker swallowtails are black and yellow. One type of female (2) resembles the male. Others (4) mimic distasteful species (5). One mimic (3) has a wing shape like the male but colors like the distasteful species. This may be a result of natural selection not having completed the change to mimic the distasteful species. (B) Mocker swallowtail females mimic numerous distasteful species.

uncontaminated areas has no resistance to metal toxicity. Since the resistant plants grow more slowly than nonresistant grasses, selective pressure operates on two traits: in contaminated soil, the resistant grasses are at an advantage; in uncontaminated soil, the faster-growing nonresistant plants are at an advantage.

Figure 23.9

In Wales, populations of bentgrass (*Agrostis tenuis*) in the neighborhoods of old mines show a striking pattern of tolerance to toxic heavy metals (lead, copper, and zinc). A. D. Bradshaw showed that populations of bentgrass growing on contaminated soil were tolerant to the metals and that a very sharp transition occurred over a few yards between tolerant and nontolerant species.

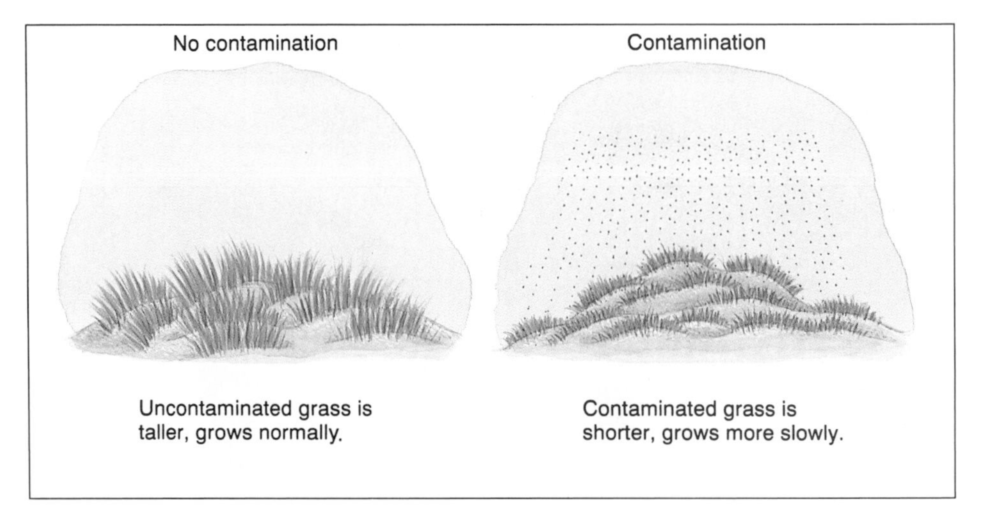

Sexual Selection

Sexual selection results in differences in the external appearance of males and females of the same species, a condition called **sexual dimorphism**. Often these differences involve sex-limited traits found in males that affect size, strength, or ornamentation. Darwin believed that a form of selection was operating among competing males, and he termed it *sexual selection*. Although Darwin treated sexual selection as a process supplementary to natural selection, the modern synthesis has included it as a special case of natural selection that explains characteristics that have a selective value affecting success in mating.

A commonly cited example is that of polygamous mammals, where the strongest male collects and guards a group of females or a territory containing females and drives off weaker males. A well-known case involves the red deer, *Cervus elaphus,* studied in Scotland. Red deer males are almost twice the size of females, and in the fall breeding season, they fight to gain control of groups of females (harems) (see Figure 23.10). The stronger, larger, and more agile males have greater reproductive success.

Finally, the striking plumage of numerous birds, like that of the peacock and the sage grouse (Figure 23.11), is thought to have resulted from the preference of females for males with those fancy feathers.

Figure 23.10

Red deer males fight during the fall breeding season to gain control of groups of females. This is an example of one type of sexual selection.

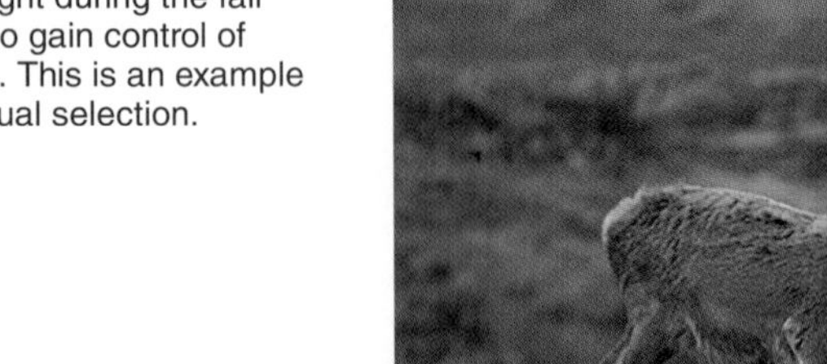

Figure 23.11

Sexual selection is thought to be responsible for some of the striking plumage of male birds, such as this peacock (A) and the sage grouse (B).

Coevolution

Coevolution occurs when two or more interacting populations of different species change mutually. The term *coevolution* was coined in 1964 by ecologists Paul Ehrlich and Peter Raven in a paper that discusses how various plants have evolved protective chemical compounds for defense against insects and how insects have subsequently evolved a tolerance of those compounds. Since that classic paper, biologists have applied the concept of coevolution to the evolution of predator-prey interactions, parasite and host relationships, and related animal-plant traits involved in pollination and seed dispersal. Two examples of coevolution are shown in Figure 23.12.

Natural selection, as applied in the modern synthesis, is the differential reproductive success of genotypes in a population. Five patterns of selection that have been identified are stabilizing selection, directional selection, disruptive selection, sexual selection, and coevolution. Each of these operates on populations and results in changes in genotypic frequencies that are reflected by alterations in observed phenotypic ratios.

Natural Selection and Adaptation

Natural selection is a complex process that involves gene fitness, gene interaction, and the interaction of organisms with their environment, including other organisms. Unlike the other forces of change that will be discussed in this chapter, natural selection not only results in changing the gene pool, or genotype frequencies in populations, but also promotes *adaptation,* a change that results in a structural, behavioral, or functional trait that promotes survival and reproduction.

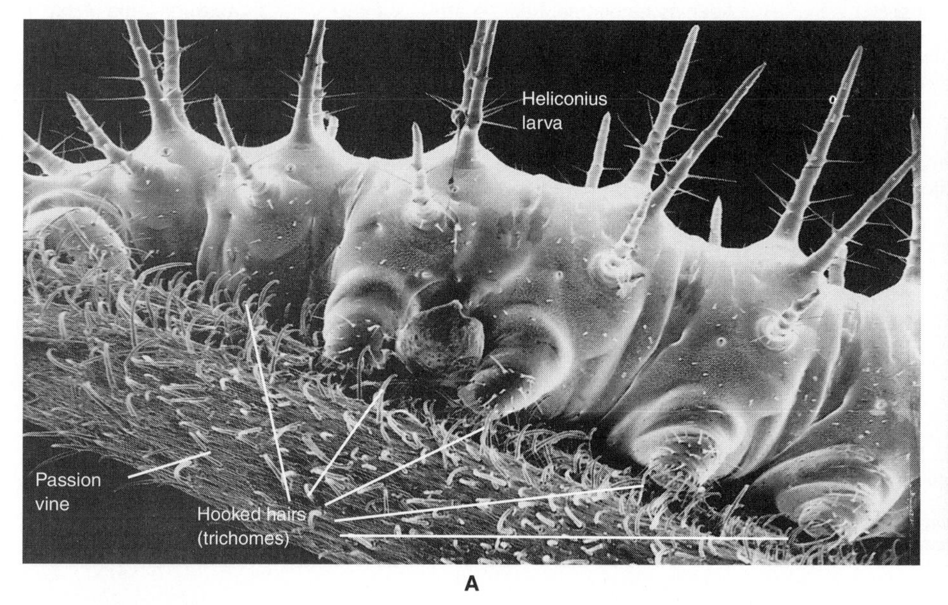

Figure 23.12

Coevolution involves mutual changes in interacting populations of different species. (A) *Heliconius* butterfly larvae feed on passion vines. They are among the few insects that can do this because passion vines contain chemicals that protect them against herbivores. Scientists believe that *Heliconius* butterflies evolved adaptations that permit them to feed on these plants. But this story continues—some species of passion vines have evolved hooked hairs (trichomes) that protect them from the *Heliconius* larvae, as this photo shows. (B) Acacia trees and ants of the genus *Pseudomyrmex* have evolved a mutual relationship that benefits both. The ants live in a tree's hollow thorns and feed on tree products that are rich in oil and protein. The ants, in turn, protect the tree from herbivores and competing plants.

It accomplishes this because individuals with certain variations are more likely to survive and leave offspring than those without them. Similarly, individuals with variations that afford them the opportunity to exploit new food sources, new defenses, or new life strategies will be selected. For this reason, modern evolutionary theory emphasizes the central role of natural selection in accounting for adaptation.

Natural selection alone does not determine how a population changes. Other processes and effects, although not directly adaptive, influence population changes. These include mutation, migration, random genetic drift, founder effects, and bottleneck effects.

Mutation

Mutation and the various causes of genetic variation were discussed earlier in this chapter (see also Chapters 14–18). Significant gene mutations are rare events, as can be seen in Table 23.1. Although mutations can be caused by various environmental factors such as ionizing radiation, ultraviolet light, heat, and chemical mutagens, mutation rates are very low, and changes associated with them do not occur frequently.

Mutations that result in a major change are usually harmful, but occasionally they increase fitness. Are these "beneficial" mutations a response by the organism to its environment, or are mutations random; that is, do they occur independently of their usefulness?

Whether or not mutations occur randomly has been a hotly debated issue

Table 23.1 Mutation Rates in Various Organisms

Organism	Mutation	Value	Units
Bacteriophage T2 (bacterial virus)	Lysis inhibition $r \rightarrow r^+$	1×10^{-8}	*Rate:* mutant genes per gene replication
	Host range $h^+ \rightarrow h$	3×10^{-9}	
Escherichia coli (bacterium)	Lactose fermentation $lac \rightarrow lac^+$	2×10^{-7}	*Rate:* mutant cells per cell division
	Histidine requirement $his^- \rightarrow his^+$	4×10^{-8}	
	$his^+ \rightarrow his^-$	2×10^{-6}	
Chlamydomonas reinhardi (alga)	Streptomycin sensitivity $str\text{-}s \rightarrow str\text{-}r$	1×10^{-6}	
Neurospora crassa (fungus)	Inositol requirement $inos^- \rightarrow inos^+$	8×10^{-8}	*Frequency* per asexual spore
	Adenine requirement $ad^- \rightarrow ad^+$	4×10^{-8}	
Drosophila melanogaster (fruit fly)	Eye color $W \rightarrow w$	4×10^{-5}	*Frequency* per gamete
Mouse	Dilution $D \rightarrow d$	3×10^{-5}	
Human (autosomal dominant)	Huntington disease	0.1×10^{-5}	
	Nail-patella syndrome	0.2×10^{-5}	
	Epiloia (predisposition to brain tumor)	$0.4\text{–}0.8 \times 10^{-5}$	
	Multiple polyposis of large intestine	$1\text{–}3 \times 10^{-5}$	
	Achondroplasia (dwarfism)	$4\text{–}12 \times 10^{-5}$	
	Neurofibromatosis (predisposition to tumors of nervous system)	$3\text{–}25 \times 10^{-5}$	

Mutation rates are extremely low.

Source: After Sager and Ryan (1961).

in the past. Many of the foremost American evolutionists of the nineteenth century believed that organisms could respond directly to environmental change either through "preadapted" or even "willed" mutations that increased their chances of surviving and leaving offspring. Many experiments have been performed to test the existence of such responses. For example, in 1960, Jack Bennett reported data from studies on *Drosophila* resistance to the pesticide DDT (see Figure 23.13). In one experiment, he selected siblings of the most resistant flies for many generations; that is, from each generation, the flies selected for breeding had *not* been exposed to DDT, but some of their siblings, used in exposure studies, had shown some resistance. By the end of 15 generations, he had created a strain of DDT-resistant flies that was descended from flies that had not been exposed to DDT. He concluded that mutations related to DDT resistance must have been present in the original population and that by selection he had isolated some of their carriers, rather than that mutations had arisen by some response to exposure on the part of the flies. Experiments like Bennett's have convinced biologists that mutations arise randomly.

Figure 23.13

Jack Bennett performed experiments on preadaptation of DDT resistance in *Drosophila melanogaster*. In the experiment shown here, a control group was later tested for DDT resistance against an experimental group. The experimental group consisted of an initial set of 46 pairs of *Drosophila* that were placed in bottles containing food and allowed to reproduce. The offspring were tested for DDT resistance by removing a female from the offspring and exposing her to DDT. Bennett selected the families that were most resistant and least resistant and allowed them to reproduce. Continual selection produced strains of flies that were more resistant and less resistant to DDT when compared to the control group.

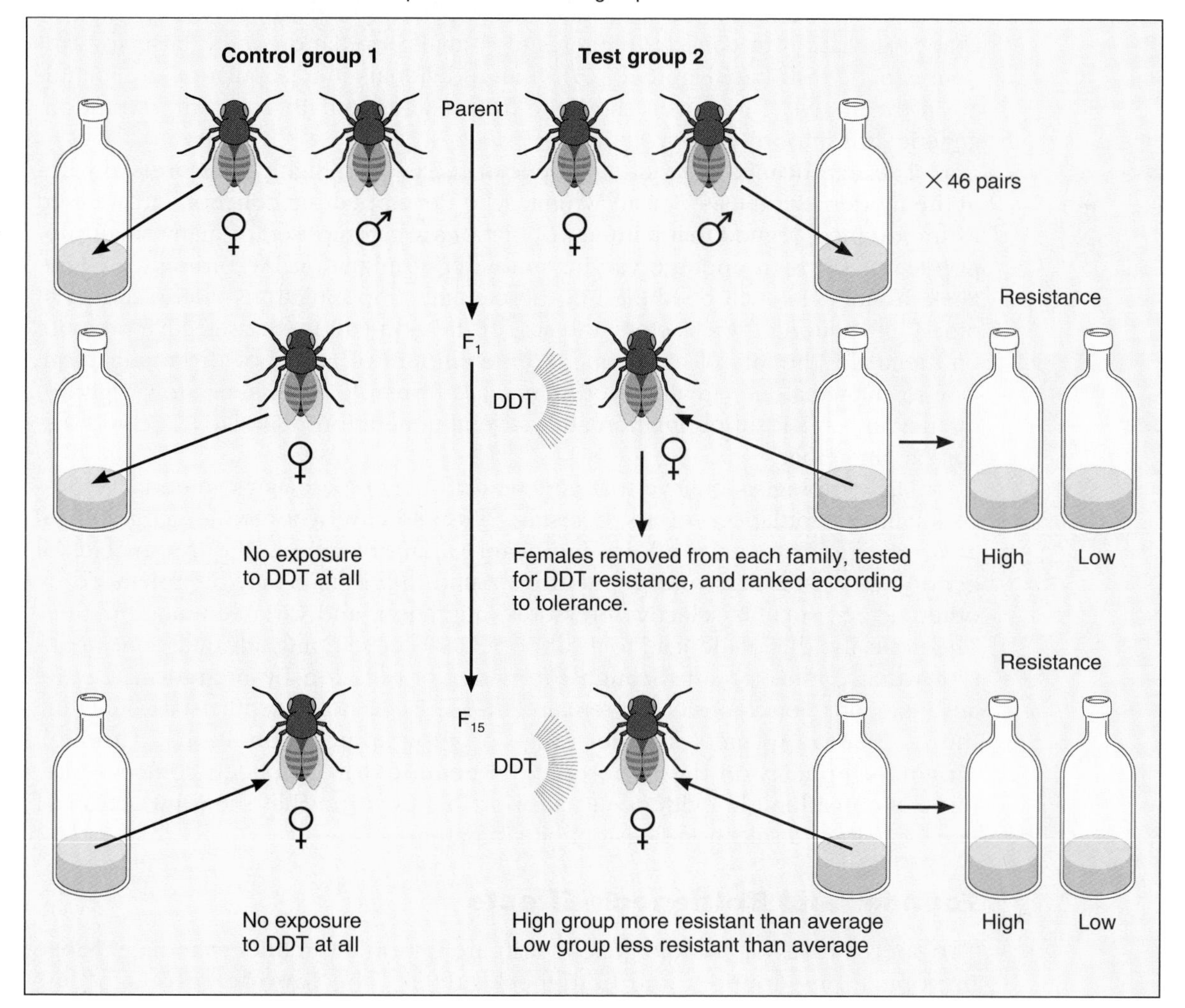

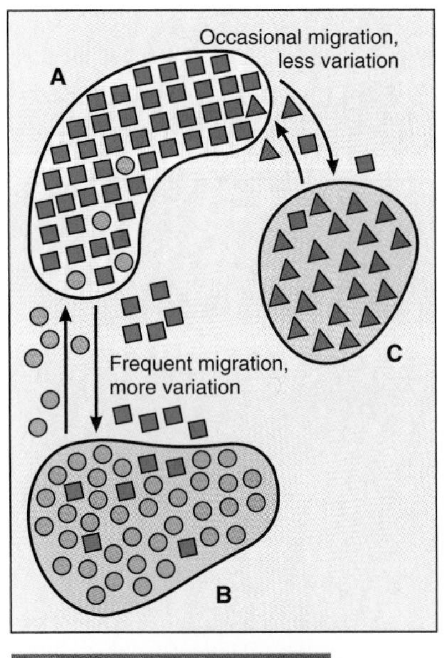

Figure 23.14

Gene flow (migration) can vary among populations. In this illustration, population A has frequent exchanges with population B, resulting in more variation for both populations. In contrast, population A has infrequent exchanges with population C, resulting in less variation for both populations. There is no gene flow between populations B and C.

Thus, although mutations arise randomly, the value of a mutation depends entirely on the environment. What may be neutral or deleterious in one setting can be advantageous in another.

Migration

When considering the genetic composition of a species, we often treat it as an isolated entity. In nature, however, species consist of many breeding populations of varying size and proximity. These groups are usually not genetically isolated; that is, some individuals interbreed with individuals from different populations. **Migration** (also called *gene flow*) is the process that describes an exchange of genes among populations (see Figure 23.14).

Migration is an important factor in evolution, for the greater the extent of migration, the greater the genetic uniformity of various breeding populations. Migration also extends successful genetic combinations and increases the genetic variability of semi-isolated populations. Alternatively, restriction of migration can lead to the formation of different races or even more significant changes that will be discussed in Chapter 24.

Random Genetic Drift

Random genetic drift refers to chance fluctuations in gene frequencies in a small population and is of importance equal to or greater than migration. In a large population, the Hardy-Weinberg law predicts that the frequency of an allele that is not subject to selective pressure will remain constant. However, this unchanging frequency actually represents a mean. If a small subset of reproducing individuals is examined, deviations from the average will be observed. This chance deviation in gene frequencies in a small population is the basis for random genetic drift (see Figure 23.15). If all individuals in a population are included in the analysis, however, the gene frequencies remain constant. Is random genetic drift important in evolution?

The significance of random genetic drift has been highly controversial within the modern synthesis. Sewall Wright, who developed the concept and stressed its importance, argued that natural populations are composed of many small subpopulations, which undergo varying amounts of migration. According to Wright, allele frequencies will be altered in these small subpopulations due to the combined interaction of four factors: the size of the population, the selective pressure on the allele, the rate of mutation, and the amount of gene flow from migration. Due to chance alone, an allele could increase from a low frequency to a high frequency in a small subpopulation after several generations. Similarly, it could disappear altogether.

Many biologists believe that genetic drift may be a major evolutionary process. Since populations of a species are often spread over a wide geographical area, genetic drift can result in significant variations among the frequencies of genotypes within these subpopulations. Such differences may be enhanced by other factors, such as selective pressures, migration, and variable mutation rates. Theoretically, this could lead to a subpopulation developing advantageous genotypes that could spread through the wider population by migration. Or the advantageous genotypes could remain in a balanced polymorphic state, allowing the population to meet a wider range of geographical conditions. Although Wright's emphasis on the importance of genetic drift has been controversial, there is considerable evidence to support his position that the joint action of genetic drift and selection does operate in nature.

Founder and Bottleneck Effects

Two special cases of random genetic drift are important in understanding change in populations: founder and bottleneck effects. The **founder effect** occurs

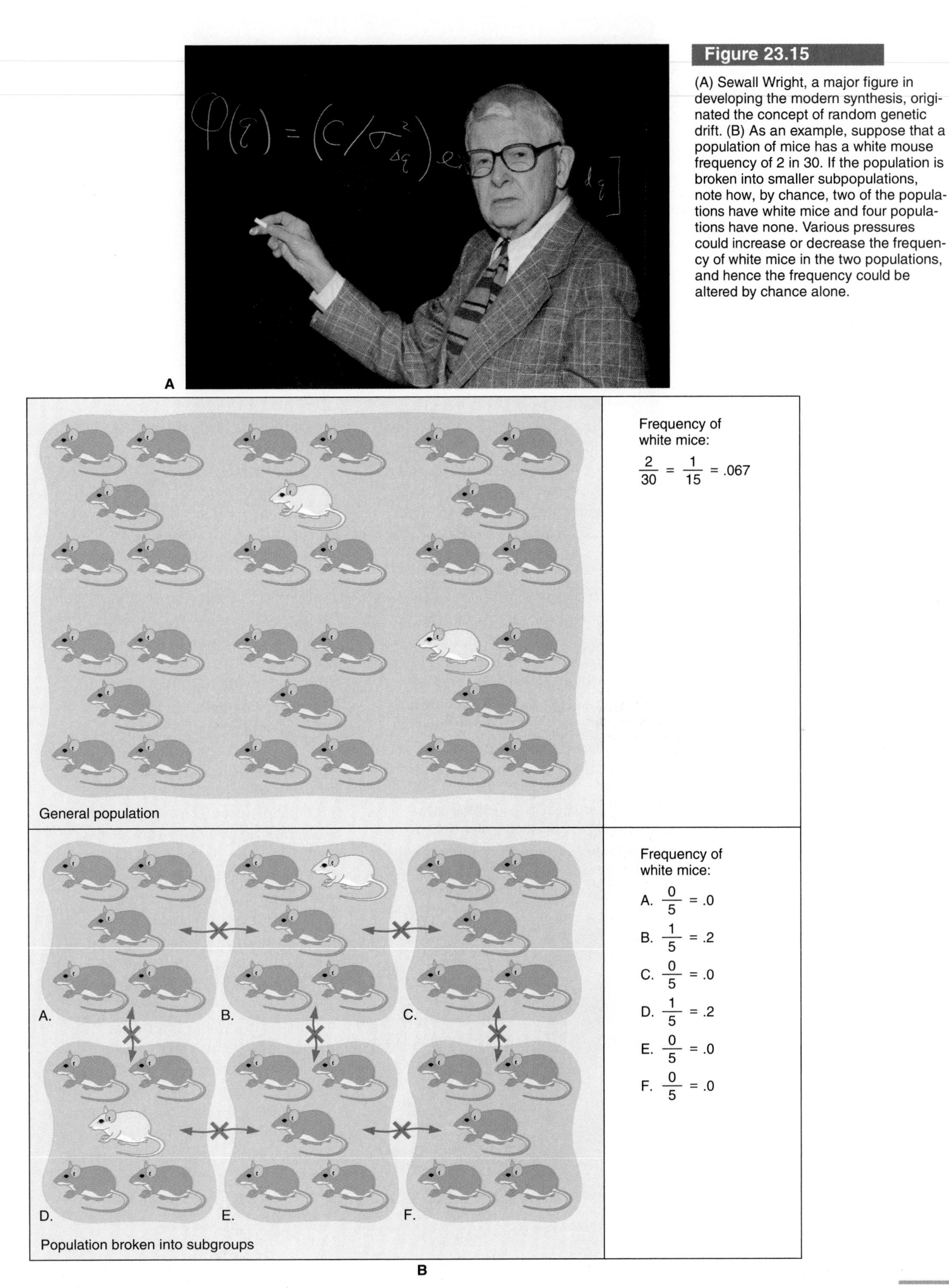

Figure 23.15

(A) Sewall Wright, a major figure in developing the modern synthesis, originated the concept of random genetic drift. (B) As an example, suppose that a population of mice has a white mouse frequency of 2 in 30. If the population is broken into smaller subpopulations, note how, by chance, two of the populations have white mice and four populations have none. Various pressures could increase or decrease the frequency of white mice in the two populations, and hence the frequency could be altered by chance alone.

Focus on Scientific Explanations

The Convergence of Ideas in Biology

The history of science provides numerous examples of main ideas that were formulated at approximately the same time by different people. In the seventeenth century, Isaac Newton and Gottfried Leibniz, two of the greatest mathematicians of their day, each invented calculus at about the same time. They fought a bitter battle over who had accomplished it "first." The history of biology also has its share of what is often called "simultaneous discovery."

In 1900, three scientists independently formulated genetic principles that Gregor Mendel had published in 1866 (see Chapter 12). After Darwin had formulated his theory of evolution, but before he had published it, he received an essay from Alfred Russel Wallace, a younger naturalist with whom he occasionally corresponded (see Figure 1). In the essay, Wallace sketched out almost exactly the theory that Darwin had constructed! Subsequently, Darwin had Wallace's essay published along with some of his own writings, so Wallace formally received equal credit for inventing the idea of evolution by natural selection.

Figure 1 Alfred Russel Wallace was coinventor of the theory of evolution.

However, since it was Darwin's book, *On the Origin of Species,* that convinced the scientific world that evolution had occurred, we hear little about Wallace today, although he was a very productive scientist and did much to support Darwin.

Several formulations of the modern synthesis were written within a short period of time. Although not as startling, perhaps, as the recovery of Mendel's laws or the independent description of the theory of evolution, it raises interesting questions: What are simultaneous discoveries, and why do they occur?

A possible explanation, and one that was popular a few decades ago, is that great ideas are "in the air" at particular times. Not like germs, but like pieces of a puzzle, the individual parts of an explanation are available, but it is necessary for someone to assemble them in the correct order. This explanation for simultaneous discovery places a high value on the earlier scientists who contributed individual pieces of the puzzle or who had described rough approximations of the completed version. Such people are called *precursors*. In Darwin's case, for instance, a likely precursor is Georges-Louis Leclerc, comte de Buffon (Figure 2A), the famous eighteenth-century French naturalist who believed in a limited form of change in plants and animals as a response to their environment. Jean-Baptiste-Pierre-Antoine de Monet de Lamarck (Figure 2B), a protégé of Buffon's, later speculated that species change and that modern species are the descendants of earlier ones.

Unfortunately, when scholars seriously analyze the writings of precursors, they usually discover that the ideas of the precursor were significantly different from

A B

Figure 2 Georges-Louis Leclerc, comte de Buffon (A), and Jean-Baptiste-Pierre-Antoine de Monet de Lamarck (B) were allegedly Darwin's precursors.

those they are supposed to have foreshadowed. Their ideas, though often very interesting, are most important for understanding the science *of their own time* rather than that of a later period. Equally problematic, when historians try to link major scientific figures with their precursors, it turns out that connections do not exist. Darwin read Lamarck and thought that his ideas were nonsense! Later, it was a source of considerable embarrassment and annoyance to Darwin that some of his conclusions were similar to those of the earlier French writer. The search for precursors usually turns out to be a sterile exercise that obscures more than it illuminates. This is especially the case when the alleged precursor is a distant or unknown figure.

A more profitable approach to understanding why simultaneous discoveries occur lies in perceiving that much scientific investigation occurs because of an enthusiasm for answering what are considered interesting or significant questions. That is, focusing on the *problem* being addressed rather than the answer that was given leads to a better understanding of the origin of ideas. Scientists, even if they work alone, rarely work in isolation. They work on problems passed down from earlier researchers or topics that have come to be regarded as especially engaging by their contemporaries. Often these activities are aimed at explaining a certain set of facts from a shared perspective, such as the search for a particulate theory of genetics. Thus it is not surprising that occasionally more than one person is involved in a simultaneous discovery.

In the case of Darwin and Wallace, both naturalists had traveled to South America and the Pacific, each was fascinated by the relationship between fossils and living forms, each was curious about the underlying causes of biogeographical distribution patterns, and both were struck by the enormous variation in nature. Finally, each wondered, in light of island life that was obviously related to, but distinct from, mainland life, how new species had come into existence. Darwin and Wallace were also interested in explaining these questions without reference to supernatural causes. Each realized that many of their questions could be solved if they assumed that species changed. But how did they change? Both had read Malthus, and it was in thinking about Malthus's discussion of the growth of unrestricted populations—which Malthus in no way connected with the idea of change of species—that each was led to the idea of natural selection.

The formulation of the modern synthesis also came about as a result of different scientists attempting to answer certain questions by synthesizing copious data that had become available through the enormous amount of research that had been conducted on evolution. Recall that Darwin, by 1870, had convinced the scientific community that evolution of life had taken place and that it was the key to understanding life on Earth. He had not, however, convinced the scientific community that natural selection was the chief factor responsible for evolution. During the next 50 years, scientists struggled to construct an alternative. Some, like the famous American paleontologist Edward Cope (Figure 3), thought that there must be

Figure 3 Edward Drinker Cope was one of America's foremost paleontologists.

a cosmic force driving the evolutionary process. Others, like the pioneering ichthyologist (a scientist who studies fish)—and later president of Stanford University—David Starr Jordan (Figure 4), thought that the solution lay in an understanding of how the environment affects species. At the turn of the twentieth century, a number of publications claimed that Darwinism—the belief that natural selection was the primary agent of species change—was dead. The most promising alternative in the early years of the century was proposed by Hugo de Vries. He claimed that he had experimental evidence indicating that individual plants occasionally mutate and give rise to individuals of a new species. This theory of a species appearing over a single generation did not last, but his research on genetics did lead to the recovery of Mendel's laws (see Chapter 12).

Biologists may have been depressed by the confusion surrounding the theory of evolution at the beginning of the twentieth century, but they were very excited about its chances for explaining the living world. The expansion of biology into university and college education and growing government support signaled a new age for the biological sciences. At that time, new disciplines such as genetics, cytology, embryology, and biochemistry were generating vast quantities of information that many thought was relevant to a broad, evolutionary understanding of life. Intensive research in traditional fields such as biogeography, anatomy, and taxonomy added even more information.

By the 1920s and 1930s, many scientists felt that the time had come to apply the body of new information to the theory of evolution. But how? The individuals mentioned in this chapter—Dobzhansky, Mayr, Huxley, and Simpson—believed that Darwin had it right to begin with; what was needed was to return to his original insight and combine natural selection with the new results from genetics, population biology, biogeography, and the other areas of research. Not everyone agreed. Nevertheless, the four books written by the leaders of what came to be called the modern synthesis convinced the biological community and served as a guiding vision for research during the past half century. All four attempted a similar task: to survey all that was known in biology from the perspective of natural selection. Their highly successful accomplishment came to be known as the modern synthesis because it synthesized our knowledge of biology from a Darwinian, evolutionary perspective. It was a powerful theory because in addition to explaining vast amounts of data, it pointed to interesting research questions. It was a theory that solved the problem of explaining the basic mechanism of evolution in a way that was consistent with the known facts of biology.

Simultaneous discovery can be understood, then, by determining the questions driving research at the time and examining the assumptions made in attempting to resolve them. Examining the common background of a problem renders its ultimate solution a little less mysterious.

Figure 4 David Starr Jordan was a leading American ichthyologist and president of Stanford University.

when a small group of individuals establishes a colony. Since the founders are only a small subset of the larger population, they may be genetically unrepresentative of the population. A certain allele may be absent in the founding population, or its frequency may be so low that it disappears before the population can increase in size.

A British study on the former inhabitants of a small island illustrates the founder effect. In 1816, British soldiers established a garrison on Tristan da Cunha, a small, isolated volcanic island in the South Atlantic midway between South Africa and Brazil. When the garrison left, one soldier and his family stayed, and they were later joined by a few other settlers. The population soon grew to 100, straining the resources of the island, and some inhabitants left. By 1961, the population had reached 267, when a volcanic eruption forced the evacuation of all island inhabitants to England.

A British geneticist, D. F. Roberts, studied these individuals and was able to chart a reproductive history of the island population. He found that only ten of the original adult settlers had living descendants in 1961 and that of the ten, two had contributed twice as much to the gene pool as the other eight. The island population in 1961 was found to have an atypical genetic composition compared to the general European population from which it was derived. For example, there were higher frequencies of genetic disorders, such as retinitis pigmentosa and genetically linked mental retardation.

A second, and similar, phenomenon, the **bottleneck effect**, occurs in populations that experience severe periodic fluctuations in size. In such circumstances, when the population is at its smallest size, random changes in gene frequency can occur. When the population then attains a greater size, the altered gene frequencies may be retained. This bottleneck effect can radically alter the gene frequencies of a large population in a relatively short period of time. Figure 23.16 illustrates the founder effect and bottleneck effects.

In addition to natural selection, the major forces of change in gene frequencies and the genotypic frequency of a population are mutation, migration, and random genetic drift. Natural selection differs from these other forces of change because it alone promotes adaptation.

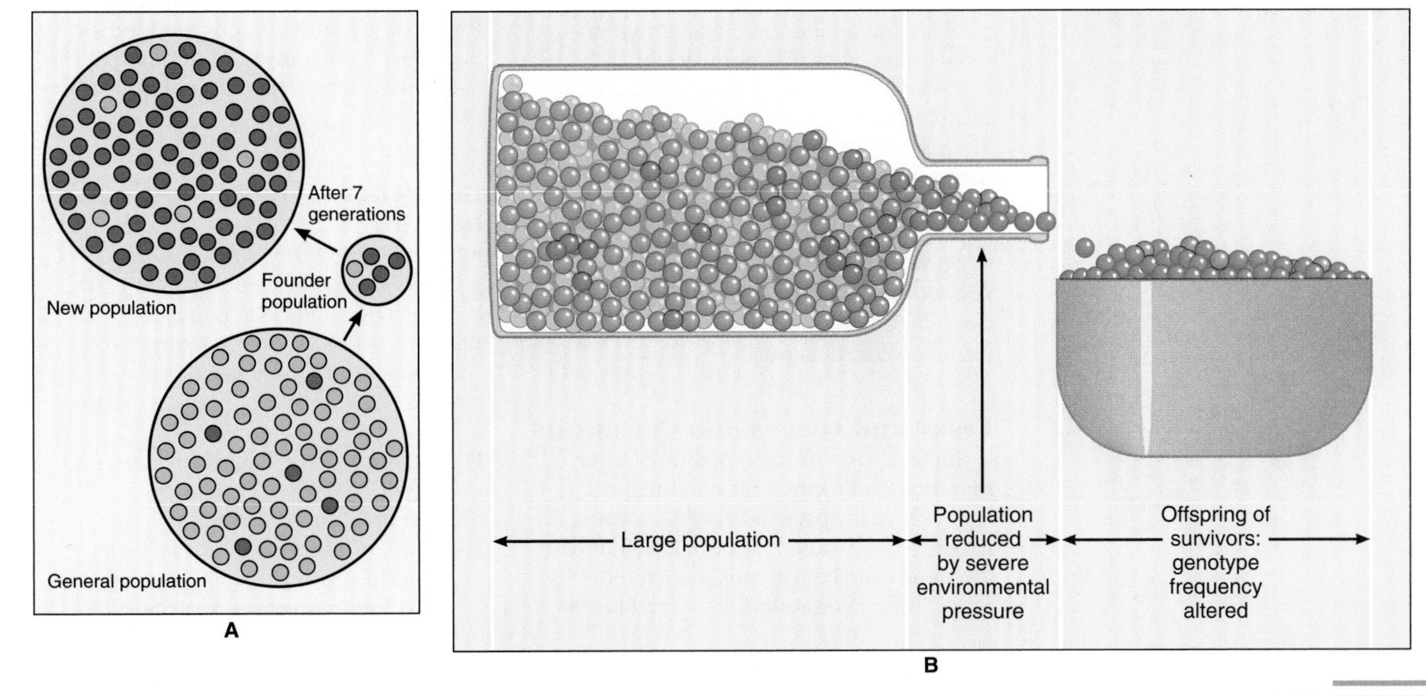

Figure 23.16

(A) A small subset of founders break off from a general population. Because, by chance, this subset is not genetically representative, the new population differs genetically from the original general population. (B) Similarly, when a population experiences a severe fluctuation in size, the survivors may, by chance, be unrepresentative of the original large population. When the offspring of the survivors reproduce, the resulting population may be different than the original large population.

JOINT ACTION OF EVOLUTIONARY FORCES

In this chapter we have examined several processes that influence the genetic composition of a population: natural selection, mutation, migration, and random genetic drift. These forces act jointly to cause population changes. Proponents of the modern synthesis stress the importance of natural selection as the main agent of evolutionary change, for it alone promotes adaptation (see the Focus on Scientific Explanations, "The Convergence of Ideas in Biology"). What are the significance and the result of population change? The next two chapters will focus on the consequences of genetic change in populations.

Summary

1. The modern synthesis states that gradual evolutionary change can be explained by the action of natural selection on small genetic changes and recombination and that the processes of species formation and of evolutionary change in groups higher than species are explicable consistent with known genetic mechanisms.
2. The study of genetics reveals a great potential for creating heritable variations among individuals of a population. Population genetics is the study of gene and genotypic frequencies in populations and how they change. The touchstone of population genetics is the Hardy-Weinberg law. This law predicts that in a large, randomly mating Mendelian population not subject to any selective pressures, after a single generation the frequency of genotypes will be in equilibrium.
3. Four related forces of change that alter the genetic composition of a population are natural selection, mutation, migration, and random genetic drift. These forces act jointly to change gene and genotypic frequency in populations.
4. Often-cited patterns of selection are stabilizing selection, directional selection, disruptive selection, sexual selection, and coevolution.

Review Questions

1. What basic assumptions does the modern synthesis make?
2. What is the study of population genetics?
3. What is the Hardy-Weinberg law?
4. What assumptions does the Hardy-Weinberg law make?
5. What forces alter the gene and genotypic frequencies of a population?
6. Describe the concepts of fitness and relative fitness.
7. Describe five patterns of selection and their effects on a population.
8. How does coevolution differ from other patterns of selection described?
9. How do mutation, migration, random genetic drift, and the founder and bottleneck effects influence the genetic composition of populations?

Essay and Discussion Questions

1. Imagine that you and several of your friends were dropped off on a planet somewhere in space to establish a colony. In 500 years, would that population have the same characteristics as the United States population today? How might it differ? How might the genotypic and phenotypic features of the population differ from the parent population?
2. Why might farmers who tried to develop the "perfect" cattle stock in the nineteenth century have become discouraged?
3. "The theory of evolution with its emphasis on the survival of the fittest has lowered the moral fiber of modern society." Based on your understanding of the modern synthesis, comment on this statement.

References and Recommended Reading

Ayala, F. J. 1982. *Population and Evolutionary Genetics*. Menlo Park, Calif.: Benjamin/Cummings.

Bennett, J. 1960. A comparison of selective methods and a test of the preadaptation hypothesis. *Heredity* 15:65–77.

Dobzhansky, T. 1937. *Genetics and the Origin of Species*. New York: Columbia University Press.

Ehrlich, P. R., and P. H. Raven. 1964. Butterflies and plants: A study in coevolution. *Evolution* 18:586–608.

Futuyma, D. 1986. *Evolutionary Biology*. Sunderland, Mass.: Sinauer.

Gilbert, L. 1971. Butterfly-plant coevolution: Has *Pasciflora adenopoda* won the selectional race with Heliconiine butterflies? *Science* 172:585–586.

Huxley, J. S. 1942. *Evolution: The Modern Synthesis*. London: Allen & Unwin.

Mayr, E. 1942. *Systematics and the Origin of Species*. New York: Columbia University Press.

———, and W. B., Provine, eds. 1980. *The Evolutionary Synthesis: Perspectives on the Unification of Biology*. Cambridge, Mass.: Harvard University Press.

Mettler, L. E., T. G. Gregg, and H. E. Schaffer. 1988. *Population Genetics and Evolution*. Englewood Cliffs, N.J.: Prentice-Hall.

Provine, W. B. 1986. *Sewall Wright and Evolutionary Biology*. Chicago: University of Chicago Press.

Sager, R., and F. J. Ryan. 1961. *Cell Heredity*. New York: Wiley.

Simpson, G. G. 1942. *Tempo and Mode in Evolution*. New York: Columbia University Press.

CHAPTER 24

The Origin of Species

Chapter Outline

Species

Speciation

Reproductive Isolating Mechanisms

The Evolution of Species

Reading Questions

1. What is a species?
2. How does speciation occur?
3. What are reproductive isolating mechanisms?
4. What is the significance of reproductive isolating mechanisms?

The number of species that have now been identified is just under 1.4 million. E. O. Wilson, one of many biologists who is worried about the alarming rate at which humans are causing species to become extinct, believes that the total number of species that inhabit the globe may be as high as 30 million. Although that figure is astonishingly high, it is relatively small compared to the number of species of plants and animals that paleontologists believe have lived on Earth at some time in the geological past, a number they estimate to be 4 billion.

How did all of these species originate? In Darwin's youth, there were far fewer species known than now, and a widely accepted idea was that God had created each species, perfectly adapted to its environment. This view was more plausible when the number of known species was only in the thousands. However, the swelling number of identified species was beginning to strain the idea that God had created each species individually. Darwin's observations in South America gave him firsthand evidence that a vast number of different plants and animals inhabited the globe. Further, his examination of the fossil record showed him that the number of extinct forms was even greater than the number that were living. The fossil record also revealed that plants and animals in the past were very different from those now inhabiting the planet and that most present species have not been around for long.

Darwin referred to the appearance of new species as "that mystery of mysteries," and, as was described in Chapter 22, he spent many years trying to understand from a scientific perspective how new species come into being. Later scientists have been equally fascinated with the processes that lead to the production of new species, and our current understanding of those mechanisms is at the center of the modern theory of evolution.

Chapter 23 reviewed the forces of change that are responsible for altering populations. Changes within species' populations are called **microevolution**. This chapter will concentrate on how species change, and Chapter 25 will examine the question of evolution above the species level.

SPECIES

In the 1940s, Ernst Mayr proposed the most widely accepted definition of *species* from the perspective of the modern synthesis: *Species are groups of actually or potentially interbreeding natural populations that are reproductively isolated from other such groups.* This definition stresses reproductive isolation. A species can occupy a wide geographical area, which allows subpopulations to be subjected to different selection pressures or to other forces of change such as genetic drift. In time, some of these subpopulations may acquire different genetic compositions and varying physical appearances. However, as long as there is migration (gene flow) between them and individuals from different subpopulations can successfully interbreed, all are considered to be members of the same species. The different subpopulations of a single species are called **races** (see Figure 24.1). If the differences in subpopulations are sufficiently great, they are called *subspecies* and given a separate taxonomic name in Latin.

According to Mayr's evolutionary definition, species are distinguished by the existence of genetic barriers to gene flow among populations, that is, **reproductive isolation**, rather than physical characteristics. The significance of reproductive isolation can be understood by considering **sibling species**. These are groups of species that are so similar, they cannot be distinguished from one another by their external characteristics. Sibling species are, however, considered to represent distinct species because of isolating mechanisms that keep them from interbreeding. An example of considerable medical importance is the European *Anopheles* mosquito. Originally this mosquito was thought to represent one species, but scientists have discovered that it is actually a group of six sibling species. It is not possible to tell them apart on the basis of their external appearance. They differ, however, in habitat preferences and egg color, characteristics

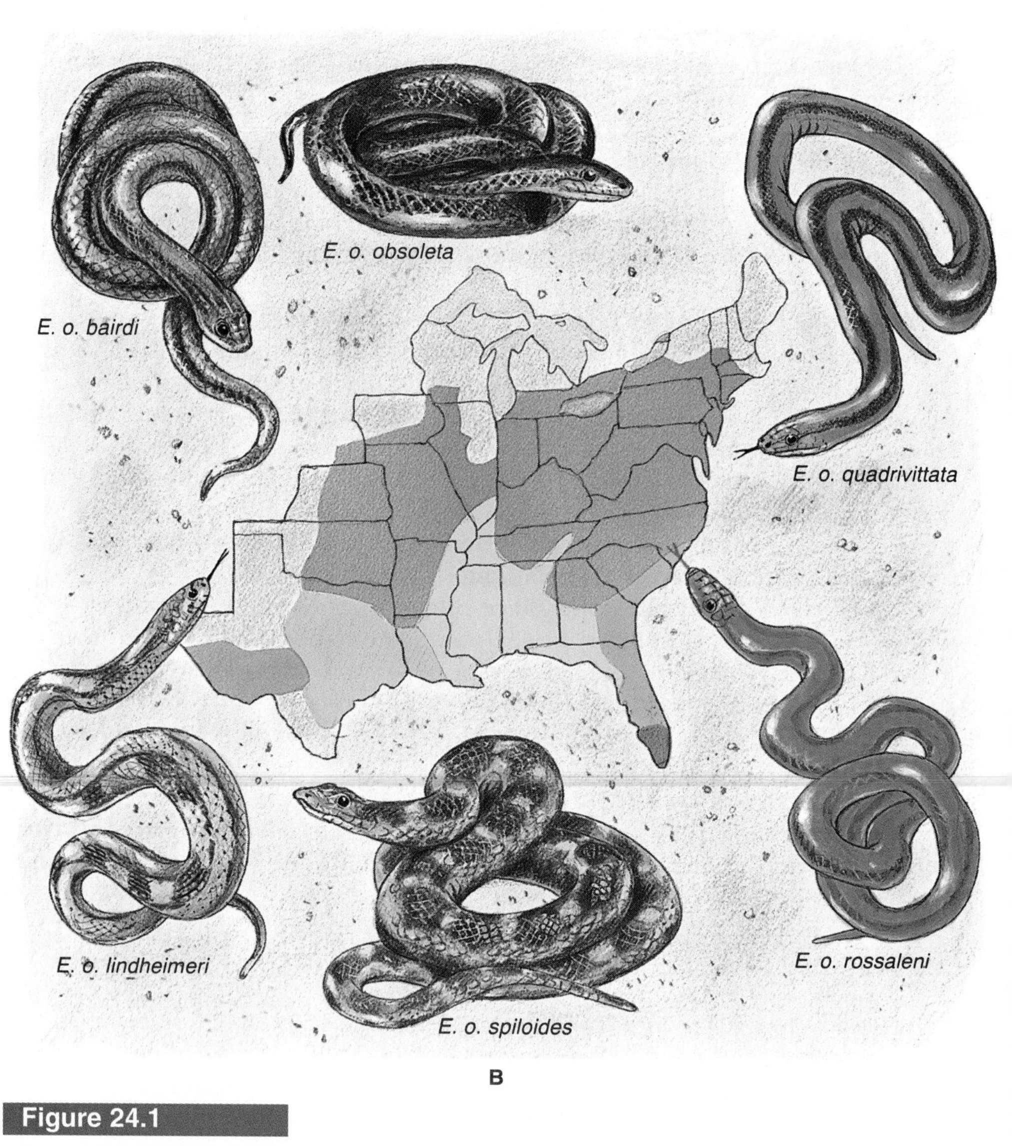

Figure 24.1

(A) The different golden whistlers (*Pachycephala pectoralis*) of the Solomon Islands constitute different races. (B) The rat snake (*Elaphe obsoleta*) of the eastern part of the United States consists of several subspecies.

that permit scientists to differentiate them. The ability to identify the sibling species of *Anopheles* is of practical value in the control of malaria, for only some of these mosquitoes transmit the disease.

Species, then, are not always as easily and simply recognized as descriptions from a bird or tree guide might lead us to believe. Whereas earlier concepts of species stressed external physical characteristics (morphological features), Mayr's newer definition is much more difficult to apply. It requires that biologists make many difficult inferences to distinguish species. As we saw in Chapter 3, taxonomists cannot always use the evolutionary concept of reproductive isolation for classifying species, and they often fall back on morphological and behavioral features.

Ernst Mayr's widely quoted definition of species is as follows: "Species are groups of actually or potentially interbreeding natural populations that are reproductively isolated from other such groups." This definition stresses reproductive isolation rather than external physical characteristics.

SPECIATION

Species change basically in one of two distinct ways: through slow, gradual change of a population (or set of populations) into a new species, a process called *phyletic evolution,* or through the splitting of a population into two or more species, a process that produces diversity and is called *cladogenesis.*

Phyletic Evolution

Phyletic evolution occurs because of the action of natural selection. The selection can act to increase the fitness of the population's genotypes in a constant environment, or it can favor certain genotypes in response to changes in the environment. In a case where phyletic evolution occurs, cumulative change in succeeding generations results in a population that is sufficiently different from the initial population that interbreeding with the original ancestral population is deemed impossible. In other words, phyletic evolution occurs when a population over time becomes so different due to natural selection that if it were possible to bring back an individual from the original population, it could not produce offspring by interbreeding with an individual of the current population.

Finding evidence of phyletic evolution is difficult. Some **paleontologists**, scientists who study fossils, estimate that the average life span of a species is probably about 3 million years. So it is not possible simply to give a science student the assignment of checking the genotypes of a species until it changes! The fossil record provides evidence of long-term change, and scientists have found remains that appear to represent sequences of forms that are consistent with the concept of phyletic evolution.

The formation of a fossil is actually a rare event. Nevertheless, paleontologists have been able to identify certain regions of the globe that have yielded impressive numbers of fossils, which permit a fragmented view of ancient events. Fossils of marine organisms have been the most useful objects for study. Drillings from the ocean floor have provided samples that are rich in the remains of microscopic organisms. Samples that are rich in foraminiferans have been particularly interesting. Foraminiferans are enclosed in hard external coverings that have been preserved in samples brought up from the ocean floor. A study shows a gradation of changes in foraminiferans during the Cenozoic era that started with *Globorotalia conoidea,* continued through three other species, and finally ended with *G. inflata* (see Figure 24.2).

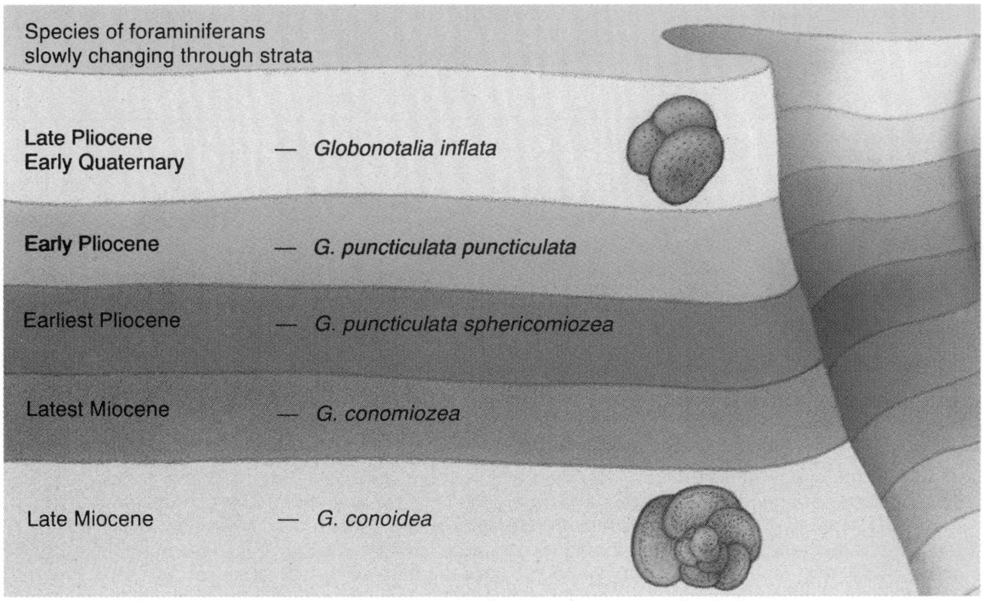

Species of foraminiferans slowly changing through strata

Late Pliocene Early Quaternary — *Globonotalia inflata*

Early Pliocene — *G. puncticulata puncticulata*

Earliest Pliocene — *G. puncticulata sphericomiozea*

Latest Miocene — *G. conomiozea*

Late Miocene — *G. conoidea*

Figure 24.2

This transitional series of species of the genus *Globorotalia* that appeared during the Cenozoic era is an example of phyletic evolution.

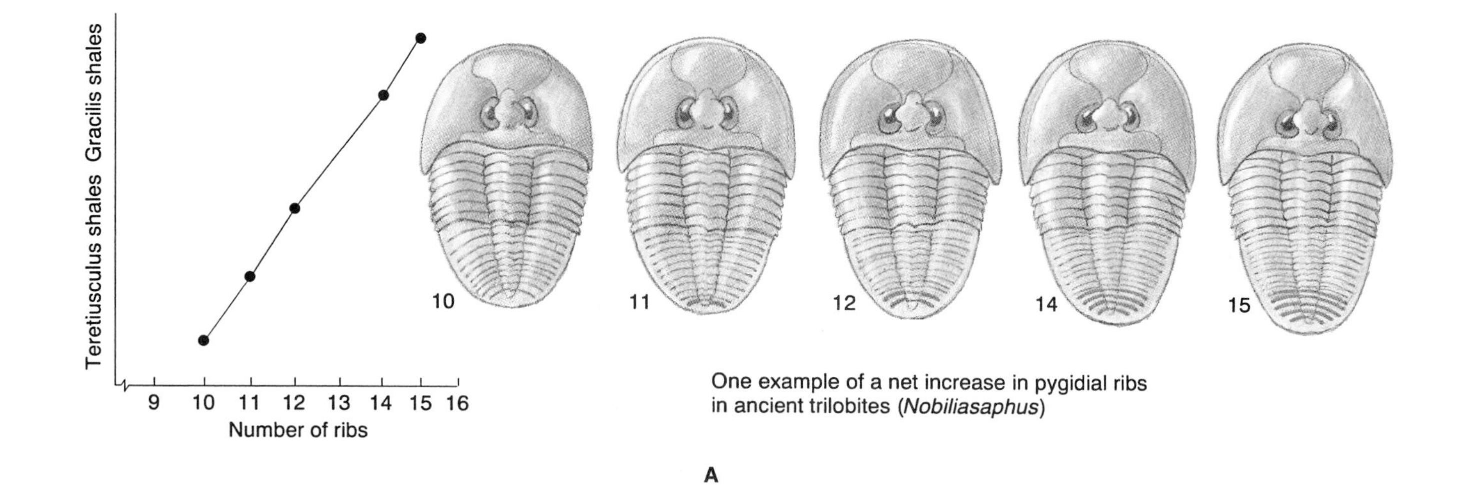

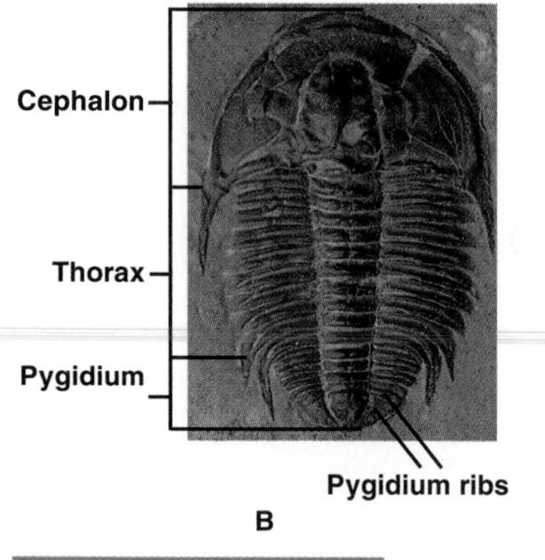

Figure 24.3

Sheldon's study of trilobites (A) showed that over a period of 3 million years, new species evolved gradually from older ones. He based his conclusions on the appearance of additional pygidial ribs (B).

Examples of phyletic evolution are not confined to microfossils. Recently, a British scientist, Peter Sheldon, examined 14,888 trilobites that had been uncovered in central Wales. These fossils (shown in Figure 24.3) are the remains of an interesting ancient group of invertebrates that have been studied extensively by numerous paleontologists. Sheldon's painstakingly exact studies showed that over a period of 3 million years, some species gradually developed an increase in pygidial ribs, a characteristic that is used to distinguish different trilobite species. This finding nicely fits the pattern expected in phyletic evolution.

Phyletic evolution is the slow, gradual change of a population or set of populations into a new species. Most of the evidence that supports the concept of phyletic evolution has come from analyzing fossils. In some cases, transitional series of fossils have been discovered that are consistent with the idea of slow, gradual change.

Cladogenesis

In contrast to the gradual transformation of one species into another over a period of time, **cladogenesis** is the splitting of one species into two or more species. Recall from Chapter 23 that many natural forces can alter the gene pool of populations. What would happen if a population were to be divided and different selective pressures acted on each of the two separate populations? Might the two populations become so different that they could no longer interbreed successfully?

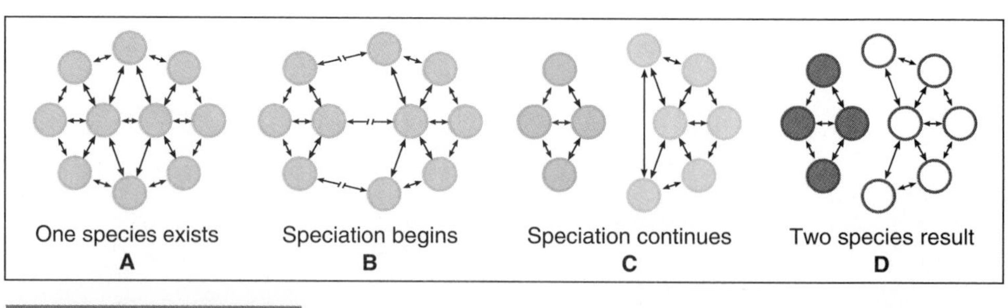

Figure 24.4

Cladogenesis is illustrated in this general model. (A) A single species consists of several local populations. (Circles represent local populations; arrows indicate gene flow.) (B) The populations become separated into two groups with no gene flow between them. The groups gradually become genetically different. In time, reproductive isolating mechanisms develop. (C) The populations reestablish contact, but due to reproductive isolating mechanisms, little gene flow occurs. Additional reproductive isolating mechanisms are favored by natural selection. (D) Speciation is complete when the two sets of populations are fully isolated reproductively; that is, the two populations coexist without any gene flow between them.

Let us consider a simple, ideal case of cladogenesis, illustrated in Figure 24.4. A single population consists of several local populations. The populations become separated into two groups. Gene flow ceases between the two groups, and over time they become genetically different, and reproductive isolating mechanisms develop. Later the populations reestablish contact, but because of reproductive isolating mechanisms, little gene flow occurs. Natural selection favors the development of additional isolating mechanisms. Ultimately, the populations become two distinct species with no gene flow at all. This ideal picture of cladogenesis is somewhat like the Hardy-Weinberg law. In nature, of course, the story is more complicated, but this model serves as a guide for investigating species origin and diversity.

When in nature do populations split? A population spread over a wide range can become broken into subpopulations due to new geographic or ecological barriers. Directional selection, random genetic drift, the founder effect, or the bottleneck effect can then act to alter the separated populations. In time, these populations might no longer be able to interbreed. Figure 24.5 illustrates two classic examples of this sort of speciation—the Galápagos finches and the Lake Victoria cichlids.

If a species arises while remaining geographically isolated, as with the Galápagos finches, the process is called **allopatric speciation**. However, a species can also originate even when it exists together with the parent species through a process called **sympatric speciation**. The occurrence of sympatric speciation is a hotly debated issue among biologists. Early work done in the modern synthesis was hostile to the idea, but recent studies have documented several possible cases. One involves a treehopper (*Enchenopa binotata*) that is thought originally to have fed on several plant species. Currently, six races of the treehopper exist, each infesting a different plant species. These races have each developed different morphological features and also show genetic differences. Female treehoppers will normally lay eggs only on their race's plant, and if raised in a cage, they mate only with individuals on the same host plant. Supporters of sympatric speciation believe that these treehopper races are on their way to becoming new species. Although several other possible cases of sympatric speciation have been recognized, allopatric speciation is still regarded as the main avenue of cladogenic speciation.

How much time is required for speciation to occur? The answer varies. Some speciation events may occur very rapidly, in a process called **quantum speciation**. The best-known form of quantum speciation is *polyploidy,* a condition often found in plants in which cells contain three or more complete sets of chromosomes. Polyploids can arise through the doubling of the same genome within a single species or when multiple sets of chromosomes result from the hybridization of individuals from different species. Polyploid individuals develop in one generation, and they are reproductively isolated from their parent population—hence they can arise by sympatric speciation. Figure 24.6 on page 458 illustrates a classic example of polyploid evolution in present-day species of bread wheat.

Cladogenesis is the splitting of one species into two or more species. This can occur when a population is split into separate subpopulations that experience different selective pressures, so that in time they diverge to the point where individuals of the different populations cannot interbreed.

REPRODUCTIVE ISOLATING MECHANISMS

Slow or fast, involving geographical isolation or not, the key to speciation is reproductive isolation. Many mechanisms lead to reproductive isolation. When two similar species live in close proximity, usually more than one isolating mechanism is operating. In some cases, the mechanisms are often not completely

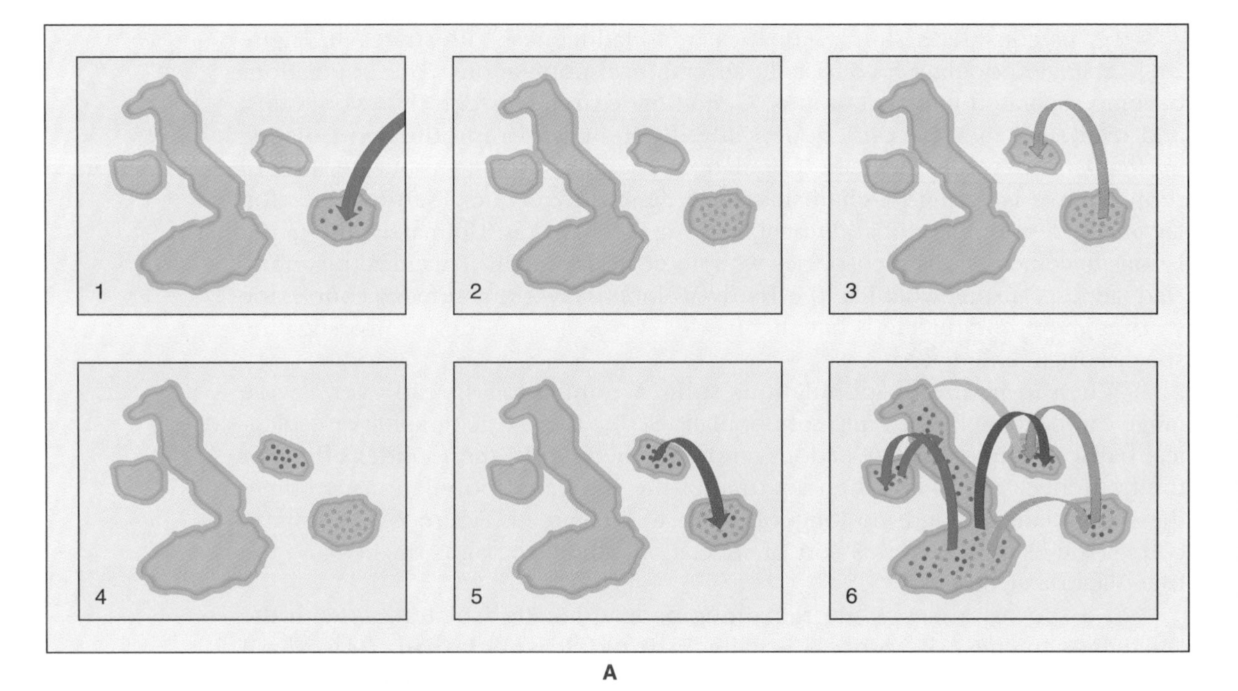

A

Figure 24.5

Two classic examples of speciation are the Galápagos finches and the cichlid fishes of Lake Victoria. This figure is a simplified version of what is known about the evolutionary story. (A) (1) In the early history of the Galápagos Islands, there were no finches. Then, a few finches from the mainland of South America reached one of the islands. (2) The number of finches increased. Due to natural selection, the finches gradually became adapted to the island's environment. (3) Some finches flew to a second island, which had a different environment. (4) These finches adapted to the new environment. (5) In time, the finches on the second island became sufficiently different so that when some of them flew back to the first island, they could not interbreed with the finches there. The populations had become different species. (6) Similar events were repeated. Now there are 13 different species of finches on the islands. Four of the finches are shown in the photos: (B) large ground finch, (C) medium ground finch, (D) woodpecker finch, and (E) warbler finch. (F) Lake Victoria, the largest lake in Africa, began to form about 1 million years ago due to movements on Earth that created some small lakes along existing rivers. (1) Initially, there were only one or a few species of cichlid fishes living in the rivers and the new lakes. (2) In time, the lakes grew and some merged. Continued earth movement altered the dimensions and borders of the lakes. In the process, many new habitats were created. The original species of cichlid fish became adapted to these new habitats. (3) When the lakes all merged, the cichlid fishes were brought together, but individuals from many different populations could no longer interbreed. The populations had become different species. (4) Lake Victoria has continued to experience changes, and new habitats have continued to come into existence, providing chances for new species to evolve. There are now over 170 species of cichlid fish in this one lake. (5) Three examples of cichlid fish (G, H, and I).

B C

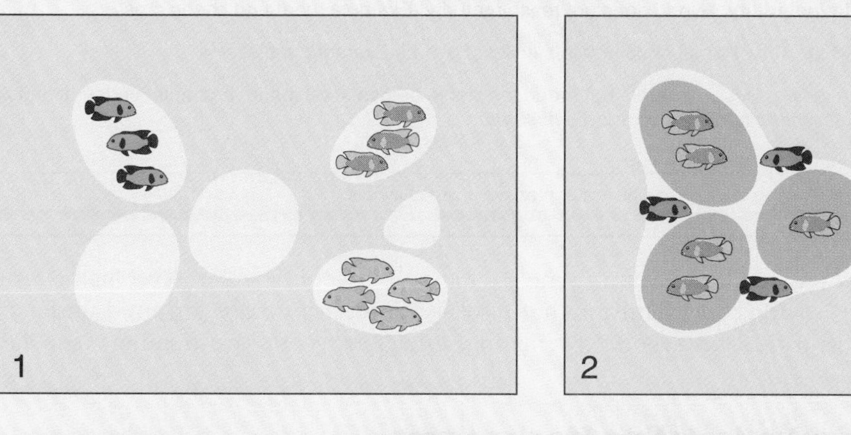

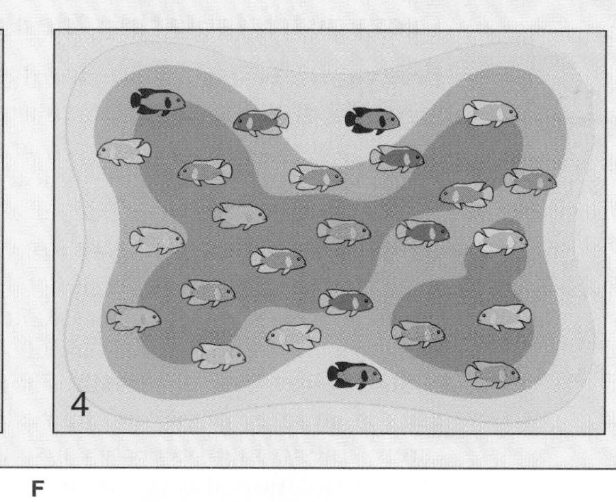

F

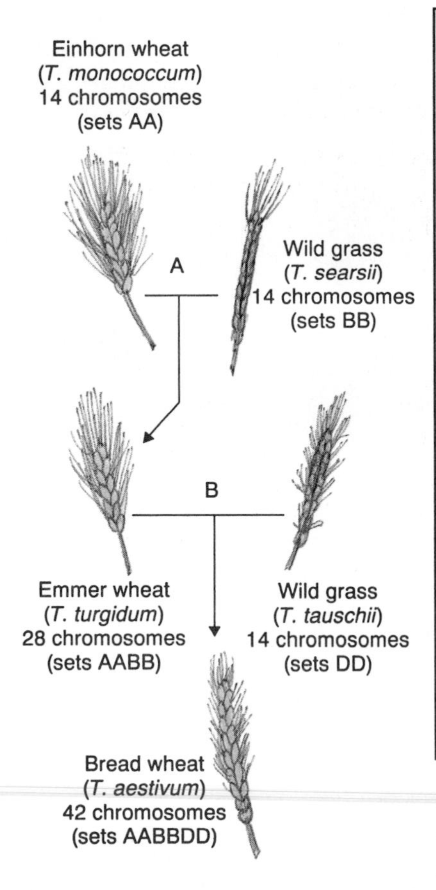

Figure 24.6

The likely origin of the wheat plant used for bread involves two hybridization events and chromosome doubling. The first hybridization was (A) between einkorn wheat (*Triticum monococcum* and a wild grass (*T. searsii*), followed by a doubling of chromosomes in the F_1 to produce emmer wheat (*T. turqidum*). The second hybridization was (B) between emmer wheat and wild grass (*T. tauschii*) followed by a doubling of the chromosomes in the F_1 to produce bread wheat (*T. aestivum*). Bread wheat contains sets of chromosomes derived from three diploid species.

Table 24.1 Reproductive Isolating Mechanisms

Prezygotic mechanisms (prevent or reduce successful interspecific crosses)

- **Habitat isolation.** Populations of different species live in different habitats within the same geographical region. Example: *Quercus velutina and Q. coccinea.*
- **Seasonal isolation.** Individuals of different species inhabit the same region but are sexually mature or breed at different times. Example: *Pinus radiata* and *P. muricata.*
- **Ethological isolation.** Individuals of different species do not mate because of differing behavior patterns. Example: courtship behavior of the crested grebe (Figure 24.7).
- **Mechanical isolation.** Individuals of different species have structural differences that prevent successful fertilization. Example: *Ficus* species pollinated by different wasps (Figure 24.8).
- **Gametic isolation.** Gametes from different species do not combine in fertilization. Example: sea urchins *Strongylocentrotus purpuratus* and *S. franciscanus* (Figure 24.9).

Postzygotic mechanisms (reduce the fitness of hybrids)

- **Hybrid inviability.** Hybrid does not reach maturity. Example: goat and sheep cross.
- **Hybrid sterility.** Partial or complete sterility of hybrid. Example: mule.
- F_2breakdown. Hybrid is fertile but produces sterile or less fit offspring. Example: cotton hybrids from *Gossypium barbadense* and *G. hirsutum.*

The major isolating mechanisms are either prezygotic or postzygotic.

effective; instead, they limit gene exchange between species. The major reproductive isolating mechanisms that have been identified are grouped under two main categories: *prezygotic* (premating) and *postzygotic* (postmating) (see Table 24.1).

Prezygotic Isolating Mechanisms

Prezygotic isolating mechanisms prevent or reduce hybridization between members of different species. Major prezygotic mechanisms include habitat, seasonal, ethological, mechanical, and gametic isolation. Examining these mechanisms allows us to understand how different species remain separated in nature.

Habitat isolation occurs when populations of different species occupy different habitats within the same general geographical region. Such isolation may be the result of some specialization that permits them to live in particular conditions. Habitat isolation is common in plants. For example, black oak (*Quercus velutina*) and scarlet oak (*Q. coccinea*) are found throughout the eastern United States. The black oak is found on well-drained soils, but the scarlet oak lives in swampy or poorly drained (acidic) soils. The two species, therefore, are rarely able to interbreed because they seldom grow in close enough proximity. Habitat isolation also occurs in animals. The sibling species of mosquitoes in the genus *Anopheles* discussed earlier live in different, although nearby, habitats.

Seasonal isolation (or *temporal isolation*) refers to species that live in the same region but whose populations are prevented from interbreeding because they are sexually mature at different time periods or seasons or they breed at different times of the day. For example, two species of *Drosophila, D. pseudoobscura* and *D. persimilis,* are found in the same regions in the western United States and breed during the same season. However, the former mates in the evening, the latter in the morning. Another classic example involves two pine species that grow on the Monterey peninsula south of San Francisco. *Pinus radiata* releases pollen in February, *P. muricata* in April; therefore, the populations remain distinct even though the individuals grow in the same area.

Ethological isolation refers to differing behavior patterns in courtship or a lack of sexual attraction between males and females of different species. It

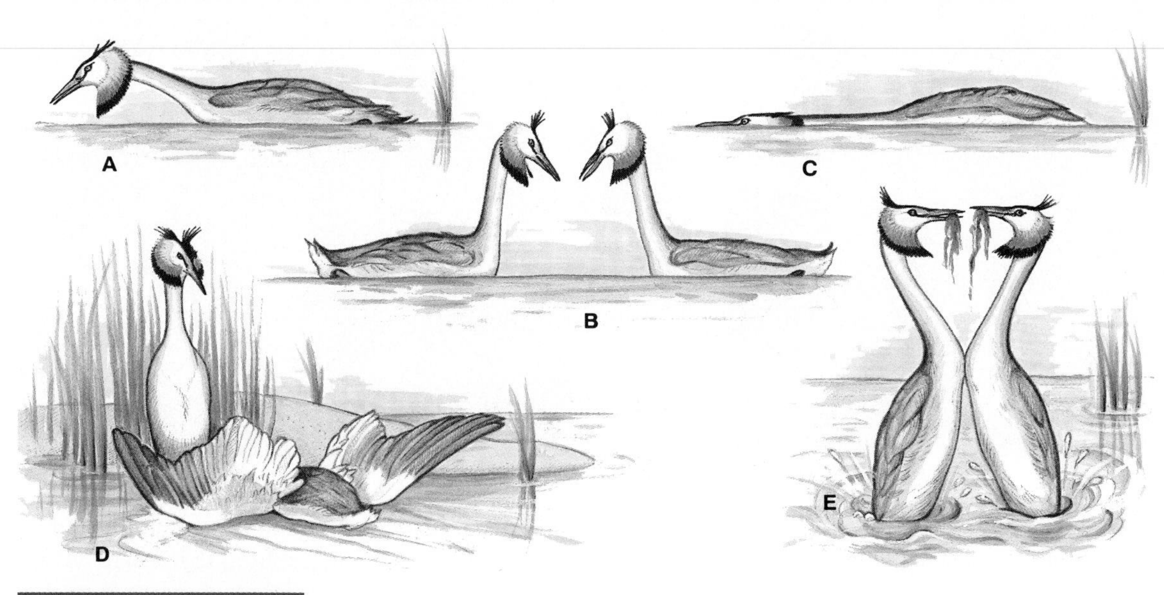

Figure 24.7

Julian Huxley, one of the architects of the modern synthesis, conducted one of the early studies (1914) on the courtship display of crested grebes (*Podiceps cristatus*). These original drawings from his paper show some of the behaviors that make up an elaborate courtship display: (A) search attitude, (B) a pair in the forward shaking attitude, (C) the passive pairing attitude, (D) part of the display ceremony, and (E) the penguin dance.

occurs only in animals and is the major prezygotic isolating mechanism in many animal groups because courtship plays such an important role in their reproduction. Courtship behavior—calls, songs, and displays—is often species-specific, and ethological isolation serves to prevent males and females of different species from attempting to interbreed. Figure 24.7 shows the elaborate courtship behavior of the crested grebe.

Mechanical isolation and gametic isolation both refer to sexual factors that inhibit or prevent fertilization. **Mechanical isolation** in animals is usually the result of incompatible sizes or shapes of genital parts. In plants, mechanical isolation results from structural differences in flowers that do not permit cross-pollination by insects or other animals. The floral structures in two populations of plants may be adapted to facilitate pollination by different animals or to prevent an animal from transferring pollen between individuals of different species. Coevolution of plants and animals also promotes mechanical isolation: most of the more than 900 species of fig (*Ficus*) are each pollinated by a different species of wasp (see Figure 24.8).

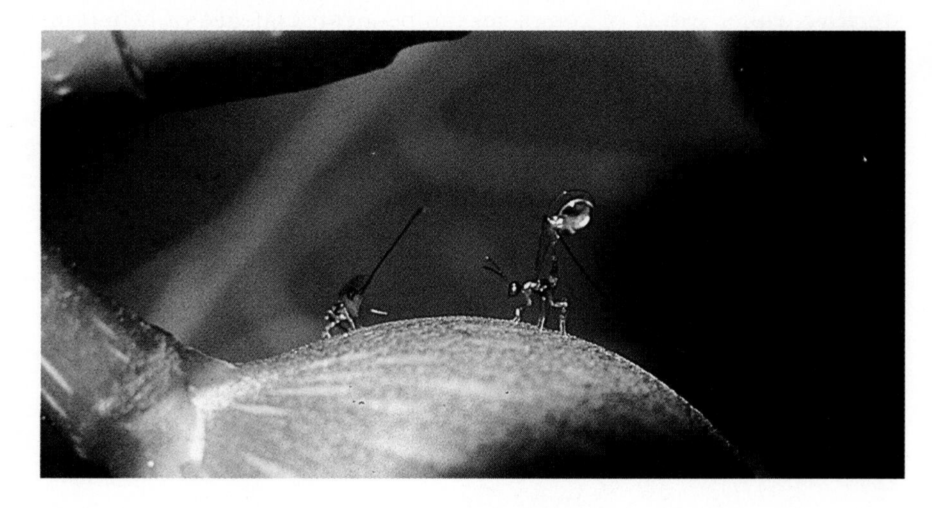

Figure 24.8

Blastophaga psenes is one of many hundreds of species of fig wasps. The 900 or so species of fig (*Ficus*) are pollinated by wasps of the same family. With just a few exceptions, each species of fig has its own species of wasp.

A

B

Figure 24.9

Stimulated to release gametes into the water, these specimens of two different species of sea urchins, *Strongylocentrotus purpuratus* (A) and *S. franciscanus* (B), will not produce viable offspring because of gametic isolation.

Gametic isolation occurs when male and female gametes cannot combine in fertilization or when pollen or sperm are rendered inviable in the female sexual structures of another species. Aquatic animals that release gametes into the water often depend on the attraction of sperm and eggs for fertilization. Sea urchins of different species, for example, display gametic isolation. In laboratory tests, male and female urchins can be stimulated to release their gametes into the water, but fertilization rarely occurs between gametes of different species (Figure 24.9).

The major prezygotic isolating mechanisms are habitat, seasonal, ethological, mechanical, and gametic isolation. Each basically ensures that a species' integrity is maintained by preventing reproduction between members of different species.

Postzygotic Isolating Mechanisms

Prezygotic isolating mechanisms prevent or reduce hybridization between members of different species. However, if such mechanisms do not exist or fail to prevent hybridization, *postzygotic isolating mechanisms* may cause the resulting offspring to die before maturity, to be sterile, or to be less fit. **Postzygotic isolating mechanisms** are processes that reduce the fitness of hybrids. They include hybrid inviability, sterility, and F_2 breakdown.

All postzygotic isolating mechanisms depend on the inability of parental genes to function effectively together in a hybrid offspring. In its most extreme form, **hybrid inviability**, the hybrid does not reach maturity. A goat and sheep cross, for example, results in an embryo that is spontaneously aborted. **Hybrid sterility**, either partial or complete, can be a consequence of hybrid inviability. It may also result from abnormal development of sex organs, incompatibility of the embryo with its food source (only in plants), abnormal segregation during meiosis, or other irregularities that prevent hybrids from reproducing. The mule is a well-known example of a sterile hybrid. Although the hybrids are vigorous, they are sterile due to faulty germ-cell formation. **F_2 breakdown** is yet another way in which postzygotic events isolate populations. In this case, the hybrids are fertile but produce sterile or less fit offspring. For example, hybrids of certain cotton species are hardy and fertile, but F_2 crosses produce plants that do not reach maturity or are less fit in other ways.

Major postzygotic isolating mechanisms are hybrid inviability, sterility, and F_2 breakdown. They prevent successful reproduction between members of different species in cases where offspring are produced by matings from individuals of different species.

Selection for Isolating Mechanisms

Isolating mechanisms (especially in combination) are effective in reducing gene flow between populations and therefore in maintaining the separation of species. Natural populations employ a wide range of isolating mechanisms. The origins of these mechanisms are correspondingly diverse. Some of them may be attributed to chance genetic variation enhanced by random genetic drift or geographical isolation. Postzygotic isolating mechanisms, for instance, are generally considered to have developed in these ways. Once partial reproductive isolation exists, however, natural selection may operate to increase that degree of reproductive isolation. An organism that mates with a member of another species and produces less fit offspring will obviously leave fewer of its own genes in the next generation than organisms that avoid such hybridization. Under these conditions, prezygotic isolating mechanisms might increase fitness, since preventing hybridization would be an advantage to the species.

Care must be taken not to oversimplify the picture. Many species that apparently do not hybridize in nature will do so under experimental conditions. By itself, the ability or inability to form hybrids cannot alone be used as a criterion for defining species.

THE EVOLUTION OF SPECIES

The origins of species are complex and varied. Some species are distributed over a wide range, while others are very localized. Some are polymorphic (see Chapter 23), some are broken into many races, and some have numerous subspecies. Species may be found living adjacent to related species, or they may have a region of overlap in which some hybridization occurs. Two related species may be found in the same region with little or no hybridization. In some cases, usually in areas disturbed by people, swarms of hybrids occur, and the boundary between separate species becomes obscured. These diverse forms of species (and we have only mentioned some obvious ones) have come about in many different ways—some forms evolve over a long period of time, and some evolve very quickly. The separation of species is maintained by a variety of mechanisms, often working together. It is little wonder that the concept of species is one of the most confusing topics among theoretical biologists.

The modern synthesis has tried to bring some order to the complexity of species by stressing underlying genetic mechanisms. This approach has proved to be very useful, but like all scientific ideas, it has its limits. For example, when we look at the fossil record, the concept of a species as a set of potentially interbreeding populations poses some formidable problems (do fossils reproduce?). Similarly, some botanists feel that the modern synthesis stresses mechanisms important for animal species formation at the expense of those, like polyploidy, that are more common in plants. Nevertheless, Mayr's new definition of a species has allowed scientists to explore populations in the field and in the laboratory. It has permitted them to bring biochemical analysis into the picture, and it has greatly expanded Charles Darwin's original idea.

Summary

1. Species are groups of actual or potentially interbreeding natural populations that are reproductively isolated from other such groups.
2. Speciation occurs in two ways: through phyletic evolution, the slow gradual change of a population or set of populations, and through cladogenesis, the splitting of one species into two or more species.
3. Allopatric speciation refers to speciation that occurs when the parent population is divided into geographically isolated areas. Sympatric speciation occurs when the new and parent species live in the same geographical area.

4. Speciation can be a slow or rapid process. When rapid, it is called quantum speciation.
5. Reproductive isolation is essential for speciation. Major known isolating mechanisms can be classed as either prezygotic or postzygotic.
6. Prezygotic isolating mechanisms include habitat, seasonal, ethological, mechanical, and gametic isolation. Postzygotic isolating mechanisms include hybrid inviability, hybrid sterility, and F_2 breakdown.

Review Questions

1. Define a species.
2. What are the two general patterns of speciation?
3. What is phyletic evolution?
4. What is cladogenesis? What is sympatric speciation?
5. What is allopatric speciation?
6. What is the importance of reproductive isolation for speciation?
7. Describe the operation of major prezygotic isolating mechanisms.
8. Describe the operation of major postzygotic isolating mechanisms.

Essay and Discussion Questions

1. Some botanists have expressed dissatisfaction with Mayr's model of speciation. Can you explain why that might be?
2. The majority of pre-Darwinian naturalists in England and the United States defined *species* as an ideal plan of God's that specified a set of unchanging essential characteristics and their relationships for each animal and plant. Darwin, by contrast, used the concept of species to refer to a population that changes over time. The radically different usage caused much confusion, and even today there is controversy over the definition of species. Should scientists come up with new terms rather than redefine old ones?
3. Since it is not possible to revive fossils and run tests with them to see if individuals can interbreed, what evidence might we use to determine that two specimens belong to the same or different species?

References and Recommended Reading

British Museum (Natural History). 1981. *Origin of Species*. London.
Dobzhansky, T., F. J. Ayala, G. L. Stebbins, and J. W. Valentine. 1977. *Evolution*. New York: Freeman.
Ford, E. B. 1975. *Ecological Genetics*. New York: Wiley.
Futuyma, D. J. 1986. *Evolutionary Biology*. Sunderland, Mass.: Sinauer.
Mayr, E. 1942. *Systematics and the Origin of Species*. New York: Columbia University Press.
Ramirez, R. 1970. Host specificity of fig wasps (Agaonidae). *Evolution* 24:681–691.
Stebbins, L. G. 1982. *Darwin to DNA, Molecules to Humanity*. New York: Freeman.
White, M. J. D. 1978. *Modes of Speciation*. New York: Freeman.
Wiebes, J. T. 1979. Co-evolution of figs and their insect pollinators. *Annual Review of Ecology and Systematics* 10:1–12.

CHAPTER 25

The Unfinished Synthesis

Chapter Outline

Reading Questions

1. What does the fossil record tell us about past life on Earth?

2. What patterns of change does the fossil record reflect?

3. How are different areas of biology related to the theory of evolution?

4. Why is the theory of evolution an "unfinished synthesis"?

The modern synthesis brings together knowledge from genetics, biogeographical distribution, and taxonomy to explain how new species come into being. Like Darwin's theory of evolution, however, the modern synthesis goes beyond explaining just the origin of species. It attempts to unify all of the current life sciences under the theory of evolution. This unification includes subjects as diverse as paleontology and physiology. Theodosius Dobzhansky epitomized this view when he commented in 1973, "Nothing in biology makes sense except in the light of evolution." How does the modern synthesis attempt to encompass all phenomena dealing with living things?

It approaches this task through various avenues. Some areas of biology, like *paleontology,* the study of fossils, include knowledge that is closely linked to evolutionary theory. Architects of the modern synthesis have used concepts and processes from evolutionary theory to explain the appearance and significance of the fossil record. In turn, the study of fossils has contributed to the development and extension of the modern synthesis. The modern theory of evolution also raises questions about the fossil record that direct research. For example, studies of ancient community evolution, population dynamics, and rates of speciation have been inspired by questions about evolution.

In other areas of biology, the relationship to evolutionary theory is less direct. *Physiology,* for example, is concerned with how organisms function. Physiologists are most interested in processes that can be studied at the present time, not their evolutionary development. Nonetheless, by comparing certain common biochemical compounds or metabolic pathways in different organisms, physiologists have thrown some light on possible evolutionary relationships among organisms. By showing the adaptive importance of various physiological processes, biologists have also related the chemical activities of an organism to its evolutionary history. Still, an examination of the major questions that now guide research in physiology does not reveal many that are directly concerned with evolution. Rather, the theory of evolution provides an intellectual framework that allows physiologists to relate the functions of organisms to other aspects of life and its history.

In this chapter, the relationship of the modern synthesis to paleontology and physiology will be discussed in order to elucidate evolution as a unifying theory. The examination of paleontology will also provide a basis for describing some of the main topics in *macroevolution,* that is, evolution above the species level.

PALEONTOLOGY AND EVOLUTION

Our consideration of paleontology begins with a general account of the fossil record. Then patterns of change revealed in the fossil record are described and explained. These topics permit an evaluation of the relationship of paleontology to the theory of evolution.

The Fossil Record

The **fossil record** is a chronicle of past life on this planet (see the Focus on Scientific Explanations, "The Fossil Record: 700 Million Years of Life"). Studies of fossils were central to Darwin in formulating his theory of evolution, and they have continued to be an important source of new ideas. Studies of the fossil record have been extremely valuable, giving scientists their only direct picture of past life on Earth. Nevertheless, the fossil record is woefully incomplete. Why is the record so fragmentary?

Fossils are remnants or traces of organisms from a past geologic age that are embedded in Earth's crust. The conditions necessary for forming fossils rarely existed. Therefore, only a fraction of the species that have lived during Earth's history left any fossils (see Figure 25.1). Moreover, since fossils are most often

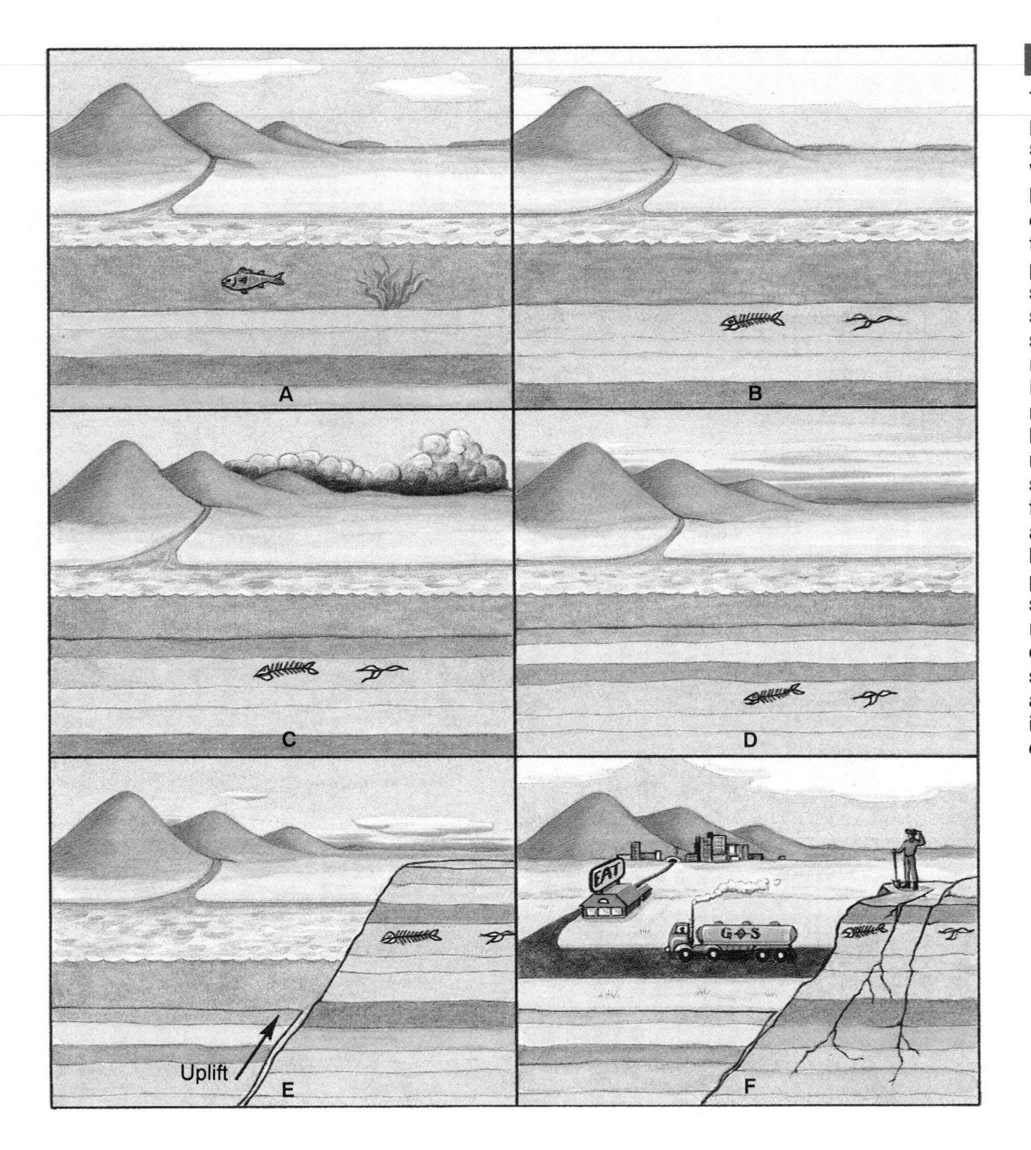

Figure 25.1

This figure describes the steps in one possible path in the formation of a fossil. (A) An animal lives in a lake. (B) When it dies, its remains sink to the bottom of the lake and in time become covered by mud and silt, which prevents their destruction. Soft tissues decompose, but bones and teeth remain. The soft parts may leave impressions in the surrounding mud. (C) As more material sinks to the lake bottom, the animal remains get buried deeper and deeper. Eventually the entire area may be submerged by more water, and the mud layer may become part of a larger sediment. (D) Elements in the surrounding sediment slowly penetrate, replace, or fill spaces in the original organic material, which may become totally replaced by a different mineral. (E) Geological processes may lead to the entire region sinking under seas to form sedimentary rock, and lakes can be uplifted to produce mountains. (F) Ultimately, the rock strata may be pushed near the surface and uncovered by erosion or by a passing traveler, research scientist, or road construction team.

formed underwater or in wet soil, the record tells us more about shallow-water marine organisms than terrestrial species. For the same reason, more has been learned about organisms that lived near lakes than those that lived on mountains. Since the soft parts of organisms have rarely been preserved, paleontologists must work with only the remains of substances like teeth, bones, shells, pollen, and wood. Entire groups of organisms (for example, jellyfish) have left practically no fossils, and even for groups that have left a record, the picture is incomplete.

The fossil record as a whole, then, is fragmentary. It tells us little about the specific history of individual species or the origins of major phyla. Remains from all the major animal phyla go back to the Cambrian, the earliest period for which we have extensive fossil remains. It is the evolution within phyla, especially animal phyla, that best shows how changes occurred in the development of life.

Despite the inherent difficulties in studying the record of past life, a great amount has been learned. The fossil record has allowed biologists to recognize many significant patterns, to trace specific lines of development, and to reconstruct a broad picture of the history of life on this planet.

Fossils are the permanent remains of organisms that lived and died in a past geologic age. They provide the clearest chronicle of life that has evolved on this planet during past geologic eras. Analyses of the fossil record have provided strong support for the theory of evolution.

A

B

Figure 25.2

Two famous "living fossils" are the coelacanth, a marine fish (A), and the ginkgo tree (B). Both have survived unchanged for millions of years. The coelacanth, once thought to be extinct, was discovered in 1938 in the waters near Madagascar. The ginkgo, a native of China, has a fossil record that extends back 60 million years.

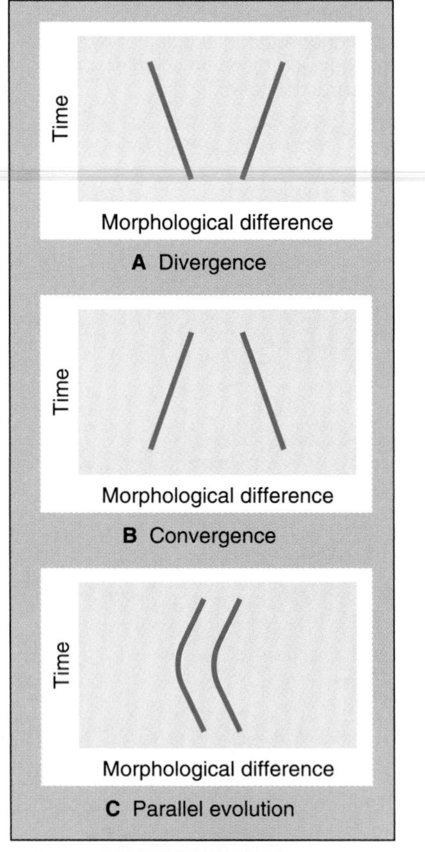

Figure 25.3

Three patterns of macroevolution as measured by morphological difference are divergence, convergence, and parallel evolution. In divergence (A) there is increased difference among separated lineages. Convergence (B) occurs when separate lineages become similar. Parallel evolution (C) resembles convergence, but occurs when the separate lineages share a common ancestor and the similarity is partly a result of the shared ancestory.

Patterns of Change

Studies of fossils cannot disclose much direct information about genetic changes in populations, the genetic similarities of different species, or the behavior of past organisms. For the most part what is revealed is morphological differences that have developed among species and higher taxa.

The fossil record shows that there are different patterns of evolutionary change. Some species appear to have remained constant for millions of years without any recognizable modification. The best-known examples are the "living fossils" such as coelacanths (an ancient marine fish species) or ginkgo trees, both shown in Figure 25.2. However, such examples are rare. More commonly, there is evidence of a gradual change without any splitting or branching. Such **lineages** (single lines of descent) are characterized by a long series of individuals that display a continuum of change. At the ends of the continuum are different species, and there may be different species in between. This gradual evolution is called *phyletic evolution*.

There are even more fossils from lineages that split over time, and several patterns of separation have been described. Divergence, convergence, and parallelism are the best documented; they are illustrated in Figure 25.3.

Divergence

Divergence describes an evolutionary pattern in which there are increased morphological differences among separated lineages. This is most striking in **adaptive radiation**, where several lineages diverge from a common ancestor. A classic example involves the evolution of Hawaiian Island honeycreepers (Drepanididae). How can the adaptive radiation of these birds be explained? Ornithologists believe that, similar to the evolution of finches on the Galápagos Islands, a South American ancestor originally colonized the oceanic islands of Hawaii and gave rise to the entire honeycreeper family, which now consists of nine genera that are divided into two subfamilies. Naturalists hypothesize that their diversity is the result of an ancestral form that invaded a new territory, enabling its descendants to exploit new sources of food. The external characteristics of honeycreepers reflect adaptations to different types of food. Walter Bock of Columbia University has described a series of radiations that start with a nectar-eating offshoot from one of the two subfamilies of honeycreepers. In time, the different genera of the second subfamily evolved. These consisted of insect, fruit, and seed eaters, each of which have very distinctive bills, as shown in Figure 25.4.

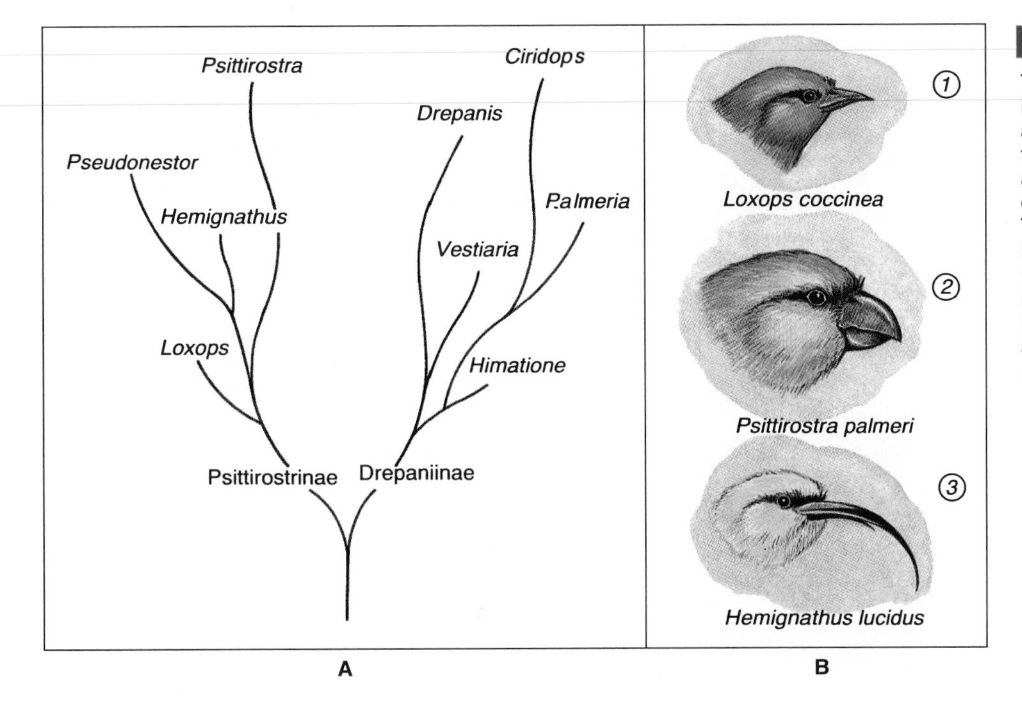

Figure 25.4

The beaks of honeycreeper birds on the Hawaiian Islands reflect great diversity and provide a classic example of the type of evolutionary divergence known as adaptive radiation. (A) A phylogenic diagram of the honeycreepers. (B) Three genera that have evolved in the Hawaiian Islands, like the six others, reflect responses to different environmental resources. *Loxops* feed on insects caught flying, *Psittirostra* on large seeds that need to be crushed, and *Hemignathus* on the nectar of flowers.

Another case of adaptive radiation involves the entire class Aves (birds). After they branched off from the reptile class during the Mesozoic, birds underwent an initial radiation, which was then followed by successive radiations during the Paleocene and Eocene epochs. These numerous radiations were related to structural and functional advances. For example, the ability to fly not only permitted birds to escape predators but also to migrate, taking advantage of mild seasons and exploiting many different food sources.

The adaptive radiation of placental mammals at the beginning of the Tertiary period is another famous example of this type of divergence (see Figure 25.5). Placental mammals, which had an efficient form of reproduction and considerable brain size, exploded in a set of radiations. They replaced the dinosaurs, which had become extinct after dominating Earth for well over 100 million years.

Divergence explains some of the striking homologies found in the living world. **Homologous structures** share a common structural, evolutionary, and embryological background. Classic examples of homologous structures are the skeletal forelimbs of humans, dogs, whales, and bats (see Figure 25.6).

Convergence

Convergence occurs when separate lineages become morphologically similar. Environmental or functional similarities often lead to convergence over time. For example, the evolution of wings in insects, birds, and mammals (bats) reflects convergent evolution related to function (see Figure 25.7). Development of the body shapes of some aquatic mammals and fishes shows convergence due to adaptations to the same environment. The independent development of mollusk and vertebrate eyes is another quite amazing example of convergence.

Convergent evolution is responsible for many of the striking *analogous* structures that can be observed, both in the fossil record and in living organisms. **Analogous structures** have similar functions but, in contrast to *homologous structures,* differ in their structure, development, and evolutionary history. For example, the wings of a butterfly and the wings of a bird are both used for flight but are analogous structures because they differ in structure, embryology, and evolutionary background.

Convergence can also occur with behavioral and physiological traits. The fossil record, however, yields little evidence of such traits, and biologists rely on comparative studies of living organisms to learn more about them.

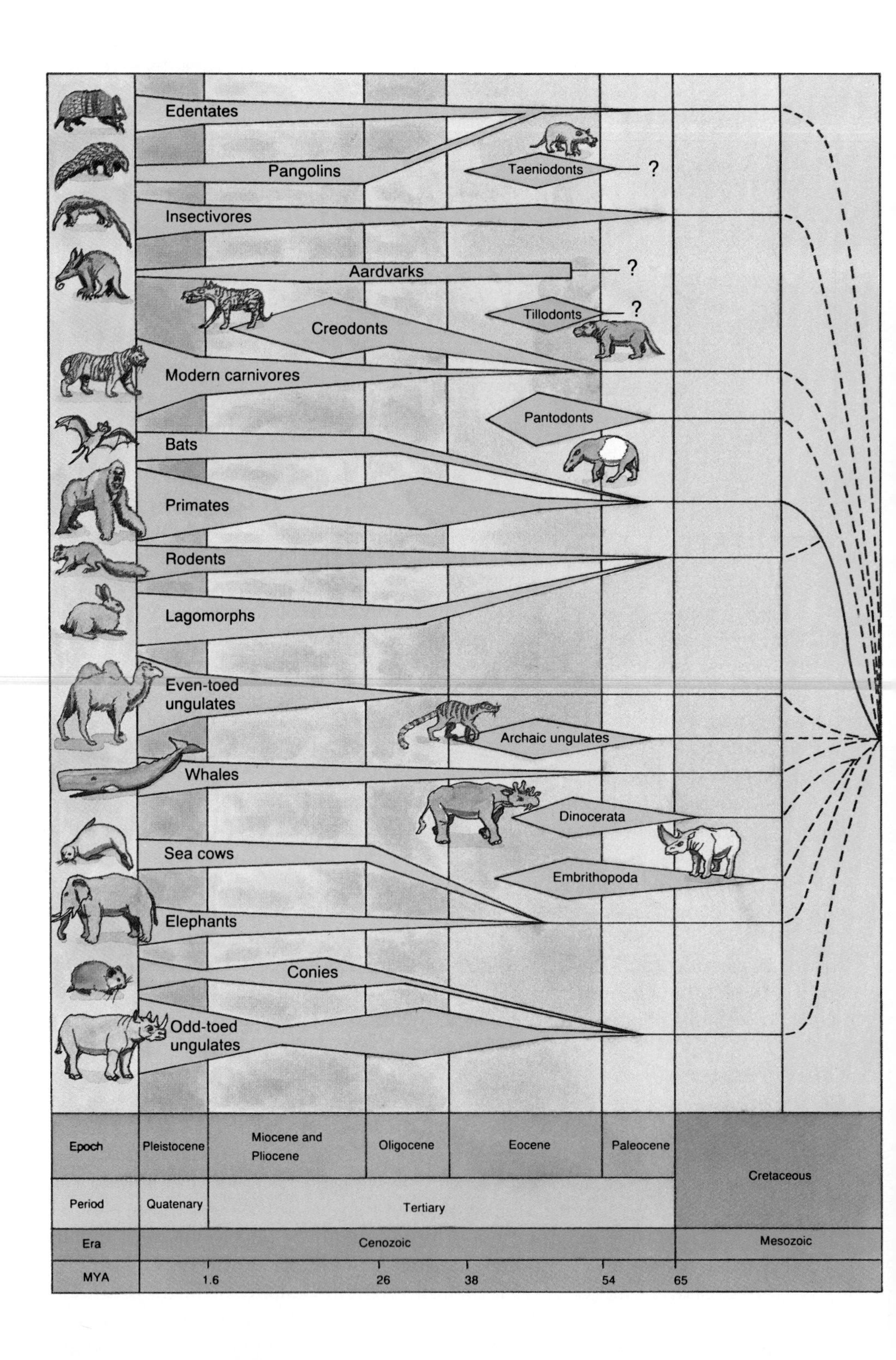

Figure 25.5

Placental mammals evolved from a primitive lineage during the Cretaceous period. Their advanced brains and offspring, which were well-developed at birth, gave this group a competitive advantage that resulted in a series of adaptive radiations. Note how most of the orders of placental mammals had evolved by the early Eocene epoch.

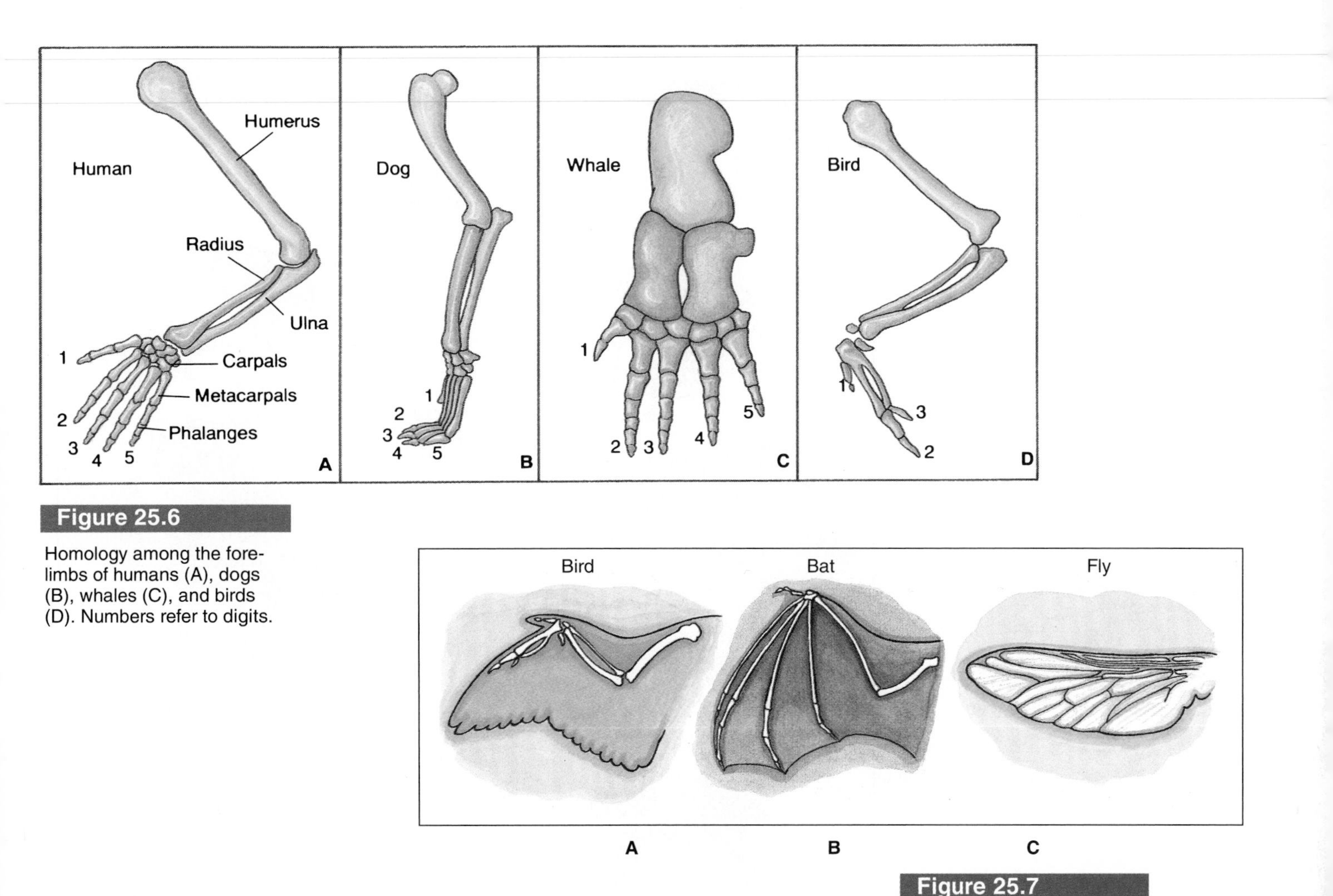

Figure 25.6

Homology among the fore-limbs of humans (A), dogs (B), whales (C), and birds (D). Numbers refer to digits.

Figure 25.7

The wings of birds (A), bats (B), and insects (C) evolved independently and thus exemplify convergent evolution.

Parallelism

Parallelism is similar to convergence. It is the development of similar characteristics in separate lineages that have a common ancestor, where the similarity is influenced by the characteristics of the shared ancestor. The evolution of placental and marsupial mammals (see Chapter 4) is an example of parallelism. Both evolved from the same subclass during the mid-Cretaceous period. Marsupial mammals evolved extensively in Australia (and in South America until the late Tertiary), in the absence of competition from the placental mammals that came to dominate other continents. Because of their similarities, marsupials evolved in Australia in parallel to mammals that evolved on other continents (see Figure 25.8).

Parallel evolution sometimes results in the production of both analogous and homologous structures. For example, the true anteater and the marsupial anteater share homologous mammalian characteristics, but their anteater characteristics—long snouts and long sticky tongues—are analogous.

The fossil record contains evidence of different patterns of evolutionary change. Although lineages remain constant for millions of years, and some show a slow, gradual change, the more common pattern is that of splitting into separate lineages.

There are different patterns of lineage splitting. In divergence, different lineages have increased morphological differences. With convergence, separate lineages become morphologically similar. Like convergence, in parallelism, lineages that share a common ancestry develop similar characteristics.

Figure 25.8

Placental mammals on the large continents evolved in parallel with Australian marsupial mammals.

Other Patterns of Change

The divergence, convergence, and parallelism of lineages are the three most common evolutionary patterns revealed by the fossil record. However, scientists recognize several other patterns. The most important of these are the emergence of *grades,* the appearance of *direction* in evolution, and the occurrence of *mass extinctions.*

Grades

Most evolutionary change involves the modification of existing structures and systems. Occasionally, however, a change occurs that results in a new functional ability. Such a development is called the creation of a new **grade**. Grades are associated with dramatic evolutionary events that lead to entirely new adaptations to the environment. They are life's accidental "inventions." Notable examples include the development of warm-bloodedness (maintaining a constant body temperature) in mammals and birds and seed reproduction in early terrestrial plants. Such "inventions," of course, do not come into existence through a single, simple change; rather, they are the result of a number of changes that confer a striking new functional ability.

Direction

The recognition of grades, particularly within the histories of higher taxonomic groups, led early paleontologists to hypothesize the existence of **progressive evolution**, an evolutionary sequence displaying a constant direction, or "goal," and coming into existence independent of selective forces. Usually, this directed evolution was thought to be the result of some unknown "internal force" that either altered embryological development or caused the organism to acquire adaptive characteristics that could be passed on to future generations. The modern synthesis has been skeptical about such forces and about the existence of progressive evolution. Although the modern theory of evolution recognizes directional selection in a population, paleontologists have argued that a critical look at the existing evidence does not support the idea of a linear, progressive evolution of major groups. Examining the evolution of ammonites provides a good model of how analyses of data have led to challenges to the claim for progressive evolution.

The subclass Ammonoida of the class Cephalopoda (mollusks) is divided into three large orders—Goniatitida, Ceratitida, and Ammonitida—that were each dominant during three different time periods (see Figure 25.9A). The goniatites lived during the last three periods of the Paleozoic, the ceratites followed them in the first period of the Mesozoic, and the true ammonites flourished during the last two periods of the Mesozoic. The three orders show a clear progression in shell complexity (Figure 25.9B). Although the functional significance of the shell complexity is not well understood, taxonomists rank the three orders as representing different grades. One might suppose that these three grades are evidence of progressive evolution.

A careful look at the ammonite fossil record, however, dispels the notion of progressive evolution. It is true that one can arrange the three orders in a chronological order that reflects the order of shell complexity, but the "higher" ceratites did not evolve from the "lower" goniatites, nor did the "highest" grade, the true ammonites, evolve from the "middle" grade ceratites. Rather, as can be seen in Figure 25.9, the ceratites arose from a different lineage, the Prolecanitida, that existed along with the "lower" goniatites and underwent a rapid radiation after the extinction of the goniatites. Similarly, the true ammonites radiated from a lineage separate from the ceratites, and did so after their extinction. To arrange the three orders of ammonites into a sequence and suggest that a linear development accounts for their evolution is to oversimplify the evidence from the fossil record.

According to recent critics of the idea of progressive evolution, it is acceptable to describe directional changes in populations or adaptive trends in the evolution of higher taxa. Such changes can be explained without resorting to alleged forces independent of natural selection that supposedly drive evolutionary development toward some goal. However, a careful examination of the fossil record casts serious doubt on the existence of long linear evolutionary sequences that display a continual advancement of grades. Progressive evolution was a hypothesis that made for nice museum displays, but it was not strong enough to stand up to the careful scrutiny of research.

Figure 25.9

The fossil record of ammonite evolution shows no direct progression. (A) The three large orders of ammonites—Goniatitida, Ceratitida, and Ammonitida—appear chronologically in the fossil record. However, the Ceratitida arose from a lineage (Prolecanitida) that existed at the same time as the Goniatitida but was not part of it. Similarly, the Ammonitida arose from a lineage that was not part of the Ceratitida. Comparing the three large orders of ammonites does reveal a progression in the complexity of the suture pattern of their shells (B), which may have misled earlier paleontologists to think that the three orders were an example of progressive evolution.

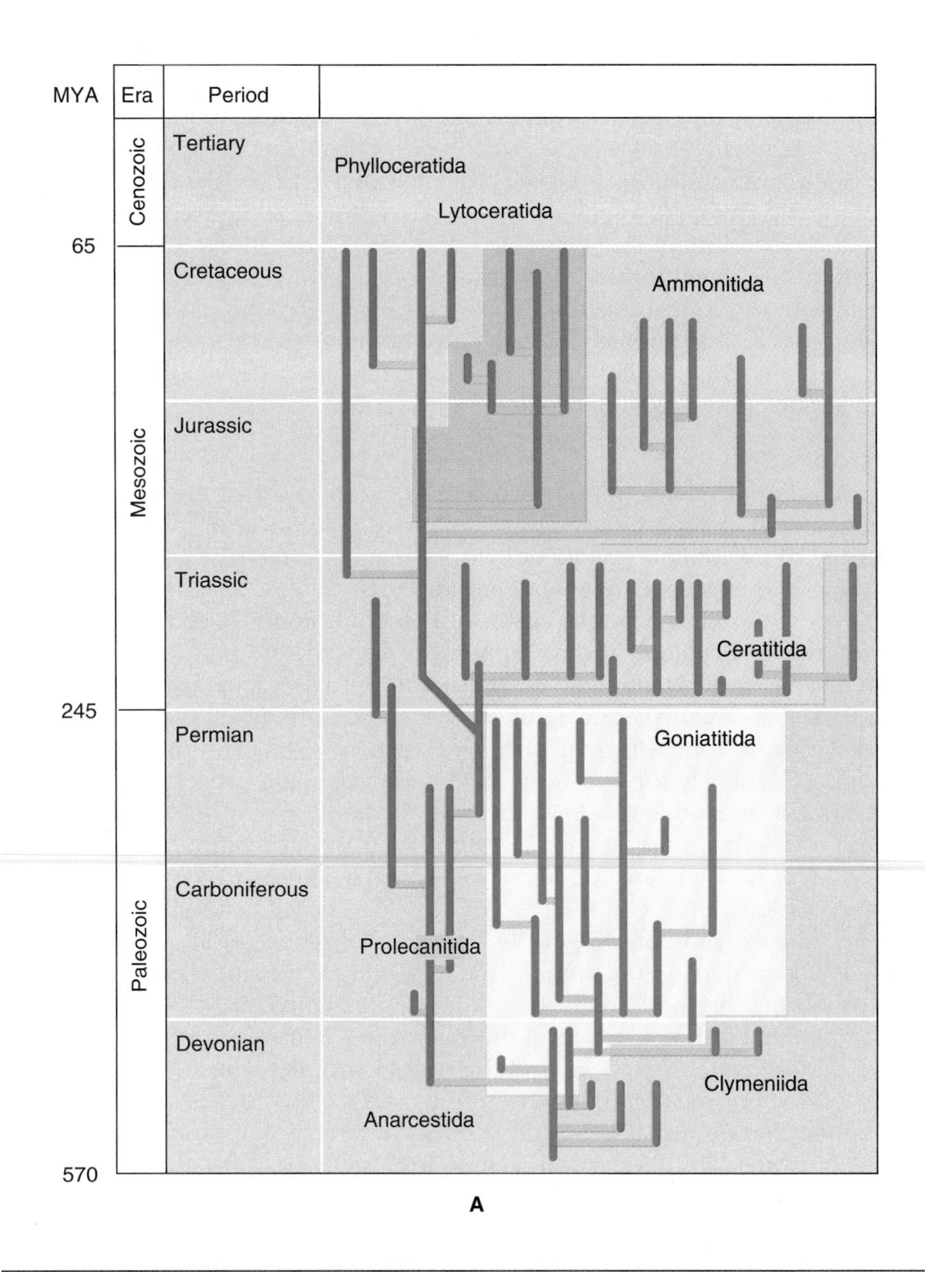

A

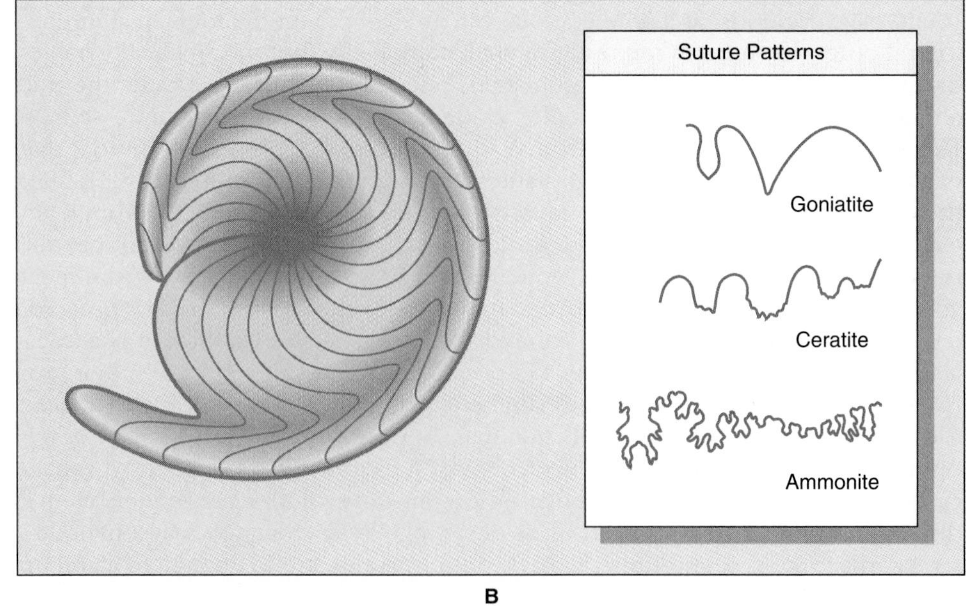

B

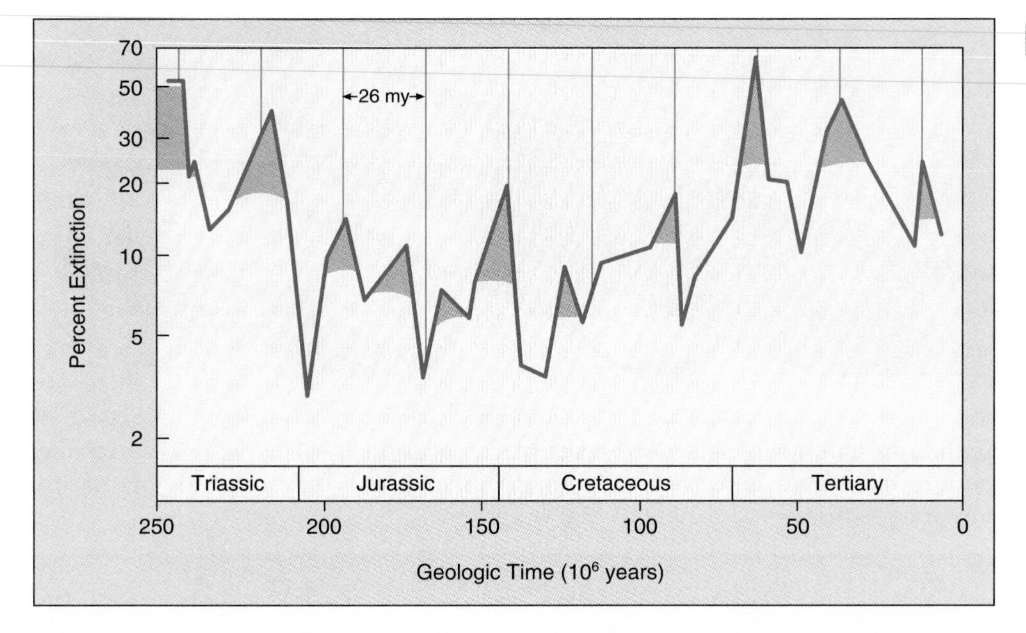

Figure 25.10

Graphing the family extinction rates of marine vertebrates, invertebrates, and protozoans for the time interval starting in the late Permian and extending through the Pliocene reveals a curve with 12 peaks. The lines above the curve mark 26-million-year intervals.

Mass Extinctions

Mass extinction refers to the destruction of a vast number of taxa; in a sense, it is the opposite of adaptive radiation. The commonly used method of dividing the geologic record into different periods is partly a reflection of mass extinctions of the past.

In 1982, two scientists, David Raup and John Sepkoski, Jr., of the University of Chicago, described five large, separate extinction events that occurred during the late Ordovician, Devonian, Permian, Triassic, and Cretaceous periods. Since most of these time periods coincide with known changes in Earth's surface and shifts in climate, scientists accept the view that general changes in the environment were responsible for the extinctions. Two years later, the same two scientists published results of another study concerned with families of marine vertebrates, invertebrates, and protozoans over the past 250 million years. Their new results, described graphically in Figure 25.10, showed 12 peaks of extinction with a regular cycle of 26 million years between the peaks. This regularity raised some very interesting questions and suggested to some scientists that there was an extraterrestrial cause of the extinctions.

The concept of a periodic (regular, recurring) astronomical cause for extinctions on Earth was not a new idea in 1982. In 1980, Luis Alvarez, his son Walter, and his colleagues at the University of California at Berkeley proposed that extinctions at the end of the Cretaceous were caused by the impact of an asteroid. They argued that the collision produced a cloud of dust that lasted long enough to suppress photosynthesis, which in turn resulted in a dramatic collapse of existing food webs. Their evidence was related to the presence of a layer of iridium, a rare metal found in asteroids, that was deposited in marine sediments from the late Cretaceous period, presumably as a result of the impact (see Figure 25.11). The Alvarez hypothesis is very controversial among paleontologists. Recent studies suggest that considerable extinction took place in the Cretaceous period *before* the iridium layer was deposited. Such evidence has weakened the Alvarez hypothesis. Nonetheless, the regularity of extinction events has led many scientists to give serious attention to nonbiological factors for explaining the fossil record.

The study of mass extinction has also suggested another feature of natural selection. Families that had a wide geographical distribution were more likely to survive than more geographically restricted families. This pattern suggests that

Figure 25.11

Scientists have found a layer of iridium in the boundary between the Cretaceous and Tertiary periods.

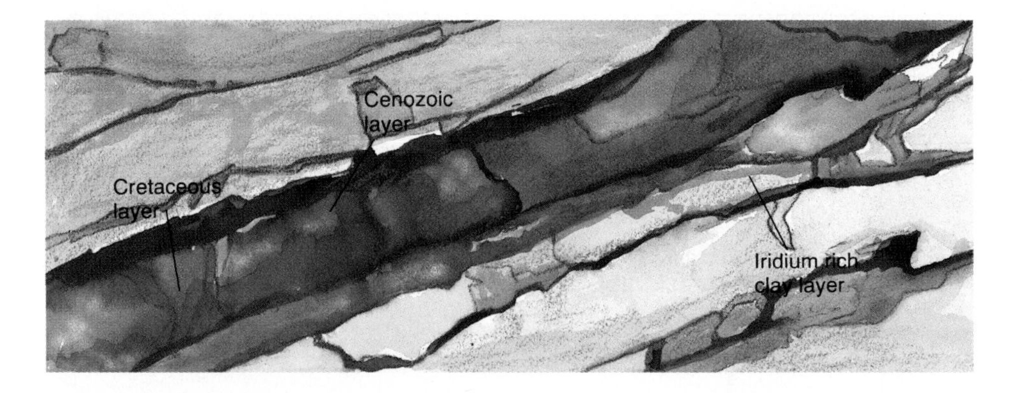

perhaps a "rare-event selection" takes place occasionally. That is, under extreme conditions, some generally overlooked adaptations may have an increased importance that is not evident during other times.

The creation of a new grade occurs when an evolutionary change results in a new and significant functional ability. The study of grades led some earlier scientists to believe in progressive evolution. More recent careful analysis of the fossil record has shown that alleged examples of progressive evolution are not valid.

Mass extinctions, the destruction of vast numbers of taxa, have occurred numerous times in the past. Evidence implies that extinctions were the result of general changes in the environment. The seemingly periodic occurrence of mass extinctions, however, has led some scientists to hypothesize an extraterrestrial cause.

Fossils and the Modern Synthesis

How does the modern synthesis explain evolutionary changes that are reflected in the fossil record? Large-scale modifications are thought to have resulted from the same factors that were responsible for microevolution. In stable environments, populations often became well adapted, and stabilizing selection worked to maintain the status quo. Under these conditions, a static evolutionary line could persist for great periods of time and result in "living fossils."

When the environment changed, populations migrated, changed, or became extinct. Various scenarios can be imagined, using information from Chapters 23 and 24. The rate of change could be slow or rapid, depending on the speed of environmental change, the size of the populations, and the amount of variation among subpopulations. A very small population might not have enough genetic variability to change in a changing environment and could become extinct. A very large population with considerable gene flow among subgroups might evolve slowly in a gradually changing environment. The result would be phyletic evolution. Such a population probably would not be able to survive a rapid change in the environment because new genetic combinations would have less of a chance of becoming established due to gene flow. However, in a large population that is broken into semi-isolated populations with limited migration, genetic drift might lead to the rapid change of some colonies. If these changes are highly adaptive, they could spread to the population at large. Or if the colonies or subpopulations were able to exploit new food sources or to exploit the environment in a different way, divergence might occur. In time, several species might branch or radiate from a common lineage, forming what we would recognize as a new genus. By extrapolating from these processes, the modern synthesis attempts to explain the origin of new genera and higher taxa (see Figure 25.12).

According to the modern synthesis, then, the origin of most higher taxa came about when either a colonizing population entered a new and relatively unoccupied territory, which permitted it to expand and diversify, or when indi-

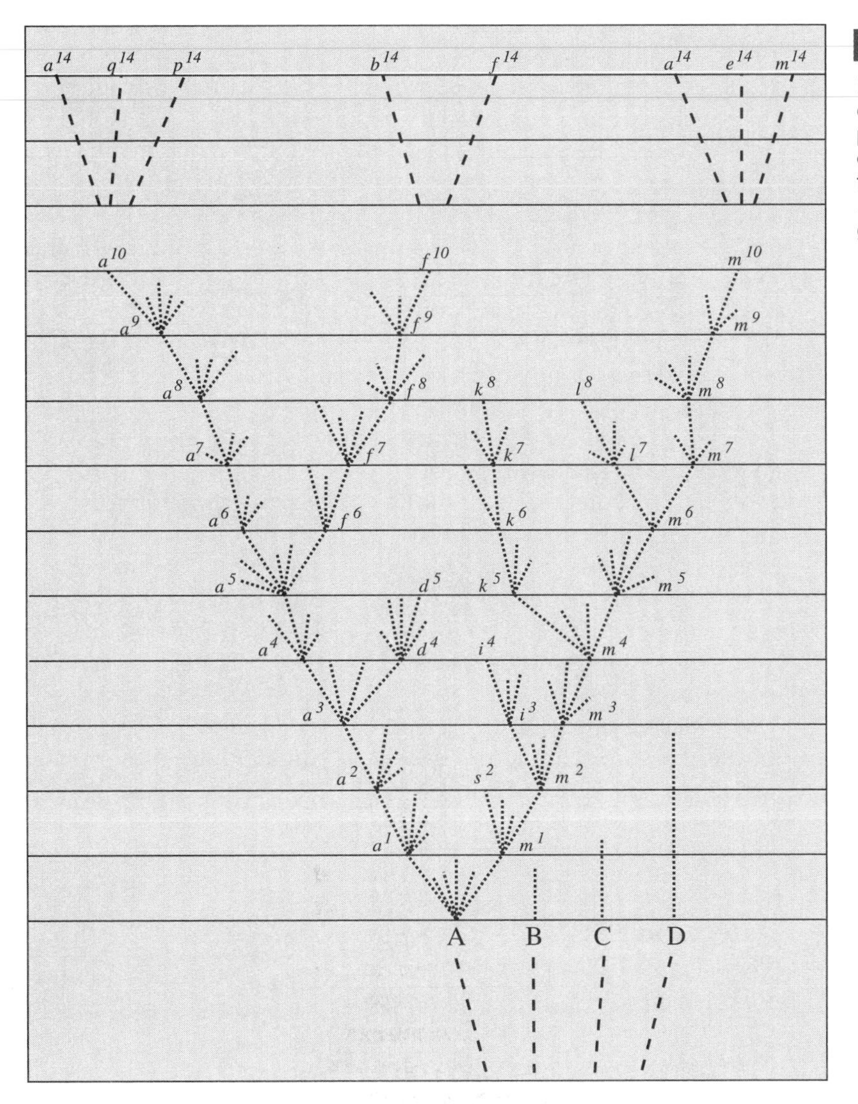

Figure 25.12

Repeated divergence coupled with extinctions of intermediate forms can produce distinctive groups of species, or genera. This is part of an illustration that Darwin used in his *Origin of Species* to convey the concept of divergence and genus formation.

viduals in a population developed some new characteristic that permitted them to exploit new habitats. In both of these circumstances, successive adaptive radiations account for the extent of evolutionary divergence.

Expanding the Modern Synthesis

In recent years, some scientists have challenged the claim that macroevolution can be explained by microevolutionary processes and have argued that the modern synthesis has to be expanded to deal adequately with the history of life on Earth as reflected in the fossil record. The best-publicized of these critiques concerns the rate of evolution. In 1972, Niles Eldredge of the American Museum of Natural History and Stephen Jay Gould of Harvard University described their **punctuated equilibrium** model of evolution, which claimed that macroevolution occurred during relatively short bursts of change, followed by long stable periods. Although this model received a lot of attention, it is not clear that it contradicts or alters the modern synthesis in any significant manner. The originators of the modern synthesis agreed that major events did occur rapidly, infrequently, and with the dramatic radiations that punctuated equilibrium describes. The debate currently centers on whether or not species generally evolved gradually or generally remained static, punctuated with occasional bursts of species formation. There are numerous examples of each in the fossil record. Some paleontol-

THE FOSSIL RECORD: 700 MILLION YEARS OF LIFE

Focus on Scientific Explanations

Chapter 2 explained the evolution of prokaryotic and eukaryotic cells. Prokaryotes make their first appearance in the fossil record about 3.5 billion years ago, and eukaryotes appeared 1.5 billion years ago. Very few fossil remains from before the Paleozoic, however, have ever been found. We have some evidence of invertebrate evolution in the period immediately before the Paleozoic. The first evidence consists merely of burrows left in sediment by primitive organisms; the earliest of these traces is 700 million years old. In South Australia, a deposit of fossils that date from 680 to 580 million years ago contains the remains of corals, segmented worms, and other marine invertebrates. Some of these are from modern phyla; others are not. Figure 1 shows some fossils from the different geological eras.

Paleozoic Era

The Paleozoic era consists of seven periods, beginning with the Cambrian and ending with the Permian.

Cambrian Period

What do we know about life in the Cambrian period? The fossil record reveals that starting about 570 million years ago, there was a sudden burst of diversity: nearly all the modern animal phyla appeared over the relatively short period of 50 million years. The fossil record contains not only all of the basic animal body plans that exist today, but also many that have become extinct. In the Burgess Shale of British Columbia, researchers discovered an unusual collection of soft-bodied fossils that contains the remains of almost a dozen phyla that have no modern counterparts. Producers do not show the same diversity and are limited to algal forms in the Cambrian.

Ordovician Period

Paleontologists have determined that late in the Cambrian and extending into the Ordovician (starting about 500 million years ago), a lineage evolved that developed an effective swimming ability. It contained organisms that came to have armor plates for protection and an axlelike skeleton with vertebrae that had lateral extensions. These were the first, primitive, jawless fishes, or *ostracoderms*. Animal life was still confined to the oceans, but during this period, much of what is now dry land was covered with shallow seas, which permitted primitive plant growth. A great increase in bivalves and other bottom-dwelling filter-feeding invertebrates can be observed in the fossil record of this period.

Silurian Period

Fossil evidence suggests that land plants first appeared in the Silurian period, which began 441 million years ago. They were simple

Figure 1A Paleozoic era: priapulid worm from the Burgess shale.

Figure 1B Paleozoic era: sea scorpion.

branching stems (up to about 12 inches long), some of which had small, scalelike structures that look like leaves. The adult plants developed spore cases that released airborne spores. Animals inhabited shallow, warm seas, and Silurian deposits reflect a diversification of fishes, as well as the presence of dramatic-looking giant sea scorpions and bottom-dwelling invertebrates.

Devonian Period

Some 413 million years ago, the Devonian period—the Age of Fish—began, according to paleontologists who have studied the fossil record. As its name suggests, a great radiation of fishes dates from this period. Jawed fish with paired fins came into existence. The fossil record contains cartilaginous fish (sharks and rays), which remained mostly marine, and bony fish that proliferated in fresh water. Among the bony fish were ray-finned fish, the ancestors of modern fish, and lobe-finned fish, some of which are thought to be the ancestors of amphibians, the first vertebrates that appeared on land. Ammonites, a group of externally shelled cephalopods that have been widely used by geologists to date strata, also made their appearance during this period. Land plants formed the first forests, which were dominated by ferns, club mosses, and horsetails. The first wingless insects are found in the late Devonian.

Carboniferous Period

Beginning 365 million years ago, the fossil record shows, great portions of the Northern Hemisphere were covered with huge swamp forests that were drowned late in the period by shallow seas, resulting in the production of massive coal deposits. Amphibians exploited many habitats during the Carboniferous and attained large body sizes. The first reptiles, which arose from a primitive amphibian lineage, appeared, as did the flying insects.

Permian Period

The Permian, starting 290 million years ago, marks the end of the Paleozoic era. Reptiles succeeded amphibians as the dominant land animals, and new insect groups emerged. Evidence indicates that great geologic changes altered the climate and topography. Gymnosperms began to increase, and conifers came to dominate many forests. The end of the Permian saw one of the most dramatic mass extinctions in the fossil record; nearly half of the known families of animals disappeared. Marine invertebrates were especially hard hit.

Mesozoic Era

The Mesozoic era, which began 245 million years ago, is conventionally divided into three periods: the Triassic, the Jurassic, and the Cretaceous. The most sensational events of this era center on the reptiles, which underwent an extensive set of radiations. Dinosaurs, that group of stars from the fossil record that intrigue not only the professional paleontologist but also grade school children and Hollywood movie makers, date from this era.

Figure 1C Mesozoic era: ammonite.

Figure 1D Mesozoic era: prawn.

The dinosaurs evolved in the Triassic and succeeded the dominant reptiles from the Permian. The dinosaurs themselves underwent several waves of extinction but each time radiated from surviving lineages and dominated Earth. At the close of the Cretaceous, they too entered the abyss of extinction, along with ammonites and some other widely distributed animals.

During the Mesozoic, the supercontinent Pangaea split apart, and by the Cretaceous period, the continents were taking on their modern shapes and positions. Earth's vegetation underwent a complete change in the last part of the Mesozoic. The first flowering plants (angiosperms) appeared roughly 130 million years ago. They underwent a spectacular radiation, which led to one of the greatest success stories in the evolutionary saga, the development of over 250,000 species. Ironically, the history of that radiation is one of the least understood chapters in evolution.

Plant and reptile evolution were not the only important Mesozoic developments. Birds and mammals arose, as did the modern orders of insects and the modern fishes.

Cenozoic Era

Some 65 million years ago, the "Age of Recent Life" began. It is divided by scientists into the Tertiary and Quaternary periods. The Tertiary comprises the bulk of the era (all but the past 2 million years) and is divided into five epochs. The Quaternary consists of two epochs, the Pleistocene and the Recent.

Depending on one's point of view, the Cenozoic can be called the "Age of Mammals," the "Age of Birds," the "Age of Insects," or the "Age of Flowering Plants." There was enormous diversification and speciation in all these groups.

Mammals are found in the fossil record to have succeeded the dinosaurs on land. They had coexisted as small, active, often nocturnal animals, but unlike the dinosaurs, some of them (including the primates) survived the mass extinctions at the end of the Cretaceous. From the several lineages that entered the Cenozoic, a strong set of radiations occurred, reaching a peak within the past 2 million years.

The Pleistocene witnessed the ice ages in the Northern Hemisphere as well as the rise of that most peculiar animal, *Homo sapiens*. Many of the large mammals became extinct before the Recent epoch, although there is evidence that humans coexisted with several of them.

Figure 1E Cenozoic era: crinoids.

Figure 1F Cenozoic era: fish.

Figure 1G Cenozoic era: snake.

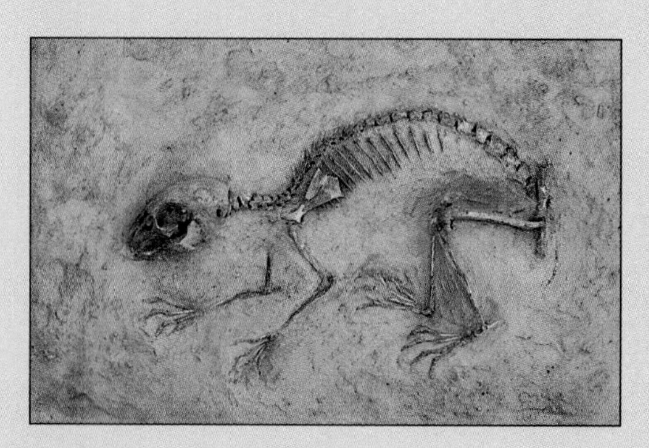

Figure 1H Cenozoic era: hare.

ogists believe that since species may evolve in several different ways, the current debate over punctuated equilibrium is not of great significance.

However, the associated claim that mere extrapolation from microevolution is not adequate to understand macroevolution is important. Eldredge and Gould, along with several other prominent evolutionists, believe that during periods of rapid radiation many new species came into being and that selection then operated on species, not just on individuals. **Species selection** assumes that certain species give rise to new species at a higher rate or last for longer intervals of time so that they produce more descendant species than others. According to species selection, some species possess adaptations that are general and therefore can give rise to many descendant species. In contrast, some new species are adapted to only a very specialized or limited set of local circumstances and may leave few or no descendant species. This view modifies the modern synthesis in having selection operate on species as well as on individuals. The notion of species selection is highly controversial today, and only further research may be able to resolve the problem.

Another possible reason to expand the modern synthesis comes from the studies of extinction patterns. Some paleontologists feel that to understand those patterns, the modern synthesis must incorporate a consideration of nonbiological factors such as astronomical effects and other rare-event selection. Some biologists also argue that to understand the fossil record better, the modern synthesis needs to integrate more fully into the theory of evolution complex biological concepts like *ecological collapse* (extensive mortalities of ecologically dependent organisms).

Debates that surround interpretation of the fossil record and its relationship to the modern synthesis show the dynamic nature of the theory of evolution. These debates do not attack the validity of the modern synthesis; rather they point to new directions in which it may be extended. Extensive research in paleontology, ecology, genetics, and population biology will ultimately establish the strength of the various ideas. At present, no new expanded version of the synthesis has met the challenge of incorporating these new ideas about the fossil record. Are we about to enter a period of punctuated equilibrium in the historical development of the theory of evolution?

PHYSIOLOGY AND EVOLUTION

Recall that physiology is the study of how organisms function. It depends heavily on chemical and physical methods. Historically, physiology was the first of the biological disciplines to develop experimental techniques. Unit IV of this text covers several areas of importance in human physiology.

Modern physiology is divided into cellular physiology, the physiology of special groups, and comparative physiology. Since at the cellular level, living organisms share many basic characteristics, **cellular physiology** is regarded by biologists as "general physiology," or the examination of fundamental functions of living organisms. The **physiology of special groups** focuses on the functioning of specific groups or species that have a special interest. For example, human physiology because of its value for medical research and plant physiology because of its agricultural importance are topics in the physiology of special groups. **Comparative physiology** examines the mechanisms that different organisms use to perform similar functions.

Relationship of Physiology to the Theory of Evolution

The leading physiologists of the mid-nineteenth century, such as Claude Bernard in France and Rudolf Virchow in Germany, were hostile to Darwin's theory of evolution when it was first proposed. Although some physiologists were inclined to believe that organisms had evolved, they rejected his theory in part because they believed that any theory that unified the biological sciences would have to be at the cellular level (Darwin didn't even mention cells in the *Origin of Species*!). Physiologists were also unreceptive because they were primarily experimental biologists, and Darwin's theory did not easily lend itself to experimental verification.

Today, physiologists have a different view of evolution. The theory serves not only as an intellectual context that relates physiology to other biological disciplines but also explains why there are similarities in basic but complex physiological processes in all cells, for example, metabolic pathways and the functions of nucleic acids (see Chapter 16). The common evolutionary history of single and multicellular organisms accounts for these similarities.

Analyses of certain biochemical compounds has contributed to evolutionary theory. Some of the best-known studies concern the protein cytochrome *c*, which is found in the cells of all animals and plants. It plays a central role in energy-generating metabolic reactions. Because the cytochrome *c* molecules of

Figure 25.13

A comparison of the amino acid sequences of cytochrome c shows a striking similarity among living organisms. Consider the amino acid sequences of cytochrome c for human, kangaroo, tuna, and wheat in this table. Each amino acid is represented by a single letter.

Amino Acid Sequences

Organism	Positions 1–49
1. Human	--------GDVEKGKKIFIMKCSQCHTVEKGGKHKTGPNLHGLFGRKTG
2. Rhesus monkey	--------GDVEKGKKIFIMKCSQCHTVEKGGKHKTGPNLHGLFGRKTG
3. Horse	--------GDVEKGKKIFVQKCAQCHTVEKGGKHKTGPNLHGLFGRKTG
4. Dog	--------GDVEKGKKIFVQKCAQCHTVEKGGKHKTGPNLHGLFGRKTG
5. Gray whale	--------GDVEKGKKIFVQKCAQCHTVEKGGKHKTGPNLHGLFGRKTG
6. Rabbit	--------GDVEKGKKIFVQKCAQCHTVEKGGKHKTGPNLHGLFGRKTG
7. Kangaroo	--------GDVEKGKKIFVQKCAQCHTVEKGGKHKTGPNLNGIFGRKTG
8. Penguin	--------GDIEKGKKIFVQKCSQCHTVEKGGKHKTGPNLHGIFGRKTG
9. Snapping turtle	--------GDVEKGKKIFVQKCAQCHTVEKGGKHKTGPNLNGLIGRKTG
10. Bullfrog	--------GDVEKGKKIFVQKCAQCHTCEKGGKHKVGPNLYGLIGRKTG
11. Tuna	--------GDVAKGKKTFVQKCAQCHTVENGGKHKVGPNLWGLFGRKTG
12. Silkworm moth	----GVPAGNAENGKKIFVQRCAQCHTVEAGGKHKVGPNLHGFYGRKTG
13. Wheat	ASFSEAPPGNPDAGAKIFKTKCAQCHTVDAGAGHKQGPNLHGLFGRQSG
14. Fungus (*Neurospora*)	----GFSAGDSKKGANLFKTRCAECHGEGGNLTQKIGPALHGLFGRKTG

Organism	Positions 50–112
1. Human	QAPGYSYTAANKNKGIIWGEDTLMEYLENPKKYIPGTKMIFVGIKKKEERADLIAYLKKATNE
2. Rhesus monkey	QAPGYSYTAANKNKGITWGEDTLMEYLENPKKYIPGTKMIFVGIKKKEERADLIAYLKKATNE
3. Horse	QAPGFTYTDANKNKGITWKEETLMEYLENPKKYIPGTKMIFAGIKKKTEREDLIAYLKKATNE
4. Dog	QAPGFSYTDANKNKGITWGEETLMEYLENPKKYIPGTKMIFAGIKKTGERADLIAYLKKATKE
5. Gray whale	QAVGFSYTDANKNKGITWGEETLMEYLENPKKYIPGTKMIFAGIKKKGERADLIAYLKKATNE
6. Rabbit	QAVGFSYTDANKNKGITWGEDTLMEYLENPKKYIPGTKMIFAGIKKKDERADLIAYLKKATNE
7. Kangaroo	QAPGFTYTDANKNKGIIWGEDTLMEYLENPKKYIPGTKMIFAGIKKKGERADLIAYLKKATNE
8. Penguin	QAEGFSYTDANKNKGITWGEDTLMEYLENPKKYIPGTKMIFAGIKKKSERADLIAYLKDATSK
9. Snapping turtle	QAEGFSYTEANKNKGITWGEETLMEYLENPKKYIPGTKMIFAGIKKKAERADLIAYLKDATSK
10. Bullfrog	QAAGFSYTDANKNKGITWGEDTLMEYLENPKKYIPGTKMIFAGIKKKGERQDLIAYLKSACSK
11. Tuna	QAEGYSYTDANKSKGIVWNNDTLMEYLENPKKYIPGTKMIFAGIKKKGERQDLVAYLKSATS-
12. Silkworm moth	QAPGFSYSNANKAKGITWGDDTLFEYLENPKKYIPGTKMVFAGLKKANERADLIAYLKESTK-
13. Wheat	TTAGYSYSAANKNKAVEWEENTLYDYLLNPKKYIPGTKMVFPGLKKPQDRADLIAYLKKATSS
14. Fungus (*Neurospora*)	SVDGYAYTDANKQKGITWDENTLFEYLENPKKYIPGTKMAFGGLKKDKDRNDIITFMKEATA-

A Alanine	F Phenylalanine	K Lysine	P Proline	T Threonine
C Cysteine	G Glycine	L Leucine	Q Glutamine	V Valine
D Aspartic acid	H Histidine	M Methionine	R Arginine	W Tryptophan
E Glutamic acid	I Isoleucine	N Asparagine	S Serine	Y Tyrosine

different species perform the same function, have similar amino acid sequences, and have comparable overall structure, they are considered homologous. When the amino acid sequences of cytochrome *c* proteins of different species are placed on a chart, their arrangement is strikingly similar (see Figure 25.13). Closely related species have amino acid sequences that are more alike than those of distantly related species. For example, the cytochromes *c* of humans and rhesus monkeys differ by only one amino acid, while those of humans and kangaroos differ by ten.

By calculating the minimum number of nucleotide substitutions in the genetic code that are necessary to produce these differences, scientists have constructed a phylogeny that agrees for the most part with those derived from the fossil record (see Figure 25.14). Other proteins have also been used in a similar manner. Not only does the comparative physiological approach verify generalizations derived from the fossil record, but it is also a valuable supplement in cases where the fossil record is weak or nonexistent.

Comparative studies of proteins and nucleic acids also suggest that base substitutions in DNA or RNA accumulate at a steady rate. Therefore, such macromolecules can act as "molecular clocks," and molecular comparisons can be used to date evolutionary divergences. The greater the dissimilarity, the older the evolutionary divergence.

Physiology and Paleontology

Although the theory of evolution provides a conceptual framework that relates physiology to other biological disciplines and provides an explanation for similar chemical processes in diverse organisms, it does not have as close a relationship to physiology as it does to paleontology. By analyzing, for example, the chemical basis of muscle contraction, hormonal regulation or reproduction, or the response of plants to airborne pollutants, physiologists can tell us how organisms

Figure 25.14

By comparing the differences among the amino acid sequences of cytochrome c in different species, scientists have been able to construct phylogenetic tables like this one that depict evolutionary relationships. The numbers on the lines are the number of nucleotide substitutions that have taken place.

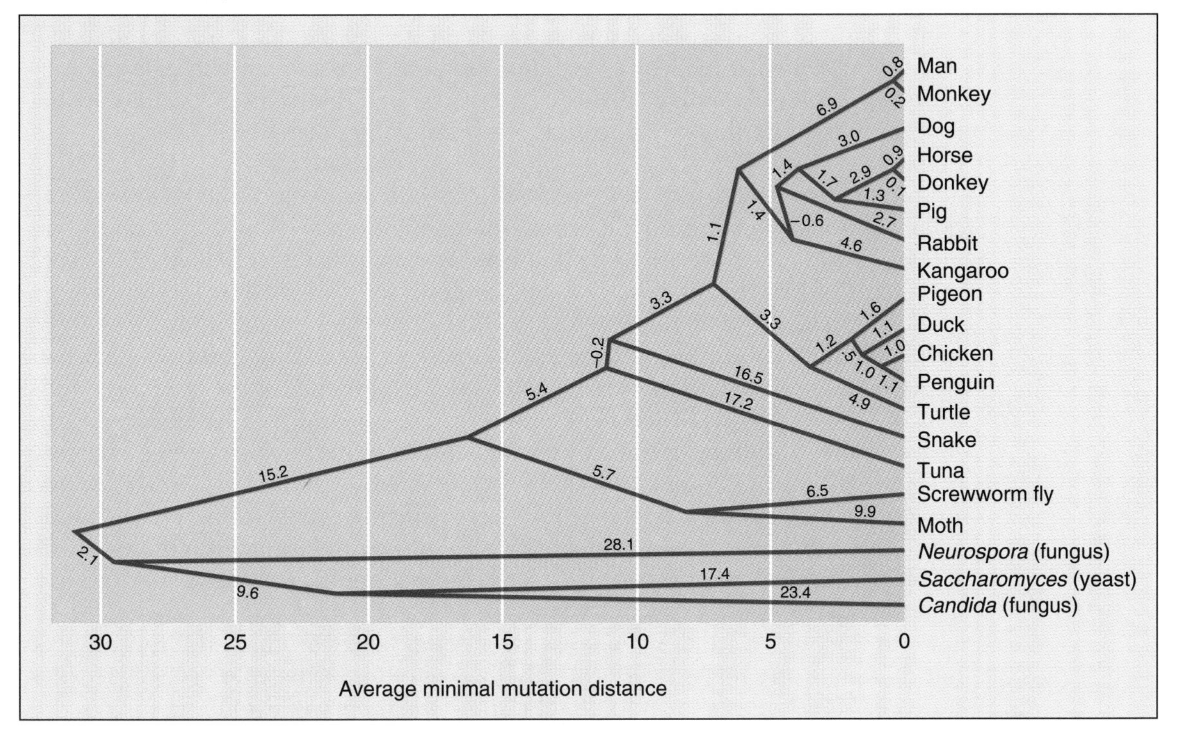

are *currently* functioning. The frame of reference is very short from an evolutionary perspective. Although the information physiologists generate occasionally has relevance for understanding evolutionary processes, most physiologists do not design their studies with evolutionary theory in mind or with an eye toward deepening our understanding of the history of life on this planet. Rather, they explore how living things work at present.

Physiology and paleontology provide a contrast in how the theory of evolution is related to the separate disciplines of biology. In paleontology, evolution not only provides the background but is central in formulating research questions that in turn have important consequences for the theory. Physiology, in contrast, looks to evolution to explain the regularities it uncovers. Although some of the specific research questions physiology explores are derived directly from the theory of evolution, most of the field has its historical roots in medicine and agriculture. Most questions that physiologists investigate relate to the functioning (or malfunctioning) of the human body or of organisms of economic importance to humans.

Physiology and paleontology provide a contrast in how the modern synthesis is related to the separate disciplines of biology. For paleontology, the theory of evolution provides a foundation and is central in guiding research. Physiology has provided important insights into evolution and can be interpreted in evolutionary terms to explain the regularities it reveals. However, physiologists are most concerned with how organisms currently function.

OTHER FIELDS OF BIOLOGY AND EVOLUTION

Other biological disciplines fall somewhere between paleontology and physiology in their relationship to evolution, depending on how directly each is guided by questions stemming from evolution. The study of classifying organisms (*taxonomy*) is close to paleontology because much of the work in taxonomy involves constructing classifications that attempt to reflect evolutionary relationships. *Ecology* studies the interaction of groups of organisms among themselves and with the environment. As a discipline related to evolution, it falls somewhere between taxonomy and physiology. Like physiology, it deals with a shorter time frame than paleontology and analyzes some interactions that can be understood without reference to their historical development. In common with paleontology, however, ecology examines issues of major concern to evolution, such as adaptation, and uses the theory of evolution in formulating many of its research questions.

Genetics, the study of heredity, also has an evolutionary as well as a nonevolutionary dimension. Questions stemming from Darwin's ideas on the origin and inheritance of variation stimulated research that eventually led to classical genetics. Investigations of population genetics were vital in formulating the modern synthesis. Modern genetics continues to explore many questions of evolutionary significance, but new areas of genetics are tackling questions that have a totally different time dimension, such as the exploration of the human genome and the development of medical genetics (see Chapter 21). As in physiology, many of these studies attempt to understand how particular systems currently operate, without reference to long-term implications. Similarly, *environmental biology,* which studies how people are disturbing ecosystems of the planet through pollution, destruction of forests, and population growth, generally involves processes that, although of monumental significance for evolution, have a shorter time period.

In Chapters 27 and 28, we discuss a topic, behavior, that until recently was not thought to be related very directly to evolution. In the past several decades, however, it has been shown to be relevant to the theory of evolution.

THE UNFINISHED SYNTHESIS

The modern synthesis has been extraordinarily successful in unifying the biological sciences. No such single theory exists for the physical sciences or the social sciences. We have seen that the modern synthesis does not relate to each of the separate biology disciplines in the same way. For some fields, like paleontology, it provides both a foundation and many research questions. For others, like physiology, it relates research results to other biological subjects and provides a conceptual framework in which to interpret the discipline as a whole.

The modern synthesis has expanded and changed considerably since it was first developed in the 1940s. Research in population genetics, molecular biology, and paleontology has resulted in lively debates and revisions of the theory. The dynamic nature of the theory proves its vitality and strength. There is no sense that the story has been completely told or that biologists have discovered all the laws regulating life. Rather most disciplines in the biological sciences pose fascinating questions that await exploration. The modern synthesis is the unifying thread that ties all of these separate studies together, gives them a coherent set of relationships, and suggests avenues of investigation and integration. That the modern synthesis is an "unfinished synthesis" is not a criticism or a weakness; rather it is an indication that the theory of evolution is one of the most exciting and dynamic products of scientific thought in the history of science and that it promises to continue to be so for a long time.

Summary

1. The modern synthesis attempts to unify all of the current life sciences. Some fields, like paleontology, are closely linked to the modern synthesis; others, such as physiology, are less so.
2. The fossil record is incomplete but does reveal certain patterns of evolution. Some species have remained stable over long periods of time. Some lineages show gradual change without splitting (phyletic evolution); others reveal splitting. Patterns of splitting include divergence, convergence, and parallelism.
3. The fossil record also displays the emergence of grades, the appearance of direction in evolution, and the occurrence of mass extinctions.
4. The modern synthesis attempts to account for macroevolution in terms of the processes of microevolution. Challenges to that view have pointed to uneven rates of evolution and have attempted to account for them by a model called *punctuated equilibrium*. In addition, some evolutionists claim that some new species have more general adaptations and can give rise to many descendant species, a process they call *species selection*.
5. Although physiology has contributed to the study of evolution, for the most part it focuses on how current organisms function. Other biological disciplines fall somewhere between paleontology and physiology in their relationship to evolution.
6. The modern synthesis is still "unfinished" but remains an exciting, dynamic theory.

Review Questions

1. What is a fossil?
2. What limits are there on what the fossil record tell us about past life on this planet?
3. What is divergence?
4. What is convergence?
5. What is parallel evolution?
6. What is a new grade in evolution?
7. Why have scientists abandoned the idea of progressive evolution?
8. What is punctuated equilibrium?
9. What is the relationship of physiology to evolution theory?
10. What is meant by the "unfinished" synthesis?

Essay and Discussion Questions

1. What would a complete, modern synthesis of evolution theory have to encompass?
2. How might you attempt to resolve a conflict between contradictory conclusions reached by paleontologists and molecular biologists about the relationship between two taxa?
3. How would you construct a display on evolution for a presentation to the general public?
4. Many examples in this chapter relied on bird models. Why might this be?

References and Recommended Reading

Alvarez, L. W., W. Alvarez, F. Asaro, and H. V. Michel. 1980. Extraterrestrial cause for the Cretaceous-Tertiary extinction. *Science* 208:1095–1108.

Alvarez, W., and F. Asaro. 1990. An extraterrestrial impact. *Scientific American* 263:78–84.

Bock, W. J. 1970. Microevolutionary sequences as a fundamental concept in macroevolutionary models. *Evolution* 24:704–722.

Eldredge, N. 1989. *Macroevolutionary Dynamics*. New York: McGraw-Hill.

———. 1985. *Unfinished Synthesis: Biological Hierarchies and Modern Evolutionary Thought*. Oxford: Oxford University Press.

———, and S. J. Gould. 1972. Punctuated equilibrium: An alternative to phyletic gradualism. In T. J. M. Schopf, ed., *Models in Paleobiology,* pp. 82–115. New York: Freeman.

Gould, S. J. 1982. Darwinism and the expansion of evolutionary theory. *Science* 216:380–387.

Mayr, E. 1988. *Toward a New Philosophy of Biology*. Cambridge, Mass.: Harvard University Press.

———, and W. B. Provine, eds. 1980. *The Evolutionary Synthesis: Perspectives on the Unification of Biology*. Cambridge, Mass.: Harvard University Press.

Raup, D. M., and J. J. Sepkoski, Jr. 1984. Periodicity of extinctions in the geologic past. *Proceedings of the National Academy of Science* 81:801–805.

Raup, D. M., and S. M. Stanley. 1978. *Principles of Paleontology*. New York: Freeman.

Stanley, S. M. 1989. *Earth and Life Through Time*. New York: Freeman.

Stebbins, G. L., and F. J. Ayala. 1981. Is a new evolutionary synthesis necessary? *Science* 213:967–971.

CHAPTER 26

Human Evolution

Chapter Outline

Human Origins

Human Evolution

Reading Questions

1. What were some major trends in the evolution of primates?
2. What is known about the biological evolution of humans?
3. How does the multiregional model compare with the single-origin model of the evolution of *Homo sapiens?*

The theory of evolution is the basic organizing theory that unifies the study of life and serves as a guiding force in directing research in many areas. Chapters 22–25 discussed the theory of evolution and its relationship to biological disciplines. This chapter describes what evolution has to say about people. Specifically, it examines an issue that has been controversial since Darwin first published *On the Origin of Species* in 1859: the origins of humans.

HUMAN ORIGINS

Darwin didn't actually discuss human evolution in the *Origin*. All that he wrote was a brief statement in the conclusion saying that in future research, "Light will be thrown on the origin of man and his history." A dozen years later, Darwin did publish a book on human evolution, *The Descent of Man,* in which he argued that people had evolved from some preexisting form. This idea was hotly contested, in part because many people thought that biology was overstepping its boundaries and delving into areas that had traditionally been the province of religion and philosophy.

Although well over a century has passed since Darwin's discussion of the origin of humans, the topic is still controversial. Creationists reject the theory of evolution itself and therefore its implications about humans, and scientists have repeatedly revised their picture of human evolution due to the continuing discovery of human fossils and other relevant new data.

HUMAN EVOLUTION

The story of human evolution is known in rough outline. It has been an area of intense research and public fascination. Like the histories of most individual species, the account contains many gaps. The incomplete nature of the fossil record precludes a complete history. Nevertheless, recent discoveries, supplemented with biochemical studies, have allowed for the development of a more detailed picture than ever.

Primates

A brief look at our place in the classification system of animals provides a starting point for the study of human history. As described in Figure 26.1, we belong to the placental mammalian order of **Primates**. The primates are one of the earliest orders of placental mammals that evolved on Earth. Fossil primates date from the late Cretaceous and early Paleocene, approximately 65 million years ago. Contemporary primates are divided into two suborders, *Prosimii,* also called the lower primates, and *Anthropoidea,* the higher primates.

The Prosimii are represented today by tree shrews, lorises, and lemurs. The Anthropoidea include monkeys, apes, and humans (see Figure 26.2). Primates share certain features, the most obvious being their adaptation for living in trees and their ability to eat many different kinds of food. The higher primates have an additional shared characteristic: they spend more time raising a smaller number of offspring.

Life in the Trees

Many animals are adapted to live in trees. However, primates show specific adaptations that distinguish them from other arboreal (tree-dwelling) creatures. Among the most significant primate adaptations are the *freely movable digits* of their hands and feet, an adaptation that permits grasping. In some primates, this has been further refined by the development of opposable thumbs and big toes (see Figure 26.3), which increases grasping and manipulatory abilities.

Primates have exceptional limb flexibility due to a *generalized limb structure.* They have retained most of the individual limb bones that characterize the

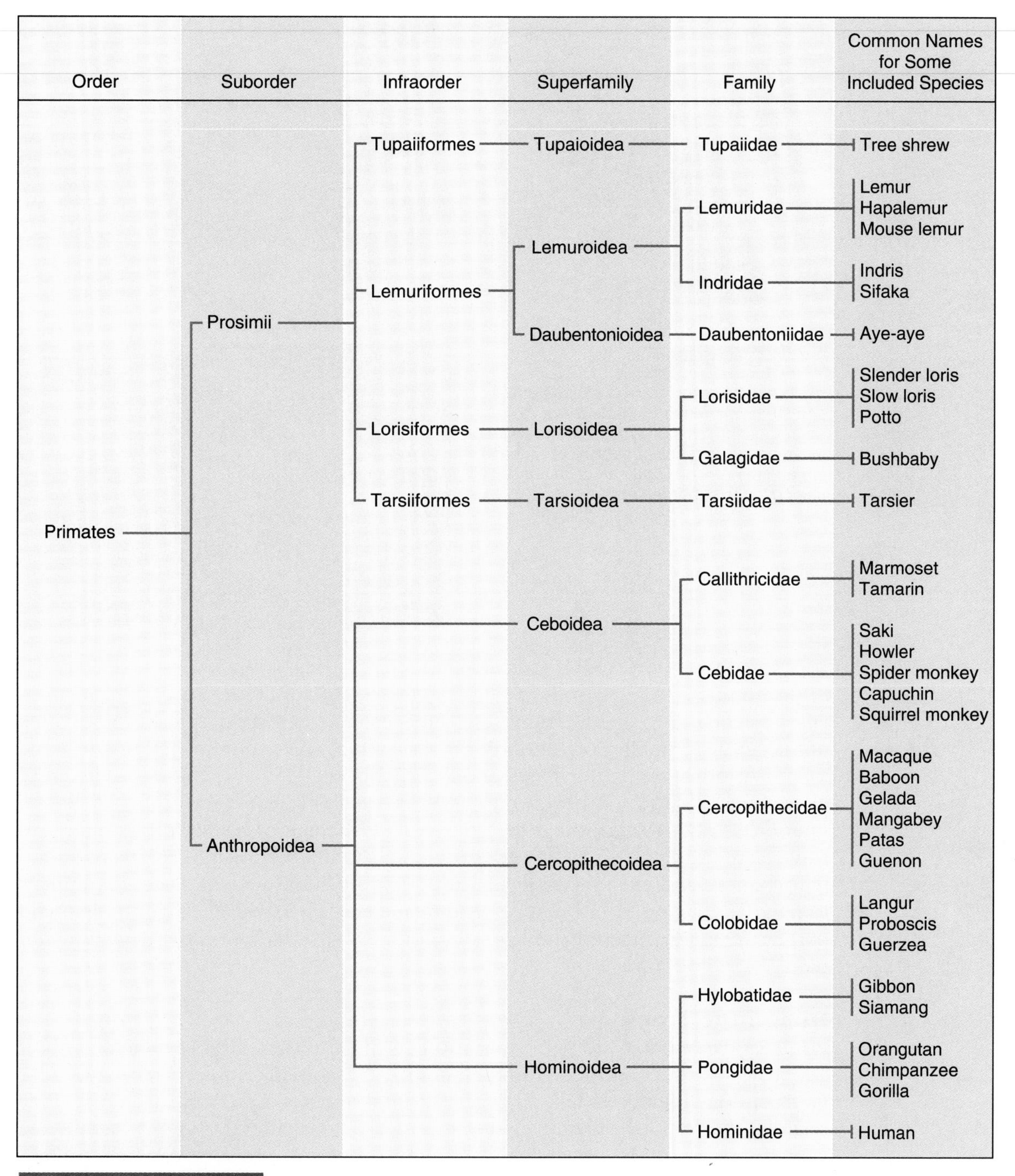

Figure 26.1

Primates are one of the 17 living orders of placental mammals. This taxonomy of the primates shows their diversity, from tree shrews to humans.

early placental mammals from which they evolved. Also, primates conserved the clavicle (collarbone), which was lost in many other mammals. The clavicle gives primates greater arm motion, permitting them to perform such actions as placing their hands behind their heads. Another feature retained in primates is two separate bones (the radius and the ulna) in the lower arm, which enhances flexibility in turning the hand.

Important changes in *sensory systems* also characterized primate evolution. Sight became the most important sense, coupled with a corresponding decline in other senses. Primates developed the ability to judge depth, to see in color, and

A

B

C

Figure 26.2

(A) The bush baby (*Galago senegalensis*) and the brown lemur of Madagascar (B) are representatives of *Prosomii*. The white-faced capuchin monkey of tropical America (C) and the African lowland gorilla (D) belong to the suborder *Anthropoidea*.

D

to discriminate objects in their field of vision. The ability to grasp and manipulate objects and to see them more clearly resulted in a reduction in importance of the sense of smell. Correspondingly, the primate nose diminished in size.

Two other adaptations associated with arboreal life were the replacement of claws by *nails* and the development of an *upright posture*. Although nails are inferior to claws for gripping, they offer the advantage of having broad, flat, sensitive tactile surfaces on their undersides. This touch sensitivity extends the ability to explore and examine objects. An upright position made it easier for animals to hang and swing from branches and to increase their visual range. It also held particular significance for the later development of walking on two legs in chimpanzees and people.

Diet in the Trees

Primates can eat a wide variety of foods. This ability is largely related to their dental structure, combined with their ability to manipulate objects. Primates have unusual dental features. Like their retention of a generalized limb structure, they retained the primitive mammalian trait of possessing different types of teeth (see Figure 26.4). Early mammals had four types of teeth—incisors, canines, premolars, and molars—each with a different purpose. Although primate evolution

A

B

Figure 26.3

Thumb and toe opposability permit greater grasping power. The gibbon (A) can examine an object with one hand, whereas other mammals, like the common mouse (B), need to use both.

shows a reduction in the number of teeth, most primates retained the different types of teeth and with them the ability to use different food sources.

Brains and Babies

Compared to other animals, higher primates invest more time raising fewer babies. As a result, offspring learn more over an extended period of time and have a greater chance of survival. The behavioral consequence of this increase in *parental investment* was an increase in the importance of social groups. Modern higher primates are almost invariably social. The increase in parental investment is also related to a general evolution of greater intelligence in the higher primates, which is reflected by increased brain size and complexity.

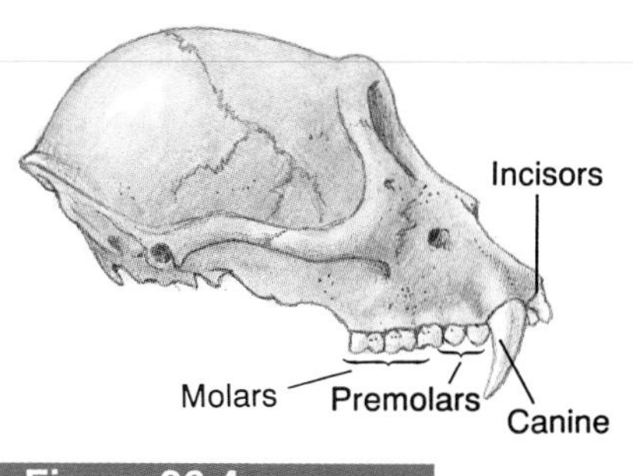

Figure 26.4

Primates have four types of teeth, each used for a different purpose.

The evolution of primates reflects adaptations for living in trees and for the ability to eat many different kinds of food. Among the most significant primate adaptations are freely movable digits of the hands and feet, generalized limb structure, changes in the sensory systems that enhance sight, replacement of claws by nails, the development of upright posture, and the ability to eat a wide range of food due to their dental structure and their ability to manipulate objects. Higher primates invest much time raising few offspring.

Hominids

To trace the evolution of **hominids**—that is, any species belonging to the human family—scientists usually begin with the fossil remains of the earliest known Anthropoidea (higher primates), which are approximately 35 million years old and come from Egyptian deposits dating from the Oligocene epoch. One of these fossils, belonging to the genus *Aegyptopithecus,* is thought to be the common ancestor of apes and hominids that diversified in the Miocene.

Higher primates underwent several radiations in the Miocene, but by the end of the epoch, approximately 5 million years ago, few descendants of these groups remained. Much of the research on these fossils has been to establish when humanlike—or ancestors immediate to humanlike forms—first appeared. One famous fossil that for many years was thought to be in a direct line with humans is *Proconsul africanus,* which was discovered by Mary Leakey in 1948. *Proconsul* species lived from 19 to 13 million years ago. They had features that are both monkeylike and apelike—short arms like a monkey, but teeth more like an ape. Another fossil that until recently was thought to represent the first example of a hominid is *Ramapithecus,* which arose about 19 million years ago in Africa and later migrated to Europe and Asia.

Recently discovered fossils and biochemical research indicate that hominids diverged from apes much later than was previously thought. The discovery of new and more complete *Ramapithecus* fossils has led paleontologists to conclude that it was more similar to orangutans than to humans and was probably ancestral to Asian great apes, including the orangutans, but not to hominids.

Hominid Features

Three features distinguish hominids—their teeth and jaws, their mode of locomotion, and their brain size and organization. Each involves both morphological and behavioral components.

Teeth and jaws differ in hominids and in apes. The most obvious difference is the lack of large projecting canines in hominid males. Male and female hominids have canines of approximately the same size that barely project above the level of the other teeth and are generally smaller than in other higher primates. They function differently as well. Hominid canines are used more like incisors for gripping, holding, and tearing rather than cutting. Figure 26.5 compares hominid teeth with those of the gorilla. Hominid permanent teeth also appear in a different sequence during maturation. Canines erupt earlier, but

Figure 26.5

Hominid teeth (A) differ significantly from gorilla teeth (B).

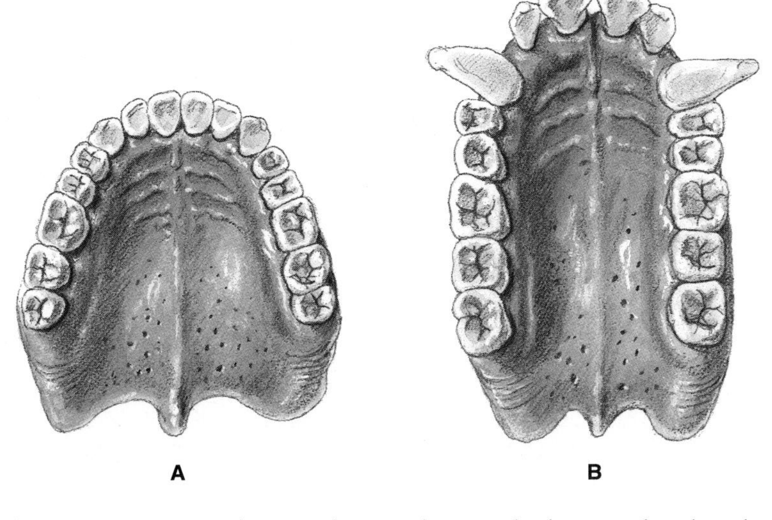

other permanent teeth erupt later in hominids than in closely related animals like chimpanzees.

Although all primates grind their food with their back teeth, hominids differ because they are able to apply more effective force during chewing. This enhanced capability is related to the size and orientation of the muscles that control the jaw.

Hominids are erect bipeds; that is, they walk erect on two feet. This *mode of locomotion* has significant advantages. It frees the hands to be used for other tasks such as carrying or manipulating objects. It also allows striding, a more energy-efficient mode of locomotion. As a result, hominids are able to cover longer distances without tiring.

Intelligence, marked by changes in *brain size* and *brain organization,* is the most prominent feature of hominids. Although hominids developed larger brains than other apes or monkeys, the importance of that increase is not clear because among humans, at least, there is no correlation between brain size and intelligence. Most biologists believe that the complexity of the human brain is more important than its size, since the highly complex human brain is capable of creating language and tools.

Hominid Fossils

The story of hominid evolution has changed considerably in recent years due to some sensational fossil discoveries. Every time textbook writers think they have the story straight, another set of fossil bones is uncovered somewhere, and everyone returns to their word processors.

At present, the earliest known hominid fossils belong to the genus *Australopithecus*. The first *Australopithecus* fossil, described by Raymond Dart in 1925, consists of the remains of a juvenile found in South Africa. At the time of its discovery, scientists were skeptical that the fossil represented a hominid, but later studies confirmed Dart's original contention. Since then, many *Australopithecus* fossils have been found, chiefly in eastern and southern Africa.

The australopithecine that has received the greatest publicity is *Australopithecus afarensis,* a species that inhabited Africa 3 to 4 million years ago. The fame that surrounds this species is related to a very unusual find in Ethiopia by Donald Johanson. In 1974, Johanson discovered the fossil remains of a small female that lived approximately 3 million years ago. What made the find so unusual was that it was 40 percent complete—the most complete specimen of such an early hominid. Nicknamed "Lucy" (after the Beatles song "Lucy in the Sky with Diamonds"), this *A. afarensis* specimen is thought by many paleontologists today to be a representative of the ancestral group that gave rise to modern humans. It has a small frame that looks humanlike but still retains an apelike head.

Australopithecus afarensis was once thought to have split into two branches, one leading to modern humans, the other, by way of the fossil discovered by Dart (*A. africanus*), to a fossil hominid with massive physical features called *A. robustus* and to a similar hominid, *A. boisei*. The *robustus* and *boisei* lines became extinct 1.2 million years ago.

A discovery in 1985 of a 2.5 million-year-old fossil in Kenya has caused the two-branch hypothesis of *Australopithecus* evolution to be revised. The newly discovered fossil has a face that is similar to the later *A. boisei* but a more primitive cranium. Several paleontologists feel that the new fossil find named *A. aethiopicus* is ancestral to *A. boisei* and represents a third branch in australopithecine evolution (see Figure 26.6).

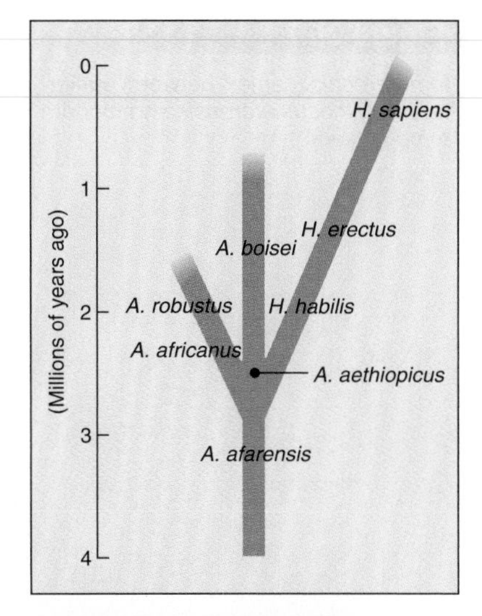

Figure 26.6

This three-pronged diagram represents current ideas on the evolution of australopithecines. *A. afarensis* split into at least three branches, one leading to *A. robustus*, one to *A. boisei*, and one leading to the the modern human, *H. sapiens*.

The lineage leading to modern humans is now thought to have proceeded from *A. afarensis* to *Homo habilis,* the earliest species to have the human genus name, *Homo*. *H. habilis* was a tool user and, like *A. afarensis,* was short, standing about 3 feet tall. A relatively brief period (in fossil record terms) of 200,000 years stands between *H. habilis* and the fossil with a distinctly modern body shape, *H. erectus*. *H. erectus* appeared approximately 1.6 million years ago and spread widely through Asia, Europe, and Africa. These early humans were the first hominids to migrate away from Africa. They had more sophisticated tools than *H. habilis,* and there is evidence that they used fire.

The evolution of the modern human, *H. sapiens,* has been highly controversial during the last several years because of different hypotheses that have been advanced by scientists who study the subject. Some scientists hold a view called the **multiregional model**. According to this hypothesis, *H. sapiens* developed from populations of *H. erectus* in several different areas of the globe, and subsequent human evolution was shaped by gene flow, natural selection, and genetic drift. Gene flow prevented populations of *H. erectus* from radiating into several different species (see Chapter 23). In between *H. erectus* and modern *H. sapiens,* therefore, are a number of "archaic" humans that are difficult to place in one species or the other because of their intermediate appearance.

The **single-origin model** of human evolution states that a population of the modern form of *H. sapiens* arose in Africa at an early date and spread rapidly across Europe and Asia. The modern form is called *H. sapiens sapiens,* to distinguish it from various "archaic" forms, and it is a subspecies of *H. sapiens* that includes both modern and archaic forms.

According to the single-origin model, the many archaic forms of *H. sapiens* found in Asia and Europe, which evolved from *H. erectus,* did not evolve into *H. sapiens sapiens*. For example, a well-known archaic human called by the subspecies name *H. sapiens neanderthalensis* lived from 130,000 to 35,000 years ago in Europe. Although for many years "Neanderthal man" was considered a direct ancestor of modern humans, recent fossils found in Africa and the Middle East suggest that modern humans (*H. sapiens sapiens*) did not evolve from populations of Neanderthals. Instead, we probably evolved from *H. erectus* in Africa 100,000 years ago but did not migrate to other parts of the globe until 40,000 years ago (see Figure 26.7).

When *H. sapiens sapiens* finally did arrive on the European scene, it was with dramatically different behavior patterns than its predecessors. Modern people planned hunts, built hearths, and fashioned complex tools. They made elaborate jewelry and buried their dead in a ritualistic manner. Art was part of their world, and they used some form of symbolic notation. Although Neanderthals buried their dead, modern humans had more elaborate preparations, and there are clear indications that they conducted final rituals (see Figure 26.8).

The changes associated with the advent of modern humans have more to do with behavior than bones. The typical male Neanderthal would probably be

Figure 26.7

Homo erectus (center) between earlier *Australopithecus afarensis* and modern *Homo sapiens.*

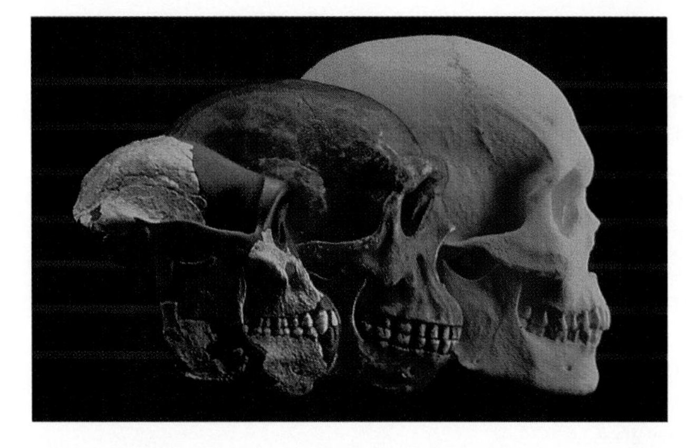

difficult to distinguish in a crowd dressed in a three-piece suit or a football uniform. Even if Neanderthals are not our direct ancestors, it is likely that some Neanderthal genes are included in the modern human gene pool (through hybridization), since Neanderthals coexisted with *H. sapiens sapiens* for thousands of years.

Neanderthals disappeared approximately 30,000 years ago, presumably because they were less fit than modern humans. What gave modern humans the selective advantage was their culture rather than any single physical trait. Careful analyses of bones indicates that modern humans lacked the brute strength of Neanderthals, but archeological remains suggest that they had a more advanced culture. Ingenuity, social organization, and an ability to plan for the future seem to have been the critical advantages. With the arrival of modern humans, biology gives way to archaeology, anthropology, and history, which all deal with the development of human cultures.

Hominids differ from apes by their teeth and jaws, their mode of locomotion, and their brain size and organization. The earliest known hominid fossil belongs to the genus *Australopithecus.*

The evolution of *Homo sapiens* remains controversial. Some scientists accept a multiregional model that holds that *H. sapiens* evolved from an original population of *H. erectus* in several areas of the globe. Other scientists accept the single-origin model, which states that a population of the modern form of *H. sapiens* arose in Africa and spread across Europe and Asia.

Figure 26.8

The beads adorning this early *Homo sapiens sapiens* skeleton bespeak an elaborate burial.

Hominid DNA

The single-origin model of *H. sapiens sapiens* evolution is strengthened by biochemical studies. By comparing DNA sequences among human populations, geneticists have discovered that relatively little variation occurs among humans, at least for nonfunctional DNA (see Chapter 17). In spite of the great morphological variation among human groups, the genetic information based on molecular clocks (see Chapter 25) suggests that there has *not* been the great divergence one would expect if many populations of humans had evolved at the same time.

A fascinating recent study of human molecular clocks has led to one of the most contentious issues in human evolution. Rebecca Cann of the University of Hawaii and Mark Stoneking and Allan Wilson of the University of California at Berkeley hypothesized from their 1987 research on *mitochondrial DNA* that the different human populations inhabiting the globe can all trace their ancestry to one particular African female who lived about 200,000 years ago. Mitochondria (small cellular organelles that provide energy to the cell) have their own unique DNA that enables them to replicate independently within the cell.

Mitochondrial DNA (mtDNA) is shorter than the DNA found in the cell's nucleus, and it is inherited only from the mother (that is, mtDNA is transmitted through the egg but not the sperm). Therefore, any changes in mtDNA must have arisen through mutations in females rather than as a result of recombination during fertilization. Since the rate of mutation defines a molecular clock, scientists contend that mtDNA is a particularly sensitive molecule for use in dating studies. Cann, Stoneking, and Wilson collected samples of placental tissue, a convenient way to obtain human tissue, from women all over the globe. (More recent research makes use of the new technique of polymerase chain reaction, described in Chapter 21, to isolate a selected portion of DNA from single hairs plucked from women and to make millions of copies of the portion required for analysis.) Comparison of mtDNA from these placental cells showed that the differences among them were extremely small. The researchers then calculated how long it would have taken these differences in mtDNA to have developed if humans diverged from a common ancestor. They concluded that approximately 200,000 years ago, a single common ancestor carried the mtDNA from which all mitochondria in the current human population originated.

This does not mean that only one female existed 200,000 years ago. It does suggest that of the females of her generation, only her descendants gave rise to a minimum of one female per generation and that only her mtDNA now exists in the human population. Presumably, along the way, descendants of other females did not leave daughters. Cann, Stoneking, and Wilson argue that this "mitochondrial Eve" lived in Africa and that a population of her descendants migrated from there to Asia and Europe, where they eventually became the dominant subspecies. Since the date of this common ancestral female is earlier than any recorded modern human fossils, it is possible that she belonged to an archaic human form. The research on mtDNA has caused debate because some scientists are not convinced that molecular clocks are accurate enough to provide very precise dates. Others, particularly those who accept the multiple-origin model, do not believe that the fossil record agrees sufficiently with the biochemical data.

Hominid Ecology

The limited number of hominid fossils makes it quite difficult to reconstruct the ecological system in which hominids evolved. If an African origin is accepted, the geologic changes that are known to have occurred there may provide clues about the original ecosystem. The early Miocene (25 to 5 million years ago) African environment consisted primarily of forests. Beginning approximately 16 million years ago, however, most of East Africa was altered by geologic changes that resulted in a drying trend and with that a replacement of forests by grasslands (see Figure 26.9). In some areas, a mosaic of forest and open areas existed,

Figure 26.9

The evolution of modern humans began as grasslands replaced forests in East Africa. This African savanna resembles the setting of hominid ecology.

and changes in the landscape created new environmental conditions. The arboreal ancestors of humans may have exploited the new opportunities and in so doing became modified. Coming down from the trees would have permitted animals to use new food sources, which might have led to the selection of altered teeth and jaws. The development of bipedal motion would have been an efficient means of moving on the grasslands. Tool use may have opened new possibilities and been an effective means of protection. New behavioral adaptations may have been dependent on a substantial amount of learned behavior, which would have resulted in more information to be learned by offspring. This emphasis on learned behavior may have added to the trend in primates toward a more complex brain.

Today, there are many different hypothetical constructions of hominid evolution. The lack of hard data makes it especially difficult to test any of the models, so their reliability is uncertain. More fossil finds and continued study of the geographical and geologic history of Earth may lead to a more detailed account.

Summary

1. Humans belong to the order Primates. Contemporary primates are divided into the Prosimii and the Anthropoidea. Primates share certain characteristics. The best known are their adaptation for living in trees and their ability to eat many different kinds of food. The Anthropoidea also spend more time raising a smaller number of offspring.
2. Among the significant adaptations of primates to tree life are their freely movable digits, generalized limb structure, and abilities to judge distance, see in color, and discriminate objects in their field of vision. Primates also replaced claws with nails and developed an upright posture.
3. Three features characterize hominids (species of the human family): teeth and jaws, mode of locomotion, and brain size and organization.
4. The earliest hominid fossils belong to the genus *Australopithecus. Australopithecus afarensis* inhabited Africa 3 to 4 million years ago. One widely held opinion is that *A. afarensis* split into three branches, one leading to modern humans, the other two to hominids that became extinct.
5. *Homo erectus* appeared approximately 1.6 million years ago and had a distinctly modern human body shape. Modern humans, *H. sapiens,* are thought to have evolved from *H. erectus*. After its beginnings, human culture soon became a primary factor in further human development.

Review Questions

1. What did Darwin have to say about human evolution?
2. What adaptations favored primates for life in the trees?
3. What features distinguish higher primates?
4. What three features distinguish hominids?
5. What genus contains the earliest known hominid fossils?
6. The lineage leading to modern humans proceeded from *Australopithecus afarensis* to what species?
7. What is the multiregional model of the evolution of modern humans?
8. What is the single-origin model of the evolution of modern humans?
9. What new conclusion has research on mtDNA supported?

Essay and Discussion Questions

1. What paleontological evidence would seriously undermine the mtDNA hypothesis on the recent origin of humans?
2. If modern humans evolved in Africa, why are there different skin colors among humans?
3. If you were writing a grant proposal to do field research on the fossils of early humans, how might you justify the overall value of the project?

References and Recommended Reading

Cann, R. L., M., Stoneking, and A. C. Wilson. 1987. Mitochondrial DNA and human evolution. *Nature* 325:31–36.

Easteal, S. 1991. The relative role of DNA evolution in primates. *Molecular Biology and Evolution* 8:115–127.

Fisher, A. 1988. Human origins. *Mosaic* 19:22–45.

Gibbons, A. 1991. Looking for the father of us all. *Science* 251:378–380.

Johanson, D. C., and M. Edey. 1981. *Lucy: The Beginnings of Humankind.* New York: Simon & Schuster.

Johanson, D. C., and J. Shreeve. 1989. *Lucy's Child: The Discovery of a Human Ancestor.* New York: Morrow.

Lewin, R. 1988. *In the Age of Mankind.* Washington, D.C.: Smithsonian Books.

Shreeve, J. 1990. Argument over a woman. *Discover* 11(8):52–59.

Stringer, C. 1990. The emergence of modern humans. *Scientific American* 263:98–104.

Wenke, R. J. 1990. *Patterns in Prehistory.* Oxford: Oxford University Press.

Wolpoff, M. 1980. *Paleoanthropology.* New York: Knopf.

CHAPTER 27

Animal Behavior

Chapter Outline

Individual Behavior

Evolution and Behavior

Reading Questions

1. What are the major types of animal behavior?
2. How do various behaviors increase the fitness of animals?
3. What are the major types of learning?
4. How does the theory of evolution explain animal behaviors?

The behavior of animals has long attracted the attention of biologists. However, in the nineteenth century, the life sciences became highly specialized, and most scientists ignored the study of behavior in favor of such areas as anatomy, physiology, embryology, and biogeography. A notable exception to this trend was Charles Darwin. Not only did he continue to do research on broad general areas, but he was also interested in behavior. In his book *On the Origin of Species,* Darwin devoted a separate chapter to behavior, or "instinct," as he called it. He attempted to demonstrate that the behavior of animals could have evolved in the same ways as their physical characteristics. Thus Darwin defined animal behavior as a field that could and should be encompassed by his new theory of evolution.

However, decades passed before animal behavior was fully integrated with evolutionary theory. Although Darwin and some of his immediate followers made observations on behavior and speculated on its evolution and development, other areas of biology continued to attract more attention. Fortunately, animal behavior was approached by scholars from other fields of science. The modern study of behavior has its roots in disciplines such as psychology, ecology, neurology, and natural history. Many of the fundamental ideas that became integrated into the discipline of behavior came from studies in these fields.

In this chapter, several basic types of animal behavior of individuals are described; social behavior will be left for Chapter 28. In both cases, evolution serves as an important unifying theme.

INDIVIDUAL BEHAVIOR

The modern theory of evolution regards the behavior of animals as adaptive, in that it allows them to survive and leave offspring. How does their behavior allow them to accomplish these ends?

Innate and Learned Behavior

During the past century, scientists approached the behavior of animals in two distinct ways. Some believed that animals inherit nervous systems that are organized in such a way that specific sets of reactions, **instincts**, occur in response to a proper stimulus. To this group of scientists, instincts have evolved, varied, and been subjected to selective pressures in the same way as anatomical features.

Other scientists held that much of the behavior that animals display does not consist of such rigid behaviors but rather reflects what they have learned in given situations. Therefore, **learning**—the change in behavior that occurs as a result of experience—has been of central interest to this second group of scientists. As with instincts, the ability to learn has been subjected to selective pressures and has adaptive value.

A few decades ago, a great amount of heat was generated by proponents of these two different positions. In the history of scientific debates, when each side has a considerable amount of verifiable data, scientists eventually come to realize that both sides are at least partially correct. That is, both instinct and learning are now thought to be important in understanding behavior. Types of animal behaviors that were originally thought to be purely instinctive have turned out to have a learned component, and vice versa. For some individual behaviors, one factor is more important than the other; in other cases, both are crucial.

Figure 27.1

The simple knee jerk is a well-known human reflex. The reflex depends on sense organs in the patellar tendon below the kneecap and on muscles in the upper leg. A sensory nerve travels up from the tendon into the spinal cord and there makes contact with a motor nerve that goes down to the responding muscle.

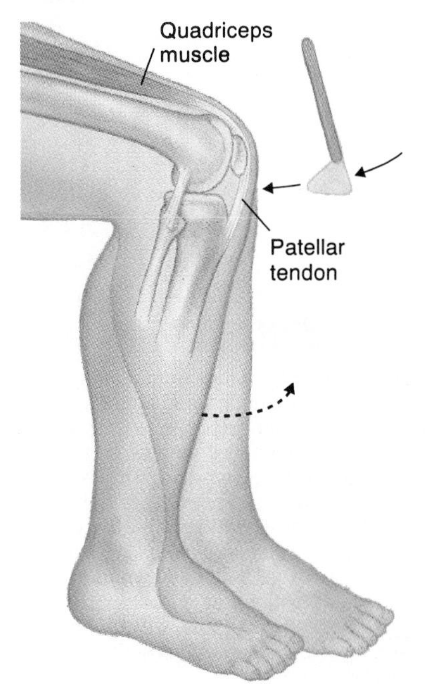

Reflexes, Kineses, and Taxes

Among the simplest forms of behavior are **reflexes**, simple reactions to external stimulation. A stimulus such as a change in light intensity or a touch can trigger an automatic, stereotyped, and involuntary response. A classic example is the human knee-jerk reflex (see Figure 27.1). Another is the scratch reflex in dogs, which was studied in the early twentieth century by the famous English physiologist Charles Sherrington.

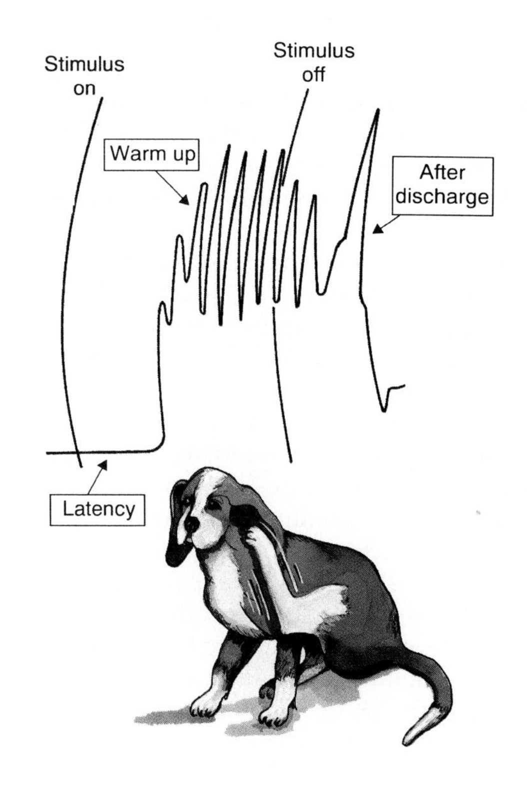

Figure 27.2

Sherrington's studies of the scratch reflex in dogs revealed various distinct stages.

Sherrington demonstrated some of the basic characteristics of reflexes in studies on dogs (see Figure 27.2). Later investigators found that several of these features apply to learned behavior patterns as well. Sherrington found that a brief period elapsed between a stimulus and the start of a response. This short delay is called a **latent period**. Similarly, there is a small delay between removal of the stimulus and the end of the response, which is called **after-discharge**. Sherrington noted that when the scratch reflex begins, it is not as strong as it becomes after a few strokes. He called this property of the reflex (to not attain its maximum strength right away) *warm-up*. He also noticed *fatigue:* when a stimulus is not removed, the response does not continue indefinitely. After a time, the response will diminish and stop altogether until the system has recovered.

Some reflexes can be performed by an animal without attention to its position in space. Most animals, however, orient themselves to the source of stimulation. If the orientation is merely an undirected change in the rate of motion in response to the intensity of the stimulus, it is called a **kinesis** (plural, *kineses*). When an animal either heads toward or away from a source of stimulation, the action is called a **taxis** (plural, *taxes*) (see Figure 27.3). Taxes are numerous. For example, animals can orient themselves to different stimuli such as light, chemicals, and gravity. They can also align themselves in different fashions, such as moving at an angle to the stimulus rather than straight at it or comparing the amount of stimulation on each side of their bodies and moving accordingly.

Figure 27.3

Kineses and taxes are simple reflexes. (A) Kinesis, an undirected change in the rate of motion in response to the intensity of a stimulus, is illustrated by these common pill bugs, which are more active in low humidity and less active in high humidity. Therefore, they become aggregated in damp places like under rocks. (B) Taxis, a motion toward or away from a stimulus, is illustrated by these housefly maggots, which after feeding will crawl away from a light source.

A

B

Figure 27.4

Pacific salmon navigate using past experience. Using their sense of smell, coho salmon can correctly choose among the forks in streams as they navigate back to their home stream.

Navigation

Navigation, the action of orienting toward a goal, is a more complex orientation behavior. Animals use a variety of mechanisms in navigating. Bees generally navigate by the sun, but when the sky is cloudy, they use a backup system that depends on polarized light in the sky. In both cases, they must compensate for the movement of the sun. Other animals use Earth's magnetic field and possibly other environmental cues. The impressive migration journeys of birds require the sun and the stars, which are used as compass points in guiding them.

The navigation of most species can be explained by innate (inborn) responses. European warblers, for example, make an annual trip from Europe to Africa by flying southwest for 40 days and then southeast for 20 to 30 days. However, the navigation of some species depends on past experience. Pacific salmon hatch in streams and rivers of western North America, and after spending one or two years in the stream, they migrate downstream to the ocean, where they spend two to four years. When salmon become sexually mature, they return to the general area where their natal river enters the ocean using the sun and other environmental cues. From there, they use their sense of smell to guide them back to the same tributary stream in which they hatched (see Figure 27.4). During the early stages of life, salmon learn the native odors of their stream. As adults, they can distinguish the odor of their native stream from that of the other waters entering the river and choose the correct direction at each fork of the river on their upstream voyage.

The study of animal behavior has roots in psychology, ecology, physiology, and natural history. Evolution theory focuses on the adaptive value of behavior, both learned and instinctive. Among the simplest forms of behavior is a reflex, which is a simple reaction to an external stimulus. Animals orient themselves to sources of stimulation and can direct themselves toward a goal, a process called navigation.

Fixed Action Patterns

Studies of reflexes initially led some scientists to hope that they could explain complex behavior in terms of chains of reflexes. Although complex behavior often incorporates reflexes, it is actually more complicated. Reflexes can usually be described very precisely and in terms of our detailed knowledge of the nervous system, whereas complex behavior involves so many systems and variables that a more comprehensive method of description is necessary. Early in the twentieth century, the concept of a *fixed action pattern* was developed and found to be useful. Although more complicated than reflexes, fixed action patterns often share some of the features of reflexes, such as latency and fatigue. Initially, it was also thought that fixed action patterns were entirely innate, but as we shall see later, this view has been modified.

The concept of fixed action patterns developed from the work of European **ethologists**, naturalists who are interested in observing behavior in the natural environment and attempt to explain it in an evolutionary context. The most famous ethologists, Konrad Lorenz in Germany and Niko Tinbergen in England, made careful observations of individual behavioral patterns and related them to specific biological situations.

Lorenz and Tinbergen: Fixed Action Patterns

If an egg rolls out of the nest of an incubating greylag goose and she notices it, a set of highly stereotyped actions will occur (see Figure 27.5). The goose fixes her attention on the egg, rises slowly, reaches her neck out over the egg, rolls the egg back into the nest with the bottom of her bill, and continues the incubation. Once started, the action continues to completion in a mechanical way. Even if the egg is removed when the bird is extending her neck, she will execute the entire sequence of actions to the end.

Lorenz and Tinbergen believed that this egg-rolling behavior was innate and stereotyped. Experiments by Tinbergen showed not only that geese would return eggs to the nest but that if they saw similar objects near the nest (or even slightly similar objects, such as metal cans or baseballs), they would automatically return these objects to the nest also. Foreign objects were rejected by the geese once they were in the nest but nonetheless were capable of triggering the action if placed near it.

Lorenz and Tinbergen called this entire sequence of actions a **fixed action pattern**. They thought that an "innate releaser" must exist that triggered such actions, and they hypothesized that an **innate releasing mechanism** could be stimulated by something egglike. The features of objects that triggered the innate releasing mechanism were called **releasers**.

Some of Tinbergen's later research identified some of these releasers. He conducted a set of classic experiments with a small freshwater fish, the three-spined stickleback, to discover which features the fish responded to. Figure 27.6 illustrates some of his basic observations. A male stickleback that has built a nest is stimulated to engage in courtship behavior by the swollen belly of a female with eggs. A male will court a featureless dummy lacking eyes, fins, spines, and a tail, so long as it has a swelling on its bottom side.

During the mating period, the breeding male stickleback has a distinctive red underside and is very aggressive toward other males that have similar markings. Tinbergen showed that the aggressive males reacted primarily to the red

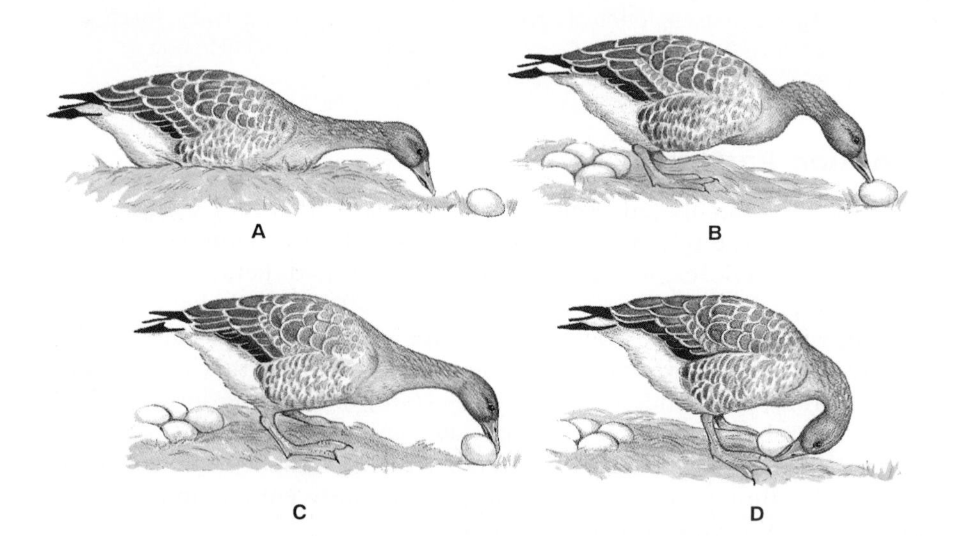

Figure 27.5

A fixed action pattern is a sequence of stereotyped behaviors that is triggered by a stimulus in the environment. The sequence of events in the egg-rolling behavior of the greylag goose is a classic example: The goose looks at the egg (A), rises and reaches beyond the egg (B), catches the egg under her bill (C), and pulls it back toward her into the nest (D). Even if the egg is removed partway through the ritual, the goose continues through all the steps.

Figure 27.6

When the male stickleback recognizes an appropriate female, he engages in a "zigzag" display. If the female responds, he leads her to his nest and shows her the entrance by poking his snout into it. The female may then enter the nest to lay eggs while the male nuzzles her tail.

marking. He did this by making models that were accurate imitations of male sticklebacks but lacked a red belly and models that barely resembled a fish but had a red belly. The breeding males reacted only to the models that had a red belly (see Figure 27.7).

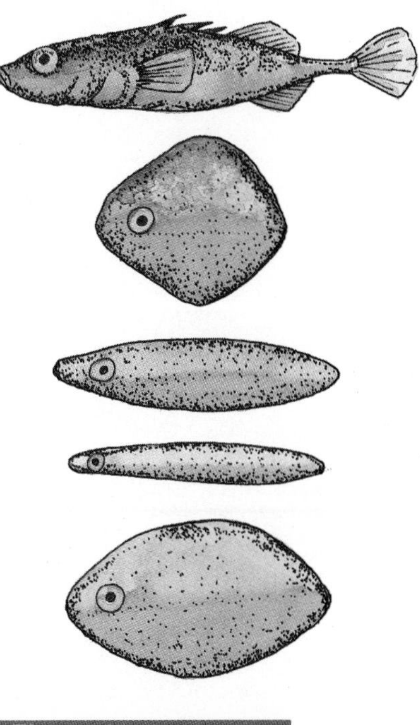

Figure 27.7

Dummy sticklebacks were used to demonstrate that the red marking of males is the releaser of aggressive behavior in breeding males.

Drives and Clocks

A male stickleback that has built a nest will react to the presence of another male that has red markings. However, males do not always have red markings, and those that are not guarding a territory in preparation for mating will not react either. Similarly, the specific egg-rolling behavior of greylag geese occurs only at certain times and with particular individuals.

Why don't animals always respond in the same way to a given stimulus? Ethologists explain this by referring to the "motivation" of the animal. Specific motivations are called **drives**; for example, thirst is a drive. A dog that has not had anything to drink for an extended period will search out a source of water, even if it is to us a seemingly inappropriate toilet bowl. Ethologists use the concept of drive in referring to the internal state of the animal that results in a tendency to organize its behavior to achieve a certain goal. The causes of that internal state are complex mechanisms that are not very clearly understood, but they are known to involve the nervous and endocrine systems. Ethologists have found it useful to describe many stereotyped behaviors in terms of drives.

Originally, ethologists spoke of animals having a number of discrete drives. However, research on these drives showed that rather than being simple forces, drives are complex states that start, stop, or modulate an animal's responsiveness to stimuli. Drives are considered only a part of a larger group of factors, both internal and external, that can stimulate overlapping behavior patterns.

One of the important elements that determines the stimulation of behavior patterns is a **biological clock**, a biochemical mechanism that is responsible for repeated patterns—rhythms—of behavior. Some of these clocks are set by external environmental factors. For example, Atlantic fireworms that inhabit the waters off Bermuda swarm during each full moon of the summer. Fifty-five minutes after sunset, the worms form breeding swarms. The reproductive cycles of most animals are correlated with specific seasons. Sticklebacks react to the lengthening days of spring by producing reproductive hormones that result in changed behavior.

By experimentally altering the conditions under which an animal lives, scientists have been able to discover which behavioral rhythms are set by external stimuli. One can, for example, shift the time of day that the marine snail, *Nassarilus festilvus,* is active by illuminating the snails at night and leaving them in darkness during the day. Under natural conditions, these snails are active at night, but under the reversed experimental conditions, they are active by day.

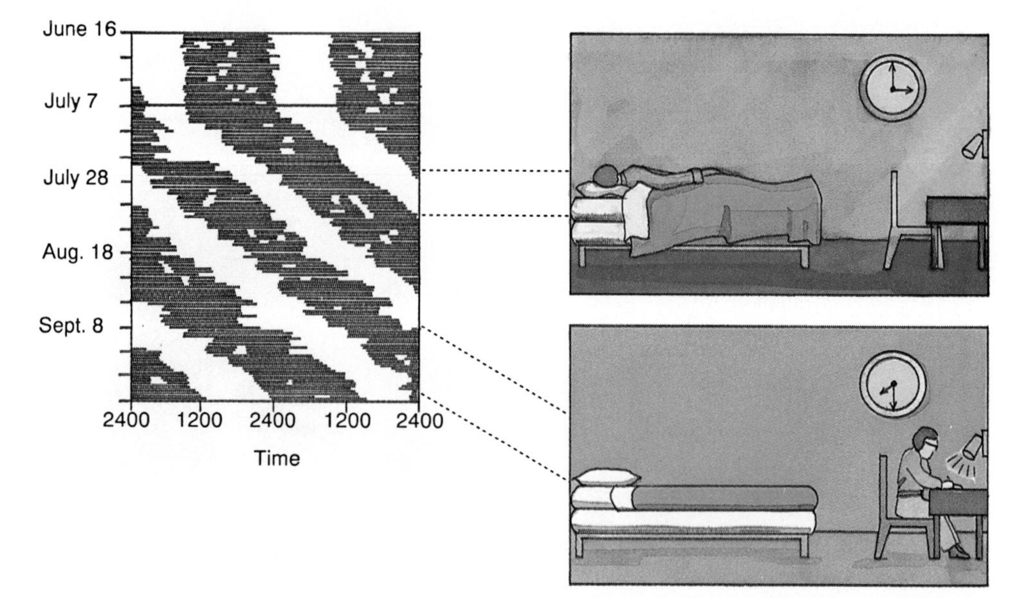

Figure 27.8

Circadian rhythms can get out of phase. In this experiment, a human was placed in isolation with a clock. At first, he tried to keep a regular schedule; then (July 7) he slept when he felt like it. Notice how his daily clock shifted.

Some biological clocks do not seem to be altered when the animal is placed in an experimental environment with constant conditions (for example, no change in light or temperature). Scientists consider these to be *internal clocks*. The best known are daily internal clocks, which reset every 24 hours, called **circadian rhythms**. Even these, however, are synchronized with the environment by external stimuli referred to as *times setters*, or **zeitgebers**. When animals are kept in isolation (that is, away from external time setters), their daily clocks will drift out of step with the environment, but if left in a normal setting, they will stay synchronized (see Figure 27.8).

Fixed Action Patterns and Learning

Early biologists who studied reflexes, kineses, taxes, and fixed action patterns assumed that these responses are inherited, unlearned, and common to all members of a species. They clearly depend on internal and external factors, but until recently, instinct and learning were considered distinct aspects of behavior. However, in some very clever experiments, Jack Hailman of the University of Wisconsin showed that certain stereotyped behavior patterns require subtle forms of experience for their development. In other words, at least some of the behavior normally called instinct is partly learned.

Hailman observed and experimented on the feeding behavior of sea gull chicks. In one species, the laughing gull, parent gulls lower their heads and point their beaks downward in front of young chicks. The chicks will then, with a pecking motion, grasp the parent's bill and stroke it downward. The parent responds by regurgitating some partly digested food (see Figure 27.9). Earlier ethologists had claimed that gull chicks used an innate "picture" of a parent gull for pecking. Hailman's experiments showed that the gull chicks indeed reacted to an innate stimulus but that learning was also important in the development of the behavior.

He first demonstrated that the accuracy of the chick's pecking rapidly increases after hatching. He then compared the accuracy of chicks raised in the dark with those of normal chicks raised in the wild. Although the chicks raised in the dark improved their accuracy with time, they never achieved the "normal" accuracy of the controls. From this experiment, Hailman concluded that visual experience was necessary for full development of the fixed action pattern of gull chick feeding.

Hailman also found that the chicks do not have a complex innate picture of a parent gull as earlier ethologists, like Tinbergen, had thought. Instead, they

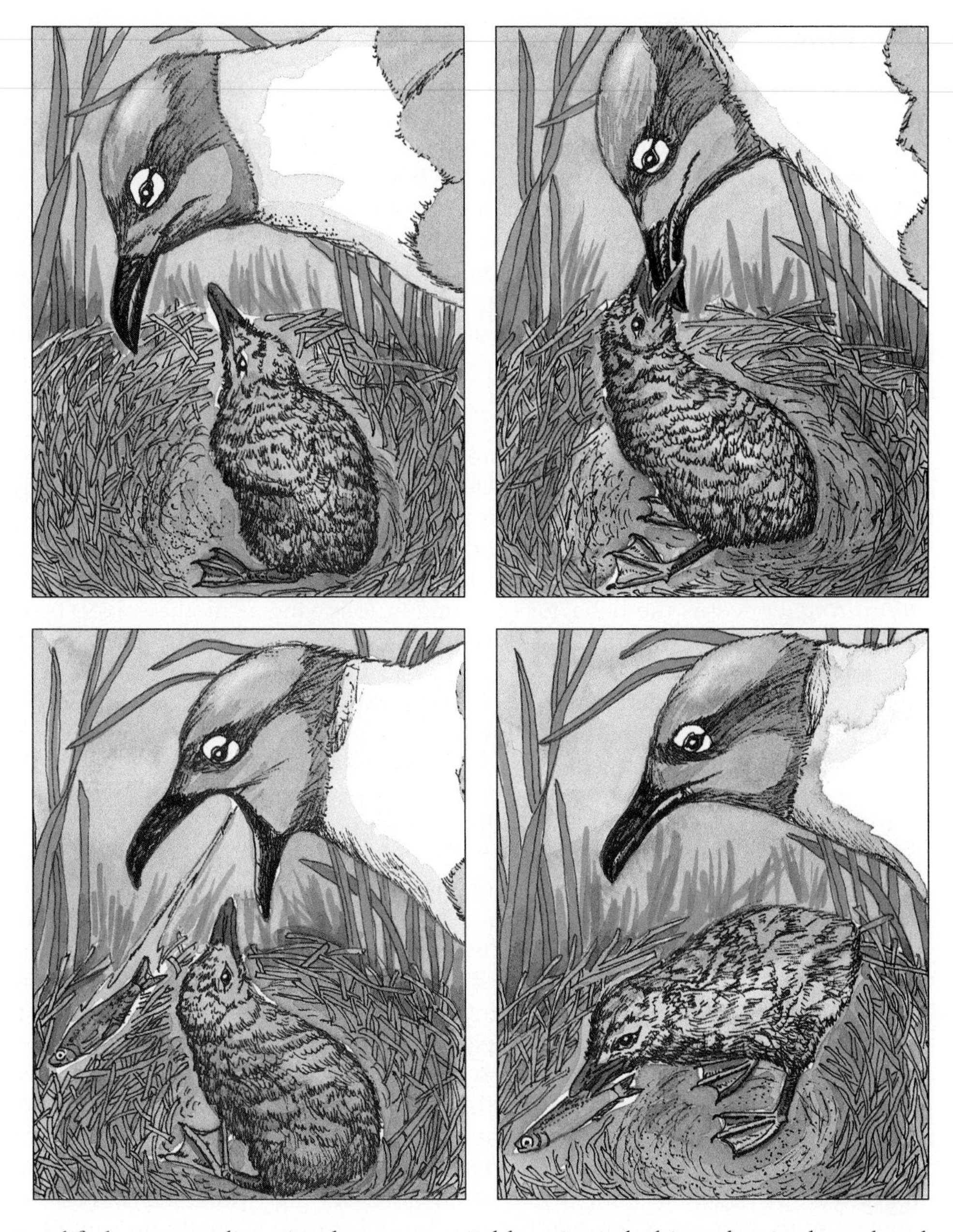

Figure 27.9

A young gull chick stimulates its parent to regurgitate partly digested food by grasping the parent's beak and stroking it downward.

start life by responding simply to any suitably oriented object that is shaped and moves like a parent's bill. Chicks respond to headless bills, but not to bill-less heads.

Much to Hailman's surprise, laughing gull chicks that had not been isolated from their natural parents at first responded similarly both to models that resembled laughing gull parents and to those that resembled the quite different-looking species of herring gulls. Within a week, however, the chicks responded only to models that closely resembled their parents. As with the increase in accuracy of the chicks' pecking, results from this experiment also suggest that some fixed action patterns have a component of learning and may not be totally innate. Only further research will show to what extent other "innate" behaviors have a learned component.

Ethologists call a sequence of innate, stereotyped behaviors triggered by a releaser a fixed action pattern. Internal states of the animal influence how animals respond to releasers and other stimuli. Originally, fixed action patterns were thought totally innate; however, some of them have been shown to have a learning component.

Learning

As stated earlier, learned behavior occurs when animals change their responses as a result of experience. A considerable amount of the early work on learning was done by psychologists. They have primarily been concerned with human learning, and even when their research has been on animals, it has been with an eye toward using animals to understand human behavior. More recently, biologists have focused directly on animal learning. Although studies on learning have been carried out on a relatively small number of species, a vast amount of information has been generated. Scientists now recognize five major categories of learning: imprinting, habituation, associative learning, latent learning, and insight.

Imprinting

Imprinting is a highly specialized form of learning. In many species, it takes place during the early stages of an animal's life when attachment to parents, the family, or a social group is critical for survival. **Imprinting** is a process whereby a young animal forms an association or identification with another animal, an object, or a class of items. The best-known type of imprinting, called *filial imprinting,* concerns the behavior of young in following a "mother object" (see Figure 27.10). During a critical *sensitive period,* a young animal is susceptible to imprinting.

Young animals are not completely indiscriminate in what they follow. For example, a mallard duckling will follow a moving object for the first two months after hatching. However, it will show a preference for yellow-green objects (the color of its parents) over others of different colors. Young may also be sensitive to sound as well as to sights. Wood ducks respond to a species-specific call in exiting from their nests.

Imprinting is an important form of learning, for it has both an effect on the parent-offspring relationship and long-term effects that become evident in adult animals. For example, lack of imprinting has been shown to result in abnormal adult social behavior in some species. Also, the breeding preferences of many birds are a consequence of early imprinting experiences. As adults, they prefer to mate with birds of their imprinted parents' color or markings. This form of imprinting, which has considerable evolutionary significance as a reproductive isolating mechanism, is called **sexual imprinting**. The phenomenon can be tested by experiments that allow a bird to be raised by foster parents of a different species. When these young birds mature, they show a sexual preference for mates with the color of their foster parents. Sexual imprinting can have some unusual outcomes, as when hand-reared birds become sexually imprinted on people.

Figure 27.10

Konrad Lorenz got goslings to imprint on him as their mother object.

Habituation

Habituation is a simple form of learning. It occurs when an animal is repeatedly exposed to a stimulus that is not associated with any positive or negative consequence. The animal ceases to respond to that stimulus, even though it will continue to respond to other stimuli.

Another way of looking at habituation is to consider that the animal has learned not to respond to meaningless stimuli. Birds will soon ignore a scarecrow that initially caused them to avoid a garden. A snail moving along a board will withdraw into its shell if the board is tapped. After a brief time, it will emerge and continue on its course. If the board is tapped again, it will withdraw again but will emerge more quickly. After six or more responses, the snail will ignore subsequent taps. Habituation is important to animals in adjusting to their environment. If animals continued to respond to meaningless (for them) stimuli, they would not be able to function effectively.

Associative Learning

Habituation is learning that results in the loss of a response that is not relevant or useful to the animal. **Associative learning**, in contrast, is the acquisition of a response to a stimulus by associating it with another stimulus. Considerable research on associative learning has taken place in the past century. The most famous of the early scientists to study associative learning was the Russian scientist Ivan Petrovich Pavlov (see Figure 27.11).

In a number of now classic experiments, Pavlov demonstrated a form of associative learning called **classical conditioning**, which results in changes in the stimuli that elicit behavior. He did this by placing meat powder in a dog's mouth and recording the amount of saliva produced. He then repeated the procedure, only this time he rang a bell before giving the dog the powder. Although at first the dog did not respond to the bell (other than by pricking up its ears), eventually the dog salivated at just the sound of the bell. That is, it had learned to associate the stimulus of the bell with the stimulus of the meat. It had learned to respond to a new stimulus, the *conditioned stimulus* (the bell), by associating it with an old one, the *unconditioned stimulus* (the meat powder). Pavlov called the "new" response a *conditioned reflex;* now we call it a *conditioned response.* Pavlov showed that he could use almost any stimulus as a conditioned stimulus,

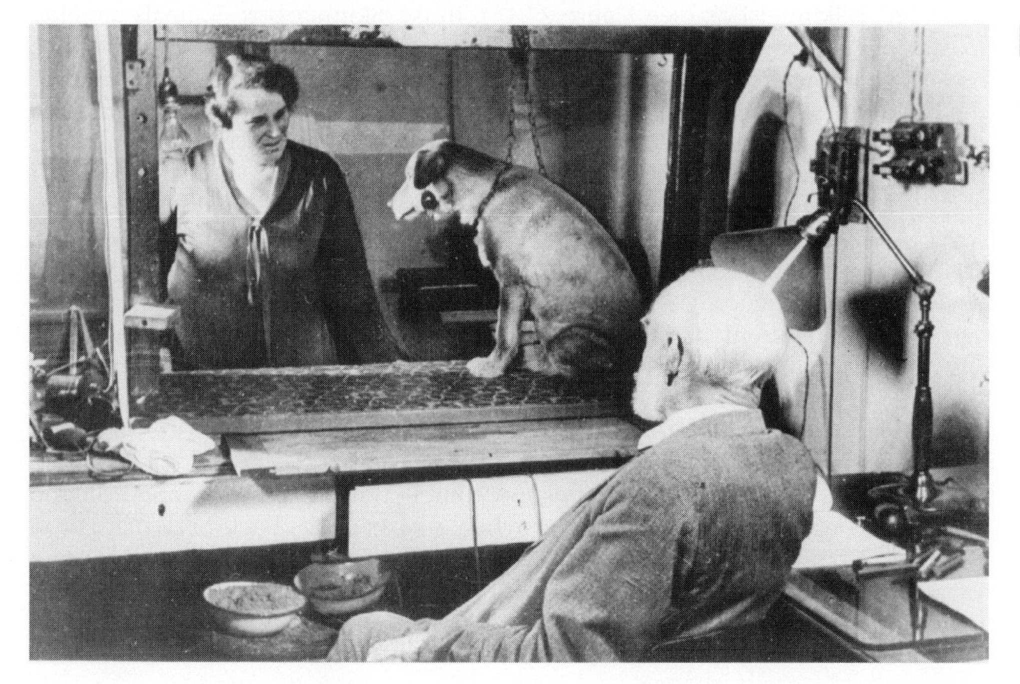

Figure 27.11

Ivan Petrovich Pavlov was one of the first investigators of associative learning, using dogs as his subjects.

Figure 27.12

E. L. Thorndike used puzzle boxes to study operant conditioning. In this version, a confined cat must learn to pull a string loop that by way of a pulley system moves a latch, which releases the door.

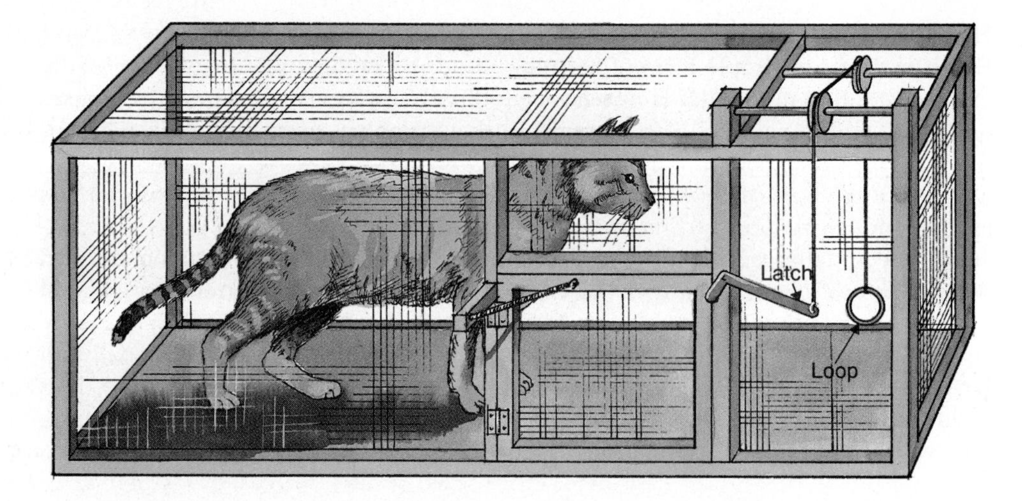

as long as it was not too strong. Although Pavlov's experiments were elegant and have been confirmed repeatedly, it is not clear to what extent classical conditioning actually occurs in nature.

A form of associative learning that has been observed in numerous animal species is **operant conditioning**, where an animal's conditioned response is instrumental in producing consequences in its environment. Early in this century, E. L. Thorndike of Columbia University constructed "puzzle boxes" (see Figure 27.12) that he used in experiments on animal learning. The puzzle box was a closed wooden box with a lever that opened the box when pushed. Thorndike placed animals in the box, observed their behavior, and found that they could learn to escape. A cat placed in the box first explored its new surroundings, but at some point, by accident, it pushed the lever (for example, by rolling over or sitting). When placed in the box again, it repeated the action that opened the box and after several trials was able to escape immediately.

Thorndike's studies stimulated considerable new research, the most famous of which was conducted by B. F. Skinner of Harvard University. Thorndike had shown that a trial-and-error form of learning occurred in animals. Skinner, who coined the term *operant,* used a refined version of the puzzle box that came to be called a Skinner box (see Figure 27.13). In the Skinner box, animals are able to manipulate some object, which will yield some reward if properly done. The reward "reinforces" the behavior, and the animal can be conditioned to repeat the behavior. Skinner and other scientists were able to train laboratory animals to perform relatively arbitrary behavior in order to obtain food. Under natural conditions, operant conditioning can modify behavior to make the animal more efficient in obtaining its goal.

Latent Learning

Associative learning stresses the importance of reward or reinforcement. Animals can associate different stimuli or situations, however, without immediate reward. The best-known cases involve the ability of animals to learn the geography of their home area or to learn from exploring new surroundings. This is called **latent learning** because at the time of learning, there is no obvious reward, and what is learned remains latent until put to use. Female sand wasps, for example, learn the features of their surrounding environment during flights to and from the nest. Their "knowledge" is used later when a wasp drags back prey that is too heavy to carry in flight. Another classic example of latent learning comes from experiments done on well-fed rats that were placed in mazes and allowed to explore. Such rats, although not reinforced for escaping the maze, were then able to choose the correct path through the maze when they were hungry.

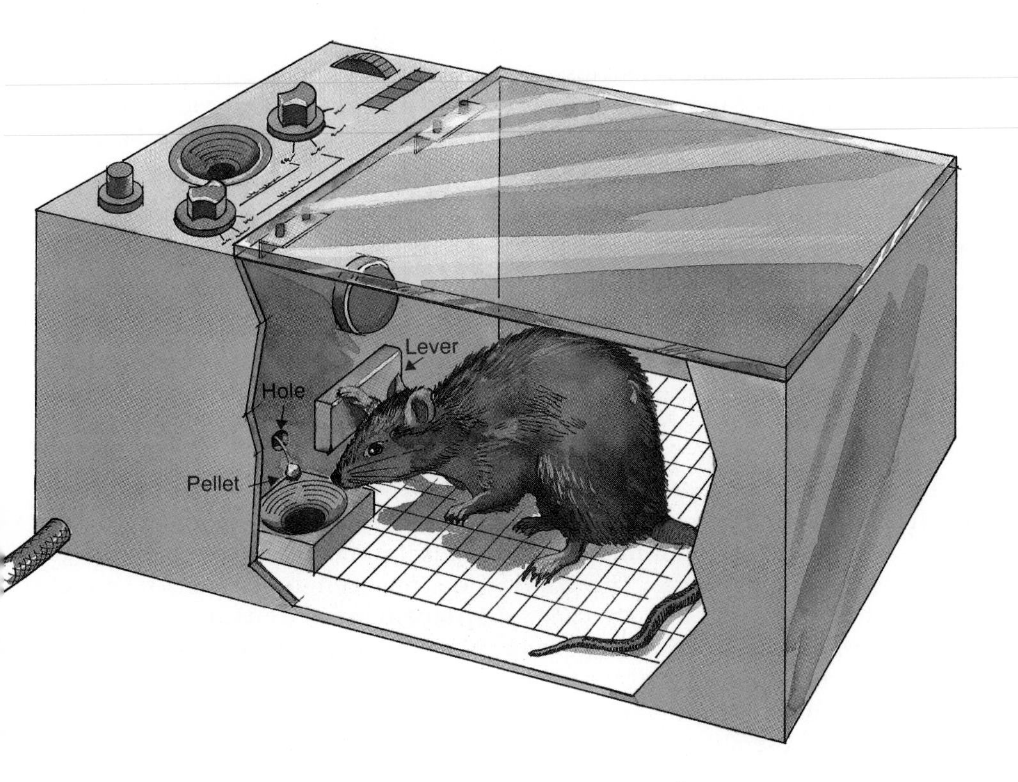

Figure 27.13

B. F. Skinner refined Thorndike's puzzle box into the famous Skinner box.

Insight

Insight refers to the seeming ability of some animals to devise new behaviors from past experiences. A chimpanzee will try to reach some bananas that are out of its reach by fitting two sticks together or by piling up boxes on which to stand. It does not arrive at the solution slowly by trial and error but rather suddenly, which suggests that it has used reasoning to solve the problem. The existence of insight in animals has been questioned by several scientists who hold that most of what passes for insight is really the result of normal behavior patterns, imitation of others, or latent learning.

Learned behavior occurs when animals change their responses as a result of experience. Learned behavior has important adaptive value. The five major categories of learning—imprinting, habituation, associative learning, latent learning, and insight—can each be understood in terms of its evolutionary significance.

Instinct and Learning

The study of individual behavior has shown that it depends on both internal and external factors. The nervous and endocrine systems respond to external stimuli or are modified by them. Behavior is also conditioned by experience. Early students of behavior thought of animals as genetically determined mechanisms with fixed action patterns that responded to specific signals. Research has shown that animal behavior is far more flexible and that animals benefit from experience. The old argument between learning and instinct is being replaced with a view that animals are born with genetically determined behavioral systems but that these systems respond to the environment in such a way that experience modifies their behavior. This ability to learn has great adaptive value, and, therefore, the study of animal behavior is closely tied to evolution.

EVOLUTION AND BEHAVIOR

It has long been recognized that inherited patterns of behavior have evolved. Unfortunately, the evolution of behavior is more difficult to study than the evolution of anatomical structures. The fossil record provides only indirect information about the evolution of behavior. However, by comparing the behavior that characterizes the individuals of related species, scientists have discovered a great deal.

Evolution of Behavior

One of the well-studied processes in the evolution of behavior is **ritualization**, the modification of behavior patterns to serve a new function, usually involving communication. The term was first used in 1914 by the English naturalist Julian Huxley in describing the evolution of the courtship behavior of crested grebes. Huxley showed that the elaborate courtship behavior of grebes (see Chapter 24) was derived from elements of their nest-building behavior.

By comparing related species, scientists have been able to infer the origin of courtship behavior by means of ritualization. A famous example involves tropical grass finches. Male zebra finches often interrupt their courtship behavior to perform a beak-wiping action. This activity does not seem to have any relevance to the situation. Such activities are called **displacement activities**, and they are often observed in circumstances where an animal has conflicting tendencies or is frustrated in obtaining its goal. In finches that are closely related to the zebra finch, such as the striated finch and the spice finch, the male performs a ritualized motion that resembles the beak-wiping action of the zebra finch (see Figure 27.14). It is likely that this bowing motion is derived from a displacement activity of beak wiping that is still observed in zebra finches. The beak wiping is a common behavior in the tropical grass finches, but in some species it has become ritualized into their courtship behavior.

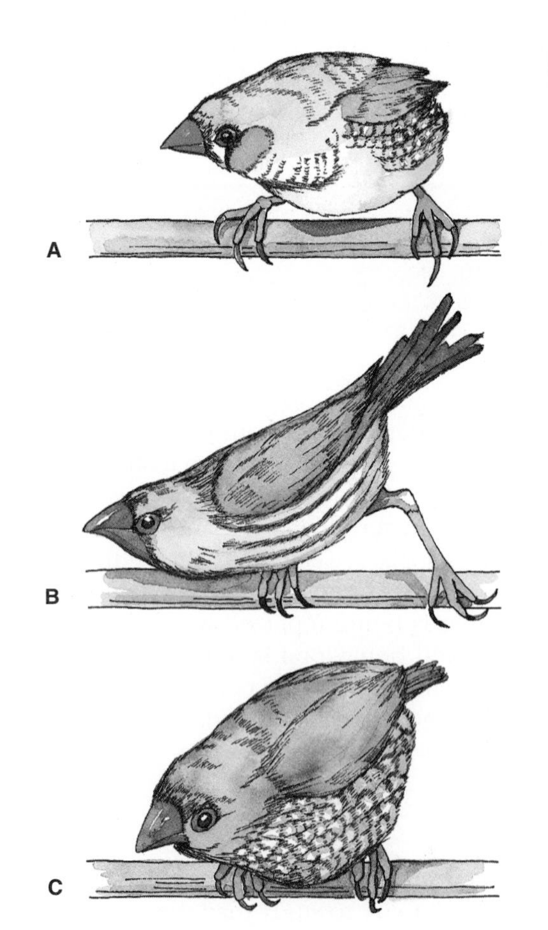

Figure 27.14

Studies of related species can lead to inferences of how specific behaviors evolved, as illustrated by this observation of beak wiping in three finches. In an unritualized movement, a zebra finch (A) prepares to wipe its beak on its perch. This movement has become ritualized in the striated finch (B) and the spice finch (C), that hold the position for some seconds during courtship so that it resembles a bow.

Behavior and the Theory of Evolution

In Chapter 25, the relationship of the theory of evolution to various biological disciplines was discussed. Some disciplines, such as paleontology, were seen to be very closely tied to the theory, whereas others, such as physiology, were less so.

Studies of individual behavior relate to the theory of evolution largely due to their emphasis on adaptation. Scientists interested in animal behavior have approached the topic primarily from the perspective of its adaptive function. Tinbergen, for example, showed the adaptive value of empty eggshell removal behavior by gulls. Breeding gulls remove broken eggshells to a distance of several meters, although when they are not breeding, they are indifferent to the shells' presence. Tinbergen thought that the bright white insides of the shells might attract the attention of predators and that if the broken shells were not removed, they would reduce the value of the well-camouflaged colored eggs. To test his hypothesis that the removal of the broken eggshells was adaptive, Tinbergen set out gull eggs in an area with broken shells nearby and gull eggs that were not close to any broken shells. Predators found two thirds of the eggs that were near the broken shells but only one fifth of the other eggs.

Emphasizing the adaptive significance of individual behavior closely ties the study of animal behavior to the theory of evolution by relating it to fitness. The study of behavior is also tied to the theory of evolution through the use of behavior in classification. Behavior is a trait, and therefore by comparing behavior, a taxonomist can use behavioral information in much the same way as anatomical information. Knowledge of the behavior of animals is also valuable in deciding if two populations are actually different species. Information about behavior can determine if the two groups interbreed in nature or if they have different sexual behaviors that would prevent them from interbreeding.

Although a considerable amount of the study of animal behavior is done with explicit reference to evolution, a significant portion of the work is also done without it. Some very exciting work being done at present concentrates on the underlying mechanisms of individual behavior. Research on neural pathways, genetic foundations, hormonal control, and learning focuses on uncovering the mechanisms involved in behavior. As with the study of physiology, evolution puts these behavioral mechanisms into a broader context and relates them to other domains of biology.

Summary

1. During the past century, scientists have approached the behavior of animals in two distinct ways. Some held that behavior is a function of inherited patterns of the nervous system that result in specific reactions known as instincts. Other scientists have held that behavior is learned.
2. Contemporary behavioral scientists realize that both instinct and learning are important in behavior.
3. The simplest forms of behavior are reflexes. Undirected orientation of animals to sources of stimulation are called kineses. Movement toward or away from a source of stimulation is called a taxis.
4. A complex form of orientation is navigation, the action of orienting toward a goal.
5. Fixed action patterns are sequences of actions that are innate and stereotyped.
6. Animals have specific motivations for behavior called drives.
7. Biological clocks are responsible for many rhythms of behavior. Daily internal clocks are called circadian rhythms.
8. There is evidence that some of what we usually regard as instinctive behavior is learned.

9. Learned behavior occurs when animals change their responses as a result of experience.
10. Five categories of behavior are imprinting, habituation, associative learning, latent learning, and insight.
11. Ethologists have demonstrated the adaptive significance of some individual behaviors.

Review Questions

1. What are instincts?
2. What is learned behavior?
3. What is a reflex? A kinesis? A taxis? How do they differ from one another?
4. What types of environmental cues are used by animals in navigating?
5. What is a fixed action pattern? Describe one.
6. How did Lorenz and Tinbergen demonstrate the occurrence of fixed action patterns?
7. What are biological clocks? Give one example.
8. How is learning related to fixed action patterns?
9. What is imprinting? Give an example.
10. What is the significance of habituation, associative learning, latent learning, and insight?
11. How is behavior related to fitness?

Essay and Discussion Questions

1. What lessons came out of the battle between scientists who stressed the importance of instinct to behavior and those who stressed learning?
2. How might an understanding of individual animal behavior relate to an understanding of individual human behavior?
3. How might an understanding of the underlying mechanisms of behavior be integrated with the theory of evolution?

References and Recommended Reading

Alcock, J. 1989. *Animal Behavior: An Evolutionary Approach*. Sunderland, Mass.: Sinauer.

Eisner, T., and E. O. Wilson. 1975. *Animal Behavior: Readings from Scientific American*. New York: Freeman.

Gould, J. L. 1982. *Ethology: The Mechanisms and Evolution of Behavior.* New York: Norton.

Grier, J. W. 1984. *Biology of Animal Behavior.* St. Louis: Times Mirror/Mosby College Publishing.

Hinde, R. A. 1970. *Animal Behavior.* New York: McGraw-Hill.

Johnson, C. H., and J. W. Hastings. 1986. The elusive mechanism of the circadian clock. *Scientific American* 74:29–36.

McFarland, D. 1982. *The Oxford Companion to Animal Behavior.* Oxford: Oxford University Press.

Tinbergen, N. 1973. *The Animal in Its World: Explorations of an Ethologist.* Cambridge, Mass.: Harvard University Press.

Waterman, T. H. 1989. *Animal Navigation.* New York: Scientific American.

CHAPTER 28

Social Behavior

Chapter Outline

Reading Questions

1. How are social interactions understood in evolutionary terms?
2. How do biologists attempt to understand social groups?
3. In what ways is behavior adaptive?

The diversity of animal forms is extraordinary. Anyone who has had to memorize part of the classification system of animals can attest to that. It will come as no surprise to learn that the behaviors of animals are equally diverse. Darwin recognized this when he was formulating his theory of evolution. He treated behavioral traits as important adaptive features that had changed over time due to selective pressures. In Chapter 27, some basic features of individual behavior were described and discussed. The adaptive value of such behavior is of considerable interest for the study of evolution. So is **social behavior**, the behavior of two or more interacting animals.

SOCIAL BEHAVIOR AND EVOLUTION

Social behavior has attracted the attention of evolutionary scientists from Darwin's time to the present. Much of the recent research on this topic has been classified as **sociobiology** after the book by Edward O. Wilson of Harvard University. Wilson, shown in Figure 28.1, published *Sociobiology: The New Synthesis* in 1975 as an attempt to integrate modern work on social behavior with that done in genetics, evolution, and ecology. He hoped to establish "the systematic study of the biological basis of all social behavior."

Fitness

The study of social behavior is currently dominated by an evolutionary perspective. Basic to that view is the idea of relative fitness. Recall that *relative fitness* refers to an individual's chance of leaving offspring compared to other individuals in the same population (see Chapter 23). It measures the relative genetic contribution of individuals to the next generation. Much of the writing on social behavior is based on the assumption that sets of genes determine particular behavioral traits. A behavior that increases reproductive success will, presumably, result in an increase in the frequency of alleles (alternate forms of a gene) associated with that behavior. Unfortunately, it is often very difficult to gauge a specific behavior's effect on reproductive success or to separate one behavior from an animal's entire behavioral repertoire. Nonetheless, biologists have attempted to assess the consequences of a behavior and to determine if it might enhance reproductive success.

Social Interactions

Social interactions can be classified, from an evolutionary perspective, by their effects on reproductive success of the *actor* (the individual displaying the behav-

Figure 28.1

E. O. Wilson wrote *Sociobiology* in an attempt to integrate research on social behavior with research on genetics, evolution, and ecology.

ior) and the *recipient* (the individual or individuals affected by the behavior). Three main behavioral categories are recognized: altruistic, cooperative, and selfish.

Altruistic Behavior

Altruistic behavior results in a benefit to the recipient at a cost to the actor. From an evolutionary point of view, one might not expect such behavior to exist, for altruistic behavior seemingly reduces individual fitness, and natural selection ought to operate against it. Why should animals give alarm calls when it places them at greater risk of predation? Why should animals help raise the young of another pair or orphaned young when they could use their energies in raising their own young?

The paradox of altruistic behavior is explained by the idea of **kinship**, that is, the degree of relatedness among organisms. Although helping another animal at one's own expense might decrease an individual's relative fitness, its genetic contribution to the next generation could be enhanced if the recipient is related to the actor. How does this occur? An individual can increase its genes in the next generation in two ways: by producing offspring itself or by helping relatives produce offspring. If the degree of relatedness is high (that is, if the parties have many genes in common) and the cost to the actor low, an altruistic act may well have value in promoting the relative fitness of the actor's genes. The altruistic act, seen from this perspective, is not so unselfish after all. It is a means by which an animal's genes are increased in the next generation, even if it occasionally comes at the individual's expense!

Figure 28.2

Belding's ground squirrels have been used in studies of kin selection. The females make calls that benefit their close relatives even though it brings them into considerable danger. This photo shows a Belding's ground squirrel giving an alarm call.

Because of the complexity of behavior, it has been difficult to document with certainty the relationship between kinship and altruistic behavior. However, many cases of reported altruistic behavior appear to involve kinship. In a famous study of Belding's ground squirrels, it was shown that these animals gave warning calls, at considerable danger to themselves, more often when they were near close relatives than when they were not. Adult females, shown in Figure 28.2, which were usually found near close relatives, give more calls than adult males, which were more widely distributed. In addition, adult females were more likely to give a warning call if they were near many close relatives than if they were not.

The evolution of an altruistic act, because of its effects on promoting the survival and reproduction of relatives, is the result of a process termed **kin selection**. Research on kin selection has led scientists to expand the idea of fitness to **inclusive fitness**, which measures both an individual's reproductive success and its effects on the reproductive success of its relatives. Since close relatives share genes, from this point of view, an altruistic individual may be enhancing the survival of his own genes by helping a relative survive and reproduce. One can visualize this with the famous quotation attributed to J. B. S. Haldane, one of the biologists who worked on the modern synthesis: "I would lay down my life for two brothers or eight cousins." He was alluding to the fact that an individual, on the average, shares half of his genes with a brother (or sister) and one eighth of his genes with a cousin. So if his two brothers or his eight cousins each survived and reproduced, the altruistic individual would have the same number of genes passed on to the next generation as if he had reproduced himself.

Cooperative Behavior

Although altruistic behavior has received considerable attention from researchers interested in theoretical issues, cooperative behavior is of greater social importance in organizing the behavior of animals. **Cooperative behavior** involves members of the same species and results in mutual benefits. In evolutionary terms, it is behavior whereby both the recipient's and the actor's fitness is increased. Examples of cooperative behavior among animals are numerous.

Figure 28.3

Cooperative defense is used by the Arctic musk-ox for protection against wolves. When threatened, the females form a defensive circle around their calves. The adult male (right) remains outside and attacks any predators, such as wolves, that approach.

Figure 28.4

Lions normally hunt small prey that can be killed by individual lions, but their hunting is more efficient when done in groups. Also, at certain times of the year, smaller prey are scarce, and lions must hunt larger, more dangerous prey. Individual lions are much less successful than groups in hunting such animals. This photo shows young lions cooperating in bringing down a wildebeest.

Mutual vigilance is one that has been studied extensively. Starlings in a flock, for example, can spend more time foraging, and do so more safely, than solitary birds. In experiments with caged starlings, scientists have shown that birds in flocks responded faster to the threat of a predator than solitary birds and that they spent more time foraging.

Cooperative defense can be more active than simple vigilance. Social ants, termites, bees, and wasps can launch massed attacks to defend their colonies. When attacked by wolves, Arctic musk-oxen form a defensive ring of adult females around the young, with the adult male on the outside to attack wolves that approach more closely (see Figure 28.3). Cooperative behavior also permits animals to be more successful in hunting, both in terms of capturing larger prey and in increasing the percentage of successful hunts. Lions have a higher percentage of success in capturing prey when they hunt in groups of two or more. Much of the increased success is attributable to cooperative ambush tactics (see Figure 28.4). Pack-hunting African wild dogs, as shown in Figure 28.5, need four to six individuals to capture most of their regular prey.

Selfish Behavior

Selfish behavior occurs when the actor benefits and there is a cost to the recipient. Natural selection will favor such acts in the actor but will also favor recipients that avoid paying the cost of the actor's selfish behavior. A well-studied example involves the langur monkey, which travels in groups of 10 to 20 adult females with one dominant adult male. Occasionally, the dominant male is displaced by another male. When this happens, the new male will often kill all of the infants in the group. This infanticide contributes to the reproductive success of the new male, for the females mate after they cease nursing. The energy they

Figure 28.5

African wild dogs cooperate to kill their prey.

Figure 28.6

Female langur monkeys attempt to protect their infants from the selfish behavior of a new dominant male. These females and their daughters traveled apart from the main troop while the daughters were infants.

would have expended on their previous infants (fathered by the earlier dominant male) will be saved for raising those of the new male. In contrast to the male, the female's reproductive success is diminished by this infanticide. As a result, female langur monkeys have developed several behaviors to counter the selfish behavior of the male. For example, after a new male has displaced the former one, females will sometimes travel alone or on the periphery of the group until their infants are older and less subject to attack (see Figure 28.6). Other females will also come to the aid of a mother in resisting attacks by the new male.

From an evolutionary perspective, social interactions may be classified by their effect on the reproductive success of the actor and the recipient. Altruistic behavior results in benefits to the recipient at a cost to the actor. The paradox of altruistic behavior is resolved by the concept of inclusive fitness, which measures both an individual's reproductive success and its effects on the reproductive success of its relatives. Cooperative behavior involves members of the same species and results in mutual benefits, whereas selfish behavior occurs when the actor benefits and there is a cost to the recipient.

SOCIAL GROUPS

Although some animals are *solitary,* or alone, for part or most of their lives, many others live in groups. These groups are often dynamic and are difficult to classify. In very broad terms, they can be placed into three general classes: aggregations, anonymous groups, and societies.

Aggregations are groups formed as a result of individuals being attracted to a stimulus or an environmental feature such as a source of food. They have little or no internal organization and are therefore the simplest form of social group. Fruit flies gathered around a mound of rotted fruit, moths around a light, and slugs in moist places, as shown in Figure 28.7, are common examples from everyday experience.

Figure 28.7

Slugs form an aggregation that is based on their attraction to moist places. Gardeners in the Pacific Northwest make use of this attraction in reducing slug damage by placing slug bait in moist spots in and around their gardens.

Figure 28.8

These caribou form an anonymous group.

Anonymous groups often exhibit coordinated movements that are more organized and may have some rudimentary division of labor. Examples are schools of fish, migratory flocks of birds, and migrating herds of caribou (see Figure 28.8). Although not highly structured, these groups still provide advantages to members in terms of mutual defense and other cooperative behaviors.

Animal *societies* have received the most attention from scientists studying animal behavior. They are very diverse and can vary with the season and are therefore difficult to characterize. Generally, **societies** consist of members of a species that show an attraction for one another, communicate with one another, exhibit cooperative behavior, and have synchronized activities. Table 28.1 lists various animal societies. (See also Figure 28.9, page 518).

Features of Social Organization

The pattern of relationships among individuals within a population at a particular time is known as **social organization**. Social organization is dynamic; it can change to meet different conditions. How can social organization be understood?

Table 28.1 Animal Societies

Type of Society	Examples
One-parent family	Polar bear (*Thalarctos maritimus*), black bear (*Ursus arctos*), hedgehog (*Erinaceus europaeus*), mallard duck, black grouse, stickleback fish, many rodents
Family	Songbirds, mute swan, golden jackal (*Canis aureus*)
Extended one-parent family	Honeybee, wasp (*Vespula vulgaris*), ants
Extended family	Termites, wolf, white-handed gibbon (*Hylobates lar*)
Harem	Wild horse (*Equus caballus*), zebra (*Equus burchelli*), vicuna (*Vicugna vicugna*), red deer, gelada baboon (*Theropithecus gelada*), patas monkey, sea lions (*Otaria*)
Female groups	African elephants (*Loxodonta africana*), wild pigs (*Sus scrofa*), red deer
Bachelor groups	Common waterbuck (*Kobus ellipsiprymnus*), Thompson's gazelle (*Gazella thomsoni*)
Multimale groups	Common baboon, rhesus monkey, spotted hyena (*Crocuta crocuta*), African hunting dogs (*Lycaon pictus*), African lion (*Panthera leo*)

Biologists usually examine several key features. First, they analyze group structure. They gather population data concerning the distribution of age, population density, reproductive rates, life spans, and participation in reproduction. **Demography** is the statistical analysis of population data collected from animal behavior study.

Next researchers look at how individuals and groups are distributed in the area they occupy and at the behavioral activity involved in that spacing. Animals do not use all of the available space in an area. What they do exploit on a daily basis is called their **home range**. Within the home range, individuals and groups are spaced in different *dispersion patterns* and show diverse behaviors related to those patterns. At one extreme is the mountain gorilla, which lives in groups that have overlapping home ranges and generally tolerates its neighbors. At the other extreme, observed in robins, is a **territory**, where the home range is defended by threats, attacks, or advertisement (with scent or song) for exclusive use of the individual or group.

The study of *territorial behavior* has long attracted interest. The evolutionary significance of that behavior has also been appreciated, since the occupation of a specific area gives access to resources such as food, nest sites, or mates that increase reproductive success.

In addition to studying population data and spacing, scientists have a critical interest in describing social interaction in groups. How do animals interact with other members of their group? How do they cooperate, compete, and communicate, and how do these activities develop? These are subjects of central interest for many scientists who study animals in the field and in the laboratory. Many behavioral interactions involve communication, which we shall discuss shortly.

Finally, biologists try to understand how social organization changes. It is constantly being modified in small ways due to the shifting relationships among individuals and also often undergoes marked changes related to season, ecological conditions, or reproductive cycles. For instance, during the red deer mating season, males form temporary harems of females, but after the season, the deer organize into male and female groups.

Communication

Communication occurs when information is passed by means of signals from one animal to another and results in a change in behavior. The study of communication is essential to understanding social organization because so much of social behavior depends on animals' ability to communicate.

Methods of Communication

Animals communicate in many ways. The simplest and perhaps oldest method is chemical. Chemical signals that convey information between animals are called **pheromones**. They are found throughout the animal kingdom and are used to convey a variety of information. Sex pheromones are species-specific and are used to attract mates. Other pheromones cause aggregation in social insects, are used for alarm signals, or are used in the recognition of territory boundaries. Most pheromones are releasers that trigger specific behaviors. Some, however, called *priming pheromones,* act by modifying the physiology of an organism over a long period of time rather than by calling forth an immediate behavioral response. Priming pheromones can synchronize the sexual maturation of animals that aggregate during reproduction, such as locusts. In social insects such as honeybees, the queen produces a priming pheromone that inhibits the sexual development of worker bees. Some researchers think that the synchronization of human female menstrual cycles, which often occurs among women living closely together (such as in college dormitories or sororities), is caused by an unknown pheromone.

Figure 28.9

Some different animal societies. (A) Mallard hen with young (one-parent family); (B) sea lions (harem); (C) African elephants (female group); and (D) Thompson's gazelle (bachelor group).

Visual and auditory signals are most commonly used in communication. Many of these signals are releasers; they can trigger stereotyped responses that are innate. Some of the responses may be conditioned by experience or modified by the motivational state of the recipient. Complex social communication—for example, that among dogs—is a rich mix of learned and innate behaviors that may involve many signals to which we humans are insensitive.

The range of communication signals is not confined to chemical, visual, and auditory cues. Tactile (touch) signals are employed by many animals in the final stages of courtship. The male garden spider, for example, signals the female by rhythmically twanging a strand of her web. One unusual mode of animal communication is electrical. Electric fish use electrical pulses to navigate in their environment. Careful experiments on one of these fish, *Brienomyrus brachyistisus,* have revealed that the male and female emit different pulses. The male can detect the species-specific signal and responds with a stereotyped electrical "rasp" (see Figure 28.10).

Information Communicated

Animals communicate information about the environment, internal states, and their identity, as well as signals that coordinate behavior. One of the best-studied and most fascinating examples of communication is the dance language of honeybees.

For a long time, it has been known that honeybees can communicate about the location of food. For decades, the Austrian biologist Karl von Frisch, shown in Figure 28.11, studied the behavior of bees and conducted experiments to comprehend their means of communication. He showed that forager bees, upon returning to the hive after finding a food source, can communicate its location to other workers. They do this by means of performing, on a vertical comb surface of the hive, one of two sets of body movements called dances: a waggle dance or a round dance (see Figure 28.12). The *waggle dance,* which follows the format

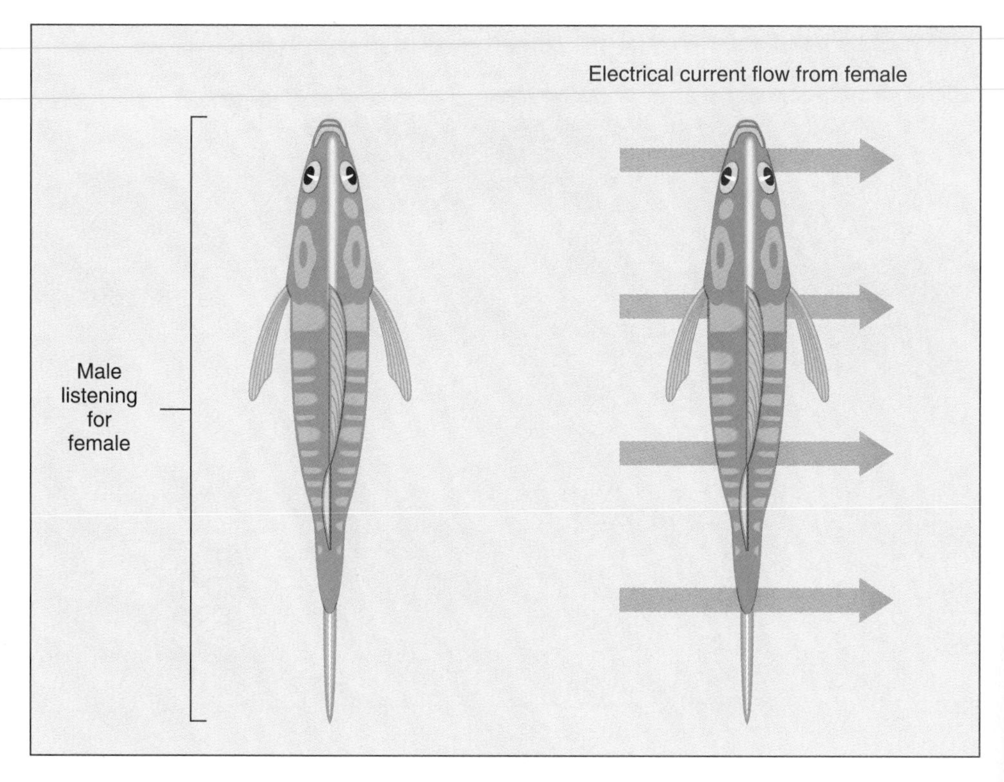

Figure 28.10

Electrical pulses of a fish, *Brienomyrus brachyistisus*, have been studied to understand electric fish communication. Males and females emit different pulses, and during the breeding season, unmated males respond to females that emit a special courtship pulse. The males detect the female's pulse through electroreceptors on both sides of their bodies. These receptors respond to current flow from one direction only, so the fish can tell when a pulse begins and ends.

of a figure eight, is done if the source is far away. The length of the dance is correlated with the distance of the food source.

The direction of the food source is also communicated by the waggle dance. As the bee dances, as described in Figure 28.13, it traces on the vertical sheets of comb a figure eight with the axis at an angle to the force of gravity. This angle matches the angle between the sun and a line that points directly at the food source. So if the food is in the same direction as the sun, the forager will perform a dance pointed straight up. If the food is in the opposite direction, it will dance pointed downward. If the food is 45° to the right of the sun, it will orient the axis of the dance 45° to the right. The *round dance,* which is merely a reduced waggle dance, is performed if the food source is close to the hive. Both dances excite foragers, which exit the hive and search until they find the food.

Figure 28.11

Karl von Frisch studying bee behavior.

Scientists study various aspects of social organization. They study population data, the spacing of individuals in an area, their methods of communication, and the information communicated. Animals communicate by chemical, visual, and auditory signals. Some use tactile signals or even electrical pulses to communicate. The information conveyed can be about the environment or about themselves and is central in coordinating behaviors.

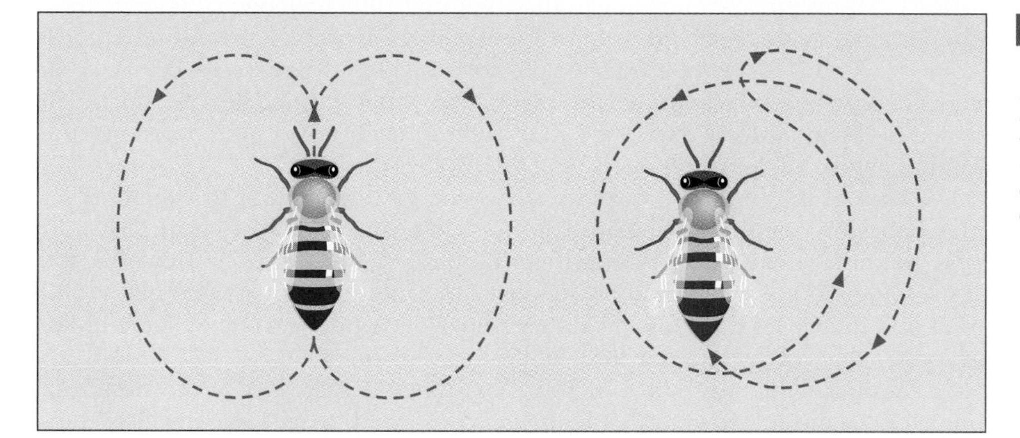

Figure 28.12

Honeybees communicate information about food sources. In the waggle dance, a figure eight is made. The bee waggles its abdomen along the straight line between the loops. In round dances, the bee traces a series of clockwise and counterclockwise circular movements.

Figure 28.13

The axis, or orientation, of the waggle dance communicates the direction of a food source. (A) If the food is located toward the sun (from the nest), the dance is straight up. (B) If the food is in a direction opposite to the sun, the axis of the dance is down. (C) If the direction of the food is 45° to the left of the sun, the axis of the dance will be 45° to the left.

THE ADAPTIVE VALUE OF SOCIAL BEHAVIOR

The evolutionary perspective by which behavior is approached stresses the value of social behavior for reproductive success. A general look at the main functions of behavior supports that position. However, explaining the detailed behavior of animals is not always easy, and often scientists are quite puzzled by their observations. All exciting science is like that. Without the challenge of unexplained phenomena, science would be a dull enterprise. However, inventing a good story to account for certain behavior is not considered good science. In 1902, Rudyard Kipling published a charming children's book, titled *Just So Stories,* in which he imaginatively explained the origin of such novel phenomena as the camel's hump, the leopard's spots, and the whale's throat.

Some of the popular books on sociobiology that attempt to explain complex behavior—especially human behavior—in terms of relative or inclusive fitness are, unfortunately, little better than Kipling's fictional stories. However, the abuse of sociobiology should not detract from the enormous value that it and evolution theory hold for understanding behavior. A brief review of some of the adaptive values of social behavior will illustrate this point.

An often-cited advantage of social existence is increased protection from predators. Many animals must expend energy to avoid being someone else's din-

Figure 28.14

A Japanese macaque "invented" potato washing, which was quickly learned by other members of the troop.

ner. Cooperative behavior in a social group allows most species to benefit from mutual vigilance. Birds, for example, as mentioned earlier, gain extra foraging time by living in flocks, for they can rely on the vigilance of other flock members.

What other benefits might an individual animal gain from group living? Researchers in behavior now recognize many advantages. For example, numerous carnivores cooperatively hunt prey that an individual would be incapable of bringing down or killing. Similarly, experiments on captive birds show that groups are more efficient in finding food than single birds. Also, birds that forage over a wide area where the food is unevenly distributed benefit by belonging to a flock, which allows them to follow successful birds.

The division of labor that is possible in social insects has important adaptive value. Social insects have colonies that consist of different **castes**, which are groups of individuals that differ in structure and function. The different castes undertake distinct tasks, which increases the efficiency of the group considerably. By dividing the work among different groups, not only is each individual task done more efficiently, but several important tasks can be undertaken at one time, rather than in a sequence. The probability of completing the task is thereby considerably increased.

Another possible advantage of social life is the transmission of learned behavior. When generations overlap, learned behavior can be transmitted to the next generation. For example, bird songs are partly learned. Male chaffinches raised in total isolation sing a song that is different from that of normal adults.

Learned behavior can be transmitted to peers as well. In studies on Japanese macaque monkeys, a particularly innovative female "invented" potato washing, which soon spread to other members of her troop (see Figure 28.14). She later discovered that she could wash wheat to rid it of sand, and this practical skill was also passed on to others and also to later generations. Compared to the evolution of human culture, this "animal cultural evolution" is limited. However, its adaptive value is nonetheless important. (See the Focus on Scientific Inquiry, "Do Animals Think?")

Social behavior is adaptive. It can increase protection from predators, help in obtaining food, make the completion of tasks more efficient, and make the transmission of learned behavior possible. These advantages can increase the reproductive success of the individuals involved.

SOCIAL BEHAVIOR, SOCIOBIOLOGY, AND HUMANS

The study of social behavior was dominated in the 1980s by the insights of sociobiology and led to fruitful research. But the field has remained controversial. The two main issues that have attracted criticism are the emphasis on the genetic basis of social behavior and the application of sociobiology to humans.

Focus on Scientific Inquiry

Do Animals Think?

Scientists have investigated the human organism from a mechanistic viewpoint for over a century. Their underlying assumption has been that humans could be viewed as elegant chemical machines that follow predictable natural laws. This approach has had stunning success. We know a great deal about the human body, we can design drugs to alleviate various aliments, and we can counter numerous conditions that cause suffering or death. Although science can tell us a great deal about our physical condition by treating the human body as a machine, no one doubts that humans are, unlike machines, conscious creatures. Our own consciousness is evident. What about animals?

To those of us who have pets, such as dogs or cats, it is difficult to think of them as machines without self-awareness—as entities more akin to our washing machines, personal computers, and blow-dryers than to our family members and friends (see Figure 1). Can it be that our clever dog, Cassie, that "comforts" us when we are down, leaps with "joy" when we return from work, and has "outsmarted" the neighbor's dog that used to "steal" her food, is simply a genetically programmed automaton? Or that ZiZi, our neighbor's cat, that would seemingly "favor" starvation to dry cat food and that, if not a connoisseur of lasagna, is known to "prefer"—very definitely—smoked salmon to canned tuna, is, in her behavior, just reflecting an idiosyncratic program rather than expressing a conscious preference?

Until shortly after World War I, it seemed obvious to scientists that animals had feelings and that they could think. Charles Darwin believed that female birds showed aesthetic preferences in their choice of mates and that sexual selection was strongly influenced by it.

Many writings done in the late nineteenth century on the animal mind, however, were uncritical and highly anthropomorphic. Human desires, fears, and attitudes were attributed to animals, and numerous stories were accepted without any careful attempts at verification. It is not surprising, then, that when we read this literature today, much of it seems comical.

Psychologists in the 1920s reacted strongly to this uncritical literature and took the position that it was not possible to verify whether or not animals could

Figure 1 (A) Can this animal think? (B) Do pet tricks reflect thinking ability?

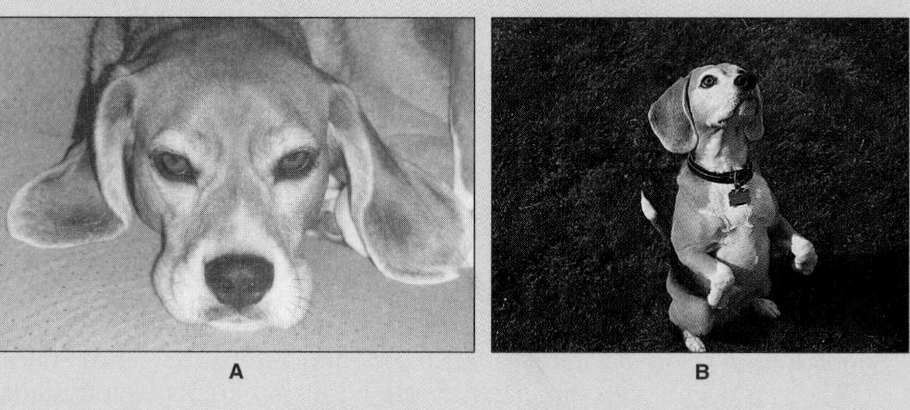

think. They concluded that the question of animal thinking was not a meaningful topic for science because it could not be tested experimentally. Instead, psychologists focused on the observable *behavior* of animals. They argued that in establishing a scientific psychology, it was irrelevant whether animals thought. They intended to establish scientific laws about how animals learn and behave that could be verified by other scientists. To psychologists like James Watson or B. F. Skinner, the private mind of the animal, if it existed, was closed to human investigation.

Ethologists who studied animal behavior were, for the most part, equally dismissive about probing the inner world of animals. A few workers were interested in how the world might "look" to animals, which have different sense organs than humans, but the primary thrust of ethology was in documenting repeatable patterns of behavior and in comparing these patterns with the object of establishing evolutionary connections.

Modern animal behavior draws on knowledge derived from psychology, ethology, and an ever-growing body of research results from genetics, ecology, neurophysiology, and neuroanatomy. Until recently, all of these areas of research had been far removed from discussions of animal thought or animal awareness.

A well-known investigator of animal behavior, Donald Griffin of Rockefeller University, argues that neglecting animal awareness and thinking is not only an overreaction to the naive acceptance of undocumented animal stories but also a blind spot that retards advances in the scientific understanding of animal behavior. Griffin believes that mental experiences in animals could have an adaptive value—the better an animal understands its environment, the better it can adjust its behavior to survive and reproduce in it. He is also interested in animal communication, which he feels can sometimes be used to convey information about objects or events that are distant in time or space. This form of information may suggest awareness.

In support of his ideas, Griffin cites various behaviors that seem to involve accurate evaluation in complex environments. For example, he refers to a classic study on the prey selection of a bird, the wagtail, found in southern England. These birds feed on fly eggs and on a number of small insects, and each day they must make several choices on where to hunt, when to move on to hunt in another area, and whether to join a flock or hunt alone. Scientists who study these wagtails have shown that they hunt with great proficiency. Although proficiency is not necessarily an indicator of awareness, Griffin argues that in cases where accurate evaluation of a changing and complex environment occurs, it is reasonable to consider that the animal is consciously thinking about what it is doing. Cooperative hunting by lions and the cultural transmission of behavior such as the potato washing done by Japanese macaques, both described in this chapter, and insight learning, described in Chapter 27, are other examples of behaviors that suggest to Griffin and others that animals are aware and can think.

At present, Griffin and analysts who agree with him are in the minority in the scientific community. How can the question of animal thinking be resolved? One way is to attempt to design experiments that might give an indication one way or the other. Psychologists in the early twentieth century were very outspoken in their rejection of animal consciousness, claiming that testing for it was impossible. Griffin has proposed that some tests may be possible and that evidence can be gathered to support his position. He argues that once we have a better understanding of the electrical signals that are correlated with conscious thinking in humans, we could search for equivalents in animals. If none were found, that would suggest that his hypothesis of animal awareness is false. The strongest supporting evidence of Griffith's hypothesis involves cases in which animal communication is active and specialized, information is exchanged, and the receiving animal responds interactively. To Griffin, such cases are compelling examples of conscious and intentional acts.

It is too early to tell what researchers of animal behavior will conclude about animal awareness. Further research on interesting phenomena, such as animal communication, will ultimately provide the results necessary for formulating a scientific conclusion. Until then, we are confident that people will continue to discuss the world with their dogs and cats.

Few experts doubt that individual and social behavior have their foundations in the genome, but such a belief is not the same as claiming that particular genes control a specific behavior. Likewise, some behavior does appear to improve fitness, but it is usually impossible to measure directly the impact of a particular behavior on the fitness of an allele. Even the influence of a behavior on the fitness of an individual is often difficult to assess, because of the multitude of factors involved in fitness. Critics argue that although genes may be responsible for the physiological and anatomical foundations of behavior, and that behavior can evolve, behavior is too complex to be understood in terms of our current knowledge of genetics. To a large extent, the resolution of the genetic issue will depend on how successful sociobiologists are in relating behavior to genes.

The issue as it relates to humans, however, is much more nebulous, for *Homo sapiens* simply cannot be studied in the same way that scientists explore the behaviors of starlings and starfish. The issue with human behavior, moreover, raises a difficult question: To what extent is human behavior controlled by genes? The idea of human behavior being under genetic control is not attractive to those who take pride in people's freedom of action, creativity, and ethical standards.

What are the implications for social change and moral choice? A few sociobiologists have argued that the study of social behavior will permit researchers to provide a scientific foundation for human values. Such a claim is not only unlikely but also confuses different domains of thought. An understanding of the biological constraints on human action is necessary for a discussion of how humans *ought to* act. Philosophers often begin their investigations of morality with a discussion of human nature. Whatever light biology could shed on that issue has always been important, but understanding the biological basis of behavior will not reveal what is good or bad behavior. Some sociobiologists have attempted to argue that behaviors that increase fitness might be the basis of morality. However, survival in itself does not have any moral value.

The study of behavior, like other branches of the biological sciences, can be approached from numerous perspectives. The study of mechanisms of individual behavior can be profitable without reference to the theory of evolution. The theory of evolution, however, provides considerable guidance for the origin of behavior and for understanding its diversity. Studies of social behavior have been especially tied to evolution and have even contributed to the enlargement of the scope of the theory. The study of social and individual behavior may provide a background for an understanding of human values but appears to be an inappropriate guide for them. The study of morality lies in the domain of the humanities. Poets tell us much about nature, but not the same sorts of things scientists do. Similarly, biologists tell us much about behavior, but not what philosophers have to say. Since all knowledge is ultimately connected, cooperative behavior among scholars will, it is hoped, provide the greatest benefit to all.

Summary

1. Sociobiology attempts to integrate research on social behavior with knowledge from genetics, evolution, and ecology. Basic to the study of social behavior is the concept of relative fitness.
2. Social interactions are characterized as altruistic, cooperative, or selfish. Study of altruistic behavior has led to the formulation of the concept of relative fitness.
3. Although some animals are solitary, many live in groups. Animal groups fall into three broad classes: aggregations, anonymous groups, and societies.
4. The pattern of relationships among individuals in a population at a particular time is known as social organization. Among the features of social organization that scientists study are group structure, population data, distribution patterns, interactions, and change.
5. Communication is the passing of information by means of signals, resulting in

a change in behavior. Animals communicate in various ways. They can use chemicals, electrical pulses, or visual, tactile, or auditory signals.
6. Animals communicate information about the environment, internal states, and their identity, as well as signals that coordinate behavior.
7. Sociobiology stresses the adaptive value of behavior. It has been a very fruitful approach but has been criticized because of its emphasis on the genetic basis of behavior and because some sociobiologists apply their techniques to the study of human behavior.

Review Questions

1. How is behavior classified from an evolutionary perspective?
2. What is altruistic behavior?
3. Altruistic behavior seemingly reduces individual fitness. How is this paradox resolved?
4. What is cooperative behavior? What are some of its benefits to the individual?
5. What is selfish behavior?
6. What are three general classes of social groups?
7. How are population data used in studies of social behavior?
8. What two aspects of animal spacing do scientists study?
9. What does *communication* mean in the study of animal social behavior?
10. What are pheromones? What are some of their functions in animal interaction?
11. What other simple communication signals do animals use?
12. What are some adaptive advantages that result from social existence?

Essay and Discussion Questions

1. A controversial issue in biology is the question of how speciation occurs. How might the study of behavior influence that discussion?
2. Some of the harshest critics of sociobiology have stated that the study of animal behavior may be of importance for ethics. What might they mean by that?
3. If it were to be established that genetically human males were more aggressive than females, would there be any implications for the formulation of social policies? What would they be? Why?

References and Recommended Reading

Alcock, J. 1989. *Animal Behavior: An Evolutionary Approach*. Sunderland, Mass.: Sinauer.

Gould, J. L. 1982. *Ethology: The Mechanisms and Evolution of Behavior.* New York: Norton.

Griffin, D. R. 1984. *Animal Thinking*. Cambridge, Mass.: Harvard University Press.

Kitcher, P. 1985. *Vaulting Ambition: The Quest for Human Nature*. Cambridge, Mass.: MIT Press.

Ristau, C. A. 1991. *Comparative Ethology: The Minds of Other Animals*. Hillsdale, N.J.: Erlbaum.

Trivers, R. L. 1985. *Social Evolution*. Menlo Park, Calif.: Benjamin/Cummings.

Wilson, E. O. 1978. *On Human Nature*. Cambridge, Mass.: Harvard University Press.

———. 1975. *Sociobiology: The New Synthesis*. Cambridge, Mass.: Harvard University Press.

Wintsch, S. 1990. Cetacean intelligence: You'd think you were thinking. *Mosaic* 21(3):34–48.

Wittenberger, J. F. 1981. *Animal Social Behavior.* North Scituate, Mass.: Duxbury Press.

UNIT IV

The Systems of Life

Through millions of years, plants and animals have evolved tissues and organs that permit them to function, survive, and leave offspring. Despite enormous diversity and variations, all organisms share common functions. What are the basic processes of life? How do organisms respond to their environment? What happens when disorders or diseases occur? Unit IV examines these questions, which are related to the systems of life.

The foremost requirement of organisms is energy. Plants are central to the capture and transformation of energy used by most organisms on Earth. Plants evolved fundamentally different strategies than animals to meet the challenges of survival and reproduction. Plant systems are superbly adapted for coping with continual stresses that characterize the various environments they inhabit.

Humans and other animals have evolved elaborate organ systems that enable them to carry out complex functions. Contemporary studies on humans and animals have led to deep levels of understanding about the normal functions of most human systems. Research has also shown scientists how certain diseases develop. There is now great hope that complicated human diseases such as coronary heart disease and cancer may yield to new approaches that emphasize prevention.

CHAPTER 29

Photosynthesis and Cellular Respiration

Chapter Outline

Reading Questions

1. What are the different forms of energy that are important for sustaining life on Earth?
2. What is photosynthesis, and why is it important?
3. What are glycolysis and cellular respiration, and why are they important?
4. How are photosynthesis, glycolysis, and cellular respiration related?
5. How can life on Earth be explained in terms of energy and energy transformation processes?

There exist on Earth complex levels of organization, such as ecosystems, that are supported by a continuous input of light (radiant) energy from the sun (Figure 29.1). Enormous quantities of light energy are essential for maintaining temperatures suitable for life on Earth and for driving the water cycle that distributes moisture over its landscapes (described in Chapter 7).

Figure 29.1

Almost all organisms on Earth depend on the sun for energy they require to sustain their lives.

ENERGY IN THE BIOLOGICAL REALM

At simpler levels of organization, trees, humans, frogs, worms, and bacteria carry on activities that require an uninterrupted supply of energy. The sun is also the ultimate source of energy for almost all of these biological systems. Only in deep-ocean hydrothermal vent systems, and perhaps a few anaerobic hot springs, are different sources of energy—hydrogen sulfide and methane—used to support life (see Chapters 2 and 9).

Plants are able to capture light energy emitted by the sun and convert it into chemical energy. Almost all organisms depend on plants to provide the chemical forms of energy required for carrying out life-supporting functions. In this chapter, we examine the energy-transforming pathways used by organisms and the importance of these reactions for sustaining life on Earth.

ENERGY IN THE PHYSICAL REALM

Exactly what is energy? Most people have an intuitive sense about energy; it is something associated with activity, movement, or power—the ability to cause change. Scientists describe energy in more technical terms. *Energy* occurs in two general states. **Potential energy** is stored or inactive energy that is capable of doing work later. A stick of dynamite, a liter of gasoline, and the vast quantities of snow deposited on mountains during winter have potential energy (see Figure 29.2). **Kinetic energy** is the energy of action or motion. A match applied to the dynamite or gasoline will explosively convert the potential energy into kinetic energy. When the warm spring sun melts snow, the water flowing downhill represents kinetic energy (Figure 29.2A). Dams harness the kinetic energy of falling water to produce electricity (Figure 29.2B).

Forms of Energy

Several forms of energy exist on Earth. **Chemical energy** is the potential energy contained in chemical bonds within compounds such as gasoline, sugar, and fat. **Chemical bonds** are forces of attraction that hold two or more atoms together in a molecule. When these bonds are broken, energy is released. **Thermal energy (heat)** is the kinetic energy of molecular motion that can be measured

Figure 29.2

(A) Snow constitutes a source of potential energy. When snow melts and flows into streams and rivers, it has been transformed into kinetic energy. (B) Dams transform the kinetic energy of moving water into electrical energy.

with a thermometer. With appropriate mechanical systems, heat can be harnessed to do certain types of work such as power an automobile or produce electricity. Within the biosphere, the movement of heat from regions of higher to lower temperature creates winds and powers the water cycle. **Electrical energy** is produced by a flow of electrons (negatively charged particles) and may occur in such diverse objects as batteries and the cells of certain organisms. Speeding race cars and flexed muscles reflect **mechanical energy**, which is expressed as motion. Energy emitted from the sun travels to Earth as **radiant energy** in the form of light waves.

The Laws of Thermodynamics

Where does energy come from? What are the relationships between the various forms of energy? Basically, all forms of energy are interrelated and can be converted from one form to another. **Energy transformations**, changing one form of energy to another, are the basis for sustaining life on Earth. However, such transformations are not very efficient since some energy is lost during each change. For example, a liter of gasoline (chemical energy) transformed to move (mechanical energy) a car 5 miles could actually move it 15 miles if the energy transformation were 100 percent efficient. Why are energy transformations so wasteful? Two laws of thermodynamics formulated by physicists help answer such questions. Recall that laws are the most powerful scientific concepts; they describe regularities that we think are invariable.

The First Law of Thermodynamics

Thermodynamics describes the relationships between heat and other forms of energy. The **first law of thermodynamics** (also called the *law of conservation of energy*) states that energy can be converted from one form to another form but can never be created or destroyed.

The first law offers this startling perspective: no organism can create the energy it requires for survival. The significance of the first law is that radiant energy from a distant source, the sun, must be converted into chemical energy that can be used by living organisms.

The Second Law of Thermodynamics

The **second law of thermodynamics** states that every energy transformation results in a reduction in the total usable energy of the system. To maintain order, organisms must transform energy every second of their lives. All usable forms of energy are ultimately dissipated as heat through energy transformations. As a result, **entropy**, defined as the useless energy of any system—whether the universe or a cell—continuously increases while order, or organization, decreases.

The ultimate biological consequences of the second law are harsh. Life on Earth will end in several billion years when the sun's energy is finally used up in accordance with the second law. Also, all organisms eventually lose their battle against entropy, biological order fades, and the organism dies.

Picture a romantic setting in which dinner is being served by candlelight. To open (and quite possibly close) the conversation, how would you explain the burning of the candles in terms of the first and second laws of thermodynamics?

An unending supply of energy is required to sustain life on Earth. This is provided by radiant energy emitted by the sun, which is captured by plants and converted into chemical energy, a form of potential energy. Organisms transform chemical energy into different forms of kinetic energy that they use for carrying out their activities. The laws of thermodynamics explain the primary consequence of energy transformations—an increase in entropy.

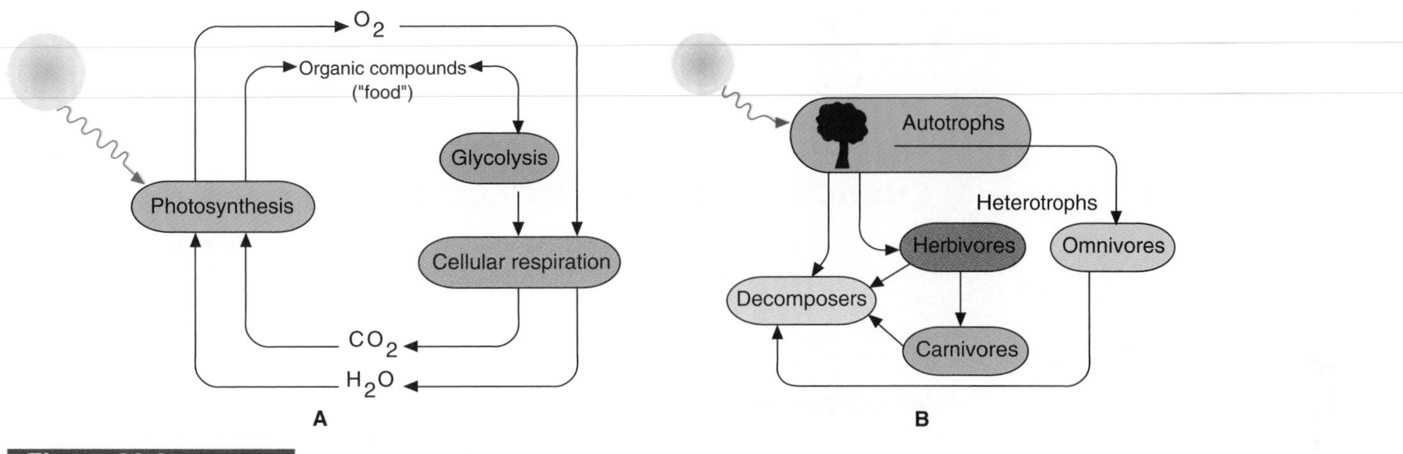

Figure 29.3

(A) Radiant energy from the sun is transformed by plant cells into organic compounds used as food through the process of photosynthesis. Plants and animals carry out the reactions of glycolysis and cellular respiration to convert food (chemical energy) into energy that can be used in the cell. Oxygen, carbon dioxide, and water are recycled through these three biological energy transformation processes. (B) Autotrophs produce food for themselves and for various heterotrophs.

BIOLOGICAL ENERGY TRANSFORMATION PROCESSES

The energy transformation process by which solar energy is captured by plants and converted into high-energy organic compounds, such as the sugar glucose, is called *photosynthesis*. Almost all organisms can convert the chemical energy in these compounds into other forms of energy necessary for their activities through two sets of reactions, termed *glycolysis* and *cellular respiration*. The latter term should not be confused with breathing, or respiring, which refers to the exchange of gases between air and lungs. Figure 29.3 describes the fundamental relationships of the energy-transforming pathways and reviews concepts introduced in Chapter 7. Using light energy, green plants are able to manufacture complex organic compounds ("food") from carbon dioxide and water during photosynthesis.

Photosynthetic organisms, including most plants, are able to use the food they synthesize to support their activities and are classified as **autotrophs** (in ecology, *producers*). **Heterotrophs** (the *consumers* of ecology) are organisms that must obtain energy and nutrients from the food molecules produced and stored by autotrophs. Heterotrophic organisms are further classified according to how they accomplish this task. *Herbivores* obtain energy-rich, carbon-containing compounds directly by eating plants or plant parts. *Carnivores* acquire products manufactured by autotrophs indirectly by consuming the flesh of herbivores and other animals. *Omnivores,* including humans, can obtain energy from either plants or flesh. *Decomposers* obtain food by breaking down dead organic materials from any source.

The reactions of photosynthesis, glycolysis, and cellular respiration are linked through an interdependence of products formed during each process. Heterotrophs and autotrophs both carry out cellular respiration, which results in the production of carbon dioxide and water. These are the raw materials of photosynthesis, which can be cycled back to plants for the synthesis of more sugar. Oxygen, a waste product of photosynthesis, is required for the reactions of cellular respiration.

The chemical reactions of photosynthesis and cellular respiration are complex and numerous. We shall limit our descriptions of chemical reactions to a few and instead concentrate on broader concepts related to the following questions:

1. How do plants harness the sun's energy?
2. What are the major products of the energy transformation pathways?
3. What are the relationships between the energy-generating and the energy-consuming pathways?

ENERGY-MANAGING MOLECULES

For energy to be produced and used in an orderly manner, certain molecules in cells are employed to "manage" the energy as it becomes available. Essentially, they serve either to store energy or to transport it between reactions until it can finally be released to the cell.

Adenosine Triphosphate

The energy used by plant and animal cells is provided by **adenosine triphosphate (ATP)**. Figure 29.4 shows the simplified molecular structure and the reactions of ATP. ATP is composed of the molecule adenosine and three linked phosphate groups (P). Energy is *released* in the cell when the terminal phosphate is removed by a specific enzyme (enzymes are discussed in Chapter 16) to produce adenosine diphosphate (ADP) and a phosphate molecule. Energy is *stored* when ADP combines with P to regenerate ATP.

The ATP-ADP system is very dynamic since the cell is a cauldron of constant chemical activities, with some releasing energy and some storing energy. To illustrate this flux, consider that a new ATP molecule is consumed within 60 seconds of its formation and that each typical plant or animal cell contains several *million* molecules of ATP-ADP. Even on the quietest day of your life, your cells require the generation of about 45 kilograms of ATP simply to keep you alive!

Electron Carriers

During the chemical reactions of photosynthesis, glycolysis, and cellular respiration, a continuous series of molecular degradation and rearrangement results in the release of energetic electrons that must be captured to prevent damage to the cell. How is this accomplished? **Coenzymes** are organic molecules that are required for the operations of enzymes, and many are able to accept the energetic electrons, serving as *electron carriers*. Three coenzymes play important roles in managing energetic electrons produced during biological energy transformations. In this chapter, we will simply use their common names. When not carrying electrons, they are referred to as $NADP^+$, NAD^+, and FAD. When they are carrying electrons, they are identified as NADPH, NADH, and $FADH_2$. ATP is produced when energy released from the electrons is captured as they are transferred to other electron carriers in later reactions (see Table 29.1).

Different biological energy transformation processes, and the organisms involved in each, are interrelated. Autotrophs convert radiant energy into chemical energy by photosynthesis. Both autotrophs and heterotrophs transform this chemical energy into forms of kinetic energy during glycolysis and cellular respiration. Several molecules—the ATP-ADP system and various coenzymes—participate in the controlled release of kinetic energy in the cell.

Figure 29.4

(A) Adenosine triphosphate is composed of an adenine base, ribose, and three phosphates. The wavy line between phosphates signifies a high-energy chemical bond. (B) ATP is able both to store and to release energy at different times. Energy is stored for future use when ADP combines with a phosphate (P) to form ATP. Energy is released to the cell when a phosphate is removed from ATP, converting it back to ADP.

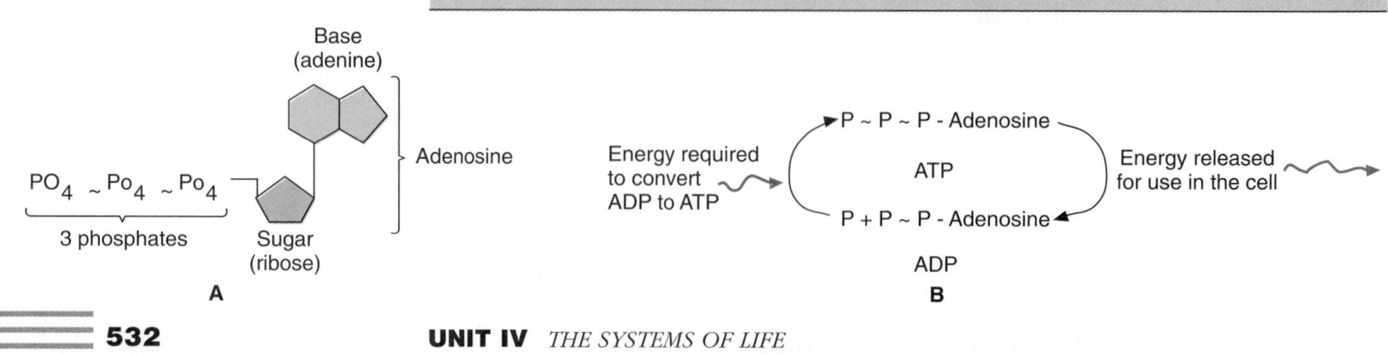

Table 29.1 Summary of Energy Reactions Occurring in Cells

Molecules	→ Molecules	Outcome
	Reactions During Photosynthesis	
ATP	→ ADP + P	Energy released for use in the cell
ADP + P	→ ATP	Energy stored for later use in the cell
$NADP^+$	→ NADPH	Energetic electron captured by coenzyme
NADPH	→ $NADP^+$	Energy from electron released to make ATP
	Reactions During Cellular Respiration	
ATP	→ ADP + P	Energy released for use in the cell
ADP + P	→ ATP	Energy stored for later use in the cell
NAD^+	→ NADH	Energetic electron captured by coenzyme
NADH	→ NAD^+	Energy from electron released to make ATP
FAD	→ $FADH_2$	Energetic electron captured by coenzyme
$FADH_2$	→ FAD	Energy from electron released to make ATP

All living things transform energy at the cellular level. Plants use all of these reactions since they carry out photosynthesis and cellular respiration; animals use only the reactions of cellular respiration.

PHOTOSYNTHESIS

Photosynthesis consists of a series of complex chemical reactions driven by energy originally supplied by the sun. What is the significance of this process for organisms inhabiting Earth? All plants and animals require food and oxygen for their survival. Plants manufacture their own food through photosynthesis. However, animals cannot make food and are ultimately dependent on the energy-rich, organic compounds produced by plants. Thus each of the millions of carbon molecules in every one of your billions of cells came from a molecule of carbon dioxide that was once present in the atmosphere. The warmth of a fire on a cold winter night represents energy that originally came from the sun tens, hundreds, or millions of years ago. Finally, most organisms require the oxygen that is released during photosynthesis. According to contemporary scientific hypotheses, if the remarkable process called photosynthesis had not evolved, there would be no oxygen and no food supply. The most familiar life forms—plants and animals—would never have appeared on Earth.

Photosynthesis occurs primarily in the cells of green plants, algae, and cyanobacteria. The process can be summarized as follows: radiant energy from the sun is used to convert carbon dioxide and water into glucose and oxygen according to the following simplified equation. These types of equations summarize the number and types of molecules *used* to the left of the arrow and the number and types *produced* to the right.

$$6CO_2 + 12H_2O + \text{radiant energy} \rightarrow \text{glucose } (C_6H_{12}O_6) + 6O_2 + 6H_2O$$

Radiant Energy

As described in Chapter 2, the sun emits radiant energy consisting of various *photons* that are each composed of a different wavelength and energy level, as shown in Figure 29.5. The shorter the wavelength, the greater the energy of the photon. High-energy photons called gamma rays and X-rays are very harmful to

Figure 29.5

Sunlight, or radiant energy, consists of a mixture of photons, each having different wavelengths that represent distinct levels of energy. Visible light includes colors with wavelengths between 400 and 750 nanometers. Photons with shorter wavelengths—gamma rays, X-rays, and ultraviolet light—possess levels of energy that can damage DNA.

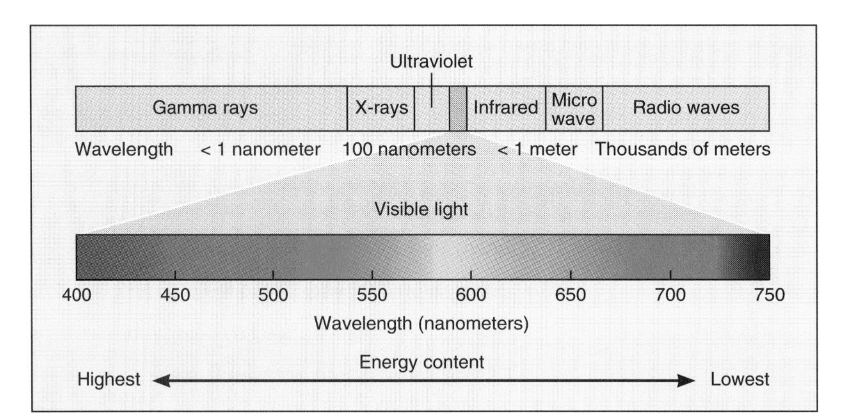

biological systems because they can cause mutations in DNA, but they rarely reach Earth's surface. Ultraviolet radiation is also harmful but is mostly blocked by gases (especially ozone) present in the upper atmosphere. In quantities emitted by the sun, lower-energy infrared radiation and radio waves are not energetic enough to cause biological effects.

Only photons that fall within the narrow *visible light spectrum* possess energy that can be captured and transformed through biological processes without causing harm. Photons with characteristic wavelengths in the visible light spectrum are perceived as distinct colors because they affect light receptor cells in the eye differently. The shortest wavelengths are seen as violet or blue colors, while the longest visible light photons appear to be red.

Light-trapping Pigments

How do plants capture the energy from sunlight? Plant **pigments** are protein molecules capable of absorbing photons and transferring energy by rearranging their own molecular structures to create an "energized state." A pigment remains energized for only a fraction of a second and then transfers its energy through a variety of processes. In photosynthesis, the two most important mechanisms are transmitting the energy to another molecule or using it directly to drive a chemical reaction.

Many types of pigments are found in plant cells. However, each pigment can absorb only light with specific wavelengths; those they do not absorb are reflected. *Chlorophyll* is the most common pigment, and it exists in different forms, the most important of which is *chlorophyll a*. Figure 29.6A shows the wavelengths of light absorbed by chlorophylls *a* and *b* and indicates that these

Figure 29.6

(A) Each type of plant pigment absorbs light of a different wavelength. Yellow, orange, and red carotenoid pigments absorb light from the violet and blue portion of the light spectrum, while chlorophylls *a* and *b* absorb blue and red wavelengths. (B) Colorful carotenoids found in leaves of many trees cannot be seen until autumn, when the green pigment, chlorophyll, is broken down prior to leaf drop.

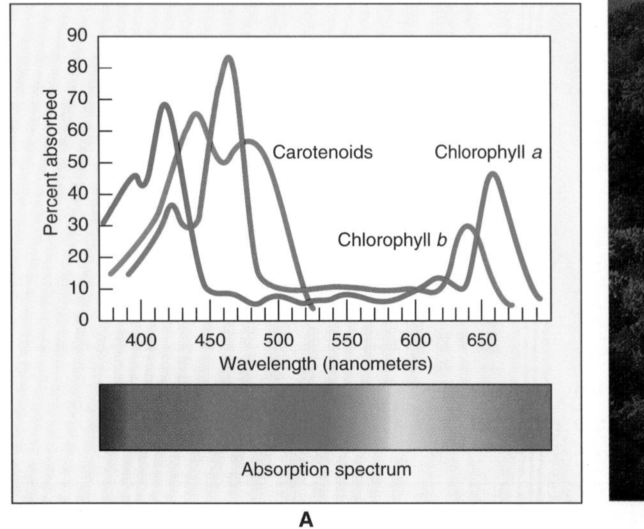

A

B

pigments absorb primarily blue and red wavelengths. Using this information, can you explain why most plants appear green?

In addition to chlorophyll, photosynthetic cells, or organisms, usually contain one or more yellow, orange, and red pigments called **carotenoids** that absorb violet and blue wavelengths but not yellow, orange, or red. The beautiful colors of autumn leaves are due to carotenoids, which are unmasked as chlorophyll is degraded prior to trees shedding their leaves (see Figure 29.6B). Aside from aesthetic considerations, is there an advantage to plants in having leaves containing several different pigments? Since each pigment absorbs light of a different wavelength, possession of multiple pigments enables a plant to use photons from a wider range of wavelengths to drive the reactions of photosynthesis.

Energy-forming Reactions of Photosynthesis

Photosynthesis consists of two distinctive sets of reactions that are light-dependent or *light reactions* and light-independent or *dark reactions*. Both take place in **chloroplasts**, specialized organelles found in plant cells that consist of an intricate network of highly organized membranes (see Figure 29.7). Chlorophyll

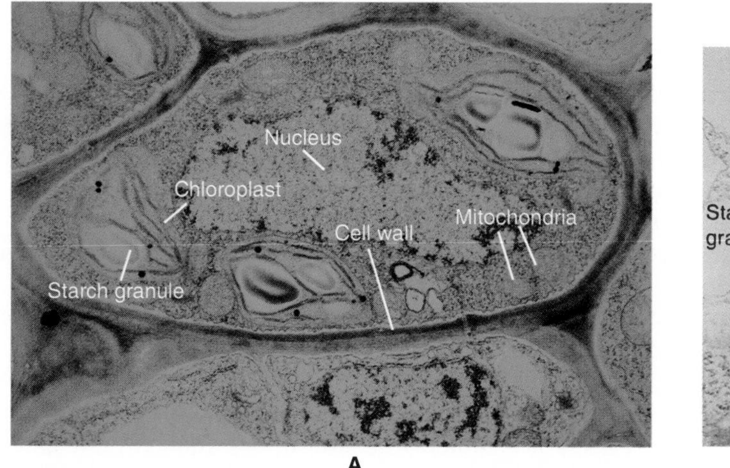

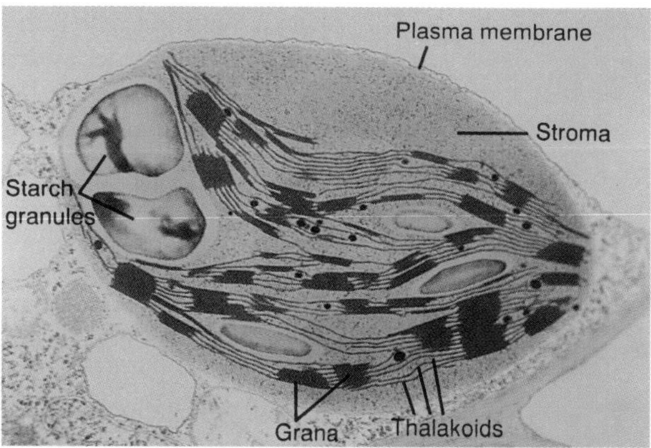

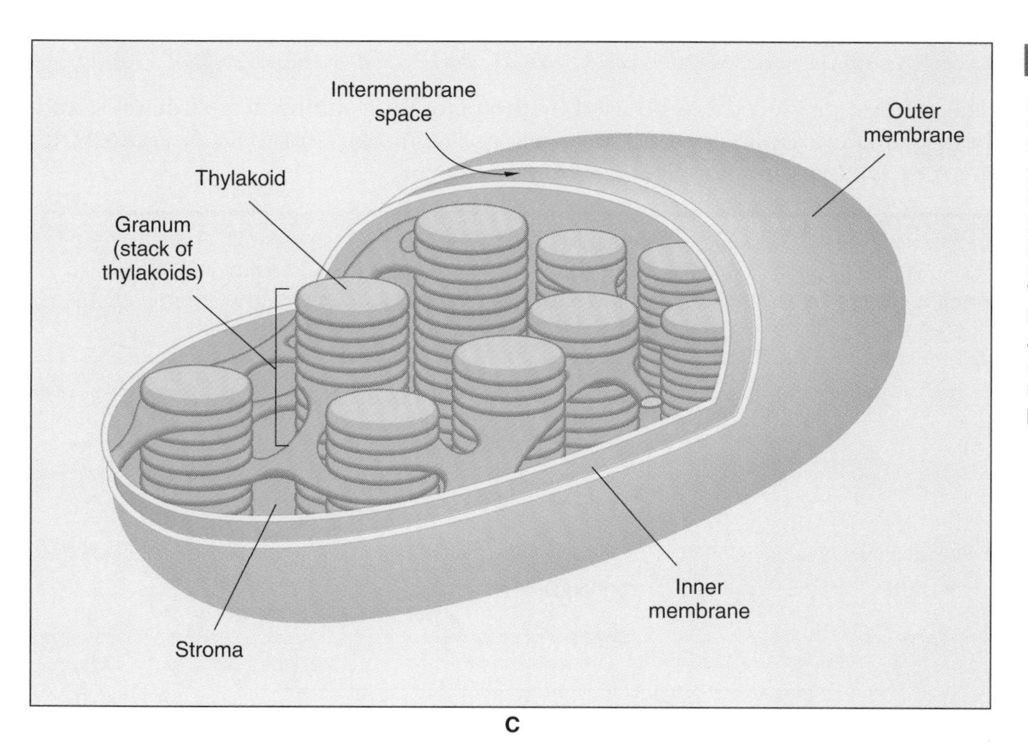

Figure 29.7

(A) Plant cells contain chloroplasts and mitochondria, the two organelles involved in photosynthesis and cellular respiration. (B) Plant chloroplasts are highly specialized organelles, composed primarily of membranes, in which photosynthesis takes place. The internal substance is called the stroma. Within the stroma are flattened membranes called thylakoids, some of which are organized in stacks called grana. (C) This diagram describes the arrangements of membranes within the chloroplast.

molecules and all of the enzymes and coenzymes required for photosynthesis are found within chloroplasts. A typical plant cell contains between 25 and 75 chloroplasts.

Light Reactions

Figure 29.8 outlines the primary products formed and broken down during the light and dark reactions of photosynthesis. The heart of the **light reactions** involves the transformation of light energy into high-energy electrons that are captured by the coenzyme $NADP^+$. During the light reactions, water is cleaved, resulting in the release of oxygen, electrons, and protons (H^+). The electrons are picked up by $NADP^+$, which is then converted to NADPH. Some radiant energy is also used to generate ATP during the light reactions. Thus during the light reactions, light energy is converted to chemical energy in the forms of NADPH and ATP. The light reactions can be summarized as follows:

$$\text{Radiant energy} + 2H_2O + 2NADP^+ + 2ADP + 2P \rightarrow O_2 + 2NADPH + 2ATP$$

Oxygen is released to the atmosphere, and NADPH and ATP are used as energy sources to drive the reactions that follow.

Dark Reactions

Dark reactions (also referred to as the *Calvin cycle,* based on the Nobel Prize-winning research on these reactions by Melvin Calvin at the University of California at Berkeley from 1946 to 1953) use energy provided by the ATP and NADPH formed during light reactions to produce glucose or other high-energy sugars. Since dark reactions are dependent on a continuous supply of chemical energy—NADPH and ATP—provided from the light reactions, they, too, take place only during daylight. For convenience, we will refer only to glucose while recognizing that several other sugars are also produced during photosynthesis. Collectively, all sugars made during photosynthesis are called **photosynthates**.

Glucose can be considered a highly efficient energy storage unit. Although glucose is a smaller molecule than ATP, when it is ultimately broken down by the cell during cellular respiration, the energy present within its molecular bonds will yield 36ATP. The essence of the dark reactions is that six molecules of carbon dioxide are systematically assembled in building one molecule of glucose. The cost of manufacturing this useful product is very high, in terms of ATP and NADPH used, as can be seen in this summary of reaction products:

$$6CO_2 + 18ATP + 12NADPH + 12H_2O \rightarrow C_6H_{12}O_6 + 18ADP + 18P + 12NADP^+ + 6H_2O$$

The glucose produced can be used by the plant for building new structures, and both plants and animals use it as an energy resource. The ADP, P, and $NADP^+$ are recycled back into the light reactions.

> In photosynthesis, pigments absorb light photons, which are used in initiating light reactions. During the light reactions of photosynthesis, water molecules react with $NADP^+$ to produce oxygen and NADPH, and some ATP is also formed. During the dark reactions of photosynthesis, energy from NADPH and ATP is used to produce glucose.

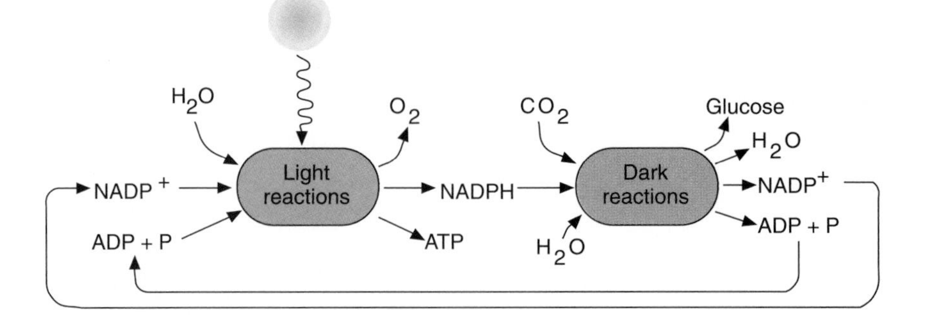

Figure 29.8

This diagram summarizes the primary products formed and broken down during the light and dark reactions of photosynthesis. During the light reactions, radiant energy from the sun is transformed into chemical energy—NADPH and ATP—that is then used to drive the dark reactions.

GLYCOLYSIS AND CELLULAR RESPIRATION

Glucose produced as a result of the large energy investment in photosynthesis ultimately yields a handsome return of ATP during the reactions of glycolysis and cellular respiration, as seen in the following equation:

$$C_6H_{12}O_6 + 6O_2 + 36ADP + 36P \rightarrow 6CO_2 + 6H_2O + 36ATP$$

Figure 29.9 summarizes the general pathways of glucose metabolism in cells. The degradation of glucose through controlled reactions liberates chemical energy that is used to generate ATP. Although we limit our discussion to glucose, fats and proteins are also broken down and fed into the cellular respiration pathways.

Glycolysis

Scientists believe that glycolysis was the first biological energy transformation process to evolve on Earth, appearing in primitive cells before much oxygen was present in the atmosphere. The universal occurrence of this process in the cells of all living organisms provides strong support for the view that all life descended from a common ancestral cell (as discussed in Chapter 2).

In the energy-releasing pathway, the reactions of **glycolysis** represent the first stage in the step-by-step breakdown of glucose to water and carbon dioxide. The reactions of glycolysis occur in the cytoplasm of cells. Since these reactions do not require oxygen, they are called **anaerobic reactions**. The enzymes required to split glucose molecules, the electron carriers, and ADP are all dissolved in the watery cytoplasm.

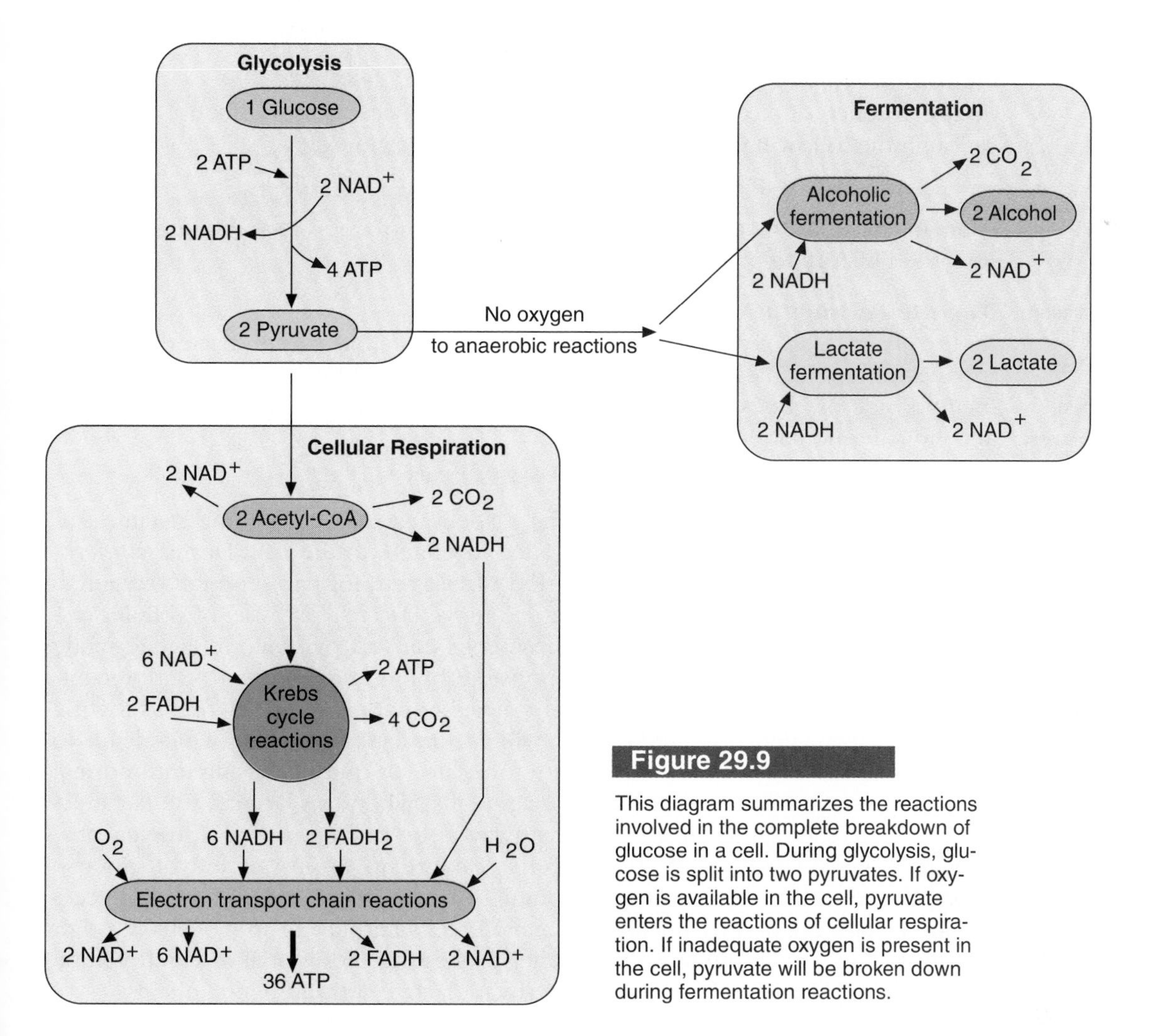

Figure 29.9

This diagram summarizes the reactions involved in the complete breakdown of glucose in a cell. During glycolysis, glucose is split into two pyruvates. If oxygen is available in the cell, pyruvate enters the reactions of cellular respiration. If inadequate oxygen is present in the cell, pyruvate will be broken down during fermentation reactions.

Initial Reactions

During glycolysis, a six-carbon glucose molecule is first split, through a series of reactions, into two molecules of **pyruvate**, a molecule that contains only three carbons. Electrons stripped off during this reaction are captured by the coenzyme NAD^+, which then becomes converted to NADH. The initial investment of energy for this step is 2ATP, but since a small amount of energy, 4ATP, is made available, the cell shows a profit of 2ATP. The reactions of glycolysis are summarized in the following equation:

$$C_6H_{12}O_6 + 2ATP + 4ADP + 4P + 2NAD^+ \rightarrow$$
$$2\text{ pyruvate} + 2ADP + 2P + 4ATP + 2NADH + H_2O$$

While there is a net gain of two ATP in glycolysis, the other two products formed, NADH and pyruvate, will eventually yield large quantities of ATP in later reactions.

Anaerobic Fermentation

What happens to pyruvate produced during glycolysis? This question reflects an active field of investigation in biochemistry, and new details are forcing reevaluations of energy management by cells and tissues. In general, the fate of pyruvate is dictated by the presence or absence of oxygen in the cell and by the type of cell that carries out the reaction. In most, but not all, eukaryotic plant and animal cells, pyruvate is shunted into the pathways of cellular respiration if sufficient oxygen is present. In these cells, if oxygen is not present, anaerobic reactions commonly called **fermentation** will take place. There are two general types of fermentation—alcoholic fermentation and lactate fermentation. In both cases, pyruvate is converted to a different type of three-carbon compound, either alcohol (two carbons) or lactate (three carbons), depending on cell type.

Alcoholic Fermentation

In the absence of oxygen, yeasts will convert pyruvate to ethyl alcohol (ethanol) through the following reaction for each molecule of glucose:

$$2\text{ pyruvate} + 2NADH \rightarrow 2\text{ ethanol} + 2NAD^+ + 2CO_2$$

Humans have taken advantage of this yeast-driven process since the beginning of recorded history to produce alcoholic beverages.

Figure 29.10

During anaerobic exercise, fermentation reactions lead to the production of lactate, a compound that causes muscle cramps unless sufficient oxygen becomes available for its breakdown.

Lactate Fermentation

In the absence of oxygen, some microorganisms such as certain bacteria convert pyruvate to molecules of lactate, a process that is used for producing yogurt and many kinds of cheeses. Lactate fermentation reactions are similar to those of alcoholic fermentation:

$$2\text{ pyruvate} + 2NADH \rightarrow 2\text{ lactate} + 2NAD^+$$

Higher animals, including humans, also are capable of forming lactate, usually if an adequate supply of oxygen is not available to the cell. During *anaerobic exercise,* such as an extended period of fast running, not enough oxygen is available in muscle cells for all pyruvate to move into the pathway of cellular respiration (Figure 29.10). As a result, these cells convert pyruvate to lactate while the individual gasps for breath—a response to obtain more oxygen. Until recently, the fate of lactate in muscle cells has been described in the following sequence: if oxygen is limited, lactate will accumulate in the cell, which leads to muscle fatigue, which forces the person to reduce or quit exercising until normal oxygen levels are restored to the deprived cells; when this occurs, the lactate is shunted into the cellular respiration pathway to be broken down. However, new research has revealed that muscle cells convert up to 50 percent of glucose metabolized to lactate no matter how much oxygen is present in the cell. Excess lactate is then apparently released by muscle cells and taken up by cells in other tissues (for example, the liver) for further processing. Additional research will be necessary to complete the picture of pyruvate-lactate processing in animals.

Anaerobic Products

Two results of fermentation reactions are significant. First, the reaction regenerates NAD^+, which is essential for glycolysis but is in short supply. Without an adequate stock of NAD^+, glycolysis cannot occur, and the cell will not be able to obtain any energy. Second, large amounts of potential energy remain in the products of fermentation. Lactate cannot be stored and is broken down when adequate oxygen becomes available. However, ethanol can either be broken down or converted into fat, a storage molecule.

Cellular Respiration

Microorganisms, certain simple plants and animals, and even some specialized cells, such as human red blood cells, can function with the meager amount of energy obtained through glycolysis. However, most cells and tissues of multicellular organisms require large amounts of ATP that are generated by the complete breakdown of glucose during the aerobic reactions of cellular respiration. **Cellular respiration** consists of two sets of reactions that result in the release and capture of energy in small steps. The first are called *Krebs cycle reactions,* and the second, *electron transport chain reactions*.

In eukaryotic cells, the many reactions of cellular respiration take place in specialized organelles called *mitochondria* (described in Chapter 13). Recall that mitochondria consist of a complex system of folded membranes that are organized to carry out aerobic respiration reactions (see Figure 29.11). All of the enzymes, coenzymes, and other molecules necessary for Krebs cycle reactions are found within the matrix of mitochondria. The cascade of events that occurs in electron transport chain reactions takes place in mitochondrial membranes.

Krebs Cycle Reactions

Following glycolysis, molecules of pyruvate move into the mitochondrial matrix. In a preparatory step, pyruvate is converted to a two-carbon molecule called acetyl-coenzyme A, or *acetyl-CoA,* for each molecule of glucose:

$$2\text{ pyruvate} + 2NAD^+ \rightarrow 2\text{ acetyl–CoA} + 2NADH + 2CO_2$$

Each acetyl-CoA molecule then proceeds through the **Krebs cycle reactions** (named after Hans Krebs, who described them in 1937 and later won a Nobel Prize for his research), where it is completely dismantled, yielding a variety of products:

$$2\text{ acetyl-CoA} + 6NAD^+ + 2FAD + 2ADP + 2P \rightarrow 6NADH + 2FADH_2 + 2ATP + 4CO_2$$

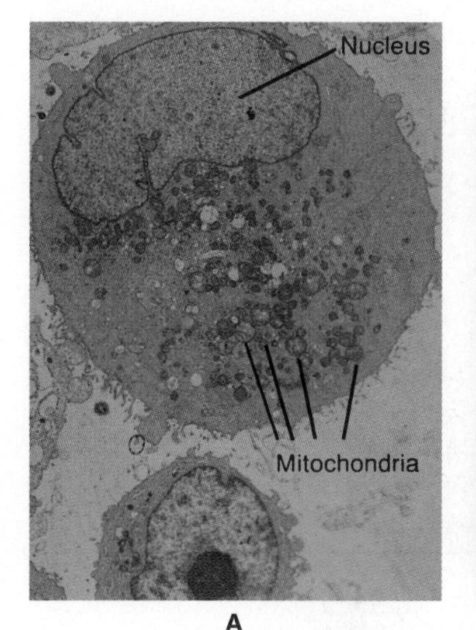

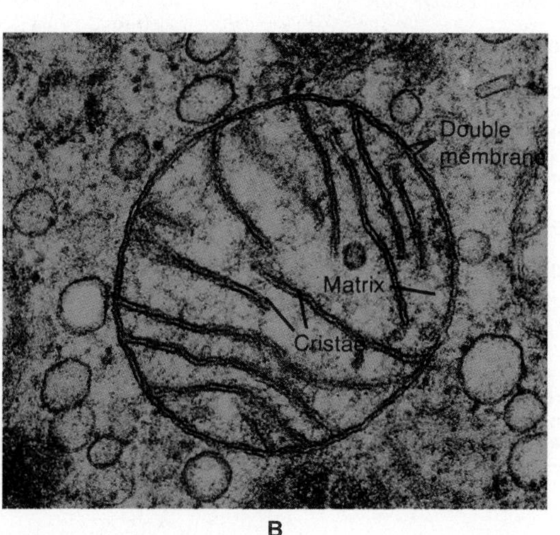

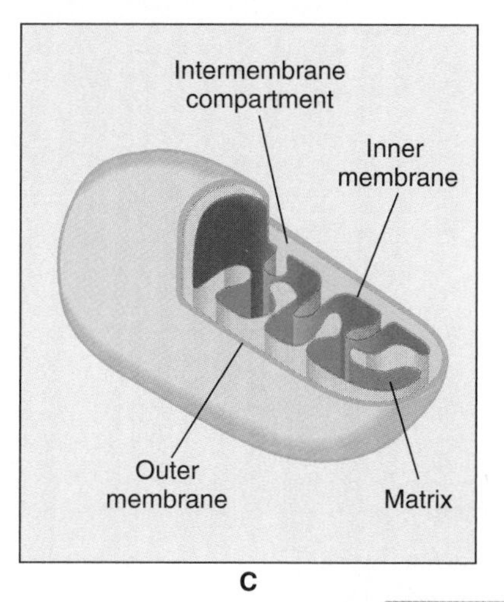

Figure 29.11

(A) An animal cell contains numerous mitochondria. (B) Mitochondria are highly specialized organelles, composed primarily of membranes, in which the reactions of cellular respiration are carried out. The inner membrane of mitochondria is folded into cristae that project into the internal matrix. (C) Krebs cycle reactions take place in the matrix, and electron transport chain reactions take place in the inner membrane.

At this point, virtually nothing remains of the original glucose molecule. Its individual atoms have met various fates. Some of the oxygen and hydrogen atoms were used to form a water molecule; some carbon and oxygen atoms were exhaled as carbon dioxide, which recycles back to the plant realm for use in photosynthesis; and the high-energy electrons released were captured by NAD^+ or FAD. The high yield of ATP finally occurs in the next set of reactions, when the electron carriers, NADH and $FADH_2$, are finally relieved of their cargo.

Electron Transport Chain Reactions

During **electron transport chain reactions**, NADH and $FADH_2$ pass their high-energy electrons on to a series of different electron carriers located in mitochondrial membranes. These electrons are passed along through a series of transfers similar to the hand-to-hand passage of a note between students in an elementary classroom. Each electron transfer results in the loss of a specific amount of energy. During some transfers, enough energy is liberated to produce a molecule of ATP. The complete processing of one NADH molecule yields two (those from glycolysis) or three ATP (those from the Krebs cycle) while one $FADH_2$ will return two ATP. When an electron finally completes its movement along the chain, it is transferred to oxygen and eventually contributes to the formation of a water molecule. About 90 percent of the oxygen we take in is used for this process. Table 29.2 summarizes the theoretical maximum (net) quantity of ATP that can be produced as a result of the different sets of cellular respiration reactions used to break down one molecule of glucose.

Glucose is ultimately converted to ATP through a sequence of chemical reactions within the cell. During glycolysis, glucose is partially broken down to pyruvate. In the absence of oxygen, pyruvate enters into fermentation reactions, where it is converted to ethanol or lactate, depending on the cell type. If oxygen is present in the cell, pyruvate moves into Krebs cycle reactions, where it is completely degraded. Electrons liberated during the Krebs cycle are processed in electron transport chain reactions. During this last set of reactions, large quantities of ATP are produced.

Table 29.2 Breakdown of Glucose

Reaction	Net ATP Gain	Other Products Formed
Glycolysis	2	2NADH
Pyruvate to acetyl-CoA	0	2NADH
Krebs cycle	2	$2FADH_2$ + 6NADH
Electron transport chain		
2NADH (from glycolysis)	4	
2NADH (from acetate)	6	
$2FADH_2$ (from Krebs cycle)	4	
6NADH (from Krebs cycle)	18	
Total ATP	**36**	

This table summarizes ATP production from the complete breakdown of one molecule of glucose during cellular respiration.

THE SIGNIFICANCE OF PHOTOSYNTHESIS AND CELLULAR RESPIRATION

Scientists have developed hypotheses to address questions about the evolution of biological energy transformation processes and the relationships among those reactions. Since photosynthesis, glycolysis, fermentation, and cellular respiration evolved billions of years ago, the following hypothesized events cannot be proved. Rather, as with all good hypotheses, they are consistent with all of the facts now available. (See the Focus on Scientific Inquiry, "The Road to Understanding Photosynthesis.")

Glycolysis and fermentation reactions are thought to have been the first biological energy transformation systems on Earth. These ancient energy-releasing pathways evolved between 3 and 4 billion years ago, when little oxygen was present in the atmosphere and the seas were filled with organic molecules that could be used for energy by the earliest organisms (see Chapter 2). Some modern organisms, such as yeasts and certain bacteria, still depend entirely on glycolysis and fermentation for energy.

Nearly all organisms have retained glycolysis as a first step in the reactions they use to release energy. The major disadvantage of a total dependency on those reactions is that only a small amount of energy (2ATP) can be obtained from each glucose molecule, and the energy-rich products formed (pyruvate, ethanol, and lactate) cannot be broken down directly to yield ATP even though they contain a great amount of additional potential energy. Consequently, organisms that remained dependent on glycolysis and fermentation for obtaining energy did not evolve into more complex forms requiring greater quantities of energy.

Between 1 and 2 billion years ago, photosynthesis evolved in primitive bacteria and, subsequently, in all plants. As a result, a new source of food became available on Earth, and great quantities of oxygen were released into the atmosphere and the seas. Following these developments, a new energy-releasing set of reactions (aerobic cellular respiration) evolved in some early eukaryotic cells. Such cells had an enormous advantage over those that remained dependent on glycolysis and fermentation because they were able to generate large amounts of ATP. Consequently, they flourished and eventually gave rise to all the multicellular plants and animals that came to inhabit Earth.

Summary

1. Different forms of energy—chemical, thermal, electrical, mechanical, and radiant—are used by organisms that inhabit Earth. All of these forms of energy are interrelated and can be converted from one form to another. The first and second laws of thermodynamics define the behavior of energy in the universe.
2. Photosynthesis is a process whereby radiant energy from the sun is transformed by plants into chemical energy that is used by almost all organisms for maintaining life.
3. In photosynthesis, beginning products, carbon dioxide and water, are converted into final products, glucose and oxygen, by means of energy provided by the sun. Pigments, such as chlorophyll, are unique light-trapping protein molecules that are able to capture radiant energy, transform it, and use it in driving the reactions of photosynthesis.
4. Cells obtain energy through the systematic breakdown of glucose into carbon dioxide and water in a series of chemical reactions. During glycolysis—a process that does not require oxygen—glucose is broken down into two smaller molecules of pyruvate. If no oxygen is present in the cell, pyruvate enters into fermentation reactions from which little energy is made available to the cell. If oxygen is present, pyruvate in a cell is completely degraded to carbon dioxide

THE ROAD TO UNDERSTANDING PHOTOSYNTHESIS

Focus on Scientific Inquiry

In science, modern understandings of many fundamental mechanisms have been derived, in part, from experiments conducted by earlier investigators. Today, the molecular events of photosynthesis are known in great detail. However, to understand the advancement of scientific knowledge, it is instructive to review some of the early, classical experiments that shed the first light on this critical biological process. An examination of these works will also reveal how results of experiments are evaluated in an existing theoretical context.

Scientific studies are based on ideas, hypotheses, and theories that prevail at the time they are undertaken. Investigations relating to photosynthesis began early in the seventeenth century. At that time, naturalists were asking two questions: What accounts for plant growth? And where does the matter used to make the branches, roots, and leaves of plants actually come from? Until about 1650, it was assumed that plant growth was due to the uptake and accumulation of materials removed from soil. Jan van Helmont conducted a simple experiment designed to test this hypothesis.

Van Helmont's Experiment

Figure 1 illustrates van Helmont's experiment. He planted a 5-pound willow tree in a vessel containing 200 pounds of dried soil. For five years, he added only rainwater to the tub of soil. At the end of that period, the tree weighed a little more than 169 pounds while the weight of the soil was almost unchanged, having lost only 2 ounces. Clearly, the new growth was not accounted for by materials removed from the soil. What was the source of the new plant matter? Van Helmont concluded that

Figure 1 Jan van Helmont conducted a study designed to determine the source of substances used in the growth of new plant tissues. Was it water, soil, or something else? He thought that the transmutation theory held the answer.

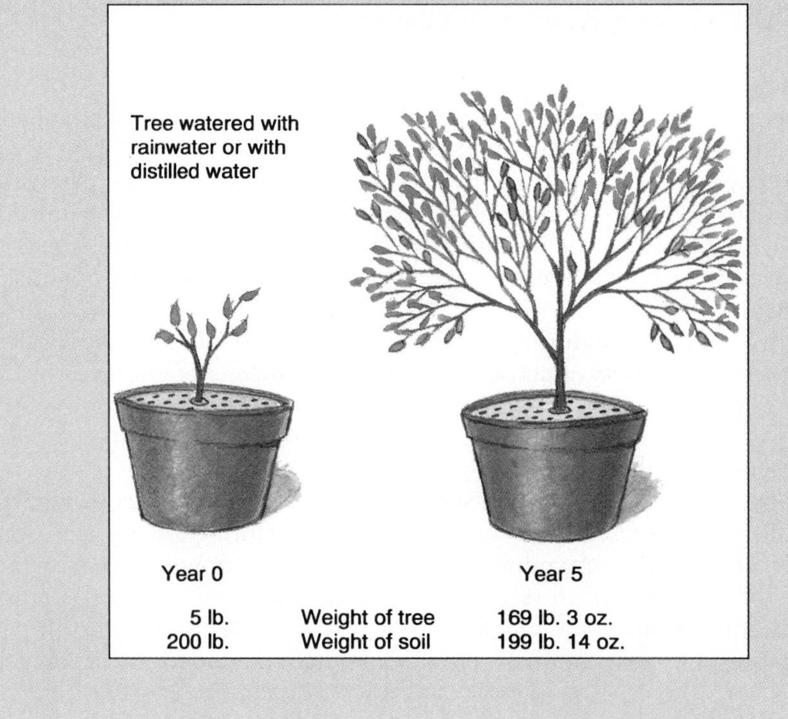

the 164 pounds of new leaves, wood, bark, and roots were "derived from water alone."

How did he arrive at this conclusion? Van Helmont evaluated his results on the basis of a popular theory that has since been discredited, the *transmutation theory*. According to this theory, one substance could be transformed into another. A familiar example was the erroneous belief that lead or some other metals could be converted into gold. Those who studied or practiced such transmutations were called *alchemists*. Even though it was incorrect, the transmutation theory, like many previously held but eventually discredited theories, was important in the development of modern science. The experiments and observations of alchemists prepared the ground from which the modern discipline of chemistry emerged. However, some of the ideas of alchemists were more fruitful than others, and some, given historical hindsight, can be seen as hindering the advancement of knowledge. Van Helmont's conclusion that new plant matter was derived from water clearly belongs in the latter category.

Hales's Experiments

Eighty years after van Helmont's experiment, Stephen Hales, an English clergyman, conducted an experiment to test his idea that plants somehow interacted with air, a possibility not considered by van Helmont. What was the basis for this hypothesis? In part, Hales was led to his experiments because of new information about plants revealed by the microscope. Using this instrument, he became aware of the existence of the very tiny openings (stomata) on the surfaces of leaves. What function did they serve? Might they act like pores in the skin of animals? Also, by that time, new knowledge had shifted the focus of chemistry. Alchemy had become chemistry, and amateurs like Hales supported a theory that assumed that the world was made up of small, hard "corpuscles of matter" that differed in size and shape. Some proponents of this *corpuscular theory* also believed that certain particles of matter were more "elastic" or active than others. Hales approached his studies on plants within the context of this corpuscular view of matter.

As shown in Figure 2, Hales's experimental system consisted of two glass containers that were filled with water and a layer of soil at the base. To one he added a peppermint plant and then covered it with a glass bell jar. The other was simply covered with the jar and served as an *experimental control,* a standard for comparison. In general, a control consists of the same components as the experimental apparatus except for the variable being tested (in this case, the peppermint plant).

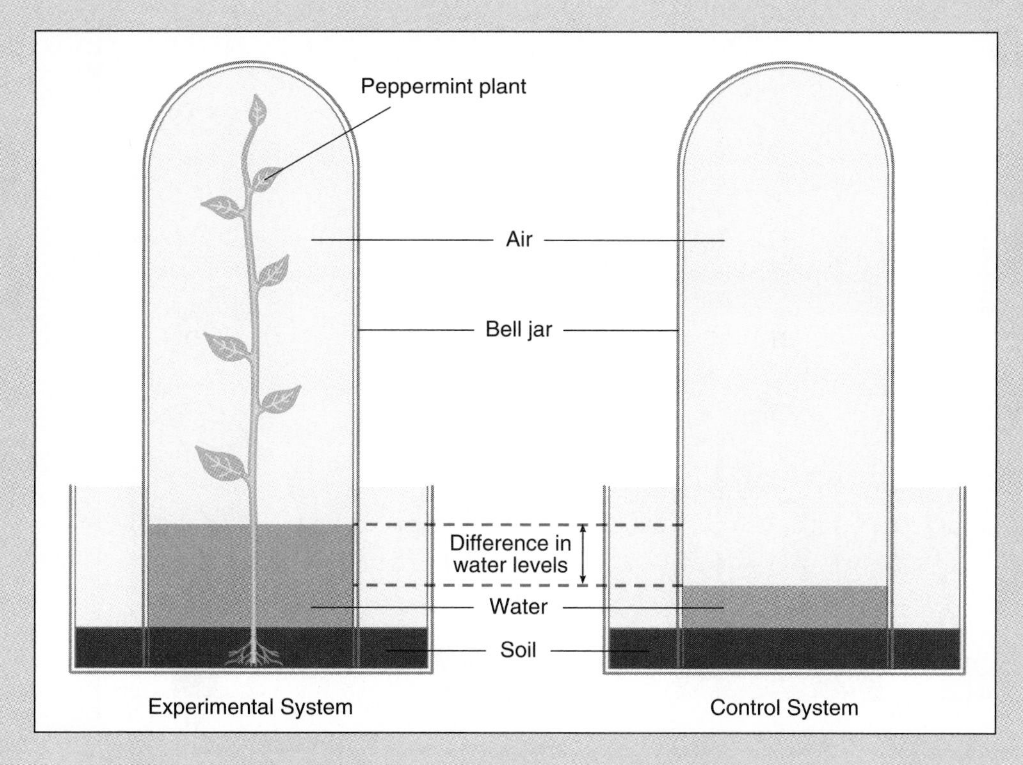

Figure 2 Stephen Hales carried out a sophisticated experiment that provided new information on the identity of substances used in plant growth. He showed that plants interacted with the air and changed its composition. He based his analyses on the corpuscular theory.

Hales observed and recorded changes in the water levels of the two systems. The control was essential because changes observed in both vessels could not be associated with the plant, but those unique to the experimental container could be considered to be caused by the plant. Note that van Helmont did not have a control in his experiment. What could have served as van Helmont's experimental control?

Hales found that the water levels fluctuated as a function of barometric pressure (the pressure or "weight" exerted by the atmosphere) and temperature. Since they fluctuated similarly in both systems, he knew that they were not associated with the plant's activities. However, after several months, Hales found that the water level in the experimental system rose significantly higher than that in the control. What could have caused this difference? Hales felt that the plant had "imbibed" or absorbed the air, thus creating a void in the jar that was "filled" by the water. He also found that there was no further reduction in air volume after two or three months. What is more, when he removed the plant from the experimental system and replaced it with a new plant, the new plant died after a few days. However, if a new plant was placed in the control system, which had been confined for the same period, the plant remained healthy. What could account for these complicated observations? Hales was unable to offer a precise interpretation of his findings. The significance of his experiment, however, was clear: plants *did* interact with the atmosphere and, in some unexplained way, modified the surrounding air.

Priestly's Experiments

The Rev. Joseph Priestly was a British chemist who conducted experiments in 1771 that significantly advanced what was then known about the relationship between plants and air (see Figure 3). Priestly established experimentally that both a burning candle and a mouse placed in a system closed to air soon expire. From

Figure 3 Joseph Priestly demonstrated that plants somehow purified air and made it fit for animals to breathe. In his experiments, burning candle (A) placed in a closed system went out within minutes (B). A mouse (C) placed in the closed system likewise died within minutes (D). But if both a candle and a plant (E) were placed in the closed system, the candle burned when lit a few days later (F), and if a plant accompanied the mouse (G), the mouse was still alive a few days later (H). Priestly based his analysis of the underlying mechanism involved on the phlogiston theory.

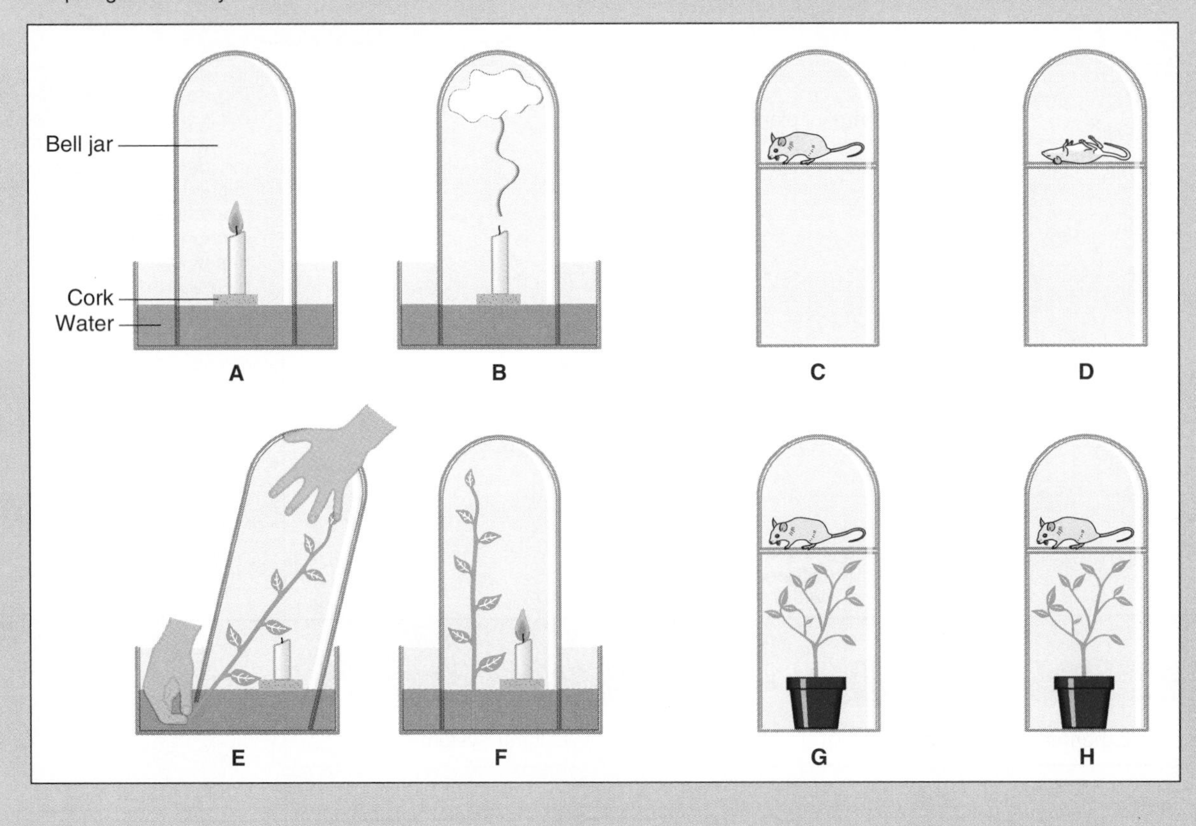

these observations, he concluded that the air had been "rendered noxious." On a broader level, he recognized that some large-scale process must exist for "rendering it fit for breathing again." Without such a mechanism, he speculated that "the whole mass of [Earth's] atmosphere would, in time become unfit for the purpose of animal life." But how was this reconditioning process accomplished?

Priestly conducted a further series of experiments that led him to conclude that he had discovered "at least one of the restoratives which nature employs for . . . restoring air which has been injured by the burning of candles. . . . It is vegetation." To test his hypothesis, he added plants to his closed experimental systems. He found that candles would burn again and mice could live in such a system. He concluded that plants served to purify the atmosphere by reversing the harmful effects caused by an animal's breathing or burning materials.

What happened chemically when the air was rendered "noxious" or when it was improved? Priestly, like most of his generation, interpreted his experiments in light of the popular chemical theory of his day, the *phlogiston theory.* Proponents of the theory held that when a substance was burned, something called *phlogiston,* "the material and principle of fire," was expelled during combustion. Conclusions formulated in using this theory seemed reasonable and consistent with existing knowledge. For example, charcoal was thought to consist of nearly pure phlogiston since it almost disappeared when burned. A candle burning in a sealed bell-shaped jar went out because the air became "phlogisticated" or "fixed" and was no longer able to support combustion or a mouse placed in the jar. "Dephlogisticated" air was capable of supporting the burning of a candle or the life of a small animal placed into the jar.

Priestly, an advocate of the phlogiston theory, very naturally interpreted his results from that perspective. Plants, according to this view, altered the air by removing phlogiston. Thus like Hales, Priestly recognized that plants change the air, but he focused on qualitative changes rather than the quantitative changes that Hales described.

Ingenhousz's Experiments

Jan Ingenhousz, a Dutch physician, confirmed and extended Priestly's experimental findings eight years later. He, too, explained his results in terms of the phlogiston theory. According to Ingenhousz, plants removed phlogiston introduced into the air by animals and burning candles and changed it into dephlogisticated air. Ingenhousz did more than just agree with Priestly. He made some other, truly significant observations. First, he noted that plants performed their "beneficial operation . . . only after the sun [appeared] above the horizon." He also observed that only the green parts of the plant performed "this office" (photosynthesis) and "that all plants contaminate the surrounding air by night, and even in the daytime in shaded places."

Soon after Ingenhousz's studies in 1779, the phlogiston theory exited from the stage of science, primarily as a result of an article published in 1786 by Antoine-Laurent Lavoisier titled *"Réflexions sur le phlogistique,"* in which he described a new theory of combustion that accurately identified the gases present in the atmosphere. Moreover, Lavoisier demonstrated that combustion and respiration were similar chemical processes.

Subsequently, a number of remaining questions about photosynthesis were soon answered. In 1796, Ingenhousz proclaimed that photosynthesis was the primary process for producing new organic materials in the biosphere. By 1845, the underlying mechanism of photosynthesis—the conversion of light energy to chemical energy—was understood. Studies conducted during the 1980s largely completed the journey that led to an understanding of photosynthesis, a trip that began nearly 300 years earlier.

The modern understanding of photosynthesis is able to draw on many scientific hypotheses that were developed in the past century and are strongly supported by research results. Like the early ideas on plants, our current knowledge of photosynthesis is rooted in broader theories of energy, chemical interactions, and biological processes.

and water by the reactions of cellular respiration. Cellular respiration results in the production of significant amounts of ATP that can be used by the cell.
5. Available evidence indicates that the first organisms to evolve on Earth used glycolysis and fermentation to obtain energy. Subsequently, certain bacteria and, later, eukaryotic cells acquired the mechanisms necessary to transform energy using the reactions of photosynthesis. Cellular respiration was the last biological energy transformation process to evolve, appearing in later eukaryotes that depended on the glucose and oxygen made available by photosynthesis.

Review Questions

1. What is energy? What are the two general energy states?
2. List and define the different forms of energy.
3. What do the first and second laws of thermodynamics tell us about energy transformations?
4. What are the major biological energy transforming processes? How are they related?
5. What are ATP, ADP, and coenzymes? What is the role of each in energy-transforming reactions?
6. What are photons? Pigments? Describe their functions in photosynthesis.
7. Describe the light and dark reactions of photosynthesis. What initial products are used and what end products are generated during each?
8. Describe the reactions of glycolysis and the different types of fermentation. What are the end products of each?
9. In which organisms do each of the following take place: glycolysis, alcoholic fermentation, and lactate formation?
10. Where in a cell does cellular respiration occur?
11. What are the end products of Krebs cycle reactions? Electron transport chain reactions?

Essay and Discussion Questions

1. Do you think it possible that unique forms of energy not present on Earth exist on other planets in other galaxies? Why?
2. Assume that life exists on other planets in the universe. Is it reasonable to believe that they use the same energy-transforming systems as their counterparts on Earth? Why? Describe a different set of transforming reactions that might have evolved.
3. Is it probable that during the next 2 billion years on Earth, one of the existing energy-transforming processes will become "extinct"? Why? Which process might be the primary candidate for elimination? Why?

References and Recommended Reading

Bassham, J. A. 1962. The path of carbon in photosynthesis. *Scientific American* 206:88–100.

Bennett, J. 1979. The protein that harvests sunlight. *Trends in Biochemical Science* 4:268–271.

Govindjee, and W. J. Coleman. 1990. How plants make oxygen. *Scientific American* 262:50–58.

Gregory, R. P. F. 1989. *Photosynthesis*. New York: Chapman & Hall.

Hendry, G. 1988. Where does all the green go? *New Scientist* 120:38–42.

Hinkle, P. C., and R. E. McCarty. 1978. How cells make ATP. *Scientific American* 238:104–112.

Krebs, H. A. 1970. The history of the tricarboxylic cycle. *Perspectives in Biology and Medicine* 14:154–170.

Loomis, W. F. 1988. *Four Billion Years: An Essay on the Evolution of Genes and Organisms*. Sunderland, Mass.: Sinauer.

Mathews, C. K., and K. E. van Holde. 1990. *Biochemistry*. Menlo Park, Calif.: Benjamin/Cummings.

Morowitz, H. J. 1970. *Entropy for Biologists*. Orlando, Fla.: Academic Press.

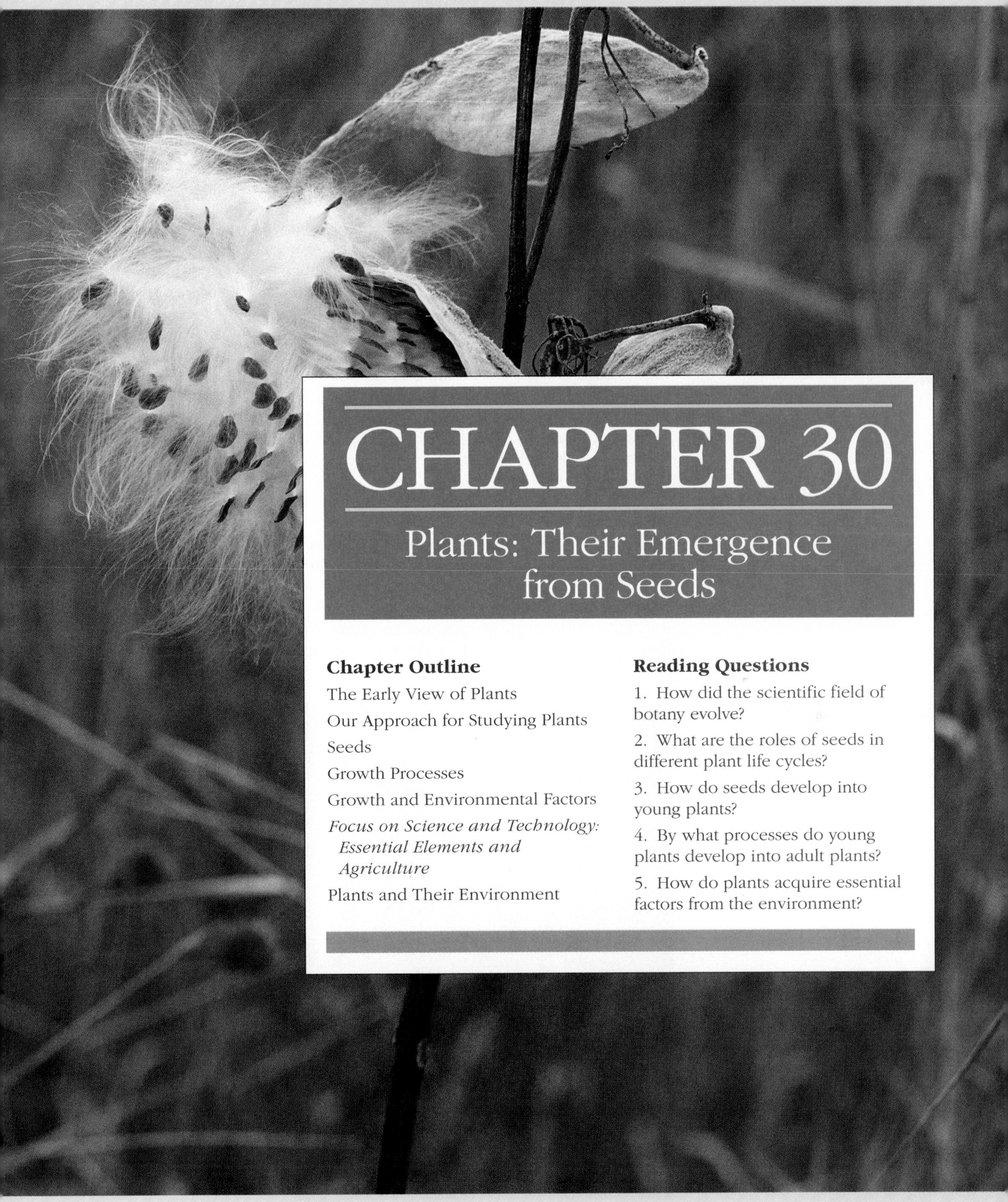

CHAPTER 30

Plants: Their Emergence from Seeds

Chapter Outline

The Early View of Plants

Our Approach for Studying Plants

Seeds

Growth Processes

Growth and Environmental Factors

Focus on Science and Technology: Essential Elements and Agriculture

Plants and Their Environment

Reading Questions

1. How did the scientific field of botany evolve?
2. What are the roles of seeds in different plant life cycles?
3. How do seeds develop into young plants?
4. By what processes do young plants develop into adult plants?
5. How do plants acquire essential factors from the environment?

Botany, the scientific study of plant life, has added significantly to our understanding and appreciation of the natural world. For centuries, plants and plant products were valued for medicine, for their beauty, for food, and for basic physical materials used in much of our culture. Botanists have described the roles of plants in creating and sustaining environments in which we live, the interactions between plants and animals, and the interesting manner in which plants develop, grow, and reproduce (Figure 30.1).

THE EARLY VIEW OF PLANTS

Scientific knowledge of plants is of recent origin. Until the nineteenth century, botany was part of natural history and had the primary objectives of naming and describing plants. Great encyclopedias of plants were created that reflected these efforts. The medical value of plants was of particular concern, and consequently, serious efforts were made to accurately identify potentially useful plants.

When early scientists contemplated plants within the total picture of life, they evaluated them in relation to the living being they knew best and valued most, the human. From the time of Aristotle and the ancient Greeks, naturalists thought that all living beings could be ranked on a scale from highest to lowest, the *chain of being* described in Chapter 3. It had what was agreed to be the highest and most perfect living being at the top (a human) and the lowest and least perfect living being at the bottom. In this scheme, plants were placed below the entire animal kingdom.

This view of plants changed in the nineteenth century as new theoretical perspectives were developed. A new general theory of physiology and the theory of evolution were of major importance. The architects of general physiology were among the first to attack the notion of a chain of being; they argued that both plants and animals shared common physiological processes such as nutrition, growth, and reproduction. Deeper investigations revealed that the chemical basis for many of these processes was the same in both plants and animals. These findings eroded the idea of an absolute separation between the two kingdoms.

Figure 30.1

Plants have long been of interest to humans, although the scientific study of plants did not begin until the nineteenth century.

The theory of evolution describes the separation of plants and animals at a very early stage in the evolution of life on Earth. It also suggests that plants and animals had separate but equal evolutionary histories and that there are no grounds for considering one kingdom "higher" or "lower" than the other. Similarities in basic functions of plant and animal cells are explained by the common evolutionary history of all life forms.

Liberated from the belief that plants were "imperfect" animals, the development of botany became one of the brilliant success stories in the history of science. Rather than being restricted to generating lists of medically valuable plants or producing field guides for identifying flowers, studies of plants have led to insights concerned with fundamental biological processes. Botany now encompasses scientific investigations of plants, animals, and environmental interactions (that is, ecology), as well as studies of fundamental principles in genetics. It has helped transform the fields of agriculture, medicine, and industry in the twentieth century.

Today, botany is one of the main fields of biology. However, until the nineteenth century, most investigations of plants were limited to identifications and descriptions. Advancements in developing scientific knowledge related to botany were hindered by a long-held view that plants were inferior to most other forms of life. The theory of evolution provided a different perspective for studying plants, and botany blossomed as a scientific discipline in the twentieth century.

OUR APPROACH FOR STUDYING PLANTS

Table 30.1 presents a simplified classification scheme for plants that will be used in Chapters 30, 31, and 32. **Nonvascular plants** have no *tissues*—aggregates of cells with a similar structure and common function—for transporting substances throughout the plant body. Nonvascular plants include mosses and liverworts; characteristics of these organisms were described in Chapter 4.

Vascular plants have specialized *vascular tissues* that transport substances such as water and nutrients throughout the leaves, shoots, and roots (see Figure 30.2). There are about 275,000 vascular plant species, which include most of the common trees, crops, garden plants, and house plants. Chapters 30–32 focus on the seed-bearing vascular plants.

In Chapters 30–32, we follow the general life history of a plant, from seed

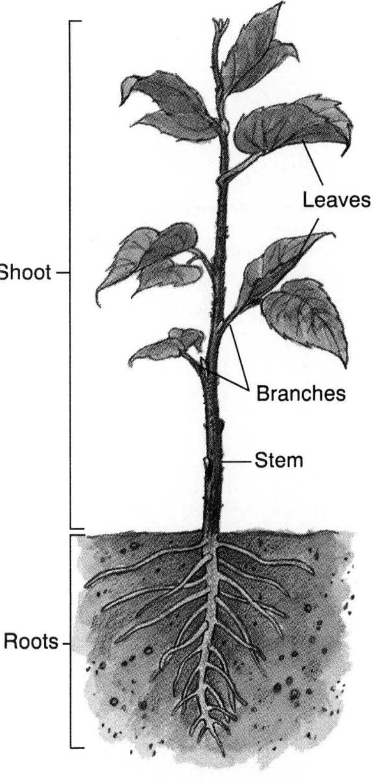

Figure 30.2

The major vascular plant structures are stems, leaves, branches, and roots.

Table 30.1 A Simplified Classification of Plants

Type of Plant	Characteristics	Examples
A. Nonvascular	No vascular tissue	Mosses, liverworts
B. Vascular	Vascular tissue, no seeds	Ferns
	Vascular tissue, seeds	Trees, flowers
1. Gymnosperms	Seeds within cones	Pine and spruce trees
2. Angiosperms	Seeds within flowers	Roses, grasses
a. Monocots	One cotyledon	Palm trees, lilies
b. Dicots	Two cotyledons	Roses, oak trees

This classification applies to our discussions of plants in Chapters 30, 31, and 32.

to adult. Our purpose is to describe the general processes that allow a plant to develop, function, and survive in the environment it comes to inhabit. A highly significant feature of a seed-bearing plant's life history is that it permanently occupies the place where it began life as a seedling. Many severe challenges have to be met between the period when a seed falls to the ground and its ultimate development into a mature, seed-producing plant. Of all the seeds produced by a plant, only a few overcome all of the obstacles and complete their life cycle. In this chapter, we begin our study of plants with the production, dispersal, growth, and establishment of seeds. How does a seed develop into a young plant? What resources does a newly established plant require? How does it acquire those resources?

In Chapter 31, we examine adaptations that enable plants to survive in the environment where they became established. Today, there is great concern about the abilities of plants to cope with human-caused environmental stresses. What types of environmental stress affect plants? What types of mechanisms allow plants to cope with such stresses? Are plant stress responses adequate for surviving in environments affected by human activities? These questions now serve as guides in a very active field of research. In Chapter 32, we consider adaptations that enable plants to withstand natural variations and selective pressures in their environment. Plants that succeed in making the transition from seed to adult may finally reproduce and form new seeds, thus completing the "seed-to-seed" span of a plant life cycle.

SEEDS

The study of seeds is a convenient point of departure to begin learning about the life cycles of higher plants. What are seeds? How are they formed? How do they become adult plants? Answers to these questions address the major events in the development and transformation of seeds into complex plants.

Production and Dispersal

Gymnosperms and *angiosperms* are the two principal groups of vascular plants that produce *seeds*—complicated structures that consist of a protective seed coat that encloses a plant *embryo* (the young, undeveloped stage), along with stored nutrients that will be used during the embryo's early growth and development. *Gymnosperms* produce seeds within **cones**, specialized reproductive structures composed mostly of modified leaves (see Figure 30.3). Seeds of *angiosperms* mature in **fruits**—structures such as apples, cherries, tomatoes, and grains—that develop from the female parts of fertilized flowers (see Figure 30.4). In this chapter, we concentrate primarily on the seeds of angiosperms.

Fruits provide support and protection for developing seeds and often play a role in their *dispersal*—scattering them away from the parent plant. In addition, some fruits, as they decay, enrich the soil, which improves the chances of a *seedling,* or young plant, becoming established. Why is seed dispersal so significant in the life history of plants? Seed dispersal acts to establish new seedlings in areas where they will not compete directly with the parent plant for limited resources. It also increases the likelihood that at least a few seeds will land in sites favorable for their growth and survival. Ultimately, seed dispersal defines the range over which a plant species exists. Plants grow in areas where their seeds have been dispersed and that are favorable for their growth and survival.

Different plant species produce an endless variety of fruits and seeds, as shown in Figures 30.3 and 30.4, and an examination of their structures will often provide some clue about their mode of dispersal. For example, *nuts,* such as acorns produced by oak trees, are large, dry, hard, one-seeded fruits that can be dispersed in various ways. They may fall and roll downhill to a new site a short distance from the parent tree, or they may be collected and carried away by

Figure 30.3

Cones of conifer species differ in size and shape. Shown are Englemann spruce (A), larch (B), Douglas fir (C), and Utah juniper (D).

squirrels and birds, which use them as food. Occasionally, squirrels may bury and forget the acorns; in so doing, they actually plant seeds, which may later grow into new oak trees that produce many more acorns. Seeds of some plants,

Figure 30.4

Seeds of angiosperms are contained within many types of fruits. Large fruits such as peaches (A) and grapes (B) protect seeds. Nuts, such as acorns (C), can be large and protect seeds but are also adapted for seed dispersal by animals or rolling away from the parent tree. Fruits produced by grains are small and may be transported by birds or animals. Small fruits with wings, such as the maple samaras (D), are dispersed by the wind.

after being consumed by animals or birds, may pass unaltered through the digestive tract and be deposited some distance from their origin.

Heavy fruits, such as acorns, are not usually dispersed as far as smaller, lighter, windblown fruits with winglike outgrowths such as those produced by elm and maple trees. Other types of fruits are also well adapted for dispersing their seeds. Those of the milkweed and dandelion mature, dry, and split open to release hundreds of small, winged seeds. These types of seeds can be dispersed by wind or become attached to animals, which then transport them to new areas.

Fruits containing seeds range from small to large, and they differ greatly in moisture content and fleshiness. Fruits produced by buttercups, buckwheat plants, and grasses contain numerous seeds that are usually small and dry. Fruits of peach trees and date palms are medium-sized and fleshy and contain only one seed. Familiar examples of fleshy fruits that contain more than one seed are tomatoes, raspberries, apples, and pears. Watermelons and squashes are examples of fruits that can reach great sizes.

Gymnosperm seeds generally take months or years to develop fully in cones. Once seeds mature, the cone leaves of many species dry and separate, allowing the seeds to be released. Some cones of several gymnosperm species, including lodgepole pine, will not open unless exposed to fire. What might be the advantages of such an adaptation? Once released, gymnosperm seeds are commonly dispersed by the wind or by birds and animals that eat them.

Seeds of Angiosperms

Based on characteristics of their seeds, shown in Figure 30.5, angiosperms are either *monocotyledonous* (*monocots*) or *dicotyledonous* (*dicots*). *Cotyledons* are leaves that are part of the embryo contained within the seed. In the seeds of some species, cotyledons are large and serve to store nutrients that will be used by the embryo. In other species, the cotyledon is a thin structure that assists in the digestion and transport of nutrients stored within the seed. *Monocot* seeds have only one cotyledon, and *dicot* seeds have two. Figure 30.5 also shows some of the other distinguishing features of monocots and dicots.

Monocots

Monocots include at least 50,000 species of plants. Monocot seeds grow into plants that have several common characteristics. These plants have flower parts, such as petals, arranged in multiples of three. Root systems of monocots consist of many small branches and are described as "fibrous." The *stem,* or above-ground axis, of monocots can become quite thick but does not develop the true wood, with seasonal growth rings, that is common in dicot trees. Finally, monocot plants have leaves with parallel *veins,* bundles of vascular tissues that run parallel to one another. Grasses, lilies, and palm trees are examples of monocots that have all of these characteristics.

Dicots

Approximately 175,000 dicot species are known to exist, including such plants as beans, peas, asters, roses, most shrubs, and deciduous trees. Dicots have flower parts that are usually arranged in fours or fives. They typically have *taproots* (a single main root with many smaller lateral branches), and the stems of woody dicots increase in size through the addition of seasonal growth rings. Their leaves usually have midribs and a netlike arrangement of veins.

Gymnosperms and angiosperms produce seeds—structures enclosing an embryo—that can develop into new plants. Seeds are usually dispersed away from parent plants through a variety of mechanisms. An angiosperm is classified as either a monocot or a dicot depending on characteristics of its seed and structural features of the adult plant.

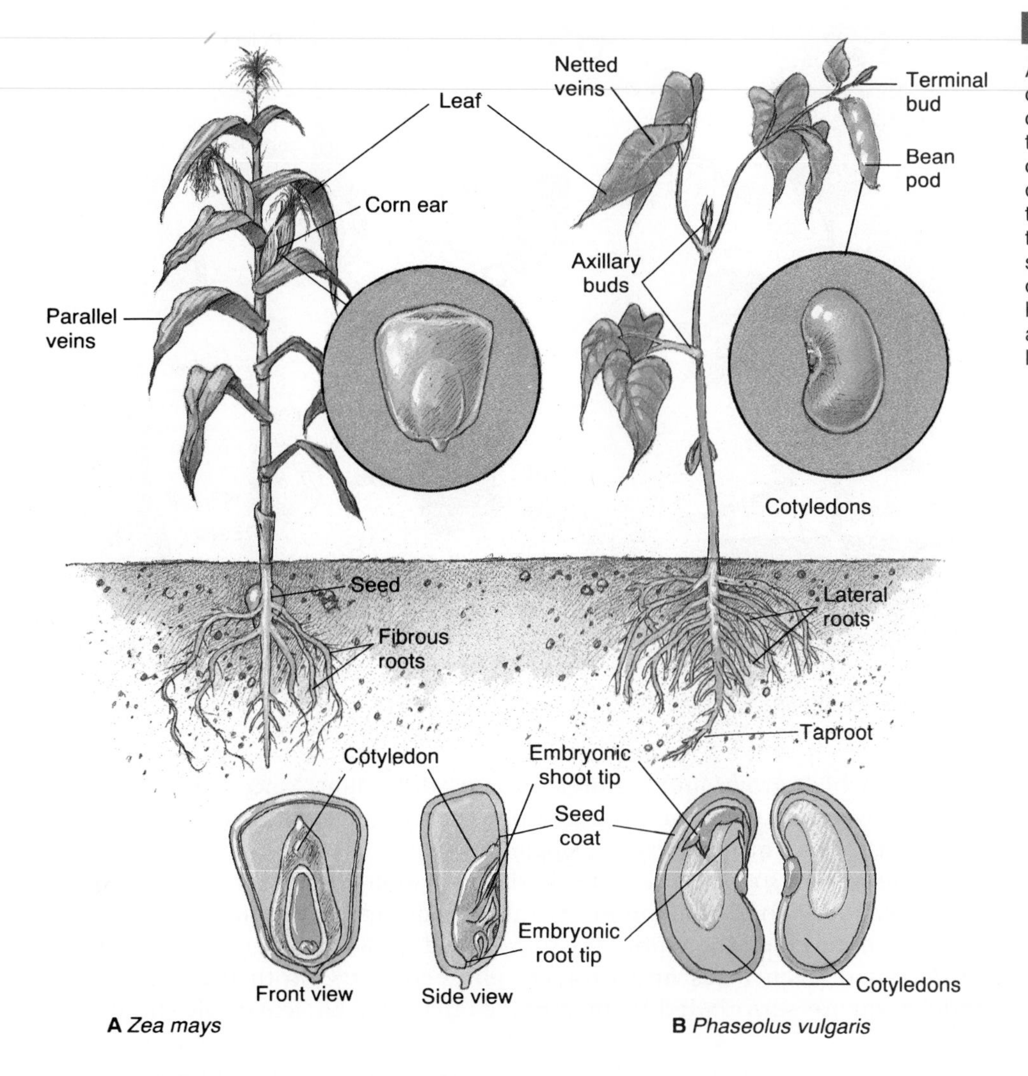

Figure 30.5

Angiosperms are classified as monocots or dicots on the basis of certain characteristics, including seed structure. The seed of a monocot such as corn, *Zea mays* (A) has only one cotyledon and an embryo. The seed coat protects the embryo. A dicot seed such as the bean, *Phaseolus vulgaris* (B) has a seed coat surrounding two large cotyledons and an embryo. Monocots also have fibrous roots and leaves with parallel veins. Dicots have a taproot and leaves with netted veins.

Seed Anatomy and Germination

Germination occurs when different structures within the seed function in an integrated way to induce development and growth of an embryo into a new plant. During germination, the seed becomes transformed from a small, inert object into a highly organized seedling. The final stage of germination is analogous to birth—after embryonic development is completed, the individual plant escapes from its protective container and begins an independent life. Besides being essential for the successful reproduction of wild plants, crops, and garden plants, germination results in the development of certain economically important products. For example, malt is produced from germinating barley seeds and has many important uses, including the brewing of beer.

Seed Anatomy

Anatomy is the study of the structures of organisms. As described in Figure 30.6, a seed starts to take shape, and the flower begins maturing into a fruit, after sperm from the pollen fertilizes an egg. *Pollen grains* each contain two sperm cells and are formed in an *anther,* the male reproductive organ. For fertilization to occur, the sperm must reach an egg that is located in the base of a *pistil,* the female reproductive organ.

The process of angiosperm fertilization involves both sperm contained within each pollen grain. One sperm combines with an egg to create a *zygote* that develops into an embryo. Embryonic cells ultimately grow and give rise to a

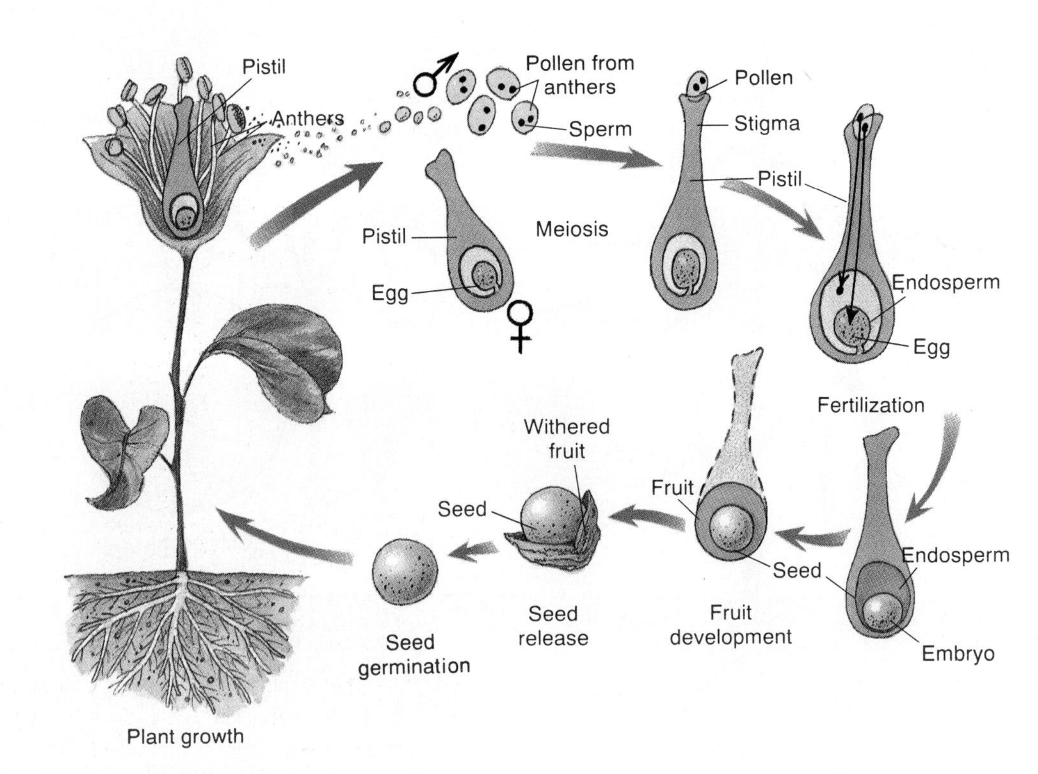

Figure 30.6

The life cycle of an angiosperm begins with a seed that germinates and develops into a seedling and then an adult plant. After the plant matures, it produces flowers that have both male and female reproductive organs. Anthers are male organs that produce pollen grains, which contain sperm. Pollen grains are transported by wind, insects, or other animals to the stigma, the top of the female organ called the pistil. The pollen grain on the stigma grows through the pistil and deposits one sperm in a position where it will fuse with the egg to form a zygote. Another sperm fuses with nuclei from female cells to form endosperm, the nutritive tissue used by the embryo. The zygote and endosperm develop into a seed within the pistil. The pistil then develops into a fruit that encloses the seed.

seedling, which grows and matures into an adult plant. Since flowers of many species have more than one pistil and pistils can have more than one egg, fertilization may occur many times in a single flower.

The second sperm participates in the formation of *endosperm.* **Endosperm** contains stored food that provides the energy and nutrients required for a seedling to develop. The embryo is surrounded by a tough outer layer called the *seed coat* that encloses and protects all seed parts. Both the embryo and endosperm are surrounded by membranes that play an active role in germination.

Besides cotyledons, the plant embryo also contains cells at its shoot tip that divide and develop into the stem and leaves. At the opposite end of the embryo is a group of cells that are destined to form the *radicle,* or embryonic root.

Seed Germination

A seed faces a demanding series of transitions before it can grow into a seedling. After formation, most seeds dry out, but there is great species variation in the events that follow. In some species, both the fruit and the enclosed seeds dry out, while in others, the seeds dry slowly after they have been released from a moist, fleshy fruit. Some plant species produce seeds that germinate soon after dispersal. The seeds of other species require passage through the digestive tract of an animal, where the seed coat is eroded by digestive juices, thereby freeing the embryo and allowing it to germinate soon after the animal eliminates it.

Once in a dried state, the seeds of some species become *dormant,* which means that their level of cellular activity is barely detectable. What is the advantage of seeds undergoing a period of dormancy? In most cases, dormancy enhances a seed's chances for survival and its eventual germination. Dormant seeds are extremely resistant to environmental stresses such as drought and extreme temperatures. Thus they can survive during periods when there is no moisture or when it is too hot or too cold for a plant to live.

The basis of dormancy differs among plant species. Seeds of many species enter into a genetically defined period of dormancy, lasting from 3 to 16 weeks, before germination can begin. Some species produce dormant seeds that also require specific intervals of cold temperature or exposure to a certain number of hours of daylight before they can germinate. Such periods of dormancy are thought to be a mechanism that prevents seeds from germinating at inappropriate times, especially during the winter. For example, if January weather was unseasonably warm, and a seed had no dormancy requirements for day length, exposure to cold, or exposure to light, it might begin to germinate, only to be killed by freezing temperatures during a February storm.

How long can dormant plant seeds survive? Botanists have found that plant species have wide ranges of viable dormancy. The seeds of some species can remain alive for only a short period of time. For example, orchid seeds may live for only a few days. More typically, weed seeds can remain dormant and viable for periods ranging up to 30 years.

Seeds of some species can remain dormant for incredibly long periods of time. In a remarkable case, Harold Schmidt, a mining engineer, made an amazing discovery in 1954. While working in the Yukon Territory of Canada, he uncovered ancient, preserved rodent burrows that contained the remains of lemmings, nest materials, and large seeds. Schmidt collected one lemming skull and several large seeds from one of the burrows and kept them for 12 years. Finally, several Canadian scientists gained possession of the materials and determined that the seeds were from an arctic tundra lupine (*Lupine articus*). Through other investigations, it was concluded that the lemming skull, burrow materials, and seeds were at least 10,000 years old. Would the seeds still germinate? (Do you suppose that the dormancy cold requirements had been satisfied?) About half of the seeds had been well preserved. When placed on wet paper in a dish, they germinated within 48 hours, and all grew into healthy plants!

Initial Events of Germination

Germination begins when a seed absorbs water and when other necessary conditions have been met (see Figure 30.7). After absorbing water, the seed swells, and many of the metabolic processes essential for plant life begin. For example, in barley seeds, the cell layer located between the seed coat and the endosperm produces enzymes needed for growth after being activated by tiny amounts of chemicals, known as **plant growth regulators**, that are released by the embryo. These enzymes are used in the processes of cellular respiration and protein synthesis. Cellular respiration reactions allow the embryo to convert carbohydrates stored in the endosperm and cotyledons into energy for growth. Protein synthesis provides the molecules that are necessary for constructing new cells and tissues during development and growth.

The first visible sign of germination is the appearance of the radicle after it expands, elongates, and then penetrates through the seed coat. As the radicle continues to elongate, it responds to gravity, turning downward and growing into the soil. As it becomes secured in the soil, the embryonic root absorbs the water and nutrients necessary for germination to continue.

Later Events of Germination

Once the radicle is established, the shoot elongates and pushes up through the soil. After the shoot breaks through the soil surface, leaves form and begin to carry out photosynthesis. This allows the new plant to produce the sugars it requires for cellular respiration and growth. When the rate of photosynthesis is great enough to satisfy the needs of the developing plant, the seedling is no longer dependent on stored energy reserves, which have usually been depleted by this time.

Signs of germination can be seen by walking through a forest, meadow, or other natural habitat during the spring. It is often possible to find newly emerged seedlings in various stages of development. Some still have remnants of their seed coats attached, others appear with their newly developed leaves, and some will have already developed many of the characteristics of the adult plant.

Figure 30.7

The germination of both monocot (A) and dicot (B) seeds generally follows the same course. After dormancy requirements are met, the root absorbs water and begins to swell. The primary root is the first part of the seedling to emerge from the seed. As the root grows down into the soil, the shoot elongates and the one or two cotyledons provide nutrients to the seedling. As the true leaves emerge, the seed coat is shed, and the cotyledons shrivel and fall away. At this point, the seedling must obtain water and nutrients from the soil and carry out photosynthesis to produce the carbohydrates it needs for continued growth and development.

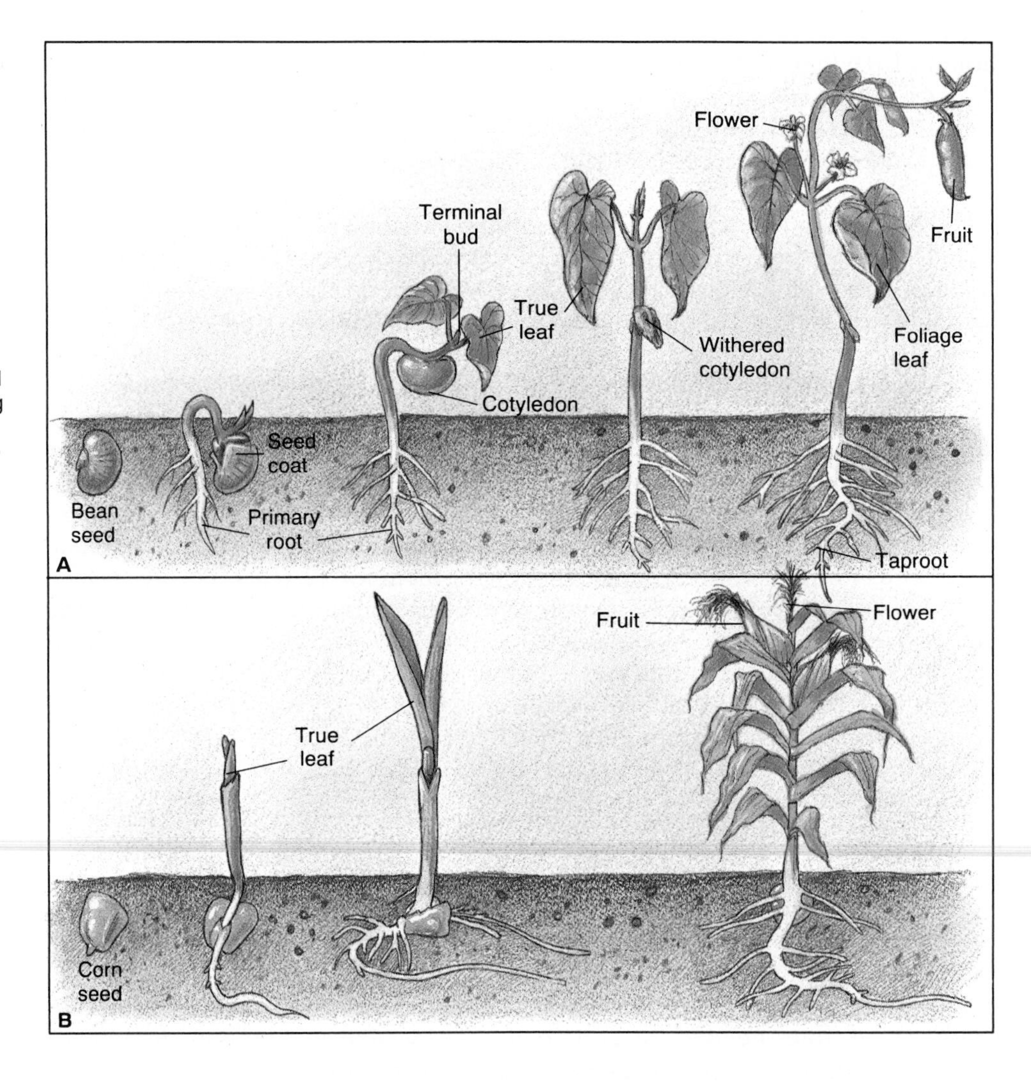

Under some circumstances, large numbers of seeds may germinate at the same time and grow at the same rate. This is most common in fields where seeds are planted at the same time, then irrigated and fertilized. Uniform germination, growth, and maturation of crop species have obvious importance for farmers, since they can harvest all plants at the same time. Seeds of wild plants may also germinate in synchrony, especially in the spring, when environmental conditions favor successful germination and seedling survival. The moist mild climate of spring normally provides the best conditions for germination and seed survival, when the plant is most sensitive to environmental stresses.

Most seeds, once formed and dispersed, enter a state of dormancy that is associated with an increased chance of survival. After the dormant period ends, germination begins when the seed absorbs water and its metabolic activities increase. The embryonic root first appears and becomes established in the soil. Subsequently, the embryonic shoot elongates, leaves form, photosynthesis begins to occur, and the new plant becomes established in its environment.

GROWTH PROCESSES

Germination marks the end of seed dormancy and the beginning of an independent existence for a new plant. Growth is an essential element in the life of a seedling because new cells must develop and increase in size before a fully mature plant arises. Growth of the root system is necessary for acquiring water

Figure 30.8

The growth of plant cells, individual plants, and populations of plants often follows a sigmoidal growth curve. Growth is initially slow, then follows a rapid phase, and finally slows or stops.

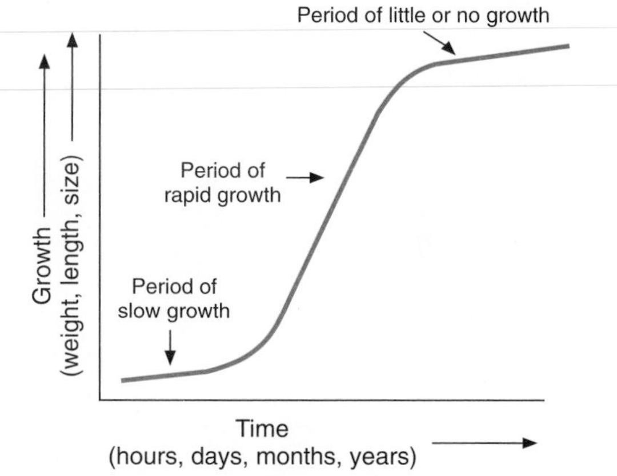

and nutrients from the soil. Similarly, the shoot of a new plant must elongate and new leaves must grow in order to carry out photosynthesis.

Patterns of Growth

Growth of plant tissues, in both the root and the shoot, often follows the pattern described in Figure 30.8. Growth is initially slow, followed by a period in which plant size increases quickly. After the period of rapid growth, there is usually little or no further increase in size. These stages of growth are characteristic of a *sigmoidal growth curve* and can be defined by plotting a specific growth characteristic, such as plant height or weight, as a function of time. Such curves are useful for describing the growth stages of a single plant, for comparing the growth of two plants, and for evaluating the effects of various environmental factors on plant growth.

Mechanisms of Growth

The growth of plant tissues is the result of two processes, the creation of new cells by *mitosis* (described in Chapter 13) and the enlargement of new or existing cells. These two mechanisms are responsible for the growth of plant roots, stems, leaves, and reproductive tissues such as flowers.

Production of New Cells

In plants, growth that is associated with mitosis is generally confined to specific regions called *meristems.* Normally, plant cells produced by mitosis are small. The initial slow phase of the sigmoidal growth curve is thought to be associated with the production of new cells. Later, the rapid growth phase may be related to the division and enlargement of cells produced during the initial phase. In the final phase of life, when there is little or no additional growth, neither cell division nor cell enlargement occurs in plant tissues.

Enlargement of Existing Cells

Plant growth can also occur through the enlargement of existing cells, as shown in Figure 30.9. For example, some sunflower daughter cells are known to increase 15-fold in size after they are produced by mitosis. Thus cell enlargement can result in substantial growth of the tissue or the entire plant.

The enlargement of existing plant cells depends, in part, on the stretching capacity of the *cell wall,* the rigid outer layer that surrounds each plant cell. Cell walls are composed primarily of *cellulose,* a complex carbohydrate molecule that can be rigid, as in wood, or flexible, as in softer plant parts. If rigid, cell walls cannot enlarge because of the unyielding restraint. However, cell walls that can bend and stretch allow a cell to enlarge in a way that is similar to expansion of a balloon's skin as it is inflated. Newly formed cell walls are generally more flexi-

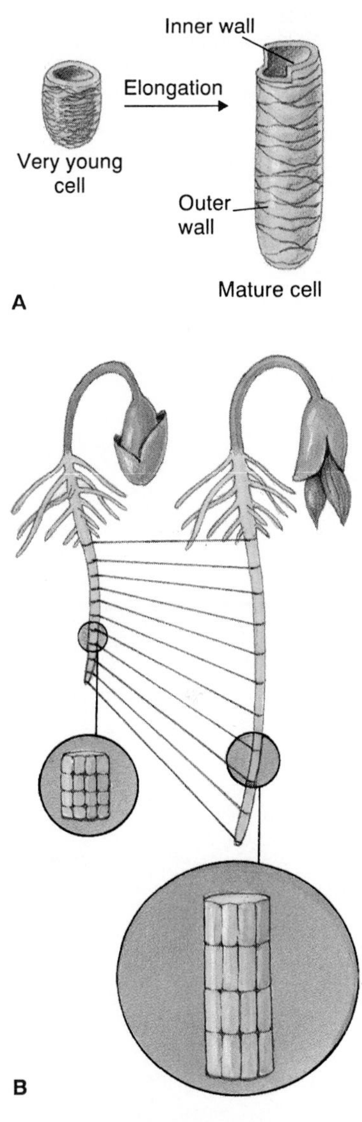

Figure 30.9

The yellow-colored regions in this figure symbolize areas—roots, stems, leaves, or flowers—where plant growth occurs as a result of cell elongation. (A) Plant cells can enlarge as a result of osmotic pressure that forces elongation. (B) Cell elongation is an important component of plant growth, as shown by the stretching zones marked on a bean root.

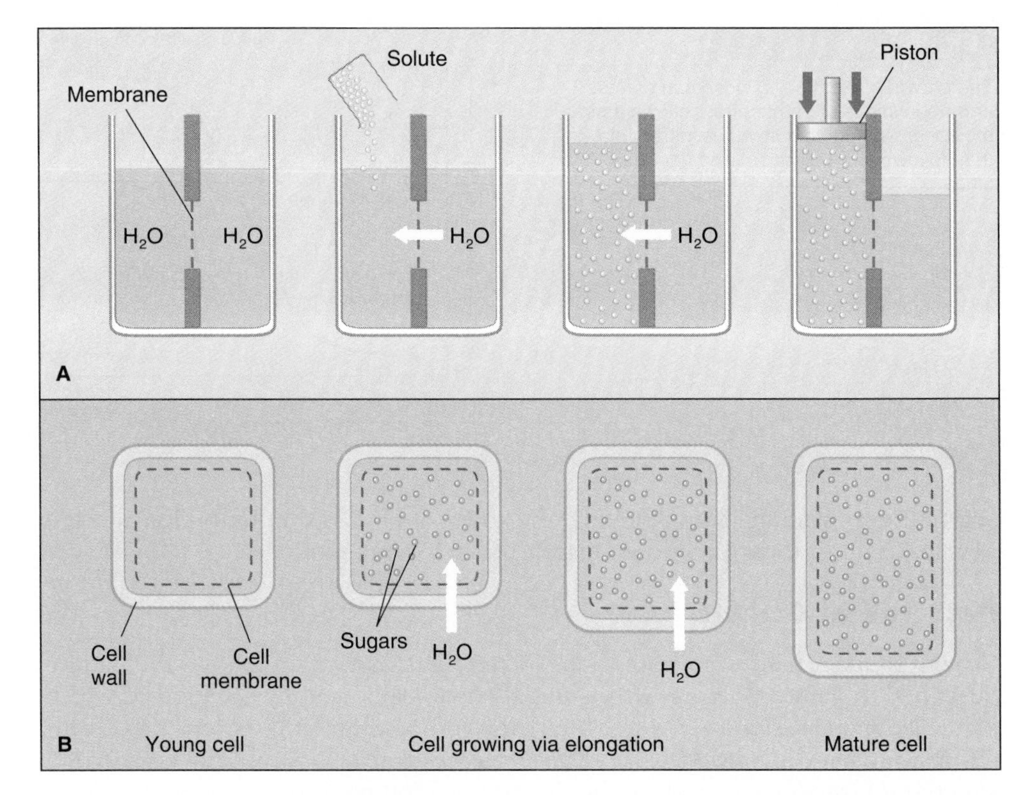

Figure 30.10

Osmotic pressure provides the force for cell elongation and for maintaining shape in mature, nongrowing cells. (A) Osmosis is the movement of water, the solvent, through a membrane, from a low to a higher concentration of solute molecules (for example, sugar). (B) When solute concentrations increase on one side (the inside) of a cell membrane, water moves into the compartment. The movement of water by osmosis may result in either a greater volume of water in the compartment where solute concentrations are higher, or, if the compartment is rigid, an increase in water pressure.

ble and elastic than the walls of mature cells. As a result, young cells are more likely to grow by enlargement than mature cells.

Plant growth due to cellular enlargement not only requires that cell wall molecules be pliable but also that pressure exerted from inside the cell causes the cell wall to stretch and expand. What is the nature of the pressure inside plant cells?

High pressure that causes the enlargement of cells is supplied by water moving between the inside and outside the cell by *osmosis*. **Osmosis** is the passage of water, a *solvent,* in which one or more substances are dissolved, through a membrane. As indicated in Figure 30.10, such movement is more rapid from a region of low *solute* (any substance dissolved in water) concentration to one of high solute concentration than in the reverse direction. This differential flow rate results in *osmotic pressure* against the membrane.

The stronger the forces that permit water to enter the cell, the greater the pressure for cellular enlargement, if the cell wall is elastic enough to allow expansion. Cellular respiration, photosynthesis, and other forms of metabolism lead to the production of dissolved substances (solutes), which plays a role in controlling the movement of water into the cell. Higher concentrations of solutes inside the cell will cause water to move into the cell from the external environment, where solute concentrations are lower (see Figure 30.11). The greater the differences between solute concentrations inside and outside the cell membrane, the stronger the force for water to move into the cell.

Water availability in the soil can also influence the expansion of cells. If moisture is abundant in the soil, sufficient water will be available for cells to enlarge. If little water is present, as in dry soils, cells will not enlarge, and the plant will be stunted and will wilt. Thus environmental factors such as moisture and heat can also influence the growth of plants by affecting the mechanism of cell enlargement.

Development

Cell differentiation is the process whereby newly formed cells become transformed into highly specialized, functional cells. During the growth phase, one of

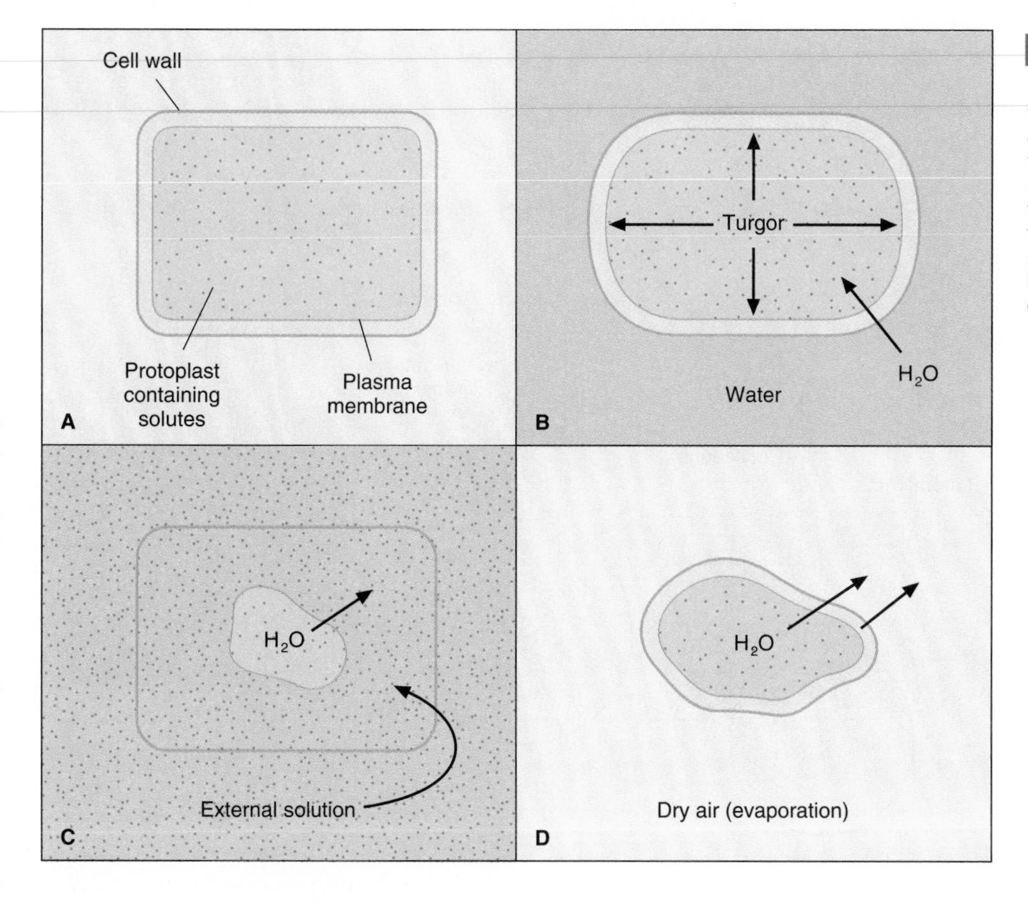

Figure 30.11

(A) High water pressure in cells affects their shape. (B) Loss of water from the cell may occur and cause changes in the cell's profile. (C) If water loss continues, the cell may lose enough volume that its cell membrane pulls away from the cell wall. (D) Extreme water loss may cause collapse of both the membrane and cell wall and result in the cell's death.

the two daughter cells created by mitosis usually remains in the meristem, does not differentiate, and will divide again. The other daughter cell differentiates, matures, and becomes capable of performing a specific function as part of a tissue.

Plants are composed of many differentiated tissues that have specific functions. Figures 30.12 and 30.13 show the anatomy of shoot and root tips. **Meristem** tissues produce new daughter cells that differentiate into the following major plant tissues. *Vascular phloem* tissue transports fluids, including water, sugars, and nutrients, from the shoot system to the root system. *Vascular xylem* tissue conducts water and nutrients absorbed from the soil by the root system to the shoot. *Parenchyma* tissue is composed of thin-walled cells that provide bulk and form to leaves, stems, branches, and roots. Parenchyma also makes up the photosynthetic cells found in leaves. All of these plant tissues originate from cells produced by meristem tissues.

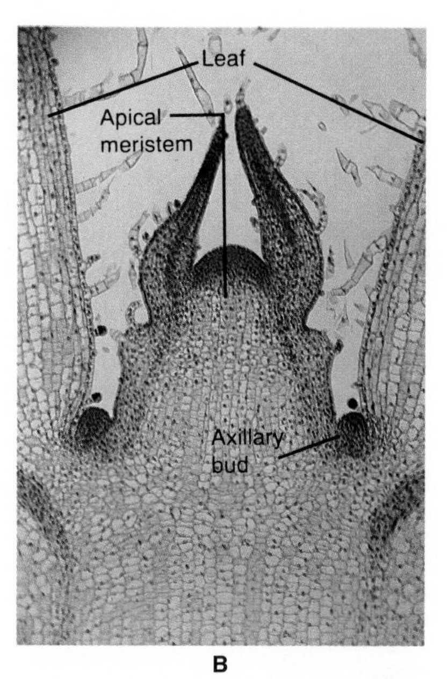

Figure 30.12

The tip of this *Coleus* shoot (A), in a magnified longitudinal section (B), shows the meristem region, developing leaves, developing buds, and cells that will eventually produce vascular tissues and other tissues.

Figure 30.13

(A) This dicot root tip shows the root cap, root hairs, and emerging lateral roots. (B) The longitudinal section shows the root cap, apical meristem, and cells destined to form various root tissues.

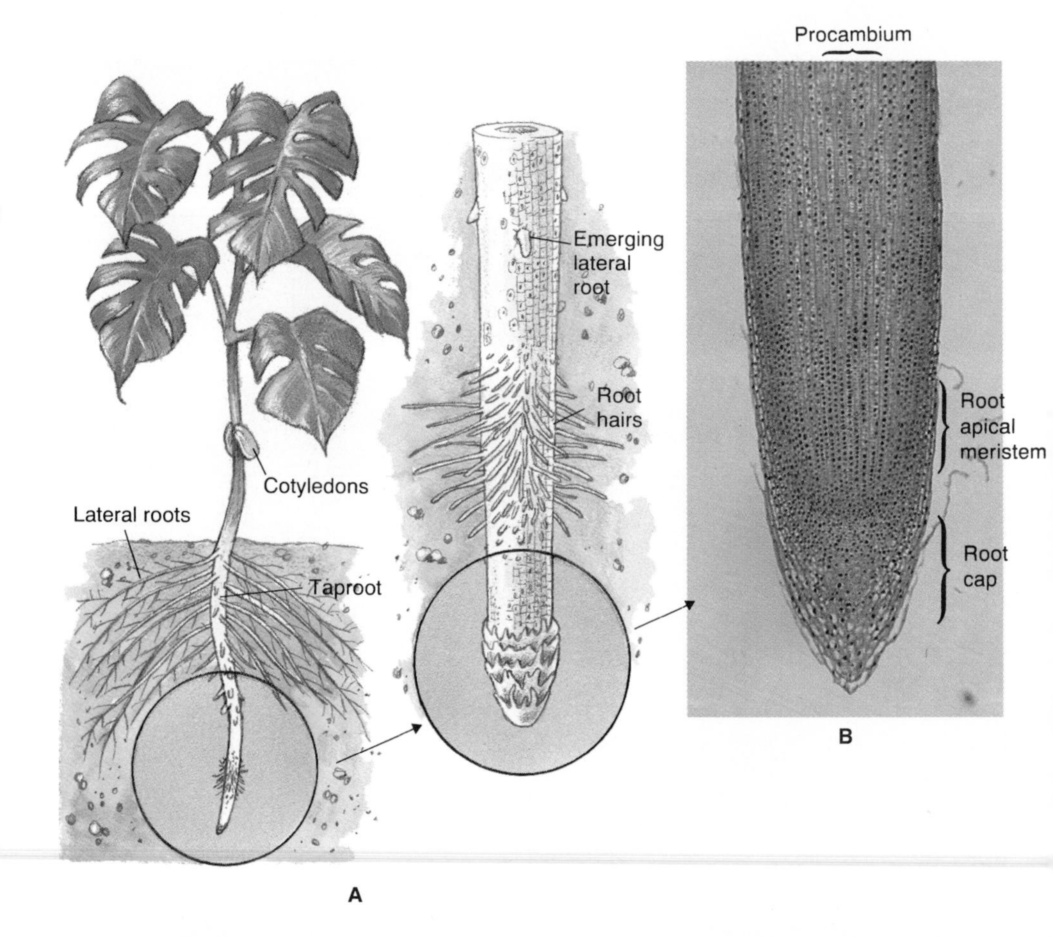

Plant growth is dependent on mitosis in meristems to produce new cells and on the enlargement of existing cells. Enlargement involves several factors, including flexible cell walls and solvent or solute movement into and out of cells. New and enlarged cells eventually differentiate and form tissues that are specialized to carry out specific functions. Meristem tissues, vascular phloem, vascular xylem, and parenchyma tissue are the primary plant tissues.

GROWTH AND ENVIRONMENTAL FACTORS

The processes of mitosis, cell enlargement, and differentiation ultimately result in the transformation of an embryo within a seed into a functional plant. The continued survival of a newly established plant depends on its abilities to acquire the resources it will need for further growth and survival. What resources does a plant require? Where do they come from? How are they obtained by the plant? Since plants are immobile, they must acquire resources from the habitat in which they grow. The nature and availability of these necessary resources and the mechanisms by which a plant acquires them define relationships between the plant and its environment.

Energy

The sun generates energy that radiates to Earth's surface, and sunlight is the primary source of energy that sustains most life on the planet. Chlorophyll molecules in plants capture solar energy, which drives the process of photosynthesis. Through complex biochemical reactions, some of the energy absorbed by chlorophyll is used to construct carbohydrate molecules from water and atmospheric CO_2. These carbohydrates, including sugars and starch, are stored in plant cells and serve as fuel for the plant to live and grow.

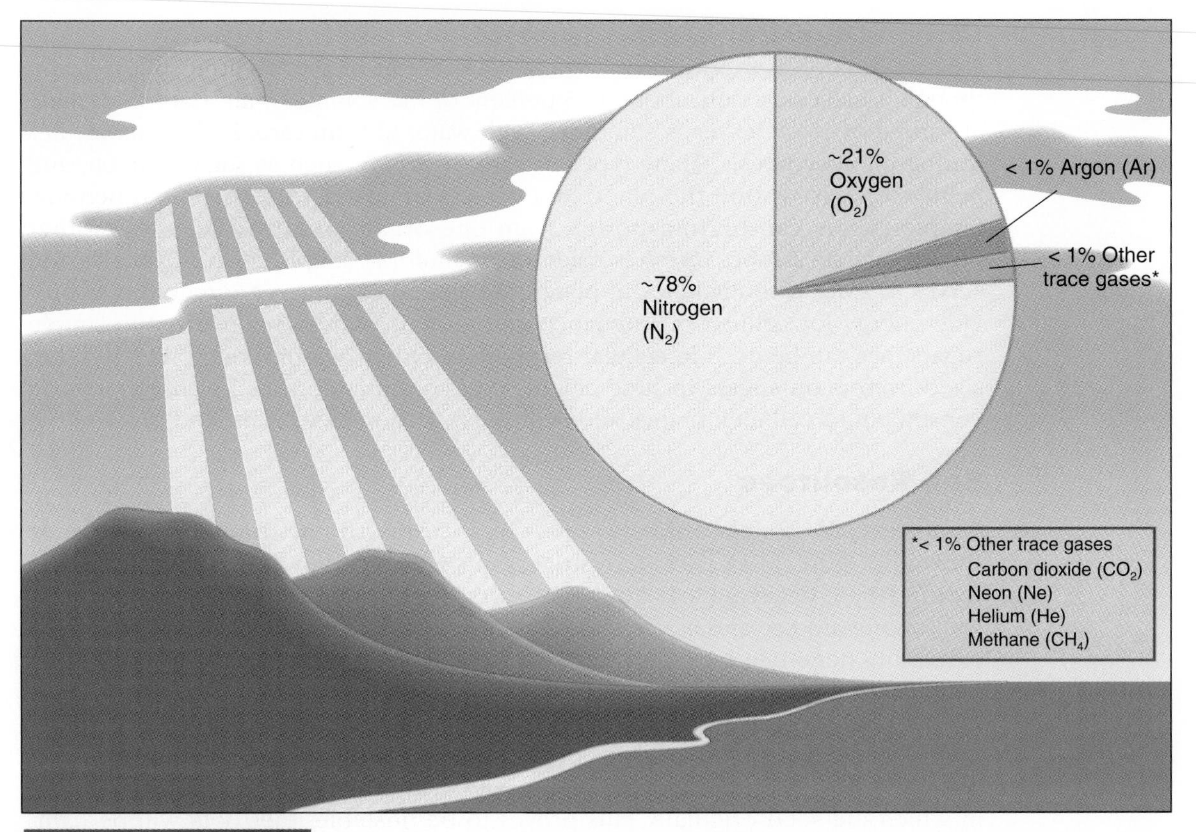

Figure 30.14

Earth's atmosphere is composed largely of nitrogen and oxygen gases. Together, they account for about 99 percent of the air. Many other gases are present in trace amounts. Carbon dioxide (CO_2) presently accounts for about 0.035 percent of Earth's atmosphere and is the source of carbon for plants. CO_2 concentrations are increasing as a result of human activities. Increasing concentrations of CO_2 and other "greenhouse" gases have the potential to cause changes in Earth's climate.

Atmospheric Resources

Plants acquire a number of essential chemicals from the atmosphere. Gases are the most critical, although atmospheric chemicals absorbed by dust, fog, rain, and snow can also be important sources of substances required for plant growth.

The chemistry of the atmosphere is extremely complex (see Figure 30.14). Unpolluted air contains more than 20 gases. Three gases, nitrogen (about 78 percent), oxygen (21 percent), and argon (1 percent) make up most of the volume of the atmosphere, and their concentrations remain nearly constant. Other gases are present in lower, variable concentrations. These include water vapor (0.7 percent) and carbon dioxide (0.03 percent), as well as traces of neon, helium, hydrogen, methane, ozone, ammonia, sulfur dioxide, and hydrogen sulfide. The gases of greatest biological importance are oxygen (O_2), carbon dioxide (CO_2), and nitrogen (N_2). These gases constitute environmental resources that are used by plants in photosynthesis, cellular respiration, and protein synthesis.

Nitrogen

Nitrogen is required by plants for synthesizing proteins. It is present as a gas in the atmosphere, but this form of nitrogen cannot be used by plants. Rather than acquiring nitrogen directly from the air, plants absorb nitrogen that has been converted from gaseous N_2 to other chemical forms, such as ammonia and nitrates. Recall from Chapter 7 that this transformation is carried out primarily by bacteria that live in close proximity to plant roots. These microbes convert atmospheric N_2 into chemical forms that can be absorbed by the roots of plants and assimilated into proteins.

Carbon Dioxide

Carbon dioxide is a critical biological component of Earth's atmosphere, even though it makes up only about 0.03 percent of the volume of air. Carbon dioxide absorbed by plant leaves is combined with water to form carbohydrate molecules during photosynthesis. Many types of carbohydrates, such as sugars, starch, and cellulose, exist within the plant, and each is involved with specific functions. Simple sugars can be transported from one organ, usually a leaf, to another organ, such as a root. *Starch* is made up of multiple simple sugar molecules and serves to store carbohydrates in plants. When carbohydrates are required to provide energy for cellular maintenance and growth, starch is converted to simple sugars that can be used in cellular respiration. Other organic compounds synthesized from plant sugars include cellulose (a component of cell walls), pectin (a constituent of cell membranes and cell walls), fats, nucleic acids, and proteins.

Soil Resources

Plants are able to take root and grow in a variety of environmental media, as shown in Figure 30.15. Most terrestrial plants are rooted in the soil, but some can root in fallen, decaying logs known as *nurse logs*. Hemlock seedlings often grow well on nurse logs, and as they continue to grow, they develop root systems that eventually penetrate into the soil. Other plants, such as Spanish moss, can grow in the tops of living trees.

Whether plants take root in other plants, logs, or soils, one universal function of the rooting medium is to provide stability so that the plant can grow in one place. Plants cannot move from one site to another, so if the rooting medium in which the seed originally falls proves to be unstable, the roots will be damaged, and the plant may die.

Water

Water is the major resource that plants must acquire from their rooting media. Even though leaves may be in direct contact with water in the form of rain, snow, or dew, most cannot absorb this water because their external surfaces are covered by a **cuticle**, which is impervious to water because it contains waxes. The cuticle is an important adaptation that allows plants to grow on land; without the cuticle, terrestrial plants would dry out and die.

Together with cell walls, water plays a central role in providing support for a plant. The osmotic pressure of water is necessary not only for cellular enlargement during growth but also for maintaining the entire structure of the plant.

Figure 30.15

Many plants, including hemlock (A), grow in soils or decomposing logs known as nurse logs. Other species, such as Spanish moss (B), root in tree canopies using branches for their support. Mistletoe, a parasitic plant, grows in the tops of pine trees (C).

A

B

C

Figure 30.16

Absorbing sufficient amounts of water allows plant cells to maintain the pressure that causes the plant to stand upright. If water is limited, plant cells lose water pressure, and the plant wilts (A). Recovery is possible if water is applied before large numbers of cells die (B).

When plant cells are fully charged with water, they are *turgid*. When insufficient water is available for cells to remain turgid, plants lose **turgor**, the capacity to sustain their shape, and they wilt (see Figure 30.16). If wilting is only temporary or not caused by severe water deficits in the soil, many plants can recover and regain their turgor when water again becomes available.

Water is also used in a plant's metabolism. It plays a critical role in the light reactions of photosynthesis and is important in other reactions carried out by the plant. Water is the primary solvent in the cell; it carries nutrients into the plant and transports chemical compounds within the plant. Just as too little water is detrimental to plants, so is an excess of water. Flooding a plant results in reduced levels of oxygen around the roots and, if prolonged, may result in death. However, many plants are adapted for growing in wetlands because they have evolved mechanisms to keep air around their roots, even when submerged. For example, water lilies can absorb air through their leaves and transport it to their roots.

Nutrients

Plants acquire the nutrient resources shown in Figure 30.17 (page 566) from soils, by root uptake, or directly from the atmosphere. Botanists generally believe that plants require at least 16 **essential elements** to grow and complete their life cycles. Plants obtain carbon from the atmosphere and the other 15 essential elements from their rooting media; the other essential elements are hydrogen, oxygen, nitrogen, potassium, calcium, magnesium, phosphorus, sulfur, chlorine, iron, boron, manganese, zinc, copper, and molybdenum. These elements are available to plants from soil because of two processes. During *decomposition,* bacteria and fungi convert organic matter in the soil into simpler chemical forms that the root system can absorb. During **mineralization**, soil particles derived from parent rock break down and release important elements. Roots usually absorb these essential elements when they become dissolved in water.

In agriculture or gardening, nutrient deficiencies may occur (see the Focus on Science and Technology, "Essential Elements and Agriculture"). These can be overcome with fertilizers. Since nutritionally deficient soils usually lack nitrogen (N), phosphorus (P), and/or potassium (K), these chemicals are most commonly contained in commercial fertilizers.

In general, plants are capable of accumulating nutrients at much higher concentrations than exist in the soils they inhabit because their roots absorb essential chemicals selectively. Nutrients absorbed by root cells are distributed

Essential Elements and Agriculture

Focus on Science and Technology

In the early nineteenth century, many American farmers in the southern and northeastern United States had become concerned about "worn-out soil." A disturbing cycle had developed. Farmers settled on new land, exploited it until fertility decreased, and then moved on to new virgin lands. Consequently, vast areas had become unsuitable for farming or habitation. However, by the middle of the nineteenth century, these barren lands were once again under cultivation, and the formerly destructive cycle had been broken. In large part this was made possible by the new field of *agricultural chemistry* that was responsible for fundamental changes in American agriculture.

The research of a German chemist, Justus Liebig, was central to the agricultural revolution in the mid-nineteenth century. His widely read book, *Organic Chemistry in Its Applications to Agriculture and Physiology,* published in 1840, went through several editions in the United States. Liebig stressed the importance of nitrogen for plants, and although his specific description of the nitrogen cycle was later replaced as new knowledge became available, he was the first to call attention to its importance. Of greater significance was his discussion of *essential elements.* Liebig was among the first scientists to realize that certain chemical elements were essential for proper plant growth. He showed that these elements were found in the tissues of all plants. Furthermore, plants either did not grow or were stunted in soils lacking essential elements.

Liebig advanced the idea that "worn-out soil" was, in fact, soil that had been depleted of essential elements. He noted that farmers could improve their fields by adding substances such as animal dung, wood ashes, ground bone, or "mineral manures" (for example, gypsum and lime). According to Liebig, these materials were effective because they replaced chemicals that had been depleted. Liebig also suggested that these chemical substances could be manufactured. This would have the advantages of providing relatively inexpensive additives that would allow agricultural experts to apply exactly the amounts of essential elements that they determined (through testing) were necessary.

Liebig's book initiated a scientific revolution in agriculture. In the United States, it began with American scientists traveling to Germany to study in laboratories such as Liebig's (see Figure 1). Later, the U.S. federal government (under President Lincoln) established the Department of Agriculture and funded land-grant colleges. These colleges, in turn, set up agricultural experiment stations that transmitted results of agricultural research to farmers throughout the country.

In the nineteenth century, some of the elements needed for agriculture (for example, magnesium) were plentiful and could be easily mined. Others were less plentiful. Recall that nitrogen, for example, although the most abundant gas in the atmosphere, cannot be used by plants in that form. Bacteria in the soil convert nitrogen into forms that a plant can metabolize. This source was supplemented with nitrates that were mined in Chile, but they were quickly exhausted. However, another German chemist, Fritz Haber, learned how to convert atmospheric nitrogen into ammo-

Figure 1 Agricultural researchers from around the world flocked to Liebig's laboratory.

nia, a product that could be used as a synthetic plant fertilizer. This discovery won Haber a Nobel Prize and was one of the key steps in the development and expansion of the huge chemical fertilizer industry.

The use of chemical fertilizers, especially since 1950, has been a mixed blessing in the United States and other developed countries throughout the world. Because of the development of new high-yield varieties of seeds (especially grains) in conjunction with the use of chemical fertilizers, agricultural yields increased dramatically, along with the ability of human populations to feed themselves. Between 1950 and 1984, farmers were able to more than double production on their lands. During this period, the annual use of chemical fertilizers increased from 14 million to 125 million tons.

Unfortunately, the production of these fertilizers consumes considerable energy. As fossil fuel reserves have become reduced, the cost of producing chemical fertilizers has increased. In addition, serious environmental problems associated with fertilizers have become recognized. Runoff of chemical fertilizers from fields has contaminated groundwater, streams, lakes, and rivers in some areas. This has caused considerable alarm over the quality of the drinking water and the ecological effects of altering food webs by accidentally "fertilizing" algae and plankton in affected rivers and lakes.

Concern over the extensive use of fertilizers and the high energy costs of agricultural production has led scientists to study new alternatives. For example, is it possible to recycle some of the United States's estimated 2 billion tons of organic wastes (such as animal manure and slaughterhouse offal) as organic fertilizer? Developing an "alternative agriculture," or "sustainable agriculture," that will reduce the use of chemical fertilizers, pesticides, and energy resources now appears to be one of the major challenges of the 1990s.

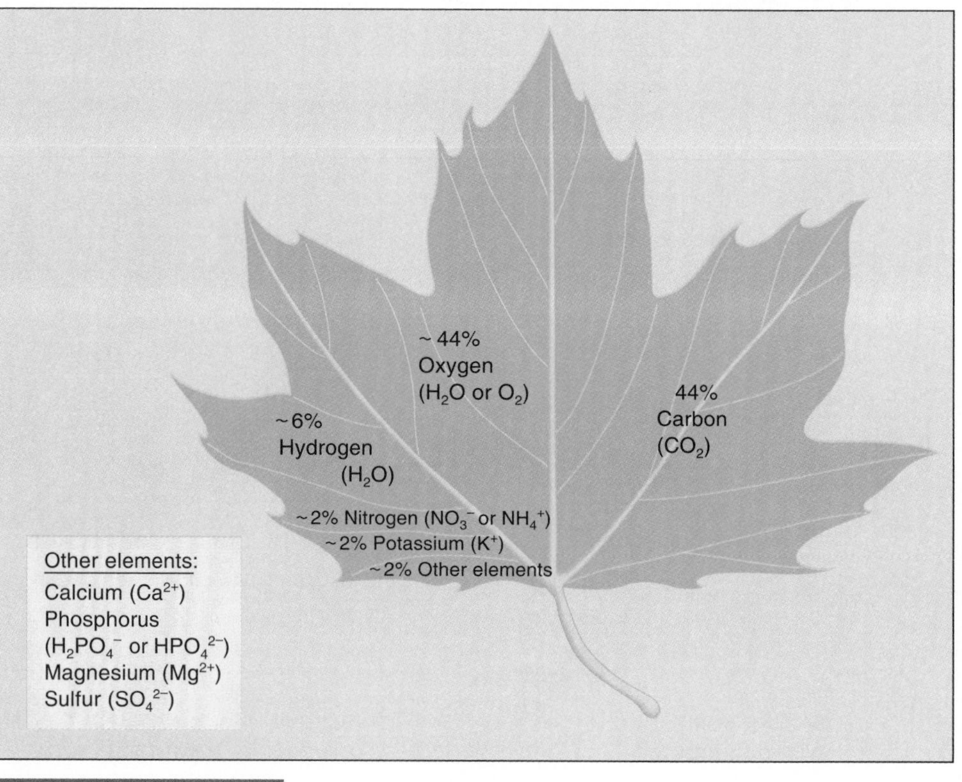

Figure 30.17

Carbon, oxygen, and hydrogen constitute about 95 percent of a plant's weight. Nitrogen usually accounts for another 2 percent or so. Many other elements needed for growth, such as potassium, calcium, and phosphorus, are present in the soil only in trace amounts.

and retained within the plant, where they may reach high concentrations. For example, potassium concentrations in plants may be several hundred times greater than the soil concentration. As much as 95 percent of the *dry weight* (the weight after water is removed) of a plant may be composed of carbon, oxygen, and hydrogen, with all other elements contributing only about 5 percent. Of that 5 percent, nitrogen may contribute as much as 2.5 percent of a plant's dry weight, and potassium and aluminum, 1 percent each. The remaining essential elements are present in only trace amounts.

To survive after germination, plants must be able to acquire essential resources from their surrounding environment. By transforming radiant energy during photosynthesis, plants are able to produce the carbohydrates they use as energy sources. Essential gases—oxygen, nitrogen, and carbon dioxide—are obtained directly or indirectly from the atmosphere. Water and essential elements are obtained from the soil through the root system and then distributed to all parts of the plant.

PLANTS AND THEIR ENVIRONMENT

In this chapter, we followed the history of seed plants from the formation of a seed to its subsequent germination, then to its development into a seedling, and, finally, to its growth into an adult plant. These events are common to many of the plants that are important to humans. Plants must absorb solar radiation and chemicals from the environment in order to grow. The processes by which they obtain resources from the habitat form the basis for the relationship between plants and their environment. The dynamic nature of this relationship demonstrates that although plants remain stationary, they must respond quickly as features of the environment change over time. We explore plant-environment relations further in Chapter 31.

Summary

1. Vascular plants have tissues that transport water and nutrients between their roots, stems, and leaves. Most trees, crop species, garden plants, and house plants are vascular plants.
2. A seed is a complex structure that contains a plant embryo and nutrients to be used by the embryo during its development. Two primary classes of vascular plants produce seeds. Gymnosperm seeds develop in cones, and angiosperm seeds form in fruits.
3. Most plant species have mechanisms that act to disperse seeds away from a parent plant. Seeds of different species are dispersed by various processes that are related to their size and shape.
4. Angiosperms are of two major types—monocots and dicots—that are distinguished by differences in number of cotyledons in their seeds, arrangement of veins in their leaves, number of floral parts, and type of root.
5. Seed formation in angiosperms is a complex process that begins when an egg in the flower is fertilized by a sperm contained in a pollen grain. Fertilization creates a zygote that develops into an embryo within the seed. In some species, a second sperm cell from the pollen grain combines with a separate female cell to form endosperm, which is also part of the seed. All of the internal seed structures are protected by a seed coat.
6. Seed germination usually occurs after the seed has dried and environmental conditions are favorable. In some plant species, seeds enter a period of dormancy that does not end until certain requirements—usually related to cold temperatures or day length—have been met.
7. During germination, the embryo constructs new cells and tissues that will be used in the development of a new plant. The radicle, or embryonic root, first emerges from the seed and becomes established in the soil. Later, the stem develops and leaves form to carry on photosynthesis. At this point, the newly formed seedling becomes independent and may eventually mature into an adult plant.
8. Growth of plant tissues depends on mitosis, cell enlargement, and differentiation. Mitosis takes place in specialized regions, called meristems. As new cells are formed by mitosis, they enlarge through the effects of osmosis and differences in water pressure inside and outside the cell. Many new cells undergo differentiation and acquire the ability to perform specialized functions.
9. Plants must carry out photosynthesis, cellular respiration, and protein synthesis in order to survive. The energy required for these biological processes comes from solar energy that is converted into sugars by photosynthesis. Nutrient resources used by plants come from the atmosphere and the soil.

Review Questions

1. How did the early concept of a chain of being influence early ideas about plants?
2. What is the basis of plant classification?
3. What is the primary difference between gymnosperm and angiosperm seed production?
4. What is the significance of seed dispersal? How is seed dispersal influenced by seed shape and fruit type?
5. What are some of the differences between monocots and dicots?
6. Describe the general anatomy of a seed.
7. What is the function of dormancy before seed germination?
8. Describe the process of seed germination.
9. What two cellular processes are associated with plant growth?
10. How is water availability related to plant cell enlargement?

11. What are the functions of vascular phloem, vascular xylem, and parenchyma?
12. How do plants acquire nitrogen and carbon dioxide?
13. Which resources do plants acquire from the soil? How do they obtain these substances?

Essay and Discussion Questions

1. Angiosperm and gymnosperm plants produce seeds during sexual reproduction. What advantages and disadvantages do seeds have as components of a plant's life history?
2. Does it seem likely that plants are more susceptible to environmental changes than animals? Why? What types of environmental changes might be especially harmful to plants or plant populations?
3. One plant reproductive strategy is to make lots of small seeds, and another strategy is to make a limited number of large seeds. What are the advantages and disadvantages of each of these strategies?

References and Recommended Reading

Bristow, A. 1978. *The Sex Life of Plants.* Fort Worth: Holt, Rinehart and Winston.

Clarkson, D. T. 1985. Factors affecting mineral nutrient acquisition by plants. *Annual Review of Plant Physiology and Plant Molecular Biology* 36:77–115.

Epstein, E. 1973. Roots. *Scientific American* 228:48–58.

Evans, M. L., R. Moore, and K. Hasenstein. 1986. How roots respond to gravity. *Scientific American* 255:112–119.

Francis, C. A., C. B. Flora, and L. D. King, eds. 1990. *Sustainable Agriculture in Temperate Zones.* New York: Wiley.

Galston, A., P. Davies, and R. Satter. 1980. *The Life of a Green Plant.* Englewood Cliffs, N.J.: Prentice-Hall.

Grace, J. B., and D. Tilman, eds. 1990. *Perspectives on Plant Competition.* Orlando, Fla.: Academic Press.

Klein, R. M. 1987. *The Green World: An Introduction to Plants and People.* New York: HarperCollins.

Raven, P. H., R. F. Evert, and S. E. Eichorn. 1986. *Biology of Plants.* 4th ed. New York: Worth.

Stern, K. 1988. *Introductory Plant Biology.* 4th ed. New York: Wiley.

Villiers, T. 1975. *Dormancy and the Survival of Plants.* London: Arnold.

Young, J. A. 1991. Tumbleweed. *Scientific American* 264:82–87.

CHAPTER 31

Plant-Environment Relations

Chapter Outline

Reading Questions

1. How do plant acquire and use carbon? Water?
2. How do plants use sugars produced by photosynthesis?
3. What are optimal environmental conditions for plant growth and survival?
4. What types of environmental stress affect plants? How do they cope with different forms of environmental stress?

Figure 31.1

Plants produce many seeds, but only a few ever develop into mature plants.

In Chapter 30, we examined the first stages in a seed-producing plant's life history. Seeds that fall into a suitable environment may develop into seedlings (Figure 31.1). For a seedling to mature into an adult plant, it has to obtain substances and energy from the surrounding environment. An adult plant, of course, must be able to do this continuously throughout its life. In this chapter, we look deeper into the following questions: What types of mechanisms allow plants to obtain and use required environmental resources? What problems do plants face in acquiring these resources? And what types of responses enable them to overcome these problems?

As shown in Figure 31.2, plant growth can be described as a multistep process in which the following events occur: (1) **acquisition**, whereby the plant obtains needed resources from the environment; (2) **assimilation**, or modification of these resources within the plant; and (3) **allocation**, or apportionment, of the products of assimilation to specific parts of the plant according to need. For example, carbon dioxide is an extremely valuable environmental resource for plants. They must acquire CO_2 from the atmosphere, assimilate it through photosynthesis, and allocate the resulting sugars to leaves, stems, or roots for cell maintenance and growth in these tissues. The same is true for water and nutrients.

At a basic level, we can describe the general processes tied to plant growth and development by examining patterns of carbon and nutrient use (see Figure 31.3). In one process, carbon, in the form of CO_2 gas in air, is taken in by leaves and converted to sugars. These sugars are either *respired* (broken down in the process of cellular respiration, as described in Chapter 29), transported to other tissues, or converted into complex carbohydrates, such as starch, that can be stored in the leaf. In a second process, plants acquire nutrients and water directly from soils through their roots. These nutrients are then transported to shoots, which ultimately allocate them to other plant tissues. Thus plants continuously obtain resources from the air and the soil and use these substances in ways that reflect the needs of the entire organism. A plant's use of environmental resources changes as a function of age, developmental state, and the availability of such resources.

The need to comprehend the nature of plant-environment relations is increasing because human activities and products are altering the physical and chemical climate of our planet (see Chapter 11). Climatic changes, regardless of cause, may directly affect the capacity of plants to acquire, assimilate, or allocate resources that they obtain from the environment and thus may eventually limit plant growth and survival. Understanding resource acquisition and use in plants can help us predict how plants will respond to environmental changes. Such predictions are essential for making wise management decisions about forests, agricultural systems, and the quality of our air, soil, and water.

Figure 31.2

Plant growth requires nutrients, water, CO_2 and light energy from the environment. These resources are assimilated into carbohydrates, proteins, and other organic molecules and are then allocated for various uses. The products of assimilation can be used immediately for metabolism, such as respiration or protein synthesis, or they can be stored. Under some conditions, assimilation products are mobilized and transported for use in other parts of the plant.

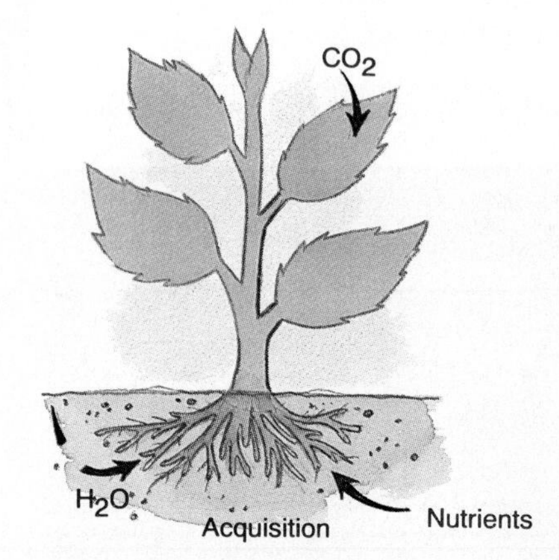

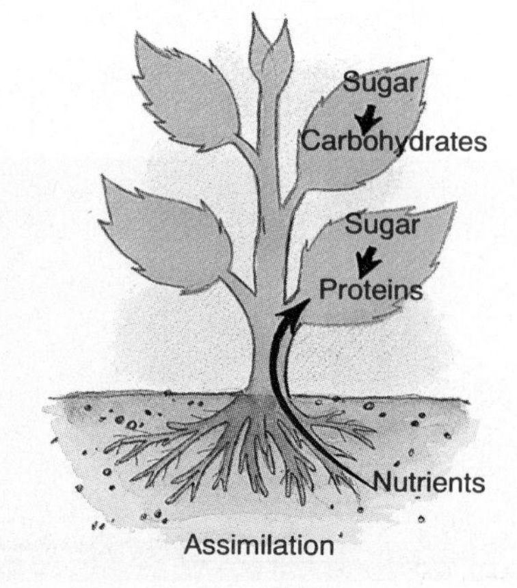

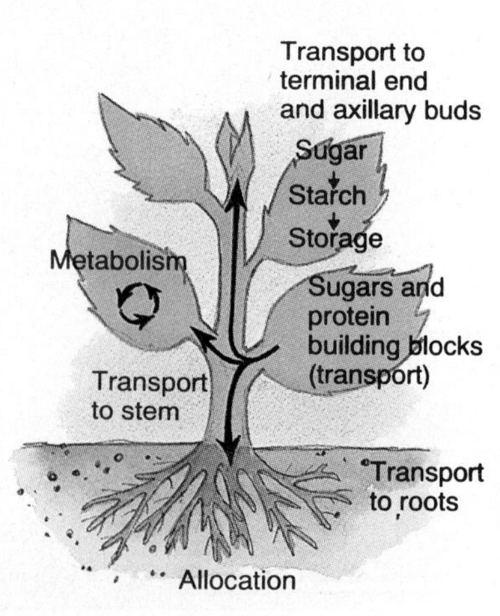

The establishment, growth, and survival of a plant are dependent on its success in acquiring necessary resources from the immediate environment. Carbon, nutrients, and water are examples of environmental resources that plants require. Plants use diverse mechanisms in obtaining these resources, assimilating and transporting them to different parts of the plant and ultimately using them for growth and other activities.

THE CARBON STORY

Carbon dioxide constitutes a small fraction of Earth's atmosphere—about 0.03 percent by volume. It is significant that this concentration is not constant; it changes seasonally in all areas of the world. In the Northern Hemisphere, CO_2 concentrations are highest in winter and lowest in summer. Also, atmospheric CO_2 concentrations are now increasing slightly each year, a factor that may be related to global warming (discussed in Chapter 11).

For convenience, most scientists express the amount of CO_2 (and other chemicals) in the air as *parts per million (ppm)* rather than percent volume. Thus air with 0.03 percent CO_2 is equivalent to 300 ppm; that is, of every 1 million liters of Earth's atmosphere, 300 liters are CO_2. During an annual cycle, atmospheric CO_2 concentrations currently range between 350 and 355 ppm.

What factors might account for seasonal fluctuations in atmospheric CO_2 concentrations? Several processes are particularly important. Some of the variation is caused by high CO_2 emissions that occur during the winter as a result of the increased use of heating fuels. Another factor is related to continuing organism respiration in the face of declines in photosynthesis during the winter. That is, CO_2-producing cellular respiration continues during the winter, but CO_2-removing photosynthesis decreases significantly. (Why?) In contrast, spring and summer peaks of photosynthesis tend to decrease atmospheric CO_2 concentrations during the growing season. Furthermore, annual mean atmospheric CO_2 concentrations increased from about 315 ppm in 1976 to about 350 ppm in 1990. This increase, and projected increases for the future, strongly influence the availability of CO_2 and the rates at which it is acquired by plants. How are plants able to control the amount of CO_2 they take into their tissues?

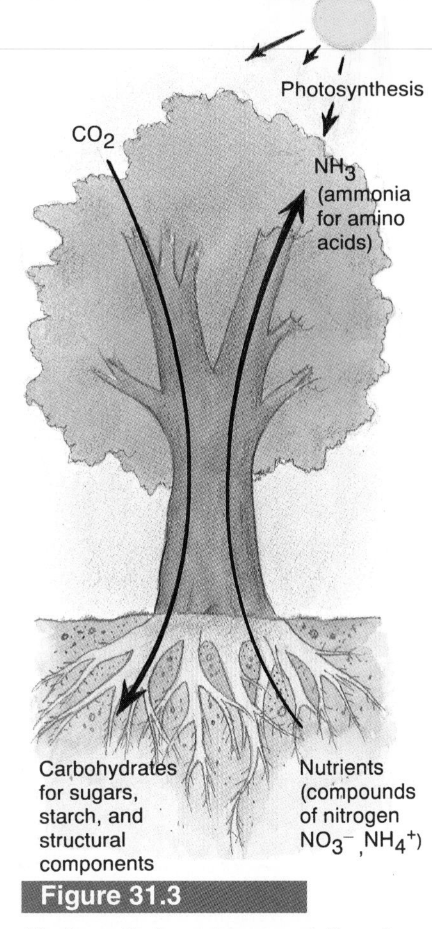

Figure 31.3

Plant growth depends on acquisition of CO_2 from the air, its assimilation into carbohydrates by photosynthesis, and allocation of carbohydrates throughout the plant where these compounds are metabolized. At the same time, plants acquire nitrogen from the soil, assimilate it into amino acids, and allocate these compounds throughout the plant where they are used for protein synthesis.

Acquisition and Regulation

All gases, including CO_2, are exchanged between the leaf and air through **stomata**, structures in the leaf surface that consist of two guard cells with a pore between them (see Figure 31.4). Since stomata can open and close—a mecha-

Figure 31.4

Stomata consist of two guard cells lying next to each other on the leaf surface. The leaf surface is covered with a waxy cuticle that is impervious to water. Physiological processes regulated by the plant affect the guard cells and cause the stomata to open (A) or close (B). Plants can transpire only when stomata are open. Stomata open when the turgor pressure of the guard cells increases. When turgor pressure rises, the guard cells increase in length and separate.

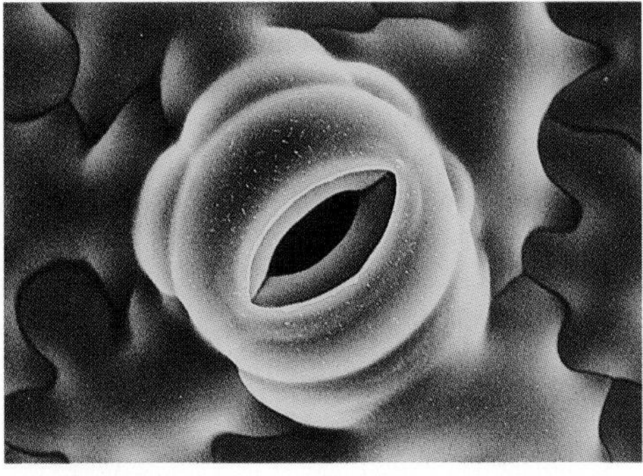

A

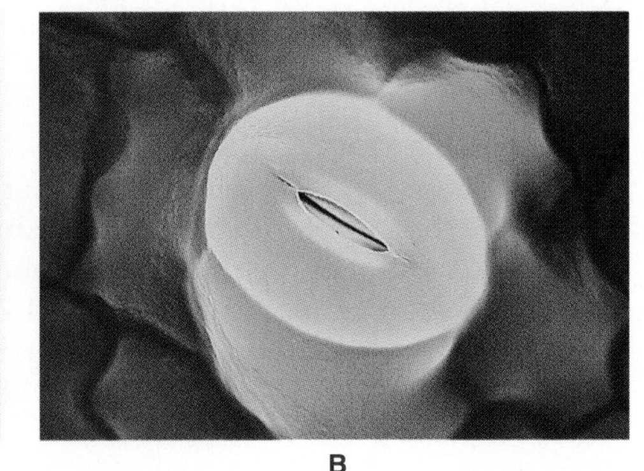

B

nism controlled by the plant—they help regulate the movement of gases into or out of leaves. The rate of CO_2 absorption is related to CO_2 concentrations in the air, leaves, and stomata.

Gas movement between the leaf and the air occurs as a result of **diffusion**, the random movement of molecules, such as CO_2 from regions of high concentration to regions of low concentration. When photosynthesis removes CO_2 from the air inside a leaf, the CO_2 concentration inside the leaf is usually lower than the CO_2 concentration in the air outside the leaf. Thus during daylight hours, when photosynthesis occurs, stomata are open, and CO_2 usually diffuses from the air into the leaf.

The magnitude of the difference in CO_2 concentration between the air and the inside of the leaf—called the **concentration gradient**—affects the rate of CO_2 acquisition by the leaf. To illustrate this effect, assume that stomata are always fully open in the following examples. If the atmospheric CO_2 concentration is 340 ppm and the concentration of CO_2 inside the leaf is 335 ppm, the concentration gradient is low, and the rate of CO_2 exchange between the leaf and air will be slow. However, if the concentration of CO_2 in the air is 340 ppm and the concentration inside the leaf is 280 ppm, the concentration gradient is high, and the exchange rate will be much faster.

Environmental factors can influence processes that affect both the CO_2 concentrations inside the leaf and the rate of CO_2 diffusion into the leaf. Light, for example, is an environmental factor that can have profound effects on CO_2 concentrations in leaves. When the sun sets and photosynthesis stops, CO_2 concentrations in leaves increase because of cellular respiration and approach those of the air whether the stomata are opened or closed.

Water Use Efficiency

Why would a plant's stomata ever close? Since stomatal closure reduces the rate of CO_2 acquisition, and therefore the quantity of photosynthetic products available to the cell, it would not seem to be beneficial to the plant. Discovering the answer to this question is essential for understanding how plants survive when environmental conditions change.

Transpiration

Water is lost from leaves by **transpiration**, the diffusion of water vapor away from the leaves and into the atmosphere. The process of water movement from roots to stem, branches, and leaves and its loss through transpiration is crucial for plant growth. For example, water movement helps distribute nutrients to all plant tissues.

The same principles that explain CO_2 regulation by leaves also apply to transpiration. Air inside the leaf is fully saturated with water vapor, whereas water vapor concentrations in the atmosphere are usually far below saturation. Figure 31.5 describes stomatal regulation of CO_2 and water vapor. The rate of transpiration depends both on the magnitude of the water vapor concentration gradient between the leaf and the air and on whether the stomata are open or closed. If transpiration rates exceed water absorption rates, the plant may dry up and die.

Controlling Water Loss

Photosynthesis and transpiration rates are both affected by changes in stomata. What is the relationship between CO_2 absorption and water loss? One key factor is that water supplies are often insufficient for plants. Also, the rate of water loss is highest when stomata are open and CO_2 absorption rates are high. Thus in terms of a cost-benefit analysis, the price plants pay for absorbing CO_2 to use in photosynthesis is the loss of water, which is an extremely valuable and often scarce resource.

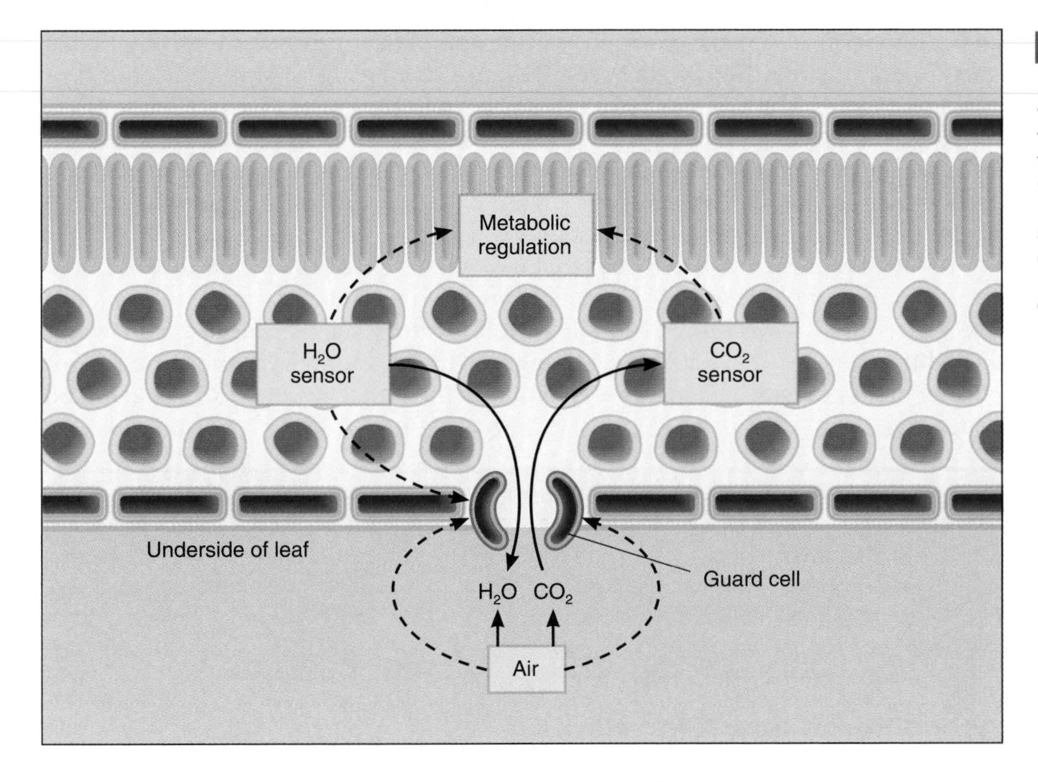

Figure 31.5

Stomata open or close in response to feedback systems that operate within the leaf. These feedback systems are thought to be controlled by water and/or CO_2 "sensors," about which little is known. When water loss rates exceed supply rates, stomata close. When environmental conditions favor high rates of photosynthesis, stomata may open.

It might seem that the most efficient way for plants to manage water intake and loss would be to open their stomata completely when photosynthesis can proceed most rapidly (see Figure 31.6). When do such periods occur? Ideally, they occur when light is high and the raw materials and enzymes used in photosynthesis are available (suitable levels of CO_2 are almost always present). Under these conditions, the amount of carbon gained per amount of water lost is high,

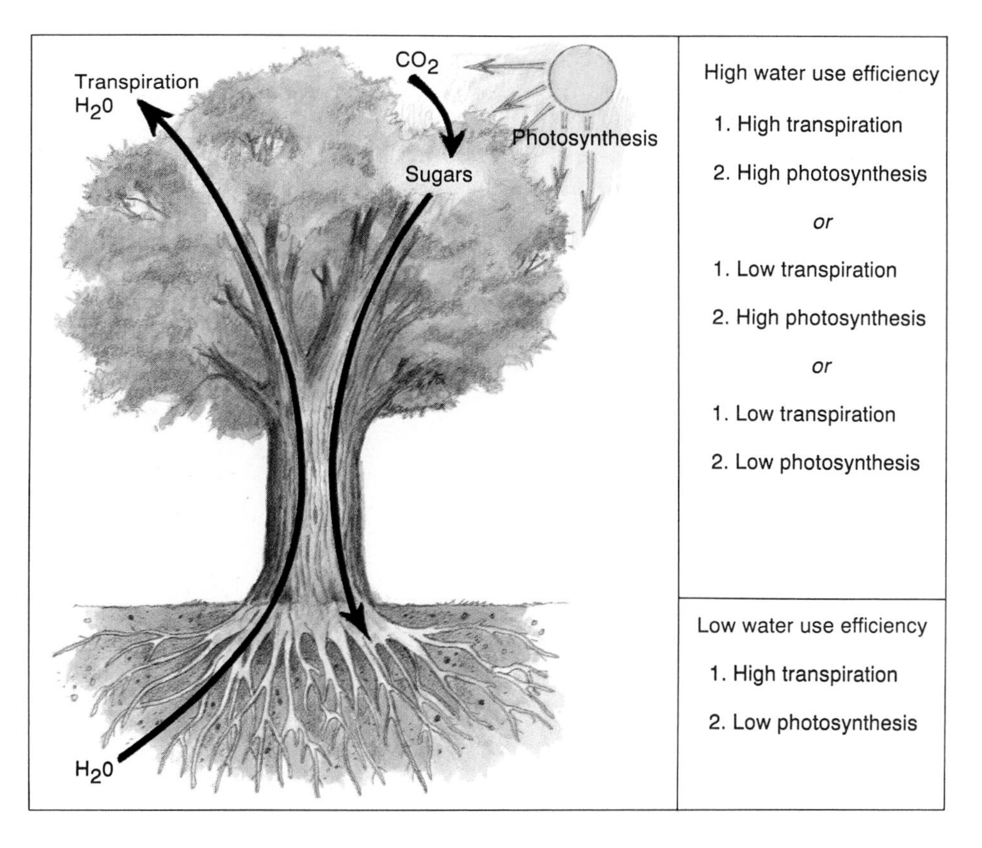

Figure 31.6

Water use efficiency—the ratio of CO_2 gain to water loss—can change with environmental conditions. High water use efficiency occurs in several circumstances. Water use efficiency is low when plants have high transpiration and low photosynthesis.

and plants are said to have a high **water use efficiency**. Some desert plants must operate with a high water use efficiency if they are to survive. Conditions can also exist in which plants have low water use efficiency. If water is abundant, a plant's stomata may remain open during the night, and transpiration rates will be high but no photosynthesis will occur. Water use efficiency is low because carbon gain is low and water loss is high. Plants growing on irrigated fields tend to slip into a regime of low water use efficiency.

Many botanists believe that plants respond to changes in their environment by opening and closing their stomata, a mechanism that optimizes their water use efficiency. Accordingly, as light levels, atmospheric humidity, soil water availability, and internal activities change, plants respond by either opening or closing their stomata. This process maximizes the ratio between carbon gain and water loss.

The carbon source for plants is atmospheric CO_2 that is converted to sugars during photosynthesis. CO_2 enters a plant through leaf stomata, structures that also have a regulatory role in the movement of gases into and out of the plant. Quantities of CO_2 that enter a plant are influenced by several factors, including CO_2 concentration gradients, light availability, and transpiration rates. Stomata play a central role in regulating carbon gain and water loss.

Assimilation and Allocation

As described in Figure 31.7, the assimilation of CO_2 by plants hinges on incorporating carbon atoms into sugars through the reactions of photosynthesis. As CO_2, water, and solar energy are converted to carbohydrates, leaves obtain the fuel needed for their growth and development. Sugars not used in cellular respiration in leaves are stored or translocated to other plant tissues.

Photosynthesis

Photosynthetic cells may produce more sugars than they can use. When conditions are optimal for photosynthesis, leaves can produce six to ten times more sugar than they use during cellular respiration. Scientists use estimates of **net photosynthesis**—the difference between the quantity of sugars produced by the leaves in photosynthesis and the quantity used in cellular respiration—as a measure of the amount available for other uses in the plant.

What is the fate of the excess sugars and other carbohydrates produced from photosynthesis? Essentially, cells that produce these substances can use

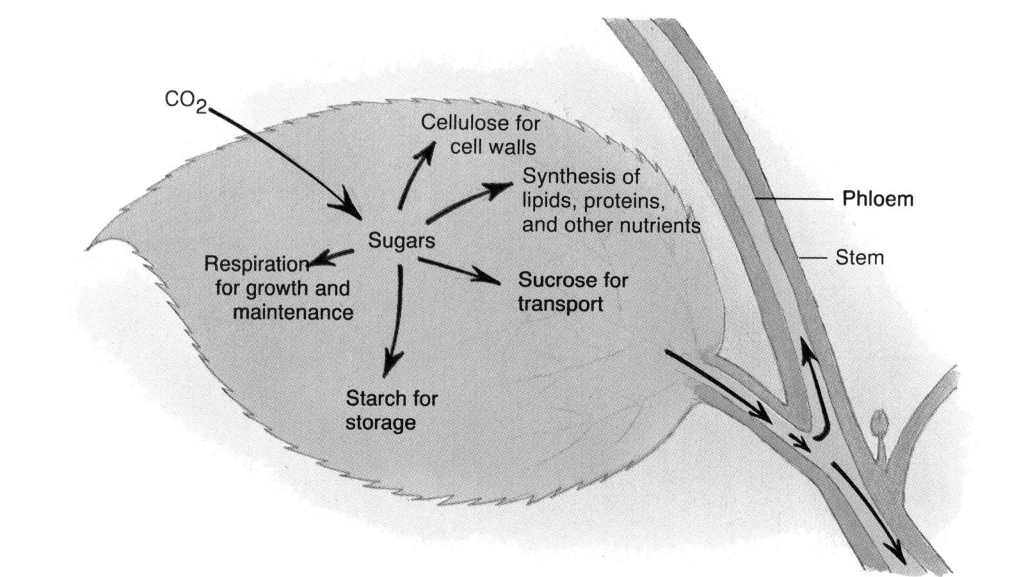

Figure 31.7

Carbon dioxide in the air is absorbed by the leaves and converted to sugars by the process of photosynthesis. Sugars can be used for respiration needed for maintenance and growth, stored as starch, used to synthesize other organic molecules, or transported from the source leaves to other plant tissues.

them immediately for respiration, store them for use at a later time, or transport them to other plant parts.

Cellular Respiration

Cellular respiration (described in Chapter 29) is the cellular metabolic process that releases the energy required for life-sustaining functions in the cells of all living organisms. Respiration has two aspects. **Maintenance respiration** involves the sugar consumption required for simply maintaining a living cell. Maintenance functions include repair of membranes and organelles and the production of new enzymes and other essential macromolecules. **Growth respiration** refers to sugar consumption that is needed for mitosis and cell enlargement. In addition, growth requires the construction of new membranes, cell walls, and organelles within cells. All sugars that are immediately available for respiration can be used either for maintenance or for growth.

Storage

Carbohydrates not used for growth or maintenance respiration can be either allocated to storage or transported for use elsewhere in the plant. Simple sugars produced by photosynthesis can be converted to more complex sugars, starch, and various forms of fats,—photosynthates—all of which can be stored. Photosynthates that are allocated to storage are generally used when the respiratory demands of cells are greater than the supply provided by photosynthesis. Over the short term, some of the starch formed during the day will be used for respiration at night, when photosynthesis cannot occur.

Storage of photosynthates is also important to plants for coping with long-term environmental fluctuations. For example, cloudy or foggy conditions may persist in coastal areas for days or even weeks. During that time, light levels may be too low for positive net photosynthesis, but the plant can draw on its stored reserves. Other environmental factors (for example, drought) can result in stomatal closure. This response may also suppress photosynthesis for prolonged periods and create conditions when plants must rely on stored carbohydrates. In both of these cases, plants can use stored photosynthates during periods when they cannot assimilate CO_2.

Translocation

Translocation occurs when photosynthates not immediately respired or stored are moved from one part of a plant to another (see Figure 31.8). Thus under ideal environmental conditions, a fully expanded leaf can produce excess sugars that can be exported to other tissues. Tissues that import and consume these sugars include immature leaves that are not yet capable of positive net photosynthesis and nonphotosynthetic tissues in flowers, fruits, seeds, branches, stems, and roots.

The complexities of sugar allocation regulation are beyond the scope of this book. Basically, sugars are allocated to various tissues in response to changes in the age of the plant, its developmental state, and environmental conditions. For instance, many plants produce only nonreproductive tissues such as roots, stems, and leaves for much of their lives. At some stage, however, a plant typically enters into a reproductive phase. At that point, most of its sugars will be allocated for use in the production of cones, flowers, pollen, fruit, or seeds.

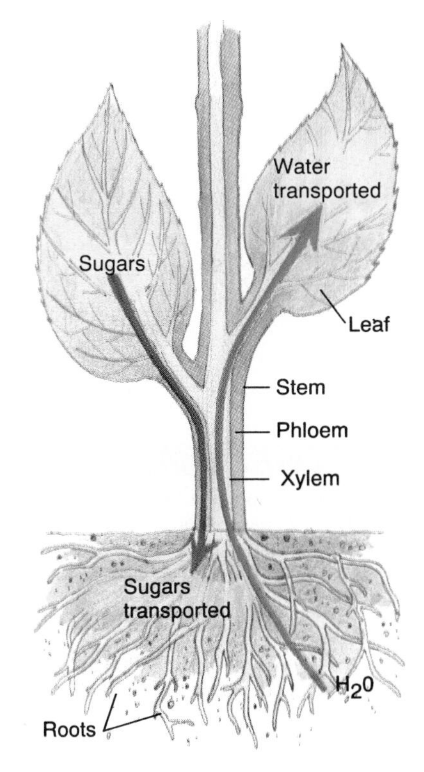

Figure 31.8

Sugars are transported in vascular phloem from leaves to other tissues that are nonphotosynthetic, such as roots, or are not capable of producing all of the sugars they need for growth and maintenance, such as young leaves. Water taken up by roots is transported to different parts of the plant through the xylem.

Plants use photosynthates in various ways. A certain amount of them is used at once in cellular respiration reactions in leaves to provide energy for maintenance and growth. Sugars not used directly in leaf cell respiration can be handled in different ways. They may be converted to complex molecules that serve to store the remaining energy for use in the future, or they may be allocated to other parts of the plant, where they are used in respiration.

THE NUTRIENT STORY

Photosynthesis and the chemical incorporation of CO_2 into various sugars are of obvious importance to the life of the plant. The mechanisms that underlie CO_2 fixation mirror the acquisition and assimilation of other plant resources.

Acquisition of Nitrogen and Sulfur

Plants acquire the nitrogen and sulfur they require for growth directly from the environment. These elements are usually absorbed by plant root cells when dissolved in the water between soil particles (see Figure 31.9). Just as carbon is absorbed by plants in the chemical form CO_2, nitrogen is generally absorbed by plants as nitrate (NO_3^-) or ammonium (NH_4^+), and sulfur is absorbed as sulfate (SO_4^{-2}). In these states, nitrogen and sulfur are of little use to the plant until they are incorporated into organic compounds such as amino acids or proteins.

Assimilation of Nitrogen and Sulfur

The patterns of nitrogen and sulfur processing reveal two common themes. First, plants acquire these elements from the environment in specific chemical forms. Second, these chemicals are modified during assimilation in leaves and other tissues, producing organic compounds of great importance to the plant. These compounds may become structural components, or they may be used in various cellular processes, including photosynthesis and respiration.

Following their absorption by roots, nutrients are transported through vascular xylem to tissues that convert them to organic molecules. The processes of assimilating nitrogen and sulfur (and carbon) are unique to plants and most clearly distinguish them from animals. Animals are ultimately dependent on plants for the energy and organic compounds they use in sustaining their lives.

ENVIRONMENTAL STRESSES

What kinds of environments are ideal for plants? Like any organism, each plant species has an optimal range of environmental conditions that provide for growth and reproduction. Given the resource requirements and mechanisms described earlier, what environmental factors would exist in a perfect plant

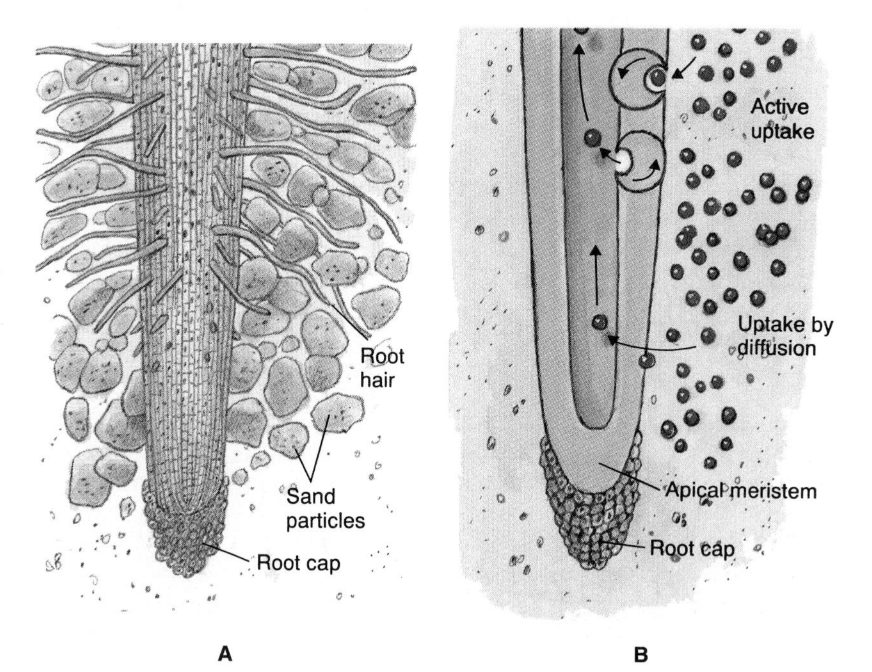

Figure 31.9

(A) Plant root tips penetrate into the soil. The root cap protects the apical meristem from being damaged by soil particles. Root hairs increase the surface area of roots and facilitate nutrient uptake. (B) Uptake of nutrients by roots occurs by diffusion and by active uptake, a process in which the cell uses energy to acquire a needed molecule.

world? Optimal conditions generally include an abundance of CO_2, water, and nutrients; high levels of light; high relative humidity; and temperatures that are neither too warm nor too cold. In such ideal conditions, plus a habitat free from insect pests, diseases, and competing plants, we might expect maximal plant growth.

Optimal conditions are closely approximated, but never achieved completely, in agricultural systems, gardens, greenhouses, and other places where environmental factors can be managed. Practices that tend to optimize environmental conditions include irrigation, fertilizer applications, planting after the last freeze, and the use of selective insecticides and herbicides to reduce predation and limit competition. Even using such expensive methods, however, may fail to provide conditions that are perfect for plant growth. Furthermore, as described in Chapter 30, human attempts to improve crop yields by supplementing water and nutrient resources and managing weeds, disease, and pests often result in more energy being invested than is recovered in increased plant yields.

In a perfect environment, plants would be readily able to obtain and assimilate CO_2 and nutrients such as nitrogen and sulfur. There would also be an abundance of water, high levels of light and relative humidity, ideal temperatures, and a scarcity of pests, disease agents, and competing plants. Although plants almost never grow in such optimal environments, they have evolved mechanisms that enable them to exist in less hospitable habitats.

The Concept of Environmental Stress

The idea that environmental conditions are never truly perfect for plant growth suggests that plants are continually subjected to stress, as depicted in Figure 31.10. It is helpful to characterize these stresses so that positive and negative environmental factors that affect plants can be identified and evaluated.

Figure 31.10

Environmental stresses always limit plant growth. Stresses in deserts (A) often include drought and excessive heat. Growth in citrus orchards (B) may be affected by freezing temperatures. Stresses even exist in agricultural fields (C), where temperature, light, water, and nutrients are never at optimal levels.

Natural Environmental Stress

An **environmental stress** is any component of the physical environment that reduces the capacity of plants to acquire, assimilate, or allocate resources needed for maintenance or growth. Relative to resource use, an example of a natural environmental stress is drought. An inadequate supply of water can limit plant maintenance, growth, and development because it results in a reduced capacity for cells to grow and can cause stomata to close, thereby decreasing photosynthesis.

Nutrient deficiencies can also be an important environmental stress. For example, inadequate levels of nitrogen in soils may lead to reduced levels of amino acids, structural proteins, and enzymes needed for all essential cellular processes. Researchers have found that photosynthetic capacity is often correlated with the nitrogen content of leaves. This finding indicates that when nitrogen levels in leaves are high, photosynthesis is not limited by a lack of enzymes or other molecules constructed from amino acids.

Light becomes a natural environmental stress factor if intensities are too low for sustaining high rates of photosynthesis. Light levels may be reduced by clouds; inanimate objects such as buildings, rocks, and mountains; or neighboring plants. Shading can also result from normal plant growth; as the plant canopy expands, self-shading occurs, and photosynthesis by leaves inside the canopy becomes limited due to insufficient light.

Light intensity levels are often correlated with temperature extremes, which also become a natural stressor to plants. For example, during periods of intense light in deserts, associated high temperatures cause reductions in rates of photosynthesis and increases in rates of respiration. Thus when plants are exposed to high temperatures, their rate of sugar production decreases while their rate of sugar consumption increases. Low temperatures can damage plants directly. If water freezes, ice crystals formed within cells may puncture membranes and cause cell death. Low temperatures may also slow down critical biological processes such as water absorption by roots or respiration throughout the plant.

Anthropogenic Environmental Stress

How have human activities affected environments in which plants live? The industries, products, and transportation devices used by human cultures have changed the chemistry of the environment, and such changes can limit the capacity of plants to survive and grow (see Figure 31.11). Human-caused environmental stresses are called *anthropogenic.* Toxic, gaseous air pollutants such as ozone (O_3), sulfur dioxide (SO_2), and nitrous oxides (NO_x) can be absorbed through stomata and moved into leaf cells, where they cause damage or death. In addition, atmospheric acid depositions, both in dry form (dust and aerosol particles) and wet form (precipitation such as rain and snow), constitute environmental stresses. These forms of air pollution can acidify soils in sensitive areas, create an unfavorable balance of nutrients, or add to the existing levels of sulfur and nitrogen already present in soils. Thus acid deposition has the potential to increase or decrease the availability of nutrients in soils and their absorption. Currently, scientists are debating the extent to which acid deposition may be affecting forest productivity in the industrialized countries.

Human activities have also resulted in increased emissions of "greenhouse" gases, including CO_2 and methane, which can have direct effects on plants. Increasing atmospheric CO_2 concentrations may cause metabolic alterations that could lead to accelerated growth in some species of trees, wild plants, and crops. Even if such an effect is beneficial, CO_2-caused increases in growth could ultimately be damaging if plants outgrow their water supply.

Multiple Stresses

Environmental stresses are interrelated and rarely occur alone. For example, light levels may become too high and lead to overheating, drought, and decreased

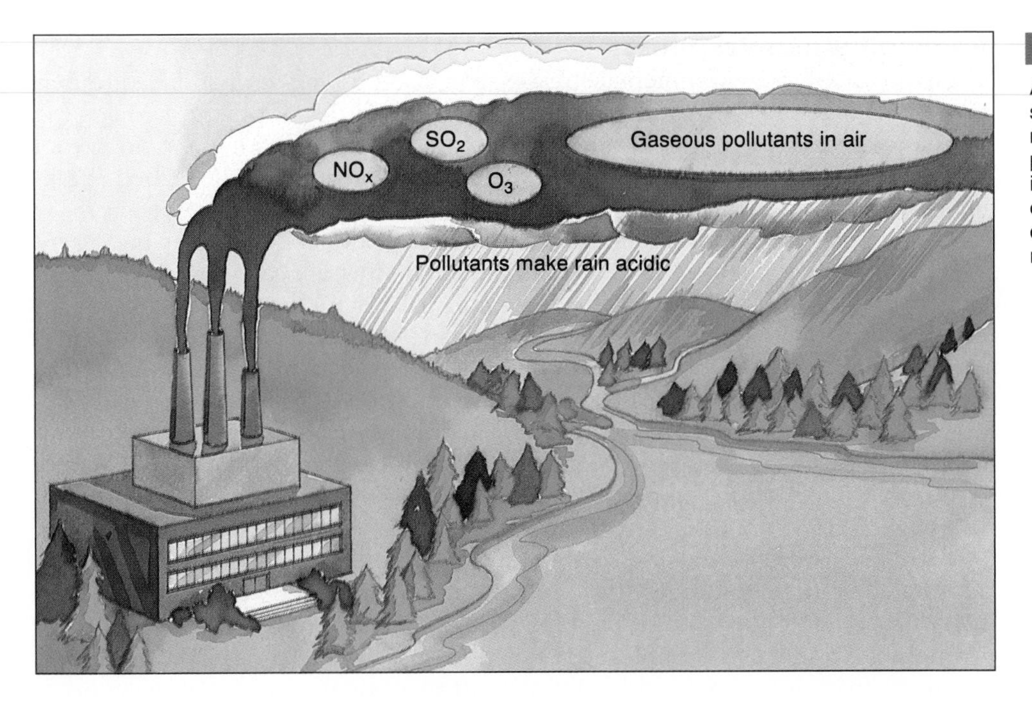

Figure 31.11

Air pollutants emitted from industrial sources can modify the chemical climate and thereby cause stress on plants. Gaseous pollutants can directly inhibit physiological processes that occur in leaves, or the atmospheric deposition of chemicals can alter the nutrient supply of plants.

photosynthesis and growth. Like single stress factors, interrelated multiple stresses can result in a decreased capacity for plants to acquire, assimilate, and allocate resources needed for growth.

Scientists are only now beginning to determine the extent to which humans have modified Earth's environment. Researchers from many disciplines including chemistry, physics, atmospheric science, oceanography, and biology are turning their attention to answering critical questions about anthropogenic environmental effects.

Plants are often subjected to environmental stresses that compromise their abilities to grow, reproduce, and survive. Nitrogen deficiencies are associated with reductions in photosynthesis and growth. Extremes in light levels and abnormal temperatures also constitute environmental stresses that lead to decreased photosynthesis and other harm to the plant. Anthropogenic stresses may now be limiting plant growth and survival on a global scale.

Stress Compensation

Given the range of stresses to which plants are subjected, how do they survive? Plants can respond to short-term environmental stresses quickly through adjustments at the cellular level. Such responses form the foundation of a dynamic relationship between plants and their environment. Plants that compensate in response to stress have both increased survival and increased growth, compared with the survival and growth of plants that do not have such compensation mechanisms.

Stomatal Responses

Stomatal responses are among the most rapid short-term reactions to an environmental stress. When soil water is limiting, stomata close and transpirational water loss is reduced in affected plants. Stomata may also close if transpiration increases in response to decreased relative humidity. Thus plants can compensate for either insufficient water supply rates or excessive transpirational water loss rates by simply closing their stomata. Stomatal responses are considered to be *compensatory* because even if growth is inhibited by environmentally caused changes in plant water status, stomatal closure will help the plant survive.

Root and Shoot Responses

Compensation for environmental stresses may also be long-term and lead to permanent changes in the patterns of plant growth and development. Perhaps the best-studied example of this sort of response involves changes in the ratio of root dry weight to shoot dry weight, or the **root-shoot ratio** that is described in Figure 31.12.

The amount of carbon allocated to roots and shoots determines the extent of growth and the biomass of these tissues. Plants that allocate carbon preferentially to shoots will use photosynthates to construct more shoots, which can gain more carbon. Such a strategy results in the plant's having a high growth rate and a low root-shoot ratio. In contrast, plants that allocate photosynthates preferentially to nonphotosynthetic tissues such as stems or roots have invested carbohydrates into tissues that have important roles but cannot be used to gain more carbon. The root-shoot ratio in such plants is high, and growth rates will be low. Why don't plants always employ the shoot-favoring strategy, since it is associated with higher growth rates?

If plants were to function solely to maximize their growth rates, they would allocate all of their resources to shoots and nothing to roots. However, the success or survival of plants depends on more than a rapid growth rate. Besides providing attachment, structural support, and the capacity to absorb water and resources from soils, roots and belowground plant tissues carry out many other vital plant functions. For example, some plants store carbohydrate reserves in belowground stems called *tubers*. Potatoes are tubers that store carbohydrates until they are needed. Sweet potatoes, beets, carrots, and radishes are plants with roots that perform a similar function.

Plant Responses to Anthropogenic Environmental Stresses

A useful way to categorize plant responses to anthropogenic stresses is to distinguish between aboveground stresses and belowground stresses. One group of aboveground stresses that acts to limit carbon gain can be broadly classified as components of the atmospheric environment (see Figure 31.13). Such stress factors include gaseous air pollutants. Plants faced with these stresses will generally have less carbon for allocation to nonphotosynthetic tissues. In this situation, plants can compensate for such stresses by shifting their carbon allocation patterns to favor shoots at the expense of roots, thereby decreasing their root-shoot

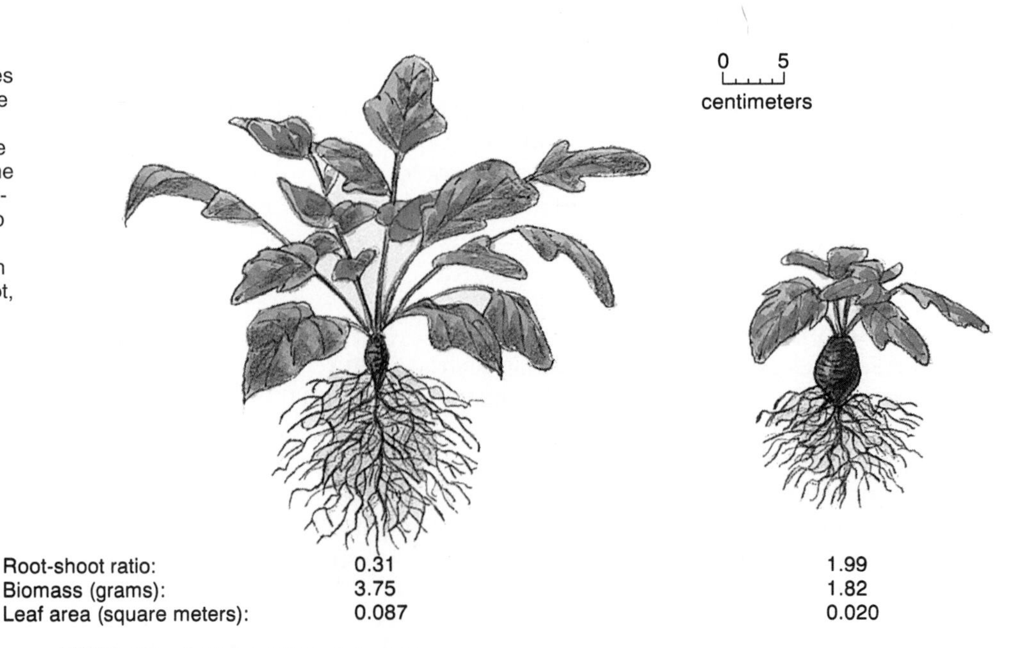

Figure 31.12

Plants with similar photosynthetic rates can have different growth rates. These two radishes are both five weeks old but show different types of growth due to different patterns of carbon use. The wild type of radish (left) used carbohydrates from photosynthesis to develop more foliage, which leads to more growth. The commercial, edible radish (right) stored carbohydrates in the root, which resulted in less growth.

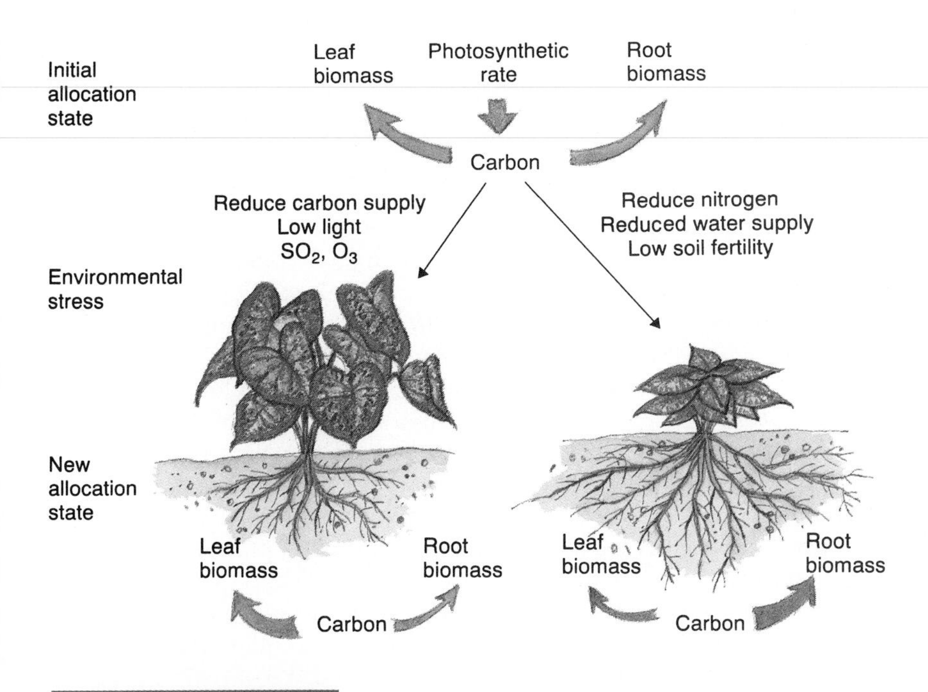

Figure 31.13

Environmental stresses can reduce plant growth and influence carbon allocation to roots and shoots. Atmospheric stresses tend to shift allocations in favor of the shoot, whereas stresses in soils tend to cause shifts in favor of the roots.

ratio. Basically, plants compensate for atmospheric stresses that limit carbon acquisition rates by increasing the proportion of their biomass allocated to acquire carbon.

Belowground anthropogenic environmental stresses, such as the presence of toxic substances, can also limit carbon gains by plants. In these cases, plants shift carbon allocation patterns by favoring belowground tissues at the expense of shoots. The shift in carbon allocation pattern in response to belowground stresses results in an increase in the root-shoot ratio, an opposite response to that elicited by atmospheric stresses.

These stress mechanisms are also used in responding to natural environmental stresses. However, it is largely because of an increasing concern about the effects of pollution on plants that emphasis is now being placed on responses to anthropogenic stresses (see the Focus on Scientific Inquiry, "Plant Physiological Ecology").

Plants are capable of both short-term and long-term responses to environmental stress. Stomatal reactions act to reduce water loss in plants when water becomes limiting. Carbon allocations to roots and shoots change in response to environmental stress. The net effect of this mechanism is to optimize the chances for long-term survival by increasing growth of either the shoot or the root, depending on the nature of the stress.

A NEW WAY OF LOOKING AT THE PLANT WORLD

As plants grow and develop over the course of a single season or over many years, the pattern of carbon allocation changes continuously. Shifts in carbon allocations can be understood from the perspective of a plant's compensation for stress, which accomplishes several purposes. Compensation results in enhanced survival and growth, at least superior to what would have occurred in the absence of compensation. Another facet of stress compensation seems to be for plants to shift structure and function so as to optimize acquisition of the most limiting resources.

Focus on Scientific Inquiry

Plant Physiological Ecology

The relationships between plants and their environment is of concern to scientists in the field of **plant physiological ecology**. **Plant physiology** is the study of processes, such as photosynthesis and cellular respiration, taking place within a single cell or an individual plant. **Plant ecology** is the study of the structure and functions of plant populations. Studies in plant physiological ecology are becoming increasingly important as more people realize that the global climate is changing. Consequently, plant scientists are challenged to describe, evaluate, and ultimately predict the effects global warming will have on individual plants, plant communities, and ecosystems. To achieve this ambitious goal, extremely large-scale experiments will be conducted.

What types of studies are carried out by plant physiological ecologists? Generally, they attempt to define physiological mechanisms and describe how they are affected by different environmental stresses. Such research is important in defining a plant's range of environmental tolerances, in understanding its ecological role in a community, and in assessing the possible consequences of environmental change. The experiments described here provide some indication of the progress in this field.

Simple plant transplant experiments have provided useful information for identifying variations in stress response within a single species. These differences can be seen through studies of a single species that inhabits a wide range of elevations. In the early 1950s, scientists at Stanford University in California studied a small plant species (*Potentilla glandulosa*) that grew along an elevation gradient ranging from 100 meters in western California to greater than 3,000 meters in the Sierra Nevada. First, they collected seeds from plants growing at three different altitudes, designated "low elevation," "moderate elevation," and "high elevation." They then grew these seeds in a common garden at a moderate elevation and compared the growth of seedlings and adult plants. The hypothesis being tested was that seeds of the same species would grow uniformly at the moderate elevation, regardless of their origin. Surprisingly, seeds collected from the high elevation produced small plants, even when grown at the moderate elevation. As shown in Figure 1, seeds from the low-elevation site also gave rise to plants smaller than those from the moderate elevation. Therefore, this experiment demonstrated that significant genetic differences existed between populations of a single species that were growing in habitats characterized by different stresses. This was one of the first scientific studies that showed in a rigorous way that plants possess an inherited ability to respond to environmental stress.

Studies conducted in the late 1950s began to show how environmental factors affected a plant's capacity to carry out photosynthesis and transpiration. Researchers wanted to know if (and when) plants growing above the timberline, at a site in the Medicine Bow Mountains of Wyoming, altered their rates of photosynthesis during the day and during the growing season. A unique feature of these studies was that experiments were done at a remote field site,

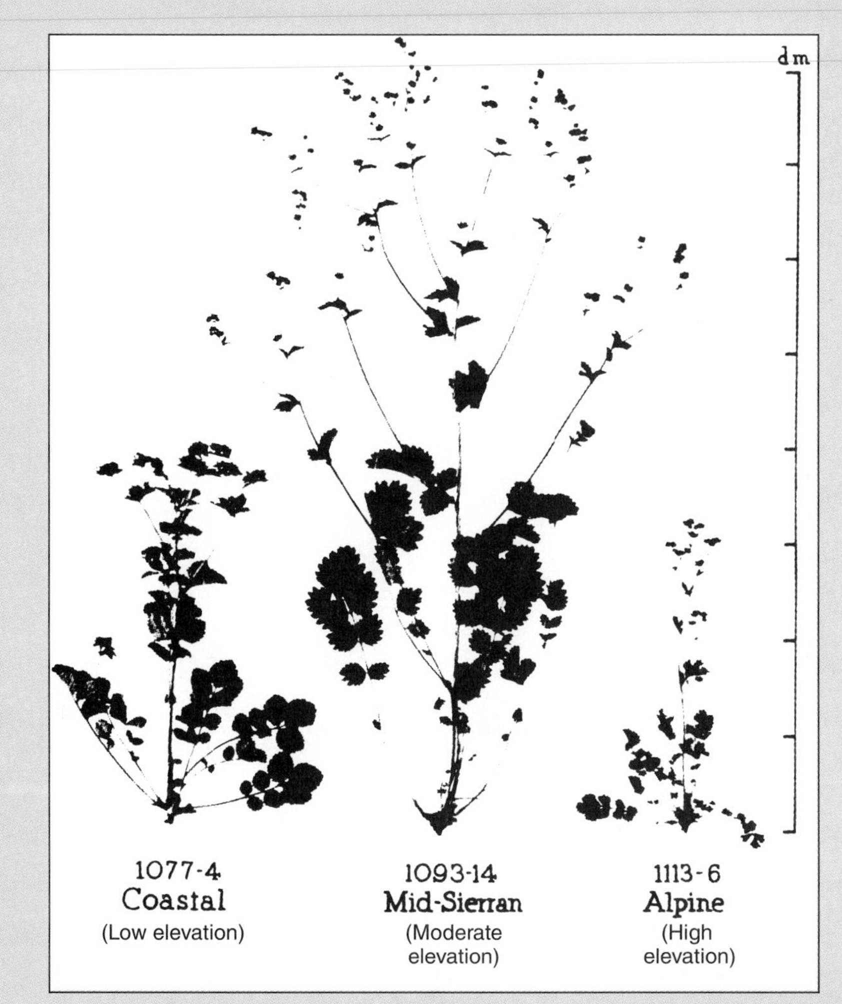

Figure 1 These *Potentilla granulosa* plants were raised from seeds collected from different habitats. These plants were raised in a single garden, and the results indicate that this species has distinct variations that represent unique adaptations to specific habitats.

using plants growing in their native environment. To make the necessary measurements, a small chamber was built that could be opened and closed around a single leaf (see Figure 2). The chamber had a hole for the *petiole,* the part of the leaf that connects it to the plant, so that the leaf could remain intact and attached to the plant. The chamber was also designed to allow airflow through inlet and outlet holes. Measurements of photosynthesis were determined by measuring the differences in CO_2 concentrations between the airstreams entering and leaving the chamber. Transpiration was measured by comparing water vapor concentrations at the chamber inlet and outlet. The investigators found that leaves became photosynthetically active very early in the spring and that rates of photosynthesis increased during the day as the leaf responded to warmer temperatures and more light. These and similar experiments conducted by plant physiological ecologists established normal patterns of plants' activities and opened the door to new research that aimed to define further the effects of environmental stresses on plant physiology and growth.

Determining the extent of anthropogenic climate change and its effects on vegetation will be difficult. One of the obstacles to unraveling plant responses to human-induced environmental changes is that many natural factors change simultaneously and affect plants in the process. For example, as summer progresses toward fall, plants may shed their leaves in response to cooler temperatures, shorter days, or other factors associated with seasonal change. How do plant physiological ecologists identify which environmental factors—natural or anthropogenic—are responsible for observed or measured changes in plant physiology and growth?

Modern, highly sophisticated experimental systems, not only

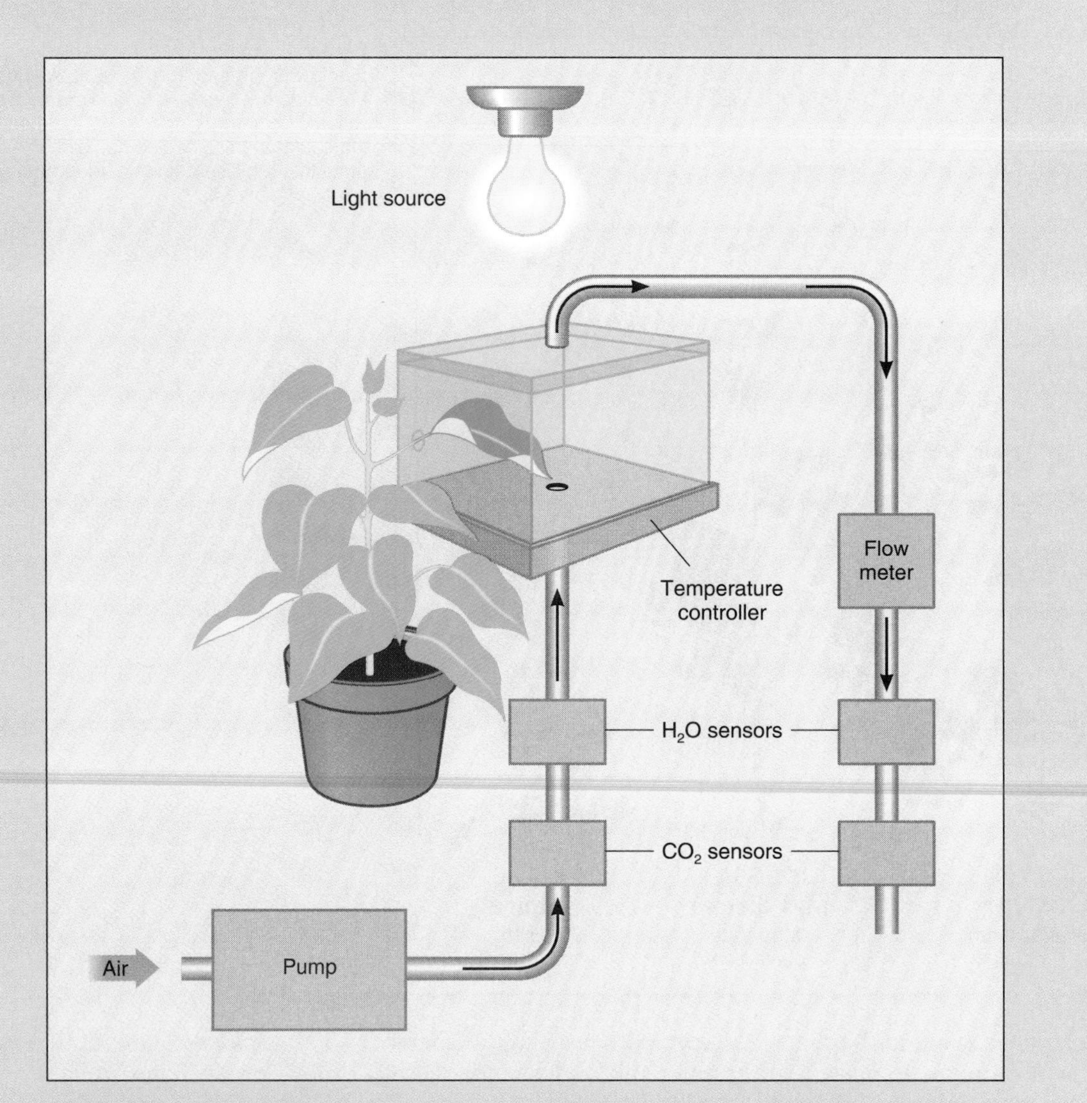

Figure 2 A chamber designed for measuring photosynthesis, respiration, and stomatal responses in single leaves has a supply of moving air and a lid that is opened to insert the leaf while it is still attached to the plant. Photosynthesis reduces the CO_2 concentration in the airstream. Transpiration increases the relative humidity of the air as it passes through the tube.

monitor important environmental factors but also allow them to be controlled. Leaf chambers have been developed that can control temperature, CO_2, and water vapor concentrations in the air flowing into the system. Special lights have also been developed, along with shading techniques, in order to control light levels. Thus temperature around the leaf can now be held constant while light levels change, or temperature can be held constant while water vapor concentration in the air is either increased or decreased. These systems can operate in the field, a greenhouse, or a laboratory and allow scientists to monitor leaf responses to systematic and controlled environmental changes. Much of the monitoring and regulation of environmental factors is done by computer systems. These new systems will be important tools for investigating the effects of changing climate on plants.

Comprehending the dynamic nature of plant-environment relations provides important insights into why plants are so successful. This helps us recognize the changes in function and growth that occur in the plants we observe each day. From this perspective, the responses of plants to natural environmental changes, such as a sudden storm, a sunset, or a period of drought, take on new meaning or pose new questions. Why do certain flowers, such as tulips, close when the sun sets and open when it rises? If you live in Buffalo, New York, why do the grasses in lawns turn brown during the winter and once again become green when spring arrives? If you live in Seattle, Washington, why is grass green all winter? The consequences of anthropogenic pollution can also be analyzed in such a way that human activities may be managed more effectively.

Analyses of plant-environment relations also reveal that not all plants respond to environmental change in the same ways. The ideas that plants have many adaptations for responding to environmental change and that these adaptations are perpetuated by generations that follow are considered in Chapter 32.

Summary

1. In order to grow and survive at the site where they take root, plants must be able to obtain resources from the air and the soil that surround them. Plants use a variety of processes to accomplish this task. Many of these same mechanisms are also used in adjusting to changing conditions and in coping with environmental stresses.
2. Carbon dioxide is a critical resource that is used by plants in producing sugars during photosynthesis. Plants acquire CO_2 from the atmosphere, and rates of intake are regulated primarily by the leaf stomata. Global climatic changes and increasing concentrations of CO_2 in the atmosphere may ultimately affect the ability of certain plant species to maintain and regulate rates of carbon fixation.
3. Water and nutrients obtained from the soil are transported continuously throughout the plant to tissues where they are required. Water reaching the leaf surface diffuses into the atmosphere, by transpiration, and is lost to the plant. The rate of transpiration and water loss is regulated by the stomata and is influenced by several environmental factors, including light, water vapor in the air, temperature, and CO_2 concentrations.
4. Under appropriate conditions, leaf cells produce excess sugars that can be transported to other plant tissues where they are metabolized or stored for future use. Stored carbohydrates represent a reserve that can be used by the plant if environmental conditions change or if the plant becomes stressed in the presence or absence of certain environmental factors.
5. Water and most nutrients required by plants, such as nitrogen and sulfur, are obtained, by the roots, from the soil. Once absorbed, they are assimilated and translocated to tissues where they are used for cell maintenance and growth.
6. Plants rarely inhabit an ideal environment. The degree to which they are able to compensate for environmental stresses determines whether or not they will be able to survive, grow, and reproduce. Inadequate amounts of water, insufficient levels of nutrients, extreme temperatures, inappropriate light intensities, and pollutants introduced by human activities are environmental stresses that plants must deal with.
7. Plants can make short-term adjustments to environmental stress. For example, if water becomes limiting, plants can decrease their rate of photosynthesis, or the stomata can respond to reduce water loss.
8. If subjected to continuing stress, plants can also make long-term adjustments. One long-term compensation mechanism involves regulating the flow of carbon and nutrients to different plant tissues, depending on the nature of the stress. If exposed to aboveground stresses, such as pollution, plants tend to invest resources in the development of stems and leaves at the expense of roots and other nonphotosynthetic tissues. Belowground stresses result in the opposite compensation pattern.

Review Questions

1. What processes are related to plant growth?
2. How do atmospheric carbon levels fluctuate during the year? Why do these variations occur?
3. What mechanisms regulate the movement of CO_2 into and out of a plant's leaves?
4. What is transpiration? How is it regulated in plants?
5. Under what conditions do plants have a high water use efficiency?
6. What plant mechanisms operate to maximize water use efficiency?
7. What are the possible fates of photosynthates produced by a plant?
8. How do plants acquire and assimilate nitrogen and sulfur?
9. What are ideal plant conditions for plant growth and survival?
10. What is environmental stress? Under what circumstances may nutrient availability, light, temperature, and human activities cause environmental stress on plants?
11. What mechanisms do plants use to compensate for environmental stress?

Essay and Discussion Questions

1. In what ways do human activities affect resource availability for plants? How do your daily activities affect Earth's physical and chemical climate?
2. Tropical plants can be made to grow in a desert. How might this be done?
3. The effects of an environmental stress on plants, such as drought, are a function of an array of factors. Which plant structures and mechanisms help define their capacity to tolerate a particular stress?

References and Recommended Reading

Bouwman, A. F., ed. 1990. *Soils and the Greenhouse Effect*. New York: Wiley.

Chabot, B. F., and H. A. Mooney. 1985. *Physiological Ecology of North American Plant Communities*. New York: Chapman & Hall.

Freedman, B. 1989. *Environmental Ecology: The Impacts of Pollution and Other Stresses on Ecosystem Structure and Function*. Orlando, Fla.: Academic Press.

Hale, M. G., and D. M. Orcutt. 1987. *The Physiology of Plants Under Stress*. New York: Wiley.

Jones, H. G., T. J. Flowers, and M. B. Jones. 1989. *Plants Under Stress*. New York: Cambridge University Press.

Katterman, F. 1990. *Environmental Injury to Plants*. Orlando, Fla.: Academic Press.

Kozlowski, T. T., P. J. Kramer, and S. G. Pallardy. 1991. *The Physiological Ecology of Woody Plants*. Orlando, Fla.: Academic Press.

Kuppers, M. 1989. Ecological significance of above-ground architectural patterns in woody plants: A question of cost-benefit relationships. *Trends in Ecology and Evolution* 4:375–379.

Mooney, H. A., B. G. Drake, R. J. Luxmoore, W. C. Dechel, and L. F. Pitelka 1991. Predicting ecosystem responses to elevated CO_2 concentrations. *BioScience* 41:96–103.

Strain, B. R. 1987. Direct effects of increasing atmospheric CO_2 on plants and ecosystems. *Trends in Ecology and Evolution* 2:18–21.

Treshow, M., and F. K. Anderson. 1990. *Plant Stress from Air Pollution*. New York: Wiley.

CHAPTER 32

Plant Adaptations and Reproduction

Chapter Outline

Adaptations

Reproduction

Reading Questions

1. What is the relationship between plant adaptations and long-term natural environmental changes?

2. What types of adaptations allow deciduous trees and conifers to survive in the environments they inhabit?

3. How do gymnosperms and angiosperms produce seeds by sexual reproduction?

Plants can respond to environmental factors in many ways and, in so doing, can increase their capacity to grow, survive, and reproduce. As described in Chapter 31, over the course of days or weeks, plants adjust their physiological and growth processes as they compensate for environmental changes or stresses (see Figure 32.1). In this chapter, we concentrate on a broader time scale. Over many generations, plants evolve adaptive characteristics as a result of natural selection. Such adaptations account for the basic shapes, structures, and physiological capacities that allow plants to survive normal environmental changes over extended time periods.

ADAPTATIONS

What is the nature of plant adaptations? Figure 32.2 shows a few examples of plant adaptations to natural environmental stresses. The most familiar aspects of a plant's appearance—the size, color, and shapes of leaves, the plant's general shape, branch pattern, and reproductive structures—reflect adaptations related to survival in the environment it inhabits. Many of these structural features, along with some of their physiological capabilities, arose in response to environmental stresses in the plant's evolutionary past. Some of these stresses persist today. For example, local and regional climate changes have occurred slowly over centuries. How do plants respond to such changes? Basically, a plant species may evolve new adaptations for coping with the new climate, or it may stop growing at sites where conditions are no longer favorable. If this happens, the plant species must be dispersed to more favorable sites, or it may disappear from a geographical region. In extreme cases, it may even become extinct. For example, conifer species became widely established throughout the United States after the last ice age ended. Now, after centuries of a general warming trend, the distribution of many conifer species is generally confined to northern regions of the country.

Two basic types of trees live in temperate zones of the world, and we focus on them to illustrate the nature of plant adaptations more closely (see Figure 32.3). **Deciduous trees** living in the Northern Hemisphere shed their leaves during the fall and grow new ones the following spring. Common deciduous species inhabiting temperate forests include oaks, elms, beeches, ashes, and many species found in fruit orchards (for example, apple trees). In contrast, most **evergreen trees** retain their leaves year round, although the leaves are regularly being shed and replaced. Pines, spruces, firs, cedars, and junipers are common evergreen species.

Many deciduous and evergreen tree species are economically important. They provide timber products, such as lumber and paper; fruits; and aesthetic beauty, an asset that is difficult to price. In addition, stands of trees are ecologically important. They play central roles in water and nutrient cycling, and they influence the chemical composition of our air. The evolutionary success of deciduous and evergreen tree species indicates that they are well adapted to many environmental stresses.

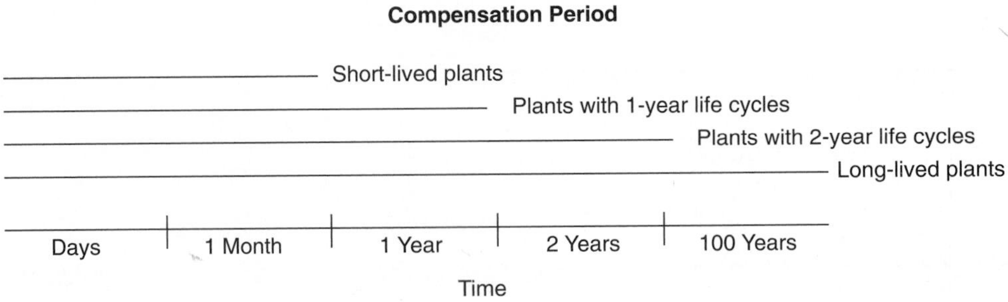

Figure 32.1

Plants use different mechanisms to respond to, and compensate for, natural environmental stresses. Short-term stress compensations take place during the life of an individual plant. Plants may also have adaptations that evolved over many successive generations in response to continuous environmental stresses.

Figure 32.2

Tropical vines (A), desert cacti (B), and flowering plants (C) represent some of the diversity found in the plant world. The diverse shapes and forms reflect adaptations that have enabled these plants to cope with environmental pressures.

Successful plants are well adapted to survive in the environments they inhabit. Adaptations are reflected in plant structures and functions. Many adaptations are related to normal environmental variations such as seasonal change. Examining the adaptations of deciduous trees and evergreen trees provides a basis for understanding normal, long-term relationships between plants and their environment.

Deciduous Trees

Deciduous trees are widespread throughout the world and can grow up to 50 meters tall. Although scattered throughout North America, forests dominated by oak, hickory, maple, beech, and elm are most common on the eastern part of the continent.

A

B

Figure 32.3

Deciduous trees, such as the maple (A), are usually broad-leaved and shed their leaves at the end of each growing season. Many familiar evergreen trees like the eastern white pine (B) have needles that are not shed annually.

Figure 32.4

Deciduous leaves are commonly arranged on branches in either an opposite (A) or alternate (B) pattern of orientation.

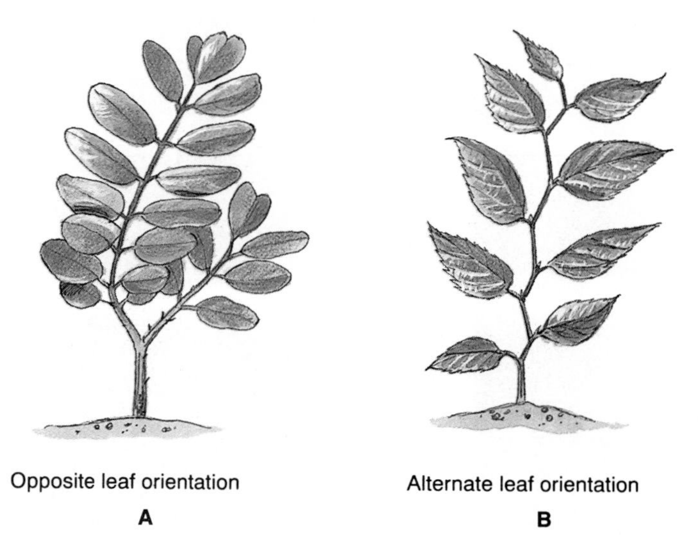

Leaves

With few exceptions, deciduous trees have *laminar* leaves, which are thin, flat, and wide. What is the adaptive significance of thin, flat leaves? The laminar shape favors the efficient absorption of solar energy. Specifically, photosynthetic components, including light-absorbing pigments and required biochemical compounds, are arranged in thin sheets and can intercept more light than those in leaves with other shapes. Thus the profile and structure of these leaves reflect adaptations that enhance photosynthesis.

Each deciduous tree species has a specific pattern of leaf arrangement, as illustrated in Figure 32.4. Most commonly, neighboring leaves are paired and positioned on opposite sides of the branch, or they are staggered on alternate sides of the branch. Leaf arrangements represent an adaptation related to minimizing self-shading. As branches grow, **buds** are produced that will eventually develop into new branches with their own leaves. As shown in Figure 32.5, older leaves that were once at the outside edge of the canopy, where they received abundant sunlight, become shaded by leaves that grow from new branches above them. As the growing season progresses, self-shading becomes more pronounced, although the pattern of leaf spacing on a branch tends to reduce this effect.

Branches

The location and arrangement of branches relative to other branches is an adaptation that also reduces self-shading. As Figure 32.6 indicates, this is particularly striking in *open-grown trees,* those that grow at a distance from neighboring trees that might compete for light, which usually have extensive branching and expansive canopies. Such trees can sustain high rates of photosynthesis because of the great amount of leaf surface area available for capturing radiant energy and absorbing CO_2.

Trees that grow in dense forests usually lose their lower branches in response to low levels of light. In this process, called *self-pruning,* once leaves on a branch are no longer able to produce sufficient photosynthates to support the branch or be transported to other tissues, resources are no longer allocated to that branch. Its leaves die, and the branch eventually dries up and falls from the tree.

Nitrogen Allocation

Nitrogen is commonly used to build structural proteins and enzymes that function in photosynthesis and other forms of metabolism. Leaves and branches are well adapted to maximize the efficiency of nitrogen use in a tree. When would

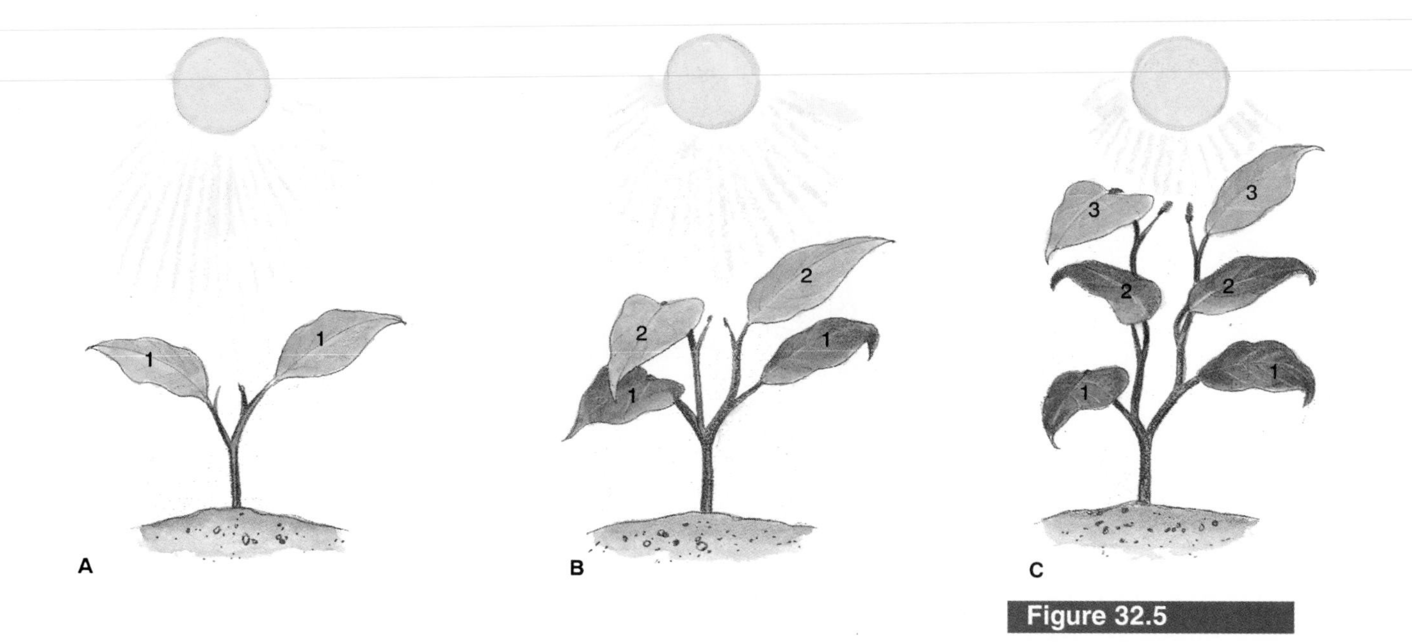

Figure 32.5

Self-shading occurs when leaves formed during the first month (1) of the growing season (A) are overgrown by leaves formed during the second month (2) of the growing season (B). During successive months (3), leaves that were formed at the outside edge of the canopy, where light was high, will become part of the inside canopy, where light intensity is lower (C).

this adaptation become important? An interesting case involves self-shading. The progression of self-shading during canopy growth leads to a situation in which leaves can no longer maintain high rates of photosynthesis because light levels are too low. As light levels decline, rates of enzyme synthesis decrease, and unused nitrogen accumulates. The excess nitrogen in these older, shaded leaves is then converted to a chemical form that can be translocated from an older "source" leaf, through the branch, to different plant structures such as younger, active leaves that are fully exposed to the sun. Once recycled to these new leaves, the nitrogen is again used to construct enzymes required for high levels of photosynthesis.

Loss of Leaves

What is the adaptive significance of deciduous trees shedding their leaves during the fall? Leaf loss during autumn represents an adaptation to natural environmental stresses associated with drought, short days, low temperatures, and other aspects of the progressing season. Loss of leaves begins with the development of a specialized layer of cells at the site where the leaf petiole joins the branch. Growth of this **abscission cell layer**, which is shown in Figure 32.7, first restricts and then severs the vascular tissues connecting leaf and branch and

Figure 32.6

Leaf distribution in the canopy of deciduous trees, such as these blue oaks, reflects a branching pattern that becomes apparent when the trees shed their foliage.

Figure 32.7

Deciduous leaves are shed after a layer of cells, the abscission layer, forms at the point where the petiole joins the branch.

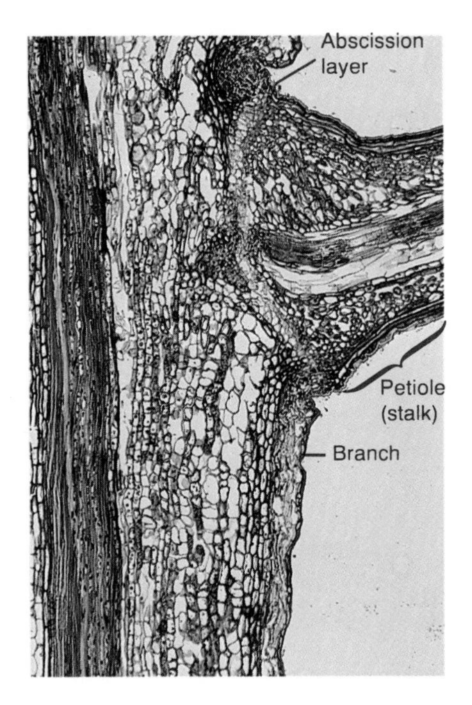

eventually causes the leaf to fall from the branch. The abscission layer often forms within a month in all leaves on a tree. During this period, nitrogen and photosynthates in leaves are converted to soluble forms that are translocated back into the shoots and roots. Thus these important resources are not lost to the tree when its leaves fall. Instead, they are stored in the tree and can be reused the following spring, when the next growing season begins.

When leaves are finally shed from a tree's canopy, they contain very little nitrogen and sugar. At this point, leaves consist mostly of structural carbohydrates, such as *cellulose,* that make up cell walls (see Figure 32.8). Leaf shedding from deciduous trees in temperate regions is considered adaptive because nutri-

Figure 32.8

Mature leaves have higher concentrations of nutrients at the peak of the growing season compared to leaves that have been depleted of nutrients prior to being shed from the tree.

Nutrient content	Live mature leaf	Dead fallen leaf
Nitrogen	3%	0.1%
Sugar	4%	0.1%
Potassium	2%	0%
Magnesium	0.5%	0%

ents and carbohydrates are not lost from foliage that might be destroyed by freezing or physical damage from winter ice and snowstorms. In other words, the tree's "currency" (nutrients and energy-rich compounds) cannot be stolen by Jack Frost during the winter because it is safe inside the vault (tree).

Deciduous trees have various adaptations that enable them to survive throughout the world. Their leaf structure promotes maximum absorption of radiant energy and high rates of photosynthesis. Leaves and branches of deciduous trees are arranged in such a way that self-shading is minimized. If leaves or branches become nonproductive, they are shed by the tree. Nitrogen allocation mechanisms ensure the most beneficial use of this resource. Deciduous trees routinely shed their leaves before winter to prevent the loss of valuable plant resources.

Evergreen Trees

Evergreen trees live throughout the world and retain their leaves for more than one growing season. Evergreen tree species have either broad leaves or needles. Evergreens with broad leaves commonly grow in the tropics. Examples of broad-leaved evergreen trees that are found in the United States are live oak, rhododendron, and holly. A few broad-leaved evergreen species are also found in hot, dry regions.

Most evergreen tree species from temperate regions are classified as *conifers* (described in Chapter 4) and have narrow, elongated, pointed leaves called *needles* (see Figure 32.9). Needles are cylindrical, two-sided (flat), or polyhedral (having several flat surfaces). Coniferous evergreens are found throughout the southeastern United States, New England, and across much of the dry, mountainous regions of western North America.

Figure 32.9

Needles of coniferous species have a wide assortment of shapes as exemplified by the ponderosa pine (A), Douglas fir (B), blue spruce (C), and juniper (D).

Needles

How do the functions and lives of needles compare with the leaves of deciduous trees? Even though most coniferous species do not shed their leaves annually, needles are not permanent parts of the tree. Some species retain their needles for only two or three years, while other species may conserve their needles for more than ten years. Needles shed from inside the conifer canopy are soon replaced by new needles produced at the outside of the canopy. Needle longevity is related to the needle's capacity to sustain positive net photosynthesis. As needles age and self-shading becomes more pronounced, the tree responds by dropping needles that are no longer self-sufficient. As with deciduous trees, nitrogen and photosynthates are reallocated to other active tissues before needles fall. As illustrated in Figure 32.10, branches of conifer trees also undergo *self-pruning*. Consequently, many conifers in dense, forested stands typically have long, straight trunks with needle-bearing branches only in the upper section of the tree, where light levels are high and photosynthesis occurs.

Summer Adaptations

Conifer needles are well adapted for conserving water during the dry summer season. Their stomata generally allow low rates of gas exchange, which results in lower water use. Also, needles are covered with a thick, prominent cuticle of wax that is impervious to water, another adaptation related to water conservation. Finally, the cylindrical shape of the needle is well suited for efficient light interception and for cooling during hot, dry summers due to air movement over the leaf surface.

Winter Adaptations

How are the leaves of evergreen trees able to survive the rigors of winter? Needles are well adapted to endure stresses imposed by winter weather and are even able to photosynthesize on warm days. Their cylindrical shape prevents the buildup of ice or wet snow that could cause a broad leaf to break. This shape also offers little resistance to high winter winds. The needle's internal structure is reinforced with cellulose and other substances that make it resistant to physical damage from wind, snow, and ice.

Both conifer needles and broad-leaved evergreen foliage have adaptations that prevent their freezing and protect them from cold damage. With the onset of cold, fall weather, needles begin the process of "hardening," which ultimately protects them from freezing temperatures. During hardening, sugars accumulate in the needles, where they function as antifreeze; their freezing point drops because lower temperatures are required to freeze needles with high sugar content than with low sugar content. If water did freeze in the cells, the expansion

Figure 32.10

In dense strands of trees, where self-pruning is common, living needles and branches occur only at the top of the tree. Needles and branches are shed from the tree after the needles' exposure to sunlight is reduced and they are no longer able to carry out a high rate of photosynthesis.

that occurs when water turns to ice would damage or tear cell membranes, organelles, and cell walls. Thus needles are usually able to survive the extreme cold temperatures of winter because of biochemical adaptations that prevent water from freezing inside their cells.

The conical tree shape is also an adaptation for winter—structures with this shape easily shed heavy snow and ice. If allowed to accumulate, the increased weight from wet snow and ice could break the branches and even the main stem of a tree.

Conifer trees have highly specialized needle-shaped leaves that are not usually shed in response to seasonal variations. As a result, conifers remain active throughout the year since their leaves are able to carry on photosynthesis. Conifer needles are also adapted for conserving water during the summer and surviving the low temperatures of winter.

REPRODUCTION

Plant adaptations are, of course, tied directly to plant reproduction. Adaptations are passed from one generation to the next through reproduction. New offspring represent products resulting from natural selection that determine whether or not individual fitness of the parent plants would be expressed in future generations. Also, plant species have adaptations that promote successful reproduction.

Reproductive periods punctuate the life histories of plants at specific intervals. During these periods, the normal patterns of resource acquisition, assimilaton, and allocation that govern the growth and development of roots, shoots, and leaves are temporarily modified. For plants to reproduce successfully, resources must be diverted to developing cones, flowers, gametes, fruits, and seeds.

Understanding plant reproduction allows humans to exploit certain plants more effectively. Desirable species can be propagated successfully, and undesirable species can be reduced or perhaps eliminated. Many industries are totally dependent on the reproductive tissues of plants. Most agricultural activities, including the production of corn, soybeans, wheat, fruit, and other crops, are directed at harvesting fruits and seeds. The landscaping and cut-flower industries rely heavily on flowering species, including roses, orchids, rhododendrons, daisies, and tulips. Products of plant reproduction provide much of the food consumed by humans, and flowers enrich our lives as well.

Growth Phases and Patterns

The life history of a plant encompasses two successive growth phases. The phase that results in the development of roots, stems, and leaves is termed the **vegetative growth phase**. During this phase, the plant normally produces and stores carbohydrates. When reproductive plant tissues begin to form, the plant enters the **reproductive phase**. The duration of these two phases and the timing of transition between them differ among species.

Three distinct patterns of vegetative growth and reproduction phases are used in classifying plants (see Figure 32.11). Angiosperms have either an *annual, biennial,* or *perennial* life history, while all gymnosperms are perennials. Each of these patterns has distinguishing qualities that include the common properties of growth and reproduction.

Annual Plants

As described in Figure 32.12A, **annual plants** complete their entire life cycle within one calendar year. Annual plants include many agricultural crops, such as corn and beans; ornamental flowers, including pansies and marigolds; and weeds, such as cocklebur and vetch. Annuals commonly have small seeds and are adapted to germinate and grow on unsettled soil.

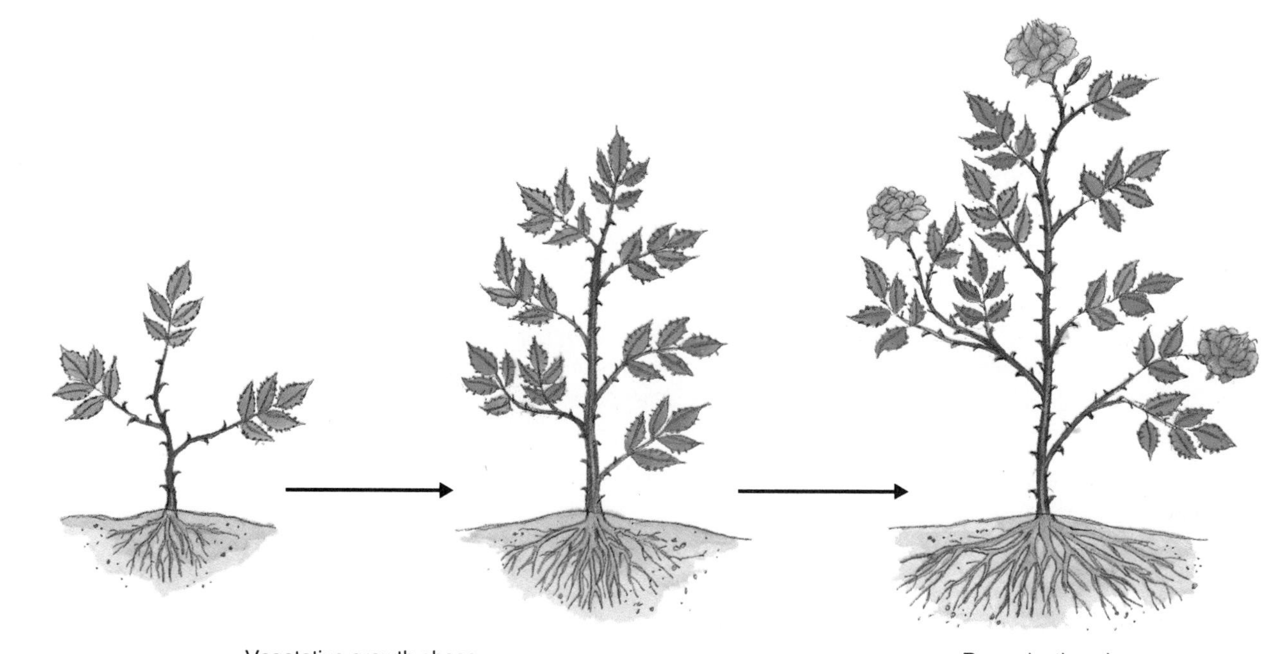

Figure 32.11

After completing a vegetative growth phase, plants enter a reproductive phase, during which they begin to produce cones or flowers.

Typically, seeds of annual plants germinate in the spring and proceed through a period of vegetative growth. They enter their reproductive phase when they reach a particular size or stage of development or when they become stimulated by an environmental cue such as a "correct" length of daylight. Once reproduction has been completed, the plant dies, leaving its seeds behind.

Some annual plants complete this life cycle very quickly. For example, desert annuals usually germinate during brief periods of spring or fall rain, grow rapidly, flower, and produce seeds before the weather becomes too hot and dry for them to survive. These plants frequently complete their entire life history within the span of two or three weeks.

Some annuals germinate in the spring, grow vegetatively, begin flowering, and persist in both growing and flowering until the plant dies from freezing in late fall. The life span of these plants may be six months or more. Familiar examples of relatively long-lived annual plants include many garden species such as beans and pansies. Even though some of these plants do not die immediately after producing seeds and can live for long periods if protected from freezing in a greenhouse, they are still considered to have annual life cycles.

Biennial Plants

Beets, cabbage, and celery are examples of **biennial plants** that complete their lives in two growing seasons (see Figure 32.12B). During the first year, the growth of biennials is usually limited to the vegetative phase. Such plants often have a large taproot, a shoot less than 2 centimeters tall, and large leaves, which grow nearly flat on the ground. Much of the nitrogen and photosynthate accumulated during the vegetative growth phase is stored in the taproot. Beets and cabbage are biennials that are usually harvested after their first year of growth.

During their second year of life, biennial plants usually enter into a reproductive phase. At the onset of reproduction, nutrients and photosynthates stored in the taproot, along with those being produced by leaves, are mobilized and sent to the shoot. When this happens, **bolting** occurs as the short, compressed shoot quickly grows to a height of half a meter or more. This process involves rapid stem elongation and may take one or two days or as long as two weeks. Once a plant has bolted, its resources are allocated to tissues forming flowers,

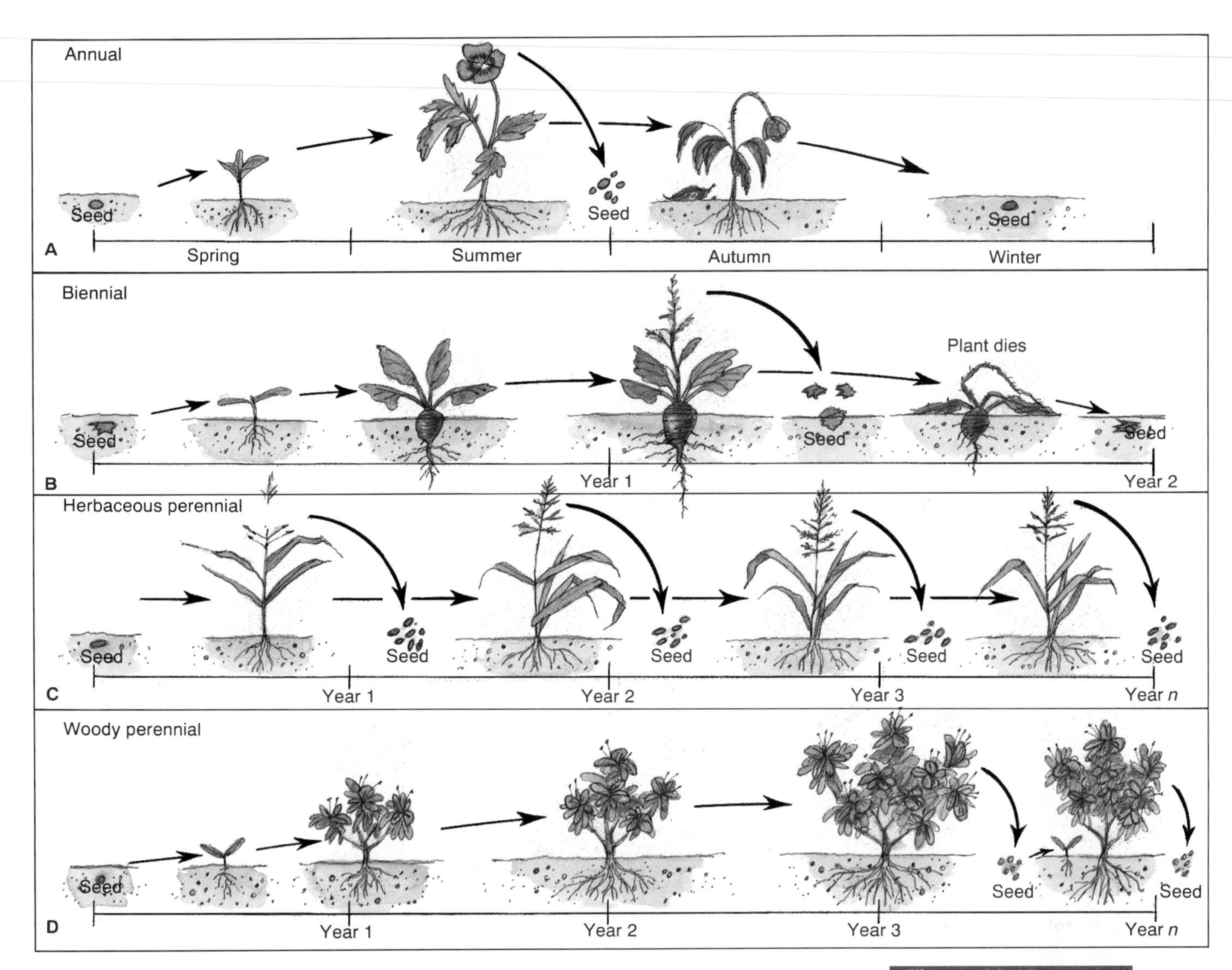

Figure 32.12

(A) Annual plants complete their life cycle in one growing season. In the temperate zone, annual plant seeds typically germinate in the spring, and flowers are produced. The plant dies after its seeds are released. (B) Biennial plants have a two-year life history. During the first year, the seed germinates, and the plant completes a vegetative growth phase. In the second year, biennial plants produce seeds during a reproductive phase and then die. (C) Perennial plants can live and produce seeds for many years. Herbaceous perennials, such as many grass species, may produce seeds during each year of growth. (D) Woody perennials, including trees and shrubs, may go through a juvenile period before they become capable of reproducing.

seeds, and fruits for the remainder of the growing season. Biennial plants die at the end of their second year of life, after they have bolted and produced seeds.

Perennial Plants

Perennial plants live more than two years and are subdivided into two groups, which are described in Figure 32.12C–D. *Herbaceous perennials* lack wood in their tissues and include asparagus, rhubarb, milkweed, lilies, and many grasses. *Woody perennials* are characterized by woody structures and include vines (for example, grapes), shrubs (rhododendron), and trees (pines and oaks).

Many perennials have vegetative growth phases that last for extended periods of time. Some species of trees, such as walnuts, may grow in a juvenile, nonvegetative phase for up to 20 years. Once perennials have matured, they enter a reproductive phase and often reproduce during each subsequent growing season. Thus once a walnut tree produces a first crop of nuts, it is likely to produce a nut crop every remaining year of its life.

The number and vigor of seeds produced by perennial plants differ from year to year, depending on environmental factors and plant age. Years of extreme environmental stress, such as a prolonged drought, will in some cases cause a great reduction in the seed production of affected perennials. In other cases, harsh conditions may actually stimulate reproduction in affected plants. The term *mast year* describes a year in which trees such as oaks, hickories, and beeches produce an abundance of seeds. Mast years are important because they increase the possibility of successful reproduction, and the prolific seed crop is a valuable supply of food for many animals such as wild squirrels and birds.

> Successful plant adaptations are passed from parent to offspring through reproduction. The life history of plants includes two growth phases—a vegetative growth phase when roots, shoots, and leaves are formed and a reproductive phase when reproductive structures develop. Three types of growth patterns occur in plants. Annual plants complete both phases within a single year. Biennial plants complete a vegetative phase during the first year and reproduce during the second year. Perennial plants have variable vegetative growth phases and generally reproduce yearly once they reach maturity.

Mechanisms of Vegetative Growth and Reproduction

Active zones of plant growth, or *meristems,* occur in plant shoots and roots. They are found in apical buds of the shoot and root, lateral buds, and vascular cambium tissues (see Figure 32.13). Many angiosperms use vegetative growth mechanisms not only in producing new tissues but also for propagating new plants from **vegetative plant tissues**—those not involved in sexual reproduction. This process is called **vegetative reproduction**, or reproduction by nonsexual tissues. Each undifferentiated plant cell contains a complete set of chromosomes and has the potential to develop into a new plant. Parenchyma cells and meristem cells (described in Chapter 30) produced at regions of active plant growth are undifferentiated cells that can generate complete new plants. In contrast, specialized cells, such as phloem or xylem, cannot develop into other cell types.

Figure 32.13

Apical buds, lateral buds, root tips, vascular cambium, and cork cambium all contain active meristem cells that divide and increase in size, resulting in growth. The cork cambium and vascular cambian lie just inside the bark of woody trees and shrubs.

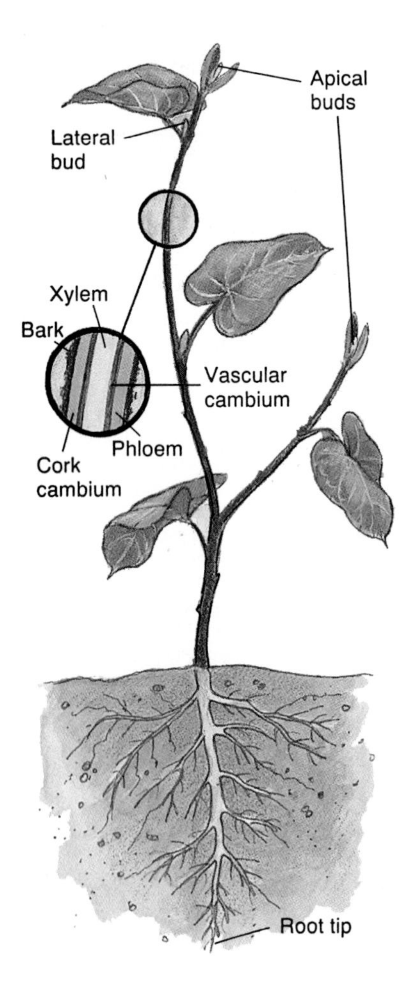

Primary Growth

Apical buds, located at the top of the plant and at the ends of branches and roots, give rise to all cells in the plant's shoots and roots. The production and growth of apical cells results in increased plant height. **Lateral buds** are located in *leaf axils,* the sites where petioles join the stem. Meristem cells of lateral buds develop and grow into branches. Growth that results in increased plant height or increased branch length is termed **primary growth**.

Apical and lateral buds generally become active in the spring, causing branches to elongate and new leaves to emerge during a primary growth phase. As the growing season ends, branches of deciduous trees stop elongating, and their leaves eventually fall to the ground. As shown in Figure 32.14, a record of these events remains on the branch in the form of leaf scars and bud scars that indicate the former locations of leaves and buds.

Secondary Growth

In woody plants, a thin cell layer, the **vascular cambium**, forms a continuous sheath of cells that surrounds the plant. It may be only two or three cells thick and is found just inside the bark (see Figure 32.15). The vascular cambium produces cells from its inner surface that ultimately form woody xylem tissue; cells originating from its outer surface differentiate into phloem tissue, the innermost layer of the bark. The continuous activity of the vascular cambium not only produces cells specialized in transporting essential resources in the plant but also causes an increase in the diameter of the stem and branches, a process known as **secondary growth**. The **cork cambium**, which lies outside the vascular cambium and phloem tissue, is a meristematic tissue that produces most of the bark of woody plants. It is also associated with secondary growth.

Vegetative Reproduction

The list of reproductive mechanisms used by plants is long and includes many types of plant tissues. Some tree species, including willow and poplar, can be vegetatively propagated by simply cutting off a young branch and planting it in the ground. Such plants have the capacity to generate roots quickly enough to allow the shoot to develop, grow, and survive. The establishment of new plants from clipped branches, or **cuttings**, depends on characteristics of the branch and environmental factors, such as sufficient warmth and water, that affect root development. Not all species of woody plants can be propagated from cuttings.

Grafting is a form of vegetative reproduction in which a bud or a branch of one plant is attached to the stem or rootstock of another, closely related plant. Nurseries commonly use this method to create one plant with superior features of two plants. For example, ornamental roses are produced by grafting buds from desirable roses to wild rose rootstock. If stems or branches that support apical or lateral buds become buried, cells of the vascular cambium can divide and differentiate into root tissues. Subsequently, an entirely new, independent plant will arise. This type of vegetative reproduction, called **layering**, is common in many tree species that grow close to the ground and are often covered by wind-blown soils and litter. It also occurs in tree species, such as high-elevation spruce, that grow where conditions are so harsh that sexual reproduction is rarely successful.

Plant tissues that grow below ground can also produce new plants by vegetative reproduction. Plants such as strawberries and spider plants have unusual stems, called **stolons**, that extend to the soil. Once on (or in) the soil, lateral buds on the stolons can develop into roots and shoots, which ultimately become transformed into independent plants. Stems of certain plants can grow into the

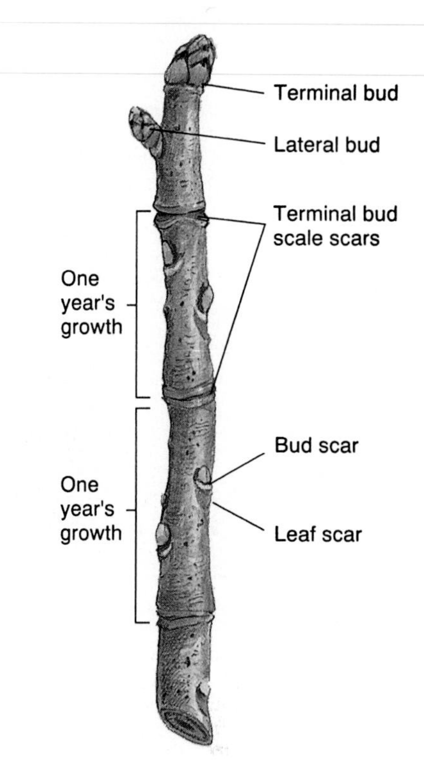

Figure 32.14

The history of leaf location, bud location, and growth during each season is recorded on branches. When apical and lateral buds open at the onset of growth, the bud scales drop, leaving a scar on the branch. The distance between apical bud scale scars represents one year of growth. Leaves also leave scars after they are shed from branches. Apical buds may be found just above the leaf scars and mark points where branches will sprout during the next growing season.

Figure 32.15

This figure shows a cross-section of a pine tree trunk with bark, annual rings, vascular cambium, phloem, and xylem.

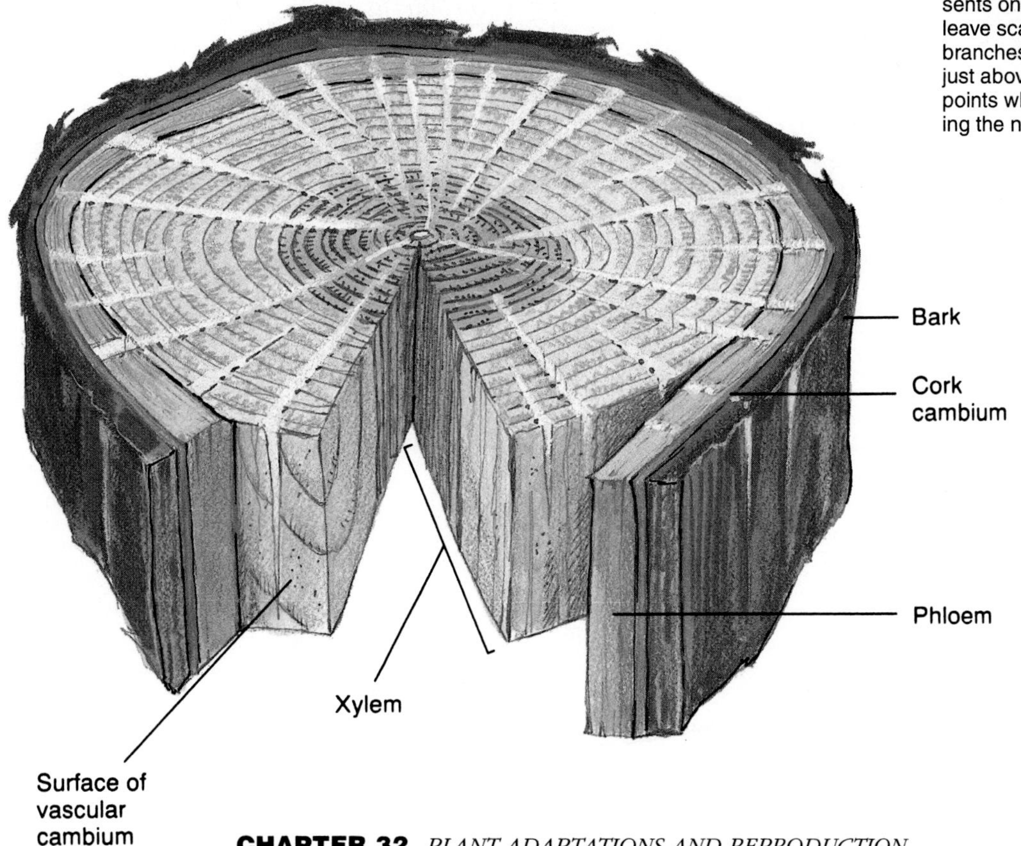

Figure 32.16

Plants such as potatoes reproduce by sprouting from underground stems, called tubers, which are also specialized to store carbohydrates.

soil surface. These stems, called **rhizomes**, produce lateral buds that contain cells capable of generating new plants. Bamboo, iris, and lily-of-the-valley are examples of common plants that reproduce vegetatively from rhizomes.

Tubers are underground stems that are highly specialized for carbohydrate storage but are also capable of vegetative reproduction (see Figure 32.16). For example, the eyes of a potato are actually the lateral buds of a tuber. Potatoes are commonly propagated by planting a segment of tissue containing an eye, or lateral bud.

Finally, even roots can produce buds that first give rise to plant structures by vegetative reproduction and then develop into new plants. Some trees, such as willows and poplars, often reproduce this way.

Vegetative growth is associated with increasing the size of most plant tissues. During primary growth, meristem regions located in apical and lateral buds produce new cells that develop and grow, resulting in an elongation of shoots, branches, leaves, and roots. In secondary growth, meristem tissues of the vascular cambium produce cells that form xylem and phloem, which results in an increase in the diameter of stems, branches, and roots. Cuttings, layering, and specialized structures—stolons, rhizomes, and tubers—are associated with vegetative reproduction.

Figure 32.17

Sexual reproduction in gymnosperms and angiosperms begins with the production of sperm and eggs by meiosis. Fertilization occurs when a sperm and an egg nuclei unite to form a zygote.

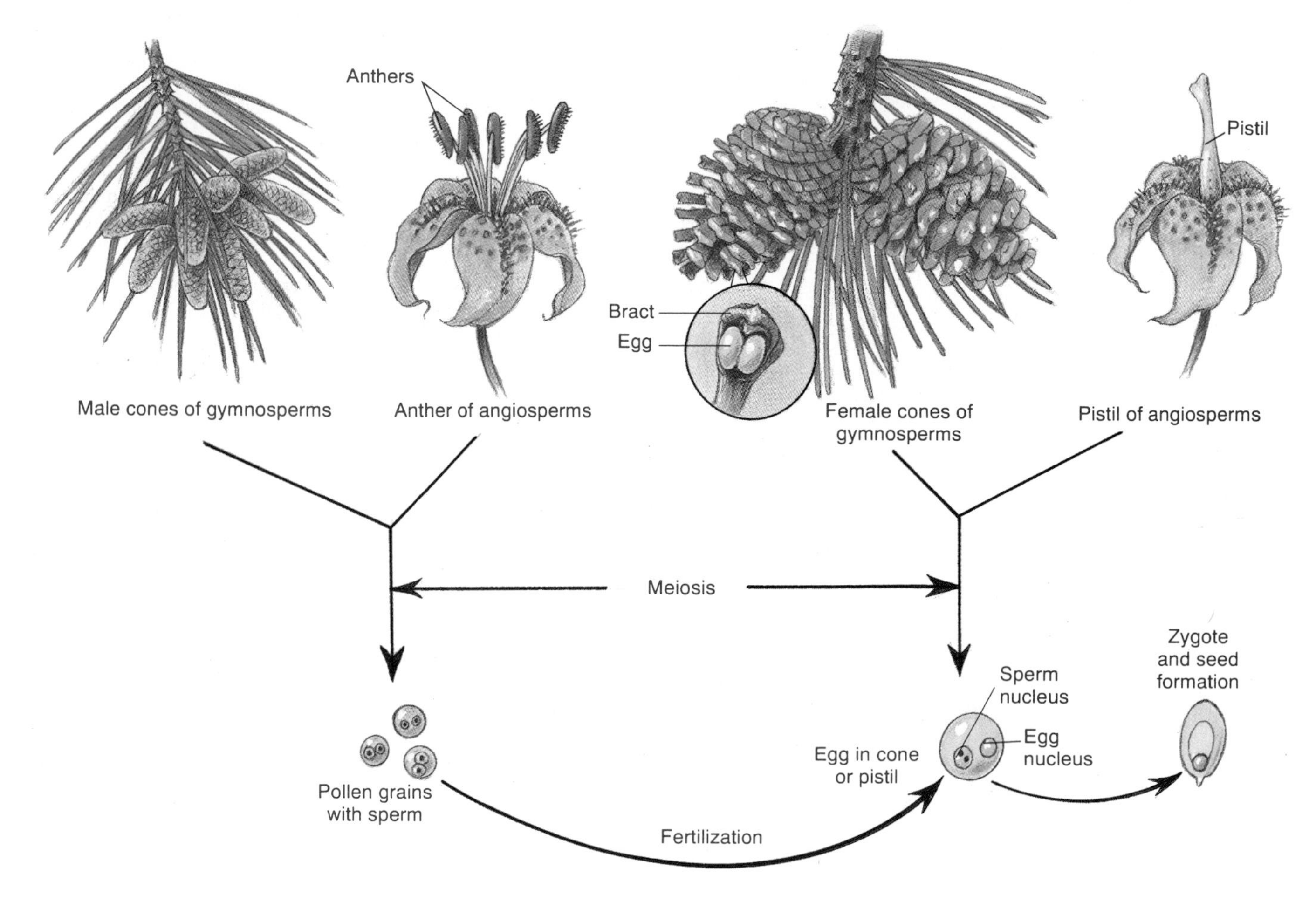

Mechanisms of Sexual Reproduction in Plants

The seeds of gymnosperm and angiosperm plants are produced through sexual reproduction. Within seed cells are genes that convey adaptations from previous generations of successful plants. These adaptations, encoded in the DNA of embryonic cells, may enable the seed to germinate successfully and proceed in its development from a seedling to a mature, seed-bearing plant.

Figure 32.17 illustrates the general process of fertilization in plants. The zygote in a seed carries chromosomes originating from a male and a female gamete. Thus a key feature of seed formation is fertilization, the process in which two gametes merge and combine their chromosomes to form a zygote. A general pattern of fertilization is common to all seed plants, but there are many variations, all of which can result in seed production.

Gymnosperm Reproduction

Coniferous evergreen trees, including pines, firs, spruces, and hemlocks, are well-known gymnosperms. Pine trees produce seeds in cones through a series of events typical of most gymnosperms. Figure 32.18 describes the pine tree life cycle. In gymnosperms, sperm-bearing pollen grains from male cones are usually transported by the wind to egg-bearing female cones. Fertilization takes place within the female cone.

Male Cones

Pines produce two different kinds of cones, one male and one female, on the same tree (this is not true for all gymnosperms). Male cones form in the late fall from buds at the ends of branches. During the spring, mature pollen grains, each containing two sperm, develop in male cones. Pine pollen grains have two small

Figure 32.18

The life cycle of a gymnosperm, such as a pine, includes seed germination, growth and maturation of the tree, and the production of male and female cones. Sexual reproduction begins with the production of sperm-containing pollen grains in male cones. Pollen grains are then carried by the wind to the female ovulate cones, where the sperm fertilize eggs. Fertilization results in the formation of a zygote and, eventually, a new seed.

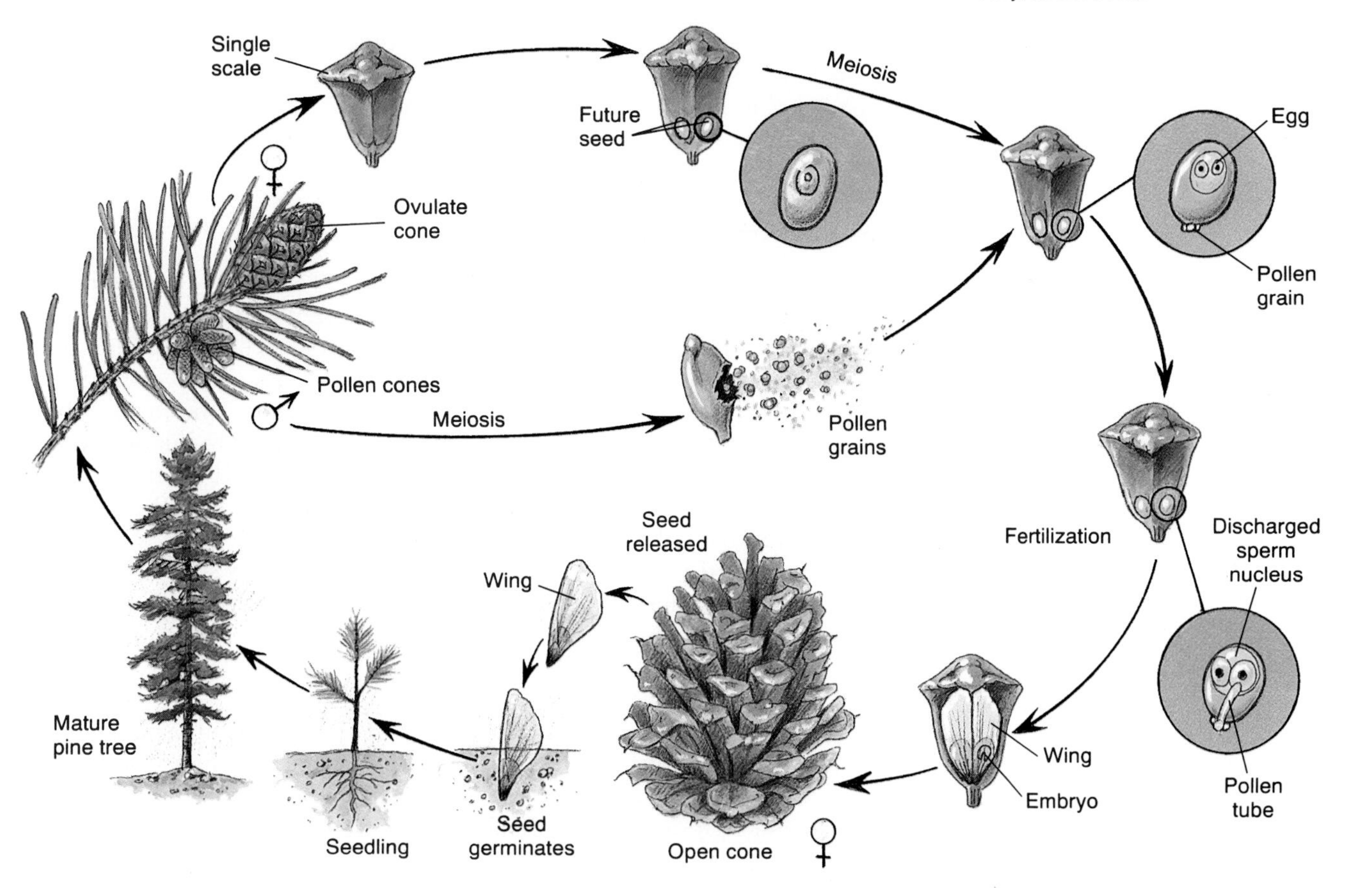

Figure 32.19

An enormous quantity of yellow pollen from pine trees is visible along the shoreline of this lake (Crater Lake, Oregon).

wings that aid in their dispersal by the wind. Pines, and other conifers, can produce pollen in such great quantities that nearby objects become covered with what appears to be yellow dust (Figure 32.19). Following the release of their pollen, male cones dry out and fall to the ground.

Female Cones

Female cones are produced on the youngest branches in the canopy and are larger than male cones. During the spring, pollen released from male cones drifts through the air and lodges between scales of the female cones. As a female cone grows, its scales close and prevent the entrance of any additional pollen. Over the next 12 to 14 months, pollen tubes emerge from the pollen grains in the female cone as the female reproductive cells undergo development and maturation.

Fertilization occurs within the female cone and leads to the formation of seeds. The seed contains the zygote that ultimately gives rise to an embryo, which has a radicle, an apical bud, and cotyledons (see Chapter 30), is surrounded by nutritive tissues, and is protected by a seed coat. The pine seed has a wing that enables it to be carried by the wind when it is released from the cone. Pine seeds are usually shed in the spring, about two years after the female cones first appear.

Angiosperm Reproduction

The angiosperm life cycle is described in Figure 32.20. In angiosperms, male gametes from pollen grains are most often transported by the wind or by insects as they move from flower to flower while foraging for nectar and pollen (see Figure 32.21). Once the pollen is in position on the stigma of the pistil, it emits a tube that grows down toward the egg located at the base of the pistil. When this pollen tube contacts the egg, one sperm from the pollen grain migrates down the pollen tube and fertilizes the egg. Another sperm moves down the pollen tube and combines with a cell from the female flower, which leads to the formation of nutritive tissues that will be used by the developing embryo.

Angiosperm flowers are produced during the spring, summer, or fall, followed by pollination and then seed dispersal. The production of flower buds and seeds typically occurs during the growing season.

Flower Structure

Angiosperms produce an incredible variety of flowers. Many species, such as orchids and roses, produce large, conspicuous flowers; in contrast, the flowers of other species may be either small or short-lived and therefore not apparent. For example, grasses produce inconspicuous flowers that the casual observer seldom notices.

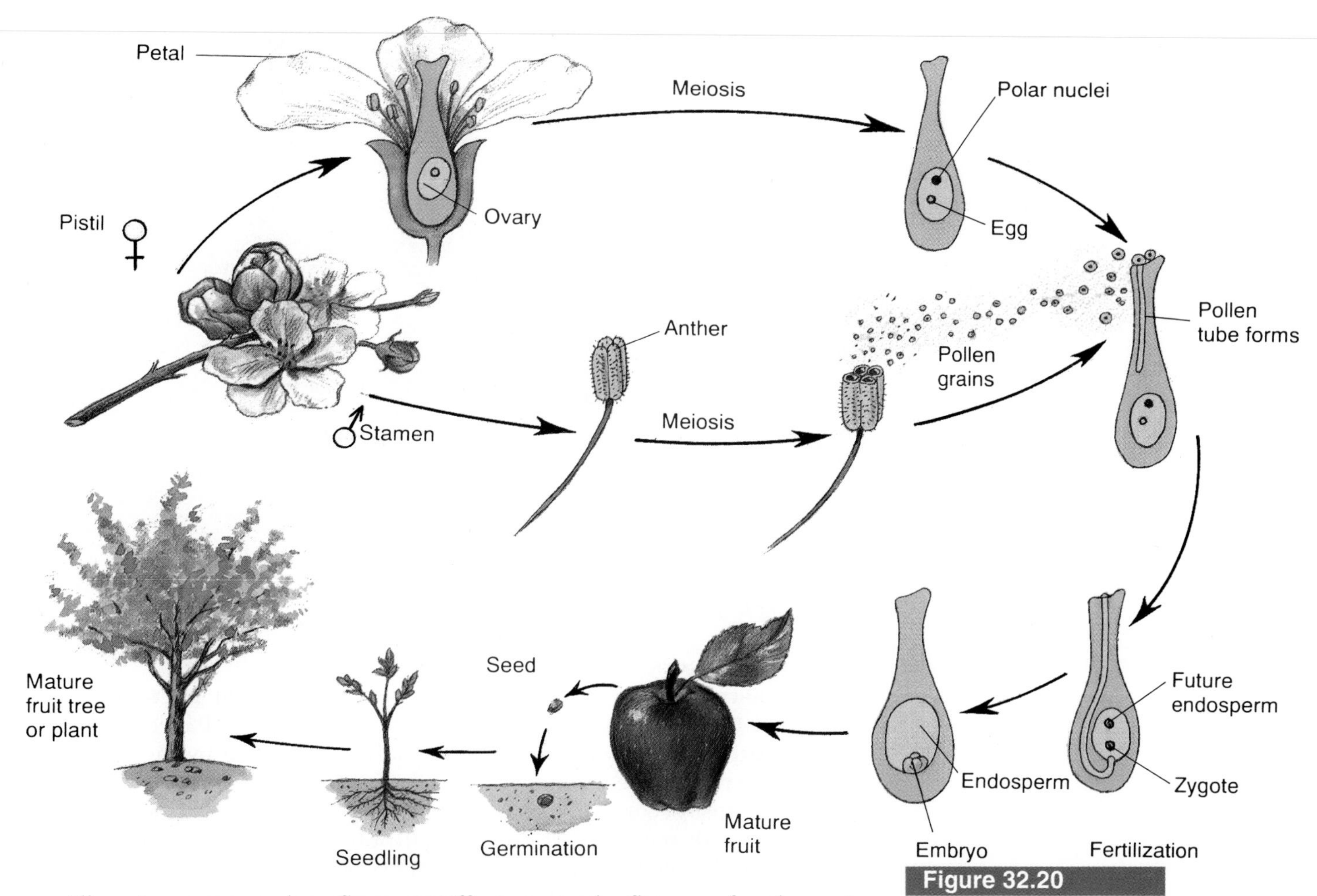

Figure 32.20

The life cycle of an angiosperm includes germination of a seed, vegetative growth, and the production of flowers that eventually bear fruit, which contains seeds. Seeds are formed when the male reproductive organs, the anthers, produce pollen that contain sperm. The pollen grains are transported by winds or animals to the stigma of the egg-bearing pistil. The sperm and egg unite to form the zygote, which becomes part of a new seed. The seed develops further within the pistil, a structure that eventually matures into a fruit.

All angiosperms produce flowers. Differences in the flowers of each species are due to variations in the color, size, and arrangement of their flower parts. Figure 32.22 illustrates the general anatomy of a flower. Flowers are attached to the plant by a stalk known as the **peduncle**. The peduncle is surrounded by **sepals**, which protect the flower while it is a bud, and also by the flower's **petals**. The sepals and petals are attached to a **receptacle**. Petals attract the insects, hummingbirds, or bats that are needed for pollination. The reproductive organs of a flower include the **stamens**, which produce pollen, and the **pistils**, which produce female gametes. After fertilization, seeds develop inside the developing fruit.

The sepals of some species are small, green, and barely visible. However, in other species such as dogwood, sepals may be quite large and take on the appearance of petals. Large, colorful flowers resulting from natural selection are

Figure 32.21

Flowers of many angiosperm species are adapted to attract animal pollinators. Certain insects, such as bees (A), birds, and mammals, such as bats that pollinate wild bananas (B), forage for nectar in flowers and consequently transfer pollen from one flower to another. The pollen of some angiosperms, such as grasses, is transported by the wind.

Figure 32.22

This is the structure of a typical flower. Sepals and petals are arranged in whorls around the axis of the flower. Male reproductive structures include the filament and the anther. The stigma at the top of the female reproductive structure, the pistil, is modified to receive pollen grains. The egg is located in the ovary at the base of the pistil.

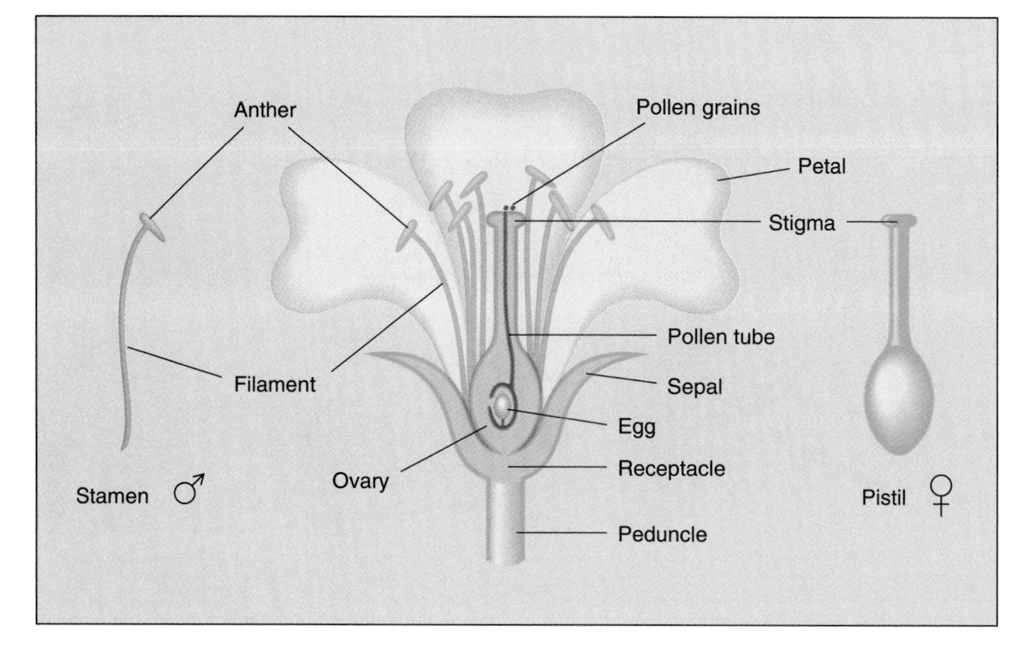

adaptations for attracting insects and other animals that transport the pollen from one flower to another, resulting in *cross-pollination*.

Types of Flowers

Flowers that contain both stamens and pistils are called **perfect flowers**; **imperfect flowers** lack one of these reproductive organs. Perfect flowers may be **self-pollinating**, which means that pollen produced by the stamen of a flower can fertilize an egg within the pistil of the same flower. Stamen and pistil maturation must occur at the same time for self-pollination to take place. Some species have adaptations that prevent self-pollination. (Why would this be an advantage to a plant?) For example, pollen of a flower may mature and be released from its stamens either before or after the period when pollen can be received by the flower's pistil.

Certain species of angiosperms, such as willows and poplars, are **dioecious**, meaning that male flowers grow on one plant and female flowers on another plant. For such species to bear fruit, male plants must grow close to female plants to ensure pollination of the female flowers. Many plants, including corn, are **monoecious**, having both male and female flowers on the same plant. Pollen from the male flowers may be either compatible or incompatible with the female flowers on the same plant. Self-fertilization can occur if flowers are compatible but not if they are incompatible.

The number of floral parts varies among species. Some species have flowers with one pistil, several stamens, and a single type of petal. In other species, flowers have many pistils, many stamens, and several types of petals and sepals. As the number, position, arrangement, and color of these structures change, so too does the appearance of the flower. Obviously, the number of possible combinations of these floral characteristics is enormous, as is reflected in the great diversity of flowers found in angiosperm species.

Sexual reproduction in gymnosperms involves pollen production by male cones and transportation of the pollen to egg-bearing female cones, where fertilization occurs and seeds develop. Sexual reproduction in angiosperms takes place after flowers are formed. Pollen from male flowers is transported to female floral parts, where fertilization occurs. Subsequently, seeds develop, and the ovary becomes transformed into a fruit. Because of the enormous diversity of flower types among angiosperm species, there are many variations in the basic process of sexual reproduction.

The events that lead to the production of seeds mark the point at which the life cycle—from seed to seed—has been completed. Whether reproduction occurs only once, as in the life of an annual plant, or many times, as in the life of a perennial plant, the release of viable seeds is the endpoint or product of all of the processes that occurred during the vegetative and reproductive phases of plant life.

Summary

1. Plants are well adapted for surviving in the environments they inhabit. Several general types of adaptations exist in plants. These include physiological mechanisms that are used in normal activities and also in making adjustments to changing environmental conditions, specialized structures and growth forms that permit survival in different geographical areas, and life history strategies that increase the likelihood of long-term survival and successful reproduction.
2. Deciduous trees have broad, flat leaves that produce great quantities of photosynthates during spring and summer. Their leaves are arranged on branches so as to receive full sunlight. The growth form and branching pattern of deciduous trees also ensure maximum exposure of leaves to sunlight. However, their leaves cannot survive cold temperatures, and they are shed from the tree during the fall. Thus deciduous trees are characterized by high levels of photosynthesis during the growing season but very low levels of activity during the rest of the year.
3. In contrast, most evergreen trees have small, needlelike leaves. Such leaves can survive low temperatures and are not shed seasonally. The growth form of evergreen trees—an elongated pyramid with sloping branches—is favorable for shedding heavy snow and ice. Hence evergreen trees are adapted for remaining active throughout the year.
4. Both deciduous and evergreen trees have other adaptations that are important to their survival. These include self-pruning and mechanisms for regulating carbohydrate and nitrogen allocation and preventing water loss.
5. Plants have two distinct growth phases during their life history. In the vegetative growth phase, plant resources are used in the development and growth of roots, stems, and leaves. During the reproductive phase, a plant invests in forming structures used in reproduction.
6. The timing and duration of the two growth phases varies with plant species. Three general types of life history pattern are used to classify plants. Annual plants complete both growth phases during a single year and then die. Biennial plants usually undergo a vegetative growth phase during their first year and a reproductive phase in their second year, after which they die. Perennial plants can live for many years, and the time required for the vegetative growth phase may last for years. Once they enter a reproductive phase, they may reproduce yearly for the rest of their lives.
7. Vegetative plant structures—stems, roots, and leaves—are produced from cells located in meristems. These cells divide and differentiate to form the various plant tissues. Primary growth leads to increases in plant height and branch length. Secondary growth—the production of xylem and phloem—occurs only in woody plants and results in an increase in branch and stem diameter.
8. Some plant species use specialized processes (layering) or structures (stolons, rhizomes, and tubers) to create new plants by vegetative reproduction.
9. Sexual reproduction in plants consists of fertilization and seed formation and development. Gymnosperms have male cones that produce pollen, which contains sperm, and female cones, which contain eggs. Fertilization leads to the formation of a zygote and a seed. The zygote develops into an embryo within the seed. Angiosperms employ flowers in their reproduction. Although the basic process of sexual reproduction is similar to that in gymnosperms, among angiosperm species there are enormous variations in flower types, flower structure, pollinators, and fertilization mechanisms.

Review Questions

1. Give some examples of plant adaptations to natural environmental changes.
2. In what ways are the leaves of deciduous trees considered to be adaptive for survival of the entire tree?
3. What is a deciduous tree's response to self-shading? Why is it considered adaptive?
4. Why is it advantageous to deciduous trees to drop their leaves in the fall?
5. In what ways are the leaves of conifers considered to be adaptive for survival of the entire tree?
6. How are conifers adapted for remaining active during the winter?
7. Describe the two basic life history phases of plants.
8. Explain the primary differences in the life histories of annual plants, biennial plants, and perennial plants.
9. Describe the different mechanisms plants use to carry out vegetative reproduction. What is the difference between primary growth and secondary growth?
10. Describe the process of sexual reproduction by gymnosperms.
11. What are the functions of flowers and the various floral parts in angiosperm sexual reproduction?

Essay and Discussion Questions

1. The life histories of evergreen and deciduous tree species differ, yet both must be well adapted to their environment in order to survive and reproduce. How can you explain the occurrence of both types of trees in a single forest?
2. Assuming that the buildup of atmospheric greenhouse gases leads to a slow rate of global warming, what types of adaptation may evolve in plants growing in your area? What might be some of the consequences for plants of a rapid rate of warming? Why?
3. Larch is a conifer species that produces new needles each spring and sheds them each fall. What are some disadvantages to this adaptation that might explain why it is rare in conifers?

References and Recommended Reading

Barbour, M. G., and W. D. Billings, eds. 1988. *North American Terrestrial Vegetation*. New York: Cambridge University Press.

Doust, J. L. 1989. Plant reproductive strategies and resource allocation. *Trends in Ecology and Evolution* 4:230–234.

Feinsinger, P. 1987. Effects of plant species on each other's pollination: Is community structure influenced? *TreRogy and Evolution* 2:123–126.

Mirov, N. T., and J. Hasbrouck. 1976. *The Story of Pines*. Bloomington: Indiana University Press.

Mlot, C. 1990. Restoring the prairie. *BioScience*. 40:804–809.

Niklas, K. J. 1987. Aerodynamics of wind pollination. *Scientific American* 257:90–96.

Primack, R. B. 1985. Longevity of individual flowers. *Annual Review of Ecology and Systematics* 16:15–38.

———. 1987. Relationships among flowers, fruits, and seeds. *Annual Review of Ecology and Systematics* 18:409–430.

Proctor, M., and P. Yeo. 1973. *The Pollination of Flowers*. New York: Taplinger.

Waring, R. H., and W. H. Schlesinger. 1985. *Forest Ecosystems: Concepts and Management*. Orlando, Fla.: Academic Press.

CHAPTER 33

Cellular Diversity and Organization in Humans

Chapter Outline

Cellular Diversity in Humans

Structural Levels of Organization in Humans

Anatomy and Physiology

Maintenance and Control

Understanding the Human Body

Reading Questions

1. What is the biological significance of the cell theory?
2. What types of functions are carried out by cells?
3. What are the major structural levels of organization in humans?
4. How are the functions of different organ systems integrated and coordinated?

In observing fellow members of the animal kingdom, it is difficult to avoid being astounded by the range of activities animals use during their lives. Collectively, animals are able to do the following: (1) inhabit all types of terrestrial and aquatic environments, some of which are characterized by extreme climatic conditions; (2) consume and extract nourishment by processing a wide assortment of foods; (3) react effectively to many different stimuli; (4) perform extraordinary feats of strength, speed, and coordination; and (5) reproduce themselves using a diverse set of strategies. All of these require structural and functional capabilities that are, of course, the end result of millions of years of evolution.

All existing animals have evolved structures enabling them to accomplish functions that are necessary for overcoming problems that constantly threaten their existence. It is beyond the scope of this text to consider the remarkably diverse ways in which different animals perform these functions. Rather, we focus on humans, the animal of greatest interest to most readers of this text.

How can the human body be understood? Begin by looking in the mirror and considering the following questions: What is inside the exterior that you observe? How does it function? To acquire a basic knowledge of how humans operate, it is helpful to begin by exploring how the human body is constructed and organized.

CELLULAR DIVERSITY IN HUMANS

The cell theory was formulated during the nineteenth century, but it has since been extended as a result of new discoveries. The contemporary cell theory encompasses several fundamental concepts in biological science, most of which have been described in earlier chapters. To review, *cells* are the basic structural units of all living organisms. That is, every organism is composed of a single cell or populations of cells. Second, all cells require energy to maintain their internal environments and perform their specific functions. This energy is derived directly from the molecule ATP (described in Chapter 29). Third, cells are the functional units of life. All of the chemical reactions necessary to maintain or reproduce an organism occur within cells. These processes are controlled through hereditary information contained in genes made of DNA. All cells translate portions of the genetic information within their DNA into specific proteins, which are synthesized by ribosomes (explained in Chapter 16). These proteins govern structure and functions in all cells. Fourth, all cells, since the origin of the first cells (traced in Chapter 2), arise from preexisting cells through the division processes of meiosis or mitosis (discussed in Chapter 13).

All cells have specific roles in all life functions in humans. Amazing diversity is reflected in the structural features and functional capabilities of cells that make up different parts of the body.

The ability of humans and other organisms to carry out functions associated with life is based on the activities of cells. The cell theory describes the general roles and requirements of cells and also explains their origin and destiny.

Cell Structure

To review briefly some key concepts, eukaryotic cells of all higher organisms, both plants and animals, have basic structural similarities (see Figure 33.1). Each cell is surrounded by a protective *plasma membrane* that regulates materials entering and leaving the cell. The interior of the cell contains *cytoplasm,* a viscous fluid containing submicroscopic organelles that are the cell's tools for accomplishing certain tasks. Within the cytoplasm of most cells is a large *nucleus,* a membrane-bound region containing hereditary information called DNA.

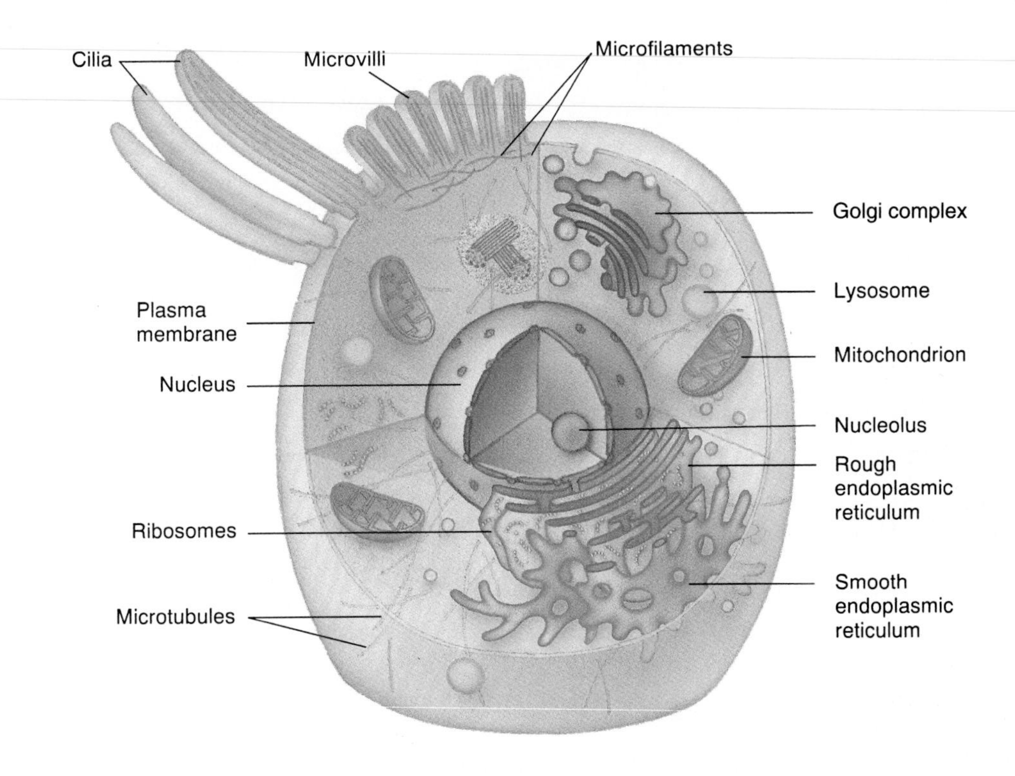

Figure 33.1

Like cells of all eukaryotic organisms, human cells are composed of basic structures and organelles.

The functions of cellular *organelles* such as mitochondria, ribosomes, endoplasmic reticulum, and Golgi complexes are summarized in Table 33.1. No cell harbors all of the organelles or structures listed in the table; rather, each contains those necessary for performing its specific cellular function. The trillions of cells in the human body are remarkably diverse. They differ in size, shape, internal components, movement, location, life span, and function.

Cell Functions

All human cells are continuously engaged in normal activities that include the conversion of food materials into energy or new materials and the removal of waste products created during these transformations. In addition, cellular components are constantly damaged or destroyed during the cell's activities. Thus **self-maintenance** involves activities used in meeting the continuous demand for repairing or replacing ribosomes, membranes, enzymes, and other structures and molecules that are lost during normal operations. It is essential for maintaining the health of the cell and hence the health of the individual.

Cellular Reproduction

Some cells in adults are capable of undergoing *cellular reproduction* in order to replace cells that normally wear out and die. Before a cell can reproduce, it undergoes a precisely controlled increase in its numbers of chromosomes, mitochondria, ribosomes, plasma membranes, proteins, and other structures and molecules. Once sufficient quantities of these elements are available, a cell divides by mitosis and gives rise to two new daughter cells, one or both of which can then continue to carry out the same function as the cell they replaced.

Specialized Functions

Most human cells are *differentiated,* which means that they have distinctive molecular and structural features that reflect a specialized function. For example, as illustrated in Figure 33.2, cells that actively synthesize proteins, such as active plasma cells, are packed with rough endoplasmic reticulum and ribosomes; those requiring great amounts of energy, such as heart muscle cells, have numerous

Table 33.1 A Review of the Major Structures and Functions of Human Cells

Structure	Functions
Plasma membrane	Encloses and protects the cell; regulates transport of substances into and out of the cell
Nucleus	Controls activities of the cell; contains DNA
Nucleolus	Assembles ribosomes used in protein synthesis
Endoplasmic reticulum	Transports substances inside the cell
Smooth ER	Synthesis of steroids and other hormones in some cells
Rough ER	Synthesis of proteins that are generally transported out of the cell
Ribosomes	Synthesis of proteins
Golgi complex	Aggregate of membranes; packages chemicals synthesized in the cell for export to other cells
Mitochondria	Provide cells with energy through the reactions of cellular respiration
Microtubules	Contribute structural support in some cells
Microfilaments	Fibrous elements made of protein; important in contraction of muscle cells
Cilia	Assist in the movement of materials outside of certain cells, such as those that line the respiratory and digestive tract
Flagella	Used in cellular locomotion; in humans, found only in sperm cells
Lysosomes	Small vesicles containing enzymes that digest cellular contents when released upon death of the cell

mitochondria. Some cells, however, remain *undifferentiated;* they do not become specialized to carry out a specific function. Such cells often have the ability to continue dividing by mitosis, thus producing new cells. These new cells commonly replace functional cells that are injured or die. To do this, they must undergo **differentiation**, the process of changing from an unspecialized cell to a cell that has a definite structure and function.

Some specific functions of differentiated human cells include conducting electrochemical impulses, forming bone, capturing and digesting bacteria that invade the body, secreting hormones, transporting oxygen, forming gametes, and storing fat. We will be looking more closely at several of these functions in the next five chapters.

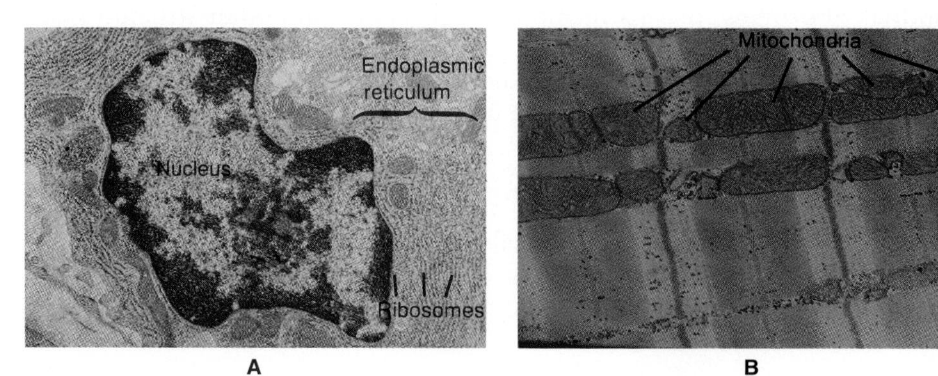

Figure 33.2

A cell's structure is closely related to its function. Plasma cells (A) contain an abundance of ribosomes and endoplasmic reticulum—organelles required for protein synthesis. Heart muscle cells (B) are full of mitochondria—organelles that produce energy.

Most human cells contain a nucleus and various organelles enclosed within a cell membrane. Throughout their lives, cells engage in self-maintenance activities to repair or replace damaged molecules and organelles. Some cells are capable of producing, by mitosis, daughter cells that replace lost or injured functional cells. These new cells become differentiated to perform specialized functions.

STRUCTURAL LEVELS OF ORGANIZATION IN HUMANS

There are four structural levels of organization in the architecture of the human body. From simplest to most complex, they are cells, tissues, organs, and organ systems. Whereas cells are able to carry out relatively few functions, organ systems are capable of accomplishing numerous tasks within the body.

Cells

Cells represent the simplest structural level of organization in all multicellular organisms. Hundreds of different cell types occur in humans. However, except for blood cells and gametes, they do not operate as independent entities. Rather, they combine with other cells to form more complex levels of organization that permit an efficient division of labor among different groups of cells.

Tissues

Tissues consist of a large assemblage of cells, primarily with the same type of structure, that are specialized to carry out a particular function. There are four primary animal tissues: epithelial, connective, muscular, and nervous. Each of these tissues is composed of cells with a characteristic appearance and functional capability.

Epithelial Tissues

Epithelial tissues (epithelium) are continuous layers of closely connected cells that cover body surfaces and line cavities inside the body. They are classified according to the shapes and surface specializations of the cells, the arrangement of cell layers within the tissue, and their functions. Several types of epithelium perform important tasks in the human body, some of which are described in Table 33.2 and shown in Figure 33.3.

Table 33.2 Selected Epithelial Tissues of the Human Body

Type	Structure	Location	Functions
Epidermis	Simple, flattened cells	Skin	Protection
Endothelium	Simple, flattened cells	Line the heart, lung, and blood vessels	Protection; allows gases and fluids to move between external and internal environments
Mesothelium	Simple, flattened cells	Line internal body cavities, hollow organs	Protection; allows fluids to move between external and internal environments
Ciliated	Flat, cuboidal, or columnar	Line some digestive tissues, parts of the lung	Assist in the movement of fluids and food
Glandular	Cuboidal or columnar	Glands	Secrete specific substances, such as hormones or milk, or excrete certain wastes

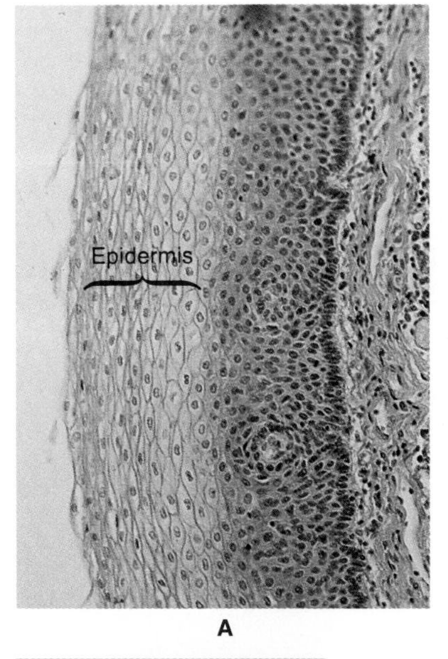

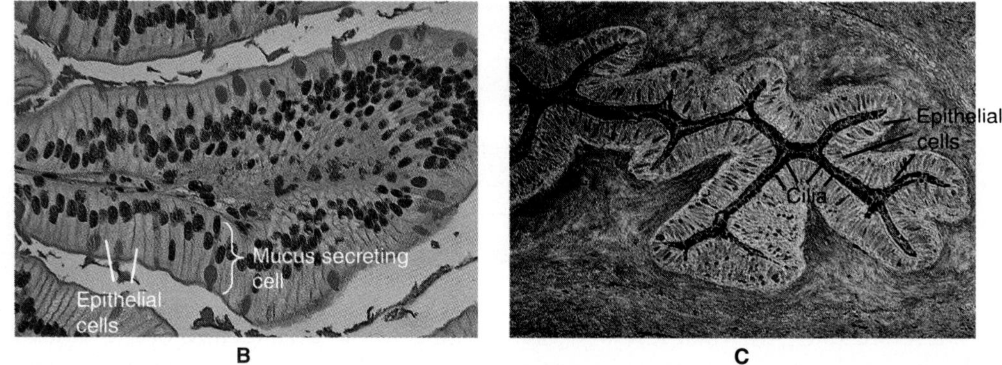

Figure 33.3

Different types of epithelial cells cover and line tissues and organs. (A) Epidermis is the outermost layer of skin and consists of flattened cells. (B) Glandular epithelium is composed of cells that are specialized to secrete substances such as mucus. (C) Ciliated epithelium is specialized to move substances, primarily in the lungs and digestive organs.

Epithelium protects underlying tissues and conducts a variety of important activities for the body, primarily in the areas of chemical synthesis and processing. Epithelial tissues absorb nutrients from the digestive tract; synthesize *mucus,* a slimy, viscous fluid, to lubricate hollow organs (for example, the lungs and the stomach); manufacture digestive enzymes and hormones; and engage in many other functions.

Connective Tissues

Connective tissues, the most abundant tissue type in the body, include a diverse group of organized cells. They function in supporting organs in their appropriate locations, binding different tissues together, storing energy reserves, and other specialized functions. Connective tissues are composed of relatively few cells that exist within an abundant *intercellular matrix*—usually a semifluid, gellike, or fibrous material made up of complex carbohydrates and proteins. Within the matrix are different types of *filamentous fibers* that are categorized according to size and molecular composition (see Figure 33.4A).

The three major types of connective tissue fibers found in humans are shown in Figure 33.4B-D. *Collagenous fibers* are the most abundant and are made up of bundles of smaller fibers, similar to the structural design of a cable. These fibers are made of a protein called *collagen* and are very strong and inelastic. *Elastic fibers* are long, thin, branching fibers made of the protein

Figure 33.4

(A) Connective tissue is composed of a chemical matrix and filamentous fibers that hold cells and tissues together. Collagenous fibers (B), which appear as irregular yellow strands with entangled red blood cells, elastic fibers (C), and reticular fibers (D), which are both stained dark blue or black, are composed of proteins, and each fiber type has properties related to its function. See Table 33.3 for details.

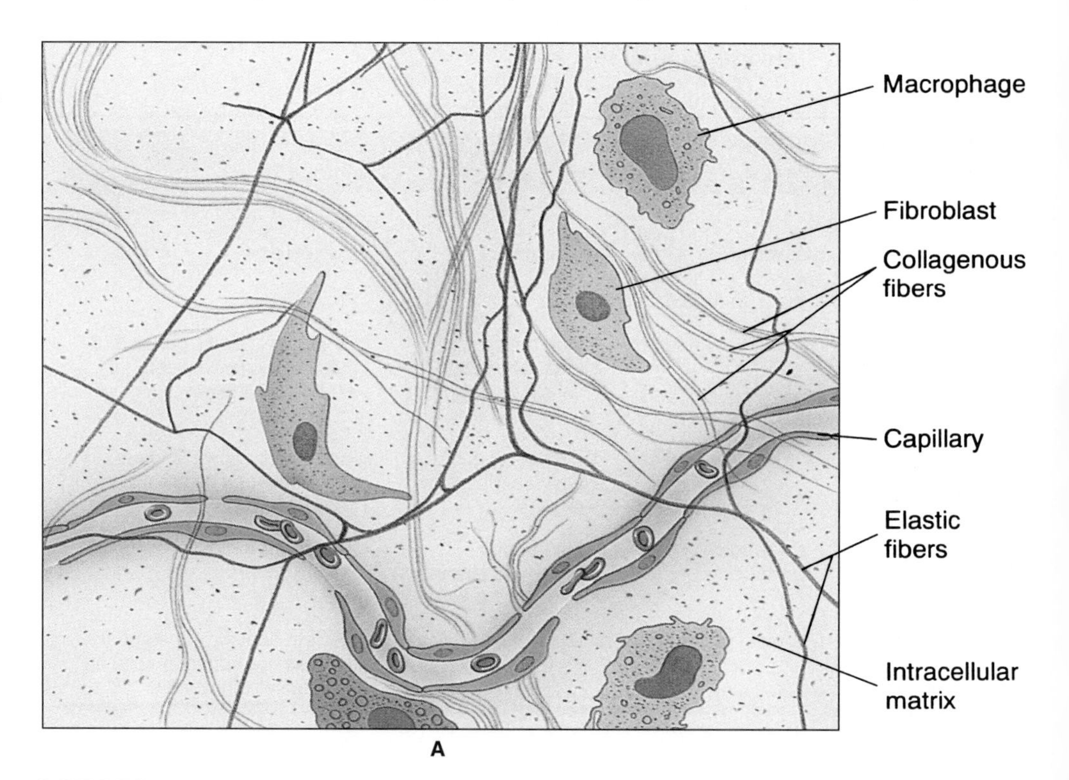

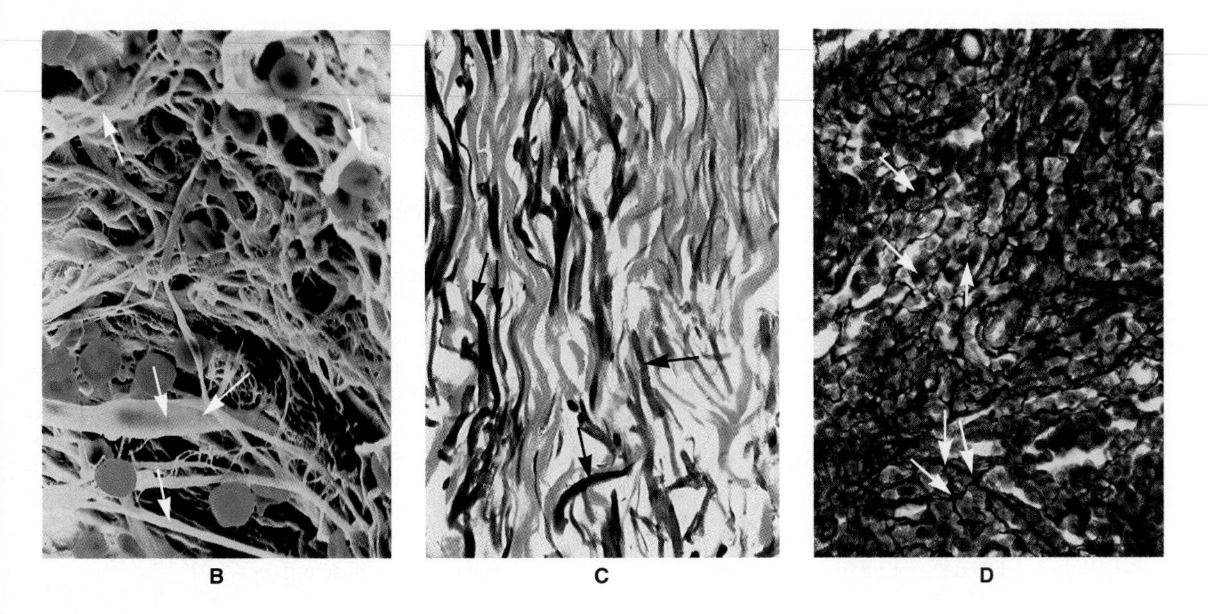

elastin (Figure 33.4C). Elastin conveys the property of resilience to the fibers, allowing them to return to their original length after they have been stretched. Loss of this property is correlated with age and accounts for the unflattering effects of gravity on the skin and external structures of older humans. *Reticular fibers* are thin, short filaments constructed of a protein called *reticulin* (Figure 33.4D). These fibers commonly form tight networks that serve as a scaffolding for the epithelial cells in tissues such as glands.

Common and important cell types associated with the connective tissues are fibroblasts and macrophages. **Fibroblasts** are spindle-shaped cells that produce the various types of connective tissue fibers. **Macrophages** are scavenger cells that wander throughout the body engulfing foreign particles, such as bacteria, and debris from dead cells. They also play a major role in immunity, which is described in Chapter 38.

Connective tissues are classified according to the character of the intercellular matrix and the types and organization of fibers within the matrix. Table 33.3 provides information about the types and functions of connective tissues in the human body. Figure 33.5 illustrates some types of connective tissues in humans. Bone and blood are classified as connective tissues because they are made up of cells embedded in an intercellular matrix. In bone tissue, the matrix becomes *mineralized;* that is, minerals such as calcium are deposited in the fibrous network and form a nearly solid substance. In contrast, the surrounding matrix of blood cells is a fluid called *plasma*. Bone and blood are discussed in Chapters 34 and 38.

Muscle Tissues

Muscle tissue is composed of elongated, thin cells commonly referred to as *fibers*. Muscle fibers are true cells and should not be confused with connective tissue fibers. Each muscle fiber (cell) contains numerous smaller fibers, called *myofibrils,* made of specialized proteins that have the ability to contract. Contraction of muscle fibers is necessary for moving body parts, changing the diameters of hollow organs within the body, moving materials throughout the body, and removing unwanted substances from the body. In addition, muscle contractions require and release energy, which results in the production of significant amounts of heat that help maintain normal body temperatures. The human body contains three types of muscle tissue, shown in Figure 33.6, which are distinguished by their structure, function, and location.

Skeletal muscle consists of extremely long fibers, masses of which are attached to bones by special connective tissues, that cause movements of the

Table 33.3 Major Types and Functions of Human Connective Tissues (CT)

Type	Location	Functions
Loose CT; unorganized collagen fibers	Widely distributed throughout the body	Attaches the skin to the underlying tissue; fills spaces between organs and holds them in place; surrounds and supports blood vessels
Dense CT; abundance of collagen fibers	Specific organs that require support or strong attachments	Provides capacity to withstand high degree of tension; support for underlying tissues of skin; forms tendons, ligaments, and some membranes
Elastic CT; majority of elastic fibers	Organs that expand and return to original size	Confers strength and elasticity to walls of arteries, trachea and bronchi, vocal cords, and lungs
Reticular CT; mostly reticular fibers	Liver, lymph nodes, and spleen	Provides a framework for support of the cells that make up these organs
Adipose CT; consists of fat cells	Underlying skin; in loose CT	Serves as a storage site for fats; insulates, pads, and protects certain areas of the body
Cartilage; modified collagen fibers	Certain skeletal structures, breathing tubes	Provides strong structural support for the embryonic skeleton and ends of bones; forms flexible structures such as the ear and nose
Bone; composed of mineralized matrix	Major portion of the adult skeleton	Provides support and protection; skeleton serves as the attachment site for muscles
Blood; cells in a fluid matrix	Propelled throughout the body, in blood vessels, by the heart	Transports oxygen, nutrients, and other materials; cellular immune functions

skeleton. Skeletal muscle also aids in manipulating facial features and moves the eyes and tongue. Skeletal muscle cells have two unique features: they contain multiple nuclei, and they have characteristic cross-striations due to the alternation of light and dark bands along the myofibrils. The bands are associated with the mechanical contraction of muscle fibers. Skeletal muscle is classified as being under voluntary, or conscious, control of the individual.

Figure 33.5

Adipose tissue (A), cartilage (B), and bone (C) are classified as connective tissues.

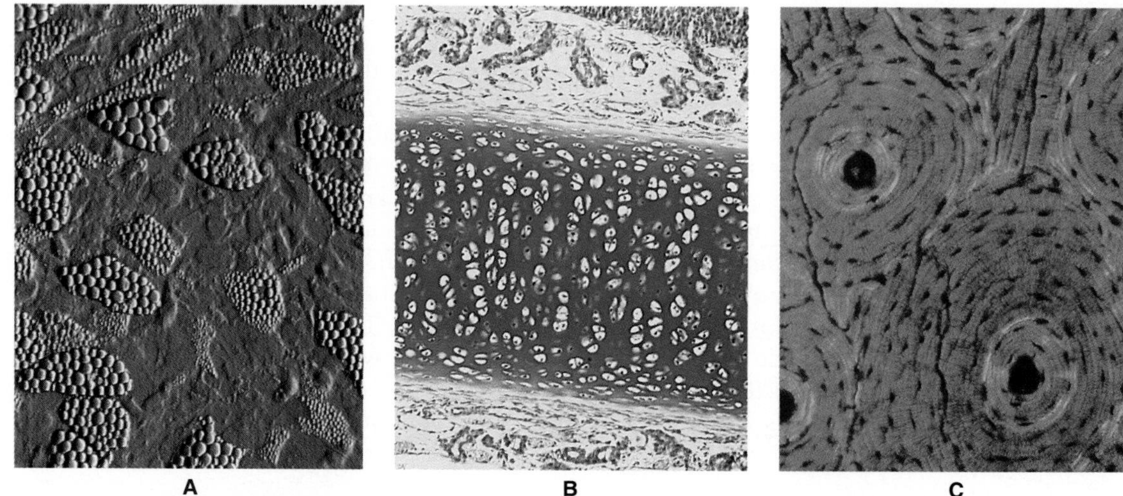

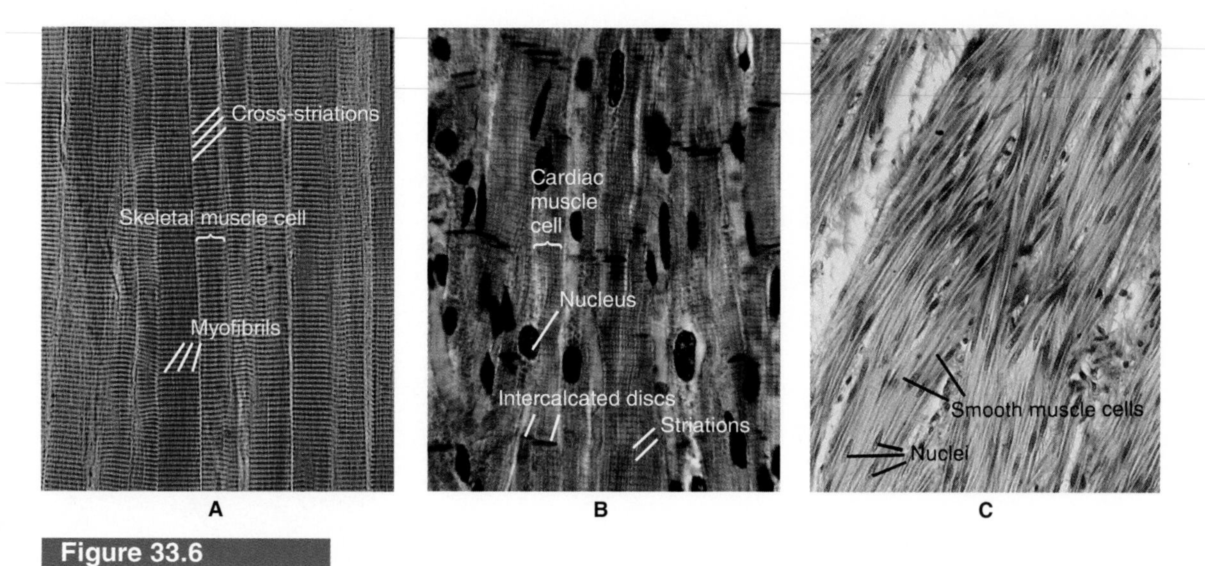

Figure 33.6

Three types of muscle tissue are found in humans. Skeletal muscle (A) is attached primarily to bones, the heart is composed of branched cardiac muscle cells (B), and smooth muscle (C) lines the walls of hollow internal organs and tubes. In smooth muscles, each nuclei represents one muscle cell.

The walls of the heart are constructed of specialized *cardiac muscle* cells, which form branching networks that impart great strength to this critical organ. Cardiac muscle cells contain a single nucleus and have striations that are similar to those in skeletal muscles. They also have unique cross-striations, called *intercalated disks,* that occur where the ends of two cells meet, end-to-end, and form junctions. Cardiac muscle is under involuntary control (how would you like to have to think about when and how often your heart should beat?) and is regulated by hormones and parts of the nervous system.

Smooth muscle, made up of cells with single nuclei and no cross-striations, is present in the walls of hollow internal organs and tubes (for example, intestines and blood vessels) and in many other locations. Contractions of smooth muscle, which serve to move materials through these organs, are also under involuntary control.

Nervous Tissue

The structural unit of **nervous tissue** is the *neuron,* a cell type specialized for the rapid conduction of electrochemical impulses. As shown in Figure 33.7, neurons come in a variety of sizes and shapes, but their basic structure consists of a cell body, which contains the nucleus, and two or more slender extensions involved in the transmission of nerve impulses. Nervous tissues are considered in greater detail in Chapter 35.

Organs

An **organ** is a distinct structure that is made up of more than one type of tissue, has a definite form, and performs a specific function in the body. Consider a familiar example, the stomach. It is a hollow organ with an interior and exterior lined with different types of epithelial tissue; its walls are composed of muscle and connective tissue; and it is regulated by nervous tissue. Each of these tissues contributes to the specific function of the stomach, the breakdown of food.

Most organs are contained within three major body cavities (see Figure 33.8). The *cranial cavity,* or skull, encloses the brain. The *thoracic cavity* contains two critical organs, the heart and lungs, as well as the esophagus, which relays food to the stomach. The *abdominal cavity* is the largest human cavity and holds the organs of digestion and others such as the kidney and liver. Further information on organs is provided in Chapters 35–38.

Figure 33.7

Neurons, the structural and functional unit of nervous tissue, have an amazing diversity of sizes and shapes.

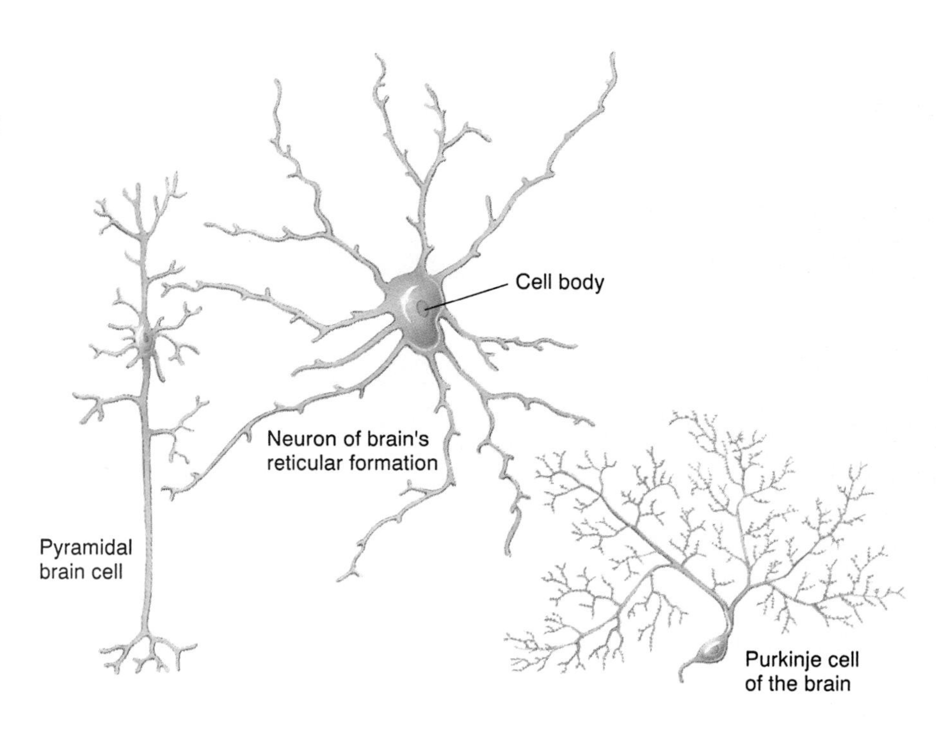

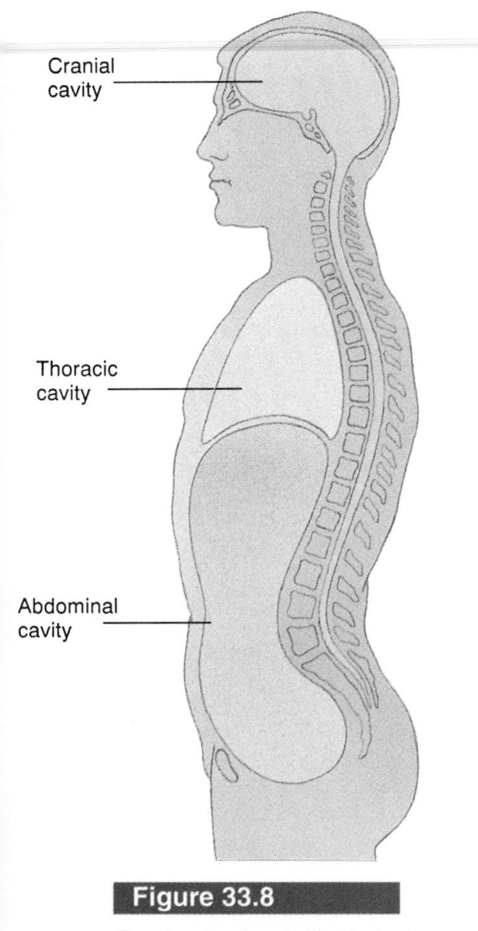

Figure 33.8

The three major cavities in the human body—cranial, thoracic, and abdominal—each contain organs of critical importance.

Organ Systems

Within individual organisms, the highest level of biological structural organization is the organ system. **Organ systems** are groups of organs operating as a unit to perform a major function or group of functions. Truly independent organ systems do not actually exist in humans. Rather, the term *organ system* is a useful way to categorize interacting structures and functions for study. The ten defined human organ systems and their major organs, structures, and functions are summarized in Table 33.4 and illustrated in Figure 33.9 on pages 618–620.

Cells are the basic structural unit of organization. A tissue is an organized collection of cells. The four major human tissues are epithelial, connective, muscle, and nervous. Organs are structures that consist of multiple tissues and perform a specific function. Organ systems are composed of several organs and represent the highest level of structural organization. Ten human organ systems—integumentary, skeletal, muscular, nervous, endocrine, circulatory, respiratory, digestive, urinary, and reproductive—perform all principal functions in the body.

ANATOMY AND PHYSIOLOGY

Anatomy is the study of the structure of an organism and the relationship among its parts. In general, anatomy focuses on whole organs, although microscopic anatomy involves the study of structures that can be seen only with the aid of a microscope. The field of anatomy has existed since the time of the ancient Greeks, when physicians in Alexandria examined dead organisms, including humans, to learn more about their internal structure. Today, it still has great value in reducing the complexity of studying the human body by describing the forms, arrangements, and relationships of organs and organ systems—in other words, how we are constructed.

Physiology is the study of the functions of an organism. It focuses on many levels of organization, ranging from functions of subcellular organelles to multicellular organs. Physiology is considered in greater detail in subsequent chapters.

Table 33.4 The Structures and Functions of Human Organ Systems

System	Major Structures	Functions
Integumentary	Skin and specialized structures such as hair, nails, and sweat glands	Covers and protects internal body structures against injury and foreign materials; prevents fluid loss; regulates temperature
Skeletal	Bones, cartilage, joints, and the ligaments connecting them	Supports and protects soft tissues and organs; provides for body movement
Muscular	Skeletal, cardiac, and smooth muscles	Movement of the skeleton and internal organs; locomotion; propels blood through body
Nervous	Brain, spinal cord, nerves, and sense organs	Primary regulatory system; regulates activities of the body; receives and interprets information from the internal and external environment
Endocrine	Hormone-secreting glands	Regulates many body functions in conjunction with the nervous system
Circulatory	Heart, arteries, veins, capillaries, lymphatic vessels, blood, and lymph	Transports oxygen, nutrients, and hormones between different cells and tissues; removes cellular waste products
Respiratory	Lungs, bronchi, windpipe, mouth, and nose	Exchange of oxygen and carbon dioxide between internal and external environment
Digestive	Mouth, esophagus, stomach, intestines, and accessory organs—liver, salivary glands, pancreas, and gallbladder	Processes foods used for energy and the synthesis of biomolecules and construction of cells and tissues
Urinary	Kidneys, bladder, and associated ducts	Regulation of blood chemistry; eliminates metabolic waste products; maintains water balance
Reproductive	Testes, ovaries, and associated structures, depending on sex	Production of gametes; secretion of hormones influencing growth, maturation, and other activities

A strong interdependence between structure and function is evident at all levels of organization. Every organism represents an integrated unit of structure and function. Examining the structure and function is a time-honored approach for studying humans that will be used in Chapters 34–38.

MAINTENANCE AND CONTROL

The trillions of cells making up the tissues and organs of the human body operate at high levels of activity during most of their existence. Not surprisingly, many of these cells wear out, are injured, or die after a genetically programmed period of service. Are cells of all tissues replaced? How? What are the consequences if they are not replaced? To answer these questions, it is necessary to begin by examining growth prior to adult maturity.

Cell Renewal

Cells of all tissues in the developing human embryo are capable of dividing by mitosis. This permits growth and repair of all tissues during that phase of life. After birth, however, cells of certain tissues become fully differentiated and lose the ability to divide. In other words, the numbers of those cells are fixed soon

Figure 33.9

There are ten major organ systems of the human body.

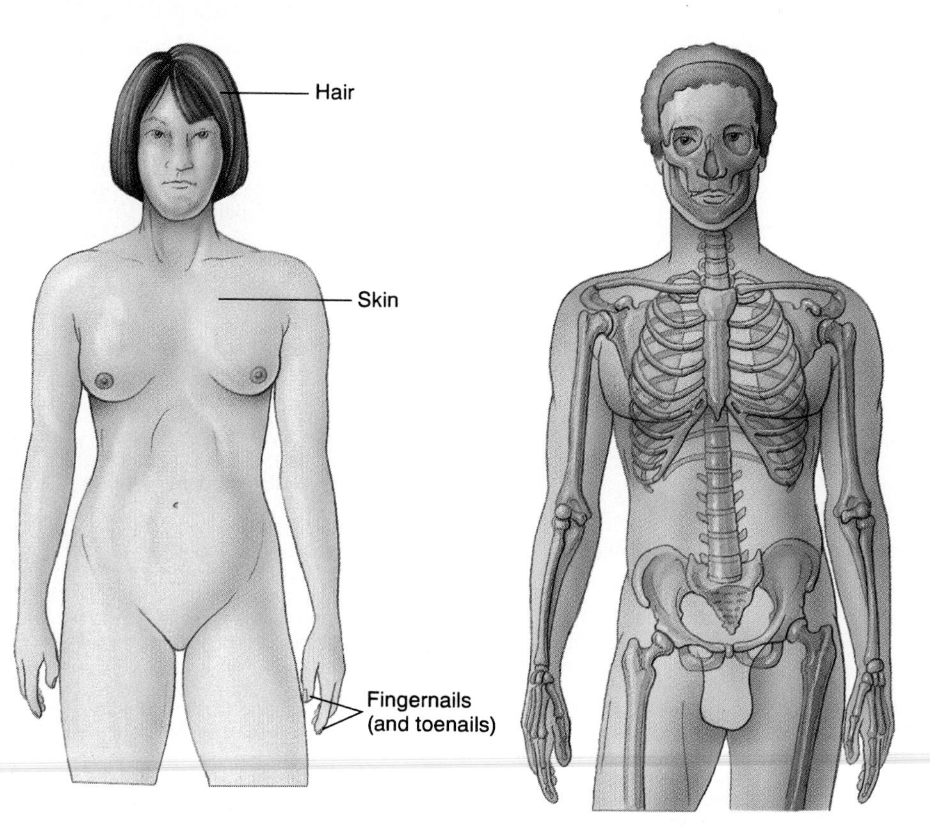

1. Integumentary System

2. Skeletal System

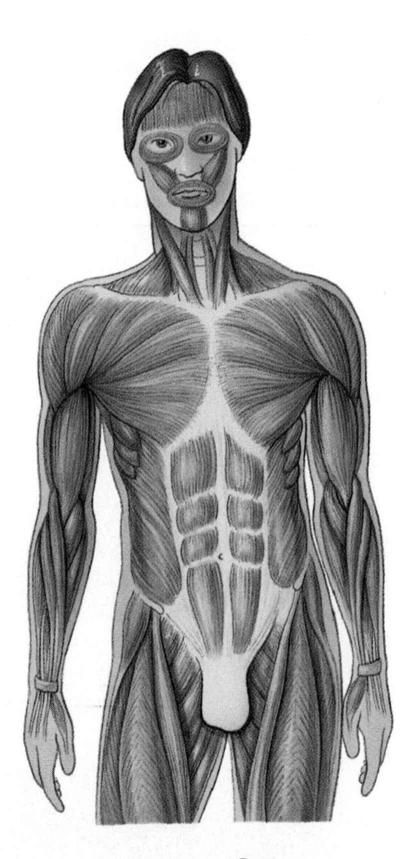

3. Muscular System

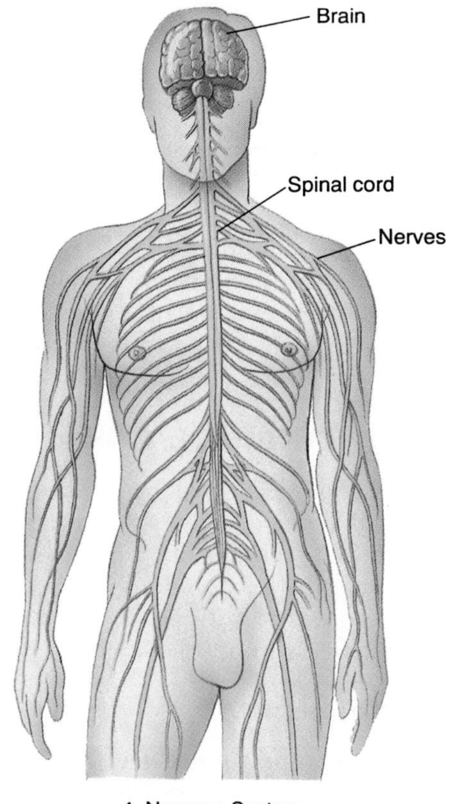

4. Nervous System

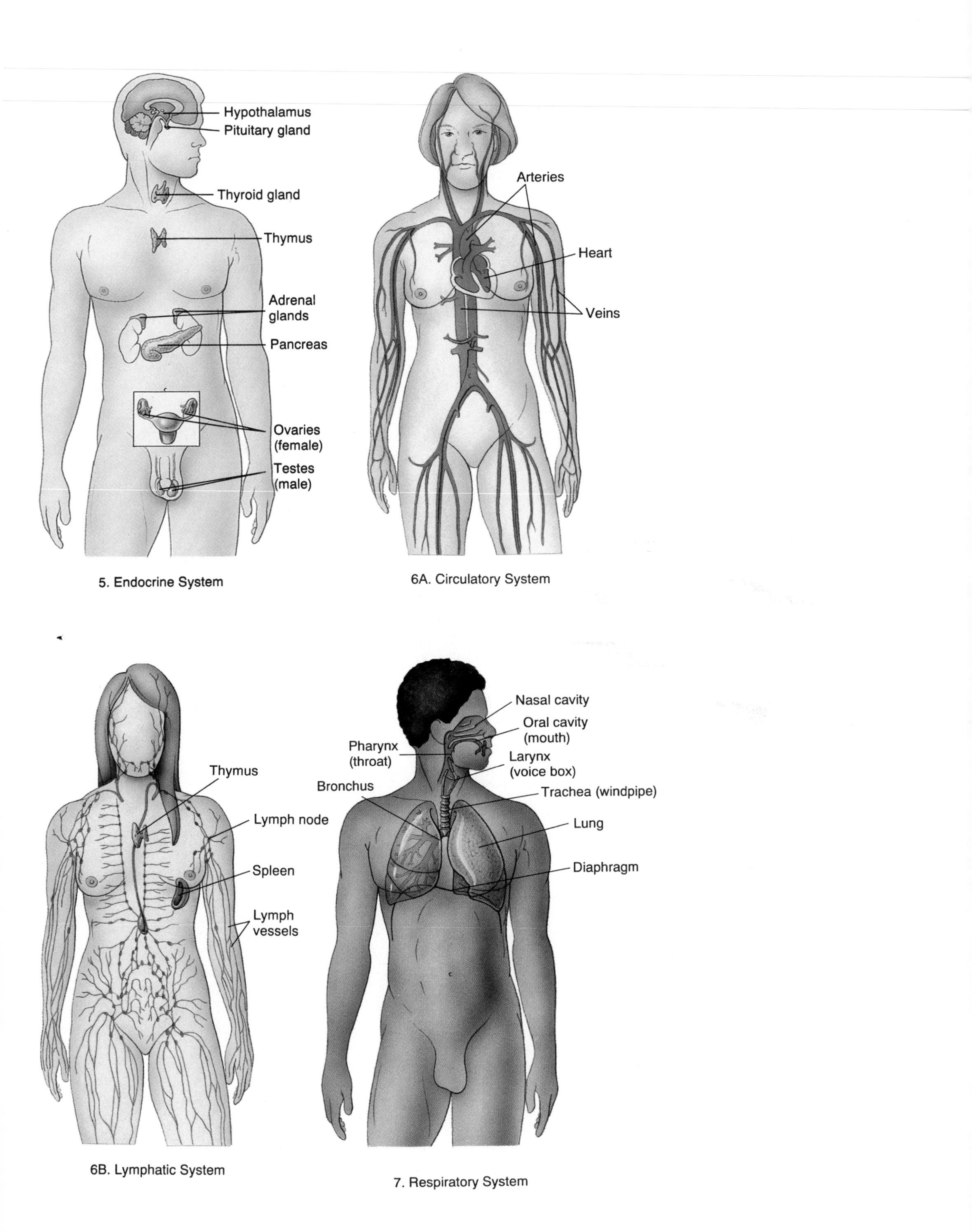

5. Endocrine System

6A. Circulatory System

6B. Lymphatic System

7. Respiratory System

Figure 33.9

(Continued)

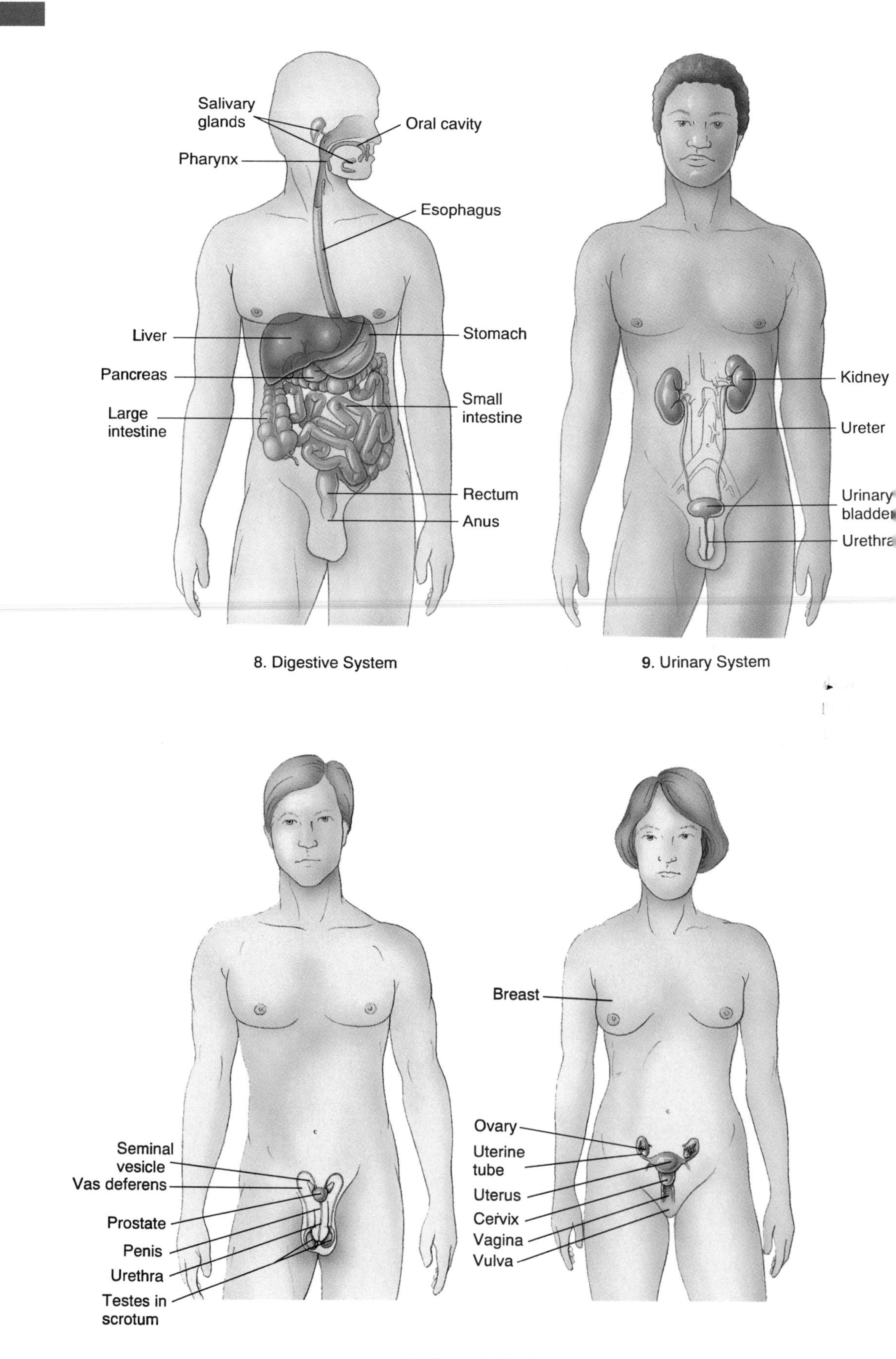

8. Digestive System

9. Urinary System

10. Reproductive System

after birth for the rest of the individual's life. Nervous and muscle tissues are two human tissues that consist of such cells. Once development of these tissues is completed within a year or two after birth, it is generally not possible to replace damaged or destroyed cells. Consequently, injuries to nerves and muscles are permanent and cannot be repaired by natural biological mechanisms.

In contrast, epithelial and connective tissues contain small populations of undifferentiated cells that retain the ability to divide and differentiate into new cells. This is important for two reasons. First, tissues that are injured can be repaired completely and with great accuracy. Second, many tissues consist of cells with genetically programmed life spans. For example, skin (several months), red blood cells (about 120 days), and cells lining the intestine (three to seven days) die and are replaced continuously throughout an individual's life. Other tissues and organs—including glands, blood, and certain digestive, respiratory, and urogenital tissues—have similar replacement systems.

To illustrate the importance of cell renewal systems, consider the normal requirement for replacing red blood cells (RBCs) in humans. The average adult human contains about 25 billion RBCs, which have a life span of about 120 days. To maintain a constant population of RBCs, it is necessary for the body to produce, by mitosis, 2.5 million new cells each second of your life! It is estimated that the total number of cells produced and differentiated by the replacement systems in the human body exceeds 10 million per second.

Homeostasis

Human cells are specialized to survive and function under relatively constant internal conditions (for example, temperature, fluid compositions, and levels of glucose) that are continuously subjected to forces of change from both the external and internal environments. **Homeostasis** refers to the condition of a relatively constant internal environment. To ensure this state, there are physiological mechanisms to resist or compensate for changes that are always occurring in the dynamic internal environment (see Figure 33.10).

Homeostatic Mechanisms

All organ systems participate in maintaining homeostasis. For example, if the internal environment becomes overheated, temperature is reduced directly through heat-dissipating operations of the skin and circulatory system (sweating). However, balance is ultimately restored through activities of the nervous, urinary, and other systems. How does the body sense changes in the internal environment? What underlying mechanisms compensate for these disruptions?

Negative Feedback Mechanisms

The body uses a regulatory mechanism known as *negative feedback* to maintain stable conditions in the internal environment. **Negative feedback** refers to a process, described in Figure 33.11, in which the following sequence of events occurs: (1) feedback from a controlled system is monitored by a receptor that (2) relays information to a processing center, which (3) causes the level of the variable to change in the *opposite* direction from the original change if it deviates from a set point value. A familiar example is a thermostatically controlled heating system. If the temperature (the variable) falls below a set point, a receptor in the thermostat (processing center) activates the heating system (controlled system), and heat will be emitted until the set point temperature is reached.

There are several thousand control systems in the human body, most of which involve negative feedback mechanisms regulated by the nervous and endocrine (hormonal) systems. Because of their critical role in many disease processes, these systems have become a very important area of research in human medicine.

Figure 33.10

The respiratory, circulatory, digestive, and urinary systems have major responsibilities in maintaining homeostasis.

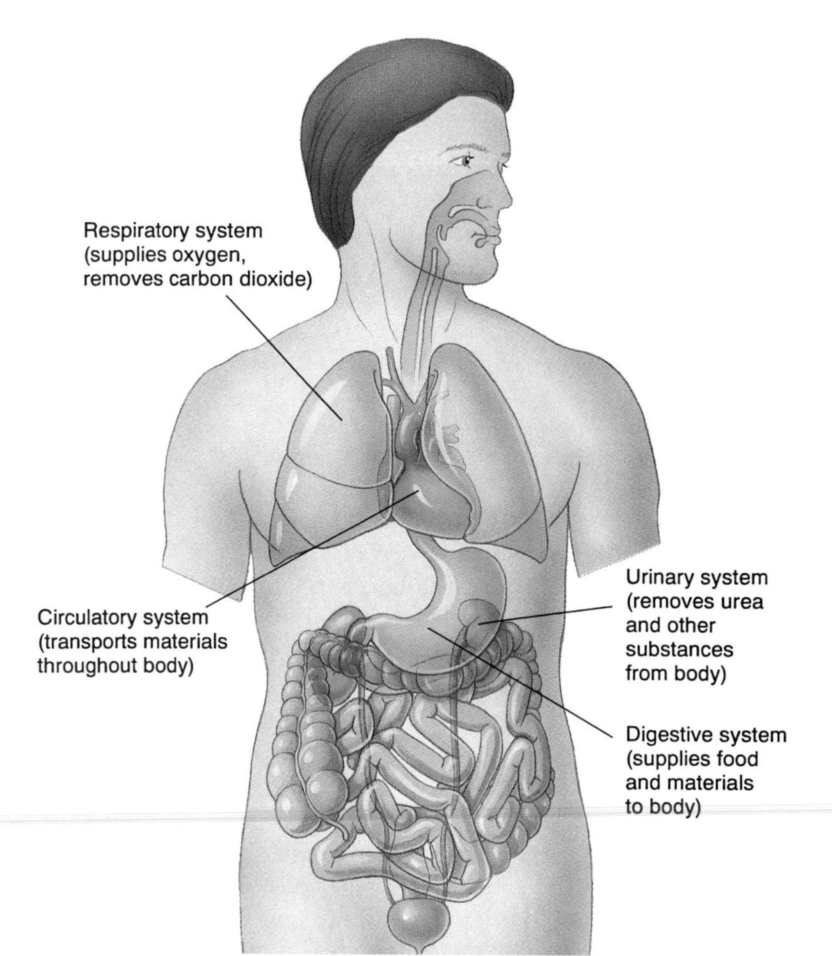

Figure 33.11

Homeostasis is maintained primarily through negative feedback systems. The systems in the human body (A) operate much like a home heating and cooling system (B).

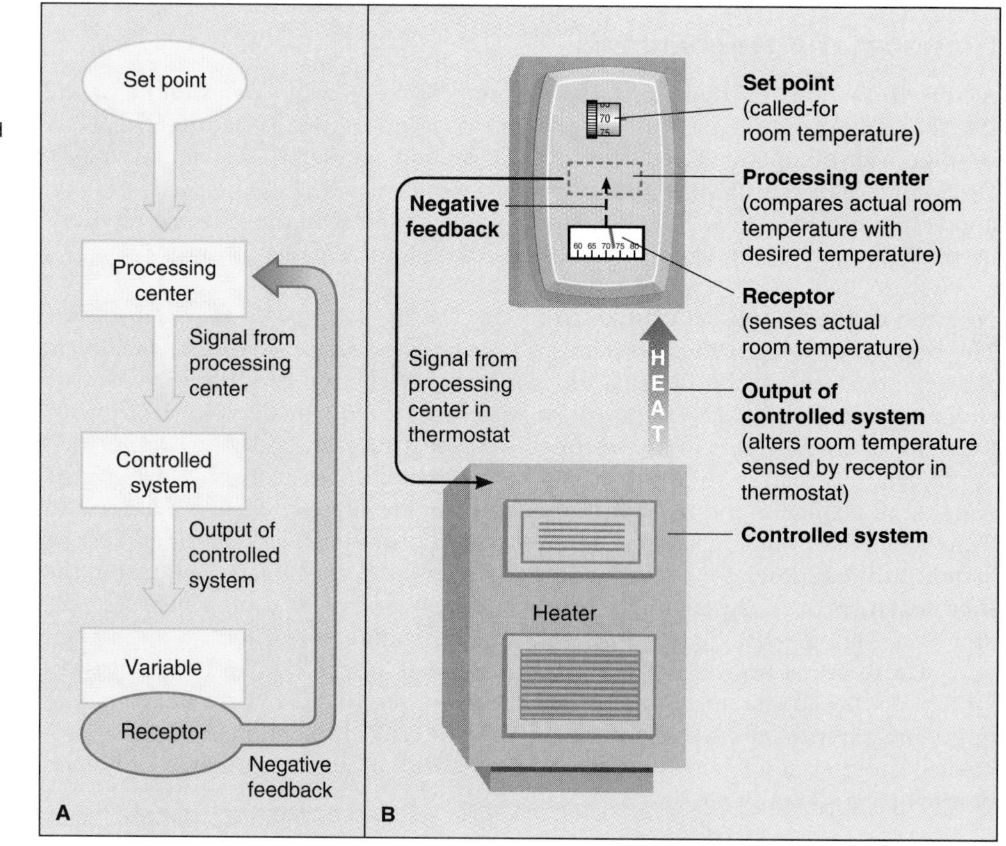

Positive Feedback Mechanisms

Positive feedback mechanisms tend to cause the level of the variable to change in the *same* direction as the initial change. Thus they do not lead to maintenance of stable internal conditions but rather tend to increase the level of a variable further. Since this mechanism does not promote internal stability, it is not commonly found in the human body. One example of a positive feedback mechanism occurs during childbirth, when the pressure of the baby's head against the birth canal stimulates further muscular contractions. In this case, positive feedback promotes expulsion of the baby at birth.

A valuable approach for studying organisms is by concentrating on anatomy and physiology—structure and function. The health and integrated control of organ systems depend on a number of mechanisms. Cells with finite lives are regularly replaced through the orderly operations of cell renewal systems. Homeostatic processes, primarily negative feedback mechanisms, act to regulate and coordinate the activities of organ systems.

UNDERSTANDING THE HUMAN BODY

The human is a wondrously complex organism. One of the most stimulating and worthwhile endeavors individuals can undertake is to learn about the operations of their bodies. Without such knowledge, it is difficult to acquire an adequate understanding of self. In the chapters that follow, we explore basic human structures and functions, the regulation and control of organs and selected physiological processes, and the consequences when failures occur at different levels of organization.

Summary

1. Human organs and tissues are composed of an enormous variety of cells that enable them to carry out specific functions.
2. All cells have the capacity for self-maintenance—the ability to repair damage and maintain health. A limited number of cell types are able to reproduce and in so doing provide a continuous supply of cells to replace those that die in the normal course of events. Most cells are highly specialized, and some carry out unique functions such as conducting an electrochemical impulse, transporting oxygen, or storing fat.
3. There are four structural levels of organization in higher organisms. Organ systems are composed of organs, organs are made up of tissues, and tissues consist of millions of cells.
4. The basic types of tissues (and their primary function) are epithelial (covering body surfaces and lining cavities), connective (binding and supporting tissues and organs), muscle (moving organs), and nervous (conducting electrochemical impulses).
5. Anatomy is the study of structure, and physiology is the study of function. These two branches of biology have long been used in the systematic investigation of organisms and how they operate.
6. Mechanisms of maintenance and control in the body ensure optimal conditions for normal cell, tissue, and organ functions. For example, new cells are normally produced at the same rate at which cells are lost, and the internal environment remains relatively constant because of homeostatic mechanisms.
7. Homeostasis, the condition of a constant environment, is maintained primarily through negative feedback mechanisms, which operate on the principle of changing the direction of the variable being monitored. If temperature increases to a dangerous level, for example, negative feedback will lead to a decrease in temperature.

Review Questions

1. What does the cell theory tell us about the structure and function of organisms?
2. What is the basic structure of eukaryotic cells?
3. What is the importance of cellular self-maintenance and cellular reproduction?
4. How do cells acquire an ability to perform specialized functions?
5. What is a tissue? Describe the basic structure and function of epithelial tissue, connective tissue, muscle tissue, and nervous tissue.
6. What is an organ? An organ system? What are the ten major human organ systems?
7. How are anatomy and physiology related?
8. What is the function of a cell renewal system?
9. What is the role of homeostatic mechanisms?
10. What is the difference between a negative and a positive feedback mechanism?

Essay and Discussion Questions

1. Do you think it is likely that the cell theory will be further modified or extended in the future? Why?
2. The endosymbiotic hypothesis explains the origin of cells (see Chapter 2). Describe a possible outline for the evolution of higher levels of structural organization in animals. Information in Chapters 3 and 4 might be useful.
3. Emphasizing anatomy and physiology is a common method for learning about organisms. What other approaches might be used in acquiring an understanding of organisms?

References and Recommended Reading

Alberts, B. M. 1989. Introduction: On the great excitement in cell biology. *American Zoologist* 29:483–486.

Barnes, R. D. 1987. *Invertebrate Zoology.* 5th ed. Philadelphia: Saunders.

Chase, R. F. 1989. *The Bassett Atlas of Human Anatomy.* Menlo Park, Calif.: Benjamin/Cummings.

Crawshaw, L. I., B. P. Moffitt, D. E. Lemons, and J. A. Downey. 1981. The evolutionary development of vertebrate thermoregulation. *American Scientist* 69:543–550.

de Duve, C. 1984. *A Guided Tour of the Living Cell.* New York: Scientific American Books.

Junqueira, L. C., J. Carneiro, and J. A. Long. 1986. *Basic Histology.* 5th ed. Los Altos, Calif.: Lange.

Porter, K. R., and J. B. Tucker. 1981. The ground substance of the living cell. *Scientific American* 244:56–67.

Ratnoff, O. D. 1987. The evolution of hemostatic mechanisms. *Perspectives in Biology and Medicine* 31:4–24.

Schmidt-Nielsen, K. 1990. *Animal Physiology: Adaptation and Environment.* 4th ed. New York: Cambridge University Press.

Tortora, G. J., and N. P. Anagnostakos. 1990. *Principles of Anatomy and Physiology.* 6th ed. New York: HarperCollins.

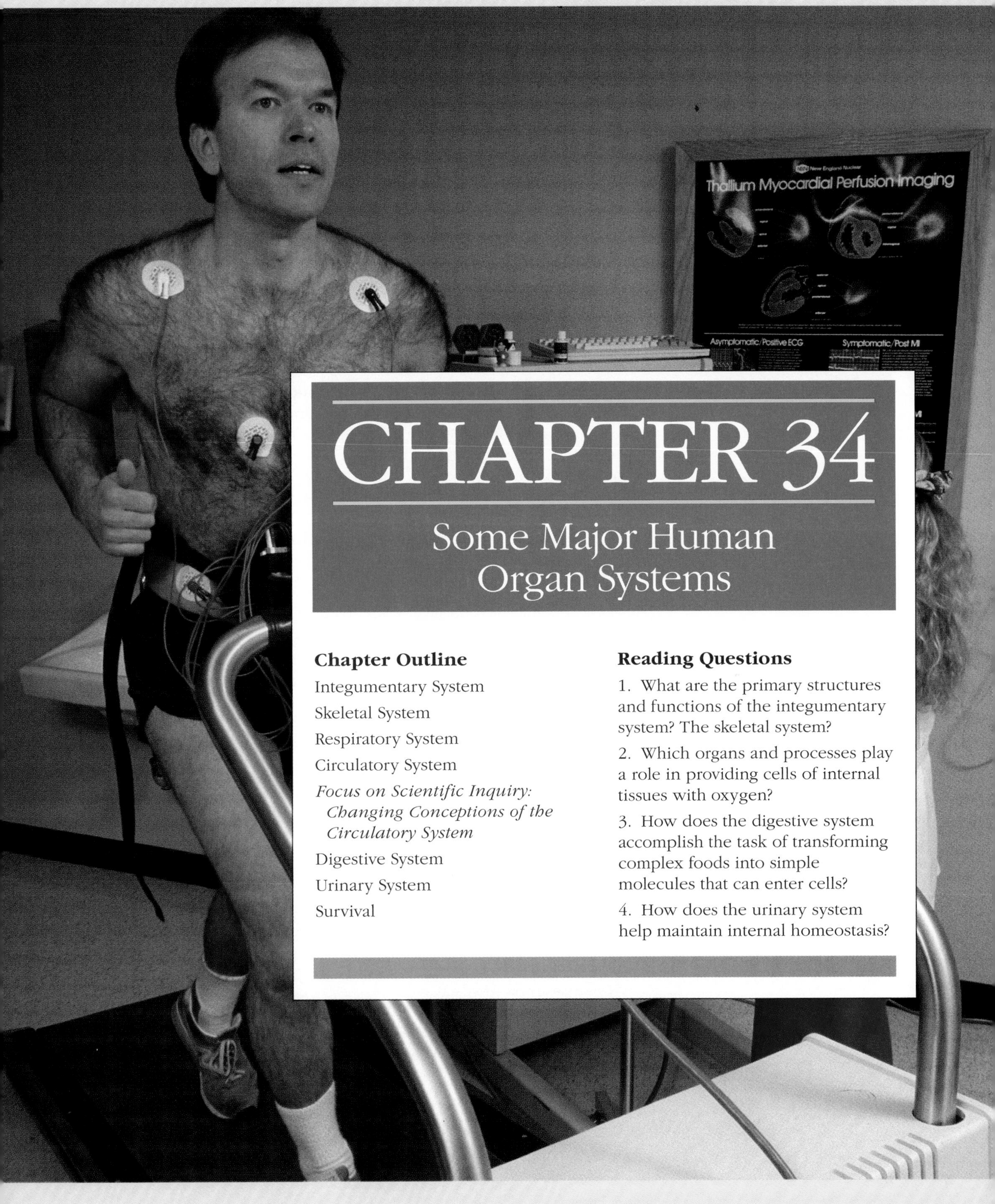

CHAPTER 34

Some Major Human Organ Systems

Chapter Outline

Integumentary System

Skeletal System

Respiratory System

Circulatory System

Focus on Scientific Inquiry: Changing Conceptions of the Circulatory System

Digestive System

Urinary System

Survival

Reading Questions

1. What are the primary structures and functions of the integumentary system? The skeletal system?
2. Which organs and processes play a role in providing cells of internal tissues with oxygen?
3. How does the digestive system accomplish the task of transforming complex foods into simple molecules that can enter cells?
4. How does the urinary system help maintain internal homeostasis?

Multicellular animals first evolved hundreds of millions of years ago. The first primitive animals, such as the unadorned sponges, had only rudimentary organizations of cells that were adapted to carry out certain essential functions, such as capturing food. Over millions of years, more complicated animals arose, and cells in them became organized into complex tissues and organs. Finally, Chordata, the taxonomic group that contains fishes, frogs, snakes, birds, and mammals (including humans), evolved. These organisms have elaborate organ systems that enable them to solve a great assortment of biological problems.

What types of biological problems do humans face, and how do we solve these problems? In this chapter, we consider some organ systems that have functions related to the following questions: What covers and protects us from the external environment? Why are we able to win the battle with gravity? How do our cells obtain the nutrients and oxygen they require? How does the body get rid of all the wastes generated during cellular activities? How are we able to maintain a constant internal environment?

INTEGUMENTARY SYSTEM

The **integumentary system** includes the **skin**, the external layer that covers the body, and structures such as nails, hair, and certain glands. The skin, shown in cross section in Figure 34.1, may be the least appreciated human organ system because of its apparent simplicity. However, microscopic examination reveals a complex arrangement of cells, tissues, and specialized structures that have many important functions. Skin is composed of two primary layers: a thin, outer layer of closely packed cells called the **epidermis** and an inner layer, the **dermis**, consisting of connective tissue and various cells and tissues.

Epidermis

The epidermis has several layers of cells, as shown in Figure 34.2A. The *basal cell layer* contains undifferentiated cells that reproduce rapidly and continuously by mitosis in order to replace cells that are sloughed from the skin's surface. The newly formed cells mature as they migrate to the surface of the skin. As maturation proceeds, cells start to become flattened and synthesize *keratin,* a tough, fibrous protein that is flexible and impermeable to water. As the flattened, keratinized cells approach the surface, they die and become closely packed to form a cell layer called the *stratum corneum*. These dead cells, which form the outer

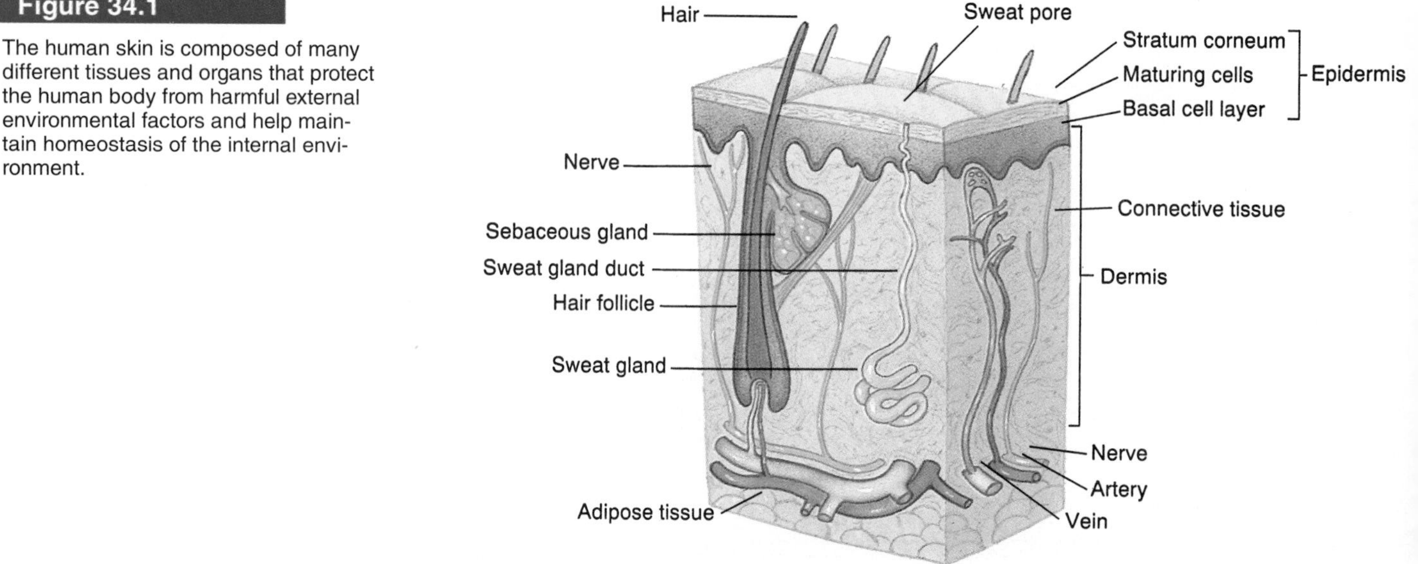

Figure 34.1

The human skin is composed of many different tissues and organs that protect the human body from harmful external environmental factors and help maintain homeostasis of the internal environment.

protective layer of the skin, are continuously sloughed from the skin's surface. The rate at which they are lost is equal to the rate of replacement from reproduction by the basal cells.

Melanocytes are epidermal cells that synthesize *melanin,* the brown pigment primarily responsible for eye, hair, and skin color (see Figure 34.2B). Exposure to sunlight, specifically the ultraviolet rays, stimulates melanocytes to enlarge and increase melanin production. Thus a suntan, an increase in skin pigmentation, is actually the end result of a biological protective response, since the function of melanin is to protect the skin from harmful irradiations of the sun. Can you explain the evolutionary advantage, for those human populations inhabiting equatorial regions of the world, of having darker skins?

Dermis

The epidermis is firmly attached to the underlying dermis. The dermis holds an assortment of specialized structures such as hair and sweat glands that are embedded in underlying connective tissues and fibers.

Hair

Like the outer epidermis, hairs are constructed of dead, keratinized cells but trace their origins to living, rapidly dividing cells in deeper tissue layers (Figure 34.2C). Two structural types of hair are produced by *hair follicles* located in the dermis. Fine, soft hairs cover much of the body, and coarser hair grows on everyone's scalp, eyebrows, and genital areas and on the face of males. *Sebaceous glands* associated with hair follicles secrete an oily substance that waterproofs the skin, prevents hair from drying, and suppresses the growth of bacteria. These glands become especially active during adolescence, and when they become obstructed or infected, pimples, the universal affliction of teenagers, develop.

Eyebrows and eyelashes offer protection against particles entering the eyes. For the most part, however, human hair is thought to be an evolutionary relic that served as insulation for preventing heat loss in our hominid ancestors. The continuing presence of scalp hair may be partly related to that function, since the rate of heat loss is greatest through the scalp. Nevertheless, the importance of a

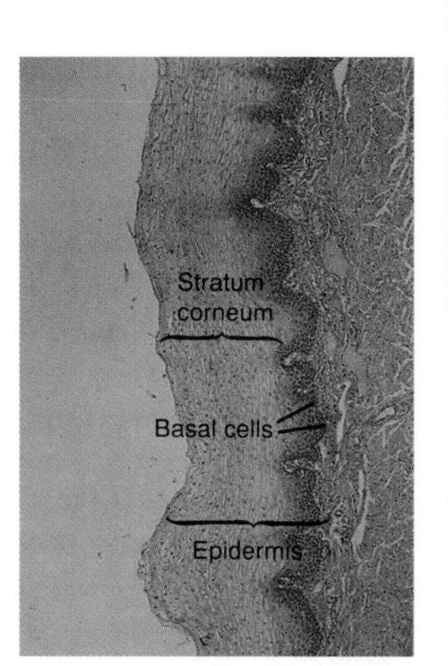

A

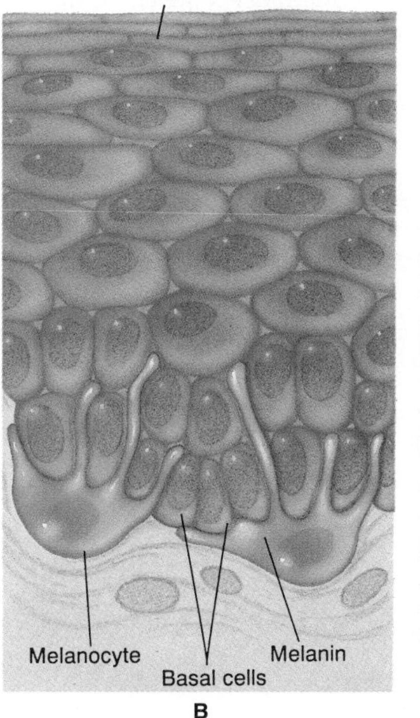

B

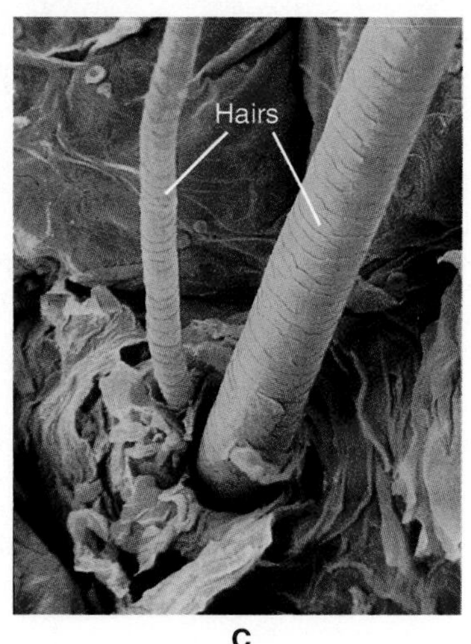

C

Figure 34.2

(A) The epidermis is composed of different types of cells. Basal cells produce new cells by mitosis. As these cells mature, they move toward the surface of the skin, become keratinized, and finally become part of the stratum corneum. (B) Melanocytes in the epidermis produce melanin, a pigment that gives hair and skin their color. Melanin also protects skin cells from the harmful effects of ultraviolet radiation from the sun. (C) Hairs project through openings in the stratum corneum.

heat-retaining hair covering has been diminished for humans by clothing and, more recently, by occupancy of heated dwellings.

The activities of hair-producing follicles in the scalp are cyclical. Normally, follicle cells are busy dividing and constructing hair at a rate in excess of 1 centimeter per month. However, follicles periodically enter a quiescent phase after months or years of activity, and hair growth is arrested for several months. Approximately 10 percent of the hair follicles are in the resting phase at any particular time, and some anxiety may occur as these hairs are shed when being combed or washed.

Sweat Glands and Heat Regulation

The normal internal, or core, temperature of humans is 37°C (98.6°F), although it fluctuates throughout the day. Heat is a by-product of metabolic reactions that occur endlessly in all living cells. Except during exposure to cold temperatures, a certain amount of body heat must be dissipated to maintain a constant internal temperature.

The skin functions as a controlled radiator that regulates heat carried by circulating blood. Ninety percent of the total excess heat is dissipated through about 2.5 million *sweat glands.* These glands are coiled structures that secrete *sweat,* a complex solution containing salts and small amounts of nitrogen-containing metabolic waste products. Most of the heat is lost when *perspiration,* the secretion of sweat onto the skin's surface, evaporates, being converted from liquid into vapor. This change requires energy, which is provided by the heat.

This homeostatic mechanism usually operates unnoticed except when excessive sweating occurs during periods of exercise or high external temperatures. When that happens, the core temperature increases beyond the normal range, and sweat glands are stimulated to increase the amount of perspiration through a negative feedback system described in Figure 34.3. This facilitates the loss of greater amounts of heat through evaporation. Evaporation is most efficient in a dry atmosphere and operates ineffectively in a humid environment. Thus special precautions are necessary when living in a hot, humid environment since overheating may have potentially serious consequences. Also, when the external temperature is cold, blood flow to the skin is reduced, and consequently, less heat is lost from the internal environment.

In summary, the epidermis is a tough, waterproof covering that serves a variety of protective functions. It acts as a flexible armor against injury, inhibits water loss, aids in regulating body temperature, prevents the entrance of harmful microorganisms, and protects the underlying tissues from excessive, harmful, ultraviolet radiation emitted by the sun. The dermis adorns the skin with hairs and plays a critical role in regulating internal body temperature.

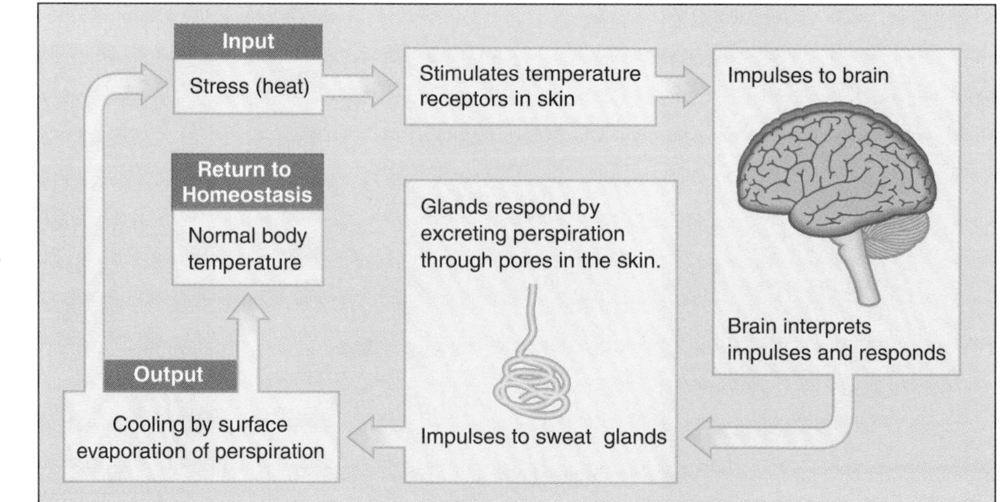

Figure 34.3

One role of the skin is to help regulate body temperature. It performs this function through a classic negative feedback system. When stimulated by heat, temperature receptors in the skin transmit impulses to the brain. Subsequently, sweat glands are stimulated to produce sweat, which evaporates from the skin's surface and results in cooling. When normal body temperature is restored, sweat production ceases.

The skin is a large, complex organ system that is the primary part of the integumentary system. Skin consists of epidermis and dermis, each having specialized functions. The basal cell layer of the epidermis generates cells that develop into the outer layer of the skin. Melanocytes produce melanin, a pigment that protects skin from harmful sunlight. The dermis contains hair follicles, sebaceous glands, and sweat glands, all of which have protective functions.

SKELETAL SYSTEM

Perhaps because of its historical association with Halloween and its traditional use in depicting Death, the misconception persists that the human skeleton is a dead structure. The living skeletal system really consists of *active* tissues and organs. The structural design and architecture of bones, cartilage, joints, and ligaments, as measured by strength and flexibility, constitute a marvel of biological engineering.

Bones and Cartilage

To the frustration of beginning medical students who are required to know such things, the skeleton of the average human adult contains 206 bones. The general structure of the human skeleton and some specific bones are shown in Figure 34.4. **Bone** is a connective tissue with cells embedded in a matrix of calcium compounds, which impart hardness, and collagen fibers, which provide strength. This chemical composition provides bones with enormous strength and makes them capable of supporting massive weights without bending or breaking.

The architecture of different bones varies according to their shape, size, and function. For example, long bones of the arms and legs have a solid, strong outer layer of *compact bone* (see Figure 34.5). The length and strength of these bones are important for mobility and as sites for muscle attachment. The inner portion of these bones consists of a network of *spongy bone* that surrounds the **bone marrow**, a type of connective tissue capable of storing fat (yellow marrow) or making several types of blood cells (red marrow). Bones are dynamic organs. They continue to grow and undergo structural modifications throughout an individual's lifetime, although most skeletal growth is completed by about 20 years of age.

Cartilage is connective tissue composed of cells, collagen, and elastic fibers embedded in a solidified chemical matrix. Cartilage is tough but flexible and is found in joints and the ends of long bones, such as those of the arms and legs, in adults. Different types of cartilage are also involved in the construction of the windpipe, backbone, and ears.

Joints and Ligaments

Joints are sites of connection between bones or between bones and cartilage. These structures are linked in such a way that skeletal movement is possible. The various joints are categorized by their structure and the types of movement that they allow. Bones in joints are connected to one another by bands of tough connective tissue called **ligaments**. Ligaments also serve to restrict bone movement, thus preventing bones from being dislocated.

Functions of the Skeleton

Besides storing blood cell-producing cells and energy reserves, the skeleton has other functions. The two most obvious are to provide support for the body and protection of critical internal organs. Different parts of the skeleton are designed to protect certain organs. The *skull* shields the brain, eyes, and ears from injury. The backbone or *vertebral column* is composed of separate small bones called

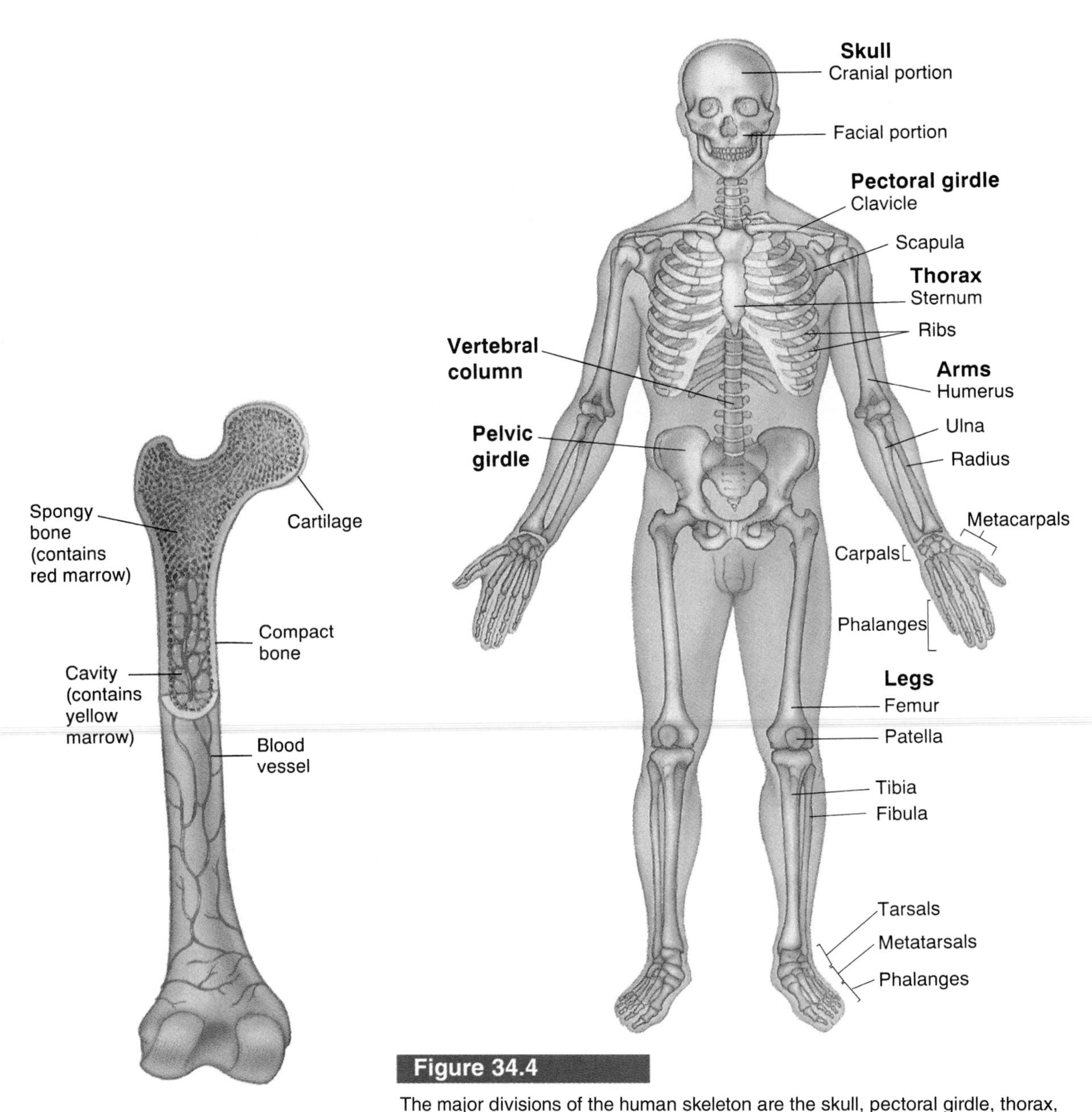

Figure 34.4

The major divisions of the human skeleton are the skull, pectoral girdle, thorax, arms, vertebral column, pelvic girdle, and legs. The names of a few of the 206 bones of the human skeleton are given.

Figure 34.5

A long bone such as the humerus consists of an outer part composed of compact and spongy bone that impart strength. The inner spongy bone contains red marrow, which produces red and white blood cells, and yellow marrow, which stores fat.

vertebrae, a type of construction that imparts great strength but also allows great flexibility. The backbone provides primary support of the skeleton and provides attachments for the skull, rib cage, and pelvic girdle. It also protects the spinal cord from injury. The lungs, heart, stomach, and liver are covered by the *rib cage,* and the *pelvic girdle* offers protection for female reproductive organs.

The skeleton contains significant quantities of many minerals, including calcium, potassium, sodium, and phosphorus; this explains why the skeleton remains intact long after death. The importance in life, however, is that the skeleton serves as a mineral storehouse within the body. When there is a demand for a specific mineral, it can be mobilized from bones and distributed by the blood to sites where it is required. This homeostatic mechanism is especially important in maintaining appropriate levels of calcium in the body. Calcium is essential for normal muscle function. Also, during pregnancy, some calcium from the mother's skeleton is transferred to the developing skeleton of the fetus. Finally, the skeleton provides sites for muscle attachment. The action of muscles imparts movement to the human skeleton.

The skeleton is constructed of bones, cartilage, joints, and ligaments. Together, these structures constitute an internal support system that has enormous strength and great flexibility. The principal functions of the skeletal system are to provide support and protection for internal organs and to permit movement of the organism. The skeleton is a dynamic system, with cells and tissues that remain active throughout a person's lifetime.

RESPIRATORY SYSTEM

All cells require a steady supply of energy for carrying out their activities, and oxygen (O_2) is required for the reactions of cellular respiration that provide such energy (see Chapter 29). Consequently, most human cells die quickly if deprived of O_2 for more than a few minutes. It is also necessary that carbon dioxide (CO_2), the waste product of energy-releasing reactions, be removed from the body. The primary gas exchange mechanism involves diffusion and differences in concentrations of CO_2 and O_2 at exchange sites. Air, containing O_2, is taken into the body from the environment, is moved to the organs of respiration, and diffuses to blood cells that deliver it to tissues. Carbon dioxide is then transported to the lungs by returning blood. The exchanges of O_2 and CO_2 between an organism and the environment is called **respiration**.

Organs of Respiration

The respiratory system, shown in Figure 34.6, consists of the **lungs**, the principal organ of respiration, and two treelike passages that lead to and from the lungs. Air entering through external openings, the *nostrils,* is filtered by small hairs that line the nasal surfaces. It then moves into hollow chambers called *nasal cavities,* where it is warmed, humidified, and filtered. Any small airborne particles present in the inhaled air are absorbed by the mucus layer that lines the cavity. The mucus and particles are continuously swept into the throat by ciliated epithelium

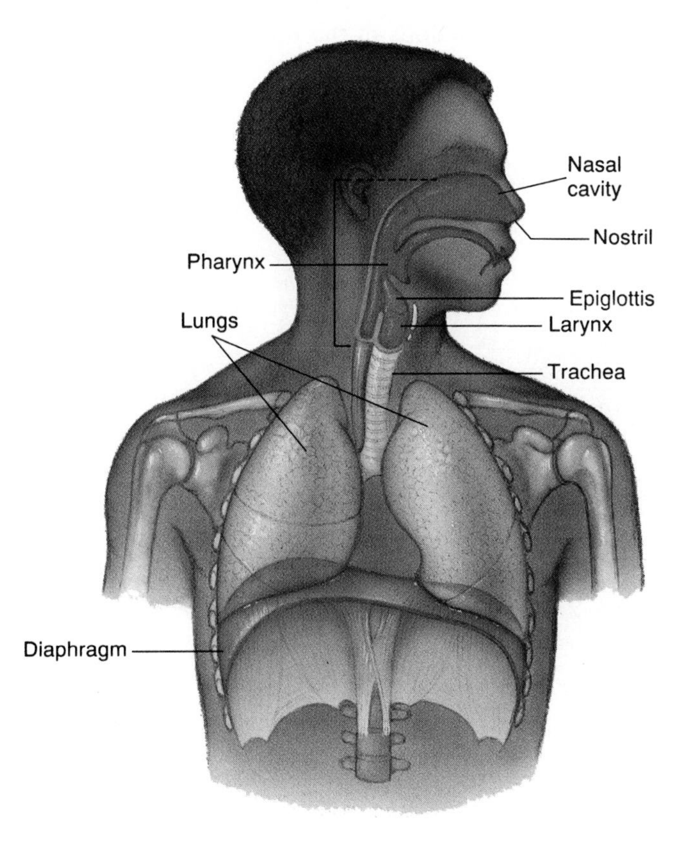

Figure 34.6

The human respiratory system provides for efficient CO_2O_2exchange.

that lines the nasal cavities, swallowed, and routed into the digestive tract, which ultimately disposes of them.

Air then progresses through a set of organs that divert it into a separate pathway from that taken by food and liquids. All these materials initially enter the *pharynx,* or throat. There they become separated by a flap of tissue called the *epiglottis,* which automatically seals the entry into the air passageways during swallowing. When there is a malfunction of this organ and food is allowed to enter the air passage, paroxysms of uncontrolled coughing result in an attempt to remove the obstacle. If this reflexive action fails and the air passage is blocked, choking ensues, and death results if the situation is not corrected.

From the pharynx, air moves through the *larynx,* commonly known as the "Adam's apple," and then into the windpipe, or *trachea,* the trunk of the respiratory tree. The structure of both the larynx and the trachea reflect their function as air conduits; outer walls containing rings of cartilage hold these tubes constantly open.

As shown in Figure 34.7A, the trachea divides into two *bronchi,* one advancing into each lung, that continue to branch into smaller *bronchioles.* These in turn branch repeatedly into tiny clusters of terminal, thin-walled sacs called *alveoli* that are the sites of gas exchange in the lung (see Figure 34.7B). Lungs, then, are not simple bags but are complex, moist, respiratory organs consisting of an enormous number of alveoli (about 750 million), with nearby capillaries, enclosed within a delicate connective tissue framework. Approximately 70 to 90 square meters of surface area is available for gas exchange in healthy lungs, an area approximately 40 times as large as the surface of the skin.

Control of Breathing

The mechanical process of taking air into the lungs and exhaling it again is **breathing**. The act of breathing is performed about 12 times each minute in an average adult, and in a 24-hour period, more than 8,000 liters of air is inhaled and exhaled.

Many organs work in partnership to carry out breathing and respiration. As with most critical functions in the human body, they are under involuntary con-

Figure 34.7

(A) Each lung contains a branching network of vessels called a *respiratory tree.* The trachea, the trunk of the tree, splits into two bronchi, which in turn branch into smaller bronchioles. (B) Bronchioles terminate in alveoli, the tissue sites where CO_2 and O_2 are exchanged. Each alveolus is serviced by lymph vessels and blood capillaries.

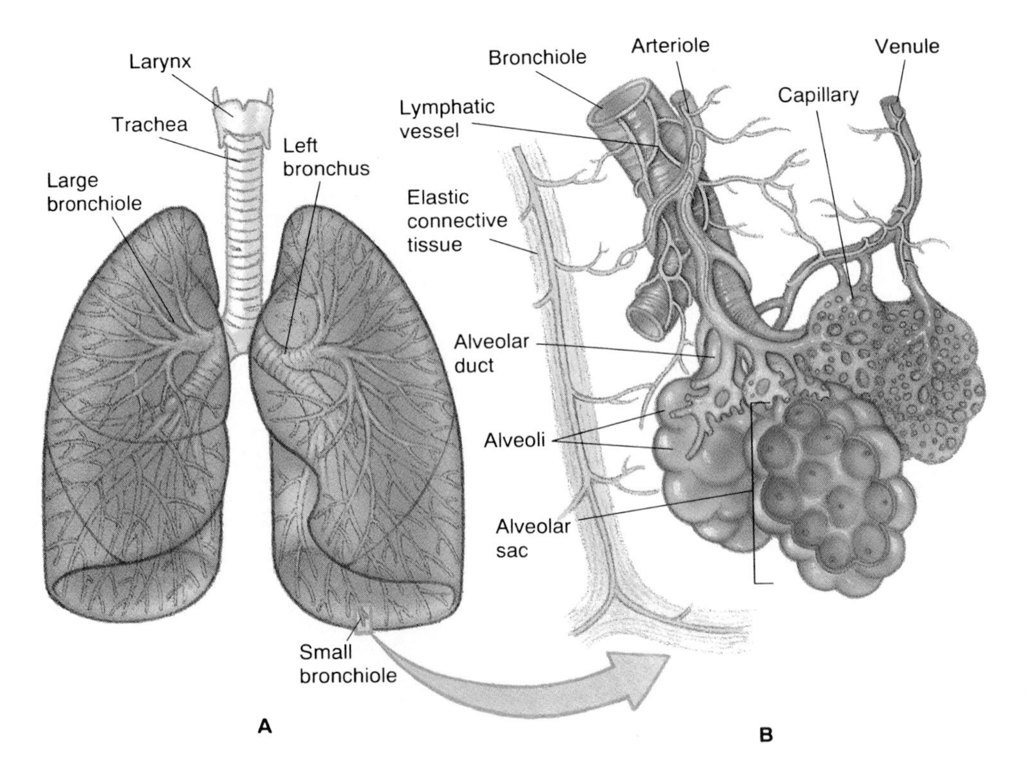

trol and do not require our attention. Yet significant differences do occur in our breathing rates, depending on whether we are exercising, reading, or attempting to perform some activity in front of an audience. For example, what regulates the rate of breathing during strenuous exercise? In contrast to many regulated functions, the primary mechanism that controls breathing during exercise is relatively uncomplicated. Within the brain, a special group of nerve cells called the *respiratory center* responds to levels of CO_2 in the blood. When levels of CO_2 become elevated, such as during exercise or stress, the breathing rate increases and becomes deeper to bring more blood, bearing CO_2, to the lungs, where it can be expelled. After excessive CO_2 is removed, homeostasis is restored, and the breathing rate returns to normal.

The main functions of the respiratory system are to supply red blood cells with O_2 and to remove CO_2 from the body. Oxygen contained in air enters the lungs, and CO_2 is exhaled, when breathing occurs. Once in the lung, air passes through a system of ever-smaller branches until it reaches the alveoli, the tissue where O_2 diffuses across membranes into red blood cells and CO_2 from the bloodstream diffuses into the lung. Breathing is controlled by a respiratory center in the brain.

CIRCULATORY SYSTEM

Cells and tissues in the human body require an incessant supply of nutrients and O_2 to survive. They also must rid themselves of the metabolic wastes generated continuously during their normal activities. These functions are accomplished largely by the **circulatory system**, an internal transport system that consists of the heart, a four-chambered muscular pump; an elaborate network of interconnecting, hollow vessels; and blood, a complex fluid containing different types of cells or cell fragments that each perform a specialized function.

The circulatory system has long been of interest to humans (see the Focus on Scientific Inquiry, "Changing Conceptions of the Circulatory System"). Discussions of this system are usually peppered with huge numbers (and exclamation marks) related to such matters as the amount of blood the heart pumps in an average 70-year lifetime (18 million barrels!), the number of miles of capillaries (greater than 60,000!), the number of red blood cells in the human body (in the neighborhood of 25 billion!), and so on. The numbers themselves are relatively unimportant, but they do give an idea of the magnitude of circulatory system activities within the human body.

The circulatory system is responsible for a range of important activities, including the transport of O_2 from the lungs to red blood cells and CO_2 from blood cells and fluids to the lungs. It is also involved in the transport of other substances. For example, metabolic wastes generated in cells are moved to organs that can process and eliminate them; also, nutrients from the digestive tract, or nutrient storage sites, are transported to cells where they are used. Blood cells defend the body against disease-causing agents. Finally, the circulatory system serves in certain homeostatic mechanisms such as regulating body temperature and maintaining fluid balance.

Blood Components

Blood is a "fluid tissue" that is transported in blood vessels and circulated by the action of the heart. It consists of a liquid ingredient, called *plasma,* that contains structures collectively called *formed elements*.

Plasma

Plasma is mainly water (about 91 percent) that contains and transports dissolved gases, proteins, hormones, nutrients, lipids, minerals such as iron, and metabolic

Focus on Scientific Inquiry

Changing Conceptions of the Circulatory System

To someone who casually observes the internal organs of an animal and sees the heart and some of the major blood vessels, it is not at all obvious that blood is circulating. It is not surprising, therefore, that the circulation of the blood within the body was not appreciated by ancient or even Renaissance physicians.

The first physicians and anatomists did realize that the heart and blood vessels were vital to living organisms. The ancient Egyptians, who were among the earliest people to study the structure of the human body, thought that the heart was the origin of all the "vessels" of the body and that through these vessels, various important substances, such as blood, semen, water, and air, traveled. Later, Greeks studied the internal anatomy of the human body for medical purposes and came to have a sophisticated knowledge of its structure (but not its functions). Some early Greeks ignored the connections between the heart and the blood vessels and assumed that the blood vessels were like irrigation ditches through which the nutrients from food flowed out to the various parts of the body to be consumed. Later Greek anatomists believed that the heart was the center of vitality and that from it an "animating heat" traveled through the arteries to the rest of the body. They also believed that digested food traveled from the stomach and liver through the veins to nourish the tissues. Famous anatomists like Galen, who lived in the Roman Empire during the second century A.D., gave accurate and detailed descriptions of the heart and the blood vessels of humans but had no conception that the blood circulated. Instead, most physicians and naturalists believed that the blood simply ebbed and flowed in the blood vessels.

Opinions on the nature and function of the heart and blood vessels did not begin to change until William Harvey, a distinguished English physician, published his observations in 1628. Harvey discovered that blood did not travel out to the tissues of the body to be consumed; rather, it traveled continuously in a closed loop. He was led to his discovery by anatomical studies on the heart and blood vessels. He was also influenced by the then-popular idea that the human body was a reflection of the cosmos. To him, the heart was like the "sun" of the body, and fluids circulated around it, similar to the way the planets traveled around the center of our solar system.

Harvey's famous 1628 book on the circulation of the blood, *Exercitatio anatomica de motu cordis et saguinis in animalibus* ("Anatomical Studies on the Motion of the Heart and Blood"), is a classic in the history of the biological sciences. He not only described the motion of the heart and the blood, but he also elegantly set out *experimental arguments* to demonstrate his points.

To show that the heart is supplied with blood from the vena cava, Harvey described observations on a living snake:

> If a live snake be cut open, the heart may be seen quietly and distinctly beating for more than an hour, moving like a worm and propelling blood when it contracts longitudinally, for it is oblong. It becomes pale in systole, the reverse in diastole. . . . The vena cava enters at the lower part of the heart, the

artery leaves at the upper. Now, pinching off the vena cava with a forceps or between finger and thumb, the course of blood being intercepted some distance below the heart, you will see that the space between the finger and the heart is drained at once, the blood being emptied by the heart beat. At the same time, the heart becomes much paler, even in distention, smaller from lack of blood, and beats more slowly, so that it seems to be dying. Immediately on releasing the vein, the color and size of the heart returns to normal.

On the other hand, leaving the vein alone, if you ligate or compress the artery a little distance above the heart, you will see the space between the compression and the heart, and the latter also, become greatly distended and very turgid, of a purple or livid color, and choked by the blood, it will seem to suffocate. On removing the block, the normal color, size, and pulse returns.

To show the one-way flow of the blood through the veins, Harvey relied on a simple but brilliant set of observations and experiments. First, he described the valves of the veins, which had been discovered by Hieronymus Fabricius of Aquapendente. These valves are positioned so that they prevent blood from flowing backward, and thus they route blood back to the heart.

Harvey then used a very elementary experiment to underscore the one-way flow of venous blood. By applying a loose tourniquet around the arm of a man, Harvey forced the veins of the arm to swell. At intervals, nodules appeared that indicated the sites of the valves of the veins, as shown in Figure 1. Then he noted:

> If you will clear the blood away from the nodule or valve by pressing a thumb or finger below it, you will see that nothing can flow back, being entirely prevented by the valve, and that the part of the vein between the swelling and the finger, disappears, while above the swelling or valve it is well distended. Keeping the vein thus empty of blood, if you will press downward against the valve, by a finger of the other hand on the distended upper portion, you will note that nothing can be forced through the valve. The greater effort you make, the more the vein is distended toward the valve, but you will observe that it stays empty below it.

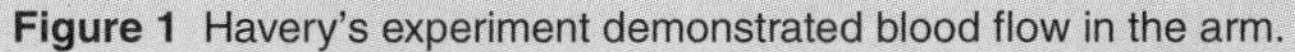

Figure 1 Havery's experiment demonstrated blood flow in the arm.

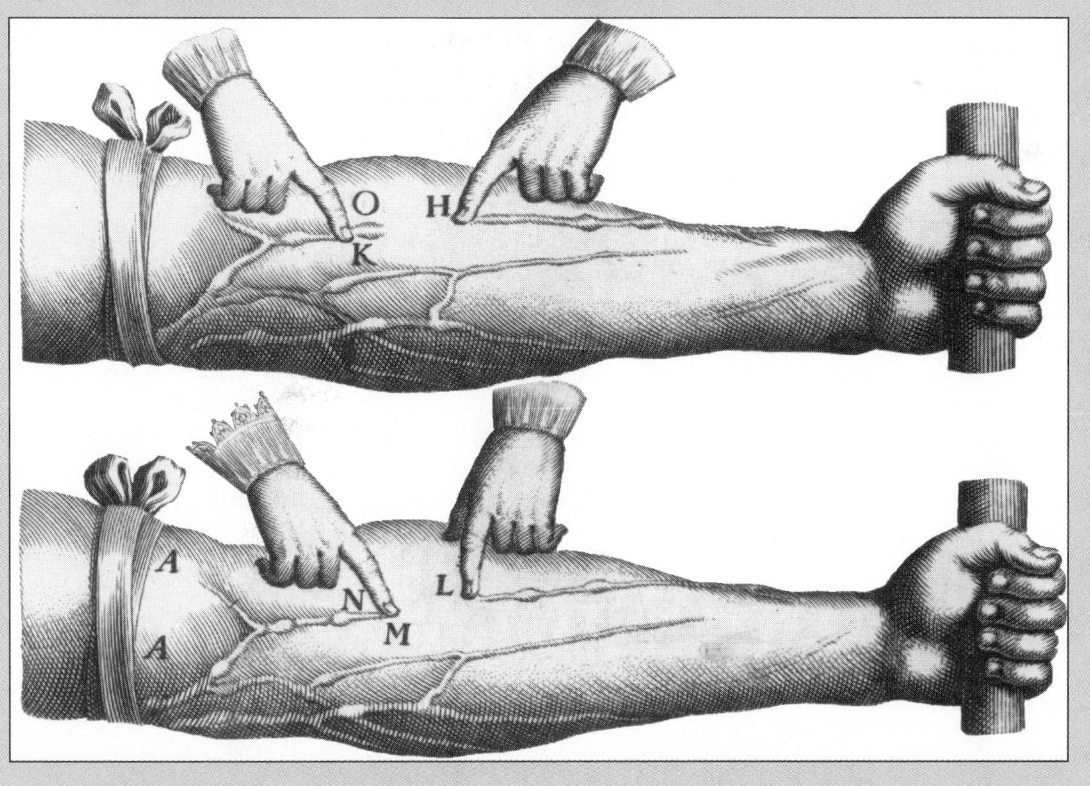

Focus on Scientific Inquiry

To make this point even more dramatically, Harvey also described a related experiment:

> With the arm bound as before and the veins swollen, if you will press on a vein a little below a swelling or valve and then squeeze the blood upward beyond the valve with another finger, you will see that this part of the vein stays empty, and that no back flow can occur through the valve. But as soon as the finger is removed, the vein is filled from below.

Harvey provided some compelling quantitative arguments as well. By calculating the volume (not very accurately, as it turned out, but reasonably close) of blood forced out of the heart into the aorta by each beat, he easily showed that the volume of blood pumped by the heart in an hour was more blood than the entire body contained. Even though Harvey did not observe the capillary system between arteries and veins (microscopes had not yet been invented), he reasonably argued that since the flow was one-way, it must circulate through a single pathway in the body.

Harvey's discovery and description of the circulation of the blood stands as a paragon in the history of the biological sciences, but what is perhaps even more fascinating than his careful experiments and reasoned arguments was the reception of his ideas. Although a group of English physicians at Oxford did appreciate the implications of Harvey's discovery, the medical community was slow to accept the concept of blood circulation. For decades anatomists and physicians rejected Harvey's findings. Why? It is difficult today to read through Harvey's short book without being totally convinced. In his own day, however, Harvey's book was revolutionary in two dramatic ways, one methodological, the other theoretical.

Harvey's use of an experimental approach now seems very ordinary to us, but in the 1620s it was very unusual. The study of living beings at that time was primarily a descriptive science. Naturalists classified animals and plants; medical researchers described certain parts of the body and speculated on their functions. Experimental investigations were being done in the physical sciences, but they were still rather novel. The idea of doing experiments on living beings was so strange and new that many serious and well-educated people found it hard to imagine that anything worthwhile could come of it. The use of powerful experiments in biology lay in the future.

The other reason that the reception of Harvey's ideas was muted was that it refuted a long-held theory. Assuming that Harvey's findings were correct meant that the accepted theory of the animal body, as it had evolved from the Greeks, through the Arabs and then medieval and Renaissance Europeans, was fatally flawed. The foundation of that theory was an idea about the uptake, distribution, and fate of food in the body. It was believed that food taken into the body first traveled to the stomach, where the "nutritious portion" was separated, then to the liver, where it was further altered and supplemented, and finally to the veins, which distributed it to the rest of the body to be absorbed. A portion of the blood passed to the arterial system, where it mixed with a small amount of air taken in by the lungs. This vital mixture then flowed out via the arteries to the entire body, where it was all consumed.

Circulation of the blood didn't fit this theoretical view at all. Harvey showed that the blood was not distributed to tissues of the body to be consumed but rather traveled in a continuous circuit. What, then, was the function of this circulation? How did it connect with nutrition? If the venous and arterial systems were *not* distinct, how were they related? Since the entire edifice of practical medicine, physiology, and anatomy was based on the earlier conception of the heart and blood vessels, physicians were hesitant to accept an idea (based on a novel method) that contradicted conventional wisdom.

Harvey's ideas were eventually accepted, of course, but not until they became integrated into a new theoretical view of the body that "made sense." This case study illustrates how important theory is to science and emphasizes the reluctance one should have in thinking of science as merely a collection of facts.

waste products. Plasma also contains a large variety of proteins that have specific structures and functions. Several of these proteins participate in the body's defense mechanisms and are discussed in Chapter 38.

Formed Elements

The **formed elements** of blood include *erythrocytes,* or red blood cells (RBCs), which are specialized for transporting O_2; various types of *leukocytes,* also called white blood cells (WBCs), which have specialized defense functions; and *platelets,* which are very small, nonnucleated cell fragments that have a critical function in blood clotting (see Figure 34.8). Leukocytes and platelets are described in greater detail in Chapter 38.

Erythrocytes are the most abundant formed element in blood. Mature RBCs are small, circular, biconcave disks that have no nuclei. Each RBC contains up to 275 million hemoglobin molecules within its cytoplasm. *Hemoglobin* is a complex, iron-containing molecule that combines with O_2 in the lung, giving blood its characteristic bright red color. After O_2 is given up to cells, hemoglobin transports some CO_2 back to the lung. RBCs are subjected to great stress during their endless journey through blood vessels. They have an average life span of 120 days and are replaced through mitosis in the bone marrow. The body has the ability to govern the number of RBCs available; regulation is associated primarily with the body's demand for O_2. For example, individuals who exercise regularly have up to twice as many RBCs as sedentary individuals.

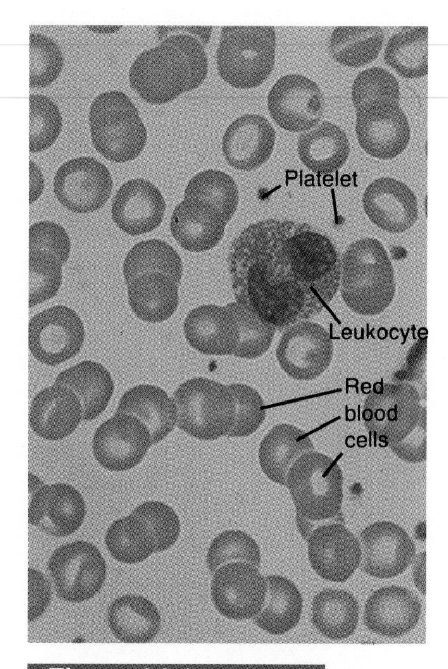

Figure 34.8

Red blood cells, leucocytes, and platelets (arrows) are the formed elements of the blood.

Vessels

Blood in the circulatory system flows in one direction, through three general types of blood vessels that are distinguished by size, structure, and function (see Figure 34.9A). **Arteries** conduct blood away from the heart and move it to separate vascular territories of the body, while **veins** return blood to the heart. **Capillaries** connect the arterial and venous systems.

Major arteries closest to the heart are strong and thick-walled, due to large amounts of muscle and elastic fibers that allow the vessel wall to be expanded by the force of blood being pumped from the heart. When the muscular artery wall rebounds after expansion, it causes movement of blood through the vessel. If you place your fingers on the inner portion of your wrist, you will be able to feel a *pulse,* which represents the wave of pressure radiated outward from the heart through the arteries. Farther along the arterial tree, after considerable branching, the size and diameter of the vessel walls become reduced and blood pressure diminishes significantly; these smaller vessels are called *arterioles.*

From arterioles, blood flows into capillaries, the smallest blood vessels, which have walls thin enough to allow the transfer of O_2, nutrients, CO_2, and wastes between cells and the blood (see Figure 34.9B). Pressure exerted from pumping by the heart has nearly dissipated by the time blood passes into capillaries. Single capillaries are about 1 millimeter long and have diameters approximately the size of RBCs. Tissues are completely infiltrated with beds of capillaries, and all cells within tissues are no more than two or three cell diameters away from a capillary. Nutrients, gases, and other materials exchanged between blood cells and tissue cells move through *tissue fluid,* composed of water, salts, and proteins, that bathes the cells.

Upon leaving the capillaries, blood flows into small *venules* that unite to form veins. Veins return blood to the heart largely through the routine movements of skeletal muscles, which squeeze the vessels and move the blood forward as described in Figure 34.9C. Backflow is prevented in larger veins by the presence of *valves,* or flaps of tissues extending from the vessel wall that close when blood attempts to flow in the opposite direction. In the absence of movement—for example, when standing rigidly at attention—blood tends to collect in veins and in extreme cases may lead to fainting because the brain becomes deprived of an adequate supply of blood.

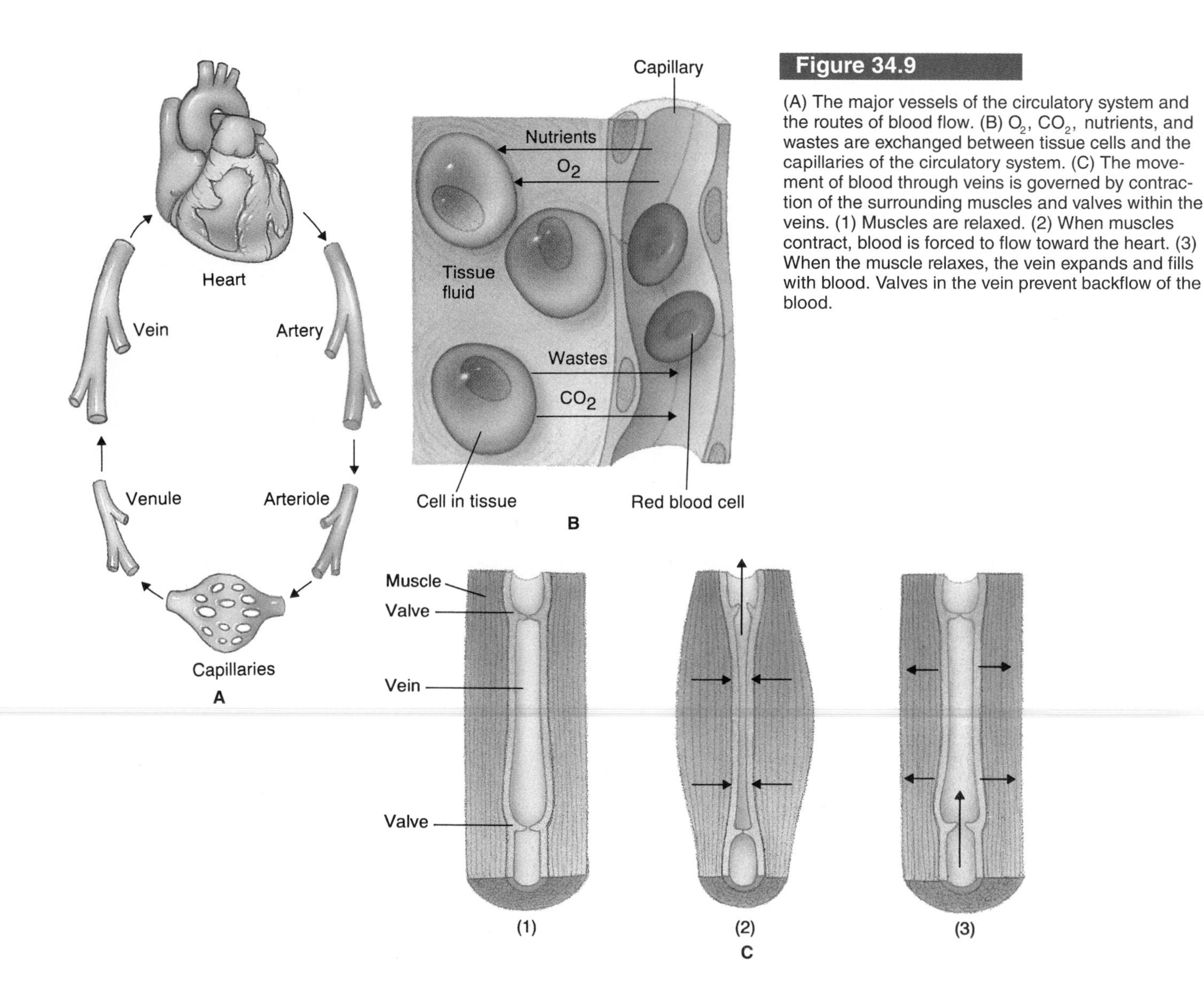

Figure 34.9

(A) The major vessels of the circulatory system and the routes of blood flow. (B) O_2, CO_2, nutrients, and wastes are exchanged between tissue cells and the capillaries of the circulatory system. (C) The movement of blood through veins is governed by contraction of the surrounding muscles and valves within the veins. (1) Muscles are relaxed. (2) When muscles contract, blood is forced to flow toward the heart. (3) When the muscle relaxes, the vein expands and fills with blood. Valves in the vein prevent backflow of the blood.

The Heart

The **heart** is a large muscular organ capable of powerful contractions that propel blood into vessels (see Figure 34.10). The human heart consists of four chambers, called the right and left *atria* and right and left *ventricles,* and blood flows through each chamber in a specific sequence. Chambers are separated by valves, which open and close automatically, to ensure that blood flows only in one direction.

Circulation of the Blood

The circulation of blood follows two specific pathways (see Figure 34.11). This design reflects a primary responsibility of transporting O_2 to the billions of cells in the human body.

Pulmonary circulation conducts blood between the heart and the lungs. Oxygen-poor, CO_2-laden blood returns through two large veins (*venae cavae*) from tissues within the body, enters the right atrium, and is then moved into the right ventricle of the heart. From there it is pumped into the pulmonary artery, which divides into two branches, each leading to one of the lungs. In the lung, the arteries undergo extensive branching, giving rise to vast networks of capillaries where gas exchange takes place, with blood becoming oxygenated while CO_2 is discharged. Oxygen-rich blood then returns to the heart via the pulmonary

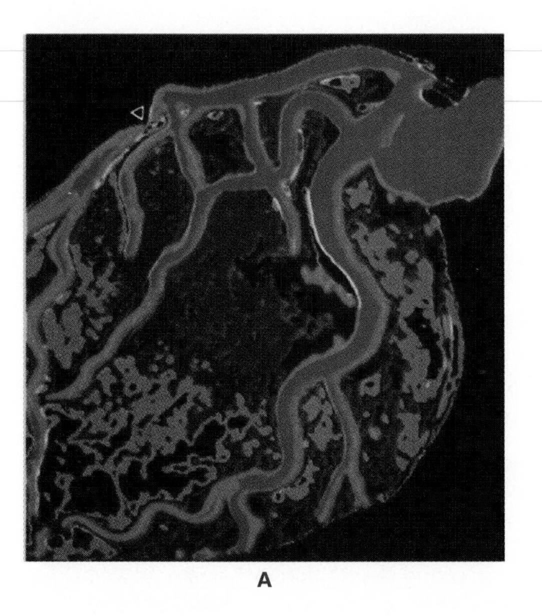

A

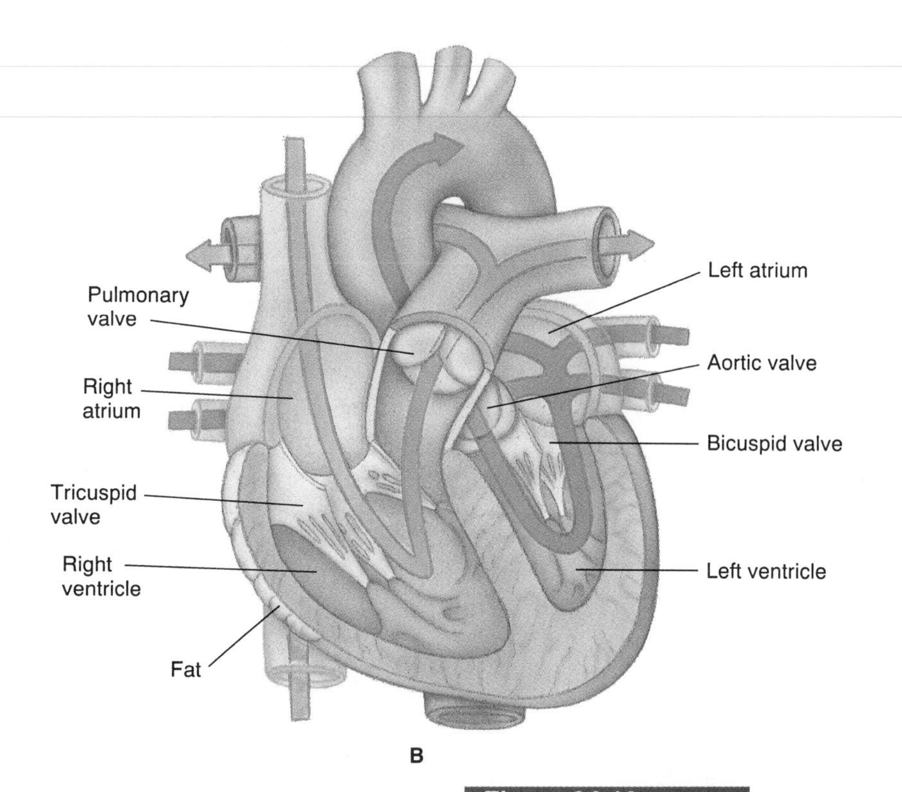

B

Figure 34.10

(A) An external view of the human heart as revealed by a new imaging technique called digital subtraction angiography. Blood vessels have been shaded red, fat tissue is green, and cardiac muscle is blue. (B) The four chambers of the human heart are separated by valves. Arrows indicate the direction of blood flow.

veins. In contrast with all other arteries and veins in the body, pulmonary arteries carry oxygen-deficient blood, and pulmonary veins convey oxygen-rich blood.

Once blood is returned from the lungs, it enters the left atrium, passes to the left ventricle, and is then distributed by **systemic circulation** to all tissues of the body. Blood emerges from the heart into the *aorta,* the largest artery in the body. The aorta branches into major arteries that lead to vascular territories that include the heart itself, the brain, the shoulder region, the intestines, the kidneys, and the legs.

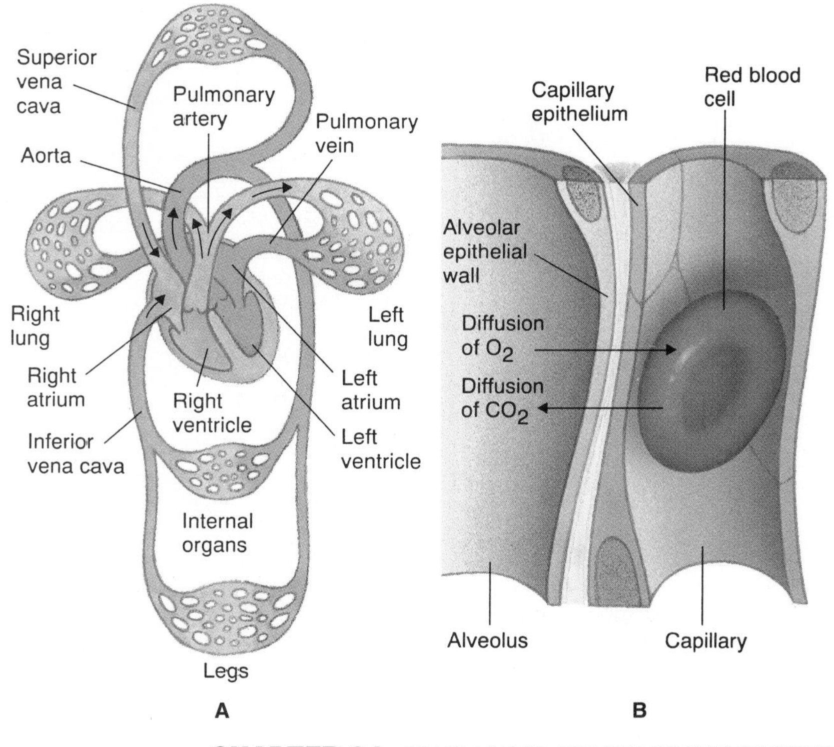

Figure 34.11

(A) Blood circulates through two general pathways in the human body. Pulmonary circulation between the heart and the lungs loads red blood cells with O_2. Oxygen-rich blood is then distributed to the head and arms, the internal organs, and the legs by systemic circulation. Oxygen-poor blood returns to the heart through the two vena cava. (B) Oxygen and CO_2 are exchanged by diffusion between red blood cells and alveoli in the lung.

The Lymphatic System

The **lymphatic system** consists of a network of vessels that run near the veins and *lymph tissue,* which is a type of connective tissue. The functions of the lymphatic system are generally related to draining excess fluids, transporting fats from the digestive system to blood, and removing foreign substances from the body. Specifically, a slight excess of tissue fluid accumulates in the body as a result of nutrient, gas, and water diffusion between tissue cells and blood cells. This fluid enters lymphatic capillaries, where it is called *lymph,* that merge to form larger vessels (see Figure 34.12). Lymph is returned to veins near the heart. Lymph tissue is organized into compact masses of connective tissue called *nodes* that filter and remove foreign particles, such as bacteria, and also harbor some of the cells that have important functions in the immune system.

The circulatory system moves blood throughout the body. Its organs include a pump, the heart, and various types of blood vessels. There are two elaborate networks of blood vessels, one associated with pulmonary circulation and the other with systemic circulation. Blood is a complex fluid composed of plasma and several formed elements, each with specific functions. The lymphatic system consists of a separate network of vessels that transport lymph and remove foreign substances from the body.

DIGESTIVE SYSTEM

The **digestive system** has many functions, all of which are generally related to delivering the nutrients contained in food to every cell in the body that requires them for energy, building new structures or molecules, or repairing damage. **Digestion** refers to the breakdown of complex food substances into simple molecules that can be absorbed into the body.

The digestive system can be perceived as a "disassembly line" where large molecules are broken down into smaller subunits. As described in Figure 34.13, it accomplishes this task by moving foods progressively through a tube called the *alimentary canal* that begins at the mouth and terminates at the *anus,* the open-

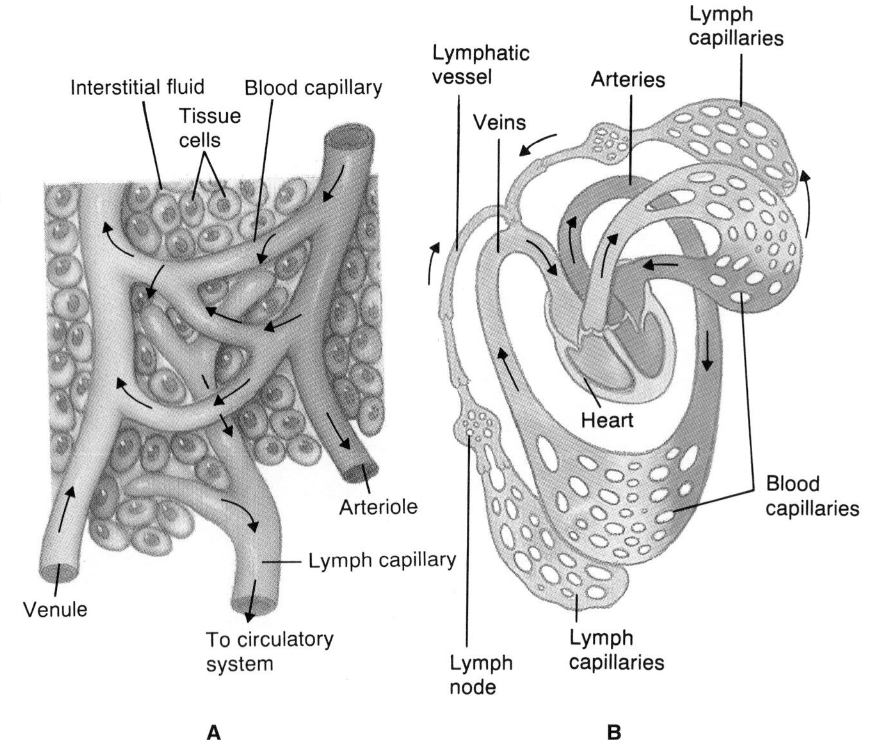

Figure 34.12

(A) Tissue fluids that surround cells diffuse into lymph capillaries. (B) Lymph capillaries flow into large lymph vessels that merge with veins near the heart, returning fluids to the blood. Lymph nodes act as filters in removing foreign particles. They also play a major role in the immune system.

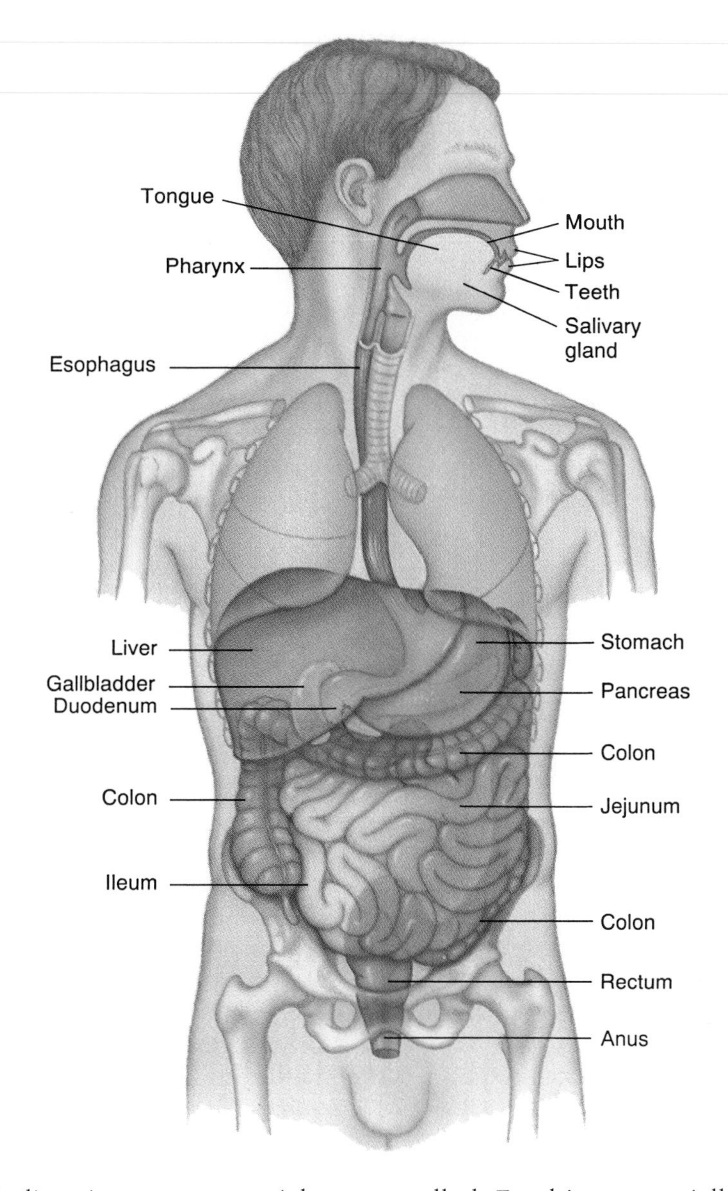

Figure 34.13

The human digestive system breaks down and delivers nutrients to all cells of the body.

ing through which digestive waste materials are expelled. Food is sequentially modified, first mechanically and then chemically, to achieve the reduction of complex molecules into simpler forms. Along the way, various organs and glands empty their products into the digestive tube at certain regions where specialized operations occur. All of these activities are regulated by the nervous and endocrine systems.

Mouth, Pharynx, and Esophagus

The human **mouth** is the opening to the exterior that receives food and begins the process of digestion using both mechanical and chemical devices. The tongue, teeth, and jaw are used to chew or grind the food into smaller pieces, which results in a greater amount of surface area being exposed to the action of enzymes. The presence of food stimulates the release of saliva through ducts from three *salivary glands* in the mouth. *Saliva* is a watery secretion consisting of mucus, which protects the lining of the mouth and lubricates the food for easier movement, and *salivary amylase,* a specific digestive enzyme that begins the breakdown of starch.

Once the food has been chewed, softened, and lubricated, it passes into the **pharynx**, a thick-walled, muscular tube. In the pharynx, the ball of pasty food is swallowed and moved to the entrance of the esophagus. The **esophagus** is an elastic tube, about 25 centimeters in length and 2.5 centimeters in diameter, that conducts food from the pharynx to the stomach. From the esophagus onward, food is forced through the alimentary canal by *peristalsis,* a series of contraction waves that ripple along the smooth muscles surrounding the tubular organs of the digestive system. From the esophagus, the food *bolus*—the rounded, softened mass of food material—moves into the stomach.

Stomach

The **stomach** is a relatively large muscular bag in the upper part of the abdomen that can stretch to accommodate about 2 liters of food or liquid. A number of chemical assaults are made on food once it enters the stomach. The epithelium that lines the interior of the stomach secretes gastric juice, a highly acidic fluid that is strong enough to dissolve nails. The powerful acid acts to dissolve the tough tissues that bind together the plant and animal cells present in food. Gastric juice also contains an enzyme called *pepsin* that initiates the breakdown of food proteins.

Given the potency of the digestive chemicals inside the stomach, why isn't the stomach lining also broken down? First, a coating of mucus secreted by epithelial cells lining the stomach offers substantial protection against the digestive actions of gastric juice. A second mechanism involves the release of an enzyme in an inactive form that becomes activated only when exposed to acid. For example, pepsin is secreted by epithelial cells in an inert form called pepsinogen. Upon exposure to the acid in gastric juice, it is converted to the active enzyme, pepsin. Many other digestive enzymes are restricted in the same manner. Despite these protective mechanisms, epithelial cells of the stomach have life spans measured in days, and they are replaced by new cells arising from mitosis in the basal epithelial layers. After spending two to six hours in the stomach, the food mass is reduced to a semifluid mixture of partially digested food, called *chyme,* that passes next into the small intestine.

Despite the complexity of its functions, the stomach is classified as a nonessential organ. In drastic circumstances such as cancer development or severe cases of ulceration, the stomach can be partly or totally removed by surgery. Removal requires eating modifications such as consuming many small meals each day rather than the three that are customary in our society.

Small Intestine

The major organ of digestion, the **small intestine**, is about 6 meters long. It consists of an upper region called the *duodenum;* a middle region, the *jejunum;* and a lower sector named the *ileum*. Proteins, fats, and carbohydrates are finally broken down to simpler molecules in the duodenum due to the actions of numerous enzymes secreted by intestinal epithelial cells. Other chemical substances released from accessory organs also aid in digesting food in the small intestine.

Following digestion, the nutrients, along with water, are absorbed into the bloodstream through the intestinal walls of the jejunum and ileum and distributed throughout the body. The architecture of the ileum is ideally suited for its absorptive responsibilities (see Figure 34.14). Folds of the inner surface are covered with small, fingerlike projections called *villi,* and the epithelial cells lining the villi are covered with thousands of *microvilli,* microscopic folds in the plasma membranes. This structural design provides an enormous amount of surface area through which molecules such as glucose and amino acids can pass into capillaries surrounding the intestine. From there, they are transported to the liver, where nutrients are removed and "processed." For example, glucose is converted to *glycogen,* a storage form of glucose. If present, toxic chemicals are also extracted and broken down by the liver before they can move elsewhere in the body.

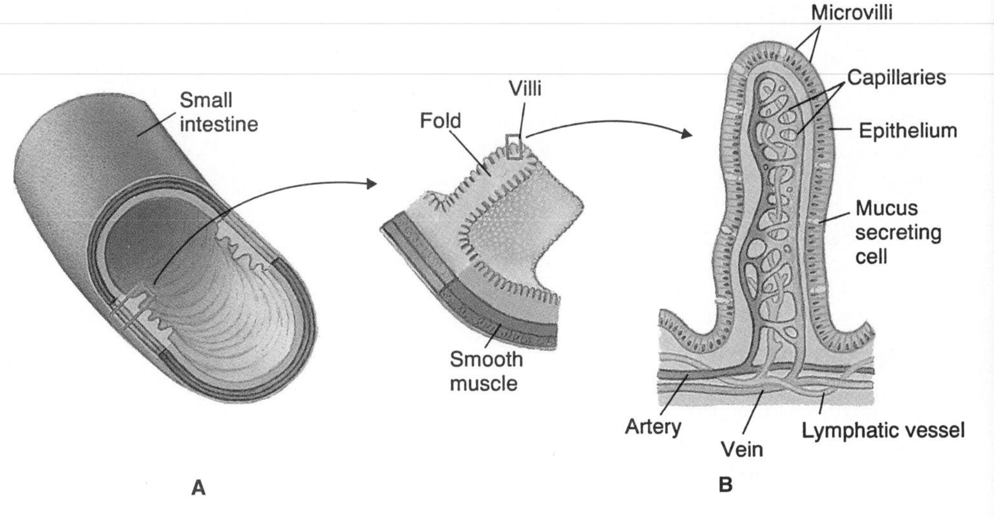

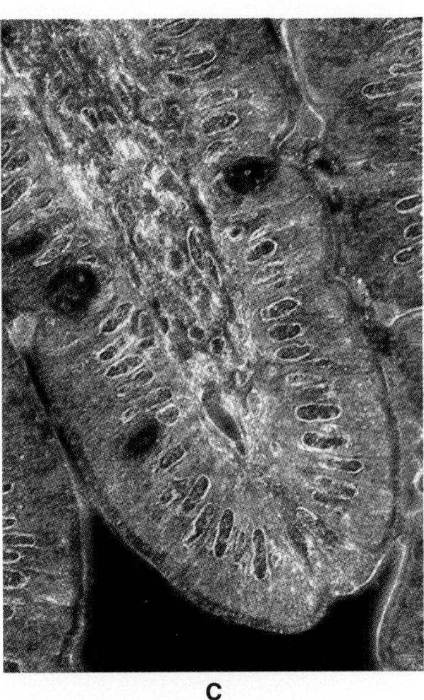

Figure 34.14

(A) The small intestine is extensively wrinkled. Millions of small projections called villi (B) line each wrinkle and are responsible for nutrient transfer between the small intestine and the circulatory system. (C) Villi of the small intestine showing details of the epithelial cells and central spaces.

Accessory Digestive Organs

Two other organs have major roles in the breakdown of food in the duodenum. They synthesize or store chemical substances that are released into the small intestine during digestion.

Pancreas

The **pancreas** carries out functions in the digestive system and also the endocrine system. It produces pancreatic juice, which contains an assortment of digestive enzymes and releases a bicarbonate solution that serves to neutralize the strong acids contained in the chyme received from the stomach.

Liver

The **liver** is an enormously complex organ that performs a wide variety of critical operations in the body. In addition to removing nutrients from circulation, converting glucose to glycogen, and detoxifying harmful compounds, it has other major digestive functions. These include breaking down hemoglobin, storing iron and some vitamins, synthesizing certain proteins, and other metabolic conversions of amino acids, carbohydrates, and fats. The liver also synthesizes *bile,* a mixture of substances that breaks down large fat globules into smaller particles. Bile is stored in the *gallbladder,* another accessory organ, and released into the small intestine as required during digestion.

Large Intestine

By the time chyme reaches the **large intestine**, or *colon,* little remains except for indigestible materials. The primary functions of the large intestine are, first, to reabsorb sodium and the remaining water that was released into the digestive tract to facilitate digestion. Reclaiming water is necessary to prevent dehydration, which would occur if the water were not recycled. Second, the colon eliminates *feces,* compacted solid wastes consisting mostly of undigested food, dead intestinal cells, and bacteria that inhabit the large intestine. The colon hosts a rich variety of bacteria that are generally harmless, including the molecular biologist's favorite, *Escherichia coli*. Bacteria are able to survive in the large intestine by extracting nutrients from certain food materials, such as cellulose, that are of no value to humans. In return, they produce small quantities of valuable vitamin K, which are absorbed by the host. Less welcome bacterial by-products—odoriferous gases such as hydrogen sulfide—are also produced in the large intestine. After one to three days, feces reach the *rectum,* the lower portion of the large intestine, where they are stored until defecation, or expulsion through the anus, occurs.

The basic function of the digestive system is to reduce food to small molecules that can be absorbed by the circulatory system and transported to all cells in the body. In the mouth, food is mechanically broken down and first exposed to digestive enzymes. Food is further degraded in the stomach by gastric juice and numerous enzymes. The final phase of digestion occurs in the small intestine, where other enzymes complete the breakdown process and absorption occurs. The pancreas, liver, and large intestine also have roles in digesting food.

URINARY SYSTEM

As we have seen, a number of human organ systems are involved in disposing of wastes produced as a result of various activities, as summarized in Figure 34.15. The **urinary system** is composed of organs, shown in Figure 34.16, that are responsible for filtering the blood and producing *urine,* a solution containing varying amounts of water and waste products that is excreted from the body. The wastes normally removed by the urinary system are excess salts, nitrogen-containing compounds formed during the chemical modification or breakdown of amino acids, proteins or nucleic acids, and other unwanted products that may be present in the blood. The primary metabolic waste that must be removed is *urea,* a nitrogen-containing product.

In addition to filtration and waste removal, the urinary system performs critical homeostatic functions. It helps maintain a constant internal environment by continuously adjusting the concentrations of water, salts, and other materials present in blood.

Organs of the Urinary System

Blood filtration and urine formation are carried out by two **kidneys**, bean-shaped organs approximately 10 centimeters long that lie on the back wall of the abdomen (see Figure 34.17A, page 646). From each kidney, urine moves down a tube called a *ureter* and into the *urinary bladder,* a hollow, muscular organ that serves as a temporary storage reservoir. From the bladder, urine is excreted through a duct, the *urethra,* that leads outside the body.

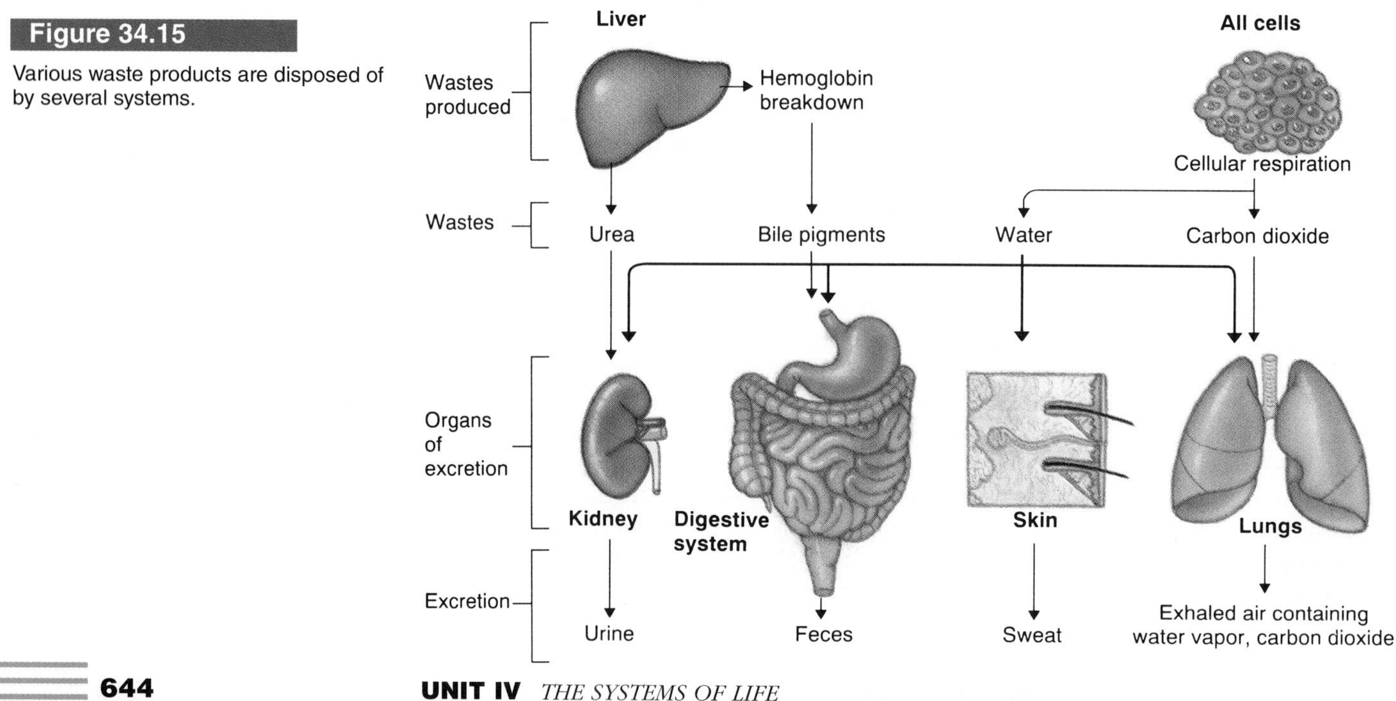

Figure 34.15

Various waste products are disposed of by several systems.

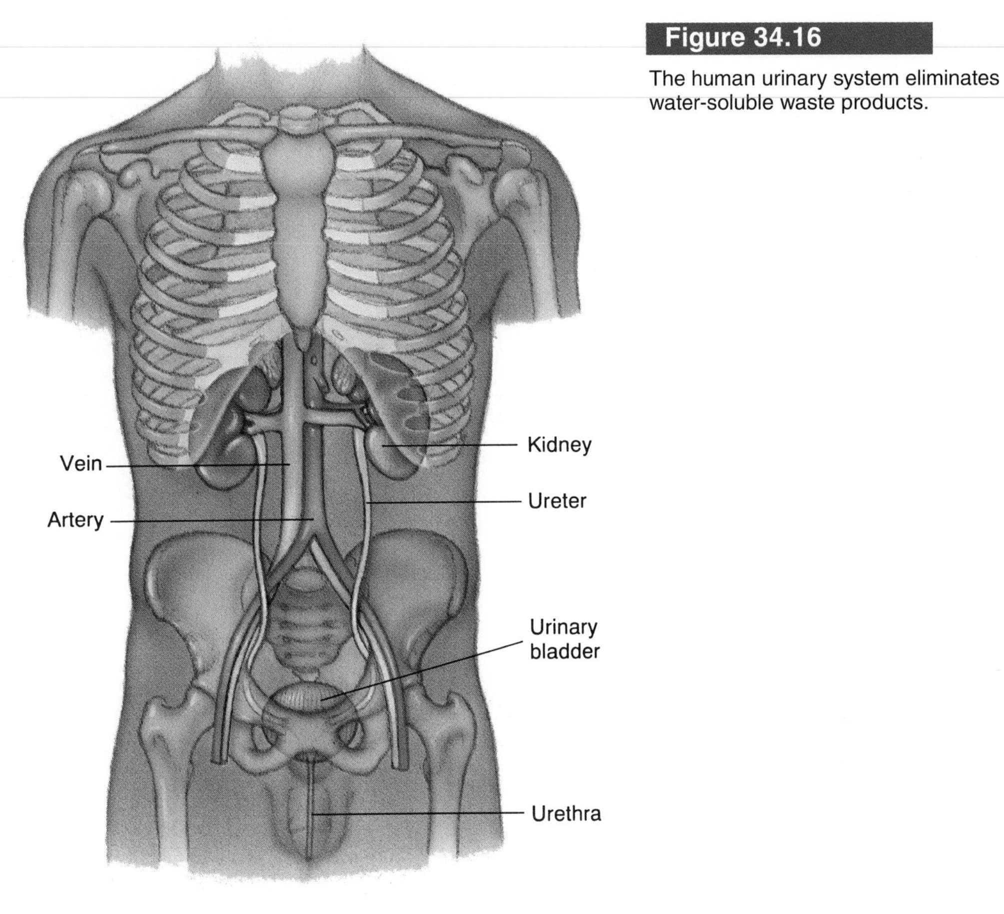

Figure 34.16

The human urinary system eliminates water-soluble waste products.

Structure and Function of the Kidney

The kidney is enclosed within a *renal capsule* that consists of a thick layer of connective tissue. The outer part of the kidney is called the *cortex,* and the inner portion is called the *medulla.*

Nephrons

Each kidney contains more than 1 million functional units called **nephrons**, which are long, curved, tubular structures of exquisite design. Nephrons regulate the composition of blood through the production of urine (see Figure 34.17B). Each nephron can be divided into two functional compartments. First, the *glomerulus,* a mass of capillaries, and associated structures filter the blood. Second, a complex of *tubules* removes valuable materials, such as glucose, and transfers them back into the bloodstream. Unwanted substances, including some excess water, proceed and are eventually excreted as urine.

Formation of Urine

To carry out the functions of removing nitrogenous wastes from the bloodstream and regulating concentrations of water and salts in the body, the kidney conducts two operations. First, wastes, salts, amino acids, glucose, some water, urea, and other substances contained in blood enter the glomerulus and are forced out of the bloodstream into the tubules, where they form a fluid called the *filtrate.* Note the unusual composition of this filtrate. Some of the materials are harmful and require elimination from the body, while others, such as glucose, are too valuable to be lost and are normally reclaimed by healthy kidneys. Thus the second step, which occurs in the tubules, involves the selective reabsorption of most of the water and other substances of value to the body. Unwanted sub-

Figure 34.17

(A) The human kidney is composed of more than 1 million nephrons that filter wastes from the blood and produce urine. (B) Each nephron consists of a glomerulus and various tubules that produce urine. (1) In the glomerulus, substances such as wastes, salts, glucose, and urea are filtered from the blood, enter the capsule, and pass into the tubule. (2) As substances move through the tubule network, some are reabsorbed by blood in surrounding capillaries. Substances not reabsorbed from the tubule move to the collecting tubule, where they become incorporated into urine. (3) Unwanted substances may also pass from capillaries directly into tubules to become part of the urine.

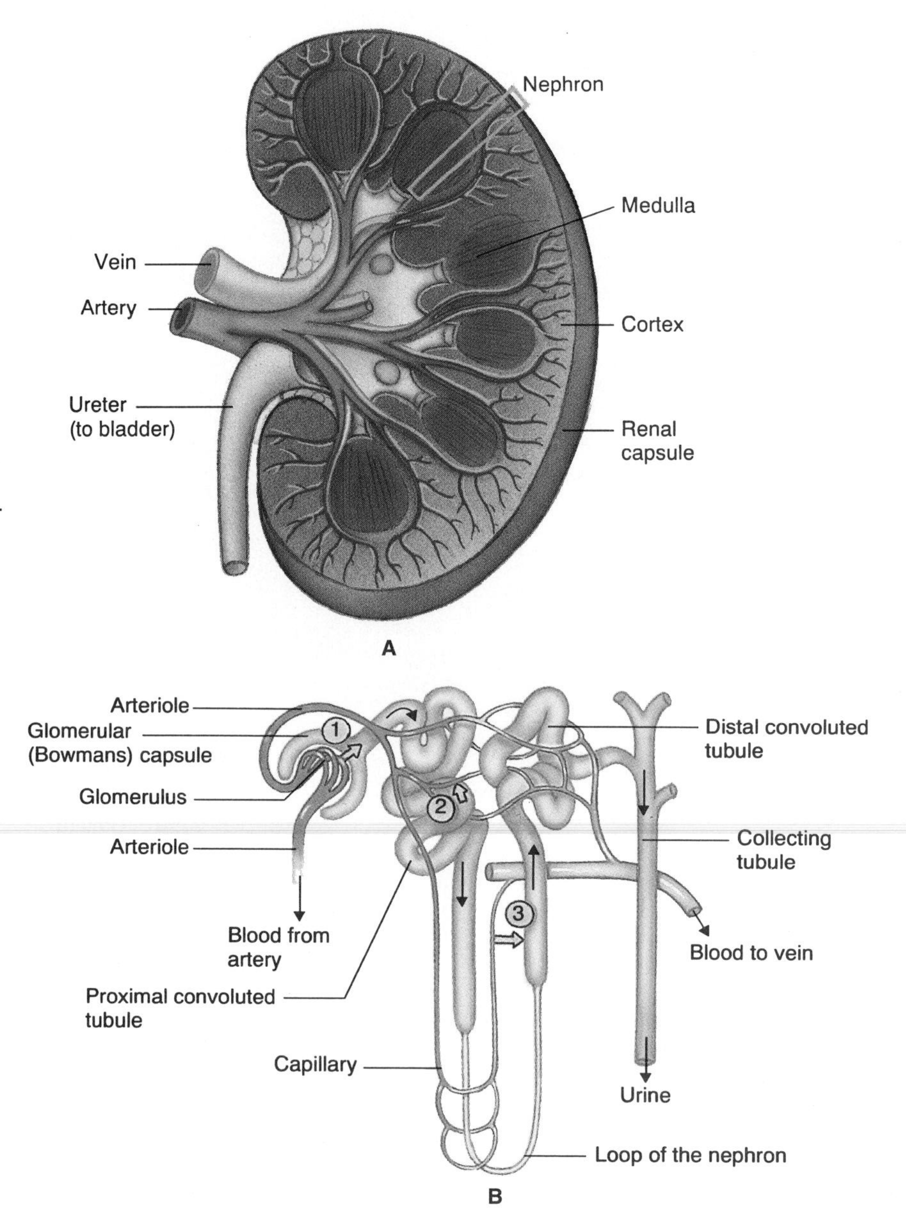

stances pass through the tubules without being reabsorbed and become part of the urine.

The workload of the kidneys is astonishing. About 1,200 liters of blood is processed daily, and almost 110 liters of filtrate is formed. More than 99 percent of the filtrate is reabsorbed from the tubules, and an average 1 liter of urine is produced and excreted per day.

Although the entire process is immensely complex, the end product, urine, contains about 96 percent water and 1 to 2 percent each of nitrogenous waste products, principally urea, and salts. However, urine composition varies depending on the state of the internal environment. For example, if an individual has consumed excessive quantities of fluids, the urine will be more dilute, since larger-than-average amounts of water will be excreted. In contrast, the urine of someone who has run 10 kilometers on a hot day will contain only the minimal amount of water necessary to form urine. Concentrations of salts also fluctuate, depending on intake and internal conditions. Some compounds, such as glucose, are reabsorbed entirely, and their appearance in urine usually signals some type of health problem. Through the precise regulation of substances reabsorbed or excreted, the kidneys play a pivotal role in maintaining a constant internal environment.

The urinary system plays a major role in maintaining internal homeostasis. Nephrons, the functional units of the kidneys, remove salts, nitrogen-containing waste products, and excess water from the blood. These substances are then incorporated into urine, which is excreted from the body.

SURVIVAL

The major structures and functions of several human organ systems have been described in this chapter. Though each has specific and distinctive responsibilities, all are interrelated and interdependent. Homeostasis is a unifying concept that is valuable for reflecting on the roles of various organ systems in the human body. Organ systems allow us to maintain the relatively constant internal environment that permits the highest level of cellular specialization. Organ systems enable us to operate with great efficiency and survive in a demanding environment. What controls or regulates the activities of each of the organ systems? That question is addressed in the next two chapters.

Summary

1. Humans are composed of a number of tissues, organs, and organ systems that allow them to solve different biological problems.
2. The major organ of the integumentary system is the skin, which consists of two cellular layers, the outer epidermis and the inner dermis. Within the epidermis and dermis are various cells and tissues—melanocytes, hair, and glands—that protect underlying tissues from potentially harmful factors present in the external environment.
3. The skeletal system is made up of bones and cartilage, which are connected at joints and held together by ligaments. The skeleton provides support for the body, protects vital internal organs, produces blood cells, and provides sites for muscle attachment.
4. Lungs are the primary organ of the respiratory system, which functions to obtain O_2 from the air and get rid of CO_2 produced during cellular respiration. Concentrations of these two gases in the body are regulated by the respiratory center.
5. The major functions of the circulatory system are to transport O_2, metabolic wastes, and nutrients from one area of the body to another; to assist in defense reactions against disease agents; and to regulate temperature and fluid balance. To accomplish these tasks, the heart pumps blood through an elaborately organized system of vessels that reaches all the tissues and cells of the body.
6. The mouth, pharynx, esophagus, stomach, small intestine, various accessory organs, and large intestine constitute the digestive system, which functions in the preparation and delivery of nutrients throughout the body.
7. To rid the body of certain wastes, the kidneys filter blood to remove unwanted substances, which are later excreted in urine. The urinary system also plays a role in the regulation of water, salts, and other substances in the blood.

Review Questions

1. What are the structure and function of epidermis?
2. What are the protective functions of hair and sweat glands?
3. How are the functions of bone related to its structure?
4. What are the major functions of the skeletal system?
5. Describe the pathway that air (oxygen) follows, beginning with the atmosphere and ending with alveoli.
6. How is breathing controlled?
7. What are the functions of the circulatory system?
8. Describe the composition of blood.
9. What are the functions of the various types of blood vessels? Of the heart?
10. What is the function of pulmonary circulation? Systemic circulation?
11. What are the structures and functions of the lymphatic system?
12. Which organs are part of the alimentary canal? What is the function of each?

13. How do the pancreas and the liver contribute to digestion?
14. What is the function of the urinary system?
15. What is the functional unit of the urinary system? Describe its structure.
16. How is urine formed?

Essay and Discussion Questions

1. What sort of adaptive changes might you predict would occur in humans if Earth's environment changed from its present state to become significantly hotter and drier?
2. There are several "essential" human organs (those absolutely required for survival). Disease or trauma can destroy any of these organs, and efforts are being made to develop artificial essential organs. How would you rank the lung, heart, liver, and kidney in terms of difficulty in development, from easiest to most difficult? Why?
3. Why do you suppose the heart became the symbol for love?
4. The lymphatic system was the last major organ system to be discovered and described. What might have accounted for this delay?

References and Recommended Reading

Bramble, D. M., and D. R. Carrier. 1983. Running and breathing in mammals. *Science* 219:251–256.

Gans, C. 1969. Functional components versus mechanical units in descriptive morphology. *Journal of Morphology* 128:365–368.

Harvey, W. (1628/1941). *Exercitatio anatomica de motu cordis et saguinis in animalibus*. Trans. C. D. Leake. 3d ed. Springfield, Ill.: Thomas.

Moore, J. A., I. Deyrup-Olsen, and W. V. Mayer. 1988. Science as a way of knowing: Form and function. *American Zoologist* 28:441–808.

Pawelek, J. M., and A. M. Kürner. 1982. The biosynthesis of mammalian melanin. *American Scientist* 70:136–145.

Radinsky, L. B. 1987. *The Evolution of Vertebrate Design*. Chicago: University of Chicago Press.

Schmidt-Nielsen, K. 1990. *Animal Physiology*. 4th ed. New York: Cambridge University Press.

Spence, A. P., and E. B. Mason. 1987. *Human Anatomy and Physiology*. 3d ed. Menlo Park, Calif.: Benjamin/Cummings.

Tortora, G. J., and N. P. Anagnostakos. 1990. *Principles of Anatomy and Physiology*. New York: HarperCollins.

Went, F. W. 1968. The size of man. *American Scientist* 56:400–413.

CHAPTER 35

Control and Regulation I: The Nervous System

Chapter Outline

Reading Questions

1. How is the structure of a neuron related to its function?

2. How are nerve impulses transmitted across a synapse?

3. How is knowledge of synaptic transmission being used in modern medicine?

4. What are the major divisions of the nervous system and the function of each?

All organisms, including humans, are constantly exposed to great quantities of "information" being transmitted from both the internal and external environments. Endless streams of signals are conveyed to internal regulatory centers from all tissues and organs in our bodies. Never-ending stimuli from the external environment—in the form of light, noise, heat, pressure, odor, color, shape, and movement—are detected and processed during every moment of our lives. Occasionally, a proper response to this incoming information must be made instantly if we are to survive or continue functioning in a normal fashion. In other cases, a more leisurely, protracted reaction may be appropriate, and sometimes no response is required.

The nervous and endocrine systems are responsible for receiving, processing, and responding to stimuli of widely different varieties and intensities. They are also central in communications that occur between cells, tissues, and organs within our bodies. Although we consider the nervous and endocrine systems in separate chapters, they are inexorably linked in governing activities that are essential to our well-being. In general, our nervous system with its extraordinary control center, the brain, allows us to formulate prompt responses with pinpoint control. In contrast, our endocrine system is characterized by slower reaction times, responses of greater duration, and more diffuse control. A common feature of both systems is that they use **chemical messengers**, specialized molecules capable of transmitting signals, in their communications.

In Chapters 35 and 36, we describe the anatomical features of these complicated systems and examine a number of functions and physiological mechanisms involved in their operation. For the most part, we concentrate on processes that can be readily visualized and those that occupy the spotlight in contemporary research efforts.

The nervous system coordinates and regulates most functions of the body. This system consists of billions of highly specialized cells, called *neurons,* that are interconnected to form vast networks of communication. The basic "wiring" of the human nervous system has been known for centuries, and more recently we have come to understand parts of its structure and function at the molecular level. Nevertheless, a great deal remains unknown about its more sophisticated operations, particularly those centered in the brain (for example, learning).

The human nervous system is often compared to a computer since both are viewed as operating in fundamentally similar ways; information is first put into the system, some form of integration occurs, and a message is then formulated. However, such correlations have only limited applications. Even the largest mainframe computers have primitive processing capabilities in comparison with our nervous system, since the latter operates as a parallel processor (two or more items can be manipulated at a time) and has unlimited storage capacity. What enables the human nervous system to operate with such great flexibility and with such a high degree of precision? The answers lie within the structure, organization, and functional capabilities of neurons.

The general function of the nervous system is to receive, process, and integrate nerve impulses caused by stimuli received from the external and internal environments. Neurons are the functional units of the nervous system. Billions of neurons are concentrated in specialized organs, such as the brain and spinal cord, and arranged in specific networks that reach all parts of the body. This organization allows for precise control of all organ systems.

NEURONS

The nervous system of all animals is composed of two principal cell types. **Neurons**, shown in Figure 35.1A, are the functional units of the nervous system. They are highly specialized cells capable of conducting electrochemical nerve

Figure 35.1

(A) Neurons are highly specialized cells that are the functional units of the nervous system. (B) Billions of small neuroglia cells, such as these astrocytes, nourish, protect, and support neurons. (C) There are many types of neurons, but all have a common structure. The basic components of a neuron are a cell body, an axon, and multiple dendrites. Arrows indicate the direction of impulse travel in a neuron.

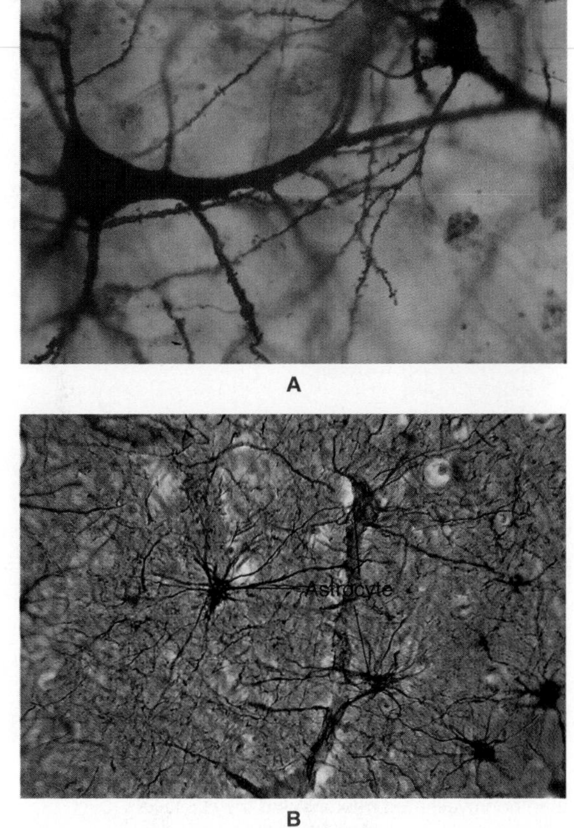

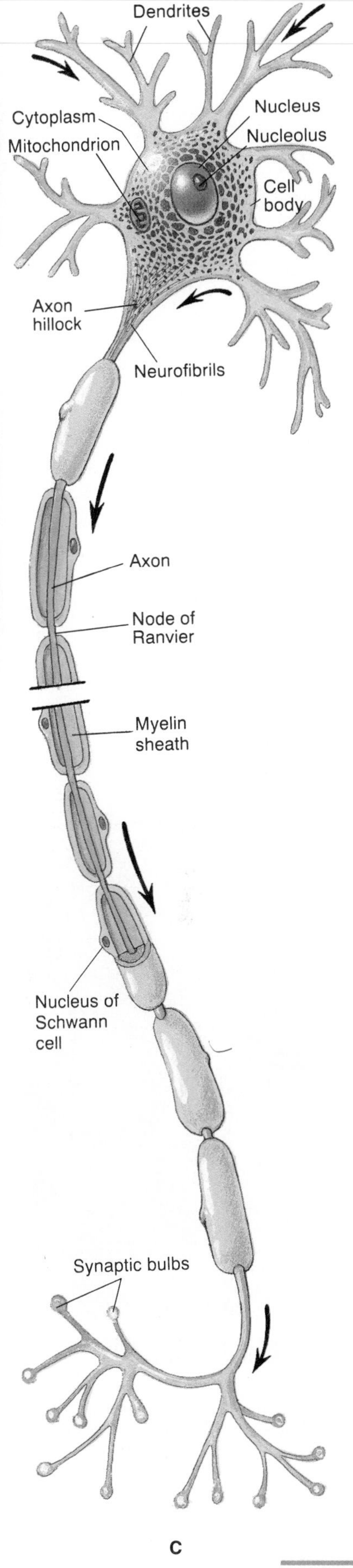

impulses. The human nervous system consists of over 100 billion neurons, most of which are located in the brain. They are supported, protected, and insulated by surrounding *neuroglia* (or *glial*) *cells,* the other major cell type of the nervous system (see Figure 35.1B). Neuroglia cells are smaller than neurons, and five to ten times as abundant. There are several kinds of neuroglia cells, and they perform different functions. Certain neuroglia cells form supporting networks by being wrapped around neurons, others bind neurons to blood vessels, and some tie nervous tissues together.

Structure of Neurons

Neurons come in hundreds of sizes and shapes, but they all have a common structure that reflects their function (see Figure 35.1C). The *cell body* contains the nucleus, cytoplasm, and organelles responsible for maintenance and for synthesizing the specific proteins required by a neuron to carry out its functions. The most prominent features of neurons are two types of stringlike cytoplasmic extensions, emanating from the cell body, that are used in conducting impulses and communicating with other cells. Most neurons contain a single **axon**, a fiber that conducts the nerve impulse away from the cell body. The length of axons in human neurons ranges from less than 1 millimeter to greater than 1 meter. The axon separates near its end into multiple small branches that each end in a tiny knob called the *synaptic bulb*. The ends of the axon's branches are called *synaptic terminals*. **Dendrites**, the second type of extension, are thinner and radiate out from the cell body. Dendrites receive messages from adjacent cells or directly from the external environment and conduct impulses toward the cell body. A neuron may have up to 200 dendrites. The cell body, axons, and dendrites contain numerous *neurofibrils,* threadlike protein structures that form delicate supporting networks within a neuron.

Axons of most large neurons are covered by an insulating layer of lipid-rich, whitish material called *myelin* that is produced through the actions of neuroglia

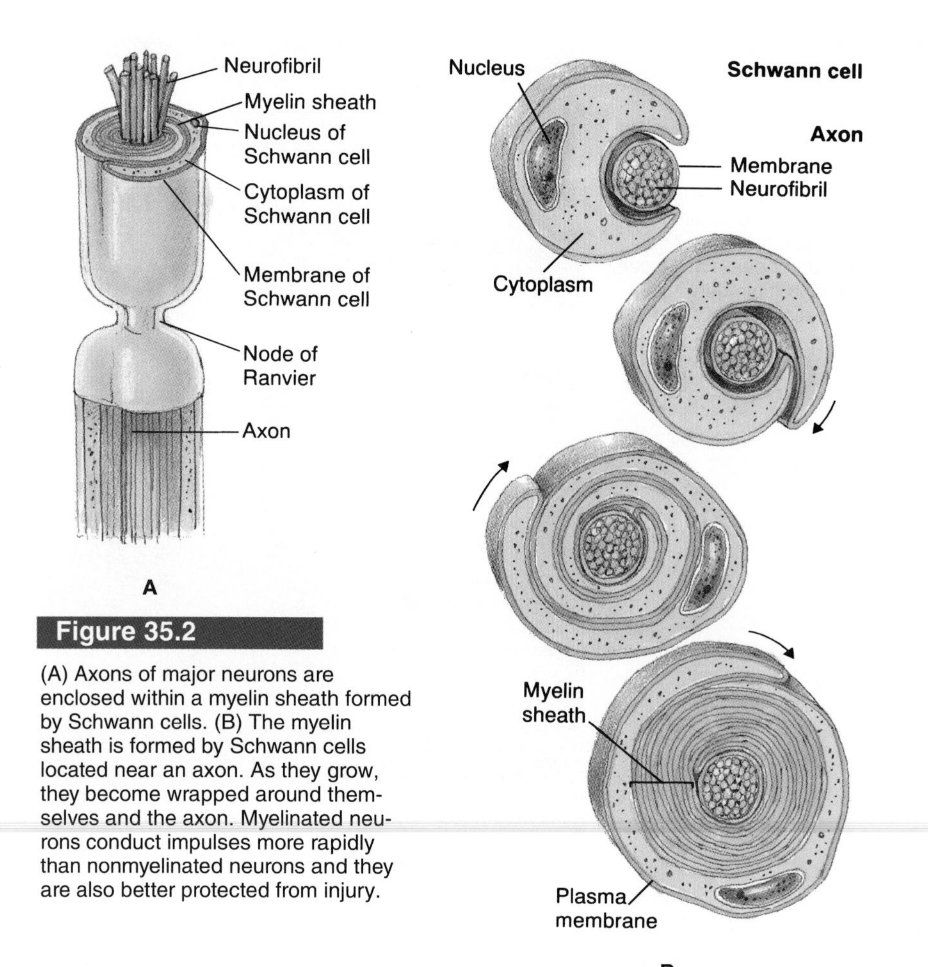

Figure 35.2

(A) Axons of major neurons are enclosed within a myelin sheath formed by Schwann cells. (B) The myelin sheath is formed by Schwann cells located near an axon. As they grow, they become wrapped around themselves and the axon. Myelinated neurons conduct impulses more rapidly than nonmyelinated neurons and they are also better protected from injury.

cells commonly known as *Schwann cells* (see Figure 35.2). During embryonic development and early childhood, Schwann cells grow in such a way that they encircle an axon several times and form a covering of overlapping membranes that is called a *myelin sheath. Nodes of Ranvier* are spaces along a myelinated neuron that are gaps between individual Schwann cells. Myelinated neurons are better protected and conduct impulses faster than nonmyelinated neurons (see Figure 35.3). If the myelin covering of an axon becomes eroded or is injured and is replaced by scar tissue, nerve function becomes impaired, and serious consequences follow. For example, *multiple sclerosis (MS)* is a disease characterized by progressive deterioration of myelin sheaths covering axons in the brain and spinal cord. Affected neurons become covered with scars, resulting in the disruption of nerve conduction pathways. Early symptoms of MS are muscular weakness, double vision, and lack of coordination. Later symptoms depend on the extent and rate of damage to other neurons. The cause of MS is not yet known, although increasing evidence suggests that a virus may be involved.

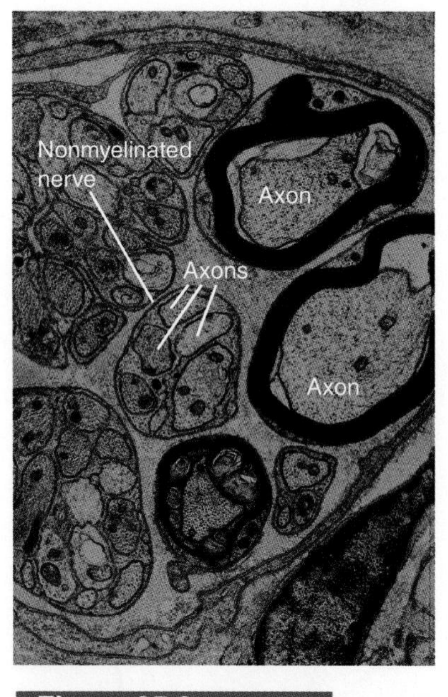

Figure 35.3

In normal myelinated neurons, myelin sheaths appear as thick, black circular structures around the axon. The axons of several nonmyelinated neurons are contained in a nerve.

Types of Neurons

Neurons are commonly classified according to their functions, which are in turn related to the routes of information flow within our nervous system (see Figure 35.4). Cells, tissues, and organs that are controlled by the nervous system through the actions of neurons are said to be *innervated*.

Sensory (afferent) **neurons** receive information from the external and internal environments and conduct it to the brain or spinal cord, where it is processed. **Motor** (efferent) **neurons** transmit impulses away from the brain or spinal cord, where the cell body is situated, toward other cells, muscles, glands, or other neurons involved in making a response. Typically, motor neurons have single, long, myelinated axons. **Association neurons**, the most abundant type (more than 99 percent of all neurons), are confined to the brain and spinal cord. There, they process incoming information by transmitting it between sensory and motor neurons. Such processing may involve transmission among a small number of neurons if a simple response occurs or millions of neurons if learning or language is involved. Brain association neurons commonly have a profusion of branched dendrites that are essential for establishing connections with other neurons.

A **nerve** is a bundle of parallel neuron fibers, both axons and dendrites, and their associated blood vessels and supporting cells, all of which are enclosed by protective connective tissue (see Figure 35.5). Each fiber within a nerve operates independently, and most nerves have both sensory and motor components. By definition, nerves are found outside the brain and spinal cord. Structures in the brain and spinal cord that have the same structural organization are called *nerve tracts*.

Functions of Neurons

Neurons have two basic properties related directly to their function. They are *excitable,* meaning that they have the capacity to respond to stimuli, and they are also *conductive,* or able to conduct a nerve impulse when they become excited. Sensory neurons are also *energy transducers,* in that they are able to convert one form of energy (for example, light, touch, or sound) into a specialized electrochemical impulse that is the means of communication between neurons and cells within the body.

Figure 35.4

(A) Information flows through the nervous system along neural pathways. Stimuli are received from the external and internal environments and transmitted through sensory neurons to association neurons in the brain and spinal cord, where they are integrated and interpreted. If a reaction is initiated, nerve impulses are transmitted to muscle and gland cells by motor neurons.

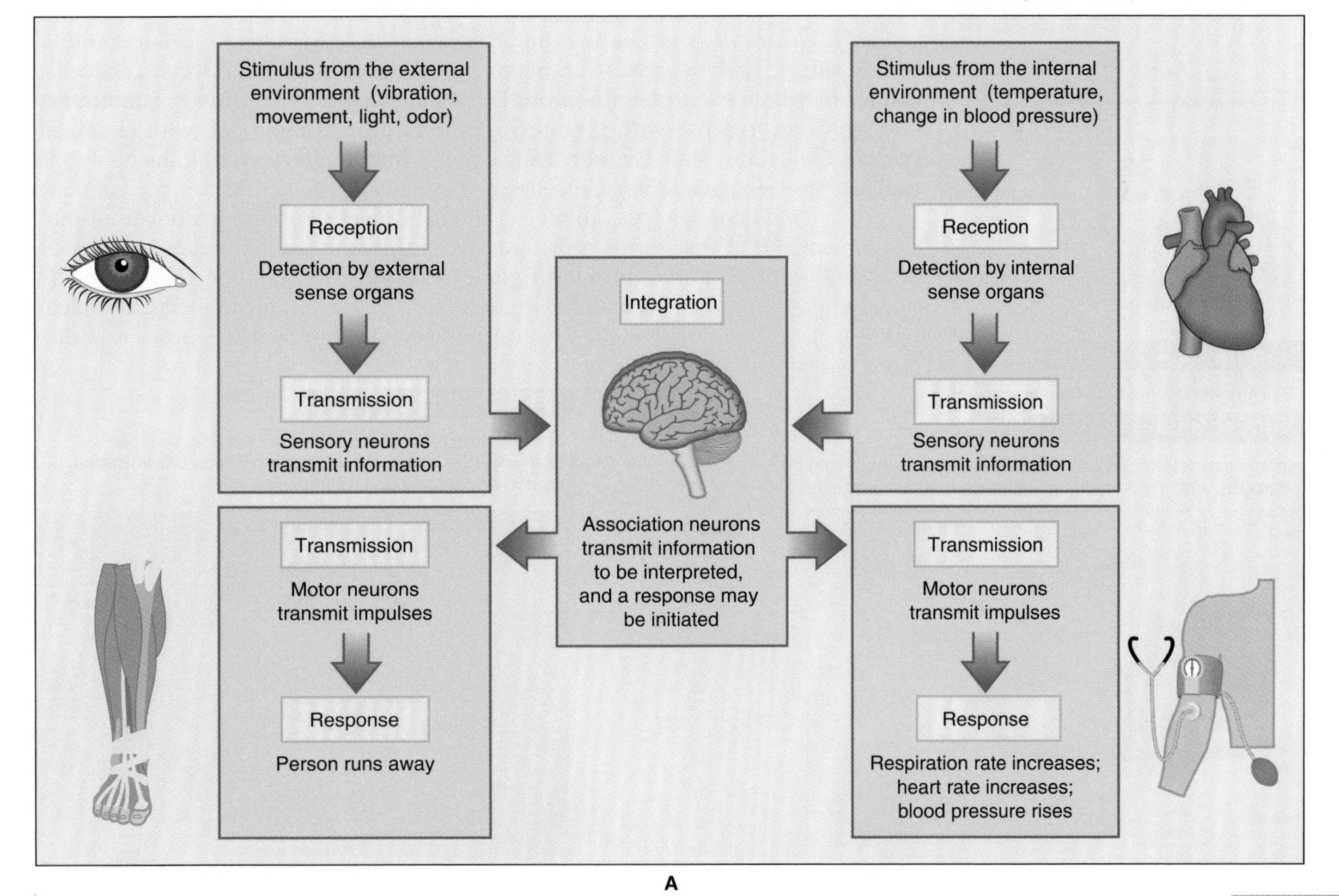

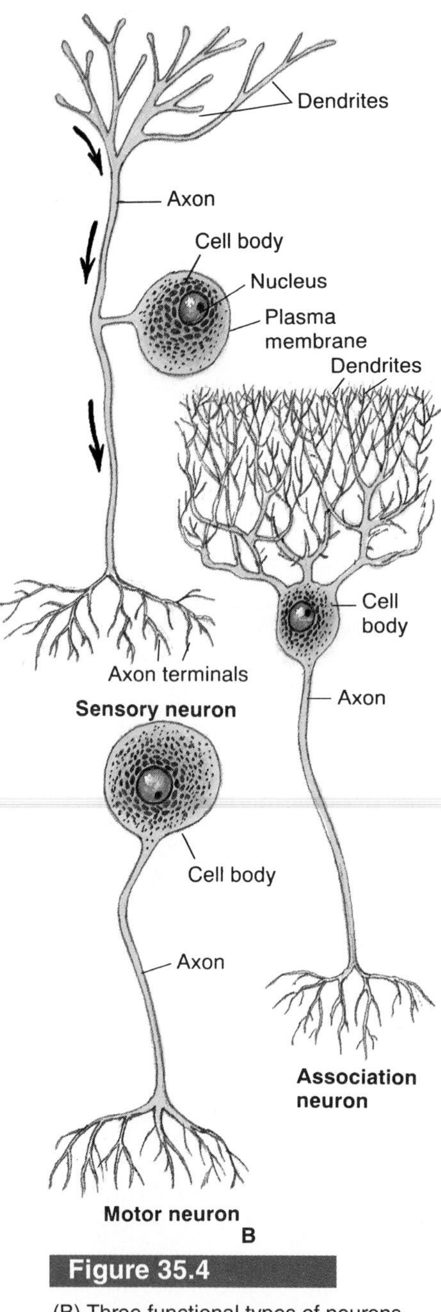

Figure 35.4

(B) Three functional types of neurons.

The nerve impulse is a complex electrochemical event involving changes in the movement of positively charged ions across a neuron's plasma membrane. The details of the electrochemical impulse are beyond the scope of this book. However, the following points about nerve impulses are of special importance: (1) they travel rather rapidly, about 100 meters per second; (2) they travel only in one direction; and (3) they follow an *all-or-none law:* a neuron conducts an impulse like a pistol fires a bullet, either at full power or not at all.

Neurons are specialized cells that are able to respond to stimuli and conduct an impulse. There are three basic neuron types, each with a specific function. Sensory neurons receive stimuli and transmit impulses to the brain or spinal cord; motor neurons transmit impulses from these organs to cells, tissues, or organs that respond to stimuli; and association neurons transmit impulses between sensory and motor neurons.

THE SYNAPSE

Synapses are narrow gaps between neurons, between a neuron and a muscle or gland cell, or between two muscle cells (see Figure 35.6). How is a nerve impulse able to cross the synapse and continue along its route? The answer to that question involves complex biochemical events that can be comprehended by examining Figure 35.7. Knowledge of this process also opens the door to understanding some interesting research questions: How do certain poisons, medical drugs, and controlled substances alter functions of the nervous system? Are there drugs that may be useful in treating different neurological disorders?

Types of Synapses

There are two basic types of synapses, electrical and chemical. In cells communicating by *electrical synapses,* the impulse travels directly from one cell to another cell. In effect, such synapses can be viewed as direct wiring between two cells, neither of which has to be a neuron. Electrical synapses permit very rapid transmission of an impulse with no interruptions, but in comparison with chemical synapses, they are relatively rare. They occur where high speed of transmission is vital, as between cardiac muscle cells.

In a *chemical synapse,* an electrochemical nerve impulse is converted into a chemical signal that forms a bridge across the synapse between neurons or between a neuron and a muscle or gland cell. This bridge allows the chemical signal to pass to the adjacent cell, where it may again become an electrochemical signal. Let us examine how this occurs and what happens if the process is disrupted or fails.

Figure 35.5

Nerves consist of bundles of neurons, blood vessels, and neuroglia, which are enclosed within connective tissue.

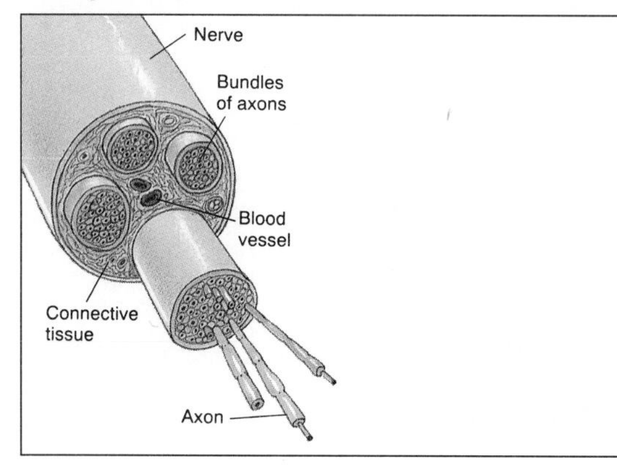

Figure 35.6

This electron micrograph shows a chemical synapse (arrows)—the space between two neurons or between a neuron and a muscle or gland cell.

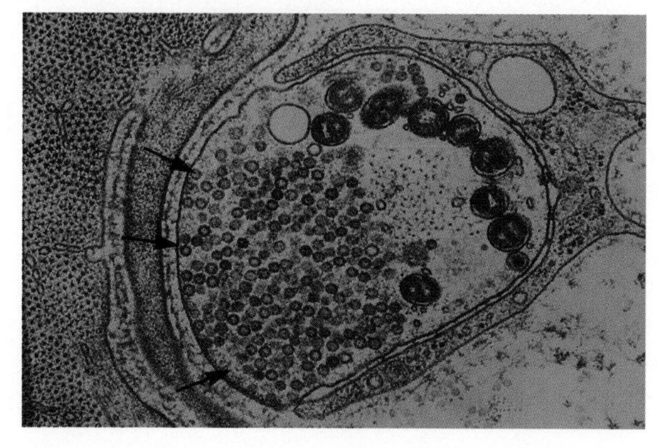

Transmission of a Nerve Impulse Across a Chemical Synapse

Figure 35.7 summarizes the events in chemical synaptic transmission. One-way transmission of an impulse is a key feature of this process. When a nerve impulse travels down the axon of a *presynaptic neuron* (the one carrying the impulse) toward the synapse, it initiates a series of events in the *synaptic bulbs,* which are located at the ends of axon terminals. Within the synaptic bulb are thousands of **synaptic vesicles**, small membrane-bound sacs filled with **neurotransmitters**, specific chemical messengers involved in nerve impulse transmission across a chemical synapse. Each synaptic vesicle contains 10,000 to 100,000 neurotransmitter molecules. There are many different neurotransmitters, but each vesicle is thought to contain only a single type. However, a synaptic terminal can harbor two or more types of vesicles that contain different neurotransmitters.

As the impulse reaches the synapse, vesicles are stimulated to fuse with the *presynaptic membrane* and release their neurotransmitter into the *synaptic cleft,* the narrow space between cells. The neurotransmitters diffuse across the cleft and bind to specific receptors located on the membrane of the *postsynaptic cell* (the one receiving the impulse). *Postsynaptic receptors* are proteins that each have a unique structure enabling them to bind with a specific neurotransmitter (that is, neurotransmitter X binds only to receptor X). When neurotransmitters are bound to receptors, chemical changes occur in the postsynaptic plasma membrane that may trigger an electrochemical impulse, thus perpetuating transmission of the original impulse.

Not all nerve impulses, however, continue to be transmitted. Neuron cell bodies, muscle cells, and gland cells are typically contacted by numerous synaptic bulbs (see Figure 35.8). Some of these belong to excitatory neurons and are likely to induce a response. Other synaptic bulbs, however, are from inhibitory neurons that tend to make a response less likely. Whether or not an impulse is transmitted depends on a final accounting by the postsynaptic cell in making a yes-or-no decision about firing. If the balance sheet tilts toward excitation, the impulse is transmitted; if toward inhibition, no impulse conduction occurs.

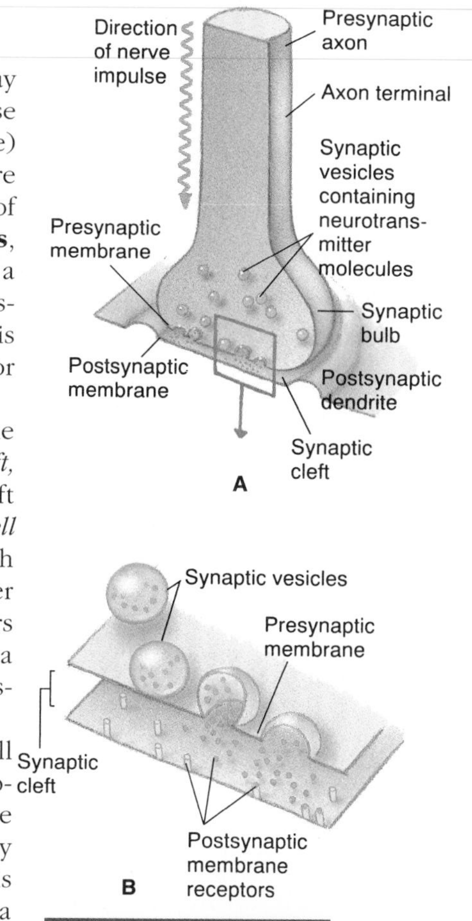

Figure 35.7

(A) Nerve impulses are transferred from a presynaptic cell to a postsynaptic cell. For the impulse to be continued in the postsynaptic cell, synaptic transmission must occur. (B) When an impulse reaches the synaptic bulb, it induces synaptic vesicles to fuse with the presynaptic membrane and release their neurotransmitters into the synaptic cleft. The neurotransmitters move across the cleft and bind with specific receptors on the postsynaptic membrane. If a sufficient number of receptors become involved, changes occur in the postsynaptic cell, which may cause an impulse to be generated. Neurotransmitters are quickly removed from the synapse once transmission is completed.

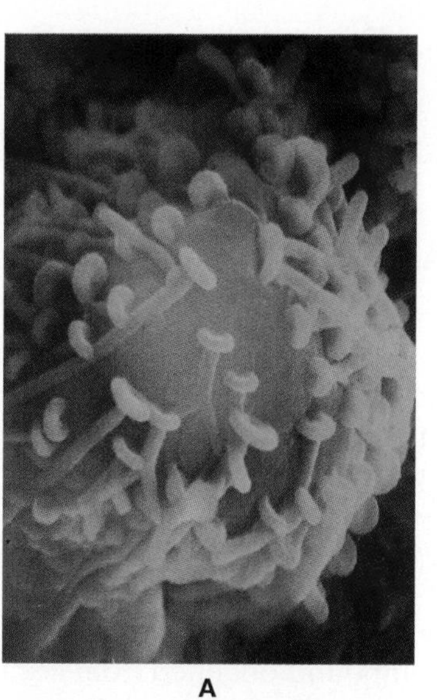

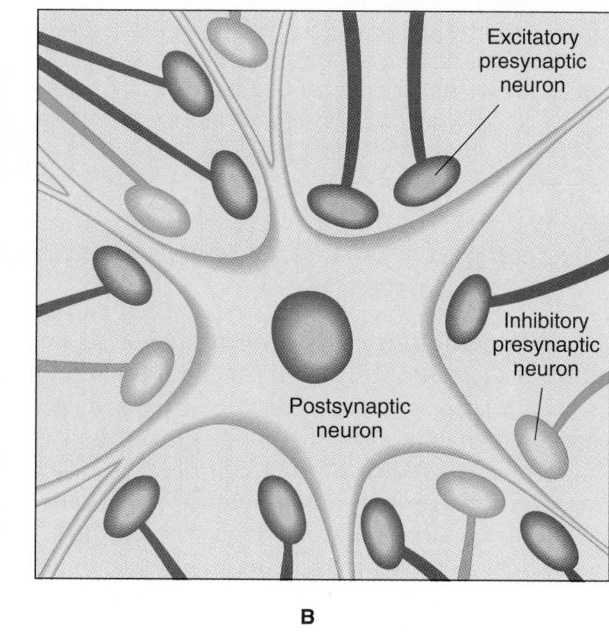

Figure 35.8

(A) This scanning electron micrograph shows numerous synaptic bulbs in contact with a neuron cell body. (B) Cells are often contacted by both excitatory and inhibitory neurons. By integrating impulses from both types of neurons, the target cell determines whether or not to fire.

Return to Normal

The action of neurotransmitters takes only a fraction of a second. What happens to them after the impulse is transmitted? How is the signaling terminated? These are critical questions, because if the synaptic cleft is not cleared of neurotransmitters, continuous firing of the excited neuron will result, which can lead to severe problems (this is the mode of action for some poisons). There are three primary means for purging the synaptic cleft of neurotransmitters: (1) specific enzymes degrade some transmitters, (2) some transmitters are taken back up by the presynaptic membrane and recycled into vesicles, and (3) some transmitters are taken back up by nearby neuroglia cells. Whatever the process, it occurs within a fraction of a second, and calm is restored at the synapse.

The Current Picture

The mechanism of chemical synaptic transmission is well understood, but there are still many open questions. The list of chemicals that serve as neurotransmitters is now lengthy but still incomplete. Also, full details about how neurotransmitters are released and interact with postsynaptic cells are known for only a handful of such substances. **Acetylcholine (ACh)**, a neurotransmitter occurring at many neuron-neuron and neuron-muscle cell synapses, has been the most extensively studied. A detailed picture has been developed for its synthesis, storage in synaptic vesicles, and release into the synaptic cleft (see Figure 35.9). Scientists have also isolated the ACh receptor found on postsynaptic membranes. The receptor is a protein, and researchers have cloned it and determined its amino acid sequence. After transmission, ACh is instantly split into choline and acetate, the two subunits forming ACh, by the enzyme called *acetylcholinesterase*. Choline and acetate are recycled back into the synaptic bulb, recombined into ACh, and reincorporated into a synaptic vesicle. This type of detailed information will likely become available for other neurotransmitters in the future and will undoubtedly lead to new pharmaceutical drugs and other compounds that can modify action at the synapse.

Most nerve impulses are transmitted from cell to cell across a chemical synapse. Impulses traveling down presynaptic neurons trigger release of neurotransmitters from synaptic vesicles into a synaptic cleft. Neurotransmitters bind to receptors on the postsynaptic cell and initiate changes that may trigger an impulse, depending on whether the cell becomes excited or inhibited. Once an impulse is transmitted, neurotransmitters are quickly removed from the synapse.

ALTERATIONS OF NORMAL SYNAPTIC TRANSMISSION

Understanding the process of normal synaptic transmission of nerve impulses enables us to recognize the potential for modifying the mechanism at different points. A large number of **drugs**—chemical agents that bring about functional or structural change in living tissues—exert their effects at the synapse. In humans, these effects include altering neurological processes involved in thinking, learning, memory, and complex behaviors. Perhaps the best-known drugs are those with questionable or no legitimate medical use. For example, *nicotine,* the active ingredient in tobacco, is a stimulant that mimics the action of ACh and causes a short-term excitatory effect. *Cocaine,* an addictive compound obtained from coca leaves, is now thought to prevent or reduce the uptake of neurotransmitters at the synapse and results in euphoria and an increased perception of alertness. *Crack,* a potent form of cocaine, disrupts the normal inactivation of neurotransmitters involved in mood and muscle coordination. In the short term, this can cause convulsions, high blood pressure, and weight loss. Chronic use may lead

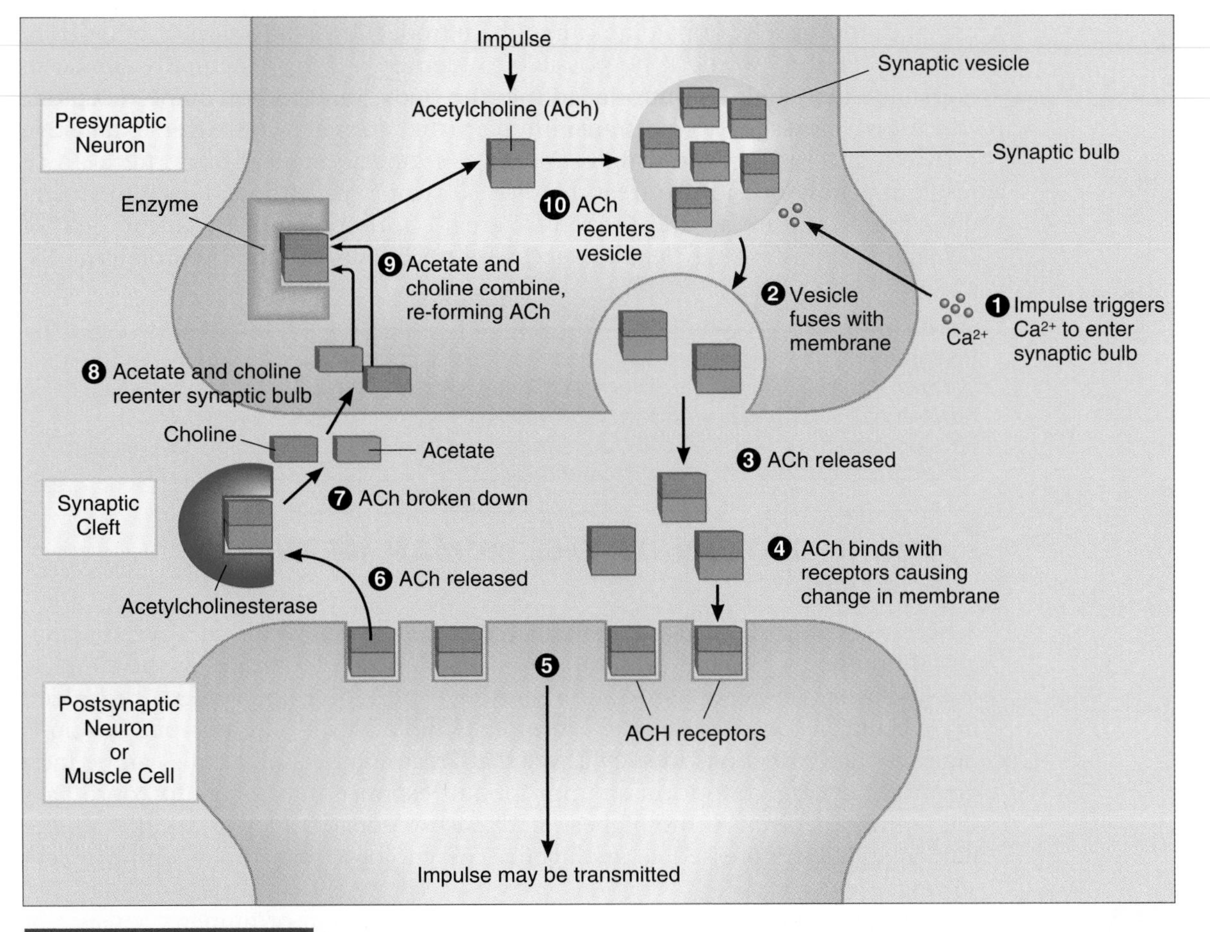

Figure 35.9

An incoming nerve impulse causes calcium ions (Ca)$^{+2}$ to enter the synaptic bulb and trigger release of acetylcholine (ACh) from synaptic vesicles into the synaptic cleft. ACh diffuses across the cleft to the postsynaptic cell, where it binds with specific ACh receptors on the membrane, causing changes in the membrane that lead to the generation of an impulse. Once this occurs, ACh is released from the receptor and broken down into acetate and choline by the enzyme acetylcholinesterase. Acetate and choline are taken up by the presynaptic cell, where they recombine to form ACh.

to a depletion of specific neurotransmitters and result in severe psychological effects such as persistent depression.

A more positive use of drugs is in treating serious neurological disorders. For example, *Parkinson's disease* is characterized by late age of onset (typically 50 years or older) and severe symptoms such as drooling, tremors, and nonfunctioning muscles. Although the cause of the disease is not known, a leading hypothesis is that exposure of certain brain neurons to toxic substances may play a role. This view was strengthened by a recent discovery involving young addicts who had taken heroin contaminated with a toxic compound known as MTTP. After entry into the body, MTTP was found to alter the same part of the brain that is affected by Parkinson's, and it caused the same effects. A major challenge to scientists is to devise ways of treating such disorders. How can the symptoms be eliminated or reduced?

The first step, always, is to identify the underlying biological mechanism that has been altered. In Parkinson's, the cause was found to be due to depletion of a neurotransmitter called *dopamine* in the brain. Consequently, neurons that require certain levels of dopamine for normal nerve transmission function abnormally, resulting in uncontrolled muscular actions. Although dopamine-requiring brain cells cannot transmit impulses normally, their dopamine receptors remain intact. Unfortunately, dopamine cannot be introduced via injection or other common procedures. Would it be possible to identify a drug structurally similar to dopamine that could be introduced into the body, bind to dopamine receptors, and allow messages to be transmitted in the normal way? After exhaustive research and testing, the drug *L-dopa* was found to meet those criteria. As a result, it is now possible to treat the disease symptoms, although such treatment does not constitute a cure. Currently, great efforts being made in the field of "synaptic chemistry" may lead to successful treatments of other neurological disorders.

Presently, intense efforts are being made to understand how specific drugs affect synaptic transmission events. There are two aspects to this research: first, to determine how harmful substances modify normal transmission mechanisms, and second, to develop new drugs that may counteract or prevent the effects of neurological disorders.

ORGANIZATION OF THE HUMAN NERVOUS SYSTEM

Schematic wiring diagrams of the human nervous system would fill several manuals and probably astound even the most experienced electrician. At the simplest level, sensory neurons, association neurons, and motor neurons are organized into circuits, as described earlier. At more complex levels, the nervous system is organized and functions in such a way that great numbers of signals can be integrated, or averaged, to form an appropriate response that serves to coordinate the entire organism. At the highest, and least understood, levels are the unique human abilities expressed as intelligence, psychology, philosophy, and problem solving.

As diagramed in Figure 35.10, the nervous system of humans consists of two primary divisions, the **central nervous system (CNS)**, which is basically the brain and the spinal cord, and the outlying **peripheral nervous system (PNS)**, which innervates all parts of the body and transmits information to and from the CNS.

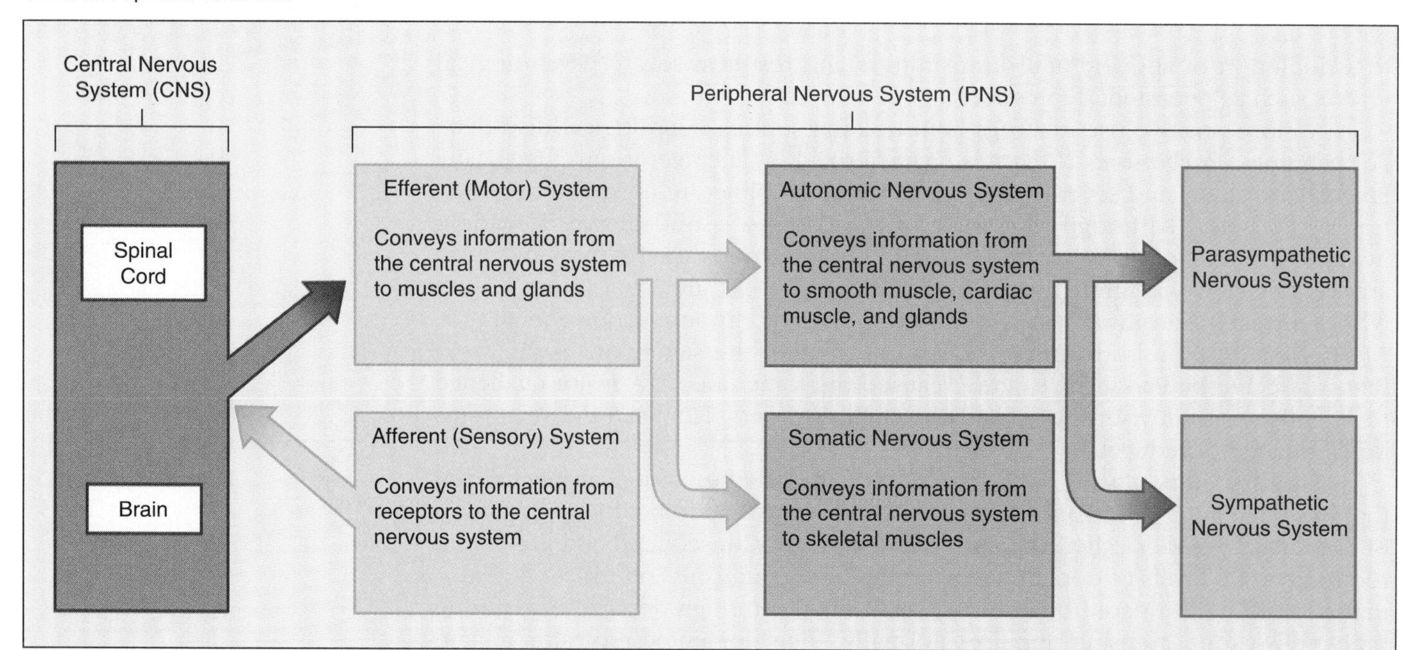

Figure 35.10

The human nervous system is composed of two principal parts, the central nervous system (CNS) and the peripheral nervous system (PNS). The PNS includes the somatic nervous system and the autonomic nervous system, which is further subdivided into the parasympathetic and sympathetic nervous systems. Each of the various divisions has specific functions.

The Peripheral Nervous System

The PNS consists of all neurons that exist outside of the CNS, and it contains 43 major nerves and a vast network of smaller nerves that connect the brain and spinal cord with all other structures in the body. The PNS has the central responsibility for maintaining homeostasis.

Of the 43 major nerves of the PNS, 12 are *cranial nerves* that originate from the lower brain, and 31 are *spinal nerves* that emerge from the spinal cord (see Figure 35.11A). Through these nerves, information is transmitted continuously between every organ and tissue and the CNS.

Cranial and Spinal Nerves

The cranial nerves lead to major **sense organs**—the eyes, ears, nose, and tongue—and to muscles located in the head. Only one cranial nerve, the critically important *vagus nerve,* serves a region outside the head. It extends to, and regulates functions of, the heart, lungs, and digestive organs. Spinal nerves innervate all other regions of the body below the neck.

Functional Classification of the PNS

The PNS is organized according to functions performed. It is first divided into the *afferent (sensory) division,* which includes nerves that relay impulses from all areas of the body to the CNS, and the *efferent (motor) division,* which is in turn subdivided into the following two parts: the **somatic nervous system**, composed of motor nerves that run directly from the CNS to skeletal muscles, and the **autonomic nervous system**, shown in Figure 35.11B, which connects motor nerves from the CNS with different internal organs and is primarily concerned

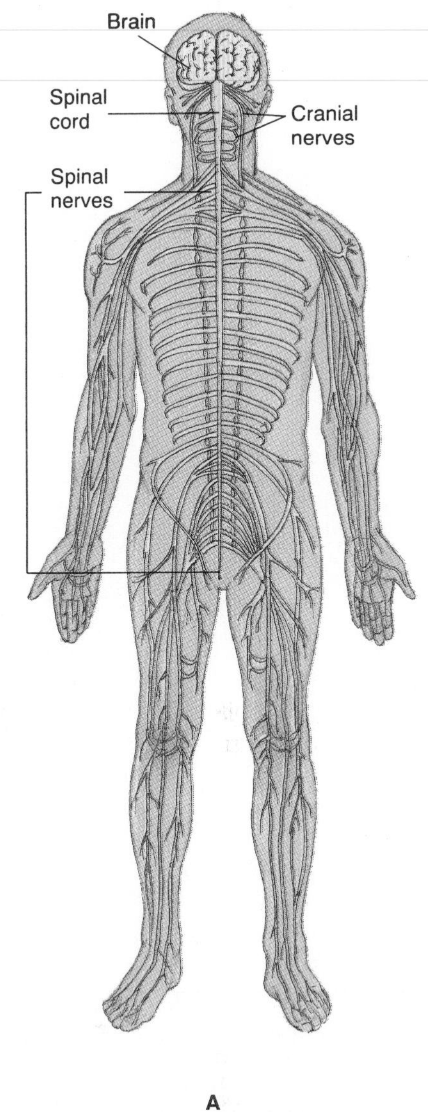

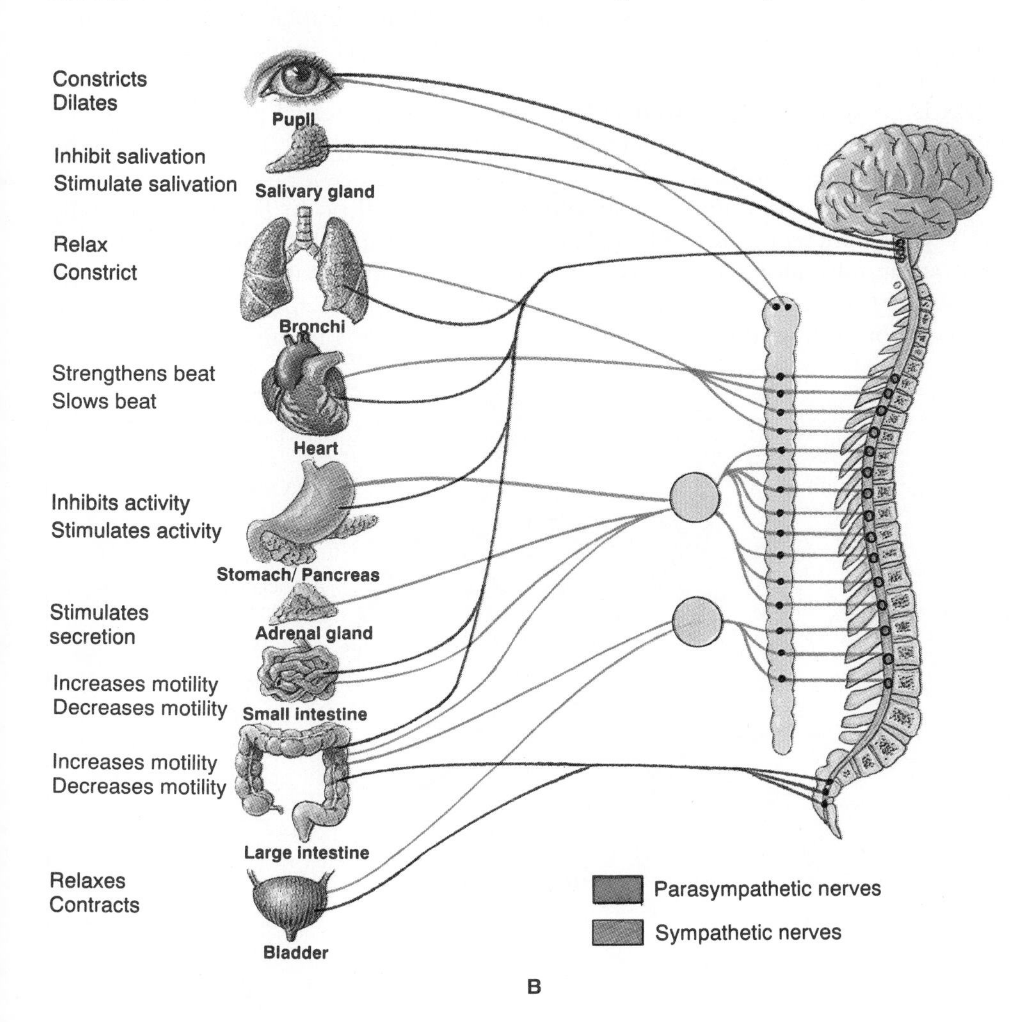

Figure 35.11

(A) The 43 nerve pairs of the peripheral nervous system emerge from the brain and spinal cord and, through branching and interconnections, reach every organ in the body. (B) The autonomic system of the PNS has two divisions, the parasympathetic and sympathetic nervous systems. Most organs are innervated by both of these systems, and they generally cause opposite effects.

with automatic functions such as the beating of the heart. Finally, there are two divisions of the autonomic nervous system, the *sympathetic* and *parasympathetic* nervous systems. Many organs of the body are affected by both of these divisions, and the two systems generally tend to cause opposite responses. Most often, the sympathetic system prepares a person for vigorous activity, for example, by increasing the heart rate. The parasympathetic system is concerned with maintaining the body under relatively calm conditions and allows us to lead tranquil lives without worrying about when it is necessary to breathe or for our hearts to beat. The dual effects of the autonomic system allow for remarkably precise control of internal functions.

The Central Nervous System

The CNS is the integrative and control center that links the sensory and motor functions of the nervous system. Sensory neurons transmit information to the spinal cord. From there, it is relayed to the brain, the coordination center of the nervous system, where thorough processing and integration take place in the association neurons and an appropriate response, involving stimulation of specific muscles or glands, occurs. The brain also analyzes sensory information, a highly complex activity that involves thousands of association neurons. Although many motor neurons originate in the brain and numerous sensory neurons terminate there, the vast majority of brain neurons are the association neurons essential to processing information and formulating complex responses.

The Spinal Cord

The spinal cord has two principal functions. First, it acts as a two-way relay system that transmits impulses between the brain and the peripheral nervous system through sensory and motor neurons. Second, it controls simple reflex actions. A reflex is an unconscious, programmed response to a specific stimulus. Each reflex is controlled by a **spinal reflex arc**, an inborn neural pathway by which impulses from sensory neurons reach motor neurons without first traveling to the brain (see Figure 35.12).

A familiar (and painful) example of a reflex will occur if you accidentally place your hand on a hot object. Pain receptors in the skin covering the hand send an impulse along a sensory nerve leading to the spinal cord. This message is switched directly, through association neurons in the spinal cord, to motor neurons controlling muscles of the arm, and the hand is instantly withdrawn. At

Figure 35.12

In a spinal reflex arc, reflexes occur automatically because they involve fixed neural pathways that travel directly from a receptor to a muscle or gland.

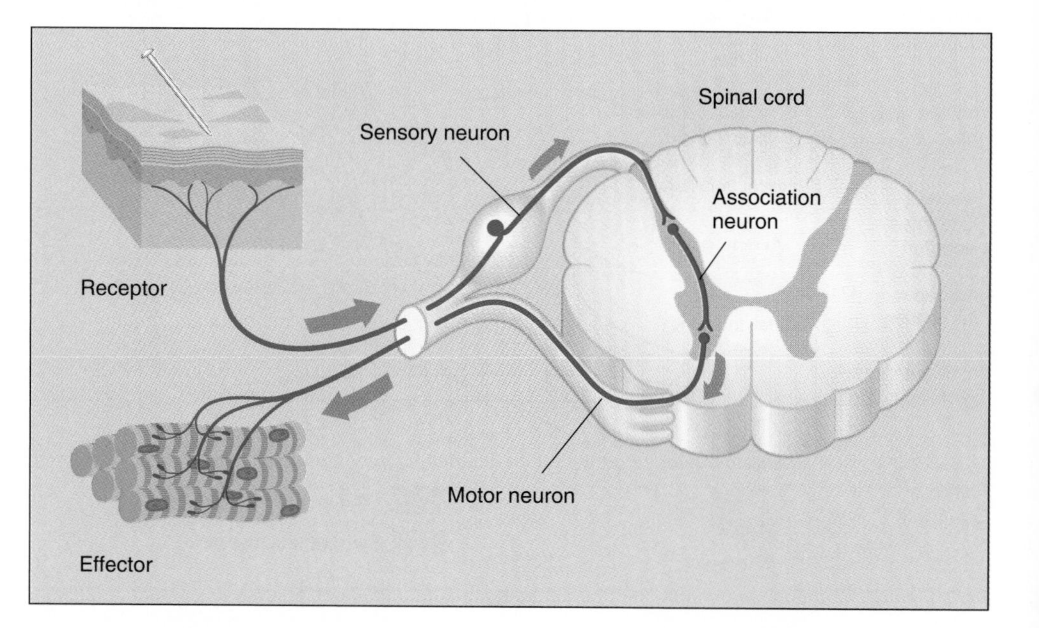

the same time, the brain also receives the message, and more complex reactions, such as vocalization, follow instantly after the reflex action. Such reflex actions are important because they permit a rapid, automatic, corrective response to a potentially dangerous stimulus.

The Brain

The three geographical regions of the human brain and their major structures and functions are described in Figure 35.13 and Table 35.1. The three primary structures of the **hindbrain**—cerebellum, medulla, and pons—are involved in maintaining homeostasis and coordinating large-scale body movements such as walking or jumping. They are also part of the *reticular formation,* a loosely organized network of nerves that receives input from several other areas of the brain and spinal cord and maintains the state of consciousness. It accomplishes

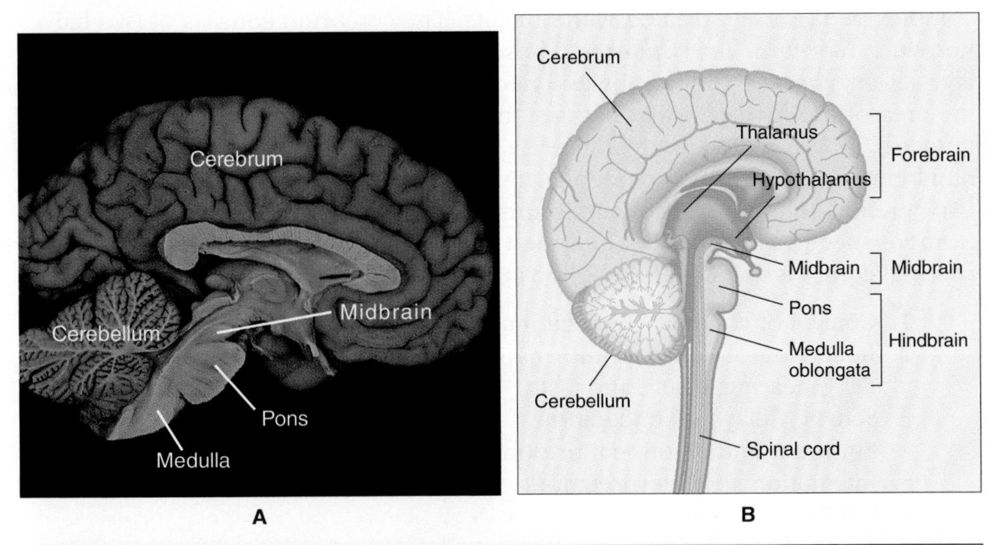

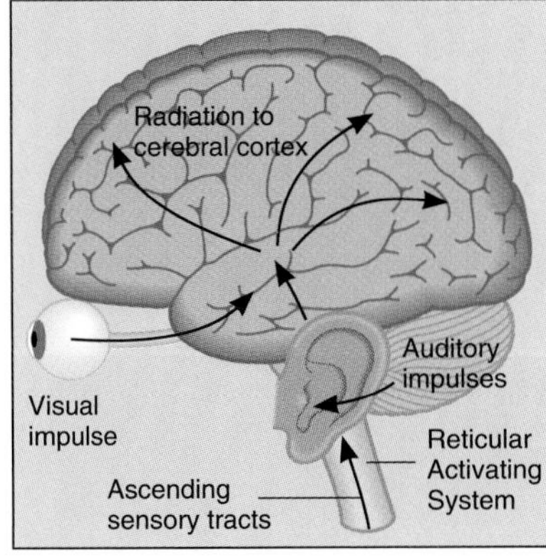

Figure 35.13

(A) The human brain contains 100 billion neurons. (B) The brain is divided into three main sections: hindbrain, midbrain, and forebrain. (C) The reticular formation is a diffuse network of nerves that receives and processes signals from the internal and external environments. Only a small number of these signals are transferred to the cerebral cortex through the reticular activating system.

Table 35.1 Major Functions of Different Structures and Regions of the Human Brain

Area	Structure	Function
Hindbrain	Cerebellum	Coordinates complex muscular movements; maintains a sense of balance.
	Medulla oblongata	Regulates several automatic functions such as breathing, heartbeat, swallowing, and digestion.
	Pons	Conducts impulses between different parts of the brain; helps regulate breathing activities.
Midbrain		Serves to connect the hindbrain and forebrain; contributes to coordination and consciousness.
Forebrain	Cerebrum	Processes and integrates information received from several sources; essential to thought, memory, consciousness, emotion, and higher mental processes.
	Thalamus	Center for relaying and coordinating sensory information; directs information to specific anatomical areas of the cerebrum.
	Hypothalamus	Main coordinating center for parts of the peripheral nervous system; for example, regulates body temperature, appetite, and emotions; it also regulates the pituitary gland, thus linking the nervous and endocrine systems.

this task by filtering the massive number of impulses received from the surrounding environment and relaying only a portion of them to the *cerebral cortex,* the conscious center of the brain. Since the reticular formation controls what reaches the cortex, it is considered to be an activating system for maintaining wakefulness and alertness and is often referred to as the *reticular activating system (RAS)*. Damage to the RAS often results in coma.

The **midbrain**, a small area between the forebrain and the hindbrain, receives and integrates several types of sensory information and relays it to specific regions of the forebrain for processing. Certain cells of the midbrain are also integrated with the reticular formation.

The most complex neural processing and integration occur in the **forebrain**. The *cerebrum* is the largest part of the entire brain and is more highly developed in humans than in any other species. This portion of the brain is essential to higher mental functions—thought, memory, consciousness, and complex behaviors associated with humans. The cerebrum consists of two halves known as *cerebral hemispheres*. These structures are covered with a delicate layer of *gray matter* (nonmyelinated nervous tissue) called the **cerebral cortex**, the area of the forebrain that has become so highly specialized and enlarged in humans. To accommodate this increase in size, our cerebral cortex consists of a number of specific folds and deep grooves that allow it to fit inside the skull. The two cerebral hemispheres contain four major *lobes,* each of which carries out one or more specific functions, as indicated in Figure 35.14. In summary, the human cerebral cortex is the site of the following:

1. Control of voluntary skeletal muscle
2. Receiving of sensory information involving touch, pain, temperature, sight, hearing, smell, and taste
3. Sensory relay and integration
4. Emotions and emotional responses
5. Integration and interpretation
6. Intellectual activities

The human brain is perhaps the most remarkable organ to have evolved in any organism. Its operations encompass a range of activities, from the simple,

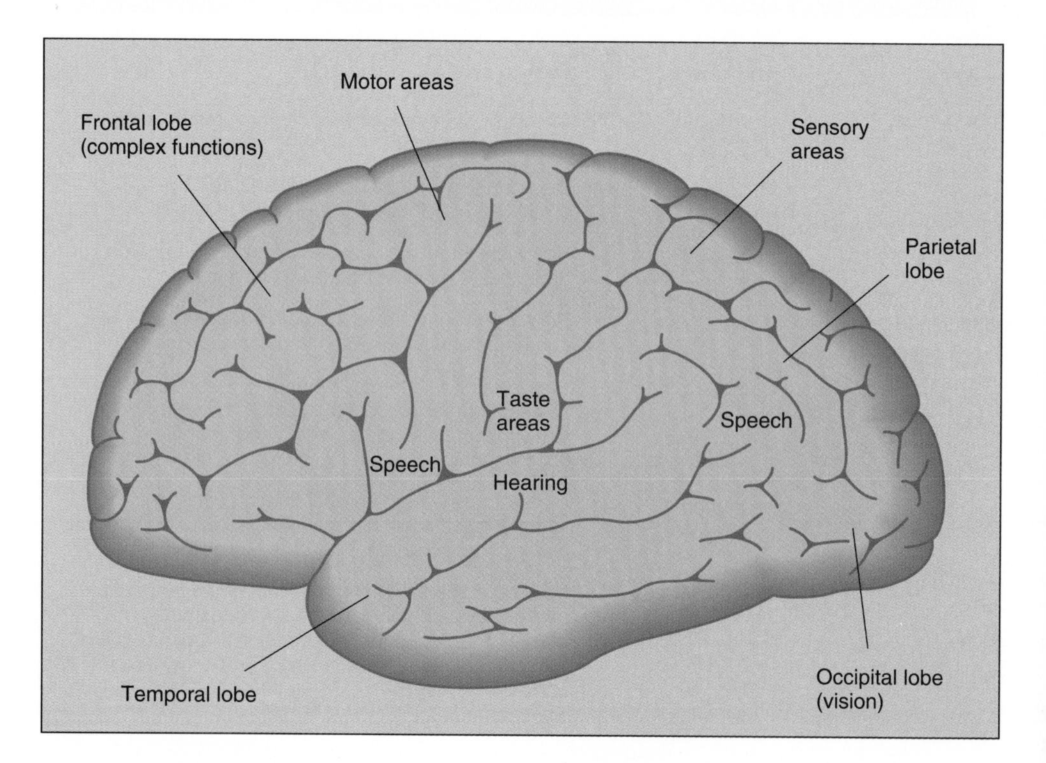

Figure 35.14

This diagram shows the four major lobes and other regions of the cerebral cortex where major functional areas, such as vision, speech, and hearing, are centered.

automatic control of muscles to the greatest intellectual achievements. Massive textbooks have been written about the structure and functional domains of the human brain. However, much remains to be learned about the most wondrous aspects, which are related to intellect, and the exploration of brain structure and function represents a frontier field of research.

The nervous system has an organizational hierarchy that is related to function. The brain and spinal cord form the CNS, which regulates most body activities. The PNS links the CNS with all other organs. The PNS is partitioned into an afferent division, which directs impulses to the CNS, and an efferent division, which is subdivided into two parts, the somatic nervous system, composed of motor nerves, and the autonomic nervous system, which regulates automatic functions.

FUTURE FRONTIERS

Interest in systems that regulate and control operations within the human body has a long history. Studies of the nervous system can be traced back to the ancient Greek and Roman cultures. As in other fields of biological research, progress accelerated during the early twentieth century and erupted in parallel with advances in molecular biology during the past four decades. As described in this chapter, much is known about the highly specialized cells, chemical messengers, and underlying mechanisms that govern our activities. Nevertheless, great voids remain in our understanding of the nervous system.

What problems remain for scientists to confront in the decade that lies ahead? Perhaps first and foremost is studying how the human brain works. Our interpretations of the outside world and how we respond to it at any given moment are dependent on brain functions. What about other processes, such as emotion, memory (see the Focus on Scientific Inquiry, "Memory"), information storage, intelligence, thought, and imagination, that make us human? How does the brain develop? How do the 100 billion neurons (give or take an order of magnitude) of the brain become so highly organized? How are different types of information processed? Neurobiologists are making headway in some of these areas, and it should be fascinating to follow their progress in the years ahead. Recognizing the importance of research on the brain, in 1990, Congress designated the 1990s as the Decade of the Brain. As Francis Crick (of DNA fame) stated in a 1984 article, "There is no scientific study more vital to man than the study of his own brain. Our entire view of the universe depends on it."

Summary

1. The nervous system receives, integrates, and responds quickly to stimuli from the internal and external environments.
2. The functional unit of the nervous system is the neuron, a highly specialized cell that forms communication networks with other neurons.
3. Neurons are classified according to their function. Sensory neurons receive information and transmit it to the central nervous system, motor neurons transmit impulses away from the central nervous system, and association neurons relay signals between sensory and motor neurons.
4. The basis of communication in the nervous system is an electrochemical impulse initiated and transmitted by neurons. An impulse can be transmitted across the synapse between cells through the complex actions of neurotransmitters released from synaptic vesicles of the neuron.
5. Once the nerve impulse has passed between two cells, neurotransmitters are removed from the synapse by different mechanisms. Drugs may disrupt normal neurotransmitter synthesis, prevent their removal from the synapse, or mimic their action. Though most drugs are ultimately harmful, several may be useful in treating neurological disorders.

MEMORY

Focus on Scientific Inquiry

Memory can be described as the store of items learned and retained as the result of an individual's activities or experience. Knowledge of a specific past event—a memory—can be recalled from storage and reproduced mentally with varying degrees of accuracy. Why are we able to remember some things and not others? Why can certain incidents in our lives, such as our first "blind date," be recalled with crystal clarity or in agonizing detail? Why do sensory stimuli, especially odors—such as the fragrance of a certain perfume, new-mown grass, or a musty attic—frequently evoke detailed memories of events in the distant past?

Scientists are very interested in answering such questions as part of their efforts to understand the brain. As much as any other trait, the ability to remember events and integrate them in intelligent ways is a hallmark characteristic of our species. Neurobiologists are concentrating on answering these and other questions. Are there different types of memory? How is information processed and transformed into a memory? Where is a memory stored? How is it accessed? How are memories formed?

As outside observers, perhaps we should ask, How can such questions even be addressed? What types of studies or experiments can be done to provide data that can be used to answer these questions? Several approaches have been used in memory research. Much of the early effort (in the 1950s and 1960s) was directed at studying patients with various types of *amnesia,* loss of memory due to a brain injury or malfunction. Certain affected individuals, usually identified by their initials, provided investigators with insights that later led to carefully designed experiments. When amnesia patients participating in a study died, their brains were carefully analyzed to determine which parts appeared damaged or abnormal and may therefore have been associated with the memory disorders.

The first and perhaps most famous case was a man referred to as H. M., an individual who had a severe form of epilepsy that could not be treated successfully in 1953. As a result, doctors made a decision to remove parts of his brain associated with the seizures. After recovery, it was discovered that the seizures were much less severe, but tragically, H. M. was no longer able to learn new facts. He was unable to remember the names of individuals he saw every day or recall what he had eaten a few moments earlier. Interestingly, he was quite proficient at solving puzzles but was unable to remember that he had done so.

What was learned by studying H. M. and other individuals with amnesias? First, it became clear that there are different types of memory linked with specific regions of the brain. Memory of past events, facts, names, dates, and places is called *declarative memory. Procedural memory,* or "habit," is acquired by repetition or continuous practice. Riding a bicycle, swinging a golf club, playing a piano, and solving puzzles are examples of procedural memory. H. M. retained procedural memories but was unable to acquire declarative memories after his operation. Second, the parts of the brain that are important in processing and transforming informa-

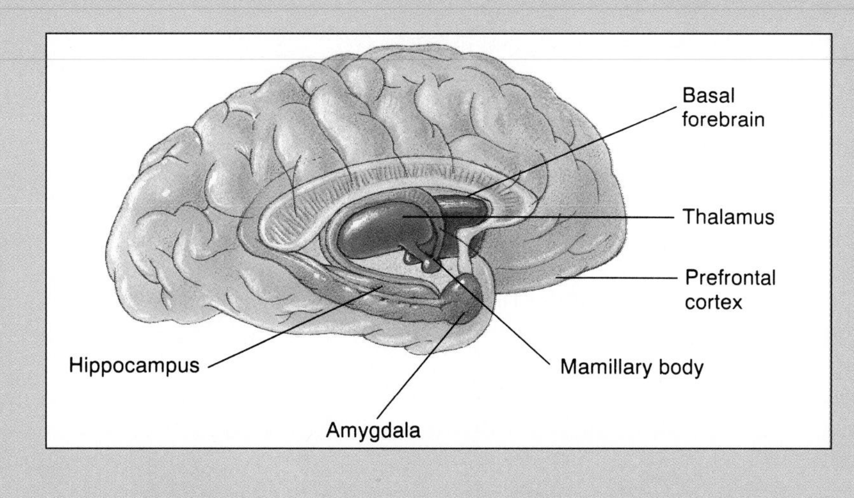

Figure 1 The indicated parts of the brain were found to be involved in memory formation.

tion into memories were identified. They included the hippocampus, the amygdala, and parts of the cortex (see Figure 1).

To identify more precisely the key parts of the brain involved in declarative memory formation and storage, investigators turned to studying memory in animal models, primarily monkeys. The usual experimental approach was to teach the animals how to play different types of "games" in which they received a reward for playing correctly. Once they learned how to play the game, small parts of their brains would be surgically removed or modified. Their memories of the game they had learned would then be evaluated using different tests.

Results from these experiments extended the knowledge gained from studies of amnesia patients. The areas of the brain that were found to be critical for memory formation and recall indicated in Figure 1 were determined to be related as shown in Figure 2. The most interesting findings were related to discoveries of the neuroanatomical pathways used in memory processing. From these experiments, conducted over a 20-year period, scientists hypothesized that perceptions are formed in areas of the brain that receive sensory information (for example, a vision, a smell, or a noise). Once this occurs, two parallel circuits, one leading to the amygdala and the other to the hippocampus, are activated. Information flow from both of these structures travels to the diencephalon (mamillary body and thalamus), the prefrontal cortex, and the basal forebrain.

How do these structures collaborate in creating a memory? This is a very complicated question, and the answer is not yet

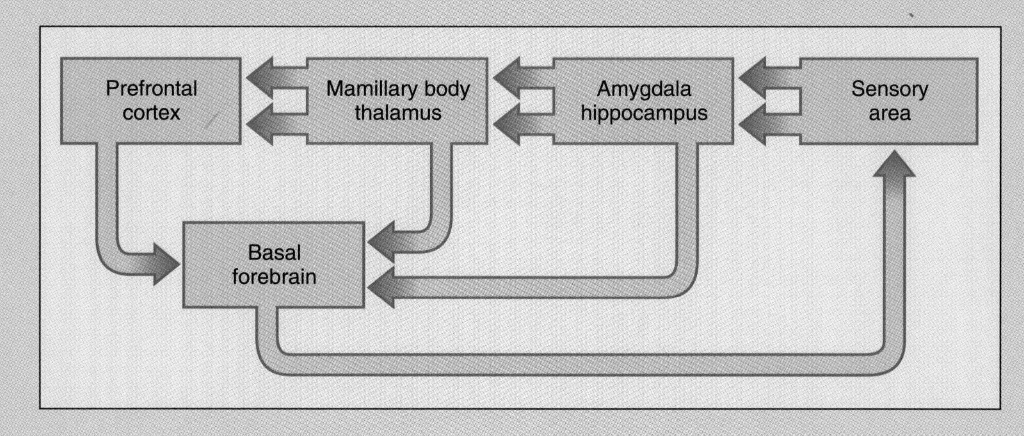

Figure 2 Experimental studies of monkeys led to this diagram of the memory system.

complete. Available evidence indicates that specific sensory areas of the brain not only process incoming signals but are also the most likely sites for memory storage. The amygdala and hippocampus are thought to be critical in determining whether or not a memory will be stored. Damage to either of these areas results in total amnesia. The amygdala is a region known to be involved in processing emotionally charged information. Thus sensory perceptions travel from sensory areas to the amygdala, which evaluates and weighs their "emotional content." "Heavy" information, such as events surrounding the first blind date, are "tagged" by the amygdala and move forward to other parts of the brain, where conscious processing ("why did I accept this blind date?") occurs. Such events are replayed constantly ("rehearsed") in the brain and are the most likely to be stored as memories. Sensory events with no emotional tag (for example, the color of the first car you saw today) are not likely to be remembered. This model seems consistent with experience, since most of our clearest memories have strong emotional content.

The hippocampus appears to play a key role in organizing neural connections between itself and the sensory areas and perhaps other parts of the brain. The establishment of new neural circuits between the sensory areas and the hippocampus is thought to be one basis for memory formation and storage. These new pathways become permanent after an appropriate number of rehearsals. This process is analogous to the creation of a network of trails in a new campground. At first, there is nothing distinctive, but as more and more people walk from campsites to different areas of the campground, a well-worn trail system develops. According to this model, when a familiar stimulus (the perfume or cologne worn by the blind date) enters the sensory area, the memory, stored as a neural circuit (the trail network), is recalled to consciousness. The exact functions of the other areas of the brain in this model are more dimly understood. However, they are also critical in information processing, since damage to each of them alone disrupts memory formation.

Many new questions arose from these studies and the hypothesized physical model of memory processing and storage. For example, what biochemical events are required for creating new neuronal circuits? What is the molecular basis for information storage and memory formation? These and other questions have guided research during the past decade, and new hypotheses about the finer aspects of memory formation have begun to emerge. Such research is intrinsically fascinating to scientists, and it is also fundamentally important. An ultimate hope is that memory can be understood at the molecular level so that it may become possible to treat conditions, such as Alzheimer's disease, that cause impairment of this extraordinary human ability.

6. The human nervous system is composed of the central nervous system—the brain and spinal cord—and the peripheral nervous system, which conveys information to and from the central nervous system.
7. The peripheral nervous system is organized according to function. Each of the divisions performs different functions, some of which are under conscious control (for example, movement of skeletal muscles) and some of which are not (for example, heartbeat). The general function of the peripheral nervous system is to maintain constant body conditions.
8. The central nervous system acts as the control center of the body. Different parts of the human brain are responsible for specific operations, such as helping maintain homeostasis, maintaining consciousness, integrating and processing sensory information, and intellectual activities.

Review Questions

1. What are the basic functions of the nervous system?
2. What are neurons? Describe their structure.
3. What are the three major types of neurons? What is the general responsibility of each type?
4. What are the functional properties of a neuron?
5. What is a synapse? What are the two types of synapses?
6. Describe the events of chemical synaptic transmission.
7. How do different drugs affect synaptic transmission? How is this knowledge being used?
8. What is the function of the peripheral nervous system? What types of nerves does it contain?
9. Describe the functional classification of the peripheral nervous system.
10. Describe the components of a reflex arc. Why are they important?
11. What are the three major regions of the brain and the function of each?
12. What are the functions of the cerebral cortex?

Essay and Discussion Questions

1. What might be the goals of research conducted on the nervous system during the 1990s? Beyond the twentieth century?
2. What might be the adaptive value of the following: (a) pain; (b) memory; (c) separation of the somatic and autonomic nervous systems?
3. What do you consider the ethical limits of doing experiments on living animals in order to obtain information and data about the nervous system?

References and Recommended Reading

Baringa, M. 1990. The tide of memory, turning. *Science* 248:1603–1605.

Black, I. B., J. E. Alder, C. F. Dreyfus, W. F. Friedman, E. F. LaGamma, and A. H. Roach. 1987. Biochemistry of information storage in the nervous system. *Science* 236:1263–1268.

Freeman, W. J. 1991. The physiology of perception. *Scientific American* 264:78–85.

Gazzaniga, M. S. 1989. Organization of the human brain. *Science* 245:947–951.

Holloway, M. 1991. Trends in pharmacology. ℞ for addiction. *Scientific American* 264:94–103.

Hooper, J., and D. Teresi. 1986. *The Three-Pound Universe.* New York: Macmillan.

Jastrow, R. 1981. *The Enchanted Loom: Mind in the Universe.* New York: Simon & Schuster.

Johnson, G. 1991. *In the Palaces of Memory.* New York: Knopf.

Kalil, R. E. 1989. Synapse formation in the developing brain. *Scientific American* 260:76–85.

Levi-Montalcini, R. 1987. The nerve growth factor 35 years later. *Science* 237:1154–1162.

Marks, J. 1990. Marijuana receptor gene cloned. *Science* 249:624–626.

Mishkin, M., and T. Appenzeller. 1987. The anatomy of memory. *Scientific American* 256:80–89.

Tetrud, J. W., and W. Langston. 1989. The effect of Deprenyl (selegiline) on the natural history of Parkinson's disease. *Science* 245:519–522.

Tulving, E. 1989. Remembering and knowing the past. *American Scientist* 77:361–367.

CHAPTER 36

Control and Regulation II: The Endocrine System

Chapter Outline

Reading Questions

1. How do modern concepts dealing with the endocrine system differ from those of classical endocrinology?

2. What are some general functions of human hormones?

3. How do hormones carry out their tasks?

4. What are some of the functions of the pituitary gland? How is it regulated?

The endocrine system, an exquisitely controlled assemblage of tissues and organs, has moved into the spotlight of biological research during the past two decades. The earlier classical view of the endocrine system was based largely on studies conducted during the first half of the twentieth century. The classical concept has value for studying this system. It holds that the **endocrine system** consists of discrete organs called **glands**, which are made up of specialized cells capable of secreting unique chemical messengers called *hormones*. Furthermore, hormones are secreted into the circulatory system and then transported to a set of distant "target cells," which they affect in some way. *Endocrinology* is the study of glands and hormones and their effects on the organism.

THE CLASSICAL VIEW OF THE ENDOCRINE SYSTEM

Figure 36.1 identifies the glands of the classical endocrine system. The endocrine and nervous systems operate together in regulating an organism's functions. The common denominator linking the two systems is the use of chemical messengers in their communications. However, in comparison to the instant responses and pinpoint control provided by neurons and neurotransmitters, the effects of hormones differ in the following ways. First, they are slow-acting. From the time they are released from cells, their effects may not occur for minutes, hours, days, or even longer. In some cases, they mediate responses measured in years (for example, sexual maturation). Second, they circulate rather randomly (and slowly)

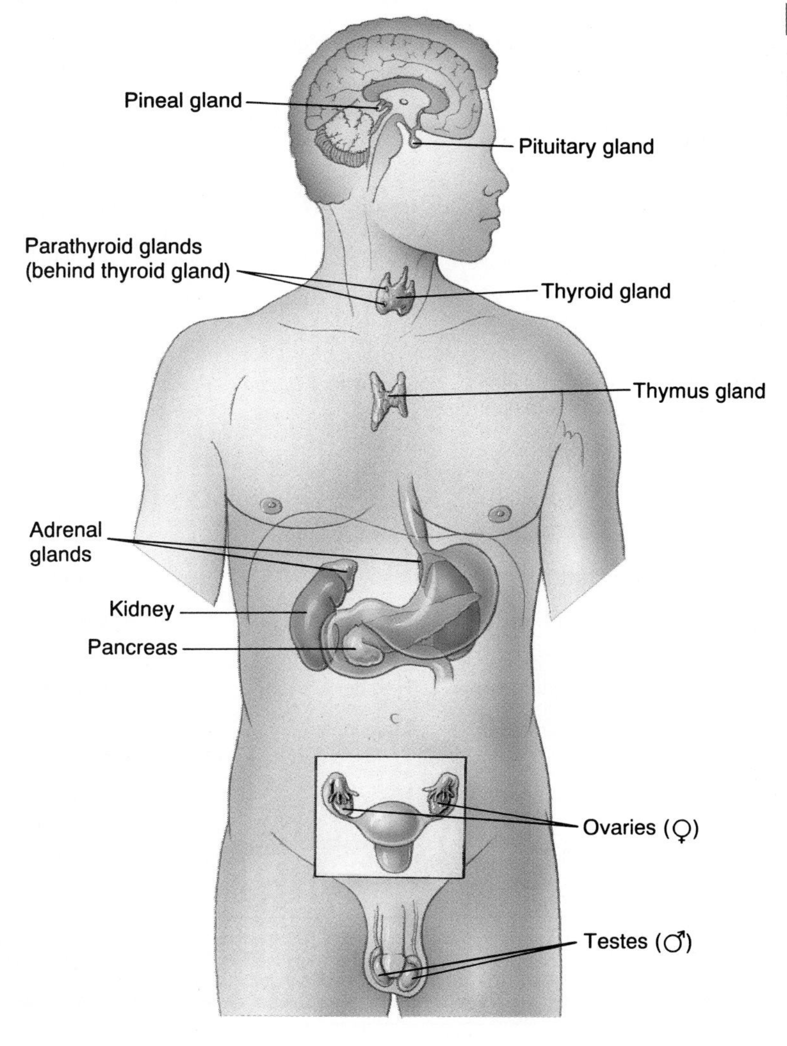

Figure 36.1

This drawing shows the major glands of the human endocrine system and their location relative to certain organs.

throughout the body while locating appropriate target cells. Third, their control is somewhat diffuse. Often, several cells or tissues may be affected by a single hormone.

In pioneering studies during the first half of the twentieth century, many important glands and hormones and their effects in the human endocrine system were described; they are summarized in Table 36.1. Hormones were found to affect development, growth, maturation, behavior, and a variety of other process-

Table 36.1 Major Glands, Hormones, and Actions of the Classical Endocrine System

Gland	Hormone[a]	Target	Primary Functions
Pineal	Melatonin	Hypothalamus	Blocks LH and FSH releasing factors
Pituitary			
Anterior	ACTH	Adrenal cortex	Synthesis of various hormones
	FSH	Ovary Testis	Female development and hormones Male development and hormones
	hGH	Bones, tissue	General growth stimulation
	LH	Ovary Testis	Stimulates ovulation, synthesis of female hormones Stimulates synthesis of male hormones
	Prolactin	Mammary gland	Stimulates milk synthesis
	TSH	Thyroid	Control of thyroid gland
Posterior	ADH	Kidney	Water conservation
Thyroid	Thyroxine	Most cells	Increases metabolic activities
	Calcitonin	Bones	Stimulates calcium uptake
Parathyroid	PTH	Bones	Increases calcium levels in blood
Thymus	Thymosin	White blood cells	Stimulates differentiation
Adrenal			
Medulla	Epinephrine, norepinephrine	Circulatory system	Prepares the body for physical action
Cortex	Steroids:		
	GC	Various cells	Increases carbohydrate metabolism
	MC	Kidneys	Regulates water and mineral balance
Pancreas	Insulin	Many cells	Stimulates glucose uptake from blood
	Glucagon	Many cells	Stimulates glucose release from cells into blood
Kidney	Erythropoietin	Red blood cells	Stimulates red blood cell production
Testis	Testosterone	Many cells	Stimulates male sexual development and maturation of sex organs
Ovary	Estrogens	Many cells	Stimulates female sexual development and maturation of sex organs
	Progesterone	Uterus	Stimulates growth of uterine lining

[a]Abbreviations: ACTH = adrenocorticotropin; FSH = follicle-stimulating hormone; hGH = human growth hormone; LH = luteinizing hormone; TSH = thyroid-stimulating hormone; ADH = antidiuretic hormone; PTH = parathyroid hormone; GC = glucocorticoid; MC = mineralocorticoid.

es by activating cellular mechanisms that were either idle or operating at a greatly reduced level or by decreasing a cell's activities. What types of studies influenced formation of the classical view of endocrinology?

The Theory of Internal Secretions

In the eighteenth and nineteenth centuries, physicians were familiar with human anatomy, including locations of the major glands, although little was known about the functions of these glands. There was considerable interest in determining experimentally the effects on animals when a certain tissue or organ was removed. This is not a terribly imaginative approach by modern scientific standards, but it led to some interesting discoveries.

Descriptions of the first such experiments on record were published in 1849 by A. A. Berthold. He found that if he removed the testes from immature male chickens, they never developed structures characteristic of mature roosters, nor did they ever crow or show any sexual interest in hens. However, if a single testis was later transplanted into the castrated male, it developed normal male structures and behaviors. What could be concluded from such experiments? Clearly, testes were necessary to the development of male characteristics in chickens. It is now known, of course, that the testes secrete a hormone, testosterone, that influences the development of males in all vertebrate species, including humans.

Little progress was made in endocrinology during the next 50 years. In 1855, the French physiologist Claude Bernard introduced the term *internal secretion* to describe, in a general way, a hypothesized mechanism responsible for controlling concentrations of glucose in the blood. However, it was not until the turn of the twentieth century that the concept of glands secreting substances captured the attention of scientists. The term *hormone* (Greek for "arouse to activity") was introduced in 1905 by Ernest H. Starling after a series of experiments established that pancreatic secretions were stimulated by a substance released from the small intestine. In 1889, it was reported that dogs developed a condition similar to human *diabetes,* a disease characterized by excess glucose in the blood, when their pancreases were removed. In a reciprocal experiment, diabetes did not occur if those dogs then received a graft of pancreas tissue. What could these results mean? Surely, the pancreas must be required for the proper regulation of glucose levels in the blood. How did the pancreas influence activities in other areas of the body? Perhaps some secretion was involved? That was confirmed in studies conducted during the next 35 years, and the pancreatic secretion (by now called a hormone) was given the name *insulin*.

Endocrinology in the Twentieth Century

After the successes of early workers, the "theory of internal secretions" continued to evolve and served as a springboard for studies on glands and hormones in the twentieth century. Similar to investigations in the field of genetics during this same period, the results achieved were stunning. As mentioned previously, many of the major hormones were isolated and identified, their functions were largely determined, a classical view of hormones and target cells emerged, and scientists in this field won a host of Nobel Prizes.

Regrettably for those of us who are fond of simple pictures, recent discoveries have made it necessary to revise the classical gland-hormone view of the endocrine system. Why must such an appealing scheme, basically summarized in Table 36.1, be extended and modified? What is known now that was not understood during the formation of classical endocrinology? Three general conclusions emerged from new research. First, we now know that organs other than classical endocrine glands secrete hormones. The gastrointestinal tract, heart, and brain all are now known to secrete hormones, although their effects are not yet clearly understood. Neurons were also found to secrete hormones that regulate activities of many organs. Second, nonhormonal substances were discovered that acted as hormones. For example, some classic neurotransmitters, such as dopamine, may function as hormones in certain cells. Third, it was determined that not all hormones were transported to their target cells by the circulatory system.

As a result of these findings, new, broader definitions have been formulated that provide greater accuracy and flexibility. A **hormone** is now considered to be any substance released by one cell that acts on another cell, near or far, regardless of the means of transport. A **neurohormone** is a hormone produced by a neuron, and a *neurotransmitter* is a neurohormone that acts at the synapse. The term *chemical messenger* can be used as a synonym for all of these terms. **Endocrinology** is the study of hormones derived from classical endocrine glands or from other cells or tissues, such as the brain or heart, and their functions.

Studies conducted between 1800 and 1950 gave rise to a classical model of endocrine system function. It explained that distinct glands secreted specific hormones, which acted on target cells and influenced processes such as development and growth. More recent research has made it clear that hormones can be produced by other organs and that substances other than classical hormones can affect the activities of target cells.

ENDOCRINE GLANDS AND HORMONES

A large and varied collection of hormones is required to regulate human physiological activities. Most hormones and their effects are probably now known. However, no one is any longer surprised by reports of a new hormone or of a hormone being produced by a previously unsuspected tissue or organ. The classical glands, hormones, and effects still provide a useful framework for studying the endocrine system. Nevertheless, discoveries in the future will continue to move the field well beyond the classical concept.

Major Classes of Hormones

There are two principal chemical classes of hormones (see Figure 36.2). **Peptide hormones** are mostly small proteins composed of linear strings of 3 to 200 amino acids. They are manufactured through the protein-synthesizing pathways and processes described in Chapter 16. Most hormones listed in Table 36.1 belong to this class.

Steroid hormones are synthesized from a large lipid molecule called *cholesterol.* Two major types of steroids are produced by the *adrenal glands,* which are located on top of the kidneys. *Glucocorticoids* play important roles in sugar metabolism, and *mineralocorticoids* are critical in maintaining water and mineral balances. The gonads produce several *sex hormones* that are essential

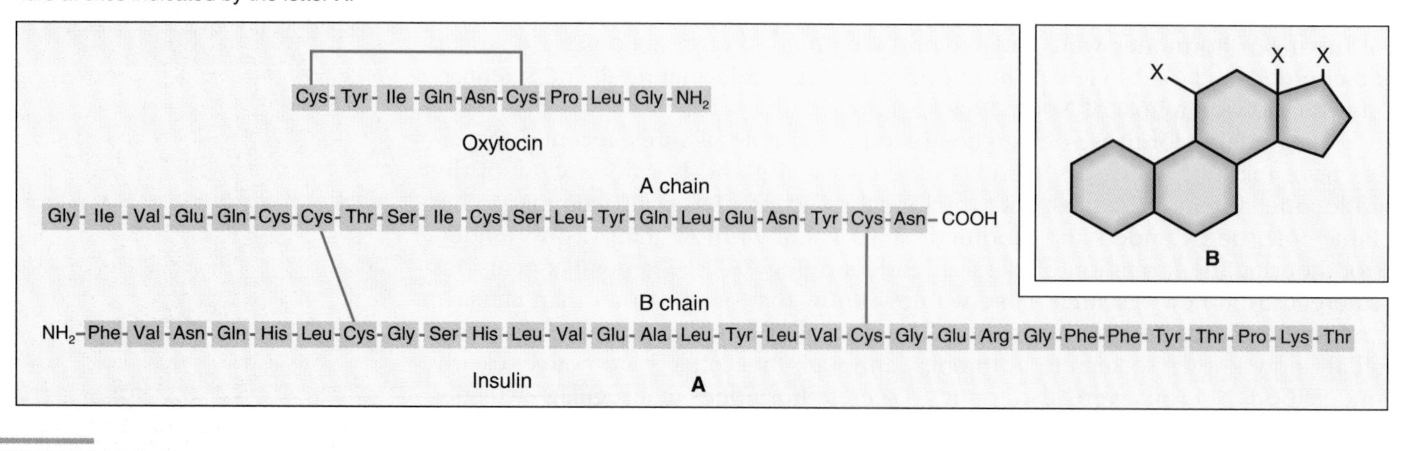

Figure 36.2

(A) Peptide hormones are small protein molecules composed of linear chains of amino acids, which are represented by rectangles. Oxytocin has only nine amino acids, whereas insulin consists of two chains, an A chain of 21 amino acids and a B chain of 30 amino acids, held together by two chemical bonds. (B) All steroid hormones are made with part of a cholesterol molecule (shown in blue), but each has a variable chemical structure at sites indicated by the letter *X*.

for reproductive development and function. The principal male sex hormones produced by the testes are *androgens,* and the female ovaries synthesize *estrogens.*

There are other types of hormones that do not fit into either major class (for example, thyroid hormones) and "candidate hormones," substances whose exact roles are unknown but that appear to function as hormones. In this latter category are chemical messengers such as dopamine, which functions as a neurotransmitter but may also act as a classical hormone in regulating the pituitary gland. Scientists have also discovered numerous *peptide growth-stimulating factors* that promote the growth or development of certain cells and tissues. Two of these include *nerve growth factor,* a chemical messenger responsible for neuron maturation during embryonic development, and *erythropoietin,* which stimulates erythrocyte (red blood cell) production in the bone marrow.

The two major classes of hormones are peptide hormones and steroid hormones. Peptide hormones are synthesized by several glands and regulate many different activities. Steroid hormones are synthesized by the adrenal glands and the gonads and play a major role in reproductive organ development and function. There are also other types of substances that act as chemical messengers.

The Lives of Hormones

Figure 36.3 summarizes the general "life cycle" of a hormone. All hormones are synthesized inside cells, and most are enclosed within secretory vesicles until released from the cell, similar to neurotransmitters of the nervous system. Once stimulated by an appropriate external or internal environmental cue, the vesicle fuses with the plasma membrane and empties the hormone outside the cell. Steroids and thyroid hormones, however, are never stored; they are secreted directly into the bloodstream as they are produced. After release, hormones reach the target cells in which they cause a response. Once the target cell's activities are completed, the hormone is removed and eliminated. Most hormone secretions are regulated by negative feedback mechanisms.

Hormone Secretion

What stimulates hormone-secreting cells to liberate their contents? Many external stimuli, such as light, sound, and temperature, may influence the release of hor-

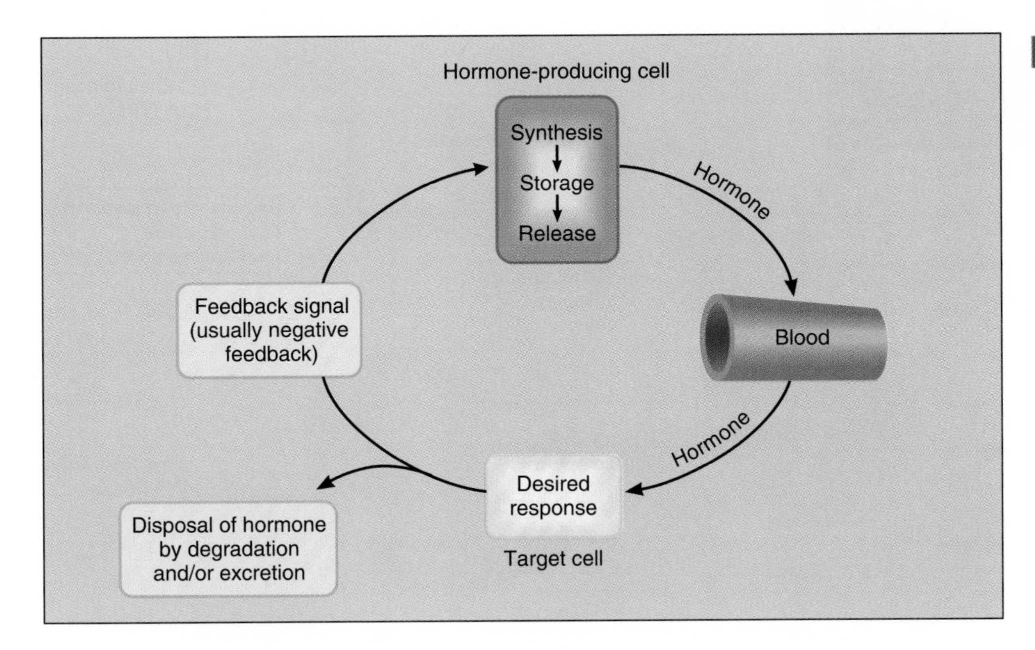

Figure 36.3

Most hormones are synthesized in cells, released into the bloodstream, and travel to target cells, where they elicit a response. Once the target cell completes its activity, the hormone is eliminated. The synthesis and release of most hormones are precisely regulated by negative feedback mechanisms involving the endocrine and nervous systems.

mones through their effects on the nervous system. For example, a loud noise interpreted as threat-related by the nervous system may provoke the release of the hormone epinephrine from the adrenal glands. Internal stimuli consist of signals from either the nervous system or the endocrine system. Most frequently, other hormones stimulate the secretion of hormones from specific glands. For example, follicle-stimulating hormone (FSH) stimulates the release of certain hormones from the gonads.

Hormone Degradation

Since hormones can cause subtle or profound changes in the activities of cells, organs, and organisms, there must be ways to inactivate them. If not, the responses they induce would not cease, and the consequences could be extremely serious. Various mechanisms ensure that the life of a hormone does not exceed the time required for its beneficial effect.

Peptide hormones have rather short lives once they are secreted from the cell. Generally, they remain intact for less than an hour before they are inactivated, usually in the liver or kidney. A smaller number of peptide hormones are broken down at the target cell area, perhaps by actions of the target cell itself.

Steroid and thyroid hormones live for considerably longer periods, some for up to several weeks. Steroids are ultimately modified and inactivated in the liver, filtered from the blood in the kidneys, and then excreted in the urine. Thyroid hormones may also follow this route.

Feedback Control of Hormone Secretion

Levels of hormones are controlled through complex feedback systems, negative feedback being the most common mechanism (see Chapter 33). An example is described in Figure 36.4. *Thyroxine* is a hormone secreted by the thyroid gland. It helps regulates metabolism and tissue growth and development, especially in

Figure 36.4

(A) The thyroid gland consists of hundreds of follicles where hormones, including thyroxine, are produced. The interior colloid material contains stored hormones and other substances. (B) A single follicle has this general structure. (C) When blood thyroxine levels are lower than normal, the hypothalamus releases TRH (thyroid-releasing hormone), which acts on the anterior pituitary gland, causing it to release TSH (thyroid-stimulating hormone). TSH stimulates the release of thyroxine from the thyroid gland. Increased levels of thyroxine in the blood inhibit further release of TRH from the hypothalamus.

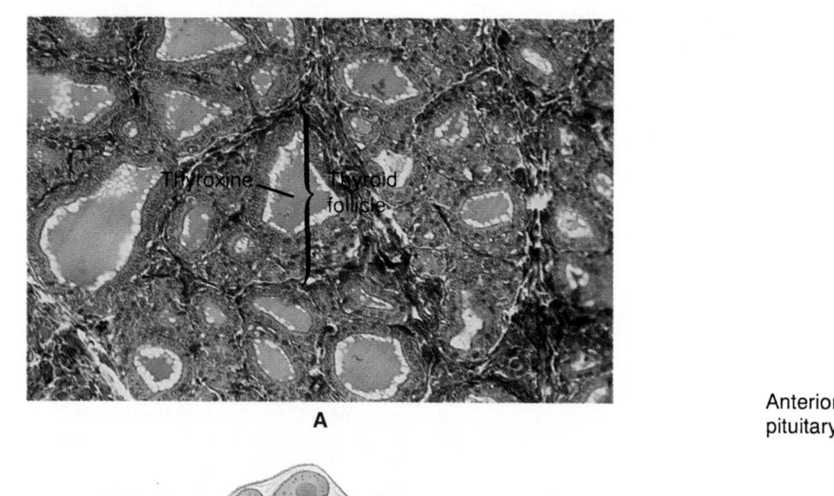

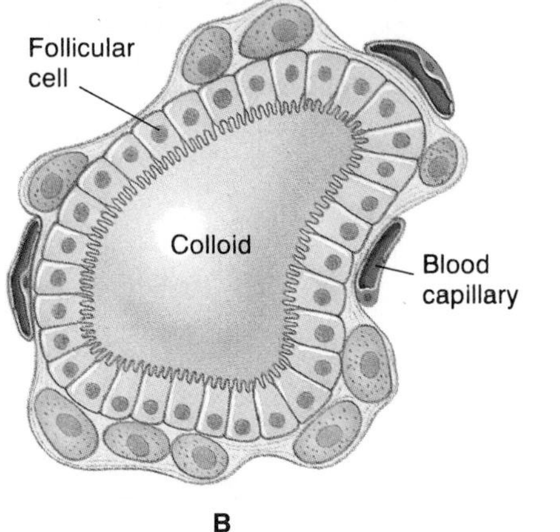

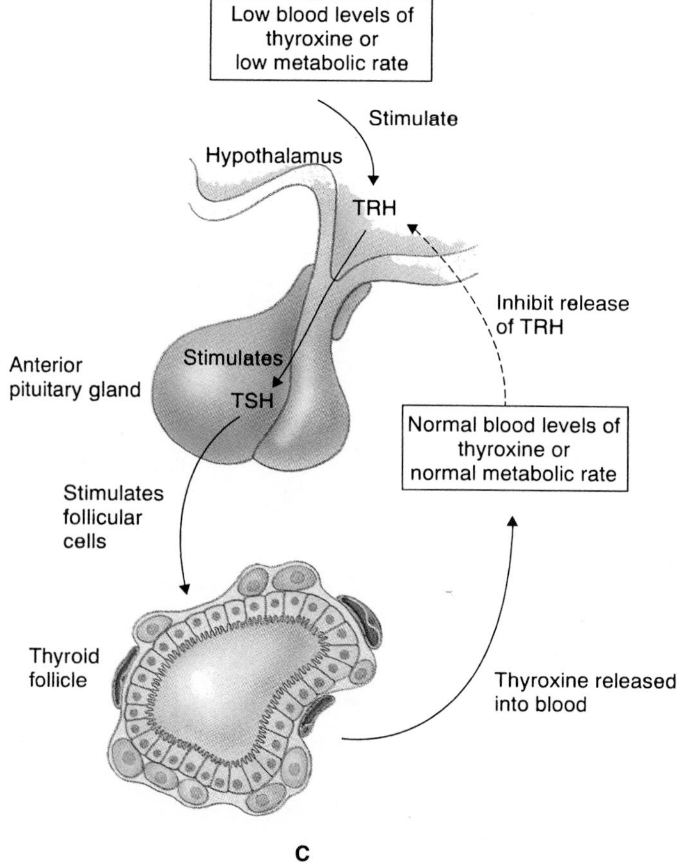

infants and children. When blood thyroxine concentrations fall below normal or metabolism is significantly reduced, the hypothalamus, a part of the brain, detects the change and secretes thyrotropin-releasing hormone (TRH) in response. TRH travels to the pituitary gland, where it stimulates the release of thyroid-stimulating hormone (TSH), which in turn causes the thyroid gland to release more thyroxine into the blood. Once metabolism and thyroxine blood levels return to normal, the hypothalamus releases a thyrotropin-inhibiting factor that ultimately causes a reduction in thyroxine secretion. Such complicated feedback mechanisms (and this example is relatively simple!), involving both the endocrine and nervous systems, result in exquisite control of related functions within the body.

Hormones are released from cells in response to signals from the nervous system or from other hormones. Hormone concentrations in the body are precisely regulated, usually by negative feedback mechanisms. Once their actions are completed, hormones are broken down, or inactivated, and removed.

Example: Regulation of Blood Glucose Levels

To conduct normal daily activities, your cells require a constant supply of energy. Most cells use glucose, a simple sugar, as an energy source. Regulating blood glucose levels is critical for maintaining homeostasis, since abnormally high or low levels of glucose in the blood can cause serious problems. *Diabetes mellitus* is a term applied to a variety of disorders characterized by elevated concentrations of glucose in the blood. How are blood glucose levels regulated? Figure 36.5 describes the organs, cells, hormones, and mechanisms involved.

Two hormones, *insulin* and *glucagon,* both secreted by the pancreas, are responsible for glucose homeostasis. Insulin acts to reduce the levels of blood glucose, and glucagon has the opposite effect, causing an increase in blood glucose. The pancreas contains different populations of secretory cells. Approximately 1 million cell clusters called *islets of Langerhans* exist in the pancreas. Each islet is composed of *alpha cells* that produce glucagon, *beta cells* that produce insulin, and *delta cells* that produce *somatostatin,* a hormone that inhibits secretion of glucagon and insulin. Digestive enzymes are secreted by pancreas cells that surround the islets.

After eating a meal, blood glucose levels increase, and within minutes, insulin is released from beta cells in the pancreas. The immediate effect (in seconds or minutes) of insulin is to increase the rate of glucose uptake from the blood into target cells—primarily liver and muscle cells. Once inside cells, glucose molecules can be metabolized or linked together to produce *glycogen,* a storage form of glucose. As blood glucose levels fall below the normal range, glucagon is released by alpha cells of the pancreas in response. This hormone primarily affects liver cells and stimulates the breakdown of glycogen and the release of glucose into circulation. Normally, the feedback regulation between these two hormones maintains blood glucose levels at optimal physiological levels.

As described in Figure 36.5C, pancreatic hormones also act in routing proteins, fats, and sugars to cells where they can be stored. Adipose cells, for example, take up excess glucose and convert it to fat for storage. They are also involved in liberating stored substances from cells if conditions warrant, as during dieting. These activities play an important role in glucose homeostasis over longer time periods.

The General Activities of Hormones

Like most matters of the endocrine system, describing the general activities of hormones is complicated by the great diversity of these substances. Some hor-

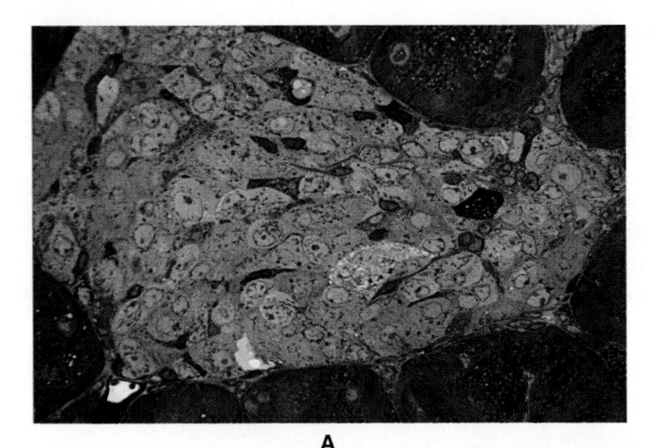

Figure 36.5

(A) Blood glucose levels are regulated by hormones released from different pancreatic cells. The pancreas contains approximately 1 million islets of Langerhans, one of which is shown here. Each islet is composed of alpha (red), beta (blue), and delta (not shown) cells; each cell type secretes a different hormone. (B) Each islet of Langerhans has this general structure. (C) When blood glucose levels increase, insulin is released from beta cells. Insulin causes several changes that all serve to remove glucose from the blood. When glucose levels are low, glucagon is released from alpha cells and stimulates activities that release glucose into the bloodstream. The complete regulatory system ensures that every cell in the body will always receive appropriate amounts of glucose.

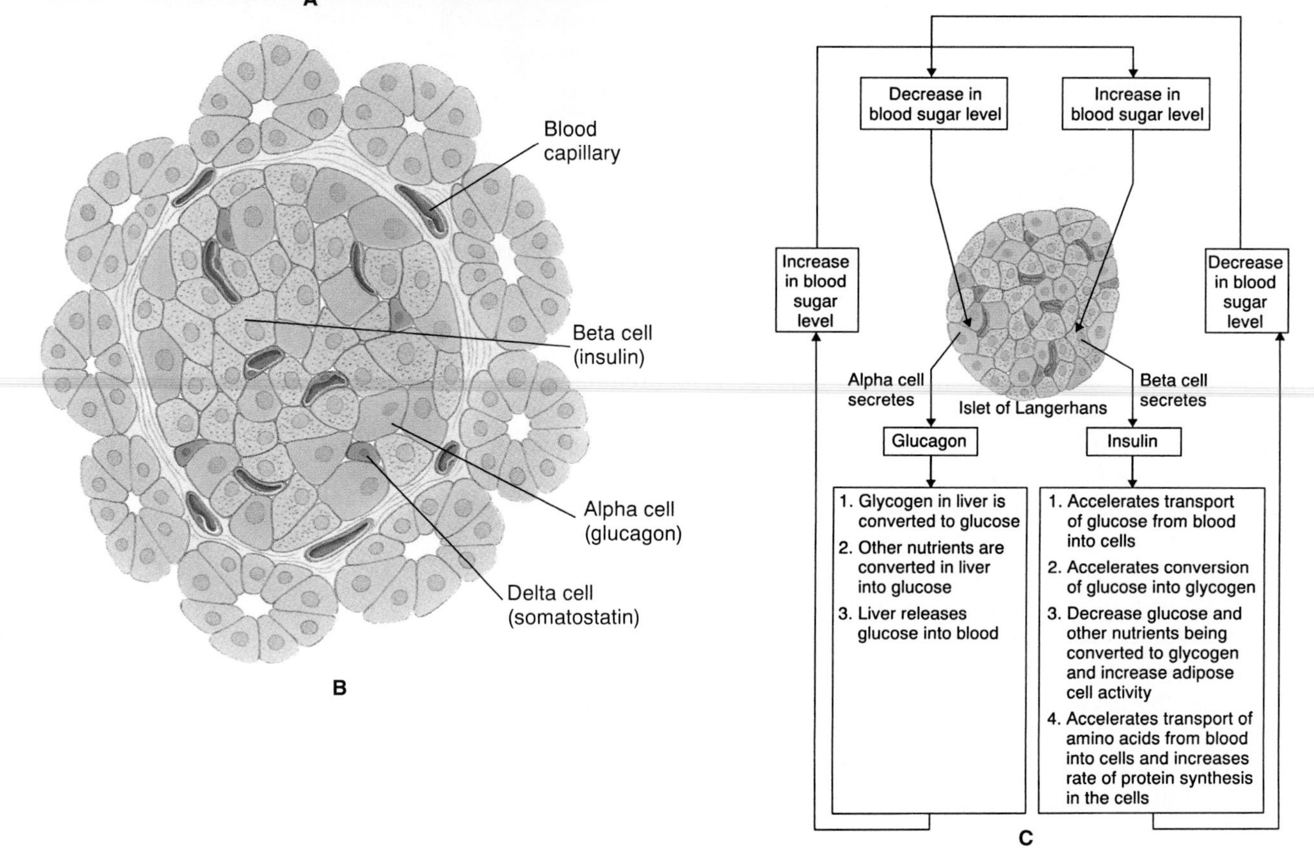

mones may never be secreted or are produced only rarely. Oxytocin, for example, is a hormone that functions only at the end of pregnancy and during the period of nursing. Others, such as insulin, are secreted in irregular on-off cycles that are related to homeostatic demands. Some, such as FSH in females, influence events for decades, but after a certain age, they play no further role in the life of the individual.

The following list attempts to categorize the vast number of physiological activities of hormones into a manageable number of general processes.

1. Hormones influence the secretions of most substances by cells in the body. These substances include, but are not limited to, other hormones, neurohormones, milk from mammary glands, enzymes and other products of the digestive tract, and sweat.
2. Hormones control both the synthesis and breakdown of fats, carbohydrates, and proteins in cells of the body.
3. Hormones stimulate or inhibit cell division (mitosis) and thus direct the growth of tissues and the organism.

4. Hormones regulate activities associated with sexual development, sexual behaviors, and the production of gametes.
5. Hormones regulate levels of minerals and other chemicals, including sodium, calcium, and potassium, in the body.
6. Hormones are thought to have a significant influence on different types of behaviors, although, except for some information on sexual behaviors, this effect is poorly understood in humans.

MECHANISMS OF HORMONE ACTION

After a gland secretes a hormone into the bloodstream, the chemical messenger reaches a target cell that then engages in some physiological process in response. Each hormone can regulate only specific target cells. How are hormones able to identify their specific target cells? What determines whether or not a cell will respond to a hormone?

Receptors

Scientists have found that each target cell possesses different **receptors**, which are complex protein molecules that "recognize" hormones. Like a neuron, gland, or muscle receptor that recognizes only a single type of neurotransmitter, a specific hormone receptor has a unique structure that matches only one hormone, and when contact occurs, the hormone and receptor fit together like lock and key. Each cell is equipped only with receptors for hormones that act on that cell. If a cell's activities are regulated by, say, two hormones, it will have one type of receptor for each hormone.

The number and even the type of receptors a cell possesses can change at different times, depending on the function of the cell and other circumstances. In general, the total number of cellular receptors is quite large. For example, target cells for steroid hormones contain 10,000 to 100,000 steroid receptors; each receptor molecule can bind with one hormone molecule. Finally, only a fraction of the receptors have to be occupied to cause a maximum physiological response by the cell. When hormone molecules are captured by receptors, what occurs that leads to changes in the cell's activities? Two different mechanisms have been identified.

Peptide Hormones: First and Second Messengers

Target cells for peptide hormones have specific receptor molecules located on their surfaces, in the plasma membrane. The exact nature and molecular composition of most hormone receptors remain to be determined. Although the attraction of peptide hormones to the surfaces of target cells was discovered relatively early, the mechanism for action long remained unclear. How could a hormone that became attached to the outside of a cell trigger chemical reactions inside the cell? The first answers came through research conducted in the 1950s by Earl Sutherland and colleagues. They discovered that a previously unknown chemical called **cyclic adenosine monophosphate (cAMP)** accumulated in "activated" target cells (cAMP is similar to ATP, but it has only one phosphate group). For discovering cAMP, Sutherland was awarded the Nobel Prize in physiology and medicine in 1971.

What was the role of cAMP in peptide hormone actions? Building on the pioneering studies of the 1950s, researchers hypothesized that after binding to surface receptors, peptide hormones triggered the synthesis of cAMP inside the cell. In turn, cAMP initiated a cascade of chemical reactions in the cell that would enable it to perform a certain function, such as synthesizing an enzyme that could break down glycogen into glucose. Subsequent research in the 1960s confirmed the basic accuracy of this hypothesis for a variety of peptide hormones.

Figure 36.6

A peptide hormone—the first messenger—interacts with its target cells by binding with receptors located on the cell's surface. Binding results in the production of cAMP—the second messenger—which causes the cell to carry out the activity specified by the hormone.

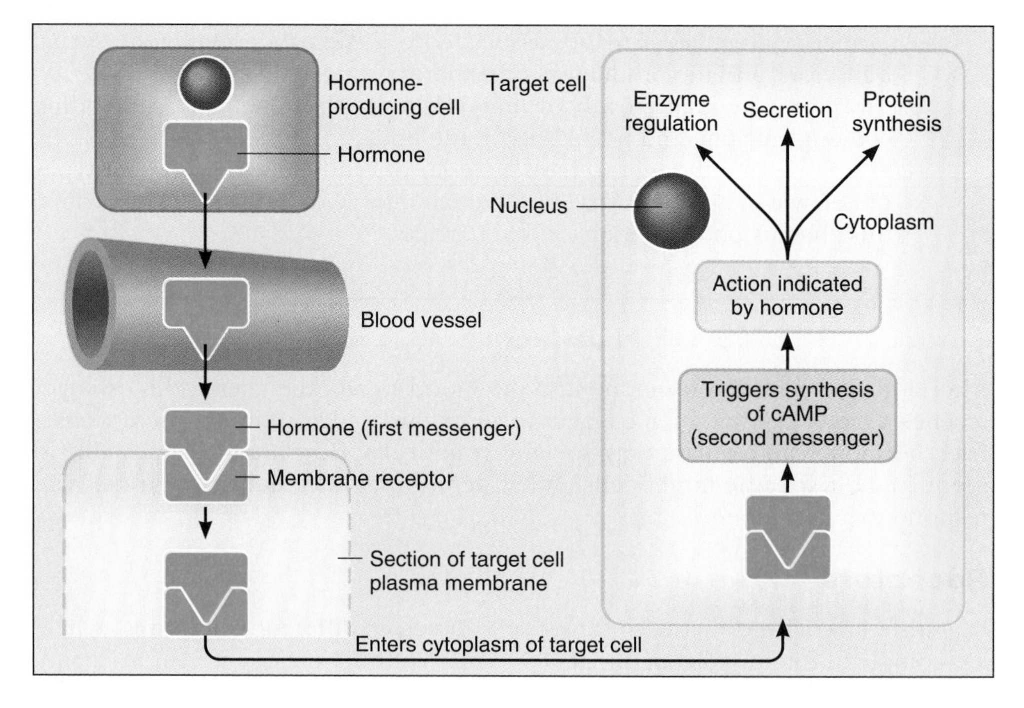

This mechanism, described in Figure 36.6, is known as a second-messenger system. The first messenger is the peptide hormone that acts with its receptor at the cell surface to initiate the synthesis of cAMP; the second messenger is the cAMP synthesized inside the cell. Increased levels of cAMP then trigger a specific molecular activity inside the cell. Other second messengers are now known to exist, but cAMP is the most common in the human endocrine system.

Models for Steroid Hormones

Steroids have the ability to pass through plasma membranes and enter the interior of most cells. However, their effects occur only in cells that contain appropriate internal receptors. Once inside a target cell, they combine with a receptor to form a receptor-steroid complex. It is still not clear whether unoccupied receptors are found in the cytoplasm (the "old model") or in the nucleus (the "new model"). Recent experimental evidence supports the latter view. Figure 36.7 illustrates the model for steroid action.

In either case, once the hormone joins its receptor, the complex migrates to chromatin within the nucleus. There it binds to the DNA at a specific sequence (gene) that codes for a certain protein (or proteins). Once that occurs, there is an accelerated synthesis of the protein "requested" by the hormone. Increased levels of this protein commonly have some effect, usually of long duration, on the growth or maturation of specific tissues. For example, the steroid sex hormones estrogen and progesterone affect the development of secondary sex characteristics, such as breast development in females. They also influence a variety of other activities, depending on the target cells, within a female.

Target cells have receptors for specific hormones. Two mechanisms are used by most hormones for exerting their effect in a target cell. In the second-messenger system, peptide hormones are the first messenger. Once bound to a receptor, they trigger synthesis of the second messenger, cAMP, which initiates an activity in the target cell. Target cells for steroid hormones have internal receptors. Once formed, the steroid-receptor complex interacts with DNA and stimulates production of a specific protein that usually affects growth or development.

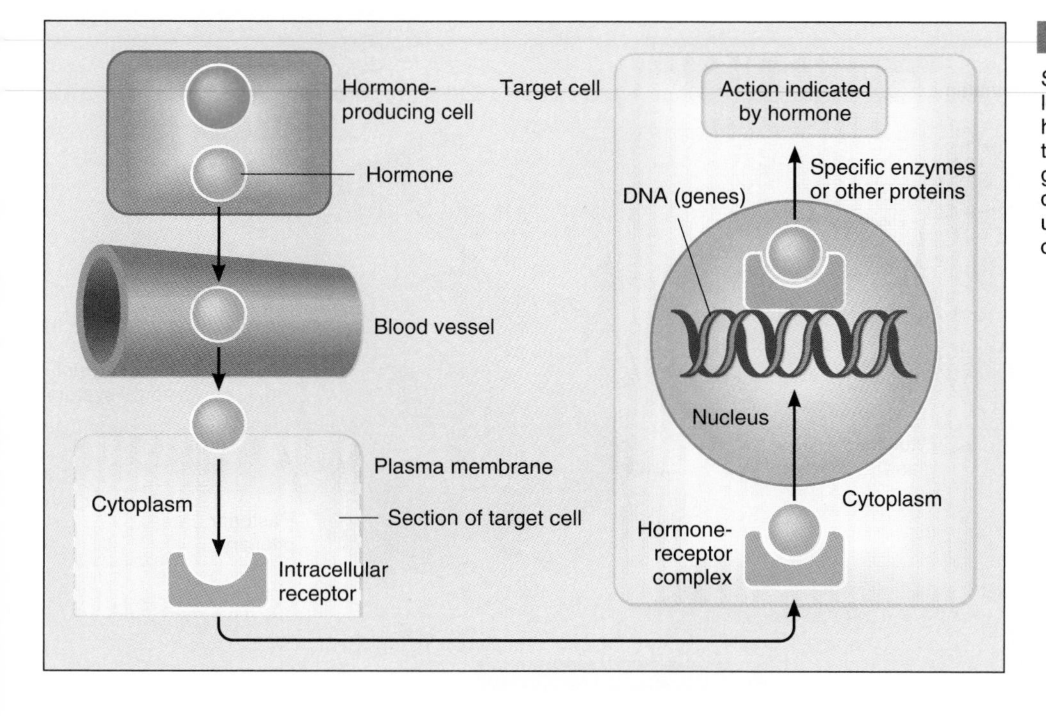

Figure 36.7

Steroid hormones bind to receptors located inside the target cell, creating a hormone-receptor complex that enters the nucleus and becomes attached to a gene. The gene becomes activated and directs the synthesis of a protein that usually affects growth or development of some tissue in the organism.

THE PITUITARY GLAND

Details about the pituitary gland illustrate the types of activities and regulatory mechanisms that characterize the human endocrine system. As shown in Figure 36.8A, this gland lies at the base of the brain and is attached to the hypothalamus region by a short stalk. The pituitary gland was known to the ancients almost 2,000 years ago. Galen, a famous Greek physician living during the Roman period, thought that the pituitary gland served to conduct waste products from the brain into the nose, where they became incorporated into nasal mucus and were then discharged. In the early nineteenth century, some 1,800 years later—and after extensive, detailed anatomical and physiological studies—a more enlightened view emerged. The pituitary was found to be an amazing organ that regulates many other glands of the endocrine system and affects diverse body activities through the secretion of several hormones. It also was discovered to have a very close relationship with the nervous system. Specialized neurons in the hypothalamus are linked to the pituitary. In addition, blood flows from the hypothalamus directly to the pituitary through a complex network of blood vessels called the *hypophysial portal system* (see Figure 36.8B). The pituitary gland consists of two lobes that are quite distinct in both form and function.

The Posterior Pituitary

The posterior pituitary releases two peptide hormones, antidiuretic hormone (ADH) and oxytocin, that are both composed of nine amino acids. *ADH,* also called *vasopressin,* helps regulate water balance within the body by promoting reabsorption of water in the kidneys. In females, *oxytocin* stimulates muscles of the uterus during childbirth and triggers secretion of milk from the mammary glands during nursing. Its function, if any, in males has not yet been determined.

Although both ADH and oxytocin are released from the posterior pituitary, these hormones are actually synthesized by specialized neurons in the hypothalamus called **neurosecretory cells** (see Figure 36.9). After being synthesized in the hypothalamus, ADH and oxytocin travel through axons of the neurosecretory cells to the pituitary, where they are released. These neurosecretory cells can

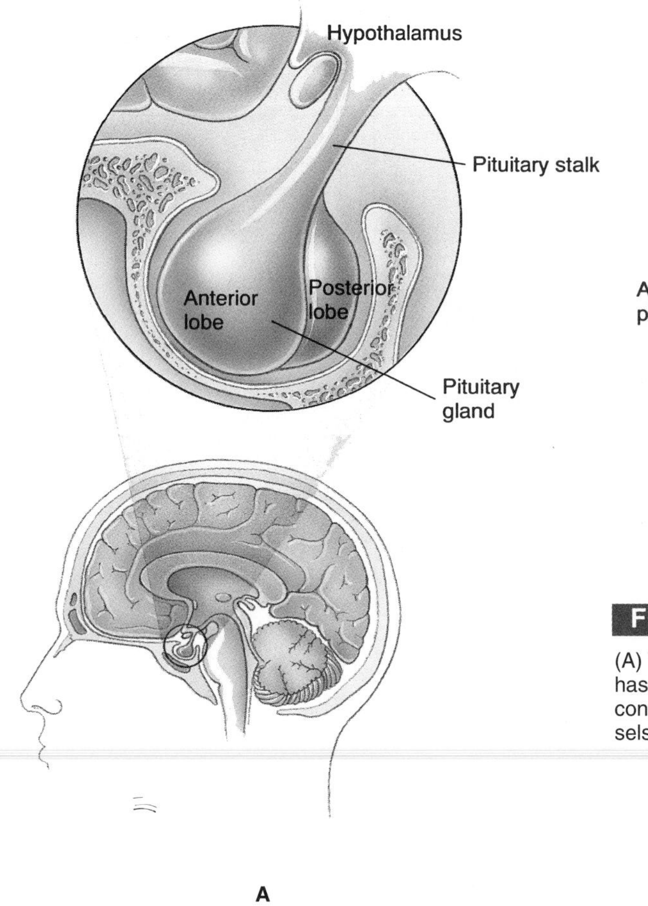

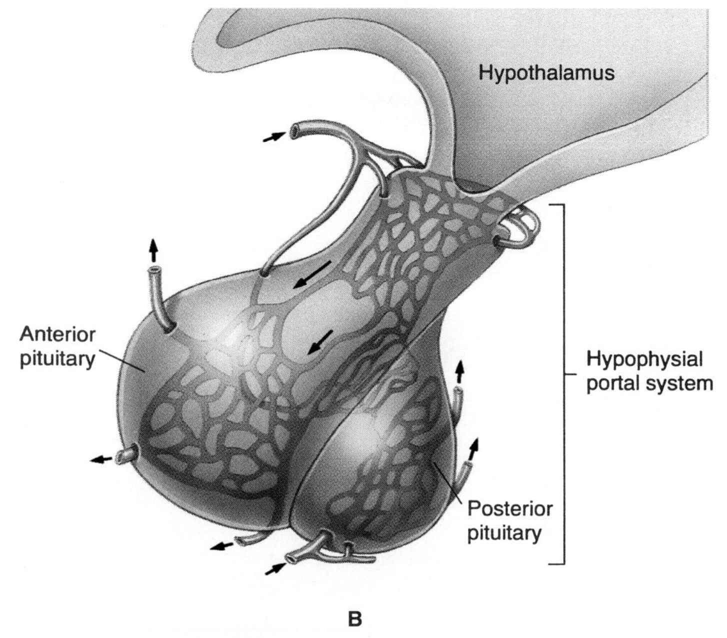

Figure 36.8

(A) The human pituitary gland is located adjacent to the brain. It has two lobes, anterior and posterior. (B) The pituitary gland is connected to the hypothalamus by a complex system of blood vessels called the *hypophysial portal system.*

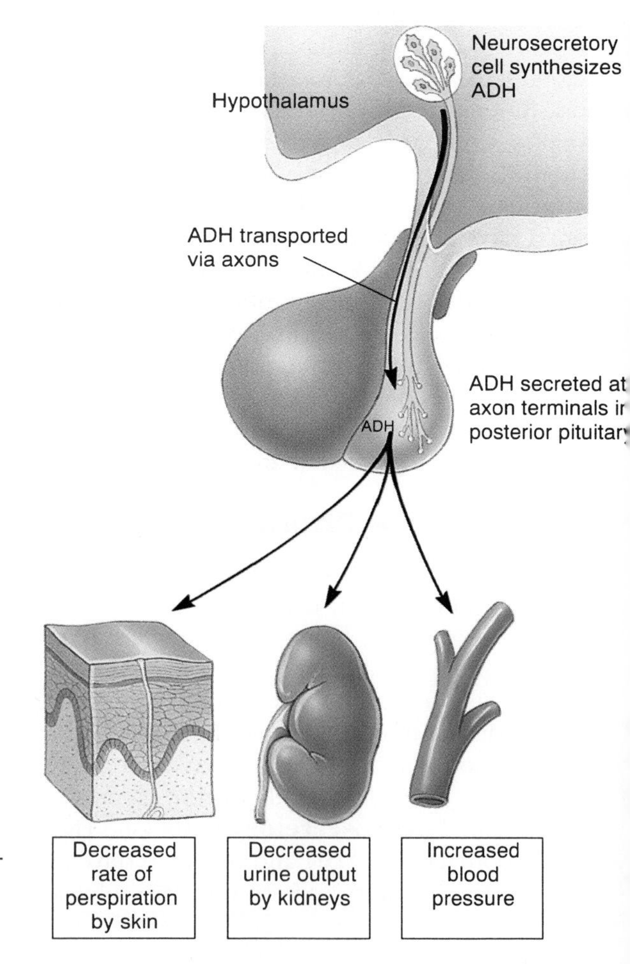

Figure 36.9

ADH (shown here) and oxytocin are synthesized in neurosecretory cells of the hypothalamus and travel to the posterior lobe of the pituitary gland, where they are released. ADH affects several organs in the body.

also receive and conduct nerve impulses. The release of oxytocin is regulated by nerve impulses from the body. Pressure on the uterus during childbirth and stimulation of the nipples during nursing trigger the release of oxytocin. ADH release is stimulated by chemical changes in the blood and is regulated by a negative feedback mechanism.

The Anterior Pituitary

The anterior pituitary produces numerous hormones that affect many activities in the body (see Figure 36.10). For this reason, and because of the neurovascular hypothesis (see the Focus on Scientific Inquiry, "The Neurovascular Hypothesis"), the anterior pituitary has been studied extensively.

Hormones

Hormones that affect the release of other hormones are called **trophic hormones**. *Follicle-stimulating hormone* (FSH) and *luteinizing hormone* (LH) are classified as **gonadotrophins** because they affect endocrine functions of the gonads (ovaries and testes). In females, FSH stimulates the functions of the ovary

Figure 36.10

The anterior pituitary secretes numerous hormones that affect many organs and processes in the human body. Endocrine activities of the anterior pituitary are regulated by the hypothalamus through the production of releasing hormones that stimulate hormone secretion. The hypothalamus also produces inhibiting hormones that suppress secretion from the anterior pituitary.

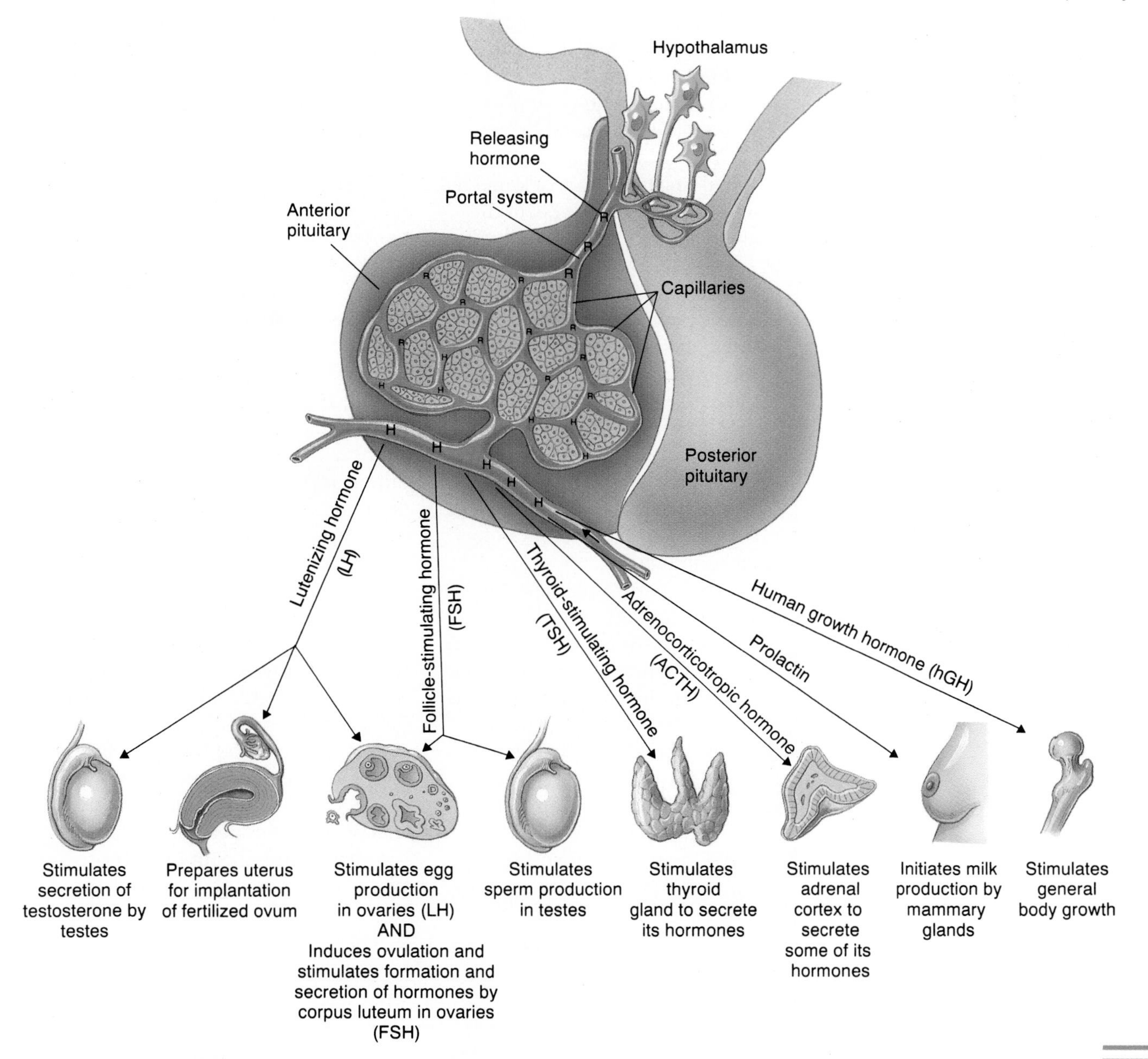

Focus on Scientific Inquiry

The Neuro-vascular Hypothesis

Confirmation of a scientific hypothesis in biology has often involved an extended period of time, brilliantly conceived experiments, and fragile resources. Verification of the existence and role of hypothalamic releasing and inhibiting hormones followed a somewhat different path that serves to illustrate the flexibility of scientists in solving difficult problems. The complete story is reported by Lawrence Crapo in his book *Hormones: The Messengers of Life*.

During the nineteenth century, experimental evidence indicated that the pituitary gland secreted several hormones essential for numerous body functions; thus the pituitary became known as the "master gland." The view persisted that regulation of hormone secretions from the pituitary involved classical negative feedback mechanisms (that is, high levels in the body eventually led to a reduction in secretion from the pituitary). In the twentieth century, however, it became clear, from many types of studies, that the nervous system was somehow involved in regulating secretions from this gland. For example, stimuli from the external environment were found to cause the release of certain hormones from the pituitary, even though this gland is incapable of detecting such stimuli.

What part of the nervous system was involved in this regulation? Several studies indicated that it was the hypothalamus. In 1894, it had been discovered that axons from a group of prominent neurons ran from the hypothalamus to the posterior pituitary. Subsequently, researchers found that these neurons synthesized ADH and oxytocin and transported them to the pituitary, where they were stored until appropriate stimuli triggered their release. These results were, of course, tremendously exciting since they offered clear evidence of integration between the nervous and endocrine systems.

But what regulated the anterior pituitary, the portion of the gland that releases the most hormones and was *not* linked to the hypothalamus by neurosecretory cells? Several lines of experimental evidence indicated that the hypothalamus was also responsible for controlling this part of the pituitary. For example, electrical stimulation of the hypothalamus caused the release of anterior-pituitary hormones, but direct stimulation of the anterior pituitary had no effect. Also, damage to the hypothalamus was found to cause decreases or increases in the secretion of different anterior-pituitary hormones. Finally, in the early 1930s, scientists discovered that the hypothalamus was directly connected to the anterior pituitary by a set of blood vessels called the hypophysial portal system.

Thus a hypothesis emerged in the 1950s: neurons in the hypothalamus synthesize substances that travel straight to the anterior pituitary, via the portal system, and stimulate or inhibit secretion of hormones from this gland. Like most notable hypotheses, this new *neurovascular hypothesis* satisfied the following criteria: (1) it was consistent with existing knowledge (no neuron connections), (2) it offered a plausible interpretation of the role of a poorly understood but obviously significant element (the portal system), (3) it made predictions (the nervous system controlled the pituitary through stimulation of the hypothalamus), and (4) it served as a guide for further research. In certain cases, scientists competed to identify the releasing factors. Some of these races lasted several decades.

How could the neurovascular hypothesis be confirmed? There seemed to be two possible lines of investigation. The first was to demonstrate that regulation of the anterior pituitary by the hypothalamus depended on the portal system. This was accomplished through several animal experiments in which surgical procedures were used. For example, the portal system between the hypothalamus and the pituitary was disconnected and hormone levels were then measured (they decreased) or hormone-dependent functions were monitored (they failed). After reconnection, normal levels and functions returned. All of these results strongly supported the view that hypothalamic control was applied through the portal system.

The second and by far more difficult task was to isolate and identify the specific releasing factors hypothesized to regulate the pituitary. That is, did releasing hormones actually exist? How could essential details be obtained for these unique but unknown substances? What were the problems?

One of the primary problems in the 1950s was that only ACTH quantities in tissues or fluids could be measured reliably; methods for analyzing other anterior-pituitary hormones were inadequate or did not exist. Thus the first efforts were aimed at isolating the ACTH-releasing hormone (termed *CRH,* for *corticotropin-releasing hormone,* since ACTH belongs to a class of hormones called *corticotropins*). For several years, numerous scientists endeavored independently to isolate and identify CRH. The key experiments involved growing pituitary cells, obtained from rats and dogs, in test tubes. At first, the cells produced ACTH, but after several days, no additional ACTH was synthesized, even though the cells remained healthy. What should be attempted next? Tissue from the hypothalamus was then added to the test tubes containing the pituitary cells, and once again they produced ACTH! This observation indicated that the hypothalamus released some factor (CRH?) that stimulated ACTH production. Results of these experiments were published in 1955, but despite serious efforts, overwhelming problems delayed the isolation and identity of CRH until 1981. The most stubborn obstacle was being able to obtain sufficient quantities of pure CRH from experimental animals. Eventually, investigators were forced to turn to slaughterhouses in order to acquire the hundreds of thousands of glands they needed to obtain adequate amounts of the desired substance.

Following the frustrating delay in identifying CRH, these same scientists (primarily two groups headed by Andrew Schally and Roger Guillemin, both of whom were to receive Nobel Prizes for their research on releasing factors) turned their attention to identifying thyrotropin-releasing hormone (TRH). To overcome the fundamental problem of having pure substances to study, both investigators used mind-boggling quantities of animal tissues. Guillemin is reported to have processed over 5 tons of hypothalamic tissue from 500,000 sheep while in Paris and the hypothalami of almost 2 million sheep after he moved his research program to Texas. Not to be outdone, Schally established a relationship with the Oscar Mayer meat-packing company that netted him donations of more than 1 million pig hypothalami. These industrial-scale quantities are not a feature shared with other studies described in earlier chapters of this book. After several false starts, some ingenious methodological developments, a few hundred thousand additional hypothalami, and seven years, both groups identified the structure of TRH, the first hypothalamus releasing hormone to be described. Thinking it had won the contest, Guillemin's group was to publish its findings on November 12, 1969. However, on November 6, the (same) results of Schally and his colleagues' work were described in another journal; they had won the race by less than a week!

The competition between these two investigators (and others) to identify new releasing hormones continued. The next hunt, centered on describing luteinizing hormone–releasing hormone (LHRH), was also won by Schally's group in 1971. Guillemin's group finally scored in 1973 when they identified the hormone that inhibits the release of growth hormone. Finally, in 1981, another group determined the structure of CRH.

Except for prolactin, the structures of all the hypothalamic releasing and inhibiting hormones are now known, and the neurovascular hypothesis has been confirmed. While less well known than the dramatic investigations of DNA, confirmation of this hypothesis was certainly as challenging and exciting.

and, along with LH, induces the production of female sex hormones. In males, FSH stimulates the development and functions of sperm-producing cells. LH stimulates the testes to produce male sex hormones and also plays a role in sperm production.

Thyrotropin (also called thyroid-stimulating hormone) stimulates the synthesis and release of hormones from the thyroid gland. FSH, LH, and thyrotropin are all glycoproteins, which are peptides combined with carbohydrate molecules. *Adrenocorticotropin* (ACTH) is a peptide hormone of 39 amino acids that stimulates the release of hormones important in the metabolism of carbohydrates from the adrenal glands.

Human growth hormone (hGH) influences a number of metabolic activities that directly or indirectly influence growth, especially of the skeleton. Its most striking effects occur during periods of rapid growth and development, but it is present throughout life. *Prolactin,* like hGH, is a relatively large peptide (191 amino acids). Its role in males is uncertain, but it functions in the production of milk in females.

Regulation of Anterior Pituitary Hormones

Hormones from the anterior pituitary regulate the activities of many other tissues and glands, but what governs the anterior pituitary? It is now known that neurosecretory cells in the hypothalamus manufacture and secrete substances called **releasing** and **inhibiting hormones** or **factors** (*hormone* if the chemical structure is known, *factor* if it is not). As described in Figure 36.10, these substances travel to the pituitary through the hypophysial portal system, where they stimulate or inhibit the release of hormones produced by the anterior pituitary. Secretion of releasing or inhibiting factors is under the control of the peripheral nervous system and the information it relays to the brain.

The pituitary gland consists of an anterior and a posterior lobe. The posterior pituitary releases ADH and oxytocin, which are synthesized by neurosecretory cells in the hypothalamus. The anterior pituitary produces many hormones, including several trophic hormones. Hormone secretions from the anterior pituitary are regulated by the hypothalamus through the actions of releasing and inhibiting hormones.

FUTURE FRONTIERS

In the 1980s, endocrinologists made enormous advances in understanding the mechanisms involved in the actions of chemical messengers. Like molecular genetics, and for many of the same reasons, endocrinology is a dynamic field in which new, exciting, and important discoveries are made almost daily. These two disciplines are able to employ state-of-the-art techniques that enable investigators to address questions related to the molecular foundations on which our existence is constructed and maintained.

Hormone receptors are a very hot research topic at present. It has become clear that a large number of human diseases are due to defects in hormone-receptor interactions. Some of these can be traced to genetic errors that may lead to mistakes in synthesizing normal proteins. Since hormones and receptors are both largely, or entirely, constructed of specific proteins, such errors result in defective hormone or receptor structure, insufficient numbers of receptors or quantities of hormones, or defects in receptor signals. In this regard, three molecular studies reported in 1989 identified specific receptor defects as causes of different types of diabetes. The root causes are genetic defects that interfere with normal receptor structure and processing because of defective protein synthesis.

The effects of specific diseases caused by biological agents may also be directed at hormone-receptor interactions. For example, cholera is a disease caused by bacteria that may result in death because of dehydration. In this case, a toxin secreted by the bacteria disrupts the regulation of cAMP production. As a result, there is continuous production of this critical messenger, which causes dangerous amounts of fluids to be lost from the intestine.

Understanding the exact molecular cause of such diseases is an essential requirement in devising an effective strategy for curing them. For hormone-related diseases, the future prospects for cures or treatment are already extremely promising. Furthermore, recent studies indicate that deeper knowledge of the endocrine system will lead to exciting discoveries related to human health and diseases. For example, new research results indicate that altered relationships between hormones and their receptors may cause certain types of cancers. This sort of startling discovery typifies the enormously exciting field of endocrinology.

Summary

1. The classical view of the endocrine system is centered on the idea that glands secrete hormones into circulation and that these "chemical messengers" are then carried to target cells, where they exert their influence. In early studies, hormones were found to affect normal development, growth, sexual maturation, and behavior.
2. Because of recent research findings, modern endocrinology has found it necessary to enlarge the simple gland-hormone picture. It is now known that many organs besides glands secrete hormones, and substances have been discovered that act in the same way as hormones but are not hormones. Neurons secrete neurohormones, and growth factors control the growth and development of certain cells.
3. The two major types of hormones are peptide hormones and steroid hormones. Other chemical messengers include thyroid hormones and growth-stimulating factors.
4. The production and release of hormones are usually regulated by complex negative feedback systems that involve both the endocrine and nervous systems. After release, peptide hormones are degraded within hours, whereas steroid and thyroid hormones may remain in the body for several weeks.
5. Peptide hormones interact with their target cell through complex protein receptor molecules in the plasma membrane of the cell. Each receptor is specific for a single hormone. Peptide hormones then trigger the synthesis of cAMP inside the cell, and this leads to a cascade of chemical reactions that cause the cell to perform a certain function. Steroid hormones enter the cell and bind to an internal receptor, forming a complex that attaches to DNA and causes the synthesis of specific proteins.
6. The pituitary gland is the major gland of the endocrine system in terms of the number of hormones it produces and the effects its hormones have on body activities. The pituitary gland is regulated by neurosecretory cells in the hypothalamus region of the brain that release substances that either stimulate or inhibit hormone synthesis in the pituitary.
7. Research on the endocrine system, specifically on hormone-receptor systems, holds great promise for the treatment of a number of important diseases.

Review Questions

1. What is the classical view of the endocrine system?
2. How did the theory of internal secretions advance the field of endocrinology?
3. How has the classical view of the endocrine system changed in the twentieth century?
4. What are two major classes of hormones? How do they differ?

5. Describe the general life cycle of a hormone.
6. Which organs, cells, and hormones play roles in regulating blood glucose levels? How is blood glucose regulated?
7. What are some general activities of hormones in the human body?
8. What are hormone receptors? What is their function?
9. What is the "second-messenger system" used by peptide hormones?
10. How are steroid hormones able to carry out their functions?
11. Which hormones are produced by the posterior and anterior lobes of the pituitary gland? How is the anterior pituitary regulated?
12. What types of endocrinology studies are now being conducted? What is the potential importance of such studies?

Essay and Discussion Questions

1. There are several types of diabetes, each with a specific cause. What might be some of the causes of diabetes? Consider the hormones involved, how they act on cells, and the general life cycle of hormones.
2. Select one of the hormones listed in Table 36.1, and devise a probable feedback mechanism that may be involved in its regulation.
3. Which control system, nervous or endocrine, do you think evolved first? Why?

References and Recommended Reading

Archer, R. 1980. Molecular evolution of biologically active polypeptides. *Proceedings of the Royal Society of London B* 210:21–43.

Berridge, M. J. 1985. The molecular basis of communication within the cell. *Scientific American* 253:142–151.

Charron, M. J., and B. B. Kahn. 1990. Divergent molecular mechanisms for insulin-resistant glucose transport in muscle and adipose cells in vivo. *Journal of Biological Chemistry* 265:7994–7999.

Crapo, L. 1985. *Hormones: The Messengers of Life*. New York: Freeman.

Evans, R. M. 1988. The steroid and thyroid hormone receptor superfamily. *Science* 240:889–894.

Hadley, M. E. 1988. *Endocrinology*. 2d ed. Englewood Cliffs, N.J.: Prentice-Hall.

Guillemin, R., and R. Burgus. 1972. The hormones of the hypothalamus. *Scientific American* 227:24–33.

Kaltenbach, J. C. 1988. Endocrine aspects of homeostasis. *American Zoologist* 28:761–773.

Snyder, S. H. 1985. The molecular basis of communication between cells. *Scientific American* 253:132–141.

Unger, R. H. 1991. Diabetic hyperglycemia. Link to impaired glucose transport in pancreatic β cells. *Science* 251:1200–1205.

Wade, N. 1978. Guillemin and Schally. *Science* 200:279–282.

CHAPTER 37

Human Reproduction and Development

Chapter Outline

Reading Questions

1. What are the major structures and functions of the male and female reproductive systems?
2. How are activities of male and female reproductive systems regulated?
3. How does a fertilized egg become transformed into a baby in nine months?
4. What types of questions are now guiding research in reproduction and development?

Reproduction is the process by which plants and animals produce new individuals. Most people are extremely interested in the way humans accomplish this task. Biologically, human reproduction mostly depends on organs, controlling systems, and strategies that first evolved in our distant vertebrate ancestors, the reptiles (see Chapter 25), and our more modern mammalian ancestors.

What are some of these inventions from evolutionary history? Like our earliest animal ancestors, we carry out *sexual reproduction,* which involves the fusion of haploid gametes to create offspring (see Chapter 13). We use **internal fertilization**, the process in which gametes fuse within the body of the female. This ancient strategy enabled reptiles to occupy terrestrial environments hundreds of millions of years ago. Finally, from our mammalian ancestors, we inherited abilities to retain the developing embryo inside the female's body and to provide it with nourishment from milk-secreting glands, practices that greatly enhanced survival of the offspring.

Humans have, of course, added a new wrinkle or two through the course of our evolutionary history. Perhaps the most interesting is that unlike all other mammalian species, which reproduce only during a specific time of the year as determined by the female's *estrous cycle* (a regular period of sexual receptivity that occurs when an egg is available to be fertilized), humans can reproduce year-round. They also engage in sexual activities when fertilization is unlikely to occur. These unique features of human reproduction have been analyzed extensively. What would be the advantage of a strategy in which sexual behavior is not necessarily linked to reproductive success or a greater chance of producing offspring? Though not convincing to all scientists, one hypothesis proposes that in early humans, this strategy may have been important in establishing long-term pair bonds. In other words, the continuous availability of pleasurable sex was the glue that held a relationship together. A durable human male-female relationship could then provide the justification for individuals to invest major amounts of time in caring for intensely dependent human offspring. Thus early human couples who engaged in such behavior would have been more likely to leave behind more offspring than those who did not.

The origins and evolutionary significance of human sexual behaviors will continue to be debated. However, there is little argument that human reproduction has been enormously successful throughout the course of history.

THE HUMAN REPRODUCTIVE SYSTEM

Males and females possess **gonads**, the *primary sex organs* that create gametes and produce hormones that influence development and functions related to reproduction. The primary sex organs of males are called **testes** and produce *spermatozoa,* or sperm, by the billions as well as male sex hormones called **androgens**. **Ovaries** are the primary sex organs of females. They produce a small number of *ova,* or eggs, during their reproductive lifetime. The ovaries also produce sex hormones, including *progesterone* and a number of *estrogens*. *Accessory reproductive organs* include all other organs involved in gamete protection, storage, and transport.

Humans can produce progeny year-round through sexual reproduction, internal fertilization, and embryonic development. All of these processes result in a high offspring survival rate. The primary sex organs of males are testes, which produce sperm. Female primary sex organs are ovaries, which produce ova. Reproductive organs and gamete production are regulated primarily by hormones.

Figure 37.1

The external organs of the male reproductive system are the penis and the scrotum. The others are internal organs.

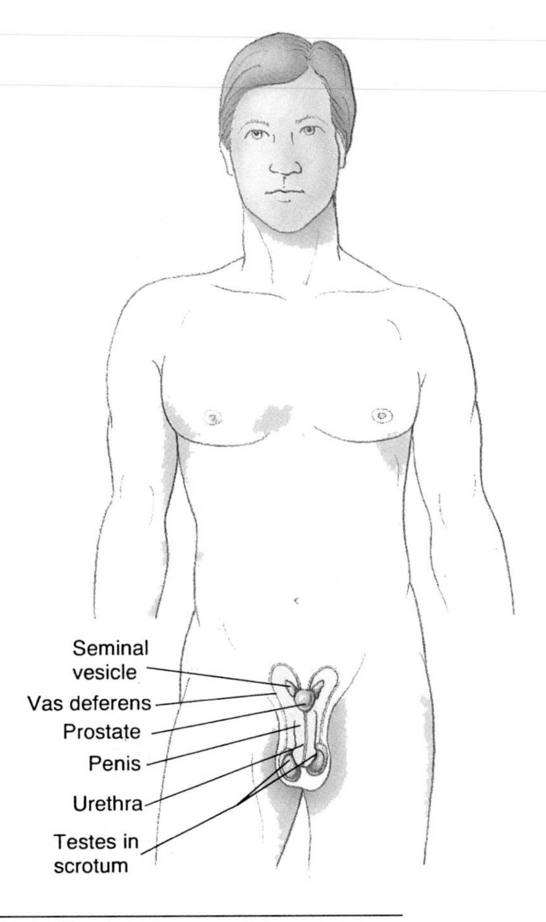

THE MALE REPRODUCTIVE SYSTEM

Organs of the male reproductive system are illustrated in Figures 37.1. The two major functions of the testes are to manufacture sperm and secrete androgens. The primary androgen is **testosterone**, a hormone that influences many activities in the male reproductive system. Various accessory organs are responsible for storing and nourishing sperm and for delivering them to the reproductive tract of a female. Regulation of all reproductive functions is carried out by pituitary hormones, with secretions managed by the hypothalamus of the brain.

Sperm Production

The testes lie outside the abdominal cavity, contained within a pouch of skin called the *scrotum*. The external location of the testes is significant because normal sperm development requires a slightly cooler temperature than that maintained in the internal environment. Each testis is surrounded and protected by a capsule of connective tissue, and inward extensions of this tissue divide the testes into about 250 compartments. Each compartment contains sperm-producing *seminiferous tubules,* highly convoluted structures that harbor male gametes in various stages of development (see Figure 37.2A). Testosterone is produced by clusters of *interstitial (Leydig) cells* located within the connective tissue between the seminiferous tubules.

Production of spermatozoa within the testes is called **spermatogenesis**. It begins in males at *puberty,* the time of life when the gonads mature and reproduction becomes possible. After puberty, spermatogenesis continues throughout life.

The manufacture of mature sperm begins in germinal cells of the seminiferous tubules shown in Figure 37.2B–C. These cells, called *spermatogonia,* divide

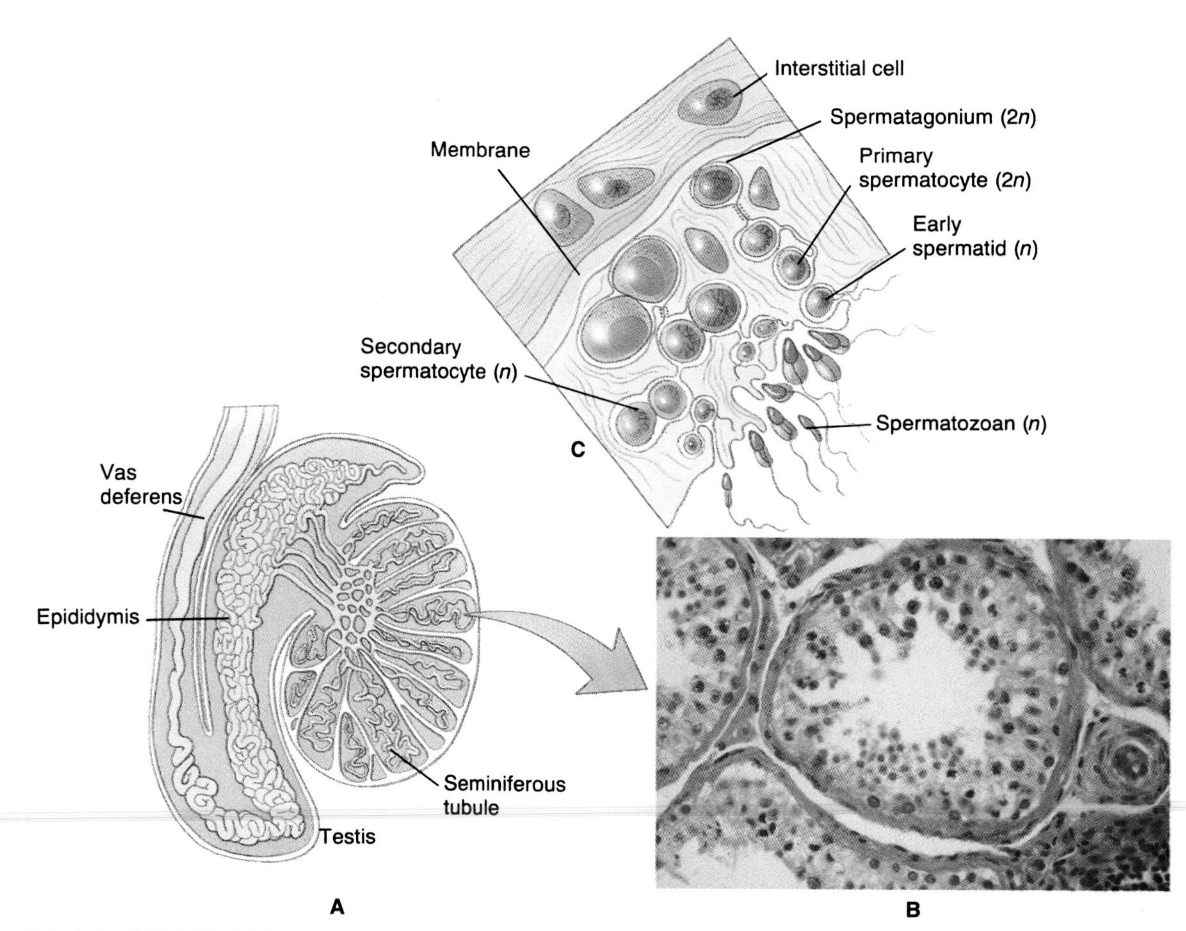

Figure 37.2

(A) Sperm are produced within seminiferous tubules of the testes. (B) This electron micrograph shows sperm in various stages of production. (C) Diploid (2n) spermatogonia proliferate by mitosis and produce cells that develop into primary spermatocytes. Primary spermatocytes undergo meiosis, giving rise to secondary spermatocytes that each contain a haploid (*n*) number of chromosomes. Secondary spermatocytes then develop into spermatids, which mature into functional spermatozoa.

by mitosis to give rise to large numbers of identical cells. Spermatogonia contain the diploid number of 23 pairs of chromosomes (44 autosomes and an X and a Y chromosome). Each day, millions of spermatogonia move toward the center of the tubule, begin to grow, and change into large developing cells known as *primary spermatocytes*. Ultimately, it is essential for each sperm to possess a haploid number of chromosomes (n = 23). How does this occur? The process is described in Figure 37.3. First, each primary spermatocyte undergoes meiosis.

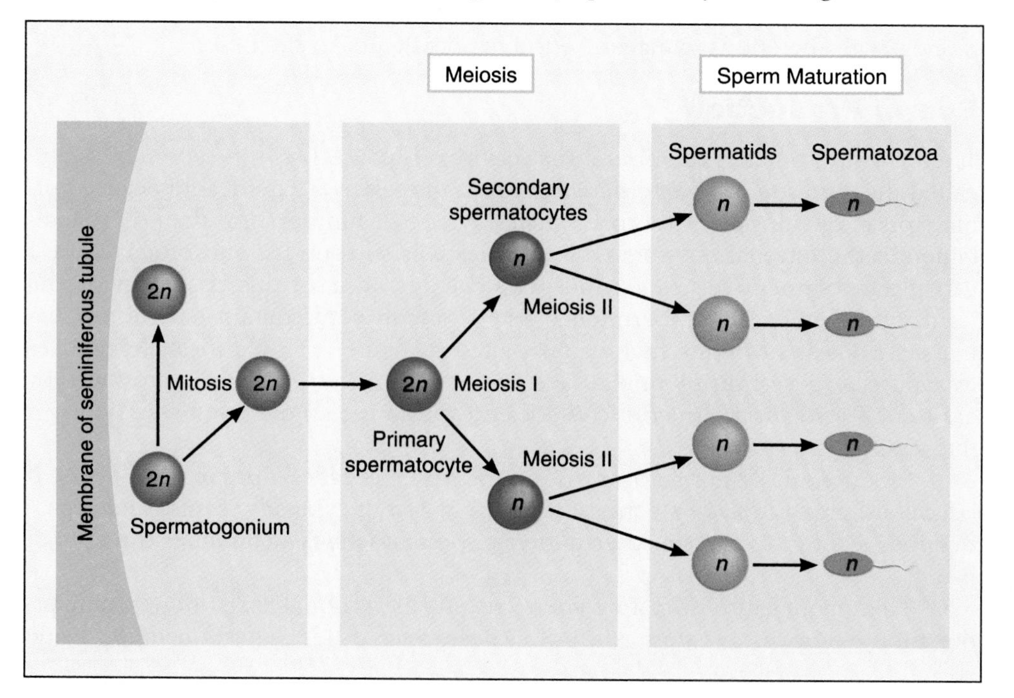

Figure 37.3

The maturation sequence from spermatogonia to spermatozoa includes a change in the chromosome number from 2*n* (diploid) to *n* (haploid) after meiosis.

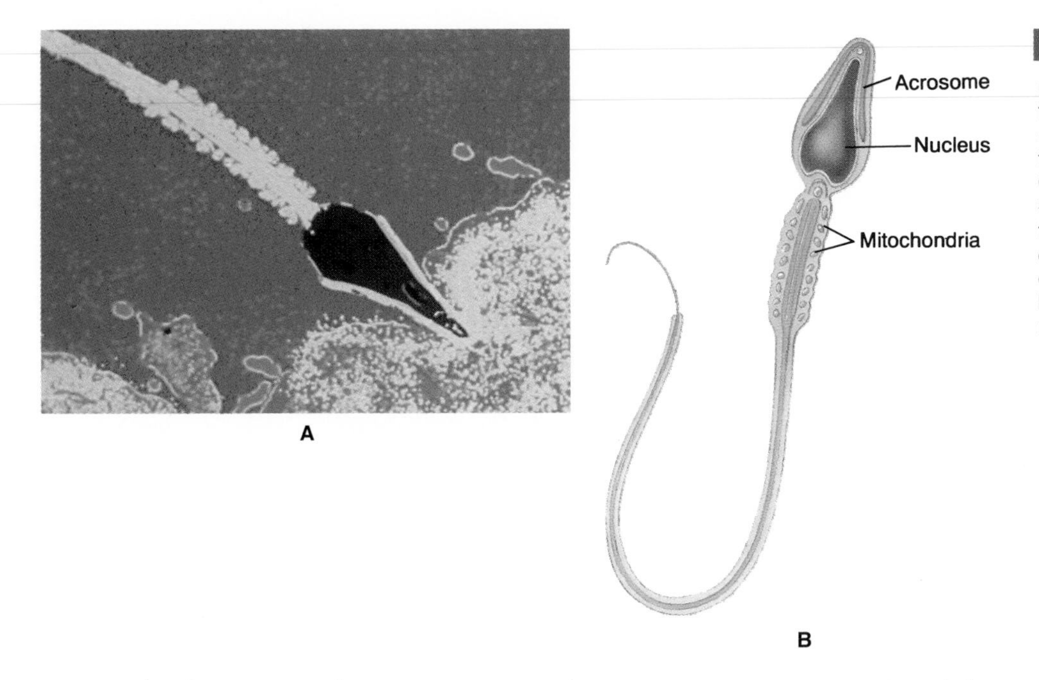

Figure 37.4

(A) A functional human sperm looks like this as it begins to penetrate an ovum. (B) The structure of a sperm is related to its function—to fertilize an ovum. Chromosomes are carried in the nucleus, and the acrosome contains fluid filled with enzymes that break down ovum membranes. Mobility is provided directly by the flagellum and indirectly by mitochondria that provide the energy required to move the sperm's "tail."

During the first meiotic division, two *secondary spermatocytes* are produced that each have 22 paired autosomes plus either a pair of X or Y chromosomes. Subsequently, during the second meiotic division, each secondary spermatocyte gives rise to two *spermatids,* each of which contains 22 autosomes and one sex chromosome. Each spermatid then undergoes a sequence of changes that leads to its final destiny, becoming a mature **spermatozoon**.

Sperm have a structure that reveals a perfect relationship with their function (see Figure 37.4). The head of a sperm contains a nucleus with 23 chromosomes and is covered by a caplike, fluid-filled structure called an *acrosome.* The acrosomal fluid contains enzymes that are released when the sperm encounters an ovum in the female reproductive tract. Apparently, these enzymes assist the sperm in dissolving membranes that surround the ovum. Another striking feature of sperm is the "tail," a single hairlike *flagellum* that provides the cell with mobility.

Sperm Storage and Transport

Once spermatozoa are formed, they move through a system of ducts, shown in Figure 37.5, until they are eventually released from the body. From the seminiferous tubules, sperm pass into highly coiled tubules of the *epididymis.* There they undergo final maturation and are stored until *ejaculation,* or discharge from the body, takes place. This occurs when a male becomes sexually stimulated and a series of physiological events takes place that finally leads to the sperm being propelled through the **penis**, the male organ that conveys sperm to the external environment.

Between the epididymis and the penis are a number of ducts (tubelike structures) and glands that play important roles in the sperm's journey. During ejaculation, sperm from both testes are propelled from the epididymis into the *vasa deferens.* These two ducts are surrounded by walls of smooth muscle that contract rhythmically during ejaculation, transporting the sperm forward. A vas deferens ascends from each testis and merges with a duct from a *seminal vesicle* to form a short *ejaculatory duct* that is located behind the urinary bladder. These ducts then pass through the *prostate gland* and enter the *urethra,* the tube that drains both the urinary and reproductive systems.

Secretions from three organs combine with sperm to form *semen,* the fluid that is ejaculated. The two seminal vesicles secrete a sticky fluid that contains fructose, a sugar that is the major energy source for ejaculated sperm, and

Figure 37.5

This median section (from the side) shows the organs involved in the transport of sperm and adjacent organs.

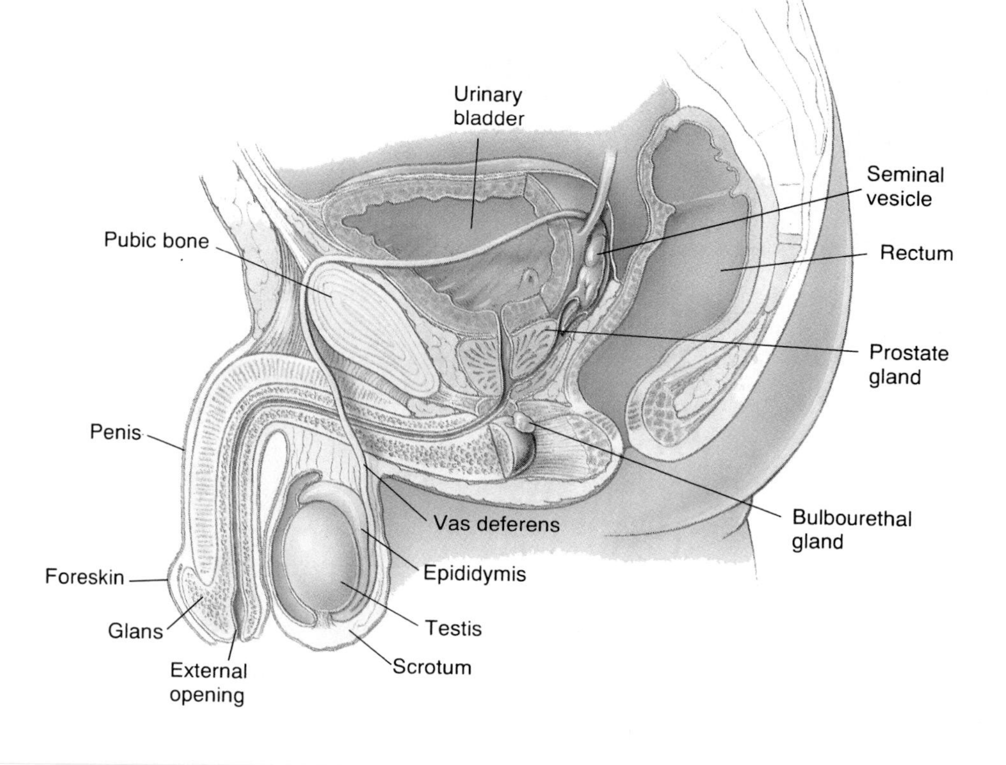

prostaglandins, substances thought to enhance the movement of sperm when deposited in the female reproductive tract. The single prostate gland secretes a milky, alkaline fluid that helps counteract the acid environment within the female reproductive tract. Finally, secretions from two *bulbourethral glands* precede ejaculation and are thought to play some role in reducing the acidity of the urethra and lubricating the penis.

Hormonal Regulation of the Male Reproductive System

Prior to the onset of puberty, which typically occurs at about 14 years of age (versus 12½ to 13 years for females), males are unable to produce mature sperm. After puberty, through the actions of testosterone, males are able to produce sperm by the billions. In addition, this hormone radically transforms several body features. Changes in the body related to sexual maturation are known as **secondary sex characteristics**. In males, these include a deepening of the voice, increased development of skeletal muscle tissue and a decrease in body fat, growth of hair on the face and other areas of the body, and enlargement of the penis, scrotum, and testes. What happens at puberty to cause these profound changes that enable males to become capable of producing offspring? The answers, of course, are tied to activities of the human control systems.

The leading hypothesis to explain the onset of puberty in humans, summarized in Figure 37.6, focuses on maturation of the part of the brain's neuroendocrine control system involved in secreting releasing hormones from the hypothalamus. Recall that specific releasing hormones from the hypothalamus are required to stimulate production of certain hormones from the anterior pituitary gland. However, the factors responsible for the onset of puberty in males are not totally understood. During childhood, the endocrine system produces only meager amounts of male sex hormones. These hormones, which are required for sexual maturation and function, consist of testosterone and two *gonadotropins* called *follicle-stimulating hormone* (FSH) and *luteinizing hormone* (LH). At puberty, the hypothalamus, in reaction to a signal from an undefined area of the brain, begins to secrete increased levels of a

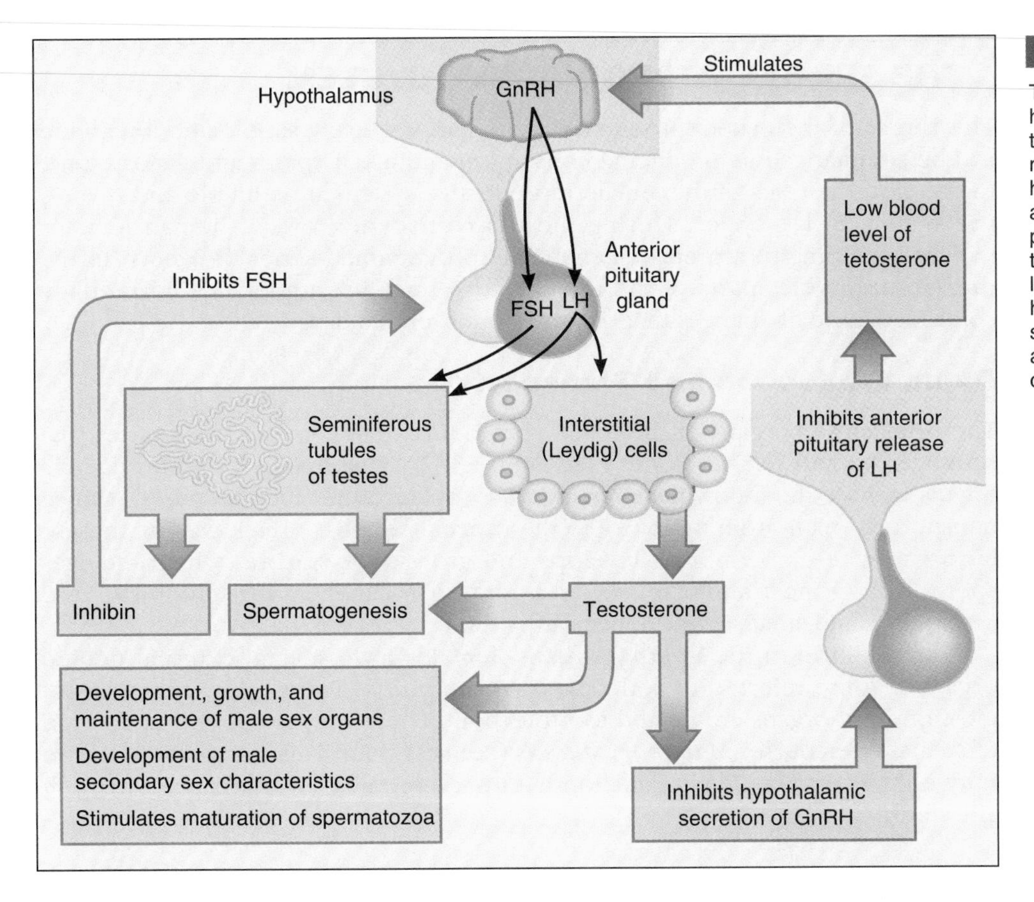

Figure 37.6

The onset of puberty and continuous hormonal regulation of sperm production in males involves several hormones. GnRH released from the hypothalamus stimulates production and release of FSH and LH from the pituitary gland. FSH activates and sustains sperm production, and LH stimulates production of testosterone, the hormone that plays a central role in spermatogenesis and the development and maintenance of secondary sex characteristics.

gonadotropin-releasing hormone (GnRH). It is still not known whether a single GnRH stimulates the production of both FSH and LH or if two different releasing hormones are necessary. All these hormones are believed to interact in classical negative feedback cycles.

Figure 37.6 indicates that GnRH stimulates production and release of FSH and LH, which are then transported by the circulatory system to the testes. FSH activates sperm production, and LH stimulates interstitial cells to secrete testosterone, which in turn triggers spermatogenesis and the development and maintenance of secondary sex characteristics.

How are levels of these various hormones regulated by control systems? The self-regulating pathways are generally understood, although some uncertainties linger. High levels of testosterone inhibit LH release through negative feedback that is thought to be directed at the hypothalamus and perhaps the pituitary. In other words, as testosterone concentrations in the blood increase, the pituitary responds by reducing production of LH.

FSH production in both sexes is hypothesized to be regulated by a hormone called *inhibin*. In males, this hormone is thought to be produced by cells in the seminiferous tubules and to exert negative feedback effects on the pituitary, the hypothalamus, or both. As testosterone levels decrease, the pituitary once again produces increased levels of FSH and LH. This hormonal cycle operates uniformly and continuously throughout the lifetime of a normal male.

Testosterone, a sex hormone, plays a major role in regulating the maturation and function of male reproductive organs and is also responsible for development of secondary sex characteristics. Spermatogenesis begins in the testes at puberty and continues throughout life. Two pituitary hormones are important in maintaining this process: FSH initiates sperm production, and LH stimulates testosterone secretion. All these hormones are regulated by negative feedback mechanisms.

THE FEMALE REPRODUCTIVE SYSTEM

The chores of the male reproductive system are relatively simple and linear and can be summarized as producing constant quantities of sperm and testosterone. In contrast, the tasks of the female reproductive system are multiple and involve two major 28-day cycles. During the **ovarian cycle**, ova of human females mature according to a precise, genetically programmed series of events. In the **menstrual cycle**, parts of the reproductive tract also undergo a hormonally controlled cycle of changes that occurs about every 28 days.

Ovum Production and Transport

The primary sex organs of the female reproductive system are two ovaries that produce ova and sex hormones (see Figure 37.7). Major accessory reproductive organs include the following: *uterine tubes* (oviducts or Fallopian tubes), which transport an ovum from the ovary to the **uterus** (womb), a thick-walled, muscular organ where an embryo develops and grows; the **vagina**, which receives sperm from a male and also serves as the birth canal during childbirth; and accessory glands that secrete a lubricating fluid.

In addition to the internal organs, females have several external genital organs that are referred to collectively as the *vulva*. The general functions of most of these organs are related to protection of sensitive tissues and facilitation of sexual intercourse. *Mammary glands* (breasts), found also in males, develop fully only in females. Their function is to provide milk for nourishing offspring after birth.

Oogenesis

In females, puberty generally begins about a year earlier than in males. Available evidence suggests that the onset of puberty in females is related to attaining a critical body weight or body composition. The beginning of puberty is marked by the first incidence of **menstruation**, a periodic (approximately monthly) loss

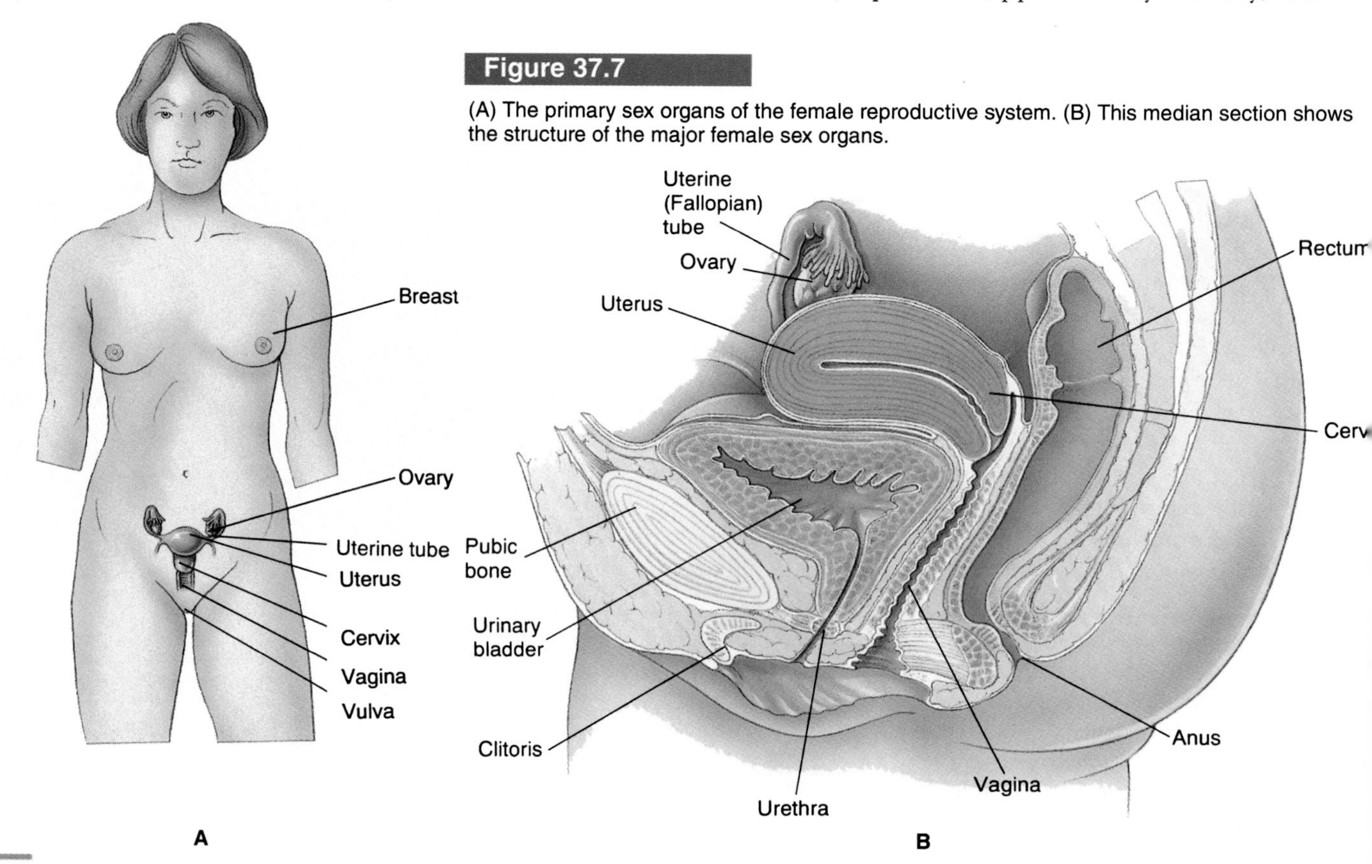

Figure 37.7

(A) The primary sex organs of the female reproductive system. (B) This median section shows the structure of the major female sex organs.

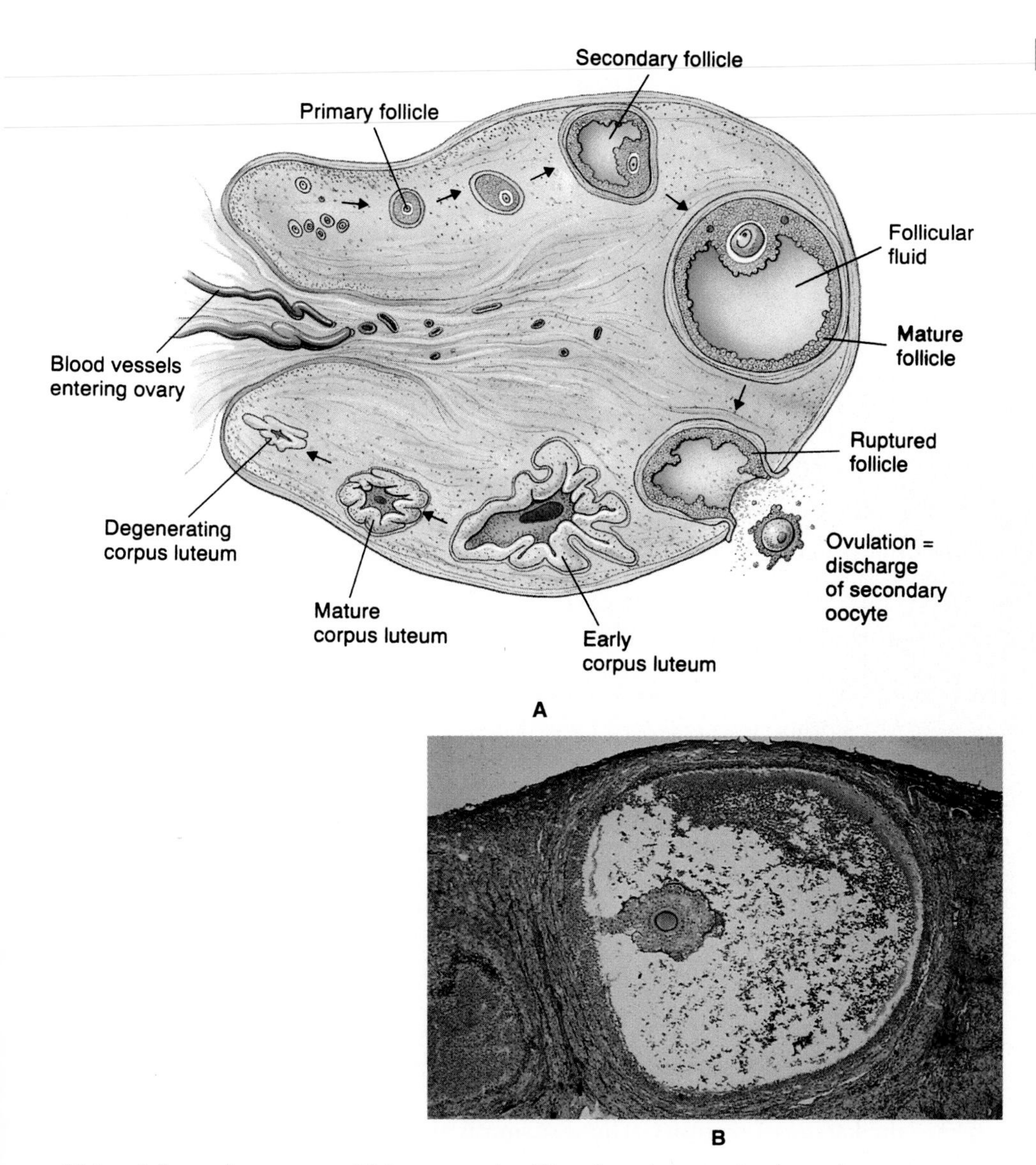

Figure 37.8

(A) Once each month, in one of the two ovaries, a primary follicle containing an oocyte develops into a secondary follicle, which releases the mature oocyte during ovulation. The follicle then becomes transformed into a corpus luteum that will degenerate if fertilization of the ovum does not occur. (B) This secondary follicle contains a maturing oocyte.

of blood from the uterus. This event signifies that the ovaries have begun to produce mature ova, a process called **oogenesis**, which is described in Figure 37.8A.

The gametogenic potential of the ovary is actually determined during embryonic development. Before birth, *oogonia,* cells in the female embryo's ovary, divide by mitosis, giving rise to about 3 million *primary oocytes* (immature ova). At birth, about 1 million oocytes remain, and by puberty, this number has been further reduced to about 250,000, of which 400 to 500 will ultimately develop into mature ova. Prior to birth, primary oocytes enter the first phase of meiosis (prophase), where they remain in arrested development until later in life.

Development of Follicles

Beginning at puberty, **ovulation**, the release of a single mature egg from one of the ovaries, occurs approximately once each month (the average cycle is 28 days). At the start of the cycle, primary oocytes are surrounded by a layer of supporting follicle cells that, together with the primary oocyte, are called a *primary follicle.* Events in the ovarian cycle begin when, under hormonal stimulation, 6 to 12 primary follicles began to grow and develop into *secondary follicles* (see Figure 37.8B). Several events occur during the maturation of a secondary follicle.

1. The first meiotic division is completed, producing two haploid cells of unequal size, as described in Figure 37.9. The largest, which may

Figure 37.9

During embryonic development, diploid oogonia (2*n*) proliferate by mitosis and produce primary oocytes (2*n*). Each month after puberty, a primary oocyte undergoes the first meiotic division, giving rise to one haploid secondary oocyte (*n*) and a haploid polar body (*n*). The first polar body, and the secondary oocyte if fertilized, may complete a second meiotic division.

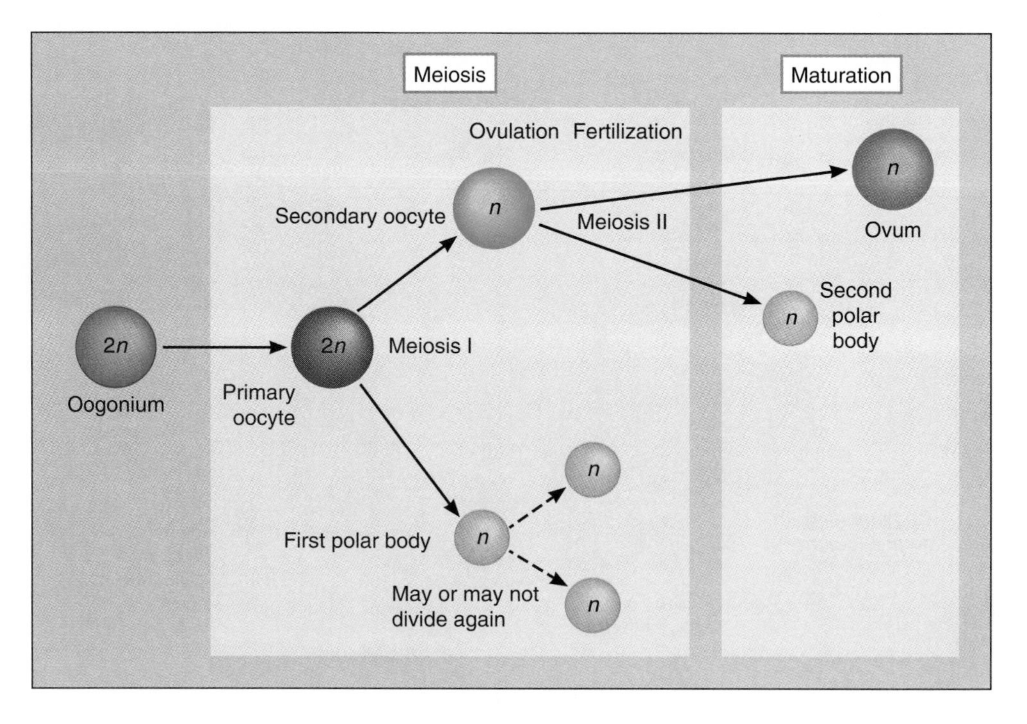

become a mature ovum, is called a *secondary oocyte,* and the smaller is referred to as the *first polar body,* a cell that has no functional future. If fertilization occurs, each of these cells may undergo a second meiotic division, giving rise to two additional polar bodies.

2. Follicle cells proliferate and form glandular cells that secrete a fluid around the developing oocyte.
3. As the oocyte approaches maturity, it becomes surrounded by a fluid-filled cavity.
4. At maturity, which occurs at 10 to 14 days into the ovarian cycle, the follicle moves toward the surface of the ovary. Although several follicles undergo initial development each month, generally only one reaches the final stage of maturity. The rest simply degenerate.

Ovulation

Upon receiving the appropriate hormonal cue, ovulation occurs as the mature follicle ruptures and the **ovum**, along with a surrounding layer of follicle cells, moves into the uterine tube. Once there, it is swept toward the uterus by the action of cilia that line the tube.

Formation of the Corpus Luteum

Soon after ovulation occurs, the ruptured follicle undergoes a series of changes, eventually becoming a cholesterol-filled tissue called a *corpus luteum*. The fate of this structure depends on whether or not the female becomes pregnant. If she does not, it degenerates about 14 days after ovulation. If pregnancy occurs, the corpus luteum undergoes further development, forming an endocrine gland that regulates several important functions during pregnancy.

Hormonal Regulation of the Female Reproductive System

Female hormonal regulating mechanisms are described in Figure 37.10. At puberty, the GnRH-FSH-LH system begins to function, and **estrogens**, female sex hormones produced by cells of the ovaries, induce the development of secondary sex characteristics. During this period, the vagina, uterus, and uterine tubes grow

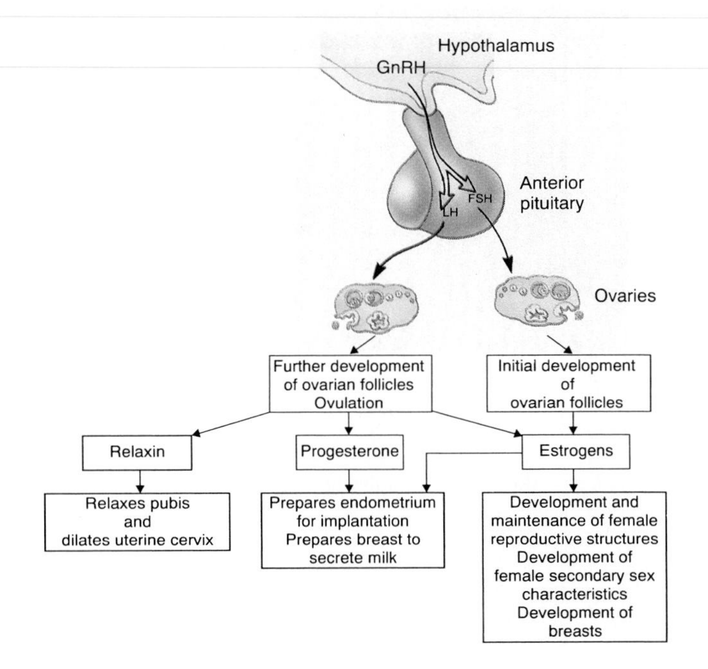

Figure 37.10

The female hormones GnRH, FSH, LH, progesterone, and estrogen are all involved in regulating the monthly ovarian and menstrual cycles in females after puberty. They also, along with relaxin, play important roles if pregnancy occurs.

and develop; fat becomes deposited in certain areas such as the hips and the breasts; and mammary growth and development occur. The female reproductive tract is prepared to receive a fertilized egg and to maintain the uterine lining if pregnancy occurs. All of these actions are induced by estrogens and a number of other hormones.

The regular ovarian and menstrual cycles continue to occur in a woman until **menopause**, the time in her life when reproduction is no longer possible. The onset of menopause, usually around age 50, has been thought to be associated with the failure of the ovary to respond any longer to FSH and LH. Consequently, the ovarian hormonal cycles cease, estrogen levels become greatly diminished, and this leads to a number of physiological and structural changes in menopausal women. Recent studies indicate that alterations in GnRH release patterns may also contribute to the onset of menopause.

The Ovarian Cycle

The events and timing of the ovarian and menstrual cycles are described in Figure 37.11A. The concentrations of different hormones in the blood during a typical 28-day period are shown in Figure 37.11B. The effects of GnRH, LH, and FSH are similar to those described for males. In females, FSH stimulates the development of ova, and an examination of Figure 37.11A–B reveals the relationships between hormone concentrations and events during development of the follicle. FSH levels decrease until midway through the cycle, probably as a result of increasing estrogen (and perhaps inhibin) concentrations. LH production is also suppressed during this period by the negative feedback inhibition of estrogens on the pituitary. At mid-cycle, estrogen levels increase, and they remain high for 36 hours or more. These high estrogen levels trigger a surge in the levels of LH and FSH. The high LH levels stimulate release of the mature ovum, which in turn results in an elevation of progesterone levels. Progesterone—which continues to increase during the latter part of the cycle—and inhibin exert negative feedback on the hypothalamus, and further FSH and LH secretion is inhibited during this time. If pregnancy does not occur, the corpus luteum degenerates, and levels of all hormones begin to fall. Consequently, negative feedback inhibition of FSH and LH secretion by estrogens and progesterone no longer occurs. As a result, FSH and LH levels begin to increase once again, new follicles will begin to develop, and the ovarian cycle is repeated.

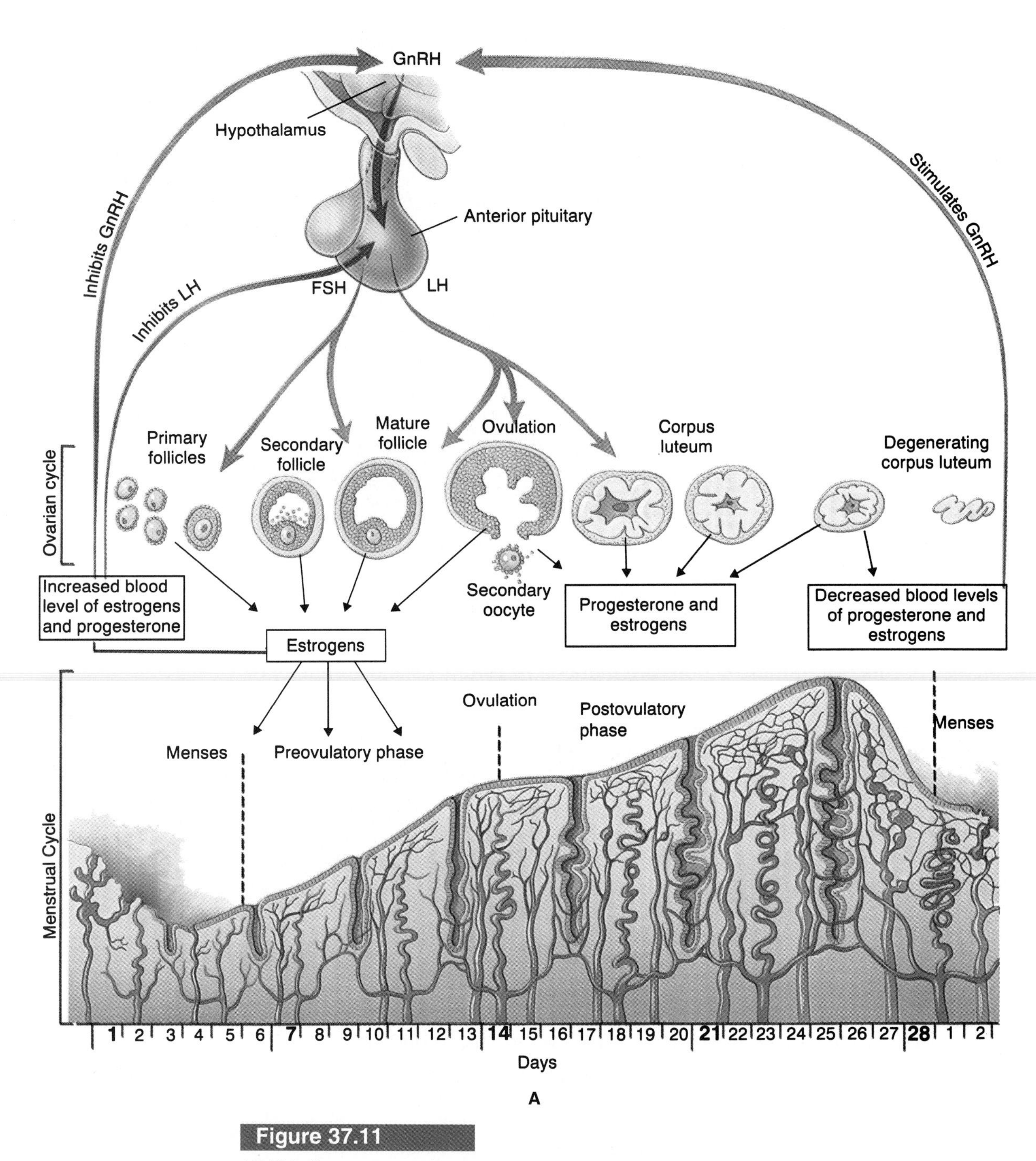

Figure 37.11

(A) This diagram traces the changes that occur during the 28-day ovarian and menstrual cycles, the hormones involved in these changes, and the relationships between the two cycles. (B) Hormone concentrations vary during a normal 28-day cycle. Inhibin concentrations are not shown, but they are apparently very low.

The Menstrual Cycle

The menstrual, or uterine, cycle involves a sequence of changes that occur primarily in the *endometrium,* the mucous glandular membrane that lines the uterus. It overlaps with the ovarian cycle since estrogens and progesterone also regulate the menstrual cycle (see Figure 37.11). During the developmental phase, which is first stimulated by estrogens, blood vessels invade the epithelium, glandular cells increase, and the endometrium becomes thicker. After ovulation, the corpus luteum secretes both estrogens and progesterone, which continue to stimulate further development of the endometrium. Essentially, it is being prepared to receive a fertilized egg. If pregnancy does not occur, hormone levels decline as the corpus luteum degenerates, and as a result, maintenance of the

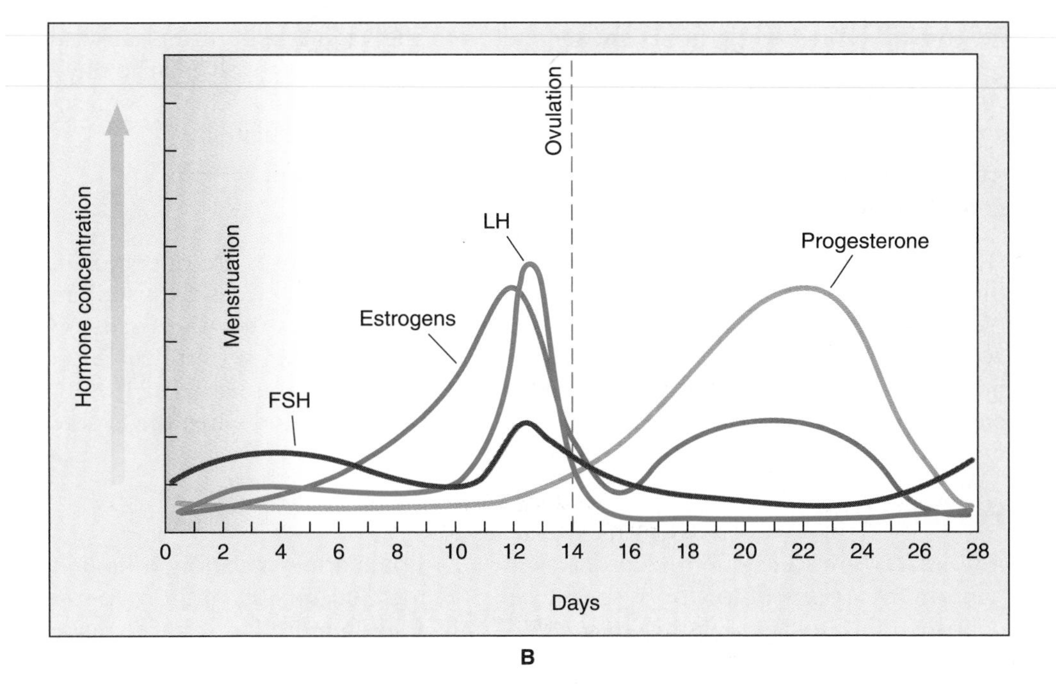

B

endometrium is discontinued. When that happens, the membrane deteriorates and cells, blood vessels, and blood are sloughed and discharged from the uterus as the menstrual flow. The endometrium is quickly repaired, and before the next cycle begins, it can again undergo the cyclical changes that characterize the menstrual cycle.

In females, mature ova are produced in a monthly cycle between the ages of puberty and menopause. Estrogens induce development of secondary sex characteristics and also play key roles in all female reproductive functions. The ovarian and menstrual cycles are connected by negative feedback mechanisms involving FSH, LH, estrogens, progesterone, and perhaps inhibin. Fluctuating levels of these hormones are responsible for the cyclical maturation and release of ova and the development and decline of the corpus luteum and the endometrium.

SEXUAL RESPONSES

Males and females are capable of responding to many types of stimuli that have sexual connotations. In males, sexual stimuli, whether visual, tactile, or psychological, may lead to an *erection* (enlarging and firming) of the penis. This is brought about by engorgement of the highly vascular tissues of the penis with blood, and it enables the penis to be inserted into the vagina. As sexual stimulation intensifies, pulse, blood pressure, and rate of breathing all increase, and *emission,* movement of the contents of sexual ducts and glands to the urethra, occurs. At this point, small amounts of semen may be emitted from the penis. Ultimately, sexual excitement reaches its peak, and sympathetic nerve impulses lead to ejaculation of semen from the urethra. Ejaculation is accompanied by a series of brief, widespread muscular contractions and pleasurable sensations that are collectively referred to as *orgasm*. Following ejaculation, regular patterns of blood flow, heart rate, and breathing are restored, and the penis returns to its normal, unaroused state. For several minutes or hours after ejaculation, most males are unresponsive to further sexual stimulation.

Sexual stimulation in women leads to many physiological responses that are similar to those in men, which is not surprising since many of the same nervous pathways are involved. The *clitoris,* a small sensitive organ that has the same embryological origin as the male penis, becomes engorged with blood, as do the breasts and the area surrounding the vaginal opening. Copious quantities of fluids to help lubricate the penis are secreted into the vagina, and further arousal leads to rhythmic stimulatory responses of all external organs involved. The pleasurable sensations and physiological changes that characterize a female's orgasm are similar to those of the male, except for the absence of ejaculation.

PREGNANCY

If all of the reproductive and hormonal functions of both sexes are operational, the various sexual responses occur, and no contraceptive devices are used, an ovum may be fertilized in one of the female's uterine tubes (see the Focus on Science and Technology, "Contraception"). The resulting *zygote* (fertilized egg) now faces a precarious series of tasks that must be accomplished if it is to emerge nine months later as a normal human being. The period when the zygote develops within the female is referred to as **pregnancy**.

Fertilization

A human ovum can be fertilized only within 24 hours after ovulation, a limited window of opportunity for a sperm cell. Of the 300 to 400 million sperm deposited in the vagina during ejaculation, only a few hundred are able to make their way to the uterine tube, where the ovum is usually located. For fertilization to occur, one of the sperm must penetrate the membrane surrounding the egg and pass through, entering into its cytoplasm (see Figure 37.12A). Following this exploit, these next events happen quickly: (1) the ovum completes the second meiotic division and becomes a true haploid cell, (2) the nuclei of the sperm and egg (called pronuclei at this stage) combine to form one nucleus and reestablish the diploid number of chromosomes (see Figure 37.12B), and (3) the outer membranes of the fertilized egg undergo changes that prevent the entry of any additional sperm. The zygote begins to undergo cell division, development, and relocation and is now called an *embryo*.

Implantation

Figure 37.13 summarizes the events that occur between fertilization and implantation of the embryo in the endometrium. Within three to four days after fertilization, the developing embryo has migrated down the uterine tube and entered the uterus. By this time, it has developed into a microscopic hollow ball of cells called a *blastocyst*. One week after fertilization, the blastocyst becomes implanted in the endometrium and begins to secrete enzymes that digest endometrial cells. This serves two purposes: the blastocyst creates a "nest" for itself in the endometrium, and the digested cells provide nutrients for its continuing development.

The Endometrium During Pregnancy

Figure 37.14 summarizes hormonal actions during pregnancy. The endometrium provides nourishment to the developing embryo for about two months. There-

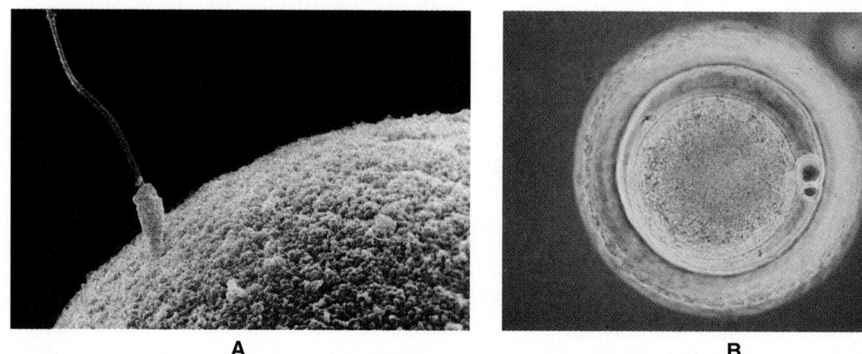

Figure 37.12

(A)This scanning electron micrograph shows a sperm penetrating the membrane of an ovum, the first step in fertilization. (B) The male and female pronuclei in this fertilized ovum are just about to fuse to form a zygote.

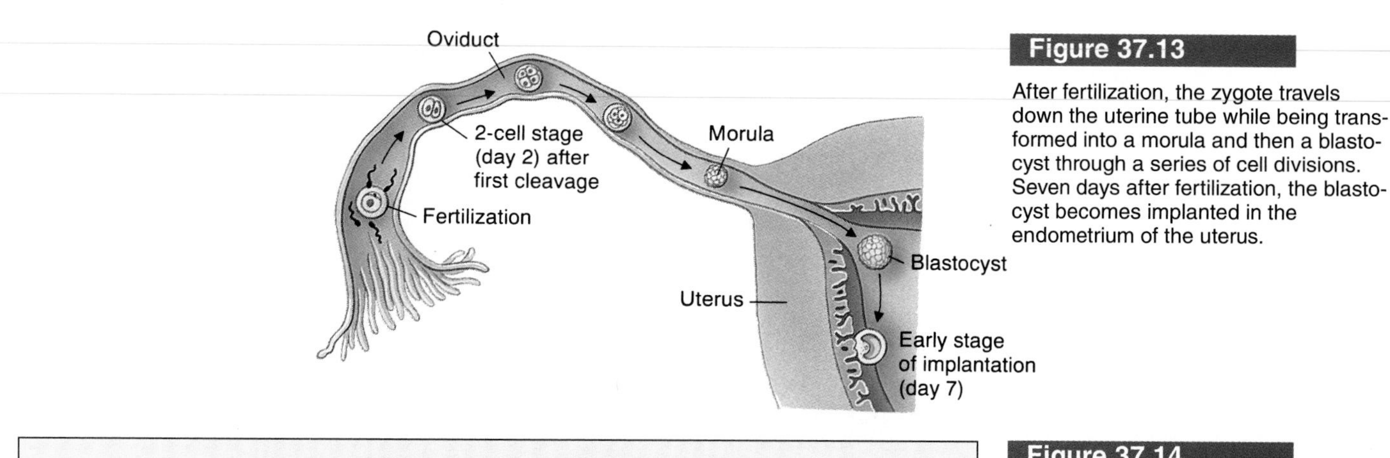

Figure 37.13

After fertilization, the zygote travels down the uterine tube while being transformed into a morula and then a blastocyst through a series of cell divisions. Seven days after fertilization, the blastocyst becomes implanted in the endometrium of the uterus.

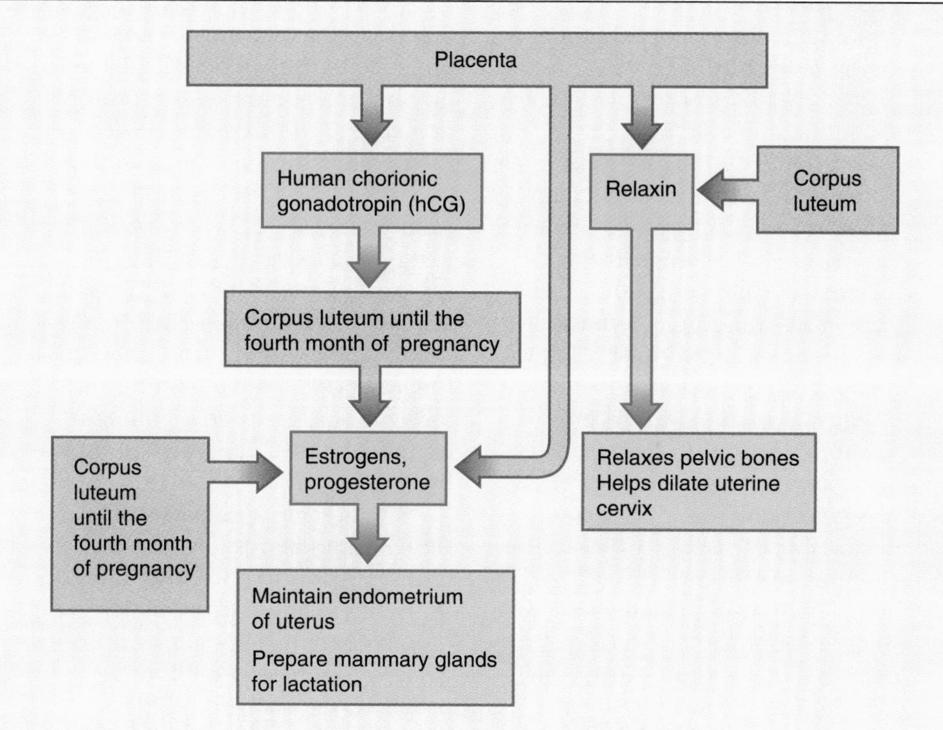

A

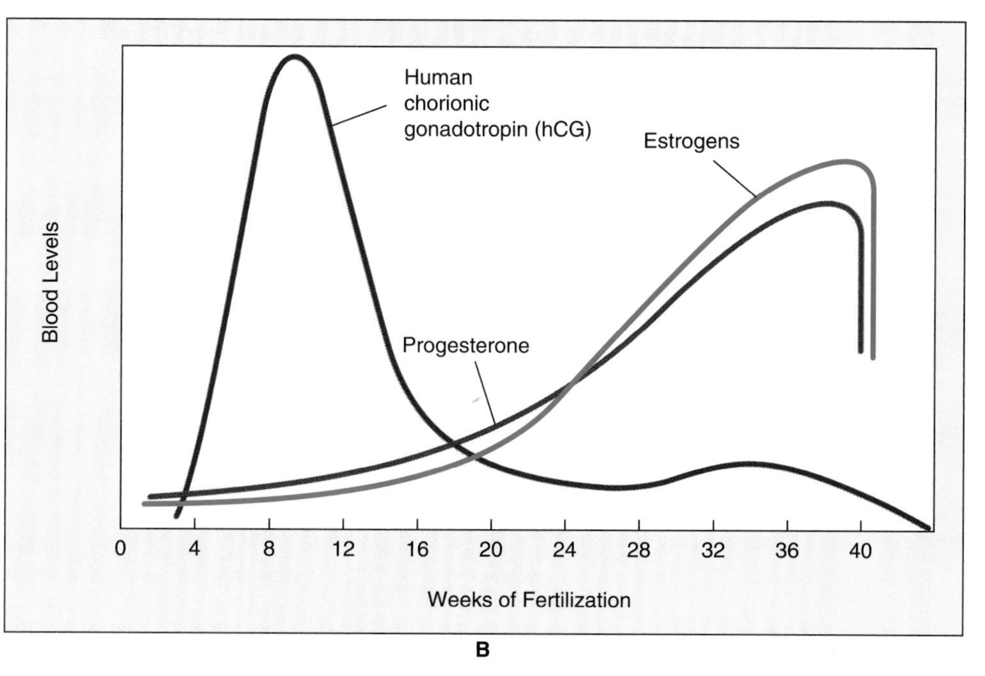

B

Figure 37.14

(A) Hormones have various functions during pregnancy. (B) High levels of estrogens and progesterone are necessary to maintain the endometrium and the placenta and to prevent the maturation of new follicles. Early in pregnancy, hCG helps maintain the corpus luteum, a source of the other two hormones. Relaxin affects certain organs just prior to childbirth.

Focus on Science and Technology

CONTRA-CEPTION

Contraception, or birth control, refers to any device or strategy that prevents fertilization of the ovum or growth of the zygote. According to this definition, abstinence qualifies as a contraceptive method. Although this is undoubtedly the most effective method, in making decisions about pregnancy prevention, many humans do not consider it an acceptable alternative. Thus through the course of history, humans have devised other methods for birth control. Contraceptive methods currently in use are summarized in Table 1.

Contraceptive methods depend on behavioral, mechanical, hormonal, or surgical maneuvers that can be categorized as follows: (1) preventing the sperm from contacting the egg in the female reproductive tract, (2) suppressing the production or release of eggs or sperm, and (3) preventing implantation. In the last category, an intrauterine device (IUD) was commonly used by women during the 1970s and 1980s to prevent implantation of the blastocyst. The IUD is a small spiral, ring, or loop made of various materials that requires insertion into the uterus by a physician. Many women suffered negative health effects from using the IUD (for example, increased menstrual bleeding, uterine infections, and allergic-type reactions to the IUD). As a result, it is no longer recommended as a method of contraception for young women who may wish to have children in the future.

Withdrawal (*coitus interruptus*) and the rhythm method are not reliable, since they require behaviors that are not always practical during passionate moments. Furthermore, the effectiveness of the rhythm method depends on accurate prediction of ovulation, which can often be irregular, and that of withdrawal depends on nondeposition of sperm, which can in fact seep out before completion of the sex act. Consequently, reliance on these methods is problematic if avoidance of pregnancy is a serious concern.

Condoms and diaphragms (used with a spermicide) are very effective barriers when used properly but may be useless if torn or defective. A new barrier device, the cervical cap, was approved for use by the U.S. Food and Drug Administration (FDA) in 1989. It is made of latex or plastic, fits over the cervix, and is used with a spermicide. Its advantages over the diaphragm are that it can be worn longer and it is tighter-fitting and rarely leaks. Both the diaphragm and the cervical cap must be fitted by a physician.

Oral contraceptives that are designed to alter regulation of hormonal cycles are currently limited to use by women. The pill commonly contains estrogens or progesterone (or both). Pills are usually taken daily for 20 days, from the fifth to twenty-fifth day of the menstrual cycle, after which their use is discontinued for about eight days. During that time, menses (uterine bleeding) occurs, and then the menstrual cycle begins again. What is the basis of this hormonal strategy, which is currently used by over 8 million women in the United States? Essentially, the pill's hormones exert negative feedback to the hypothalamus, and perhaps the pituitary, that suppresses both LH and FSH secretions. Thus there is

Table 1 Methods of Contraception Most Commonly Used in Developed Countries

Method	Procedure	Effectiveness[a]	Advantages	Disadvantages
Sperm Prevented from Contacting Egg				
Withdrawal	Male withdraws penis from vagina prior to ejaculation	75–80%	No costs; no side effects	Unnatural action required; ineffective
Rhythm	No intercourse for several days around time of ovulation	65–85%	No costs; no side effects; no action required during intercourse	Requires careful studies and record keeping to establish pattern of ovulation; periods of abstinence required; relatively ineffective
Condom	Rubber or plastic sheath fitted over penis; sperm trapped, cannot enter the vagina	85–90%	No side effects; simple to use; prevents spread of sexually transmitted diseases; not permanent	Expensive; must be applied before intercourse; disrupts sexual activities; somewhat ineffective
Diaphragm (with spermicide)	Plastic or rubber cup inserted to cover entrance to uterus; sperm unable to enter uterus; kills sperm	90–95%	No side effects; reversible	Must be applied before intercourse; disrupts sexual activities; must be inserted correctly
Vaginal spermicide alone	Kills most sperm	60–80%	No side effects; simple to use	Must be inserted just before intercourse; may disrupt sexual activities; ineffec tive
Vaginal douche	Kills or inhibits sperm	50–70%	Inexpensive; simple; can be used after intercourse	Must be done immediately following intercourse; ineffective
Suppression of Sperm or Egg Production or Release				
Oral contraceptive ("pill")	Prevents follicle maturation; suppresses ovulation	95–100%	Effective; reversible; simple	Expensive; increased risk of blood clots, heart disease; must take daily; possible short-term effects including weight gain, nausea
Vasectomy	Vas deferens cut and tied; prevents release of sperm	99–100%	Effective; simple	Usually irreversible
Tubal ligation	Uterine tubes cut and tied; ovum fails to reach uterus	99–100%	Effective	Usually irreversible

[a]The percentage of sexually active women who did not become pregnant when using the method for one year.

no LH-FSH surge, and follicle maturation and ovulation are prevented. Oral contraceptives have been linked to an increased risk of certain vascular (blood-clotting) disorders and perhaps other side effects. However, in women under 35, the risk is about 25 percent or less than the risk of death due to complications of pregnancy and parturition.

Is it possible to devise a hormonal contraceptive for males? Before 1990, most research directed toward this goal involved attempts to lower testosterone levels or block their spermatogenic effect. Unfortunately, this caused unacceptable effects on other target organs that resulted in decreasing muscle mass and a loss of sexual urge. However, in 1990, results of a study were published in which males had been injected weekly with a synthetic variant of testosterone. Consequently, sperm production diminished, presumably because of negative feedback. Only one pregnancy occurred among 157 couples who participated in the study, and no serious side effects were observed. Additional studies involving this synthetic hormone will be done in the next few years.

The two surgical methods, vasectomy in males and tubal ligation in females, are considered 100 percent effective, although a rare pregnancy may result in cases where the tubes are not completely separated or reattach spontaneously. They have no effect on sexual desire or intercourse, and both are now considered to be relatively risk-free operations. The major drawback is that they are usually irreversible. However, new surgical techniques have had modest success in restoring the severed ducts.

Another form of birth control is induced abortion, which entails removing the embryo from the uterus. Written records over 4,000 years old indicate that abortion has been used for centuries to terminate pregnancies. The use of this technique as a birth control measure is very controversial, even though in some parts of the world it is the predominant method of birth control. The moral issues surrounding abortion continue to be debated, and in the United States, no consensus is expected soon.

Several questions are relevant to individuals, or couples, in evaluating the various methods of contraception available. Is the method effective in preventing pregnancy? How effective? Is it safe for the individual using the method, or are risks involved? What are the risks? Is it reversible? Is it convenient, or does it interfere with the enjoyment and spontaneity of sexual activities for one or both partners? No existing method provides the desired answers for all these questions. Thus individuals must use personal criteria, balancing advantages and disadvantages against the possibility of an unwanted pregnancy, in selecting an appropriate method for their own use.

As described in an important 1990 report from the National Academy of Sciences, the consensus among health officials is that the United States has remained in the Stone Age when it comes to the development and use of advanced birth control technologies. In countries throughout the world, new forms of contraception have been developed and used. For example, women in Finland, Sweden, and several other European countries have access to a process in which a contraceptive (hormone-releasing) device is implanted under the skin of the upper arm. A similar under-skin implant system (Norplant) was finally approved by the FDA in 1991 for use in the United States. The implant method protects against pregnancy for up to five years. Germany, China, and Mexico manufacture an injectable contraceptive that prevents pregnancy for one to two months. The French have RU 486, which is described in the next paragraph. All of these advanced methods have been found to be convenient, safe, effective, and relatively inexpensive. The NAS report also stressed that more contraceptive choices for Americans would result in fewer abortions. Available data indicate that contraceptive failure now accounts for at least half of all abortions in the United States each year.

In 1986, a new birth control substance, called *RU 486,* was introduced in France. The development of this compound dramatizes how advances in biomedical research can create new social problems. The standard treatment, which usually begins within ten days after a missed menstrual period, consists of a woman taking three RU 486 pills, followed by a small amount of progesterone 48 hours later. RU 486 binds to progesterone receptors of the uterus. Without progesterone, the endometrium deteriorates and, along with the embryo, is expelled. Since studies indicated that RU 486 was very effective, caused few side effects, and did not impair future fertility, its use was approved in France. Subsequently, considerable controversy has arisen over its use in France and its possible approval for use in other countries.

The primary social issue associated with the use of RU 486 is familiar but involves new definitions of terms. Does this drug induce menstruation or abortion? In general, groups in the United States and other countries that oppose abortion do not support the drug's approval for general use under any circumstances. These groups have exerted considerable pressure on the manufacturer of RU 486 and also on researchers interested in studying all potential uses of the drug. Groups and individuals for whom abortion is an option for terminating pregnancy favor allowing access to RU 486 as quickly as possible. Both groups are currently using economic threats against the parent company, and it is not yet clear if, or when, RU 486 will be made available in other countries. Most developed countries, such as the United Kingdom and the United States, require extensive testing and clinical trials prior to approving any drug for use. Such requirements have not yet been completed in the United States. Given the intense debate over abortion in this country, approval for use from the FDA is not expected soon. It remains to be seen whether or not other new contraceptive technologies will be approved for use.

Whether new contraceptive technologies are developed or not, their use will still be subject to moral discussion. Even for people who approve of them, their use will be dependent on knowledge of all the available alternatives and how to gain access to them. The United States has the highest rate of unplanned pregnancies among the industrialized countries of the West. It also lags behind those other countries in the dissemination of relevant information in a formal way, through the existing educational system.

fore, it cannot be allowed to degenerate as it normally does during the regular menstrual cycle. Can you hypothesize about how this might be accomplished, knowing that hormones play a key role? High levels of estrogens and progesterone are required to maintain the endometrium and also to suppress FSH and LH production, which prevents development of any new follicles during pregnancy. For the first few weeks, supplemental quantities of these hormones are secreted by the corpus luteum, a temporary endocrine gland that remains intact if pregnancy occurs. Two other unique, temporary tissues also secrete hormones that are critical during pregnancy. The *chorion* is the outer membrane of the embryo, and it secretes *human chorionic gonadotropin* (hCG), a hormone that helps sustain the corpus luteum during the first months of pregnancy. The presence of this hormone in urine can be detected and is used in simple tests to confirm pregnancy.

The Placenta

The *placenta* is a vascular organ that connects mother and fetus and secretes estrogens and progesterone throughout pregnancy (see Figure 37.15). (An embryo becomes a **fetus** after the second month of development.) It is a complex organ that serves as a lifeline for the developing fetus from the third month of pregnancy until birth. The placenta is a combination of maternal and fetal tissues. During placental formation, the chorion develops numerous fingerlike extensions called *chorionic villi*. As these villi develop, they erode limited areas of the endometrium, creating microscopic lagoons of maternal blood. Eventually, tiny arteries and veins from the fetus grow into the villi, each of which is surrounded by a pool of maternal blood. The structure containing fetal arteries and veins that connects the fetus with the placenta is called the **umbilical cord**. Blood vessels from the mother are also connected with these blood pools in the placenta, but they are not in direct contact with the blood vessels of the fetus. The pools serve as common reservoirs through which waste substances from the fetus, such as carbon dioxide, and nutrients and oxygen contained in the mother's blood can be exchanged.

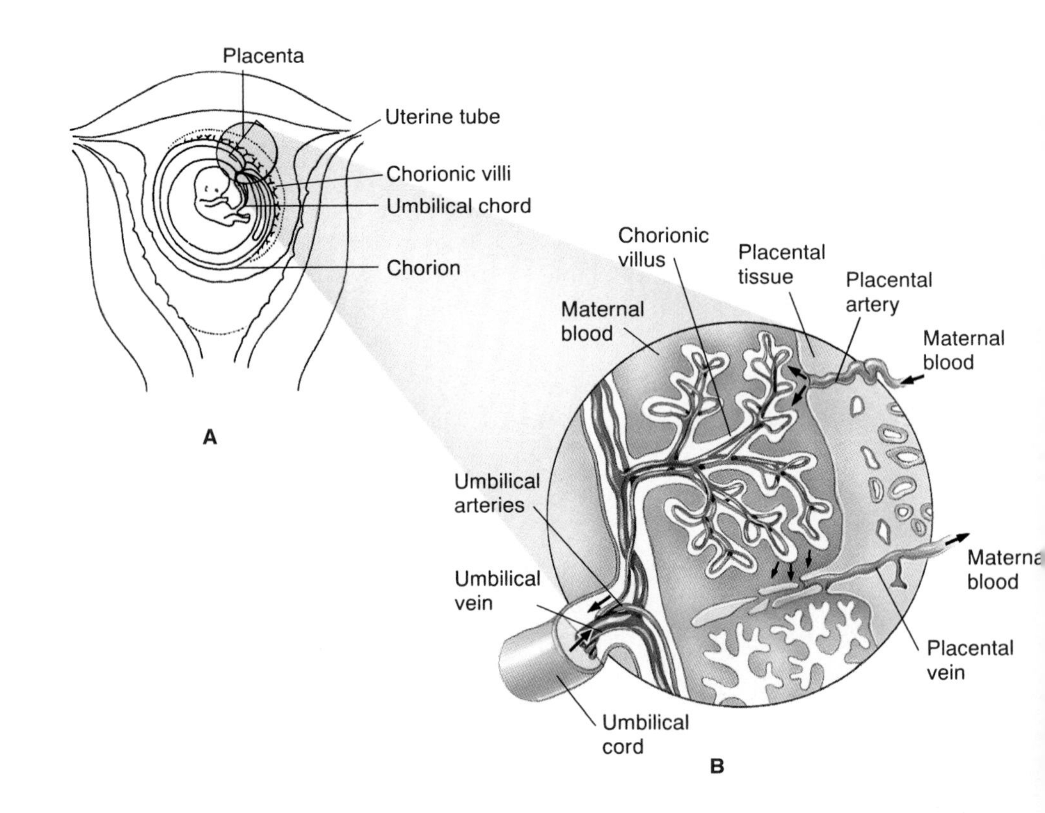

Figure 37.15

(A) The umbilical cord contains arteries and veins to and from the fetus and connects the fetus with the placenta. (B) In the placenta, nutrients, waste products, carbon dioxide, and oxygen are exchanged between blood in vessels from the mother and fetal vessels in the umbilical cord.

During pregnancy, the placenta produces progesterone, which is important in maintaining the uterus, inhibiting GnRH secretion, and stimulating mammary development. The placenta also produces estrogen, but its synthesis and secretion are exceedingly complex, requiring substances from both the fetus and the mother. The full details about this process are not yet known.

After fertilization takes place, the resulting zygote begins a series of transformations that ends nine months later when the final product, a baby, is born. Within one week, the zygote develops into a blastocyst that becomes implanted in the endometrium, where it will remain for two months and develop into an embryo. After two months, the fetus becomes connected with the placenta, and the mother, through the umbilical cord. Hormones control the development and maintenance of all these structures.

DEVELOPMENT

Development is a general concept that encompasses all irreversible changes that occur throughout an individual's lifetime. The major developmental phases in a human life are commonly classified as embryonic development, infancy, childhood, and adulthood, this last of which is a lengthy stage with few profound changes that ends with the death of the individual.

In this chapter, *development* refers to the systematic increase in complexity that occurs from the time of fertilization until birth. How does the zygote, a single cell, give rise to a new human being consisting of about 200 billion highly organized cells within a period of only nine months? This question can be broken into three parts. First, what cellular events occur during development? Second, how are these events initiated and regulated at the molecular level? Third, how do cell aggregates form specialized tissues and organs? The answers to the first and third questions are quite well understood. Extensive, ongoing research is slowly enabling scientists to close in on the answers to the second question.

Embryonic Development

The events that occur during the first three months of development are staggeringly complex. At the most fundamental level, a single-celled zygote undergoes cell divisions (by mitosis), a process that eventually produces millions of cells. Through controlled—but poorly understood—organizational processes, these cells become the normal tissues and organs that make up a human being. There are three stages of embryonic development, referred to as *cleavage, gastrulation,* and *organogenesis.*

Cleavage

Cleavage is a series of rapid mitotic divisions that gives rise to a sequence of structures (see Figure 37.16). Cleavage begins almost immediately after the zygote is formed and continues during its passage down the uterine tube to the uterus. During cleavage, the embryo does not increase in size; rather, the cells produced simply do not grow after division. Thus after mitosis, the resulting two daughter cells are half the size of the mother cell.

By day 4 after fertilization, a solid ball of cells called the *morula* is formed. As further divisions occur, additional cells accumulate in the morula. By the sixth day, some of these cells become rearranged, forming a hollow, fluid-filled *blastocyst* consisting of an *inner cell mass,* a cluster of cells at one end, and a single layer of surrounding cells called the *trophoblast.* After one week, the trophoblast of the developing embryo burrows into the wall of the uterus, where it ultimately forms the embryo's portion of the placenta.

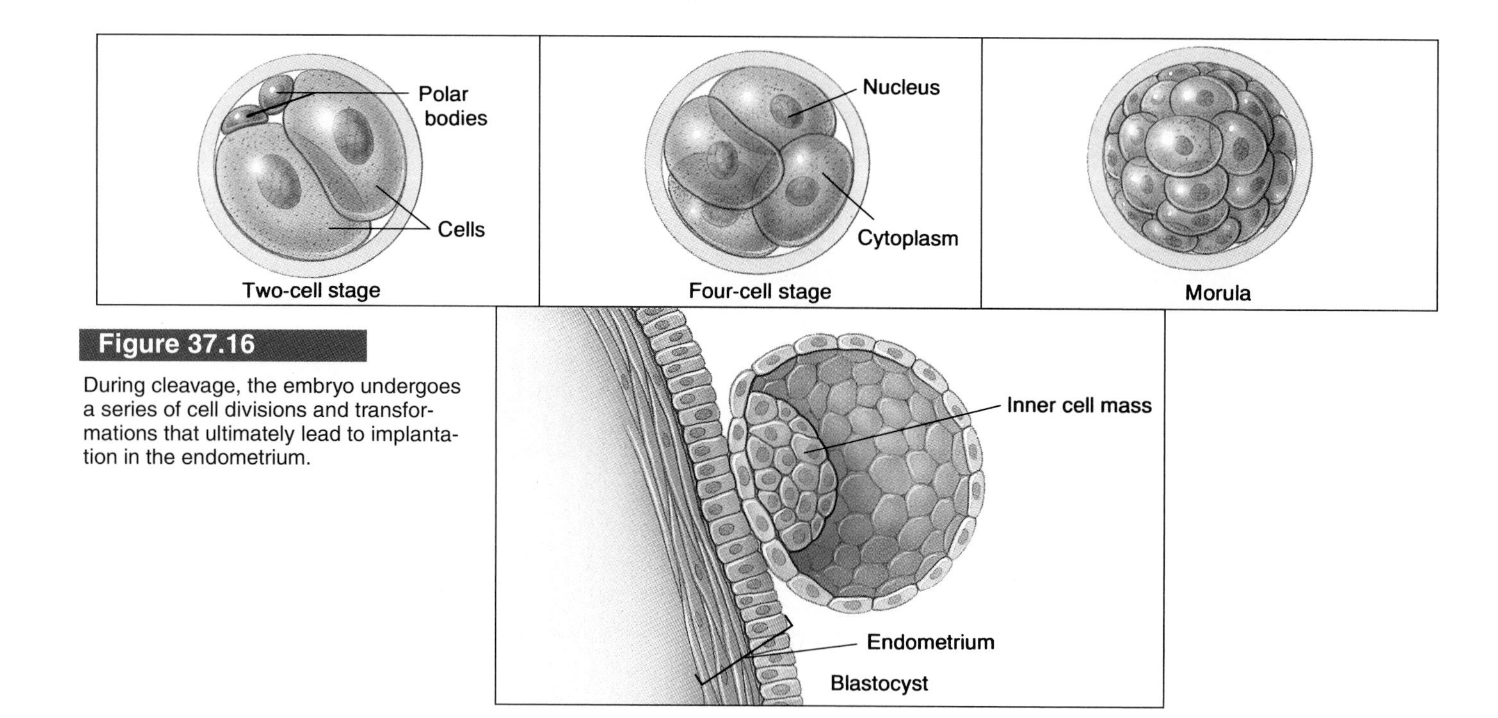

Figure 37.16

During cleavage, the embryo undergoes a series of cell divisions and transformations that ultimately lead to implantation in the endometrium.

Gastrulation

By the third week, **gastrulation** occurs, and the cells undergo a series of migrations that lead to a dramatic remodeling of the embryo, described in Figure 37.17. The inner cell mass becomes the *embryonic disk,* which will form three layers of cells, the *ectoderm, mesoderm,* and *endoderm*. These cell layers are called the *primordial germ layers,* and during the next few weeks, each gives rise to a specific set of tissues and organs.

During this period, before the placenta has developed, four *extraembryonic membranes* form in the developing embryo. The membrane surrounding the cavity between the trophoblast and the inner cell mass is the fluid-filled *amnion*. This structure eventually surrounds the fetus, and the fluid it encloses (the "water" released during birth) cushions the developing fetus. The *yolk sac* emerges as a small cavity below the embryonic disk and eventually becomes incorporated into the umbilical cord. Despite its name, it provides no nutrients to the developing human embryo (although this is its major function in nonmammalian vertebrates). This membrane is the source of germ cells that will later migrate to the developing gonad and give rise to gametes. The *allantois,* a third embryonic membrane, is also found in the umbilical cord and may be involved in the transport of wastes from the embryo and oxygen and nutrients from the mother. The fourth membrane, the *chorion,* develops from the trophoblast and completely surrounds the embryo. As described previously, this membrane secretes hormones essential for maintaining the corpus luteum.

Organogenesis

The process through which primary organs are formed in the developing embryo is called **organogenesis**. During this period, which lasts for about five weeks, or until the embryo reaches its eighth week, cells of the primordial germ layers differentiate, becoming specialized cells that form major tissues and organs (see Figure 37.18). Perhaps most dramatically, the sex of the offspring is established about six or seven weeks after fertilization. In humans and other mammals, current evidence indicates that the embryo develops into a male due to the action of a gene called *SRY* located on the Y chromosome (see Chapter 17). This gene is thought to encode a protein that transforms sexually neutral embryonic tissues into the testes and other organs that characterize the male phenotype. If this pro-

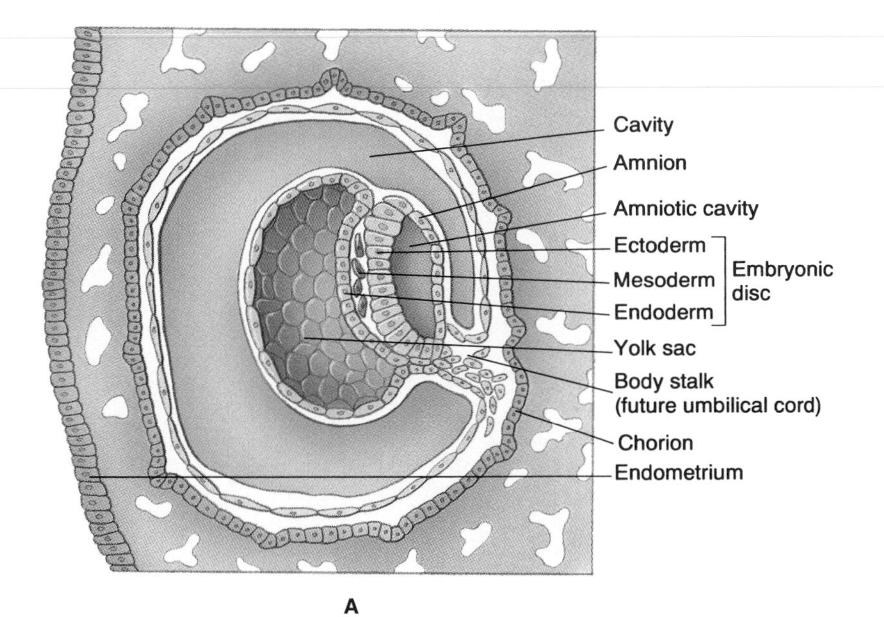

A

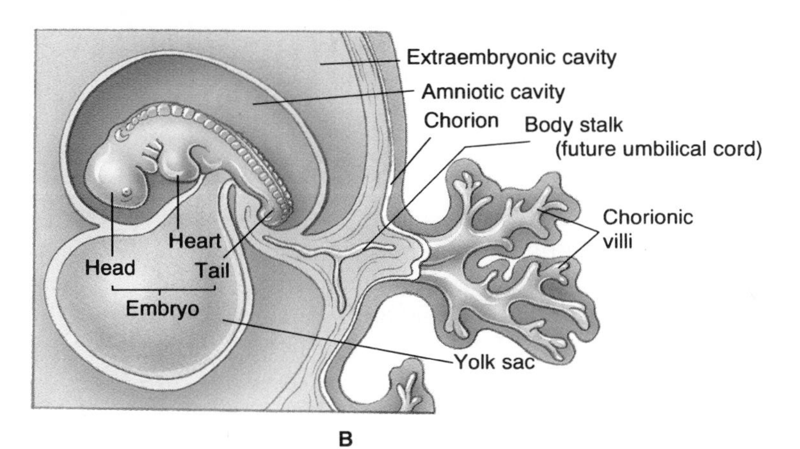

B

Chorionic villi
Allantois
Umbilical cord
Yolk sac
Amniotic cavity
Amnion
Chorion
Cavity of uterus

C

Figure 37.17

(A) By the third week of development, gastrulation has taken place, and three primordial germ layers—the ectoderm, endoderm, and mesoderm—have originated in the embryo. Each of these three germ layers will give rise to specific tissues and organs as organogenesis continues during the next few weeks. (B) The developing embryo looks animallike three to four weeks after fertilization. (C) Four extraembryonic membranes—the amnion, yolk sac, allantois, and chorion—develop during gastrulation.

Figure 37.18

The three embryonic cell layers—ectoderm, mesoderm, and endoderm—each give rise to different tissues and organs during development.

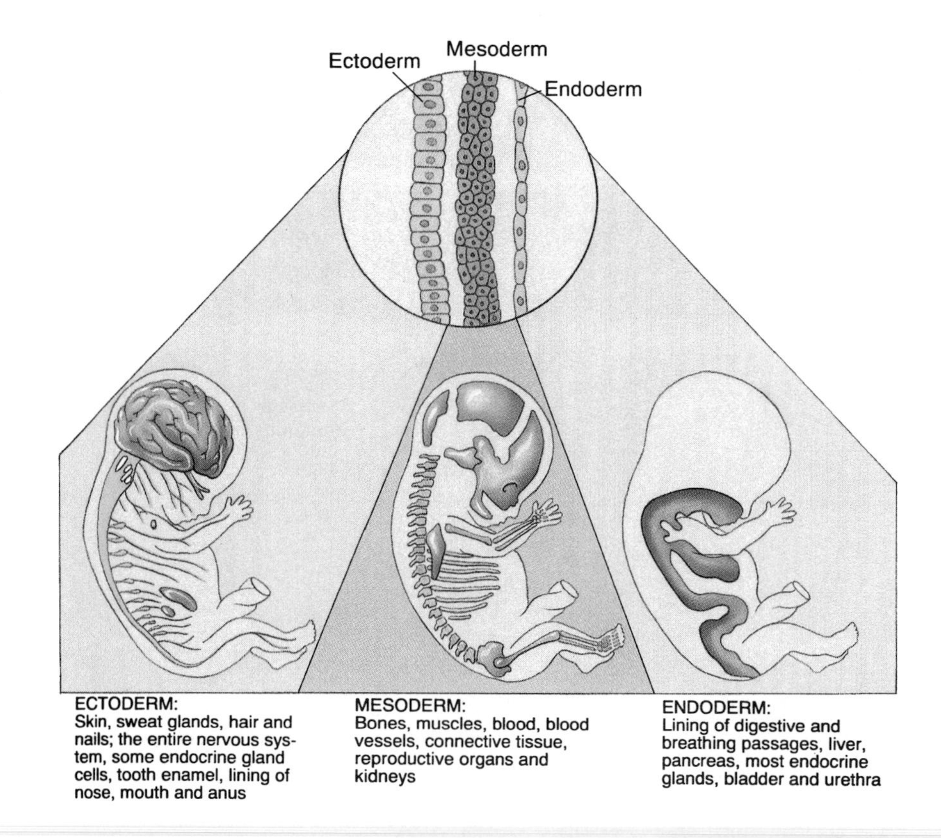

tein is not present, the embryo follows an alternate, longer pathway that leads to the development of a female phenotype between the thirteenth and sixteenth weeks. By the sixteenth week, the brain, spinal cord, eyes, limbs, heart, and most other organs can be identified (see Figure 37.19).

Fetal Development

The second and third trimesters are characterized by spectacular growth of the fetus. By the fourth month, a mother can detect movements of the fetus. The respiratory and circulatory systems have appeared, although they will not be functional for another month or two, which severely limits the chances of survival if birth were to occur before the sixth month. By the sixth month, the fetus is about 5 centimeters long and weighs 700 grams. Scalp hair is visible, the fetus may suck its thumb, and the first bones (ribs) develop. By the seventh month, the fetus weighs around 1 kilogram. The nervous system has developed to the point where controlled breathing can occur if birth takes place, although the chances of survival would still be low. During the eighth and ninth months, all essential organs become fully developed, and the weight of the fetus doubles.

Embryonic development occurs during the first three months of life. In the first three weeks, the fertilized egg undergoes cleavage. Gastrulation then occurs, characterized by the formation of ectoderm, mesoderm, and endoderm, the three cell layers that ultimately give rise to all tissues and organs by the third month of pregnancy. During the last six months of pregnancy, all fetal organs undergo rapid growth.

BIRTH

Parturition, the process by which the fetus is ousted from the uterus about 265 days after fertilization, is a complicated process. As in other reproductive events, a number of different hormones influence this dramatic separation of the fetus from the mother.

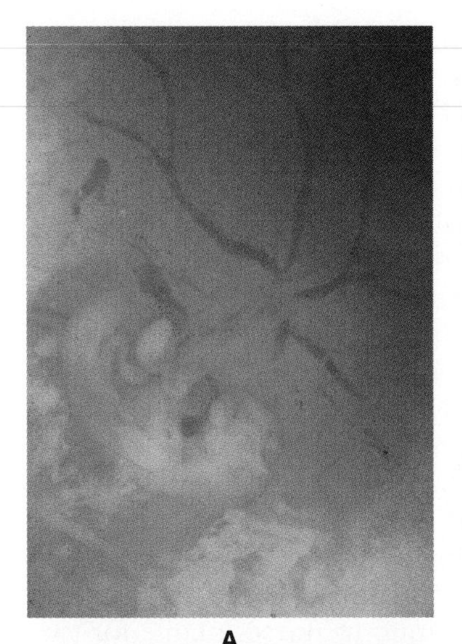

A

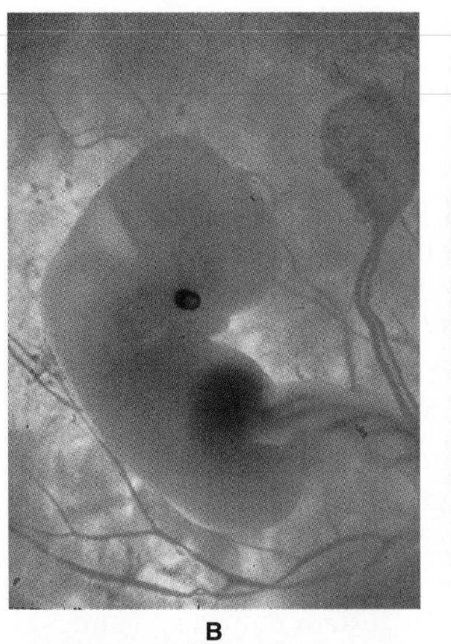

B

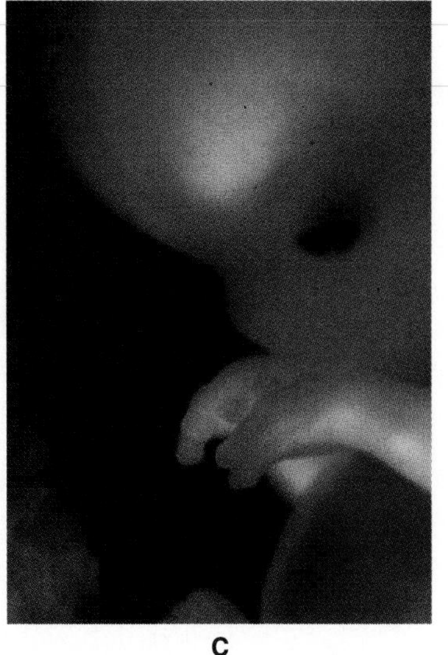

C

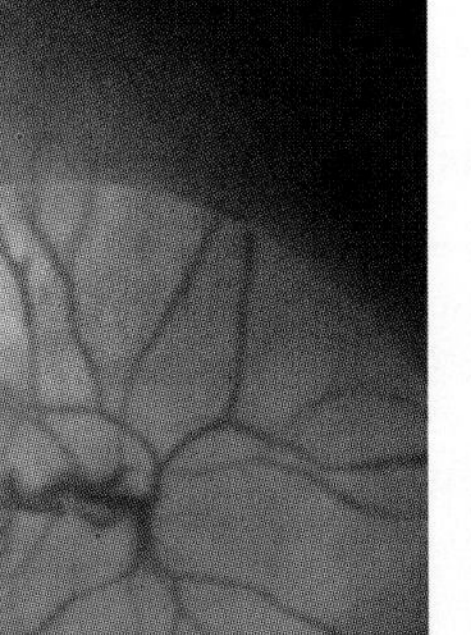

D

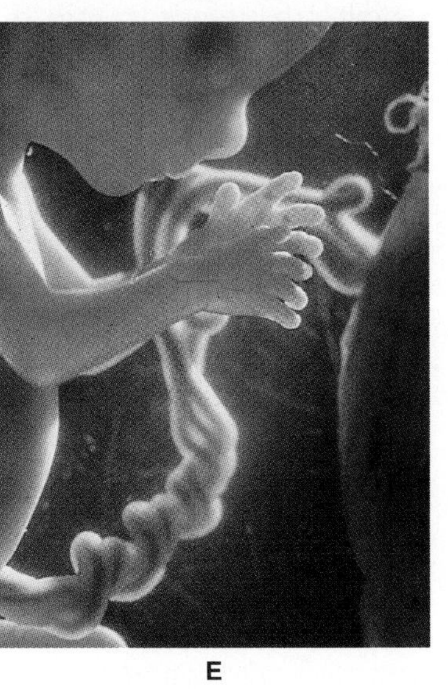

E

Figure 37.19

The sequence of human development during first four months of pregnancy. An embryo 4 weeks after fertilization (A), at 6–7 weeks (B); a fetus at 10 weeks (C), at 12 weeks (D), and at 4 months (E).

The onset of *labor,* the period when strong uterine contractions ultimately lead to expulsion of the fetus from the uterus, is thought to be triggered by three hormones. The fetus initiates the "contraction cascade" by releasing the hormone *oxytocin,* which travels through the umbilical cord and binds to receptors on the placenta and uterus. Oxytocin induces the placenta to produce prostaglandins that act to stimulate muscle contractions in the uterus. As uterine contractions continue, moving the fetus down the birth canal, nerve impulses are transmitted from the uterus to the hypothalamus. These impulses lead to the release of more oxytocin, this time from the mother's pituitary gland, which further increases uterine contractions. The cycle continues to escalate until the fetus is finally ejected from the birth canal. This cyclical activity, involving oxytocin, uterine contractions, and hypothalamus-pituitary stimulation, is a rare example of a positive-feedback mechanism.

Before birth, *relaxin,* a hormone whose origin is unclear, acts to increase the flexibility of joints between bones around the birth canal. This allows enlargement of the birth canal and enables the baby to pass through, usually head first. Since relaxin is present in the blood as early as the fourth week of pregnancy, it has been hypothesized that relaxin is produced by the corpus luteum and then stored within certain blood cells until release near the time of parturition.

At birth, several changes take place in the fetus that enable it to make a successful transition from living in an aquatic environment to becoming an air-breathing animal. Most notably, blood enters capillaries in the newborn infant's lungs and causes them to expand and begin functioning (confirmed by a crying response). A few minutes after delivery, the uterus contracts, and the placenta separates from the uterine wall and is expelled as the *afterbirth*. The uterus and its membranes then undergo repair and, if the mother does not nurse the child, menstrual cycles usually resume after six weeks. Nursing inhibits resumption of the menstrual cycle through the actions of *prolactin,* a pituitary hormone secreted during pregnancy that has diverse reproductive roles. This hormone stimulates mammary gland growth and development during pregnancy, and after birth, it induces *lactation,* the secretion of milk. In addition, it inhibits release of LH and probably FSH, which in turn leads to repression of ovulation and the menstrual cycle. In certain cultures, women continue to nurse infants for two years or longer, a practice that tends to spread the ages of siblings, especially if no other form of birth control is used. However, this natural method of birth control is highly unreliable since the length of time that the menstrual cycle remains repressed varies greatly among individuals.

Parturition begins when oxytocin and prostaglandins initiate uterine contractions, which first cause the fetus to move into the birth canal and then force its entry into the outside world. After birth, the placenta is expelled, and the uterus and other maternal tissues are restored. The ovarian and menstrual cycles begin six weeks later if the mother does not nurse the child. Otherwise, they will resume soon after nursing is discontinued.

CELLS AND DEVELOPMENT

As in many fields of the life sciences, the cardinal problems of developmental biology are now being successfully investigated using modern molecular methods. To date, much of what has been learned has come from studies of animal models including fruit flies (*Drosophila*), frogs, and rats. For much of the past century, the central question has been, How does a single-celled zygote give rise to an organism that consists of billions of cells, each with distinct capabilities, which are organized perfectly into different tissues and organs?

Much is now known about the development of *cell specificity,* or how a cell ultimately becomes specialized to perform a precise activity or function. The processes by which a single cell, the fertilized egg, ultimately produces an entire organism by mitosis and cell specialization are known, at least in outline form. For example, genes that determine the course of development of certain groups of cells have been identified in some animals. The pathways of embryonic development, including the genes responsible for creating an adult organism, have been determined in *Drosophila*. However, important questions about molecular events during embryonic development remain open. How are signals transmitted between embryonic cells during development? How do genes or gene products control the development of cells with defined sizes, shapes, and capabilities?

At the other end of the life span, why do cells ultimately fail, causing death? What causes aging? Evidence generally supports a *genetic clock* hypothesis, the view that all species contain unique genetic instructions that dictate its course of aging. Every carefully studied species has been found to have a defined average life span, and all appear to follow an unalterable, programmed timetable that leads to death. For example, mice normally live about 2 years, beagle dogs 10 to 15 years, humans 75 to 100 years, and Galápagos turtles for perhaps hundreds of years. Why does the effectiveness of our tissues and organs begin to decline at about age 30? During the autumn of our lives, do "aging genes" become activated? Do they produce a protein, an aging substance, that disrupts cell function or prevents cell division? Evidence suggests that this does occur. Cell development and mechanisms of aging are fascinating, challenging fields that will continue to occupy scientists well into the next century.

Summary

1. Humans produce offspring by sexual reproduction. Distinguishing features of human reproduction include internal fertilization, year-round sexual receptivity, unique behaviors, and complex endocrine mechanisms that are responsible for sexual maturity and the regulation of gamete production.
2. Testes are the primary sex organ of males. Once a male becomes sexually mature, the testes continuously produce sperm and androgens (hormones). Testosterone, the most important androgen, initiates the development of male secondary sex characteristics and sustains continuous production of sperm. Spermatogenesis is regulated by releasing factors (hormones) and hormones from the hypothalamus-pituitary complex.
3. Ovaries are the primary sex organs of females. At puberty, the ovaries begin releasing mature ova and various sex hormones that influence the development of secondary sex characteristics. Gamete production in females is characterized by monthly cycles. Ovulation occurs about once each month and is regulated by hormones secreted by the ovary and numerous releasing factors (hormones) and hormones from the hypothalamus-pituitary complex. Females continue to produce ova, on a monthly cycle, for about 35 years.
4. If the egg is fertilized and pregnancy occurs, the resulting embryo becomes implanted in the endometrium. By the third month of pregnancy, the fetus becomes linked with the mother via the placenta. From the third month of pregnancy until birth, various substances are exchanged between the mother and the fetus through the placenta. The placenta is maintained by the actions of various hormones.
5. During the first three months of pregnancy, major organs arise in the developing embryo through the processes of cleavage, gastrulation, and organogenesis. During the second and third trimesters, organs undergo rapid growth and become part of fully functional organ systems.
6. Pregnancy ends about nine months after fertilization. Parturition, and associated events such as labor, are controlled and coordinated by various hormones.

Review Questions

1. What are the significant features of human reproduction?
2. What are the major organs of the male reproductive system?
3. How are sperm produced in the male?
4. How is the male reproductive system regulated by hormones?
5. What are the major organs of the female reproductive system?
6. How are mature ova produced by the female?
7. Describe the major events in the ovarian and menstrual cycle. How are these two cycles regulated by hormones?
8. List the events that occur between fertilization and the embryo's becoming connected with the uterus.
9. What occurs during the following stages of embryonic development: cleavage, gastrulation, and organogenesis?
10. Which hormones are involved in parturition? What are their functions?
11. What questions are now guiding research in developmental biology?

Essay and Discussion Questions

1. Women often stop menstruating when placed under great stress or when not receiving adequate amounts of food. What might be the adaptive significance of this response?
2. In certain cultures (ancient and modern), people do not believe that sexual intercourse is responsible for producing children. Can you identify any biological facts that might account for such a belief?

3. If you could redesign the human body, what would you do to improve the human reproductive system?
4. Which human type would probably be most harmed by being exposed to mutagens (chemicals or physical agents that cause mutations): adult male, adult female, embryo, or fetus? The least sensitive? Why?

References and Recommended Reading

Baulieu, E. 1989. Contragestion and other clinical applications of RU 486, an antiprogesterone at the receptor. *Science* 245:1351–1357.

Edelman, G. M. 1988. *Topobiology: An Introduction to Molecular Embryology*. New York: Basic Books.

Gehring, W. J. 1985. The molecular basis of development. *Scientific American* 253:153–162.

Hull, R. T. 1990. *Ethical Issues in the New Reproductive Technologies*. Belmont, Calif.: Wadsworth.

Katchadourian, H. 1979. *Human Sexuality: Sense and Nonsense*. New York: Norton.

Kline, D. 1991. Activation of the egg by the sperm. *BioScience* 41:89–95.

Konner, M. 1982. She and he. *Science 82* (September): 54–61.

Marx, J. L. 1988. Sexual responses are—almost—all in the brain. *Science* 241:903–904.

National Academy of Sciences. 1990. *Developing New Contraceptives: Obstacles and Opportunities*. Washington, D.C.: NAS Press.

Rowe, J. W., and R. L. Kahn. 1987. Human aging: Usual and successful. *Science* 237:143–149.

Rubin, G. M. 1988. *Drosophila melanogaster* as an experimental organism. *Science* 240:1453–1459.

Ulmann, A., G. Teutsch, and D. Philibert. 1990. RU 486. *Scientific American* 262:42–48.

CHAPTER 38

Human Defense Systems

Chapter Outline

Reading Questions

1. What are the two general human defense systems?

2. How do nonspecific defense mechanisms prevent infection and respond to injury?

3. What is the significance of self and nonself in immunity?

4. What is the function of cellular and humoral immune responses? When do they occur? How do they operate?

5. What problems associated with the immune system are of current interest to researchers?

Throughout their lives, all organisms are constantly threatened by biological, chemical, and physical agents from the surrounding environment. Most commonly, these external agents are microorganisms attempting to colonize an environment that can provide them with nutrition and a suitable place to grow and multiply. The cells and tissues in a multicellular organism's body provide a nearly ideal habitat in which bacteria, viruses, fungi, and *parasites* (protozoans and worms) can thrive. However, through the course of evolution, potential host organisms have developed mechanisms for defending themselves against invaders. In this chapter, we examine defense systems that usually keep us safe from such unwelcome guests.

DEFENSE SYSTEMS

Human defense systems are separated into two general categories that are summarized in Table 38.1 and Figure 38.1 on page 718. *Nonspecific defense mechanisms* always operate in the same way when presented with a challenge. They include mechanical barriers, secreted products, and the inflammation response. In contrast, *specific immune responses* involve a complicated adaptive system composed of various cells, cell products, and control mechanisms. The immune system has great flexibility and responds in a specific way to each unique provocateur that enters the body.

The immune system is fascinating to explore. It embodies intricacies that offer a unique perspective on the development and success of an advanced system that evolved in higher organisms in response to natural selection pressures. Of all animals, only birds and mammals have the elaborate immune system described in this chapter.

Human defenses consist of nonspecific defense mechanisms that are inborn and conventional, reacting in the same way to all initiating agents, and specific immune responses that create a customized reply to each challenge.

NONSPECIFIC DEFENSE MECHANISMS

Figure 38.2 on page 718 identifies the major nonspecific defense mechanisms. Most materials (microorganisms, chemicals, and physical agents such as splinters) that pose a threat to an individual's internal homeostatic mechanisms gain access to the body from the external environment.

Body Surfaces

Routes of entry by which potentially dangerous substances may pass from the external to an internal environment include penetration or absorption through the skin, eyes, or ears and the digestive, respiratory, urinary, and reproductive systems. Our first line of defense consists of organs, tissues, and cell products that form barriers and prevent or limit access at these portals.

Skin and Mucous Membranes

The basic structure of the skin was described in Chapter 34. The skin is a complex organ that covers the external surface of the body. *Keratin,* the tough protein found in the skin's outer cell layers, is resistant to water, some chemicals, and enzymes secreted by certain bacteria. Even though it is subjected to considerable abuse (for example, exposure to sun, soaps, and cosmetics) and is frequently damaged, it has a remarkable ability to repair itself. Because of this property, it generally remains intact and thus constitutes a highly effective, dependable defense barrier.

Table 38.1 Human Defense Systems

Type	Components	Functions
Nonspecific Defense Mechanisms		
Body surfaces	Skin	Barrier to invasion by external agents
	Sweat and sebaceous glands	Secrete substances toxic to many bacteria
	Mucous membranes	Barrier to invasion by external agents
	Secreted products	Mucus traps small particles; tears in eyes contain an enzyme with antimicrobial activity; gastric juice in stomach destroys bacteria and bacterial toxins
Inflammation	Leukocytes	
	Basophils (mast cells)	Phagocytosis of foreign materials; release histamine
	Eosinophils	Phagocytosis; important in parasite infections and allergic responses
	Macrophages (monocytes)	Phagocytosis; release pyrogens
	Neutrophils	Phagocytosis; produce factors that attract macrophages; release enzymes that digest foreign materials
	Natural killer cells	Kill foreign cells or modified self cells
	Chemical mediators	
	Histamine	Causes dilation of blood vessels
	Complement system	Mediates events in inflammation
	Chemical factors	
	Pyrogens	Increase body temperature
Specific Immune Responses		
Cellular immunity	Macrophages	Initiate immune responses
	T lymphocytes	
	T helper cells	Assist in antibody formation
	T cytotoxic cells	Kill cells with nonself antigens
	T suppressor cells	Diminish immune response
	T memory cells	Possess "memory" of a specific antigen
Humoral immunity	B lymphocytes	
	Plasma cells	Produce antibodies
	B memory cells	Possess "memory" of a specific antigen
	Antibodies	React with antigens

Mucous membranes are composed of an outer epithelial cell layer and an inner connective tissue layer. Many of the epithelial cells have cilia, and some secrete *mucus,* a substance that covers the surfaces of mucous membranes. Mucous membranes line body cavities and the surface of organs or tissues that open to the external environment. The digestive, reproductive, respiratory, and urinary systems and the eyes are all lined by these moist membranes. Like the skin, they form a barrier between the external and internal environments.

Figure 38.1

Nonspecific defense mechanisms generally prevent infections by blocking the entry of disease-causing agents into the body. If this first line of defense fails, disease results, and the specific immune system is activated. Specific immune responses ultimately eliminate foreign invaders, and recovery occurs. If the same agent reinfects the body, an immune memory process ensures that it will not cause disease a second time.

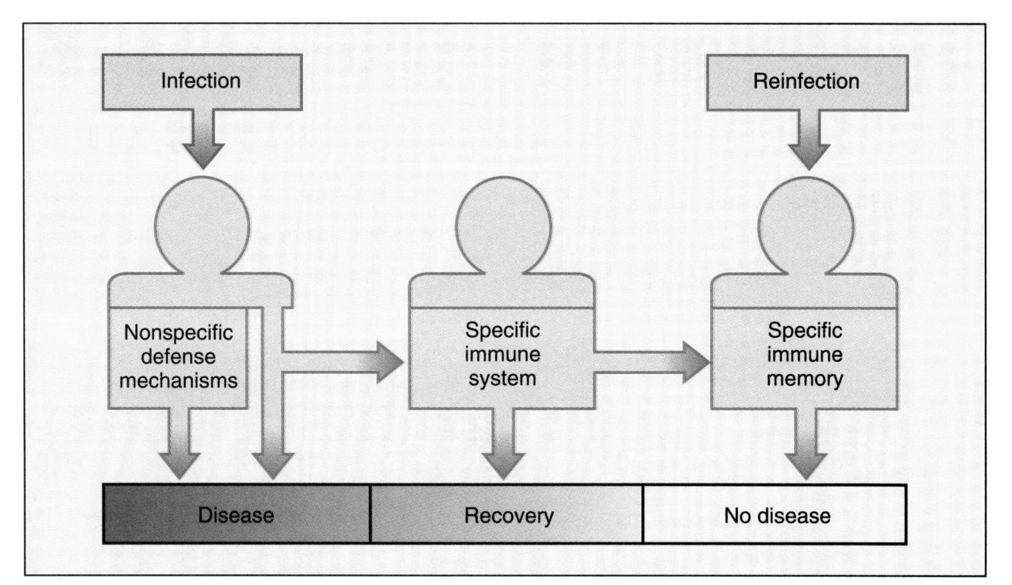

Figure 38.2

Nonspecific defense mechanisms include barriers to penetration (skin and mucous membranes) and chemical secretions and enzymes that are harmful to the infecting agent.

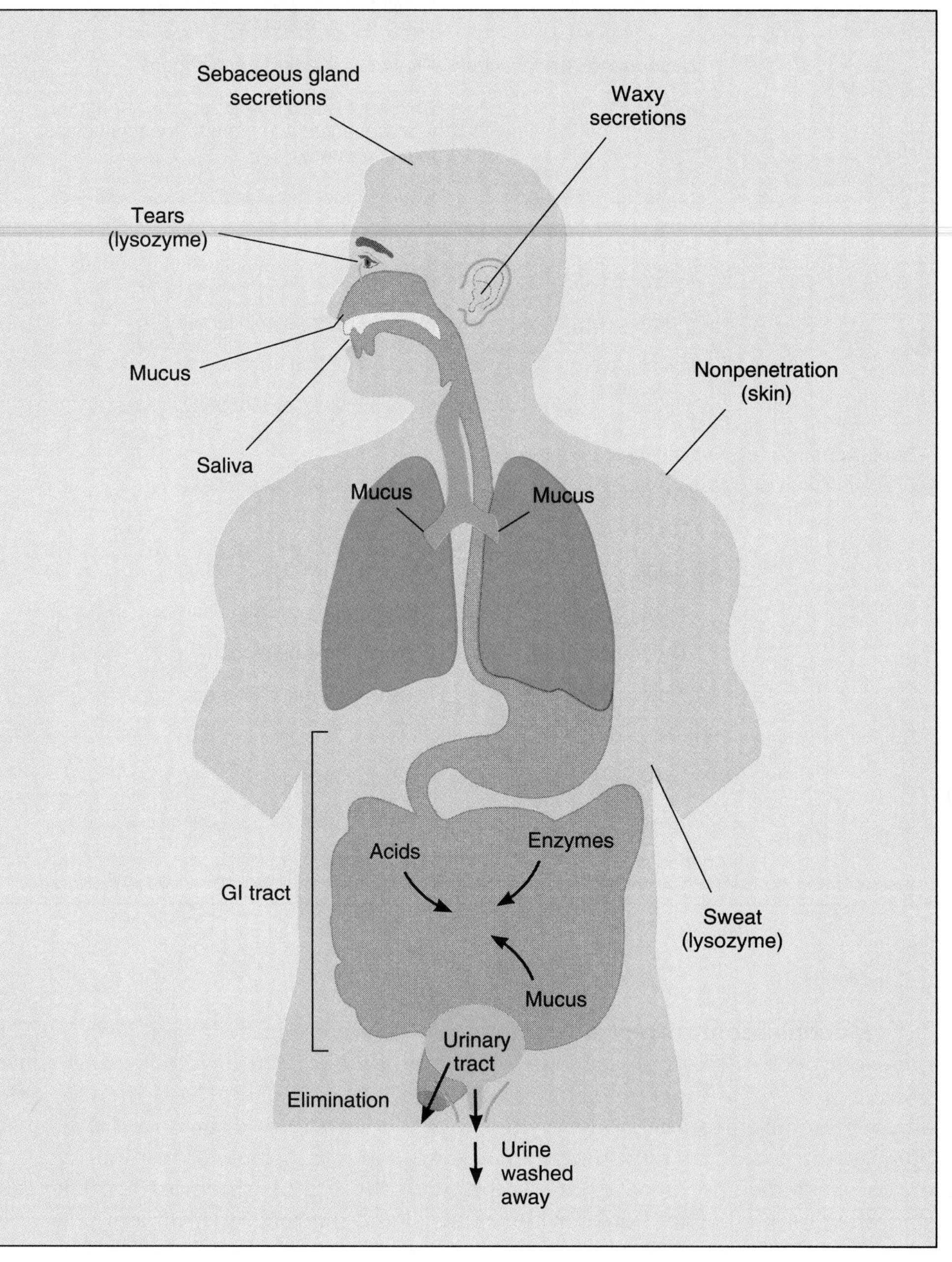

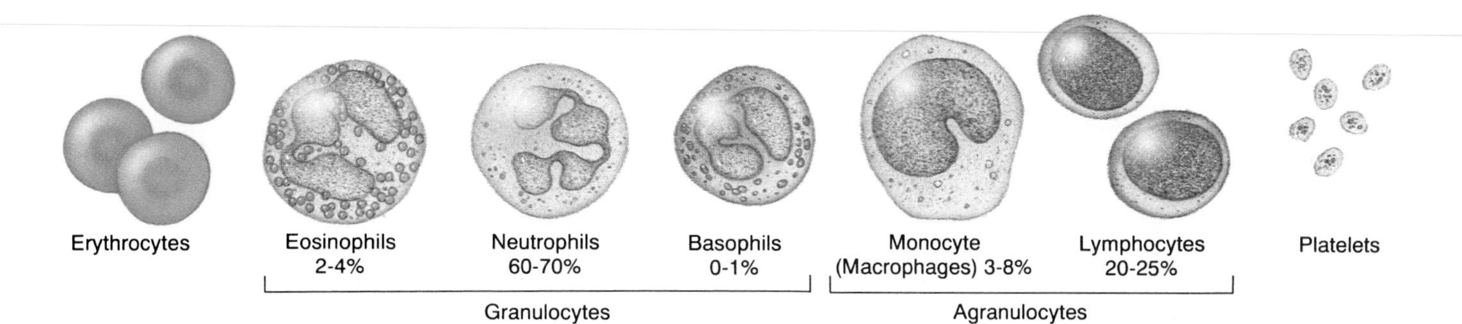

Figure 38.3

Platelets, erythrocytes, granulocytes, and agranulocytes are blood cells that play roles in nonspecific defense responses. Monocytes, which develop into macrophages, and lymphocytes are important in specific immune responses. The percentages indicate normal ranges of abundance for each cell type in blood.

Secreted Products

The skin contains glands that secrete substances that play a role in defense. *Sebaceous glands* secrete an oily substance (sebum) that prevents hair and skin from drying and is also toxic to many types of bacteria. *Sweat glands,* in addition to producing sweat, also secrete antimicrobial substances such as *lysozyme,* an enzyme that can break down the outer cell wall of many bacteria.

Mucus entraps particles that can then be swept away by cilia or removed by white blood cells capable of engulfing them. Saliva rinses foreign materials from surfaces in the mouth and also contains antimicrobial substances. Enzymes (lysozymes) are present in tears, which bathe the eyes, and in the gastric juices of the digestive tract; they undoubtedly destroy millions of bacteria every day of our lives.

Despite the remarkable effectiveness of the skin, mucous membranes, and special substances in their defense roles, they are often breached by unwanted materials. What protects us when that happens?

Blood Cells and Nonspecific Defense

Once an organism or foreign substance gains entry into the body, a marvelous array of blood cells and chemical regulators acts to repel and eliminate the invader. Under the control of chemical regulating substances, some blood cells react immediately, and others join in at different stages of the response as circumstances dictate. Before considering the responses, it will be helpful to identify the participants. Table 38.1 contains information on different types of blood cells, and they are shown in Figure 38.3.

Red blood cells (*erythrocytes*) were discussed in Chapter 34. **Platelets** are small cell fragments that are important in clotting reactions (see Figure 38.4). Humans have two general types of white blood cells (**leukocytes**) that can be

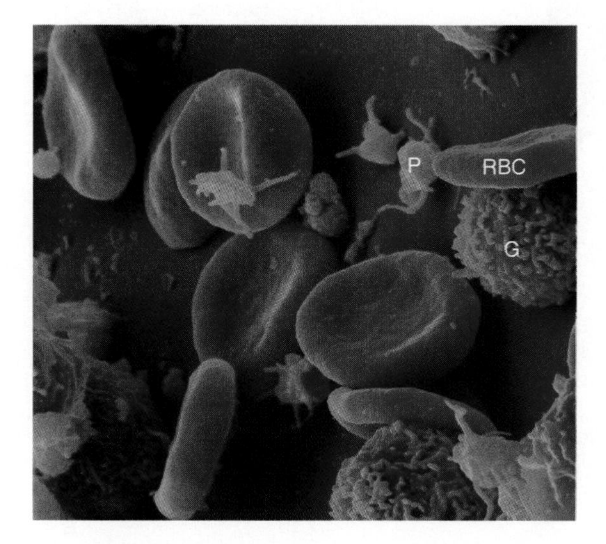

Figure 38.4

This scanning electron micrograph shows erythrocytes (RBC), platelets (P), and granulocytes (G).

distinguished microscopically as *granulocytes* and *agranulocytes*. **Granulocytes** possess granules in their cytoplasm. *Granules* are microscopic organelles that contain substances related to function. Three types of granulocytes can be distinguished by the color of their granules after they are stained with biological dyes and by the shape of their nuclei. *Neutrophils* have granules that contain enzymes used to digest microorganisms, *eosinophil* granules hold various substances used in defense reactions, and *basophil* granules contain factors that play a role in reactions against parasites and in allergic responses initiated by *mast cells,* a type of basophil found in some tissues.

In contrast to granulocytes, the cytoplasm of **agranulocytes** appears clear under the microscope. There are three types of agranulocytes that differ in size: *monocytes* develop into larger *macrophages,* and both are bigger than *lymphocytes*. A curious class of lymphocytes is referred to as *natural killer (NK) cells*. The full capabilities of NK cells are not yet known, but they play an active role in destroying virus-infected cells, fungi and parasites, and cancer cells that arise within the body.

One outstanding characteristic of granulocytes and macrophages is that they can carry out *phagocytosis,* a process of engulfing and destroying foreign particles (see Figure 38.5). This is the primary process by which microorganisms, dead cells, and cell fragments are removed from the body. Neutrophils and macrophages are voracious phagocytes, but basophils and eosinophils are only mildly phagocytic.

Most leukocytes have short lives, from a few days to a week or two. The relative abundance of leukocytes in the blood is used in diagnostic tests. When disease or infection occurs, these values change, and sharp increases of certain leukocytes may be used as the first indicator that something is wrong.

Nonspecific defense mechanisms act to reduce the chance of tissues being damaged by physical, chemical, or biological agents. The skin, mucous membranes, and secreted substances prevent or inhibit foreign substances from entering the body. Many types of blood cells carry out activities aimed at eliminating or killing invading agents that do enter the body.

Inflammation

Inflammation is an innate protective and defensive response that eliminates or prevents the spread of foreign materials at sites of injury and prepares the damaged tissue for repair. What comes to mind when you think of inflammation? The

Figure 38.5

(A) A macrophage's capture of a bacterium is the first step in phagocytosis. (B) Once ingested after capture, the bacterium is enclosed within a membrane vesicle that fuses with a lysosome. The lysosome releases enzymes into the vesicle that digest the bacterium but not the vesicle. The remaining material may then be released if the vesicle empties its contents to the exterior.

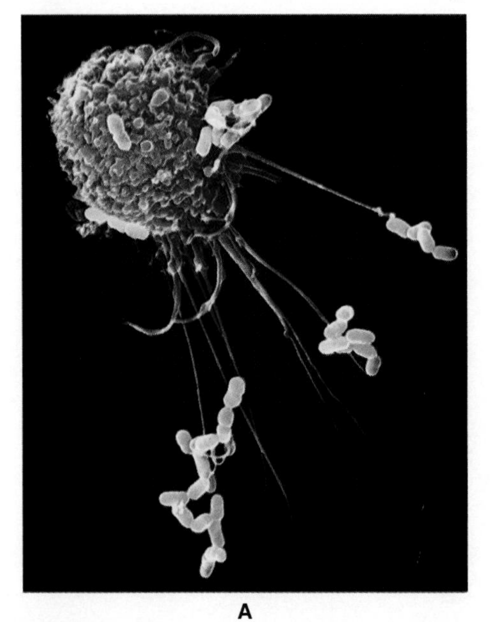

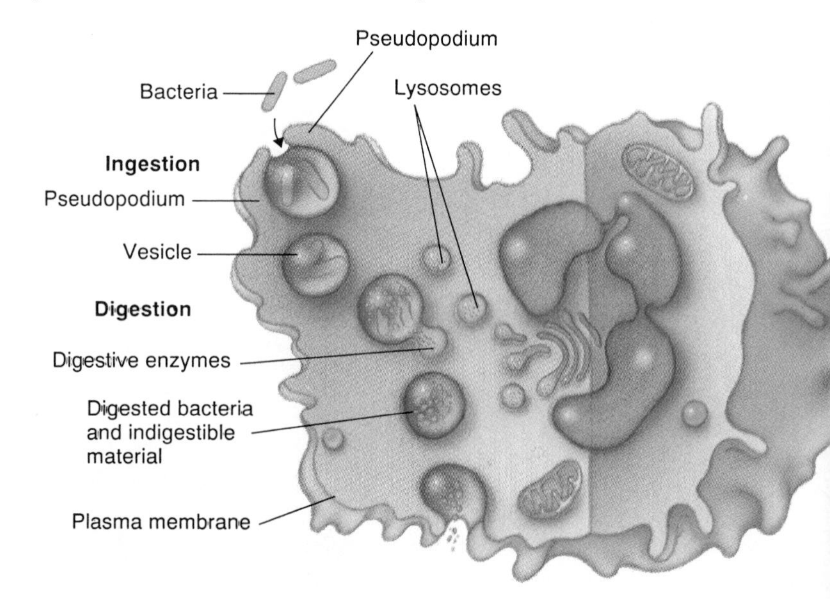

four cardinal signs of acute inflammation were described as long ago as 25 B.C. by Celsus, a famous Roman encyclopedist. They are redness, heat, swelling, and pain. A fifth classic sign, loss of function, was added by Virchow in the nineteenth century.

The inflammatory response consists of two phases. First, blood cells and plasma proteins enter tissues that have been injured or infected and act to limit damage. Second, tissue damage or wounds are cleaned up and later repaired. Thus being scratched by a cat, bruised in a fall, or infected by a bacterium may all initiate the following sequence of events:

1. Blood flow to the affected area increases (redness and heat).
2. Fluid accumulates at the site of injury (swelling).
3. Neutrophils invade the affected site (pain).
4. Lymphocytes and macrophages infiltrate.
5. The normal architecture of the affected tissue is restored (wound repair), if possible; function is lost, if not.

All of the inflammatory processes are controlled by interrelated feedback mechanisms. The magnitude of the responses varies according to the severity of the trauma. Hence only increased blood flow may occur, for example, if the skin is lightly scratched, or the entire inflammatory response may occur if the damage is serious.

The first three events on the list are part of the **acute inflammation response**. There are two key features of acute inflammation: it occurs in exactly the same way in all humans, no matter what the cause, and it happens quickly, beginning within seconds of the initiating event. Although a large number of chemical and cellular activities are involved in inflammation, our discussion will focus on the major events of an inflammatory response.

The Inflammatory Response: Minutes to Hours

Immediately following injury, blood vessels in the surrounding area dilate, and their permeability increases (see Figure 38.6A). As a result, blood cells flow quickly into the area.

When a serious injury occurs, a number of mechanisms, collectively called the **coagulation system**, are activated by chemical mediators to arrest bleeding. *Chemical mediators* are small molecules that either attract other molecules or cells to a site or signal them to perform some function. Platelets arrive on the scene and begin to plug leaks in small vessels of the vascular system. Blood *clotting* is a very complex process that requires a number of proteins and other factors present in the plasma. The most important mechanism involves two proteins; the first, *fibrinogen,* is converted into the second, *fibrin.* Fibrin is a threadlike protein that fashions a network that entraps erythrocytes, platelets, and other materials to create a blood clot. Formation of a blood clot, shown in Figure 38.6B, halts bleeding and walls off the damaged area. Later, the fibrin complex is dissolved by enzymes so that the area can be repaired.

Hours to Days

Increased blood flow results in sharp increases in the delivery to the site of injury of blood cells, plasma, fluids, and other circulating substances. Also, damaged and dead cells release their contents into this mixture. The **complement system** consists of a set of blood proteins—chemical mediators—that function during acute inflammation and participate in a cascade of reactions. They participate in breaking down invading organisms and dead cells, dilating blood vessels, attracting various leukocytes, and enhancing phagocytosis by leukocytes.

Neutrophils are generally the first leukocytes to migrate into damaged tissues. They are attracted by a number of chemical mediators released from basophils at the site and also by substances released from dead or dying tissue

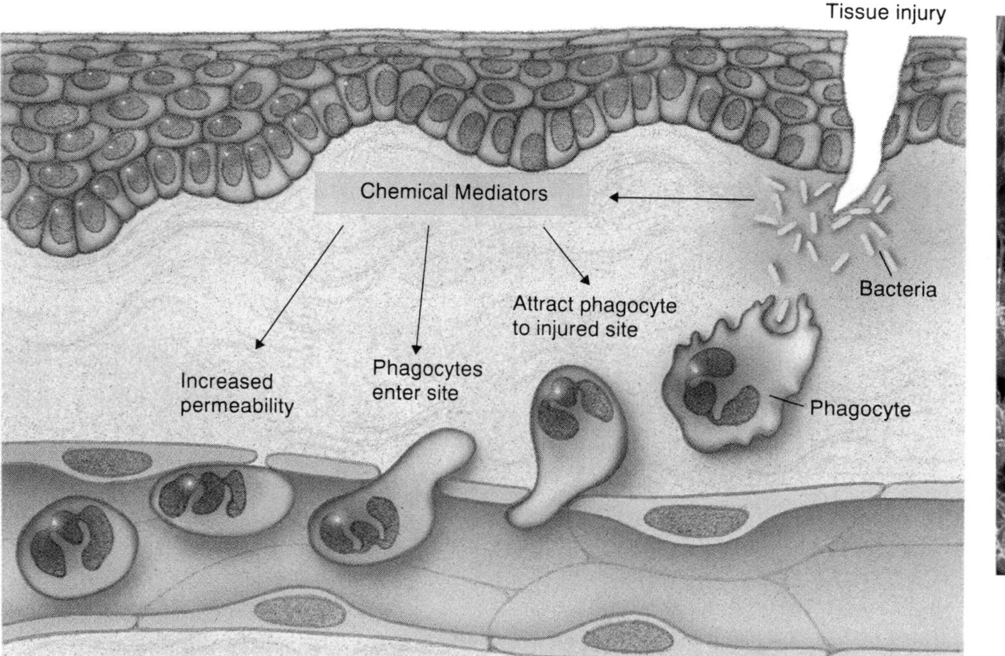

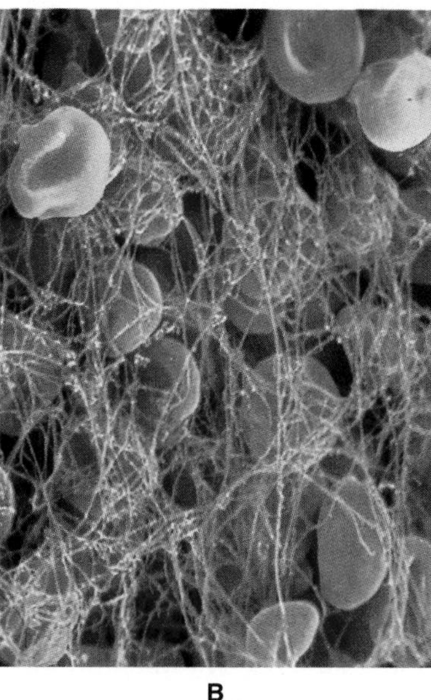

Figure 38.6

(A) Immediately after injury, affected tissues release chemical mediators that increase blood flow to the area and induce phagocytic cells to migrate to the damaged site. First neutrophils and later macrophages phagocytose bacteria and cell debris and prepare the site for repair. (B) If necessary, early in the inflammatory process, a fibrin network (pink threads) is thrown up to trap red blood cells to form a clot that prevents further loss of blood.

cells or bacterial cells. Neutrophils phagocytose foreign substances and the remains of dead cells. In so doing, they release enzymes that degrade dead cells or tissues. Neutrophils carry out these processes by the thousands or even millions, and many are killed or incapacitated as a result.

Macrophages are large phagocytic cells that are critically important in inflammation and also in specific immune responses. Probably because of their size, they are slower-moving than the smaller neutrophils and arrive at the inflammatory site a few hours later, attracted by several chemical mediators. Macrophages (and some other inflammatory cells) release substances called *pyrogens,* which cause fever, or elevation of the body temperature. What does this accomplish? The elevated temperature increases the efficiency of phagocytic cells and also retards multiplication of some viruses or bacteria that may be present.

Macrophages put the finishing touches on an acute inflammatory response. They clean up dead cells and tissue fragments, eliminate any remaining microorganisms, and clear the site of other inflammatory products. Depending on the severity of the injury, the inflammatory process may be resolved by these actions.

Days to Weeks

The goal of an acute inflammatory response is to limit injury, prevent invasion by microorganisms, clear the inflamed area of debris, and allow the tissue to return its normal state (*resolution*). Sometimes, however, tissue damage is too extensive, and the inflammatory response must be extended in time, or become *chronic,* in order to repair the tissue. When resolution cannot be accomplished, a wound repair response will occur, with damaged tissues being replaced by collagen, a tough, fibrous protein produced by fibroblasts. This results in the formation of a scar and loss of function in the tissue replaced.

> Inflammation is a nonspecific defense mechanism that occurs when cells and tissues are damaged. During an acute inflammatory response, blood flow into the injured site increases, the coagulation and complement systems are activated, and neutrophils and macrophages carry out phagocytosis and perform other functions that lead to resolution. In cases of extensive injury, damaged tissues are replaced by collagen, forming a scar.

SPECIFIC IMMUNE RESPONSES

All animals, from the simplest invertebrates to the most complicated mammals, use inflammation and wound repair responses in maintaining the integrity of their bodies when injury or infection occurs. However, phagocytosis, the primary antimicrobial response of inflammation, is relatively sluggish and inefficient. A highly sophisticated protective system evolved in higher vertebrates that is much more effective in the continual campaign to prevent colonization by bacteria, viruses, other microorganisms, and parasites.

Overview

The **immune system** consists of many different cells and organs that participate in a highly coordinated mission to inactivate or destroy foreign organisms or substances. **Antigens** are substances that induce formation of specialized protein molecules called *antibodies* as part of complex responses that lead to *immunity*. **Immunity** is the state of being able to resist a particular disease-causing agent or the substances that such an agent may produce. **Immunology** is the study of the immune system, immune responses, and immunity. Table 38.2 defines terms used frequently in this field.

A critical feature of immune responses is that the immune system is able to recognize *self* molecules and does not normally react against them. Thus the goal of an immune response is to eliminate any foreign, or *nonself*, antigen present in the body. Research by molecular biologists has shown that the complete immune picture is extremely complicated, and much remains to be learned. The critical feature of modern immunology is that immune cells are able to communicate with one another through chemical mediators and specific cell surface *receptors*.

When challenged by an antigen, the immune system responds in two basic but interrelated ways. **Cellular immunity** involves the actions of several classes of *T lymphocytes* (*T cells*), each of which carry out specialized immune reactions

Table 38.2 Terms Used in Immunology

Term	Definition
Antibodies:	Specialized protein molecules (immunoglobulins) produced in response to exposure to an antigen.
Antigen:	A substance that induces a specific immune response.
Immunity:	State of protection from a foreign agent or organism due to previous exposure.
Cellular immunity:	Immunity mediated by populations of lymphoid cells called T lymphocytes.
Humoral immunity:	Immunity mediated by antibodies produced by plasma cells, a type of B lymphocyte.
Lymphocytes:	A population of leukocytes with similar appearance but different immune functions.
B lymphocytes (B cells):	Lymphocytes that differentiate into plasma cells that produce antibodies.
T lymphocytes (T cells):	Thymus-derived lymphocytes consisting of several cell subsets that participate in cellular immune responses.
Major histocompatibility complex (MHC) molecules:	Self molecules that are unique for each individual's cells.
Self versus nonself:	The *self* is a person's own cells as determined by the presence of specific MHC surface molecules; the *nonself* is foreign antigens or cells whose MHC molecules differ from those of self cells.

that are directed toward the destruction of nonself cells or antigens. **Humoral immunity** consists of responses characterized by the formation of antibodies. Antibodies are manufactured by *plasma cells,* a type of *B lymphocyte* (*B cell*). Antibodies bind to a foreign (nonself) antigen, which leads to its being inactivated, destroyed, or targeted for removal by phagocytic cells. A key feature of antibodies is their *specificity*. An antibody will recognize and bind to only one type of antigen; however, the immune system is capable of generating billions of different antibodies, each with a different specificity.

Where are B and T cell populations found in the body? Several organs, collectively referred to as *lymphoid organs,* contain dense populations of lymphocytes and have critical functions relative to the immune system (see Figure 38.7). **Lymph nodes** are complex structures containing specific tissue territories composed of either B or T lymphocytes, with a few phagocytic leukocytes mixed in. They are found in areas where lymph drains in the body and act as filters for fluids in lymphatic vessels. Lymph nodes are the sites where most of the activities of the immune response actually occur. The *spleen* filters circulating blood and also contains populations of B cells, T cells, and macrophages. It is especially important in clearing the blood of pathogenic microorganisms. The *thymus gland* is involved in the production and maturation of T cells. Local collections of diffuse, unencapsulated *lymphoid tissues* are found in tissues underlying the gastrointestinal tract (the appendix), the airways leading to the lung (tonsils), and the urinary and reproductive tracts.

Specific immunity involves a set of cellular responses to a foreign antigen. Underlying immune system actions is an ability to distinguish between self and nonself substances. Cellular immunity consists of coordinated responses by T cell populations that are directed at destroying nonself cells or unique antigens. Humoral immunity is based on plasma cells producing antibodies that react against antigens. Most T and B cells live in lymphoid organs and tissues.

Antigens

Since most scientists believe that the immune system evolved in response to selection pressures exerted by *pathogens* (organisms that cause disease), antigens are classically considered to consist of nonself proteins and other molecules that are recognized as nonself. In general, an immune response is directed at a limited region of the antigen called the **antigenic determinant**, rather than at the entire antigen (see Figure 38.8A). A specific pathogen may have many different antigenic determinants on its surface, but individual antibody molecules can recognize only one.

The immune system also recognizes self cells that come to possess unique, nonself antigens. Cancer cells, for example, may be killed by the immune system if they have antigens on their surface that immune cells recognize as foreign. If the immune system fails to recognize such cells, a cancer may develop.

How does our immune system recognize self cells? The cells of every individual possess a unique set of self molecules called *major histocompatibility complex* (*MHC*) proteins. In humans, these self molecules are encoded in a cluster of genes—the MHC—located on chromosome 6. Cells possessing MHC molecules on their surface will not normally be destroyed by the immune system.

Antibodies

Antibodies are specialized proteins that have the capacity to bind specifically to an antigen or antigenic determinant (see Figure 38.8B). Antibodies may be either attached to the surfaces of B cells as receptors or unattached when circulating in the blood, lymph, and tissue fluids. Antibodies belong to a family of proteins known as *immunoglobulins* (Ig) and are secreted from plasma cells. An antibody

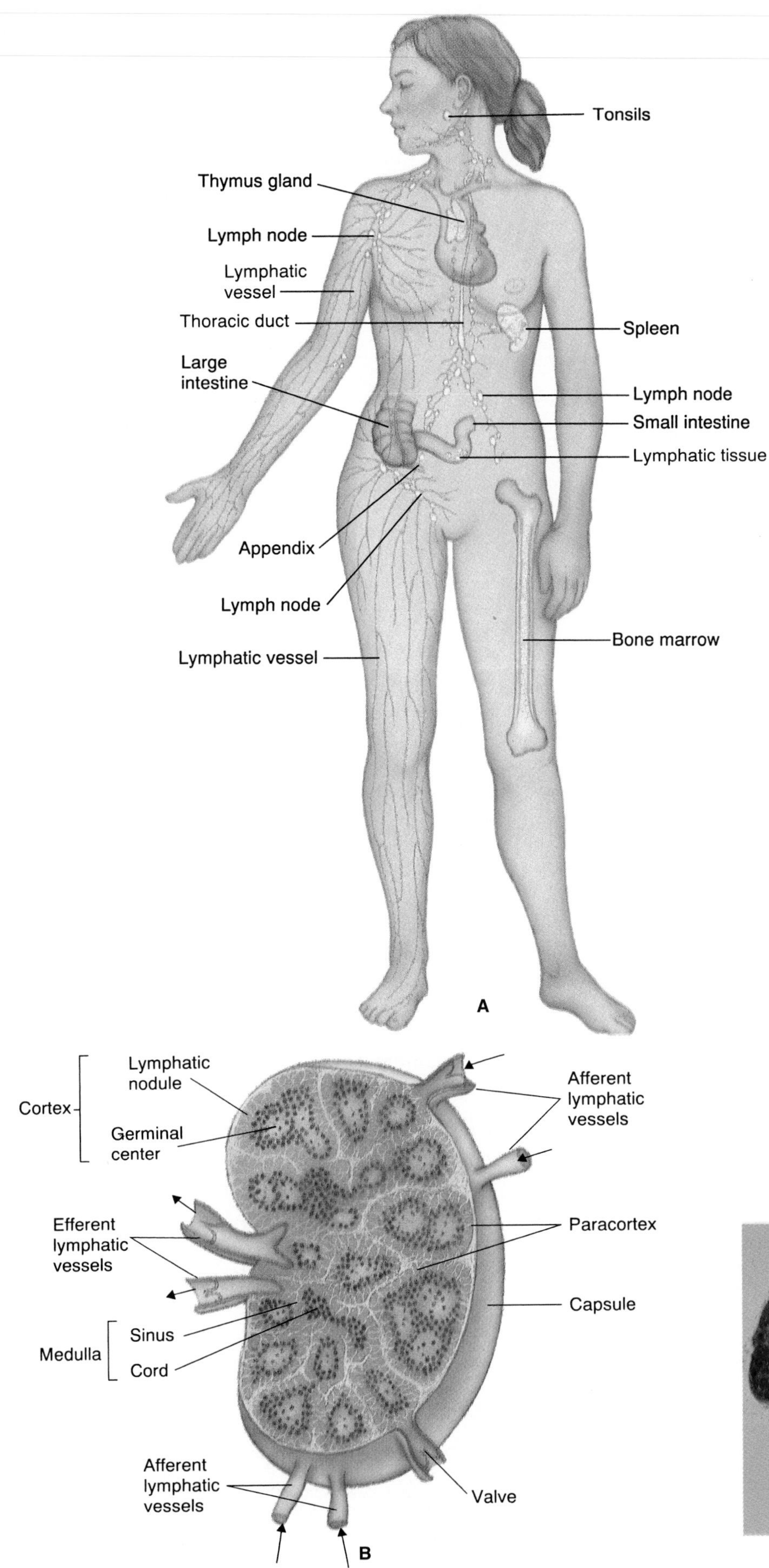

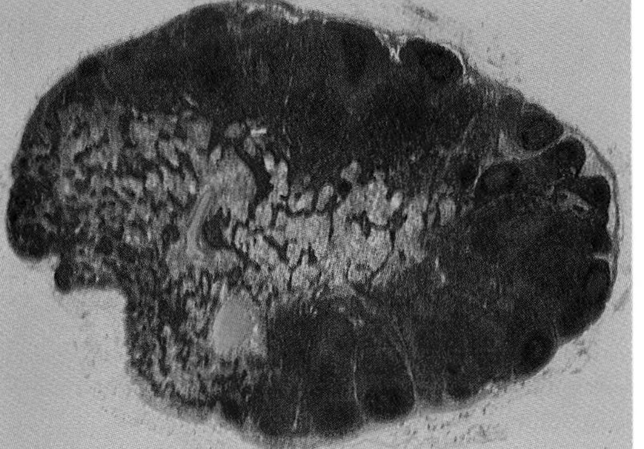

Figure 38.7

(A) The human lymphoid system consists of various organs and tissues. (B) Lymph nodes are enclosed by a thick collagen capsule and contain clusters of smaller nodules. A node's architecture consists of different functional regions. Sinuses are lined by phagocytic cells that remove and process any foreign materials that may be present in lymph entering from the afferent vessels. A nodule cortex contains B cells and germinal centers where new B cells are produced. The paracortex contains T cells and antigen-processing cells. The medulla contains T and B cells, and the cord region contains plasma cells that produce antibodies. Lymphocytes leave the node through efferent vessels. (C) This photomicrograph shows a cross section of a lymph node.

Figure 38.8

(A) Antigens have a three-dimensional structure in which specific regions, antigenic determinants, are the target for antibodies. (B) Antibodies have sites that bind with an antigenic determinant much as a key fits a lock. (C) Binding results in the formation of an antigen-antibody complex.

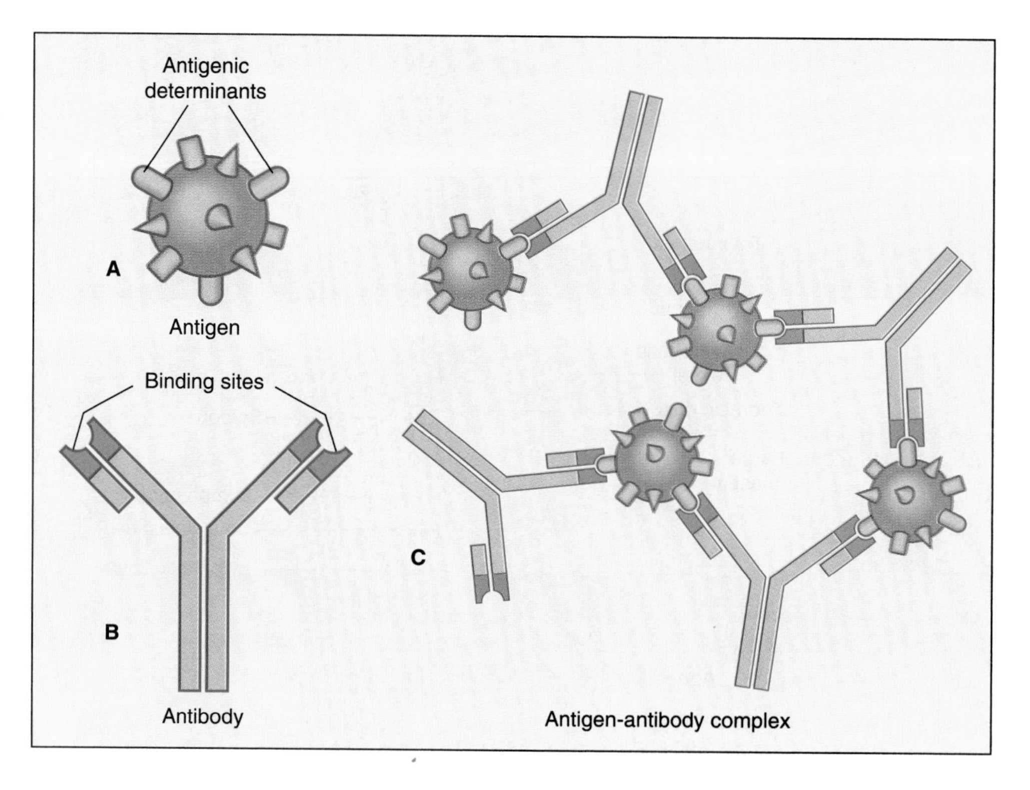

is able to combine with a specific antigen in a lock-and-key fashion to form an *antigen-antibody complex* (see Figure 38.8C).

Classification and Functions

Five major classes of immunoglobulins, each with different biological properties, are recognized in humans: IgG, IgA, IgM, IgD, and IgE. The structure and function of each are shown in Figure 38.9.

Structure

Antibodies have two critical roles in immunity that are reflected in their structure. The first is *recognition,* a process in which an antibody receptor on the surface of a B lymphocyte "recognizes" and binds to an antigen. Once that happens, a generalized response follows that leads to large-scale production of the antibody.

Antibody molecules have a fundamental structure that is described in Figure 38.10. They are Y- or T-shaped and consist of two identical long protein chains, called *heavy (H) chains,* and two identical shorter protein chains, called

Figure 38.9

This figure presents schematically the general structure, size, and function of the five types of immunoglobulins.

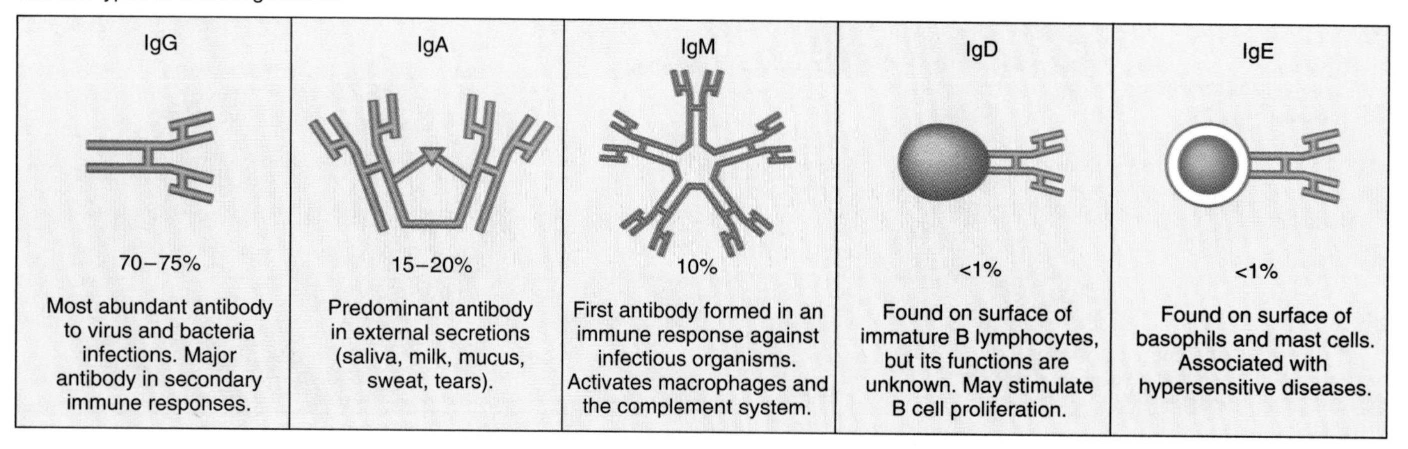

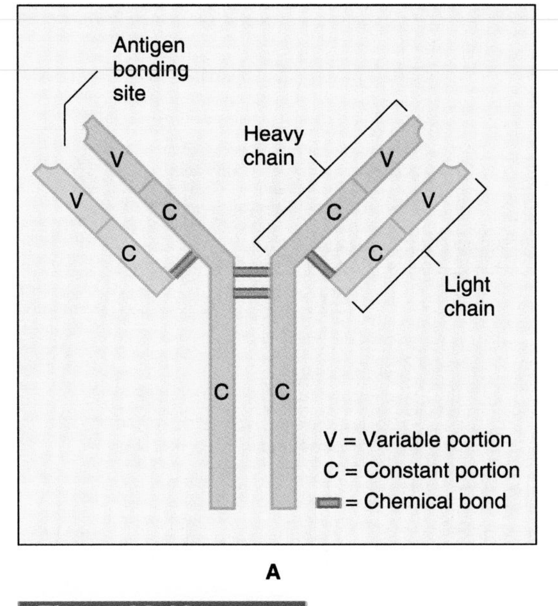

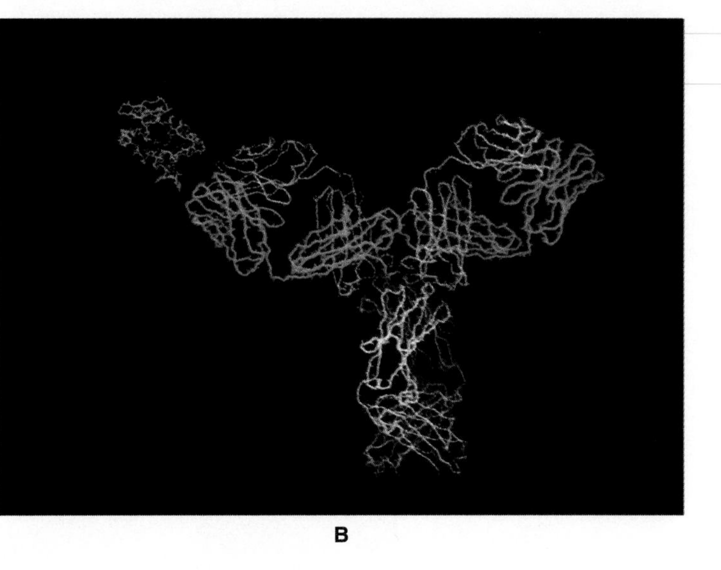

Figure 38.10

(A) In an antibody molecule, both H (heavy) chains and both L (light) chains have identical amino acid sequences, but the H and L sequences differ. The V (variable) regions of each antibody are responsible for binding to an antigen. Each V region of an antibody has a different amino acid sequence. The hinge portion provides flexibility and allows the two binding sites to act independently. (B) A computer-generated antibody-antigen (bright green) model.

light (L) chains. The amino acid sequence of much of the H and L chains does not vary, and these parts of the antibody chains are referred to as the *constant (C) region.* At the end of each chain are small *variable (V) regions* with amino acid sequences that are unique to each antibody. The V portions of one H and one L chain that form an arm of the antibody interact with, and bind to, a specific antigen.

Self cells have MHC molecules that are recognized by immune cells. Any substance without such MHC molecules is recognized as a nonself antigen, and antibodies will be formed against it. Antibodies are immunoglobulin molecules. Humans have five major antibody classes, each with a different function in an immune response. Antibodies consist of two H chains and two L chains. Each chain has constant structural regions and a variable, antigen-binding region.

Cells of the Immune System

Several types of lymphocytes have specific responsibilities in immune responses. All lymphocytes have a similar appearance, but several functional classes are known to exist. In healthy adults, about 20 percent of the lymphocytes circulate in the blood, while the remainder reside in lymph nodes, the spleen, the thymus gland, the tonsils, and the appendix. When first formed in the bone marrow, lymphocytes have no special immune functions. After maturation, they are capable of performing some critical roles when challenged by an appropriate antigen. The production, maturation, and functions of lymphocytes are described in Figure 38.11.

T Lymphocytes

One class of lymphocytes produced in the bone marrow must first circulate to the thymus gland, where they are "processed" before being released to the circulation and distributed to lymphoid organs and tissues. These are known as

Figure 38.11

T and B cells are produced in the bone marrow. T cells then move to the thymus gland, where they are surveyed for having appropriate receptors. The site of B cell processing in humans is not known. Once processing is completed, T and B cells migrate to lymphoid organs, where they may participate in immune reactions.

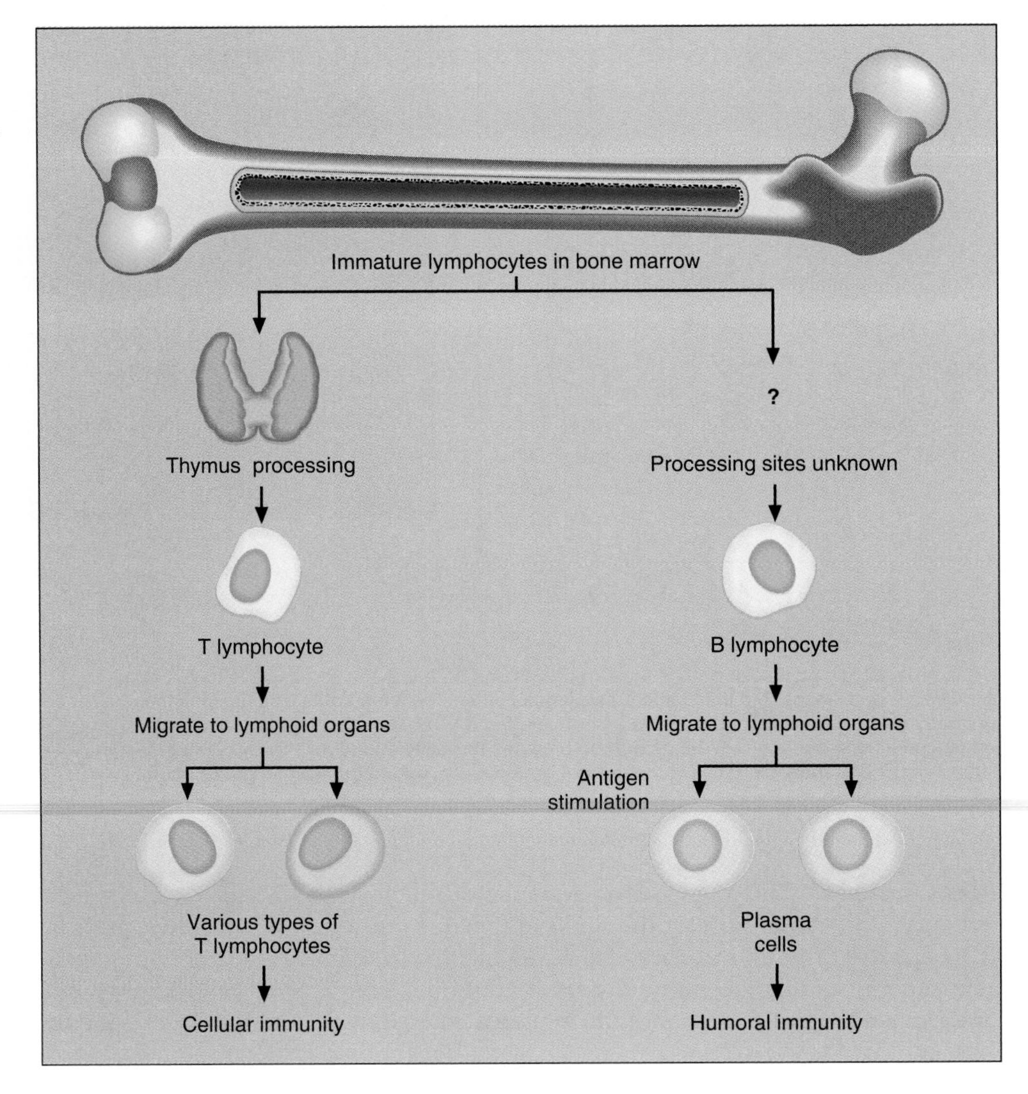

T lymphocytes, or T cells. There are several T cell subpopulations; each has a specific function in a cellular immune response.

During their development in the bone marrow, T cells acquire, on their plasma membrane, unique antigen-binding molecules of a specific configuration called *T cell receptors*. What is the nature of T cell receptors? Their structure is similar to, but distinct from, that of immunoglobulins. Like antibodies, T cell receptors can bind with an antigen. Each T cell type bears unique receptors that can recognize and react with only one type of nonself antigen.

T cell receptors recognize and react with antigenic determinants that are "presented" on the surfaces of other self cells, usually macrophages. *Antigen presentation* is a process whereby certain self cells display a foreign antigen on their surfaces in a form that is recognized by T cell receptors. Basically, the nonself antigen must be bound to an MHC (self) molecule before it will be recognized by a T cell receptor (see Figure 38.12A). The fundamental structure of the MHC-presenting molecules is now known and is shown in Figure 38.12B. Its shape reveals that a nonself antigen becomes bound to a MHC molecule because it fits into a "binding pocket" in the MHC protein.

Why is it necessary for immature T cells to be processed in the thymus gland? What happens to them in this organ? Evidence suggests that T cells are subjected to a brutal selection process in the thymus. Only T cells that have receptors capable of recognizing a nonself antigen in combination with a self

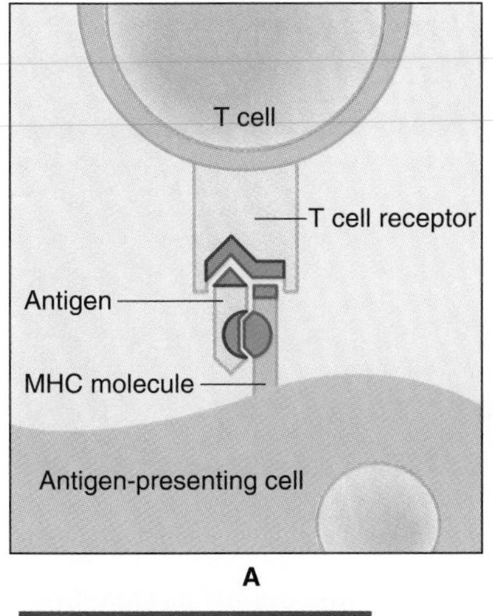

A

Figure 38.12

(A) Processed T cells are able to recognize foreign, nonself antigens when they are combined with a self MHC molecule and presented on the surface of an antigen-presenting cell. The basis for recognition is the presence of a specific receptor on the T cell that matches the shape of the nonself antigen when combined with a self MHC molecule. (B) MHC-presenting molecules have a structure that matches their function. Foreign antigens fit into the binding pocket on MHC proteins.

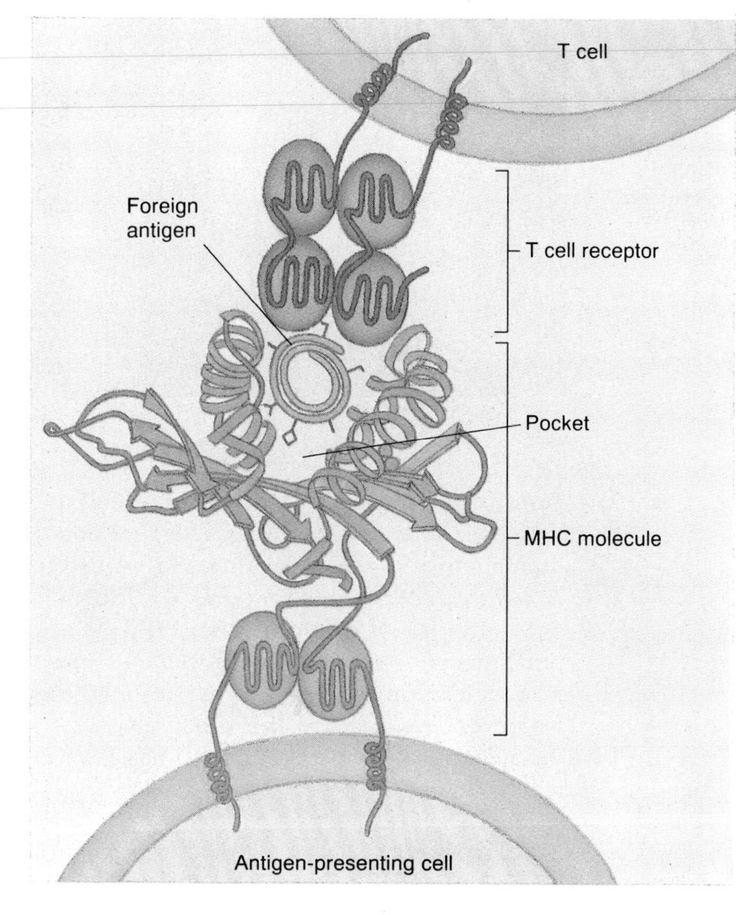

B

MHC molecule are allowed to survive. T cells with only nonself antigen receptors, only self MHC molecules, or no receptors are destroyed, probably by macrophages. Thus full-fledged T cells that survive the thymic selection process are equipped with the appropriate receptors for performing their function. Surviving T cells can recognize only one type of foreign antigen in combination with an MHC molecule.

T Cell Subpopulations

Four T cell subpopulations are important in normal cellular immune responses. Sensitized *cytotoxic T cells* (*T_C cells*; also known by the more colorful name *T killer cells*) destroy cells possessing foreign antigens on their surface by puncturing them (see Figure 38.13). Such cells include virus-infected self cells and cancer cells. Only modified self cells are killed by T_C cells. Normal host cells are not killed because they do not display foreign (viral or cancer cell) antigens on their surfaces. Activated *helper T cells* (*T_H cells*) stimulate B cells and other types of T cells involved in the total immune response and stimulate proliferation of T_C cells. *Suppressor T cell* (*T_S cells*) inhibit the production of T_C cells and help control and limit humoral immune responses. Finally, *T memory cells* (*T_M cells*) recognize previously encountered antigens.

B Lymphocytes

B lymphocytes arise in the bone marrow, but their course of processing or maturation in humans is not yet understood. After they are formed, B cells migrate to lymph nodes and other lymphoid organs and tissues. Populations of inactive B cells have many antibody receptors on their surfaces, but all receptors on a single B cell have the same structure and can recognize only a single type of anti-

Figure 38.13

(A) Cytotoxic T lymphocytes constitute the primary natural defense mechanism against the formation of malignant tumors. T cells also react against virus-infected cells. In this photograph, T_C cells are shown attacking a much larger cancer cell. (B) This cancer cell has been punctured and will soon die.

gen. If the surface antibody receptor cell does come in contact with its matching antigen, the B cell becomes activated and gives rise to plasma cells capable of manufacturing and secreting the same type of antibody (see Figure 38.14). *B memory cells* are also produced during humoral immune responses.

Millions of different T and B cells are produced daily, each with unique receptors and recognition capabilities. These cells live only a few days, and only a small fraction of them ever come in contact with their corresponding antigens. However, because of the great diversity and the continuous production of unique lymphocytes, an antigen has little chance of escaping detection for any length of time.

Macrophages

Several other cells have significant functions in specific antibody responses, including macrophages found in lymphoid organs and tissues. As described earlier, macrophages are the primary phagocytic cells and are particularly important in cleaning up damaged tissues by ingesting dead cell fragments and any bacteria or other microorganisms present. This latter function serves to bridge the inflammatory response, which occurs first, and the immune responses, which follow if foreign antigens are present.

The uptake of nonself antigens by macrophages usually constitutes the first step leading to an immune response. In a sense, at least one member from an invading horde has been tracked down and trapped. We shall see what happens next.

Figure 38.14

(A) B cells become activated to make a specific antibody after they contact their matching antigen. In this photograph, B cell antibody receptors have contacted chlamydia bacteria that contain nonself antigens. (B) Once activated, the B cell proliferates and gives rise to identical cells that develop into plasma cells (shown here) that produce antibodies against the initiating antigen.

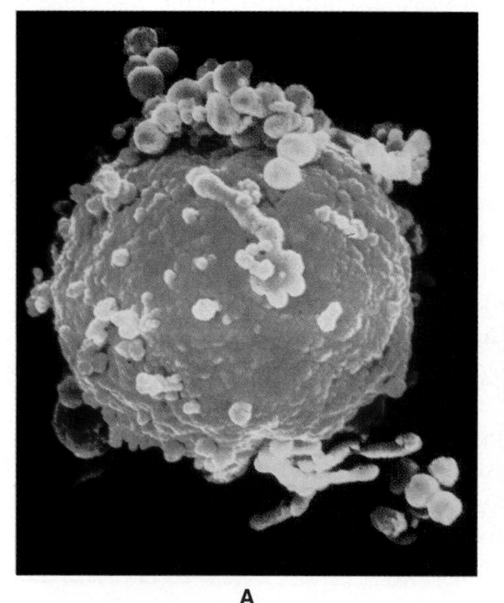

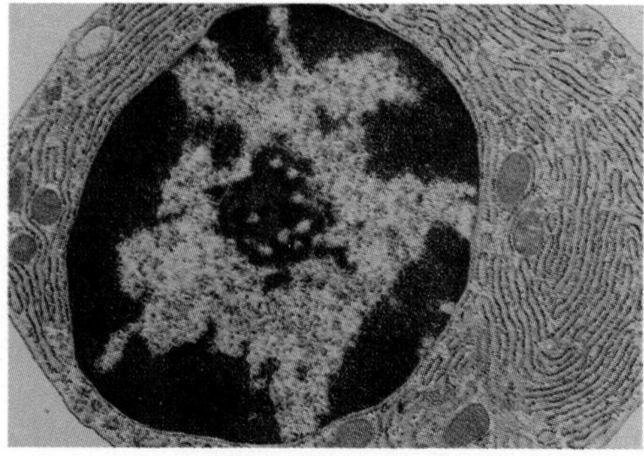

T cells are produced in the bone marrow and then enter the thymus gland, where they survive only if they have receptors capable of recognizing an MHC molecule in combination with a foreign antigen. B cells are also produced in the marrow, but their site of processing is not known. Mature B and T cells migrate into lymphoid organs and tissues. Four classes of T cells—T_C, T_H, T_S, and T_M—take part in cellular immune responses. B cell populations that participate in humoral immune responses are plasma cells and B memory cells. Macrophages initiate specific immune reactions by capturing and processing a foreign antigen.

Primary Immune Responses

To assist in conceptualizing the immune response, it is commonly divided into cellular responses and humoral responses. Although such a separation of functions facilitates understanding, it must be emphasized that both types of immune response occur simultaneously, in the same tissues, and are regulated by overlapping mechanisms. The first response to an antigen not previously encountered by the immune system is called a **primary immune response**. Lymphocytes are dormant until stimulated by an antigen. When an immune response is initiated in response to an antigen, a series of cellular interactions occurs that leads to the creation of active T cell subpopulations and the production of antibodies by B plasma cells.

T Cells

Figure 38.15 summarizes the major events in a cellular immune response. A primary immune response is initiated after a nonself antigen is first "processed" by a macrophage. For example, a protein antigen is broken down to smaller segments (10 to 20 amino acids), at least one of which acts as an antigenic determi-

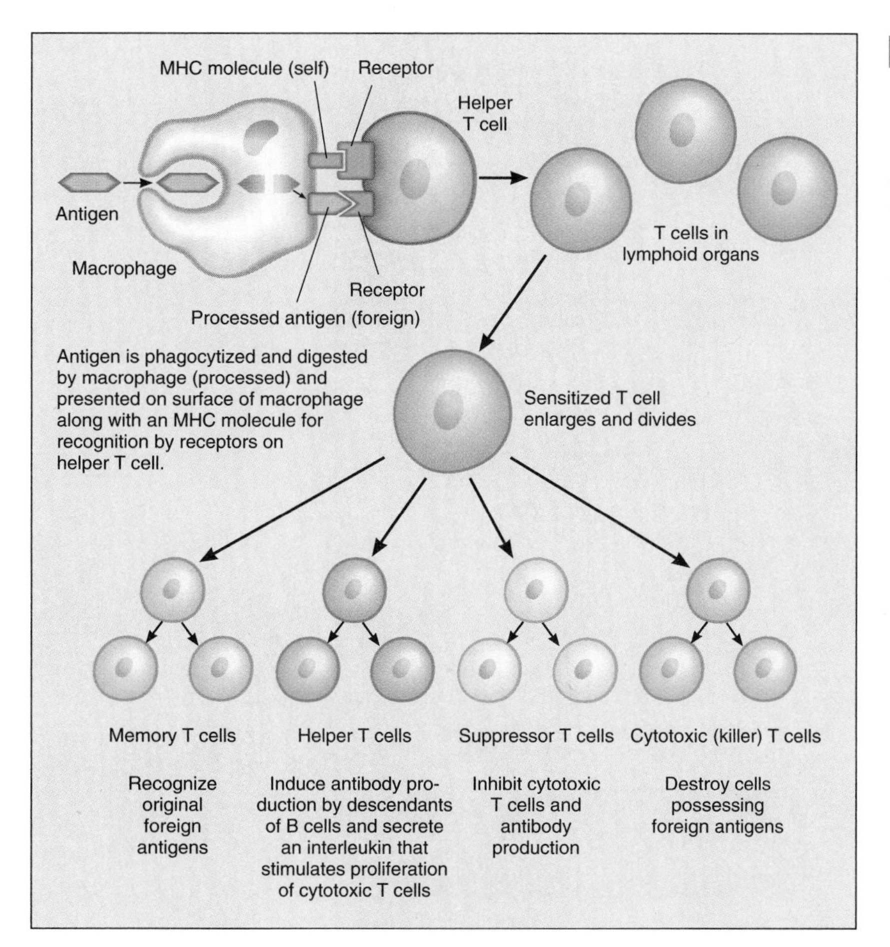

Figure 38.15

After a macrophage phagocytoses and processes a foreign antigen, it presents an antigenic determinant along with an MHC molecule on its surface. A helper T cell with matching receptors binds with the presented molecules, becomes activated, and induces proliferation and maturation of the various T cell populations that play roles in the immune response.

nant. Once this task is completed, the antigenic determinant is bound to a self MHC molecule, and both are displayed on the surface membrane of the activated macrophage. In addition, the presenting macrophage secretes a substance called *interleukin 1* that attracts and activates T cells. **Interleukins** are hormones that act on lymphocytes, and they are critical for the activation, maturation, proliferation, and functioning of all cells involved in immune responses. Following these events, a T_H cell with receptors for the specific nonself antigen-MHC protein complex is attracted to, and reacts with, the presenting macrophage and becomes activated.

Activated T_H cells induce the growth, proliferation, and development of other functional T cell strains, including T_C, T_S, and T_M cells. Following activation, the immune response could cause the body to becoming overloaded with immune cells and antibodies. What regulates T and B cell proliferation and antibody production? Several factors are probably involved, but the actions of T_S cells are extremely important. T_S cells are thought to limit quantities of T cells and antibodies through the action of a suppressor factor they release that retards further production of both B cells and T_H cells.

B Cells

B cells can become activated by two processes (see Figure 38.16). A "naive" B cell (one that has not yet been activated) possessing a matching antibody receptor for the specific nonself antigen may be stimulated directly by the antigen, which leads to its proliferating and producing large numbers of identical cells, or clones. This process is known as **clonal selection** (see the Focus on Scientific Inquiry, "Antigen Recognition and Antibody Specificity"). Alternatively, T_H cells formed in response to an antigen may activate a B cell with matching receptors and stimulate its proliferation, growth, and maturation.

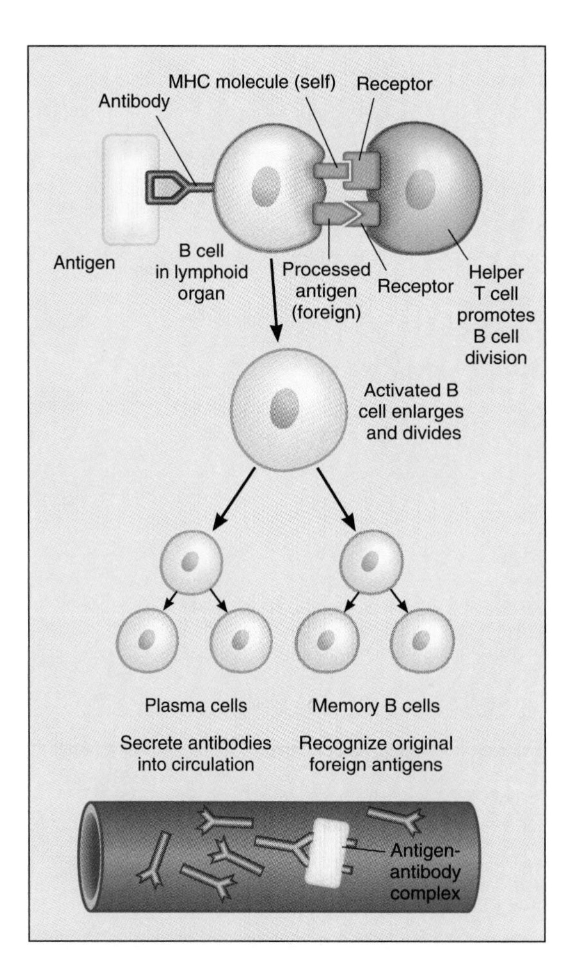

Figure 38.16

A naive B cell becomes activated by one of two processes. Its antibody receptors may contact a fitting antigen, or a helper T cell activated by the same antigen stimulates B cell proliferation. The resulting B cells develop either into plasma cells that secrete antibodies specific for the initiating antigen or into B memory cells.

The B cells produced from clonal selection or T_H cell activation develop into plasma cells capable of producing only a single type of antibody—the type that was responsible for its development. Plasma cells remain in lymphoid tissues, but antibodies are released and are present in plasma and tissue fluids. Antigen-antibody reactions generally occur in lymphoid organs and tissues.

Memory Cells

In the primary immune response just described, activated B and T cells proliferate and develop into different subsets of functional immune cells that have the capacity to react directly with the specific, initiating nonself antigen. During this first exposure to the antigen, B and T cells also produce an additional type of cell, called a memory cell. B or T **memory cells** are specific for (retain a memory of) the antigenic determinant that provoked their development. That is, like their antecedents, they possess receptors that bind with only one antigen. Memory cells have very long lives (up to 20 years) and do not mature into active immune cells unless challenged by the same antigen that led to their creation.

Secondary Immune Responses

As illustrated in Figure 38.17, if the immune system encounters an antigen for the second time, a more intense **secondary immune response** occurs quickly because it is not necessary to activate naive cells. Memory cells can give rise to active B and T cells directly. Memory cells and secondary immune responses provide the basis for long-lasting immunity. Once an individual has been exposed to a foreign antigen and has mounted a successful immune response, he or she will usually not suffer from the effects of that antigen at a later time. If the original antigen is encountered again, memory cells ensure that it will be eliminated before it can cause any noticeable symptoms.

A primary immune response begins when a T_H cell contacts a presenting macrophage. Activated T_H cells induce production and maturation of other T cell populations. B cells are activated to proliferate and develop into antibody-secreting plasma cells by T_H cells or through a process of clonal selection. Ultimately, the initiating foreign antigen is eliminated through the actions of plasma cells and T cells. B and T memory cells remain in the body and can trigger a secondary response if the antigen reappears at a later time.

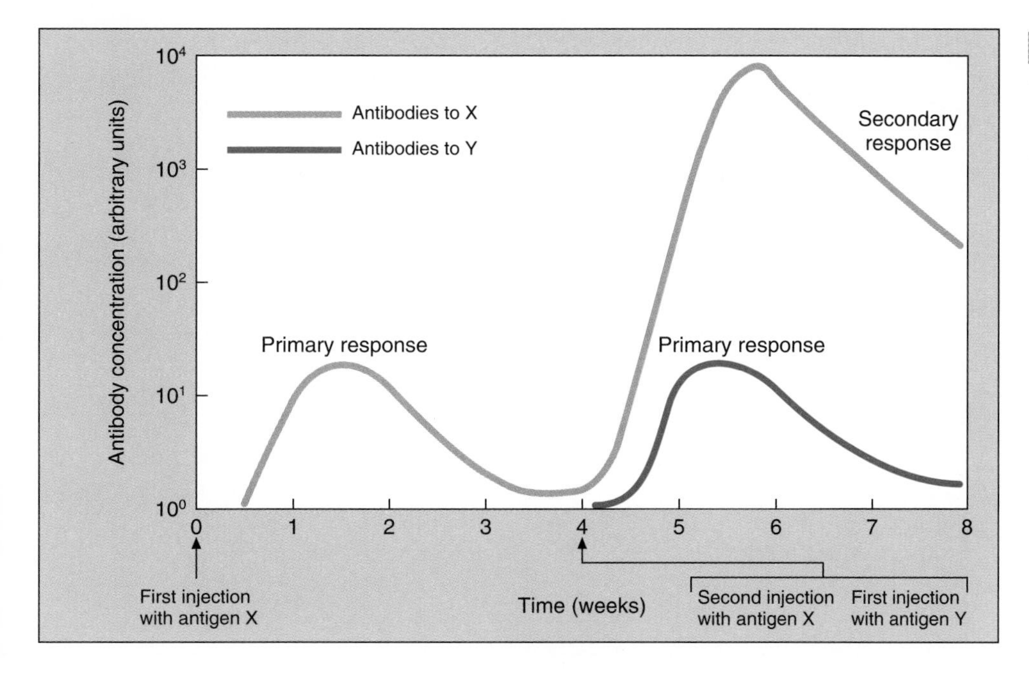

Figure 38.17

The primary immune response to antigen X leads to antibody production resulting in a certain antibody concentration. If reinfection to antigen X occurs, a rapid secondary immune response results in much higher antibody concentrations.

Focus on Scientific Inquiry

Antigen Recognition and Antibody Specificity

We now know that the human immune system can recognize and react against an almost indefinite number of antigens. We also understand much about the underlying cellular and genetic mechanisms that provide us with this capability. Like most major intellectual advances in biology, knowledge was derived from creating hypotheses and then designing and conducting experiments to test them. Early hypotheses about antigen recognition were formulated and provided a framework for further advances. As new information became available, they were modified or gave way to new hypotheses. The evolution of hypotheses about antibody formation illustrates this dynamic scientific process.

In common with other fields of physiology, much of the experimental work in immunology was (and still is) done on animals, especially mice. Since these fellow mammals have a similar evolutionary history, it is presumed that results and conclusions from such studies can be extrapolated to humans. Except for minor details, this has generally proved to be the case. Modern immunology terms are used to simplify the story.

By the late nineteenth century, scientists recognized that animals were capable of reacting against foreign materials. If an animal was infected with bacteria that produced a toxin, "antitoxins" would soon appear in the blood serum of the host. This was illustrated by the formation of clumps when fresh serum from the host was mixed with the bacterial toxin. E. A. von Behring, a German bacteriologist, referred to the host's clumping substances as *antibodies*. Paul Ehrlich, a German physician, was an active researcher in immunology. During the 1890s, he developed a method for measuring amounts of antibodies produced in response to a bacterial infection. This quantitative technique was to prove especially valuable in forming early pictures of the antibody response in animals. Its use made clear that exposure to an infectious agent resulted in an irruptive increase in responding antibodies. How did these agents cause such a reaction?

In 1897, Ehrlich proposed the first coherent hypothesis about antibody formation (see Figure 1). He postulated that white blood cells had special chemical groups ("side chains") on their surface that normally functioned in the cell's metabolic processes. However, these groups could also serve as receptors for antigens that entered the body. Once a side chain was linked to an antigen, though, the side chain's normal functioning would be disrupted. To counteract this effect, the cell would be stimulated to synthesize new side chains to replace those that no longer functioned. Ehrlich claimed that "the antitoxins represent nothing more than the side-chains reproduced in excess during regeneration and are therefore pushed off from the protoplasm—thus to exist in a free state." Note that this early hypothesis was essentially correct in two respects: antigens are captured by surface receptors, and cells do produce antibodies that appear in the circulation after exposure to an antigen. However, the hypothesis was too vague and was based, in part, on incorrect assumptions. For example, Ehrlich felt that cells naturally made side chains that could bind to all antigens. Thus specificity could not be explained by this hypothesis.

Karl Landsteiner, an immunologist who spent most of his career at the Rockefeller Institute for Medical Research, demonstrat-

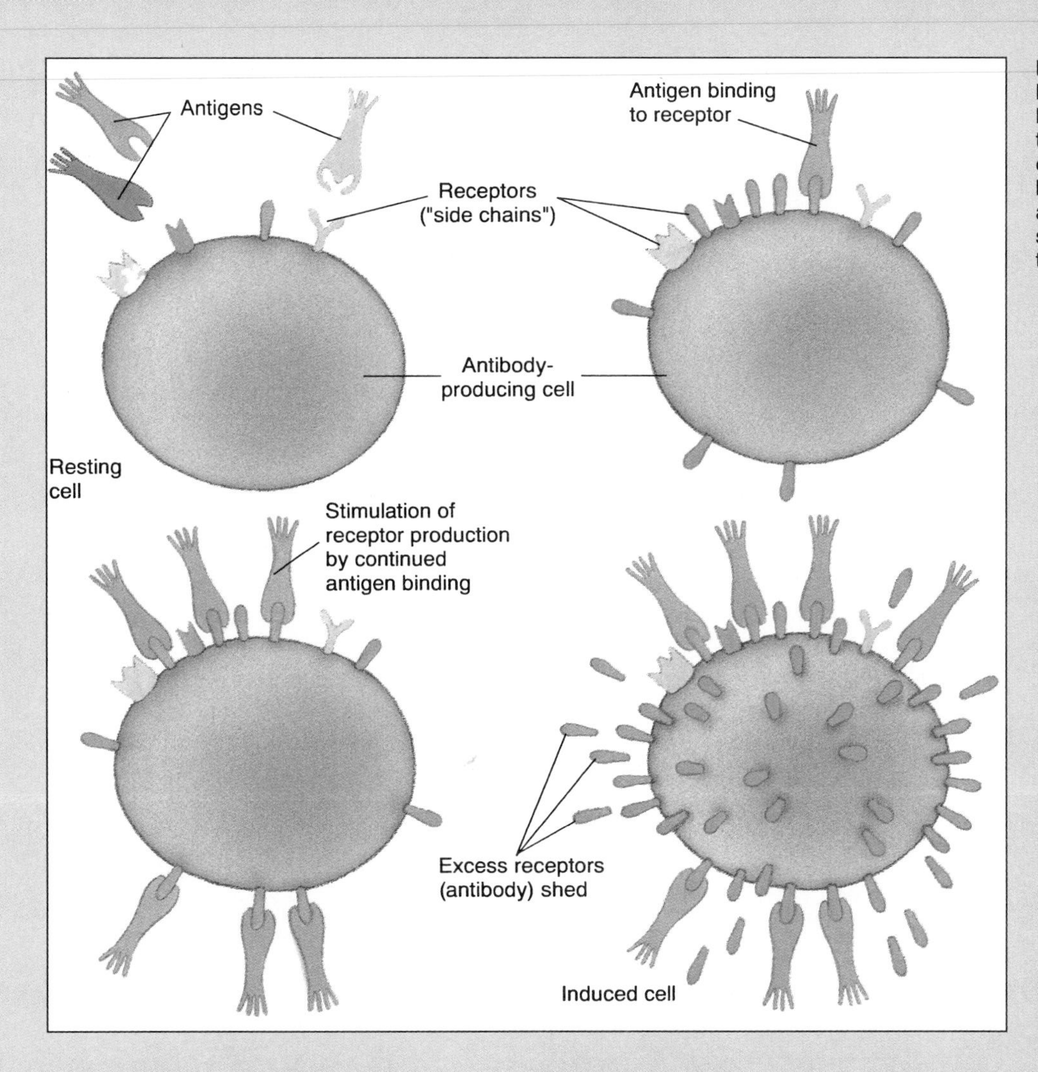

Figure 1 Ehrlich's illustration of his side-chain hypothesis, published in 1897. Using modern terms, Ehrlich thought that a foreign antigen, indicated in red, became bound to a cell receptor and stimulated the cell to synthesize and release identical receptors, which acted as antibodies.

ed that specific antibodies were formed in response to specific antigens. In his 1945 book *The Specificity of Serologic Reactions,* he summarized the view of immunology at that time:

> The immune antibodies all have in common the property of specificity, that is, they react as a rule only with the antigens that were used for immunizing or with similar ones, for instance, with proteins or blood cells of one species, or particular bacteria, and closely related species.

Further, "the specificity of antibodies . . . constitutes one of the two chief theoretical problems, the other being the formation of antibodies." In other words, although the concept of specificity had become clearly established, neither the basis for specificity nor the mechanisms of antibody formation were known.

How could an antigen induce the formation of a specific antibody? In the early 1930s, several hypotheses were advanced to answer this question, but one claimed the attention of most scientists for the next 25 years. The "direct template" or "instructive" hypothesis of antibody formation maintained that an antigen entered the antibody-forming cell and acted as a template against which the antibody molecule would shape itself in a way complementary to the antigen, as shown in Figure 2. Most of the arguments in favor of this hypothesis were based more on conceptual ideas than on experimental data. Professor Linus Pauling published a detailed paper in 1940 ("A Theory of the Structure and Process of Formation of Antibodies") that attempted to explain how the antibody might fold, or be shaped, by the antigen. The basis of complementary folding was related to chemical bonding phenomena, one of his fields of expertise. However, one of his assumptions was that all antibody molecules contained the same amino acid structure, and hence specificity could not be related to variations

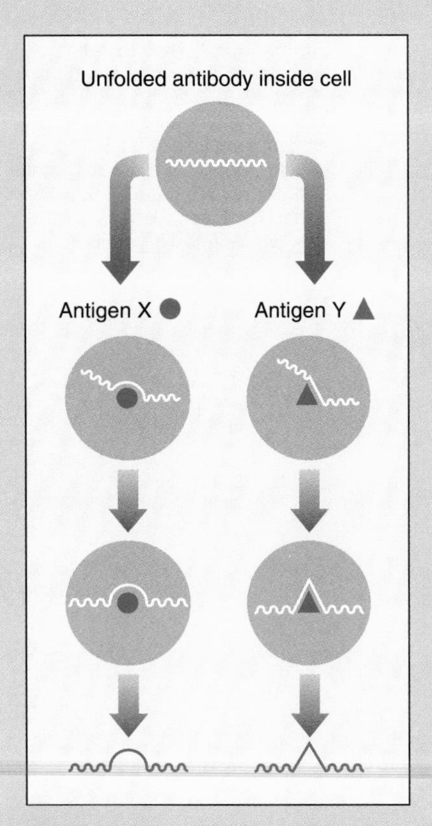

Figure 2 According to the instructive hypothesis, nonself antigens (X and Y) served as templates for antibodies that would then react against them.

in an antibody's protein chains. This, of course, turned out to be incorrect, but that was not determined until years later.

By 1950, the template hypothesis was crumbling. A number of scientists, including F. Macfarlane Burnet, an Australian immunologist, identified several shortcomings of the hypothesis. For example, it could not account for the early rapid rise in antibody concentrations following infection. How could antibodies so quickly outnumber the antigen templates? The hypothesis also failed to explain secondary responses (well understood by the 1930s), continuing antibody production even in the absence of an antigen, or immunological tolerance (the concept of self-toleration became known in the 1940s). Obviously, the fall of the template hypothesis could be anticipated as soon as a new hypothesis arose that could address these criticisms.

A 1955 paper published by a Danish immunologist, Niels Jerne, provided a foundation for a new hypothesis that was to emerge in 1957. In this paper, Jerne formulated a "natural selection" or "selective" hypothesis of antibody formation that proposed the following scheme:

1. Animals contain at least one circulating antibody that can react to every antigen.
2. After an antigen becomes bound to the antibody, it will interact with a lymphoid cell to trigger an immune response.
3. It does this by stimulating the cell to generate and release large quantities of the specific antibody.

Figure 3 The clonal selection theory, formulated in 1957, proved to be essentially correct and is used today to account for the humoral immune response.

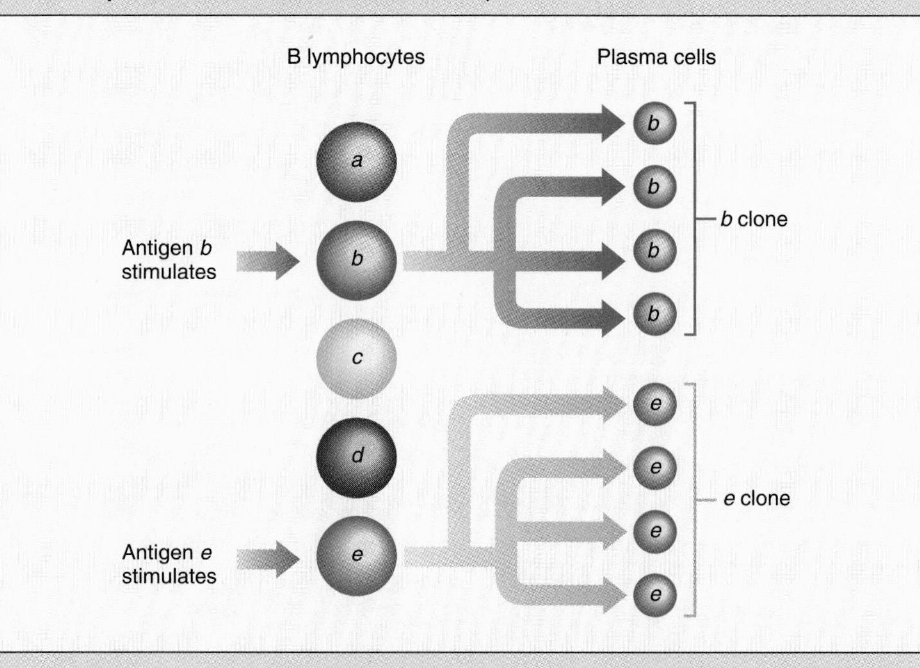

Note the similarity of this theory with Ehrlich's earlier side-chain theory. The primary flaw in Jerne's theory was his assumption that the body contained preexisting antibodies for every conceivable antigen.

In 1957, David W. Talmage, of the University of Colorado, added to Jerne's ideas. He proposed that in addition to circulating antibodies, replicating cells possessing antibodies (receptors) were involved in the immune response. Such cells would then multiply after their antibody (receptor) was "selected" by the antigen. He further suggested that single cells might produce only one type of antibody. In the same year, Burnet, independently of Talmage, published his "clonal selection" theory of antibody formation. Though his theory was very similar to that proposed by Talmage, Burnet is usually given credit for its development.

The *clonal selection theory,* described in Figure 3, has been called one of the great intellectual constructs of modern biology. It explained how antigens stimulated antibody production and why there had to be an enormous number of potential antibody-producing B cells in the body. The theory made the following assumptions:

1. Naive B cells are produced continuously in the body.
2. Each B cell has only a single type of antibody receptor.
3. An antigen is capable of reacting with ("selecting") only a B cell that has the "correct" receptor.
4. Once selected, the B cell proliferates and creates a population of identical cells ("cloning").
5. Each of the clones secretes the same antibody.

This description of immune events remains valid today, even though scientists now recognize that numerous other factors are involved.

Thus from Ehrlich's first revolutionary concepts, the modern theory of antibody formation evolved. As with all major theories, there were many false starts, blind alleys, and nonproductive ideas that have not been described here. And the story is still not completed. Much remains to be learned about the induction and preservation of antibody diversity as well as the intricate regulating features of the entire immune process. All of the participants in this story, except Linus Pauling, later received Nobel Prizes for their studies in immunology. The great Pauling went on to receive two Nobel Prizes for his other works.

TOLERANCE

Immunological **tolerance** is the process whereby lymphocytes do not respond to self molecules. It is quite obvious that it would not be beneficial for your immune system to produce many antibodies or T cells that react against your own proteins or cells. What prevents or limits our immune system from reacting against self molecules? Normally, an individual develops tolerance during embryonic and fetal development. However, since T and B cells are produced throughout life, tolerance must also be developed in new lymphocytes produced by adults.

How does the immune system distinguish between self and nonself? This is one of the major mysteries in immunology, but it is beginning to be unraveled. Two main hypotheses on tolerance induction have been advanced. The first (referred to as *clonal deletion*) involves the idea that all developing lymphocytes, but primarily T cells, that react against MHC (self) molecules during maturation will be killed. The second hypothesis (known as *clonal anergy*) proposes that immature B or T lymphocytes that have antiself potential receive a signal that retards or abolishes their ability to react against a self molecule. This mechanism is thought to operate mainly on B cells. A third hypothesis is that during an immune response, a network of T_H and T_S cells is established that somehow prevents reactions against self molecules. Evidence suggests that if such a mechanism exists, it plays only a minor, backup role to clonal deletion and clonal anergy.

THE OTHER EDGE OF THE SWORD

Given the enormous complexity of immune responses, the number of chemical and genetic factors involved, and the coordination of events required for success, it is perhaps not surprising that many problems can, and do, occur. Autoimmune responses and allergies are examples of abnormal immune responses. In some cases, the efficiency of our immune system causes frustrating problems for the medical community. For example, it is extremely difficult to transplant organs from one individual into another if the immune system is operating normally. What accounts for autoimmune diseases, allergies, and organ transplant rejections?

Autoimmune Diseases

In 5 to 7 percent of humans, self-tolerance mechanisms fail, and an **autoimmune response** occurs in which T cells attack self cells or antibodies react against self molecules. More than 40 diseases have been characterized by the presence of destructive self antibodies, including multiple sclerosis, rheumatoid arthritis, and insulin-dependent diabetes mellitus. In some cases, the autoimmune response is directed against specific self molecules; in others, the response is directed at molecules present in the nucleus or cytoplasm of cells. Autoimmunity can, of course, have disastrous consequences. Why do these harmful immune responses occur? Their causes are poorly understood, although genetic mutations, chemicals, viruses, bacteria, and drugs have been implicated in some cases.

Intense research is currently being conducted to identify the molecular targets of autoimmune responses and to determine the underlying causes of the reactions. Are self-reactive T cells involved? Is there a failure in the mechanisms that normally suppress self-reactive T cells? Some evidence suggests that autoimmune diseases involve genes encoding MHC proteins. How does this fit into the autoimmune picture? Answering these questions presents a major challenge to immunologists.

Allergies

Hypersensitivity is a heightened immune response that may be harmful to the body. If a person is *allergic* to a substance (called an *allergen,* in this case), exposure to that allergen will cause a hypersensitive reaction. About 35 to 40 million Americans suffer from allergies. Common allergens include dust, feathers, pollens, penicillin, strawberries, and lobster.

One type of hypersensitivity is mediated by the antibody IgE (see Figure 38.18A). During the first exposure to an allergen, the susceptible individual becomes sensitized, and his or her B cells produce IgE antibodies that bind to mast cells. In subsequent exposures, the allergen binds to these IgE antibodies in such a way that it causes mast cells to release great quantities of chemical mediators such as histamine, which activates various defense reactions (see Figure 38.18B). When this occurs, breathing becomes difficult because smooth muscles lining the air passages constrict; excess mucus is released (runny nose); and there is much coughing, sneezing, and sometimes itching. Drugs called *antihistamines* can sometimes be effective in neutralizing histamine and relieving some of these effects.

In the allergies just described, the reactions are concentrated in the nasal area. In some allergies, particularly those involving injected substances, such as insect stings or penicillin shots, the immune reactions are more generalized and affect the whole body. Severe reactions involving the respiratory and circulatory systems may occur and can be fatal if not treated immediately. Epinephrine, a hormone, is commonly injected in affected individuals to counteract the effects of histamine and other mediators. Attempts are now being made to develop new types of therapies for preventing or treating allergies.

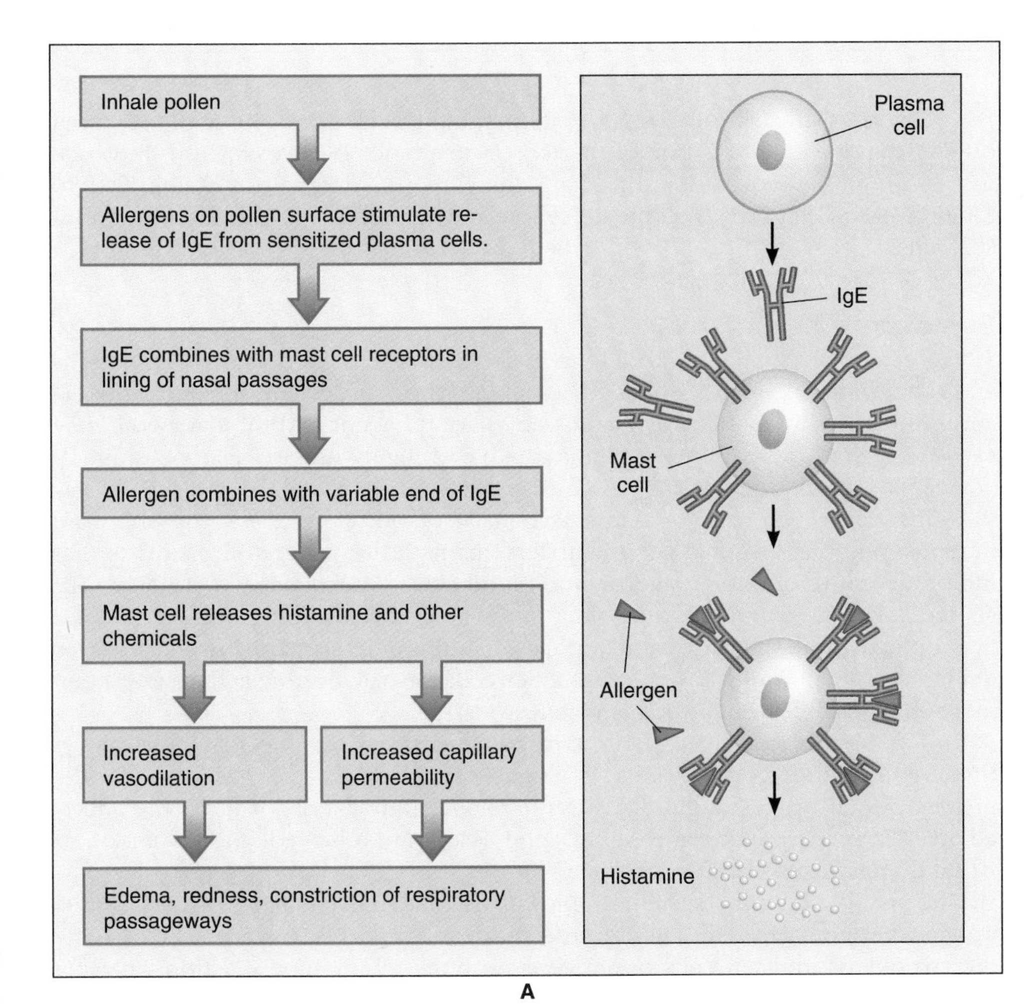

A

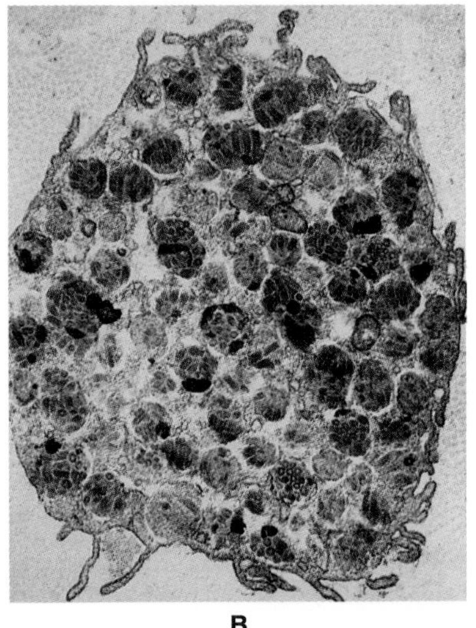

B

Figure 38.18

(A) In allergic responses, basophils are stimulated to release great quantities of histamine when an appropriate antigen, such as pollen, is encountered. Histamine is a chemical mediator that triggers allergic reactions. (B) This mast cell contains numerous granules (stained dark) that are filled with histamine and other immune systems.

Organ Transplants

Beginning in the 1960s, it became technically possible to transplant organs from one individual to another. There was initially great enthusiasm over the possibility of providing new organs to patients with hopeless heart or kidney ailments. However, disappointment followed a short time later when it became clear that transplanted organs tended to be rejected because of immune system reactions against foreign MHC molecules.

Since then, higher success rates have been achieved through the use of precise tissue typing and drugs that suppress immune responses. Great efforts are made to match the MHC molecules between patient and donor as closely as possible. In most cases, close relatives now provide the transplanted organ or tissue. The use of drugs to suppress the immune system is problematic, since it may lead to serious infections from pathogenic microorganisms. Drugs are usually used for only a short period of time or in quantities that reduce but do not abolish the immune response. New approaches may soon be developed. For example, it may be possible to prevent nonself antigens from binding to self MHC presentation molecules, thus preventing a full-blown immune response to the transplanted tissue.

Many important questions about immunity remain unanswered. The basis of immunological tolerance has not yet been defined, although two attractive hypotheses—clonal deletion and clonal anergy—have been advanced. If tolerance mechanisms fail, autoimmune diseases may result, but their causes are generally unknown. Being able to diminish immune responses may be desirable in two cases: extreme immune responses that occur in allergic reactions and organ transplants that have poor success rates because of immune rejections.

Given its great importance in maintaining the health of humans, research on the immune system is one of the most active fields of investigation. In Chapters 39 and 40, we consider what happens when this marvelous system fails to repel disease agents or to eliminate self cells that become transformed into nonself cells.

Summary

1. Humans have two general defense systems that are used against pathogens and other harmful agents. Nonspecific defense mechanisms are innate and operate in the same inflexible way against any harmful agent and the damage it may cause. The specific immune system is capable of highly precise responses directed against a single agent.
2. Nonspecific defense mechanisms consist of body surfaces—the skin and mucous membranes—that act as barriers against entry, secreted products that inhibit the entry of foreign agents, and blood cells that react against foreign substances or organisms that gain entry into the body. Granular leukocytes and macrophages remove various materials, including microorganisms and dead or dying cells, by phagocytosis. Natural killer cells destroy host cells that have been infected by viruses or have become cancerous.
3. Inflammation is a nonspecific two-phase process that occurs after injury. Immediately after cells, tissues, or organs are damaged, various blood cells invade the site and, through the action of chemical mediators, wall off the affected area. Subsequently, destroyed cells and tissues are removed, and resolution or wound repair takes place.
4. The specific immune system employs two basic mechanisms to react against nonself antigens. In cellular immune reactions, various T lymphocytes react directly against nonself cells or antigens. In humoral immune reactions, plasma

cells, a type of B lymphocyte, produce specific antibodies that respond to a single antigen or antigenic determinant. Self cells are protected from being destroyed by the immune system because of the presence of MHC molecules on their surfaces.
5. B and T lymphocytes are produced in the bone marrow. T cells then undergo processing in the thymus, while B cells are apparently processed elsewhere. When mature, they move to lymphoid organs and tissues, where they carry out their immune functions.
6. In a primary immune response, a nonself antigen is first captured and processed by an antigen-presenting cell, usually a macrophage. Subsequently, a nonself antigen is joined with a self MHC molecule, and both are presented on the surface of the macrophage. A helper T cell with specific receptors for the combination interacts with the presenting macrophage and becomes activated. B cells are activated directly by reacting with the antigen or indirectly by helper T cells. Once activation occurs, various T cells that react against the nonself antigen, against foreign cells, or against affected host cells are produced, along with plasma cells capable of producing specific antibodies against the nonself antigenic determinant.
7. A legacy of the primary immune response is the presence of memory B and T cells—cells that have specific receptors for the initiating antigenic determinant. These cells have very long lives, and if the original antigen reappears, a secondary immune response quickly results in its elimination.

Review Questions

1. What are some fundamental differences between the two general human defense systems?
2. What are the defense functions of skin, mucous membranes, and various secretions?
3. Describe the types of blood cells that participate in nonspecific defense responses. What is the role of each?
4. What is inflammation? Why is it important? What are the major events in an inflammatory response?
5. How is the concept of self used in immunology? Why is it important?
6. What is an antigen? An antibody? How are they related? How do their structures reflect this relationship?
7. Describe the life cycle of a T cell and a B cell. What type of immune response is associated with each cell?
8. What are the general functions of T cells, B cells, and macrophages in immune responses?
9. How does a primary immune response occur? What does this type of immune response accomplish?
10. What is the significance of a secondary immune response? How does it differ from a primary immune response?
11. What explains the development of immune tolerance?
12. What is the relationship between immune responses and autoimmune diseases? Allergies? Organ transplants?

Essay and Discussion Questions

1. What might be some of the consequences of removing the thymus gland at birth?
2. Some physicians who did pioneering immunology research believed that understanding immune processes could lead to curing all human disease. What are some inherent limits to this idea?
3. Explain the possible evolutionary significance of complicated specific immune systems being limited to birds and mammals.
4. Allergy sufferers rely on drugs such as antihistamines to control symptoms. What other strategies might be used to control allergies in the future?

References and Recommended Reading

Abbas, A. K., A. M. Lichtman, and J. S. Pober. 1991. *Cellular and Molecular Immunology.* Orlando, Fla.: Saunders.

Ada, G. L., and G. Nossal. 1987. The clonal-selection theory. *Scientific American* 257:62–69.

Alt, F. W., K. Blackwell, and G. D. Yancopoulos. 1987. Development of the primary antibody repertoire. *Science* 238:1079–1086.

Blackman, M., J. Kappler, and P. Marrack. 1990. The role of the T cell receptor in positive and negative selection of developing T cells. *Science* 248:1335–1341.

Burnet, F. M. 1959. *The Clonal Selection Theory of Immunity*. London: Vanderbilt and Cambridge University Press.

Nilsson, L. 1987. *The Body Victorious*. New York: Dell/Delacorte Press.

Ramsdell, F., and B. J. Fowlkes. 1990. Clonal deletion versus clonal anergy: The role of the thymus in inducing self tolerance. *Science* 248:1342–1348.

Sprent, J., E. Gao, and S. R. Webb. 1990. T cell reactivity to MHC molecules: Immunity versus tolerance. *Science* 248:1357–1363.

Till, J. E. 1981. Cellular diversity in the blood-forming system. *American Scientist* 69:522–527.

Unanue, E. R., and P. M. Allen. 1987. The basis for the immunoregulatory role of macrophages and other accessory cells. *Science* 236:551–557.

Yancopoulos, G. D., and F. W. Alt. 1988. Reconstruction of an immune system. *Science* 241:1581-1583.

CHAPTER 39

Human Diseases

Chapter Outline

Reading Questions

1. Which human diseases are of greatest importance in the United States?
2. How does the concept of balanced pathogenicity explain the relationship between a disease agent and its host?
3. What is cancer? What are vascular diseases?
4. Why are individuals being asked to assume control in managing disease risk factors in their lives?

As a result of impressive strides in producing antibiotics during the first half of the twentieth century, progress in understanding the causes of human diseases actually slowed. Many people came to think that all diseases were due to invasion of the body by *microbes* (single-celled microorganisms and viruses). They also thought that the way to combat such disease agents was to destroy them with antibiotics, vaccines, and drugs.

There was considerable hope in the 1950s that all human disease would someday be conquered. However, further developments soon showed that such optimism was misguided. The discovery of numerous human viruses revealed a class of disease agents that was much more difficult to control than bacteria. It also became clear that many human diseases were caused not by microbes but by other factors.

DISEASES AND MODERN MEDICINE

Concepts of human disease have changed with time (see the Focus on Scientific Inquiry, "Changing Concepts of Disease"). Today we generally think of **disease** as any abnormal condition of the body that impairs normal functioning. This is a very broad definition that includes mental disease, social maladaptations, and trauma, as well as the more common meaning, which includes heart disease, cancer, and **infectious diseases** caused by microorganisms such as bacteria, viruses, protozoans, fungi, and parasitic worms. Other causes of disease are also becoming better understood. Nutritional diseases have been recognized for nearly a century, and our national preoccupation with vitamins and "health foods" reflects widespread concern. Genetic disorders, discussed in Chapter 18, form a long list of complex diseases. Modern medicine is based on the concept that disease impairs normal physiological functioning. This reflects the mechanical view of the body that has been the foundation of Western medicine for over 100 years.

The Role of Lifestyles

The medical community has become increasingly aware of the influences of lifestyle and environment on the individual. Chemical and psychological stresses have been found to cause harmful effects on the body (see Figure 39.1). Public health officials consider cigarette smoking the single greatest health risk factor, accounting for nearly one third of all deaths due to vascular diseases, cancer, and respiratory diseases. Other environmental factors such as preservatives, charcoal-broiled food, drinking water contaminated with chemicals, fouled air, pesticide residues, and a host of other agents may be involved in some diseases, although their contributions have not been definitively established.

Figure 39.1

Until the mid-twentieth century, it was thought that infectious agents accounted for most diseases. Now it is clear that many other factors, such as stress, nutrition, and lifestyle, contribute to disease.

The Future

Perhaps a lesson from the nineteenth century has relevance for controlling diseases of modern civilization. Before the discovery of antibiotics, several of the most potent infectious diseases, such as cholera, were effectively controlled by the development of public health programs. Sanitation and clean water ended some of the worst epidemics in industrial Europe well before the real causes (bacteria) were discovered. Thus it is not altogether a novel idea that control of environmental and lifestyle diseases may involve engineering, social workers, nutritionists, education, and modification of our attitude toward exercise.

Modern medicine also appreciates the extent to which different categories of disease often affect the same individual. Invading organisms, genetics, lifestyle, and environmental conditions are often different facets of problems faced by the modern physician. The patient who complains of a chronic sinus infection undoubtedly harbors an invading microbe. However, effective treatment may involve more than just prescribing an antibiotic if the patient smokes heavily, lives in a seriously polluted atmosphere, is a perfectionist who works 80 hours a week, or thinks a balanced meal is a hamburger with fries and a milkshake.

In fact, major campaigns designed to raise public awareness of health risk factors have achieved some notable successes in the past decade or two. For example, the National High Blood Pressure Education Program, initiated in 1972 by the National Heart, Lung and Blood Institute, is credited with reducing deaths due to strokes by 55 percent by 1984. Such education programs are likely to become increasingly important in the future.

DISEASES OF CONTEMPORARY IMPORTANCE IN DEVELOPED COUNTRIES

Advances in modern medicine have greatly benefited citizens living in developed countries; most obviously, we live longer (see Figure 39.2). Antibiotics, effective sanitation practices, public health measures, clean living environments, and education have diminished the importance of infectious diseases as a cause of death. For example, deaths due to cholera, typhus, scarlet fever, and many other diseases of historical importance are now rare. Much of that success is due to childhood immunizations against dangerous infectious diseases. The picture is less cheering for inhabitants of developing countries, where infectious diseases still claim a frightful toll.

Although mortalities have diminished, the frequency of infectious diseases in all citizens of the world has probably not decreased significantly in the twentieth century. Rather, some new diseases, and some diseases that have been around for a long time, have become more common because of new opportunities for infecting the dense, mobile human populations that now exist on Earth.

Figure 39.2

The life expectancy of men and women in the United States has increased steadily for nearly 100 years. This is due largely to improved health care, nutrition, sanitation, and education.

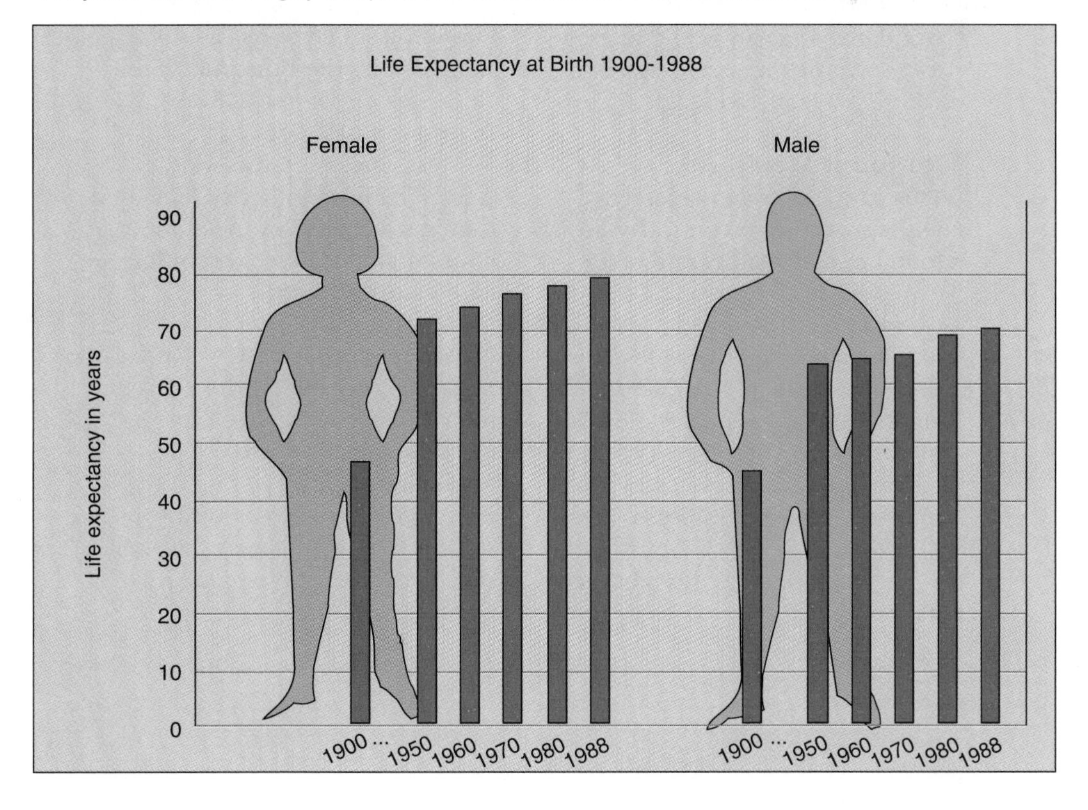

CHANGING CONCEPTS OF DISEASE

Focus on Scientific Inquiry

Early Human Cultures

The concept of disease has a long history and has changed along with culture. In very early cultures, the human worldview was basically magical. No distinctions were made between living and nonliving things, and all objects were thought to have a potential for affecting other objects. In such a setting, illnesses were believed to be a consequence of someone's anger or spite. They could be brought about from a spell cast by a jealous rival, by a person who had been mistreated, or by a demon or spirit that was offended or was seeking punishment for some offense. The symptoms of any illness were of little use in determining its cause. The primary concerns were discovering who had caused the problem and why. All early human societies had individuals whose special function was to investigate the causes of illness. These medicine men (and women), or *shamans,* used various methods of divination and careful interviews to uncover the force, and perhaps the motive, behind the problem. Once understood, the shaman would attempt to resolve the issue through exhortation, prayers, and magic.

Rational Medicine

This early belief in magic was widespread, and even today it is found in many cultures that are classified as "primitive." The Greeks first formulated a system of rational medicine (as well as the first systematic rational worldview). The early Greeks' view of disease was expressed in the fifth century B.C. and survives in a set of books called the *Hippocratic Corpus*.

Greek doctors thought that the body was composed of substances they termed *humors*. There were four humors: black bile, yellow bile, phlegm, and blood. Health was the optimal body state when the four humors were in balance, and illness was caused by an imbalance, most often an excess of one of the humors. The body also had an internal force known as its *vital heat* that maintained the balance of humors. If there was an excess of a humor, the vital heat "cooked" it, which allowed the body to expel it.

Many factors could influence the balance of humors. Diet, age, weather, daily habits, and season were among the most important. To correct humor imbalances (disease), rest and proper diet were recommended to help the body heal itself. Greek doctors recognized many specific diseases and carefully described their symptoms. They believed that humor imbalances could be best treated at certain periods and that many diseases had critical periods when the patient would either recover or die.

The Greek humor theory endured for centuries. The Romans inherited the humor theory and passed it to the Arabs, who in turn transmitted it back to Western Europe in the Middle Ages. The humor theory remained dominant in the European medical community until the end of the eighteenth century.

Nineteenth-Century Medicine

Starting at the end of the eighteenth century, Europe was transformed by the industrial revolution. As part of that revolution, science came to occupy a greater role in all aspects of society. Medicine was particularly affected, and consequently, the concept of disease was fundamentally revised. In place of a disease being caused by an altered balance of the body humors, disease came to be seen as a specific entity in itself. The new concept of disease arose from three changes in medical science: localism, cell theory, and germ theory.

Localism was formulated in Paris by a group of physicians who began to carefully perform autopsies with an eye toward correlating internal changes with external symptoms of disease. This led to the concept of *localism,* the idea that local disturbances in specific organs were responsible for disease.

Localism was extended by the cell theory in the 1830s. The *cell theory* claimed that the body was composed of cells, which were organized into tissues and organs. Rudolf Virchow, the greatest theoretician of the cell theory, maintained that a local disturbance in a cell or group of cells could spread through the body and affect other cells. The cell theory replaced the humor theory as the major conception of the body and of disease. Thus disease was considered to be caused by a local disturbance of a group of cells that could undermine their proper functioning.

The great French chemist Louis Pasteur (see Figure 1) carried the cell theory even further. In a set of brilliant experiments, he showed that many diseases

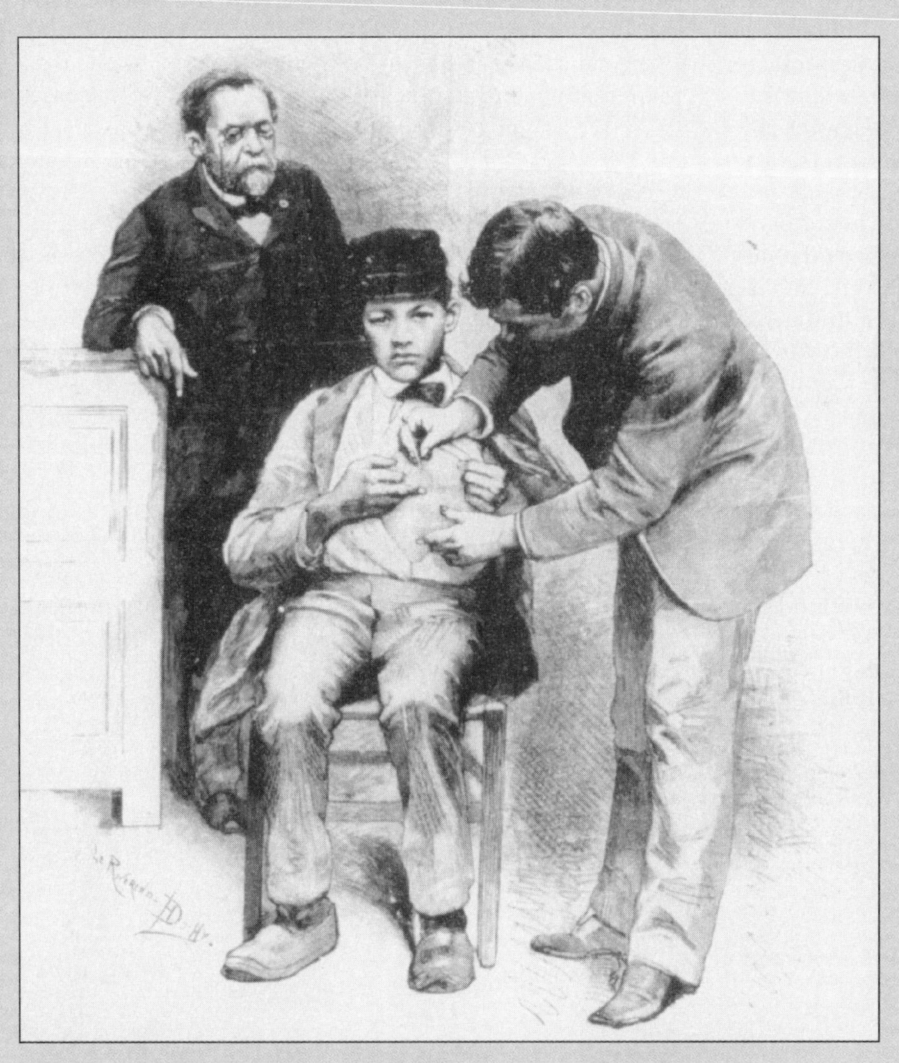

Figure 1 Louis Pasteur developed the germ theory—the idea that diseases were caused by microbes that invaded the body. This drawing, from a nineteenth-century French magazine, shows Pasteur (with glasses) supervising a rabies vaccination.

were caused by microbes, such as bacteria, that invaded healthy bodies. His new theory was called the *germ theory*. His ideas came from studying *germs* that spoiled beer, milk, and other natural products. To counteract these germs, he invented the process of pasteurization (or partial sterilization) to keep milk and beer from spoiling.

Obviously, humans cannot be gently boiled to keep germs from growing in them. Pasteur also discovered that some diseases could be combated by *vaccination*. In modern terms, the basic approach involves exposing a potential host to an appropriate disease-agent antigen (a *vaccine*). This will lead to an immune response and result in permanent immunity because of memory cells (explained in Chapter 38). Vaccinations were powerful preventives but were not well understood or of much value after an individual had contracted a disease. For these reasons, much of the medical research in the second half of the nineteenth century involved attempts to identify and to kill microbes.

Joseph Lister appreciated the finding that germs spoiled organic matter. He quickly realized that germs were probably responsible for causing infections of wounds. Shortly thereafter, he found that carbolic acid could be used to kill germs that caused infections in wounds and surgical incisions. Through his use of antiseptic surgery and dressings, the mortality rate in surgery was drastically reduced. Indeed, before Lister, surgery was a last-resort procedure because the survival rate was abysmally low.

Finding a means for combating infectious diseases was more difficult. Paul Ehrlich, the German physician mentioned in Chapter 38, had the idea of trying to find chemicals that would adhere to microbes but not to human cells. Ehrlich reasoned that if one could attach a poison, like arsenic, to such a chemical, it would kill the microbe but not the host. He referred to these hypothetical agents as "magic bullets." After many discouraging years of research, such a drug was found. Called *salvarsan,* it was the first of the sulfa drugs that were effective in treating certain diseases.

Antibiotics and Vaccines

An even more dramatic discovery occurred early in the twentieth century, when it was found that certain microbes produce chemicals that inhibit the growth of other microorganisms. These chemicals are called *antibiotics,* and they revolutionized medicine in that they made it possible to treat bacterial infections after they occurred. Antibiotics disrupt vital metabolic pathways in the microbe. For example, the most famous antibiotic is produced by a mold, *Penicillium*. When used in making an extract that can be injected into the human body, it prevents bacterial cell wall synthesis in susceptible species (and thus kills them) but does not damage human cells.

Viruses, however, cannot be treated with antibiotics because they use the host cell's metabolic machinery. To control modern viral diseases, emphasis was first placed on developing new vaccines. Highly effective virus vaccines have been developed for measles, polio, mumps, and rabies, and in one of the great triumphs of modern medicine, smallpox was eradicated by a vaccine.

Although the development of new vaccines remains an important goal, much attention is now being directed toward identifying or creating antiviral drugs. *Drugs* are chemical substances used to control diseases. Successful antiviral drugs are now appearing that are targeted at a specific feature of the virus life cycle. By disrupting, or abolishing, an essential step in viral synthesis, the disease may never develop. For example, the drug acyclovir is a specific inhibitor of replication by certain herpesviruses. It prevents the normal synthesis of virus-directed DNA and hence new virions cannot be produced. It does not, however, eliminate the virus. Despite these advances in treating many diseases, there are several important diseases for which no effective antibiotic or vaccine is available (for example, malaria and herpes). Treatment of these diseases remains a challenge for the future.

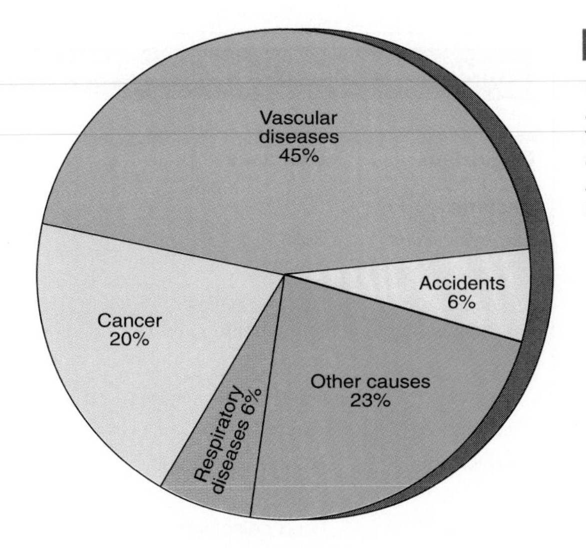

Figure 39.3

Each year, approximately 2 million Americans die of cancer, diseases of the vascular and respiratory systems, accidents, and "other" causes, such as AIDS, pneumonia, diabetes mellitus, murder, and suicide.

Two consequences of the reduction in mortalities caused by infectious diseases are that we now live to an older age and we die of other diseases. What are the primary causes of death in the United States? Figure 39.3 indicates that vascular (circulatory) diseases are the leading cause of death, and cancer is second. Infectious diseases fall in the "other" category, with no specific disease accounting for as many deaths as any of the leading causes.

Why do our grandparents and parents now tend to die of circulatory diseases and cancer? Which infectious diseases are serious at the present time and likely to be important in the twenty-first century?

It once seemed possible that human diseases could be eradicated by the use of antibiotics. It is now recognized that many diseases have causes that cannot be addressed with antibiotics. In developed countries, infectious agents still cause serious diseases. However, nutrition and lifestyle factors are associated with circulatory diseases and cancer, the major causes of death in the United States.

INFECTIOUS DISEASE

Humans live in an environment filled with microorganisms. Vast numbers of beneficial or harmless bacteria normally inhabit our skin, our mouth, and all of our bodily openings. Also, different viruses routinely enter and exit our body during the course of our lives. Fortunately, we serve as hosts to only a few bacteria, viruses, and other microorganisms that cause disease.

Infectious Disease–causing Agents

Diseases can be caused by microorganisms at all levels of classification, but we shall confine our discussion to bacteria (kingdom Prokaryotae, subkingdom Eubacteria) and viruses. Viruses are not considered to be living by many scientists since they are unable independently to carry out the activities that define living cells or organisms. Table 39.1 contains information about a few important human bacterial and viral diseases. More than 400 microorganisms are known to infect humans. Many protozoans (kingdom Protoctista), fungi (kingdom Fungi), and parasitic worms (kingdom Animalia) also cause important human diseases.

Bacteria

Bacteria are a diverse group of single-celled organisms with varying shapes and sizes whose most important biological roles are decomposition, nitrogen fixa-

Table 39.1 Some Bacteria and Viruses That Cause Diseases in Humans

Organism	Diseases	Features
Bacteria		
Staphylococcus[a]	Boils	Common inhabitant of skin; may infect skin cells, causing pimples or boils
Streptococcus	Dental cavities, pneumonia	Regular inhabitant of mouth; causes plaque Normal inhabitant of lung epithelium; causes infections in damaged lungs
Neisseria	Gonorrhea	Sexually transmitted disease; infects epithelium lining urogenital tract
Clostridium	Tetanus, gangrene, botulism	Widely distributed in soil and intestines; opportunistic infections cause disease
Salmonella	Food poisoning	Caused by an exotoxin that affects cells of the intestinal epithelium
Pseudomonas	Urinary tract and wound infections	Common human intestinal bacteria; resistant to many antibiotics; opportunistic infections
Legionella	Legionnaire's disease	Respiratory pathogen; often acquired from contaminated air-conditioning units
Viruses		
Herpesvirus	Herpes	Two forms: type 1 infects salivary glands; type 2 the genital tract. Both are persistent and cannot be eliminated
Papovavirus	Warts	Infects skin cells
Orthomyxovirus	Influenza types A, B, and C	Respiratory infections
Coronavirus	Common cold	More than 100 antigenetically distinct cold viruses
Retrovirus	Leukemia and other cancers	A retrovirus—HIV—causes AIDS

[a]For bacteria, the genus name is included in this table. There are several species of most of these bacteria that cause disease.

tion, and disease agents (see Figure 39.4). All bacterial cells are surrounded by a *cell wall,* composed of protein and carbohydrate molecules, that provides rigidity and protection. Many bacteria have a surrounding *capsule* that is gelatinous and lies outside the cell wall. The capsule provides protection and aids in attaching bacteria to cells they can infect. Molecules of the cell wall and capsule often act as antigens in initiating a host immune response. Bacteria may also have external flagella and pili, structures that originate inside the cell. Flagella permit movement, and *pili* are used to attach bacteria to host cells and in bacterial reproduction. Finally, a few bacteria, such as *Clostridium,* can produce *spores* that are resistant to most environmental stresses (for example, heat, cold, and drying). New bacteria develop from spores when favorable conditions occur.

One outstanding characteristic of many bacteria is that they reproduce by cell division every few minutes when they inhabit a favorable environment. Thus a single bacterium can give rise to a population that numbers in the millions in less than 24 hours. Bacteria that normally colonize external areas of the human body cause little or no harm, other than distinctive odors from inhabited areas such as feet, mouth, and underarms, unless they have an opportunity to gain entry through wounds. It is impossible to eradicate these normal occupants, but

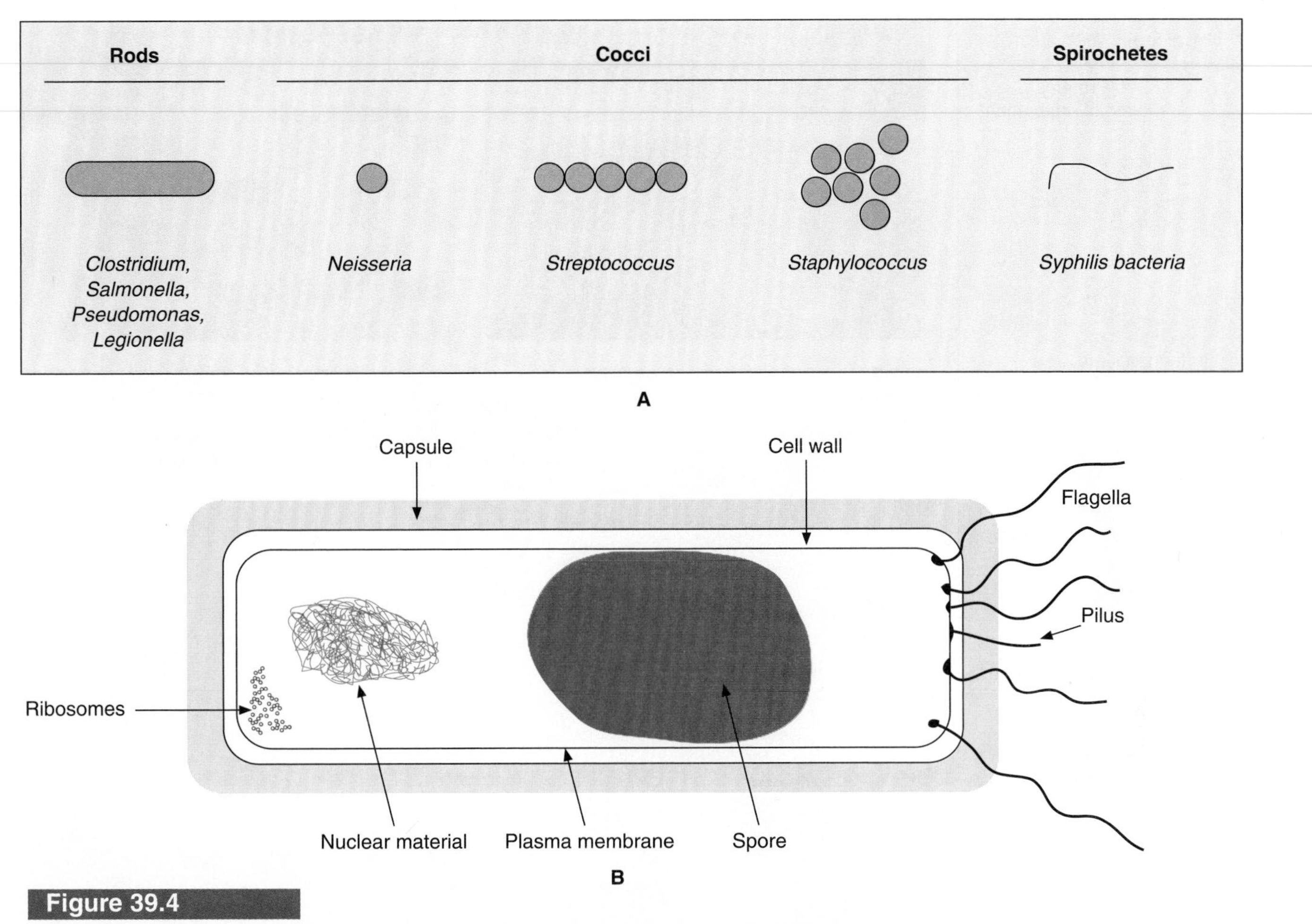

Figure 39.4

(A) Bacteria are commonly classified according to shape—rod, coccus (sphere), or spirochete (coil)—and whether or not they occur singly or together in chains. (B) Bacterial cells are enclosed by a cell wall and a plasma membrane. Within the cell are the nuclear material and ribosomes for protein synthesis. Different bacterial species may also have a capsule, flagella, or pili and the ability to form spores.

proper personal hygiene serves to keep their numbers within a tolerable range. Traditional pathogenic bacteria, by contrast, may cause disease if they evade the immune system and become established on or in the human body. The mechanisms by which they cause damage vary, depending on the bacteria and tissues affected. Bacterial infections can generally be treated with antibiotics.

Viruses

Viruses occupy a world between living and nonliving. They can replicate only inside a host cell, since they are unable to carry on metabolic activities by themselves. The structure of different virus particles varies (see Figure 39.5A). In general, a virus is a submicroscopic particle consisting of a nucleic acid *genome* (that is, all of its genetic material) surrounded by a protein coat called a *capsid*. Some viruses, such as the influenza *virion* (an infectious virus particle), have an outer *envelope* consisting of lipids and viral proteins. Unlike all living cells, which contain both DNA and RNA, viruses contain either RNA or DNA but not both. The genome of viruses is limited in size, ranging between 3 and 200 genes. The origin of viruses is steeped in mystery, but they are clearly stupendously successful pathogens that cause many serious human diseases.

A general viral growth cycle is described in Figure 39.5B. A viral infection begins when a virion comes in contact with a host cell, becomes attached, penetrates the plasma membrane, and introduces its DNA or RNA into the cytoplasm.

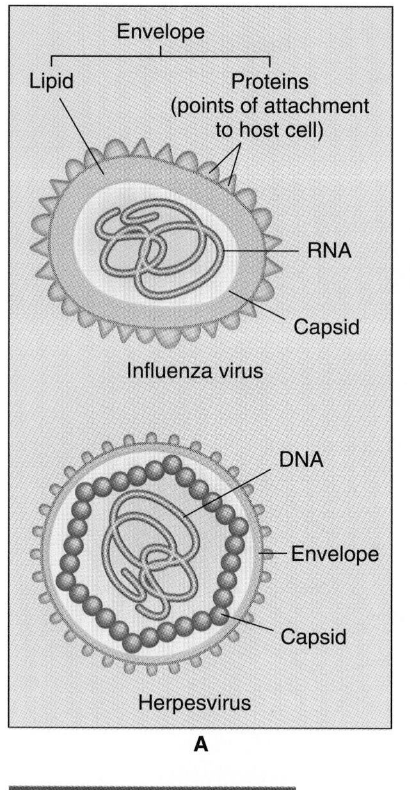

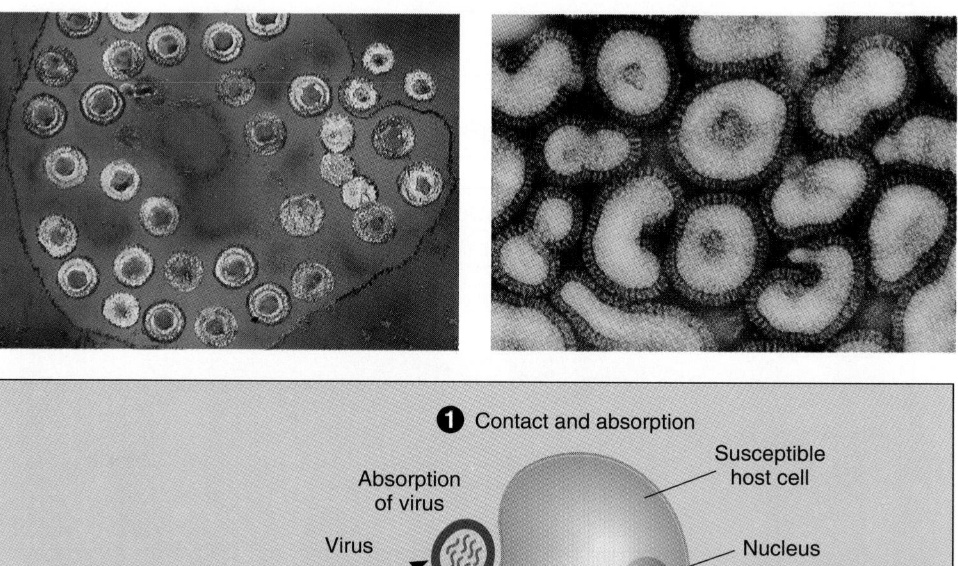

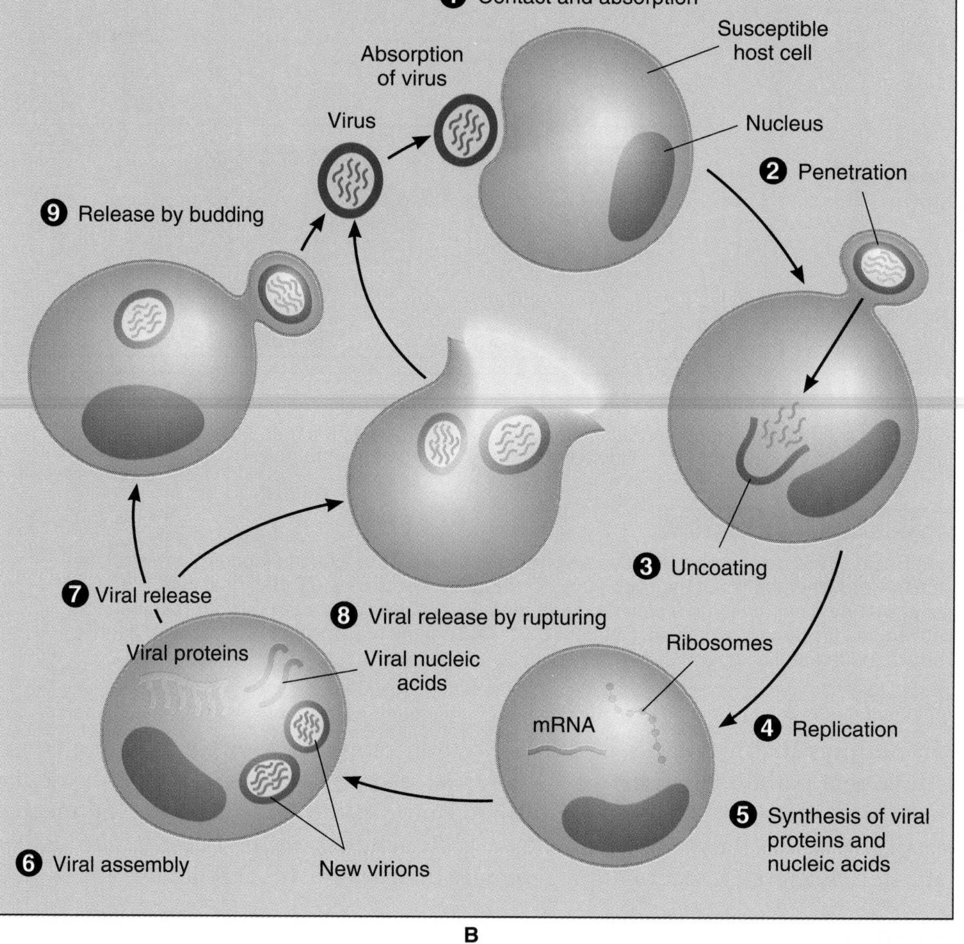

Figure 39.5

(A) Viruses are extremely small particles that infect cells and cause many human diseases. Their basic structure includes an outer envelope composed of lipid and protein, a protein capsid, and genetic material that is enclosed within the capsid. The photo on the left shows herpesviruses (yellow and green spheres), and the photo on the right shows human influenza viruses. (B) Viral replication can take place only within host cells and occurs in a stepwise fashion. A virion first contacts a host cell and enters the cytoplasm, where it sheds its envelope and capsid and releases its genetic material. The viral DNA (or RNA) directs the host cell's protein and DNA-synthesizing machinery to produce viral proteins and DNA. These viral molecules are then used in assembling new virions. Once formed, they are released after rupturing the host cell or by budding from the cell's surface. Released virions may then infect other cells.

A DNA virion's nucleic acid then enters the nucleus, where it takes over the host cell's protein- and DNA-synthesizing machinery. RNA viruses usually operate in the cytoplasm, where conditions for RNA synthesis are optimal. Regardless of location, the virion nucleic acid "orders" the host cell's different RNAs, enzymes, and ribosomes to synthesize virion proteins and nucleic acids. These viral proteins and nucleic acids are then used in assembling new virions. The time required for most viruses to produce 100 to 1,000 new virions in a cell is less than one hour. The normal activities of the host cell often cease once the virion takes control. When the infected cell becomes full of new virions, it may rupture, releasing its contents all at once; it may simply disintegrate and release the virions more slowly; or the virions may "bud" from the cell's membrane. The effect of these processes is to kill the host cell and create new virions that can go on to infect fresh cells.

Other viruses, including some DNA "tumor viruses" that cause cancer and the extremely important RNA retroviruses that cause cancer and AIDS (discussed in Chapter 40), use a different strategy. When these virions infect a cell, their nucleic acid can be stably integrated into the host's DNA molecules, where it remains "silent" until turned on by a certain signal. The effects of these agents are described in Chapter 40.

The treatment of viral diseases is generally complex. Many of the early symptoms of viral disease—fever and inflammation—are the direct result of non-specific defense responses to the virus and virus-infected cells. However, much to everyone's frustration, there is no quick cure for many viral diseases. Why can't the symptoms of a common cold or the flu be alleviated by getting a shot or taking a pill? The viral life cycle holds the answer. Drugs or antibiotics intended to kill a virus directly would have to kill host cells, an unacceptable approach. Thus the strategy commonly employed ("get lots of rest and drink as much fluid as possible") is simply to wait until the immune system catches up with the virus. This eventually occurs during the virus's passage from one host cell to another, when appropriate immune cells are able to capture antigenic virions. For several of the most important viral diseases, such as polio and smallpox, effective vaccines and drugs have been developed.

Bacteria, viruses, fungi, protozoans, and parasitic worms cause infectious diseases. Both harmless and pathogenic bacteria may cause disease if they enter the body. Bacterial infections can usually be treated with antibiotics. Viruses have unique life cycles. Host cells are required to synthesize viral genetic material and proteins and to assemble new virions. Since they live inside host cells, viruses cannot be reached with antibiotics.

Evolutionary Considerations

What are the basic requirements for a microbe that causes disease? They must include infection of a host, survival, successful reproduction, and at some point, transmission to a new host. If any of these is not accomplished, the microbe will not be an evolutionary success. The consequences of these requirements are interesting and enable us to understand disease better.

At the heart of a microbial pathogen's success is an extraordinary rate of evolution in comparison with that of their human hosts. As discussed in Chapter 38, people have developed highly effective immune defenses against disease agents. However, the rate at which new generations of bacteria and viruses are produced is measured in minutes or hours, not decades. Thus there are many more chances for microbes to benefit from the genetic events (mutations, recombination) that lead to new opportunities. As a result, successful disease microbes are usually a step ahead in the race with our immune system.

Microbe Survival

To survive and cause disease, an individual microbe must, at some stage of its life cycle, be able to finesse the human immune system. Different microbes accomplish this in a number of distinct ways. Some microbes are only weakly antigenic, so our immune system fails to mount an effective response against them. Others prevent or suppress immune reactions (to their antigens) in the infected host. For example, bacteria use several mechanisms to avoid being phagocytosed and processed by macrophages (see Figure 39.6). Intracellular infectious agents, such as the herpesviruses that live inside nerve cells, are not usually exposed to immune cells or circulating antibodies. Also, a few disease agents, especially certain viruses, frequently undergo changes in the molecular composition of their antigens (see Figure 39.7). As a result, the immune system does not initially recognize them when infection occurs, and they are able to become established in the

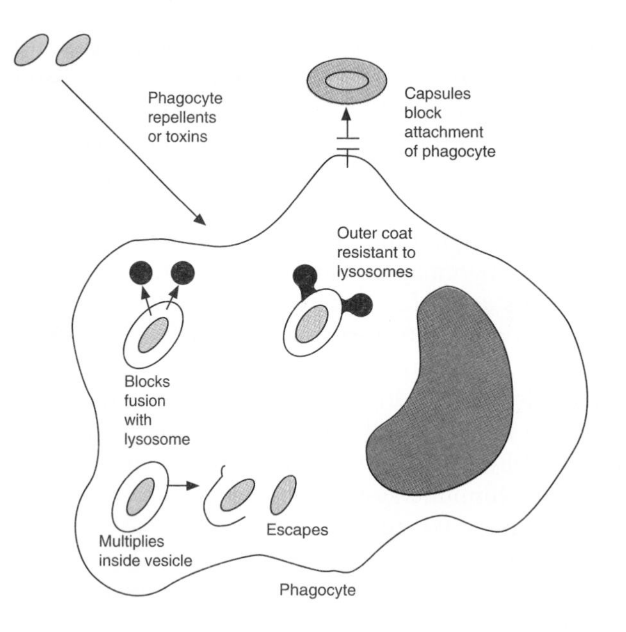

Figure 39.6

Some bacterial species have mechanisms to avoid being killed by phagocytes, among them secreting substances that kill or deflect phagocytes, capsules that block phagocyte attachment or fusion with lysosomes, or capsules that are resistant to the action of lysozymes. Also, some bacteria can multiply inside phagocytes without being destroyed.

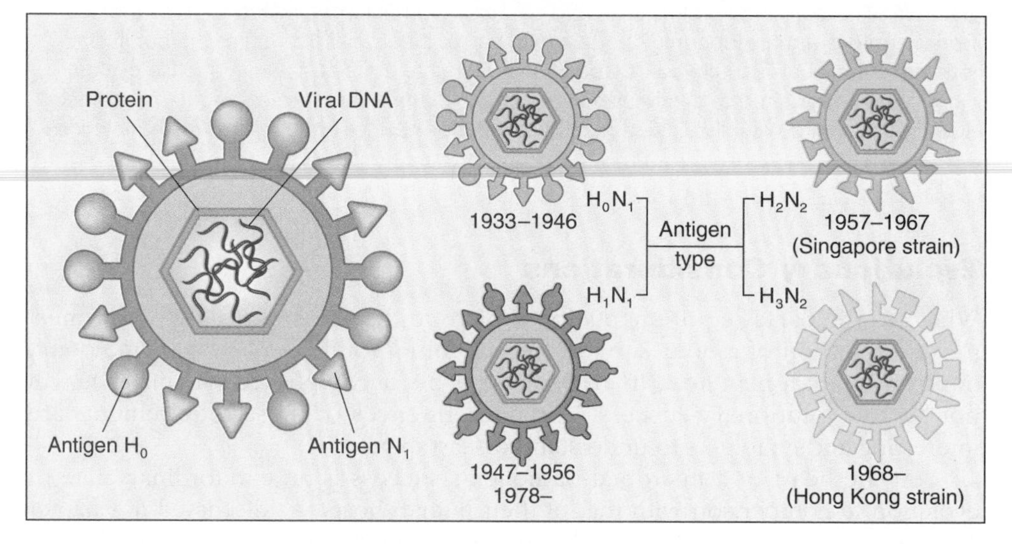

Figure 39.7

Influenza viruses occasionally undergo changes in their envelope proteins (antigens) that make them resistant to host immune responses. (Why?) Strain A of the influenza virus has two major antigenic proteins on its envelope—antigen H is associated with cell attachment, antigen N with cell penetration. The "standard" strain A antigens are designated H_0N_1. In 1947, a new strain, H_1N_1, with different H antigens, appeared. This strain remained in the population for ten years and then subsided, only to reappear in 1978. Other changes are indicated in the figure. The Hong Kong strain was particularly virulent and remains in the population today.

body. Finally, many microbes (for example, those causing colds and sexually transmitted diseases) are successful because their "offspring" pass quickly and efficiently to a new host before an effective immune response that kills them in the original host can be initiated.

Disease Transmission

Assuming successful reproduction, how are microbes that require transmission passed on to other human hosts? The two types of transmission are described in Figure 39.8. **Horizontal transmission** occurs when one individual infects another, directly or indirectly. **Vertical transmission** occurs when the microbe is transferred from parent to offspring via infected sperm, ova, or milk or through the placenta. Certain cancer-causing viruses are known to be transmitted by vertical spread, and a number of other diseases (for example, multiple sclerosis) are suspected to be spread by this route.

In developed countries, horizontal transfers via infected feces, food, and urine have decreased in significance during the past century, while two other routes have become increasingly important. Viruses that spread by *respiratory transmission* (such as sneezing, coughing, talking, and kissing) and *sexual*

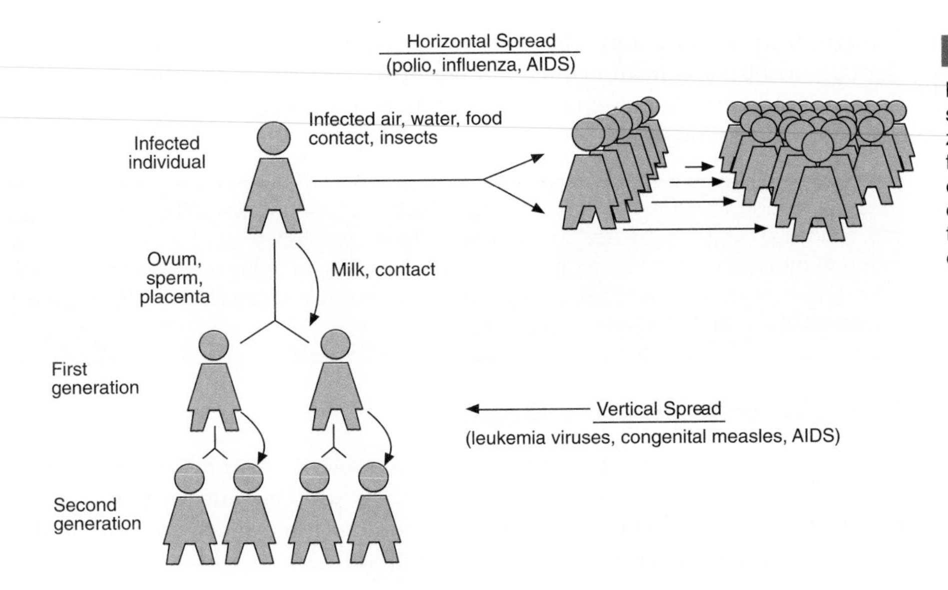

Figure 39.8

Most infectious disease organisms are spread from person to person by horizontal transmission—through air, water, food, or personal contact. Some diseases can be transmitted from parent to offspring by vertical transmission—through the placenta or infected sperm or ova.

transmission have entered their "golden age"—the first because of increasing population size and mobility, which allows individuals to contract infections from all parts of the world, and the second largely because of a greater degree of promiscuity and increased numbers of sex partners. Some disease agents, such as the AIDS virus, are transmitted both horizontally and vertically.

The Concept of Balanced Pathogenicity

Can a disease microbe be successful if it kills its human host? In general, the answer must be no, because transmission cannot occur as a result of that effect. Therefore, from an evolutionary viewpoint, it is in the best interest of the microbe to cause as little damage to the host as possible. Disease agents that have infected humans for thousands of years have generally succeeded in establishing a balance between causing harmful (*pathological*) effects in the host and being able to survive, reproduce, and be transmitted (see Figure 39.9). This is known as the concept of **balanced pathogenicity**, and it helps us understand certain aspects of infectious disease. If an especially virulent disease, such as AIDS or a new flu strain, appears suddenly in humans, it is assumed to be a new disease or a disease to which we have never been previously exposed. No immune resistance has been developed for such diseases. This also explains why North American Indians were initially devastated by diseases (such as smallpox)

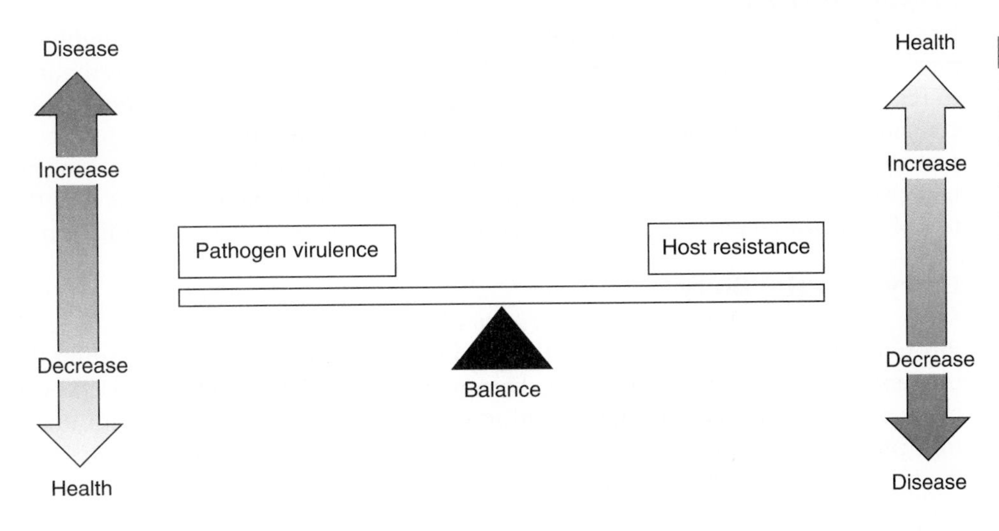

Figure 39.9

The concept of balanced pathogenicity explains that most long-term relationships between a pathogen and a host result in a balance between host resistance and pathogen virulence. In these cases, the pathogen does not usually cause serious harm because of host immunity that has developed over a long time period. However, when new pathogens enter a host, dangerous diseases may result because no secondary immune response will occur. Likewise, if the host's immune system is deficient, an old pathogen may cause serious disease.

brought from Europe and why diseases (such as rabies) that are transmitted accidentally to humans from normal animal hosts cause severe, often fatal effects. In these cases, no balance exists because there has been no long-term experience with the disease organism.

Microbial pathogens must be able to infect hosts, survive in the host's environment, reproduce, and be transmitted to new hosts. To accomplish these tasks, they depend on different mechanisms for eluding the immune system and being transferred to a new host. Disease agents that have a long evolutionary relationship with humans usually cause relatively little damage to the host.

The Life of a Disease Organism

Bacteria, viruses, and other disease-causing agents face a number of supreme challenges in their effort to become successful. They must gain entry into a human host, survive once they arrive, find a suitable place to grow and reproduce, and finally, succeed in being transmitted to another human.

In Search of a Host

Disease-causing microbes commonly enter and leave the host from one of the body surfaces. The tissue of exit is generally related to the means by which the pathogen will be transmitted to and infect a new host.

Respiratory Tract

Microorganisms in the respiratory passages and mouth can become incorporated into fluid droplets (aerosols) and expelled to the external environment by coughing, sneezing, talking, or kissing. For example, a person with a common cold can emit up to 20,000 virus-containing droplets during one sneeze. Given the crowded environments in which most people spend at least part of the day (classrooms, buses, offices), little can be done to prevent the spread of diseases via respiratory transmission.

Intestinal Tract

Microbes that infect organs of the digestive tract are expelled in feces, and for many diseases, the process is accelerated by diarrhea. Numerous pathogens can be transmitted indirectly from feces to the mouth through contaminated drinking water. Many of the great disease epidemics in our history (for example, cholera) involved this route of transmission. Today, in developed countries at least, waterborne diseases are rare because of effective public health measures (proper sewage disposal and treatment, purified water supplies).

Urogenital Tract

Most disease organisms of the urinary and reproductive tracts are transmitted to new hosts through direct mucous membrane contact during sexual intercourse. In recent decades, there has been a tremendous increase in the *prevalence* (percentage of the population infected) of sexually transmitted diseases such as gonorrhea, syphilis, and genital herpes. The underlying causes of this continuing increase are complex but include changes in social and sexual customs and the use of oral contraceptives rather than condoms that block disease transmission.

Entry into the Body

Some microbes, such as *Staphylococcus* bacteria, can invade the body only through breaks in the surfaces of the skin or in mucous membranes lining the urogenital tract, respiratory tract, or digestive tract (see Figure 39.10A). However, many microorganisms, through evolution, have developed mechanisms for entering into or between cells of mucous membranes lining specific organs (see Figure 39.10B). For example, flu and cold viruses enter cells lining the respiratory

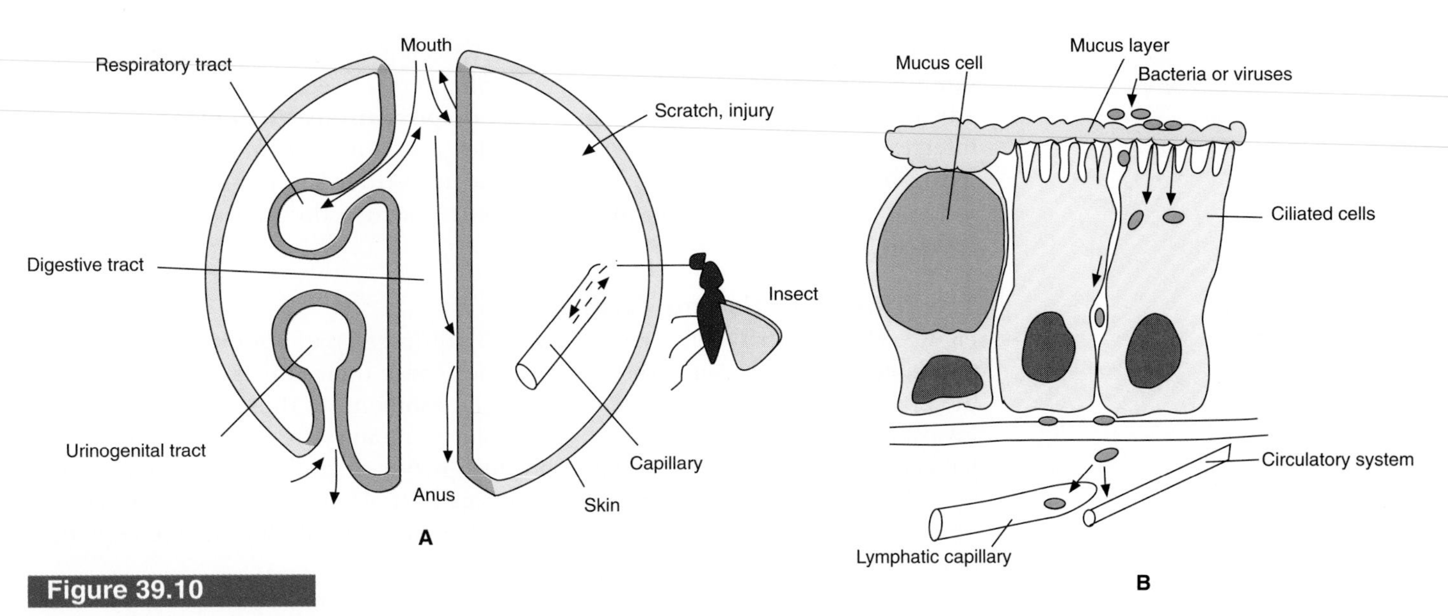

Figure 39.10

(A) Many pathogens enter and leave the human body through the digestive, respiratory, and urogenital tracts. Others enter through wounds in the skin or are injected into the body by mosquitoes or other insects. (B) Viruses and bacteria that cause disease often invade through or between cells of mucous membranes that line the major tracts.

tract, *Neisseria* (gonorrhea bacteria) infect cells of the urinary tract, and *Salmonella* (food poisoning bacteria) pass through cells lining the intestinal tract. Finally, for certain diseases, such as malaria, the agent (a protozoan) is injected through the skin by a mosquito, an insect that carries and transmits the agent.

Entry Through the Respiratory Tract

The air we breathe is filled with microorganisms, which exist as suspended particles. It has been estimated that we inhale about 10,000 of these each day, most of which are harmless. One of the roles of the respiratory system is to remove such particles, and it achieves this largely by trapping them in mucus and moving them to the back of the throat by ciliary action, where they are swallowed and later destroyed by acids and enzymes in the digestive system. Smaller particles that escape this "mucociliary elevator" reach the lung and are phagocytosed by macrophages present in the epithelium. Some viruses (flu, common cold) are able to enter epithelial cells directly. Others (measles virus, *Streptococcus*) cause abrasions and enter through the damaged sites.

Entry Through the Intestinal Tract

The lower intestinal tract is a site of intense bacterial activity. It has been estimated that normal individuals provide housing for 10^{14} bacteria in their digestive tract (versus 10^{12} on their skin and 10^{10} in their mouths)! Normal bacterial inhabitants (such as *E. coli,* the favorite of molecular geneticists; *Streptococcus;* and *Pseudomonas*) are intimately associated with epithelial cells lining the intestinal tract; they cause no harm under ordinary conditions, and some may even be beneficial (for example, as a source of vitamin K). Pathogenic digestive tract bacteria and viruses have mechanisms for attaching to epithelial cells. This allows them to enter the cell or penetrate the epithelial surface, which may result in cell damage and inflammation.

Opportunistic Infections

We all harbor billions of bacteria in or on various areas of our body. Like any guests, they are normally considerate of their hosts and do not cause trouble.

Sometimes, however, the host may suffer from a wound or a reduced level of immune resistance, and these normal inhabitants are usually the first to take advantage and invade. The result is an **opportunistic infection**, and such infections are quite common. For example, *Pseudomonas aeruginosa* are free-living bacteria that are also found in the intestine. They now cause serious opportunistic infections in many hospitals because they are widespread and resistant to many antibiotics.

Infection of Cells and Tissues

Figure 39.11 describes the routes taken by different pathogenic microbes once they enter the body. Many disease microbes cause their effects by multiplying in the epithelial surface at the site of their establishment. In these cases, there is no invasion of underlying tissues. For example, respiratory flu viruses infect epithelial cells of the respiratory tract and produce new virions that quickly infect nearby cells. This hit-and-run process continues for a few days, with new viruses being shed, until the immune system catches up or until all available cells have been infected.

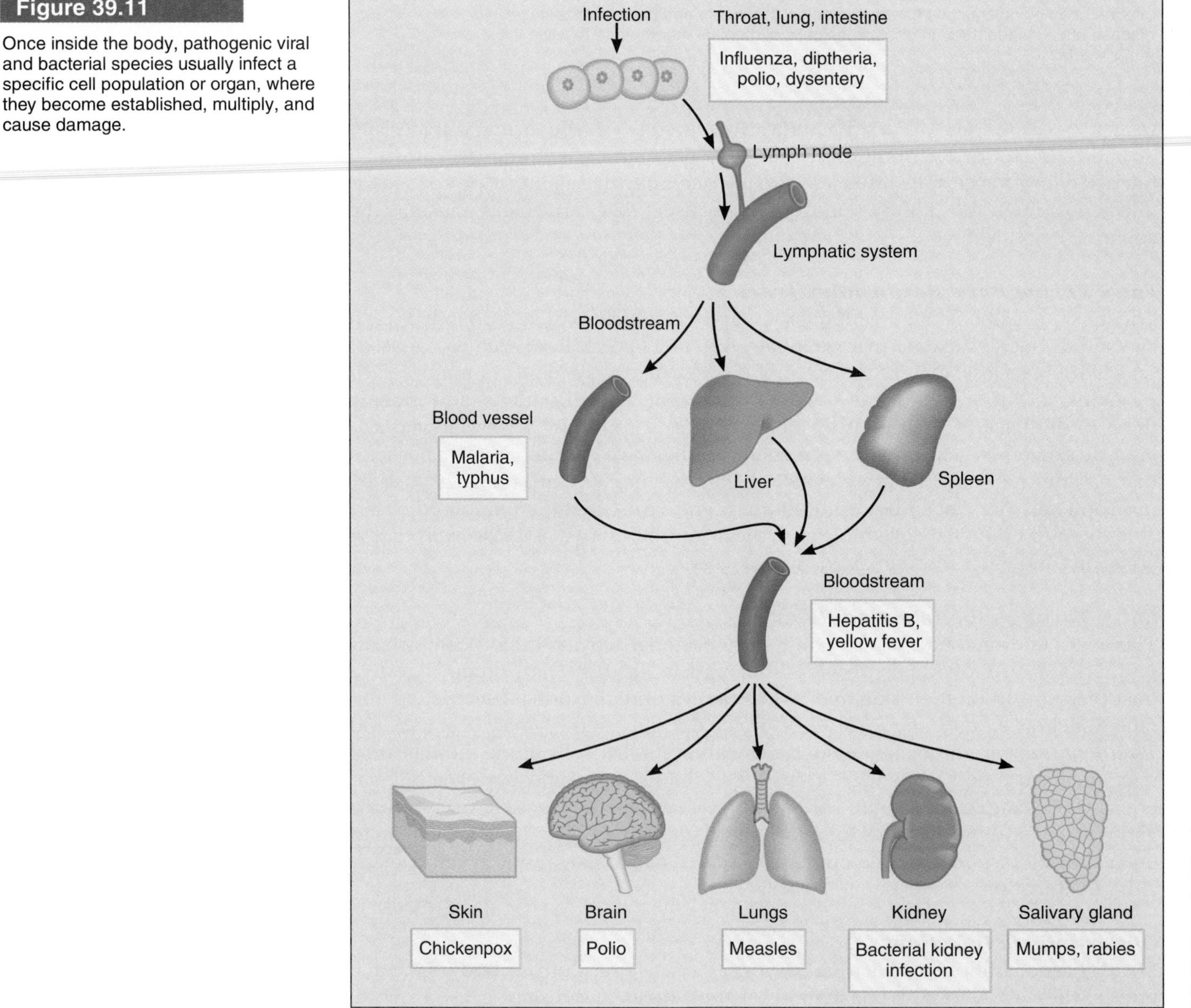

Figure 39.11

Once inside the body, pathogenic viral and bacterial species usually infect a specific cell population or organ, where they become established, multiply, and cause damage.

Gonorrhea bacteria live in the epithelial surface of the urogenital tract, but they sometimes infect underlying tissues, which causes an inflammatory response. This leads to a characteristic yellowish discharge consisting of epithelial cells, leukocytes, bacteria, and inflammatory substances.

Once they enter the internal environment, most microbes are filtered out, inactivated, or destroyed in lymph nodes. This is the reason why swollen, sore nodes indicate that a disease agent–immune system battle is under way. Microbes that escape destruction by this system can then spread to other parts of the body in plasma, blood cells, or both. Components of the immune system continue to remove such microbes, but if they avoid these forces or replicate more rapidly than they are removed and reach their target cells or tissues where they can grow and multiply, disease occurs.

Cell and Tissue Damage

All human cells, tissues, and organs can be occupied and affected by specific disease-causing microbes. There is an obvious relationship between the tissue affected and the severity of a disease. Thus infections of the heart and brain that cause pathological changes and disease are usually much more serious than those of the liver or intestine. Pathogenic microbes cause damage through several mechanisms.

Direct Damage

Viral replication usually results in the death of the infected host cell, and tissue damage occurs when a large number of cells have been killed. For example, one of the common cold viruses infects epithelial cells lining the nasal cavity. Infected cells die, become detached, and are carried away in a stream of fluid (a "runny nose"). Bacteria that infect cells also destroy their host cells.

Dental cavities are perhaps the most common disease of developed countries. They occur as a result of direct damage caused by a complex of bacteria, which colonize the surfaces of teeth, cause plaque (a film of bacteria and saliva) to form, and ultimately, as shown in Figure 39.12, create a cavity in the tooth. These bacteria metabolize sugars in the mouth as an energy source, producing acids that act to decalcify the tooth, and a cavity results. This action can be reduced by removing or preventing plaque and reducing the dietary intake of sugars.

Figure 39.12

Bacteria that colonize the teeth can cause cavities by releasing acids that penetrate the tooth enamel.

Bacterial Toxins

Toxins released from multiplying bacteria are called *exotoxins*. These substances are for the most part proteins, and some are among the most powerful biological poisons known. For example, tetanus and botulism are often fatal diseases caused by bacterial exotoxins. Many bacteria, including species of *Streptococcus, Staphylococcus,* and *Legionella,* secrete exotoxins that kill or damage cells. Toxic shock syndrome is a very dangerous disease that occurs most commonly in menstruating women. In some cases, tampons have been contaminated by an exotoxin-secreting *Staphylococcus*. Alternatively, since menstrual fluids are a good growth medium for bacteria, tampons may scratch the vaginal lining, allowing *Staphylococcus* entry into the body.

Susceptibility and Risk Factors

Infectious diseases, like most diseases, do not tend to strike individuals randomly. Many factors influence the chances of an individual suffering from a certain disease. **Risk factor** is a term applied to anything that increases the probability of developing a disease. For example, some risk factors for sexually transmitted diseases are fairly obvious: sexual intercourse with multiple partners, intercourse with an infected person, intercourse with no condom, and so on. There are two general types of risk factors: those that are subject to some degree of individual control (use of a condom, cigarette smoking) and those that are not (genetic disorders, age). Scientists attempt to identify specific risk factors for a disease and assign a quantitative value to them that reflects their relative importance. This approach has emerged as an extremely valuable tool that can be used by enlightened individuals in making decisions about their personal behavior or lifestyle relative to disease.

Some general risk factors for infectious diseases are immune depression, age and sex of the host, malnutrition, hormonal factors, fatigue, and stress. Which of those are "manageable" and which are not? There are, of course, different risk factors for specific diseases (not brushing your teeth, attending crowded lectures during flu season).

Microbial infections usually depend on delivery to a new host through the respiratory, digestive, or urogenital tract. Once inside the body, a microbe often lives in specific cells or tissues, which may be damaged or killed. Through knowledge and management of risk factors, individuals can greatly reduce their chances of getting some infectious diseases.

CANCER

The word *cancer* is known to evoke powerful feelings in many individuals. Fear is one of the most common responses, perhaps because cancer is perceived to strike individuals at random and to have no cure. For the most part, however, cancer is a disease of old age, as indicated in Figure 39.13. The major reason for this age association is that cancer takes a long time to develop, but there are also other considerations. The ravages of a lifetime of exposure to cancer-causing substances, the age-related decline in immune system effectiveness, and an inability to repair cell damage may explain, in part, why cancer most often affects older individuals.

Because of the importance of this disease, in terms of numbers of people affected directly and indirectly, enormous amounts of money and research effort have been directed toward solving the cancer problem. Certain efforts, such as the "war against cancer" proclaimed in the 1960s, were based on political concerns and naive assumptions. Underlying this effort was a simplistic belief that cancer was a distinct disease and that a cure could be found, given enough time and money.

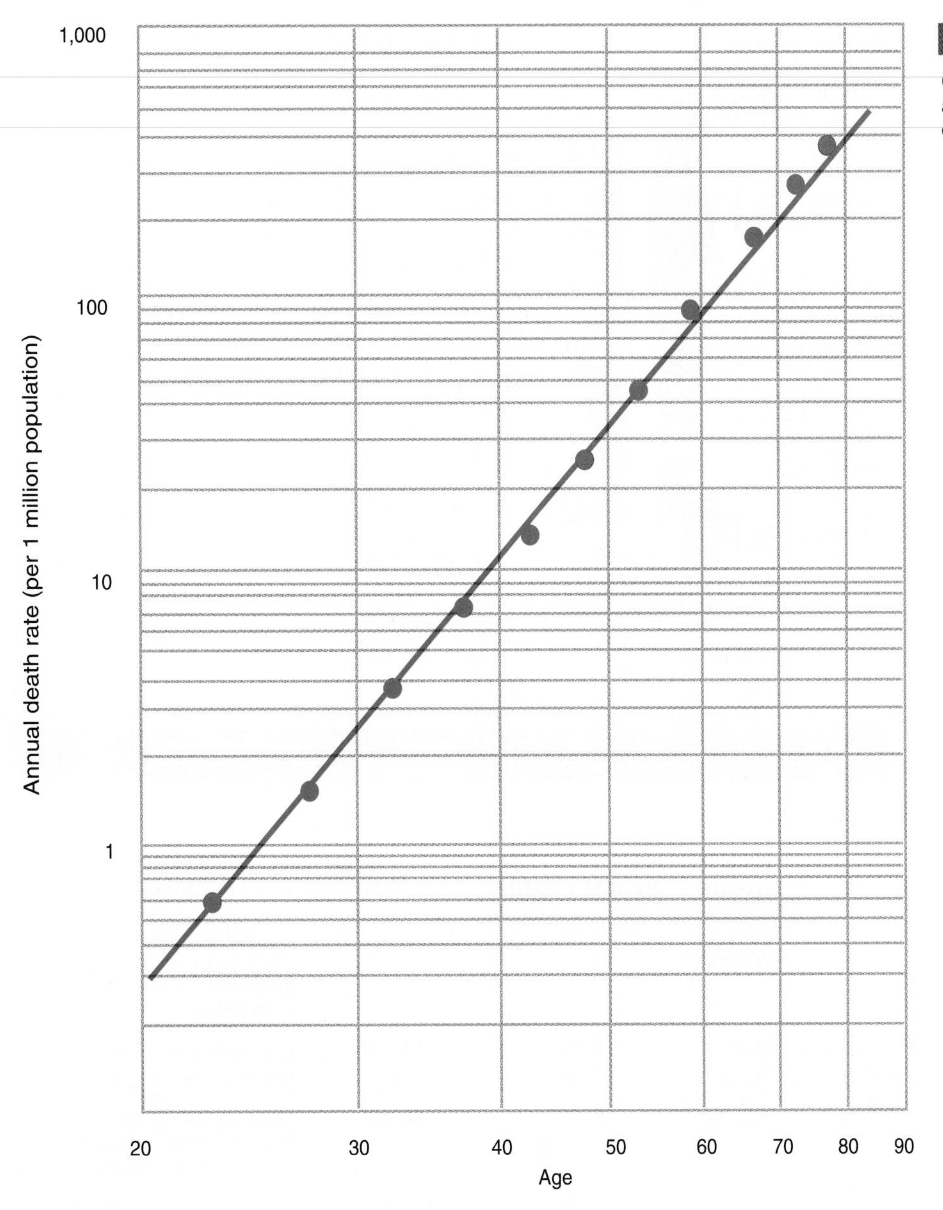

Figure 39.13

Cancer is primarily a disease of old age and is the second leading cause of death in the United States.

It is now recognized that understanding cancer was an extremely difficult challenge. Yet phenomenal progress was made in cancer research during the 1980s, and certain facets of this disease have rapidly come into focus. At least partial answers are now available for many fundamental questions. What is cancer? What causes cancer? Who gets cancer? Can cancer be cured? What does the future hold?

Cancer in the United States

The American Cancer Society, using data from national registries that record cancer information in the United States, estimates that approximately 1 million new cases of cancer will be diagnosed each year in the 1990s. About 30 percent of all Americans will develop cancer during their lifetime and, at the present time, 40 percent of those people will be treated successfully (that is, patients will survive for at least five years after treatment). Each year, just under 500,000 Americans die of cancer. Cancers of various organs have declined or remained nearly constant for over 50 years; the exception is lung cancer, which has increased. Figure 39.14 and Table 39.2 provide information about cancers of selected tissues in U.S. citizens.

Figure 39.14

Cancer deaths in 1990 were estimated to break down as indicated here.

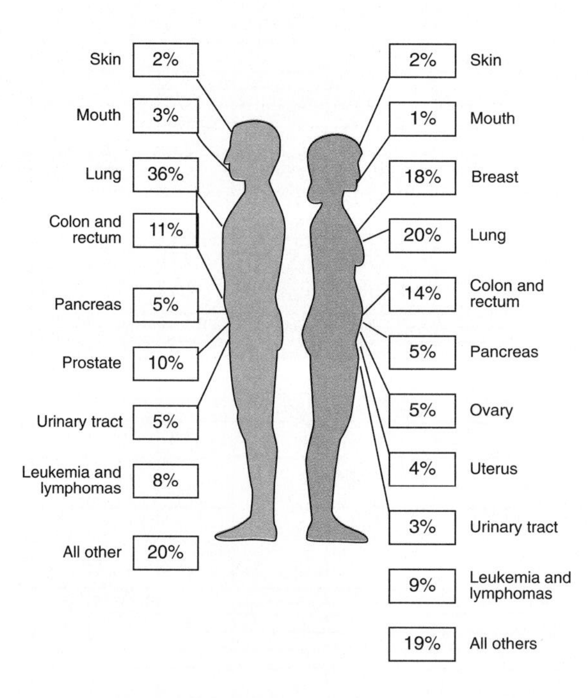

Table 39.2 Information on Selected Cancers of Various Organs

Cancer Site	Risk Group	Percentage of All Cancer Deaths (Rank)[a]	Percentage Survival[b]	Risk Factors
Lung	Men Women	36 (1) 20 (1)	13	Cigarette smoking; smoking 20 or more years; exposure to certain industrial substances; exposure to radiation
Colon and rectum	Men Women	11 (2) 14 (3)	35–80	Family history of these cancers or polyps in the colon or rectum
Breast	Women	18 (2)	60–90	Over age 50; personal or family history of breast cancer; never had children; first child after age 30
Prostate	Men	10 (3)	83	Incidence increases with age (80 percent in men over 65)
Leukemia	Men Women	8 (5) 9 (4)	33	Causes of most cases unknown; linked with excessive exposure to radiation and certain chemicals
Uterus and cervix	Women	4 (7)	85	Cervix: early age at first intercourse; multiple sex partners. Uterus: history of infertility; prolonged estrogen therapy and obesity
Skin	Men Women	2 (8) 2 (9)	90	Excessive exposure to sun; fair complexion; occupational exposure to certain chemicals (coal, tar, creosote); severe sunburn in childhood
Oral	Men Women	3 (7) 1 (10)	32–90	Cigarette, cigar, and pipe smoking; use of smokeless tobacco; excessive use of alcohol

[a]Rank indicated is specific for sex.
[b]If a range is given, higher number is for cancers detected early, lower number for cancers discovered after they have spread.

Source: American Cancer Society (NCI SEER Program).

The direct causes of cancer involve changes in gene structure or function (see Chapter 40). However, the occurrence of many cancers is thought to be related to an individual's environment, personal habits, and lifestyle. The dominant cancer risk factor is the use of tobacco; it is estimated to cause 30 percent of all cancers. However, it is impossible to quantify the level of risk associated with most other factors or even to identify specific risk factors with certainty. Those listed in Table 39.2 have been shown to be associated with an increased probability of developing the cancer indicated. The exact relationship between diet and cancer is unknown, although diet is suspected to be associated with 20 to 35 percent of all cancer deaths. There is good evidence that cholesterol is linked with cancers of the digestive tract. A great deal of research is now focused on defining the role of diet and nutrition in causing cancer.

Normal Cell Growth

Tissues and organs are composed of cells that carry out specific functions. In many tissues, cells carry out their activities for a genetically prescribed length of time, die, and are replaced by new cells, which are produced at a constant pace. The rate of cell replacement varies among organs. For example, intestinal cells and most leukocytes exist for only a few days; red blood cells live three to four months; liver cells seldom die yet can be rapidly replaced if serious loss occurs; and cells of the central nervous system are never replaced. In adults, a steady state exists between cell birth and cell death. To maintain this balance, cell growth and reproduction are exquisitely regulated, primarily by hormones and growth factors. Different regulatory substances exert their control either by inhibiting or stimulating cell growth or proliferation. What happens if regulation is lost?

Abnormal Cell Growth

Rarely, a cell (or more) escapes from the control system and grows and divides at an accelerated rate, well beyond the normal replacement requirements of a tissue. If such a cell gives rise to a clone of cells that also does not respond to regulatory substances, a cell mass eventually develops that is called a **tumor**.

Tumors

Tumors often arise but are rarely a problem because they remain localized, grow slowly, and become surrounded by connective tissue. These types of growths are called **benign tumors**, and they can usually be removed by surgeons. Tumors that consist of rapidly reproducing cells that invade surrounding tissues and spread throughout the body are called **malignant tumors**. **Cancer** is a large group of diseases that are characterized by malignant tumors.

Malignant Cells

We refer to cells of a malignant tumor as *malignant* (or cancer) *cells*. Why do malignant cells fail to respond to normal control mechanisms? A simple (and incomplete) answer is that their DNA regulating mechanisms become permanently altered through some process, and as a result they no longer receive, recognize, or respond to regulating signals. Such changes cause alterations in the growth and developmental properties of a cell, a process called **transformation**. Transformed cells, such as those shown in Figure 39.15, specialize in proliferation, and this is reflected in their structure; they have an undifferentiated appearance that is characteristic of rapidly reproducing cells. The frequency at which cells are transformed is not known, but it is thought to occur repeatedly. Most transformed cells are killed by cells of the immune system. For unknown reasons (except for immune depression), malignant cells that cause a cancer have escaped such detection or elimination.

Figure 39.15

Cancerous cells are transformed normal cells. Once transformation occurs, cancerous cells are able to proliferate and grow quickly, since they are not regulated by host control mechanisms. This ability is reflected in their structure. These breast cancer cells are poorly differentiated, vary in size, and are abnormally large.

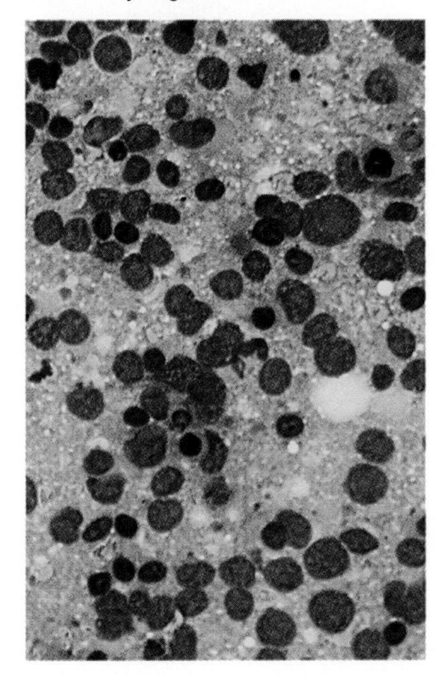

Malignant cells have other distinguishing characteristics. Most notably, they have abnormal numbers of chromosomes, they can penetrate and invade surrounding tissues, and they can be transported through the circulatory system to distant sites in the body, where they create new malignant growths. This last process is called **metastasis** (see Figure 39.16).

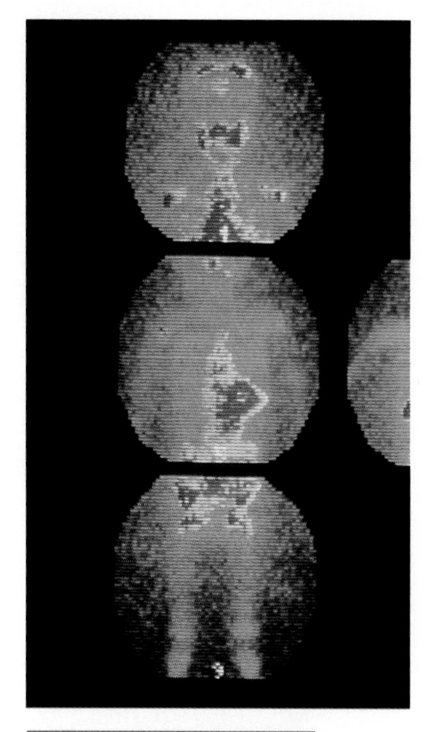

Figure 39.16

Cancer cells often metastasize to new organs. Specialized imaging technology shows that cancerous growths (in red) have become established in the brain, lungs, liver, and marrow of the pelvic girdle.

Effects on Untreated Patients

If unchecked, the hallmark characteristics of malignant cancer—uncontrolled proliferation, invasiveness, and metastasis—will lead to the death of the individual. Because their rate of growth vastly exceeds replacement requirements, malignant cells displace or overwhelm normal cells. Invasiveness and metastasis result in the same effect, uncontrolled growth and displacement of normal tissues over a widespread area (see Figure 39.17). Death caused by cancer is related to depletion of the host's nutritional reserves due to the growth requirements of malignant cells or loss of function in the organs or tissues affected.

Understanding these effects enables one to appreciate the critical importance of early treatment with drugs, radiation, surgery, or a combination of these therapies. Rarely does cancer regress without some form of medical intervention. The earlier this whole progression of events—loss of regulation, transformation, proliferation, malignant tumor formation, invasion, metastasis—is halted, the better the chance for survival. Once invasion and metastasis have occurred, the probability of successful treatment is very low.

Causes of Cancer

To inquire about the cause of cancer is to ask, "What causes cell transformation?" Until very recently, most efforts to answer this question were directed at identifying specific types of **carcinogens** (cancer-causing agents). From research conducted over the past half century, certain viruses, chemicals, and radiation have been implicated in causing human cancers, although there is no clear understanding of the numbers of cancers attributable to each. Causes of cancer are considered in greater detail in Chapter 40.

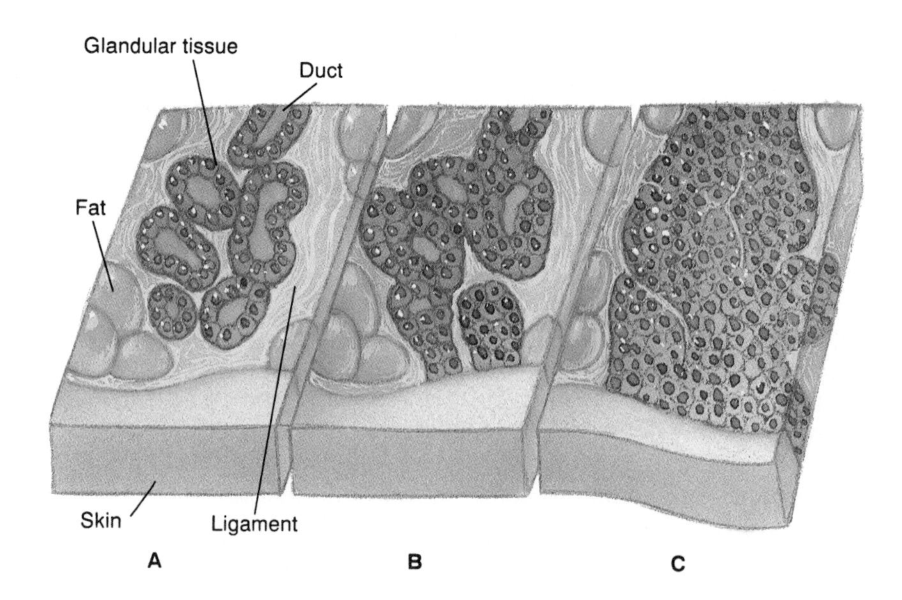

Figure 39.17

The increase in malignant cell populations is characterized by two processes: a rapid rate of mitosis and invasiveness, in which cells grow and penetrate into surrounding tissues. This drawing illustrates a breast tumor that arose in the glandular tissue (A), grew rapidly (B), and invaded the surrounding tissues (C). A lump would be detectable when a tumor had formed (B).

Cancer occurs when normal cells become transformed into malignant cells that are not killed by the immune system. The growth of malignant cell populations cannot be regulated by the host; consequently, they increase rapidly and may metastasize to new sites. If treatment is delayed, death usually occurs. The major risk factor for cancer is tobacco products.

VASCULAR DISEASES

Cardiovascular diseases are malfunctions of the heart and blood vessels. **Cerebrovascular diseases** occur when the blood supply to the brain is either temporarily or permanently interrupted. Given the critical functions of the circulatory system and these two organs, it is not surprising that failures of this system have serious, often fatal, consequences.

Cardiovascular Disease

The major cause of coronary heart disease is *atherosclerosis,* the buildup of fibrous, fatty deposits, called *plaques,* on the inner walls of major arteries. The development of plaques consists of a complex series of interactions in the artery lining involving cholesterol, macrophages, and cell growth factors (see Figure 39.18). Atherosclerosis results in loss of artery elasticity and a thickening of arterial walls, which can diminish or block blood flow to major organs, including the heart and the brain. Blood flow dynamics around plaques are altered in such a way that blood clots may also form on the arterial wall. If these break loose, they may block a smaller artery downstream.

When the flow of blood to the heart is hindered because of plaques or blocked by a blood clot, muscle cells of the heart are deprived of oxygen and

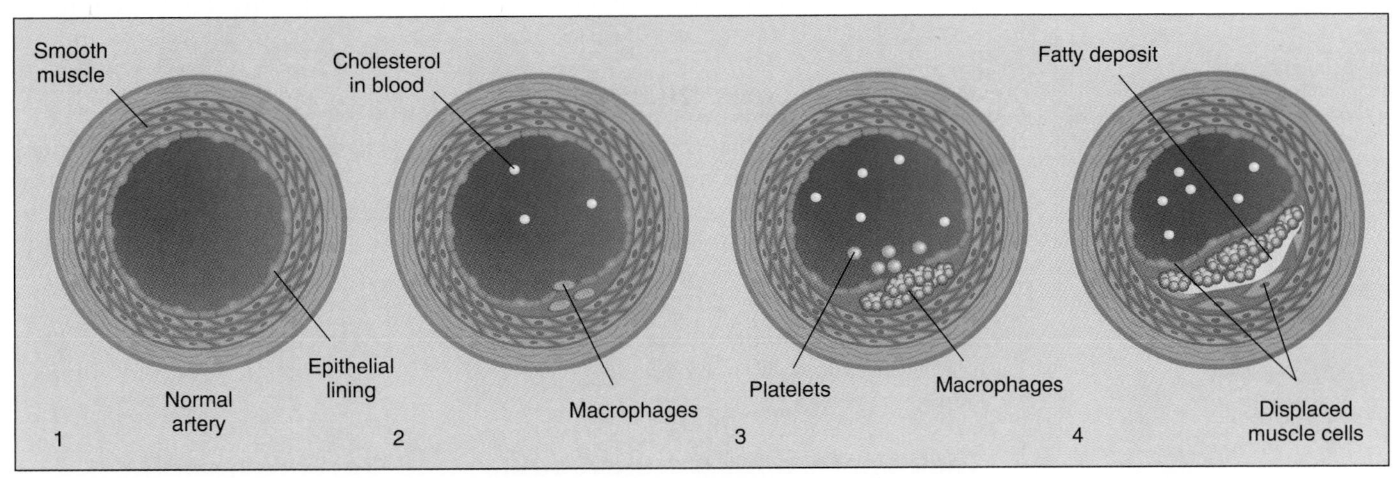

A

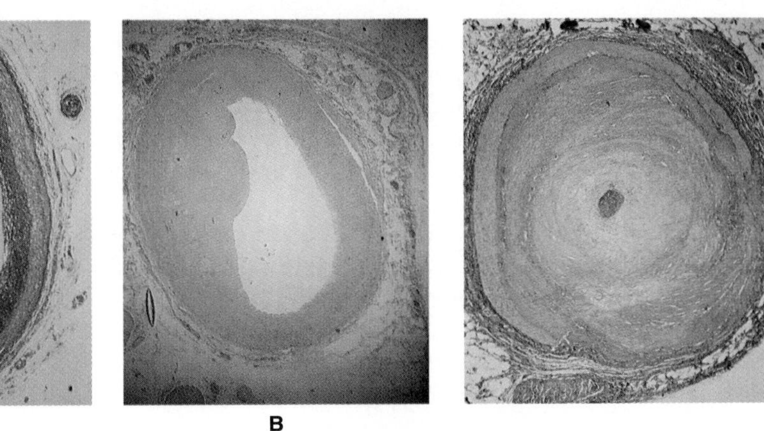

B

Figure 39.18

(A) Plaque formation in blood vessels begins when excess cholesterol becomes incorporated into the lining of blood vessels and macrophages attempt but fail to remove it. Subsequently, fat deposits at the site increase, underlying muscle and connective tissue cells proliferate, and the entire complex eventually becomes a plaque. (B) The photographs show a normal artery (left), a partially obstructed artery (center), and an artery with advanced atherosclerosis (right).

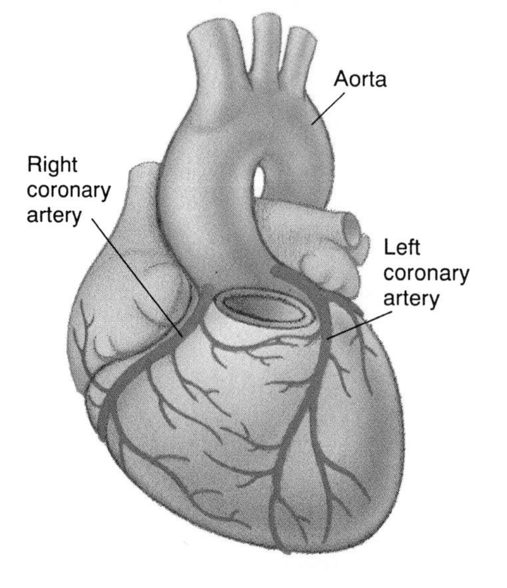

Figure 39.19

The severity of heart attacks is related to the site where circulation is blocked. If blockage occurs at the beginning of a major coronary artery, death usually results.

glucose. When this occurs, the cells die, and consequently, there is some loss of heart function, and a heart attack follows. The degree of damage is related to the location of the obstruction and the number of heart cells lost. If the block occurs at the beginning of a coronary artery, the heart attack will be severe and most likely fatal (see Figure 39.19). If the blood supply is reduced or lost in a more distant part of the artery, the effect may be less severe.

There have been a number of important developments in preventing and treating heart attacks during the past decade. New drugs, such as blood clot-dissolving enzymes, and new surgical procedures (for example, to remove or bypass plaques) are now available that are highly effective in extending the lives of individuals suffering from heart disease.

Cerebrovascular Disease

A **stroke** is a sudden, severe event causing irreversible damage to the brain due to an interference in normal blood circulation. Such interruptions occur if a blood clot blocks one of the arteries leading to the brain (*cerebral thrombosis*) or if an artery in the brain ruptures (*cerebral hemorrhage*). The severity of the stroke generally depends on the part of the brain affected and the extent of the damaged area. There is a broad range of possible effects, from relatively minor (a speech impediment) to major (permanent loss of consciousness, death). The common causes of stroke are atherosclerosis and *hypertension* (high blood pressure).

Risk Factors for Cardiovascular and Cerebrovascular Diseases

Risk factors for vascular diseases as determined through various studies are listed in Table 39.3. The basic approach used in identifying risk factors is to collect and analyze relevant data on individuals who were affected by a cardiovascular disease. Two basic questions are asked: Is there a correlation between a risk factor of interest and the disease under study, and what is the strength of the relationship?

Typically, results from hundreds of studies are analyzed by scientists representing agencies associated with different diseases (for example, the National Cancer Institute; the National Heart, Lung, and Blood Institute; and the American Heart Association), and a consensus is formed about the degree of risk. *Major* risk factors are those that are most commonly found to be correlated with the disease and that most scientists agree are likely to be associated with the disease.

Table 39.3 The Five Leading Causes of Death in the United States and Associated Risk Factors

Cause of Death	Risk Factors
Cardiovascular disease	Tobacco use, elevated serum cholesterol, high blood pressure, obesity, diabetes, sedentary lifestyle
Cancer	Tobacco use, improper diet, alcohol, occupational or environmental exposures to carcinogens
Cerebrovascular disease	High blood pressure, tobacco use, elevated serum cholesterol
Accidental injuries	Safety belt noncompliance, alcohol or substance abuse, reckless driving, occupational hazards, stress or fatigue
Chronic lung disease (such as emphysema)	Tobacco use, occupational or environmental exposures

Source: McGinnis (1988–1989), pp. 46–52.

It must be emphasized, however, that these do not represent *proof* in the classic scientific use of the term. What would it take to prove that a specific risk factor caused a disease? The simplest study required would be to compare the occurrence of the disease in two groups of people that differed only in the risk factor being investigated (for instance, one group smoked, and the other did not). Because of the great diversity of individual lifestyles, such a study would be impossible. Nevertheless, the risk factors listed in Table 39.3 are thought by most scientists to be involved in the development of cardiovascular diseases.

Circulatory diseases of the heart and brain are the number one cause of death in the United States. Atherosclerosis is the principal cause of cardiovascular disease and, along with hypertension, cerebrovascular disease. Major risk factors for these diseases are tobacco use, diet, and lack of exercise.

RISK FACTORS AND LIFESTYLE MANAGEMENT

Each year, about 11 percent of the U.S. gross national product ($2,000 to $2,500 for every man, woman, and child) is spent on health, with the largest share by far going for treatment—not prevention—of disease. Since it is estimated that more than 60 percent of all Americans die prematurely (not of "old age"), it is widely believed that much more of that amount should be directed toward preventing disease.

Detection and treatment are important for all diseases, but preventive actions addressed at changing individual behaviors could be much more productive. Researchers estimate that more effective management of major risk factors for vascular diseases and cancer (poor diet, use of tobacco, abuse of alcohol, lack of exercise) could prevent between 40 and 70 percent of all premature deaths. In comparison, the use of advanced medical treatments is not likely to reduce premature death by more than 10 or 15 percent.

Understanding risk factors is helpful to individuals wishing to optimize their chances of not getting a particular disease. A straightforward analysis suggests that a person who makes decisions not to smoke, to eat foods with little or no fat and cholesterol, and to lower blood pressure through exercise and careful nutrition would greatly reduce the risk of coronary heart disease and stroke. In fact, data indicate that such lifestyle changes reduce the chances of suffering from one of these diseases by a factor of 5 to 10.

This approach recognizes the complexities of modern diseases. It also enables individuals to take primary responsibility in making decisions about managing their lives in such a way that they reduce the likelihood of falling victim to a disease.

Summary

1. Until the 1950s, it was widely believed that most diseases were caused by infectious agents that could be controlled if the proper antibiotic was discovered. Since that time, other disease-causing agents have been found that modified this view of disease. Today, disease is considered to be any abnormal condition that impairs functioning. Causes of disease include infectious microbes, nutritional deficiencies, genes, and other agents, as well as lifestyle factors.
2. Many infectious diseases of historical importance have largely been eliminated in developed countries because of antibiotics, sanitation practices, and public health measures. As a result, citizens in these countries generally die of diseases commonly associated with old age—cancer and cardiovascular disease.
3. Bacteria and pathogenic viruses can gain entry to the interior of body through openings in the external surface or through the digestive, respiratory, or urogenital system. If they gain entry, they can become established, grow, and reproduce in various cells and tissues. As a result, they may kill or damage the cells or tissues they infect and other tissues if they are able to spread to different areas of the body. Bacterial infections can often be treated with antibiotics without harming host cells. Viruses infect cells and cannot be killed by antibiotics without affecting host cells. For some viral diseases, vaccines and drugs have been developed that result in destruction of the infective virus. Most bacterial and viral infections are eventually terminated through actions of the immune system.
4. Cancer encompasses a large number of diseases characterized by the unregulated and excessive growth of a malignant cell population that arises from the transformation of a normal cell. If not brought under control, the growth of malignant cell populations will usually deplete host resources, displace normal cells, disrupt normal organ function, and cause the death of the host.
5. Vascular diseases are a group of circulatory system diseases that cause the deaths of about 1 million Americans each year. The major cause of coronary heart disease—the primary cause of death—is atherosclerosis, a complex process that leads to a failure of blood vessels leading to the heart, which results in heart attacks. Strokes occur if blood flow to the brain is disrupted.
6. Risk factors for various diseases and other causes of death have been identified and can be used for preventive purposes by informed individuals.

Review Questions

1. What is disease? What are some of the causes of disease?
2. Why are infectious diseases less important in developed countries than undeveloped countries? What types of diseases are most serious in developed countries?
3. How do viruses cause disease?
4. What are the two general processes of disease transmission?
5. Explain the concept of balanced pathogenicity.
6. How do bacteria and viruses enter the body?
7. How do bacteria and viruses affect cells and tissues after infection?
8. What are risk factors? How can they be used in health protection?
9. What is cancer? How does cell transformation occur? What is the significance of cell transformation?
10. Why is prompt treatment of cancer necessary for increasing the chances of survival?
11. What is cardiovascular disease? Cerebrovascular disease? What are the major risk factors for these diseases?
12. Why is there increasing emphasis on lifestyle management in reducing disease?

Essay and Discussion Questions

1. If you were the minister of health in a developing country, what recommendations would you make when preparing your budget to address the nation's health problems? How would they differ from those made for a developed country?
2. Assume that you are asked to create a lifestyle management program for yourself. Which diseases concern you most? What risk factors would you emphasize? Why?
3. Drugs, antibiotics, and lifestyle management are now used to combat diseases in the United States. What new approaches might be used in the future? Explain.
4. A significant number of people die from disease. What factors, other than disease, are important in determining the theoretical maximum life expectancy of humans?

References and Recommended Reading

Amler, R. W., and H. B. Dull. 1987. *Closing the Gap: The Burden of Unnecessary Illness*. New York: Oxford University Press.

Aral, S. O., and K. K. Holmes. 1991. Sexually transmitted diseases in the AIDS era. *Scientific American* 264:62–69.

Cohen, L. A. 1987. Diet and cancer. *Scientific American* 257:42–48.

Dalen, J. W. 1991. Detection and treatment of elevated blood cholesterol: What have we learned? *Archives of Internal Medicine* 324:303–317.

Duffy, J. 1990. *The Sanitarians: A History of American Public Health*. Urbana: University of Illinois Press.

Feldman, M., and L. Eisenbach. 1988. What makes a tumor cell metastatic? *Scientific American* 258:60–68.

Finkle, B. 1988. New medicines from industry. *Journal of Chemical and Biotechnology* 43:313–327.

Hirsch, M. S., and J. C. Kaplan. 1987. Antiviral therapy. *Scientific American* 255:76–85.

Kolberg, R. 1990. Molecular biology takes cholesterol to heart. *Journal of NIH Research* 2:65–69.

McEvedy, C. 1988. The bubonic plague. *Scientific American* 258:118–123.

McGinnis, J. M. 1988–1989. National priorities in disease prevention. *Issues in Science and Technology* (Winter): 46–52.

Readings from Scientific American. 1986. *Cancer Biology*. New York: Freeman.

Sanford, J. P., and J. P. Luby. 1981. *Infectious Diseases,* vol. 8. Orlando, Fla.: Grune & Stratton.

CHAPTER 40

Contemporary Research on Human Diseases

Chapter Outline

Reading Questions

1. How did the existence of a new disease, AIDS, become known?
2. How does HIV-1 cause AIDS?
3. What does the future hold for AIDS in the United States and the world?
4. What is the concept of cellular homeostasis? How is it related to cancer?
5. How are oncogenes and tumor suppressor genes related to cancer?

Beginning in the early 1950s, rapid progress was made in learning about all aspects of human disease, primarily through the use of powerful tools developed in molecular biology. New knowledge was obtained about host biology, the immune system, and the underlying molecular causes of "simple" human diseases—those caused by a single disease agent (for example, polio) or a single gene (for example, muscular dystrophy).

STUDYING COMPLEX HUMAN DISEASES

In the decade of the 1990s, the principal focus of human disease research will shift to "complex" diseases—those with complicated or multiple causes, such as AIDS, cancer, vascular diseases, and mental disorders such as schizophrenia. Investigators will continue to concentrate on the molecular level because complex disorders often involve sequences of molecular changes. Also, understanding diseases at the molecular level is the first step leading to the development of treatments or cures.

Several relevant questions now guide research on complex diseases. What causes the disease? Which cellular processes are disrupted and why? What approaches might be used to treat or cure the disease? In this chapter, we describe research on two of the most important human diseases—AIDS and cancer. Case studies of these two diseases illustrate the complexities of modern disease research. We emphasize that great gaps remain in our understanding of AIDS and cancer. The intent of this chapter is to present the most current information available and to provide a background that will enable you to follow progress in the years ahead.

AIDS: A NEW DISEASE

Beginning in 1981, it slowly became evident that a new disease, which became known as *AIDS,* was causing mortalities in the United States, Europe, and Africa. This new disease was to become one of the great *pandemics* (a disease that occurs worldwide) of the twentieth century. How did scientists determine that a new disease was affecting people and that it was caused by an infectious agent?

Discovery of the Causative Agent

Epidemiology is the field of science dealing with the relationships of various factors that influence the frequencies and distributions of a disease in a human community or population. Epidemiologists conduct investigations designed to accomplish the following:

Determine the occurrence of different diseases
Identify patterns of mortality in different geographical areas of a country or the world
Define methods of transmission
Identify specific groups that may be at risk
Determine risk factors
Create strategies to prevent and control disease

In the United States, the Centers for Disease Control (CDC), located in Atlanta, have the primary responsibility for conducting such investigations. Scientists at the CDC rely on data forwarded from local and state health departments. Such data provided the first indications that a new disease was present in the United States and also contributed clues about the nature of the disease.

Early Epidemiological Investigations

In 1981, workers at the CDC became aware that over an eight-month period, five new cases of an extremely rare type of pneumonia caused by a protozoan (*Pneu-*

mocystis carinii) had occurred in homosexual men from Los Angeles. At approximately the same time, 26 cases of an unusual type of cancer (Kaposi's sarcoma) were reported in homosexual men in New York and California. Many other severe opportunistic infections were also reported in the early 1980s. Recall that opportunistic diseases occur in individuals whose immune systems have become impaired for some reason.

Occurrence and Transmission of AIDS

Most of the early cases of AIDS in the United States occurred in male homosexuals living in New York and California. Was the disease actually caused by an infectious agent? Or was some lifestyle factor responsible?

In comprehensive studies of homosexuals with AIDS, CDC investigators established a strong link between the occurrence of AIDS and the pattern of sexual contacts. Patients with the disease tended to have frequent sex with multiple partners in comparison with control subjects (homosexuals without AIDS). Such a finding was consistent with the view that AIDS was caused by an infectious agent and that it was transmitted through sexual contacts between males.

By 1982, AIDS cases were being reported in individuals who were not homosexuals. People with hemophilia and recipients of blood transfusions had contracted AIDS, apparently from injections of blood or blood products. AIDS was also reported in people who injected themselves intravenously with controlled drugs using shared hypodermic needles. Finally, AIDS was reported in two females who were sexual partners of male intravenous drug abusers. Therefore, AIDS could also be transmitted via heterosexual intercourse. Somewhat later, AIDS-infected mothers were shown to have infected their infants during pregnancy or nursing (see Figure 40.1). All of these findings, of course, emphasized the crucial importance of developing tests to identify infected individuals or to detect the AIDS agent, whatever it was, in blood.

The results of these various studies led scientists to conclude that AIDS was being transmitted by the injection of blood from an affected person or from body fluids exchanged during sexual contact. Further, since AIDS could be transmitted, it must be caused by an infectious agent. In 1982, the CDC identified individuals with the new disease syndrome (a *syndrome* is a set of symptoms or characteristics that occur together) as having **acquired immune deficiency syndrome (AIDS)**. Thus the combination of rare opportunistic infections along with a seriously deficient immune system became known as AIDS. Data analysis eventually revealed that there was a latent period of several years between the time of initial infection and the appearance of AIDS. Therefore, infections in AIDS patients living in the United States had actually occurred in the 1970s.

A New Retrovirus

What relevant information about a possible AIDS-causing agent was available to researchers in 1982, when the search for such an agent began in earnest? AIDS

Figure 40.1

Volunteers at the Birk Childcare Center in Brooklyn, New York, caring for infants with AIDS.

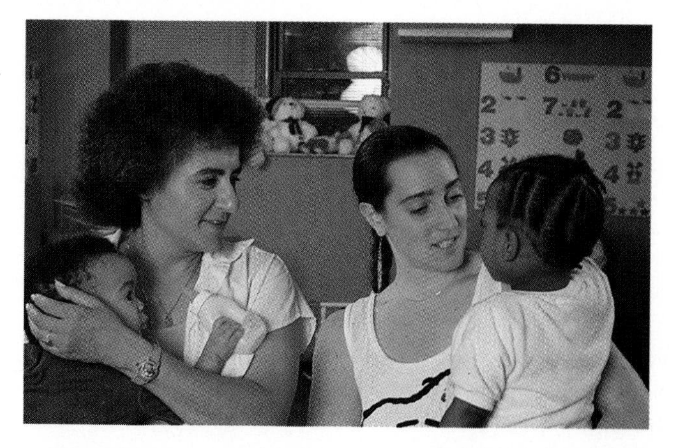

was considered to be a new disease, apparently caused by an agent that was present in different body fluids including blood, blood products, plasma, and semen. Transmission could occur through sexual contact, blood transfusions, or injections using shared needles containing blood or plasma. Two additional clues were particularly significant. First, before plasma is transfused into individuals with hemophilia, it is filtered, a process that removes bacteria and other microbes but not the much smaller viruses. Second, immunodeficient patients with AIDS were found to have very low levels of helper *T4 lymphocytes*. This type of T cell regulates a variety of immune functions and appeared to be the target of the AIDS agent. Also, one type of T lymphocyte cancer (leukemia) was already known to be caused by a retrovirus.

The problem of identifying the AIDS agent had attracted top scientists in Europe and the United States, and they accomplished their goal during 1983 and 1984. Researchers in France and the United States identified a new retrovirus that was responsible for causing AIDS. During this same period, diagnostic blood tests were also developed for detecting the presence of the virus. These tests were important for identifying infected individuals and also for screening blood donated for transfusions. Use of these tests enhanced the safety of blood supplies .

In the early 1980s, epidemiologists discovered that a new disease, characterized by opportunistic infections, was causing mortalities. People initially affected in the United States were mostly male homosexuals and recipients of blood transfusions. Later, other population groups were found to have the disease. In 1982, the CDC designated the new disease syndrome as AIDS. By 1984, the cause of AIDS had been determined to be a new retrovirus.

HIV-1

Progress in learning more about the AIDS virus continued at a rapid pace following its discovery. In 1985, a virology nomenclature committee named the new AIDS virus *human immunodeficiency virus, type 1,* now known as **HIV-1** (see Figure 40.2A). Throughout the rest of this chapter, we use HIV-1 in reference to the established AIDS virus. Subsequently, a second, different HIV (HIV-2) was isolated; it is thought to cause a milder form of AIDS. To many people, HIV-1 has become synonymous with AIDS. However, HIV-1 causes a broad range of disease effects. AIDS, as defined by the CDC, occurs late in the progression of infection with HIV-1.

The Molecular Structure of HIV-1

The molecular structure of HIV-1 is described in Figures 40.2B–D. The *virion,* or virus particle, is enclosed by an *envelope* studded with knobs made of two glycoproteins (gp) (proteins containing carbohydrate side chains) called *gp120* and *gp41*. A protein (p) called *p18* lies inside the envelope. Within the protein (*p24*) core are two identical strands of RNA, the genetic information of retroviruses. The RNA is surrounded by structural proteins and reverse transcriptase, the enzyme that enables HIV-1 to make DNA corresponding to its RNA sequence. This retro, or backward, flow of genetic information—RNA converted to DNA—is the distinguishing characteristic of retroviruses. The newly synthesized DNA contains viral genes required for replicating new HIV-1 virions inside the host cell.

HIV-1 Life Cycle

Like all viruses, HIV-1 cannot be replicated or cause harmful effects until it enters an appropriate host cell. What happens once HIV-1 virions gain entry into the body through sexual contact, injection, or other means? How does HIV-1 find and enter an appropriate cell? What events follow that lead to the cell's being killed? The HIV-1 life cycle is described in Figure 40.3.

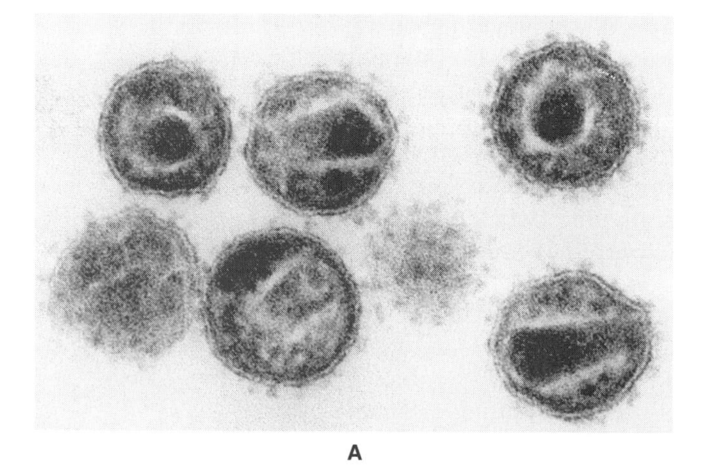

A

B

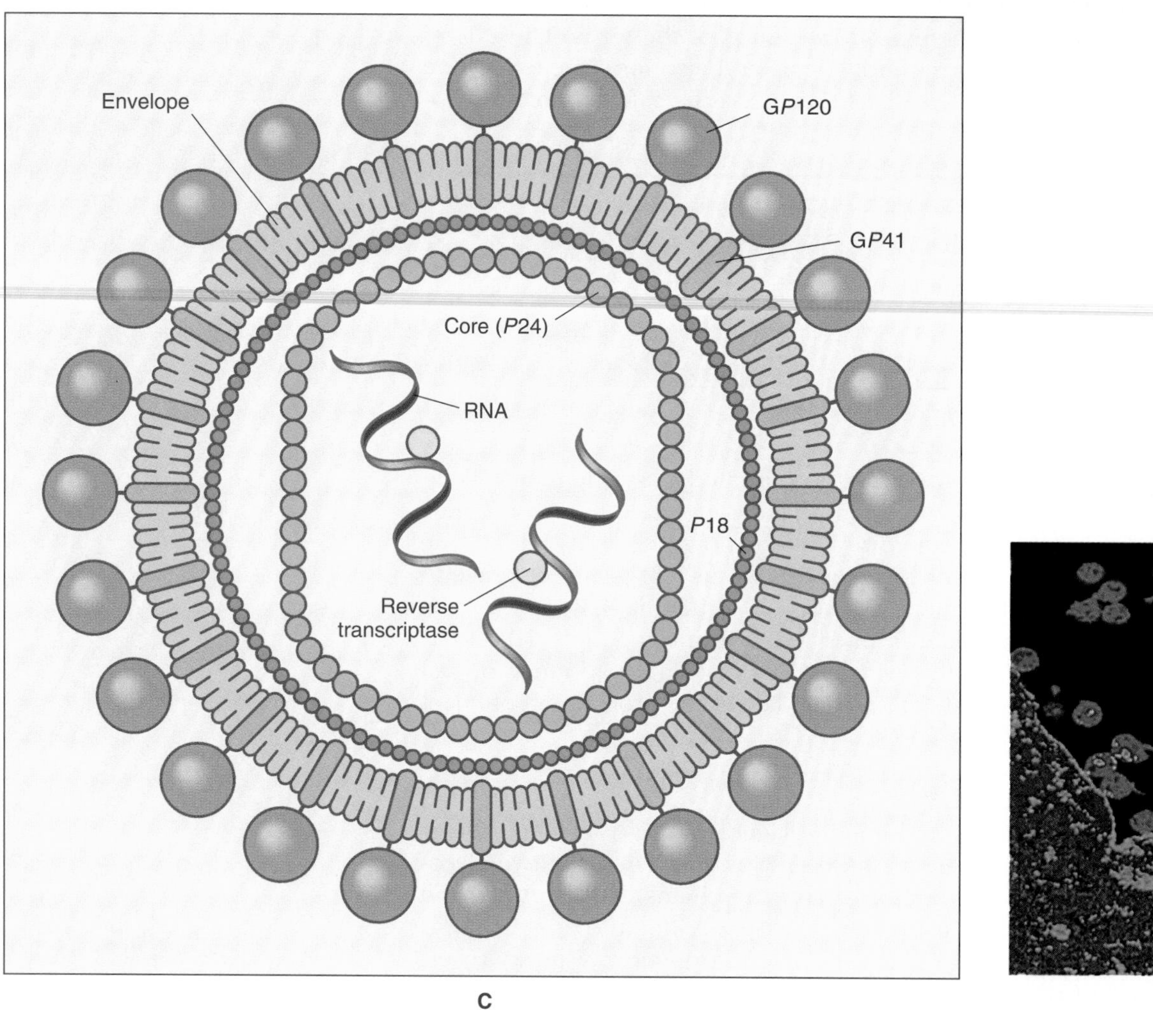

C

D

Figure 40.2

(A) This electron micrograph shows HIV-1. (B) This is a three-dimensional model of HIV-1. (C) This cross section shows the structure of HIV-1. The knobs are composed of two glycoproteins: gp120 attaches to CD4 receptors of T4 lymphocytes prior to infection of the cell, and gp41 is embedded in the viral envelope. The genetic material, RNA, is enclosed within a protein (p24) core. The enzyme reverse transcriptase transcribes viral RNA into DNA that becomes integrated into a host cell chromosome. (D) This scanning electron micrograph (SEM) image shows HIV-1 RNA (red), glycoproteins (blue), and proteins (green).

Entering the Host Cell

After HIV-1 virions enter the circulatory system, they eventually contact, bind to, and then enter specific host ("target") cells. HIV-1 binds to host cells that have

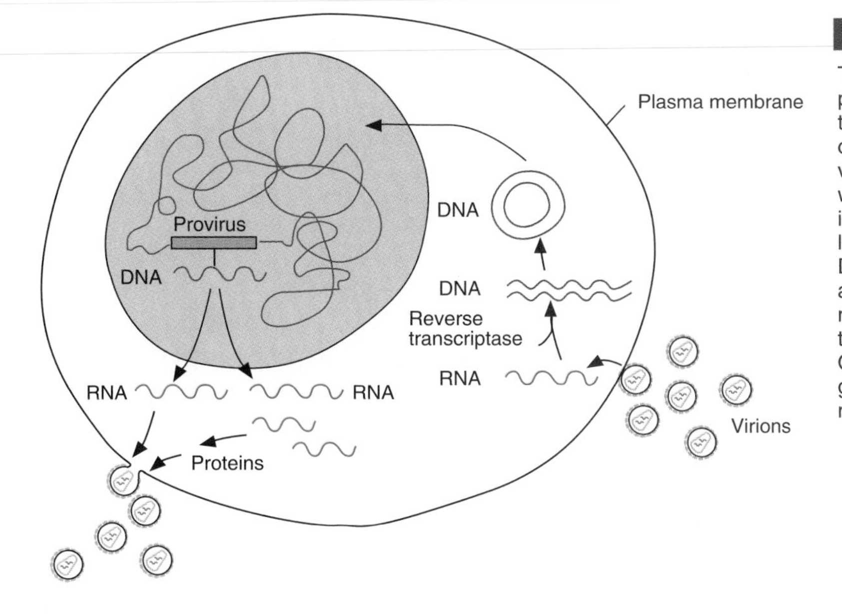

Figure 40.3

The HIV-1 life cycle consists of four phases: infection, transcription, integration, and viral production. Infection occurs when gp120 knobs of an HIV-1 virion bind to CD4 receptors; they fuse with the cell membrane, and the core is inserted into the cell. Transcription follows when viral RNA is copied into DNA, by reverse transcriptase, to create a provirus. The provirus enters the nucleus and becomes integrated into the DNA of the host cell's chromosome. Once activated by exposure to an antigen, the infected T4 cell's metabolic machinery will produce HIV-1 virions.

CD4 (T4) molecules, or receptors, protruding from their surface membrane. CD4 is found primarily on helper T4 cells, where its functional role is related to interactions between immune cells or molecules involved in immune responses. It is also found on monocytes, macrophages, and apparently certain brain cells, intestinal cells, and bone marrow cells, which can also become infected.

The binding of HIV-1 to the host cell membrane can be described as a molecular lock-and-key fit; specifically, gp120 binds with the CD4 receptor on a susceptible cell. Following the initial binding event, the viral core is injected into the host cell.

RNA Transcription and Provirus Integration

Once inside the host cell, the virion is rapidly uncoated and the viral RNA genome is copied by reverse transcriptase into a double-stranded DNA molecule using host cell nucleotides. Once synthesized, *integration* occurs—the viral DNA becomes inserted into a host cell chromosome, where it remains inactive for weeks, months, or years, until triggered to make new virus particles. In this state, the integrated viral DNA is known as a *provirus*. When an integrated provirus becomes activated, the consequences are devastating.

Production of New HIV-1 Virions

The production of new HIV-1 virions is linked to activation of the cells they occupy. Thus when an HIV-1-infected T4 cell is stimulated by the presence of a foreign antigen, it begins to produce substances, including DNA for creating new T4 cells, necessary to carry out its immune function. Unfortunately, the proviral DNA is also activated, and its genes, essentially instructions for building new HIV-1 virions, are expressed. This results in the viral takeover of the T4 cell's metabolic machinery. Instead of building new cellular proteins used in normal immune responses, the host cell becomes an instrument for assembling the agents of its own destruction.

Proviral DNA initially directs host cell enzymes to transcribe the integrated HIV-1 DNA back into RNA. The specific viral proteins encoded by genes in the RNA, along with more genomic RNA molecules, are then synthesized by the host cell. The proteins include structural molecules for building new virions, enzymes necessary to assemble them, and regulatory substances that control their life cycle.

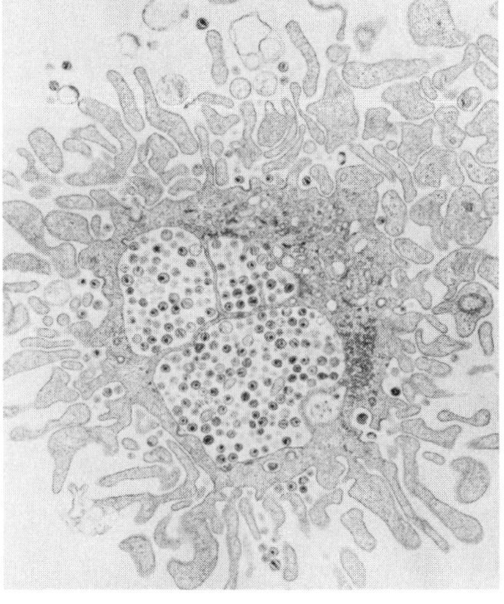

A

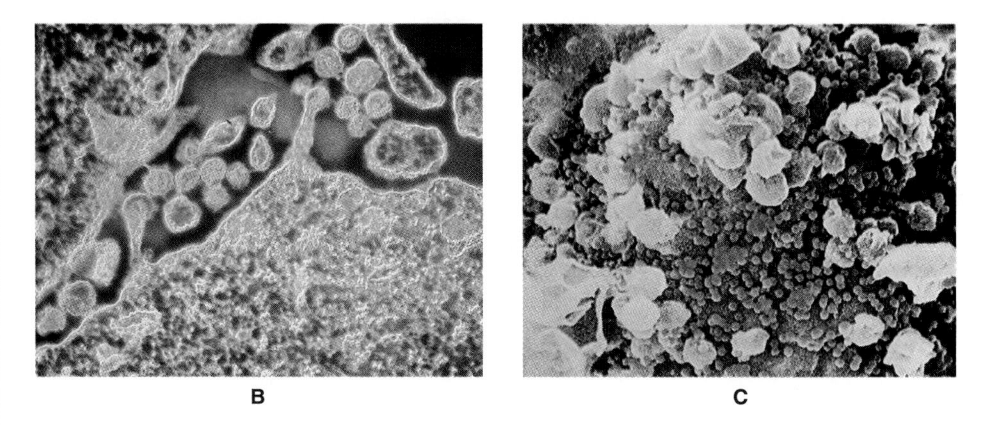

B C

Figure 40.4

(A) This electron micrograph shows HIV-1 assembled within an infected cell. (B) This electron micrograph shows HIV-1 virions (orange/green) in different stages of budding from the plasma membrane of an infected cell. (C) This SEM shows HIV-1 virions (green) as they bud from a T4 lymphocyte.

As viral proteins and genomic RNA molecules become available, new virions are assembled at the plasma membrane (see Figure 40.4A). The plasma membrane, which is now destined to become the new viral membrane, begins to constrict around the new HIV-1 particle and forms a bud at the surface of the cell. When the virus is assembled, the membrane pinches off, and a new HIV-1 is shed from the cell, as shown in Figures 40.4B–C.

Death of T4 Cells

The hallmark of HIV-1 infection is depletion of the helper T4 cell population. It has been found, however, that only a relatively small percentage of T4 cells are actually infected with HIV-1 at any given time. What accounts for the eradication of the T4 cell population? In general, T4 cells are killed by one of two processes. First, HIV-1-infected cells may be killed by immune cells (NK cells and T_C cells, described in Chapter 38) or through other immune reactions. Second, when new virions bud off from the cell, holes may be torn in the plasma membrane, which causes the loss of structural integrity and eventually the death of the cell.

Macrophages, monocytes, and other cells that pick up HIV-1 during early immune reactions are apparently not killed by the virus. These cells are thought to act as *reservoirs,* sites from which viable virions may be spread to other, uninfected cells.

The Genome of HIV-1

Scientists have succeeded in determining the entire molecular structure of HIV-1, including its genetic repertoire. The genome of HIV-1 is extremely limited, consisting of 9,749 nucleotides (compared to 3 to 4 billion in the human genome) and only nine genes, which are described in Figure 40.5. Three of the genes (*GAG, POL,* and *ENV*) encode viral proteins. The other six have roles in different phases of the HIV-1 life cycle. Remarkably, these six genes guide all HIV-1 molecular operations concerned with proviral integration, construction of new HIV-1 virions, regulation of the HIV-1 life cycle within the infected cell, and the escape and infection of fresh cells. The nine viral genes are bounded at each end by stretches of DNA known as long terminal repeats (LTRs). The LTRs do not encode any protein but somehow initiate construction of new HIV-1 in the infected cell by directing host cell enzymes to make viral RNA from proviral DNA.

HIV-1 virions consist of an outer envelope and an inner core containing RNA, their genetic material. During infection, the RNA is inserted into a host cell bearing T4 receptors. Reverse transcriptase then converts viral RNA into DNA that becomes integrated into a host cell chromosome. When the host cell is activated, proviral genes are replicated that, once expressed, direct the host cell to produce HIV-1 virions. The new virions leave the original host cell and infect fresh cells.

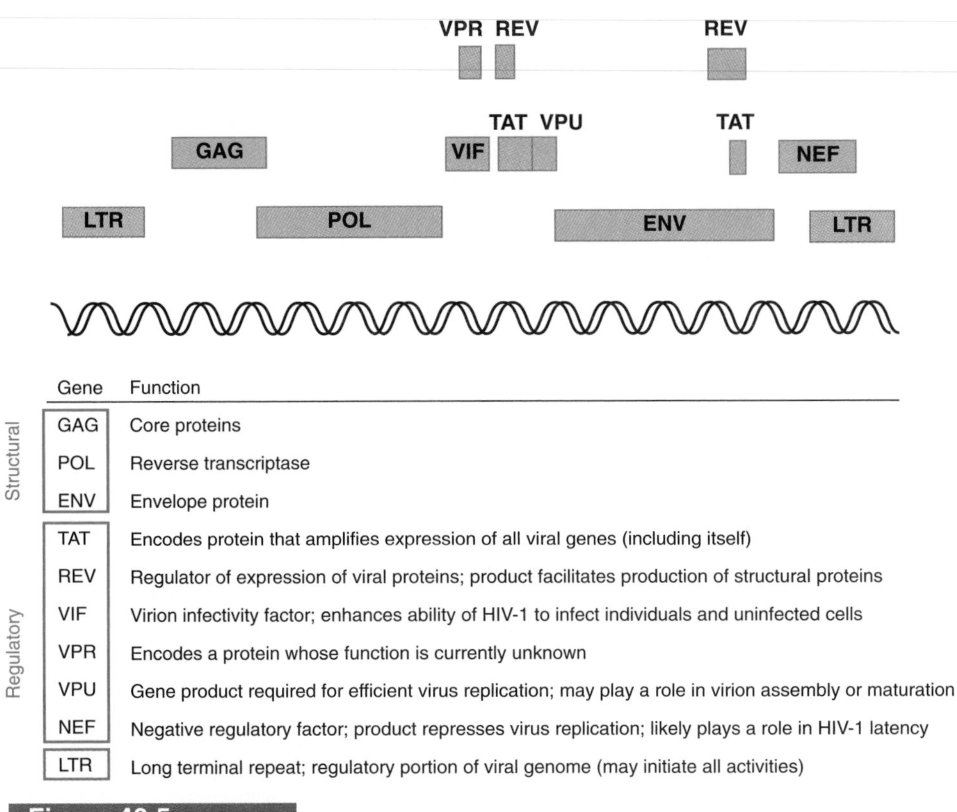

	Gene	Function
Structural	GAG	Core proteins
	POL	Reverse transcriptase
	ENV	Envelope protein
Regulatory	TAT	Encodes protein that amplifies expression of all viral genes (including itself)
	REV	Regulator of expression of viral proteins; product facilitates production of structural proteins
	VIF	Virion infectivity factor; enhances ability of HIV-1 to infect individuals and uninfected cells
	VPR	Encodes a protein whose function is currently unknown
	VPU	Gene product required for efficient virus replication; may play a role in virion assembly or maturation
	NEF	Negative regulatory factor; product represses virus replication; likely plays a role in HIV-1 latency
	LTR	Long terminal repeat; regulatory portion of viral genome (may initiate all activities)

Figure 40.5

The genome of HIV-1 consists of nine genes. Three are structural genes that encode the core and envelope proteins and reverse transcriptase. The other six genes encode proteins that play a role in the production and assembly of HIV-1 virions within the infected cell.

HIV-1 and Human Diseases

The consequences of HIV-1 infection and the resulting immune deficiency are numerous and varied. The disease syndrome called AIDS represents the cumulative effects of immune dysfunction that become evident only during the terminal phase, several years after infection. What happens during the period between infection and the appearance of AIDS?

General Aspects of HIV-1-induced Diseases

Most HIV-1-induced diseases are associated with gradual depletion of the T4 lymphocyte population. As these cells are lost, viruses and opportunistic bacteria, fungi, and protozoans are able to establish themselves. Beside infectious diseases, individuals may also develop cancers and disorders of the central nervous system. There is considerable variation in diseases and disease progression among infected patients.

The Walter Reed Classification System

In 1984, physicians at the Walter Reed Army Institute of Research in Washington, D.C., created an objective classification system that categorizes patients according to their stage of infection as determined by different indicators of immune impairment. The system was created to facilitate research and treatment strategies because most individuals were found to follow a predictable progression of disease once infection occurred.

The scheme begins with Walter Reed stage 0 (WR0), before infection is confirmed, and progresses through six stages, from WR1 to the last stage, WR6, which is AIDS. The essential distinguishing elements of the stages are the numbers of T4 cells per cubic millimeter of blood and the functional capabilities of the patient's immune system. The different states are summarized as follows:

WR0: No confirmed infection; T4 cell counts are normal, about 800 cells per cubic millimeter. A poorly understood but vigorous humoral (antibody) immune response is usually mounted against HIV-1 within weeks or a few months after infection occurs. HIV-1 antibody presence is the basis of the test used to identify infected individuals.
WR1: A positive HIV-1 test reveals infection; there may or may not be any symptoms, such as fatigue, fever, and headaches during the next year or two.
WR2: Development of swollen lymph nodes, thought to be caused by high levels of B cell activity in response to HIV-1 antigens. Stage 2 usually lasts between three and five years and is characterized by generally good health.
WR3: T4 cell counts drop below 400 cells per cubic millimeter; within one to two years, there are further indications of immune impairment.
WR4: Poor response to special antigens injected under the skin, which indicates impending loss of humoral immunity. This stage lasts for several months.
WR5: Immune system functions continue to deteriorate until there is no measurable immune response. The patient may then develop an uncontrolled fungal (*Candida*) infection in the mouth, a disease called *thrush,* that marks the loss of cellular immunity. There may be other serious opportunistic viral and bacterial infections of the skin and mucous membranes.
WR6: WR6 is synonymous with AIDS, the stage defined by numerous opportunistic infections and complete collapse of the immune system that usually occurs six to ten years after infection with HIV-1. Some of the diseases that commonly occur during this stage are *Pneumocystis carinii* and other protozoan infections; fungal infections of the nervous system, liver, bone, and other tissues; bacterial infections of the respiratory and digestive systems; and viral infections of the digestive tract. About 75 percent of AIDS patients develop significant neurological disorders whose causes are poorly understood. Most patients die within two years of entering WR6.

Tragically, most patients studied have followed this general course of disease progression, and no one has yet survived once the infection becomes established. That may change as scientists find new ways to treat the diseases caused by HIV-1. The greatest hope, of course, is that vaccines or drugs will be discovered that will prevent infection or cure infected individuals (see the Focus on Scientific Inquiry, "Can We Prevent or Cure AIDS?").

HIV-1 infection causes a progressive destruction of the immune system that is associated with different diseases. The Walter Reed classification system categorizes the sequential pattern of diseases in affected individuals. Health generally remains good three to five years after infection. Then there is a loss of humoral and cellular immunity, and opportunistic infections occur. In the final stage, AIDS, many diseases may develop. There is no cure for HIV-1 infections.

Epidemiology

The first epidemiological studies of AIDS played a pivotal role in identifying the new disease and set events in motion that led to the discovery of HIV-1. Since then, numerous large investigations have been conducted in the United States and throughout the world.

United States

What is currently known about the status of HIV-1 in the United States? How many people have been or will be infected? Who are they? Why did they become infected with HIV-1? The answers to these questions are based on data collected mostly by the CDC. These data have allowed scientists to formulate some important conclusions about mortality rates, ways in which HIV-1 is transmitted, and populations at risk for becoming infected.

Numbers of Infected People

In the 1980s, 132,000 cases of AIDS were diagnosed in the United States, and nearly half of those patients had died by 1990. In 1991, the U.S. Public Health Service estimated that between 1 and 2 million people were infected with HIV-1 in the United States. Of those infected with HIV-1, it is expected that 50,000 to 60,000 new cases of AIDS will occur in 1992 and, beginning in 1993, new cases will increase to 70,000 to 90,000 each year. The higher rate of increase is expected to continue for at least several more years because of the long latent period between infection and the onset of AIDS, an average span of eight years.

Who Gets AIDS?

Figure 40.6 describes the population groups in which AIDS cases occurred in the United States prior to 1990. During that period, 89 percent of the cases were associated with two risk groups, homosexual or bisexual men and intravenous drug users. Most of the AIDS cases linked to blood transfusions occurred before 1985. Since that time, all blood donations have been tested for HIV-1, and very few new infections have been traced to contaminated blood in the United States. Most of the heterosexual transmission involved sexual contact with a partner in one of the two primary risk groups. At present, the greatest rate of increase in AIDS cases is occurring in newborn children. Children born to mothers infected with HIV-1 have about a 50 percent chance of becoming infected. The infections in these cases occur during or shortly after pregnancy, and most are due to intravenous drug use by one or both parents.

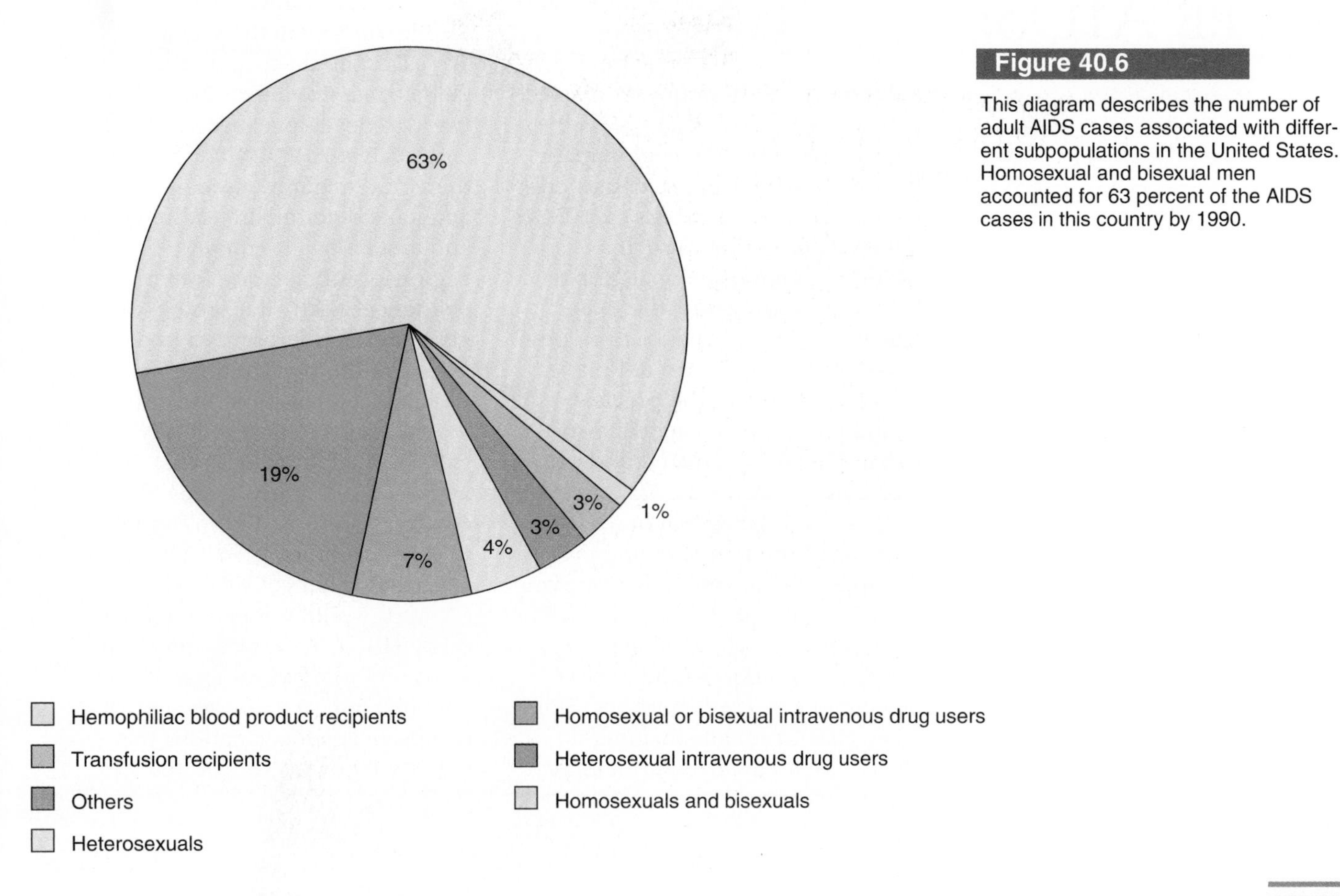

Figure 40.6

This diagram describes the number of adult AIDS cases associated with different subpopulations in the United States. Homosexual and bisexual men accounted for 63 percent of the AIDS cases in this country by 1990.

Can We Prevent or Cure AIDS?

Focus on Scientific Inquiry

In 1984, after AIDS was shown to be caused by HIV-1, investigators in laboratories around the world began to search for ways to prevent infections and treat or cure the disease. Two basic approaches are used to counteract disease agents: vaccines and drug therapy. *Vaccines* are produced from dead or modified forms of a disease-causing virus, bacterium, or protozoan that, when introduced into the body, stimulate an immune response. This results in the production of immune memory cells that prevent future infections. *Drug therapy* involves creating drugs that attack the disease agent once it has infected the host. Thus drug therapy is used to treat and cure diseases after they occur.

HIV-1 Vaccines

Whereas there have been celebrated successes in developing vaccines against the viruses that cause smallpox, polio, mumps, measles, yellow fever, and other important human diseases, no effective HIV-1 vaccines have yet been produced. A common question asked by concerned citizens is, Why, given the economic and human costs of AIDS, can't scientists create a vaccine that works on HIV-1? Research to develop such a vaccine is taking place on a global scale. Unfortunately, there are three major obstacles to overcome.

First, there is no definitive understanding of the precise immune protective mechanism that must be stimulated to destroy or inhibit HIV-1. Second, once HIV-1 becomes integrated as a provirus, the immune system is helpless in locating and destroying it. Thus even if a vaccine led to the development of HIV-1 memory cells, integration would likely occur before they could act. Third, key HIV-1 proteins that might serve as antigenic determinants change continuously because of extraordinarily rapid mutation rates in the viral genome. An infected individual may have several different *strains* of HIV-1, varieties that differ only slightly in their protein structure or genetic content. As many as 100 HIV-1 strains have been isolated from a single individual. Consequently, the virion represents an elusive target for vaccines, which by their nature are designed to represent a precise, unchanging part of the viral structure.

There are also a number of other problems in the development of HIV-1 vaccines: difficulties in conducting tests to evaluate the effectiveness of a candidate (new) vaccine, the need to resolve certain ethical questions (for example, who should receive the vaccine when it is being tested?), and possible liability suits stemming from tests of candidate vaccines. What if someone should become infected with HIV-1 through receiving the vaccine?

Despite these and other difficulties, massive efforts are being made to develop a successful vaccine against HIV-1. What part of the virion might be used as an antigenic determinant for initiating and sustaining an immune response sufficient for conferring resistance? Each viral protein or glycoprotein represents a potential target. At the present time, most research is focused on the envelope glycoproteins, especially gp120. The rationale for this approach is that if viral gp120 could be prevented from binding to CD4 receptors, then T4 lymphocytes and other cells with that receptor would not become infected by HIV-1. To date, results from studies of test vaccines are not encouraging, perhaps because of extensive genetic variations that are now known to occur in the *env* gene. Thus gp120 vaccines may be "obsolete" soon after production. Despite the lack of significant progress, there is considerable optimism in the scientific community that a successful vaccine will be produced, given sufficient time and resources.

HIV-1 Drug Therapies

How does a therapeutic agent, or drug, act to cure a disease? Basically, it must either kill the disease-causing agent or prevent it from reproducing (as in bacteria and protozoa) or replicating (as in viruses), without irreparably harming host cells. To carry out its task, a drug targets a biochemical pathway that is unique to the agent and disrupts its function. Consequently, some critical substance may not be available to the pathogen, or it may not be able to complete its life cycle.

Because viruses live inside host cells, they are a difficult target for a drug to reach. This problem becomes even more severe when dealing with integrated HIV-1 proviruses. The fundamental problem is how to eliminate the virus without also killing the host cell. Unfortunately, to some degree, all drugs are harmful to at least some cells. Thus a fine balance must be maintained between effects on the pathogen and effects on the host. For nonthreatening diseases such as colds, a conservative approach is taken, and drugs having little effect on the host are prescribed. In contrast, for life-threatening diseases such as AIDS, radical therapies that have substantial side effects may be employed.

How do scientists identify possible targets of opportunity in the case of HIV-1? A general approach is to study the virus's life cycle and define points of vulnerability. Examination of the HIV-1 life cycle indicates several possible points of attack by drugs (see Figure 1). Research is being conducted to identify drugs capable of acting in these different stages. Three examples are considered that represent varying degrees of success.

In thinking about disrupting the life history of HIV-1, perhaps the most obvious question that might be asked is how its binding to CD4 receptors can be prevented. Successful efforts were made (using recombinant DNA technology and the CD4 gene) to create a soluble form of "plain" CD4, called *recombinant soluble CD4* (*rsCD4*), that can bind to gp120. The general idea was that if enough rsCD4 molecules were present in the body, viral gp120 binding sites would become saturated, and the HIV-1 could not infect T4 cells. In laboratory tests using cultured T4 cells, it was found that binding was blocked by a drug consisting of rsCD4, a

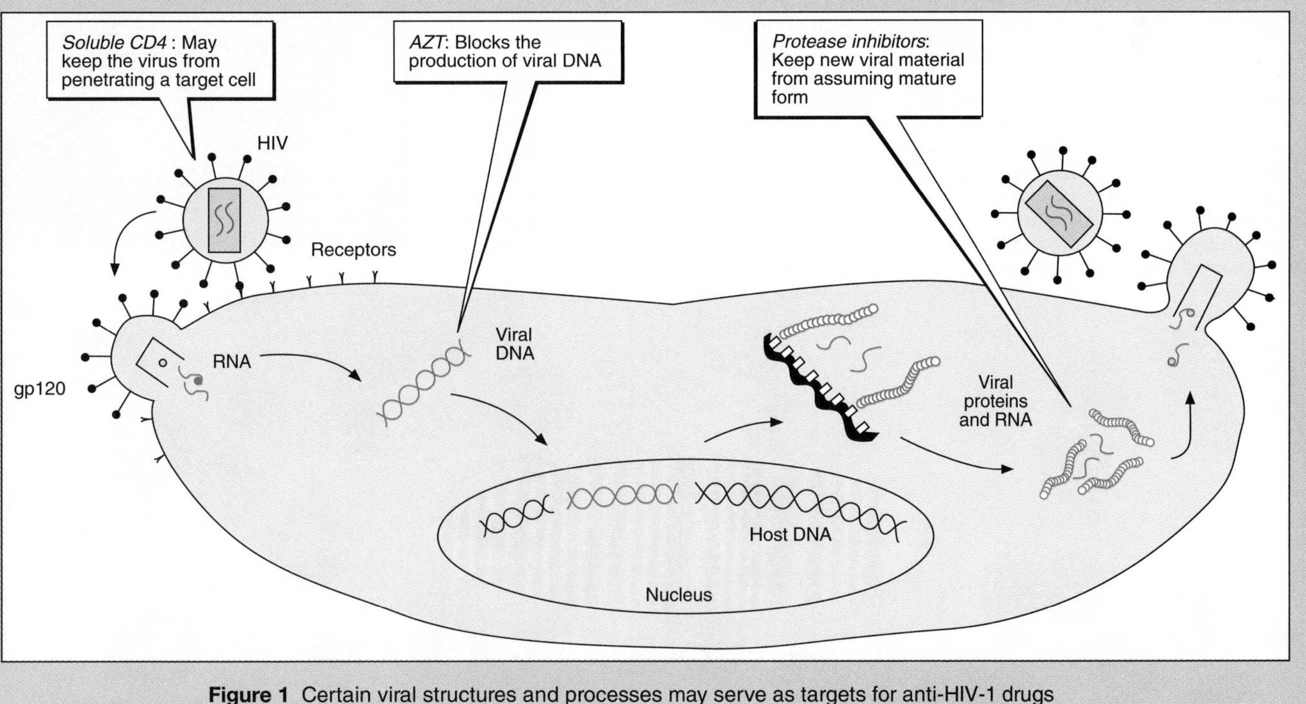

Figure 1 Certain viral structures and processes may serve as targets for anti-HIV-1 drugs currently under development or being tested.

Focus on Scientific Inquiry

promising start. Unfortunately, tests on humans indicated that there was no significant enhancement of the immune response. Research to modify rsCD4 is now under way. For example, attempts are being made to combine rsCD4 molecules with potent protein toxins that can kill HIV-1-infected cells.

In 1984, when HIV-1 was identified as the causal agent of AIDS, scientists at the National Cancer Institute began testing known antiviral drugs to determine if any of them might react against the new virus. It was discovered that one of them, 3'-azido-2'3'-dideoxythymidine, or *AZT* (also called *azidothymidine* or *zidovudine*), was a powerful inhibitor of HIV-1 growth in T4 cells maintained in culture. Initially, it did not appear to be harmful to cells, although it is now known to be toxic to bone marrow cells at high doses. Intensive efforts led to the development of an AZT drug that could be tested in AIDS patients. By 1986, it had been demonstrated that AZT extended the survival of AIDS patients and improved the quality of their lives. Further, the earlier the treatment occurred after infection, the greater the positive effect. The U.S. Food and Drug Administration approved AZT for use in 1987.

How does AZT react against HIV-1 or prevent T4 cells from being affected by the virus? AZT mimics thymidine (T), one of the nucleotides used to construct DNA. When reverse transcriptase converts viral RNA into DNA, it may mistakenly incorporate AZT instead of T into the chain being synthesized. When that happens, the next DNA base cannot be added, viral DNA synthesis is aborted, and no provirus is manufactured.

AZT is not the final solution, but it remains the most effective drug yet identified and tested. However, it has some serious drawbacks. In many patients, HIV-1 becomes resistant to AZT. Also, AZT causes moderate to severe side effects, primarily because of its bone marrow toxicity, which often leads to anemia. At present, there is great excitement about other new drugs that are also directed at disrupting reverse transcriptase.

Finally, researchers have identified the three-dimensional structure of one of the HIV-1 enzymes (protease) required for viral replication. It appears to have a specific site that may be vulnerable to attack by appropriate drugs (see Figure 2). If this enzyme were disabled, new virions could not be assembled in infected cells. Clinical trials using protease inhibitors are expected to be under way by 1992.

Many other drugs are now being developed and tested. In fact, the number of new drugs is now beginning to exceed the government's capacity to conduct clinical tests of their safety and effectiveness. Given the large number of viral targets, the enormous effort being made by researchers, and the support provided by various funding agencies, there is great hope that additional effective, or more effective, drugs will soon be discovered.

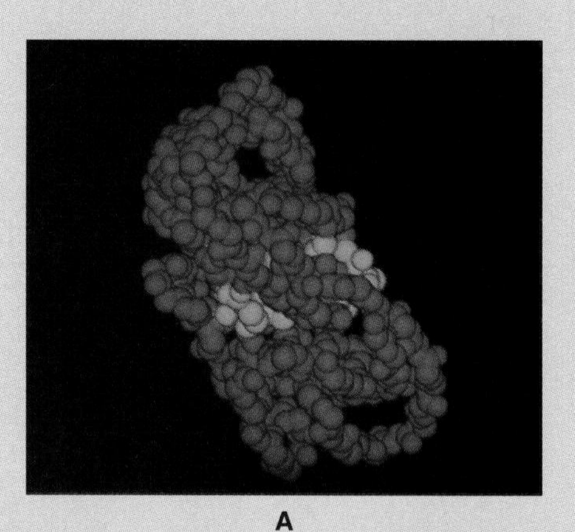

A

Figure 2 A new approach for treating HIV-1 infections is now being studied. The basic concept involves inactivating the HIV-1 protease enzyme, which is necessary for viral replication, with a drug. (A) This model shows how a drug (green) binds to the protease molecule (red) at a specific site, thus disabling it. In a recently completed experiment, infected cells were treated with a protease inhibitor. (B) HIV-1 in untreated cells (the experimental control) replicated normally as indicated by either a bullet-shaped or circular central core (arrows). (C) In treated cells, most virions failed to replicate normally as indicated by the absence of a central core and the dense plaque under the membrane (arrows).

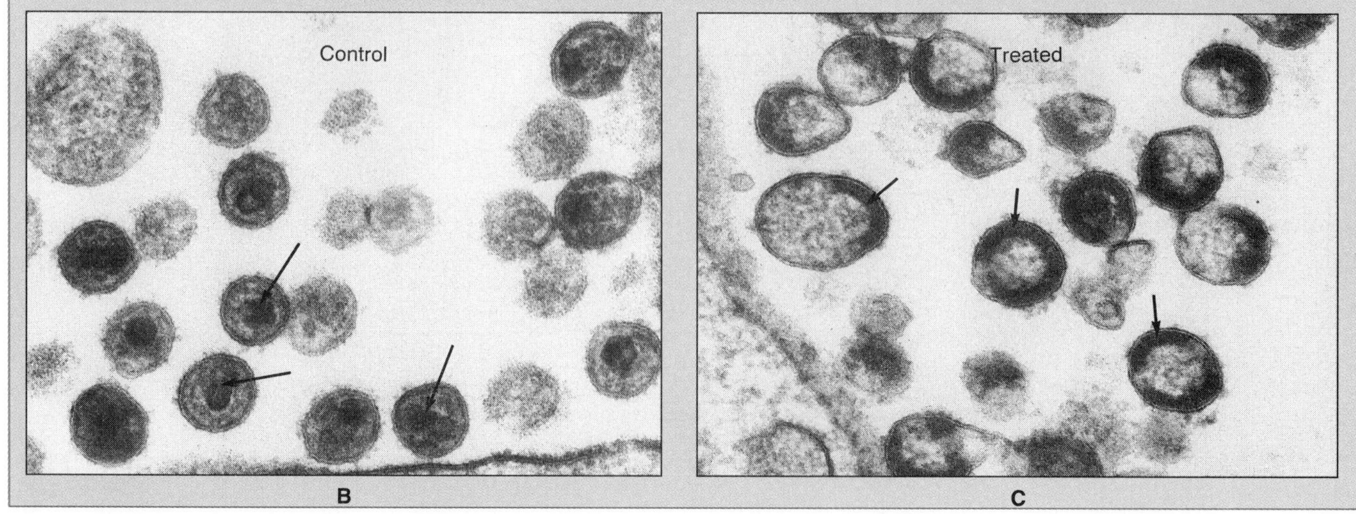

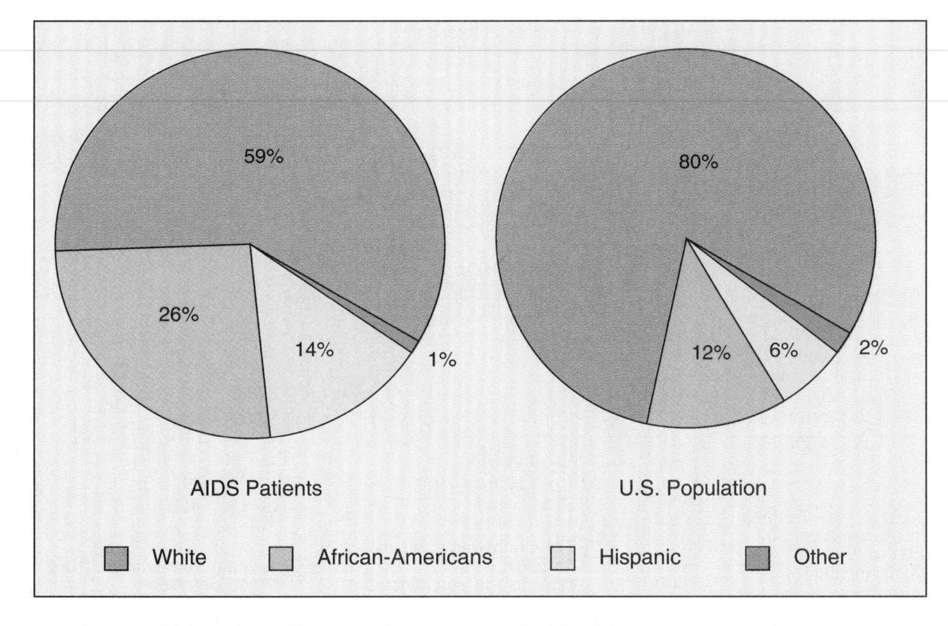

Figure 40.7

African-Americans and Hispanics account for a greater percentage of AIDS cases in the United States than their percentage of the population.

As shown in Figure 40.7, there is a unbalanced percentage of AIDS cases among African Americans and Hispanics. African-Americans make up 12 percent of the total population, but they account for 26 percent of the adult AIDS cases and 53 percent of the pediatric cases. For Hispanics, these numbers are 6.5 percent of the total population, 14 percent of adult cases, and 23 percent of pediatric cases. These percentages are related primarily to the higher numbers of reported intravenous drug users in these two groups.

It is now well established that HIV-1 is transmitted mainly through sexual contact, exposure (via needles) to contaminated blood or blood products, from mother to offspring during pregnancy, or while nursing. Risk factors for homosexual men include number of sexual partners, frequency of unprotected anal intercourse, and the presence of sexually transmitted diseases (STD), especially syphilis and genital herpes, in either partner. Available evidence indicates that use of a condom significantly reduces the risk of contracting HIV-1 from an infected sexual partner.

Who Doesn't Get AIDS?

A great deal of research has been directed at determining the risk of transmission between members of a household when one of the members is infected with HIV-1, the risk of contracting the virus from an infected individual at the workplace or school, and the possibility of contracting the virus via an insect vector. Except for sexual partners of an infected individual and a few workers accidentally injected in hospitals or research labs, no one has yet been infected with HIV-1 in any of these circumstances. Also, convincing evidence indicates that HIV-1 is not transmitted by insects or from food, coughing, sneezing, toilet seats, holding hands, or kissing.

Worldwide

The World Health Organization (WHO) has assumed primary responsibility for studying HIV-1 and AIDS outside the United States. Figure 40.8 describes WHO data for HIV-1 infections in different parts of the world in 1991. AIDS data from developed countries are quite reliable, but information from developing nations is very poor. For many countries, few or no data are available on the numbers of AIDS cases. WHO estimates that 1 million cases of AIDS have already occurred in the world population. Between 8 and 10 million people are currently infected with HIV-1, and at least 1 million new AIDS cases will develop during the next five years. By the year 2000, it is predicted that a minimum of 15 to 20 million people will be infected.

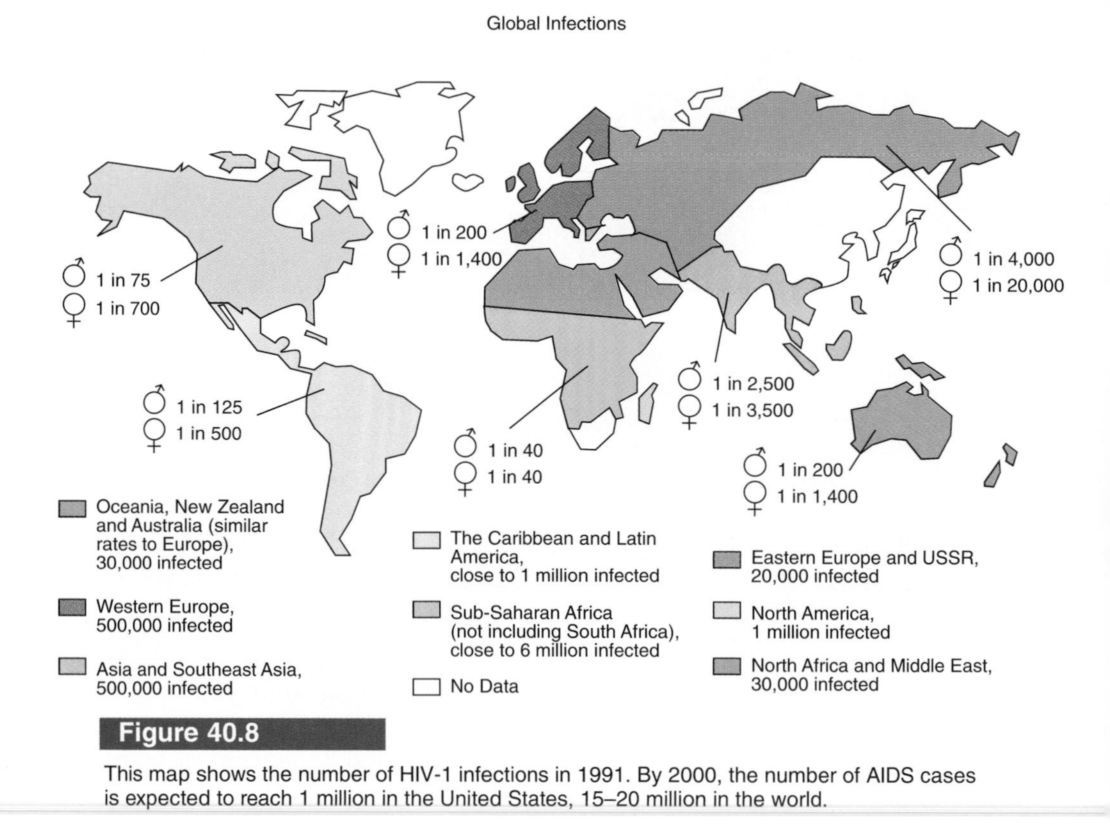

Figure 40.8

This map shows the number of HIV-1 infections in 1991. By 2000, the number of AIDS cases is expected to reach 1 million in the United States, 15–20 million in the world.

The epidemiological pattern (time of appearance, risk groups, risk factors) seen in the United States also occurs in Canada, Mexico, South America, Australia, New Zealand, and many Western European countries. A much different pattern has been described for Africa and, increasingly, parts of Central America and the Caribbean. In these countries, AIDS also began to spread in the late 1970s, but the major route of transmission was through heterosexual intercourse, and drug use was not a major risk factor. Many countries face grave risks because of unsafe blood products and a scarcity of sterile needles. For example, in Romania and the USSR, HIV-1 has been spread by poor medical practices and a lack of money required to buy the equipment necessary to test blood and sterilize syringes.

Africa has been the continent most seriously affected by HIV-1 infections. Available data indicate that 6 to 7 million Africans have already been infected. AIDS is now a leading health problem in many central and eastern African countries. WHO estimates that between 5 and 20 percent of sexually active people in urban centers in these countries are now infected, and that figure is expected to increase. Given the lack of health and economic resources in these areas, the current picture in Africa is devastating.

Epidemiological data have enabled scientists to identify risk groups and risk factors and to determine approximate numbers of people infected. Between 1 and 2 million people in the United States are now estimated to be infected with HIV-1. Most are homosexual or bisexual men or intravenous drug users, but the numbers of infected women and children are increasing. Major risk factors involve unsafe sex practices and drug abuse. In underdeveloped countries, HIV-1 infections are spread through heterosexual intercourse and unsafe medical practices. Several African populations have very high rates of HIV-1 infection.

Some Other Aspects

Before AIDS, the last great pandemic occurred just after World War I, when influenza killed more than 20 million people. A complete AIDS accounting, in terms of mortalities and economic costs, will not be known for decades. However, it is already evident that the final costs of HIV-1 will be enormous, and for some countries and population groups, they will be absolutely disastrous.

Origin of HIV-1

How did such a virus originate? Where did it come from? Definitive answers to these questions will probably never be known, but certain clues have given rise to a hypothesis. Analyses of blood stored in different countries for decades revealed that a sample from Zaire (Africa), collected in 1956, contained antibodies against HIV-1. Thus it appears that the virus was present in that country at least 20 years before it appeared elsewhere.

Scientists hypothesize that HIV-1 existed in isolated pockets in central Africa until the 1950s. At that time, rural Africans began migrating into large cities and transported the HIV-1 with them. There the virus was transmitted to larger numbers of individuals, presumably through heterosexual intercourse and transfusion of contaminated blood. By the 1960s and 1970s, technology in the form of passenger jets and global distribution of blood for medical uses had led to the dissemination of HIV-1 throughout the world. AIDS then appeared in countries to which the virus had been conveyed.

The Present Situation

A major question that remains unanswered involves the level of risk in acquiring HIV-1 through heterosexual intercourse with no other risk factors involved. No large-scale studies of heterosexual behaviors have been conducted, so few data are available from which to draw definitive conclusions. From the data that are available, the CDC has identified the major risk factors associated with heterosexual intercourse. They are number of sexual partners, the chance of the partner being infected with HIV-1, and not using a condom during intercourse. Of these, multiple partners and use of a condom are risk factors subject to individual control.

Data collected between 1989 and 1991 indicate that there has been a startling rise in the number of teenagers developing AIDS. This increase began in cities such as New York and Miami, where there were already large numbers of HIV-1 infections, but higher numbers are now evident in other areas of the United States. The increasing number of teenage AIDS cases is of great concern to epidemiologists because of the means of transmission (thought to be primarily heterosexual intercourse) and the potential for large numbers of young people to become infected.

The total number of pediatric cases continues to expand in direct relation to the increased number of women becoming infected. The incidence of infection among homosexual men has dropped significantly because of education efforts in this group.

Scientists can provide the type of technical information contained in this chapter. However, it falls to all educated citizens not only to understand the nature of these biological problems but also to plot a course of action that may lead to solutions of social and cultural problems. How can the spread of AIDS be stopped? Clearly, education about safe sex, risk factors, and the importance of supporting research on AIDS will be critical. However, what is the best means of educating people about HIV-1 and AIDS? Also, whereas scientists can describe risk groups and risk behaviors precisely, only individuals can modify their own risky behaviors. Finally, how are we going to pay for the medical bills of AIDS patients, now expected to be tens of billions of dollars in the 1990s? Perhaps the most important consideration for all citizens is to recognize that AIDS has not disappeared and will not within our lifetimes. It remains to be seen whether or not, and to what extent, it will become established in the general population.

Available evidence indicates that HIV-1 originated in rural Africa by the 1950s and then spread to large African cities. Subsequently, it was dispersed throughout the world. For the heterosexual population, major risk factors for becoming infected with HIV-1 are multiple partners, infected partners, and failure to use condoms. HIV-1 is now spreading more rapidly among teenagers than in the past, presumably by heterosexual intercourse. There are major problems in trying to prevent the spread of HIV-1.

A UNIFYING HYPOTHESIS OF CANCER CAUSATION

Cancer, the disease, was described in Chapter 39. It begins when fundamental molecular changes lead to unregulated cellular proliferation and loss of cellular homeostasis. Before the 1970s, cancer researchers placed great emphasis on identifying the causes of cancer in experimental animals and evaluating the behavior of transformed, malignant cells.

In the past 20 years, however, new discoveries have opened doors to deeper, molecular levels of understanding cancer and its underlying causes. Currently, there is great excitement in the cancer research community because of new conceptual breakthroughs. Scientists are now attempting to develop a *unifying hypothesis*—a synthesis of existing knowledge to create a single series of statements or ideas—that will explain the causes of cancer. The general principle that gene mutations are a prerequisite for transforming normal cells into cancer cells led to several new, critical questions that guided related research during the 1980s. Exactly what types of genetic damage lead to the transformation of normal cells? Why is cancer the consequence of cell transformation?

The Concept of Cellular Homeostasis

Figure 40.9 illustrates the concept of **cellular homeostasis**: a relatively constant number of cells makes up tissues and organs in the body, and this number is maintained throughout an individual's life. The four major types of tissues have two different patterns of cellular homeostasis. In general, cells of nervous and muscle tissues cannot be replaced if they are damaged or destroyed. Thus their numbers normally decline slowly with age. In contrast, cells of most epithelial and connective tissues are systematically and continuously replaced through the operations of highly regulated **cell renewal systems**. These systems consist of four "compartments," each of which describes a sequential stage in the life of a cell: *proliferation* (mitosis), *growth and differentiation, function,* and *senescence* (death). Complex regulatory mechanisms operate in each compartment of the system. For example, the frequency of mitosis, the rate of growth and differentiation, the length of time a cell will function, and the life span of a cell are all controlled by positive and negative regulatory proteins produced by genes. What is the nature of these genes? What happens if they malfunction? The answers to these questions have their roots in a famous discovery made early in the twentieth century.

Oncogenes

In 1910, Peyton Rous demonstrated that he could transmit cancer (specifically, a *sarcoma*—a type of connective cell cancer) from an affected chicken to one that was not affected. He did this by first removing cancerous cells, crushing them, and then filtering the material to remove all cellular fragments, bacteria, and other substances. When the filtered fluid was injected into healthy chickens, they developed sarcomas! What was the agent that caused these cancers? Like those of Mendel (Chapter 12) and Garrod (Chapter 16), Rous's discovery was scientifically premature—the agent and its significance were not understood until decades later. It is now known that Rous had discovered the first oncogenic (cancer-caus-

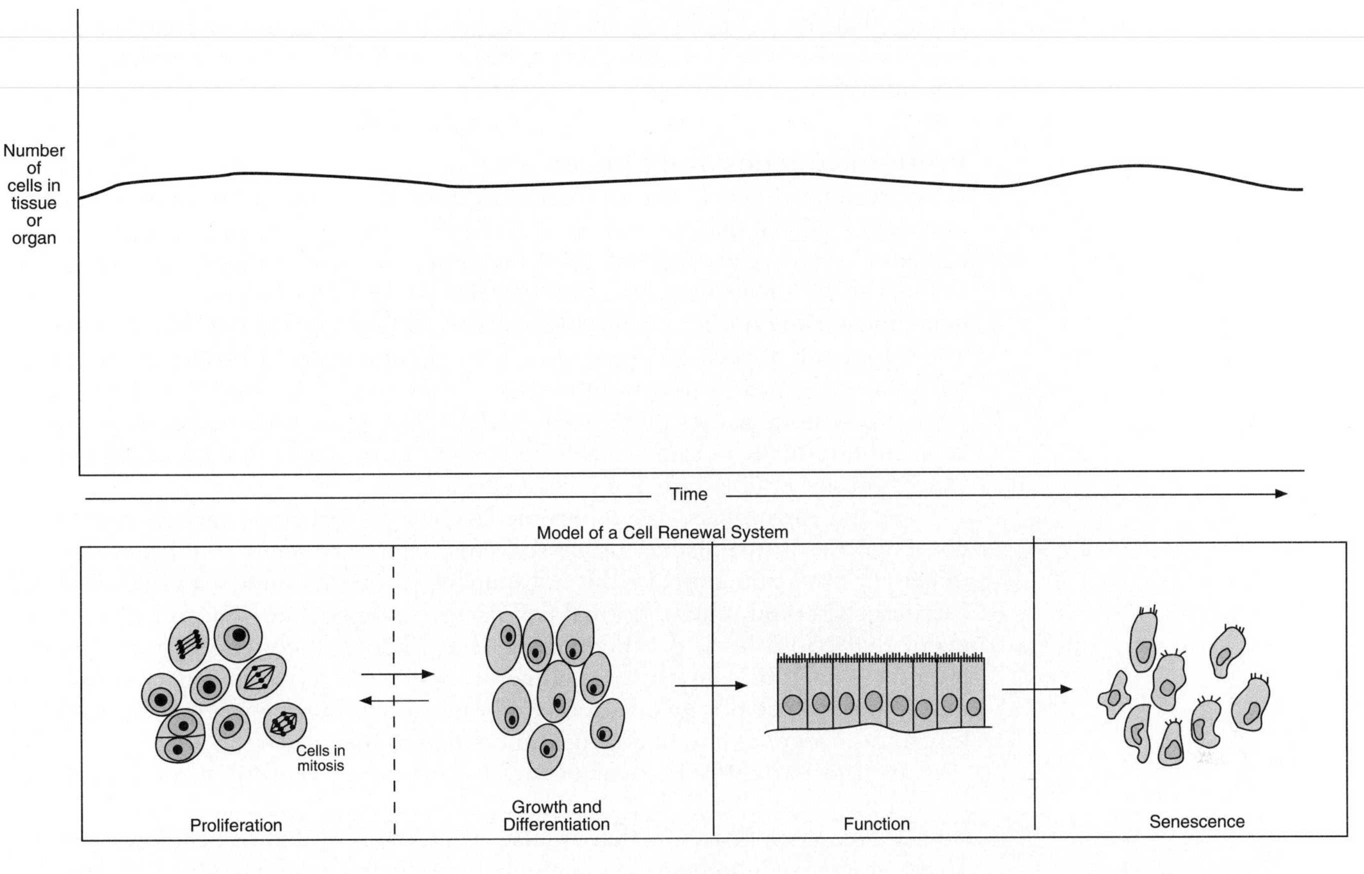

Figure 40.9

Cells of different tissues arise through mitosis. They then undergo a period of differentiation when they become specialized to perform a specific function in the body. Such cells are functional for a genetically defined period of time, after which they die. In normal tissues, the number of new cells produced by mitosis is approximately equal to the number lost. If cells fail to differentiate, they may continue to proliferate—a common characteristic of cancerous growth.

ing) retrovirus (now called the *Rous sarcoma virus*). Further, this virus was found to have only four genes. Three of them are used to replicate the virus in infected cells, and the fourth is a gene that causes cancer—an **oncogene** known as *src*. For his discovery, Rous finally received a Nobel Prize in 1966, when he was 87 years old.

Retroviruses proved to be fascinating to study. Beginning in the 1950s, they were shown to cause cancers in several types of animals. Ultimately, more than 20 retrovirus oncogenes were discovered. As a result, in 1969, George Todaro and Robert Huebner of the National Cancer Institute proposed the *oncogene hypothesis* to explain how cancer developed in humans. Their idea was that because of infections that occurred early in the evolutionary history of humans, viral oncogenes existed within the chromosomes of our normal cells, and cancer developed when they became activated by chemical carcinogens or radiation. In other words, viral oncogenes were integrated in the normal genome, and malignant growths arose in humans when they were turned on. What type of research finding would support or confirm the oncogene hypothesis? Investigators began by searching for the *src* gene in various cells. These efforts led to a very surprising discovery.

Proto-oncogenes

In 1976, J. Michael Bishop and Harold Varmus found *src* sequences in the normal DNA of birds and later in mammals, including humans. However, the *src* gene in these cells is not a viral oncogene! Rather, in normal DNA, it is a cellular gene that encodes an ordinary structural protein found in the plasma membrane. Thus in normal cells, *src* is not an oncogene; consequently, it was classified as a **proto-oncogene**—a gene that has a specific function but when damaged or mutated has the potential to become a *cellular oncogene*. For their crucial discovery that normal cells contain proto-oncogenes that can cause cancer if they become transformed into oncogenes, Bishop and Varmus were awarded a Nobel Prize in 1989.

Proto-oncogenes and Cancer

How are retroviral oncogenes related to normal cellular proto-oncogenes and oncogenes? Recall that the retroviral genome (provirus) becomes integrated into cellular chromosomes after infection. According to current theory, at some point in their evolutionary history, a provirus picked up part of a cellular proto-oncogene, most likely during a replication phase, and incorporated it into its genome. The frequency of such an event over many million years of evolutionary history must have been extremely rare. Evidence from molecular studies indicates that pirated proto-oncogenes underwent modification in the provirus replication process and eventually became established as *viral oncogenes* that can cause certain cancers in susceptible bird and animal species.

By the early 1980s, it had become established that some cancers could be a consequence of two distinct processes. First, viral oncogenes could cause cancer in certain birds and animals (but not humans) if they became activated. Second, carcinogens could mutate normal cellular proto-oncogenes into oncogenes, an event that could lead to cancer. Thus the original oncogene hypothesis was incorrect—the ultimate cause of human cancer (and most cancers of other animal species) does not involve retroviral oncogenes. What proteins are encoded by proto-oncogenes and oncogenes, and what do they do in the cell?

By the early 1990s, over 60 proto-oncogenes had been identified. All known proto-oncogenes and oncogenes encode either growth factors, their cell surface receptors, or factors that regulate gene transcription or cell development. These products all function in the regulation of cellular proliferation, growth, differentiation, or senescence.

Conversion to Oncogenes

How are proto-oncogenes converted to oncogenes? Different types of damage have been identified—translocations between chromosomes or within the same chromosome, deletions of tiny segments of specific chromosomes, mutations within proto-oncogenes, and abnormal increases in portions of a chromosome, a process called *gene amplification*. When a proto-oncogene becomes an oncogene, three types of malfunctions can occur. First, the oncogene or its product cannot be regulated, and increased, uncontrolled cell activity may result. Second, gene amplification may lead to *overexpression,* the production of excess quantities of the encoded protein, or the production of an abnormal protein. Third, mutations can alter the action of the protein produced.

Any of these events may result in a loss of ability to regulate cell differentiation and proliferation, a step that can lead to cancer. For example, the first oncogene protein was described in 1988. This protein is produced by a member of the *ras* oncogene "family" (one of a number of genes with similar structure and function) and has a role in regulating cell growth. The difference between the normal *ras* proto-oncogene protein and the *ras* oncogene protein usually involves a change in only a single nucleotide base. Yet the consequence of this change is that the oncogene encodes a protein that is associated with uncontrolled proliferation of the affected cell. Active *ras* genes are the most common oncogenes isolated from human cancers.

Thus from research conducted during the 1980s, it became known that there were links between oncogenes and certain cancers in humans. In general, oncogenes produce active products that lead to a loss of control at some point in a normal cell renewal system. New questions continued to arise. Can single oncogenes cause cancer, or are many active oncogenes required? Are there other genes involved?

Tumor Suppressor Genes

Throughout the 1980s, evidence accumulated that another type of regulatory gene played a role in the genesis of many cancers. Such genes became known as **tumor suppressor (TS) genes**. In contrast to oncogenes, they influenced cancer induction if they were inactivated or lost from the chromosome. TS genes have been implicated in many cancers. At present, the picture of TS genes is poorly developed, but it is known that their encoded proteins tend to inhibit cellular renewal processes rather than driving them forward as oncogenes do. In addition, some TS genes regulate DNA repair—an essential mechanism for maintaining chromosome stability. How are these genes lost or inactivated? Basically, by the same processes involved in proto-oncogene transformation—translocations, deletions, and mutations. Figure 40.10 illustrates the two general types of mutations linked to loss of normal cell regulation and the development of cancer.

A unifying hypothesis on cancer causation is now evolving. Cancer develops when there is a loss of regulated cellular homeostasis. This can occur when proto-oncogenes encoding regulatory proteins are converted to oncogenes. Consequently, the oncogenes may not be regulated, or their encoded proteins may be abnormal or overexpressed. Any of these outcomes may lead to cancer. Proteins encoded by tumor suppressor genes are also involved in cellular homeostasis. The loss or inactivation of these genes have also been implicated in some cancers.

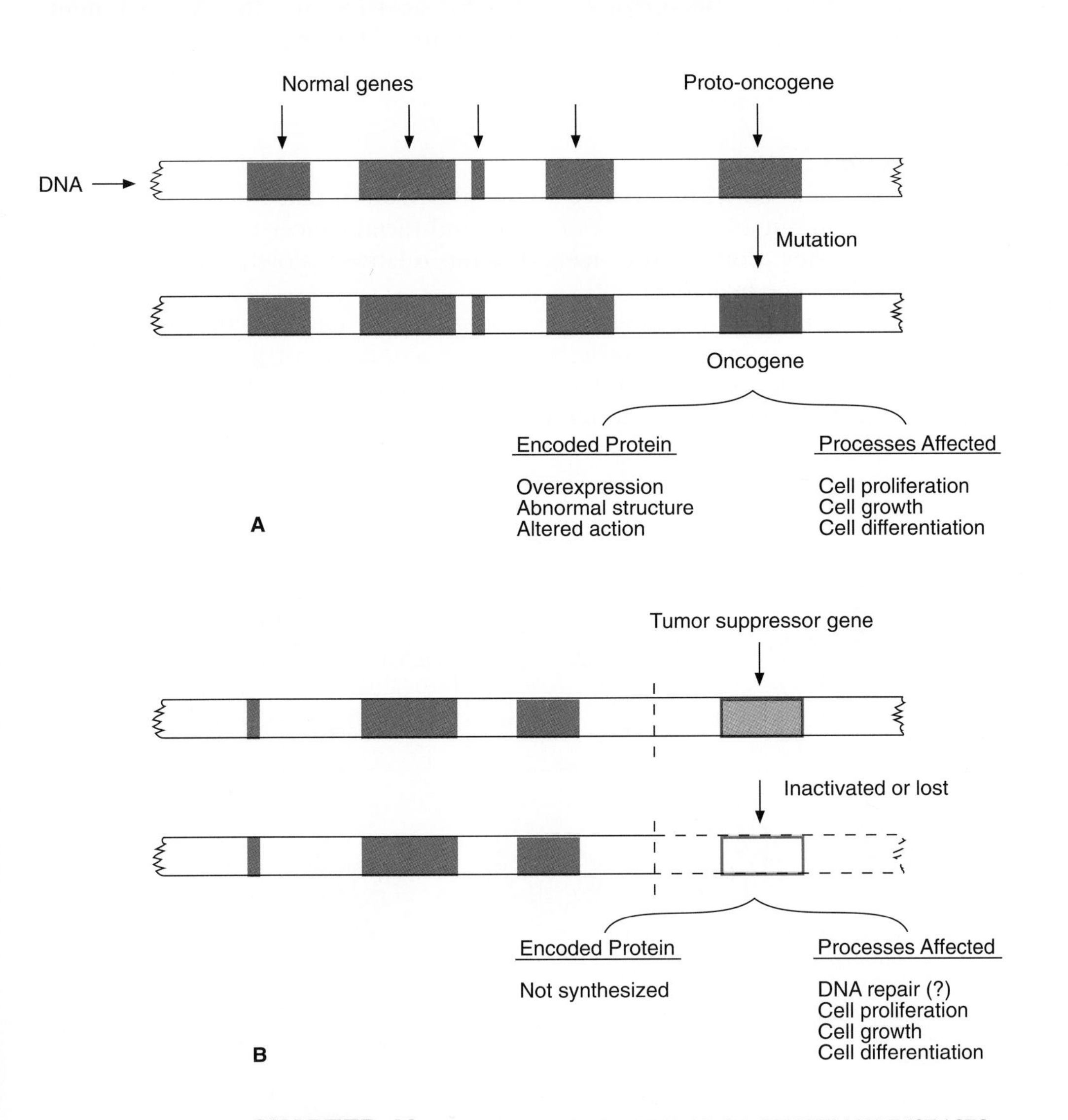

Figure 40.10

Two general types of mutations lead to loss of cell regulation and cancer. (A) A proto-oncogene is converted to an oncogene, a change that leads to alterations in levels or types of proteins involved in normal cell regulation. (B) Loss or inactivation of a tumor suppressor gene results in abnormal cell regulation because of the absence of an essential protein.

Two Case Studies

Two case studies will illustrate the types of research now under way. Some very exciting research results were published between 1989 and 1992 that may have great relevance in creating effective treatments for two important human cancers.

Breast and Ovarian Cancers

Malignancies of the breast and ovary account for one third of all female cancers and are responsible for one fourth of cancer-related deaths in women. Several oncogenes and growth factors have been linked with breast cancer. In one large study of breast cancer patients, investigators found that amplification of a specific oncogene known as *erb*B2 occurred in 25 to 30 percent of the cases. *erb*B2 encodes a protein with a structure that is consistent with that of a growth factor receptor. In addition, there was evidence of direct correlations between having multiple copies of *erb*B2, which resulted in overexpression, and the appearance of tumors, and between *erb*B2 alterations and prognosis for survival. Similar results were also reported in analyses of ovarian tumors.

The significance of this research is threefold. First, one specific mechanism of both breast and ovarian cancers may have been identified. Second, extra copies of *erb*B2 is predictive for clinical outcome; the greater the degree of amplification or overexpression, the less likely the patient will survive. In these cases, the most aggressive forms of therapy are to be recommended. Third, it indicates that *erb*B2 analysis should be used in the diagnosis of breast and ovarian cancer and may also provide a focus for devising an effective treatment. Another relevant feature is the extreme importance of a single oncogene in these cancers, although others may be involved, and tumor suppressor genes may also be found to fit into the final equation.

Colorectal Cancer

Figure 40.11 is a model for colorectal tumor formation that has emerged in the early 1990s. About 150,000 cases of colon and rectal cancer occur each year in the United States. Colorectal cancers develop relatively slowly, taking years or decades to become malignant.

The evolution of this cancer, beginning with benign growth and progressing to malignancy and metastasis, requires a sequence of genetic alterations involving both oncogenes and tumor suppressor genes. According to this model, tumor suppressor genes are sequentially lost from chromosomes 5, 18, 17, and

Figure 40.11

Colorectal cancer develops over many years or decades and requires a sequence of mutations or other types of genetic damage. Each of these events is associated with a change in the dynamics of the cell renewal system. Initially, there is increased growth, followed by the development of a series of benign tumors (adenomas). Eventually, cells acquire the ability to grow invasively into underlying tissues (carcinoma) and metastasize to other sites in the body.

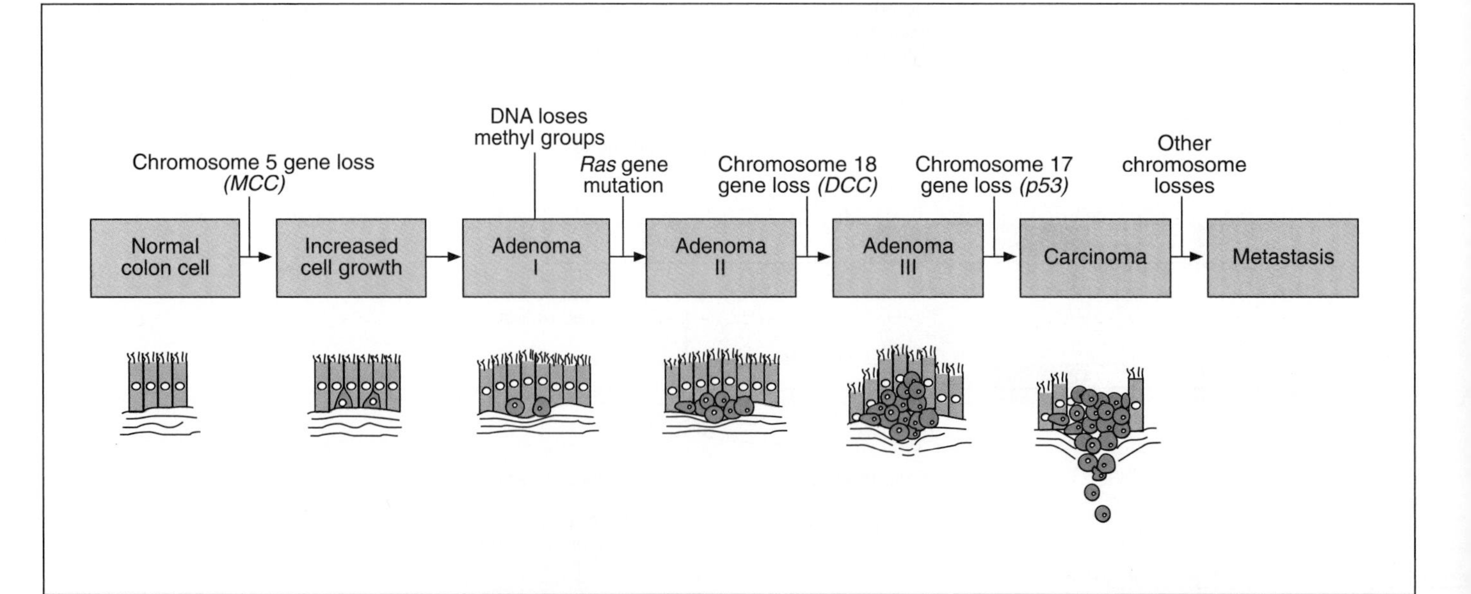

perhaps others at a later stage, just prior to metastasis. Activated *ras* oncogenes are apparently the only oncogenes involved in the progressive development of colorectal cancer. What is the identity of the tumor suppressor genes and their encoded proteins? At present, the tumor suppressor genes identified in the colorectal model are a gene identified as *MCC* (*m*utated in *c*olorectal *c*ancer), on chromosome 5, the *DCC* gene (*d*eleted in *c*olon *c*ancer), located on chromosome 18, and the *p53* gene, located on chromosome 17. Loss or mutation of the p53 gene also occurs in many other types of human cancer. Thus it is being intensely studied, but its regulatory function in human cells has not yet been established. One leading hypothesis is that normal protein p53 activates the transcription of genes that suppress cell proliferation.

This model is extremely useful for several purposes. First, it offers an explanation for the relatively late age of onset of colorectal cancer—it takes years or decades for the multiple mutations to occur. Second, it may be possible to detect the tumor in an early, benign stage and remove the growth before additional mutations accumulate and it becomes malignant and metastasizes. Finally, it provides points of reference for devising treatments, or even cures, for colorectal cancer in the future.

Two types of cancer exemplify the power of the developing hypothesis. Many breast and ovarian cancers are associated with *erb*B2, an oncogene. Increased numbers of this oncogene and overexpression of its encoded protein are negatively correlated with chances of survival. Colorectal cancer develops as a result of sequential changes in a proto-oncogene and several tumor suppressor genes.

Current Ideas About Cancer Causation

The emerging hypothesis of cancer causation suggests that cancer occurs when some part of a complex cell renewal regulatory system fails because of an accumulation of genetic changes within a single cell. Consequently, the affected cell continues to proliferate—but no longer differentiates or becomes functional—or dies, according to a genetically defined timetable. These cells may eventually invade surrounding tissues and metastasize to other sites in the body. The underlying elements that lead to this loss of control include active oncogenes and their proteins and lost tumor suppressor genes and their missing proteins.

Many human cancers have now been studied, and most are thought to occur because of a combination of activated oncogenes and the loss of tumor suppressor genes. In some cases, a specific oncogene has been linked to several different cancers. The same is also thought to be true for tumor suppressor genes; many cancers are characterized by the loss of a specific tumor suppressor gene. The model for colorectal cancer—beginning with normal epithelium and progressing to benign tumors (*adenomas*) and finally to the malignant, invasive stage of the disease—strongly supports the emerging hypothesis of cancer causation.

Much remains unknown, and years of additional research will be required to put the complete puzzle together. Nevertheless, understanding cancer at the level described in this chapter offers great hope for eventually being able to prevent or retard its development. For example, it may become possible to disrupt the sequence of genetic events that leads to the final, malignant stage of the disease. Someday it may become practical to use gene therapy approaches (discussed in Chapter 21) for replacing defective or missing genes or proteins involved in cancer progression.

LOOKING AHEAD

This chapter has focused on HIV-1 and cancer research being conducted at the molecular level of biological organization. It is worth noting that the speed with which many of the puzzles associated with HIV-1 were solved is simply phenomenal. Equally rapid headway is now being made in unraveling the causes of cancer. You will recognize that this progress was due in large part to the advances made in molecular biology discussed in earlier chapters.

Despite this progress, much remains to be learned about HIV-1 and cancer. Perhaps the most promising feature of the research now under way lies in the future development of treatments, or even cures, for these two dreaded diseases. For example, in 1991, clinical trials began using experimental therapies designed to counter the *erb*B2 gene in breast cancer patients. It should be very exciting to follow research advances in AIDS and cancer in the 1990s.

Summary

1. During the 1990s, complex human diseases such as AIDS and cancer will be investigated at the molecular level. This approach will enable scientists to determine the basis for pathogenic effects caused by these diseases and to establish rationales for diagnosis, therapy, and eventual cures for these disorders.
2. AIDS was identified as a significant disease in the United States and other parts of the world in 1981. AIDS is characterized by loss of immune function and the inability to resist opportunistic infections.
3. Through epidemiological investigations conducted by the CDC and other agencies, it became obvious by 1982 that AIDS was caused by an infectious agent, not some unique lifestyle factor. In 1983 and 1984, the cause of AIDS was discovered to be a new retrovirus called HIV-1.
4. HIV-1 infects helper T4 lymphocytes and becomes integrated into host cell chromosomes. When the T4 cell becomes activated by contact with a foreign antigen, it replicates new virions rather than substances and cells required for a normal immune response. Subsequently, the virions bud from the cell, enter circulation, and infect other T4 cells. The original T4 cell dies during the budding process.
5. The eventual depletion of T4 cell populations accounts for the characteristic immune deficiency of AIDS and is the indirect cause of death of infected individuals.
6. The major groups infected with HIV-1 in the United States are homosexual and bisexual men, intravenous drug users, sexual partners of individuals in these two groups, people who received blood transfusions, and babies of infected mothers. Relatively few data are available for defining the risk of heterosexual intercourse involving members of the general population.
7. It is now hypothesized that cancer arises because of sequential mutations in genes and chromosomes. These changes modify the structure and function of proto-oncogenes, converting them into oncogenes, and cause the loss of tumor suppressor genes.
8. Both proto-oncogenes and tumor suppressor genes play key roles in the complex regulating mechanisms of cell renewal systems. The combination of active oncogenes and lost tumor suppressor genes results in an inability to regulate normal cell proliferation and differentiation, which leads to cancer.

Review Questions

1. Why is modern disease research emphasizing changes at the molecular level?
2. What is epidemiology? What are the objectives of epidemiological research?
3. What types of results from early studies indicated that AIDS was caused by an infectious agent?
4. What is AIDS? What causes AIDS?
5. Describe the HIV-1 life cycle.
6. What is the general sequence of diseases that occur in people infected with HIV-1? What factors influence the development of different diseases?
7. Who is most likely to be infected with HIV-1 in the United States? Why?
8. Why is the AIDS picture so bleak for Africa?
9. What are proto-oncogenes? Oncogenes? How are they related to cellular homeostasis?

10. What are tumor suppressor genes? What is their normal function?
11. Explain one cause of breast and ovarian cancers and one cause of colorectal cancer.
12. Summarize the developing hypothesis of cancer causation.

Essay and Discussion Questions

1. Based on current epidemiological information in the United States, what would you recommend to halt the spread of HIV-1? What problems might affect implementation of your recommendations? How could they be overcome?
2. New data indicate that AIDS is spreading in the heterosexual population, the rise in teenage AIDS cases is alarming, and high school and college students have not altered their sexual behaviors. What might be done to change the attitude and behaviors of this risk group? Should anything be done? Why?
3. Whether or not HIV-1 tests should be required for all marriage license applicants remains an issue in the United States. An analysis published in 1987 (*JAMA* 258:1757–1762) included the following data: applicants for marriage licenses, 3.8 million; true positives (actual cases identified), 1,219; false positives (cases misdiagnosed), 382; false negatives (cases missed), 129; HIV-infected births prevented, 250; cost of screening, $100 million. Do you favor marriage applicant screening? Why? What arguments support or do not support HIV-1 screening?
4. Describe the types of studies and data that you think will be necessary to confirm the unifying hypothesis of cancer causation. Do you believe that this hypothesis will explain the origin of all cancers? Why? How do the risk factors described in Chapter 39 fit into the hypothesis?

References and Recommended Readings

Baum, R. B., and R. Dagani. 1990. AIDS vaccine, drug research advancing on several fronts. *Chemical & Engineering News* 68:7–15.

Bishop, J. M. 1991. Molecular themes in oncogenesis. *Cell* 64:235–248.

Greene, W. C. 1991. The molecular biology of human immunodeficiency virus type 1 infection. *The New England Journal of Medicine* 324:308–317.

Hollingsworth, R. E., and W. Lee. 1991. Tumor suppressor genes: New prospects for cancer research. *Journal of the National Cancer Institute* 83:91–96.

Kotloff, K. L., C. O. Thacket, J. D. Clemens, S. S. Wasserman, J. E. Cowan, M. W. Bridell, and T. C. Quinn. 1991. Assessment of the prevalence and risk factors for human immunodeficiency virus type 1 (HIV-1) infection among college students using three survey methods. *Journal of Epidemiology* 133:2–8.

Lippman, M. E. 1991. Growth factors, receptors, and breast cancer. *The Journal of NIH Research* 3:59–62.

Marshall, C. J. 1991. Tumor suppressor genes. *Cell* 64:313–326.

Marx, J. 1990. Oncogenes evoke new cancer therapies. *Science* 249:1376–1378.

Mitsuya, H., R. Yarchoan, and S. Broder. 1990. Molecular targets for AIDS therapy. *Science* 249:1533–1544.

Quackenbush, M., M. Nelson, and K. Clark. 1988. *The AIDS Challenges: Prevention Education for Young People*. Santa Cruz, Calif: Network Publications.

Slamon, D. J., W. Godolphin, L. A. Jones, J. A. Holt, S. G. Wong, D. E. Keith, W. J. Levin, S. G. Stuart, J. Udove, A. Ullrich, and M. F. Press. 1989. Studies of the HER-2/*neu* proto-oncogene in human breast and ovarian cancer. *Science* 244:707–712.

Sluyser, M., ed. 1990. *Molecular Biology of Cancer Genes*. New York: Ellis Norwood.

Stanbridge, E. J. 1990. Identifying tumor suppressor genes in human colorectal cancer. *Science* 247:12–13.

Touchette, N. 1990. Tumor suppressors: A new arena in the war against cancer. *The Journal of NIH Research* 2:62–66.

Turner, C. F., H. G. Miller, and L. E. Moses, eds. 1989. *AIDS: Sexual Behavior and Intravenous Drug Use*. Washington, D.C: National Academy Press.

Weiss, R. 1989. A national strategy on AIDS. *Issues in Science and Technology, Volume V:* 52–59.

What science knows about AIDS (A single-topic issue). 1988. *Scientific American* 259:40–134 (ten articles).

CHAPTER 41

Biology in Future Decades

Chapter Outline

Reading Questions

1. Why is the integration of biological knowledge considered to be an important goal for the coming decades?
2. Of what scientific or practical value is the study of individual taxonomic groups?
3. What are some ways that our biological knowledge may be applied in the near future?
4. What are currently active areas of biological research.?

Throughout this book, we have emphasized the dynamic nature of biology, discovering that studies of life have changed significantly in the past century. Biology has shifted focus on its central questions, expanded its domain to incorporate what were formerly separate disciplines, and developed new methods, technologies, concepts, and theories for understanding nature. There is little reason to doubt that biology will continue to evolve in the coming decades. What might we look forward to?

At best, crystal ball gazing is a perilous task. One of the primary lessons from the history of science is that biology has often veered in unexpected directions. New ideas, discoveries, and challenges have altered the course of biological research in unpredictable ways. Nevertheless, it is possible to reflect on probable directions, likely areas of development, or desirable paths that biology might follow in the coming years.

How does one attempt to answer the question, What direction will, or should, biology take in coming decades? One obvious method is to consider the fastest-growing and most exciting areas in biology. Throughout this book, we have described developments in rapidly progressing fields. These "hot areas" are recognized by funding agencies, experts in the field, and, through the media, the general public. Some of these prominent areas of research are described here.

There are also broad, general ways in which the biological sciences might change or in which serious researchers would like it to move. These general changes include the overall manner in which scientific ideas relate to one another, the basic perspectives that guide research, and the makeup and structure of the scientific community. Forecasting such changes is an uncomfortable activity for scientists, but what Ernst Mayr, one of the architects of the modern theory of evolution and an active participant in evolutionary thinking for several decades, calls "wishful hoping" is not. Many authorities, like Professor Mayr, have given serious thought to how they would like to see biology develop in the coming decades. We shall describe several major themes, related to the future of biology, that leading scientists, historians of science, and philosophers of science have identified.

ACTIVE AREAS OF BASIC RESEARCH

Several active areas of basic research, because of their importance and momentum, will remain prominent in the coming decade. Neurobiology is one of these. Research on the immune system and human diseases is another field where considerable work will be done.

Many active areas of research were described in earlier chapters. From these we developed a list of "hot questions" for the next decade. We challenge you to formulate other questions.

How are different ecosystems integrated?
How is the human population affecting ecosystems?
What are the causes of global climate change? Can global warming be reversed?
What is the molecular basis for various human genetic disorders?
What is the best use of genetic engineering of microbes, plants, and animals?
How can genetic technology be used in treating or curing human disorders?
Is the process of macroevolution different from microevolution?
What is the genetic basis of behavior?
Will there be a new synthesis of the modern theory of evolution?
How is the endocrine system related to other systems?
What is the molecular basis of human diseases such as AIDS and cancer?

GENERAL TRENDS

The list of ways in which biology might change is much shorter than the inventory of active areas for research. Three major items on the list of potential changes are (1) integrating knowledge, (2) shifting from studying life in terms of levels of biological organization to studying taxonomic groups of organization, and (3) applying biological knowledge.

Integrating Knowledge

The most commonly expressed concern about change in biology involves integrating scientific knowledge. During the past century, biology has progressively become more specialized. There once was a time when scientists could envision writing a set of volumes that would encompass all knowledge of the living world. That time is long past. Now many societies and journals concentrate on finely defined areas of research, and scientists find it increasingly difficult to keep abreast of significant research in their own area of interest. As a consequence, more and more information is becoming available, but it is not being connected or integrated into existing bodies of knowledge.

Why is this a problem? One reason is related to efficiency. On a practical level, the proliferation of information—what has often been referred to as an information explosion—creates a situation where similar problems arise in different fields but are attacked independently rather than jointly. Also, time and money may be lost in "reinventing the wheel" because scientists are often ignorant of relevant developments in other fields.

The enormous number of new research articles published each year tells us a great deal about pieces of the story of life, but it leaves great gaps in developing the overall picture. Ernst Mayr has stated:

> My own attitude is perfectly clear. In science, we have to do analysis first so that we know what the pieces are with which we are working. However, in many areas we have now reached the place where we have to try to put the systems together that we have analyzed. The two areas—and almost everybody seems to agree in this—that are in particular need of this are the genotype as a whole and the functioning of the central nervous system.*

Mayr is referring to the great concern among biologists about our current lack of understanding the functional genotype—how it becomes expressed in forming a phenotype, how it has evolved, and its inherent constraints. His comment on the nervous system points to a hot area of research, neurobiology. Many scientists are now attempting to understand the brain and the nervous system. Notable figures, like Francis Crick, who pioneered our understanding of DNA's structure and function, are now exploring the brain. Although scientists have uncovered much information about genotype and the nervous system, they have not produced comprehensive explanations of either (Figure 41.1).

Other examples illustrate the need to integrate knowledge in addressing scientific issues. Margaret Davis, of the University of Minnesota and past president of the Ecological Society of America, recently stated that she was convinced that global change can be understood only if research is coordinated at all levels of ecological organization. The scientific problems of global change, she contends, touch every subfield of the discipline and present the most exciting intellectual challenge that has confronted ecologists in decades. But the answers we seek can be reached only by coordinating and integrating research.

In Chapter 25, we considered why the theory of evolution is the broadest integrating theory in the biological sciences and how it is more closely related to

*All quotations in this chapter are from personal communications.

Figure 41.1

Research being conducted with this relatively simple sea hare (*Aplesia california*) is advancing our knowledge of neurobiology.

some fields than others. For many biologists, the intellectual challenge of the 1990s is to integrate all biological disciplines within an evolutionary perspective. Ledyard Stebbins, a leading evolutionary botanist associated with applying the modern synthesis to studying plants, feels this way and agrees with Theodosius Dobzhansky's dictum that "nothing makes sense except in the light of evolution." Professor Stebbins also recognizes that other unifying concepts, such as energy transfer and biological organization, may be useful in drawing together all information about the living world. David Hull, of Northwestern University and one of the foremost philosophers of biology in the United States, hopes

> to see advances in our understanding of the general process of evolution. We know a lot about a few processes, but when it comes to understanding how different areas in biology relate evolutionarily, we are still mostly in the dark. We don't even understand how ecology and population biology go together.

Biodiversity and the New Natural History

Biologists have recently questioned the current focus on studying distinct levels of organization. E. O. Wilson, father of sociobiology, the world's authority on ants, and curator of entomology at the Museum of Comparative Zoology at Harvard University, has recently argued that it is time to shift from a biology that concentrates on levels of organization to a biology that concentrates on different taxonomic groups. Since the 1950s, biological research has concentrated on searching for generalizations on different levels: the organism level, the cellular level, or the molecular level. Wilson argues that it is time to emphasize "the study of particular groups of organisms across all levels of organization." He is not advocating a return to old-fashioned descriptive natural history but rather, he stresses, in this new natural history, the value of studying particular taxa with tools that have been developed in molecular and cell biology.

Why pursue the study of individual taxa? In part, Wilson says, the search for new and interesting generalizations that hold true across all the species at any one level of organization is yielding fewer and fewer successes. Although we continue to discover many interesting truths, they often turn out to apply to just the group in which they were discovered, not all groups. The deep study of different taxa will, it is hoped, reveal new unifying generalizations.

Of equal importance is an appreciation of the inherent interest and value of individual groups. Another way of stating this idea is that future emphasis will be placed on exploring diversity. The theory of evolution provides a basic framework for understanding the phenomenon of diversity itself. Wilson and others hope to expand on that general understanding. The diversity of life, *biodiversity,* is the storehouse of all that exists in the biological world. A detailed understand-

Figure 41.2

Many species in this tropical rain forest might be endangered due to the loss of its native habitat.

ing of diversity can provide information about which organisms are of benefit to humans, which are endangered, and which are useful for different types of research. This approach is particularly relevant at the present time because scientists have documented the alarming rate at which the biodiversity of Earth is being dissipated by destruction of natural habitats (Figure 41.2).

Applications

Another common theme that will run through discussions of biology in the coming decades concerns the enormous potential for greater practical applications of knowledge. Which areas of active research will lead to applications in the 1990s and the twenty-first century? Recent discoveries associated with medicine and medical technology will undoubtedly lead the way (Figure 41.3). Marvin Druger, a geneticist and science educator at Syracuse University, states:

> This is an especially exciting time for biology. Technological advances have given us the tools to explore living things in ways that were not possible before. Biologists can use these tools in clever and creative ways that benefit humankind. New technologies will continue to be developed and refined, and they will be applied to making the environment cleaner, curing diseases, predicting disease risks, improving agriculture, and learning more about organism and cell functions. The entire pharmaceutical industry will be transformed by genetic engineering technology, and most major body chemicals will be produced by microorganisms. Also, all sorts of new creatures will be designed by genetic engineers, combining the desirable qualities of different species. Organ transplants will become more common, and we may have centers where arms, legs, eyes, kidneys, and other parts are generated from cell cultures *in vitro*. Cures will be found for AIDS, cancer, and other menacing diseases. . . . Chemical substances will be available to prevent deterioration that occurs with aging.

Linus Pauling, a distinguished biochemist and twice a Nobel Prize laureate, is one of many experts who are optimistic about what future research may hold for *human health*. For example, he believes that research on nutrition will be particularly fruitful, especially in determining

> the properties of vitamins and other orthomolecular substances (substances normally present in the human body, usually required for life). I foresee that much information will be gathered about the optimum levels of these substances in the blood, or perhaps I should say optimum levels in the body. . . . I think that it will be recognized that the greatest progress in the control of disease, improvement of health and well-being, and extension of the life span will be made during the coming decade by studies in this direction.

Figure 41.3

These flasks in a National Cancer Institute laboratory contain interferon, a substance produced in large quantities through genetic engineering.

Figure 41.4

The short-term effects of this oil spill near Galveston, Texas, were very dramatic. Long-term effects, which are often less obvious, are now being studied by scientists.

A concern for the *health of life on Earth* also occupies a prominent position in biological research and will continue to be important in the coming decade. Scientists are exploring the effects of pollution on various environments and attempting to understand alterations in ecosystems that will follow if predicted global climate changes occur (Figure 41.4). Biologists are also concerned with the management of natural resources, such as forests, and with restoration of areas that have been severely altered by humans.

Scientists now understand that processes affecting the environment operate over different time scales. Long-term effects are sometimes "invisible" in the short term and may not be noticed until substantial damage has occurred. Until recently, much ecological research was focused on relatively short time scales. The current work on global climate change is concerned with very long periods. Recently, the National Science Foundation established a new program, the Long-Term Ecological Research Program (LTERP), that bridges the gap between short and very long term effects and also attempts to look at problems on an intermediate spatial scale.

The Ecological Society of America, in a recent report titled *The Sustainable Biosphere Initiative: An Ecological Research Agenda for the Nineties,* urges scientists to investigate topics that are important for science and society, such as preparing for global change, protecting biodiversity, and maintaining a sustainable biological ecosystem.

ORGANIZATION OF BIOLOGY

Numerous scientists deplore the overspecialization—the narrow attention of researchers to their own area—that seems to characterize biology today. One way to reduce or eliminate overspecialization is to reform education and have it become broader and more integrative. Another way is to encourage more group research and collaborative efforts. Much research is now being done by teams, especially in some fields. For example, the problems associated with global climate change are so complex that teams of atmospheric scientists, ecologists, physiologists, geographers, and oceanographers will be required to provide some answers.

BIOLOGY AND SOCIETY

If the nineteenth century was the "century of physics," the twentieth may be labeled by historians as the "century of biology." It has been a period of enormous expansion in the quantity and complexity of the questions asked, the solutions developed, and the impact of biology on our lives. Biology is now a part of

Figure 41.5

This genetically-engineered cantaloupe plant is being field tested to see if it is resistant to two common viruses that often damage it. It is hoped that such genetic engineering will replace the need for chemicals currently used to control infestations.

big business, the educational establishment, the recreation industry, health care, government policy, and government budgets. We see or hear about biology on comics pages, on television, and even during congressional debates (Figure 41.5). In the coming decade, this interest and significance will increase. The number of scientists studying the living world will also increase, but probably in a somewhat different way. Women and minorities, who now constitute a small fraction of the biological research community, will increase proportionately to the white males who now make up the vast majority of biologists. The shift will come about because of a concerted effort by the scientific and educational communities to recruit and retain women and minorities who are interested in pursuing careers in biology. Biology will see many new faces in the coming decade in more than one way, for it will interface increasingly with other intellectual disciplines, such as philosophy and medical ethics. New insights will provide us with a deeper understanding of humans, relationships between all forms of life, and the planet Earth. Because biology is and will continue to be concerned with problems that affect so many aspects of human life, it will increasingly have a political dimension. Issues concerning land use, the protection of endangered species, and the direction of medical research are examples of topics that have scientific, ethical, and political facets. Scientists will have to work closely with other specialists in the coming decades to face the challenges that now confront us.

References and Recommended Reading

Crick, F. 1988. *What Mad Pursuit: A Personal View of Scientific Discovery*. New York: Basic Books.

Davis, M. 1989. Insights from paleoecology on global change. *Bulletin of the Ecological Society of America* 70:222–228.

Franklin, J., C. Bledsoe, and J. Callahan. 1990. Contributions of the Long-Term Ecological Research Program. *BioScience* 40:509–523.

Hull, D. 1988. *Science as a Process: An Evolutionary Account of the Social and Conceptual Development of Science*. Chicago: University of Chicago Press.

Lubchenco, J. et al. 1991. The sustainable biosphere initiative: An ecological research agenda. *Ecology* 72(2):371–412.

Mayr, E. 1988. *Toward a New Philosophy of Biology: Observations of an Evolutionist*. Cambridge, Mass.: Harvard University Press.

Wilson, E. O. 1989. The coming pluralization of biology and the stewardship of systematics. *BioScience* 39:242–245.

Appendix

Classification of Organisms: The Five Kingdoms, Their Phyla, and One Genus of Each

	Genera	Common Names
KINGDOM: PROKARYOTAE		
Subkingdom: Archaebacteria		
Division: Mendosicutes (methanogenic bacteria)		
Methanocreatrices	*Methanobacterium*	Methanogenic bacteria
Unnamed (1988)	*Thermoplasma*	Hot springs bacteria
Subkingdom: Eubacteria		
Division: Tenericutes (lack rigid cell walls)		
Aphragmabacteria	*Mycoplasma*	Pneumonia bacteria
Division: Gracilicutes (gram-negative bacteria)		
Spirochaetae	*Treponema*	Syphilis bacteria
Thiopneutes	*Desulfovibrio*	Sulfate bacteria
Unnamed (1988)	*Rhodomicrobium*	Purple bacteria
Cyanobacteria	*Anabaena*	Blue-green algae
Chloroxybacteria	*Prochloron*	Prochlorophytes
Unnamed (1988)	*Azotobacter*	N_2-fixing aerobes
Pseudomonads	*Pseudomonas*	Pseudomonas
Omnibacteria	*Escherichia*	Escherichia
Unnamed (1988)	*Nitrobacter*	Chemoautotrophs
Myxobacteria	*Simonsiella*	Mucus bacteria
Division: Firmicutes (gram-positive bacteria)		
Unnamed (1988)	*Streptococcus*	Lactic-acid bacteria
Aeroendospora	*Bacillus*	Bacillus
Micrococci	*Micrococcus*	Micrococcus
Actinobacteria	*Streptomyces*	Strep bacteria
KINGDOM: PROTOCTISTA		
Caryoblastea	*Pelomyxa*	Giant amoeba
Dinoflagellata	*Gonyaulax*	Gonyaulax
Rhizopoda	*Amoeba*	Amoeba
Chrysophyta	*Dinobryon*	Golden algae
Haptophyta	*Prymnesium*	Marine heterophyte
Euglenophyta	*Euglena*	Euglena
Cryptophyta	*Cryptomonas*	Cryptophytes
Zoomastigina	*Trichonympha*	Flagellates
Xanthophyta	*Ophiocytium*	Zoospore algae
Eustigmatophyta	*Vischeria*	Eustigs
Bacillariophyta	*Diploneis*	Diatoms
Phaeophyta	*Postelsia*	Sea palm
Rhodophyta	*Corallina*	Corallin algae
Gamophyta	*Spirogyra*	Spirogyra
Chlorophyta	*Codium*	Dead man's fingers
Actinopoda	*Acanthocystis*	Actinopods
Foraminifera	*Globigerina*	Forams
Ciliophora	*Paramecium*	Paramecium
Apicomplexa	*Plasmodium*	Malaria sporozoa
Cnidosporidia	*Ichthyosporidium*	Fish microsporidian
Labyrinthulomycota	*Labyrinthula*	Slime nets
Acrasiomycota	*Dictyostelium*	Cellular slime molds
Myxomycota	*Echinostelium*	Slime molds

Plasmodiophoromycota	*Plasmodiophora*	Plasmodia
Hyphochytridiomycota	*Hypochytrium*	Hypochytrids
Chytridiomycota	*Blastocladiella*	Chytrids
Oomycota	*Saprolegnia*	Fish fungus
KINGDOM: FUNGI		
Zygomycota	*Rhizopus*	Black bread mold
Ascomycota	*Saccharomyces*	Yeast
Basidiomycota	*Amanita*	Fly agaric
Deuteromycota	*Penicillium*	Penicillium
Mycophycophyta	*Lobaria*	Lungwort lichen
KINGDOM: PLANTAE		
Bryophyta	*Sphagnum*	Peat moss
Psilophyta	*Psilotum*	Whisk ferns
Lycopodophyta	*Lycopodium*	Club moss
Sphenophyta	*Equisetum*	Horsetail
Filicinophyta	*Pteridium*	Bracken fern
Cycadophyta	*Zamia*	Cycads
Ginkgophyta	*Ginkgo*	Ginkgo
Coniferophyta	*Pinus*	Pine
Gnetophyta	*Welwitschia*	Welwitschia
Angiospermophyta	*Rosa*	Roses
KINGDOM: ANIMALIA		
Placozoa	*Trichoplax*	Trichoplax
Porifera	*Leucilla*	Urn-shaped sponge
Cnidaria	*Hydra*	Hydra
Ctenophora	*Pleurobrachia*	Cat's eye combjelly
Acoelomate Phyla		
Mesozoa	*Dicyema*	Mesozoans
Platyhelminthes	*Planaria*	Planaria
Nemertina	*Nemertea*	Ribbon worms
Gnathostomulida	*Gnathostomaria*	Gnathostomulids
Gastrotricha	*Tetranchyroderma*	Gastrotrichs
Pseudocoelomate Phyla		
Rotifera	*Brachionus*	Rotifers
Kinorhyncha	*Kinorhynchus*	Kinorhynchs
Loricifera	*Nanaloricus*	Loriciferans
Acanthocephala	*Acanthocephalus*	Spiny-headed worms
Entoprocta	*Barentsia*	Entoprocts
Nematoda	*Trichinella*	Trichinosis worms
Nematomorpha	*Gordius*	Horsehair worms
Protostome Coelomate Phyla		
Ectoprocta	*Plumatella*	Ectoprocts
Phoronida	*Phoronis*	Phoronid worms
Brachiopoda	*Terebratulina*	Lamp shells
Mollusca	*Octopus*	Octopus
Priapulida	*Tubiluchus*	Priapulids
Sipuncula	*Phascolosoma*	Peanut worms
Echiura	*Urechis*	Spoon worms
Annelida	*Lumbricus*	Earthworms
Tardigrada	*Echiniscus*	Water bears
Pentastoma	*Linguatula*	Tongue worms
Onychophora	*Peripatopsis*	Velvet worms
Arthropoda	*Homarus*	Lobsters
Deutrostome Coelomate Phyla		
Pogonophora	*Riftia*	Beard worms
Echinodermata	*Pisaster*	Sea star
Chaetognatha	*Sagitta*	Arrow worms
Hemichordata	*Saccoglossus*	Acorn worms
Chordata	*Odocoileus*	Deer

Answers to Selected Problems and Questions

Chapter 12

1. a. All yellow pods
   b. 3 yellow to 1 green
2. The following has two traits, YYyy and TTtt. You may substitute any other two.

Dihybrid cross: Yellow pod (Y) Tall (T) ♂ Green pod (y) Short (t) ♀

Homozygous dominant parent YYTT x Homozygous recessive parent yytt

F_1

♀ \ ♂	YT	YT
yt	YyTt	YyTt
yt	YyTt	YyTt

YyTt x YyTt

F_2

♀ \ ♂	YT	Yt	yT	yt
YT	YYTT	YYTt	YyTT	YyTt
Yt	YYTt	YYtt	YyTt	Yytt
yT	YyTT	YyTt	yyTT	yyTt
yt	YyTt	Yytt	yyTt	yytt

3. Mendel could have crossed the F_1 plants with a homozygous recessive. If the cross yielded a ratio of 1 dominant: 1 recessive, he would have shown that the monohybrid cross was heterozygous.

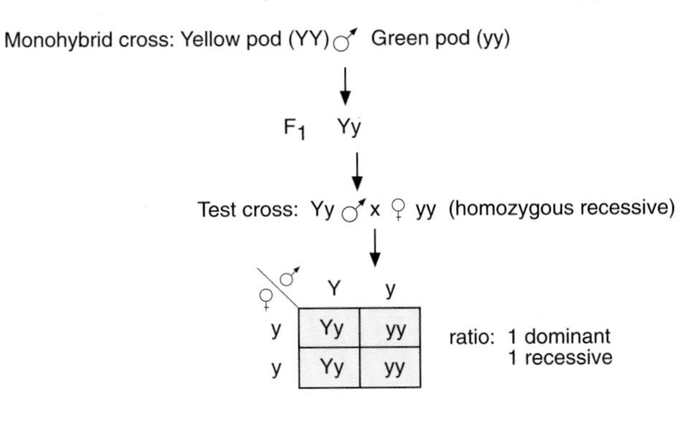

Chapter 14

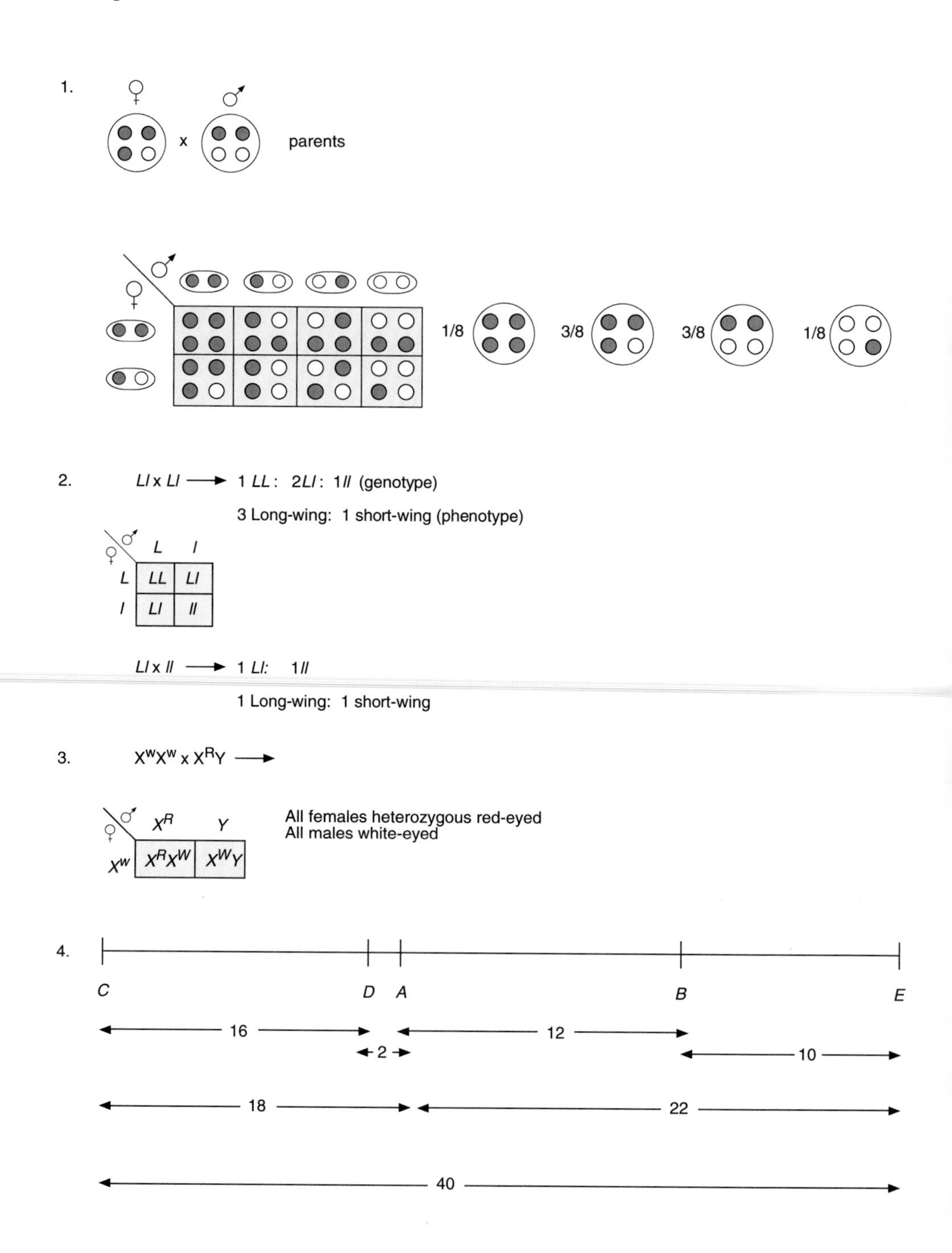
1.
x
parents
1/8
3/8
3/8
1/8
2.
Ll x Ll ⟶ 1 LL : 2Ll : 1ll (genotype)
3 Long-wing: 1 short-wing (phenotype)
L
l
L
LL
Ll
l
Ll
ll
Ll x ll ⟶ 1 Ll: 1ll
1 Long-wing: 1 short-wing
3.
X^wX^w x X^RY ⟶
X^R
Y
X^w
X^RX^W
X^WY
All females heterozygous red-eyed
All males white-eyed
4.
C
D
A
B
E
16
12
2
10
18
22
40

Chapter 18

4. a. Autosomal recessive
   b. Autosomal dominant: also, autosomal recessive
   c. Sex-linked recessive

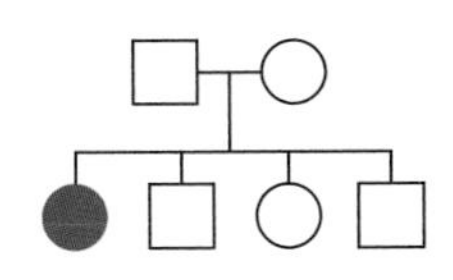

(a) Autosomal recessive

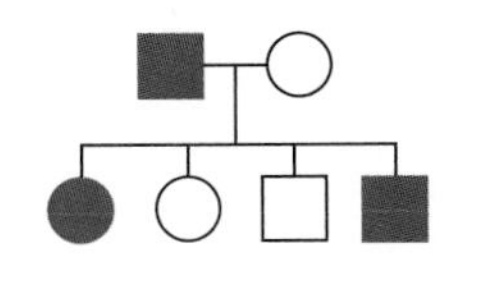

(b) Autosomal dominant; also, autosomal recessive

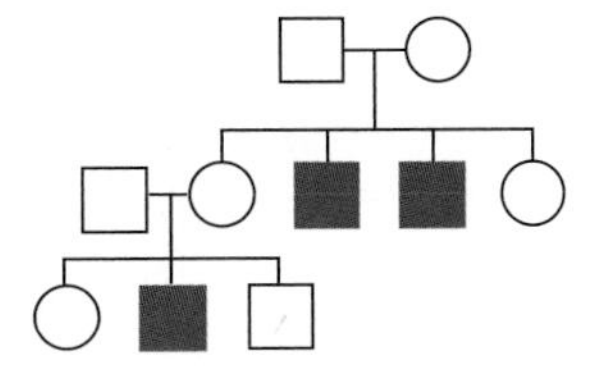

(c) Sex-linked recessive

Chapter 21

4. a. Numbers 14 pedigree
   b. 1
   c. A, C (the 2 grandfathers)
   d. 2
   e. G, I, L

Glossary

AAET See *average annual evapotranspiration.*

abdominal (ab-DOM-i-nul) **cavity** The largest human body cavity; contains the organs of digestion and the kidneys, liver, spleen, and pancreas.

abiotic (ā´bī-OT-ik) **factors** All environmental factors not directly associated with living organisms; physical, chemical, and temporal (time) components of the environment.

aboveground net primary production (ANPP) The amount of plant biomass (leaves and stems) produced after subtracting that used in plant respiration per unit area per unit of time.

abscission (ab-SIZH-un) **cell layer** Layer of cells connecting the leaf to the branch; changes in it in autumn cause the leaf to fall.

abyssal (uh-BIS-ul) **zone** Ocean region 1,500 to 6,000 meters below the surface.

accessory reproductive organs All organs other than testes or ovaries involved in gamete protection, storage, and transport.

acetylcholine (uh-sēt´ul-KŌ-lēn) A neurotransmitter released at many neuron-neuron and neuron-muscle cell synapses.

acid deposition Process whereby acidic substances are delivered from the atmosphere to the earth's surface.

acid-neutralizing capacity The capacity of alkaline substances in water to neutralize a certain quantity of acid.

acoelomates (ā-SĒ-luh-māts) The most primitive bilateral phyla; do not develop a coelom.

acquired immune deficiency syndrome See *AIDS.*

acrosome (AK-ruh-sōm) A caplike, fluid-filled structure that covers the head of a sperm.

acquisition Process in which a plant obtains required resources such as carbon dioxide from the environment.

ACTH See *adrenocorticotropin.*

acute inflammation response An initial response to injury, characterized by redness, heat, swelling, and pain.

adaptation (1) In ecology, the process of adjustment of an individual organism to environmental stresses. (2) In evolution, a change that results in a structural, functional, or behavioral trait that promotes survival and reproduction.

adaptive radiation Evolutionary divergence in which several lineages arise from a common ancestor.

adenosine triphosphate (uh-DEN-uh-sēn trī-FOS-fāt) **(ATP)** An energy-liberating molecule used by plant and animal cells, composed of the molecule adenosine and three linked phosphate groups.

ADH A hormone that helps regulate water balance in the body by promoting reabsorption of water in the kidney; also called *vasopressin.*

adrenal (uh-DRĒN-ul) **glands** Two glands that produce certain steroid hormones; located on top of the kidneys.

adrenocorticotropin (uh-drē´nō-kor´ti-kō-TRŌ-pin) **(ACTH)** A peptide hormone of 39 amino acids that stimulates the release from the adrenal gland of other hormones that are important in carbohydrate metabolism.

aerobic (e-RŌ-bik) **cellular respiration** A set of oxygen-requiring chemical reactions that release energy stored in sugar for use by an organism.

afferent (AF-uh-runt) **(sensory) division** Part of the peripheral nervous system that includes nerves that relay impulses from all areas of the body to the central nervous system.

afterbirth The placenta expelled after childbirth.

after-discharge The brief delay between removal of a stimulus and the end of a response.

aggregation (ag´ruh-GĀ-shun) Group formed by individuals that are attracted to a stimulus or some feature of the environment.

Agnatha (AG-nuh-thuh) The class of the jawless fishes.

agranulocytes (ā-GRAN-yoo-lō-sīts) White blood cells without cytoplasmic granules; macrophages and lymphocytes.

AIDS Acquired immune deficiency syndrome, caused by the virus HIV-1; characterized by a combination of opportunistic infections and a seriously deficient immune system.

AIDS-related complex (ARC) Various symptoms, including fever, diarrhea, and weight loss, that occur in AIDS patients.

albinism (AL-buh-niz´um) The condition of having no pigment in certain cells.

alchemists People who studied or attempted transmutations such as turning base metals into gold.

alimentary canal Digestive tube that begins at the mouth and terminates at the anus.

alkaptonuria (al-kap´tuh-NYOOR-ē-uh) A metabolic disorder characterized in children by urine turning black after it is exposed briefly to air; causes arthritis in adults.

allantois (uh-LAN-tuh-wis) A membrane involved in the transport of wastes from the embryo and oxygen and nutrients from the mother.

alleles (uh-LĒLZ) Alternate forms of a gene; one of two or more forms that can exist at a single gene locus.

allergen (AL-ur-jin) A substance (antigen) to which some people have a hypersensitive reaction.

allergy Hypersensitive immune reaction to an allergen.

allocation Process in which plant metabolism products are apportioned to specific tissues and organs according to need.

allopatric speciation (al´uh-PAT-rik spē´shē-Ā-shun) Process by which a species arises in geographic isolation.

alpha cells Pancreas cells that produce glucagon.

alpine zones Habitats on mountains located between timberline and snow line, with environmental features similar to those of tundra biomes.

alternation of generations Plant life cycle feature characterized by two different structural forms, *gametophytes* and *sporophytes*, produced alternatively.

alternative hypothesis A statement that is accepted if a null hypothesis is rejected.

altitudinal zonation A change in community structure as a function of changing elevation.

altruistic (al´troo-IS-tik) **behavior** Behavior that results in a benefit to the recipient at a cost to the actor.

alveoli (al-VĒ-uh-lī) Tiny clusters of thin-walled air sacs that are sites of gas exchange in the lung.

amino (uh-MĒ-nō) **acids** Small building-block molecules that make up protein.

ammonification (uh-mon´uh-fuh-KĀ-shun) Process whereby fungi and bacteria convert organic nitrogen from plants and animals into ammonia; the release of ammonia from nitrogenous organic matter.

amniocentesis (am´nē-ō-sen-TĒ-sus) Procedure in which a long, thin needle is inserted through the mother's abdominal wall and into the amniotic cavity to withdraw cells and fluids that are analyzed to determine if genetic abnormalities are present in a fetus.

amnion (AM-nē-on) Fluid-filled membrane that first surrounds the cavity between the trophoblast and the inner cell mass and later becomes a membranous sac holding the fetus.

amniote (AM-nē-ōt) **egg** Type of leathery or hard-shelled egg produced by reptiles or birds that is waterproof and contains a fluid-filled membrane.

amoebocytes (uh-MĒ-buh-sīts) Blood cells that can move through the cell or tissue layers with their pseudopodia.

anabolism (uh-NAB-uh-liz-um) Chemical reactions involved in the synthesis of molecules.

anaerobic (an´e-RŌ-bic) **exercise** Exercise in which not enough oxygen is available in muscle cells for all pyruvate to move into the pathway of cellular respiration.

anaerobic (an´e-RŌ-bic) **reaction** A reaction that does not require oxygen.

analogous (uh-NAL-uh-gus) **structures** Structures that have similar functions but differ in their structural, embryological, and evolutionary background.

analysis of data Numerical evaluations conducted to determine whether or not the results obtained in an experiment are significant.

anaphase (AN-uh-fāz) The fourth stage of mitosis, in which paired chromatids separate at their centromeres and appear to be pulled toward the poles of the parent cell.

anaphase I Stage of meiosis in which homologous chromosomes are separated.

anaphase II Stage of meiosis in which sister chromatids are separated.

anatomy (uh-NAT-uh-mē) Study of the structure of an organism and the relationship among its parts.

androgens (AN-druh-jinz) Principal male sex hormones produced by the testes; stimulate development and maintenance of the male reproductive system and secondary sex characteristics.

anemia (uh-NĒ-mē-uh) A low number of red blood cells.

aneuploid (AN-yoo-ploid) Containing an abnormal number of chromosomes.

angiosperms (AN-jē-uh-spurmz) Plants that produce seed-containing fruits from flowers.

animal model An animal species with a disease that serves as a model for learning about the nature of a similar or identical condition in humans.

annual plants Plants such as tomatoes, beans, pansies, and lettuce that complete their entire life cycle within one calendar year.

anomaloscope (uh-NOM-uh-luh-skōp) An instrument that projects light of different colors onto a screen.

anonymous group Loosely structured group that sometime exhibits coordinated movements and cooperative behaviors.

ANPP See *aboveground net primary production.*

anthropogenic (an´thruh-puh-JEN-ik) Tracing its origin to humans.

antibiotics (an´tē-bī-OT-iks) Chemicals that inhibit the growth of microorganisms.

antibodies (AN-ti-bod´ēz) Specialized proteins that have the capacity to bind specifically to an antigen or antigenic determinant, leading to neutralization or destruction of the antigen.

anticodon (an´tē-KŌ-don) A particular three-nucleotide sequence in transfer RNA that is complementary to the three base pairs of a specific codon.

antigen (AN-ti-jin) A foreign macromolecule that induces an immune response.

antigen-antibody complex Structure formed when an antibody combines with a specific antigen in a lock-and-key fashion.

antigenic determinant A region of an antigen that is the site of antibody attachment.

antigen presentation A process by which certain self cells, usually macrophages, display a foreign antigen on their surfaces in a form that is recognized by T cell receptors.

antihistamine (an´tē-HIS-tuh-mēn) Drug used in neutralizing histamine and relieving some of the effects of allergic reactions.

antiparallel Of DNA strands, running in opposite directions relative to their base sequence.

anus (Ā-nus) The opening through which digestive waste materials are expelled.

aorta (ā-OR-tuh) The largest artery in the body; emerges from the left ventricle of the heart.

apical (Ā-pi-kul) **buds** Buds located at the top of the plant and at the ends of branches and roots that give rise to all cells in the plant's shoots and roots.

aquatic ecosystem An ecosystem that occurs in water; includes both fresh and marine bodies of water.

ARC See *AIDS-related complex.*

archaebacteria (ar´kē-bak-TĒR-ē-uh) The oldest of two major lineages of bacteria; the subkingdom Archaebacteria in the kingdom Prokaryotae includes the methanogenic, halophilic, and thermoacidophilic bacteria.

arteries Vessels that conduct blood away from the heart.

arterioles (ar-TĒR-ē-ōlz) Small blood vessels that branch off arteries and deliver blood to capillaries.

artificial insemination The manual placement of sperm into the female reproductive tract.

artificial system A classification system that is used primarily for retrieving information and makes no claims about the relationships among the objects classified.

asci (AS-kī) Microscopic reproductive structures in fungi that appear as tubular spore sacs filled with ascospores.

asexual reproduction Reproduction in which a single parent cell or individual gives rise to two or more identical (or nearly identical) offspring.

assimilation (uh-sim´uh-LĀ-shun) Modification of molecules by means of cellular metabolism.

association neuron Neuron in the brain and spinal cord that processes incoming information by transmitting it between sensory and motor neurons.

associative learning Learning in which the subject acquires a response to a stimulus by associating it with another stimulus.

atherosclerosis (ath´uh-rō-skluh-RŌ-sus) Blockage of arteries that supply the heart due to the buildup of fibrous, fatty deposits on their inner walls.

atmosphere The gaseous envelope surrounding Earth.

atmospheric cells Circular zones where air rises and moves north or south, then descends in cooler regions.

atom The fundamental chemical unit composed of neutrons, protons, and electrons; the most common atoms in biological structures are carbon, hydrogen, nitrogen, and oxygen.

ATP See *adenosine triphosphate.*

autoimmune response Response in which T cells attack self cells or antibodies react against self molecules.

autonomic (ot´uh-NOM-ik) **nervous system** Part of the nervous system that connects motor neurons from the central nervous system to various internal organs; primarily concerned with automatic functions such as heartbeat.

autosomes (OT-uh-sōmz) The first 22 pairs of the chromosomes that are identical in human males and females; chromosomes not involved in sex determination.

autotroph (OT-uh-trōf) An organism that synthesizes organic molecules from inorganic molecules using sunlight (photosynthesis) or the energy released from the oxidation of chemically reduced inorganic compounds (chemosynthesis).

average annual evapotranspiration (AAET) The average amount of water that evaporates from soil plus that transpired from vegetation per year.

axon (AK-son) A process of a neuron that conducts a nerve impulse away from the cell body.

bacteria (bak-TĒR-ē-uh) Organisms composed of a single prokaryotic cell without a nucleus or organelles.

bacteriophage (bak-TĒR-ē-uh-fāj) A virus that infects bacteria.

balanced pathogenicity (path´uh-juh-NIS-uh-tē) The idea that a balance exists between disease agents and their hosts; the agent is able to survive, reproduce, and be transmitted without the host's suffering serious harm.

balancing selection Selection that favors an equilibrium based on more than one genotype for a given trait.

bands Sections of the chromosome that differ from other areas because of its lighter or darker staining intensity.

Barr body A densely staining structure in the nucleus that represents an inactivated X chromosome.

basal cell layer Deep cell layer in the epidermis of the skin, containing undifferentiated cells that reproduce rapidly and continuously by mitosis in order to replace cells that are sloughed from the skin's surface.

base A chemical that forms part of one of the linked nucleotides of a nucleic acid chain; adenine (A), thymine (T), guanine (G), cytosine (C), or uracil (U).

base pair Two nucleotides held together by weak hydrogen bonds.

basophils (BĀ-suh-filz) Granulocytes containing factors that react against parasites and in allergic responses initiated by mast cells.

bathyal (BATH-ē-ul) **zone** Ocean region between 200 and 1,500 meters below the surface.

B cells See *B lymphocytes*.

benign (bē-NĪN) **tumor** A tumor that remains localized, grows slowly, and becomes surrounded by connective tissue.

benthic barrel A device used to obtain samples of gases discharged from deep-ocean-vent water.

benthic zone The bottom of lakes, where sediments accumulate and most bacterial decomposition occurs, releasing nutrients required by surface producers.

beta cells Pancreas cells that produce insulin.

biennial (bī-EN-ē-ul) **plant** Plant that completes its life cycle in two growing seasons or two calender years.

big bang theory The theory that the universe originated 10 to 20 billion years ago in an infinitely dense, infinitely hot point in space.

bile Secretion of the liver containing a mixture of substances that break down large fat globules.

binary fission (BĪ-nuh-rē FISH-un) A type of cell division that occurs in prokaryotes.

binomial (bī-NŌ-mē-ul) **system** The use of a genus and species name for identifying each kind of organism.

biogenesis (bī´ō-JEN-uh-sus) Concept that cells arise from preexisting cells through the division processes of meiosis or mitosis.

biogenic (bī´ō-JEN-ik) **origin** Derivation from organic matter, as of nutrients released in the decomposition of sediments accumulating in subduction zones.

biogeochemical cycles Cyclical systems in which important chemical elements such as nitrogen, phosphorus, and carbon are transferred between abiotic and biotic components of the biosphere.

biogeographical realms Large-scale distribution patterns of plants and animals on Earth.

biological clock A biochemical mechanism responsible for repeated patterns of behavior.

biological magnification The increase in body burdens of metals or pesticides in each successive trophic guild of an affected food chain.

biology The scientific study of life.

biomes (BĪ-ōmz) Large regions of Earth with characteristic assemblages of plants and animals.

biosphere (BĪ-us-fēr) Portion of Earth in which all known life forms exist, consisting of a thin envelope of air, water, and land.

biotic Pertaining to living organisms or their products.

biotic ecosystem models Traditional systems that emphasize the amount of energy and materials moving through different trophic levels.

biotic factors Living plants, animals, and microbes that affect other organisms.

biotic hierarchy A level of biological organization within the biosphere; for example, biome, ecosystem, community, population, deme, individual.

bivalves Most of the marine and freshwater mollusks known as *shellfish*; clams, oysters, and other mollusks that have two shells.

blastocyst (BLAS-tuh-sist) In animal embryonic development, a microscopic hollow ball of cells that becomes implanted in the endometrium shortly after fertilization.

blastopore (BLAS-tuh-por) The opening in a developing gastrula that will develop into the mouth in protostome animal species, into the anus in deuterostomes.

blastula (BLAS-chuh-luh) The developmental stage of an animal embryo between cleavage and before gastrulation; usually a hollow ball made of a single layer of cells.

blending theory of inheritance The erroneous theory that hereditary materials from the male and female are mixed, or blended, in producing offspring.

blood A fluid tissue that is transported in blood vessels and circulated by the action of the heart.

B lymphocytes (LIM-fuh-sīts) Lymphocytes that develop into plasma cells or memory cells.

bolting Rapid elongation of a shoot that often indicates the end of vegetative growth and the onset of reproduction.

bone A type of connective tissue composed of cells embedded in a matrix of calcium compounds, which impart hardness, and collagen fibers, which provide strength.

bone marrow A type of connective tissue capable of storing fat (yellow marrow) or making several types of blood cells (red marrow).

boreal (BOR-ē-ul) **woodlands** Areas of the taiga where white and black spruce grow, permafrost does not occur, and deciduous trees are restricted to margins along watercourses where deeper soils have developed from past sedimentation.

botany (BOT-uh-nē) The scientific study of plant life.

bottleneck effect A special case of random genetic drift in which a severe fluctuation in population size results in random changes in gene frequency.

breakdown Change of a chemical into different products ($x \rightarrow y + z$).

breathing The mechanical process of inhaling and exhaling air from the lungs.

broad-spectrum insecticides Pesticides that kill most insect species.

bronchi (BRON-kī) Tubular structures leading from the trachea to the lungs.

bronchiole (BRON-kē-ōl) Smaller tubular structures that branch off the bronchi into lung alveoli.

buds Plant structures that will eventually develop and grow into new branches, leaves, or flowers.

buffering capacity The acid-neutralizing capacity of alkaline substances in water.

bulbourethral (bul´bō-yoo-RĒ-thrul) **glands** A pair of glands in the male reproductive system that secrete seminal fluid that reduces acidity and lubricates the penis.

callus (KAL-us) A mass of unspecialized plant cells.

Calvin cycle The dark reactions of photosynthesis.

cancer A large group of diseases characterized by abnormal cell proliferation and malignant tumors.

capillaries (KAP-uh-ler-ēz) Microscopic blood vessels that connect arterioles and venules, through which materials pass between blood and cells.

capsid (KAP-sid) A protein coat surrounding a virus.

capsule A gelatinous structure surrounding the cell walls of some bacteria species.

carbohydrates Various sugar molecules, such as glucose and starch.

carbon sinks Locations in the biosphere where large amounts of carbon are stored; for example, in trees, fossil fuels, and calcium carbonate rock strata.

carcinogen (kar-SIN-uh-jin) A cancer-causing agent.

cardiac (KAR-dē-ak) **muscle cells** Muscle cells that form the heart wall; their branching networks impart great strength to this critical organ.

cardiovascular (kar´dē-ō-VAS-kyuh-lur) **diseases** Disorders of the heart and blood vessels.

carnivores (KAR-ni-vorz) Flesh-eating animals; organisms that feed in nonherbivore trophic guilds.

carotenoids (kuh-ROT-uh-noidz) Photosynthetic pigments that absorb violet and blue photons but not yellow, orange, or red photons.

cartilage (KAR-tuh-lij) Connective tissue composed of cells, collagen, and elastic fibers embedded in a solidified chemical matrix.

castes (kasts) Groups of individuals of the same species that differ in structure and function.

catabolism (kuh-TAB-uh-liz-um) Chemical reactions by which a cell breaks down molecules.

cDNA See *complementary DNA*.

cell The fundamental structural and functional unit of life, bounded by a membrane that separates the internal and external environments and characterized by self-production and sets of chemical reactions associated primarily with protein synthesis and energy transformation; two basic cell types—prokaryotic and eukaryotic—make up all living organisms.

cell body Part of a neuron that contains the nucleus, cytoplasm, and organelles; responsible for maintenance and for synthesizing specific proteins required by a neuron to carry out its functions.

cell division The division of a parent cell into two daughter cells.

cell renewal systems Highly regulated systems in which cells of most epithelial and connective tissues are systematically and continuously replaced throughout life.

cell specificity The specific function of a differentiated cell.

cell theory Theory that holds that (1) cells are the basic structural units of life, (2) all cells require energy, (3) cells are the functional units of life, and (4) all cells arise from preexisting cells.

cellular homeostasis (SEL-yuh-lur hō´mē-ō-STĀ-sus) Maintenance of a relatively constant number of cells in tissues and organs in the body.

cellular immunity Result of the actions of several classes of T lymphocytes, each of which carry out specialized immune reactions directed toward the destruction of nonself cells or antigens.

cellular reproduction The process whereby a cell divides by mitosis and cytokinesis, giving rise to two identical daughter cells.

cellular respiration Two sets of chemical reactions that result in the production of ATP from glucose and other organic molecules.

cellulose (SEL-yuh-lōs) The structural carbohydrate of plant cell walls.

cell wall A tough, extracellular matrix composed of cellulose and other substances that encloses cells of bacteria, certain protoctists, fungi, and plants, providing rigidity and protection.

central dogma The belief that genetic information flowed from DNA to RNA to protein and that information flow in the reverse direction, from protein to RNA to DNA, was not possible. Later it was discovered that retroviruses make DNA from RNA.

central nervous system Major part of the nervous system, consisting of the brain and the spinal cord.

centrioles (SEN-trē-ōlz) Two small cylindrical structures in animal cells, composed of microtubules, that appear during cell division and help to organize microtubules that move chromosomes.

centromeres (SEN-truh-mērz) Constricted regions where sister chromatids are joined.

cephalization (sef´uh-luh-ZĀ-shun) The concentration of specialized nervous tissues in the head of more advanced animals.

cephalopods (SEF-uh-luh-podz) Class that includes the largest and most complex protostome coelomate mollusks.

cerebral cortex (suh-RĒ-brul KOR-teks) Area of the forebrain that has become highly specialized and enlarged in humans; the conscious center of the brain.

cerebral hemispheres The two halves of the cerebrum, or forebrain.

cerebral hemorrhage Rupture of an artery in the brain.

cerebral thrombosis A blood clot blocking an artery leading to the brain.

cerebrovascular (suh-rē´brō-VAS-kyuh-lur) **disease** Ailment that occurs when the blood supply to the brain is temporarily or permanently interrupted.

cerebrum (suh-RĒ-brum) See *forebrain*.

chain of being An early scale of life forms that progressed from the smallest, "simplest" one-celled organism up to what were considered the most "perfect" animals, humans.

chaparral (shap´uh-RAL) A minor biome that receives significant winter rains but has summers typical of a desert.

Chargaff's rule Rule that states that in DNA, the amount of adenine always equals the amount of thymine (A = T) and the amount of guanine equals the amount of cytosine (G = C).

chemical bonds Forces of attraction that hold two or more atoms together in a molecule.

chemical energy Potential energy contained in chemical bonds of compounds such as gasoline, sugar, and fat.

chemical evolution hypothesis The idea that organic molecules were first formed in shallow oceans on Earth and ultimately gave rise to prebionts.

chemical factors Water and all nutritional elements required by living organisms.

chemical mediators Small molecules that either attract other molecules or cells to a site or signal them to perform some function.

chemical messengers Specialized molecules capable of transmitting signals between cells, such as hormones and neurotransmitters.

chemical synapse Site where an electrochemical nerve impulse is converted into a chemical signal that forms a "bridge" that allows the signal to pass to an adjacent cell.

chemosynthetic (kē´mō-sin-THET-ik) **bacteria** Autotrophic bacteria that use energy-rich molecules like hydrogen sulfide or methane for synthesizing complex organic molecules.

chlorophyll (KLOR-uh-fil) In plants, green pigments located in chloroplasts that function in photosynthesis.

chlorophyll a The predominant form of chlorophyll.

chlorophytic (klor´uh-FIT-ik) **green algae** Phylum of green algae that live in both freshwater and marine environments, distinguished from the gametophytic algae phylum by the presence of undulipodate gametes, which fuse during conjugation.

chloroplast (KLOR-uh-plast) Organelles that function in photosynthesis and are found only in eukaryotic plant and algae cells.

cholesterol (kuh-LES-tuh-rōl) A large lipid molecule that is an important component of animal cell membranes and is also used in the synthesis of steroid hormones.

cholinesterase (kō´li-NES-tur-ās) An enzyme that splits acetylcholine into choline and acetate.

Chondrichthyes (kon´DRIK-thē-ēz) The cartilaginous sharks and rays.

Chordata (kor-DĀ-tuh) Phylum containing the most advanced animals.

chorea (kuh-RĒ-uh) Constant and uncontrollable body movements.

chorion (KOR-ē-on) The outer membrane surrounding an embryo.

chorionic villi (kor´ē-ON-ik VIL-ī) Fingerlike extensions of the chorion that contain fetal blood vessels.

chorionic villus sampling (CVS) Removing a plug of tissue from a developing fetus with a small tube inserted through the mother's cervix and analyzing the cells removed to determine if genetic abnormalities are present in the fetus.

chromatin (KRŌ-muh-tin) The dispersed substance of chromosomes between cell divisions; consists of DNA and proteins.

chromosomal (krō´muh-SŌ-mul) **mutations** Changes in chromosome structure, function, or number.

chromosome (KRŌ-muh-sōm) **maps** Diagrams that show the location of gene loci and their exact linear order along the length of a chromosome.

chromosomes (KRŌ-muh-sōmz) Filamentous structures composed of protein and tightly coiled DNA; the genetic material.

chromosome set In each species, the group of different chromosomes that carries the complete collection of genes.

chromosome theory The theory that inheritance patterns can be explained by assuming that genes are located on chromosomes.

chronic (KRON-ik) Extended in time, as of an inflammation or a disease.

cilia (SIL-ē-uh) Short undulipodia used for motility in some protozoans and for the movement of particles or fluids in certain cells of more advanced eukaryotic organisms.

circadian (sur-KĀ-dē-un) **rhythms** Daily internal clocks that are synchronized to the environment.

circulatory (SUR-kyuh-luh-tor´ē) **system** An internal transport system consisting of the heart, a network of blood vessels, and blood.

cladogenesis (klad´uh-JEN-uh-sus) The splitting of one species into two or more species.

class A taxon that includes similar and related orders; the major subdivision of a phylum.

classical conditioning Associative learning that results in changes in the stimuli that elicit behavior.

cleavage (KLĒ-vij) A series of rapid cell divisions that gives rise to a sequence of structures in a developing embryo.

climax community A permanent community of organisms that is stable through time; the final stage or sere resulting from ecological succession in a given biotic community.

clitoris (KLIT-uh-rus) A small, sensitive female organ that has the same embryological origin as the male penis.

clonal (KLŌ-nul) **selection** Process in which a single B lymphocyte with the potential for producing antibodies against a specific antigen encounters that antigen, proliferates, and gives rise to large numbers of identical cells (clones).

clonal selection theory An attempt to explain how antigens stimulate antibody production by the process of clonal selection.

cloning (KLŌ-ning) The process of asexually producing a group of cells (clones), all genetically identical to the original ancestor.

cloning vector Usually a plasmid or a virus that can enter a living cell where replication can occur under appropriate conditions; used to transfer a DNA fragment from a test tube into a living cell.

closed-canopy forests Forests where mature trees shade understories of small shrubs, herbs, and moss-covered ground, limiting the penetration of sunlight in the lower strata and on the forest floor.

clotting A change in plasma from a liquid to a fibrous mass that entraps blood cells; requires a number of proteins and other factors present in the plasma.

coagulation (kō-ag´yuh-LĀ-shun) **system** Several mechanisms that are activated by chemical mediators to arrest bleeding.

cocaine (kō-KĀN) An addictive compound obtained from coca leaves.

codominance (kō´DOM-uh-nuns) Equal expression of both alleles in a heterozygote.

codon (KŌ-don) A section of DNA, three nucleotides in length, that codes for a single amino acid or acts as a start-or-stop signal in protein synthesis.

coelom (SĒ-lum) A body cavity that develops within the mesoderm of more advanced animals (coelomates).

coenzymes (kō´EN-zīmz) Protein molecules that are required for the functioning of enzymes; many are able to accept energetic electrons emitted from chemical reactions in cells.

coevolution The mutual evolution of two or more interacting populations of different species.

colchicine (KOL-chuh-sēn) A chemical that stops cell division at metaphase.

cold deserts Deserts distinguished by their low average yearly temperatures, usually occurring at high latitudes in the rain shadows of mountain ranges.

collagenous (kuh-LAJ-un-nus) **fibers** The most abundant connective tissue in humans, made of bundles of smaller fibers composed of a strong, inelastic protein called *collagen*.

collectors Consumer guilds in lotic ecosystems that feed on fine particulate organic matter.

colon (KŌ-lun) A division of the large intestine.

communication Transmission of information by means of signals from one animal to another that results in a change in behavior.

compact bone A solid, strong outer layer of bone tissue found in long bones of the arms and legs.

comparative method Looking at similarities and differences to discover regularities.

complementary base pairs In DNA, the pairs of bases that are held together by weak chemical bonds; adenine (A) bonds with thymine (T), and guanine (G) bonds with cytosine (C).

complementary DNA (cDNA) DNA that is synthesized from messenger RNA.

complement system A set of chemical mediators that function during acute inflammation and other defense reactions.

computerized thermal imagery A technique used to create colorized visual images based on temperature variations that result in emissions of different heat patterns.

concentration gradient The difference in concentration of a substance between two points.

conductance values The capacity for gases to move between the interior of a leaf and the surrounding air.

conductive Able to conduct a nerve impulse when excited.

cones (1) Color-sensing cells located in the retina of the eye. (2) Specialized reproductive structures of gymnosperms, consisting of modified leaves.

congenital (kun-JEN-uh-tul) **malformations** Deformities that appear during fetal development.

conidia (kuh-NID-ē-uh) Sexual spores produced in fruiting bodies of mature fungi from which new fungi arise.

coniferous forest biome (kuh-NIF-uh-rus) Forest region in which coniferous trees predominate, including the boreal forests and the forests of the Rocky Mountains and the Pacific Northwest.

conifers (KON-uh-furz) Evergreen trees that have narrow, elongated, pointed leaves called needles; include fir, pine, spruce, cedar, and larch.

conjugation (kon´juh-GĀ-shun) Process of reproduction in ciliated protozoans that involves meiosis and the exchange of haploid nuclei to produce new individuals with unique genetic combinations.

connective tissue The most abundant of four tissue types in the body, consisting of cells and intercellular substances that bind and support.

constant region Portion of antibody H and L chains with similar amino acid sequences.

consumers Organisms that obtain their nutritional requirements by feeding on producers or their products.

continental upwelling The movement of cold, nutrient-laden, deep ocean water up to the surface along some western continental coasts where the combined effects of surface winds and Earth's rotation toward the east cause a net offshore movement of surface water.

contraception (kon´truh-SEP-shun) Prevention of fertilization of an ovum or growth of a zygote.

contractile vacuoles (kun-TRAK-tul VAK-yoo-ōlz) Specialized structures used by ciliated protozoans for expelling excess water from their cells.

control A sample that is not manipulated—a basic feature of experiments.

control site A section of DNA that includes a promoter and an operator.

convergence (kun-VUR-juns) An evolutionary pattern whereby separate lineages become morphologically similar over time.

cooperative behavior Behavior among members of the same species that results in mutual benefit.

Coriolis (kor-ē-Ō-lus) **forces** Forces that result from Earth's rotation and cause circular air and water currents to rotate clockwise in the Northern Hemisphere and counterclockwise in the Southern Hemisphere.

cork cambium (KAM-bē-um) A living tissue that lies outside the vascular cambium and produces cells that form bark.

corpuscular (kor-PUS-kyuh-lur) **theory** An erroneous early theory that certain particles of matter were more "elastic" or active than others.

corpus luteum (KOR-pus LOO-tē-um) A temporary cholesterol-filled tissue formed from a ruptured ovarian follicle that acts as a gland, secreting estrogens, progesterone, and relaxin.

cortex (KOR-teks) Outer part of an organ; commonly used in reference to the brain and the kidney.

cosmology (koz´MOL-uh-jē) The field of science that studies the origin and evolution of the universe.

cotyledon (kot´uh-LĒD-un) An embryonic seed leaf that contains stored food for young angiosperms; monocots have one and dicots have two.

crack A potent form of cocaine.

cranial (KRĀ-nē-ul) **cavity** The cavity formed by the skull that encloses the brain.

cranial (KRĀ-nē-ul) **nerves** Twelve pairs of nerves that originate from the lower brain and innervate the head, neck, and limb attachments.

crossing-over The exchange of corresponding chromosome parts between duplicated homologs; a process that can introduce genetic variation in sexually reproducing organisms.

cross-pollination The transport of pollen from the stamen of one plant to the stigma in the flower of another plant.

crown gall A harmful tumorous enlargement caused by infecting bacteria that contain Ti plasmids.

cultural eutrophic (yoo-TRŌ-fik) **lakes** Lentic systems where excessive nutrients generated by human activity are introduced into the water, resulting in production that exceeds 1,000 grams of carbon per square meter per year; decomposition of the excessive production results in oxygen depletion and fish mortality.

culture Attributes of human societies that include language, social structure, value systems, and the development of tools and their use in agriculture.

cuticle (KYOO-ti-kul) A waxy, waterproof covering found on upper and lower leaf surfaces.

cuttings Clipped branches that can be used to propagate new plants.

CVS See *chorionic villus sampling.*

cyclic adenosine monophosphate (cAMP) (SIK-lik uh-DEN-uh-sēn mon´uh-FOS-fāt) A second messenger in initiating the actions of peptide hormones in cells.

cysts Tough, enclosed resting stages formed by single-celled amoebas during some stage of their life cycles.

cytogeneticist (sī´tō-juh-NET-uh-sist) Scientists who study chromosome structure and behavior.

cytokinesis (sī´tō-kuh-NĒ-sus) The division of the cytoplasm of a parent cell into two daughter cells.

cytology (sī-TOL-uh-jē) The study of cells.

cytoplasm (SĪ-tuh-plaz´um) A viscous fluid enclosed within the plasma membrane that surrounds submicroscopic organelles and the cell nucleus.

cytoskeleton (sī´tuh-SKEL-uh-tun) A delicate weblike structure within the cytoplasm composed of microfilaments and microtubules.

cytosol (SĪ-tuh-sol) The gelatinlike material in the interior of the cell that consists primarily of chemical substances dissolved in water.

cytotoxic (sī´tuh-TOK-sik) **T cell** A T cell that destroys cells possessing foreign antigens on their surface.

dark reactions The second set of chemical reactions in photosynthesis that use energy provided by ATP and NADPH formed during light reactions to produce glucose or other high-energy carbohydrates.

daughter cells Two cells produced by cell division of a parent cell.

deciduous (dē-SID-yuh-wus) **forests** Forests found in three temperate regions of the Northern Hemisphere in which deciduous trees predominate.

deciduous (dē-SID-yuh-wus) **trees** Trees that shed their leaves at the end of a growing season before entering a dormant period.

declarative memory Memory of past events, facts, names, dates, and places.

decomposers (dē´kum-PŌ-zurz) Certain fungi and bacteria that obtain food by breaking down nonliving organic materials from any source.

deforestation (dē-for´uh-STĀ-shun) Clearance of vast areas of forest for lumber, planting subsistence crops, or grazing cattle.

deletion The loss of a chromosome segment.

delta cells Pancreas cells that produce somatostatin.

deme (dēm) An interbreeding population that exists in a limited geographical area; a tier in the biotic hierarchy between populations and individuals.

deme density The number of individuals of the same species in a unit area or volume.

demography (di-MOG-ruh-fē) The study of populations, especially of growth rates and age structure.

dendrites (DEN-drīts) Processes of neurons that transmit nerve impulses toward the cell body.

denitrification (dē-nī´truh-fuh-KĀ-shun) The process in which anaerobic bacteria convert nitrates (NO^-_3) to nitrites and ammonia.

density-dependent factors Environmental factors that affect populations as a function of changes in deme density; generally retard population growth as density increases or enhance growth as density decreases.

density-independent factors Environmental factors, usually abiotic, that cause changes in the number of individuals per unit area or in a deme.

dermis (DUR-mus) The inner layer of skin, consisting of dense connective tissue and various other cells and tissues.

desert Regions of environmental extremes with dry areas usually located in the doldrums of 30° north and south latitude and receiving less than 25 centimeters of annual rainfall.

deuterostomes (DYOO-tuh-ruh-stōm) Animal phyla in which the blastopore becomes the anus.

development All irreversible changes that occur throughout an individual's lifetime.

developmental totipotency (tō-TIP-uh-tun-sē) The potential to regenerate a whole organism from a single cell.

diabetes mellitus (dī-uh-BĒ-tēz MEL-uh-tus) A variety of hereditary disorders characterized by elevated concentrations of glucose in the blood.

dicots (DĪ-kots) Angiosperms that develop from germinated seeds with two cotyledons and grow into plants with netted leaf venation and concentric vascular systems that produce flowers with parts in fours or fives or multiples of four or five.

differentiated (dif´uh-REN-chē-āt-ud) Of cells, having acquired distinctive molecular and structural features associated with a specific function.

differentiation (dif´uh-ren´chē-Ā-shun) Process of changing from an unspecialized cell to a cell that has a specific structure and function.

diffusion (dif-YOO-zhun) Process in which molecules in a gas or a solution move from regions of high concentration to regions of low concentration.

digestion The breakdown of complex food substances into simple molecules that can be absorbed into the body.

digestive system System performing numerous functions related to delivering nutrients to every cell in the body.

dihybrid cross (1) In Mendelian genetics, a cross involving individuals with two different pairs of traits. (2) A cross that involves differences at two gene loci.

dinitrogen molecules The most stable form of nitrogen (N_2).

dioecious (dī´Ē-shus) Having male flowers on one plant and female flowers on another plant.

diploid (DIP-loid) Of a cell, having two complete sets of homologous chromosomes; of an organism, having two chromosome sets in each of its cells.

directional selection Selection that shifts the mean (average) in a distribution of certain characteristics in response to a change in the environment.

disease Any abnormal condition of the body that impairs normal functioning.

displacement activities Behavioral activities characterized by irrelevance to the situation, usually in a response to frustration.

disruptive selection Selection that acts to favor two or more traits simultaneously; typically occurs when a population is subjected to separate selective pressures in different occupied areas.

divergence In evolution, a pattern in which there are increased morphological differences among separated lineages.

DNA Deoxyribonucleic acid; the material of which genes are composed; a double chain of linked nucleotides having deoxyribose sugars.

DNA polymerase (POL-uh-mur-ās) An enzyme that replicates DNA by joining complementary bases to a parental strand, which serves as a template.

DNA replication The process of making DNA.

dominant trait (1) In Mendelian genetics, the trait that dominates in the first generation of a monohybrid cross. (2) The allele in a heterozygous genotype that is expressed in the phenotype.

dopamine (DŌ-puh-mēn) A neurotransmitter in the brain; depletion may lead to Parkinson's disease.

dorsal (DOR-sul) Located near the back or upper surface of most animals; opposite the ventral surface.

double helix The shape of a DNA molecule, much like a ladder twisted about its long axis.

doubling time The time it takes a population growing at a given rate to double in size.

Down syndrome An abnormal human phenotype usually caused by a trisomy of chromosome 21; characterized by mental retardation and various physical abnormalities; more common in babies born to older mothers.

drift Dissolved materials, suspended particulates, living insects, and other organisms that move seaward with the flowing water in lotic ecosystems.

drive In classical ethology, the internal state of an animal that results in a tendency to organize its behavior to achieve a certain goal.

drugs (1) Chemical agents that bring about a functional or structural change in living tissues. (2) Chemical substances used to control disease.

drug therapy The use of drugs to attack a disease agent after it infects a host.

dry deposition Acid deposition from atmospheric particle fallout.

Duchenne muscular dystrophy (DMD) (doo-SHEN MUS-kyuh-lur DIS-truh-fē) A genetic disorder associated with an X-linked, recessive allele that encodes a protein, dystrophin, necessary for normal muscle cell development; occurs in at least one in 3,500 males born.

duodenum (dyoo´uh-DĒ-num) Upper region of the small intestine.

duplication An increase in the number of genes carried by a chromosome.

dystrophin (dis-TRŌ-fin) A large protein necessary for normal muscle cell development.

ECC See *environmental carrying capacity.*

echinoderm (ē-KĪ-nuh-durm) Coelomate, deuterostome marine organism in the phylum Echinodermata; includes starfish and sea urchins.

ecological succession The predictable, gradual transformation of a site into different communities (seres) with the passage of time; ends with formation of a climax community.

ecology (ē-KOL-uh-jē) The scientific study of interrelationships that exist between organisms and their environments.

ecosystem (Ē-kō-sis´tum) (1) All organisms and physical factors that form an ecological unit. (2) A system of physical and biological processes in a space-time unit of any magnitude, such as the biotic community plus its abiotic environment.

ecosystem-centered school An informal group of early ecologists who considered ecosystems, consisting of both abiotic and abiotic factors, to represent the fundamental ecological level of organization. Followers were often placed in a specific group; those who agreed with F. E. Clements or those who agreed with H. C. Cowles.

ectoderm (EK-tuh-durm) Outermost of three primordial germ layers in a developing animal embryo; gives rise to the nervous system and epidermis.

efferent (EF-uh-runt) **(motor) division** Division of the peripheral nervous system composed of the somatic nervous system and the autonomic nervous system.

eggs Haploid gametes produced by females.

ejaculation (ē-jak´yuh-LĀ-shun) Reflexive discharge of semen from the penis.

ejaculatory (ē-JAK-yuh-luh-tor´ē) **duct** A tube that transmits sperm from the vas deferens to the urethra.

elastic fibers Long, thin, branching fibers made of the protein elastin that are found in certain connective tissue cells.

electrical energy A form of energy produced by a flow of electrons (negatively charged particles).

electrical synapse Site where a nerve impulse travels directly from one cell to another cell.

electromagnetic radiation Beams of light of different wavelengths emitted from the sun.

electron carriers Coenzymes that accept energetic electrons.

electron transport chain reactions A set of stepwise chemical reactions in which high-energy electrons from NADH and $FADH_2$ pass through a series of electron carriers located in mitochondrial membranes, producing large quantities of ATP.

embryo (EM-brē-ō) A diploid organism in its early stages of development; in humans, the first two months after fertilization.

embryonic (em´brē-ON-ik) **disk** Part of the developing embryo that gives rise to the three primordial germ layers.

emigration (em´uh-GRĀ-shun) The movement of individuals out of a deme, usually to a deme in a different area.

emission Movement of the contents of sexual ducts and glands to the urethra in males.

endocrine (EN-duh-krin) **system** A system consisting of discrete organs called glands, made up of specialized cells capable of secreting hormones, as well as other organs that secrete hormones and nonhormonal substances that act as hormones.

endocrinology (en´duh-kruh-NOL-uh-jē) The study of hormones derived from classical endocrine glands or from other cells or tissues, such as the brain or heart, and their functions.

endoderm (EN-duh-durm) Innermost of the three primordial germ layers in a developing animal embryo; gives rise to the stomach, intestines, urinary bladder, and respiratory tract.

endometrium (en-duh-MĒ-trē-um) The mucous glandular membrane that lines the uterus.

endoplasmic reticulum (en-duh-PLAZ-mik rē-TIK-yuh-lum) **(ER)** An intracellular network of interconnected membranous tubules where proteins and lipids are synthesized in the cell.

endosperm (EN-duh-spurm) A tissue containing stored food in seeds, which provides the energy and nutrients required for seedling development.

endosymbiotic (en´dō-sim´bī-OT-ik) **hypothesis** A hypothesis stating that modern eukaryotic cells arose as a result of unions between at least two types of eubacteria (prokaryotic cells) that came to live within ancestral eukaryotic cells.

energy The capacity to do work or cause change; occurs in two general states, potential and kinetic.

energy transducer A neuron that is able to convert one form of energy, such as light, into an electrochemical impulse.

energy transformation A change from one form of energy to another.

entropy (EN-truh-pē) The useless energy of any system, whether the universe or a cell; a quantitative measure of randomness or disorder.

envelope A structure surrounding some viruses, composed of lipids and proteins.

environmental carrying capacity (ECC) The density of an animal deme that can be sustained in each stage of ecosystem succession.

environmental stress Any component of the physical environment that reduces the capacity of plants to acquire, assimilate, or allocate resources needed for growth.

enzymes (EN-zīmz) Catalysts that participate in most chemical reactions within cells; usually composed of proteins.

eosinophil (ē´uh-SIN-uh-fil) A granulocyte whose granules hold various substances used in defense reactions.

epidemiology (ep´uh-dē´mē-OL-uh-jē) Field of science dealing with the relationships of various factors that influence the frequencies and distributions of a disease in a human community or population.

epidermis (ep´uh-DUR-mus) The skin's thin, outer layer, composed of closely packed epithelial cells.

epididymis (ep´uh-DID-uh-mus) Small mass of tubules in which sperm undergo final maturation.

epiglottis (ep´uh-GLOT-us) A flap of tissue in the throat that automatically seals off the air passageways during swallowing.

epithelial (ep´uh-THĒ-lē-ul) **tissue** Continuous layers of closely connected cells that cover body surfaces, form glands and part of the skin, and line cavities inside the body.

ER See *endoplasmic reticulum.*

erection Enlarged, firm temporary state of the penis or the clitoris.

erythrocytes (ē-RITH-ruh-sīts) Red blood cells; specialized for transporting oxygen.

erythropoietin (ē-rith´rō-POI-uh-tin) A chemical messenger that stimulates erythrocyte production in red bone marrow.

esophagus (i-SOF-uh-gus) Tube that conducts food from the pharynx to the stomach.

essential elements Chemical elements essential for normal plant growth.

estrogens (ES-truh-jinz) Female sex hormones produced by the ovaries that affect development and maintenance of female reproductive organs and secondary sex characteristics.

estrous (ES-trus) **cycle** A regular period of sexual receptivity in female mammals, except higher primates; occurs when an egg is available to be fertilized.

ethological (ē´thuh-LOJ-i-kul) **isolation** Isolating mechanism that occurs due to differing behavior patterns in courtship or a lack of sexual attraction between males and females of different species.

ethologist (ē-THOL-uh-jist) A scientist who studies animal behavior.

ethology (ē-THOL-uh-jē) The scientific study of animal behavior.

eubacteria (yoo´bak-TĒR-ē-uh) Prokaryotic cells that can live in diverse environments, including most common groups of contemporary bacteria; the more recent of two main lineages of bacteria.

eugenicists (yoo-JEN-uh-sists) Scientists concerned with the implications of evolution for human heredity; they believe that knowledge from both genetics and evolution can be used for medical and social improvement.

eukaryotic (yoo´kar-ē-OT-ik) **cell** The more complex of two cell types, with complicated internal structures including a nucleus defined by a nuclear envelope and numerous membrane-bound organelles; all protoctists, fungi, plants, and animals are composed of eukaryotic cells and are commonly called *eukaryotes.*

euphotic (yoo-FŌ-tik) **zone** Lake and ocean region to a maximum depth of about 200 meters, the limit of light penetration.

eutrophic (yoo-TRŌ-fik) **lakes** See *Natural eutrophic lakes.*

evapotranspiration The total water loss from direct evaporation and plant transpiration.

evergreen rain forests Forests containing a variety of conifer trees such as spruce, fir, pine, and hemlock, occurring in the coastal plains and mountains of the Pacific Northwest.

evergreen trees Broad-leaved and coniferous trees and shrubs that retain leaves year round, although the leaves are regularly being shed and replaced.

evolution (ev´uh-LOO-shun) Changes that occur in populations of living organisms over long periods of time.

excitable Able to respond to a stimulus.

exons (EK-sonz) The protein-coding sequences of a gene; the parts of a gene that are transcribed as RNA.

exotoxins (ek´so-TOK-sinz) Toxic substances released from multiplying bacteria.

experimental method The use of experiments as a tool of investigation.

experimentation In science, applying a set of procedures designed to accomplish certain goals.

exponential (ek´spuh-NEN-chul) **growth** Growth that can be described mathematically by the formula 10^x, where x (the exponent) is some computed number that defines the rate of growth; also called *geometric growth*.

expression vectors Plasmids that contain specific nucleotide sequences used to direct protein synthesis.

extraembryonic membranes Four membranes formed early in embryonic development that later become the amnion, yolk sac, allantois, and chorion.

familial hypercholesterolemia (fuh-MIL-yul hī´pur-kuh-les´tuh-ruh-LE-mē-uh) **(FH)** A genetic disorder characterized by high levels of cholesterol in the blood; the most frequent cause of inherited heart disease.

family A taxon that includes similar and related genera; the major subdivision of an order.

feces (FĒ-sēz) Compacted solid digestive wastes discharged through the anus.

fermentation (fur´men-TĀ-shun) Process whereby microorganisms synthesize useful products from nutrients and raw materials in the absence of oxygen; specifically, an anaerobic reaction in which pyruvate is converted to a different three-carbon compound, either alcohol or lactate, depending on cell type, producing a small amount of ATP.

fertilization (fur´tuh-luh-ZĀ-shun) Process in which haploid gametes unite to form a diploid zygote.

fetoscopy (fē-TOS-kuh-pē) A technique for viewing an embryo that requires insertion of a fetoscope into the amniotic sac through a small surgical opening in the abdomen.

fetus (FĒ-tus) A developing human from the third month of pregnancy to birth.

FH See *familial hypercholesterolemia*.

fibrin (FĪ-brin) A threadlike protein that forms a network entrapping erythrocytes, platelets, and other materials to create a blood clot.

fibrinogen (fī-BRIN-uh-jin) A protein that is converted to fibrin during blood-clotting reactions.

fibroblasts (FĪ-bruh-blasts) Spindle-shaped connective tissue cells that produce collagenous and elastic fibers and intracellular substances.

filamentous fibers Structures that make up part of the intercellular matrix.

filopodia (fil´uh-PŌ-dē-uh) Needlelike pseudopodia that protrude from pores in the tests of foraminifers and are used in movement and feeding.

filtrate Substances, including salts, amino acids, glucose, water, and urea, that are removed from blood as it enters the kidneys.

first law of thermodynamics Law stating that energy can be converted from one form to another form but can never be created or destroyed (also called the *law of conservation of energy*).

first-order consumers Herbivores or herbivore guilds.

first-order streams The smallest headwater streams.

first polar body The smaller of two haploid cells formed during the first meiotic division of an oocyte.

fitness A measure of success among genotypes that is calculated as the net reproductive rate (the average number of offspring produced by individuals) times the probability that the individuals will survive to reproductive age.

fixed action pattern A species-specific set of stereotyped, coordinated behavioral activities.

flagella (fluh-JEL-uh) Long, thin, whiplike structures used in eukaryotic cell mobility; undulipodia.

flowers Plant reproductive structures containing male and female reproductive parts that produce male and female gametes.

flux The movement of material between trophic levels of an ecosystem in a unit of time; the amount of any substance that moves from one place to another in a unit of time.

foliar (FŌ-lē-ur) **uptake** The process by which trees take up substances from the air through their leaves or needles.

follicle-stimulating hormone (FSH) The hormone that in females initiates development of ova and stimulates the ovaries to secrete estrogens and in males initiates sperm production.

food chain The sequence of energy or food transfers from one trophic level to another.

food cycles Early conceptual diagrams that incorporated both energy flow and biogeochemical cycles in studies of communities.

food web The complex, interlocking series of food chains in a community.

forebrain Part of the brain where most complex neural processing and integration occur; also called the *cerebrum.*

forest decline Condition characterized by trees' developing yellow leaves or needles and by a decrease in tree growth per unit of ground area; often associated with acid deposition.

forest dieback The mortality of select species within a forest stand or even of an entire stand; often associated with acid deposition.

formed elements Solids in blood, including erythrocytes, various types of leukocytes, and platelets.

fossil record All fossils, which collectively form a chronicle of past life on Earth.

fossils Remnants or traces of organisms from past geological ages embedded in Earth's crust.

founder effect A special case of random genetic drift in which a small group of individuals establishes a genetically unique colony.

fourth-order consumers Trophic guilds that feed on third-order consumers; in lentic ecosystems, larger trout, bass, or pike.

fourth-order streams Streams that form at the junction of two third-order streams.

frameshift mutation The insertion or deletion of nucleotides, which disrupts translation.

freshwater ecosystems Aquatic systems other than marine ocean ecosystems, including lotic systems characterized by free-flowing streams and rivers and lentic systems with lakes and ponds.

fruit Seed-containing structure of angiosperms that originates from the female parts of fertilized flowers.

FSH See *follicle-stimulating hormone.*

F_2 breakdown An isolating mechanism that results in a hybrid producing sterile or less fit offspring.

functional classification system A way of classifying organisms in the five kingdoms of life, in which their mode of nutrient acquisition is used to categorize them as being producers, consumers, or decomposers.

functional ecosystem models Models based on the rate at which energy and materials move through the trophic guilds of an ecosystem.

fusion gene A gene constructed by combining DNA segments from different organisms.

Galápagos (guh-LOP-uh-gus) **Islands** A group of Pacific islands near the equator about 600 miles off the coast of South America.

gallbladder An accessory digestive organ that stores bile.

gametes (guh-MĒTS) Specialized cells that contain the haploid number of chromosomes; in humans, sperm and egg cells.

gametic (guh-MET-ik) **isolation** Mechanism that occurs when male and female gametes cannot combine in fertilization or when pollen or sperm are rendered inviable in the female sexual structures of another species.

gametophyte (guh-MĒT-uh-fīt) The gamete-producing generation of plants; plants that grow from germinated haploid spores.

gaseous cycles Biogeochemical element cycles in which the atmosphere or hydrosphere is a major reservoir.

gastrula (GAS-truh-luh) Developmental stage of an animal with two cell layers and a blastopore that develops from a blastula.

gastrulation (gas´truh-LĀ-shun) Process during which cells undergo a series of migrations that lead to a dramatic remodeling of the embryo and the establishment of primordial germ layers.

GCMs See *general circulation models*.

gel electrophoresis (JEL ē-lek´truh-fuh-RĒ-sus) Technique used to separate protein fragments by size.

gene The fundamental physical and functional hereditary unit, composed of a DNA segment.

gene amplification Abnormal increases in portions of a chromosome.

gene augmentation Providing cells with added genes to produce the missing protein associated with a genetic disorder; a form of gene therapy.

gene cloning The central process of genetic engineering, using microorganisms to produce millions of identical copies of one gene, usually one that codes for a desirable protein.

gene dosage compensation Process in which X chromosome inactivation results in both males' and females' producing the same amount of X chromosome gene proteins.

gene expression Process whereby genes are converted into structures; all the steps necessary to transpose a genotype to a phenotype.

gene families A set of genes descended from an ancestral gene; genes that encode similar proteins.

gene library A collection made from a set of overlapping DNA fragments representing the entire genome of an organism.

gene mutation A change in a single gene from one allelic form to another.

gene pool All genes that are distributed among all individuals of a population.

gene substitution Removal of a mutant gene and replacement with a normal gene; a form of gene therapy.

general circulation models (GCMs) Computer models used to make predictions about long-term changes in global atmospheric conditions.

gene therapy The replacement, correction, or augmentation of existing genes.

genetic code The sequence of nucleotides, coded in mRNA triplets, that determines the amino acid sequence of a protein.

genetic counseling Communication process concerning the risks of occurrence of a genetic disorder in a family; involves an attempt to help the person or family comprehend the medical facts, appreciate the hereditary nature and recurrence risks in specific relatives, understand the options for dealing with the risk, choose the most appropriate course of action, and make the best possible personal decision.

genetic disorders A predictable set of consequences associated with a specific gene or chromosomal mutation.

genetic engineering The manipulation of genetic information to alter the characteristics of an organism.

genetic linkage map A map of the relative position of genetic loci on a chromosome; determined on the basis of how often the loci are inherited together.

genetic polymorphisms (pol-ē-MOR-fiz-umz) Variations in the nucleotide sequences in a specific segment of an individual's DNA.

genetic recombination New combinations of genes or chromosomes that are created by any process.

genetics (juh-NET-iks) The scientific study of heredity and the transmission of characteristics from parents to offspring.

genome (JĒ-nōm) The complete collection of genes in one organism or in a chromosome set.

genotype (JĒ-nuh-tīp) The specific alleles contained in a cell or an individual.

genus (JĒ-nus) A taxon that includes similar and related species; the major subdivision in a family.

geological time scale The various eons, eras, periods, and epochs that occurred in Earth's 4.5 billion–year history.

germination (jur´muh-NĀ-shun) Growth of an embryo within a seed that develops into a new plant.

germline therapy Form of gene therapy in which genes are modified in cells that produce sperm and eggs or in early embryonic cells.

germ theory Theory that many diseases are caused by microbes, such as bacteria, that invade healthy bodies.

glands Organs composed of specialized epithelial cells that secrete hormones.

glomerulus (gluh-MER-yuh-lus) A rounded mass of capillaries (or nerves) involved in blood filtration, especially in the kidneys.

glucagon (GLOO-kuh-gon) Hormone secreted by the pancreas involved in glucose homeostasis; acts to increase glucose levels in blood.

glucocorticoid (gloo´kō-KOR-tuh-koid) A steroid hormone that plays an important role in sugar metabolism.

glycogen (GLĪ-kuh-jin) A storage form of glucose in animals.

glycolysis (glī-KOL-uh-sus) The first chemical reaction in the step-by-step breakdown of glucose to water and carbon dioxide; occurs in all living cells.

GnRH See *gonadotrophin-releasing hormone.*

Golgi (GOL-jē) **complex** Organelle in which proteins from the endoplasmic reticulum are modified and stored prior to export from the cell.

Golgi (GOL-jē) **sacs** Flattened membranes that make up a Golgi complex.

gonadotrophin-releasing hormone (GnRH) Hormone (possibly more than one) that stimulates production of FSH and LH.

gonadotrophins (gō-nad´uh-TRŌ-finz) Hormones that regulate the endocrine functions of the gonads (ovaries and testes).

gonads (GŌ-nadz) Reproductive organs of animals.

grade An evolutionary change resulting in a new functional ability.

gradients Changes in the rate of variables over distance; for example, increasing or decreasing air temperature with changing elevation or latitude.

grafting A form of vegetative reproduction in which a bud or a branch of one plant is attached to the stem or rootstock of another, closely related plant.

Gram stain A specific biological staining agent used to distinguish different types of bacteria; named after the Danish microbiologist Hans Christian Gram.

granules Microscopic organelles in granulocytes that contain substances related to specific defense functions.

granulocytes (GRAN-yuh-lō-sīts) Several types of white blood cells that possess granules in their cytoplasm.

grassland biomes Areas dominated primarily by grasses and grasslike plants where about one fourth of the total area is covered by vegetation, found in a number of different latitudinal zones and, depending on elevation, interspersed within tropical and temperate forest biomes; the largest of four major natural vegetation formations covering Earth and largest biome type in North America.

gray matter Nonmyelinated nervous tissue.

grazers Insect guilds in lotic ecosystems that feed on algae attached to benthic surfaces.

greenhouse effect Warming of the atmosphere that occurs when concentrations of CO_2, water vapor, and other gases increase and absorb more of the sun's longer (infrared) wavelengths radiated from Earth's surface.

green revolution An agricultural movement built using chemical fertilizers and insecticides and on the selective propagation of high-yielding varieties of cereal grains such as wheat and rice.

gross production The total assimilation of organic matter by an organism, population, or trophic unit per unit time per unit volume or area.

growth respiration The amount of metabolic energy used for mitosis and cell enlargement in plants.

gymnosperms (JIM-nuh-spurmz) Tracheophytes such as conifer trees that produce seeds within cones.

habitat isolation Isolating mechanism that occurs when populations of different species occupy different habitats within the same general geographical region.

habituation (huh-bich´uh-WĀ-shun) A simple form of learning in which an animal is repeatedly exposed to a stimulus not associated with any positive or negative consequence and eventually ceases to respond.

hadal (HĀ-dul) **zone** Deep region of the ocean between 6,000 and 11,000 meters below the surface.

hair follicles Structures in the dermis that produce hairs.

halophilic bacteria (hal-uh-FIL-ik) "Salt-loving" bacteria that can thrive in saline habitats.

haploid (HAP-loid) Having one chromosome set; having only one homolog of each chromosome type.

Hardy-Weinberg law Law that predicts genotypic frequencies of succeeding generations of a population on the basis of an initially defined distribution of genotypes.

HB See *hydrological balance.*

hCG See *human chorionic gonadotrophin.*

HD See *Huntington disease.*

heart A large hollow muscular organ capable of powerful contractions that propel blood through arteries.

heat See *thermal energy.*

heavy chains Two identical long protein chains that form part of an antibody molecule.

helper T cell T cell that stimulates B cells and other T cells involved in the total immune response.

hemoglobin (HĒ-muh-glō-bin) An iron-containing protein molecule of red blood cells that carries oxygen from the lungs to the internal tissues.

herbaceous perennials (hur-BĀ-shus puh-REN-ē-ulz) Plants that lack wood in their tissues and live more than two years, such as asparagus, rhubarb, milkweed, lilies, and many grasses.

herbicides (HUR-buh-sīdz) Pesticides that kill plants.

herbivores (HUR-buh-vorz) Members of trophic guilds that feed on primary producers, plants, and algae.

hereditary factor Mendel's term for the hereditary material responsible for individual traits.

herpesviruses (hur´pēz-VĪ-ruh-siz) Viruses that normally live inside nerve cells.

heterotrophs (HET-uh-ruh-trōfs) Organisms that must obtain energy and nutrients from the molecules produced and stored in autotrophs.

hGH See *human growth hormone.*

High Arctic Tundra at higher latitudes north of the Low Arctic; considered a desert where annual precipitation is low.

hindbrain Part of the brain that includes the cerebellum, medulla, and pons; largely involved in maintaining homeostasis and coordinating large-scale body movements.

HIV-1 Human immunodeficiency virus, type 1; the retrovirus that causes AIDS.

holdfasts Structures that anchor kelp to rocky shores and allow it to survive in surf zones.

homeostasis (hō´mē-ō-STĀ-sus) The condition of maintaining a relatively constant internal environment.

homeothermic (hō´mē-ō-THUR-mik) Having internal mechanisms to control body temperature.

home range The space or area that animals exploit on a daily basis.

hominid (HOM-uh-nid) A family of the Primate order that includes the human species.

homolog (HŌ-muh-log) A member of a pair of homologous chromosomes.

homologous (huh-MOL-uh-gus) **pair of chromosomes** A pair of chromosomes that are identical in size, shape, and gene composition, one homolog from each parent.

homologous (huh-MOL-uh-gus) **structures** Structures that share a common evolutionary, embryological, and structural background.

homology (huh-MOL-uh-jē) (1) Similarity between different genes. (2) Structures that have a similar embryological and evolutionary history.

horizontal transmission Direct or indirect transfer of a disease agent from one individual to another.

hormone (HOR-mōn) Any substance released by one cell that acts on another cell anywhere in the body.

hot deserts Deserts distinguished by a high average yearly temperature.

human chorionic gonadotrophin (kor´ē-ON-ik gō-nad´uh-TRŌ-fin) **(hCG)** A hormone that helps sustain the corpus luteum during the first months of pregnancy.

human growth hormone (hGH) A hormone that influences a number of metabolic activities that directly or indirectly influence growth, especially of the skeleton.

humoral immunity (HYOO-muh-rul im-YOO-nuh-tē) Part of the immune response, characterized by the formation of antibodies.

Huntington disease (HD) An inherited genetic disorder caused by a late-acting dominant allele that results in harmful nervous system effects.

hybrid (HĪ-brid) (1) The heterozygous offspring in crosses between parents having different forms of a single trait. (2) The offspring of a cross between individuals of different species or different varieties.

hybrid inviability An isolating mechanism that results in a hybrid's failure to reach maturity.

hybrid sterility An isolating mechanism that results in partial or complete sterility of a hybrid.

hydrological (hī´druh-LOJ-i-kul) **balance (HB)** The amount of precipitation that penetrates a canopy and reaches the ground surface minus the total water loss from evaporation and from transpiration by the foliage.

hydrologic (hī´druh-LOJ-ik) **cycle** The total water cycle of the biosphere, including evaporation, condensation, precipitation, and the elevational gradient-induced flow of water.

hydrosphere (HĪ-drus-fēr) All the water on Earth's surface.

hypersensitivity A heightened immune response that may be harmful to the body.

hypertension High blood pressure.

hyphae (HĪ-fē) Fungal structures that develop from conidia and usually become multicellular.

hypophyseal portal (hī-pof´uh-SĒ-ul POR-tul) **system** A complex network of blood vessels through which blood flows from the hypothalamus directly to the pituitary gland.

hypothesis (hī-POTH-uh-sus) An explanation derived by scientists from careful observations and supported by results from experiments and other evidence; less certain than a theory or a law.

ileum (IL-ē-um) Lower section of the small intestine.

immigration The movement of individuals into a deme, usually from a deme in another area.

immune system System composed of many types of cells and organs that participate in highly coordinated processes to inactivate or destroy abnormal self cells, foreign organisms, or antigens.

immunity (im-YOO-nuh-tē) The state of being able to resist a particular disease-causing agent or the antigens of such an agent.

immunoglobulins (im´yuh-nō-GLOB-yuh-linz) A family of proteins that includes antibodies.

immunology (im´yuh-NOL-uh-jē) Study of the immune system, immune responses, and immunity.

imperfect flowers Flowers that lack one of the reproductive organs (stamens or pistils).

imprinting Process whereby a young animal forms an association or identification with another animal or an object.

inclusive fitness The measure of an individual's reproductive success and the effects of that success on the reproductive success of its relatives.

incomplete dominance Situation when offspring have a range of intermediate phenotypes from crosses between different homozygous parents.

incomplete penetrance Situation when a trait is not evident in all individuals who have the relevant genotype.

inducer An agent that activates transcription from an operon.

infectious diseases Diseases caused by microorganisms such as bacteria, viruses, and protozoans, and by fungi and parasitic worms.

inflammation (in´fluh-MĀ-shun) A nonspecific defense mechanism that eliminates or prevents the spread of foreign materials at a site of injury and prepares the damaged site for repair.

inhibin (in-HIB-in) A male sex hormone thought to be produced by cells in the seminiferous tubules that exerts negative feedback effects on the pituitary, the hypothalamus, or both and inhibits FSH release.

inhibiting hormones (inhibiting factors) Substances that travel through the hypophyseal portal system and inhibit release of hormones produced by the anterior pituitary.

initiation site The site in a DNA molecule where replication begins.

innate releasing mechanism In classical ethology, a hypothesized mechanism responsible for triggering a fixed action pattern in response to a stimulus.

inner cell mass A cluster of cells at one end of a blastula.

innervated (IN-ur-vā´tid) Controlled by the nervous system through the actions of neurons.

insecticides (in-SEK-tuh-sīdz) Pesticides that kill insects.

insertion Relocation of one or more genes to a different section of the same chromosome or to a different chromosome; a type of translocation.

insight The ability of some animals to devise new behaviors based on past experiences.

instinct An innate set of reactions that occurs in response to a stimulus.

insulin (IN-suh-lin) Hormone secreted by the pancreas, required by humans for normal processing of sugar in the blood.

integumentary (in-teg´yuh-MEN-tuh-rē) **system** System that includes the skin and structures such as nails, hair, and certain glands.

intercalated (in-TUR-kuh-lā´tid) **disks** Unique cross-striations in cardiac muscle cells that occur where two cells meet end to end and form junctions.

intercellular matrix A semifluid, gel-like, or fibrous material made up of complex carbohydrates and proteins.

interkinesis (in´tur-kuh-NĒ-sus) A period in meiosis when chromosomes fade and new nuclear envelopes form.

interleukin 1 (in´tur-LOO-kin) A hormone that attracts and activates T cells.

interleukins (in´tur-LOO-kinz) Hormones that act on lymphocytes, affecting activation, maturation, proliferation, and functioning of all cells involved in immune responses.

internal fertilization Process in which gametes fuse inside the body of the female.

interphase The first and longest stage of cell division, in which chromosomes are functionally active and genes are transcribed and translated; also, the stage in which each chromosome is duplicated and the cell acquires enough materials to survive division into two daughter cells.

interspecies interactions Interactions between demes of two or more different species.

interstitial (Leydig) (in´tur-STISH-ul; LĪ-dig) **cells** Cells that produce testosterone, located within the connective tissue between seminiferous tubules in the testes.

intraspecies interactions Interactions among individuals of a species within a deme or between members of the same species from different demes.

intrinsic control Factors that regulate parameters affecting individuals, groups, or systems; for example, biotic factors in a deme that affect the number of individuals per unit area or volume.

introns (IN-tronz) The intervening sequences between exons; noncoding, nontranscribed sections of a gene.

inversion Situation when portions of chromosomes break, rotate 180°, and become reinserted at the same position in the chromosome.

islets of Langerhans (LONG-ur-honz) Clusters of alpha, beta, and delta cells in the pancreas.

isopleth (Ī-suh-pleth) A line on a graph or map that connects points of equal or corresponding values.

jejunum (ji-JOO-num) Middle region of the small intestine.

joints Sites of connection between bones or between bone and cartilage.

karyotype (KAR-ē-uh-tīp) Photomicrograph showing a complete set of chromosomes, each of which has a characteristic size, shape, and staining pattern.

kelp Large forms of brown algae, some of which attain a length of nearly 100 meters.

keratin (KER-uh-tin) A tough, fibrous protein that is flexible and impermeable to water, found in the hair, nails, and epidermis.

kidneys Two bean-shaped organs on the back wall of the abdomen that regulate the composition of the blood and produce urine.

kinesis (kuh-NĒ-sus) An undirected change in the rate of motion in response to the intensity of a stimulus.

kinetic (kuh-NET-ik) **energy** The energy of action or motion that results in change.

kingdom The most inclusive taxon of living organisms in the modern classification system; includes similar and related phyla.

kin selection Selection of behavior that promotes the survival and reproduction of relatives.

kinship Degree of relatedness among organisms.

Krebs cycle reactions A stepwise set of chemical reactions in which an acetyl-coA molecule is completely dismantled, yielding a variety of products that can enter into electron transport chain reactions.

krill Small crustaceans that are abundant in some marine ecosystems, especially in arctic and antarctic waters.

labor The period when strong uterine contractions ultimately lead to expulsion of the fetus from the uterus.

lac operon (LAK OP-uh-ron) In bacteria, three consecutive regions of DNA consisting of structural genes (Z, Y, and A) that code for enzymes needed to metabolize lactose.

lactation (lak-TĀ-shun) The secretion of milk from mammary glands.

LAI See *leaf area index*.

laminar (LAM-uh-nur) Thin, flat, and wide.

landmarks Consistent features—position of the centromere, position of major bands, or the ends of chromosome arms—used in identifying a specific chromosome.

large intestine Digestive organ between the small intestine and the anus that reabsorbs sodium and water.

larynx (LAR-inks) The Adam's apple or voice box; passageway between the pharynx and the trachea.

latent heat of evaporation The amount of heat required to change water from its liquid state to its gaseous state.

latent heat of melting The amount of heat required to change water from its solid state (ice) to its liquid state.

latent learning Learning in which there is no obvious reward at the time of learning and what is learned remains latent until put to use.

latent period The brief period of time between a stimulus and the start of a response.

lateral buds Meristematic zones where leaf petioles join the stem and grow to form branches.

law A generalization that has stood the test of time and is continuously confirmed by new evidence; laws have the highest level of certainty.

law of independent assortment Mendel's second law, stating that traits are transmitted independently of one another.

law of segregation Mendel's first law, stating that hereditary traits are caused by a pair of factors and that during gamete production these factors separate, with only one member of the pair going to any single gamete.

layering A form of vegetative reproduction in which meristems, stems, or branches give rise to a new and independent plant.

LDL See *low-density lipoprotein.*

leaf area index (LAI) The ratio of total plant leaf surface area to the total ground surface area covered by the leaves.

leaf axils The sites where petioles join the stem.

learning A change in behavior that occurs as a result of experience.

lentic ecosystems (LEN-tik Ē-kō-sis´tumz) Ecosystems that occur in standing bodies of fresh water, including lakes and ponds.

lethal mutation A change in a gene that results in death of the organism.

leukocytes (LOO-kuh-sīts) White blood cells; of two general types, granulocytes and agranulocytes.

LH See *luteinizing hormone.*

lichens (LĪ-kinz) A phylum of fungi consisting of two different species, one a fungus and the other a cyanobacterium or chlorophytic alga, living symbiotically.

life zone A habitat and the plants and animals that live there.

ligaments (LIG-uh-munts) Tough tissues that connect bones in joints.

light chains Two identical short protein chains that form part of an antibody molecule.

light reactions The first set of chemical reactions in photosynthesis in which light energy is transformed into ATP or high-energy electrons that are captured by the coenzyme $NADP^+$, which then becomes converted to NADPH.

limnetic (lim-NET-ik) **zone** Part of a lake, excluding the littoral zone, in which sufficient light energy penetrates to depths where photosynthesis can occur.

lineage (LIN-ē-ij) A single line of evolutionary descent.

linkage The proximity of two or more genes or markers on a chromosome; the closer together the markers are, the lower the probability that they will be separated during meiosis and the greater the probability that they will be inherited together.

linkage analysis A process for locating the approximate sites of genes on chromosomes based on analyses of the inheritance of linked genes.

linkage group All the genes present on a single chromosome.

lipids (LIP-idz) Various fat molecules; key components of cellular membranes.

lithosphere (LITH-uh-sfēr) The rigid crustal plates of Earth; distinguished from the atmosphere and the hydrosphere.

littoral (LIT-uh-rul) **zone** An area of transition between riparian zones along the shore and the water, extending to a depth of about 10 meters in lentic ecosystems.

liver Complex organ that removes nutrients from circulation, converts glucose to glycogen, and detoxifies harmful compounds.

lobe One of four main areas in the cerebral hemispheres; each carries out one or more specific functions.

localism An early idea that local disturbances in specific organs were responsible for disease.

locus Gene locus; the specific site on a chromosome where a gene is located.

lotic ecosystems Ecosystems that occur in flowing bodies of fresh water, streams, and rivers.

Low Arctic Area south of the High Arctic; a treeless zone at the edge of boreal woodlands that extends north to areas where low temperature and available moisture limit the growth of vegetation.

low-density lipoprotein (LĪ-pō-prō´tēn) **(LDL)** Protein that helps to transport cholesterol to the cells.

lungs Principal organs of the respiratory system in which oxygen and carbon dioxide are exchanged between the internal and external environments.

luteinizing (LOO-tē-uh-nī´zing) **hormone (LH)** Hormone that stimulates ovulation and progesterone secretion by the corpus luteum in females and testosterone secretion in males.

lymph Fluid in lymph vessels, composed of water and substances collected from various body tissues.

lymphatic system System composed of networks of vessels that run near veins and lymph tissue.

lymph nodes Compact masses of connective tissue organized from lymph tissue; filter and remove foreign particles and certain B and T cells and macrophages that have important functions in the immune system.

lymphocytes (LIM-fuh-sīts) Types of leucocytes; two general classes, T cells and B cells, function in specific immune responses.

lymphoid organs Organs that contain dense populations of lymphocytes; important in immune responses.

lymphoid tissues Tissues containing lymphocytes; underlie the gastrointestinal tract, airways leading to the lung, and the urinary and reproductive tracts.

lymph tissue Connective tissue organized into nodes containing cells that remove foreign particles and participate in immune responses.

lysozyme (LĪ-suh-zīm) An enzyme that can break down the cell wall of many bacterial species.

macroevolution (mak´rō-ev-uh-LOO-shun) Evolutionary change in taxonomic groups higher than species.

macromolecule (mak´rō-MOL-uh-kyool) A large molecule of living matter, composed of subunits.

macronucleus (mak´rō-NYOO-klē-us) The larger of two nuclei types present in ciliate protozoans.

macrophages (MAK-ruh-fāj-iz) Phagocytic cells that engulf foreign particles, such as bacteria, and debris from dead cells and also play a major role in immunity in processing foreign antigens and activating T helper cells.

macroproducers Plants, mosses, and some algae in freshwater ecosystems and large macroalgae in marine ecosystems.

maintenance respiration The energy from carbohydrates that is used in maintaining a living cell.

major histocompatibility complex proteins A unique set of self molecules found on cells of an individual that are the basis for immune tolerance.

malignant cells Cells of malignant tumors.

malignant tumors Tumors composed of rapidly reproducing cells that invade surrounding tissues and spread throughout the body.

mammary glands Breasts; develop fully only in females to provide milk for nourishing offspring after birth.

marine ecosystems Ecosystems in open ocean systems, intertidal systems, deep ocean systems, and bays and estuaries.

marsupials (mar-SOO-pē-ulz) Animals such as opossums, koalas, and kangaroos in which fertilization and early development occur internally but later development takes place in an external pouch where the embryo feeds on milk provided through the nipple of a mammary gland.

mass extinction The sudden destruction of a vast number of taxa.

mast year A year in which trees such as oaks, hickories, and beeches produce an unusually large seed crop.

mechanical energy Energy expressed as motion.

mechanical isolation Isolating mechanism that occurs due to incompatible reproductive structures, such as incompatible genital parts in animals or features that prevent cross-pollination in plants.

mechanical philosophy The attempt to explain all phenomena in terms of matter following the laws of physics.

medulla (muh-DUL-uh) The inner portion of an organ such as the kidney.

meiosis (mī-Ō-sus) Process consisting of two consecutive cell divisions that result in the formation of four gametes in animals and spores in plants and fungi.

melanin (MEL-uh-nin) A brownish pigment that gives color to skin, hair, and eyes.

melanocytes (muh-LAN-uh-sīts) Epidermal cells of the skin that synthesize melanin.

memory The store of information learned and retained somewhere in the brain as a result of an individual's activities or experience.

memory cells T and B lymphocytes that are specific for (retain a memory of) the antigenic determinant that provoked their development.

Mendelian (men-DĒ-lē-un) **population** A population of interbreeding, sexually reproducing individuals.

menopause The time in a woman's life after which reproduction is no longer possible, marked by the cessation of menstrual cycles.

menstrual (MEN-stroo-ul) **cycle** A hormonally controlled cycle of changes in the endometrium that occurs about every 28 days.

menstruation (men´stroo-Ā-shun) A loss of blood, mucus, and cells from the uterus that usually lasts about five days and recurs monthly.

meristems (MER-i-stemz) Tissues in plant shoots and roots that produce unspecialized cells that differentiate and become new tissues such as phloem, xylem, and epidermis.

mesenchyme (MEZ-un-kīm) A gelatinous layer in which amoebocytes and spicules can be found in sponges.

mesoderm (MEZ-uh-durm) Middle layer of the primordial germ layers in animal embryos; gives rise to blood and blood vessels, muscles, and connective tissues.

mesotrophic (mez´uh-TRŌ-fik) **lakes** Lentic ecosystems in which the production ranges between 10 and 70 grams of carbon per square meter per year.

messenger RNA (mRNA) The type of RNA used to transmit information from the DNA of a gene to a ribosome, where the information is used to make a protein.

metabolic (met´uh-BOL-ik) **pathway** A complete sequence of chemical reactions in which a molecule becomes progressively modified.

metabolism (muh-TAB-uh-liz´um) All chemical reactions involved in the synthesis or breakdown of molecules within a living cell.

metabolites (muh-TAB-uh-līts) The chemical products or by-products of metabolism.

metallothionine (muh-tal´uh-THĪ-uh-nēn) A protein that binds to metals and prevents metal poisoning; expressed in all tissues, especially the liver.

metaphase (MET-uh-fāz) The third stage of mitosis, in which sister chromatids are aligned at the center of the cell.

metaphase I Stage of meiosis in which paired homologous chromosomes become aligned at the equatorial plane of the cell.

metaphase II Static phase of meiosis in which sister chromatid pairs are aligned at the equatorial plane of each daughter cell.

metastasis (muh-TAS-tuh-sus) Process whereby malignant cells can be transported through the circulatory system to distant sites in the body, where they create new malignant growths.

methanogenic (muh-than´uh-JEN-ik) **bacteria** Organisms that use simple organic molecules as food sources; most use CO_2 and hydrogen as an energy source and produce methane and water as by-products.

microbes (MĪ-krōbz) Single-celled microorganisms and viruses.

microevolution The changes within species' populations.

microfilaments Components of the cytoskeleton; thin fibers composed of globular protein subunits that function in moving organelles around the cell and in contraction movements in cells specialized for that activity.

microhabitat A small, specialized habitat; the immediate surrounding environment of an organism.

microinjection A technique of inserting cloned genes or DNA strands generated by rDNA techniques into fertilized eggs or cells of an embryo using finely engineered instruments.

micronucleus The smaller of two nucleus types in ciliated protozoans.

microproducers Diatoms, desmids, and other small species of algae in aquatic ecosystems.

microtubules Components of the cytoskeleton; made of globular, beadlike subunits organized into hollow, cylindrical tubes that form a skeletal fiber network.

microvilli (mī´krō-VIL-ī) Microscopic folds in the plasma membranes of small intestinal cells.

midbrain A small area between the forebrain and the hindbrain that receives and integrates several types of sensory information and relays it to specific regions of the forebrain for processing.

migration (1) The exchange of genes among populations; also called *gene flow*. (2) The cyclic annual movement of populations between different distant regions.

mineralization The process by which soil particles formed from parent rock break down and release important chemical elements.

mineralized Characterized by minerals such as calcium deposited in a fibrous network, forming a nearly solid substance, such as bone or shell.

mineralocorticoids (min´uh-ruh-lō-KOR-tu-koidz) Steroid hormones involved in maintaining water and mineral balances.

mitochondria (mī´tuh-KON-drē-uh) Membranous organelles found in all eukaryotic cells that supply energy to the cell through reactions of cellular respiration.

mitochondrial (mī´tuh-KON-drē-ul) **DNA (mtDNA)** DNA found in mitochondria and transmitted to offspring through an egg but not a sperm.

mitosis (mī-TŌ-sus) Process in nuclear division whereby each daughter cell receives the same number and kind of chromosomes as the parent cell.

mixed prairies Grasslands composed of both shortgrass and tallgrass vegetation.

model A structure developed to describe or predict a cause-and-effect relationship in nature; more generally, a structure that describes change in a system through time.

modern synthesis The modern theory of evolution stating that gradual evolutionary change can be explained by the action of natural selection on small genetic changes and that the processes of species formation and evolutionary changes in higher taxonomic groups are explainable in terms that are consistent with known genetic mechanisms.

modification A stepwise change in chemical structure ($M \rightarrow N \rightarrow O \rightarrow P$).

molecular biology The field of study based on physics, biochemistry, microbiology, and genetics.

molecular genetics The study of molecules involved in heredity.

molecule A group of certain kinds of atoms held together in a specific arrangement; for example, a water molecule has one atom of oxygen held between two atoms of hydrogen.

monocots (MON-uh-kots) Angiosperms that develop from germinated seeds having a single cotyledon and grow into plants with parallel leaf venation, scattered vascular bundles, and floral parts that occur in threes or multiples of three.

monoculture Growth of a single-species crop.

monocyte (MON-uh-sīt) Agranulocyte that develops into a macrophage.

monoecious (muh-NĒ-shus) Having both male and female reproductive structures.

monohybrid cross (1) In Mendelian genetics, a cross that involves two forms of a single trait. (2) A cross involving different alleles at a single gene locus.

monoploid (MON-uh-ploid) Having the haploid number of chromosomes in any cell.

monosomic (mon´uh-SŌ-mik) Lacking one chromosome from a homologous pair.

montane (mon´TĀN) Cool, moist, mountainous habitat where evergreen trees grow.

montane forests Forests in the mountains of western North America.

mortality Death rate of a population expressed as a percentage or fraction.

morula (MOR-yuh-luh) A solid ball of cells that forms a few days after fertilization.

motor neuron Neuron that transmits impulses away from the brain or spinal cord toward other cells, muscles, glands, or other neurons involved in making a response.

mouth The opening to the exterior that receives food and begins the process of digestion using both mechanical and chemical processes.

mRNA See *messenger RNA.*

mtDNA See *mitochondrial DNA.*

mucous (MYOO-kus) **membranes** Membranes composed of an outer epithelial cell layer and an inner connective tissue layer that line body cavities and the surface of organs or tissues that open to the external environment.

mucus (MYOO-kus) A viscous fluid secreted by mucous glands that covers the surfaces of mucous membranes.

multiple alleles Several alleles that can occur at a single gene locus.

multiple cropping An agricultural approach in which several different crops are grown within a given area.

multiple sclerosis (skluh-RŌ-sus) Disease characterized by progressive deterioration of myelin sheaths covering axons in neurons of the brain and spinal cord.

multiregional model The hypothesis that *Homo sapiens* developed from populations of *H. erectus* in several areas of the globe and that subsequent human evolution has been shaped by gene flow, natural selection, and genetic drift.

muscle tissue Specialized tissue that produces motion, composed of elongated, thin cells commonly referred to as *fibers*.

mutant An organism possessing a trait resulting from a mutation.

mutation (myoo-TĀ-shun) A permanent structural change in a DNA molecule that may result in a new form of an expressed trait.

mutualism (MYOO-choo-ul-iz´um) A relationship in which two organisms of different species live together in an intimate association that benefits both.

mycelia (mī-SĒ-lē-uh) Mats of tissue formed by fungal hyphae.

myelin (MĪ-uh-lin) Lipid-rich, whitish material that covers and insulates axons of most large neurons; produced by neuroglia cells.

myelin (MĪ-uh-lin) **sheath** Overlapping membranes covering an axon.

myofibrils (mī´ō-FĪ-brilz) Threadlike structures in muscle cells made of specialized proteins that have the ability to contract.

nasal cavities Hollow chambers inside the nose where air is warmed, humidified, and filtered.

natality (nā-TAL-uh-tē) Birth rate; the number of live births per female in a deme or population per unit of time.

natural eutrophic (yoo-TRŌ-fik) **lakes** Lentic ecosystems in which productivity ranges between 70 and 400 grams of carbon per square meter per year; lakes where excessive organic production, resulting from nutrients of natural origin, is decomposed by bacteria, depleting oxygen concentrations to levels that cause fish mortality.

natural history The general study of nature, stressing description, classification, and interrelationships.

natural killer (NK) cells Lymphocytes that play an active role in destroying virus-infected cells, fungi and other parasites, and cancer cells.

natural selection (1) In modern evolution theory, the nonrandom, differential reproductive success of genotypes. (2) More generally, the nonrandom, differential survival or reproduction of individuals or sets of individuals that possess advantageous characteristics for survival in specific environments.

natural system A classification system that reflects an existing order in nature.

navigation The action of orienting toward a goal.

negative feedback Process in which feedback from a controlled system is monitored by a receptor that relays information to a processing center, which causes the level of the variable to change in the direction opposite that of the original change if it deviates from a set point value; governs most actions of the endocrine and nervous systems.

nephrons (NEF-ronz) Functional units of the kidney that regulate the composition of blood through the production of urine.

neritic (nuh-RIT-ik) **waters** Portions of oceans that include the littoral zone in near-shore waters adjacent to coasts, bays, estuaries, and waters extending over part of the continental shelf.

nerve A bundle of parallel neuron fibers and their associated blood vessels and supporting cells, all of which are enclosed by protective connective tissue.

nerve growth factor A chemical messenger responsible for neuron maturation during embryonic development.

nerve tracts Collections of nerves in a bundle; found in the brain and spinal cord.

nervous tissue Tissue composed of neurons specialized for initiating and conducting electrochemical impulses.

net photosynthesis The difference between total gross photosynthesis and cellular respiration.

net plant production The gross production of an organism, trophic guild, or community less that consumed during cellular respiration.

net productivity The amount of organic material produced by plant photosynthesis less that used during plant respiration.

neural (NYOO-rul) **tube defect** An abnormal development of the neural tube, the structure that gives rise to the nervous system in vertebrate embryos.

neurofibrils (nyoo´rō-FĪ-brilz) Thread-like proteins that form delicate supporting networks within a neuron.

neuroglia (nyoo-RŌ-glē-uh) **cells** One of two major cell types of the nervous system; most are specialized to function as connective tissue cells.

neurohormone (nyoo´rō-HOR-mōn) A hormone produced by a neuron.

neuron (NYOO-ron) One of two major cell types of the nervous system, consisting of a cell body, dendrites, and an axon and specialized for the rapid conduction of electrochemical impulses.

neurosecretory (nyoo´rō-si-KRĒ-tuh-rē) **cells** Specialized neurons in the hypothalamus that produce hormones.

neurotransmitters (nyoo´rō-tranz-MIT-urz) Chemical messengers that allow nerve impulse transmission across a chemical synapse.

neurovascular (nyoo´rō-VAS-kyuh-lur) **hypothesis** Hypothesis stating that neurosecretory cells in the hypothalamus produce substances that regulate hormone secretions of the anterior pituitary gland.

neutrophils (NYOO-truh-filz) Granulocytes whose granules contain enzymes used to digest microorganisms.

niche (nich) The ecological role a species plays within a community.

nicotine The highly addictive active ingredient of tobacco; mimics acetylcholine.

nitrification (nī´truh-fuh-KĀ-shun) The conversion of ammonia (NH^+_4) into nitrite (NO^-_2) and nitrate (NO^-_3).

nitrogen fixation The process of converting atmospheric nitrogen (N_2) into ammonia (NH^+_4).

nitrogen oxides Molecules, such as nitrites and nitrates, composed of nitrogen and oxygen.

NK cells See *natural killer cells.*

nodes See *lymph nodes.*

nodes of Ranvier (RON-vē-ā´) Spaces along a myelinated neuron that represent gaps between individual Schwann cells.

nondisjunction The failure of paired homologous chromosomes to separate during anaphase of meiosis I or meiosis II.

nonspecific defense mechanisms Defense mechanisms that always operate in the same way when presented with a challenge; include mechanical barriers, secreted products, and the inflammation response.

nonstructural genes Genes that code for tRNA and rRNA.

nonvascular bryophytes Plants that have no vascular tissues and require that substances diffuse from the environment into their tissues; mosses are the most common type.

nonvascular plants Algae, fungi, mosses, and lichens that lack true phloem and xylem tissues.

notochord (NŌT-uh-kord) An internal, cartilaginous rod present in all chordates during some developmental stage; it is replaced by the developing vertebral column in higher vertebrates.

nuclear (NOO-klē-ur) **envelope** The double membrane surrounding the nucleus in eukaryotic cells.

nuclear (NOO-klē-ur) **pores** Openings in the surface of a nuclear envelope.

nucleic (noo-KLĒ-ik) **acids** Large molecules such as DNA and RNA that are composed of nucleotides.

nuclein (NOO-klē-in) Nineteenth-century term for what turned out to be DNA.

nucleoid (NOO-klē-oid) The region in a prokaryotic cell, mitochondrion, or chloroplast where DNA is concentrated.

nucleolus (noo-KLĒ-uh-lus) A prominent structure in the nucleus of most eukaryotic cells that is active in the synthesis of ribosomes.

nucleoplasm (NOO-klē-uh-plaz´um) Substance in the interior of a nucleus.

nucleotide (NOO-klē-uh-tīd) A subunit of DNA or RNA that consists of a base (adenine, guanine, thymine, or cytosine in DNA; uracil instead of thymine in RNA), a phosphate molecule, and a sugar molecule (deoxyribose in DNA, ribose in RNA); thousands of nucleotides are linked to form a DNA or RNA molecule.

nucleus (NOO-klē-us) A membrane-enclosed structure that contains genetic material and regulates many cell activities in eukaryotic cells.

null hypothesis A statement formulated before performing an experiment that the experimental condition is due to chance alone; to be tested and accepted or rejected.

numerical mutations Mutations that increase or decrease the number of whole chromosomes without changing the structure of individual chromosomes.

nutrient deserts Areas in oceans where there are no mechanisms to cycle nutrients from the bottom back into the euphotic zone.

occult deposition Direct acid deposition from fog or cloud droplets onto surfaces.

oceanic waters All areas of ocean zones, excluding neritic waters.

oligotrophic (OL-i-gō-trō´fik) **lakes** Lakes in which productivity is low, ranging between 0.1 and 10 grams of carbon per square meter per year.

omnivore (OM-ni-vor) A consumer of both plants and animals; an organism that feeds in both carnivore and herbivore guilds.

oncogene (ON-kuh-jēn) A gene that can cause cancer.

one-gene, one-enzyme hypothesis Hypothesis that originally stated that each biochemical reaction was catalyzed by a single enzyme and that each enzyme was specified by one gene; as more became known about genes and proteins and their structure, the hypothesis was extended to become the *one-gene, one-polypeptide hypothesis.*

one-gene, one-polypeptide hypothesis See *one-gene, one-enzyme hypothesis.*

oocyte An immature ova (egg).

oogenesis (ō´uh-JEN-uh-sus) Process by which ovaries produce mature ova.

oogonia (ō´uh-GŌ-nē-uh) Cells in the ovaries of a female embryo that ultimately give rise to oocytes.

open-grown trees Trees growing far from neighboring trees that might compete for light and other needed resources.

open woodlands Areas where short trees fail to form a closed canopy.

operant conditioning A form of associative learning whereby an animal is conditioned to repeat a behavior by reinforcement.

operator A DNA segment capable of interacting with a repressor in controlling the function of an adjacent region.

operculum (ō-PUR-kyuh-lum) A structure that covers and protects the gill slits of bony fishes.

operon (OP-uh-ron) A set of adjacent structural genes whose mRNA is synthesized in one piece, plus the adjacent regulating regions that affect transcription of the structural genes.

opportunistic infection Infection usually caused by bacteria or fungi that normally exist in or on the body without causing harm; occurs when a host suffers from a wound or a reduced level of immune resistance.

order A taxon that includes similar and related families; the major subdivision of a class.

organ A distinct structure that is made up of more than one type of tissue, has a definite form, and performs a specific function in the body.

organelles (or´guh-NELZ) Microscopic membranous structures in a eukaryotic cell that are responsible for carrying out specific functions; include mitochondria, ribosomes, endoplasmic reticula, and Golgi complexes in animal cells.

organic chemicals Carbon-containing molecules essential for life, including proteins, sugars, lipids, DNA, and RNA.

organogenesis (or´guh-nō-JEN-uh-sus) The formation of primary organs during embryonic development.

organ system A group of organs integrated into a unit for performing a major function or group of functions.

orgasm A series of brief, widespread muscular contractions and pleasurable sensations that occur in sexual intercourse; accompanied by ejaculation in males.

osmosis (oz-MŌ-sus) The passage of a solvent through a cell membrane.

outcrossing Process whereby offspring are produced from gametes contributed by two different parents.

ova Mature egg cells.

ovarian cycle A monthly cycle in human females in which an ovum matures.

ovary Primary sex organ of females that produces ova and hormones, including estrogens, progesterone, and relaxin.

overexpression The production of excess quantities of a protein or an abnormal protein.

ovulation (ov´yuh-LĀ-shun) The release of a single mature egg from one of the ovaries.

oxidation (ok´suh-DĀ-shun) Process in which an atom or molecule loses one or more electrons.

oxytocin (ok´si-TŌ-sin) Hormone that stimulates muscles of the uterus during childbirth and triggers secretion of milk from the mammary glands during nursing.

ozone (Ō-zōn) Gas composed of three oxygen atoms (O_3) that forms a layer in the outer atmosphere that absorbs ultraviolet radiation emitted by the sun and protects life from the harmful effects of this highly reactive form of energy.

paleoclimatic (pā´lē-ō-klī-MAT-ik) **indicators** Records that indicate ancient climates.

paleontologists (pā´lē-on-TOL-uh-jists) Scientists who study the fossil record.

paleontology (pā´lē-on-TOL-uh-jē) The study of fossils and ancient life.

pancreas (PAN-krē-us) An organ that carries out functions in the digestive system by secreting pancreatic juice and in the endocrine system by secreting the hormones insulin, glucagon, and somatostatin.

pandemic (pan-DEM-ik) A disease that occurs worldwide.

parallelism The development of similar characteristics in separate lineages that have a common ancestor.

parasites (PAR-uh-sīts) Organisms such as protozoans and worms that obtain substances from a host.

parasitism (PAR-uh-suh-tiz´um) A relationship in which two organisms of different species live together in an intimate association in which only one of the organisms benefits while the other is harmed.

parasympathetic nervous system Part of the nervous system that maintains the body under relatively calm conditions.

parent cell A cell that divides, giving rise to two daughter cells.

Parkinson's disease Disease of the nervous system, characterized by late age of onset and severe symptoms such as drooling, tremors, and nonfunctioning muscles.

parturition Process in which a human fetus is expelled from the uterus about 265 days after fertilization.

pathogens (PATH-uh-jinz) Organisms or biological agents that cause disease.

pedigrees (PED-uh-grēz) Simple genetic diagrams used in tracing the inheritance of genetic traits and disorders in families.

peduncle (PĒ-dun-kul) A stalk that attaches the flower to the plant.

pelvic girdle The portion of the skeleton that protects the reproductive organs.

penis (PĒ-nis) The male copulatory organ that conveys sperm into the female vagina.

pepsin (PEP-sin) An enzyme in gastric juice that initiates the breakdown of food proteins.

peptide bond The type of chemical linkage that joins two amino acids.

peptide growth-stimulating factors Substances that promote the growth or development of certain cells and tissues.

peptide hormones Mostly small proteins composed of linear strings of 3 to 200 amino acids that affect target cells through the operation of a second messenger system.

perennial (puh-REN-ē-ul) **plants** Plants that may live more than two years, grow continuously, and produce seeds once a year on several occasions; examples are cattails, rhododendrons, and redwood trees.

perfect flowers Flowers that have both stamens and pistils.

peripheral nervous system (PNS) Part of the nervous system that innervates all parts of the body and transmits information to and from the central nervous system.

periphyton (puh-RIF-uh-ton) Plants or algae adhering to rock substrates in streams and rivers.

peristalsis (per´uh-STOL-sus) A series of contraction waves that ripple along the smooth muscles surrounding the tubular organs of the digestive system.

permafrost The subsurface layer of tundra ground that is permanently frozen to a depth of 400 to 600 meters.

perspiration Substance produced by sweat glands that is secreted onto the skin's surface; its evaporation helps maintain body temperature.

pesticides (PES-tuh-sīdz) Substances that kill unwanted or harmful organisms.

petals Parts of a flower that are usually colored to attract insects, hummingbirds, or bats needed for pollination.

petiole (PET-ē-ōl) The stalk that connects the leaf to a plant.

phage (fāj) See *bacteriophage.*

phagocytosis (fag´uh-suh-TŌ-sus) Process in which cells engulf and destroy foreign particles.

pharyngeal (fuh-RIN-jē-ul) **gill slits** Openings that allow water to flow through the mouth, over the gills, and out the gill slits of fishes.

pharynx (FA-rinks) The throat; a thick-walled, muscular tube.

phenetic (fi-NET-ik) **system** A classification system based on computer tabulations and analyses of the number and degree of similarities that exist among organisms.

phenotype (FĒ-nuh-tīp) The physical trait or traits that occur as a result of a specific genotype.

phenylketonuria (fen´ul-kē-tuh-NYOO-rē-uh) **(PKU)** A disorder distinguished by abnormally high levels of an amino acid, phenylalanine, in the blood of newborn babies; can result in severe mental retardation.

pheromones (FER-uh-mōnz) Chemical signals that convey information between animals.

phlogiston (flō-JIS-tun) **theory** An erroneous early theory stating that when a substance was burned, something called *phlogiston,* the "material and principle of fire," was expelled.

photons (FŌ-tonz) Different increments of light or radiant energy emitted by the sun, each having a specific wavelength and amount of energy.

photoperiod The length of the day between sunrise and sunset.

photosynthates (fō´tō-SIN-thātz) All sugars or other carbohydrates made during photosynthesis.

photosynthesis Process in which photosynthetic cells in plants, algae, and certain cyanobacteria use radiant energy from the sun to convert carbon dioxide and water into energy-rich sugars that are the source of energy for most organisms on Earth, releasing oxygen as a by-product.

photosynthetic bacteria Autotrophic prokaryotes that use light as an energy source for converting simple molecules into complex organic molecules.

phyletic (fī-LET-ik) **evolution** The slow, gradual change of a population, or set of populations, into a new species.

phylogeny (fī-LOJ-uh-nē) The evolutionary development of a species or of a larger taxonomic group of organisms.

phylum (FĪ-lum) A taxon that includes similar and related orders; the major subdivision of a kingdom.

physical factors Gravity and light and heat energy; all environmental factors that are not biotic.

physical map A map showing the actual location of genes and the distances between them.

physiognomy (fiz´ē-OG-nuh-mē) The appearance or characteristic features of vegetation present in a community.

physiology (fiz´ē-OL-uh-jē) The study of functions of an organism.

pigments Protein molecules capable of absorbing photons and transferring energy by rearranging their own molecular structures to create an energized state.

pili (PĪ-lī) Bacterial structures used in attachment to host cells and in bacterial reproduction.

pioneer species Organisms that successfully colonize new ground or areas that have sustained recent perturbations.

pistil (PIS-tul) The plant reproductive organ that produces female gametes.

PKU See *phenylketonuria.*

placenta (pluh-SEN-tuh) Structure through which materials are exchanged between the mother and the embryo or fetus; secretes estrogens and progesterone throughout pregnancy.

placental (pluh-SEN-tul) **mammals** Animals that complete their development in the female uterus.

plant ecology The study of the structure and functions of plant populations and their environments.

plant growth regulators Substances that are produced in one tissue, transported to another part of the plant, and have a physiological effect at a different time.

plant physiological ecology The study of the relationships between individual plants and their environment.

plant physiology The study of processes, such as photosynthesis and respiration, taking place in a single cell or in an individual plant.

plant respiratory biomass The amount of stored energy required by plants to conduct their normal activities.

plaques (plaks) Fibrous, fatty deposits in blood vessels.

plasma (PLAZ-muh) The extracellular fluid that surrounds blood cells, containing and transporting dissolved gases, proteins, hormones, nutrients, lipids, certain minerals such as iron, and metabolic waste products; blood without the formed elements.

plasma cell A B lymphocyte that secretes antibodies.

plasma membrane The membrane surrounding a cell; responsible for regulating substances entering and leaving the cell.

plasmids (PLAZ-midz) Small, circular, double-stranded DNA molecules that reside inside prokaryotic cells such as bacteria.

platelets (PLĀT-lits) Very small, nonnucleated cell fragments that have a critical function in blood clotting.

plate tectonics The field of science that studies the continuous, slow movement of plates on which Earth's landmasses and ocean basins rest.

PNS See *peripheral nervous system.*

poikilothermic (poi´kuh-lō-THUR-mik) Having no internal mechanisms for controlling body temperature.

polygenic inheritance A trait determined by many genes at different loci; the numbers, chromosomal locations, and degree of expression of all the different genes are usually not known.

polymerase (POL-uh-mur-ās) **chain reaction** A method in which short, unique pieces of DNA can be amplified.

polymorphism (pol´i-MOR-fiz-um) The existence within a population of two or more genotypes for a given trait that results from balancing selection.

polypeptide A chain of amino acids that may be smaller than a complete protein.

polyploid (POL-i-ploid) An organism that has three or more complete sets of chromosomes.

population A group of individuals of the same species that live in the same place.

population-community school An informal group of early ecologists who considered populations to represent fundamental levels of organization. They identified factors such as competition and predation and studied their influence on populations.

population genetics The study of the genetic constitution of populations and how it changes.

positive feedback A process in which the level of a variable changes in the same direction as the initial change.

postsynaptic cell Part of a cell that receives a nerve impulse from another cell; transmits impulses away from the cell.

postsynaptic receptors Proteins on the membranes of a postsynaptic cell; of many types, each with a unique structure enabling it to bind with one type of neurotransmitter.

postzygotic isolating mechanisms Mechanisms that reduce the fitness of hybrids; they include hybrid inviability, sterility, and F_2 breakdown.

potential energy Stored or inactive energy that is capable of doing work or creating change at a later time.

prebionts (PRĒ-bī´onts) Hypothesized prebiological systems composed of organized organic molecules that gave rise to forms capable of conducting processes associated with life.

prebiotic phase The time before organisms were present on Earth.

predictive relationship A correlation in which a data set can be used to predict the value for one variable, given a value for the other.

preformation An erroneous explanation of reproduction that held that successive generations were preformed and encapsulated in previous generations like nested boxes.

pregnancy The period of development between fertilization and birth.

prenatal diagnosis A process used to determine the genetic status of a developing fetus.

presynaptic membrane The membrane of a presynaptic neuron at the site of a synapse.

presynaptic neuron The part of a neuron that transmits a nerve impulse toward the synapse.

prevalence (PREV-uh-luns) The percentage of a population affected by a disorder.

prezygotic isolating mechanisms Mechanisms that prevent or reduce hybridization between members of different species, including habitat, seasonal, ethological, mechanical, and gametic isolation.

primary follicle A primary oocyte and surrounding follicle cells.

primary growth Growth from apical meristems that results in increased length of roots, stems, and branches.

primary immune response The first specific immune response to an antigen not previously encountered.

primary oocytes (Ō-uh-sīts) Immature ova.

primary producers Organisms that carry out photosynthesis.

primary sex organs Organs that produce gametes and secrete hormones involved in development and functions related to reproduction.

primary spermatocytes (spur-MAT-uh-sīts) Large developing cells that give rise to mature sperm.

primary succession The progression of different species that come to inhabit new ground where life forms have not previously existed.

primates (PRĪ-māts) Order of placental mammals that includes tree shrews, lorises, lemurs, monkeys, apes, and humans.

primordial (prī-MOR-dē-ul) **germ layers** Three embryonic cell layers—ectoderm, mesoderm, and endoderm—that give rise to all tissues and organs during embryonic development.

primordial (prī-MOR-dē-ul) **origin** Source of prebiotic substances created when Earth was young, some of which are still being released from deep hydrothermal zones where oceanic plates are separating.

probes Small single-stranded segments of radioactive DNA or RNA that are used to identify bacterial colonies with cloned genes.

procedural memory Memory acquired by repetition or continuous practice; habit.

producer See *autotroph*.

profundal (prō-FUN-dul) **zone** The deep part of lakes where light does not penetrate.

progenote (PRŌ-juh-nōt) The hypothesized precellular stage of chemical organization that gave rise to both prokaryotic and eukaryotic cells.

progesterone (prō-JES-tuh-rōn) Sex hormone produced by the ovaries that helps prepare the endometrium for implantation of the embryo.

progressive evolution An evolutionary sequence displaying a constant direction, or *goal,* and coming into existence independently of selective forces.

prokaryotic (prō´kar-ē-OT-ik) **cell** A cell that lacks a membrane-bound nucleus and has no membranous organelles; a bacterium.

prolactin (prō-LAK-tin) Pituitary hormone secreted during pregnancy that stimulates mammary growth and development and after birth induces lactation; its role in males, if any, is not known.

promoter A short DNA nucleotide sequence where RNA polymerase binds and initiates transcription.

prophase (PRŌ-fāz) The second stage of mitosis, in which chromosomes become shorter and thicker.

prophase I Stage of meiosis in which chromatin coalesces and sister chromatids become visible.

prophase II Stage of meiosis in which newly formed nuclear envelopes of the two daughter cells disappear as chromosomes re-form.

prostaglandins (pros´tuh-GLAN-dinz) Substances thought to enhance the movement of sperm when deposited in the female reproductive tract.

prostate (PROS-tāt) **gland** A gland that empties into the urethra in males and secretes a substance that contributes to sperm mobility.

proteins (PRŌ-tēnz) A diverse group of organic molecules composed of chains of amino acid molecules.

proto-oncogene (prō´tō-ON-kuh-jēn) A gene that has a specific function but when damaged or malfunctioning has the potential to become an oncogene.

protoplast fusion Process in which two protoplasts from different plant species may be stimulated to fuse and form a somatic-cell hybrid.

protoplasts Plant cells with their outer cell wall removed by digestion with certain enzymes.

protostomes Animal phyla in which the blastopore becomes the mouth.

provirus The state in which viral DNA is integrated into a host chromosome.

pseudocoelomates (soo´dō-SĒ-luh-mātz) Animal phyla that develop a body cavity between the mesoderm and the endoderm that is often called a *false coelom.*

pseudogenes (SOO-dō-jēnz) Genes that are no longer functional; that is, they are not transcribed; derived from an ancestral cell.

pseudopodia (soo´dō-PŌ-dē-uh) Specialized, temporary membrane extensions formed by single-celled amoeba-like cells and used in locomotion or engulfing food.

puberty (PYOO-bur-tē) The time of life when gonads mature and reproduction becomes possible.

pulmonary circulation The flow of deoxygenated blood between the right ventricle and the lungs and the return of oxygenated blood to the left atrium.

pulse The wave of pressure radiated outward from the heart through the arteries.

punctuated equilibrium Evolutionary sequences in which macroevolution occurs in short bursts of change, followed by long stable periods.

Punnett (PUN-it) **square** A diagram for illustrating all possible combinations of gametes in crosses.

pure-breeding lines Lines that when self-pollinated or cross-pollinated with a member from the same line produce only offspring that are identical to the parents.

purine (PYOO-rēn) A base; adenine and guanine in DNA and RNA.

pyramids of biomass Representations of the amount of living mass in each trophic level of an ecosystem, usually expressed as grams of dry weight per area or volume.

pyramids of energy Representations of the amount of energy in each trophic level of an ecosystem, usually presented as kilocalories per unit area or volume.

pyramids of numbers Representations of the density of organisms in each trophic level, expressed as numbers per unit area or volume.

pyrimidine (pī-RIM-uh-dēn) A base; cytosine and thymine in DNA, cytosine and uracil in RNA.

pyrogens (PĪ-ruh-jinz) Substances that cause an elevation of body temperature.

pyruvate (pī-ROO-vāt) A molecule containing only three carbon atoms that is formed by the splitting of a glucose molecule during glycolysis.

quantitative methods Methods scientists use to measure, count, and calculate results.

quantum speciation Speciation events that occur very rapidly.

races Different subpopulations of a single species.

radiant energy Energy emitted from the sun in the form of light waves or photons.

rain shadow The warm, dry area on the leeward side of a mountain range that results from air losing its moisture as it rises up and over the windward side and then becoming warmer as it descends.

random genetic drift Chance fluctuations in the gene frequencies of a small population.

rDNA See *recombinant DNA.*

receptacle The part of a flower attached to the petals and sepals.

receptors Different classes of complex protein molecules that recognize hormones, neurotransmitters, or antigens; found on the surface of the plasma membrane or inside the cell.

recessive trait (1) In Mendelian genetics, a trait that disappears in the first generation in a monohybrid cross. (2) The allele in a heterozygous genotype that is not expressed in the phenotype.

recognition Process in which an antibody receptor on the surface of a B cell identifies and binds to an antigen.

recombinant (rē-KOM-buh-nunt) An individual with traits determined by genes on different homologous chromosomes.

recombinant DNA (rDNA) A DNA molecule containing two or more segments of DNA from two different genes or species.

rectum The lower portion of the large intestine.

reduced nitrogen Molecules such as ammonia, ammonium, and various forms of organic nitrogen that are composed of nitrogen and hydrogen.

reduction Process in which an atom or a molecule gains one or more electrons.

reflex (RĒ-fleks) An unconscious, rapid response of the nervous system to a specific stimulus.

regulatory gene Gene encoding a repressor protein that controls transcription of structural genes.

regulatory sequences Specific DNA sequences that control the production of a protein.

relative fitness A comparative measure of the fitness of different genotypes in a population, calculated by dividing the net reproductive rate of a genotype by the reproductive rate of the genotype with the greatest fitness.

relaxin (ri-LAK-sin) Female hormone that acts to increase the flexibility of joints between bones around the birth canal, facilitating birth.

releaser In classical ethology, any aspect of an external stimulus that triggers an innate releasing mechanism.

releasing hormones (releasing factors) Substances that travel from the hypothalamus to the pituitary gland through the hypophyseal portal system and stimulate release of hormones from the anterior pituitary.

renal capsule A thick layer of connective tissue that encloses a kidney.

repetitive DNA Regions of DNA that contain repeated sequences of varying lengths.

replication (rep´luh-KĀ-shun) DNA synthesis.

repressor A protein molecule that binds to an operator and prevents transcription of an operon.

reproduction The origination of new organisms from preexisting ones; also, the formation of cells used in growth, repair, and replacement.

reproductive isolation Genetic barriers to gene flow among populations.

reproductive phase Part of the plant life cycle in which reproductive tissues begin to form, leading to the formation of seeds.

reproductive technologies Strategies involving controlled breeding or the direct manipulation of sex cells; widely used to improve livestock herds.

reservoirs (1) Places where large amounts of a chemical element or compound are found. (2) Sites in the body from which viruses may spread to uninfected cells.

resin (REZ-in) A viscous, sticky substance secreted by plants, especially conifers.

resistance The ability of survivors of pesticide applications to give rise to new populations that can tolerate pesticide applications in ever-increasing concentrations; also used in reference to bacteria and antibiotics.

resolution The final stage of inflammation, in which an injured tissue returns to its normal state.

respiration The exchange of oxygen and carbon dioxide between an organism and its environment.

respiratory (RES-pur-uh-tor-ē) **center** A special group of nerve cells that responds to levels of carbon dioxide in the blood and regulates the rate of respiration.

respiratory (RES-pur-uh-tor-ē) **transmission** Transmission of a disease agent by sneezing, coughing, talking, or kissing.

respired Broken down in the process of cellular respiration.

restriction enzymes Enzymes that cut DNA molecules at specific nucleotide sequences.

restriction fragment length polymorphisms (RFLPs) Inherited variations in the size of DNA fragments produced when a defined piece of DNA is cut with a specific restriction enzyme.

reticular (ri-TIK-yuh-lur) **activating system** A network of branched neurons in the brain; when stimulated, the individual becomes awake and alert.

reticular (ri-TIK-yuh-lur) **fibers** Thin, short filaments constructed of the protein *reticulin*.

reticular (ri-TIK-yuh-lur) **formation** A loosely organized network of nerves that receives input from several other areas of the brain and spinal cord and maintains the state of consciousness.

retroviruses (ret´ruh-VĪ-rus-iz) A class of viruses that can direct the synthesis of DNA from RNA.

reverse transcriptase (tran-SKRIP-tās) An enzyme that synthesizes DNA from RNA.

RFLPs See *restriction fragment length polymorphisms.*

rhizoids (RĪ-zoidz) Rootlike structures of bryophytes and ferns that are composed of elongated single cells or multicellular filaments that anchor the plant and absorb water and mineral nutrients.

rhizomes (RĪ-zōmz) Underground plant stems from which new plants can sprout.

rib cage The portion of the skeleton that encases the lungs, heart, stomach, and liver.

ribosomal (rī´buh-SŌ-mul) **RNA (rRNA)** RNA molecules used in ribosomal structure and function.

ribosomes (RĪ-buh-sōmz) Structures composed of rRNA and proteins that appear as small particles in the cytoplasm or attached to rough endoplasmic reticulum; sites where mRNA is translated into an amino acid sequence.

riparian (ri-PER-ē-in) **zones** Areas along stream or riverbanks that are influenced by flowing water.

risk factor Anything that increases the probability of developing a disease or suffering harm.

ritualization The modification of behavior patterns to serve a new function, usually involving communication.

RNA Ribonucleic acid; single-chain molecules formed by nucleotide links; different RNAs include messenger RNA (mRNA), ribosomal RNA (rRNA), and transfer RNA (tRNA); retroviruses use RNA as their genetic material.

RNA polymerase (POL-uh-mur-ās) An enzyme that attaches to a specific nucleotide sequence on the DNA molecule, causes it to separate partly into two strands, and then moves through a gene, making an RNA molecule complementary to the gene DNA.

rods Photoreceptors in the eye containing rhodopsin, a pigment that functions in dim light.

root-shoot ratio The ratio of root dry weight to shoot dry weight.

rough ER The part of the endoplasmic reticulum that is studded with ribosomes, where proteins are synthesized.

rRNA See *ribosomal RNA.*

rule of parsimony (PAR-suh-mō´nē) Refers to the idea that in living systems, the simplest "answer" evolved in response to a biological problem.

saliva (suh-LĪ-vuh) A watery secretion from the salivary glands that consists mainly of mucus, which protects the lining of the mouth and lubricates the food for easier movement, plus salts and enzymes.

salivary amylase (SAL-uh-ver´ē AM-uh-lās) A specific enzyme in saliva that begins the breakdown of starch.

salivary (SAL-uh-ver´ē) **glands** Three pair of glands near the mouth that secrete saliva.

sample size An adequately large number of subjects necessary to ensure that an experimental result is not due to chance.

saprophytic (sap´ruh-FIT-ik) Obtaining nutrients by absorbing of breakdown products from dead plants and animals.

sarcoma (sar-KŌ-muh) A cancer involving connective tissue cells.

savannas (suh-VAN-uz) Modified tropical grasslands characterized by alternating wet and dry seasons.

scale of analysis The relative size of different biotic hierarchies in the biosphere; for example, in moving from biome to ecosystem to community to population to deme to individual, the total number of variables that may influence an observed pattern becomes greater.

scanning electron microscope (SEM) A microscope that makes it possible to see the three-dimensional appearance of a cell at high magnifications.

Schwann (shwon) **cells** Neuroglia cells that produce myelin.

scientific serendipity (ser´un-DIP-i-tē) The making of an important scientific discovery by chance or accident.

scrotum (SKRŌ-tum) Pouch of skin containing the testes, located outside the abdominal cavity.

seasonal isolation Isolating mechanism that occurs when species that live in the same region have populations that are prevented from interbreeding because they are sexually mature at different time periods or seasons.

sebaceous (si-BĀ-shus) **glands** Glands in the dermis that secrete an oily substance (sebum) that prevents hair and skin from drying and is also toxic to many types of bacteria.

secondary follicles An intermediate developmental stage leading to the formation of a mature ovum.

secondary growth An increase in stem and branch diameters due to new cells produced by the vascular and cork cambiums.

secondary immune response Immune response to a previously encountered antigen that occurs quickly because of the presence of memory cells specific for the antigen.

secondary oocytes (Ō-uh-sīts) The larger of the two types of haploid cells formed during the first meiotic division of a secondary follicle.

secondary sex characteristics Changes in the body that are related to sexual maturation, including changes in distribution of body hair and body fat, voice pitch, and muscle development.

secondary spermatocytes (spur-MAT-uh-sīts) Cells produced during the first meiotic division of a primary spermatocyte that give rise to spermatids.

secondary succession The progression of a community, after disruption, through a number of predictable types (seres) that eventually culminates in a new climax community.

second law of thermodynamics Law stating that every energy transformation results in a reduction in the total usable energy of the system.

second-order consumers Large crustacean zooplankton, insect nymphs, and other small arthropods in guilds that feed on herbivores in lentic ecosystems.

second-order streams Streams that originate when two first-order streams meet.

sedimentary (sed´uh-MEN-tuh-rē) **cycles** Biogeochemical cycles in which rocks, soil, or sediments act as primary reservoirs.

seeds The reproductive structures formed following fertilization of a cone or a flower that contain a protective coat, stored nutrients, and an embryo.

selfish behavior Behavior that benefits the actor at a cost to the recipient.

self-maintenance The set of activities used in meeting the continuous demand for repairing or replacing ribosomes, membranes, enzymes, and other cellular structures and molecules that are lost during normal functioning.

self-pollination Fertilization of an egg by pollen from the same flower.

self-pruning The shedding of unproductive branches by trees.

SEM See *scanning electron microscope.*

semen (SĒ-min) Fluid ejaculated by a male during sexual intercourse that consists of sperm and secretions from various glands.

semiconservative Descriptive of DNA replication in which each new double-stranded molecule is composed of one parental strand and one newly formed daughter strand.

seminal vesicles (SEM-uh-nul VES-i-kulz) A pair of structures that secrete a component of semen into ejaculatory ducts.

seminiferous tubules (sem´uh-NIF-uh-rus TOO-byoolz) Highly convoluted structures in the testes that harbor male gametes in various stages of development.

sense organs The eyes, ears, nose, skin, and tongue.

sensory neuron A neuron that receives information from the external and internal environments and conducts it to the brain or spinal cord, where it is processed.

sepals (SĒ-pulz) Flower parts that surround a bud.

septa Connecting cross walls in fungi.

sequence hypothesis The hypothesis that the sequence of bases in DNA and RNA molecules specifies the sequence of amino acids in proteins.

sex chromosomes Chromosomes that determine the sex of an individual.

sex hormones Steroid hormones produced by the gonads that are essential for reproductive development and function.

sexual dimorphism (dī-MOR-fiz-um) Differences in the external appearance of males and females of the same species.

sexual imprinting The breeding preferences of many birds as determined by early imprinting experiences.

sexual reproduction The fusion of haploid gametes from two parents to create offspring.

sexual selection Selection for characteristics that have value for success in mating.

sexual transmission The transfer of disease agents through sexual intercourse.

shift The insertion of portions of a chromosome in a different region of the same chromosome.

shortgrass prairies Plains grasslands in which drought-resistant shortgrass species constitute the climax community; occur over vast areas of the Great Plains of North America.

shredders Lotic insect guilds that consume coarse particulate organic matter and obtain most of their nutrients from digesting bacteria and fungi that colonize leaves and other organic debris entering streams.

sibling species Morphologically similar species that cannot be distinguished from one another by external characteristics but have isolating mechanisms that keep them from interbreeding.

sickle-cell anemia A disorder in which red blood cells become sickle-shaped, resulting in a loss of function; caused by an autosomal recessive gene.

sickle-cell disorder A disorder that occurs in individuals who are homozygous recessive for a particular gene, characterized by severe anemia.

sickle-cell trait A disorder that occurs in individuals who are heterozygous for a particular gene, characterized by mild anemia.

single-origin model The hypothesis that a population of the modern form of *Homo sapiens* arose in Africa at an early date and spread rapidly across Europe and Asia.

sister chromatids (KRŌ-muh-tidz) Replicated homologs joined by a centromere that exist in mitosis and meiosis.

skeletal muscle A tissue specialized for contraction that consists of extremely long fibers, masses of which are attached to bones by special connective tissues; causes movements of the skeleton.

skin A complex organ forming the external surface layer of the body.

small intestine The major organ of digestion, consisting of a long tube from which nutrients are absorbed into the blood and lymph.

smooth ER Part of endoplasmic reticulum where lipids are synthesized; has no ribosomes on its surface.

smooth muscle A tissue specialized for contraction that is made up of cells with single nuclei and no cross-striations, present in the walls of hollow internal organs and tubes.

social behavior Behavior involving two or more animals.

social organization The pattern of relationships among individuals within a population at a particular time.

societies Groups consisting of members of a species that are attracted to one another, communicate with one another, exhibit cooperative behavior, and engage in synchronized activities.

sociobiology The study of the biological basis of social behavior.

softwood Lumber from conifers.

solar system Our sun and the planets and other celestial bodies that circle it.

somatic (sō-MAT-ik) **cell hybrid** A cell formed by fusing protoplasts of two different species.

somatic (sō-MAT-ik) **cell therapy** A type of gene therapy in which genes in somatic cells are modified.

somatic (sō-MAT-ik) **nervous system** Part of the nervous system composed of motor nerves that run directly from the central nervous system to skeletal muscles.

somatostatin (sō-mat´uh-STAT-in) Hormone secreted by the pancreas, involved in glucose homeostasis; inhibits secretion of glucagon and insulin.

spacer DNA DNA that connects remote islands of structural genes.

spatial patterns The distributions of specific plant populations in space.

speciation (spē´she-Ā-shun) The processes of species formation.

species (SPĒ-shēz) Groups of actual or potentially interbreeding natural populations that are reproductively isolated from other such groups; the major subdivision of a genus.

species diversity The number of different species present in an area.

species selection Evolutionary change where species are the units of selection.

specific heat The amount of heat required to raise the temperature of any substance one unit; the standard is the amount of heat required to raise the temperature of 1 gram of water 1°C.

specific immune responses A complicated adaptive defense system that functions in response to nonself antigens, involving various cells, cell products, and control mechanisms.

specificity (spes´uh-FIS-uh-tē) The ability of an antibody to recognize and bind to only one type of antigen.

sperm Haploid gametes produced by males; spermatozoa.

spermatids (SPUR-muh-tidz) Haploid cells arising from secondary spermatocytes; in humans, each contains 22 autosomes and one sex chromosome.

spermatogenesis (spur-mat´uh-JEN-uh-sus) Production of spermatozoa in the testes.

spermatogonia (spur-mat´uh-GŌ-nē-uh) Diploid germinal cells of the seminiferous tubules that divide by mitosis to produce cells that ultimately develop into mature sperm.

spermatozoon (spur-mat´uh-ZŌ-on) A mature sperm cell.

spicules (SPIK-yoolz) Skeletal elements of sponges, radiolarians, and sea cucumbers composed of calcium carbonate or silicates produced by amoebocytes.

spina bifida (SPĪ-nuh BĪ-fuh-duh) A developmental anomaly characterized by a defect in the bony encasement of the spinal cord.

spinal nerves Thirty-one pairs of nerves that emerge from the spinal cord and innervate most organs below the neck.

spinal reflex arc An inborn neural pathway by which impulses from sensory neurons reach motor neurons without first traveling to the brain.

spindle fibers Specialized microtubules that move chromosomes during eukaryotic cell division.

spleen An organ that filters circulating blood and also contains populations of B cells, T cells, and macrophages.

spongy bone The network of inner bone surrounding the marrow.

spore (1) Haploid structure that can germinate directly into the gametophyte stage of a plant. (2) Inert form of a bacterium or a fungus enclosed in a thick protective wall that is resistant to most environmental stresses.

sporophyte (SPOR-uh-fīt) The spore-producing generation in plant life cycles; a diploid plant stage that produces haploid spores through meiosis.

stabilizing selection Selection against extreme variants in a population, with the result that a standard phenotype is favored.

stamens (STĀ-menz) Male reproductive parts that produce pollen in flowers.

standing crop The number or amount of organisms present per unit of area or volume at any one time.

statistically significant Shown by mathematical analysis to be associated with a low probability of error.

statistical methods The mathematical analyses and interpretations of data (numerical facts) that enable scientists to make an objective appraisal about the reliability of conclusions based on the data.

steroid (STĒ-roid) **hormones** Hormones synthesized from cholesterol that usually cause target cells to manufacture proteins through a mechanism involving internal receptors.

stolons (STŌ-lunz) Stems that extend to the soil, from which lateral buds can develop into roots and shoots.

stomach A relatively large, muscular organ of the digestive system in the upper abdomen that can stretch to accommodate about 2 liters of food or liquid.

stomata (STŌ-muh-tuh) Pores in the leaf surface that can open and close in response to metabolic control mechanisms.

stomatal (STŌ-muh-tul) **conductance** The capacity for stomata to allow diffusion of gases between the leaf and air; the degree to which stomata are open.

strains Varieties of a specific virus or bacterium that differ only slightly in their protein structure or genetic content.

stratum corneum (STRĀ-tum KOR-nē-um) A layer of dead skin cells that forms the outer protective layer of the skin and is continuously sloughed.

stroke Sudden, severe trauma causing temporary or permanent loss of consciousness or sensation, resulting from an interference in normal blood circulation.

structural genes Genes that code for proteins.

structural mutations Mutations involving a rearrangement of genes or sets of genes along specific chromosomes.

subduction The downward movement of oceanic plates when continental plates ride up and over them.

subspecies (SUB-spē´shēz) A subpopulation of a single species that is sufficiently distinct to be given a separate taxonomic name.

subtropical forests Forests found in northern and southern portions of the Torrid Zone and at higher elevations on mountains within rain forests, resulting from a decrease in total available moisture or a seasonal distribution of rain that is less than that received by tropical forests.

succession See *ecological succession.*

supercooled Water that cools rapidly and fails to freeze at 0°C.

suppressor T cell T cell that inhibits production of certain immune cells and antibodies and helps control and limit humoral immune responses.

sweat A complex solution containing salts and small amounts of nitrogen-containing metabolic waste products.

sweat glands Glands in the skin that secrete sweat and antimicrobial substances such as lysozyme.

symbionts (SIM-bī-onts) Organisms of different species participating in a close living association (symbiosis); for example, algae and fungi in a lichen.

sympathetic nervous system Part of the nervous system that prepares for vigorous activity, for example, by increasing the heart rate.

sympatric speciation The process by which a species originates in the same geographical region as its parent species.

synapse (SIN-aps) A narrow gap between neurons or between a neuron and a muscle or gland cell.

synapsis (suh-NAP-sus) A close pairing of homologs during prophase I of meiosis.

synaptic (suh-NAP-tik) **cleft** The narrow space between two cells at a synapse.

synaptic (suh-NAP-tik) **terminals** The ends of an axon's branches.

synaptic vesicles (suh-NAP-tik VES-i-kulz) Small membrane-bound sacs filled with neurotransmitters that are found in presynaptic neurons.

syndrome (SIN-drōm) A set of symptoms or characteristics that occur together.

synthesis (SIN-thuh-sus) The combining of simple chemical building blocks into a larger molecule.

systemic (sis-TEM-ik) **circulation** The circulation of oxygenated blood from the left ventricle to all tissues of the body and the return of deoxygenated blood to the right atrium.

taiga (TĪ-guh) The northern coniferous forest biome located about 50° north latitude that ends at the lower limit of the permafrost.

tallgrass prairies Plains grasslands receiving ample rainfall, once covered with tall grasses but now used to grow corn, soybeans, and other crops.

taxa (TAK-suh) The various levels in the classification system.

taxis (TAK-sis) A motion toward or away from a source of stimulation.

taxonomy (tak-SON-uh-mē) The science of classifying organisms; also called *systematics.*

T cells See *T lymphocytes.*

TDF See *testis-determining factor.*

telophase (TEL-uh-fāz) The fifth and last stage of mitosis, in which daughter nuclei form, nuclear envelopes and nucleoli appear, and chromosomes disappear.

telophase I Stage of meiosis in which the separated homologous chromosomes cluster at each pole and cytokinesis takes place.

telophase II Stage of meiosis in which each pole receives one set of chromosomes.

TEM See *transmission electron microscope.*

temperate forest biome Forest biome that occurs in the Temperate Zones, where increasing latitude results in greater seasonal extremes, with lower average temperatures and less precipitation than in biomes of the Torrid Zone.

temperate forests Forests in latitudes of the westerlies, between 5° and 10° north and south of the 40th parallel.

temporal factors (1) Normal changes that occur throughout the life of an organism. (2) Gradual environmental changes in a site over long periods of time.

territory A home range that is defended with threats, attacks, or advertisement (via scent or song) for exclusive use by an individual or group.

test cross A cross between an individual with a dominant trait and one with a recessive phenotype to determine the genotype of the dominant individual.

testis The male gonad or reproductive organ that produces sperm and sex hormones.

testis-determining factor (TDF) Gene on the Y chromosome that initiates the development of a male in embryonic development. The gene has recently been identified as *SRY.*

testosterone (tuh-STOS-tuh-rōn) The male sex hormone that influences the growth and development of male sex organs, secondary sex characteristics, and sperm.

tests Hard coverings of foraminiferans composed of mineral and organic molecular complexes; the internal skeletons of echinoderms.

tetranucleotide (tet´ruh-NOO-klē-uh-tīd) **hypothesis** The erroneous belief that DNA was a static molecule, composed of equal amounts of adenine, guanine, thymine, and cytosine.

T4 lymphocytes (LIM-fuh-sīts) T cells that regulate a variety of immune functions; targets of HIV-1.

theory A systematic set of concepts that explain and relate data, answer questions, and guide research.

thermal energy Kinetic energy of molecular motion that can be measured with a thermometer; heat.

thermoacidophilic (thur´mō-uh-sid´uh-FIL-ik) **bacteria** Literally, "heat-and-acid-loving" prokaryotes; bacteria that thrive in habitats with high temperatures and high concentrations of acids.

thermodynamics The study of relationships between heat and other forms of energy.

third-order consumers Lentic guilds of minnows and young trout, bass, or pike that feed on second-order consumers and are in turn consumed by fourth-order consumers.

third-order streams Streams that arise where two second-order streams meet.

thoracic (thuh-RAS-ik) **cavity** The body cavity that contains the heart, lungs, and esophagus.

thrush A fungal infection in the mouth that is common in AIDS patients.

thymus (THĪ-mus) **gland** Gland involved in the production and maturation of T cells.

thyrotropin (thī´ruh-TRŌ-pin) Hormone that stimulates synthesis and release of hormones from the thyroid gland (also called *thyroid-stimulating hormone*).

thyroxine (thī-ROK-sēn) Hormone secreted by the thyroid gland that helps regulate metabolism and tissue growth and development.

timberline The upper limit of tree growth; varies from 4,200 meters in the southwestern forests to about 3,500 meters in Glacier National Park, in Montana, near the Canadian border.

Ti plasmid (TĪ-PLAZ-mid) A segment of DNA that disrupts normal plant cell function, resulting in the formation of a tumor; exists in a bacterium, *Agrobacterium tumefaciens.*

tissue fluid The fluid that bathes cells, composed of water, salts, and proteins.

tissues Large assemblages of similar cells that are specialized to carry out a particular function.

T lymphocytes (LIM-fuh-sīts) One of two general types of lymphocytes that circulate to the thymus gland, where they are processed before being released to the circulation and distributed to lymphoid organs and tissues; develop into T helper cells, cytotoxic T cells, T suppressor cells, or T memory cells.

T memory cell T cell that recognizes previously encountered antigens.

tolerance The capacity of lymphocytes not to respond to self molecules.

top carnivores Flesh-eating organisms in the final consumer guild of a food chain.

Torrid Zone The area of Earth between the Tropics of Cancer and Capricorn; divided by the equator.

trachea (TRĀ-kē-uh) The windpipe, major trunk of the respiratory tree.

transcription Synthesis of mRNA from a sequence of DNA; so called because information coded in the sequence of DNA nucleotides is copied (transcribed) into RNA nucleotides; the first step in gene expression.

transfer RNA (tRNA) RNA molecules with triplet nucleotide sequences that are complementary to the triplet coding sequences of mRNA; tRNAs bond to amino acids and transfer them to ribosomes.

transformation Alteration in the normal growth and developmental properties of a cell.

transforming principle An unknown substance that changed one form of *Diplococcus* bacteria into a different form; later determined to be DNA.

transgenic animal An animal that developed from embryos in whose chromosomes, genes, or DNA from a different species had been integrated.

transgenic plant A plant with cells containing genes from a different species.

translation The process of producing a protein whose linear amino acid sequence is derived from the mRNA codon specified by a gene.

translocation (1) The insertion of genes or portions of chromosomes in nonhomologous chromosomes. (2) The movement of substances from one part of a plant to another.

transmission electron microscope (TEM) A microscope that can magnify objects up to 300,000 times.

transmutation theory An erroneous early theory that one substance could be transformed into another.

transpiration The loss of water vapor through a membrane or pore of an organism; usually associated with water loss from plants during photosynthesis or cooling.

transposons (trans-PŌ-zonz) Short DNA sequences capable of migrating and becoming inserted into a different chromosomal site.

trichocysts (TRIK-uh-sists) Specialized structures used by ciliated protozoans for capturing food and for defense.

trisomy (TRĪ-sō-mē) The presence of a third copy of a homologous chromosome in cells.

tRNA See *transfer RNA*.

trophic dynamics The transfer of energy within ecosystems.

trophic guilds Groups of species that exploit the same class of trophic resources in a similar way.

trophic hormone A hormone that stimulates the release of other hormones.

trophic levels Successive steps of a food chain, each of which has less energy available than the previous level; the levels are often referred to as producers; primary, secondary, tertiary (and higher) levels of consumers; and decomposers.

trophoblast (TRŌ-fuh-blast) A single layer of cells surrounding a blastocyst.

tropical forests Forests that occur in the equatorial portion of the Torrid Zone, where more than 240 centimeters of annual rainfall combined with an average annual temperature greater than 17°C has resulted in the most productive forests on Earth.

TS genes See *tumor suppressor genes*.

tubers (TOO-burz) Large, fleshy underground stems, like that of the potato, that are highly specialized for carbohydrate storage and are also capable of vegetative reproduction.

tubule (TOO-byool) A small tube.

tumor (TOO-mur) A cell mass that develops when cells no longer respond to substances that normally regulate their growth, differentiation, or reproduction.

tumor suppressor (TS) genes Genes that normally suppress rates of cell growth and development; may play a role in cancer induction if they become inactivated or removed from a chromosome.

tundra (TUN-druh) A northern biome that begins at the southern limit of the permafrost and extends north to permanent ice fields.

turgor (TUR-gur) The capacity of plants to sustain their shape with water pressure.

turnover The process by which cold surface water sinks and is replaced by deep water from near the lake bottom; occurs when the surface water reaches its greatest density at 4°C.

ultrasound scanning A visualization procedure that employs high-frequency sound waves to provide a profile of the structural features of the fetus.

umbilical (um-BIL-i-kul) **cord** The structure containing fetal arteries and veins that connects the fetus with the placenta.

undifferentiated Not specialized to carry out a specific function.

undulipodia (un´joo-luh-PŌ-dē-uh) Specialized structures used in mobility by some members of the kingdom Protoctista; a cellular structure used in classifying some biota into specific kingdoms.

unifying hypothesis A synthesis of existing knowledge to create a single explanation that can be tested.

upwelling The vertical transport of nutrients from ocean depths to the surface.

urea (yoo-RĒ-uh) The main nitrogen-containing waste product excreted in urine.

ureter (YOO-ruh-tur) One of two tubes from the kidney to the urinary bladder.

urethra (yoo-RĒ-thruh) A tube from the urinary bladder to the exterior; transports urine in females and urine and semen in males.

urinary (YOO-ruh-ner-ē) **bladder** A hollow muscular organ that serves as a temporary storage reservoir for urine.

urinary (YOO-ruh-ner-ē) **system** System composed of organs that function in filtering blood and producing urine.

urine (YOO-rin) A fluid containing varying amounts of water and waste products that is excreted from the body.

urkaryote (oor´KAR-ē-ōt) The hypothesized first eukaryote to have evolved from the progenote.

uterine (YOO-tuh-rīn) **tube** One of two tubes that transport an ovum from the ovary to the uterus; also called the *Fallopian tube* or *oviduct.*

uterus (YOO-tuh-rus) A thick-walled, muscular organ in females where an embryo develops and grows.

vaccination (vak´suh-NĀ-shun) Process of exposing a potential host to a disease-agent antigen to stimulate immunity to the antigen.

vaccine (vak-SĒN) Substance produced from a dead or modified form of a disease-causing virus, bacterium, or protozoan that stimulates an immune response after vaccination.

vacuole (VAK-yuh-wōl) Intracellular cavity in plant cells that is surrounded by a single membrane and stores water, sugars, proteins, or waste products.

vagina (vuh-JĪ-nuh) The female organ that receives sperm from a male and also serves as the birth canal during childbirth.

vagus (VĀ-gus) **nerve** The cranial nerve that extends to and regulates the functions of the heart, lungs, and digestive organs.

valves Flaps of tissues in blood vessels that prevent blood from flowing in the opposite direction; in the heart, they regulate blood flow between chambers.

variable A characteristic that changes.

variable expressivity The appearance of a trait differently in people that apparently have the same genotype.

variable regions Amino acid sequences in antibody protein chains that are unique for each antibody.

vascular cambium (VAS-kyuh-lur KAM-bē-um) A cylinder of meristematic cells that surround the stem, branches, and roots of woody plants and give rise to phloem and xylem cells.

vascular (VAS-kyuh-lur) **plants** Plants with phloem and xylem to transport substances such as water and nutrients.

vas deferens (VAS DEF-ur-enz) Either of two male reproductive ducts surrounded by walls of smooth muscle that contract rhythmically during ejaculation to transport sperm forward.

vegetative growth phase Part of the plant life cycle in which roots, stems, and leaves develop, but not reproductive tissues.

vegetative plant tissues Plant tissues that are not involved in sexual reproduction.

vegetative reproduction Reproduction by nonsexual plant tissues.

vein (vān) A blood vessel that returns blood from tissues to the heart.

vena cava (VĒ-nuh KĀ-vuh) Either of two large veins through which blood returns to the right atrium of the heart from tissues within the body.

venules (VĒN-yoolz) Small blood vessels that transport blood from capillaries to veins.

vertebral (vur-TĒ-brul) **column** The backbone or spine, composed of separate small bones called *vertebrae* that impart great strength but also allow great flexibility.

vertical transmission The transfer of a disease agent from parent to offspring by infected sperm, ova, or milk or through the placenta.

vesicles (VES-i-kulz) Small membrane-bound sacs in which proteins are stored.

villi (VIL-ī) Small, fingerlike projections located in the folds of the inner surface of the ileum that contain blood vessels and a lymphatic vessel; sites where food products of digestion are absorbed.

virion (VĪ-rē-on) An infectious virus particle.

viruses (VĪ-rus-iz) Nonliving disease agents that can replicate only inside host cells.

visible light spectrum Formed by photons with wavelengths between 400 and 700 nanometers that are detectable as various colors by the human eye.

vulva (VUL-vuh) The external genital organs of females.

water use efficiency The amount of carbon gained per amount of water lost.

wet deposition Acid deposition on surfaces from precipitation.

woody perennials Plants that live for two or more years and are characterized by woody structures; include vines (such as grapes), shrubs (rhododendrons), and trees (pines and oaks).

X chromosome inactivation Process in which one of two X chromosomes in cells of females becomes inactivated during embryonic development.

X-ray crystallography (kris´tuh-LOG-ruh-fē) A technique used in determining molecular structure.

yolk sac A small cavity below the embryonic disk; nonfunctional in humans' development.

zeitgeber (TSĪT-gā-bur) An external stimulus that sets a biological clock.

zygote (ZĪ-gōt) A fertilized egg formed from the union of haploid female and male gametes (egg and sperm).

Credits and Acknowledgments

Photo Credits

Detailed Contents
P. vii, detail of Chapter 3 opener: Flagellate protozoan, *Trichomonas,* David M. Phillips/Visuals Unlimited. P. vii, detail of Chapter 5 opener: Extravehicular activity performed by astronaut above Earth's surface, NASA. P. viii, detail of Chapter 7 opener: Beaver Meadow Falls in autumn, St. Hubert's Adirondack Park, New York, Sharon Beris/Tom Stack & Associates. P. viii, detail of Chapter 12 opener: Flowering pea plants, Larry Lefever/Grant Heilman Photography. P. ix, detail of Chapter 13 opener: Lily Meiosis I, early prophase, John D. Cunningham/Visuals Unlimited. P. ix, detail of Chapter 17 opener: Similarities in looks of family members, Will & Deni McIntyre/Photo Researchers. P. x, detail of Chapter 20 opener: Aerial view of corn harvest, Grant Heilman from Grant Heilman Photography. P. x, detail of Chapter 22 opener: Galápagos Islands tortoise in cactus forest, David Cavagnaro. P. xi, detail of Chapter 25 opener: Katydid preserved in amber, John Cancalosi/Peter Arnold, Inc. P. xi, detail of Chapter 26 opener: Petroglyphs found in Rainbow Park, Utah, SUPERSTOCK. P. xii, detail of Chapter 30 opener: Milkweed pod releasing its seeds, Stephen P. Parker/Photo Researchers. P. xiii, detail of Chapter 35 opener: Giant multi-polar neurons, Stan Elems/Visuals Unlimited. P. xiii, detail of Chapter 36 opener: Diver in pike position, David Madison/Bruce Coleman Inc. P. xiii, detail of Chapter 37 opener: Beautiful babies, Michel Tcherevkoff/The Image Bank. P. xiv, detail of Chapter 39 opener: Electron micrograph of spirochaetes bacteria, CNRI/SPL/Photo Researchers.

UNIT I
Pp. xxii–1, Unit I opener: Satellite composite view of cloudless Earth, Tom VanSant/The GeoSphere Project, Santa Monica, CA.

Chapter 1
P. 2, Chapter 1 opener: White-tailed deer crossing stream of water, Thomas Ketchin/Tom Stack & Associates. P. 3, Figure 1.1: (A) Charles Heidecker/Visuals Unlimited; (B) Polaroid, R. Oldfield/Visuals Unlimited. P. 4, Figure 1.3: (A) John Cancalosi/Tom Stack & Associates; (B) Steve McCutcheon/Visuals Unlimited. P. 5, Figure 1.4: *The Snake Charmer* by Henri Rousseau/Musee du Louvre, Paris. P. 5, Figure 1.5: Art Resource, NY. P. 5, Figure 1.6: Courtesy of The Newberry Library, Chicago. P. 5, Figure 1.7: By permission of the Houghton Library, Harvard University from Otto Brunfel's *Living Portraits of Plants,* c. 1530, woodcut illustration adapted from drawing by Hans Weiditz. P. 6, Figure 1.8: (A) SIU/Visuals Unlimited; (B) Dan McCoy/Rainbow; (C) Science VU/Visuals Unlimited; (D) New York Hospital/Peter Arnold, Inc. P. 7, Figure 1.9: John D. Cunningham/Visuals Unlimited.

Chapter 2
P. 10, Chapter 2 opener: Sunset on Oregon Coast, Bandon State Park, Oregon, Willard Clay. P. 11, Figure 2.1: NASA. P. 16, Figure 2.6: NASA. P. 16, Figure 2.7: NASA. P. 18, Figure 2.10: (A) Paul Miller/Black Star; (B) Greg Vaughn/Tom Stack & Associates. P. 19, Figure 2.12: Smithsonian News Service. P. 28, Figure 2.17: (A) John Cancalosi/Tom Stack & Associates; (B) South Australia Museum.

Chapter 3
P. 31, Chapter 3 opener: Flagellate protozoan, *Trichomonas,* David M. Phillips/Visuals Unlimited. P. 32, Figure 3.1: Keith I. King. P. 33, Figure 3.2: (A) Michael S. Quinton/Visuals Unlimited; (B) Charles Seaborn/Odyssey Productions, Chicago. P. 33, Figure 3.3: Keith I. King. P. 36, Figure 3.6: Helmut Gritsch-

er/Peter Arnold, Inc. P. 37, Figure 3.7: (A) K. G. Murti/Visuals Unlimited; (B) John D. Cunningham/Visuals Unlimited. P. 38, Figure 3.8: Keith I. King. P. 39, Figure 3.9: Brian Parker/Tom Stack & Associates. P. 39, Figure 3.10: M. Abbey/Visuals Unlimited. P. 39, Figure 3.11: Arthur M. Siegelman/Visuals Unlimited. P. 39, Figure 3.12: M. Abbey/Visuals Unlimited. P. 40, Figure 3.13: Kevin Schafer/Tom Stack & Associates. P. 40, Figure 3.14: Veronika Burmeister/Visuals Unlimited. P. 40, Figure 3.15: T. E. Adams/Visuals Unlimited. P. 41, Figure 3.16: Jeff Foott/Tom Stack & Associates. P. 41, Figure 3.17: Jeff Foott/Tom Stack & Associates. P. 41, Figure 3.18: D. Gotshall/Visuals Unlimited. P. 41, Figure 3.19: Greg Vaughn/Tom Stack & Associates. P. 43, Figure 1: By permission of the Houghton Library, Harvard University, from Otto Brunfel's *Living Portraits of Plants,* c. 1530, woodcut illustration adapted from drawing by Hans Weiditz. P. 44, Figure 3.20: Keith I. King. P. 45, Figure 3.21: John Shaw/Tom Stack & Associates. P. 45, Figure 3.22: John D. Cunningham/Visuals Unlimited. P. 45, Figure 3.23: Kerry T. Givens/Tom Stack & Associates. P. 45, Figure 3.24: (A) Kerry T. Givens/Tom Stack & Associates; (B) George Knaphus/Visuals Unlimited. P. 46, Figure 3.25: William H. Amos.

Chapter 4

P. 47, Chapter 4 opener: Epiphytic orchids in blossom, Allan Roberts. P. 49, Figure 4.2: (A) Rod Planck/Tom Stack & Associates; (B) John Logan/Visuals Unlimited. P. 49, Figure 4.3: (A) John D. Cunningham/Visuals Unlimited; (B) Forest W. Buchanan/Visuals Unlimited; (C) William H. Amos. P. 50, Figure 4.4: (A) John D. Cunningham/Visuals Unlimited; (B) G. Ziesler/Peter Arnold, Inc.; (C) Brooking Tatum/Visuals Unlimited. P. 50, Figure 4.5: Milton Rand/Tom Stack & Associates. P. 51, Figure 4.7: (A) Jacques Jangous/Peter Arnold, Inc.; (B) W. H. Hodge/Peter Arnold, Inc.; (C) Dora Lambrecht/Visuals Unlimited. P. 51, Figure 4.8: Keith I. King. P. 51, Figure 4.9: (A) Milton Rand/Tom Stack & Associates; (B) Thomas Kitchin/Tom Stack & Associates. P. 52, Figure 4.10: Keith I. King. P. 52, Figure 4.11: Keith I. King. P. 52, Figure 4.12: (A) Daniel W. Getshall/Visuals Unlimited; (B) David M. Dennis/Tom Stack & Associates. P. 54, Figure 4.14: Fred Bavendam/Peter Arnold, Inc. P. 54, Figure 4.15: (A) William H. Amos; (B) Dave B. Fleetham/Tom Stack & Associates. P. 54, Figure 4.16 (A, B): William H. Amos. P. 55, Figure 4.17: (A) Robert DeGoursey/Visuals Unlimited; (B) Brian Parker/Tom Stack & Associates. P. 55, Figure 4.18: (A) Kjell B. Sandved/Visuals Unlimited; (B) Thomas Kitchin/Tom Stack & Associates; (C) Allan Roberts. P. 56, Figure 4.19: (A) Don and Esther Phillips/Tom Stack & Associates; (B) C. P. Hickman/Visuals Unlimited; (C) John Gerlach/Tom Stack & Associates; (D) William H. Amos; (E) Adrian Wenner/Visuals Unlimited; (F) John Gerlach/Tom Stack & Associates. P. 57, Figure 4.20: (A) Jeff Rotman/Peter Arnold, Inc.; (B) Daniel W. Gotshall/Visuals Unlimited. P. 57, Figure 4.21: (A) Denise Tackett/Tom Stack & Associates; (B) William C. Jorgensen/Visuals Unlimited. P. 58, Figure 4.22: (A) Gary Bell/The Wildlife Collection; (B) Gary Milburn/Tom Stack & Associates. P. 59, Figure 4.23: (A) David M. Dennis/Tom Stack & Associates; (B) David M. Dennis/Tom Stack & Associates; (C) Glenn Oliver/Visuals Unlimited. P. 59, Figure 4.24: Joe McDonald/Tom Stack & Associates. P. 59, Figure 4.25 (B): Allan Roberts. P. 59, Figure 4.26: (A, B) Allan Roberts. P. 60, Figure 4.27: (A) Dave Watts/Tom Stack & Associates; (B) William H. Amos; (C) John D. Cunningham/Visuals Unlimited. P. 60, Figure 4.28: (A) Allan Morgan/Peter Arnold, Inc.; (B) John Cancalosi/Peter Arnold, Inc.; (C) Kjell B. Sandved/Visuals Unlimited; (D) Thomas Kitchin/Tom Stack & Associates.

Chapter 5

P. 65, Chapter 5 opener: Extravehicular activity performed by astronaut above Earth's surface, NASA. P. 69, Figure 5.5: (A) John Gerlach/Visuals Unlimited; (B, C) Keith I. King. P. 70, Figure 1: Keith I. King. P. 75, Figure 5.9: (A) Link/Visuals Unlimited; (B) Greg Vaughn/Tom Stack & Associates. P. 77, Figure 2: George

Hebren/Visuals Unlimited. P. 78, Figure 5.10: (A) Keith I. King. P. 80, Figure 5.12: Keith I. King.

Chapter 6

P. 84, Chapter 6 opener: Arctic pack ice, Baffin Bay, Atlantic Ocean, E. R. Degginger. P. 85, Figure 6.1: William H. Amos. P. 87, Figure 6.2: (A, B, C, D) Bill Beatty/Visuals Unlimited. P. 88, Figure 6.3: (A) Greg Vaughn/Tom Stack & Associates; (B) Glenn Oliver/Visuals Unlimited. P. 89, Figure 6.4: Steve Kaufman/Peter Arnold, Inc. P. 89, Figure 6.5: Doug Sokell/Visuals Unlimited. P. 89, Figure 6.6: (A) Larry Brock/Tom Stack & Associates; (B) Joe McDonald/Visuals Unlimited. P. 90, Figure 6.7: Joe McDonald/Visuals Unlimited. P. 91, Figure 6.8: (A) Keith I. King; (B) John S. Flannery/Visuals Unlimited. P. 91, Figure 6.9: Steve McCutcheon/Visuals Unlimited. P. 91, Figure 6.10: Tom J. Ulrich/Visuals Unlimited. P. 92, Figure 6.11: Steve McCutcheon/Visuals Unlimited. P. 94, Figure 6.13: (A) Steve McCutcheon/Visuals Unlimited; (B) Tom J. Ulrich/Visuals Unlimited. P. 94, Figure 6.14: Keith I. King. P. 95, Figure 6.15: (B) Keith I. King. P. 96, Figure 6.16: Keith I. King. P. 96, Figure 6.17: (A) William E. Ferguson Photography; (B) E. R. Degginger. P. 97, Figure 6.18: Bruce Davidson/ANIMALS ANIMALS. P. 97, Figure 6.19: Nigel Smith/Earth Scenes.

Chapter 7

P. 99, Chapter 7 opener: Beaver Meadow Falls in autumn, St. Hubert's Adirondack Park, New York, Sharon Beris/Tom Stack & Associates. P. 100, Figure 7.1: (B) Thomas Kitchin/Tom Stack & Associates. P. 104, Figure 1: E. C. Williams/Visuals Unlimited. P. 110, Figure 7.7: Glenn Oliver/Visuals Unlimited. P. 112, Figure 7.10: Steve McCutcheon/Visuals Unlimited. P. 114, Figure 7.12: David J. Books/Visuals Unlimited. P. 115, Figure 7.14: (B) C. P. Vance/Visuals Unlimited. P. 117, Figure 7.15: John D. Cunningham/Visuals Unlimited.

Chapter 8

P. 119, Chapter 8 opener: Canyon De Chelly, Montana, Don and Pat Valenti. P. 121, Figure 8.2: (A) Jeff Foott/Tom Stack & Associates. P. 124, Figure 8.6: Breck P. Kent/Earth Scenes. P. 125, Figure 8.7: (B) Barbara Gerlach/Visuals Unlimited; (C) Ron Spomer/Visuals Unlimited. P. 129, Figure 8.9: (A) John D. Cunningham/Visuals Unlimited; (B) Keith I. King. P. 132, Figure 1: Jeff Henry/Peter Arnold, Inc. P. 133, Figure 3: Richard Thom/Visuals Unlimited. P. 135, Figure 8.12: (A, B) Keith I. King. P. 135, Figure 8.13: John Gerlach/Visuals Unlimited. P. 136, Figure 8.14: (A, B) E. R. Degginger.

Chapter 9

P. 138, Chapter 9 opener: Coral reef, Cindy Baroutte/PDS/Tom Stack & Associates. P. 139, Figure 9.1: (A) Ron Spomer/Visuals Unlimited; (B) Milton Rand/Tom Stack & Associates. P. 144, Figure 9.5: Keith I. King. P. 144, Figure 9.6: NASA. P. 146, Figure 9.8: (B) J. R. Williams/Earth Scenes; (C) A. Gurmankin/Visuals Unlimited; (D) Doug Sokell/Tom Stack & Associates. P. 146, Figure 9.9: Kirtley-Perkins/Visuals Unlimited. P. 148, Figure 9.12: Noah H. Poritz/Visuals Unlimited. P. 149, Figure 9.13: (A, B) Keith I. King. P. 152, Figure 9.16: (A) Denise Tackett/Tom Stack & Associates. P. 154, Figure 9.18: Gerry Ellis/The Wildlife Collection. P. 154, Figure 9.19: Thomas Kitchin/Tom Stack & Associates. P. 154, Figure 9.20: E. R. Degginger. P. 155, Figure 9.22: Brian Parker/Tom Stack & Associates. P. 156, Figure 1: La Verne D. Kulm. P. 157, Figure 2: La Verne D. Kulm. P. 158, Figure 3: La Verne D. Kulm. P. 159, Figure 9.23: (A) Tom J. Ulrich/Visuals Unlimited; (B) David M. Doody/Tom Stack & Associates.

Chapter 10

P. 161: Chapter 10 opener: Bicycle riders in Beijing, China, Peter Turnley/Black Star. P. 162, Figure 10.1: Joe McDonald/Tom Stack & Associates. P. 170, Figure

10.7: Science VU/Visuals Unlimited. P. 171, Figure 10.9: (A) Stephen W. Kross/Visuals Unlimited; (B) Ted Levin/ANIMALS ANIMALS. P. 174, Figure 1: (A) D. Wilder/Tom Stack & Associates; (B) Bruce Iverson/Visuals Unlimited. P. 175, Figure 3: Keith I. King.

Chapter 11

P. 178, Chapter 11 opener: Ripples in sand dunes, Death Valley National Monument, California, Willard Clay. P. 182, Figure 11.3: Thomas Kitchin/Tom Stack & Associates. P. 183, Figure 11.4: (B) Keith I. King; (C) Jack Swenson/Tom Stack & Associates; (D) Andrew Holbrooke/Black Star. P. 188, Figure 11.7: Thomas Kitchin/Tom Stack & Associates. P. 192, Figure 11.11: (A) William Campbell/TIME Magazine; (B) Miriam Austerman/Earth Scenes. P. 193, Figure 11.12: (A) Goddard Space Flight Center/NASA. P. 194, Figure 11.13: NASA. P. 195, Figure 11.14: Goddard Institute for Space Studies/NASA. P. 197, Figure 1: (A) Official U.S. Navy photograph; (B) Geoffrey Wheeler/NOAA.

UNIT II

Pp. 200-201: Unit II opener: Computer-generated image of natural DNA, Lawrence Berkeley Laboratory/Courtesy University of California.

Chapter 12

P. 202: Chapter 12 opener: Flowering pea plants, Larry Lefever/Grant Heilman Photography. P. 204, Figure 12.1: (A, B, C) From *Micrographia,* by Robert Hooke, first edition published by The Royal Society, 1665. P. 205, Figure 12.2: Scott, Foresman. P. 206, Figure 12.4: The Bettmann Archive. P. 206, Figure 12.5: Derek Fell. P. 206, Figure 12.6: (A, B) Mendelianum of the Moravian Museum, Brno, Czechoslovakia. P. 215, Figure 1: From *Memoir on Heat,* 1783, Messers Lavoisier & De La Place, Royal Academy of Sciences.

Chapter 13

P. 217, Chapter 13 opener: Lily Meiosis I, early prophase, John D. Cunningham/Visuals Unlimited. P. 218, Figure 13.1: From *Micrographia,* by Robert Hooke, first edition published by The Royal Society, 1665. P. 220, Figure 13.3: Dwight R. Kuhn. P. 221, Figure 13.4: (A) David M. Phillips/Visuals Unlimited. P. 221, Figure 13.5: Martha J. Powell/Visuals Unlimited. P. 222, Figure 13.6: (A) Don Fawcett/Photo Researchers. P. 223, Figure 13.7: (A) Don W. Fawcett/Visuals Unlimited. P. 223, Figure 13.8: (A) Science VU/R. Bolender/Don W. Fawcett/Visuals Unlimited. P. 224, Figure 13.9: (A) David M. Phillips/Visuals Unlimited. P. 225, Figure 13.11: Don W. Fawcett/Visuals Unlimited. P. 225, Figure 13.12: (A) Jeremy Burgess/Science Source Library/Photo Researchers. P. 226, Figure 13.13: (A) M. Schliwa/Visuals Unlimited. P. 228, Figure 13.15: (A) David M. Phillips/Visuals Unlimited; (B) E. White/Visuals Unlimited. P. 229, Figure 13.17: (A–F) William H. Amos.

Chapter 14

P. 239, Chapter 14 opener: Chromosomes of fruit fly, *Drosophila,* Ed Reschke/Peter Arnold, Inc. P. 240, Figure 14.1: John D. Cunningham/Visuals Unlimited. P. 242, Figure 14.2: AP/Wide World. P. 246, Figure 1: Runk/Schoenberger/Grant Heilman Photography. P. 254, Figure 14.6: (A) Barry L. Runk/Grant Heilman Photography.

Chapter 15

P. 261, Chapter 15 opener: DNA filaments from chromosomes, J. R. Paulsen/U.K. Laemmli/D. W. Fawcett/Visuals Unlimited. P. 262, Figure 15.1: (A) K. G. Murti/Visuals Unlimited; (B) David Phillips/Visuals Unlimited. P. 266, Figure 15.5: Ed Reschke/Peter Arnold, Inc. P. 270, Figure 15.8: (A) Les Simon/Stammers/Science Source/Photo Researchers. P. 273, Figure 15.10: A. D. Barrington Brown photo, from *The Double Helix,* by J. D. Watson, 1968, Atheneum, New York, Cold Spring Harbor Laboratory Archives, p. 215, New York. P. 274, Figure 15.11: A. D. Barrington Brown photo, from *The Double Helix,* by J. D. Watson, 1968, Atheneum, New York, Cold Spring Harbor Laboratories Archives, p. 215, New York. P. 274, Figure 15.12: Anne Sayre. P. 276, Figure 1: David M. Dennis/Tom Stack & Associates. P. 280, Figure 15.16: Leonard Lessin/Peter Arnold, Inc.

Chapter 16

P. 283, Chapter 16 opener: RNA (orange) and DNA (blue) ribbons of a newt oocyte during lampbrush chromosome stage, Zheng'an Wu, Christine Murphy, Joseph Gall, Carnegie Institution of Washington. P. 283, Figure 16.1: John Watney/Science Source/Photo Researchers. P. 284, Figure 16.2: (A) James W. Richardson/Visuals Unlimited.

Chapter 17

P. 304, Chapter 17 opener: Similarities in looks of family members, Will & Deni McIntyre/Photo Researchers. P. 306, Figure 17.1: L. Lisco/D. W. Fawcett/Visuals Unlimited. P. 307, Figure 17.2: Courtesy, Children's Memorial Hospital, Chicago, IL. P. 311, Figure 17.5: AP/Wide World. P. 312, Figure 17.6: (A) Custom Medical Stock Photo; (B) Bill Beatty/Visuals Unlimited; (C) Carolina Biological Supply Company; (D) Carolina Biological Supply Company. P. 315, Figure 17.8: (A–C) Courtesy FBI. P. 316, Figure 17.9: George Wilder/Visuals Unlimited. P. 318, Figure 1: Runk/Schoenberger/Grant Heilman Photography. P. 324, Figure 2: From *The Eugenic Marriage,* by Grant Hague, M.D., 1914.

Chapter 18

P. 327, Chapter 18 opener: Normal and sickle cell blood cells, David M. Phillips/Visuals Unlimited. P. 334, Figure 18.5: (A, B) Stanley Fiegler/Visuals Unlimited. P. 336, Figure 18.6: Steve Uzzell/Hereditary Disease Foundation, Santa Monica, Ca. P. 339, Figure 18.9: (A, B) National Heart, Lung and Blood Institute, National Institutes of Health. P. 341, Figure 18.10: Howard Sochurek. P. 345, Figure 2: William McCoy/Rainbow.

Chapter 19

P. 348, Chapter 19 opener: Plasmids, self-replicating circles of DNA, K. G. Murti/Visuals Unlimited. P. 355, Figure 19.3: (A) K. G. Murti/Visuals Unlimited. P. 356, Figure 19.4: (A) Tom Broker/Rainbow. P. 357, Figure 19.5: (A) David M. Phillips/Visuals Unlimited; (B) Elmer Koneman/Visuals Unlimited. P. 359, Figure 19.8: Dan McCoy/Rainbow.

Chapter 20

P. 266, Chapter 20 opener: Aerial view of corn harvest, Grant Heilman from Grant Heilman Photography. P. 367, Figure 20.1: John Colwell/Grant Heilman Photography. P. 368, Figure 20.2: (A, B) C. P. Vance/Visuals Unlimited. P. 371, Figure 20.4: (B) Sinclair Stammers/SPL/Photo Researchers; (C) Plantek/Photo Researchers. P. 373, Figure 20.5: (A) Dan McCoy/Rainbow. P. 375, Figure 20.6: Grant Heilman Photography. P. 376, Figure 20.7: (A) GRANADA. P. 378, Figure 20.8: Dr. R. L. Brinster, Laboratory of Reproductive Physiology, University of Pennsylvania. P. 382, Figure 2: Science VU/Jackson Laboratory/Visuals Unlimited. P. 383, Figure 20.9: (B) Dr. J. B. Gurdon/Wellcome Institute for the History of Medicine/CRC Institute, University of Cambridge.

Chapter 21

P. 386, Chapter 21 opener: Scientist compares autoradiographs of DNA bands, S.I.U./Visuals Unlimited. P. 388, Figure 21.1: Bill Pierce/Rainbow. P. 388, Figure

21.2: Hank Morgan/Rainbow. P. 393, Figure 21.5: Matt Meadows/Peter Arnold, Inc. P. 397, Figure 21.9: Margot Bennet/Cold Spring Harbor Laboratory.

UNIT III

Pp. 406-407, Unit III opener: The Galápagos Islands, Galen Rowell/Peter Arnold, Inc.

Chapter 22

P. 408, Chapter 22 opener: Galápagos Islands tortoise in cactus forest, David Cavagnaro. P. 409, Figures 22.1 and 22.2: Courtesy of Mr. G. P. Darwin, by permission of the Darwin Museum, Down House, Royal College of Surgeons of England. P. 410, Figure 22.4: From *Historia Fiscia y Politica de Chile.* P. 412, Figure 22.6: William E. Ferguson Photography. P. 412, Figure 22.7: Kjell V. Sandved/Visuals Unlimited. P. 413, Figure 22.8: Breck P. Kent/ANIMALS ANIMALS. P. 418, Figure 1: The Burndy Library. P. 420, Figure 3: By permission of the Darwin Museum, Down House, The Royal College of Surgeons of England. P. 421, Figure 22.12: The Bettmann Archive. P. 423, Figure 1: UPI/Bettmann. P. 424, Figure 2: UPI/Bettmann.

Chapter 23

P. 427, Chapter 23 opener: Numerous varieties of squash, David Cavagnaro. Figure 23.1: (A, B) AP/Wide World; (C) The Bettmann Archive; (D) AP/Wide World. P. 436, Figure 23.7: (A) Tweedie/Bruce Coleman Inc.; (B) Breck P. Kent/ANIMALS ANIMALS. P. 438, Figure 23.10: Warren & Genny Garst/Tom Stack & Associates. P. 439, Figure 23.11: (A) M. Long/Visuals Unlimited; (B) Diana L. Stratton/Tom Stack & Associates. P. 439, Figure 23.12: (A) Lawrence Gilbert, Dept. of Zoology, University of Texas, Austin; (B) D. Wilder/Tom Stack & Associates. P. 443, Figure 23.15: (A) James F. Crow, University of Wisconsin, Madison Laboratory of Genetics. P. 444, Figure 1: The Bettmann Archive. P. 445, Figure 2: (A, B) Farber Collection. P. 445, Figure 3: The Bettmann Archive. P. 446, Figure 4: UPI/Bettmann.

Chapter 24

P. 450, Chapter 24 opener: Trilobite fossils, Albert Copley/Visuals Unlimited. P. 454, Figure 24.3: (B) John Cancalosi/Tom Stack & Associates. P. 456, Figure 24.5: (B) Tui A. DeRoy/Bruce Coleman, Inc.; (C) Jack Couffer/Bruce Coleman, Inc. P. 457, Figure 24.5: (D) Jack Couffer/Bruce Coleman, Inc.; (E) William E. Ferguson; (G) Jane Burton/Bruce Coleman, Inc.; (H) E. R. Degginger/Animals Animals; (I) E. R. Degginger. P. 459, Figure 24.8: Kjell B. Sandved/Visuals Unlimited. P. 460, Figure 24.9: (A) Brian Parker/Tom Stack & Associates; (B) Daniel W. Gotschall/Visuals Unlimited.

Chapter 25

P. 463, Chapter 25 opener: Katydid preserved in amber, John Cancalosi/Peter Arnold, Inc. P. 466, Figure 25.2: (A) L. L. T. Rhodes/ANIMALS ANIMALS; (B) John D. Cunningham/Visuals Unlimited. P. 476, Figure 1A: Albert C. Copley/Visuals Unlimited. P. 477, Figure 1B: Albert C. Copley/Visuals Unlimited; Figure 1C: John Cancalosi/Tom Stack & Associates; Figure 1D: Albert C. Copley/Visuals Unlimited. P. 478, Figures 1E–H: John Cancalosi/Tom Stack & Associates.

Chapter 26

P. 485, Chapter 26 opener: Petroglyphs found in Rainbow Park, Utah, SUPERSTOCK. P. 488, Figure 26.2: (A) Norman Tomalin/Bruce Coleman Inc.; (B) Douglas T. Cheeseman, Jr./Peter Arnold, Inc.; (C) Jack Swenson/Tom Stack & Associates; (D) Nancy Adams/Tom Stack & Associates. P. 492, Figure 26.7: Kevin O'Farrell/Institute of Human Origins. P. 492, Figure 26.8: G. Shlionsky/SOVFOTO. P. 494, Figure 26.9: Gerry Ellis/The Wildlife Collection.

Chapter 27
P. 496, Chapter 27 opener: Affectionate yellow-bellied marmots, Don and Pat Valenti. P. 498, Figure 27.3: (A) Phil Degginger; (B) Bill Beatty/Visuals Unlimited. P. 499, Figure 27.4: Daniel W. Gotshall/Visuals Unlimited. P. 504, Figure 27.10: Nina Leen/LIFE Magazine © Time Warner Inc. P. 505, Figure 27.11: Tass from SOVFOTO.

Chapter 28
P. 511, Chapter 28 opener: Queen bee surrounded by drones, FPG. P. 512, Figure 28.1: Courtesy, Joe Wrinn, Harvard News Office, Harvard University. P. 513, Figure 28.2: John Gerlach/Visuals Unlimited. P. 514, Figure 28.3: Tom J. Ulrich/Visuals Unlimited. P. 514, Figure 28.4: Y. Arthus-Bertrand/Peter Arnold, Inc. P. 514, Figure 28.5: Joe McDonald/Tom Stack & Associates. P. 515, Figure 28.6: Zig Leszczynski/ANIMALS ANIMALS. P. 515, Figure 28.7: Donald Specker/ANIMALS ANIMALS. P. 516, Figure 28.8: Steve McCutcheon/Visuals Unlimited. P. 518, Figure 28.9: (A) G. Ziesler/Peter Arnold, Inc. (B) Steve McCutcheon/Visuals Unlimited; (C) Douglas T. Cheeseman Jr./Peter Arnold, Inc.; (D) Jim Tuten/ANIMALS ANIMALS. P. 519, Figure 28.11: Nina Leen/LIFE Magazine Time Warner Inc. P. 521, Figure 28.14: Albert C. Copley/Visuals Unlimited. P. 522, Figure 1: (A, B) Michael Mix.

UNIT IV
Pp. 526-527, Unit IV opener: Peruvian woman in potato field, Lynn Johnson/Black Star.

Chapter 29
P. 528, Chapter 29 opener: Sunlight penetrating cattail leaves, John Gerlach/Visuals Unlimited. P. 529, Figure 29.1: Jonathan T. Wright/Bruce Coleman Inc. P. 529, Figure 29.2: (A) Jay Labov/Visuals Unlimited; (B) Science VU/Visuals Unlimited. P. 534, Figure 29.6: (B) John D. Cunningham/Visuals Unlimited. P. 535, Figure 29.7: (A) Martha Powell/Visuals Unlimited; (B) Biophoto Associates/Science Source/Photo Researchers. P. 538, Figure 29.10: Jean Marc Barey/Agence Vandystadt/Photo Researchers. P. 539, Figure 29.11: (A) E. H. White/Visuals Unlimited; (B) K. G. Murti/Visuals Unlimited.

Chapter 30
P. 547, Chapter 30 opener: Milkweed pod releasing its seeds, Stephen P. Parker/Photo Researchers. P. 548, Figure 30.1: Willard Clay. P. 551, Figure 30.3: (A) Doug Sokell/Tom Stack & Associates; (B) John D. Cunningham/Visuals Unlimited; (C) Doug Sokell/Visuals Unlimited; (D) R. J. Erwin/Photo Researchers. P. 551, Figure 30.4: (A) Thomas Kitchin/Tom Stack & Associates; (B) Link/Visuals Unlimited; (C) Walter H. Hodge/Peter Arnold, Inc.; (D) W. Ormerod/Visuals Unlimited. P. 559, Figure 30.12: (B) Ed Reschke/Peter Arnold, Inc. P. 560, Figure 30.13: (B) R. Knauft/Photo Researchers. P. 562, Figure 30.15: (A) Milton Rand/Tom Stack & Associates; (B) Jack Dermid/Photo Researchers; (C) Kathy Merrifield/Photo Researchers. P. 563, Figure 30.16: (A, B) John D. Cunningham/Visuals Unlimited. P. 565, Figure 1: J-L Charmet/SPL/Photo Researchers.

Chapter 31
P. 569, Chapter 31 opener: Stomata of leaf epidermis, P. Dayanandan/Photo Researchers. P. 570, Figure 31.1: Gerry Ellis/The Wildlife Collection. P. 571, Figure 31.4: (A, B) Dr. Jeremy Burgess/SPL/Photo Researchers. P. 577, Figure 31.10: (A) John D. Cunningham/Visuals Unlimited; (B) M. Timothy O'Keefe/Tom Stack & Associates; (C) William Munoz/Photo Researchers. P. 580, Figure 31.12; Adapted from Winner, William E., Mooney, Harold A., Williams, Kimberly, von Caem-

merer, Suzanne, from *Measuring and Assessing SO² Effect on Photosynthesis and Plant Growth*, from *Sulfur Dioxide and Vegetation: Physiology, Ecology and Policy Issues,* Stanford Press, CA, 1985. P. 583, Figure 1: From *Experimental Studies on the Nature of Species: I: Effect of Varied Environments on Western North American Plants,* Carnegie Institute of Washington, publication 520, by Jesn Clausen, David D. Keck, and William M. Hiesey, 1940.

Chapter 32

P. 587, Chapter 32 opener: Field of yellow poppies along California coast, Joe McDonald/Visuals Unlimited. P. 589, Figure 32.2: (A) G. Ziesler/Peter Arnold, Inc.; (B) Allan Morgan/Peter Arnold, Inc.; (C) John Shaw/Tom Stack & Associates. P. 589, Figure 32.3: (A) John Gerlach/Visuals Unlimited; (B) E. R. Degginger. P. 591, Figure 32.6: Walt Anderson/Visuals Unlimited. P. 592, Figure 32.7: E. J. Cable/Tom Stack & Associates. P. 593, Figure 32.9: (A) Patti Murray/Earth Scenes; (B) Milton Rand/Tom Stack & Associates; (C) Michael P. Gadomski/Photo Researchers; (D) John Sohlden/Visuals Unlimited. P. 594, Figure 32.10: Galen Rowell/Peter Arnold, Inc. P. 600, Figure 32.16: Kenneth D. Whitney/Visuals Unlimited. P. 602, Figure 32.19: Marty Cooper/Peter Arnold, Inc. P. 603, Figure 32.21: (A) Hans Pfletschinger/Peter Arnold, Inc.; (B) Merlin Tuttle/Bat Conservation International/Photo Researchers.

Chapter 33

P. 607, Chapter 33 opener: Skeletal muscle fibers, Ed Rescshke/Peter Arnold, Inc. P. 610, Figure 33.2: (A) David M. Phillips/Visuals Unlimited; (B) Don Fawcett/Visuals Unlimited. P. 612, Figure 33.3: (A) Fred Hossler/Visuals Unlimited; (B) John D. Cunningham/Visuals Unlimited; (C) E. R. Degginger. P. 613, Figure 33.4: (B) CNRI/SPL/Photo Researchers; (C) Biophoto Associates/Science Source/Photo Researchers; (D) R. Calentine/Visuals Unlimited. P. 614, Figure 33.5: (A) Cecil Fox/Science Source/Photo Researchers; (B) Biophoto Associates/Photo Researchers; (C) Eric V. Grave/Photo Researchers. P. 615, Figure 33.6: (A) Eric V. Grave/Photo Researchers; (B) John D. Cunningham/Visuals Unlimited; (C) Biophoto Associates/Photo Researchers.

Chapter 34

P. 625, Chapter 34 opener: Treadmill fitness examination, John Coletti/Stock Boston. P. 627, Figure 34.2: (A) Bruce Iverson/Visuals Unlimited; (C) CNRI/SPL/Photo Researchers. P. 635, Figure 1: Countway Library, Harvard University. P. 637, Figure 34.8: Fred Hossler/Visuals Unlimited. P. 639, Figure 34.10: (A) The Sterling Company. P. 643, Figure 34.14: (C) Cecil Fox/Science Source/Photo Researchers.

Chapter 35

P. 649, Chapter 35 opener: Giant multipolar neurons, Stan Elems/Visuals Unlimited. P. 651, Figure 35.1: (A) Dan McCoy/Rainbow; (B) Don Fawcett/Visuals Unlimited. P. 652, Figure 35.3: David M. Phillips/Visuals Unlimited. P. 654, Figure 35.6: T. Reese & D. W. Fawcett/Visuals Unlimited. P. 655, Figure 35.8: (A) Science VU/E. R. Lewis/T. E. Everhart/Y. Y. Zeevi/Visuals Unlimited. P. 661, Figure 35.13: (A) Dr. Colin Chumbley/SPL/Science Source/Photo Researchers.

Chapter 36

P. 668, Chapter 36 opener: Diver in pike position, David Madison/Bruce Coleman Inc. P. 674, Figure 36.4: (A) Astrid & Hanns-Frieder Michler/SPL/Photo Researchers. P. 676, Figure 36.5: (A) David M. Phillips/Visuals Unlimited.

Chapter 37

P. 687, Chapter 37 opener: Beautiful babies, Michel Tcherevkoff/The Image Bank. P. 690, Figure 37.2: (B) VU/SIU/Visuals Unlimited. P. 691, Figure 37.4: (A)

C. Edelmann/La Villette/Photo Researchers. P. 695, Figure 37.8: (B) Biophoto Associates/Photo Researchers. P. 700, Figure 37.12: (A) D. W. Fawcett/Science Source/Photo Researchers; (B) Biophoto Associates/Photo Researchers. P. 711, Figures 37.19: (A–D) Petit Format/Nestle/Science Source/Photo Researchers.

Chapter 38

P. 715, Chapter 38 opener: Leukocytes ingesting bacteria, David M. Phillips/Visuals Unlimited. P. 719, Figure 38.4: Lennart Nilsson/Boehringer Ingelheim Zentrale GmbH. P. 720, Figure 38.5: Lennart Nilsson/Boehringer Ingelheim Zentrale GmbH. P. 722, Figure 38.6: (B) CNRI/SPL/Photo Researchers. P. 725, Figure 38.7: (C) John D. Cunningham/Visuals Unlimited. P. 727, Figure 38.10: (B) Petit Format/I.P.R.P/Photo Researchers. P. 730, Figure 38.13: (A, B) Andrejs Liepins/SPL/Science Source/Photo Researchers. P. 730, Figure 38.14: (A) Lennart Nilsson/Boehringer Ingelheim Zentrale GmbH; (B) Don W. Fawcett/Visuals Unlimited. P. 739, Figure 38.18: Dr. Rosalind King/SPL/Photo Researchers.

Chapter 39

P. 743, Chapter 39 opener: Electron micrograph of spirochaetes bacteria, CNRI/SPL/Photo Researchers. P. 744, Figure 39.1: Kim Newton/Woodfin Camp & Associates. P. 747, Figure 1: Pasteur Institute, Paris. P. 752, Figure 39.5: (A, left) CDCI/Science Source/Photo Researchers; (A, right) K. G. Murti/Visuals Unlimited. P. 759, Figure 39.12: KROPPENS FORSVAR/Lennart Nilsson/Bonnier Fakta. P. 763, Figure 39.15: Cecil Fox/Science Source/Photo Researchers. P. 764, Figure 39.16: Philippe Plailly/SPL/Photo Researchers. P. 765, Figure 39.18: (B) National Institutes of Health.

Chapter 40

P. 770, Chapter 40 opener: Non-invasive diagnostic technique using nuclear magnetic resonance, David M. Dennis/Tom Stack & Associates. P. 772, Figure 40.1: Hank Morgan/Science Source/Photo Researchers. P. 774, Figure 40.2: (A) Hans Gelderblom, Robert Koch Institute, Berlin; (B) Courtesy, Dupont; (D) Phototake. P. 776, Figure 40.4: (A) Hans Gelderblom, Robert Koch Institute, Berlin; (B) CNRI/SPL/Photo Researchers; (C) NIBSC/SPL/Photo Researchers. P. 782, Figure 2: (A) Leonard Lessin/Peter Arnold, Inc.; (B, C) T. K. Hart, P. J. Bugelski, Dept. of Experimental Pathology, SmithKline Beecham.

Chapter 41

P. 794, Chapter 41 opener: Genetically engineered human insulin, SIU/Visuals Unlimited. P. 797, Figure 41.1: Daniel W. Gotshall/Visuals Unlimited. P. 798, Figure 41.2: Dr. Nigel Smith/Earth Scenes. P. 798, Figure 41.3: National Cancer Institute, National Institutes of Health. P. 799, Figure 41.4: Matt Bradley/Tom Stack & Associates. P. 800, Figure 41.5: Upjohn Company.

Text Credits

P. 796, quotation by permission of Ernst Mayr. P. 797, quotation by permission of David L. Hull. P. 797, summary and quotation by permission of E. O. Wilson. P. 798, first quotation by permission of Marvin Druger. P. 798, second quotation by permission of Linus Pauling.

Index